Novel protein and lipid sources in animal nutrition

Novel protein and lipid sources in animal nutrition

Applications in aquaculture, livestock feed and pet food

Edited by

Cristina Murcia

BRILL | WAGENINGEN ACADEMIC

Kemin Industries sponsored this book to support open-access education and the free dissemination of scientific knowledge, without any commercial interest. The sponsor had no influence on the editorial process or scientific content, which remain the sole responsibility of the editor and authors.

ISBN 978-90-04-69281-7
ISBN 978-90-04-69282-4 (PDF)
DOI https://doi.org/10.1163/9789004692824

Library of Congress Control Number: 2026939536

Wageningen Academic and Koninklijke Brill BV are part of De Gruyter Brill.
www.degruyterbrill.com

Questions about General Product Safety Regulation: productsafety@degruyterbrill.com

Cover illustration: Vane Julián

One should not be afraid to say 'I don't know' or 'I don't understand'.
To continue even when things appear to be impossible;
to stand alone or to be different;
and not to be afraid to make and admit mistakes,
for only those who dare to fail greatly can ever achieve greatly.

Margaret Hamilton, Lead Software Engineer, Apollo Space Program (NASA)

Contents

PART 4
Applications in animal nutrition

PART 7
Market insights in feed and pet food

PART 8
Case studies from industry

PART 9
Conclusions and broader impact

Acknowledgements

This book has been a journey of science, collaboration, and shared commitment. I would like to express my sincere gratitude to all co-authors for their trust, generosity, and intellectual openness. Your willingness to share knowledge across disciplines, sectors and borders transformed an ambitious idea into a collective achievement.

My sincere thanks go to Wageningen Academic Publishers, and specially to Mascha Sappok, for believing in this project from the very beginning. Four years ago, after reading my article "Novel Ingredients in Pet Food", Mascha encouraged me to develop an initial concept into a full book. Her confidence in the project and in my ability to bring together scientific rigor with a broader societal perspective on sustainability and social responsibility was the catalyst that made this possible.

I am deeply grateful to Kemin Industries for supporting the open-access publication of this book. While this project originated from an independent initiative, Kemin recognized its value and chose to support its dissemination without imposing any editorial restrictions. Throughout the process, I retained full independence in defining the scope, structure and contributors, and the final content was shared with the sponsoring team only shortly before publication. Such support represents a precious example of integrity and respect for scientific independence.

My appreciation also extends to the external peer reviewers whose critical insights and constructive feedback strengthened the scientific rigor and balance of the chapters. Their contributions were essential in ensuring the quality and credibility of this volume.

I would also like to acknowledge those authors who went beyond their own chapter by reviewing additional manuscripts and offering thoughtful feedback to fellow contributors. This collegial engagement greatly enriched the book.

I am equally grateful to all industry contributors, including representatives of different and sometimes competing companies, for demonstrating that collaboration is possible when the common objective extends beyond individual interests. The openness and professionalism surrounding this project have been inspiring.

To those who initially questioned the feasibility of such an undertaking, suggesting it was too ambitious or too complex, I am grateful as well. Those doubts served as a reminder that progress often begins with ideas that challenge existing expectations.

Finally, from my heart, I thank my family in Spain and Germany for their support throughout this long process. And to my two daughters, three and two years old when this book was completed, thank you for your patience, your smiles, and for making this long journey much happier than it would otherwise have been. This project required many hours away from you, but it also represents a commitment to the values of knowledge, responsibility and care and to the kind of world we hope to help shape for future generations.

Cristina Murcia
Editor

Figures and tables

Figures

Tables

Notes on contributors

Cristina Murcia

PhD in Chemistry (University of Bonn). Currently Senior Technical Service Manager at Kemin Pet Food and Rendering. She specializes in ingredient stability, oxidation control, shelf-life management, and food safety in pet food and rendered products. Her work bridges science and industry to support more sustainable feed and pet food systems..

Raluca Alexa

PhD in Food Science and Technology (University College Dublin & Teagasc). Food scientist specializing in hydrocolloid applications for wet pet food, focusing on texture optimization, stability, and cost-effective formulations.

Cláudia Aragão

PhD in Aquaculture (University of Algarve). Researcher at CCMAR with over 20 years of experience in aquaculture nutrition, developing sustainable feeding strategies that enhance fish resilience, circularity, and environmentally responsible aquaculture systems.

André Barreto

PhD in Aquatic Sciences – Biology and Ecology (University of Porto). He is Junior Researcher and Operational Manager at Riasearch, specializing in fish and shrimp nutrition and health across different life stages.

Ana Filipe Basto dos Santos

PhD in Animal Science (University of Porto). She is a researcher at CIIMAR–University of Porto and has worked in fish nutrition research for over ten years focusing on sustainable aquaculture, functional aquafeeds, fish robustness, and flesh quality.

Arnaud Bouxin

Agronomist (Institut National Agronomique Paris-Grignon). He is Director for Feed Safety and Regulatory Affairs at the European Feed Manufacturers' Federation (FEFAC).

Guido Bosch

Associate professor at Wageningen University. He is specialized in the nutrition of animals with various research focus areas including the alternative ingredients and foods, appetite regulation and feed/food manufacturing.

Ana Rita Cabrita
Associate Professor with Habilitation in Animal Nutrition at ICBAS, University of Porto. Her research focuses on animal nutrition and sustainable feeding systems.

Anne Carlson
CEO of Jiminy's, a leader in insect-based sustainable dog food. A former Big Heart Pet Brands VP and founder of Secant, focusing on sustainable pet nutrition innovation.

Sebastián Castro Reyes
Veterinarian (University of Chile), specializing in laboratory research on nanoemulsions, protein isolates, microencapsulation, and sustainable alternative protein sources.

Antonios Chalaris
PhD in Aquaculture (Stirling, UK). He is an aquaculture consultant with 15+ years of commercial and technical experience and board member at the European Aquaculture Society.

Kelly Condon
Senior Lecturer in the Aquaculture Discipline, at James Cook University, specialising in Aquatic Animal Health and is interested multidiscipline approaches to reduce the impact of disease in aquaculture.

Mark Digby Cyrus
has a PhD in Aquaculture (University of Cape Town) and is a Senior Lecturer at James Cook University and lead of the JCU PhycoLab, focusing on sustainable aquaculture development, macroalgae integration and new species development.

Antonella Dalle Zotte
Professor at the University of Padova specializing in Aquaculture, Poultry, and Rabbit Science with research on sustainable livestock production and insects-based feed.

Simon Kumar Das
Research Fellow at the Tropical Aquafeed Innovations Lab, James Cook University, Australia. Specializing in grouper nutrition and aquaculture sustainability.

Sofia Engrola
Principal Researcher at CCMAR and Coordinator of the Aquaculture Research Group, specializing in fish nutrition, protein metabolism, and functional aquafeeds.

António José Mira da Fonseca
PhD in Animal Science from UTAD. He is a Full Professor of Animal Science at ICBAS, University of Porto, and an integrated researcher at LAQV/REQUIMTE.

José García Alvarez
graduated in Veterinary Medicine (Complutense University of Madrid) and MSc. in Animal Nutrition and Health from ESIC. After working in feed formulation and technical support at Kelmer Internacional, he joined Visán in 1997, where he currently leads the Veterinary Nutrition Department.

Lorraine Giver
PhD in Molecular Cell and Developmental Biology (Indiana University). Chief Scientific Officer at Windfall Bio and former Chief of Science and Sustainability Officer at Calysta Inc., focusing on development and scale up of their protein production platform.

Ana Teresa Gonçalves
PhD in Marine Biosciences and Fish Health Management (University of Tokio). She is a functional nutrition expert with over 17 years of experience in fish health. Currently senior researcher at SPAROS and GreenCoLab, focusing on algae-derived ingredients and host-microbiota interactions.

Florian Henne
PhD in Chemistry (University of Dresden). He is a senior Scientist and independent consultant in feed, aquaculture, and pet nutrition, specializing in sustainable ingredient strategies, nutrient stabilization, and novel feed raw materials.

Xun Li
MSc in Animal Science (Wageningen University). Regulatory affairs expert specializing in global feed and pet food regulatory strategy, product registration, and innovation enablement. Currently Regulatory Manager at dsm-firmenich.

Matt Longshaw
PhD in Veterinary Pathology and Parasitology (University of Plymouth). Science Director at Aseno Ltd, providing consultancy across livestock, pet and aquaculture feeds industry. He previously worked in the feed ingredients, and pharmaceutical industries, and for the UK government.

Eike Luedeling
Professor of Horticultural Sciences at the University of Bonn. His research focuses on horticulture, climate change adaptation, agroforestry, and decision analysis for sustainable agricultural development and resilient production systems.

Elda Madai Melchor Martínez
Research Professor at Universidad Autonoma de Nuevo Leon. Her research is focused on algal biotechnology and natural products chemistry. She is a National Research Level I (Mexico) researcher.

Ally Motta
MSc in Animal Nutrition (University of Nottingham). She is a specialist in companion animal nutrition and Senior Product and Application Specialist at MicroHarvest Unipessoal Lda., focusing on sustainable protein development and companion animal nutrition research.

Sara Patrícia Pinto Moutinho
PhD in Biology (University of Porto). Currently researcher at CIIMAR, focusing on sustainable and innovative feed ingredients and strategies to enhance fish growth and welfare.

Elena Murcia Garcia
is a PhD student at UCAM Catholic University San Antonio Murcia, investigating on obesity-related molecular mechanisms and bioactive compounds identification.

Leo Nankervis
Associate Professor at James Cook University, leading the Tropical Aquafeeds Innovation Lab, focusing on aquaculture nutrition, metabolism and feed innovation. He is a global expert in aquaculture nutrition, driving industry-relevant research across diverse species and international collaborations.

Rachel Neil
PhD in Coral Aquaculture (James Cook University). Currently working at the Tropical Aquafeed Innovations Lab at James Cook University as a project officer, researching aquaculture sustainability and improving linkages between industry and the research sector.

Belén Marisol Padilla Ramírez
Veterinarian (University of Chile), specializing in palatability, acceptability, digestibility, and fecal score tests at the Pet Nutrition and Food Research Center (CINAM), Faculty of Veterinary and Animal Sciences, University of Chile.

Bianca Palumbo
PhD in Animal and Food Science (University of Padova). Currently researcher in Aquaculture, Poultry, and Rabbit Science at the University of Padova, working on sustainable livestock production, meat quality, and innovative feeding strategies, including insects as alternative feed ingredients.

Roberto Parra Saldívar
Professor and Chair of Sustainable Mycoremediation at Cranfield University. He holds a PhD in Biotechnology from Cranfield. His research focuses on bioprocessing and bioengineering. He is a National Research System Level III (Mexico) researcher.

Nathan Pelletier
Associate Professor (University of British Columbia). He manages the Food Systems Priority Research for Integrated Sustainability Management (PRISM) Lab. He is an industrial ecologist and ecological economist whose research establishes baselines and assesses technological and management interventions to improve sustainability outcomes in industrial food systems.

Valentina Pinilla Jara
MSc in Animal and Veterinary Sciences (University of Chile), specializing in animal nutrition and insect-based feed innovation for sustainable feed and food systems.

Omar J. Portillo Castillo
Professor at Universidad Autónoma de Nuevo León. He holds a PhD in Biomedical Chemistry, specializing in chromatographic separations and deep eutectic solvents. He is a National Research System Level I (Mexico) researcher.

Alexandra Rankovic
PhD in Companion Animal Nutrition (University of Guelph). Comparative nutritionist specializing in fat metabolism and nutrient utilization in companion animals and horses, with an emphasis on advancing evidence-based approaches to diet formulation and metabolic health.

Marta Reshetiuk
MSc in Business, Management, Marketing and Related Support Services (University WSB Merito Wrocław). She currently serves as Supply Chain Manager at Calysta GmbH specializing in sustainability, commercial strategy, and supply chain management in the feed and pet food sectors

Ana Gisela Reyes Alvarado
PhD in Biotechnology (UAM-Iztapalapa). Currently researcher at CIBNOR specializing in the biotechnological valorization of plant and marine resources, sustainable bioprocesses and circular economy principles.

Lucia Rettenbeck
PhD in Veterinary Medicine (LMU Munich). She serves as Veterinary Scientist at Interquell GmbH working on pet food development and quality assurance, focusing on formulation optimization and product innovation.

Verónica Mayela Rivas Galindo
Full Professor (Universidad Autonoma de Nuevo Leon). She is the Head of Department of Analytical Chemistry in the School of Medicine specializing in natural products extraction and chemical characterization by Nuclear Magnetic Resonance.

Rui Jorge Miranda Rocha
Assistant Professor (University of Aveiro), Integrated Member of CESAM, and Director of Innovation & Development at Riasearch. He specializes in aquaculture production optimization and academia–industry collaboration.

Zahra Salimi
MSc in Biological Chemistry (University of Stavanger). She serves as Product Developer and Laboratory Lead at Passion4Food, specializing in product development and technical innovation in pet food.

Karen Schaich
Professor at Rutgers University. She spent 14 years in the Medical Dept. at Brookhaven National Laboratory investigating free radical reactions in toxicology. Now studies and teaches mechanisms of lipid oxidation, protein co-oxidation, and antioxidant action in foods at Rutgers University.

Susanne Scherber
Veterinarian (Free University of Berlin). She previously worked at Mera Tiernahrung GmbH as specialist in research and development, specializing in companion animal nutrition, focusing on product development and formulation quality control. Currently Head of research and development at Pets Nature GmbH.

Anna Kate Schoveller
Professor in the Department of Animal Biosciences (University of Guelph). She is a comparative nutritionist whose research focuses on companion animal nutrition, specifically protein and energy metabolism, the effects of diet on behaviour and welfare, and alterative protein and fat ingredients.

Shaiyan Siddique
Shaiyan Siddique is a PhD student in Sustainability, working under the supervision of Dr. Nathan Pelletier in the Food Systems PRISM Lab at the University of British Columbia – Okanagan in Kelowna, Canada.

Rita Teodósio

PhD in Biological Sciences (University of Algarve). Researcher at CCMAR specializing in aquaculture fish nutrition, development of functional feeds that enhance antioxidant capacity, immune function, and gut health to improve the performance and resilience of marine fish in aquaculture.

Milka Kosanovic Tesla

MSc infeed Technology (Norwegian University of Life Scieces, NMBU). She is Head of R&D at Passion4Food AS and Founder of Nordic Pet Food Events AS, specializing in pet food nutrition, formulation, production, and processing.

Ekaterina Tretyakova

Msc in Economics (Moscow State University). She is Senior Market Insights Analyst at Innova Market Insights, specializing in Pet Food industry analytics and market research.

Luisa M.P. Valente

Full Professor at the School of Medicine and Biomedical Sciences (ICBAS), University of Porto focusing on feed ingredients, functional feeds, muscle growth and flesh quality, fish robustness, environmental impact, circular bioeconomy and One Health.

Carolina Paz Valenzuela Venegas

Veterinary professor and PhD from the University of Chile with experience in research and development in animal nutrition. She specialises in food ingredient encapsulation and applied biotechnologies, leading scientific projects and promoting innovation in the field of R&D.

Kaiti Valm

MSc in Environmental and Life Sciences (Trent University) and MBA (University of Alberta). She works in the Regulatory Sciences group at SGS Nutrasource and has extensive experience in animal sciences including nonclinical research, wildlife research, and veterinary health technologies.

Valeria Sofía Villanueva Villanueva

Veterinary doctor from the University of Chile with experience in research and development in the field of animal nutrition. She specializes in food ingredient encapsulation and applied biotechnologies, leading scientific projects and promoting innovation in the field of R&D.

Alexandra Wesker
MSc in Animal Nutrition (Wageningen University). She is currently the Technical Director at Aseno, guiding innovative startups and established companies. As a writer and animal scientist, she connects science with market dynamics. Alexandra authored a book on equine nutrition and regularly writes for pet trade magazines.

Cory Whitney
PhD in Agricultural Sciences (University of Kassel) with habilitation in Agroecology and Sustainable Development (University of Bonn). Human ecologist developing probabilistic decision models that translate local knowledge into actionable forecasts, supporting agricultural strategies that improve food security and sustainability.

Janina Wimmers
BSc in Marketing and Digital Media (FOM University of Applied Sciences). She is Head of Product Management at Mera Tiernahrung, focusing on strategic and operational direction of the product management department and drives innovation and brand development with great passion.

Derya Yıldız
MSc in Journalism (University of Istanbul). She is Editor-in-Chief of Feed & Additive, Alternative Proteins, and Yem & Katkı magazines. She specializes in technical, industry-focused editorial leadership and has over 25 years of experience in food, agriculture, and livestock publishing.

Franziska Zorn
MSc in Agricultural Sciences (University of Kiel) with focus on animal nutrition. She has experience in developing dog and cat food and is now Head of New Business at Saturn Petcare, responsible for third-party products.

Notes on external reviewers

María del Carmen Cerón García

Professor in Chemical Engineering (University of Almeria), Spain. She specializes in microalgae bioprocessing, valorization of microalgal biomass enhancing high value products extraction (carotenoids, fatty acids, and bioactives), and antibiofouling materials.

Aulus Cavalieri Carciofi

Professor (School of Agricultural and Veterinary Sciences, UNESP) Brazil. Coordinator of the Residency Program in Nutrition and Clinical Nutrition of Dogs and Cats. Board member of the Brazilian College of Animal Nutrition and current President of the European Society of Veterinary and Comparative Nutrition.

Marco Cullere

Associate Professor at the Department of Animal Medicine, Production, and Health (University of Padova). His research topics include genetic, nutritional, and post-mortem strategies to improve the quality of meat and meat products from poultry, rabbit, and other minor species.

Naouel Gharbi

PhD in Molecular Biology (University of Bergen). Former Research Director of the Fish Biology and Aquaculture group at NORCE Norwegian Research Centre (2022–2025). Her work focuses on fish health/welfare, aquafeed innovation, circular production systems, integrated environmental, social and economic sustainability.

Iago Teles

PhD in Bioprocess Engineering (Wageningen University). Currently research manager of the Bioprocess Engineering Group at Wageningen University focusing on microalgal biotechnology, innovation mapping and scientific communication.

Francisco Hidalgo

Research Professor at CSIC (Spanish National Research Council), specializing in lipid oxidation chemistry in foods and their impact on quality and safety. Author of 150+ food science papers and Editor of *Food Chemistry*.

Rosario Zamora

Professor at CSIC (Spanish Research Council) specializing in the characterization and quality of lipids.

Jörg Bartelt

PhD in Animal Nutrition (Humboldt University Berlin). Experienced animal nutrition specialist with extensive expertise in feed additives. Currently Technical Sales Manager at Vetagro S.p.A. (D-A-CH region). He served for several years as a postdoctoral researcher at Humboldt and Free Universities of Berlin.

Anca Becze (Naghiu)

PhD in Veterinary Medicine (University of Agricultural Science and Veterinary Medicine, Cluj-Napoca). She is currently senior researcher at INCDO-INOE 2000, ICIA Cluj-Napoca, Romania, specializing in food safety, bioeconomy, and sustainable valorisation of agro-industrial byproducts with focus on alternative proteins.

Paul-Alexandru Uiuiu

Assistant Professor (University of Agricultural Sciences and Veterinary Medicine, Cluj-Napoca). His research interests include animal anatomy and physiology, fish farming, and aquaculture. His primary work focuses on fish welfare and the protection of natural aquatic ecosystems.

Guan Gong

PhD in Animal Nutrition (Chinese Academy of Agricultural Sciences). She has extensive experience in product standards, feed ingredients and additives regulations. Currently Regulatory Approvals Manager at Royal Canin focusing on compliance and safety evaluation of novel ingredients in pet nutrition.

PART 1

Introduction and foundations

CHAPTER 1

Introduction: the case for novel ingredients in feed and pet food

C. Murcia

Kemin Nutrisurance, EMEA, Via della Tecnica 11, 37040, Veronella (VR), Italy

The growing global population of humans and animals, including the rapidly expanding number of companion animals, is generating unprecedented pressure on our planet's resources and on the Earth system as a whole. Animal feed and pet food systems sit at the core of this challenge. They link agricultural production and resource use with the nutritional needs of livestock, aquaculture species, and companion animals, while at the same time influencing greenhouse gas emissions, biodiversity loss, and disruptions of the nitrogen cycle. How we choose ingredients, formulate diets, and source raw materials therefore matters far beyond the feed mill or the pet food factory.

This is placing agrifood systems in urgent need of innovation to optimize water, land and energy use while maintaining productivity and resilience (FAO, 2024). A key aspect of this challenge requires finding ways to meet the rising demand for proteins and lipids, which supply essential amino acids and fatty acids that cannot be synthesized by animals, without overwhelming ecosystems or compromising socio-economic stability. That is why the search for novel protein and lipid sources for feed, aquaculture, and pet food has become a central focus in the global transition toward more sustainable and climate-smart production systems.

Traditional protein and lipid sources for animal diets are soybean meal, fish meal, and poultry fat among other plant- and animal-derived ingredients. However, the increasing demand for these materials is associated with growing environmental pressures, including deforestation and overfishing, and more broadly with the exceedance of key planetary boundaries to which animal feed production is a major contributor (Ineichen et al., 2023; van Riel et al., 2023; Mottet et al., 2017). While these ingredients remain nutritionally valuable and indispensable in many contexts, it is increasingly clear that relying on the same sources alone will not be enough in a world facing demographic growth, climate uncertainty, and finite natural resources.

This book does not argue for abandoning conventional ingredients. Instead, it emphasizes the need to expand the toolbox available to formulators and nutritionists. Novel ingredients can complement existing feedstocks and provide nutritional diversity while contributing to the efficiency and sustainability of animal diets. When used strategically, they can help mitigate the environmental footprint of feed production and at the same time ensure that the growing nutritional needs of animals and humans are met (Makkar, 2017). The integration of novel ingredients into animal nutrition also reflects visible trends in sustainability, consumer awareness, and innovation in a society that is more and more conscious of food and feed supply challenges and the impacts on climate change (Andre et al., 2024; Pew Research Center, 2021). In pet food in particular, purchasing decisions are not driven solely by nutritional adequacy or environmental impact, but also by emotional factors linked to the human–animal bond, trust, and perceptions of care. As a result, the acceptance of novel ingredients in pet food often follows different dynamics than in livestock or aquaculture, where decisions are more strongly driven by performance and cost efficiency.

Livestock producers, aquaculture businesses, and pet food companies increasingly recognize the need for solutions that align with environmental goals and growing consumer expectations. From my own experience working in companies that supply livestock, aquaculture, and pet food industries, I have seen this shift taking shape first-hand and this awareness is no longer limited to sustainability reports or marketing strategies, but it is now driving real innovation throughout the value chain. In the chapters dedicated to market trends in this book, these market developments are discussed in more detail.

The aim of the book *Novel protein and lipid sources in animal nutrition* is to offer a balanced and evidence-based exploration of these emerging ingredients. It is designed to serve as both a tool and a reference guide for university students, industry professionals, policy makers and anyone interested in the future of animal and pet nutrition.

This book focuses exclusively on novel lipid and protein sources as potential new ingredients in feed and pet food formulations due to their central role in meeting nutritional requirements and because proteins and lipids are typically among the most environmentally impactful feed components, particularly when derived from land- and resource-intensive systems. For these reasons, the investigation of new and more sustainable sources of lipids and proteins has strong potential to reshape future animal diets, improving efficiency while reducing environmental pressure.

At the same time, the integration of novel ingredients inevitably involves trade-offs across environmental, economic, and social dimensions, highlighting the need for evidence-based, case-specific evaluation rather than one-size-fits-all solutions.

1 Scope of the book

Novel protein and lipid sources in animal nutrition is organized into thematic sections. Each addresses critical aspects of novel ingredients for animal and pet nutrition. Together, they provide a coherent narrative: from understanding nutritional value, to applications across species, to sustainability, regulation, and future perspectives.

- **Understanding novel ingredients:** the book begins by providing an overview of the nutritional composition of novel ingredients, where we include their fatty acid and amino acid profiles as well as their proximate analyses. These metrics are compared with those of conventional ingredients, providing the reader with the ease of a quick comparison and a clear understanding of their potential roles in animal diets. However, it is important to remember that many of these ingredients are genuinely novel, and their nutritional composition can vary considerably. For instance, in insects, the protein and lipid content can change depending on the rearing substrate, resulting in ingredients from the same species showing very different nutritional values. Similarly, proteins derived from fermentation technologies may vary according to the specific production or processing methods used. To the best of our knowledge, there are still no official or harmonized sources for the nutritional composition of these novel ingredients, which makes research, formulation, and regulatory comparison more complex. This is precisely why a careful, science-based evaluation is essential when integrating them into feed and pet food systems.
- **Oxidation and shelf life:** another essential part of this book focuses on the critical topic of lipid and protein oxidation, a key factor influencing the shelf life and quality of feed and pet food products. Oxidation is the primary degradation process that affects lipids leading to the formation of undesirable compounds that can compromise nutritional value, palatability, and safety. In addition, an innovative chapter about co-oxidation of proteins is broadening the knowledge of important consequences of radical change reactions affecting not only lipids but also proteins and other nutrients such as vitamins present as key nutrients in feed and pet food diets.
- **Application in aquaculture, pet food and livestock:** the core of the book is structured around three main applications: aquaculture, pet food and livestock feed. Each section examines the suitability of various novel ingredients for these sectors, focusing on their impact on animal health, growth performance and acceptance among other important factors. Updated research is presented to highlight the current investigations, limitations and promising developments in each field.
- **Sustainability and the circular economy:** a dedicated chapter examines how novel feed ingredients can advance the principles of sustainability and the cir-

cular economy. It explores how low-opportunity-cost biomass (LOCB) and alternative protein and lipid sources could contribute to more resource-efficient and lower-emission feed systems, while also reducing waste and overall environmental footprints. By integrating environmental and economic assessments such as Life Cycle Assessment (LCA), the chapter highlights potential pathways toward more resilient and responsible feed and pet food systems. In the context of aquaculture, a separate chapter addresses the interaction between feed formulation, nutrient flows and production system design, including concepts such as Integrated Multi-Trophic Aquaculture (IMTA) and sustainability certification frameworks.

- **Holistic decision modeling:** this forward-looking chapter applies holistic decision models to explore how novel ingredients can transform animal nutrition under resource limitations, climate pressures, and demographic change. It guides readers through the shift toward feed systems that are efficient yet resilient, fair to people and animals, and respectful of planetary boundaries. By linking disciplines that rarely intersect, from agroecological modeling to animal nutrition and processing, it offers a clear and integrated way to see how choices at feed formulation level can impact broader socio-environmental systems.
- **Regulatory and market perspectives:** another part of this book examines the regulatory landscape surrounding novel ingredients in different regions, including the European Union, Canada, the United States of America and China. Additionally, market trends and consumer acceptance are discussed, providing insights into the economic viability of these ingredients and the challenges faced not only regarding availability and resources use but also human (consumer) perception and acceptance. In the case of pet food, these market dynamics are closely linked with the emotional relationship between pet owners and their animals, which can accelerate or hinder the adoption of novel ingredients independently of their technical merits.
- **Successful cases from the industry:** to bridge science and practice, the book presents case studies of companies that have successfully incorporated novel ingredients into their commercial products. These examples highlight the challenges and opportunities associated with adopting novel ingredients, including formulation complexities, and consumer acceptance.
- **Summary and reflections:** finally, the book concludes with a summary of key learnings and open questions to the reader, highlighting the importance of volumes like this one for fostering constructive dialogue among stakeholders. It also links the topics discussed to the relevant United Nations Sustainable Development Goals (SDGs).

2 Climate urgency, systemic change, and the role of feed innovation

Ten years have passed since the adoption of the Paris Agreement, a milestone in international climate diplomacy that marked a unified global commitment to limit global warming to well below 2 °C, and ideally to 1.5 °C above pre-industrial levels. For the agriculture and food sector, and particularly for animal feed and pet food systems, this commitment underscores the urgency of rethinking how key inputs such as proteins and lipids are produced, sourced, and used, given their contribution to greenhouse gas emissions, biodiversity loss, and pressure on planetary boundaries. This agreement, celebrated as a turning point, was given powerful momentum when the United States and China ratified it jointly on September 3rd, 2016, a moment I experienced firsthand from the United Nations Framework Convention on Climate Change (UNFCCC) in Bonn, where I was contributing to the communications department. That Saturday, I was called to the headquarters to help prepare digital content for this historic event and the atmosphere was electric; a sense of global cooperation and hope was palpable!

A decade later, that sense of unity feels more fragmented. This paradox between awareness and action has been widely discussed across scientific, policy, and public spheres in recent years. While the scientific evidence for climate change is stronger than ever, and the physical impacts more visible and severe, there is an undeniable sense of disconnection between what is known and what is being done. However, public awareness about climate change has never been higher. Across many regions, a clear majority of the population agrees that governments should implement more ambitious climate policies (Andre et al., 2024; Fagan et al., 2025). Surveys confirm a widespread recognition of the anthropogenic causes of climate change and the urgent need for action. Nevertheless, despite this growing awareness and support, climate change often fades into the background of political and media agendas, displaced by other global crises such as geopolitical conflicts, inflation, and energy insecurity.

This discrepancy between awareness and action is especially concerning because the pace of climate change is accelerating, and its impacts are no longer proportional to the rise in emissions (IPCC, 2023). We are entering a *nonlinear phase* of the climate crisis. The Earth system is beginning to react to warming through feedback loops: thawing permafrost releases methane, deforestation reduces carbon sinks, and ocean heat disrupts weather patterns and marine life. These processes amplify one another, pushing us dangerously close to tipping points with profound and potentially irreversible consequences. One of the most striking examples of this acceleration is what is happening in the Amazon rainforest. For decades it was seen as one of the planet's greatest carbon sinks, helping to absorb part of the CO_2 we emit every year. But today, large areas of this unique ecosystem have started to lose that function. In some regions, deforestation and repeated droughts have pushed the system beyond its limit, turning parts of the forest from a carbon sink into a carbon source. A clear signal that the Earth's natural stabilizing mechanisms are starting to fail. (Rockström et al., 2023; Rockström, 2024).

We are running out of buffering capacity. Forests and oceans, which for decades have absorbed more than half of our emissions, are now showing cracks in their resilience. The window to keep global warming within safe limits is rapidly closing, and the longer we delay systemic change, the more difficult and expensive it becomes to mitigate and adapt (IPCC, 2023; FAO, 2022). The agriculture and food sector, in particular animal feed and pet food systems, sits at the crossroads of this transformation. It is both a source of greenhouse gas emissions and one of the sectors most vulnerable to climate impacts such as drought, resource scarcity, and supply chain disruptions (Pelletier et al., 2010; FAO, 2022; FAO, 2024).

This book was conceived as a response to that reality. It brings together a diverse group of researchers, practitioners, and policy experts to explore how novel protein and lipid sources can contribute to reshaping the feed and pet food industry. From insects, algae, and fermentation-derived proteins to oilseeds, co-products, and underutilized aquatic species, these ingredients are not just nutritional innovations, they also represent an opportunity to move toward greater resilience, circularity, and climate-smart agriculture (Makkar, 2017; Pam et al., 2020).

Beyond the technical dimension, this volume also examines the enabling conditions that will determine the success of novel ingredients: regulatory clarity, public and pet owner acceptance, ethical frameworks, and fair market access. Innovation alone is not enough. We must integrate these changes within systems that respect ecological boundaries while meeting nutritional and socioeconomic needs.

Ten years after the world came together in Paris, the challenge before us is not simply to remember that moment but to reclaim its spirit. The future of animal nutrition must be part of the broader climate solution. With this book, we hope to contribute a small but meaningful piece to that collective effort.

3 Objective of the book

In light of these accelerating challenges, the intention behind this book becomes clearer. The objective of this book is to contribute to the awareness of novel ingredients as valuable additions to the existing repertoire of feed and pet food components. Each ingredient, whether conventional or novel, has unique strengths and limitations, and its suitability depends on the specific dietary requirements of different animal species.

By adopting a balanced approach, this book aims to foster understanding and collaboration among all stakeholders, including those who may be skeptical of novel ingredients. It emphasizes that the integration of these materials should be driven by evidence, practicality, and the overarching goal of sustainability.

Novel protein and lipid sources in animal nutrition is published in an open-access format, reflecting a strong belief that knowledge should be freely accessible. Accessibility is essential to advancing education, innovation, and more sustainable practices. By removing barriers to information, the book aims to empower researchers, students, and professionals from different regions and backgrounds to participate in the development and application of novel ingredients in animal nutrition.

By highlighting the role of innovative protein and lipid sources in addressing critical challenges in animal nutrition, this book contributes to ongoing efforts to build more sustainable feed and pet food systems. It aligns with global initiatives such as the United Nations Sustainable Development Goals, particularly those related to responsible production and consumption, climate action, and life below water and on land.

At the same time, the book underscores the importance of innovation in facing complex problems. Novel ingredients represent a frontier of research and development, offering exciting opportunities to reimagine how we produce and consume food. Through its interdisciplinary approach, this book serves as a bridge between academic research, industry practice, and societal needs.

References

Andre, P., Boneva, T., Chopra, F. and Falk, A. (2024). Globally representative evidence on the actual and perceived support for climate action. *Nat. Clim. Chang.* 14, 253–259 https://doi.org/10.1038/s41558-024-01925-3

Fagan, M. and Leong, G.M. (2025). People in Middle-Income Countries Say Climate Change Is Affecting Their Community. Pew Research Center. https://pewrsr.ch/40ksDHb

FAO (2022). *The future of food and agriculture – Drivers and triggers for transformation. The Future of Food and Agriculture*, no. 3. Rome. https://doi.org/10.4060/cc0959en

FAO (2024). *The State of Food and Agriculture 2024.* https://doi.org/10.4060/cd2616en

Ineichen, S.M., Zumwald, J., Reidy, B., Nemecek, T. (2023). Feed-food and land use competition of lowland and mountain dairy cow farms, *Animal, Volume 17, Issue 12*, 101028, https://doi.org/10.1016/j.animal.2023.101028

IPCC (2023). Climate Change 2023: Synthesis Report. Contribution of Working Groups I, II and III to the Sixth Assessment Report of the Intergovernmental Panel on Climate Change [Core Writing Team, H. Lee and J. Romero (eds.)]. IPCC, Geneva, Switzerland, 184 pp., https://doi.org/10.59327/IPCC/AR6-9789291691647

Makkar, H.P.S. (2017). Feed demand landscape and implications of food-not feed strategy for food security and climate change. *Animal*, 12(8), 1744–1754. https://doi.org/10.1017/S175173111700324X

Mottet, A., & Tempio, G. (2017). Global poultry production: current state and future outlook and challenges. *World's Poultry Science Journal*, 73(2), 245–256. https://doi.org/10.1017/S0043933917000071

Pam, I., Senaratne-Lenagala, L., Stube, A., & Brackenridge, A. (2020). Protein demand: review of plant and animal proteins used in alternative protein product development and production. *Animal Frontiers*, 10(4), 53–63. https://doi.org/10.1093/af/vfaa040

Pelletier, N., & Tyedmers, P. (2010). Forecasting potential global environmental costs of livestock production 2000–2050. *PNAS*, 107(43), 18371–18374. https://doi.org/10.1073/pnas.1004659107

Pew Research Center (2021). *"In Response to Climate Change, Citizens in Advanced Economies Are Willing To Alter How They Live and Work".* https://pewrsr.ch/3EgOx08

Rockström, J., Gerten, D., Wang-Erlandsson, L. (2023). Safe and just Earth system boundaries. *Nature*, 619(7968), 458–468. https://doi.org/10.1038/s41586-023-06083-8

Rockström, J. (2024). *The Tipping Points of Climate Change — and Where We Stand.* TED Talk, July 2024. Available at: https://www.ted.com/talks/johan_rockstrom_the_tipping_points_of_climate_change_and_where_we_stand

van Riel, A-J, Nederlof, M.A.J., Chary, K., Wiegertjes, G.F., de Boer, I.J.M. (2023). Feed-food competition in global aquaculture: Current trends and prospects. *Rev Aquac.* 15(3): 1142-1158. https://doi.org/10.1111/raq.12804

CHAPTER 2

Reader's guide

C. Murcia

Kemin Nutrisurance, EMEA, Via della Tecnica 11, 37040, Veronella (VR), Italy

1 How to read this book

This book brings together a wide range of perspectives on novel protein and lipid sources in livestock feed, aquaculture and pet food. It is designed as a reference resource that can be approached in different ways, depending on the reader's background, interests and objectives. Across the volume, novel proteins and lipids are examined from multiple angles, including their production, nutritional value, technological performance, stability and shelf life, regulatory status, sustainability implications and practical use in commercial diets. Rather than promoting specific solutions, the book aims to provide evidence-based insights that support informed evaluation and decision-making.

The chapters are written by contributors from universities, research institutes, regulatory organizations and industry. Each chapter is self-contained and can be read independently. At the same time, the book follows a structured progression, guiding the reader from foundational concepts and ingredient characterization to applications in animal nutrition, regulatory frameworks, sustainability considerations, market perspectives and industrial implementation.

This Reader's Guide is intended to help different types of readers navigate the structure of the book and identify the sections most relevant to their interests and professional needs.

2 How the book is organized

2.1 Part 1: Introduction and foundations

This part sets the stage:
- Why diversification of protein and lipid sources is needed in livestock feed, aquaculture and pet food
- How climate pressures and resource constraints challenge current systems
- What this book aims to achieve and how the individual chapters connect

The Reader's Guide itself is part of this introductory framework.

© 2026 THE AUTHOR(S) | HTTPS://DOI.ORG/10.1163/9789004692824-003

2.2 Part 2: Emerging protein and lipid sources

This section presents the major categories of novel ingredients:

- **Novel proteins:** insects, microalgae, fungi, bacteria, plant-based proteins and proteins derived from fermentation and upcycled agricultural by-products
- **Novel lipids:** algal oils rich in DHA, camelina oil, and insect lipids

The reader will find a clear overview of the key nutritional features of novel ingredients, particularly essential amino acid and fatty acid profiles, and how these compare with traditional raw materials. The section also highlights the natural variability of novel sources due to origin, processing and other factors helping readers to understand both their potential and their limitations. Overall, Part 2 provides the conceptual foundation needed to navigate the diversity of new proteins and lipids before moving on to species-specific applications in later chapters.

2.3 Part 3: Stability, oxidation and shelf-life

This part explains:

- Mechanisms of lipid oxidation
- How lipid oxidation leads to protein co-oxidation, off-flavors, reduced palatability, and potential safety issues
- Why stability is a critical factor for formulators, quality assurance teams, and industry adoption

Rather than focusing on specific species or inclusion scenarios, this section addresses the fundamental chemical mechanisms that govern oxidation and shelf-life across feed and pet food systems. It provides the scientific foundations needed to understand how different raw materials behave during processing and storage, and why oxidative stability remains a limiting factor for both novel and conventional ingredients.

Building on the amino acid and fatty acid profiles introduced in Part 2, this section equips the reader with tools to evaluate oxidative risk, shelf-life implications, and quality outcomes based on ingredient composition. These principles are broadly applicable across feed and pet food formulations, making Part 3 a key reference not only for novel ingredients, but for formulation and quality decisions more broadly.

2.4 Part 4: Applications in animal nutrition

This is the most extensive part of the book, organized by species, reflecting how nutrition, research and formulation are performed in practice.

Aquaculture
This section focuses on the application of novel ingredients in aquaculture systems, addressing nutritional performance, lipid and protein utilization, and sustainability considerations across commercially relevant fish species. It includes chapters on:
- Camelina oil as a potential alternative to fish oil in aquafeeds
- Insect-based ingredients used as protein, lipid or/and functional sources in fish nutrition
- Micro- and macroalgae with antioxidant, immunological, and microbiome-related effects
- Bacterial proteins and their role in commercial aquafeeds
- Sustainable certification schemes, circularity frameworks, and IMTA (Integrated Multi-Trophic Aquaculture)

Pet food
This section addresses topics that are less extensively covered in scientific literature:
- Palatability and sensory attributes of insects for dogs and cats
- Nutritional implications of plant-based proteins (pulses) in cat and dog food
- Microbial proteins for pet nutrition
- Edible fungi as emerging protein sources
- Underutilized aquatic ingredients for canine diets

It uniquely blends science, consumer perception, industrial processing, and premium formulation.

Livestock feed
This section covers insect-based ingredients, microalgae, agricultural by-products, fermentation-derived proteins, and camelina. It explores:
- Inclusion levels and digestibility in monogastrics and ruminants.
- Effects on performance, health, antioxidant status, and product quality.
- Opportunities and limitations for replacing conventional proteins and lipids.

Part 4 is particularly useful for feed formulators, R&D departments, and professionals working at the interface of nutrition and sustainability.

2.5 Part 5: Regulatory framework

Regulation is often the determining factor for the adoption of novel ingredients. This part provides:
- EU legislation on circularity, feed materials and restrictions for novel ingredients
- Approval pathways in the US and Canada
- Regulatory conditions and authorization processes in China

This section is highly practical, offering clear guidance for industry, policymakers and regulatory professionals.

2.6 Part 6: Sustainability and decision-making

This section contains two key contributions:

- A chapter examining the sustainability implications of novel feed ingredients for feed and pet food production, with a focus on life cycle assessment, resource-use efficiency, and nutrient circularity
- A conceptual and methodological chapter on holistic decision modelling, integrating predictive tools, systemic thinking, behavioral aspects and socio-ecological considerations

Together, they provide a strategic framework of how sustainability can be assessed and integrated into decisions.

2.7 Part 7: Market insights

This section places scientific advances within real-world market dynamics. It includes:

- Trends in pet food innovation and emerging ingredients
- A global overview of the compound feed market, segmented by species and region
- Key drivers and barriers to adoption of novel raw materials

This is particularly valuable for industry professionals, analysts and educators.

2.8 Part 8: Industry case studies

This part brings together practical examples from companies that have already adopted novel ingredients:

- Insect-based dog food formulations
- Fermentation-derived proteins in premium pet food
- Microbial proteins as sustainable alternatives
- Collaborative pathways for innovation and scaling

These case studies illustrate how scientific advances translate into commercial solutions.

2.9 Part 9: Conclusions and broader impacts

The final part:
- Summarizes the main messages of the book
- Aligns the content with the Sustainable Development Goals
- Identifies knowledge gaps and future research directions

It provides a forward-looking perspective and highlights the transformative potential of novel ingredients.

3 Intended audience

This book is designed for a broad audience within the animal nutrition and sustainability sectors:
- **Undergraduate and graduate students** seeking a structured, comprehensive introduction
- **Researchers and lecturers** looking for updated scientific references
- **Feed and pet food professionals** in formulation, R&D, quality, procurement, regulatory affairs and corporate sustainability
- **Start-ups and innovators** developing new raw materials
- **Regulators and policymakers** requiring clarity on authorization pathways and safety considerations

Each chapter is autonomous but embedded within a shared conceptual framework.

4 How to navigate the book according to your interests

The structure of the book allows flexible reading depending on the reader's background and objectives. Table 2.1 provides recommended entry points for different audiences.

TABLE 2.1 Suggested reading pathways based on reader interests

If your main interest is...	Start with...
Understanding novel ingredients	Parts 2 and 3
Species-specific nutrition	Part 4
Regulation and authorization	Part 5
Sustainability and decision-making	Part 6
Market and industry dynamics	Parts 7 and 8
Teaching or curriculum design	Suggested sequence: Parts 1 → 4 → 6 → 9

5 Consistency across chapters

Despite the diversity of authors, a common structure was encouraged to support comparability:
- Context and rationale
- Composition or mechanism
- Applications in specific species
- Challenges and limitations
- Future perspectives

This consistency makes the book suitable as a teaching tool and as a reference for industry and research.

PART 2

Sources of novel proteins and lipids

CHAPTER 3

From soybean to insects: nutritional characterization of emerging protein sources

F.D. Henne

Pet Food Ingredient Specialist, Applications and Palatability, Verona, Italy
Correspondence: florian.d.henne@science4climate.org

Abstract

The continuous growth of the population results in an increasing demand for high-quality proteins and essential lipids. Innovative approaches are required to secure sustainable feed and pet food supply. Conventional protein sources (soybean meal, fish meal, meat and bone meal, poultry by-product meal) remain the benchmark for animal and aquaculture feed, however, their production is increasingly constrained by environmental, economic, and ethical concerns. Novel ingredients from insects, algae, fungi, fermentation, oilseeds, and food by-products have emerged as promising alternatives.

This chapter provides a comparative overview of the nutritional characteristics of conventional and emerging protein sources used in animal feed and pet food. Special emphasis is placed on crude protein content, proximate composition and essential amino acid (EAA) profiles in relation to the requirements of livestock, companion animals and aquaculture species. Animal-derived protein meals generally contain high protein concentrations and balanced EAA profiles, while plant-based and novel proteins show greater variability depending on species, substrate and processing conditions. Fish and poultry meals are typically rich in lysine, methionine, and valine. Plant meals are typically low in sulfur-containing amino acids (methionine, cystine) but provide favorable lysine concentrations. Novel sources such as insects are particularly high in lysine and valine, whereas meals of algae offer EAA profiles comparable to animal meals.

By comparing conventional and novel protein ingredients side by side, this chapter highlights opportunities and limitations for their inclusion in sustainable feed formulations and supports the development of nutritionally balanced diets across animal species.

© 2026 THE AUTHOR(S) | HTTPS://DOI.ORG/10.1163/9789004692824-004

Keywords

dietary proteins – essential amino acid – animal nutrition – sustainability – novel ingredients – insects – algae – marine species

1 Introduction

The United Nations (UN) stated in its *World Population Prospects 2022* report that the global human population is expected to reach 8.5 billion in 2023 (8.3 billion currently, rounded, Worldometer, 2025), 9.7 billion in 2050 and around 10.4 billion at the end of this century. Main driving forces are improved health situations and decreasing mortality which results in increasing average life span (72.8 years in 2029, 77.2 years in 2050), combined with slightly decreasing but still relatively high fertility (average 5 births per woman in 1950, 2.3 births per woman in 2021, 2.1 births per woman in 2025). As a large share of future global population growth is projected to occur in eight countries across Asia and Africa, including countries facing major development and food security challenges, ensuring resource availability and improving resource efficiency will be critical (UN, 2022).

Currently, more than one-third of the global land area (4.8 billion hectare) is used in agriculture (cropland, permanent meadows and pastures) to feed the existing 8.3 billion people and its animals (FAO, 2024). A growing population increases the demand for food (e.g. high-quality proteins and essential fatty acids) and puts pressure on the land-dependent agriculture for classical proteins from cereals (57% of global supply), meat (18%), dairy (10%), fish and shellfish (6%) and other animal products (9%) due to limited land available (Henchion et al., 2017). In addition, the production of proteins from conventional sources brings challenges such as high environmental footprints, resource-intensive production systems, decline in agricultural land, and growing health concerns linked to saturated fat intake. However, this paves the way towards alternatives such as proteins from insects, algae, other marine species, plants, food waste, single-cell proteins and proteins from fermentation technologies (microbial proteins), that may offer more sustainable and nutritionally comparable options. These ingredients may also provide bioactive lipids, such as polyunsaturated fatty acids (PUFAs), which are often limited in conventional protein sources. Although they are unlikely to fully replace traditional ingredients in all applications, they can complement or partially substitute them, enabling more balanced and diversified protein formulations..

The nutritional value of a feed ingredient is mainly defined by its proximate composition, essential amino acid (EAA) profile and essential fatty acid (EFA) profile. These parameters, among others, define the capacity of a feed to meet the protein and energy demand of animals. The origin, processing conditions and biochemical composition of plant-, animal-, and novel protein sources lead to significant variability among these parameters.

This chapter aims to compare nutritional parameters (proximate composition) across conventional and novel ingredients for livestock feed, aquaculture and pet food. It further goes into detail by highlighting the protein sources and their EAA profiles to provide a critical insight into the potential of each ingredient. By comparing nutritional profiles side by side and in detail, this chapter seeks to highlight both opportunities and limitations of novel ingredients. It provides a valuable resource for understanding the nutritional potential, limitations, and practical relevance of novel ingredients in sustainable feed production..

2 Conventional and novel protein sources

In general, for humans and animals, a nutrient is any compound or element that must be supplied through the diet. Nutrients usually have a defined physiological role for maintenance, growth, health and reproduction and include water, minerals, vitamins, specific amino acids, and fatty acids (FAO, 2001; McDonald et al. 2011). Macronutrients such as carbohydrates (e.g. sugars, starch, fiber), lipids, and proteins serve primarily as energy substrates. Proteins are primarily used for protein synthesis. While their use as an energy source is possible, it is extremely inefficient compared to other macronutrients. Micronutrients (minerals and vitamins) act as cofactors, electrolytes and structural modulators. However, nutrient requirements are not only expressed by macronutrient classes but also by subcomponents of the macronutrients: EAAs and EFAs. Essential amino acids (including histidine, methionine, and phenylalanine) and essential fatty acids (α-linolenic acid (ALA), eicosapentaenoic acid (EPA), and docosahexaenoic acid (DHA) among others) cannot be synthesized endogenously and must be supplied through a balanced diet (FAO, 2002; NRC, 2021).

Animal species such as cattle, pigs, poultry and fish (white fish and salmon) and crops such as soybean, are the benchmark of global protein production, since they provide high digestible proteins with a relatively good essential amino acid profile and contribute significantly to dietary lipid intake: saturated fatty acids (SAFs) from red meat, unsaturated fatty acids (UFAs) from soybean and polyunsaturated fatty acids (PUFAs; e.g. long-chain omega-3 PUFAs) from fish (Sauvant et al. 2004, Figure 3.1). The above-mentioned animal species are mainly kept and raised for human consumption.

After slaughtering and removal of edible portions, certain parts of the animals are not consumed or used as human food. The residual materials or by-products from slaughterhouses, such as heads, carcasses, bones, fishbones, viscera, feathers, skin, hooves and trimmings are collected and transported to rendering facilities. In the rendering process, the materials are thermally processed and converted into

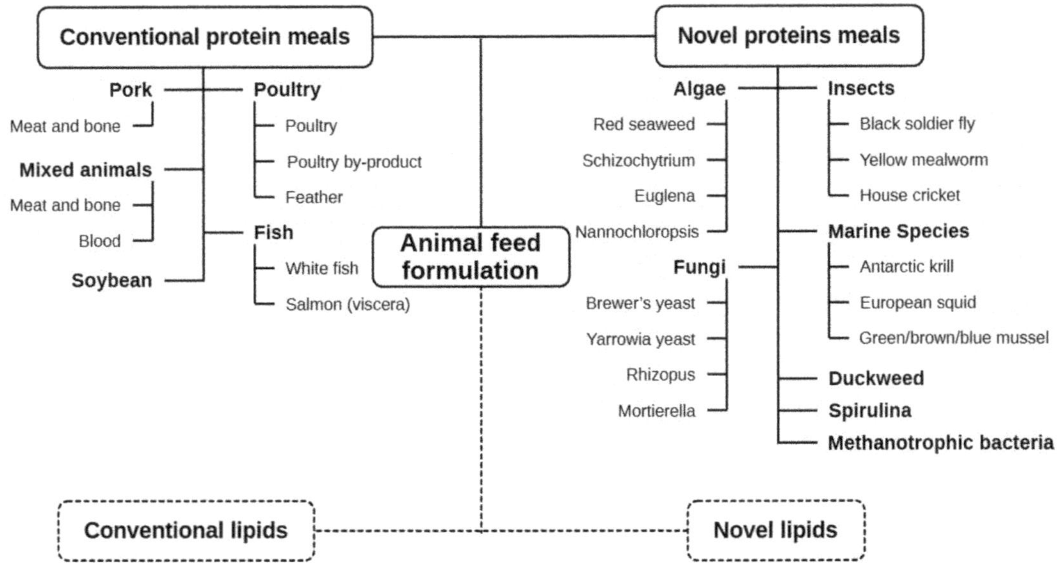

FIGURE 3.1 Overview of conventional and novel protein meals used in animal feed formulation, including animal-derived from pork, poultry, and fish, plant proteins, algae, insects, fungi, and other emerging ingredients.

high-value ingredients for feed and pet food applications. During the cooking step, heat treatment (120 °C–140 °C / 250 °F–280 °F) combined with moisture removal under pressure (up to 4 bar / 60 psi) separates the liquid (fat), solid (protein-rich), and aqueous fractions (Woodard and Curran, 2006). A subsequent press or centrifuge step separates fat from the protein fraction. After drying and grinding, the solid fraction can be processed into products such as meat and bone meal, feather meal, poultry by-product meal, or fish meal, depending on the type of animal by-products used. The recovered fat fraction can be utilized without further purification and, depending on its origin, may be marketed as tallow, lard, poultry fat, or fish oil (Ashilenje et al., 2026; Perez et al., 2010).

Rendering is recognized as a recycling and bio-conversion process. Animal by-products, that would otherwise end up as waste, are transformed into safe, stable and nutritionally rich ingredients. Since the animals were originally raised for human food production, rendered meals and fats are often considered to have a relatively low environmental impact. In contrast, the rendering industry makes a major contribution to the circular economy through the further processing of animal by-products. In addition, modern rendering meets strict biosafety, hygiene and traceability standards. It ensures that all pathogens are inactivated and the resulting meals are microbiologically safe to use (Pandey et al., 2020; Nath et al., 2023). Furthermore, rendering also reduces greenhouse gas emissions by diverting organic residues from landfill or incineration and making the recovered nutrients available to use again in agriculture and feed industry. Overall, the rendering industry is a cornerstone of the animal feed and pet food sectors because it converts by-products from meat, poultry, and fish processing into nutritionally valuable and microbiologically safe ingredients.

Over the last years, a growing number of novel protein sources gained attention (Figure 3.1, Table 3.1). Insects, algae, duckweed (aquatic plant), marine species, spirulina (cyanobacteria), camelina (oilseed crop), fungi, edible mushrooms and microbial proteins are among those, providing high protein content and unique amino acid profiles. In addition, it is claimed that their production brings environmental advantages using less water and land, and producing less greenhouse gas emissions compared to conventional sources (Henchion et al., 2017). In this chapter, insects will be represented by black soldier fly larvae (BSFL, *Hermetia illucens*), meal worm larvae (MWL, *Tenebrio molitor*), and house cricket (HC, *Acheta domesticus*). Algae will be divided into macroalgae, a multicellular organism that can grow to large sizes (red, brown, and green seaweed) and microalgae (phytoplankton), a small unicellular organism (*Schizochytrium* sp., *Euglena gracilis*, and *Nannochloropsis* sp.). Selected marine species are mussels (*Perna viridis, Modiolus modulaides*, and *Mytilus edulis*), krill (*Euphausia superba*), and squid (*Loligo vulgaris*). Within

TABLE 3.1 Organism considered as novel protein sources: common name, scientific name and classification in the chapter

Common name	Scientific name	Classification in chapter
Black soldier fly	*Hermetia illucens*	Insect
Yellow mealworm	*Tenebrio molitor*	Insect
House cricket	*Acheta domesticus*	Insect
Red seaweed	*Asparagopsis taxiformis*	Macroalgae
Schizochytrium	*Schizochytrium* sp.	Microalgae
Euglena	*Euglena gracilis*	Microalgae
Nannochloropsis	*Nannochloropsis* sp.	Microalgae
Duckweed	*Lemna minor*	Aquatic plant
Spirulina	*Arthrospira platensis*	Cyanobacteria (blue-green algae)
Methanotrophic bacteria	*Methylococcus capsulatus*	Bacteria (singlecell protein)
Antarctic krill	*Euphausia superba*	Marine species
European squid	*Loligo vulgaris*	Marine species
Asian green mussel	*Perna viridis*	Marine species
Brown mussel	*Modiolus modulaides*	Marine species
Blue mussel	*Mytilus edulis*	Marine species
Brewer's yeast	*Saccharomyces cerevisiae*	Yeast
Yarrowia yeast	*Yarrowia lipolytica*	Yeast
Rhizopus (filamentous fungus)	*Rhizopus oryzae*	Microfungus
Mortierella	*Mortierella alpina*	Microfungus

the fungal category, this chapter discusses single-cell proteins derived from yeasts such as *Saccharomyces cerevisiae* and *Yarrowia lipolytica*, as well as microfungi including *Rhizopus oryzae* and *Mortierella alpina* (Figure 3.1; Table 3.1). Macrofungi (edible mushrooms) are covered separately in a multispecies overview, alongside proteins derived from pulses.

Rendered meals are rich in highly digestible proteins and lipids, supplying essential amino acids, energy and minerals. Poultry by-product meal and fish meal typically contain protein levels between 60%–70%, with relative moderate lipid levels (10%–20%) and low fiber concentration. The ash content is often elevated (up to 30%) due to the residual bone and mineral fractions and thus a good source of highly digestible phosphorus. In contrast, soybean meals contain less crude protein (40%–50%) and relatively low levels of lipids but significantly higher levels of fiber (up to 10%). The rendered lipids or obtained from pressing seeds, are pure energy and concentrated essential fatty acid sources.

This chapter is all about exploring nutritional values of various conventional and novel sources, comparing and providing a way for fast references. More detailed information on further protein and lipid sources, their stability and respective oxidation, production, health benefits, and practical application might be found in the respective chapters within this book.

3 Nutrient composition

Tables 3.2 (conventional) and 3.3 (novel) summarize the major nutritional fractions including protein, fat, ash, fiber and moisture of selected animal- and plant-derived meals. In general, animal-derived meals contain higher protein concentrations and lower fiber levels compared to plant-based alternatives. The presented values are representative but in no way claim to be complete or to cover the entire spectrum of available sources, since nutritional values and proximate composition vary depending on factors such as the species itself, source animal's diet, and processing and storage of the raw materials (FAO, 2004). However, it can be concluded that the selected novel protein sources from insects, algae, other marine species, fungi and plant-based such as pulses and camelina contain overall protein ranges from <5% (specific mushroom) – 70% (cricket), covering the range of conventional protein meals (Table 3.3). The big difference can be explained by the different water content.

The meal's fat content varies from <1.0% (yeast) up to 50% in certain insect meals (black soldier fly) but certainly depends on species, developmental stage, the production and processing parameters and specially on the rearing substrate and defatting process (Fuso et al., 2021; Isnaini and Andri, 2025; Schiavone et al., 2017). The fat content in the selected conventional protein meals ranges from <1.0% (blood meal) up to 15% in poultry meal. The fiber fraction ranges from <1.0% in

TABLE 3.2 Nutrient composition (protein, lipid, ash, fiber and moisture) from conventional protein sources (pork, poultry, fish, and soybean)

Conventional sources	Protein [%]	Lipid [%]	Ash [%]	Fiber [%]	Moisture [%]
Pork					
Meat and bone meal[1]	56.7	10.6	26.4	-	9.7
Poultry					
Poultry meal[2,3]	26.9–50.2	13.9–15.5	6.6	10.1	8.2
Poultry by-product meal[4]	66.5	13.6	12.6	-	5.5
Feather meal[5] (n=123)	79.0	6.8	0.3	-	7.9
Mixed animals					
Meat and bone meal[#5] low fat (<7.5%, n=114) high fat (>7.5%, n=1230)	24.5–55.6	3.9 (low)–14.4 (high)	30.0–33.6	1.8–9.6	3.7–8.1
Blood meal[#5] (n=129)	89.7	0.8	2.8	-	5.7
Fish					
White fish meal[6,7,8]	53.0–71.9	4–10	5.3–25.0	<0.2	4.0–9.5
Salmon (viscera) meal[9]	74.0	11.0	8.5	-	7.0
Soybean meal[5,7,10]	40.0–54.9	1.1–3.5	5.4–7.0	3.0–7.0	9.8–12.4

Sources: [1]Bolarinwa et al., 2012, [2]Bednar et al., 2000, [3]Karapanagiotidis et al., 2018, [4]Yu et al., 2023a, [5]Sauvant et al., 2004, [6]alaska pollock FAO, 1997, [7]anchovy Cho and Kim, 2011, [8]tuna, sardine, and herring Lizárraga-Hernández et al., 2024, [9]Bayrakli and Yildiz, 2024, [10]Banaszkiewicz, 2011; [#]dried blood of slaughtered, warm-blooded animals; if no range is reported, it is one set of data from a specific meal resource or average in case of (n=xx); if no value is presented, it is either because there is not such nutrient in the source or the original reference did not mentioned.

spirulina up to 35% in camelina meals and are significantly higher compared to conventional animal or plant-based protein sources (≤10%, Table 3.2). Rather than being nutritionally disadvantageous, such fibers, particularly chitin, β-glucans, or complex polysaccharides, can act as functional nutrients and can bring additional benefits to diets. The ash content is highly specific to the raw materials used. For animal-based protein meals commonly used in feed, the more bones are used in the rendering process, the higher is the ash content. Ash refers to the total mineral residue left following complete combustion, consisting primarily of hydroxyapatite ($Ca_5(PO_4)_3OH$), a calcium phosphate mineral, together with magnesium and various trace elements incorporated into its crystal structure (Coutand et al., 2008). In novel protein sources, the inorganic ash fraction originates from structural or metabolic components. These include chitin-based exoskeletons in insects, intracellular or intercellular mineral deposits in algae and yeast, mineral-rich seed fractions in oilseed crops, and chitin- and glucan-based cell wall matrices of molds and edible mushrooms. Moisture of a meal is mainly the result of press settings and the drying time and can therefore vary significantly depending on the process. More detailed

TABLE 3.3 Nutrient composition (protein, lipid, ash, fiber and moisture) from novel protein sources (insects, algae, aquatic plant, marine species, camelina, fungi, and fermentation)

	Protein (%)	Lipid (%)	Ash (%)	Fiber (%)	Moisture (%)
Insects					
Black soldier fly larvae meal (*Hermetia illucens*)[1]	41.0–43.0	17.0–34.0	15.0–27.0	4.0–10.0	-
Mealworm larvae meal (*Tenebrio molitor*)[1]	48.0–57.0	32.0–40.0	2.0–4.0	2.0–5.0	-
House cricket larvae meal (*Acheta domesticus*)[1]	58.0–69.0	11.0–23.0	3.0–8.0	6.0–8.0	-
Marcoalgae					
Red seaweed meal[2] (*Asparagopsis taxiformis*)	13.3	2.3	58.3	22.0	-
Microalgae					
Spirulina meal[3] (*Arthrospira platensis*)	58.2	2.6	-	0.8	11.9
Euglena gracilis meal[2]	26.1	17.7	4.4	0.6	-
Aquatic plant					
Duckweed meal (*Lemna minor*)[4]	36.1	8.5	21.4	-	-
Marine species					
Green mussel (*Perna viridis*, n=10)*[5]	11.4	2.1	2.7	-	81.8
Brown mussel (*Modiolus modulaides* (n=10)*[5]	9.2	2.1	2.8	-	83.6
Blue mussel (*Mytilus edulis*)[6]					
Squid meal[7]	75.9	2.9	9.7	-	6.3
Shrimp hydrolysate[7]	63.3	9.0	14.5	-	3.8
Antarctic krill meal[8]	65.7	8.6	14.2	-	11.9
Plants					
Camelina meal[9] (solvent extraction)	35.7–43.0	1.2–5.8	5.6–5.9	18.1–37.5	6.0 –8.6
Camelina meal[9] (expeller pressed)	20.3–37.3	9.7–24.2	4.2–5.8	20.2–34.4	6.8–13.5
Pulses (7 species)[10]	14.5–46.1	0.5–11.9	2.2–9.3	1.2–11.9	2.7–22.9
Yeast					
Brewer's yeast meal (*Saccharomyces cerevisiae*, n=57)[11]	46.5	3.9	7.1	1.9	6.7
Rhizopus oryzae[6] (filamentous microfungus)					
Mushroom meal[12] (8 edible species)	4.5–37.4	1.0–4.3	4.3–9.7	-	-
Microbial proteins					
Methylococcus capsulatus[13] (methanotrophic gram-negative bacterium)	74.7	8.5	7.2	<0.1	-

Sources: [1]Valdés et al., 2022, [2]Ahmed et al., 2023, [3]Alvarenga et al., 2011, [4]Chakrabarti et al., 2018, [5]Nuñal et al., 2023, [6]Vidakovic et al. 2015, [7]Guilherme-Fernandes et al., 2023, [8]Yu et al., 2025, [9]Delver and Smith, 2024, [10]data were extracted from Chapter 16, based on INRAE-CIRAD-AFZ, 2025 and Government of Canada 2024 food composition databases, [11]Sauvant et al., 2004, [12]Valverde et al., 2015, [13]Longshaw et al., 2025; *data expressed on a wet weight basis; if no range is reported, it is one set of data from a specific meal resource or average in case of (n=xx); if no value is presented, it is either because there is not such nutrient in the source or the original reference did not mentioned.

information about the applications in animal nutrition may be found in the respective book chapters: insects (7; 13; 18), algae (9; 10; 20), camelina (11, 22), pulses (16), mushrooms (17), bacterial and microbial proteins from fermentation (8, 14, 19), underutilized marine species (15) and utilization of agricultural waste (21).

4 Protein sources and their amino acid profiles

While crude protein gives mainly information about the total nitrogen, the real biological value of a protein emerges when its amino acid composition is considered. The proportion and balance of especially essential amino acids (EAAs) are of high importance for diet and animal-specific feed. Amino acids (AAs) are considered essential when a species cannot synthesize them. Livestock, companion animals and fish require the same ten EAAs as humans: arginine, histidine, isoleucine, leucine, lysine, methionine (cystine), phenylalanine (tyrosine), threonine, tryptophan, and valine. Moreover, cats, because of their carnivorous metabolism, have additional requirements for taurine. Taurine is not an amino acid but a non-proteinogenic amino sulfonic acid and mentioned because of its importance for cats. A deficiency can cause dilated cardiomyopathy and central retinal degeneration (Hayes et al., 1975; Pion et al., 1992). As a result, if not supplied separately, cats must have some animal-based protein sources in their diet because plant-based diets do not contain taurine. Alternatively, taurine can also be added as a supplement, though this is not common practice. Pigs have a high demand for lysine, methionine, threonine, tryptophan, and valine. Corn, the primary grain in swine diets, is markedly deficient in lysine and tryptophan. In addition, lysine is susceptible to heat during drying or feed processing (e.g. pelleting) and therefore often the first limiting amino acid in swine diets (NRC, 2012). Dairy cattle, especially during early lactation, have a high demand for lysine and methionine, which are also the first limiting EAAs in typical dairy cattle diets (NRC, 2021). Poultry have a relatively high requirement for cysteine and methionine (Babazadeh and Ahmadi Simab, 2022). Furthermore, carnivorous fish such as salmonids have a relatively high need for arginine, lysine, and methionine (Wilson, 1986; Yu et al., 2023b). Table 3.4 shows the AA requirements in g/100 g protein for livestock (cattle, swine, poultry), pets (dog, cat) and aquaculture (e.g. salmonids). The presented values are representative but in no way claim to be complete or to cover the entire spectrum, since the amino acid requirements vary with age, life stage, and health status of animals as well as activity level, temperature and protein quality of the diet. All presented mammals (livestock, dogs, cats) seem to have comparable EAA requirements in g/100 g protein. In contrast, the requirements for salmonids seem to be lower compared to those of the mammals.

Knowing the essential amino acid requirements for specific animals is one part but selecting the right ingredients able to provide the required EAAs for the for-

TABLE 3.4 Essential amino acid requirements for livestock, dogs, cats and salmonids (g/100 g protein)

Amino acid	Livestock[1]	Dogs[2]	Cats[2]	Salmonids[3]*
Arginine (Arg)	4.0–6.0	3.50	4.85#	3.0–6.5
Histidine (His)	1.5–2.0	1.85	1.65#	1.0–2.0
Isoleucine (Ile)	4.0–5.0	3.75	2.70#	1.0–3.0
Leucine (Leu)	7.0–8.0	6.75	6.40#	3.0–5.0
Lysine (Lys)	6.0–7.0	3.50	1.70	3.5–6.0
Methionine (Met)	2.0–3.0	3.25	0.85	2.0–4.0
Met + Cystine (Cys)	3.5–4.5	6.50	1.70	2.0–5.0
Phenylalanine (Phe)	4.0–5.0	4.50	2.50#	-
Phe + Tyrosine (Tyr)	6.0–7.0	7.40	9.60#	4.0–6.5
Threonine (Thr)	4.0–5.0	4.25	3.25#	2.0–3.5
Tryptophan (Trp)	1.0–1.5	1.40	0.85#	0.5–1.5
Valine (Val)	5.0–6.0	4.90	3.20#	2.0–4.0
Taurine	-	-	0.05–0.1% diet	-

Sources: [1]NRC, 2012 and NRC, 2021; [2]NRC, 2006; [3]Lall and Anderson, 2005; *further fish in FAO, 2025a and NRC, 2011. #for cats adequate intake (#) replace minimum requirement where data were not available.

mulation of nutritionally adequate diets can be challenging. Feedstuffs differ considerably in their amino acid composition. Factors such as the origin of the animal species, the tissue type, and the functionality of the proteins, however, influence the amino acid requirements for protein synthesis. Furthermore, external factors such as processing methods, heat exposure, and storage conditions, as well as the rearing substrate in the case of insects, affect the amino acid profile.

The concentrations of essential amino acids in selected conventional protein meals from soybeans, feathers, fish, meat and bones, poultry by-products, blood, and pork are summarized in Table 3.5. Table 3.6 shows the amino acid composition of novel protein meals from insects, algae, other marine species, plant-based, and fungi. Plant-derived protein sources such as soybean meal provide favorable lysine (Lys: 2.7–2.9 g/100 g protein) concentrations but a limited amount of sulfur-containing amino acids methionine and cystine (Met: 0.6 g/100 g protein; Cys: 0.7 g/100 g protein; Table 3.5). In contrast, animal-derived protein sources contain a relatively balanced EAA profile, and a higher digestibility compared to plant-based proteins (Berrazaga et al., 2019). Plant-derived meals often contain antinutritional factors such as phytates, protease inhibitors, tannins, oligosaccharides, glucosinolates and other secondary metabolites (alkaloids) (Salim et al., 2023). The most complete EAA profile is usually found in fish meals containing high concentrations of Lys and valine (Val). Poultry by-product meal is also a very valuable protein source considering the EAA quality and quantity. However, the ingredients used to produce a by-product meal can vary significantly and therefore the full amino acid profile. Novel protein sources (insects, algae, camelina, fungi, microbial proteins) are promising candidates to complement or even replace conventional plant- and animal-based protein sources. Because of numerous pulse and mushroom species, their EAA profiles are not discussed here but can be found in detail in the respective chapters (16 and 17). Camelina meal for example can exceed the amount of Met but has slightly lower concentrations of Lys compared to soybean meal (Met: max. 1.1 g/100 g protein camelina meal vs. 0.6 g/100 g protein soybean meal; Lys: max. 2.1 g/100 g protein camelina meal vs. max. 2.9 g/100 g protein soybean meal; Table 3.5 and 3.6). Insect meal is gaining attention as alternative to conventional animal-based protein sources because of its high concentration of Lys and Val (Lys: max. 8.4 g/100 g protein insect meal vs. max. 5.3 g/100 g protein fish meal vs. max 6.4 g/100 g protein poultry by-product meal; Val: max. 6.8 g/100 g protein insect meal vs. max. 3.5 g/100 g protein fish meal vs. max. 4.5 g/100 g protein poultry by-product

TABLE 3.5 Essential amino acid composition of selected conventional protein sources (g/100 g protein)

	Arg	His	Ile	Leu	Lys	Met	Met + Cys	Phe	Phe + Tyr	Thr	Trp	Val
Soybean meal												
46% protein (n=316)	3.2	1.2	2.0	3.2	2.7	0.6	1.3	2.2	3.6	1.7	0.6	2.1
48% protein (n=10409)	3.4	1.2	2.1	3.3	2.8	0.6	1.3	2.3	3.8	1.8	0.6	2.2
50% protein (n=1453)	3.5	1.3	2.2	3.5	2.9	0.7	1.4	2.4	3.9	1.8	0.6	2.3
Pork												
Meat and bone meal[2]	3.6	1.3	2.0	3.8	3.3	1.0	1.6	2.1	3.5	1.9	0.5	2.7
Poultry												
Feather meal (n=123)	5.7	0.7	4.2	6.8	1.8	6.0	4.3	4.1	6.3	4.0	0.5	6.2
By-product meal[1]	5.6	2.4	4.2	5.7	6.4	2.4	3.4	4.3	8.1	3.7	-	4.5
Fish meal*												
62% protein (n=297)	3.8	1.6	2.5	4.5	4.6	1.6	2.2	2.4	4.3	2.6	0.6	3.0
65% protein (n=1198)	3.9	1.6	2.7	4.7	4.9	1.8	2.3	2.6	4.5	2.7	0.7	3.2
70% protein (n=431)	4.1	1.7	3.0	5.0	5.3	2.0	2.6	2.7	4.9	2.9	0.7	3.5
Meat and bone meal#												
<7.5% fat (n=114)	3.8	1.1	1.6	3.4	2.8	0.7	1.4	1.9	3.2	1.9	0.4	2.5
>7.5% fat (n=1230)	3.6	1.0	1.5	3.1	2.6	0.7	1.2	1.7	2.9	1.7	0.3	2.3

Sources: Sauvant et al. 2004, [1]Bolarinwa et al., 2012, and [2]Yu et al., 2023a; *no specification on fish species; #derived from slaughtered, warm-blooded animals

TABLE 3.6 Essential amino acid profiles of selected novel protein sources (g/100 g protein)

Amino acids (AA)	Arg	His	Ile	Leu	Lys	Met	+ Cys	Phe	+ Tyr	Thr	Trp	Val
Insects												
Black soldier fly larvae[1] (n=6)	2.1	1.3	1.9	3.0	2.6	0.7	1.0	1.7	4.4	1.9	0.6	2.8
Mealworm larvae[2] (n=40)	5.3	2.9	4.5	7.8	8.4	-	-	3.4	9.0	4.9	-	6.8
House cricket larvae[3] (n=10)	8.4	2.6	4.1	7.4	5.6	1.6	2.6	3.2	8.7	3.7	0.8	5.5
Macroalgae												
Red seaweed[4] (n=3)	18.4	1.2	3.2	6.2	4.8	2.0	2.3	3.6	5.9	3.4	2.2	6.7
Brown seaweed[4] (n=2)	3.9	2.0	2.8	5.3	4.0	1.8	2.9	3.5	5.0	3.4	0.8	4.1
Green seaweed[4] (n=2)	11.8	3.1	3.5	6.8	4.4	2.1	3.3	5.5	8.6	5.0	0.3	5.1
Microalgae												
Euglena gracilis[5]	7.0	2.7	4.2	8.7	7.3	2.3	3.9	4.8	9.0	5.0	1.9	6.7
Spirulina[6]	4.0	1.0	3.1	4.8	4.8	2.0	2.7	2.5	5.1	2.8	-	3.3
Aquatic plant												
Lemna minor[7] (Duckweed)	3.1	0.9	2.0	4.1	2.7	0.9	1.3	2.6	4.5	1.9	0.4	2.7
Other marine species												
Green mussel[8] (n=10)	0.5	2.7	3.4	7.9	0.2	2.9	3.9	2.7	5-8	4.2	0.4	3.9
Brown mussel[8] (n=10)	14.4	2.0	2.3	5.5	5.7	2.3	2.3	2.2	4.5	3.1	0.4	3.3
Blue mussel[9]	5.6	1.6	3.4	5.1	5.9	2.9	5.1	2.8	6.1	3.3	-	3.8
Squid meal[10]	7.2	1.7	3.2	5.6	6.7	3.7	4.2	3.4	10.4	3.4	-	4.1
Shrimp hydrolysate[10]	4.8	1.3	2.7	4.6	5.9	2.5	2.9	3.7	8.2	2.7	-	4.6
Antarctic krill meal[11]	5.6	2.0	4.0	6.8	5.6	4.9	5.6	4.8	8.6	3.5	-	4.2
Fungi												
Saccharomyces cerevisiae[12] (yeast)	4.7	4.2	3.7	6.3	6.5	1.4	2.3	3.3	6.8	4.8	0.9	5.3
Yarrowia lipolytica[12] (yeast)	4.8	2.6	4.4	6.8	7.0	1.2	2.3	4.0	15.3	4.8	4.7	5.3
Rhizopus orzyae[9] (microfungus)	2.2	1.9	3.5	4.6	5.5	0.5	0.7	2.9	4.5	1.7	-	4.3
Oilcrop seed												
Camelina[13] solv. extracted (n=10)	3.8	1.9	2.1	3.1	2.1	1.1	2.0	2.3	3.9	1.4	0.6	2.8
Camelina[13] expelled (n=3)	3.0	0.9	1.3	2.3	1.7	0.6	1.2	1.5	2.5	1.4	0.5	1.8
Microbial protein												
Methylococcus capsulatus[14]	3.9	1.3	2.7	4.6	3.4	1.6	2.0	2.5	4.1	2.6	0.8	3.5

Amino acids from left to right: arginine, histidine, isoleucine, leucine, lysine, methionine, cystine, phenylalanine, tyrosine, threonine, tryptophan, valine. Sources: [1]Lu et al., 2022, [2]Kröncke et al., 2023, [3]Oonincx and Finke, 2020, [4]Fleurence, 2004, [5]Ahmed et al., 2023, [6]Alvarenga et al., 2011; [7]Chakrabarti et al., 2018; [8]Nuñal et al., 2023, [9]Vidakovic et al., 2015, [10]Guilherme-Fernandes et al., 2023, [11]Yu et al., 2025; [12]Jach et al., 2022; [13]Delver and Smith, 2024; [14]Longshaw et al., 2025 meal). Please remember that data collected in Tables 3.5 and 3.6 are selected samples and do not represent the whole range of meals available. Therefore, the result of comparing different meals can be significantly different by choosing other meals. Algae meals including microalgae (e.g. spirulina) and macroalgae (e.g. seaweed) represent an interesting category of novel protein sources resembling animal-based EAA compositions (high in Lys: max. 4.8 g/100 g protein; Val: max. 6.7 g/100 g protein, Table 3.6), while lacking most of the classical antinutritional factors typically associated with plant-based protein sources (Bleakley and Hayes 2017). Single-cell proteins from yeast, as well as mushroom proteins can show very consistent EAA profiles and high digestibility, which makes them a good alternative for specialized feed. However, the inclusion levels of novel proteins might be constrained by cost (typically higher than conventional plant- or animal-based proteins) and regulatory situations.

Both conventional and novel protein sources can contribute to meeting the essential amino acid (EAA) requirements of livestock, companion animals, and aquaculture species, depending on their inclusion level, amino acid profile, digestibility, and the composition of the final diet. The greater challenge lies in selecting the right ingredients to ensure nutrient quality while maintaining compliance with the relevant regulatory framework. Nutrient quality can be strongly influenced by formulation choices and processing conditions, including temperature, pressure, and moisture during operations such as cold pressing, pelleting, extrusion, spray drying, or wet pet food processing. In particular, heat and pressure may promote undesirable Maillard reactions between sugars and amino acids, thereby reducing amino acid availability. In practice, such limitations are often addressed through supplementation with commercially available amino acids. Beyond amino acid supply, the nutritional value of protein ingredients is closely linked to their associated lipid fractions, which contribute energy and essential fatty acids but are also susceptible to degradation during processing and storage (Chapter 4; Moughan and Rutherfurd, 2008; Rao and Artz, 1989).

5 Conclusion

By comparing the nutritional potential of novel proteins with conventional sources, the chapter clearly highlights the growing relevance of novel proteins in feed and pet food. From promising applications to reduce methane emissions in livestock using algae toward already established formulations with insect ingredients in pet food or in aquaculture, the future feed will most likely be looking different. Conventional animal meals such as fish meal and poultry by-product meal are likely to remain as highly digestible and amino acid-balanced protein sources. However, their availability is constrained by increasing competition with human food and by price volatility, which justifies the growing exploration of novel proteins. Soybean meal, the most important plant-based protein source, is large-scale supplied but contains antinutritional factors and has negative environmental connotations related to deforestation, water pollution, monoculture and soil degradation (Barona et al., 2010; FAO, 2018; Gibbs et al., 2010; Voora et al., 2020).

Insects, algae, marine species, fungi, and ingredients from fermentation can bridge the gap between nutritional quality and sustainability. Insect meals offer a high protein concentration with favorable lysine (Lys) and valine (Val) contents. Macro- and microalgae as plant sources provide both high EAA concentrations in leucine, Lys, and Val, and long chain polyunsaturated fatty acids such as EPA and DHA. Yeast and filamentous fungi provide highly digestible single-cell proteins but vary in their amino acid profiles. Knowledge on how the essential nutrient profiles can be manipulated via selecting growth media and process optimization gives an advantage toward ingredients serving species-specific requirements. These single-cell proteins have the potential to become the most versatile ingredients in the toolbox for tailored nutritional design.

The nutritional quality of both conventional and novel protein sources is influenced by origin, processing conditions and inclusion levels, resulting in considerable variability in crude protein content and essential amino acid composition. While many novel proteins demonstrate favorable amino acid profiles comparable to animal-derived meals, their practical use may be constrained by cost, regulatory approval and processing-related effects on nutrient availability.

The stability of proteins is highly sensitive to processing and storage conditions. Managing oxidation through understanding of ingredient compositions, control of manufacturing parameters, inclusion of suitable antioxidants and selection of protective packaging is essential to maintain the nutritional value throughout the feed value chain up to the final feed product (Chapter 6).

While conventional protein sources will likely continue to play a central role in livestock, companion animal and aquaculture nutrition, emerging protein alternatives are gaining relevance and may increasingly complement, or in some cases partially replace, traditional ingredients as their functionality, availability and scalability improve. The goal must be a stepwise shift from utilizing novel proteins as complementary ingredients toward their integration as core components of modern feed systems. However, more work needs to be done before these meals from insects, algae, fungi, fermentation and oilseeds are a permanent part of the toolbox from which nutritionists and feed producers can choose. Starting with reducing barriers in consumer perception for the inclusion of novel ingredients and ingredient availability as well as standardized quality on a global scale up to legal regulations for their safe use (EFSA, 2022; Lähteenmäki-Uutela et al., 2021; Sogari et al., 2023; VanRaamsdonk et al., 2023).

Overall, strategic combinations of conventional and emerging protein sources allow feed formulators to meet species-specific amino acid requirements while improving sustainability and resource efficiency. For the formulation of nutritionally balanced diets, protein quality must be considered alongside the energy and lipid contributions of the ingredients.

References

Ahmed, E.; Suzuki, K.; Nishida, T. 2023. Micro- and Macro-Algae Combination as a Novel Alternative Ruminant Feed with Methane-Mitigation Potential. *Animals* 13:796. https://doi.org/10.3390/ani13050796

Alvarenga, R. R.; Rodrigues, P. B.; de Souza Cantarelli, V.; Zangeronimo, V. G.; da Silva Júnior, J. W.; da Silva, L. R.; dos Santos, L. M.; Pereira, L. J. 2011. Energy values and chemical composition of spirulina (*Spirulina platensis*) evaluated with broilers. *Revista Brasileira de Zootecnia (R. Bras. Zootec.)* 40:992-996. https://doi.org/10.1590/S1516-35982011000500008

Ashilenje, D.; Ashour, F.; Barz, M.; Belandria, V.; Borello, A.; Bostyn, S.; Boushaki, T.; Branciari, R.; Bwapwa, J. K.; Cerza, E.; Dell'Orto, A.; El Bari, H.; Govender, M.; Habchi, S.; Karouach, F.; Khashaba, N. H.; Kouisni, L.; Marconi, O.; Marrocchi, A.; Saber, M.; Sarh, B.; Trois, C.; Willenbacher, M.; Wohlgemuth, V. 2026. A literature review of slaughterhouse waste valorisation: Techniques, environmental, and economic implications. *Resources, Conservation and Recycling* 224:108571. https://doi.org/10.1016/j.resconrec.2025.108571

Babazadeh, D.; Ahmadi Simab, P. 2022 Methionine in Poultry Nutrition: A Review. *Journal of World's Poultry Science (J. World's Poult. Sci.)* 1: 1-11. http://doi.org/10.58803/jwps.v1i1.1

Banaszkiewicz, T. 2011. Nutritional value of soybean meal. *Soybean and Nutrition*, 12:pp.1-20. https://doi.org/10.5772/23306

Barona, E.; Ramankutty, N.; Hyman, G.; Coomes, O. T. 2010. The role of pasture and soybean in deforestation of the Brazilian Amazon. *Environmental Research Letters (Environ. Res. Lett.)* 5:024002. https://doi.org/10.1088/1748-9326/5/2/024002

Bayrakli, B.; Yildiz, S. 2024. Comparative Analysis of Nutritional Values of Fishmeals Produced From Whole Anchovy and Sprat and Farmed Salmon Viscera in the Black Sea Region. *Acta Natura Et Scientia (Act. Nat. Sci.)* 5:150-159. https://doi.org/10.61326/actanatsci.v5i2.299

Bednar, G. E.; Murray, S. M.; Patil, A. R.; Flickinger, E. A.; Merchen, N. R.; Fahey Jr., G. C. 2000, Selected animal and plant protein sources affect nutrient digestibility and fecal characteristics of ileally cannulater dogs. *Archiv für Tierernährung (Arch. Tierernahr.* 53: 127-140. https://doi.org/10.1080/17450390009381942

Berrazaga, I.; Micard, V.; Gueugneau, M.; Walrande, S. 2019. The Role of the Anabolic Properties of Plant- versus Animal-Based Protein Sources in Supporting Muscle Mass Maintenance: A Critical Review. *Nutrients* 11:1825. https://doi.org/10.3390/nu11081825

Bleakley, S.; Hayes, M. 2017. Algal Proteins: Extraction, Application, and Challenges Concerninc Production. *Foods* 6:33. https://doi.org/10.3390/foods6050033

Bolarinwa, O. A.; Olukosi, O. A.; Adeola, O. 2012. Metabilozable energy value of porcine meat and bone meal for broiler chickens. *Canadian Journal of Animal Science (Can. J. Anim. Sci.)* 92: 73-78. https://doi.org/10.4141/cjas2011-103

Chakrabarti, R.; Clark, W.D.; Sharma, J.G.; Goswami, R.K.; Shrivastav, A.K.; Tocher, D.R. 2018. Mass Production of Lemna minor and Its Amino Acid and Fatty Acid Profiles. *Frontiers in Chemistry (Front. Chem.)* 6:479. https://doi.org/10.3389/fchem.2018.00479

Cho, J. H.; Kim, I. H. 2011. Fish meal – nutritive value. *Journal of Animal Physiology and Animal Nutrition (J. Anim. Physiol. Anim. Nutr.*) 95: 685-692. https://doi.org/10.1111/j.1439-0396.2010.01109.x

Coutand, M.; Cyr, M.; Deydier, E.; Guilet, R.; Clastres, P. 2008. Characteristics of Industrial and Laboratory meat and bone meal ashes and their potential applications. *Journal of Hazardous Materials (J. Haz. Mat.*) 11: 522-532. https://doi.org/10.1016/j.jhazmat.2007.04.133

Delver, J. J.; Smith, Z. K. 2024. Opportunities for Camelina Meal as a Livestock Feed Ingredient. *Agriculture* 14: 116. https://doi.org/10.3390/agriculture14010116

European Food Safety Authority (EFSA). James, K.; Millington, A.; Randall. N. 2022. Food and feed safety vulnerabilities in the circular economy. *EFSA Supporting Publications* 19:EN-7226.112. https://doi.org/10.2903/sp.efsa.2022.EN-7226

Food and Agriculture Organization of the United Nations (FAO). 1997. Feed ingredients and fertilizers for farmed aquatic animals – Sources and composition. FAO Fisheries Technical Paper No. 375/2. Available at: https://www.fao.org/4/i1142e/i1142e04.pdf

Food and Agriculture Organization of the United Nations (FAO). 2001. Human vitamin and mineral requirements. Report of a Joint FAO/WHO Expert Consultation, Bankok. Food and Nutrition Division, FAO, Rome. Available at: https://www.fao.org/4/y2809e/y2809e00.htm

Food and Agriculture Organization of the United Nations (FAO). 2002. Food energy – methods of analysis and conversion factors. FAO Food and Nutrition Paper No. 77, Report of a Workshop, Rome, 3–6 December 2002. Rome. ISBN 92-5-105014-7. Available at: https://www.fao.org/4/y5022e/y5022e00.htm

Food and Agriculture Organization of the United Nations (FAO). 2004. Animal Production and Health. Assessing Quality and Safety of Animal Feeds. Volume 160. Rome. ISBN: 92-5-105046-5. Available at: https://openknowledge.fao.org/server/api/core/bitstreams/437d6eb2-0ced-4a55-a98b-e1efb9047115/content

Food and Agriculture Organization of the United Nations (FAO). 2018. Mateo-Sagasta, J.; Zadeh, S. M.; Turral, H. More people, more food, worse water? A global review of water pollution from agriculture. Available at: https://openknowledge.fao.org/server/api/core/bitstreams/686ea465-7847-428e-b599-b236f2240e47/content

Food and Agriculture Organization of the United Nations (FAO). 2024. Land statisics 2001–2022 – Global, regional and country trends. FAOSTAT Analytical Briefs, No 88. Rome. https://doi.org/10.4060/cd1484en

Fleurence, J. 2004. Seaweed Proteins. Chapter 9 in Proteins in Food Processing. *Woodhead Publishing Series in Food Science, Technology and Nutrition*, 197pp. https://doi.org/10.1533/9781855738379.1.197

Fuso, A.; Barbi, S.; Macavei, L. I.; Luparelli, A. V.; Maistrello, L.; Montorsi, M.; Sforza, S.; Caligiani,

A. 2021. Effect of the Rearing Substrate on Total Protein and Amino Acid Composition in Black Soldier Fly. *Foods* 10:1773. https://doi.org/10.3390/foods10081773

Government of Canada, 2024. Canadian Nutrient File (CNF), Government of Canada.

Guilherme-Fernandes, J.; Aires, T.; Fonseca, A. J. M.; Yergaliyev, T.; Camarinha-Silva, A.; Lima, S. A. C.; Maia, M. R. G.; Cabrita, A. R. J. 2023. Squid meal and shrimp hydrolysate as novel protein source for dog food. *Frontiers in Veterinary Science (Front. Vet. Sci.)* 11:1360939. https://doi.org/10.3389/fvets.2024.1360939

Hayes, K. C.; Carey, R. E.; Schmidt, S. Y. 1975. Retinal degeneration associated with taurine deficiency in the cat. *Science* 188:949-51. https://doi.org/10.1126/science.1138364

Henchion, M.; Hayes, M.; Mullen, A. M.; Fenelon, M.; Tiwari, B. 2017. Future Protein Supply and Demand: Strategies and Factors Influencing a Sustainable Equilibrium. *Foods*, 6:53. https://doi.org/10.3390/foods6070053

INRAE-CIRAD-AFZ, 2025. INRAE-CIRAD-AFZ Feed tables, INRAE-CIRAD-AFZ

Isnaini, N.; Andri, F. 2025. Amino Acid and Fatty Acid Profiles of Black Soldier Fly Larvae Meal Reared on Rice Bran, Milk Waste, and Slaughterhouse Waste Combined Substrate Under Different Processing Methods. *F1000 Research* 14:676. https://doi.org/10.12688/f1000research.163839.1

Jach, M.E.; Serefko, A.; Ziaja, M.; Kieliszek, M. 2022. Yeast Protein as an Easily Accessible Food Source. *Metabolites*, 12: 63. https://doi.org/10.3390/metabo12010063

Karapanagiotidis, I. T.; Psofakis, P.; Mente, E.; Malandrakis, E.; Golomazou, E. 2018. Effect of fishmeal replacement by poultry by-product meal on growth performance, proximate composition, digestive enzyme activity, haematological parameters and gene expression of gilthead seabream (*Sparus aurata*). *Aquaculture Nutrition (Aquacult. Nutr.)* 00:1-12. https://doi.org/10.1111/anu.12824

Kröncke, N.; Wittke, S.; Steinmann, N.; Benning, R. 2023. Analysis of the Composition of Different Instars of *Tenebrio molitor* Larvae using Near-Infrared Reflectance Spectroscopy for Prediction of Amino and Fatty Acid Content. *Insects* 14: 310. https://doi.org/10.3390/insects14040310

Lall, S. P.; Anderson, S. 2005. Amino acid nutrition of salmonids: Dietary requirements and bioavailability. In: *Mediterranean fish nutrition*. Zaragoza : CIHEAM, p. 73-90 (Cahiers Options Méditerranéennes; n. 63). Available at https://om.ciheam.org/om/pdf/c63/05600068.pdf

Lähteenmäki-Uutela, A.; Marimuthu, S. B.; Meijer, N. 2021. Regulations on insect as food and feed: a global comparison. *Journal of Insects as Food and Feed (J. Insect. Food Feed)* 7:849-856. https://doi.org/10.3920/JIFF2020.0066

Lizárraga-Hernández, F. D.; Soberanes-Yepiz, M. L.; Cortés-Jacinto, E.; Rangel-Dávalos, C.; Cientuegos-Martínez, K. 2024. Comparative analysis of the chemical quality of fishmeal produced on the Northwest coast of Mexico. *Agro Productividad,* 17: 177-184. https://doi.org/10.32854/agrop.v17i4.2870

Longshaw, M.; Quest, B.; Miller, W.; Oba, P. M.; Swanson, O. R.; Swanson, K. S.; Miller, K. 2025. The Safety of FeedKind Pet® (Methylococcus capsulatus, Bath) as a Cultured Protein

Source in the Diet of Adult Dogs and Its Effect on Feed Digestibility, Fecal Microbiome, and Health Status. *Animals*, 15:1975. https://doi.org/10.3390/ani15131975

Lu, S.; Taethaisong, N.; Meethip, W.; Surakhunthod, J.; Sinpru, B.; Sroichak, T.; Archa, P.; Thongpea, S.; Paengkoum, S.; Purba, R. A. P.; Paengkoum, P. 2022. Nutritional Composition of Black Soldier Fly Larvae (Hermetia illucens L.) and Its Potential Uses as Alternative Protein Sources in Animal Diets: A Review. *Insects*, 13:831. https://doi.org/10.3390/insects13090831

McDonald, P.; Edwards, R. A.; Greenhalgh, J. F. D.; Morgan, C. A.; Sinclair, L. A.; Wilkinson, R. G. 2011. *Animal Nutrition*, Seventh Edition. Pearson Education. Available at: https://eliasnutri.wordpress.com/wp-content/uploads/2020/07/animal-nutrition-7th-edition.pdf

Moughan, P. J.; Rutherfurd, S. M. 2008. Available Lysine in Foods: A Brief Historical Overview. *Journal of AOAC International (J. AOAC Int.),* 91:901-906. https://doi.org/10.1093/jaoac/91.4.901

Nath, P. C.; Ojha, A.; Debnath, S.; Sharma, M.; Nayak, P. K.; Sridhar, K.; Inbaraj, B. S. 2023. Valorization of Food Waste as Animal Feed: A Step towards Sustainable Food Waste Management and Circular Bioeconomy. *Animals*, 13:1366. https://doi.org/10.3390/ani13081366

National Research Council (NRC). 2006. Nutrient Requirements of Dogs and Cats. Washington, DC. The National Academies Press. https://doi.org/10.17226/10668

National Research Council (NRC). 2011. Nutrient Requirements of Fish and Shrimp. Washington, DC: The National Academies Press. https://doi.org/10.17226/13039

National Research Council (NRC). 2012. *Nutrient Requirements of Swine: Eleventh Revised Edition*. Washington, DC: The National Academies Press. https://doi.org/10.17226/13298

National Research Council (NRC). 2021. *Nutrient Requirements of Dairy Cattle: Eighth Revised Edition*. Washington, DC: The National Academies Press. https://doi.org/10.17226/25806

Nuñal, S. N.; Parcon, R. B.; Logronio, S. B.; Cartago, N. H. N.; Muegue, M. F. S. 2023. Nutritional Evaluation of Green Mussel Perna viridis (Linnaeus, 1758) and Brown Mussel Modiolus modulaides (Röding, 1798) From Panay Island, Philippines. *Asian Fisheries Science (Asian Fish. Sci.),* 36:152-163. https://doi.org/10.33997/j.afs.2023.36.3.004

Oonincx, D. G. A. B.; Finke, M. D. 2020. Nutritional value of insects and ways to manipulate their composition. *Journal of Insects as Food and Feed (J. Insects Food Feed*), 7:639-660. https://doi.org/10.3920/JIFF2020.0050

Pandey, P.; Vidyarthi, S. K.; Vaddella, V.; Venkitasamy, C.; Pitesky, M.; Weimer, B.; Pires, A. F. A. 2020. Improving Biosecurity Procedures to Minimize the Risk of Spreading Pathogenic Infections Agents After Carcass Recycling. *Frontiers in Microbiology (Front. Microbiol.*), 11:623. https://doi.org/10.3389/fmicb.2020.00623

Peréz-Calvo, E.; Castrillo, C.; Baucells, M. D.; Guada, J. A. 2010, Effect of rendering on protein and fat quality of animal by-products. *Journal of Animal Physiology and Animal Nutrition (J. Anim. Physiol. Anim. Nutr.*), 94:.154-e163. https://doi.org/10.1111/j.1439-0396.2010.00998.x

Pion, P. D.; Kittleson, M. D.; Thomas, W. P.; Skiles, M.L.; Rogers, Q. R. 1992. Clinical findings in cats with dilated cardiomyopathy and relationship of findings to taurine deficiency. *Journal of the American Veterinary Medical Association (J. Am. Vet. Med. Assoc.*), 201:267-74. https://doi.org/10.2460/javma.1992.201.02.267

Rao, S. K.; Artz, W. E. 1989. Effect of Extrusion on Lipid Oxidation. *Journal of Food Science (J. Food Sci.*), 54:1580-1583. https://doi.org/10.1111/j.1365-2621.1989.tb05164.x

Salim, R.; Nehvi, I. B.; Mir, R. A.; Tyagi, A.; Ali, S.; Bhat, O. M. 2023. A review on anti-nutritional factors: unraveling the natural gateways to human health. *Frontiers in Nutrition (Front. Nutr.*), 10:1215873. https://doi.org/10.3389/fnut.2023.1215873

Sauvant, D.; Perez, J.-M. and Tran G. 2004. Tables of Composition and Nutritional Values of Feed Animals. Wageningen Academic Publishers, Wageningen, the Netherlands, 192-284. https://doi.org/10.3920/978-90-8686-668-7. (English transation of the French original work)

Schiavone, A.; De Marco, M.; Martínez, S.; Dabbou, S.; Renna, M.; Madrid, J.; Hernandez, F.; Rotolo, L.; Costa, P.; Gai, F.; Gasco, L. 2017. Nutritional value of a partially defatted and a highly defatted black soldier fly larvae (*Hermetia illucens* L.) meal for broiler chickens: apparent nutrient digestibility, apparent metabolizable energy and apparent ileal amino acid digestibility. *Journal of Animal Science and Biotechnology (J. Anim. Sci. Biotechnol.*), 8:51. https://doi.org/10.1186/s40104-017-0181-5

Sogari, G.; Bellezza Oddon, S.; Gasco, L.; van Huis, A.; Spranghers, T.; Mancini, S. 2023. Review: Recent advances in insect-based feeds: from animal farming to the acceptance of consumers and stakeholders. *Animal (The International Journal of Animal Biosciences*), 17:100904. https://doi.org/10.1016/j.animal.2023.100904

United Nations Department of Economic and Social Affairs (UN). 2022. World Population Prospects 2022 Summary of Results. 27th edition of official UN population estimates and projections. Available at: https://www.un.org/development/desa/pd/sites/www.un.org.development.desa.pd/files/wpp2022_summary_of_results.pdf

Valdés, F.; Villanueva, V.; Durán, E.; Campos, F.; Avendaño, C.; Sánchez, M.; Domingoz-Araujo, C.; Valenzuela, C. 2022. Insects as Feed for Companion and Exotic Pets: A Current Trend. *Animals,* 12:1450. https://doi.org/10.3390/ani12111450

Valverde, M. E.; Hernández-Pérez, T.; Paredes-López, O. 2015. Edible Mushrooms: Improving Human Health and Promoting Quality Life. *International Journal of Microbiology (Int. J. Microbiol.*) 2015:376387. https://doi.org/10.1155/2015/376387

Van Raamsdonk, L. W. D.; Meijer, N.; Gerrits, E. W. J.; Appel, M. J. 2023. New approaches for safe use of food by-products and biowaste in the feed production chain. *Journal of Cleaner Production (J. Clean. Prod.*), 388:135954. https://doi.org/10.1016/j.jclepro.2023.135954

Vidakovic, A.; Langeland, M.; Sundh, H.; Sundell, K.; Olstorpe, M.; Vielma, J.; Kiessling, A.; Lundh, T. 2015. Evaluation of growth performance and intestinal barrier function in Arctic Charr (*Salvelinus alpinus*) fed yeast (*Saccharomyces cerevisiae*), fungi (*Rhizopus oryzae*) and blue mussel (*Mytilus edulis*). *Aquaculture Nutrition (Aquacult. Nutr.)*, 12344. https://doi.org/10.1111/anu.12344

Voora, V.; Larrea, C.; Bermudez, S. 2020. Global Market Report: Soybeans. Sustainable Commoditites Marketplace Series 2019. International Institute for Sustainable Development. State of Sustainability Initiatives. Available at: https://www.iisd.org/system/files/2020-10/ssi-global-market-report-soybean.pdf

Wilson, R. P. 1986. Protein and amino acid requirements of fishes. *Annual Reviews of Nutrition (Annu. Rev. Nutr.)*, 6:225-244. https://doi.org/10.1146/annurev.nu.06.070186.001301

Worldbarometer, available at https://www.worldometers.info/world-population/ access: February 2026.

Woodard & Curran, Inc. 2006. Wastes from Industries (Case Studies) in Industrial Waste Treatment Handbook (second edition), Butterwotrh-Heinemann, pp 409-496. https://doi.org/10.1016/b978-075067963-3/50012-6

Yu, H.; Li, M.; Yu, L.; Ma, X.; Wang, S.; Yuan, Z.; Li, L. 2023a. Partial Replacement of Fishmeal with Poultry By-Product Meal in Diets for Coho Salmon (*Oncorhynchus kisutch*) Post-Smolts. *Animals*, 13: 2789 https://doi.org/10.3390/ani13172789

Yu, L.; Yu, H.; Yuan, Z.; Zhang, J.; Li, L.; Ma, C.; Kong, W. 2023b. Dietary L-Lysine Requirements of Coho Salmon (*Oncorhynchus kisutch*) Alevins. *Animals*, 12:3670. https://doi.org/10.3390/ani13233670

Yu, H.; Yu, L.; Rahman, A.; Govindharajan, S.; Zhang, Y.; Wang, G.; Li, L. 2025. Impact of dietary supplementation with Antarctic krill meal on growth performance, biochemical parameters, and antioxidant status in Coho Salmon (*Oncorhynchus kisutch*) post-smolts. *Frontiers in Marine Science (Front. Mar. Sci.)*, 12:1553629. https://doi.org/10.3389/fmars.2025.1553629

CHAPTER 4

Lipids in animal nutrition: sources, fatty acid profiles and practical considerations

F.D. Henne

Pet Food Ingredient Specialist, Applications and Palatability, Verona, Italy
Correspondence: florian.d.henne@science4climate.org

Abstract

Dietary lipids are essential components of animal feed, providing concentrated energy and essential fatty acids required for growth, health, reproduction and product quality. Conventional lipid sources such as animal fats, fish oil and vegetable oils are widely used in feed and pet food formulations. However, sustainability concerns, price volatility and supply limitations have stimulated interest in alternative lipid sources.

This chapter is designed as a practical reference tool for feed and pet food formulators. Rather than providing an exhaustive review of lipid metabolism or nutritional physiology, it compiles and compares fatty acid profiles of conventional and emerging lipid sources commonly considered in feed formulations. The objective is to support ingredient selection by enabling fast comparison of fatty acid composition across sources in relation to species-specific requirements. Particular attention is given to key polyunsaturated fatty acids relevant for animal nutrition, including linoleic acid, α-linolenic acid, arachidonic acid, eicosapentaenoic acid (EPA) and docosahexaenoic acid (DHA). The chapter further addresses the role of pure fats and oils in feed formulation, highlighting their importance for energy supply, palatability, processing performance and nutritional precision.

Overall, this chapter demonstrates how conventional and emerging lipid sources can be strategically combined to meet species-specific energy and fatty acid requirements while supporting sustainable feed formulation practices.

Keywords

dietary lipids – essential fatty acids – fatty acid profiles – feed formulation – pet food – aquaculture – sustainability – novel lipid sources – novel ingredients – insects – algae – camelina

1 Introduction

Proteins form the structural and functional foundation of animal nutrition by supplying essential amino acids required for growth, maintenance and metabolic activity (Wu et al., 2014). However, nutritionally balanced diets cannot be achieved by protein supply alone as dietary lipids are equally indispensable. They serve as the most concentrated source of energy and provide essential fatty acids (EFAs) that animals cannot synthesize endogenously (Kerr et al., 2015).

While protein meals inherently contribute small and variable lipid fractions, targeted lipid supplementation is required to meet species-specific energy demands and fatty acid requirements in livestock, companion animals and aquaculture species (Becker et al., 2023). Lipids are closely interconnected with protein metabolism, influencing nutrient utilization, membrane structure, hormone synthesis and immune function (Zeng et al., 2023). In addition, lipid composition directly affects feed palatability, processing behavior and oxidative stability, thereby linking nutritional quality with technological performance (Kerr et al., 2015).

Conventional lipid sources such as animal fats and fish oils coming from rendering processes and vegetable oils obtained from extraction or press processes have long been used to fulfill these roles. However, similar to protein ingredients, their availability and sustainability are increasingly challenged. As a result, novel lipid sources derived from insects, algae, fungi, and oilseed crops have emerged as promising alternatives, often characterized by unique fatty acid profiles rich in polyunsaturated and long-chain ω-3 fatty acids.

Building on the protein-focused evaluation presented in the previous chapter (Chapter 3), this chapter examines the nutritional quality, fatty acid composition and functional relevance of conventional and emerging lipid sources. By comparing essential fatty acid requirements with the profiles of lipid fractions in protein meals as well as pure lipids and fats, this chapter aims to support informed and sustainable feed formulation across animal production systems.

2 Lipid fractions in protein meals and their fatty acid profiles

Fatty acids are components of triglycerides and can be distinguished based on the length of their carbon atoms into short chain (SC; <C6), medium chain (MC; C6–C12) and long chain (LC; >C12). Fatty acids, comparable to amino acids in proteins, determine the quality of lipid fractions and play an important role in defining nutritional and functional values. Lipids and proteins are metabolically interconnected and are key energy sources, co-operate in membranes and serve as signaling molecules (Escribá et al., 2008). The lipid fraction of a meal or pure fats and oils provide essential fatty acids (EFAs) required for optimal growth, immune function and reproduction in livestock, companion animals and aquatic species (Carr, 2023; Lenox,

2015; Moallem, 2018; Palmquist and Jenkins, 2017; Spain, 2002; Tocher, 2015). As it is for essential amino acids (EAAs), a fatty acid is considered essential when it cannot be synthesized in the body and must be delivered with the diet. According to the FA chain length definition of Food and Agriculture Organization (FAO, 2010) of the United Nations, all of the EFAs for livestock, companion animals and aquaculture are long-chain polyunsaturated fatty acids such as linoleic acid (C18:2ω-6, LA), α-linolenic acid (C18:3ω-3, ALA), but also arachidonic acid for cats (C20:4ω-6, ARA), eicosapentaenoic acid (C20:5ω-3, EPA) and docosahexaenoic acid (C22:6ω-3, DHA; Glencross, 2009; Tocher, 2015). EPA and DHA are essential for aquaculture species but are also recently discussed as complementary FAs for dogs and cats to support growth and reproduction (Dzanis, 2015; FEDIAF, 2024a). It is important to keep in mind, just as discussed for amino acids, that the lipid quality and fatty acid profile can vary significantly during manufacturing and application in feed production because of its sensitivity towards thermal, mechanical, and oxidative stress. The rendering process involves several hours of exposure at elevated temperatures (up to 140 °C / 280 °F) and often steam-pressure cooking (up to 4.1 bar / 60 psi) for manufacturing animal-based fats and meals (Woodard and Curran, 2006). Extrusion as an example for a final feed production process also requires high temperatures (150 °C / 300 °F) and mechanical pressure induced by one or more screws. However, the combination of high temperature and pressure especially applied during a short time can rather generate favorable lipid/starch and protein/starch complexes than decompose the fatty acids and amino acids (Rokey et al., 2010). To preserve the quality of lipids and proteins, it is not only essential to consider manufacturing and processing conditions but also to manage stability during storage and transportation via adequate antioxidant systems and packaging (FAO, 1991; Tian et al., 2013).

Table 4.1 shows fatty acid requirements for livestock (swine, poultry), companion animals (dogs and cats) and aquaculture species (e.g. salmonids). The presented values are representative but in no way claim to be complete or to cover the entire spectrum, since the fatty acid requirements vary with age, life stage, and health status of animals as well as activity level, environmental temperature and the quality and oxidative stability of dietary fats and oils.

Swine and poultry have fatty acid requirements for LA (swine: 0.1% of diet; broiler: 0.2% of diet; laying hen: 0.3% of diet). ALA is an EFA for swine but has no defined requirements. However, the requirements are usually met by fat present in natural dietary ingredients (NRC, 2012). For poultry, ALA is not considered an essential or limiting fatty acid under practical feeding conditions (Ravindran, 2013; Zornig et al., 2001). Ruminant diets are usually low in total fat content (3% total fat), especially unsaturated fatty acids (UFAs). It is known that UFAs, especially polyunsaturated fatty acids (PUFAs: LA, ALA, EPA, and DHA) at certain dosages have negative effects on the activity of *butyrivibrio fibrisolvens*, a rumen bacterium assisting to digest fiber and metabolize lipids (biohydrogenation of UFAs into saturated fatty acids (SFAs); Maia et al., 2010). If PUFAs are added to ruminant feed, then only

TABLE 4.1 Essential fatty acid requirements for swine, poultry, dogs, cats and salmonids (% of dry matter basis)

Fatty acid	Swine[1]	Poultry[2] – broiler / – laying hen	Dogs[3]	Cats[3]	Salmonids[4]*
Linoleic acid (LA) (C18:2ω-6)	0.1	0.2 / 0.3	1.3	0.6	-
αLinolenic acid (ALA) (C18:3ω-3)	Yes, but not defined	Not critical	#	#	1.0–2.0 (ALA + EPA + DHA)
Arachidonic acid (ARA) (C20:4ω-6)	-	-	-	0.02	-
Eicosapentaenoic acid (EPA) (C20:5ω-3)	-	-	#	#	0.5–1.0 (EPA + DHA)
Docosahexaenoic acid (DHA) (C22:6ω-3)	-	-	#	#	0.5–1.0 (EPA + DHA)

Sources: [1]NRC, 2012, [2]Ravindran, 2013, [3]FEDIAF, 2024a, [4]NRC, 2011; *further fish in NRC, 2011; #not essential but complementary during growth and reproduction; Ruminants are not included because dietary EFA requirements are generally not specified in requirement tables, due to ruminal biohydrogenation and endogenous synthesis. Nevertheless, fatty acid supply can be relevant for targeted outcomes such as reproduction and milk fat composition.

in protected form, to prevent hydrogenation in the rumen. One way of doing this is, for example, to add protected linseed oil (Chikunya et al., 2004; Fabjanowska et al., 2023; Kim et al., 2020). Dogs and cats, such as swine and poultry, have dietary requirements for LA (dog: up to 1.3% of dry matter; cat: up to 0.6% of dry matter; Dzanis, 2015; FEDIAF, 2024a). In addition, cats have special requirements for arachidonic acid (C20:4ω-6, ARA; 0.02% of dry matter), another omega(ω)-6 fatty acid. Higher unsaturated ω-3 fatty acids such as ALA, EPA and DHA are not essential but recently discussed as complementary during growth and reproduction. Since fat is the main energy source, dog foods typically contain between 5% and 15% fat, whereas the fat content in cat food can range between 5% and 40% (dry matter basis; Dzanis, 2015; FEDIAF, 2024a). However, as dietary fat content increases, maintaining adequate protein concentration becomes important, especially considering health status and age of the animal (Buff et al., 2014; Laflamme, 2005; Vecciato et al., 2022). For Salmonids and many marine fish EPA and DHA, besides ALA, are essential fatty acids for membrane function, growth and reproduction (Bou et al., 2017; NRC, 2011). It is recommended to have a combined inclusion of ω-3 fatty acids (ALA + EPA + DHA) at 1.0–2.0% or at least EPA + DHA at 0.5–1.0% as baseline in the diet, while higher levels up to 3% or more are needed depending on life stage and objective. ALA is considered as a primary essential fatty acid for salmonids, because it serves as the metabolic precursor for the synthesis of the longer-chain ω-3 polyunsaturated fatty acids EPA and DHA (NRC, 2011).

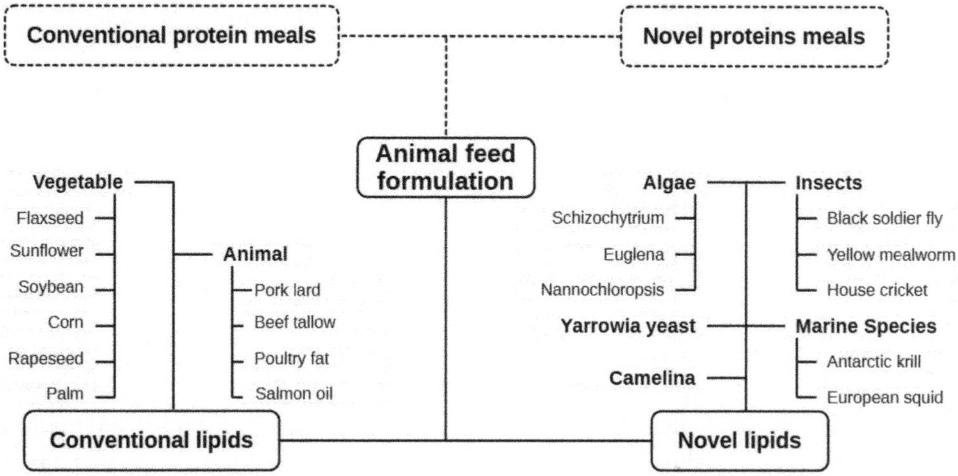

FIGURE 4.1 Overview of conventional and novel lipid sources used in animal feed formulation, including animal-derived from pork, poultry, and fish, plant-based from flaxseed, sunflower, and soybean, algae, insects, yeast, and other emerging ingredients.

TABLE 4.2 Organisms considered as novel lipid sources in this chapter: common name, scientific name and classification

Common name	Scientific name	Classification in chapter
Black soldier fly	*Hermetia illucens*	Insect
Yellow mealworm	*Tenebrio molitor*	Insect
House cricket	*Acheta domesticus*	Insect
Red seaweed	*Asparagopsis taxiformis*	Macroalgae
Schizochytrium	*Schizochytrium* sp.	Microalga
Euglena	*Euglena gracilis*	Microalga
Nannochloropsis	*Nannochloropsis* sp.	Microalgae
Duckweed	*Lemna minor*	Aquatic plant
Antarctic krill	*Euphausia superba*	Marine species
European squid	*Loligo vulgaris*	Marine species
Brewer's yeast	*Saccharomyces cerevisiae*	Yeast
Yarrowia yeast	*Yarrowia lipolytica*	Yeast
Mortierella	*Mortierella alpina*	Microfungus

A growing number of novel lipid sources have gained attention as complements to conventional options (Figure 4.1). These include lipids derived from insects, micro- and macroalgae, aquatic plants (duckweed), selected marine species (krill, squid), oleaginous yeasts and fungi (Table 4.2) as well as oil crops such as camelina.

In practical feed formulation, fatty acids originate from two complementary inputs: the lipid fraction intrinsically present in lipid-rich meals (e.g., fish meal, poultry by-product meal, insect meals) and added pure fats and oils used to adjust energy density and essential fatty acid supply. Because the final dietary fatty acid profile is the weighted outcome of both inputs, this chapter first summarizes representative fatty acid profiles of lipid fractions in selected meals and then provides profiles of commonly used pure fats and oils. This structure allows formulators to rapidly compare sources, anticipate the combined dietary profile, and select ingredients consistent with species-specific requirements, oxidative stability constraints, palatability targets, and sustainability objectives.

Tables 4.3–4.5 summarize representative fatty acid profiles of the lipid fraction present in selected protein-rich ingredients. Table 4.3 includes conventional protein meals (soybean meal, feather meal, fish meal, meat and bone meal and poultry by-product meal), whereas Tables 4.4 and 4.5 cover selected novel protein sources such as insects, algae, mushrooms, camelina and yeast.

The presented values are representative and intended to provide general orientation rather than exhaustive coverage. Fatty acid composition may vary depending on species origin, growth and feeding conditions, processing, manufacturing and storage (FAO, 2004). Accordingly, reported averages or ranges may include uncertainties and profiles should be interpreted together with ingredient inclusion level and fat content. Further detailed information on specific novel lipid sources may be found in the respective book chapters on applications in animal nutrition: insects (7; 13; 18), algae (9; 10; 20), camelina (11, 22), pulses (16), mushrooms (17), bacterial and microbial proteins from fermentation (8, 14, 19), underutilized marine species (15) and utilization of agricultural waste (21). Without mentioning every fatty acid profile in detail, soybean meal shows a high concentration of PUFA (60%), mainly due to linoleic acid (18:2ω-6, LA; 52% of the 60%). The poultry by-product meal contains 28% PUFA, mainly LA (20% of the 28%). Another representative of PUFA-rich animal meal is fish meal, containing 26% PUFAs, mainly fatty acids with four (C20:4ω-6, ARA), five (C20:5ω-3, EPA; C22:5ω-3, DPA) and six (C22:6ω-3, DHA) double bonds, besides 27% SFA and 34% monounsaturated fatty acids (MUFAs). Further details on the influence of the number of double bonds on stability and oxidation may be found in the respective lipid oxidation chapter (Chapter 5). Feather as well as meat and bone (MBM) meal show a high degree of saturated fatty acids (feather: 50% SFAs, MBM: 51% SFAs) and nearly the same concentration of MUFA (feather: 46%, MBM: 44%), mainly oleic acid (C18:1, OA; feather: 40% of the 46%, MBM: 41% of the 44%). Camelina meal, an oilseed plant-based ingredient, shows a PUFA concentration comparable to soybean (camelina: 56%, soybean: 60%, Table 4.3 and 4.5) with the advantage to be relatively equal rich in linoleic acid (24%) and α-linolenic acid (30%, C18:3ω-3, ALA).

Lipids from algae, depending on the source, are naturally rich in highly polyunsaturated fatty acids (red and brown seaweed: ARA and EPA; green seaweed: ALA, Table 4.4) but contain also considerably high amounts of SFAs (21–31%) and are

TABLE 4.3 Fatty acid profile of selected conventional protein meal sources [% of total FA]

Fatty acid	Soybean Meal# – 46% (n=316)/ – 48% (n=10409)/ – 50% protein (n=1453)	Feather meal^ (n=123)	Fish*# – 62% (n=297)/ – 65% (n=1198)/ – 70% protein (n=431)	Meat and bone meal^# – <7.5% (n=114)/ – >7.5% fat (n=1230)	Poultry by-product meal[1]
Palmitic (C16:0) PA	10.5	34.8	17.8	27.5	23.3
Palmitoleic (C16:1)	0.2	6.2	7.2	3.7	7.4
Stearic (C18:0)	3.8	13.8	3.6	19.2	5.7
Oleic (C18:1) OA	21.7	39.9	12.3	40.7	41.0
Linoleic (C18:2ω-6) LA	53.1	3.3	2.1	3.6	19.3
α-Linolenic (C18:3ω-3) ALA	7.4	-	1.9	0.9	1.0
Eicosenoic (C20:1)	-	-	6.6	-	-
Arachidonic (C20:4ω-6) ARA	-	-	2.4	-	-
Eicosapentaenoic (C20:5ω-3) EPA	-	-	9.0	-	-
Erucic (C22:1)	-	-	7.7	-	-
Docosapentaenoic (C22:5ω-3) DPA	-	-	2.6	-	-
Docosahexaenoic (C22:6ω-3) DHA	-	-	6.6	-	-
ΣSFA	14.3	48.6	21.4	46.7	29.0
ΣMUFA	21.9	46.1	33.8	44.4	48.4
ΣPUFA	60.5	3.3	22.2	4.5	20.3

Sum (Σ) SFA: saturated fatty acid, MUFA: monounsaturated fatty acid, PUFA: polyunsaturated fatty acid; sources: Sauvant et al., 2004 and [1]Nates and Suresh, 2011, *no specification on fish species; ^derived from slaughtered, warm-blooded animals; #the different concentrations in protein (soy bean meal, fish meal) or fat (meat and bone meal) have no influence on the fatty acid composition in % fatty acid of total fat; because only selected FAs are presented, the sum of SFA + MUFA + PUFA ≠100.

therefore promising alternatives for both satisfying plant-based and animal-based fatty acid requirements. Lipid profiles in fungi (mushroom and yeast) show a broad variety of SFA, MUFA, and PUFA concentrations. What makes these novel ingredients so unique is the fact that protein and lipid profiles can be modulated using specific feeding and growth media, resulting in ingredients able to provide highly desirable amino acid or fatty acid profiles.

A good example is the single-cell protein saccharomyces cerevisiae, also known as brewer's yeast, which is extremely low in PUFAs and contains 70% MUFAs, mainly palmitoleic acid (C16:1, 51%) and oleic acid (19%). Overall, novel ingredients can become promising replacements for conventional ingredients considering fatty acid profiles. Some of them (insects and algae) are already established in feeding concepts (Allen et al., 2019; Delver and Smith 2024; Dias et al. 2025; Kaur and Torrecillas, 2025; Sánchez-Muros et al., 2014). However, the same applies here as for proteins, that the inclusion levels of the novel ingredients might be constrained by cost and regulatory situation.

TABLE 4.4 Fatty acid profiles of selected novel protein meal sources [% of total FAs]

Fatty acid	**Insects** – Black soldier fly larvae[1] (n=6) – Mealworm larvae[2] (n=40) – House cricket larvae[3] (n=3)	**Macroalgae[4]** **24 species*** – Red Seaweed (n=11) – Brown seaweed (n=5) – Green seaweed (n=8)	**Microalgae** – *Euglena gracilis*[5] – *Schizochytrium* sp[6]	**Aquatic Plant** – Duckweed[7]
C10 + C12	29.2* / - / -	- / - / -	7.9∞ / -	-
Myristic (C14:0)	12.0 / 3.3 / 0.7	3.0 / 3.8 / 1.1	41.0 / 3.9	1.1
Palmitic (C16:0)	16.7 / 18.9 / 26.1	27.3 / 16.2 / 20.8	8.8 / 51.5	19.1
Palmitoleic (C16:1)	0.8 / 1.4 / 0.4	0.6 / 2.4 / 1.9	1.7 / -	5.6
Stearic (C18:0)	4.0 / 4.2 / 9.1	1.0 / 1.1 / 0.5	1.5 / 1.7	1.1
Oleic (C18:1)	13.7 / 33.3 / 23.9	5.3 / 7.9 / 6.7	3.9 / 0.1	2.2
Linoleic (C18:2ω-6) LA	11.8 / 37.9 / 35.1	1.3 / 6.0 / 4.5	1.7 / -	16.9
α-Linolenic (C18:3ω-3) ALA	1.1 / 1.0 / 1.5	0.4 / 9.1 / 19.8	1.1 / 0.1	46.4
Eicosenic (C20:1)	-	-	- / -	-
Arachidonic (C20:4ω-6) ARA	-	23.1 / 12.0 / 2.2a	2.9 / -	-
Eicosapentaenoic (C20:5ω-3) EPA	-	28.8 / 13.2 / 3.6	0.5 / 7.8	0.1
Erucic (C22:1)	-	-	-	-
Docosapentaenoic (C22:5ω-3) DPA	-	- / - / 2.5b	0.9 / -	-
Docosahexaenoic (C22:6ω-3) DHA	- / - / 0.4	- / - / 0.5c	- / 34.3	-
ΣSFA	62.4 / 26.4 / 35.9	31.3 / 21.1 / 22.4	59.2 / 57.1	21.3
ΣMUFA	14.5 / 34.7 / 24.3	5.9 / 10.3 / 8.6	5.6 / 0.1	7.8
ΣPUFA	12.9 / 38.9 / 37.0	53.6 / 40.3 / 57.1‡	7.1 / 42.2	63.4

Sum (Σ) SFA: saturated fatty acid, MUFA: monounsaturated fatty acid, PUFA: polyunsaturated fatty acid; sources: [1]Srisuksai et al., 2024, [2]Kröncke2023, [3]Tzompa-Sosa et al., 2021, [4]Vaskovsky et al., 1996; [5]Ahmed et al., 2023, [6]Nobre et al., 2025, [7]Chakrabarti et al., 2018; a n=8, b n=6, c n=3; because only selected FAs are presented, the sum of SFA + MUFA + PUFA ≠ 100; *0.8% capric acid (C10:0) + 28.4% lauric acid (C12:0); ‡ including 15.7% hexadecatetraenoic acid (C16:4ω-3) + 8.3% stearidonic acid (C18:4ω-3); ∞0.5% C10:0 + 7.4% C12:0.

TABLE 4.5 Fatty acid profiles of selected novel protein meal sources [% of total FAs]

Fatty acid	Camelina[1] – solvent-extracted (n=10) – expelled (n=3)	Yeast[2] – *Saccharomyces cerevisiae* – *Yarrowia lipolytica*	Mushroom[3]* (edible, 11 species)	Filamentous fungus – *Mortierella alpina*[4]
C10 + C12	-	13.8† / -	-	-
Myristic (C14:0)	-	1.9 / -	-	-
Palmitic (C16:0)	8.3	12.7 / 12.0	9.0 – 20.0	19.0
Palmitoleic (C16:1)	0.3	51.2 / 17.2	0.3 – 1.7	-
Stearic (C18:0)	2.4	1.0 / 0.8	0.7 – 34.6	9.8
Oleic (C18:1)	20.3	18.5 / 42.9	2.5 – 57.8	24.0
Linoleic (C18:2ω-6) LA	23.9	- / 23.0	19.8 – 78.0	7.3
α-Linolenic (C18:3ω-3) ALA	29.5	-	0.3 – 0.7	5.5
Eicosenic (C20:1)	10.7	-	0.1 – 1.5	-
Arachidonic (C20:4ω-6) ARA	-	-	-	21.0
Erucic (C22:1)	1.8	-	-	-
ΣSFA	10.7	29.4 / 12.8	12.9 – 44.3	28.8
ΣMUFA	33.1	69.7 / 60.1	3.6 – 58.5	24.0
ΣPUFA	53.4	- / 23.0	19.8 – 78.8	33.8

Sum (Σ) SFA: saturated fatty acid, MUFA: monounsaturated fatty acid, PUFA: polyunsaturated fatty acid; * % of dry weight; sources: [1]Delver and Smith, 2024, [2]SAYDIUC, 2023, [3]Ayaz et al., 2011, [4]Takeno et al., 2005, because only selected FAs are presented, the sum of SFA + MUFA + PUFA ≠ 100; †6.2% capric acid C10:0 + 7.6% lauric acid C12:0. No EPA, DPA and DHA reported.

3 Pure lipid and fat sources and their fatty acid profiles

Protein meals inherently contribute to the total dietary fatty acid profile. However, their lipid fraction is often limited and variable in composition. To achieve both energetic and essential fatty acid requirements in animals, the inclusion of pure lipids (fats and oils) become standard practice in feed formulation. Compared to the residual fat present in meals during rendering, the refined lipids provide a concentrated, controllable and digestible source of energy. They represent the most efficient way to meet animals' caloric and essential fatty acid needs. Fats and oils do not only supply EFAs such as LA, ALA, EPA and DHA, but also enhance nutrient utilization and palatability by improving feed intake, feed conversion ratio, and overall performance (Attia et al., 2020). In addition, lipids can also reduce dustiness and improve the physical quality of pelleted feeds, important in both terrestrial and aquatic feed manufacturing (FAO, 1977).

Economically, fats and oils can offer a cost-effective energy density. However, the choice among vegetable oils, animal fats, and marine oils must balance price volatility, supply security, and sustainability. Vegetable oils are widely available but linked to land use and deforestations (Chiriacò et al. 2024; Fearnside, 2001; Meijaard et al., 2020, Savilaakso et al., 2013). Animal fats are valuable by-products improving circularity (FEDIAF, 2024b). Marine oils provide long-chain highly-PUFAs but face resource limitations (Glencross et al., 2025; Miller et al., 2008).

Consequently, optimizing the lipid inclusion is not only with respect to a nutritionally balanced diet but can also come with environmental and economic decisions directly aligned with the overall goal of a sustainable feed production. Novel lipid sources can play a key role here. If the range of suitable and legally approved conventional lipids can be expanded to include several new sources, it is possible to reduce the pressure, particularly on certain vegetable and marine oils (Tables 4.6–4.8).

Table 4.6 represents conventional vegetable fats and oils whereas Table 4.7 shows animal fats. Soybean oil has the highest global production among the listed oils and ranks among the top 3 lipid sources across livestock, pet food, and aquaculture, owing to its balanced fatty acid profile and cost-effectiveness (Gaffield et al., 2024; Voora et al., 2024). Palm oil is also a highly cost-effective energy source and is therefore widely used in livestock and tropical aquaculture (Ng, 2002; Ravindra et al., 2016; Tomkins and Drackley, 2010). It is rather limited in pet food due to digestibility and consumer perception (Plasek et al., 2021; Yuangklang et al., 2016). Rapeseed oil is an important ingredient in Europe and Canada and especially used in salmonids aqua feed (ACFG, 2025). Because of its high MUFA content (nearly 59%; 58.0% of it oleic acid, C18:1, OA; Table 4.6), it is relatively stable against oxidation. Sunflower oil is a high linoleic acid (C18:2ω-6, LA) source and especially beneficial for skin and coat in pets (Akkaya, 2018; Popa et al. 2018). Sunflower is a highly relevant crop in Europe and North America and is moderately priced (FAO, 2025). Corn oil is also high in linoleic acid but mainly produced for the human food market and therefore not commonly used as low-cost bulk lipid in the feed industry (Boukid and Rosentrater, 2024). Flaxseed oil is relatively high in α-linolenic acid (53%, C18:3ω-3, ALA) and shows health benefits in pets but because of its high amount of PUFA, it is also difficult to stabilize against oxidation (Panasevich et al., 2022; Purushothaman et al., 2011; Purushothaman et al., 2014; Richards et al., 2023). It is also an expensive ingredient for energy use and is usually used, in premium pet food and experimental aquafeeds (Figueiredo et al. 2017; Menoyo et al., 2005).

Lard is widely available in pork-producing regions and easily digestible by monogastrics (poultry, swine) but rarely used in pet food due to consumer perceptions. It has a high OA content (42% out of 46% MUFA, Table 4.7) but also a relatively

TABLE 4.6 Fatty acid profile of selected conventional vegetable oils [% of total FAs]

Fatty acid	Flaxseed[1]	Sunflower	Soybean	Corn[1]	Rapeseed	Palm
Palmitic (C16:0)	5.0	6.3	10.5	11.0	4.2	43.0
Palmitoleic (C16:1)	-	0.4	0.2	-	0.4	0.2
Stearic (C18:0)	4.0	4.3	3.8	2.0	1.8	4.4
Oleic (C18:1)	20.0	20.3	21.7	28.0	58.0	37.1
Linoleic (C18:2ω-6) LA	17.0	64.9	53.1	58.0	20.5	9.9
α-Linolenic (C18:3ω-3) ALA	53.0	0.3	7.4	1.0	9.8	0.3
Eicosenoic (C20:1)	-	-	0.2	-	-	-
Erucic (C22:1)	-	-	0.3	-	0.4	-
ΣSFA	9.0	10.6	14.3	12.0	6.0	47.4
ΣMUFA	20.0	20.7	22.4	28.0	58.8	37.3
ΣPUFA	70.0	65.2	60.5	59.0	30.3	10.2

Sum (Σ) SFA: saturated fatty acid, MUFA: monounsaturated fatty acid, PUFA: polyunsaturated fatty acid; Sources: Sauvant et al., 2004 and [1]ISEO, 2016; because only selected FAs are presented, the sum of SFA + MUFA + PUFA ≠ 100.

high PUFA (nearly 12%) (Ravindra et al., 2016). Because of negative effects of fats in general to microbial activity in the rumen, lard is not suitable as primary animal fat source for ruminants. Tallow in contrary, because of its low PUFA content (3%, Table 4.7) shows a higher rumen stability and is therefore a better energy source for cattle diets (Glasser et al. 2008). It may seem surprising that tallow is permitted as feed for ruminants, but tallow is a category 3 (low risk) animal by-product and therefore a protein-free source of fat, it is permitted in cattle feed if the insoluble impurities are <0.15% (AEAN, 2003; OJEU, 2011). It offers high OA (38%) and overall high saturated fat content (45%) and is considered as ecological source of energy in dairy nutrition because it acts as a bypass fat supporting milk production (Rabiee et al., 2012). Tallow shows a lower digestibility in young pigs and poultry. Poultry fat is a valuable product of the poultry rendering and the main energy source in poultry feed as well as pet food. Due to its high oleic acid content (43%) and its overall balanced fatty acid content (28% SFA, 48% MUFA, 22% PUFA, Table 4.7), it is highly digestible for swine and poultry, provides excellent palatability benefits in pet food as top coating and experiences an increased use as fish oil replacement in aquaculture (Klinmalai et al., 2025). Fish oil (e.g. salmon oil) is the only natural conventional major source of EPA (12%) and DHA (14%). It is essential in aquaculture to maintain growth, immunity and fillet quality. However, because of its price and high oxidation sensitivity (31% PUFA) it is rarely used as energy source and only strategically for specific purposes.

TABLE 4.7 Fatty acid profile of selected conventional animal fats and oils [% of total FAs]

Fatty acid	Lard*	Tallow#	Poultry fat	Salmon oil‡
Palmitic (C16:0) PA	24.4	25.3	21.0	10.2
Palmitoleic (C16:1)	2.9	3.4	5.0	8.7
Stearic (C18:0)	14.4	19.2	7.1	4.7
Oleic (C18:1) OA	42.3	37.5	41.7	18.6
Linoleic (C18:2ω-6) LA	9.2	2.8	20.6	1.2
α-Linolenic (C18:3ω-3) ALA	0.9	0.6	1.6	0.6
Eicosenoic (C20:1)	0.9	-	0.7	8.4
Arachidonic (C20:4ω-6) ARA	1.7	0.2	0.1	0.9
Eicosapentaenoic (C20:5ω-3) EPA	-	-	-	12.0
Erucic (C22:1)	-	-	-	5.5
Docosapentaenic (C22:5ω-3) DPA	-	-	-	2.9
Docosahexaenoic (C22:6ω-3) DHA	-	-	-	13.8
ΣSFA	38.8	44.5	28.1	14.9
ΣMUFA	46.1	40.9	47.4	41.2
ΣPUFA	11.8	3.6	22.3	31.4

Sum (Σ) SFA: saturated fatty acid, MUFA: monounsaturated fatty acid, PUFA: polyunsaturated fatty acid; source: Sauvant et al., 2004, *rendered pork fats; # heat-treated fat trimmings from ruminants obtained before the carcass is split; ‡ FA profiles of further fish (e.g. anchovy, capelin, cod liver, herring, menhaden, redfish, sardine) in Sauvant et al., 2004; because only selected FAs are presented, the sum of SFA + MUFA + PUFA ≠ 100.

Table 4.8 summarizes selected novel lipid sources. Oils from black soldier fly larvae (*Hermetia illucens*, BSFL), mealworm larvae (*Tenebrio molitor*, MWL) and house cricket (*Acheta domesticus*, HC) represent sources from insects. Depending on the rearing substrate, insects such as BSFL can show substantial variation in their lipid content and fatty acid composition (Oonincx et al., 2020; Veldkamp and Bosch, 2015). However, it can be determined that BSFL oil is rich in saturated fatty acids (66%, SFA), especially lauric (C12:0) which represents 40% of the total 66%. The fatty acid profiles of MWL and HC oil revealed a relatively high concentration in C18 fatty acids (MWL: 70%; HC: 60%), distributed in OA (C18:1; MWL: 38%, HC: 26%) and LA (C18:2ω-6; MWL: 32%, HC: 34%). In addition, HC oil is also rich in palmitic acid (30%, C16:0, PA; Table 4.8).

From the source of microalgae, fatty acid profiles of *Schizochytrium* sp., *Euglena gracilis*, and *Nannochloropsis* sp. are shown. *Schizochytrium* species are especially rich in polyunsaturated fatty acids (PUFAs, 53%) with a high concentration of docosahexaenoic acid (43%, DHA) and eicosapentaenoic acid (8%, EPA, Table

TABLE 4.8 Fatty acid profile of selected novel lipid sources [% total FAs]

Fatty acid	Insects[1] – Black soldier fly larvae (n=66)/ – Mealworm larvae (n=23)/ – House cricket (n=8)	Microalgae[2] – *Schizochytrium* sp.[2]/ – *Euglena gracilis*[3]/ – *Nannochloropsis* sp.[4]	Marine species – Antarctic Krill[5]/ – European Squid[6]	Camelina oil[7]	Yeast – *Yarrowia lipolytica*[8]
C10 + C12	39.8* / 0.3* / -	1.6 / 15.9 / -	-	-	-
Myristic (C14:0)	8.2 / 2.5 / 0.7	- / 41.0 / 4.3	5.0 - 13.0 / 5.5	-	-
Palmitic (C16:0)	15.0 / 17.0 / 27.9	19.5 / 8.8 / 24.6	17.0 - 24.6 / 7.5	7.5	10.4
Palmitoleic (C16:1)	-	- / 1.7 / 30.2	2.5 - 9.0 / 7.8	0.2	14.8
Stearic (C18:0)	3.0 / 3.7 / 8.1	1.5 / 1.5 / 1.1	- / 4.6	4.1	9.6
Oleic (C18:1) OA	12.8 / 38.1 / 26.1	18.3 / 3.9 / 11.0	6.0 – 14.5 / 13.9	14.2	37.1
Linoleic (C18:2ω-6) LA	12.0 / 32.8 / 33.8	1.4 / 1.7 / 1.9	0.1 - 3.0 / 3.1	24.9	8.4
α-Linolenic (C18:3ω-3) ALA	1.3 / 1.5 / 1.1	- / 1.1 / -	0.1 – 4.7 / 3.4	23.8	0.3
Eicosenic (C20:1)	- / - / -	- / 0.2 / 1.5	- / -	11.6	-
Arachidonic (C20:4ω-6) ARA	- / - / -	- / 2.9 / 21.8	- / 2.8	-	-
Eicosapentaenoic (C20:5ω-3) EPA	- / - / -	8.0 / 0.6 / -	14.3 – 28.0 / 6.5	-	-
Erucic (C22:1)	- / - / -	- / - / -	0.1 - 1.5 / -	2.9	-
Docosapentaenoic (C22:5ω-3) DPA	- / - / -	- / 0.2 /	0.1 – 0.7 / 4.5	-	-
Docosahexaenoic (C22:6ω-3) DHA	- / - / -	42.6 / - /	7.1 – 15.7 / 4.4	-	-
ΣSFA	66.0 / 22.7 / 36.7	13.3 / 67.2 / 30.0	22.0 – 37.6 / 15.6	11.6	20.0
ΣMUFA	12.8 / 38.1 / 26.1	22.6 / 5.8 / 42.7	8.6 – 25.0 / 21.7	18.9	51.9
ΣPUFA	13.3 / 34.3 / 34.9	52.0 / 6.5 / 23.7	21.7 – 42.1 / 24.7	48.7	8.7

Sum (Σ) SFA: saturated fatty acid, MUFA: monounsaturated fatty acid, PUFA: polyunsaturated fatty acid; sp = species unspecified; sources: [1]Toral et al., 2025; [2]Fedorova-Dahms et al., 2011; [3]Ahmed et al., 2023; [4]Ma et al., 2016; [5]FAO, 2017; [6]Asadpour, 2016; [7]Rodríguez-Rodríguez et al., 2021; [8]Gao et al., 2020; * lauric acid (C12:0); because only selected FAs are presented, the sum of SFA + MUFA + PUFA ≠ 100.

4.8) and noteworthy amounts of OA (18%). *Euglena gracilis* is rich in SFAs (C12:0, C14:0). *Nannochloropsis* sp., however, shows a relatively balanced fatty acid profile (SFA: 30%, MUFA: 43%, PUFA: 24%) with high concentrations in PA (25%) and arachidonic acid (22%; C20:4ω-6, ARA). Oil from other marine species, such as krill (*Euphausia superba*) and squid (*Loligo vulgaris*) are rich sources in PUFA (krill: 14%–28% EPA and 7%–14% DHA; squid: 20% ARA + EPA + DPA + DHA, Table 4.8). Camelina oil is a naturally rich source of essential PUFAs (49: 25% LA; 24% ALA) as well as OA (14%), whereas the yeast *Yarrowia lipolytica* is rich in MUFAs (37% OA, 15% palmitoleic acid, C16:1).

4 Conclusion

Dietary lipids are indispensable components of animal feed formulations, providing a highly concentrated source of energy and supplying essential fatty acids (EFAs). EFAs are required to support growth, health, reproduction and product quality across livestock, companion animals and aquaculture species. Beyond their nutritional role, lipids contribute to feed palatability, physical quality and processing performance, thereby linking nutritional adequacy with technological functionality. Consequently, the selection and inclusion of lipid sources represent a critical aspect of modern feed formulation.

Conventional lipid sources derived from animal fats, fish oil and vegetable oils continue to dominate commercial feed production due to their availability, well-characterized fatty acid profiles and established nutritional value. However, these sources differ substantially in their content of saturated, monounsaturated and polyunsaturated fatty acids, as well as in their contribution of essential fatty acids such as linoleic acid, α-linolenic acid and long-chain ω-3 fatty acids. Matching these profiles with species-specific requirements remains essential, particularly in aquaculture and companion animal nutrition, where long-chain polyunsaturated fatty acids play a central role.

In parallel with developments in protein ingredients, increasing attention is being directed toward novel lipid sources originating from insects, algae, fungi and oilseed crops. Many of these emerging ingredients exhibit unique and, in some cases, highly favorable fatty acid compositions, including elevated levels of polyunsaturated and ω-3 fatty acids. Their integration into feed formulations offers opportunities to reduce reliance on finite marine resources and to improve the sustainability of animal production systems. At the same time, variability in lipid composition, production scale, regulatory acceptance and cost currently limits their widespread adoption, highlighting the need for continued research and technological development.

While conventional lipids such as rapeseed oil, fish oil, poultry fat, tallow and lard, provide essential fatty acids (LA, ALA, EPA, DHA) and are highly valuable energy sources, novel lipids have the potential to complement or even partially substitute them. Insect oils are rich in saturated fatty acids (SFAs) and C18 fatty acids (OA and LA). Oils from microalgae and other marine species such as krill and squid oil can provide essential PUFA in high concentrations (ARA, EPA, and DHA). Camelina oil is naturally rich in LA and ALA, whereas *Yarrowia lipotytice* yeast provides OA and palmitoleic acid.

The nutritional and functional value of dietary lipids is strongly influenced by processing conditions, storage and formulation strategies. Lipids are particularly susceptible to oxidative degradation, which can compromise fatty acid integrity, reduce palatability and negatively affect animal performance. Therefore, careful consideration of lipid source selection, antioxidant inclusion and process control is required to preserve feed quality and ensure consistent nutritional outcomes. These technological considerations are especially relevant when incorporating lipid-rich novel ingredients or highly unsaturated fats and oils.

Overall, the strategic use of both conventional and emerging lipid sources enables feed formulators to tailor dietary energy density and essential fatty acid supply to the specific needs of different animal species and production stages. When combined with appropriate protein sources and balanced amino acid profiles, dietary lipids contribute to the development of nutritionally balanced, economically viable and environmentally responsible feed formulations. As sustainability considerations continue to shape the future of animal nutrition, lipids will remain a key component in the optimization of feed efficiency, animal health and resource use across global feed and pet food industries.

Literature

Animals, England Animal Health (AEAH). 2003. The Animal By-Products Regulations. No.1482, Part 3: Restrictions on access to animal by-products and their use. https://www.legislation.gov.uk/uksi/2003/1482/2005-01-01/data.pdf

Ahmed, E.; Suzuki, K.; Nishida, T. 2023. Micro- and Macro-Algae Combination as a Novel Alternative Ruminant Feed with Methane-Mitigation Potential. *Animals*, 13:796. https://doi.org/10.3390/ani13050796

Akkaya, M. R. 2018. Prediction of fatty acid composition of sunflower seeds by near-infrared reflectance spectroscopy. *Journal of Food Science and Technology (J. Food Sci. Technol.)*, 55:2318-2325. https://doi.org/10.1007/s13197-018-3150-x

Allen, K. M.; Habte-Tsion, H.-M.; Thompson, K. R.; Filer, K.; Tidwell, J. H.; Kuma, V. 2019. Freshwater microalgae (*Schizochytrium* sp.) as a substitute to fish oil in shrimp feed. *Scientific Reports*, 9:6178. https://doi.org/10.1038/s41598-019-41020-8

Aquaculture Feed Guide (ACFG). 2025. Canola Meal for Aquaculture. Available at: https://www.canolacouncil.org/canolamazing/about/aquaculture-feed-guide. Accessed January 2026.

Asadpour, Y. A. 2016. Squid (*Loligo loligo*): The new soure to extract omega-3 and omega-6 rich marine oils. *Iranian Journal of Fisheries Sciences (Iran. J. Fish. Sci.)*, 15:100-107. http://jifro.ir/article-1-1411-en.html

Ayaz, F. A.; Chuang, L. T.; Torun, H.; Colak, A.; Sesli, E.; Presley, J.; Smith, B. R.; Glew, R. H. 2011. Fatty acid and amino acid composition of selected wild-edible mushrooms consumed in Turkey. *International Journal of Food Sciences and Nutrition (Int. J. Food Sci. Nutr.)*, 62: 328-335. https://doi.org/10.3109/09637486.2010.533160

Becker, S. L.; Humphrey, D. C.; Karriker, L. A.; Brown, J. T.; Skoland, K. J.; Greiner, L. L. 2023. The effects of dietary essential fatty acid ratios and energy level on growth performance, lipid metabolism, and inflammation in grow-finish pigs. Journal of Animal Science (J. Anim. Sci.), 101:1–17. https://doi.org/10.1093/jas/skad151

Bou, M.; Berge, G. M.; Baeverfjord, G.; Sigholt, T.; Ostbye, T.-K.; Ruyter, B. 2017. Low levels of very-long-chain n-3 PUFA in Atlantic salmon (Salmo salar) diet reduce fish robustness under challenging conditions in sea cages. Journal of Nutritional Science (J. Nutr. Sci.), 6:e32. https://doi.org/10.1017/jns.2017.28

Boukid, F.; Rosentrater, K. A. 2024. Edible corn oil: A holistic exploration from processing to market dynamics. European Journal of Lipid Science and Technology (Eur. J. Lip. Sci Technol.), 126:e2400022. https://doi.org/10.1002/ejlt.202400022

Buff, P. R.; Carter, R. A.; Bauer, J. E.; Kersey, J. H. 2014. Natural pet food: A review of natural diets and their impact on canine and feline physiology. Journal of Animal Science (J. Anim. Sci.), 92:3781-3791. https://doi.org/10.2527/jas2014-7789

Carr, I.; Glencross, B.; Santigosa, E. 2023. The importance of essential fatty acids and their ratios in aquafeeds to enhance salmonid production, welfare, and human health. *Frontiers in Animal Science (Front. Anim. Sci.*), 4:1147081. https://doi.org/10.3389/fanim.2023.1147081

Chakrabarti, R.; Clark, W.D.; Sharma, J.G.; Goswami, R.K.; Shrivastav, A.K.; Tocher, D.R. 2018. Mass Production of Lemna minor and Its Amino Acid and Fatty Acid Profiles. *Frontiers in Chemistry (Front. Chem.*), 6:479. https://doi.org/10.3389/fchem.2018.00479

Chikunya, S.; Demirel, G.; Enser, M.; Wood, J. D.; Wilkison, R. G.; Sinclair, L. A. 2004. Biohydrogenation of dietary n-3 PUFA and stability of ingested vitamin E in the rumen, and their effects on microbial activity in sheep. *British Journal of Nutrition (Brit. J. Nutr.*), 91:539-550. https://doi.org/10.1079/bjn20031078

Chiriacò, M. V.; Galli, N.; Latella, M.; Rulli, M. C. 2024. Pressure on Global Forests: Implications of Rising Vegetable Oils Consumption Under the EAT-Lancet Diet. *Global Change Biology (Glob. Chang. Biol.)*, 31:e70077.77. https://doi.org/10.1111/gcb.70077

Dzanis, D. A. 2015. The revised AAFCO dog and cat food nutrient profile, part 2. Available at: https://www.petfoodindustry.com/pet-food-market/article/15460851/the-revised-aafco-dog-and-cat-food-nutrient-profiles-part-2. Original document: Association of Americal Feed Control Officials (AAFCO). 2015. https://www.aafco.org/wp-content/uploads/2023/01/Model_Bills_and_Regulations_Agenda_Midyear_2015_Final_Attachment_A.__Proposed_revisions_to_AAFCO_Nutrient_Profiles_PFC_Final_070214.pdf

Delver, J. J.; Smith, Z. K. 2024. Opportunities for Camelina Meal as a Livestock Feed Ingredient. *Agriculture,* 14: 116. https://doi.org/10.3390/agriculture14010116

Dias, M.; Marmelo, I.; António, C.; Rodrigues, A. M.; Marques, A.; Diniz, M. S.; Maulvault, A. L. 2025. Asparagopsis taxiformis Feed Supplementation as a Tool to Improve the Resilience of Farmed Diplodus sargus to Marine Heatwave Events—A Metabolomics Approach. Fishes, 10:350. https://doi.org/10.3390/fishes10070350

Escribá, P. V.; González-Ros, J. M.; Goñi, F. M.; Kinnunen, P. K. J.; Vigh, L.; Sánchez-Magraner, L.; Fernández, A. M.; Busquets, X.; Horváth, I.; Barceló-Coblijn, G. 2008. Membranes: a meeting point for lipids, proteins and therapies. *Journal of Cellular and Molecular Medicone (J. Cell. Mol. Med.*), 12:829-875. https://doi.org/10.1111/j.1582-4934.2008.00281.x

Fabjanowska, J.; Kowalczuk-Vasilev, E.; Klebaniuk, R.; Milewski, S.; Gümü, H. 2023. N-3 Polyunsaturated Fatty Acids as a Nutritional Support of the Reproductive and Immune System of Cattle—A Review. *Animals*, 13:3589. https://doi.org/10.3390/ani13223589

Fearnside, P. M. 2001. Soybean cultivation as a threat to the environment in Brazil. *Environmental Conservation (Environ. Conserv.)*, 28:23-38. https://doi.org/10.1017/s0376892901000030

Fédération Européenne de l'industrie des Aliments pour Animaux Familiers, European Pet Food Industry Federation (FEDIAF). 2024a. Nutritional Guidelines For Complete and Complementary Pet Food for Cats and Dogs. Brussels. Available at: https://europeanpetfood.org/wp-content/uploads/2024/09/FEDIAF-Nutritional-Guidelines_2024.pdf

Fédération Européenne de l'industrie des Aliments pour Animaux Familiers, European Pet Food Industry Federation (FEDIAF). 2024b. Category 3 animal fat – an important pet food ingredient. FEDIAF Factsheet. Brussels. Available at: https://europeanpetfood.org/pet-food-facts/fact-sheets/environment-and-sustainability/category-3-animal-fat-an-important-pet-food-ingredient

Fedorova-Dahms, I.; Marone, P. A.; Bauter, M.; Ryan, A. S. 2011. Safety evaluation of DHA-rich Algal Oil from *Schizochytrium* sp. *Food and Chemical Toxicology (Food Chem. Toxicol.*), 49:3310–3318. https://doi.org/10.1016/j.fct.2011.08.024

Figueiredo, P. S.; Candido, C. J.; Jaques, J. A.; Nunes, A. A.; Caires, A. R.; Michels, F. S.; Almeida, J. A.; Filiú, W. F.; Hiane P. A.; Nascimento V. A.; Franco O. L.; Guimarães R. C. 2017. Oxidative stability of sesame and flaxseed oils and their effects on morphometric and biochemical parameters in an animal model. *Journal of the Science of Food and Agriculture (J. Sci. Food. Agric.*), 97:3359-3364. https://doi.org/10.1002/jsfa.8186

Food and Agriculture Organization of the United Nations (FAO). Rumsey, G. L. 1991. Antioxidants in Coumpoud Feeds, Chapter 10 in Fish Feed Technology (ADCP/REP/80/11). Rome. ISBN: 92-5-100901-5. Available at: https://www.fao.org/4/x5738e/x5738e0b.htm

Food and Agriculture Organization of the United Nations (FAO). 2004. Animal Production and Health. Assessing Qualaity and Safety of Animal Feeds. Volume 160. Rome. ISBN: 92-5-105046-5. Available at: https://openknowledge.fao.org/server/api/core/bitstreams/437d6eb2-0ced-4a55-a98b-e1efb9047115/content

Food and Agriculture Organization of the United Nations (FAO). 2010. Fats and fatty acids in human nutrition: Report on an expert consultation. FAO Food and Nutrition Paper 91, pp19. Rome. Available at: https://www.fao.org/3/i1953e/i1953e.pdf

Food and Agriculture Organization of the United Nations (FAO) and World Health Organization (WHO). 2017, amended in 2021 and 2024. Standard for Fish Oils. Codex Alimentarius International Food Standards, No. CXC 329-2017. Codex Alimentarius Commission. Rome. Available at: https://www.fao.org/fao-who-codexalimentarius/sh-proxy/es/?lnk=1&url=https%253A%252F%252Fworkspace.fao.org%252Fsites%252Fcodex%252FStandards%252FCXS%2B329-2017%252FCXS_329e.pdf

Food and Agriculture Organization of the United Nations (FAO). 2025. Land and Water Databases and Software. Crop Information. Sunflower. Available at: https://www.fao.org/land-water/databases-and-software/crop-information/sunflower/en/

Gaffield, K. N.; Goodband, R. D.; DeRouchey, J. M.; Tokach, M. D.; Woodworth, J. C.; Denny, G.; Gebhardt, J. T. 2024. A review of soybean processing byproducts and their use in swine and poultry diets. *Translation Animal Science (Translat. Anim. Sci.*), 8;txae063. https://doi.org/10.1093/tas/txae063

Gao, R.; Li, Z.; Zhou, X.; Bao, W.; Cheng, S.; Zheng, L. 2020. Enhanced lipid production by Yarrowia lipolytica cultured with synthetic and wastederived highcontent volatile fatty acids under alkaline conditions. *Biotechnology for Biofuels (Biotechnol. Biofuels*), 13:3. https://doi.org/10.1186/s13068-019-1645-y

Glencross, B. D. 2009. Exploring the nutritional demand for essential fatty acids by aquaculture species. *Reviews in Aquaculture (Rev. Aquac.),* 1:74-124. https://doi.org/10.1111/j.1753-5131.2009.01006.x

Glencross, B. D.; Bachis, E.; Betancor, M. B.; Calder, P.; Liland, N.; Newton, R.; Ruyter B. 2025. Omega-3 Futures in Aquaculture: Exploring the Supply and Demands for Long-Chain Omega-3 Essential Fatty Acids by Aquaculture Species. *Reviews in Fisheries Science & Aquaculture (Rev. Fish. Sci. Aquac.*), 33:167-216. https://doi.org/10.1080/23308249.2024.2388563

Glasser, F.; Schmidely, P.; Sauvant, D.; Doreau, M. 2008. Digestion of fatty acids in ruminants: a meta-analysis of flows and variation factors: 2. C18 fatty acids. Animal, 2:691-704. https://doi.org/10.1017/S1751731108002036

Institute of Shortening and Edible Oils Technical Committee (ISEO). 2016. Food fats and oils. Nineth Edition. Washington, DC. Available at: https://edibleoilproducers.org/wp-content/uploads/2025/03/foodfatsoils2016.pdf

Kaur, K.; Torrecillas, S. 2025. Impact of krill products on the growth, health, and fillet quality of farmed non-salmonid fish. *Aquaculture,* 608:742723. https://doi.org/10.1016/j.aquaculture.2025.742723

Kerr, B. J.; Kellner, T. A.; Shurson, G. C. 2015. Characteristics of lipids and their feeding value in swine diets. *Journal of Animal Science and Biotechnology (J. Anim. Sci. Biotechnol.*), 6:30. https://doi.org/10.1186/s40104-015-0028-x

Kim, T.-B.; Lee, J.-S.; Cho, S.-Y.; Lee, H.-G. 2020. In Vitro and In Vivo Studies of Rumen-Protected Microencapsulated Supplement Comprising Linseed Oil, Vitamin E, Rosemary Extract, and Hydrogenated Palm Oil on Rumen Fermentation, Physiological Profile, Milk Yield, and Milk Composition in Dairy Cows. *Animals,* 10:1631. https://doi.org/10.3390/ani10091631

Klinmalai, P.; Sodsai, J.; Promhuad, K.; Srisa, A.; Laorenza, Y.; Kovitvadhi, A.; Areerat, S.; Seubsai, A.; Harnkarnsujarit, N. 2025. Modern Palatant Strategies in Dry and Wet Pet Food: Formulation Technologies, Patent Innovations, and Market Evolution. *Foods*, 14:2824. https://doi.org/10.3390/foods14162824

Kröncke, N.; Wittke, S.; Steinmann, N.; Benning, R. 2023. Analysis of the Composition of Different Instars of Tenebrio molitor Larvae using Near-Infrared Reflectance Spectroscopy for Prediction of Amino and Fatty Acid Content. *Insects,* 14: 310. https://doi.org/10.3390/insects14040310

Laflamme, D. P. 2005. Nutrition for Aging Cats and Dogs and the Importance of Body Condition. Veterinary Clinics of North America. *Small Animal Practice (Vet. Clin. North Am. Small Anim. Pract.*), 35:713-742. https://doi.org/10.1016/j.cvsm.2004.12.011

Lenox, C. E. 2015. An overview of fatty acids in companion animal medicine. *Veterinary Medicine Today (Vet. Med. Today*), 246:1198-1202. https://doi.org/10.2460/javma.246.11.1198

Ma, X.-N.; Chen, T.-P.; Yang, B.; Liu, J.; Chen, F. 2016. Lipid Production from *Nannochloropsis. Marine Drugs (Mar. Drugs*), 14:61. https://doi.org/10.3390/md14040061

Maia, M. R. G; Chaudhary, L. C.; Bestwick, C. S.; Richardson, A. J.; McKain, N.; Larson, T. R.; Graham, I. A.; Wallace, R. J. 2010. Toxicity of unsaturated fatty acids to the biohydrogenating ruminal bacterium, Butyrivibrio fibrisolvens. *BMC Microbiology,* 10:52. http://www.biomedcentral.com/1471-2180/10/52

Menoyo, D.; López-Bote, C.; Obach, A.; Mautista, J. M. 2005. Effect of dietary fish oil substitution with linseed oil on the performance, tissue fatty acid profile, metabolism, and oxidative stabillity of Atlantic salmon. *Journal of Animal Science (J. Anim. Sci.*), 83:2853-2862. https://doi.org/10.2527/2005.83122853x

Meijaard, E.; Brooks, T. M.; Carlson, K. M.; Slade, E. M.; Garcia-Ulloa, J.; Gaveau, D. L. A.; Lee, J. S. H.; Santika, T.; Juffe-Bignoli, D.; Struebig, M. J.; Wich, S. A.; Ancrenaz, M.; Koh, L. P.; Zamira, N.; Abrams, J. F.; Prins, H. H. T.; Sendashonga, C. N.; Murdiyarso, D.; Furumo, P. R.; Macfarlane, N.; Hoffmann, R.; Persio, M.; Descals, A.; Szantoi, Z.; Sheil, D. 2020. The environmental impacts of palm oil in context. *Nature Plants (Nat. Plants.*), 6:1418-1426. https://doi.org/10.1038/s41477-020-00813-w

Miller, M. R.; Nichols, P. D.; Carter, C. G. 2008. n-3 oil sources for use in aquaculture – alternatives to the unsustainable harvest of wild fish. *Nutrition Research Reviews (Nutr. Res. Rev.*), 21:85-96. https://doi.org/10.1017/S0954422408102414

Moallem, U. 2018. Invited review: Roles of dietary n-3 fatty acids in performance, milk fat composition, and reproductive and immune systems in dairy cattle. *Journal of Dairy Science (J. Dairy Sci.*), 101:8641-8661. https://doi.org/10.3168/jds.2018-14772

Nates, S. F.; Suresh, A. V. 2011. Fat components, energy in poultry byproduct meal provide valuable nutrients to shrimp. Global Seafood Alliance 16. Available at: https://www.globalseafood.org/advocate/fat-components-energy-in-poultry-byproduct-meal-provide-valuable-nutrients-to-shrimp/

National Research Council (NRC). 2011. *Nutritient Requirements of Fish and Shrimp*. The National Academies Press , Washington, DC, United States. https://doi.org/10.17226/13039

National Research Council (NRC). 2012. Nutrient Requirements of Swine: Eleventh Revised Edition. Washington, DC: The National Academies Press. https://doi.org/10.17226/13298

Ng, W.-K. 2002. Potential of palm oil utilisation in aquaculture feeds. Asia Pacific *Journal of Clinical Nutrition (Asia Pac. J. Clin. Nutr.*), 11:473-476. https://doi.org/10.1046/j.1440-6047.11.s.7.7.x

Nobre, A. D.; Mendonça, R. V.; de S. Farias, A. B.; Yamamoto, F. Y.; Gonçalves, L. U. 2025. Dietary Schizochytrium sp. Meal Enhances the Fatty Acid Profile in Pirarucu (Arapaima gigas) Fillets with No Effect on Growth Performance and Health Status. *Animals,* 15:712. https://doi.org/10.3390/ani15050712

Official Journal of the European Union (OJEU). 2011. Commission Regulation (EU) No. 142/2011 of 25 February 2011 implementing Regulation (EC) No. 1069/2009 of the European Parliament and of the Council laying down health rules as regards animal by-products and derived products not intended for human consumption and implementing Council Directive 97/78/EC as regards certain samples and items exempt from veterinary checks at the border under that Directive. 26.2.2011. https://www.eurl.craw.eu/wp-content/uploads/2019/10/142_2011-en.pdf

Oonincx, D. G. A. B.; Finke, M. D. 2020. Nutritional value of insects and ways to manipulate their composition. *Journal of Insects as Food and Feed (J. Insects Food Feed)*, 7:639-660. https://doi.org/10.3920/JIFF2020.0050

Palmquist, D. L.; Jenkins, T. C. 2017. A 100-Year Review: Fat feeding of dietary cows. *Journal of Dairy Science (J. Dairy Sci.)*, 100:10061-10077. https://doi.org/10.3168/jds.2017-12924

Panasevich, M. R.; Daristotle, L.; Yamka, R. M.; Frantz, N. Z. 2022. Dietary Ground Flaxseed Increases Serum Alpha-Linolenic Acid Concentrations in Adult Cats. *Animals*, 12:2543. https://doi.org/10.3390/ani12192543

Plasek, B.; Lakner, Z.; Badak-Kerti, K.; Kovács, A.; Temesi, A. 2021. Perceived Consequences: General or Specific? The Case of Palm Oil-Free Products. *Sustainability*, 13:3550. https://doi.org/10.3390/su13063550

Popa, I.; Watson, A. L.; Solgadi, A.; Butowski, C.; Allaway, D.; Portoukalian, J. 2018. Linoleate-enriched diet increases both linoleic acid esterified to omega hydroxy very long chain fatty acids and free ceramides of canine stratum corneum without effect on protein-bound ceramides and skin barrier function. *Archives of Dermatological Research (Arch. Dematol. Res.)*, 310:579. https://doi.org/10.1007/s00403-018-1845-5

Purushothaman, D.; Brown, W. Y.; Wu, S.-B.; Vanselow, B. A. 2011. Evaluation of breed effects on n-3 PUFA metabolism with dietary flaxseed oil supplementation in dogs. *British Journal of Nutrition (Brit. J. Nutr.)*, 106:S139-S141. https://doi.org/10.1017/S0007114511000523

Purushothaman, D.; Brown, W. Y.; Vanselow, B. A.; Quinn, K.; Wu, S.-B. 2014. Flaxseed oil supplementation alters the expression of inflammatory-related genes in dogs. *Genetics and Molecular Research (Genet. Mol. Res.)*, 13:5322–5332. https://doi.org/10.4238/2014.July.24.11

Rabiee, A. R.; Breinhild, K.; Scott, W.; Golder, H. M.; Block, E.; Lean, I. J. 2012. Effect of fat addition to diets of dairy cattle on milk production and components: A meta-analysis and meta-regression. *Journal of Dairy Science (J. Dairy Sci.)*, 95:3225-3247. http://dx.doi.org/10.3168/jds.2011-4895

Ravindran, V. 2013. Poultry feed availability and nutrition in developing countries. Food and Agriculture Organization of the United States (FAO). Available at: https://www.fao.org/4/al703e/al703e00.pdf

Ravindran, V.; Tancharoenrat, P.; Zaefarian, F.; Ravindran, G. 2016. Fats in poultry nutrition: Digestive physiology and factors influencing their utilisation. *Animal Feed Science and Technology (Anim. Feed Sci. Technol.)*, 213:1-21. https://doi.org/10.1016/j.anifeedsci.2016.01.012

Richards, T. L.; Burron, S.; Ma, D. W. L.; Pearson, W.; Trevizan, L.; Minikhiem, D.; Grant, C.; Patterson, K.; Shoveller, A. K. 2023. Effects of dietary camelina, flaxseed, and canola oil supplementation on inflammatory and oxidative markers, transepidermal water loss, and coat quality in healthy adult dogs. *Frontiers in Veterinary Science (Front. Vet. Sci.)*, 9:1085890. https://doi.org/10.3389/fvets.2023.1085890

Rodríguez-Rodríguez, M. F.; Moreno-Pérez, A. J.; Makni, S.; Troncoso-Ponce, M. A.; Acket, S.; Thomasset, B.; Sánchez, R.; Venegas-Calderón, M.; Garcés, R.; Martínez-Force, E.; Salas, J. J. 2021. Lipid profiling and oil properties of *Camelina sativa* seeds engineered to enhance the production of saturated and omega-7 fatty acids. *Industrial Crops & Products (Ind. Crop. Prod.)*, 170:113765. https://doi.org/10.1016/j.indcrop.2021.113765

Rokey G. J.; Plattner, B.; de Souza, E. M. 2010. Feed extrusion process description. *Revista Brasileira de Zootecnica (R. Bras. Zootec.)*, 39:510-518. https://doi.org/10.1590/S1516-3598201000 1300055

Sánchez-Muros, M.-J.; Barroso, F. G.; Manzano-Agugliaro, F. 2014. Insect meal as renewable source of food for animal feeding: a review. *Journal of Cleaner Production (J. Clean. Pro.)*, 65:16-27. https://doi.org/10.1016/j.jclepro.2013.11.068

Sauvant, D.; Perez, J.-M. and Tran G. 2004. Tables of Composition and Nutritional Values of Feed Animals. Wageningen Academic Publishers, Wageningen, the Netherlands, 192-284. https://doi.org/10.3920/978-90-8686-668-7. (English transation of the French original work)

Savilaakso, S.; Laumonier, Y.; Guariguata, M. R.; Nasi, R. 2013. Does production of oil palm, soybean, or jatropha change biodiversity and ecosystem functions in tropical forests. *Environmental Evidence*, 2:17. https://doi.org/10.1186/2047-2382-2-17

Spain, J. N. 2002. Essential of Specific Fatty Acids in Reproductive Performance of High Producing Dairy Cows. *Advances in Dairy Technology,* 14:185.192. No DOI

Srisuksai, K.; Limudomporn, P.; Kovitvadhi, U.; Thongsuwan, K.; Imaram, W.; Lertchaiyongphanit, R.; Sareepoch, T.; Kovitvadhi, A.; Fungfuang, W. 2024. Physicochemical properties and fatty acid profile of oil extracted from black soldier fly larvae (*Hermetia illucens*). *Veterinary World (Vet. World),* 17:518-526. https://www.doi.org/10.14202/vetworld.2024.518-526

Takeno, S.; Sakuradani, E.; Tomi, A.; Inohara-Ochiai, M.; Kawashima, H.; Ashikari, T.; Shimizu, S. 2005. Improvement of the Fatty Acid Composition of an Oil-Producing Filamentous Fungus, Mortierella alpina 1S-4, through RNA Interference with 12-Desaturase Gene Expression. *Applied and Environmental Microbiology (Appl. Environ. Microbiolo.)*, 71:5124-5128. https://doi.org/10.1128/AEM.71.9.5124-5128.2005

Tian, F.; Decker, E. A.; Goddard, J. M. 2013. Controlling lipid oxidation of food by active packaging technologies. *Food & Function (Food Funct.)*, 4:669. https://doi.org/10.1039/c3fo30360h

Tocher, D. R. 2015. Omega-3 long-chain polyunsaturated fatty acids and aquaculture in perspective. *Aquaculture,* 449:94-107. https://doi.org/10.1016/j.aquaculture.2015.01.010

Tomkins, T.; Drackley, J. K. 2010. Applications of Palm Oil in Animal Nutrition. *Journal of Oil Palm Research (J. Oil Palm Res.)*, 22:835-845. https://palmoilis.mpob.gov.my/publications/joprv22dec10-Trevor.pdf

Toral, P. G.; Renna, M.; Frutos, P.; Gasco, L.; Hervás, G. 2025. Insect fat as feed: Potential to modify fatty acid composition of animal-derived foods. *Animal Nutrition (Anim. Nutr.)*, 22:179-190. https://doi.org/10.1016/j.aninu.2025.03.016

Tzompa-Sosa, D. A.; Dewettinck, K.; Provija, P.; Brouwers, J. F.; de Meulenaer, B.; Oonincx, D. G. A. B. 2021. Lipidome of cricket species used as food. *Food Chemistry,* (Food Chem.), 349:129077. https://doi.org/10.1016/j.foodchem.2021.129077

United Nations Department of Economic and Social Affairs (UN). 2022. World Population Prospects 2022 Summary of Results. 27th edition of official UN population estimates and projections. Available at: https://www.un.org/development/desa/pd/sites/www.un.org.development.desa.pd/files/wpp2022_summary_of_results.pdf

Vaskovsky, V. E.; Khotimchenko, S. V.; Xia, B.; Hefang, L. 1996. Polar lipids and fatty acids of some marine macrophytes from the yellow sea. *Polychemistry*, 42:1347-1356. https://doi.org/10.1016/0031-9422(96)00117-3

Vecchiato, C. G.; Schwaiger, K.; Biagi, G.; Dobenecker, B. 2022. From Nutritional Adequacy to Hygiene Quality: A Detailed Assessment of Commercial Raw Pet-Food for Dogs and Cats. *Animals,* 12:2395. https://doi.org/10.3390/ani12182395

Veldkamp, T.; Bosch, G. 2015. Insects: a protein-rich feed ingredient in pig and poultry diets. *Animal Frontiers*, 5:45-50. https://doi.org/10.2527/af.2015-0019

Voora, V.; Bermudez, S.; Le, H.; Larrea, C.; Luna, E. 2024. Global Market Report: Soybean prices and sustainability. International Institute for Sustainable Development (IISD). State of Sustainability Initiatives (SSI). Available at: https://www.iisd.org/system/files/2024-02/2024-global-market-report-soybean.pdf

Woodard & Curran, Inc. 2006. Wastes from Industries (Case Studies) in Industrial Waste Treatment Handbook (second edition), Butterwotrh-Heinemann, pp 409-496. https://doi.org/10.1016/b978-075067963-3/50012-6

Wu, G.; Bayer, F. W.; Dai, Z.; Li, D.; Wang, J.; Wu, Z. 2014. Amino Acid Nutrition in Animals: Protein Synthesis and Beyond. *Annual Review of Animal Biosciences (Annu. Rev. Anim. Biosci.*), 2:387-417. https://doi.org/10.1146/annurev-animal-022513-114113

Yuangklang, C.; Vasupen, K.; Wongsuthavas, S.; Beynen, A. C. 2016. Digestibility of a Structural Fat Consisting of Stearic and Palmitic Acid in Dogs. *Journal of Mahanakorn Veterinary Medicine (J. Mahanakorn Vet. Med.*), 11:47–56. Available at: https://li01.tci-thaijo.org/index.php/jmvm/article/view/142699

Zeng, X.; Li, S.; Liu, L.; Cai, S.; Ye, Q.; Xue, B.; Wang, X.; Zhang, S.; Chen, F.; Cai, C.; Wang, F.; Zeng, X. 2023. Role of functional fatty acids in modulation of reproductive potential in livestock. *Journal of Animal Science and Biotechnology (J. Anim. Sci. Biotechnol.*), 14:24. https://doi.org/10.1186/s40104-022-00818-9

Zornig, W. O.; Pesti, G. M.; Bakalli, R. I. 2001. The essential fatty acid requirements of broilers. *Journal of Applied Poultry Research (J. Appl. Poultry Res.*), 10:45. https://doi.org/10.1093/japr/10.1.41

PART 3

Stability concerns and shelf-life of novel ingredients

CHAPTER 5

Lipid oxidation as a quality factor in food: reactions and products, degradation effects, and analyses

K.M. Schaich

Dept. of Food Science, Rutgers University, 65 Dudley Rd., New Brunswick, NJ 08901-8520, USA
Correspondence: schaich@sebs.rutgers.edu

Abstract

Lipid oxidation in animal feed and pet foods is gaining new attention as higher consumer and legal standards for quality, nutrition, and safety of animal food products are being established. Though well-established as a free radical chain reaction, lipid oxidation has remained difficult to control, produces many products at times not predicted by traditional theory, and induces some effects that cannot be explained. In addition, lipid oxidation in foods is now being implicated as a potential factor in inflammatory bowel diseases for both animals and humans. Thus, there are compelling reasons to better understand lipid oxidation and use new knowledge to improve stabilization of foods against the process.

In traditional theory, lipid peroxyl (LOO·) and alkoxyl (LO·) radicals drive oxidation forward via hydrogen atom (H) abstractions alone. This chapter presents a new view of lipid oxidation which adds alternate competing reactions of these radicals to H abstractions to more fully account for the broad range of products found in rancid materials and appearing early and throughout the oxidation process, not just in late stages, as well as to explain why products change in type as well as quantity under different conditions. Overlooking these alternate pathways in the past probably contributed in major ways to defects in control strategies and failure to maintain food qualities and nutritional value.

To show why lipid oxidation is so important, one section provides an overview of lipid oxidation impacts and consequences in foods, from changes in flavor, odor, and texture to loss of nutritional value and potential health effects. The chapter closes with consideration of chemical and instrumental analyses that can be combined to document oxidation activity in the multiple pathways to more completely and accurately account for lipid oxidation in foods.

Keywords

epoxides – oxylipins – product profiles

1 Introduction

Although stabilizing foods to protect physical, sensory, and nutritional quality during shipping and storage has always been a front-line focus for human foods, it received less attention for animals until recently. Now, new standards for animal nutrition that set specific baselines for nutrient composition of feed for different animals, requirements for nutritional labeling, and increased recognition that animals, like humans, have sensitive palates and their health is compromised by low quality feed are forcing industry to rethink how they formulate and stabilize food materials for animals.

That animals are sensitive to tastes and textures in their food is no surprise to pet owners, farmers, and even researchers who find that animals reduce consumption or outright reject foods with off-flavors, off-odors, or strange textures. In addition, animals frequently can sense the presence of toxins in foods, and then they do not consume them (Skelhorn and Rowe, 2010). Here, the issue is not rotted food, but materials that have changed chemistry during processing, transport, or storage, allowing production of molecules that trigger animal sensors. Chief among chemical degradation modes (as distinguished from microbial spoilage) in any food is oxidation of unsaturated fatty acids in triacylglycerols, phospholipids, glycolipids, and other complex lipids that are critical components in structure and function of animal and plant tissues.

Recognized by consumers as 'rancidity' with characteristic off-odors and flavors, lipid oxidation is a free radical reaction that, once started, is difficult to stop and leads to accumulation of many kinds of products that mediate considerable deterioration of food properties.

Controlling lipid oxidation has always presented hurdles, but meeting new requirements for assuring nutritional value and chemical safety in addition to desirable sensory properties will require much deeper understanding of chemistry involved in lipid oxidation processes than were used to move animal feed and pet foods to their current standards.

To provide a strong base of working knowledge of oxidation chemistry that can be applied to interpret degradation patterns in animal food materials as well as to develop more effective stabilization approaches for maintaining their sensory qualities, nutritional value, and safety, this chapter will outline the complex multiple lipid oxidation pathways that cause oxidative damage in foods, review degradation of food qualities caused by lipid oxidation, and briefly discuss analyses of multiple oxidation products that most labs can apply to track active pathways and determine the state and extent of lipid oxidation.

To be sure, the view of lipid oxidation chemistry presented here is relatively new, but the incorporation of multiple alternate pathways explains many aspects of oxidation that have presented quandaries in the past. The goal of this chapter is to encourage developers, producers, and researchers of animal food materials to "think chemistry", analyze products broadly, and apply new information thus gained to solve our continuing problems with maintaining stability, sensory quality, nutritional value, and safety of foods for our animals.

2 Lipid oxidation: basic reactions and generation of products

2.1 Traditional free radical chain

For more than 80 years, lipid oxidation has been recognized as a chain reaction in which an initiator such as light, heat, or metals generates an initial lipid free radical (L·) localized on a carbon atom in the acyl chain. Oxygen almost immediately adds to this radical to form a peroxyl radical (LOO·), which then abstracts a hydrogen atom H from a neighboring lipid molecule to form a relatively stable lipid hydroperoxide, LOOH, plus a new radical on the target lipid (L_1·). LOOH can accumulate until light, heat, or transition metals decompose them to alkoxyl radicals (LO·) plus hydroxyl radicals (HO·) and/or hydroxide anions (HO^-). LO· then extend the radical chain by also abstracting H atoms from neighboring molecules, and since their abstraction rates are greater than by LOO·, the rate of oxidation increases. At some point, various lipid radicals recombine to form a variety of final non-radical products. A simple version of this process is shown in Figure 5.1. More details about the reactions are available in (Frankel, 1984; 1985; 1991; Frankel, 2012; Schaich, 2005; Schaich *et al.*, 2013).

Since the 1950s, this free radical chain process has been the main guide to following and controlling lipid oxidation in both food and non-food materials. Indeed, nearly all current analysis and protection protocols derive from this traditional mechanism. Hydroperoxides are the major product measured. Initiators start the radical process so extensive efforts have been focused on chelating metals in for-

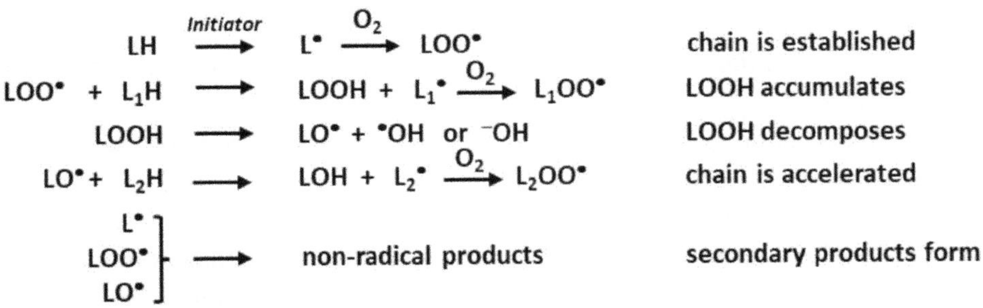

FIGURE 5.1 Steps in traditional free radical chain of lipid oxidation

mulations, blocking light during processing and via packaging, and reducing heat in processing and storage. Oxygen converts relatively unreactive L• radicals into very active LOO• and LO• radicals, so inert gas headspaces and vacuum packaging are applied extensively to limit O_2 access. Lipid radicals extend reaction chains by abstracting H atoms from other lipids, so phenolic compounds and other antioxidants are added as competitors to quench lipid radicals without forming new reactive ones.

Despite its extensive application, the traditional reaction scheme has not been able to explain all lipid oxidation patterns, products, and kinetics observed in food systems. As the traditional scheme is written, all products derive from hydroperoxides, accumulate only *after* LOOH decomposition, and form only via radical recombination. These requirements defy common observations of secondary products early in some oxidations, even when LOOH are still actively increasing, they fail to account for generation of aldehydes and other secondary products that are major sources of off-odors and flavors, and the radical recombination products prescribed are not the dominant compounds observed in lipids from oxidized oils and foods. Thus, the simple radical chain reactions as typically written, e.g. in Figure 5.1, are not incorrect, but they are incomplete and should not be used alone to track oxidation progression, calculate kinetics, or plan stabilization strategies.

2.2 Competing alternate pathways of lipid oxidation

What is missing in the traditional oxidation reactions? Free radicals are very unstable and as soon as they form, they seek a way to immediately pair the odd electron. If abstractable hydrogens are not close by, the free electron finds another route to reduce its energy. Substantial research on oxygen free radical reactions has documented that both alkoxyl and peroxyl radicals undergo multiple alternate reactions such as double bond additions, rearrangements, bond scissions, and disproportionations in addition to and in competition with hydrogen abstractions to form hydroperoxides (Gardner, 1989; Howard *et al.*, 1968; Ingold, 1967; Ingold, 1969; Ingold, 1978; Kochi, 1962; 1973a; b). Occurrence of such alternate reactions in oxidizing lipids could significantly change the direction, kinetics, and products of lipid oxidation and thus explain

- products that were previously ignored or considered unimportant,
- detection of previously-considered secondary products such as epoxides and aldehydes at all stages of oxidation, both early and late,
- generation of different patterns of products with different kinetics in different materials, and
- how lipid oxidation is a dynamic process that changes with conditions.

One alternate reaction actually replaces hydrogen abstraction and bypasses hydroperoxides; other alternate reactions supplement hydrogen abstraction and generate product mixes that are more complex than expected from traditional oxidation.

Most critically, when alternate reactions are active, determining oxidation kinetics and progress only by conjugated dienes and hydroperoxides, even perhaps with hexanal included, can seriously underestimate the extent and breadth of lipid oxidation.

The reaction scheme shown in Figure 5.2 provides one view of how alternate competing reactions of LOO· and LO· radicals can be integrated into the traditional free radical chain reaction (Schaich, 2005; Schaich, 2020a). The core process – the free radical chain of hydrogen abstractions to generate hydroperoxides – is shown vertically down the center of the scheme. Then added to this are competing reactions that divert peroxyl and alkoxyl radicals to yield different intermediate products and secondary radicals at new sites. When drawn this way, it is clear just how extensively alternative reaction paths increase the complexity of both the kinetics and the product distributions of lipid oxidation. Products from these pathways must be included with LOOH to accurately determine the extent and direction of lipid oxidation and they must be considered when developing all stabilization protocols to build in controls of all deterioration modes.

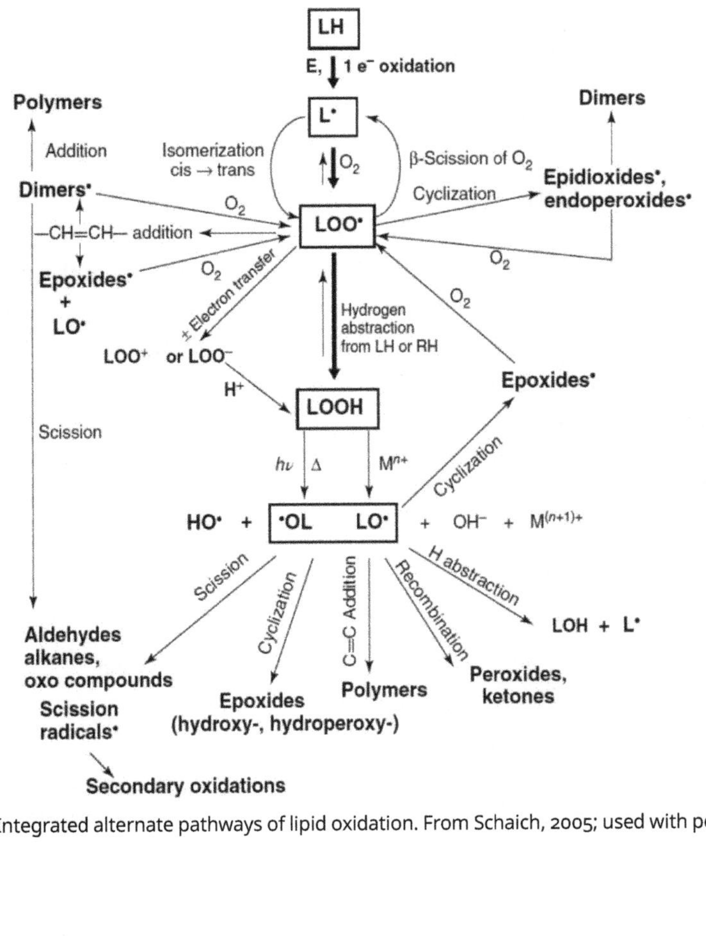

FIGURE 5.2 Integrated alternate pathways of lipid oxidation. From Schaich, 2005; used with permission.

Figures 5.1 and 5.2 show the flow of reactions in lipid oxidation. The following discussion outlines how each of the reactions in the multiple pathways modify lipids. Due to space limitations, initiations and LOOH decompositions to LO· are considered givens, then the basic chemistry of the remaining reactions is explained, starting with hydrogen abstractions that drive the radical chain reactions, then proceeding in sequence through the competing reactions shown in Figure 5.2.

Readers are referred to (Schaich, 2005; Schaich, 2020a; 2023) for additional details, background, and references on all the alternate reactions.

3 Establishing the radical chain

3.1 Hydrogen abstractions by LOO• and LO• propagate oxidation chains and generate LOOH and LOH

Once a lipid radical L· is formed, oxygen adds to it almost instantaneously to form LOO· radicals, then these radicals and derivative LO·, *not* the initiators, drive free radical chains forward by transferring radicals serially to up to several hundred other lipid molecules via hydrogen abstractions. Each lipid oxyl radical removes an H atom from a neighboring lipid molecule to form a non-radical hydroperoxide (Reaction 1) or alcohol (Reaction 1a) and a new radical on the second lipid.

$$LOO\cdot + L_1H \rightarrow LOOH + L_1\cdot \quad (k = 62\ LM^{-1}s^{-1}) \tag{1}$$

$$LO\cdot + L_1H \rightarrow LOH + L_1\cdot \quad (k = 10^{7-8}\ LM^{-1}s^{-1}) \tag{1a}$$

Hydrogen abstraction is not random but occurs at carbons with the lowest C-H bond energies: doubly-allylic carbons between two double bonds (65 kcal/mol) or allylic carbons next to single double bonds (74 kcal/mol) (Kerr, 1966; Scott, 1965) (Figure 5.3). The most susceptible positions for fatty acids with one, two, and three double bonds are marked in Figure 5.4. The extra doubly allylic carbons are one reason why oxidation increases disproportionately with the number of fatty acid double bonds.

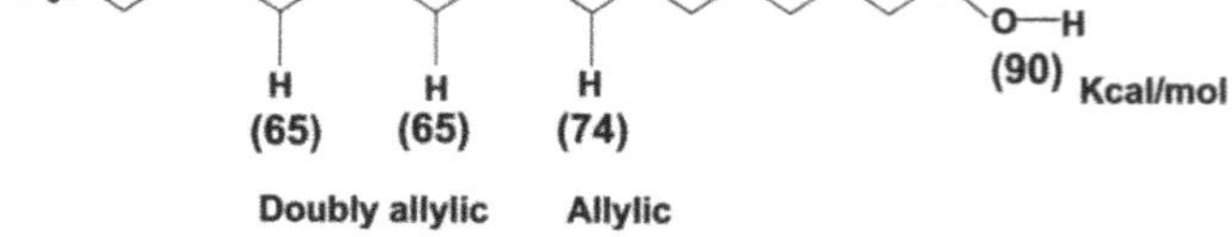

FIGURE 5.3 Hydrogen abstraction occurs preferentially at acyl carbons with lower energy, i.e. next to and between double bonds. From Schaich (2013b), used with permission.

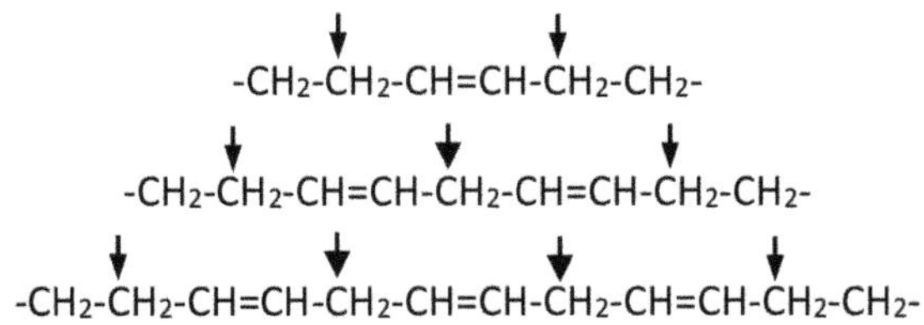

FIGURE 5.4 Dominant carbon positions for H abstractions from lipid acyl chains. Heavy arrows indicate highest preference for abstraction.

A unique feature of lipid oxidation is that double bonds are not destroyed – they either remain in place or migrate. After H abstraction next to the double bond, the free electron e^- distributes from its source carbon over the double bond to form a resonance system where electron density concentrates in the center, thus creating electron-deficient exterior carbons to which oxygen adds (Figure 5.5). When oxygen adds to a resonance carbon originally in the double bond, the double bond migrates one carbon over and reforms while inverting from *cis* to *trans*, as shown for 1,4-dienes in Figure 5.5 (Farmer *et al.*, 1943).

This pattern of H abstraction and double bond migration is responsible for two critical features in lipid oxidation. First, regardless of the number of double bonds in polyunsaturated fatty acids, the dominant position for hydroperoxide formation is on the outside carbons of the double bond system, e.g. C9 and C13 in linoleic acid, regardless of the number of double bonds (Figure 5.6). Hydroperoxides at internal positions are minor products in autoxidation, but become dominant products in singlet oxygen photosensitized oxidation, which provides a means of distinguishing light catalysis in samples. Second, double bond migration in polyunsaturated fatty acids generates conjugated dienes (-CH=CH-CH=CH-), the first chemical indicator that oxidation has started. Conjugated diene structures are retained when

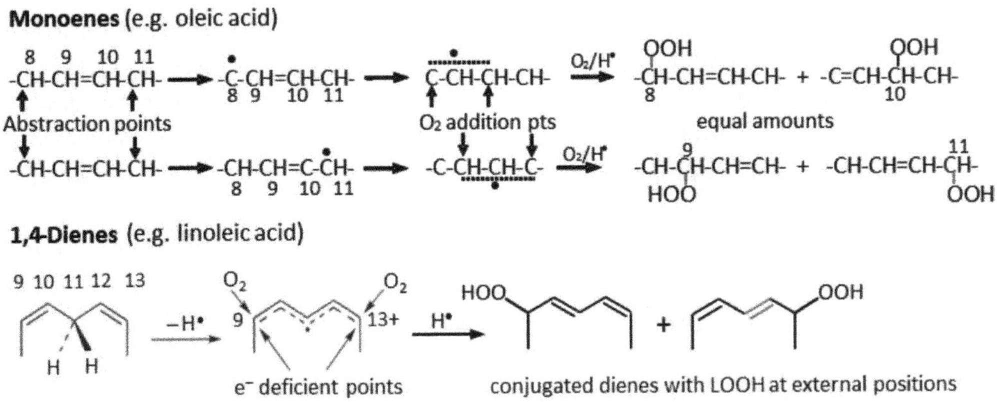

FIGURE 5.5 Electron delocalization in resonance structures determines preferred positions of hydroperoxide formation in oxidizing lipids

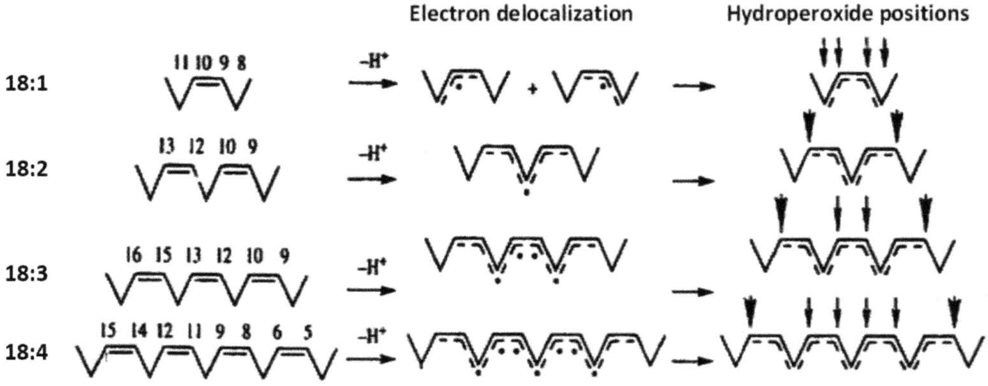

FIGURE 5.6 Dominant positions of hydroperoxide formation in oxidized polyunsaturated fatty acids. From Schaich (2005), used with permission.

hydroperoxides decompose to LO•, so are perhaps even more important indicators than hydroperoxides for detecting early oxidation.

One additional important H abstraction site must be noted. Once radical chains are established, LOO• and LO• also abstract H atoms from lipid hydroperoxides in a lesser known but faster reaction (Reactions 2 and 2a).

$$\mathrm{LOO\cdot + L_1OOH \rightarrow LOOH + L_1OO\cdot} \tag{2}$$

$$\mathrm{LO\cdot + L_1OOH \rightarrow LOH + L_1O\cdot} \tag{2a}$$

Abstractions of H from LOOH play critical roles in maintaining the progress of oxidation by recycling static LOOH and LOH to their reactive radical forms, thus eliminating the need for other catalysts to decompose the hydroperoxides to expand the radical chain.

How do H abstractions by LOO• and LO• differ? LOO• are the main chain carriers in early oxidation. With lower energy, their H abstractions are slower ($k = 62\ LM^{-1}s^{-1}$) and more selective, mainly from doubly allylic -CH_2- groups between two double bonds. LOO• get the chain started and established. LO• are much stronger oxidizers, react faster ($k=2.5 \times 10^8\ M^{-1}s^{-1}$) and less selectively, so they abstract H from all allylic carbons and become the dominant reaction driver after LOOH decomposition (Schaich, 2005). These actions give them a key role in increasing oxidation rates, mediating chain expansion and branching once the main radical chain is established, as shown in Figure 5.7, and directing final product pathways in competition with cyclization to epoxides and scission to aldehydes as will be discussed later. In reality, though, new chains are constantly being initiated as earlier ones are in secondary stages, so practical oxidations in food materials involve both LOO• and LO• simultaneously, each contributing uniquely and independently to the progression and outcome of lipid oxidation.

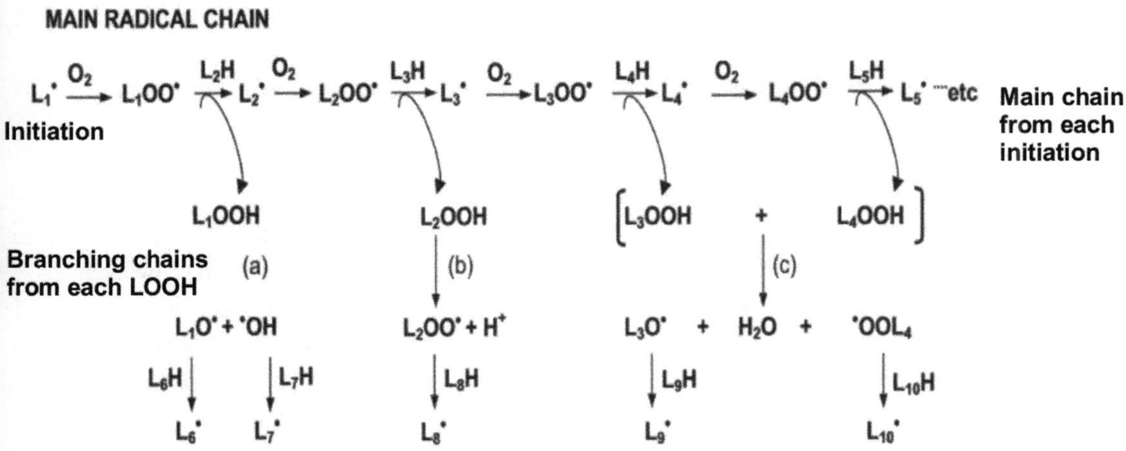

FIGURE 5.7 Differentiating propagation of the main radical chain by LOO• from chain expansion and branching by LO• in lipid autoxidation. Both processes are driven by H abstractions. Modified from Schaich 2005), used with permission.

4 Going from LOO˙ to LOOH… Reactions that compete with H abstraction during propagation

4.1 β-Scission of O_2 from LOO• to regenerate L• and trans double bonds

The rapid addition of O_2 to initial L· to generate LOO· is not a static process. If LOO· do not contact abstractable H atoms fast enough, the oxygen is released from L· by β-scission, leaving an alkyl radical behind (Reaction 3).

$$LOO\cdot \rightarrow L\cdot + O_2 \tag{3}$$

β-scission of oxygen has two significant consequences. Continuously repeating O_2 scission contributes to the Induction Period, a time when oxidation is barely measurable by oxygen consumption or hydroperoxides (Campbell *et al.*, 1974). Previously, this initial oxidation lag was attributed to ineffective initiation or presence of endogenous antioxidants that block radical propagations (Labuza, 1971). Importantly, β-scission of O_2 from LOO· also mediates positional and *cis,trans* isomerization of double bonds during oxidation. Without OO on the carbon radical, the carbon bond in the acyl chain can rotate, allowing the free radical to migrate within the full double bond resonance system of the fatty acid. When oxygen adds again, radicals may be at a new carbon position or chains may be in a different geometric configuration. The process stops with reformation of a hydroperoxide and resumes if the LOOH hydrogen is abstracted (Porter *et al.*, 1981; Porter *et al.*, 1980; Porter and Wujek, 1984; Porter and Wujek, 1987).

β-elimination is most important when abstractable hydrogens are limited (at low lipid concentrations) and at elevated temperatures. Its activity can be deduced by the presence of high conjugated dienes with very low levels of other products, including LOOH.

4.2 Addition of LOO• to double bonds to generate epoxides

LOO· add readily to conjugated double bonds (Sims and Hoffman, 1962), so for decades the reaction was connected only to later stages of oxidation. However, critical roles of LOO· addition to double bonds in early oxidation have now been demonstrated. LOO· add to isolated or non-conjugated double bonds faster than they abstract H atoms, forming a peroxo dimer on one carbon and a radical on the second carbon of the double bond (Reaction 4) (Mayo, 1968). The free radical induces scission of the O-O bonds in the dimers, then the LO· on the base lipid adds to the radical to form an internal epoxide while LO· from the adding lipid is released (Reaction 4a) (Gardner, 1989). Here, the epoxide is the stable product and the LO· radical chain takes over expansion of oxidation by propagating the oxidation chain (Reaction 4b), forming epoxides by rearrangement (Reaction 4c and subsection below), or undergoes scission to generate aldehydes and other compounds (Reaction 4d and subsection below).

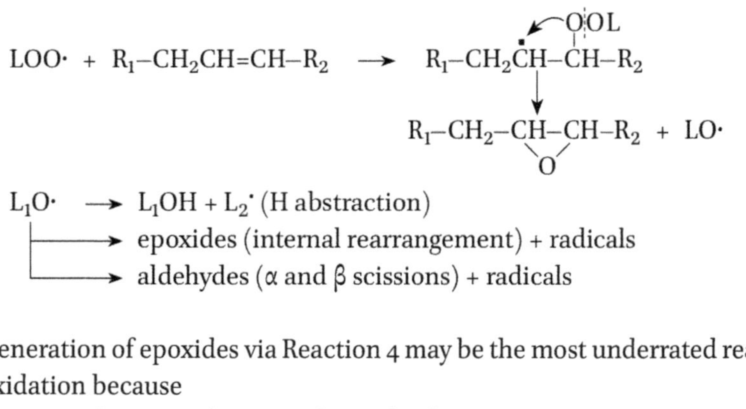

(4)

(4a)

$L_1O\cdot \rightarrow L_1OH + L_2^{\cdot}$ (H abstraction) (4b)

$\rightarrow$ epoxides (internal rearrangement) + radicals (4c)

$\rightarrow$ aldehydes (α and β scissions) + radicals (4d)

Generation of epoxides via Reaction 4 may be the most underrated reaction of lipid oxidation because

- it provides an oxidation pathway that bypasses LOOH,
- it diverts LOO· products from expected LOOH to two sources of epoxides, one that forms early in parallel with LOOH but is seldom analyzed, and another that forms via rearrangement of released alkoxyl radicals continuously throughout the tenure of oxidation,
- it effectively accelerates oxidation by releasing faster-reacting LO· without forming LOOH first, and
- it generates epoxides and aldehydes directly at the very beginning of oxidation without hydroperoxides as intermediates (Xie, 2015).

4.3 Internal cyclization of LOO• to epidioxides

When abstractable hydrogen atoms are not immediately available, internal peroxyl radicals can find pairing electrons by adding to downstream *cis* double bonds in the same acyl chain to form cyclic peroxides called epidioxides (Reactions 5 and 5a). To have sufficient space, LOO· add to the first carbon of cis double bonds two carbons away to generate the epidioxides and form free radicals on the distal carbons of the double bonds (Taube, 1965). Addition of oxygen generates new peroxyl radicals that may either propagate radical chains by hydrogen abstraction or undergo any of the other alternate reactions of peroxyl radicals discussed in this chapter.

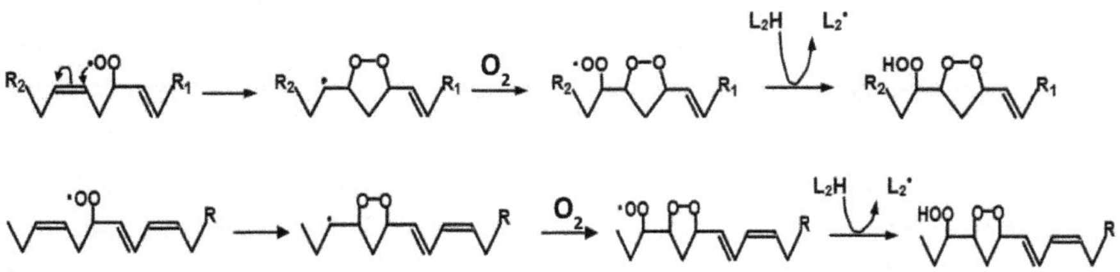

(5)

(5a)

In fatty acids with four or more double bonds, acyl chains have hair pin-like configurations that bring double bonds from opposite ends into close proximity that facilitate LOO· internal additions. In such arrangements, epidioxides can form linearly, as in Reactions 5 and 5a or cross-chain to form endoperoxides and prostaglandin-like structures as shown in Reactions 6 and 6a (Funk *et al.*, 1975; Porter, 1986). These compounds have pharmacological action, so their potential formation when fish meals or oils are ingredients in animal feed should be given serious attention.

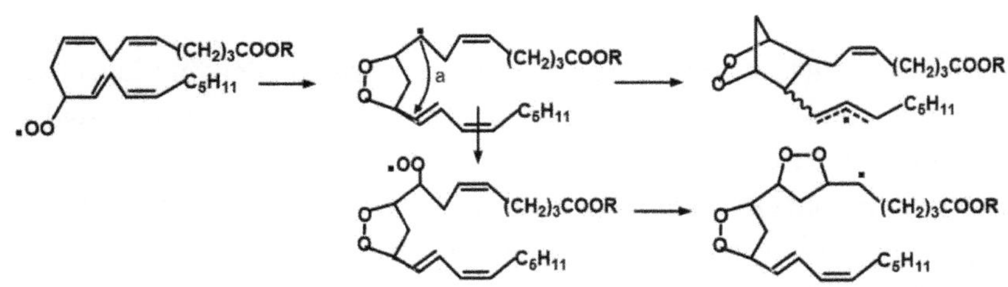

(6)

(6a)

Epidioxides are also major products in $^1O_2$-photosensitized oxidations (Chan *et al.*, 1980; Neff *et al.*, 1988). Indeed, internal cyclic products are characteristic markers distinguishing autoxidation from $^1O_2$-photosensitized oxidation.

4.4 Disproportionation of LOO•

When high concentrations of peroxyl radicals are generated or accumulate in a localized region that brings them into close contact, two peroxyl radicals can join and decompose in a process called disproportionation (self-reaction). Here, the peroxyl radicals join to form a tetroxide which then decomposes to release oxygen and produce two alkoxyl radicals (Reaction 7, most common) or a peroxide, LOOL (Reaction 7a).

$$L_1OO^\bullet + L_2OO^\bullet \longrightarrow [L_1OOOOL_2] \longrightarrow L_1O^\bullet + {}^\bullet OOOL \longrightarrow L_1O^\bullet + O_2 + {}^\bullet OL_2 \quad (7)$$

$$[L_1OOOOL_2] \longrightarrow L_1OOL_2 + O_2 \quad (7a)$$

Disproportionation is a favored reaction only in lipids oxidizing as pure oils, for example in oils being stored or thermally processed as ingredients. In aqueous systems such as emulsions, other reactions are faster. Even so, disproportionation has important consequences. Converting two relatively slow-reacting peroxyl radicals to two alkoxyl radicals that are 1000x more reactive as hydrogen abstractors greatly accelerates propagation and oxidation (Lindsay *et al.*, 1973; Traylor and Russell, 1965). At lower temperatures or high viscosity that slow LO· movements, the radicals do not separate but join to create an alkyl peroxide (L_1OOL_2) in a termination reaction (Reaction 7a). This pathway forms peroxides early and without hydroperoxide involvement. Finally, both pathways of disproportionation release oxygen back into the headspace or sample to provide fuel to maintain the radical chain. Such action is especially important when oxygen is limited in samples, e.g. due to vacuum packaging or inert gas headspace.

5 Reactions that compete with chain propagation by LO˙ in secondary oxidation stages

Anticipating LO· reactions and products is particularly challenging because, although H abstraction to extend and expand the radical chain is considered paramount, several reactions compete to change reaction direction and product distributions. Rate constants show that all of the alternate reactions of LO· are competitive under conditions in most foods and oils (Table 5.1). The balance between alternate pathways changes with the dominant fatty acids, the conditions in the materials, and the environment, so all pathways need to be accounted for when assessing lipid oxidation.

TABLE 5.1 Rate constants for alternate competing reactions of t-butoxyl radicals with linoleic acid*

LO•	Target	k_{Habs} L $M^{-1}s^{-1}$	k_{cycliz} s^{-1}	k_{addn} $M^{-1}s^{-1}$	$k_{scission}$ $M^{-1}s^{-1}$	$k_{disprop}$ $M^{-1}s^{-1}$
t-BuO•	L	8.8×10^6	10^4-10^5	10^4-10^8	10^4-10^5	10^9-10^{10}

* Data excerpted from Schaich (2005)

5.1 Scission of LO· to generate aldehydes, alkanes, and other products

LO· scissions play key roles in secondary propagation but also termination and generation of products in lipid oxidation. Scissions of the C-C bond occur on either side of the alkoxyl radical group (α is on the acid end, β is on the distal end, Reaction 8), so each LO· position on a fatty acid theoretically generates two scission pairs (four scission products). Each scission pair has one carbon chain fragment with a terminal C· radical that either abstracts a hydrogen atom to form stable alkanes of alkenes (Reaction 8a) while starting a new radical chain, or adds O_2 directly to generate a new peroxyl radical on the now shorter acyl chain (Reaction 8a). This radical continues the original chain (Reaction 8b) and forms additional downstream volatile compounds and flavor components that provide undertones and secondary notes that round out flavors. Recombination of the fragment radicals produces dimers and polymers that increase viscosity. The second product from each scission contains the alkoxyl group and two free electrons (Reaction 8); this rearranges to the aldehyde products that are associated with major off-odors and off-flavors of lipid oxidation (Reaction 8a).

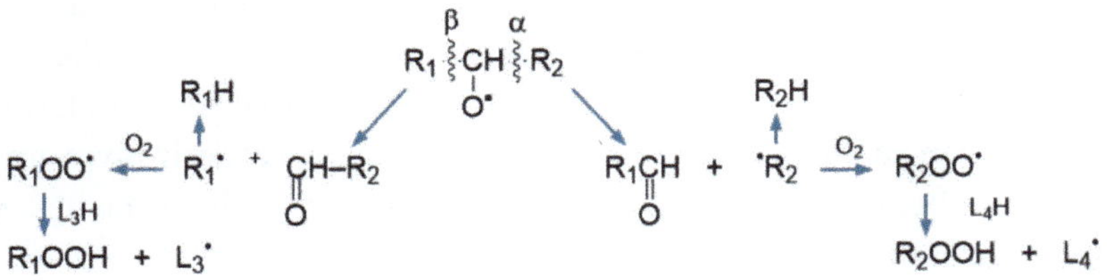

(8)

(8a)

(8b)

Considering the large number of products that can result from scission of a single LO·, it is easy to see how product mixtures can become quite complex when alkoxyl radicals form and react at multiple positions on polyunsaturated fatty acids.

Scissions are not all equivalent in either chain position of LO· or scission point relative to LO·. For example, 9- and 13-LO· in linoleic acid do not give the same levels or distributions of scission products. The final product mixture can be all α products, all β products, or a variable mixture of the two depending on the LO· structure and the reaction conditions. Still, the consistency and reproducibility of the patterns of aldehydes deriving from specific LO· on major fatty acids allow individual aldehydes to be used as markers for activity of scission reactions. See the Termination and Effects of Lipid Oxidation sections below for more details.

In terms of competition, LO· scission is somewhat slower than H-abstraction, with rate constants on the order of 10^5 s^{-1} in neat lipids or aprotic solvents, but rates increase in water and polar solvents such as alcohols ($k=10^6$-10^7 s^{-1}) where strong H bonding and charge stabilization of the polar products drive the reaction forward (Erben-Russ *et al.*, 1987). Thus, scission ranges from negligible to a minor process (less than 5–10% of total products) in oxidized fatty acids and oils but become more important in polar solvents or when water is present, as in emulsions. This difference provides one explanation for variations in "rancid" off-flavor and odor intensities in different types of foods, for example pet food kibbles vs wet patés.

5.2 Internal cyclization of LO• to epoxides

Scission of LO· must compete with not only H abstraction and chain propagation but also rapid cyclization to epoxides for contributions to lipid oxidation. Internal cyclization of LO· to epoxides (Reaction 9) occurs when abstractable H atoms are not immediately available within the reaction radius, and the alkoxyl radical instead adds to an allylic double bond to form an epoxide with the first carbon of the double bond and a radical on the distal carbon. The reaction is very fast ($k=10^4$-$10^5 s^{-1}$); it can even exceed rates of H abstraction by LO· under some conditions (Wilcox and Marnett, 1993) and is comparable to scission.

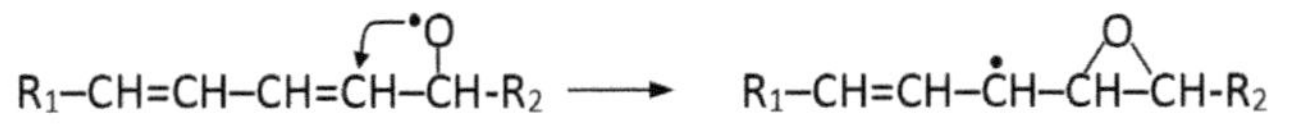

(9)

Importantly, the allylic radical now becomes the chain carrier– it adds oxygen to form peroxyl radicals, hydroperoxides, and alkoxyl radicals next to the epoxide (Reaction 10). As with other alkoxyl radicals, the epoxy LO· then can either abstract an H to extend the chain further and form hydroxy epoxide products(Chan *et al.*, 1980), or can undergo scission to form oxo epoxides or terminal epoxides (Reaction 9a), both of which have higher potential toxicity than the internal epoxides formed by LOO· via Reaction 4a (Wade *et al.*, 1978). These secondary oxy derivatives are of particular concern because LO· continue their generation throughout the lifetime of lipid oxidation, from early to late.

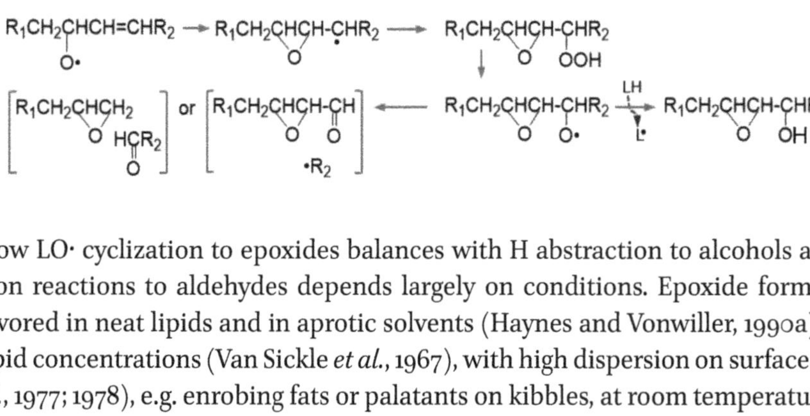

(10)

How LO· cyclization to epoxides balances with H abstraction to alcohols and scission reactions to aldehydes depends largely on conditions. Epoxide formation is favored in neat lipids and in aprotic solvents (Haynes and Vonwiller, 1990a), at low lipid concentrations (Van Sickle *et al.*, 1967), with high dispersion on surfaces (Wu *et al.*, 1977; 1978), e.g. enrobing fats or palatants on kibbles, at room temperature (Bors *et al.*, 1984) and at low oxygen pressures (Mayo *et al.*, 1958; Van Sickle *et al.*, 1967). However, α and β scissions are facilitated in polar solvents, as in extracts (Kochi, 1962c) and H abstractions are faster in high water systems (Bors *et al.*, 1984), as in food emulsions.

5.3 Addition of LO• to double bonds

Like LOO·, LO· add to cis double bonds in neighboring acyl chains (Reaction 11), although not readily due to competition with hydrogen abstraction (Reaction 1a) and cyclization to epoxides (Reaction 9). The addition itself creates a dimer with a radical that adds oxygen and continues the oxidation chain. Repeat additions of downstream LO· generate polymers (Mounts *et al.*, 1970).

(11)

LO· additions are most active in secondary stages of oxidation when allylic hydrogens are less available for abstraction and double bonds are conjugated, conditions that develop as oxidation progresses. Even under these conditions, though, addition occurs only to *cis* double bonds (Walling and Thaler, 1961). LO· additions are most important in oils rather than emulsions (Bors *et al.*, 1984), and also increase with temperature (Mounts *et al.*, 1970), as in processing and frying.

6 Terminating the chain: radicals → products

In lipid oxidation, "termination" is almost a misnomer since lipid oxidation never fully stops. A *specific radical* may be terminated and form some product, but usually there are still radicals left behind on other lipids *so the chain reaction continues.* Net oxidation slows down when radical quenching processes exceed the rate of new chain production, and the momentum of oxidation shifts from radical propagation and chain expansion to generation of stable products. Thus, 'termination' should be thought of as converting an individual lipid radical to a product, not stopping the overall reaction. The cumulative effect on any food system is determined by the number of radical chains being terminated and the products generated therein.

Lipid free radicals are terminated, i.e. form non-radical products, by four major mechanisms.

6.1 Radical recombination to form products

The seemingly limitless number of variations possible for radical recombination certainly contributes to the broad range of oxidation products detected in lipid oxidation. Nonetheless, the distinct patterns of favored recombinations identified, as shown in Reactions 12–18, add to the types of products generated in lipid oxidation but are not the dominant products usually detected or known to affect food qualities.

Alkyl radicals

$$R_1^\bullet + R_2^\bullet \longrightarrow R_1\text{-}R_2 \qquad \text{alkane polymers} \tag{12}$$

Peroxyl radicals

$$2\ R_1CH(OO^\bullet)R_2 \longrightarrow R_1(R_2)HC\text{-OOOO-}CH(R_1)R_2 \longrightarrow R_1CH(OH)R_2 + R_1C(=O)R_2 + O_2 \tag{13}$$

$$2\ ROO^\bullet \xrightarrow{-O_2} R^\bullet \xrightarrow{+ROO^\bullet} ROOR \tag{14}$$

Alkoxyl radicals

$$R_1O^\bullet + R_2O^\bullet \longrightarrow R_1OOR_2 \quad \text{peroxides} \qquad (15)$$

$$R_1O^\bullet + R_2^\bullet \longrightarrow R_1OR_2 \quad \text{ethers} \qquad (16)$$

$$R_1\text{-}\underset{\displaystyle O^\bullet}{\underset{|}{C}}H\text{-}R_2 + R^\bullet \longrightarrow R_1\text{-}\underset{\displaystyle O}{\underset{||}{C}}\text{-}R_2 + RH \quad \text{ketones, alkanes} \qquad (17)$$

$$R_1\text{-}\underset{\displaystyle O^\bullet}{\underset{|}{C}}H\text{-}R_2 + RO^\bullet \longrightarrow R_1\text{-}\underset{\displaystyle O}{\underset{||}{C}}\text{-}R_2 + ROH \quad \text{ketones, alcohols} \qquad (18)$$

Radical recombinations depend strongly on reactions conditions, particularly oxygen levels and temperature. L· reactions dominate under low oxygen (pO_2 = 1 to about 80–100 mm Hg). Because oxyl radicals LOO· and LO·are more reactive than alkyl radicals L·, chain termination by recombination increases with oxygen pressure and extent of oxidation, and the dominant recombinations involve increasing percentages of LO· then LOO· (Figure 5.8, left). LO· contributions to the product mix dominate when LOOH or LOO· decompositions are faster than their formation, e.g. in secondary stages of oxidation and at moderate temperatures and oxygen pressures. At high pO_2, LOO· may form rapidly, but with high LOO·, recombinations must compete with faster additions (Reaction 4) or disproportionation (Reaction 6). Perhaps contrary to expectations, recombinations involving oxyl radicals decrease with temperature (Figure 5.8, right). Lower oxygen solubility at higher temperatures has been offered as an explanation (Labuza, 1971) but increased transformation of oxyl radicals by other reactions at elevated temperatures is more likely.

6.2 Scission reactions of alkoxyl radicals

Alkoxyl radical scissions were introduced under alternate reactions above, noting how the fragment radicals thus formed contributed to chain propagation (Reaction 8), but these same scissions are even more significant as a termination pathway. In termination, alkoxyl radical fragments generate the aldehydes responsible for

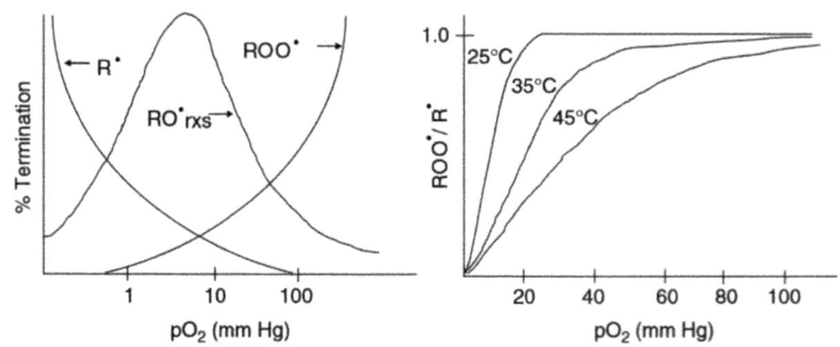

FIGURE 5.8 Effect of oxygen and heat on radical dominance in recombinations (left) and on rates of termination by radical recombinations (right). Adapted from Labuza (1971) by Schaich (2005); used with permission.

the rancid odors and off-flavors most recognized by consumers. The alkane fragments, too, contribute products – alkanes by H abstraction, dimers and polymers by radical recombinations, and eventually downstream oxidation products that contribute flavors and browning potential. Significance of these products will be discussed further under the section on effects of lipid oxidation: generation of off-flavors and odors.

One closely-related reaction that occurs downstream of LO· scissions and is often overlooked or not recognized is the oxidation of scission aldehydes to carboxylic acids whenever oxygen is present during handling, heating, and storage (Repko-Reader, 2014; Tian, 2013). This is shown with hexanal as an example in Reaction 19.

$$CH_3(CH_2)_4CHO \rightarrow CH_3(CH_2)_4COOH \quad (19)$$

Aldehyde oxidation to acids is the reaction that drives Rancimat and Oxygen Stability Index analyses. Because these short chain acids from oxidation are detected along with normal fatty acids in acid testing, their presence is often misinterpreted as deriving from fatty acid hydrolysis.

6.3 Group eliminations

HO^- and HOO^- can be eliminated from LOOH, yielding ketones and products with additional double bonds (Reaction 20) (Bothe *et al.*, 1978; Terao *et al.*, 1975). These are not major reactions but do account for some lipid oxidation products, especially in the presence of metals.

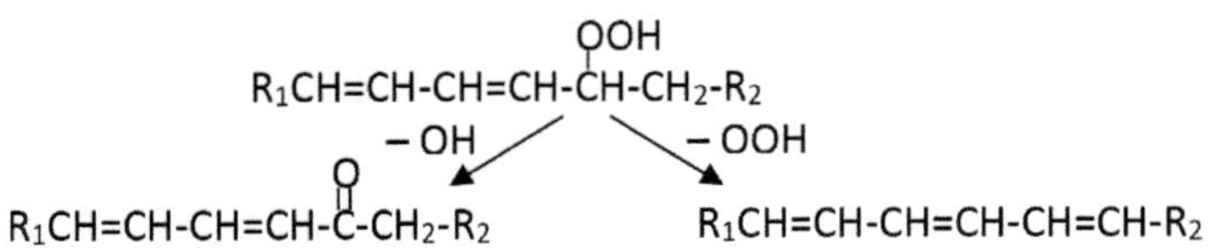

(20)

6.4 Co-oxidation of non-lipid molecules

The three termination reactions just discussed are important because they stop individual chains and slow the overall progress of lipid oxidation. Indeed, their direct effects are limited to lipids. However, the last termination reaction (co-oxidation) has far greater significance and impact on overall food quality, nutritional value, and safety because it extends and broadcasts oxidation from lipids to other molecules so that lipid oxidation becomes system oxidation with far greater degradation in quality.

Lipids transfer radicals to any other molecules that have groups with abstractable H atoms or double bond sites for addition, as shown in Reactions 21 and 21a, respectively, where T is any target molecule such as proteins, vitamins, antioxidants, pigments, and also starches in food materials. This stalls oxidation in lipids but activates oxidation in other molecules, leading to broader damage in the materials.

$$LOO^{\bullet} + TH \longrightarrow LOOH + T^{\bullet} \xrightarrow{O_2} TOO^{\bullet} \xrightarrow{RH} TOOH \quad (21)$$

$$LOO^{\bullet} + \text{-C=C-C=C-} \longrightarrow \text{-C=C-}\dot{\text{C}}\text{-}\underset{\text{OOL}}{\text{C}}\text{-} \longrightarrow \text{additional oxidation} \quad (21a)$$

Radicals alone drive lipid oxidation forward, but they are not the only reactive species generated in oxidizing lipids. Epoxides, in particular, and also aldehydes react with other molecules, particularly proteins, antioxidants, and vitamins to form adducts and sometimes crosslinks (Schaich, 2008). Although co-oxidations via epoxides and aldehydes make one product at a time without radical chains, their impact on quality, nutrition, and safety nevertheless is substantial, as will be discussed in detail for proteins in the accompanying chapter on Protein Co-oxidation Reactions, Products, and Analyses.

Co-oxidations have been largely disregarded until recently, but this must change if problems with human and animal food quality and safety are to be solved. When lipid radicals are not regenerated and their oxidation products are removed by reaction, in analyses lipid oxidation may seem to progress less rapidly or not at all. Nevertheless, at the same time, shifts from tell-tale rancid odors and flavors to increased browning, changes in textures and functional properties, as well as formation of different off-flavors and odors that are not associated directly with lipids indicate that product quality is deteriorating. Too often such degradation is misinterpreted as arising from physical stresses such as heat, shear, and moisture loss during processing, and focus for control and prevention is misdirected. To be sure, processing stresses cause damage to food molecules and qualities, including lipid oxidation, and damage from physical forces needs to be distinguished from chemical damage to develop appropriate and effective controls. Much more research is needed in this area.

Take-home message: co-oxidation products must be determined in addition to lipid oxidation to track true extent of oxidation in complex systems.

7 Why must alternate reactions and co-oxidations be integrated with traditional radical chain reactions to understand and control lipid oxidation in food materials?

Until recently, the alternate reactions described above were considered minor contributors that only generated secondary products, and co-oxidations were largely disregarded. However, competing alternate reactions of oxidizing lipids and co-oxidations of other molecules have several critical impacts on food quality, nutrition, and safety:

1. Multiple LOO· and LO· reaction pathways occur simultaneously and in competition with each other to form mixtures of products at all stages, not just at

end of chains. These competing reactions of LOO· and LO· alter propagation sequences and thus change kinetics, dominant products, detectability, and effects of lipid oxidation dramatically and in complex ways when oxidation occurs under varied conditions.

2. A combination of product analyses (as many as possible) is needed to fingerprint all alternate pathways and accurately measure the full extent of lipid oxidation especially in early stages. If the dominant pathway is not being monitored, an inaccurate picture of the rate, extent, and character of lipid oxidation is generated and oxidation levels are misinterpreted. In particular, peroxide values alone under conditions favoring cyclization or scission may miss much of the lipid change altogether.
3. Lipid oxidation does not proceed by the same pathways, just at different rates, at all temperatures. Major changes in oxidation mechanism and product distributions occur at 40, 60, 100, 120, and 150 °C due to shifts in the activation energies and the balance between the alternate pathways. Analyzing and interpreting degradation by only a few products will not reflect these differences accurately, underestimate degradation tremendously, and fail to detect reactions and products that cause important damage.
4. All changes caused by lipid oxidation (flavors, odors, browning, texture alterations, color loss, etc) derive from specific lipid oxidation products. Shifts away from H abstraction means aldehydes, epoxides, dimers or other products are formed in place of alcohols, and this alters food qualities and stability patterns markedly. Understanding how lipid oxidation affects food properties thus requires analysis of a full range of lipid oxidation products.
5. Toxicity and safety issues. Off-odors from volatile scission products are the consumer's first clue that foods are rancid and should not be eaten. Shifts from scission to internal rearrangement and addition reactions may reduce some off-odors, but the epoxides thus generated are more toxic and react more rapidly with DNA and proteins, while dimers have reduced digestibility and some evidence of toxicity. More importantly, these products have no sensory warning signals for consumers.
6. Finally, developing the most effective strategies for inhibition of lipid oxidation requires information about dominant and active propagation pathways in each food system under different conditions, then targeting antioxidant approaches to specific reactions. For example, phenolic antioxidants quench lipid radicals but also shift pathways away from addition and cyclization to LOOH formation. This may seem like a positive effect, but in polar systems where LOOH decomposition is fast and LO· scission is dominant, the net effect (especially at high antioxidant concentrations) may be enhanced production of off-flavors and odors. To achieve long-term stability, therefore, antioxidant approaches must be tailored specifically to control all active propagation pathways.

8 Effects of lipid oxidation on quality degradation of human and animal food quality

As noted in the Introduction, lipid oxidation is the chemical process responsible for the greatest degradation of quality in human and animal foods, and the complex chemical reactions outlined in the previous section help explain why. However, it is the impact of the reactions and their products, rather than the pathways per se, that consumers perceive. Here, lipid oxidation is just as complicated and extensive in its effects (Figure 5.9) as in its pathways because lipid radicals and nearly all lipid oxidation intermediates and products contribute to deterioration of food attributes in distinctive ways. The brief overview here provides merely a minimal justification for controlling lipid oxidation in food, addressing food properties affected by lipid products directly and by co-oxidations, and also health effects mediated by both.

8.1 Degradation of sensory properties

Off-odors and flavors. Best known of lipid oxidation quality effects in food are off-odors and flavors. Consumers may not know reaction mechanisms or responsible products, but they certainly recognize characteristic oily rancid odors and off-flavors in foods. Animals experience taste sensations similar to humans (Yarmolinsky *et al.*, 2009) so, like humans, they also detect rancid off-odors and flavors and reject or reduce intake of foods exuding them (Becker and Lawrence, 2025; Kimura *et al.*, 1984; Skelhorn and Rowe, 2010). Largely responsible for these smells and flavors are aldehydes, ketones, and other products generated in α and β scissions of LO· and other secondary reactions. Typical flavors or odors associated with specific compounds, well-documented by flavor studies, link changes in foods to lipid oxidation and even make it possible to track oxidation pathways in some cases (Waltking and Goetz, 1983). More often, however, hundreds of compounds intermix to generate complex major notes and undertones that account for overall flavors, and contributions of most secondary oxidation compounds in these mixes remain to be elucidated. Table 5.2 lists flavors/odors that are known for common lipid oxidation products as examples, along with their thresholds for perception in oil and water.

Two key features must be noted in Table 5.2. First, the threshold values for smelling and tasting lipid oxidation products are very low – ppm for taste and ppb levels or lower for smell. Humans and animals can readily detect even trace levels of oxidation products, so food is rejected with less than 1% of all lipids oxidized (Labuza, 1971). Thus, tiny differences in oxidation have huge impacts on sensory qualities of foods, even for animals. Second, the threshold for detection is much higher in oils or oil phases, due to impeded volatilization and because oils interfere with taste receptors on the tongue. In contrast, thresholds for perception of oxidation products are orders of magnitude lower in water (Belitz and Grosch, 1986), so oxidation in milk, salad dressings and other products with mixed phases is tasted and smelled much more rapidly than in the oils from which they were prepared. The same pattern should apply to animal feed materials, e.g. oxidation will be more readily detectable in fresh, wet canned foods than dry feed and kibbles.

Notable also in Table 5.2 is the *absence* of flavors or odors contributed from lipid epoxides and alcohols. Lipid alcohols are known to be bitter, as shown in stored oats (Doehlert *et al.*, 2010), yet lipid alcohol contents and notation of bitter flavors are rarely reported in foods, Whether this is due to low LOH formation or high thresholds for perception is not known. For epoxides, neither flavors nor odors have yet been reported. Epoxide adhesives and reagents have strong odors, so it is somewhat surprising that related odors have not been detected in oxidizing foods. Maybe no odors are released because epoxides bind readily to proteins. Considering the toxic and physiochemical potential of lipid epoxides and alcohols, respectively, identifying flavors and odors associated with these two lipid oxidation products would seem to be an important protection and cue for food rejection.

Off-flavors that derive from co-oxidations of proteins and starches and cause food rejection will be discussed in the next chapter on Protein Co-oxidation Reactions, Products, and Analyses.

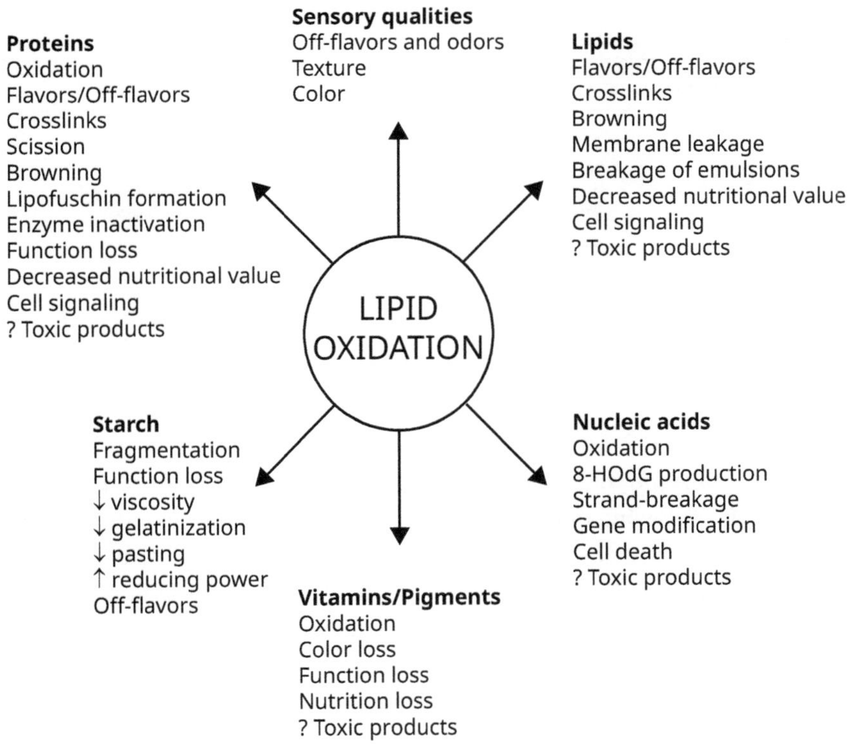

FIGURE 5.9 Molecular and property degradation mediated by oxidizing lipids. From Schaich (2013b), used with permission.

TABLE 5.2 Odor/flavor characteristics and perception thresholds of selected lipid oxidation products

		Threshold value (ppm)[a]		
		In oil		In water
Compound[a]		**Odor**	**Taste**	**Odor**
Aldehydes				
Pentanal	Sharp, bitter almond	0.24	0.15	0.012
Hexanal	Green-fruity, bitter almond	0.32	0.08	0.008
Octanal	Fatty, soapy-fruity	0.32	0.04	0.0007
Nonanal	Tallowy, soapy-fruity	13.5	0.2	0.001
Decanal	Orange peels	6.7	0.7	0.0001
Nonenal (3c)	Green cucumber	0.25	0.03	
Nonenal (2t)	Tallowy, starch-glue	3.5	0.04	0.0008
Nonadienal (2t,4t)	Fatty, oily	2.5	0.46	
Nonadienal (2t,6c)	Cucumbers	0.01	0.0015	
Nonadienal (2t,6t)	Tallowy, green	0.21	0.018	
Decadienal (2t,4c)	Frying odor		0.02	
Decadienal (2t,4t)	Deep-fried	2.15	0.1	
Ketones and furans				
1-Pentene-3-one	Sharp, fishy	0.003		0.001
1-Octen-3-one	Metallic	0.0001		0.0001
1-Octen-3-ol	Moldy, mushroomy	0.077	0.0001	0.00009
2-Pentylfuran	Buttery, beany	2		
2-(1-pentenyl)furan[b] (cis and trans)	Licorice	2 to 6		

Excerpted from Belitz and Grosch, 1986: [a]p.166, [b]p.168.

Yellowing, browning and fading. Three pathways change color of foods with oxidizing lipids. Yellowing derives from condensations of lipid carbonyl products, as can be seen in oils as they age (Kumarathasan *et al.*, 1992). Browning derives from Maillard and Michael addition reactions with proteins and amino acids in foods (Hidalgo and Zamora, 2000; Zamora and Hidalgo, 2005), and fading results from radical or carbonyl oxidation of pigments, especially carotenoids (Anguelova and Warthesen, 2000) and myoglobin (Faustman *et al.*, 1999; Faustman *et al.*, 2010).

Texture changes. Textures in foods are created by molecular interactions, and modification of these alters structures in multiple directions. As just a few examples, disruption of acyl chain associations in triacylglycerols decreases melting points and disorganizes crystal structures (Yang *et al.*, 2024a), shifting β to β' or α crystals. Lipid co-oxidation of proteins and starches degrade macromolecular starches to break down native structures but also create crosslinks that harden and stiffen structures (Schaich, 2008). These reactions will be discussed in detail in the accompanying chapter on Co-oxidation. Oxidation also changes emulsion interfaces and lipid phase viscosity, which predisposes emulsions to destabilization (Ghelichi et al., 2023).

8.2 Health impacts of lipid oxidation and co-oxidations

Loss of nutritional value. Perhaps the first major impact of lipid oxidation in foods is loss of essential unsaturated fatty acids because oxidation decreases levels of linoleic, linolenic, and higher n-3 fatty acids (Kara, 2021), as well as co-oxidized proteins, fat-soluble vitamins like A, D, E, and K, and antioxidants. While this can be tolerated for a single meal or short-term challenge, continued deprivation can cause oxidative stress and deficiencies that reduce health and lifespan.

Reduced food intake and lipid digestibility. Off-flavors and odors in oxidized foods lead to reduced consumption or outright food rejection by both animals and humans (Hayam *et al.*, 1995). In animals, this usually results in weight loss due to caloric restriction, to decreased digestion and absorption of consumed lipids (Márquez-Ruiz *et al.*, 1992), and to blocked synthesis of new lipids (Eder, 1999; Eder *et al.*, 2003). Decreased digestibility results from impaired lipase recognition and release of fatty acids, lack of absorption of lipid dimers and polymers, and disruption of micellar organization involved in fatty acid uptake (Marquez-Ruiz *et al.*, 2008).

Are oxidized lipids consumed in foods toxic? There is no question that lipid radicals, epoxides, and many of its aldehydes are cytotoxic and/or mutagenic when formed in vivo or exposed directly to cells in culture (Alghazeer *et al.*, 2008; Cortesi and Privett, 1972; Eckl and Bresgen, 2017; Esterbauer, 1993; Esterbauer *et al.*, 1991; Greene *et al.*, 2000; Moran *et al.*, 1997; O'Brien *et al.*, 2005; Sevanian, 1990). However, whether oxidized lipids are toxic when they are consumed in foods is a complex and controversial question because it is so difficult to track digestion, absorption, and distribution of individual classes of molecules without complications. In-depth analysis of published studies (Schaich, 2020b) showed that types and severity of toxic responses in animals vary significantly with levels of lipids and oxidation products consumed as well as with adequacy of essential fatty acids and glutathione or its amino acid precursors glutamine and cysteine in the diet. Many studies have reported systemic toxicity of consumed oxidized lipids, but at the same time, substantial evidence documents that animals and humans effectively detoxify lipid oxidation products in consumed foods and that only LOH are absorbed. Most 'toxicity' reported resulted from dietary lipid or oxidation product overload.

Two issues of concern arise from this evidence. The question of long-term effects of low-level oxidation exposure in animal feed was raised three decades ago (Engberg *et al.*, 1996), yet answers still are not available. Clearly, studies are needed to determine levels of oxidation tolerable in animal and human diets.

Second, what causes observed changes in physiological chemistry if lipid oxidation products are not absorbed? Here, accumulating evidence indicates that oxidizing lipids consumed in food actively interact with mucosal proteins and other molecules in early sections of the gastrointestinal tract. These interactions have two opposing effects. Some protein changes mediate signal transduction to upregulate production of detoxification enzymes and genes (Sulzle *et al.*, 2004) and mobilize liver and plasma antioxidants (Eder *et al.*, 2003) that direct protective and adaptive responses in tissues. While such changes have been interpreted as toxicity, they really reflect detoxification responses. Countering these are molecular changes that induce inflammation, degrade mucosal membranes, and impair physiological functions, as will be discussed in the following section. Factors controlling balance between protection and damage need to be explored.

Readers are referred to (Schaich, 2020b) for an extensive review of the complex issues involved in this controversy, including effects in animals.

Impairment of physiological function. Mucosa cells in the stomach and upper small intestine are directly exposed to lipid oxidation products in foods as they pass through the gastrointestinal tract. This makes them major targets for reaction and damage, and indeed, modifications of these cells and signal transduction initiated by them explain many pathologies reported for dietary oxidized lipids (Schaich, 2020b). The organization and differentiation from crypt to epithelium tip cells in intestinal villi are very sensitive to thiol redox balance (Aw, 1999). Oxidizing lipids disrupt this, causing deranged metabolism and differentiation in the cells. Inflammation and release of peptides that activate signal transduction are always active. In the epithelium, repair dominates at low oxidation with increased apoptosis (epithelial denuding) and proliferation to replace damaged cells, while damage and necrosis take over at high oxidation and long-term exposure, (Circu and Aw, 2012). Levels of aberrant crypts increase, providing pre-cancerous foci (Yang *et al.*, 1998). Such changes are likely contributors to the current marked rise in digestion disturbances and inflammatory bowel syndrome reported for both humans (Almario *et al.*, 2018) and animals (K.L. Rogers, DVM, personal communication). Also in the epithelium, alteration of Peyer's patches impairs immune function, e.g. by reducing serum immunoglobulin A and IL-6 that impedes B-cell differentiation and antibody production in chickens (Liang *et al.*, 2015) and increasing transcription factors that degrade protective responses of neutrophils and monocytes in dogs (Turek *et al.*, 2003).

A common effect when animals consume oxidized lipids in foods is loss of body weight but increased liver weight. Weight loss was originally attributed to reduced consumption due to off-flavors, but more recent studies demonstrated that oxidized lipids are recognized as abnormal and signal transduction shuts down enzymes involved in de novo fatty synthesis, leading to weight loss, while increasing antioxidant and protective enzymes (Brandsch and Eder, 2004; Eder *et al.*, 2003). Adaptive increase in membranes to accommodate production of the enzymes adds to liver weight.

More details about how dietary oxidized lipids mediate systemic responses without absorption are provided in (Schaich, 2020b).

Toxic lipid oxidation products in foods. Oxyenals such as hydroxynonenal (HNE), oxynonenal, and hydroxyhexenal immediately come to mind as toxic lipid oxidation products because their cytotoxic and genotoxic actions *in vivo* are well-known (Eckl and Bresgen, 2017; Esterbauer, 1993). However, their effects when consumed in food remain in question. For now, we will accept detailed evidence that very toxic HNE and similar 1,2-enals are efficiently metabolized, detoxified, and excreted after consumption in foods, except under conditions of overload (Alary et al., 2003).

More critical for animal feed and economy pet foods are the high levels of aldehydes, cyclic monomers, dimers, and oxypolymers formed in frying oils that are common lipid sources in kibbles and other dry animal food forms. Consumption of heated oils in foods can cause diarrhea and interfere with growth (Crampton *et al.*, 1951a; 1951b; Crampton *et al.*, 1953; Perkins and Kummerow, 1959), induce embryonic abnormalities (Huang *et al.*, 2014) by altering metabolic and neuroendocrine environments of the fetus (Kabaran and Besler, 2015), enhance systemic oxidative stress and a wide range of derivative pathologies, and create an inflammatory environment in the intestines that enhances development of inflammatory bowel disease and colon cancer (Zhang *et al.*, 2019). Part of these effects result from essential fatty acid and antioxidant deficiencies caused by the diet. Levels of used frying oils required for these effects depend on many factors, including protein composition of the diet and levels of glutathione provided. The complex issues of heated oil toxicity are discussed in detail by (Schaich, 2020b).

9 Analyses of lipid oxidation processes and products

Monitoring lipid oxidation to determine rates of progression and identify products throughout the process is always challenging because product concentrations are generally low on a molar basis (<1% of the lipid weight), products are altered during handling, and accurate, sensitive analyses are not available for all products. Given space limitations, this chapter restricts discussion to issues important for tracking low levels of multiple products accurately to determine reaction pathways that are active and require control, with focus on methods that detect and quantitate single classes of oxidation products using instrumentation available in most laboratories. Results of multiple analyses can then be combined to provide full product distributions under varying conditions. Readers are referred to (Schaich, 2013a; 2016) for more details regarding chemistry and protocols of specific analyses. Capabilities of contemporary instrumental methods to determine structures and provide profiles of oxidation products are described, but protocols are too complex to be covered here.

9.1 First, measure oxidation products generated by treatments, not by handling

Analyses of oxidized lipids are not for the faint in heart because intense effort is required to assure that compounds detected are the products and levels that actually occurred in the food sample and were not generated during handling. Such accuracy requires that derivatization, handling, and storage (even frozen) be minimized, that all solvents in analyses be saturated with argon (higher solubility and density than nitrogen), and that all samples be manipulated, transferred, and reacted under argon to prevent degradation and transformation of initial products as well as formation of new products. In addition, while it may seem mundane, rigorous glassware washing is particularly critical in lipid oxidation analyses to first saponify traces of lipids adhering to the glassware surface and then to remove metal contaminants in the glass with acid. Maintain laboratory temperature at 20 °C or lower during extractions and analyses and keep lights off or low to prevent LOOH decomposition and photochemical formation.

Samples all have at least low levels of dissolved oxygen in their tissues, so oxidation and product transformation can continue even during frozen storage under vacuum, Thus, avoid storage of samples after extraction, incubation, or reaction. Most accurate results are obtained when samples are analyzed immediately after processing or reaction.

Finally, to facilitate accurate comparisons when the goal of analyses is to track pathway dominance and shifts as well as product levels, all analysis results should be expressed on molar bases rather than mixed unitless numbers, equivalents, moles, etc.

9.2 Combine class analyses of multiple products to follow pathways and determine extent of lipid oxidation

Analyses described in this section typically are reacted in test tubes or vials then analyzed by spectrophotometry or HPLC with optical detection, they detect all compounds with a given functional group collectively but do not distinguish individual molecules. Performing separate assays for multiple products in samples may be time-consuming but it quantitates oxidation while tracking activity and balance of alternate pathways under different conditions of formulation, processing, and storage without requiring advanced instrumentation and omics capabilities. Choose and combine results for assays of as many products as possible based on equipment and facilities available.

Oxygen consumption. The classic monitoring of oxygen uptake by manometers or oxygen electrodes, sensors, or bombs remains valuable for determining kinetics of lipid oxidation but has two serious limitations. First, unless the electrode or sensor is in direct contact with the sample, the reported oxygen consumption is a confounding of endogenous oxygen reacting in the sample, diffusion of oxygen into the sample, and changes of oxygen pressure in the headspace. Because more than enough oxygen to drive lipid oxidation is dissolved in all samples and cannot be completely removed even under vacuum, oxidation chemistry can progress substantially before active oxygen consumption appears to begin. Second, the method is product blind so does not contribute to delineating active reaction pathways. Such information requires one or more of the secondary analyses of individual (e.g. specific product assays) or collective products (e.g. instrumental analyses that detect multiple products simultaneously) that are outlined below.

Conjugated dienes (CD). The formation of conjugated dienes is the first change that occurs in the oxidation of polyunsaturated fatty acids and analysis of these products is relatively simple, so this assay belongs in every lipid oxidation analysis toolbox along with cautionary notes. Oils or lipid extracts are dissolved in solvent and optical absorbance is determined variably in the 231–235 nm range. Specific isomer information can be obtained by recording absorbance at 235 nm for *cis-cis,* 233 nm for *cis-trans, trans-cis,* 231 for *trans-trans* isomers. Non-lipid components absorbing near 233 nm, especially antioxidant phenols interfere, so non-specific absorbance (as indicated by very high absorbance) is sometimes corrected for by measuring ratio of absorbance at 233 and 215 nm (oxidation index). Standardized procedures such as AOCS method Ti 1a-64 typically report conjugated dienes as % lipid weight to eliminate differences in lipid composition, but when detailed differences between samples must be determined, results must be converted to molarity using Beer's Law and calculations must be corrected for density differences between oils. Finally, all solvents contain dissolved oxygen. To obtain accurate CD results, use solvents with lower oxygen solubility (e.g. chloroform or alcohols), sparge solvents with argon (not N_2) before analysis, and analyze samples immediately after mixing with solvent.

Peroxide values (LOOH). As the assay most applied to lipid oxidation, by far, analysis of hydroperoxides has many variations, controversies, and cautionary notes.

Iodometric titration with thiosulfate. Although this assay has been much maligned, it is chemically accurate, quantitates absolute rather than relative concentrations, and depends entirely on the peroxide functionality and is independent of the R group of the hydroperoxides. Because of this accuracy, iodometric titration has been recommended for primary quantification of hydroperoxide standards used in other assays. Iodometric titration is also the only assay that can handle highly oxidized lipids without dilution, and it can be performed in even minimal labs, requiring only an inert gas supply, a good magnetic stirrer, and a burette.

Iodometric titration has two major limitations: difficult handling and low sensitivity (minimum detection about 0.5 meq/kg lipid). The assay is notoriously "empirical", requiring particular care and exact conditions to yield reproducible results.

Scrupulous oxygen removal is required to prevent oxidation of I^- and generation of erroneously high peroxide values. This can be achieved by thoroughly sparging solvents with argon before reaction and maintaining active argon sparging throughout the titration. The reaction should take place in the dark or in diffused daylight to prevent decomposition of hydroperoxides. The reaction has two immiscible phases – lipids in chloroform (or iso-octane):methanol and potassium iodide in dilute acetic acid. For reaction with iodide ions to occur, the hydroperoxides must partition from the organic to the aqueous phase, and this requires active stirring or agitation (e.g. magnetic stirring). Gentle hand-swirling is inadequate. Adsorption or addition of iodine to unsaturated bonds of fatty acids, which decreases peroxide values, can be avoided by rapid titration. Autotitrators facilitate handling large numbers of samples but argon-sparging, active stirring, and protection from light still must be maintained (Steltzer, 2012).

Low sensitivity in this reaction derives from difficulty in visually distinguishing the color change marking the actual end point determination. The purple color fades gradually in many shades rather than colored to clear instantaneously, so reactions are often over-titrated. Visual endpoint detection may be obscured altogether with colored lipids or extracts. As an alternative, potentiometric titration can take the guesswork out of endpoint detection and reduce the concentration of peroxides detectable to nanoequivalents, although deaeration requirements here are even more rigid than for the standard assay. Nevertheless, given the chemical accuracy of the KI reaction with hydroperoxides and the greatly increased sensitivity of electrochemical endpoints, this endpoint detection certainly merits more attention.

Finally, it has been claimed that this assay requires large samples. If following the standard AOCS method, five grams of sample are recommended for peroxide values below 5, and about 1 gram of sample for peroxide values greater than 10, just to be able to see the endpoint clearly. However, smaller quantities of lipids are routinely analyzed with standard procedures, especially in research laboratories, and the method is still valid. Less than 10 mg of lipid are required when potentiometric endpoints are employed.

Xylenol orange (XO) and Ferric thiocyanate (FeSCN). These two assays were developed to decrease hydroperoxide detection levels to micro-/nano- moles and reduce sample size requirements to µl. Ferrous ions in an acidic medium reduce hydroperoxides (Reaction 22), and the resulting ferric ions are complexed by the dye to produce colored complexes: blue-purple complexes with absorbance maximum at 550–600 nm for xylenol orange (Reaction 22a) and red complexes absorbing at 500 nm for FeSCN (Reaction 22b). Hydroperoxide concentrations are determined by comparison of absorbance values to a standard curve and normalization to lipid weight in the sample. Commercial kits are available for these assays.

$$LOOH + Fe^{2+} \rightarrow Fe^{3+} + LO\cdot + {}^{-}OH \qquad (22)$$

$$Fe^{3+} + XO \quad \text{orange} \rightarrow (Fe^{3+}...XO) \text{ complex} \quad \text{purple} \tag{22a}$$

$$Fe^{3+} + SCN \quad \text{colorless} \rightarrow (Fe^{3+}...SCN) \text{ complex} \quad \text{deep red} \tag{22b}$$

Absolute quantitation of hydroperoxides is not possible with either assay because they depend on accurate extinction coefficients or comparison to standard curves for quantitation. Here, extinction coefficients vary with the solvent, reagent proportions, and hydroperoxide being analyzed. Color responses vary with hydroperoxide structure and dye batch so separate standard curves must be generated for the specific conditions of each test system and each dye batch. Standard hydroperoxides should be analyzed independently by iodometric titration before plotting quantitation curves.

A second drawback of both assays is that reaction stoichiometry and accurate quantitation are both complicated by oxidation of Fe^{2+} by oxygen and by solvent radicals generated by LO· (Bou *et al.*, 2008; Mihaljevic *et al.*, 1996). These side reactions result in stoichiometry that varies with conditions and is greater than 2 moles XO-Fe^{3+} complex per mole of LOOH, particularly with FeSCN (Steltzer, 2012). To reduce inaccuracies, reactions should be run with all solutions carefully sparged with argon and blanks of reagent iron must be run with each analysis to account for oxidation of ferrous iron over time.

Perhaps the greatest practical limitation of these assays is that, at high oxidation levels, excess hydroperoxides react with the Fe-dye complexes and bleach the color, making it appear that little or no reaction occurred with highly oxidized lipids. In samples with very low absorbance readings, bleaching can be distinguished from low peroxide levels by diluting samples sequentially and re-analyzing. Increase in absorbance with sample dilution verifies bleaching from high levels of hydroperoxides, in which case the assay must be repeated with samples of smaller size or higher dilution (Steltzer, 2012).

Finally, solvent purity is critical for these assays. Isopropanol must be peroxide-free for the XO assay and chloroform stabilized with ethanol rather than amylene must be used for the FeSCN assay.

***Triphenylphosphine* (*TPP*).** The triphenylphosphine assay, a solvent trace hydroperoxide analysis recently adapted to lipids, eliminates the variations of XO and FeSCN assays in stoichiometry and reactions with different hydroperoxides while maintaining nanomole sensitivity (Zeng, 2025). The reaction involves reduction of LOOH in lipids or lipid extracts by TPP, then oxidation of TPP to TPP oxide (TPPO) by -OH, with release of LOH (Reaction 23) (Akasaka and Ohrui, 2000).

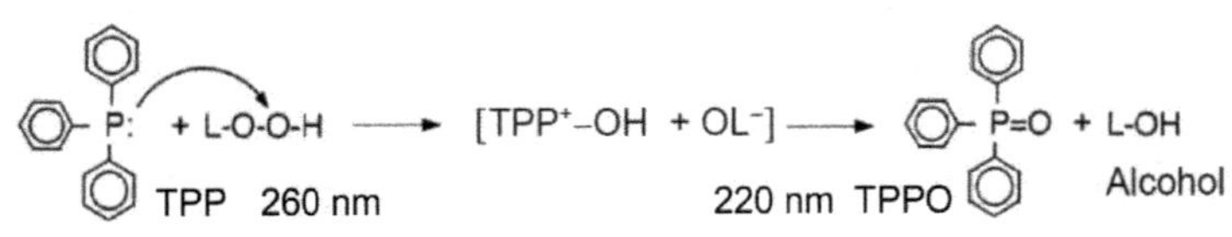

(23)

Reaction stoichiometry of 1 mol TPPO generated per LOOH is independent of LOOH structure, hence TPPO monitored at 220 nm provides accurate LOOH quantitation and all lipid samples give the same response curves as TPPO (Zeng, 2025). This simplifies quantification tremendously. With high TPP excess over LOOH concentrations, reaction is immediate at room temperature, eliminating side reactions that occur with heating and incubation. Extensive spectral overlap between TPPO and lipid LOOH make spectrophoto-metric assays unfeasible, but all four reactants separate cleanly by normal phase HPLC on pentafluorphenyl columns. Separation of LOOH and LOH is particularly useful for assessing reaction completeness. Inclusion of butylated hydroxytoluene blocks continued lipid oxidation and thereby stabilizes TPPO while samples are in the analysis queue. The TPP assay is simple, fast, and offers a promising highly-sensitive (sub-nanomoles) but accurate alternative to iodometric titration. Protocols are available from author until publication.

Epoxides. Lipid epoxide analysis has long been hampered by lack of sensitive, accurate assays. This situation remains currently and is partially responsible for deficient data on lipid epoxides in food materials.

Hydrogen bromide. AOCS standard method Cd 9-57 quantitates epoxides by titration with hydrobromic acid (HBr) (Reaction 24) (AOCS, 1996). Epoxide concentrations are calculated as % Oxirane oxygen = [ml HBr * N * 1.6] / mass oil (g), where N is the normality of the HBr.

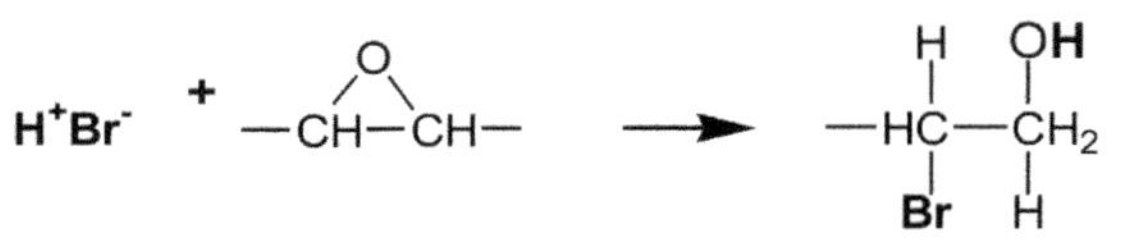

(24)

Developed for analysis of industrial epoxidized lipids, this assay targets high epoxide levels (0.0075 M to 0.1 M) and lacks the sensitivity required for detecting epoxides in food materials (nM to μM, mM max). Like the iodometric titration for LOOH, it is quite variable without rigorous handling controls and the reaction endpoint is difficult to detect. Perhaps most problematic for analyzing oxidized lipids in foods is that the HBr reaction rate and stoichiometry varies with epoxide structure, creating uncertainty for quantitation and comparisons between oils and extracts of differing composition, and it is not specific for epoxides. β-unsaturated ketones, conjugated dienols, and other oxygenated secondary products of lipid oxidation compete with epoxides (Liao, 2013).

Diethyldithiocarbamate. A new epoxide analysis with nanomole detection sensitivity and ability to quantitate total and individual epoxides uses diethyldithio-carbamate to complex epoxides (Reaction 25) then the complexes are separated by HPLC and quantitated by comparison of sample peak areas to standard curves of pure epoxides.

This assay was adapted for lipid epoxides (Liao, 2013) and used to follow methyl linoleate oxidation (Xie, 2015) and has been tested extensively to determine controlling factors and product stability (Praveen, 2024). Importantly for preserving native epoxide content, this assay requires no pre-separation or modification, and oils or extracts can be analyzed without hydrolysis. DETC is inert to other lipid oxidation products but does chelate metals, so metal-free reagents, glassware, and equipment (including HPLC columns) are essential for accuracy. Although currently still in publication stages, DETC offers the best opportunities for sensitive and accurate detection of lipid epoxides with simple instrumentation. Protocols are available from the author.

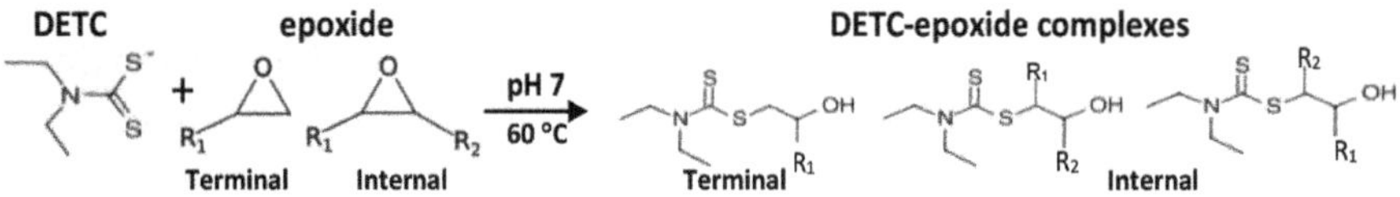

(25)

9.3 Carbonyls

Dinitrophenyl hydrazine. Carbonyls (both ketones and aldehydes) react with DNPH in acidic alcohol to form dinitrophenylhydrazones that are red wine-colored as ions (Reaction 26) and can be detected optically at 432 nm for saturated aldehydes and 458 nm for unsaturated aldehydes (Allen, 1930; Brady and Elsmie, 1926).

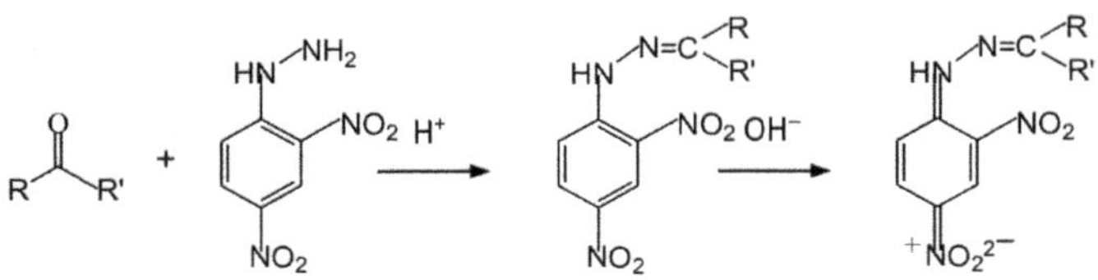
(26)

Although this assay has been used for decades, poor reproducibility and sensitivity limited its solution applications in quantitating lipid oxidation. However, recent re-evaluation of the assay identified the key problem as excess acid that converted too much DNPH to inactive $DNPH_2$, greatly reducing yields of hydrazones; it also degraded hydrazones that did form (Kandrac, 2018). Maintaining reaction pH at 2.5 catalyzes carbonyl reactions while stabilizing DNPH and product hydrazones, thus enabling accurate quantitation of sub-micromoles of carbonyls. With this critical correction in the assay, accurate carbonyl levels can be determined in oils and extracts and compared on a molecular level with other products to determine active oxidation pathways. As revised, the DNPH assay now supports class assays by test tube reactions, separation and quantitation of hydrazones by HPLC, and antibody detection of hydrazones in Elisa assays. With HPLC in particular, both total carbonyl quantitation and identification of reacting carbonyls in fatty acids and oils, lipid extracts, and phospholipids can be achieved (Kandrac, 2018; Yao, 2015). These improved capabilities offer promise for use in research and quality control analyses.

Anisidine. Unsaturated aldehydes (2-alkenals and 2,4-alkadienals) in animal fats and vegetable oils react with *p*-anisidine in glacial acetic acid to form a yellowish product that absorbs at 350 nm (White, 1995). Easy to run, this assay is often the preferred method to detect secondary products in industrial quality control, especially when thermal degradation is involved.

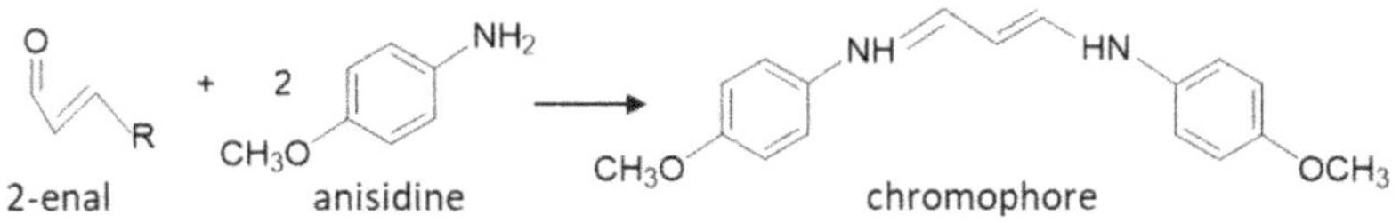

(27)

Anisidine values are only semiquantitative and unitless, being defined as 100 times the optical density change at 350 nm before and after reaction in 1.0 cm cell of a solution containing 1.0 g of oil in 100 mL of mixture of solvent and anisidine reagent (0.25 in glacial acetic acid). To account for both early and later stages of lipid oxidation, anisidine values (AV) are often combined with double peroxide values (PV) to give Totox Values (Totox value = AV + 2PV), which are used extensively for monitoring and regulating in frying operations. The degree of divergence from fresh oil conditions (AV <2.0 and Totox <4) provides a scale for assessing extent of oxidation. Totox of 20 is the reject or discard point. Standardized procedures are available for both anisidine value (AOCS Cd 18-90, ISO 6885) and Totox value (AOCS Cg 3-91) (AOCS, 2011a; IUPAC, 1987).

Chemically, anisidine has three major limitations. First, it is highly toxic to blood regardless of route of exposure and should be put on regulatory "Do Not Use" lists. Second, the anisidine reaction is neither quantitative nor specific. All aldehydes react but unsaturated aldehydes have higher color response than saturated, so results can give only relative, not absolute, aldehyde concentrations. Anisidine also reacts slowly with hydroperoxides so the assay should not be used to analyze oils with peroxide values greater than 5 (although this seems to be a common practice). The method cannot be used with highly colored oils, particularly with carotenoids, which also absorb in the 350 nm wavelength range. Third, the reaction mixture must be water-free (reactions are not fully completed in presence of water) and use carbonyl-free solvents to avoid general interferences. These limitations argue for replacement of anisidine values with safer more accurate methods such as DNPH-HPLC (Schaich, 2016).

Malonaldehyde or Thiobarbituric acid assay (TBA). Malonaldehyde is a minor secondary degradation product derived from internal LOOH in polyunsaturated fatty acids with three or more double bonds, so it is relevant mostly in animal products with substantial levels of arachidonic acid or higher fatty acids. Nominally, it can be detected by heating samples with thiobarbituric acid in acidic solutions to form a pink-colored adduct that absorbs at 532–535 nm and fluoresces at 515 nm excitation, 553 nm emission (Reaction 28) (Tsikas, 2017). Results of TBA analyses are usually expressed as µmoles TBARS / g sample to cover all reactants, or mg malonaldehyde / kg sample.

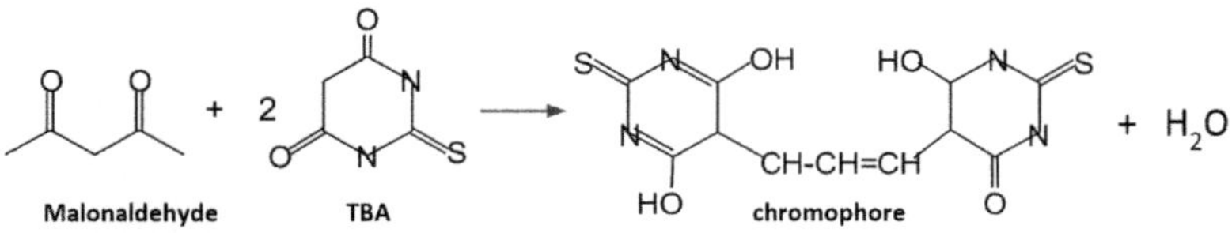

(28)

The TBA reaction remains intensely controversial for two reasons: lack of specificity and questionable relevance of malonaldehyde in following lipid oxidation. The simplicity of this reaction makes it very attractive for analyses – it can be performed by heating ground or dispersed samples in TBA without extracting lipids. However, its carbonyl groups make TBA reactive with many compounds, including alkanals, alkenals, alkadienals, and perhaps other oxidation products as well as malonaldehyde in lipids plus carbonyls sugars and starches, amines in proteins, and ascorbic acid, so its reactions report much more than malonaldehyde and lipid oxidation or underreport oxidation because the TBA remains bound to food proteins and starches and forms the chromophore there (Kwon *et al.*, 1965). Such side reactions are particularly a problem in food materials that give positive TBA responses but lack the requisite unsaturated fatty acids. What then is being measured? Too many studies look for easy procedures and ignore this issue.

The question of malonaldehyde relevance arises because it is a minor product (<0.1%) at best, and strong evidence suggests that malonaldehyde is not endogenous but is generated during the reaction heating. Thus, why not monitor products that are more prevalent and appear in more materials? Not being able to present definitive data on relevant products makes this assay questionable for determining extent of oxidation and useless when active oxidation pathways need to be distinguished (Frankel, 2012).

To improve the specificity and sensitivity, modifications of the TBA assay include reacting extracted lipids rather than complete foods and/or isolating and quantitating individual aldehydes, with and without TBA complexation, by HPLC (Esterbauer *et al.*, 1984). Analysis of lipid extracts eliminates the time advantage of the method but avoids side reactions with non-lipid reactants and provides most detailed information about oxidation in the lipid. For carbonyl detection in lipid extracts, TBA may offer some advantages over DNPH in stability, but sensitivity and specificity are both lower.

Several excellent reviews on TBA chemistry, procedures, and limitations are available for interested readers (Hoyland & Taylor, 1991; Fernández et al., 1997; Guillén-Sans & Guzmán-Chozas, 1998).

Rancimat/OSI. These two tests are mentioned here because they are used frequently in industry and anyone using them needs to understand that they measure oxidizability of lipids, not their level of oxidation. Rancimat and Oxidative Stability Index (OSI) are fundamentally the same procedure just in instruments from different manufacturers – Rancimat (Metrohm) and OSI (Omnion). In these analyses, oils are bubbled with air or oxygen while being heated to ≥100 °C. This oxidizes the oils and carries volatile secondary oxidation products into a tube with distilled water where they are condensed. Conductivity is constantly recorded and plotted against time to produce oxidation curves. Products responsible for conductivity are not the oxidation products originally in the oil; rather they are carboxylic acids generated from oxidation of aldehydes during heating. As long as the oils resist oxidation, conductivity remains low. When antioxidants are consumed or resistance is overcome, conductivity rises rapidly. The inflection point in the conductivity curve marks the end of the induction period. Results are expressed as time (minutes or hours) to the inflection point.

Although these two analyses are staples in industrial analysis of lipid oxidation, it must be stressed that they do not reflect specific oxidation products or pathways. Rather, they force formation of tertiary products that can be monitored. Thus, these methods are useful for comparing sensitivity of different oils or the same oil with different antioxidants, but they provide no information about the chemistry occurring in the oils.

Gas chromatography (GC) – headspace vs SPME vs purge and trap. Volatile reaction products are critical components of all lipid oxidation analysis toolboxes because they are what consumers smell and associate with rancidity so provide critical warning signals. At the same time, it can be difficult to collect volatiles accurately and quantitatively. Volatiles must first be released from samples, usually by mild heat, then collected and transferred to a gas chromatography column for separation and quantitation. Three modes for collecting volatiles directly from lipid samples are:

- static headspace where headspace gas over samples in sealed containers is removed by a syringe and transferred to GC; volatiles collected are limited by their relative volatilities and headspace solubility limits; least sensitive and representative.
- solid phase microextraction (SPME) where a polymer fiber is inserted into the headspace and volatiles are adsorbed and concentrated on it before desorbing onto GC; volatiles bound are determined by selectivity of the fiber material and time of adsorption.
- purge and trap / thermal desorption where volatiles released from samples are constantly swept onto a trap by inert gas, then concentrated there until desorbed onto a GC column for analysis; eliminates disadvantageous of headspace equilibria and saturation that limit product evaporation – emptying the headspace supports continuous release of volatiles and concentration increases detection limits; most sensitive method, collects largest number of products.

All three methods present two distinct advantages: collecting and analyzing multiple volatile compounds (sometimes hundreds) directly from any kind of sample (oil or solid food) and separating multiple classes of products in a single analysis. At the same time, analyzing lipid oxidation of complex foods this way is complicated because only volatile products are detected, the same products are not detected by all methods, and no clues are available to distinguish sources of products, e.g. triacylglycerols, phospholipids, glycolipids, or non-lipid compounds. Thus, supplementary pre-separations and additional analyses may be required to obtain full and definitive information.

A common practice for increasing sensitivity is first hydrolyzing fatty acids from triacylglycerols or phospholipids then derivatizing for specific product detection or esterification before direct analysis by GC. While useful in concept, these modifications also modify or transform lipid oxidation products in the process so do not provide an accurate oxidation picture. For most accurate analysis of oxidation, use procedures that minimize handling and modification of lipid samples.

The capability of measuring large numbers of products by GC is a huge untapped advantage for tracking multiple oxidation pathways and their shifts under different conditions. Previous practice identified and quantitated only products important to flavors or odors; others were ignored even though they held clues to active reactions. Instead, all products possible should be identified, quantitated, and connected to conditions for generation. New statistics and software for data mining now make it possible to organize products into reaction patterns and reconstruct the active oxidation pathways. Such information can then be focused to tailor control of specific pathways.

More details about these methods, their applications, advantages, and limitations are available in (Schaich, 2016).

9.4 Use instrumental analyses of products to detect and identify individual products

The rapid advancement of instrumental analyses offers exciting potential to put class assays under the microscope and identify individual components as well as see multiple types of components at the same time. These capabilities shift focus from limited quantitation of single products to following full oxidation progress while also determining all the products and pathways involved – i.e. detailed mapping of lipid oxidation. Contemporary instrumental analyses are now changing how we see lipid oxidation.

With space limitations, only a few key features can be presented here.

Monitoring molecules: HPLC- or GC-MS/MS with lipidomics. When it is important to identify individual products, e.g. to determine potential sources of off-flavors or toxicity or to track alternate oxidation pathways, instrumental analyses that combine product separation with quantitation and identification of individual components provide tremendous capabilities. Gas chromatography (GC, for volatile products) and high pressure liquid chromatography (HPLC, for non-volatile components) separate lipid components, tentatively identify them by comparison of retention times to those of standards, and quantitate amounts present by peak areas. Definitive identification of components then requires a mass spectrometer (MS) connected directly or in separate analysis. Simplistically, in MS analytes are ionized by hard or multiple soft methods, sorted, and fragmented; molecular structures are then determined from the molecular fragment patterns (Wu *et al.*, 2020).

These techniques may seem to be straightforward, but in practice they are challenging because lipid samples are complex mixtures of mainly unoxidized lipids with low levels of oxidation products in multiple classes. Thus, separation of lipid class fractions in extracts (Yuan *et al.*, 2018) or hydrolysis of fatty acids from acylglycerols and phospholipids (Quehenberger *et al.*, 2018) is usually necessary before analysis, which adds time and risk of oxidation product modification to the process. Lipidomics has developed to eliminate these hurdles and simplify generating molecular maps or profiles of materials to replace multiple chemical analyses. Although early in its applications to lipid oxidation, use of omics can sort and identify specific components from a complex mixture (e.g. oxidized products from unmodified lipids), determine profiles of components in sorted groups (e.g. distribution of all detected lipid oxidation products), and quantitate selected components (e.g. a suspected toxic product), even at trace levels.

Very briefly, lipidomics accomplishes this in two approaches: 1) Global or untargeted (often called shotgun) analyses of total lipid extracts without prior purification to separate and identify components molecular ions and generate total profiles, and 2) targeted analyses in which one or more molecules or groups of molecules are separated from complex mixtures by HPLC, then only these are analyzed by MS. After data collection, both approaches require application of advanced informatics to organize and interpret the data (Li *et al.*, 2014; Wu *et al.*, 2020; Yang and Han, 2016).

As one example, untargeted screening successfully profiled oxidation products in heated oils compared to native oils (Gruneis *et al.*, 2018), then targeted analyses of just the thermal products detected epoxides and hydroperoxides at micromolar levels, with epoxides being dominant (Grüneis *et al.*, 2019). Applying the same molecular ion information facilitated detection of epoxides in margarine and canola oil.

Lipidomics has also been applied to analysis of oxylipins, oxidation products of highly unsaturated ω-6 and ω-3 fatty acids (Strassburg et al., 2012; Yang et al., 2024b). While oxylipin focus on physiological oxidations, profiling of oxylipins in fish and algae oils demonstrates that the methodology can be very useful in tracking oxidation of fish, high PUFA components, and ω-3 additives in animal foods (Emami et al., 2020).

Cautionary note: although lipidomics has attracted intense interest for broader, more sensitive analysis of lipid oxidation, applications are slow to develop because the technique is complicated and requires considerable experimental expertise, plus the required instrumentation is costly. Even so, the potential power of lipidomics to elucidate complex chemical processes like lipid oxidation is undisputable and will continue to develop. This is one technique that would benefit from labs with lipidomics expertise being shared by animal and pet food manufacturers and researchers.

Useful references: overview of MS-based lipidomics with oxidation applications (Kang and Yeo, 2025), some complexities of managing lipidomic analyses of oxidized lipids (Criscuolo *et al.*, 2022; Krettler *et al.*, 2020). A database of lipid ions in MS is available at https://lipidmaps.org/tools/ms/.

Monitoring functional groups by nuclear magnetic resonance (NMR) and Fourier transform infrared spectroscopy (FT-IR). Both NMR and FT-IR detect functional groups of specific oxidation products rather than individual molecules. Thus, when the analytical goal is to follow formation of specific product classes such as epoxides or hydroperoxides or hydroxyalkenals, and full product structures are not needed, NMR and FT-IR can be useful tools.

^{1}H-NMR. Simplistically, when exposed to a fixed radiofrequency, functional groups of detected nuclei separate and are identified by their chemical shift (τ or ppm) in a magnetic field relative to the standard trimethylsilane, which has a shift of zero. Chemical shifts increase with polarity and presence of electronegative nuclei, so oxidation products of lipids adventitiously appear substantially downfield from their alkyl components, e.g. epoxides (2–3 ppm), LOH (~3–4 ppm), LOOH (4.25 ppm), aldehydes RCHO (9.7 ppm), COOH (>10 ppm) versus CH_3 (0.9 ppm), CH_2 in chain (1.3 ppm), C=C (5–6 ppm) (Gunstone and Knothe, 2010)

NMR analyses are simple and can be powerful, but have limitations. Due to cost and facilities requirements, NMR is more likely to be available in specialized instrumentation labs of universities, companies, or research consortia, and only least sensitive 300 MHz versions are found in standard analytical labs. Oxidation products can only be analyzed in oils or extracts dissolved in deuterated solvents so that interactions of solvent protons with the field do not overwhelm the sample spectra. Spectra require long times (up to an hour) to obtain and detection sensitivity is limited to mM or slightly less (Boerkamp *et al.*, 2022).

Nevertheless, research laboratories specializing in NMR have successfully applied ^{1}H-NMR to profile lipid oxidation products in food emulsions (Merkx *et al.*, 2018), determine oxidation products in frying oils (Hwang *et al.*, 2017), follow oxidation of food oils (Moya Moreno *et al.*, 1999; Silwood and Grootveld, 1999; Wanasundara and Shahidi, 1993), quantitate epoxides in oils (Kaur *et al.*, 2021; Xia *et al.*, 2015), and track oxidation during digestion of oils (Alberdi-Cedeño *et al.*, 2020; Nieva-Echevarría *et al.*, 2017a; Nieva-Echevarría *et al.*, 2017b). Details regarding NMR applications and methods for analyzing lipid oxidation products are available in (Hwang, 2017; Hwang and Bakota, 2016).

Note must also be made about the introduction of NMR oxylipidomics, powerful new applications of high frequency (600 to 950 MHz) multidimensional ^{1}H-NMR to profile lipox products, (Boerkamp *et al.*, 2025; Boerkamp *et al.*, 2022). Individual oxidation products are quantified by single dimension ^{1}H-NMR (Merkx *et al.*, 2018) for hydroperoxides and aldehydes, two dimension HSQC (heteronuclear single quantum correlation) for epoxides, hydroxides, ketones, and endoperoxides and at a substructure level (e.g. cis-trans isomers) by 2D HSQC. Overall, for oleic, linoleic, and linolenic triacylglycerols, 42 oxidation products comprising LOOH, LOH, epoxides, ketones, and aldehydes were structurally identified and quantitated with a detection limit of 0.5 mmol kg^{-1} triacylglycerol. Hence, the technique offers exciting new opportunities for tracking oxidation pathways, so bears close watching as it develops further. Like MS in lipidomics, this will be another technique that would benefit from availability of labs with specific expertise being shared by animal and pet food manufacturers and researchers

FT-IR. IR detects and distinguishes individual functional groups by their various vibrational modes – e.g. stretching, bending, rocking, wagging, scissoring, twisting. Each functional group and vibration mode absorbs energy at independent frequencies, providing a means of identification. To detect all the functional groups in a molecule or sample, IR spectrometers scan the full frequency and energy range (4000–400 cm^{-1}), then plot the absorption intensity as a function of frequency (wave numbers, cm^{-1}) in a spectrum. Peaks appear at frequencies where groups present in the molecule(s) absorb energy, with peak intensities proportional to the number of groups present. Peak frequencies and intensities provide a basis for identification of structure in pure molecules (the traditional application) or for following changes in materials during chemical reactions or treatments, as is most often used with processing and lipid oxidation (Workman Jr, 2001). Frequencies associated with lipid oxidation products are listed in Table 5.3.

FT-IR offers several important advantages in analyses. Recording full spectra is rapid and requires only ml of sample, attenuated reflectance modes for sampling makes it possible to analyze intact food samples directly without lipid extraction (Schaich, 2016), disposable Teflon sampling cards facilitate oil analyses and shelf-life studies (Schaich, 2016), and FT-IR instruments are affordable and can be accommodated in normal quality control and research labs. Important from a practical standpoint, a large database of methods and food applications is available for analysis of lipid hydroperoxides in oils (Guillen and Cabo, 2002; Kontogianni and Gerothanassis, 2022; Ma *et al.*, 1998; Ma *et al.*, 1997), lipid hydroperoxides with triphenylphosphine (Deyrieux *et al.*, 2018), lipid carbonyls (Dubois *et al.*, 1996; Guillen and Cabo, 2002), oxidation of edible oils (Muik *et al.*, 2007) and olive oil (Lerma-Garcia *et al.*, 2011; Maggio *et al.*, 2009), and lipid alcohols (LOH) (Marengo *et al.*, 2004).

A troublesome disadvantage of IR analyses, especially with mixed food matrices, is that too many key absorptions (–OH from alcohols, hydroperoxides, free fatty acids,

TABLE 5.3 Mid-Infrared frequencies for detection of lipid oxidation products

Frequency (cm^{-1})	Functional group	Molecular action	References
3610–3640	Hydroperoxides, LOOH[a]	O-H stretching	Guillén & Cabo, 2000; Borchman & Sinha, 2002
3520	LOOH-epidioxide	O-H stretching	Neff, 1982
3435	LOOH, PUFA	O-H stretching	Guillén & Cabo, 2000
3444–3450	LOOH, isolated	O-H stretching	Guillén & Cabo, 2000
3470	C=O, glyceride ester	C=O overtone	Guillén & Cabo, 2000
3468	Hydroxides (alcohols), LOH	O-H stretch	Guillén & Cabo, 2000
3006–3012	Conjugated dienes (cis)	C=C stretch	Guillén & Cabo, 2000
1728–1743	C=O, sat'd aldehydes	C=O stretch	Guillén & Cabo, 2000; Vlachos et al., 2006
1720	C=O in free acid	C=O stretch	Guillén & Cabo, 2000; Borchman & Sinha, 2002
1700-1730	C=O, ketones	C=O stretch	Stuart, 2004
1650-1654	C=C, conjugated	C=C stretch	Guillén & Cabo, 2000; Hayati et al., 2005
1630	C=O, α,β-unsaturated		van de Voort et al., 1994
950-815	epoxide (C-O)	asymmetric ring stretch	ChemistryBlog, 2012
880-750	epoxide (C-O-C)	symmetric ring stretch	ChemistryBlog, 2012
987	trans, trans conjugated diene	C-H bend	Guillén & Cabo, 2000
982-983	cis, trans conjugated diene	C-H bend	Guillén & Cabo, 2000

[a]LOOH, lipid hydroperoxide; LOH, lipid alcohol

amino acids, sugars; carbonyls from fatty acids and aldehydes) overlap so interpretation of spectra can be difficult and accurate absolute quantitation is not possible.

10 Summary

The traditional free radical chain of lipid oxidation outlined a simple series of reactions that generated a narrow range of products. Time has demonstrated that this scheme cannot fully explain either kinetics or product distributions of lipid oxidation, and food stabilization strategies based on radical quenching and measuring hydroperoxides alone frequently don't stop oxidation. To provide a base for rethinking how lipids oxidize, this chapter has presented a more complete view of lipid oxidation that includes alternate reactions that compete with hydrogen abstraction to produce epoxides at all stages of oxidation and account for a broad range of aldehydes and other products. The balance between pathways changes with conditions, generating different product distributions and altering food qualities, nutritional value, and potential safety. Thus, to fully account for the lipid oxidation and accurately determine its potential impact, all the alternate reactions need to be followed when lipids in any type of food product oxidize during processing and storage. The task of monitoring many products is not simple. Single product chemical assays that can be used in any lab and multiproduct instrumental assays that can provide total or selected product profiles were described.

Acknowledgements

This work was supported by the New Jersey Agricultural Experiment Station under Grant NJ10195.

References

Akasaka, K. and Ohrui, H., 2000. Development of phosphine reagents for the high-performance liquid chromatographic-fluorometric determination of lipid hydroperoxides. Journal of Chromatography A 881: 159-170.

Alary, J., Gueraud, F. and Cravedi, J.P., 2003. Fate of 4-hydroxynonenal in vivo: Disposition and metabolic pathways. Mol Aspects Med 24: 177-187. 10.1016/s0098-2997(03)00012-8.

Alberdi-Cedeño, J., Ibargoitia, M.L. and Guillén, M.D., 2020. Study of the in vitro digestion of olive oil enriched or not with antioxidant phenolic compounds. Relationships between bioaccessibility of main components of different oils and their composition. Antioxidants (Basel) 9(6): 543. 10.3390/antiox9060543

Alghazeer, R., Gao, H. and Howell, N.K., 2008. Cytotoxicity of oxidised lipids in cultured colonal human intestinal cancer cells (Caco-2 cells). Toxicology Letters 180: 202-211. 10.1016/j.toxlet.2008.06.859

Allen, C.F.H., 1930. The identification of carbonyl compounds by use of 2,4-dinitrophenylhydrazine. Journal of the American Chemical Society 52: 2955-2959. 10.1021/ja01370a058

Almario, C.V., Ballal, M.L., Chey, W.D., Nordstrom, C., Khanna, D. and Spiegel, B.M.R., 2018. Burden of gastrointestinal symptoms in the United States: Results of a nationally representative survey of over 71,000 Americans. American Journal of Gastroenterology 113: 1701-1710. 10.1038/s41395-018-0256-8

Anguelova, T. and Warthesen, J., 2000. Degradation of lycopene, α-carotene, and β-carotene during lipid peroxidation. Journal of Food Science 65: 71-75. 10.1111/j.1365-2621.2000.tb15958.x

AOCS, 1996. Oxirane oxygen: Method cd9-57. AOCS Press Champaign, IL.

AOCS, 2011a. *p-Anisidine* value: AOCS official method cd 18-90 (11). In: D. Firestone (Ed.), Official methods and recommended practices of the American Oil Chemists' Society, 6th edition, second printing. AOCS Press, Champaign, IL.

Aw, T.Y., 1999. Molecular and cellular responses to oxidative stress and changes in oxidation-reduction imbalance in the intestine. American Journal of Clinical Nutrition 70: 557-565. 10.1093/ajcn/70.4.557

Becker, E. and Lawrence, N.S., 2025. Disgust and distaste – differential mechanisms for the rejection of plant- and animal-source foods. Appetite 212: 108033. 10.1016/j.appet.2025.108033

Belitz, H.-D. and Grosch, W., 1986. Chapt. 3. Lipids, Food Chemistry. Springer-Verlag, Berlin, pp. 158-247.

Boerkamp, V.J.P., Hennebelle, M., Vincken, J.-P. and van Duynhoven, J.P.M., 2025. Comprehensive quantitative profiling of vegetable oil oxidation products by NMR-based oxylipidomics. Food Research International 202: 115612. 10.1016/j.foodres.2024.115612

Boerkamp, V.J.P., Merkx, D.W.H., Wang, J., Vincken, J.-P., Hennebelle, M. and van Duynhoven, J.P.M., 2022. Quantitative assessment of epoxide formation in oil and mayonnaise by ^{1}H-^{13}C HSQC NMR spectroscopy. Food Chemistry 390: 133145. 10.1016/j.foodchem.2022.133145

Bors, W., Tait, D., Michel, C., Saran, M. and Erben-Russ, M., 1984. Reactions of alkoxyl radicals in aqueous solutions. Israel Journal of Chemistry 24: 17-24. 10.1002/ijch.198400003

Bothe, E., Schuchmann, M.N., Schulte-Frohlinde, D. and von Sonntag, C., 1978. $Ho^{2\cdot}$ Elimination from a-hydroxyalkylperoxyl radicals in aqueous solution. Photochemistry Photobiology 28: 639-644. 10.1111/j.1751-1097.1978.tb06984.x

Bou, R., Codony, R., Tres, A., Decker, E.A. and Guardiola, F., 2008. Determination of hydroperoxides in foods and biological samples by the ferrous oxidation-xylenol orange method: A review of the factors that influence the method's performance. Analytical Biochemistry 377: 1-15. 10.1016/j.ab.2008.02.029

Brady, O.L. and Elsmie, G.V., 1926. The use of 2:4-dinitrophenylhydrazine as a reagent for aldehydes and ketones. Analyst 51: 77-78. 10.1039/AN9265100077

Brandsch, C. and Eder, K., 2004. Effects of peroxidation products in thermoxidised dietary oil in female rats during rearing, pregnancy and lactation on their reproductive performance and the antioxidative status of their offspring. British Journal of Nutrition 92: 267-275. 10.1079/bjn20041195

Campbell, I.M., Caton, R.B. and Crozier, D.N., 1974. Complex formation and reversible oxygenation of free fatty acids. Lipids 9: 916-920. 10.1007/BF02532619.

Chan, H.W.-S., Matthew, J.A. and Coxon, D.T., 1980. A hydroperoxy-epidioxide from the autoxidation of a hydroperoxide of methyl linolenate. Journal of the Chemical Society Chemical Communications: 235-236. 10.1039/C39800000235

Circu, M.L. and Aw, T.Y., 2012. Intestinal redox biology and oxidative stress. Seminars in Cell & Developmental Biology 23: 729-737. 10.1016/j.semcdb.2012.03.014

Cortesi, R. and Privett, O.S., 1972. Toxicity of fatty ozonides and peroxides. Lipids 7: 715-721. 10.1007/bf02533120

Crampton, E.W., Common, R.H., Farmer, F.A., Berryhill, F.M. and Wiseblatt, L., 1951a. Studies to determine the nature of the damage to the nutritive value of some vegetable oils from heat polymerization I. The relation of autoxidation to decrease in the nutritional value of heated linseed oil. The Journal of Nutrition 43: 533-539. 10.1093/jn/43.4.533

Crampton, E.W., Common, R.H., Farmer, F.A., Berryhill, F.M. and Wiseblatt, L., 1951b. Studies to determine the nature of the damage to the nutritive value of some vegetable oils from heat treatment II. Investigation of the nutritiousness of the products of thermal polymerization of linseed oil. The Journal of Nutrition 44: 177-189. 10.1093/jn/44.1.177

Crampton, E.W., Common, R.H., Farmer, F.A., Wells, A.F. and Crawford, D., 1953. Studies to determine the nature of the damage to the nutritive value of some vegetable oils from heat treatment III. The segregation of toxic and non-toxic material from the esters of heat-polymerized linseed oil by distillation and by urea adduct formation. The Journal of Nutrition 49: 333-346. 10.1093/jn/49.2.333

Criscuolo, A., Nepachalovich, P., Garcia-del Rio, D.F., Lange, M., Ni, Z., Baroni, M., Cruciani, G., Goracci, L., Blüher, M. and Fedorova, M., 2022. Analytical and computational workflow for in-depth analysis of oxidized complex lipids in blood plasma. Nature Communications 13: 6547. 10.1038/s41467-022-33225-9

Deyrieux, C., Villeneuve, P., Baréa, B., Decker, E.A., Guiller, I., Michel Salaun, F. and Durand, E., 2018. Measurement of peroxide values in oils by triphenylphosphine/triphenylphosphine oxide (TPP/TPPO) assay coupled with FTIR-ATR spectroscopy: Comparison with iodometric titration. European Journal of Lipid Science and Technology 120: 1800109. 10.1002/ejlt.201800109

Doehlert, D.C., Angelikousis, S. and Vick, B., 2010. Accumulation of oxygenated fatty acids in oat lipids during storage. Cereal Chemistry 87: 532-537. 10.1094/CCHEM-05-10-0074

Dubois, J., Van de Voort, F.R., Sedman, J., Ismail, A.A. and Ramaswamy, H.R., 1996. Quantitative Fourier transform infrared analysis for anisidine value and aldehydes in thermally stressed oils. Journal of the American Oil Chemists' Society 73: 787-794.

Eckl, P.M. and Bresgen, N., 2017. Genotoxicity of lipid oxidation compounds. Free Radical Biology and Medicine 111: 244-252. 10.1016/j.freeradbiomed.2017.02.002

Eder, K., 1999. The effects of a dietary oxidized oil on lipid metabolism in rats. Lipids 34: 717-725. 10.1007/s11745-999-0418-0

Eder, K., Suelzle, A., Skufca, P., Brandsch, C. and Hirche, F., 2003. Effects of dietary thermoxidized fats on expression and activities of hepatic lipogenic enzymes in rats. Lipids 38: 31-38. 10.1007/s11745-003-1028-6

Emami, S., Zhang, Z. and Taha, A.Y., 2020. Quantitation of oxylipins in fish and algae oil supplements using optimized hydrolysis procedures and ultra-high performance liquid chromatography coupled to tandem mass-spectrometry. Journal of Agricultural and Food Chemistry 68: 9329-9344. 10.1021/acs.jafc.0c02461

Engberg, R.M., Lauridsen, C., Jensen, S.K. and Jakobsen, K., 1996. Inclusion of oxidized vegetable oil in broiler diets. Its influence on nutrient balance and on the antioxidative status of broilers. Poult Sci 75: 1003-1011.

Erben-Russ, M., Michael, C., Bors, W. and Saran, M., 1987. Absolute rate constants of alkoxyl radical reactions in aqueous solution. Journal of Physical Chemistry 91: 2362-2365. 10.1021/j100293a033

Esterbauer, H., 1993. Cytotoxicity and genotoxicity of lipid-oxidation products. American Journal of Clinical Nutrition 57: 779S-785S; discussion 785S-786S. 10.1093/ajcn/57.5.779S

Esterbauer, H., Lang, J., Zadravec, S. and Slater, T.F., 1984. Detection of malonaldehyde by high-performance liquid chromatography, Methods in Enzymology. Academic Press, pp. 319-328. 10.1016/S0076-6879(84)05041-2

Esterbauer, H., Schaur, R.J. and Zollner, H., 1991. Chemistry and biochemistry of 4-hydroxynonenal, malonaldehyde and related aldehydes. Free Radical Biology and Medicine 11: 81-128. 10.1016/0891-5849(91)90192-6

Farmer, E.H., Koch, H.P. and Sutton, D.A., 1943. The course of autoxidation reactions in polyisoprenes and allied compounds. VII. Rearrangement of double bonds during autoxidation. Journal of the Chemical Society: 541-547. 10.1039/JR9430000541

Faustman, C., Liebler, D.C., McClure, T.D. and Sun, Q., 1999. a,β-Unsaturated aldehydes accelerate oxymyoglobin oxidation. Journal of Agricultural and Food Chemistry 47: 3140-3144. 10.1021/jf990016c

Faustman, C., Sun, Q., Mancini, R. and Suman, S.P., 2010. Myoglobin and lipid oxidation interactions: Mechanistic bases and control. Meat Science 86: 86-94. 10.1016/j.meatsci.2010.04.025

Frankel, E.N., 1984. Lipid oxidation: Mechanisms, products, and biological significance. Journal of the American Oil Chemistry Society 61: 1908-1917. 10.1007/BF02540830

Frankel, E.N., 1985. Chemistry of free radical and singlet oxidation of lipids. Progress in Lipid Research 23: 197-221. 10.1016/0163-7827(84)90011-0

Frankel, E.N., 1991. Recent advances in lipid oxidation. Journal of Science of Food and Agriculture 54: 495-511. 10.1002/jsfa.2740540402

Frankel, E.N., 2012. Lipid Oxidation, second edit. Woodhead Publishing, Cambridge, UK. 10.1016/B978-0-9531949-8-8.50021-7

Funk, M.O., Isaac, R. and Porter, N.A., 1975. Free radical cyclization of unsaturated hydroperoxides. Journal of the American Chemical Society 97: 1281-1282. 10.1021/ja00838a074

Gardner, H.W., 1989. Oxygen radical chemistry of polyunsaturated fatty acids. Free Radical Biology and Medicine 7: 65-86. 10.1016/0891-5849(89)90102-0

Ghelichi, S., Hajfathalian, M., Yesiltas, B., Sørensen, A.-D.M., García-Moreno, P.J. and Jacobsen, C., 2023. Oxidation and oxidative stability in emulsions. Comprehensive Reviews in Food Science and Food Safety 22: 1864-1901. 10.1111/1541-4337.13134

Greene, J.F., Newman, J.W., Williamson, K.C. and Hammock, B.D., 2000. Toxicity of epoxy fatty acids and related compounds to cells expressing human soluble epoxide hydrolase. Chemical Research in Toxicology 13: 217-226. 10.1021/tx990162c

Grüneis, V., Fruehwirth, S., Zehl, M., Ortner, J., Schamann, A., König, J. and Pignitter, M., 2019. Simultaneous analysis of epoxidized and hydroperoxidized triacylglycerols in canola oil and margarine by LC-MS. Journal of Agricultural and Food Chemistry 67: 10174-10184. 10.1021/acs.jafc.9b03601

Gruneis, V., Popovic, N., Zehl, M., Konig, J. and Pignitter, M., 2018. Non-targeted screening for oxidized lipids in foods, 2018 AOCS National Meeting, Minneapolis, MN.

Guillen, M.D. and Cabo, N., 2002. Fourier transform infrared spectra data versus peroxide and anisidine values to determine oxidative stability of edible oils. Food Chemistry 77: 503-510.

Gunstone, F. and Knothe, G., 2010. NMR spectroscopy of fatty acids and their derivatives. Available at: http://lipidlibrary.aocs.org/nmr/nmr.html. accessed April 14, 2013.

Hayam, I., Cogan, U. and Mokady, S., 1995. Dietary oxidized oil and the activity of antioxidant enzymes and lipoprotein peroxidation in rats. Nutrition Research 15: 1037-1044. 10.1016/0271-5317(95)00065-Q

Haynes, R.K. and Vonwiller, S.C., 1990a. Iron(III) and copper(II) catalyzed transformation of fatty acid hydroperoxides: Efficient generation of peroxy radicals with Cu(II) trifluoromethane sulfonate. Journal of the Chemical Society Chemical Communications: 1102-1104. 10.1039/C39900001102

Hidalgo, F. and Zamora, R., 2000. The role of lipids in nonenzymatic browning. Grasas y Aceites 51. 10.3989/gya.2000.v51.i1-2.405

Howard, J.A., Schwalm, W.J. and Ingold, K.U., 1968. Absolute rate constants for hydrocarbon autoxidation. VII. Reactivities of peroxy radicals toward hydrocarbons and hydroperoxides. In: F.R. Mayo (Ed.), Oxidation of organic compounds, Adv. Chem. Series vol. 75 American Chemical Society, Washington, D.C., pp. 6-23. 10.1021/ba-1968-0075.ch002

Huang, C.-F., Lin, Y.-S., Chiang, Z.-C., Lu, S.-Y., Kuo, Y.-H., Chang, S.L.-Y. and Chao, P.-M., 2014. Oxidized frying oil and its polar fraction fed to pregnant mice are teratogenic and alter mRNA expressions of vitamin a metabolism genes in the liver of dams and their fetuses. The Journal of Nutritional Biochemistry 25: 549-556. 10.1016/j.jnutbio.2014.01.005

Hwang, H.-S., 2017. Advances in NMR spectroscopy for lipid oxidation assessment. Springer, Cham, Switzerland. 10.1007/978-3-319-54196

Hwang, H.-S. and Bakota, E.L., 2016. NMR spectroscopy for evaluation of lipid oxidation. In: A.-u.-R. Rahman and M.I. Choudhary (Eds.), Applications of NMR spectroscopy, vol. 4. Applications in food sciences. Bentham Science, Sharjah, UAE, pp. 62-95. 10.2174/9781681081434116040004

Hwang, H.-S., Winkler-Moser, J.K. and Liu, S.X., 2017. Reliability of ^{1}H NMR analysis for assessment of lipid oxidation at frying temperatures. Journal of the American Oil Chemists' Society 94: 257-270. 10.1007/s11746-016-2945-z

Ingold, K.U., 1967. Rate constants for some reactions of oxy radicals. Pure and Applied Chemistry 15: 49-68. 10.1351/pac196715010049

Ingold, K.U., 1969. Peroxy radicals. Accounts of Chemical Research 2: 1-9. 10.1021/ar50013a001

Ingold, K.U., 1978. Some applications of free radical rearrangements, Organic free radicals. ACS Symposium series. American Chemical Society, pp. 187-207. 10.1021/bk-1978-0069.ch011

IUPAC, 1987. Determination of the *p*-anisidine value, method 2.504, IUPAC standard methods for the analysis of oils, fats and derivatives (7th ed.). Alden Press, Oxford, pp. 210-211.

Kabaran, S. and Besler, H.T., 2015. Do fatty acids affect fetal programming? Journal of Health, Population, and Nutrition 33: 14. 10.1186/s41043-015-0018-9

Kandrac, M., 2018. Factors affecting the 2,4-dinitrophenylhydrazine reaction with lipid carbonyls. M.S., Food Science, Rutgers University, New Brunswick, NJ.

Kang, J. and Yeo, J., 2025. Critical overview of mass spectrometry-based lipidomics approach for evaluating lipid oxidation in foods. Food Science and Biotechnology 34: 837-849. 10.1007/s10068-024-01726-6

Kara, K., 2021. Effect of stocking conditions on fatty acid composition and oxidation capacities of different class and type dog food. Italian Journal of Animal Science 20: 1042-1053. 10.1080/1828051X.2021.1939805

Kaur, A., Bhardwaj, N., Kaur, A., Abida, K., Nagaraja, T.P., Ali, A. and Prakash, R., 2021. Proton nuclear magnetic resonance-based method for the quantification of epoxidized methyl oleate. Journal of the American Oil Chemists' Society 98: 139-147. 10.1002/aocs.12439

Kerr, J.A., 1966. Bond dissociation energies by kinetic methods. Chemical Reviews 66: 465-500. 10.1021/cr60243a001

Kimura, T., Iida, K. and Takei, Y., 1984. Mechanisms of adverse effect of air-oxidized soy bean oil-feeding in rats. Journal of Nutritional Science and Vitaminology 30: 125-133. 10.3177/jnsv.30.125

Kochi, J.K., 1962. Chemistry of alkoxyl radicals: Cleavage reactions. Journal of the American Chemical Society 84: 1193-1197. 10.1021/ja00866a026

Kochi, J.K., 1973a. Free radicals, Vol. I. John Wiley and Sons, New York.

Kochi, J.K., 1973b. Oxygen radicals. In: J.K. Kochi (Ed.), Free radicals, vol. 2. John Wiley, New York, pp. 665-710.

Kontogianni, V.G. and Gerothanassis, I.P., 2022. Analytical and structural tools of lipid hydroperoxides: Present state and future perspectives. Molecules 27. 10.3390/molecules27072139

Krettler, C.A., Hartler, J. and Thallinger, G.G., 2020. Identification and quantification of oxidized lipids in LC-MS lipidomics data. Studies in Health Technology and Informatics 271: 39-48. 10.3233/shti200072

Kumarathasan, R., Rajkumar, A.B., Hunter, N.R. and Gesser, H.D., 1992. Autoxidation and yellowing of methyl linolenate. Progress in Lipid Research 31: 109-126. 10.1016/0163-7827(92)90005-4

Kwon, T.-W., Menzel, D.B. and Olcott, H.S., 1965. Reactivity of malonaldehyde with food constituents. Journal of Food Science 30: 808-813. 10.1111/j.1365-2621.1965.tb01845.x

Labuza, T.P., 1971. Kinetics of lipid oxidation in foods. CRC Critical Reviews in Food Science and Nutrition 2: 355-405. 10.1080/10408397109527127

Lerma-Garcia, M.J., Simo-Alfonso, E.F., Bendini, A. and Cerratani, L., 2011. Rapid evaluation of oxidised fatty acid concentration in virgin olive oil using Fourier-transform infrared spectroscopy and multiple linear regression. Food Chemistry 124: 679-684.

Li, L., Han, J., Wang, Z., Liu, J., Wei, J., Xiong, S. and Zhao, Z., 2014. Mass spectrometry methodology in lipid analysis. Int J Mol Sci 15: 10492-10507. 10.3390/ijms150610492

Liang, F., Jiang, S., Mo, Y., Zhou, G. and Yang, L., 2015. Consumption of oxidized soybean oil increased intestinal oxidative stress and affected intestinal immune variables in yellow-feathered broilers. Asian-Australasian Journal of Animal Sciences 28: 1194-1201. 10.5713/ajas.14.0924

Liao, C.-H., 2013. Comparison of chemical assays used to determine epoxides in oxidized lipids. M.S., Food Science, Rutgers University, New Brunswick, NJ.

Lindsay, D.A., Howard, J.A., Horswill, E.C., Iton, L., Ingold, K.U., Cobbley, T. and Li, A., 1973. The bimolecular self-reaction of secondary peroxy radicals: Product studies. Canadian Journal of Chemistry 51: 870-880. 10.1139/v73-131

Ma, K., van de Voort, F., Ismail, A. and Sedman, J., 1998. Quantitative determination of hydroperoxides by Fourier transform infrared spectroscopy with a disposable infrared card. Journal of the American Oil Chemists' Society 75: 1095-1101. 10.1007/s11746-998-0119-3

Ma, K., van de Voort, F., Sedman, J. and Ismail, A., 1997. Stoichiometric determination of hydroperoxides in fats and oils by Fourier transform infrared spectroscopy. Journal of the American Oil Chemists' Society 74: 897-906. 10.1007/s11746-997-0001-8

Maggio, R.M., Kaufman, T.S., Carlo, M.D., Cerretani, L., Bendini, A., Cichelli, A. and Compagnone, D., 2009. Monitoring of fatty acid composition in virgin olive oil by Fourier transformed infrared spectroscopy coupled with partial least squares. Food Chemistry 114: 1549-1554. 10.1016/j.foodchem.2008.11.029

Marengo, E., Bobba, M., Robotti, E. and Lenti, M., 2004. Hydroxyl and acid number prediction in polyester resins by near infrared spectroscopy and artificial neural networks. Analytica Chimica Acta 511: 313-322. 10.1016/j.aca.2004.01.053

Marquez-Ruiz, G., Garcia-Martinez, M.C. and Holgado, F., 2008. Changes and effects of dietary oxidized lipids in the gastrintestinal tract. Lipids Insights 2: 11-19. 10.4137/LPI.S904

Márquez-Ruiz, G., Pérez-Camino, M.C. and Dobarganes, M.C., 1992. Digestibility of fatty acid monomers, dimers and polymers in the rat. Journal of the American Oil Chemists' Society 69: 930-934. 10.1007/BF02636346

Mayo, F.R., 1968. Free-radical autoxidation of hydrocarbons. Accounts of Chemical Research 1: 193-201. 10.1021/ar50007a001

Mayo, F.R., Miller, A.A. and Russell, G.A., 1958. The oxidation of unsaturated compounds. IX. The effects of structure on the rates and products of oxidation of unsaturated compounds. Journal of the American Chemical Society 80: 2500-2507. 10.1021/ja01543a034

Merkx, D.W.H., Hong, G.T.S., Ermacora, A. and van Duynhoven, J.P.M., 2018. Rapid quantitative profiling of lipid oxidation products in a food emulsion by ^{1}H NMR. Analytical Chemistry 90: 4863-4870. 10.1021/acs.analchem.8b00380

Mihaljevic, B., Katusin-Razem, B. and Razem, D., 1996. The reevaluation of the ferric thiocyanate assay for the lipid hydroperoxides with special consideration of the mechanistic aspects of the response. Free Radical Biology and Medicine 21: 53-63. 10.1016/0891-5849(95)02224-4

Moran, J.H., Weise, R., Schnellmann, R.G., Freeman, J.P. and Grant, D.F., 1997. Cytotoxicity of linoleic acid diols to renal proximal tubular cells. Toxicology and Applied Pharmacology 146: 53-59. 10.1006/taap.1997.8197

Mounts, T.L., McWeeny, D.J., Evans, C.D. and Dutton, H.J., 1970. Decomposition of linoleate hydroperoxides: Precursors of oxidative dimers. Chemistry and Physics of Lipids 4: 197-202. 10.1016/0009-3084(70)90050-2

Moya Moreno, M.C.M., Mendoza Olivares, D., Amezquita Lopez, F.J., Peris Martonez, V. and Bosch Reig, F., 1999. Study of the formation of carbonyl compounds in edible oils and fats by ^{1}H-NMR and FTIR. Journal of Molecular Structure 482-483: 557-561. 10.1016/S0022-2860(98)00938-7

Muik, B., Lendl, B., Molina-Diaz, A., Valcarcel, M. and Ayora-Cañada, M.J., 2007. Two-dimensional correlation spectroscopy and multivariate curve resolution for the study of lipid oxidation in edible oils monitored by FTIR and FT-Raman spectroscopy. Analytica Chimica Acta 593: 54-67. 10.1016/j.aca.2007.04.050

Neff, W.E., Frankel, E.N. and Fujimoto, K., 1988. Autoxidative dimerization of methyl linolenate and its monohydroperoxides, hydroperoxy epidioxides and dihydroperoxides. J. Amer. Oil Chem. Soc. 65: 616-623. 10.1007/BF02540690

Nieva-Echevarría, B., Goicoechea, E. and Guillén, M.D., 2017a. Behaviour of non-oxidized and oxidized flaxseed oils, as models of omega-3 rich lipids, during in vitro digestion. Occurrence of epoxidation reactions. Food Research International 97: 104-115. 10.1016/j.foodres.2017.03.047

Nieva-Echevarría, B., Goicoechea, E., Manzanos, M.J. and Guillén, M.D., 2017b. ^{1}H NMR and SPME-GC/MS study of hydrolysis, oxidation and other reactions occurring during in vitro digestion of non-oxidized and oxidized sunflower oil. Formation of hydroxy-octadecadienoates. Food Research International 91: 171-182. 10.1016/j.foodres.2016.11.027

O'Brien, P.J., Siraki, A.G. and Shangari, N., 2005. Aldehyde sources, metabolism, molecular toxicity mechanisms, and possible effects on human health. Critical Reviews in Toxicology 35: 609-662. 10.1080/10408440591002183

Perkins, E.G. and Kummerow, F.A., 1959. The nutritional effect of polymers isolated from thermally oxidized corn oil. The Journal of Nutrition 68: 101-108. 10.1093/jn/68.1.101

Porter, N.A., 1986. Mechanisms for the autoxidation of polyunsaturated lipids. Accounts of Chemical Research 19: 262-268. 10.1021/ar00129a001

Porter, N.A., Lehman, L.S., Weber, B.A. and Smith, K.J., 1981. Unified mechanism for polyunsaturated fatty acid autoxidation. Competition of peroxy radical hydrogen atom abstraction, β-scission, and cyclization. Journal of the American Chemical Society 103: 6447-6455. 10.1021/ja00411a032

Porter, N.A., Weber, B.A., Weenan, H. and Kahn, J.A., 1980. Autoxidation of polyunsaturated lipids. Factors controlling the stereochemistry of product hydroperoxides. Journal of the American Chemical Society 102: 5597-5601. 10.1021/ja00537a032

Porter, N.A. and Wujek, D.G., 1984. Autoxidation of polyunsaturated fatty acids, an expanded mechanistic study. Journal of the American Chemical Society 106: 2626-2629. 10.1021/ja00321a022

Porter, N.A. and Wujek, J.S., 1987. Allylic hydroperoxide rearrangement: β-scission or concerted pathway? Journal of Organic Chemistry 52: 5085-5089. 10.1021/jo00232a004

Praveen, A., 2024. Development and testing of diethyldithiocarbamate – HPLC method for detecting and quantitating epoxides in oxidized lipids. B.S., George H. Cook Honors Thesis, Dept of Nutrition, Rutgers University, New Brunswick, NJ.

Quehenberger, O., Dahlberg-Wright, S., Jiang, J., Armando, A.M. and Dennis, E.A., 2018. Quantitative determination of esterified eicosanoids and related oxygenated metabolites after base hydrolysis. Journal of Lipid Research 59: 2436-2445. 10.1194/jlr.D089516

Repko-Reader, I., 2014. Thermal degradation vs autoxidation of frying oils: Evidence from volatile products. M.S., Food Science, Rutgers University, New Brunswick, NJ.

Schaich, K.M., 2005. Lipid oxidation in fats and oils: Theoretical aspects. In: F. Shahidi (Ed.), Bailey's industrial fats and oils, 6th edition. John Wiley, New York, pp. 2681-2767. 10.1002/047167849X.bio067

Schaich, K.M., 2008. Co-oxidations of oxidizing lipids: Reactions with proteins. In: A. Kamal-Eldin and D.B. Min (Eds.), Lipid oxidation pathways. Volume 2. CRC Press, Boca Raton, FL, pp. 183-274.

Schaich, K.M., 2013a. Chapter 2 - challenges in analyzing lipid oxidation: Are one product and one sample concentration enough? In: A. Logan, U. Nienaber and X. Pan (Eds.), Lipid oxidation: Challenges in food systems

AOCS Press, pp. 53-128. 10.1016/B978-0-9830791-6-3.50005-9

Schaich, K.M., 2013b. Lipid oxidation. In: N.A.M. Eskin and F. Shahidi (Eds.), Biochemistry of foods, 3rd edit. . Elsevier, Amsterdam, pp. 419-478. 10.1016/B978-0-12-242352-9.00011-4

Schaich, K.M., 2016. Chapter 1. Analysis of lipid and protein oxidation in fats, oils, and foods. In: M. Hu and C. Jacobsen (Eds.), Oxidative stability and shelf life of foods containing oils and fats. AOCS Press, pp. 1-131. 10.1016/B978-1-63067-056-6.00001-X

Schaich, K.M., 2020a. Lipid oxidation: New perspectives on an old reaction. In: F. Shahidi (Ed.), Bailey's industrial oil and fat products, 7th edition. Wiley, Hoboken, NJ, pp. 1-72. 10.1002/047167849X.bio067.pub2

Schaich, K.M., 2020b. Toxicity of lipid oxidation products consumed in the diet. In: F. Shahidi (Ed.), Bailey's industrial oil and fat products, 7th edit. John Wiley & Sons, Ltd., Hoboken, NJ, pp. 1-88. 10.1002/047167849X.bio116

Schaich, K.M., 2023. Epoxides: An underestimated lipid oxidation product. Free Radical Research: 1-48. 10.1080/10715762.2023.2277142

Schaich, K.M., Shahidi, F., Zhong, Y. and Eskin, N.A.M., 2013. Lipid oxidation. In: N.A.M. Eskin and F. Shahidi (Ed.), Biochemistry of foods, third edition. Elsevier, Amsterdam, pp. 419-478. 10.1016/B978-0-08-091809-9.00011-X

Scott, G., 1965. Atmospheric oxidation and antioxidants. Elsevier, London. 10.1016/C2009-0-15759-8

Sevanian, A., 1979. Epoxides as products of lipid autoxidation in rat lungs. Lipids 14(7):634–643. 10.1007/BF02533449.

Silwood, C.J.L. and Grootveld, M., 1999. Application of high-resolution, two-dimensional ^{1}H and ^{13}C nuclear magnetic resonance techniques to the characterization of lipid oxidation products in autoxidized linoleoyl/linolenoylglycerols. Lipids 34: 741. 10.1007/s11745-999-0421-5

Sims, R.P. and Hoffman, W.H., 1962. Oxidative polymerization. In: W.O. Lundberg (Ed.), Autoxidation and antioxidants vol. II. Interscience, London, pp. 629-694.

Skelhorn, J. and Rowe, C., 2010. Birds learn to use distastefulness as a signal of toxicity. Proc Biol Sci 277: 1729-1734. 10.1098/rspb.2009.2092

Steltzer, E., 2012. Evaluation of chemical assays for determining hydroperoxides levels in oxidized lipids. M.S., Food Science, Rutgers University, New Burnswick, NJ.

Strassburg, K., Huijbrechts, A.M.L., Kortekaas, K.A., Lindeman, J.H., Pedersen, T.L., Dane, A., Berger, R., Brenkman, A., Hankemeier, T., van Duynhoven, J., Kalkhoven, E., Newman, J.W. and Vreeken, R.J., 2012. Quantitative profiling of oxylipins through comprehensive LC-MS/MS analysis: Application in cardiac surgery. Analytical and Bioanalytical Chemistry 404: 1413-1426. 10.1007/s00216-012-6226-x

Sulzle, A., Hirche, F. and Eder, K., 2004. Thermally oxidized dietary fat upregulates the expression of target genes of PPAR alpha in rat liver. J Nutr 134: 1375-1383.

Taube, H., 1965. Mechanisms of oxidation with oxygen. Journal of General Physiology 49: 29-50. 10.1085/jgp.49.1.29

Terao, J., Ogawa, T. and Matsushita, S., 1975. Degradation process of autoxidized methyl linoleate. Agricultural and Biological Chemistry 39: 397-402. 10.1080/00021369.1975.10861616

Tian, X., 2013. Factors affecting stability of frying oils. PhD Dissertation, Food Science, Rutgers University, New Brunswick, NJ.

Traylor, T.G. and Russell, C.A., 1965. Mechanism of autoxidation. Terminating radicals in cumene autoxidation. Journal of the American Chemical Society 87: 3698-3706. 10.1021/ja01094a029

Tsikas, D., 2017. Assessment of lipid peroxidation by measuring malondialdehyde (MDA) and relatives in biological samples: Analytical and biological challenges. Analytical Biochemistry 524: 13-30. 10.1016/j.ab.2016.10.021

Turek, J.J., Watkins, B.A., Schoenlein, I.A., Allen, K.G., Hayek, M.G. and Aldrich, C.G., 2003. Oxidized lipid depresses canine growth, immune function, and bone formation. J Nutr Biochem 14: 24-31. 10.1016/s0955-2863(02)00221-8

Van Sickle, D.E., Mayo, F.R., Gould, E.S. and Arluck, R.M., 1967. Effects of experimental variables in oxidations of alkenes. Journal of the American Chemical Society 89: 977-984. 10.1021/ja00980a040

Wade, D.R., Airy, S.C. and Sinsheimer, J.E., 1978. Mutagenicity of aliphatic epoxides. Mutation Research 58: 217-223. 10.1016/0165-1218(78)90012-5

Walling, C. and Thaler, W., 1961. Positive halogen compounds, III. Allyl chlorination with *t*-butyl hypochlorite. The stereochemistry af allylic radicals. Journal of the American Chemical Society 83: 3877-3884. 10.1021/ja01479a033

Waltking, A.E. and Goetz, A.G., 1983. Instrumental determination of flavor stability of fatty foods and its correlation with sensory flavor responses. CRC Critical Reviews in Food Science and Nutrition 19: 99-132. 10.1080/10408398309527371

Wanasundara, U.N. and Shahidi, F., 1993. Application of NMR spectroscopy to assess oxidative stability of canola and soybean oils. Journal of Food Lipids 1: 15-24. 10.1111/j.1745-4522.1993.tb00231.x

White, P.J., 1995. Conjugated diene, anisidine value, and carbonyl value analyses. In: K. Warner and N.A.M. Eskin (Eds.), Methods to assess quality and stability of oils and fat-containing foods. AOCS Press, Champaign, IL, pp. 159-178.

Wilcox, A.L. and Marnett, L.J., 1993. Polyunsaturated fatty acid alkoxyl radicals exist as carbon-centered epoxyallylic radicals: A key step in hydroperoxide-amplified lipid peroxidation. Chemical Research in Toxicology 6: 413-416. 10.1021/tx00034a003

Workman Jr, J., 2001. Review of near-infrared and infrared spectroscopy, The handbook of organic compounds, Vol. 1. Academic Press, Burlington, pp. 79-129. 10.1016/B978-012763560-6/50015-5.

Wu, G.-S., Stein, R.A. and Mead, J.F., 1977. Autoxidation of fatty acid monolayers adsorbed on silica gel: II. Rates and products. Lipids 12: 971-978. 10.1007/BF02533320

Wu, G.-S., Stein, R.A. and Mead, J.F., 1978. Autoxidation of fatty acid monolayers adsorbed on silica gel: III. Effects of saturated fatty acids and cholesterol. Lipids 13: 517-524. 10.1007/BF02533590

Wu, Z., Bagarolo, G.I., Thoröe-Boveleth, S. and Jankowski, J., 2020. "Lipidomics": Mass spectrometric and chemometric analyses of lipids. Advanced Drug Delivery Reviews 159: 294-307. 10.1016/j.addr.2020.06.009

Xia, W., Budge, S.M. and Lumsden, M.D., 2015. New ^{1}H NMR-based technique to determine epoxide concentrations in oxidized oil. Journal of Agricultural and Food Chemistry 63: 5780-5786. 10.1021/acs.jafc.5b01719

Xie, J., 2015. Evidence for multiple oxidation pathways from non-volatile products of methyl linoleate. PhD Dissertation, Food Science, Rutgers University, New Brunswick, NJ.

Yang, C.M., Kendall, C.W.C., Stamp, D., Medline, A., Archer, M.C. and Bruce, W.R., 1998. Thermally oxidized dietary fat and colon carcinogenesis in rodents. Nutrition and Cancer 30: 69-73. 10.1080/01635589809514643

Yang, D., Lee, Y.Y., Lu, Y., Wang, Y. and Zhang, Z., 2024a. Internal factors affecting the crystallization of the lipid system: Triacylglycerol structure, composition, and minor components. Molecules 29. 10.3390/molecules29081847

Yang, K. and Han, X., 2016. Lipidomics: Techniques, applications, and outcomes related to biomedical sciences. Trends in Biochemical Sciences 41: 954-969. 10.1016/j.tibs.2016.08.010

Yang, L., Choi, J., Maier, C.S. and Stevens, J.F., 2024b. Profiling of oxylipins as markers of oxidative stress in biological samples. Current Protocols 4: e992. 10.1002/cpz1.992

Yao, L., 2015. Development of new methods for analyzing lipid oxidation: Accelerated solvent extraction and HPLC-DNPH analyses of carbonyl oxidation products. PhD Dissertation, Food Science, Rutgers University, New Brunswick, NJ.

Yarmolinsky, D.A., Zuker, C.S. and Ryba, N.J., 2009. Common sense about taste: From mammals to insects. Cell 139: 234-244. 10.1016/j.cell.2009.10.001

Yuan, Z.-X., Majchrzak-Hong, S., Keyes, G.S., Iadarola, M.J., Mannes, A.J. and Ramsden, C.E., 2018. Lipidomic profiling of targeted oxylipins with ultra-performance liquid chromatography-tandem mass spectrometry. Analytical and Bioanalytical Chemistry 410: 6009-6029. 10.1007/s00216-018-1222-4

Zamora, R. and Hidalgo, F.J., 2005. Coordinate contribution of lipid oxidation and maillard reaction to the nonenzymatic food browning. CRC Critical Reviews in Food Science and Nutrition 45: 49-59. 10.1080/10408690590900117

Zeng, C., 2025. Triphenylphosphine assay for nanomolar detection of lipid hydroperoxides. M.S., Food Science, Rutgers University, New Brunswick, NJ.

Zhang, J., Chen, X., Yang, R., Ma, Q., Qi, W., Sanidad, K.Z., Park, Y., Kim, D., Decker, E.A. and Zhang, G., 2019. Thermally processed oil exaggerates colonic inflammation and colitis-associated colon tumorigenesis in mice. Cancer Prevention Research 12: 741-750. 10.1158/1940-6207.CAPR-19-0226

CHAPTER 6

Protein co-oxidation in food and feed: degradation of quality, reactions and products, and analyses

K.M. Schaich

Dept. of Food Science, Rutgers University, 65 Dudley Rd., New Brunswick, NJ 08901-8520, USA
Correspondence: schaich@sebs.rutgers.edu

Abstract

Protein oxidation initiated by reactive oxygen species such as hydroxyl radical, HO·, and superoxide anion, O_2^{-}·, in living tissues has been the focus of intense research because of its ability to cause damage in disease and pathologies. In contrast, protein co-oxidation caused by reactive oxygen species of oxidizing lipids in either cells or foods has been largely ignored due to perceived slow kinetics, irrelevance, or inability to recognize or analyze effects. However, as consumer demands for higher quality and safer products as well as regulatory oversight of foods have increased, so has pressure to more fully understand details of oxidative degradation in feed and foods in order to maintain nutritional and sensory quality and to design more effective controls.

Such challenge requires recognition of protein co-oxidation in its many guises in foods, knowledge of lipid reactions that cause the multiple changes, accurate quantitative and qualitative measurements of protein changes in parallel with lipid oxidation pathways, and ability to connect the lipid and protein chemistry and property data into a meaningful picture of system oxidation.

To support recognition of protein co-oxidation as a critical degradative process in foods, show how it occurs, and encourage adoption of protein co-oxidation analyses in shelf-life and quality control standard protocols, this chapter presents first an overview of how protein co-oxidation alters foods so readers will have flags for recognizing its presence. The second section then describes how reactions of lipid radicals, hydroperoxides, epoxides, and carbonyls degrade proteins and generate products that are responsible for observed food behaviors. The last section provides information on analyses used to track physical and chemical changes in co-oxidized proteins, with focus on modifications needed to accommodate changes inherent in oxidation and on connecting results to food qualities.

Keywords

crosslinking – adducts – radicals – protein carbonyls – pyrroles – pyridines

1 Introduction

The previous chapter introduced lipid oxidation as the chemical reaction most damaging to food and feed qualities during processing and storage, and degradation of sensory properties, food functionalities, and nutritional value were described along with potential for generation of toxic products. The reactions and products of lipid oxidation are indeed extensive and directly cause significant deterioration of food qualities, yet if these were the only effects, assigning the descriptor of 'most damaging' might be arguable. What elevates lipid oxidation to the top of the damage chart is that lipids do not oxidize alone and in isolation. Nearly all intermediates and products of lipid oxidation react with other food molecules in parallel to their own oxidation, and these reactions broadcast oxidation, expanding it beyond lipids to become *system oxidation* (Figure 6.1). It is here that the molecular classes involved and the types and extent of degradation that develop are multiplied many times to become a tsunami of damage that earns the label 'most damaging'.

This process mediated by oxidizing lipids, called co-oxidation, was first reported for proteins more than 70 years ago (Desai and Tappel, 1963) and since then has been observed also with fat-soluble vitamins, antioxidants, starches, and pigments. Until recently, the idea that lipid co-oxidations occurred and were important met with some resistance, especially in the food industry, even while degradation patterns in human and animal foods could not be explained by lipid oxidation alone or controlled by radical scavengers alone. However, as animal owners now demand the same quality and accountability for their animal feed as human consumers do for their foods, and as legal requirements to meet quality and labeling standards in animal feeds increase, interest in co-oxidations as a process, the molecules and reactions involved, damage caused, and how to control it has resurfaced.

Part of the controversy over co-oxidations is that, unlike lipid oxidation where rancid off-flavors and odors are distinctive and characteristic, co-oxidation in foods is unfamiliar and has been difficult to characterize and connect to specific chemistry due to the complexity of food structures and composition. Indeed, documentation of co-oxidation effects, reactions causing them, and markers for analytical identification in food materials remain limited. For simplicity and chemical clarity, most understanding of co-oxidation processes has been generated in model systems to follow reactions of individual lipid components directly and identify products. This information was then expected to be applied to explain food behaviors where co-oxidation was suspected, but progress has been slow because co-oxidation was not viewed as having sufficient importance or relevance. Co-oxidation as a field thus remains in a state where chemistry and food behaviors are not well-connected and questions remain on both sides of the issue.

To encourage a change of state, this chapter presents support for lipid co-oxidation of proteins that play major roles in structure, function, and nutrition of both human and animal foods. It first delineates properties already associated with protein co-oxidation to lay a rational foundation for a second section that outlines responsible reactions. Overall goals are to connect cause and effect and demonstrate how accounting for protein oxidation is critical for explaining and developing effective preventions for oxidative degradation in animal feed and pet foods. The chapter ends with a brief overview of analytical methods useful for tracking protein co-oxidation and its effects on food qualities.

2 Protein and food properties changed by lipid co-oxidation

Protein co-oxidation has antioxidant effects on the lipid radical chain reactions as it redirects and broadcasts oxidation to proteins and other biomolecules, and mediates degradation of food texture, flavor, color, and nutrition in ways that may

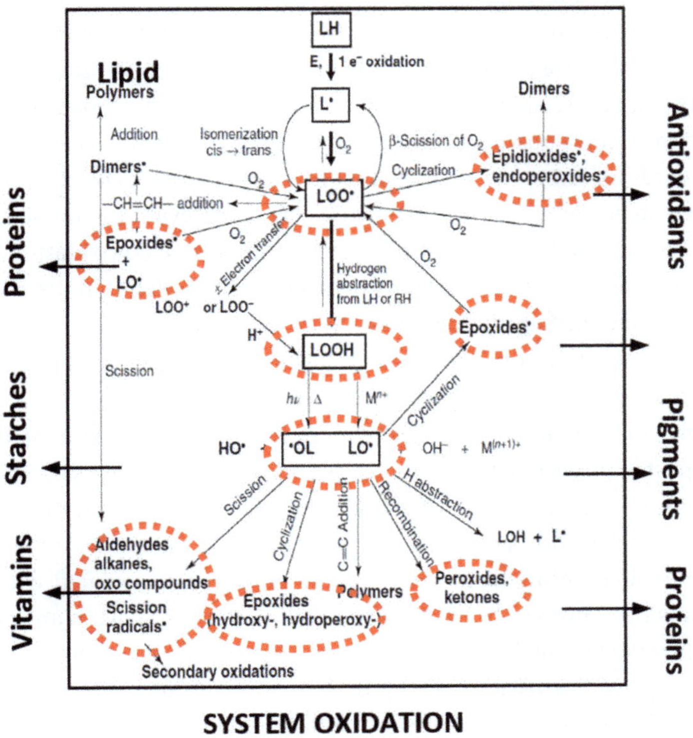

FIGURE 6.1 Nearly all lipid oxidation intermediates and products react with proteins → co-oxidation, broadening lipid oxidation to system oxidation. Adapted from Schaich (2005).

not be obviously connected to lipid oxidation. For example, peanut butter hardens without dehydration as it oxidizes in sealed laminate packages, baked and fried tortilla chips both harden but baked chips also become friable, and flour loses bread-making ability over time without development of typical rancid odors and flavors (Schaich, 2014). These changes all result from protein co-oxidation by lipids. Warmed-over flavors in meats, which have never been explicitly connected to specific lipid oxidation products, probably also derive from protein co-oxidation. Thus, when quality loss is evident in complex foods even when measures of lipid oxidation are low, protein co-oxidation is always a likely explanation.

What functional and quality degradations are associated with protein co-oxidation?

2.1 Co-oxidation effects on food quality

The off-odors and flavors associated with rancidity are quite distinctive, so consumers recognize lipid oxidation readily. In contrast, what does protein oxidation look like, smell like, or taste like? Certainly, protein co-oxidation is not in the lexicon of consumers, but neither has it connected with food producers, at least in part because they have not known what to look for as indicators of co-oxidation in food products. Related to this is that most, if not all, of the protein changes in co-oxidation resemble damage caused by heat, shear, and other stresses during processing and by dehydration during storage, so the possibility that oxidation might be involved as a cause has simply not been considered important for decades. The situation is finally changing and new information about all pathways leading to degradation of food quality, nutrition, and safety is urgently needed to meet current standards for animal feed and pet foods. Hence, protein co-oxidation is moving from being an academic curiosity to being actively relevant to practice.

The following quality degradations occur with protein co-oxidation in foods. Although they have not always been recognized or attributed to proteins in the past, the degradations in food quality and chemistry described below should become more familiar as attention is focused on protein co-oxidation, then they can serve as markers that oxidation has progressed past lipids to proteins.

Off-flavors and odors. Off-flavors and odors have been documented and recognized as problems with native proteins, especially plant proteins (Vatansever et al., 2024), but not from oxidized proteins. Descriptors, when offered, are usually rather general such as 'off', not fresh, stale, cardboard, bland, musty, etc., or lack of normal flavor (Chanadang et al., 2016) and specific off-flavors associated with individual compounds have not yet been characterized. Protein off-flavors are not the same as those from oxidizing lipids, but they are intermixed with them and thus are often misinterpreted as arising from lipids. Interestingly, many desirable meat flavors arise from Maillard reactions between proteins and carbonyls in sugars such as ribose in meats, and meat proteins have been shown to interact with carbonyls in oxidized lipids to generate distinctive off-flavors (Farmer and Mottram, 1990; Farmer et al., 1989). This supports the likelihood that warmed-over flavor in reheated muscle foods as well as other protein off-flavors are protein co-oxidation products and that lipid oxidation products are precursors rather than direct sources of the flavors.

Browning, fading, and discoloration. Browning is a common occurrence when lipids and proteins oxidize together in any kind of food because lipid carbonyl products readily replace sugars in Maillard reactions (Hidalgo and Zamora, 2000a). In addition, a different browning as well as graying or fading also occurs in fresh (Faustman, 2014; Faustman *et al.*, 1999) and intermediate moisture (Obanu et al., 1980) meat products where lipid radicals, hydroperoxides, and aldehydes all react with the meat pigment myoglobin to oxidize the iron (→ brown color) or the apo-protein (→ more intense brown or gray-brown color). Unusual red colors form when free α-amino acids condense with aldehydic products of arachidonic acid from meats (Nakamura and Hama, 1988) and ω-3 fatty acids from fish (Nakamura et al., 1998).

Texture changes. A variety of texture changes occur with protein co-oxidation, with the exact type varying with the form of food, as might be expected. For example, hardening dominated in baked and extruded pet food kibbles (K.M. Schaich, unpublished), in both baked and fried tortilla chips (Dong, 2011; Schaich, 2014), in protein bars made with pea, rice, or whey proteins (Dietrich *et al.*, 2025), and in semisolid peanut butter stored in laminate packaging (Schaich, 2014; Wan Ibadullah, 2013). Toughening occurred in fresh meats (Utrera *et al.*, 2012) and fish (Castell, 1971; Ramadhan *et al.*, 2025) undergoing co-oxidation during refrigerated or frozen storage, but the opposite occurred in intermediate moisture meats where tissue integrity degraded (Obanu et al., 1980). Lipids oxidized concurrently with texture changes in all these systems. Hardness, chewiness, and gumminess in cooked ground pork increased in parallel with lipid oxidation while springiness and cohesiveness decreased (Skowyra *et al.*, 2015). Loose protein aggregation occurred in soy proteins oxidized by linolenic acid hydroperoxides (Huang *et al.*, 2006). Textural nuances change with the food system, the nature of the protein, and the type of protein co-oxidation occurring; reasons for this will be explained in the next section on co-oxidation reactions of proteins causing degradation of proteins.

Decreased digestibility and bioavailability. In animals, protein oxidation is notably manifested in decreased overall wellness and decreased growth due to impaired protein digestibility (Levine et al.,1990; Lund et al., 2011) and availability (Nielsen et al., 1985). These effects result from protein modifications that block peptide recognition and hydrolysis by proteases so that amino acids are not released (Morzel et al., 2006; Sante-Lhoutellier et al., 2007) and by oxidation of digestive enzyme themselves so their activity is reduced (Matsushita et al., 1970; Roubal and Tappel, 1996).

Formation of toxic products. That oxidizing lipids form toxic products alone or with proteins is unquestioned (Eckl and Bresgen, 2017; Eder *et al.*, 1993), but effects of these products formed *in situ* in cells where they are compartmentalized must be differentiated from the same compounds formed in model systems in direct contact with cells or proteins or in foods where they are subjected to rigorous detoxification and elimination as the food passes through the gastrointestinal tract (Schaich, 2008). Many of the products that will be described in the Reactions section have toxicity when tested in cell culture, but whether they escape detoxification, are absorbed from foods, and cause toxic effects in animals remains hotly debated (Schaich, 2020).

One product of specific concern, however, is acrylamide (Zamora *et al.*, 2016), a carcinogen and neurotoxin generated from asparagine via two pathways. Best known is the reaction of asparagine with the carbonyl group of glucose in the Maillard/Strecker degradation pathway, but di-unsaturated carbonyl products of oxidized lipids (Hidalgo *et al.*, 2009; Hidalgo and Zamora, 2008) and oxidized oils themselves (Capuano *et al.*, 2010) can replace sugars in the Maillard reaction that produces acrylamide. A second pathway for formation of acrylamide in foods utilizes reaction of acrolein (formed from thermal oxidation of glycerol in triacylglycerols) with amino acids. This pathway may be particularly relevant to animal feed because it is facilitated at low moisture levels whereas the Maillard pathway requires water, and elevated temperature requirements for acrolein and acrylamide formation are provided during processing. Acrylamide formation may be a particular risk in feeds incorporating used frying oils as ingredients. Acrylamide levels of 39.6, 11.0, and 10.7 ng/g concentrations detected in dry, retort, and canned dog food, respectively, translate to a canine daily intake of 590 ng/kg/day (Sugita *et al.*, 2021). This is below the acrylamide LD_{50} for rats, cats, and monkeys (Le Roux-Pullen and Lessing, 2011), but high enough to mediate low level toxicity with long-term daily exposure. Thus, the potential for acrylamide formation should be considered in all feed formulation and processing designs, and quantitation of acrylamide should be part of regular quality control.

Acrylamide formation pathways, toxicity, and daily intake mitigation strategies in food have recently been reviewed (Govindaraju *et al.*, 2024).

2.2 Co-oxidation effects on protein physical and functional properties

Altered solubility versus extractability. One of the first changes noticed in co-oxidized proteins is modified apparent solubility in water and normal buffers. Most often, the effect is decreased solubility due to protein crosslinking or aggregation as well as formation of lipid adducts that reduce protein surface polarity (Kanazawa *et al.*, 1987; Lea, 1957; Smith *et al.*, 1990). However, protein native amino acid composition and conformations can reverse this pattern. In baked and fried tortilla chips based on cornmeal, for example, apparent (measured) solubility increased because denaturation of proteins during mixing and processing exposed buried tryptophan residues to the Coomassie Blue detecting reagent and polar oxygenated groups were added to very hydrophobic amino acids to increase interactions with buffers.

More importantly for co-oxidized proteins, however, measurement of solubility is controlled by the specificity of the reagent used to quantitate the amount of protein that dissolves and stays in solution. As will be discussed in the later section on analyses of co-oxidized proteins, most colorimetric reagents used to quantitate protein in solution react with the same amino acids – Trp, His, Arg, Lys, and Cys – that are primary oxidation targets (Schaich, 2008; Schaich, 2016; Schaich and Karel, 1976). When those residues have been blocked or modified, less dye is bound and concentrations of oxidized proteins in solution lower than the actual mass are reported. However, if the solubilized protein has salts, detergents, and other contaminants removed, then is dried, the dry mass weight can provide an unbiased measure of protein that was soluble and extracted in the designated solvent.

Altered surface properties. As a chemical property, hydrophobicity is not visible to consumers, but it is manifested in protein conformations and protein–protein and proteins–solvent interactions. Consequently, it strongly affects physical properties such as solubility, gelling and foaming, emulsification, water holding capacity, solution turbidity, aggregation, and cohesion (Huang et al., 2006). Hydrophobicity both increases and decreases with protein co-oxidation depending on the protein and oxidation conditions.

Impaired gluten formation by bread flour. Commercial and home bakers alike have long recognized that as flour ages, it develops musty, cardboardy odors, and breads baked with it have lower loaf heights and denser textures. These changes result from lipid oxidation and its co-oxidation of gluten proteins (Lea, 1957; Swindler, 2013).

3 Co-oxidation reactions causing degradation of protein qualities

Co-oxidation occurs by multiple mechanisms as long as lipids are oxidizing, early to late, but the complexity of degradation increases with oxidation:

- Peroxyl and alkoxyl radicals transfer radicals to proteins, leading to molecular scissions, formation of protein carbonyls, and crosslinking / polymerization.
- Lipid hydroperoxides (LOOH) hydrogen bond to amines and thiols on proteins, leading to induced decomposition of the LOOH *in situ* with subsequent formation of protein radicals and lipid radical addition to protein sites.
- Epoxides bind to proteins, forming adducts.
- Carbonyl products, particularly aldehydes, add to proteins to form linear and cyclic adducts, crosslinks, fluorescent products, and browning.

Because of the dynamic nature of lipid oxidation, these four classes of lipid oxidation products react independently, sequentially, and, in rapidly oxidizing or extensively oxidized systems, even simultaneously. Because the lipid products 'disappear' when they react with proteins, it is often difficult to link lipid oxidation directly to specific food or tissue damage by correlating protein effects to extent of lipid oxidation. Instead, elucidating the reactions of each lipid oxidation product with proteins makes it possible to establish causality backwards by correlating protein products from specific pathways with missing lipid oxidation products.

3.1 Transfer of free radicals from oxidizing lipids to proteins

Lipid radicals can be considered initiators of protein oxidation, just as other radicals are initiators of lipid oxidation, and subsequent protein transformations are also similar.

Lipid radicals are transferred from oxidizing lipids to proteins mainly by LOO· or LO· abstracting hydrogen atoms from side chains of amino acids with abstractable N and S functional groups, i.e. tryptophan, histidine, arginine, lysine, and cysteine, leaving proteins radicals in the process. Electrons from some of these side-chain radicals migrate to α-carbons and/or nitrogen atoms on the peptide backbone, tending to localize at glycine residues. Aliphatic side chains may similarly serve as electron sinks, leading to free radical oxidation of hydrophobic residues (Figure 6.2, upper section). Peptide backbone radicals are detected in the central lines of electron paramagnetic resonance signals and localization is distinguished by the g-values (Figure 6.2, lower section). Side chain radicals appear as downfield broadening of the main signal or additional downfield peaks (lower section), and can also be identified by their g-values (Schaich, 1980; 2008; Schaich and Karel, 1976).

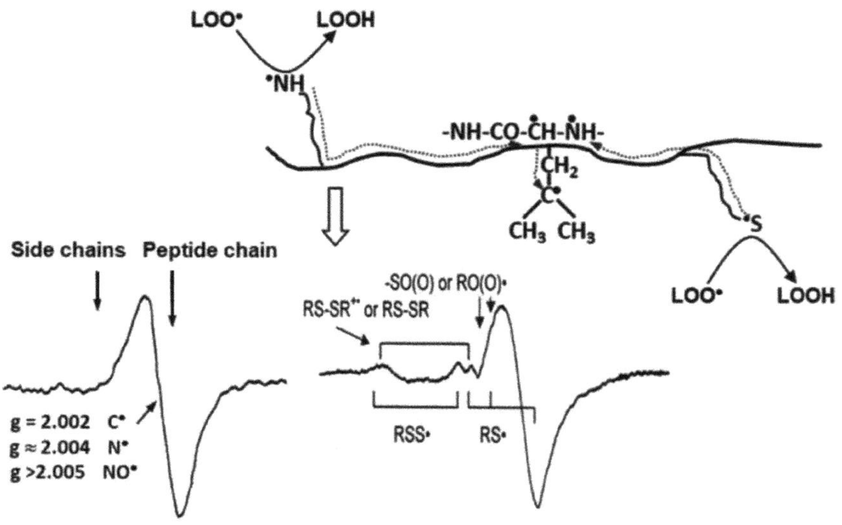

FIGURE 6.2 Oxidizing lipids transfer radicals by hydrogen abstraction from amino acid side chains of proteins. Migration of free electrons to α-carbon and/or nitrogen atoms on the peptide backbone as well as to alkyl side chains of hydrophobic amino acids (upper section) can be detected by electron paramagnetic resonance spectroscopy (lower section).

Once formed, protein radicals (P·) behave just like lipid radicals – they add oxygen to form protein peroxyl radicals (POO·), which then abstract hydrogen atoms from proteins and potentially other molecules to form protein hydroperoxides (POOH). POOH in turn decompose to protein alkoxyl radicals (PO·) when exposed to metals, heat, or light (Reaction 1) (Davies *et al.*, 1995). Critically important in this process is that P·, POO·, POOH, and PO· are all reactive and go on to damage other molecules (though not in chain reactions) and to start system degradation with formation of protein carbonyls, fragments, and polymers.

$$LOO^{\bullet}/LO^{\bullet} + PH \longrightarrow LOOH/LOH + P^{\bullet} \xrightarrow{O_2} POO^{\bullet} \xrightarrow[P^{\bullet},\, R^{\bullet}]{PH,\, RH} POOH \longrightarrow PO^{\bullet} \qquad (1)$$

Protein alkoxyl (PO·) radicals undergo α- and β-scission reactions that parallel those of lipid alkoxyl (LO·) radicals (Reaction 2) to generate protein dicarbonyls and carbonyls, new protein radicals, and peptide fragments with amide and amine end groups (Reaction 2a) (Schaich, 2008; Stadtman and Levine, 2003).

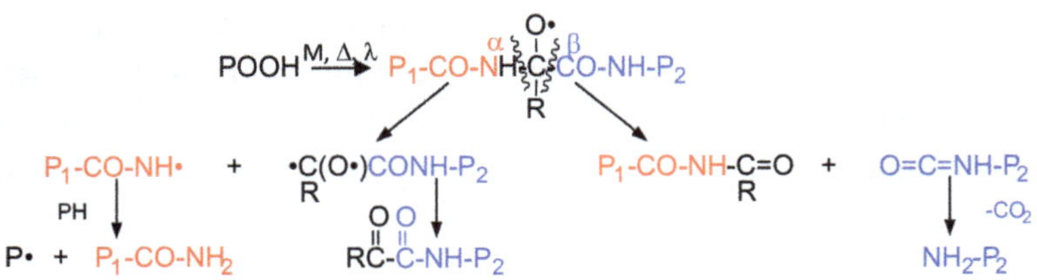

(2)

(2a)

The new protein radicals (P·) formed in the scissions can repeat the oxidation process or, along with POO· and PO· radicals, can generate direct and oxy dimers and polymers by combining with protein radicals, either on backbone or side chain (Reaction 3) or with lipid radicals (Desai and Tappel, 1963) (Reaction 4).

$$\begin{array}{ccccccc} PH & \longrightarrow & P^{\bullet} & \longrightarrow & POO^{\bullet} & \longrightarrow & PO^{\bullet} \\ & & \downarrow & & \downarrow & & \downarrow \\ & & P\text{-}P & & P\text{-}OOP & & P\text{-}OP \\ & & P\text{-}P\text{-}P & & P\text{-}O\text{-}P & & \\ & & P\text{-}P\text{-}P\text{-}P, \text{etc.} & & & & \end{array} \qquad (3)$$

$$\begin{array}{rcl} LOO^{\bullet} + P & \longrightarrow & {}^{\bullet}LOOP \\ {}^{\bullet}LOOP + O_2 & \longrightarrow & {}^{\bullet}OOLOOP \\ {}^{\bullet}OOLOOP + P & \longrightarrow & {}^{\bullet}POOLOOP \end{array} \qquad (4)$$

Reactions 1 through 4 show that a significant amount of damage can be inflicted on proteins early in lipid oxidation mostly without incorporation of lipid molecules and while lipid oxidation has been slowed by the radical transfer to proteins. Under such circumstances, it is difficult to connect the protein changes causally to lipid oxidation except by protein radicals detected by EPR. This inability to connect the processes is one reason the field has been very slow to progress.

When the radical transfer localizes mostly on the peptide backbone, Reactions 1 through 4 translate into the following system properties:

- peptide scission which disrupts food textures and makes them more friable,
- peptide crosslinking which makes food textures harder, tougher, and chewier,
- browning through Maillard reactions of the dicarbonyls,
- increased surface polarity through the terminal amides, amines, and carbonyls, with associated changes in intermolecular associations necessary for functional properties,
- production of off-flavors from the aldehydes themselves or via Maillard reactions with protein amine groups; such off-flavors can contribute to warmed-over flavors of meats or other subtle off-flavors that develop as foods age.

Radicals that remain localized on amino acid side chains lead to amino acid oxidation without incorporation of lipids. The amino acids for which products have been identified are lysine, arginine, histidine, tryptophan, and methionine. Transformations that have critical effects on protein nutrition and surface properties include deamination, decarboxylation, and transformation of polar essential to non-polar non-essential amino acids. Due to space limitations, in lieue of detailed reactions with structures, amino acid degradation reactions and products mediated by lipid oxyl radicals are compiled in Table 6.1 to provide compounds to watch for as oxidation markers in amino acid analyses.

3.2 Hydroperoxides

Lipid hydroperoxides, LOOH, are not radicals themselves, but they hydrogen bond to proteins (P) to form transient complexes that decompose to release protein radicals, P·, and lipid alkoxyl radicals, LO· (Schaich and Karel, 1976) (Reaction 5). In early oxidation, both radicals propagate oxidation; as oxidation progresses they contribute also to formation of volatile compounds and protein polymers (Pokorný et al., 1988).

$$\mathrm{LOOH + PH \rightarrow [LO\text{-}OH...H\text{-}P] \rightarrow LO\cdot + P\cdot + H_2O} \quad (5)$$

LOOH are strong oxidizing agents capable of converting tertiary amines to nitric oxides (Reaction 6) (Harris & Olcott, 1966) and sulfhydryls on cysteine side chains to mainly sulfur oxides (RSOxH) such as sulfenic -SOH, sulfinic -SO2H, and sulfonic -SO3H acids and lesser amounts of disulfides RSSR (Reaction 6a) (Little & O'Brien, 1968). Products of LOOH in these reactions have not been identified.

$$\mathrm{R_3N + LOOH \rightarrow [R_3N...HOOR'] \rightarrow R_3NO + R'O\cdot} \quad (6)$$

$$\mathrm{2\,RSH + LOOH \rightarrow RSOxH + RSSR} \quad (6a)$$

TABLE 6.1 Amino acid oxidation products generated by reaction with lipid oxyl radicals

Amino acid	Reaction	Products	References
Cysteine	radical recombination	cystine	Little and O'Brien, 1967; Little and O'Brien, 1968
	R-S scission	alanine, H_2S, cystine	Roubal and Tappel, 1966b
	oxidation	alanine sulfinic acid, cysteic acid	Lewis and Wills, 1962; Little and O'Brien, 1968
		cystine disulfoxide, cystine thiosulfinate, cystinethiosulfonate, cysteine sulfoxides, cysteine sulfinic and sulfonic acids	Finley and Lundin, 1980
Lysine	decarboxylation	diaminopentane	Karel *et al.*, 1975
	α-deamination/cyclization	pipecolic acid	Karel *et al.*, 1975
	ε-deamination/rad recombination	1,10-diamino-1,10-dicarboxydecane	Karel *et al.*, 1975
	ε-deamination/oxidation/ decarboxylation	α-amino adipic acid, aspartic acid, alanine, glycine	Karel *et al.*, 1975
Histidine	deamination	imidazole lactic acid, imidazole acetic acid	Yong and Karel, 1978; 1979
	decarboxylation, scission	histamine, ethylamine	Roy and Karel, 1973
	ring scission	aspartic acid, asparagine, valine	Karel *et al.*, 1975; Roy and Karel, 1973
Tryptophan	H abstraction, peroxidation	kynurenine, N-formyl kynurenine	Krogull and Fennema, 1987; Yong *et al.*, 1980
	addition	dioxindole-3-alanine	Yong *et al.*, 1980
	ring scission	indoleacetic acid and indolelactic acid	Yong *et al.*, 1980
Tyrosine	H abstraction, radical dimerization	dityrosine	Kikugawa *et al.*, 1991

Amine groups in proteins also add to lipid hydroperoxides (Fruebis *et al.*, 1992), shown below for linoleic acid. One O from the hydroperoxide is transferred to the distant double bond to either a) oxidize it, release R'CHO by scission, and form a cyclic fluorescent adduct on the protein (Reaction 7), or b) form an epoxide across it, followed by ring scission and cyclization to pyrrolines or tetrahydropyridines (Reaction 8). The cyclic adduct increases surface hydrophobicity on proteins, while pyrrolines and tetrahydropyridines contribute flavors, good and bad, that vary with their structure (Vernin, 1982). Tetrahydropyridines also have the potential to be bioactive if absorbed (Gedeon *et al.*, 2023), so their formation has recently become a target of concern.

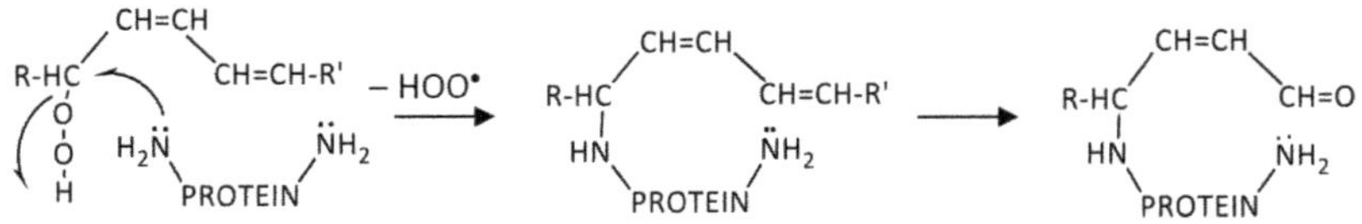

(7)

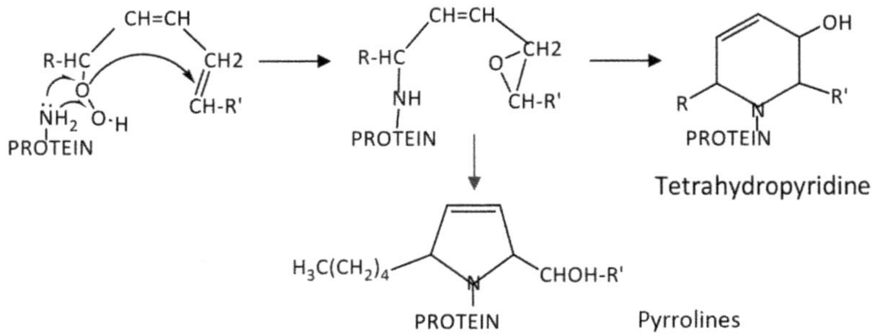

(8)

Additional details on LOOH reactions with proteins can be found in (Schaich, 2008).

3.3 Epoxides

Epoxides are probably the most underrated lipid oxidation product. Until recently they were considered minor products so were seldom measured, but when measured, they were often detected in trace quantities or not at all (Schaich, 2023). One reason may be that epoxides react very rapidly with proteins, which removes them from products detectable in lipid extracts. Under such conditions, protein damage is attributed to aldehydes or radicals and epoxide reactions are overlooked. Involvement of epoxides in protein co-oxidation therefore must be determined by looking for characteristic epoxide-protein products in the overall mixture of modified proteins.

Epoxides react with proteins to form a variety of adducts bound to the protein surface. In early oxidation, linear epoxide-protein adducts with unique structures are aminols. This special structure forms when the epoxide is adjacent to a double bond, and -N or -S nucleophilic groups on protein side chains add to the epoxide carbon connected to the double bond. The epoxide ring opens, the oxygen stays on the less highly substituted carbon, and transfer of a proton from the amino acid generates the alcohols (Lederer, 1996). Aminol structures such as those formed by lysine, histidine, and cysteine (the most active protein targets, Reaction 9) serve as specific markers for epoxide co-oxidations.

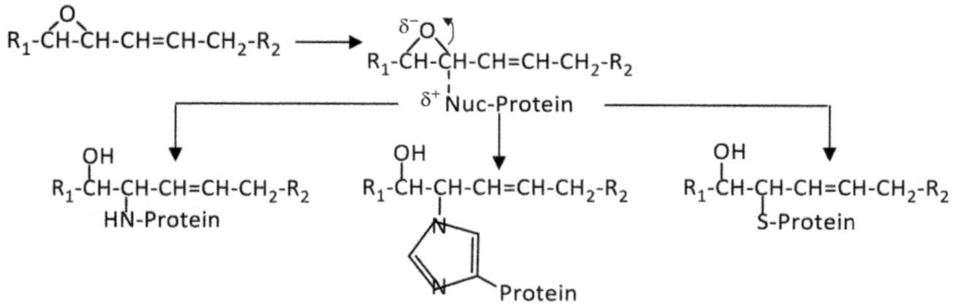

(9)

(9a)

Aminol formation is most important under dry conditions and in lipophilic or aprotic solvents. The presence of water prevents protein-epoxide associations, so no additions occur. Instead, the epoxide rings undergo hydrolysis to form lipid dihydroxy products (Lederer *et al.*, 1999) (Reaction 10) which are also characteristic markers of prior epoxide presence, even if not protein products.

(10)

As oxidation progresses, lipid products form with both epoxides and aldehydes on the same molecule. Now proteins have two targets for protein complexing, and linking the amine between the epoxide and aldehyde generates several types of cyclic adducts, outlined below, that remain bound to the protein surface. The specific reactions that dominate and the structures of the adducts formed vary with the amino acid side chains available on the protein, the epoxide structure, temperature, solvent, and pH (Zamora *et al.*, 2006; Zamora and Hidalgo, 2005). Pyrroles are major products under most conditions.

a N-substituted hydroxyalkyl pyrroles and pyrroles plus a new short chain aldehyde are the main products in buffer (Hidalgo and Zamora, 2000b; Zamora and Hidalgo, 1994; 1995). Amines first add to the terminal carbonyl and the chain bends around for amine addition to the first carbon of the epoxide to form the hydroxyalkyl ring structure (Reaction 11). Release of the hydroxyalkyl side chain forms the aldehyde and pyrrole (Reaction 11a).

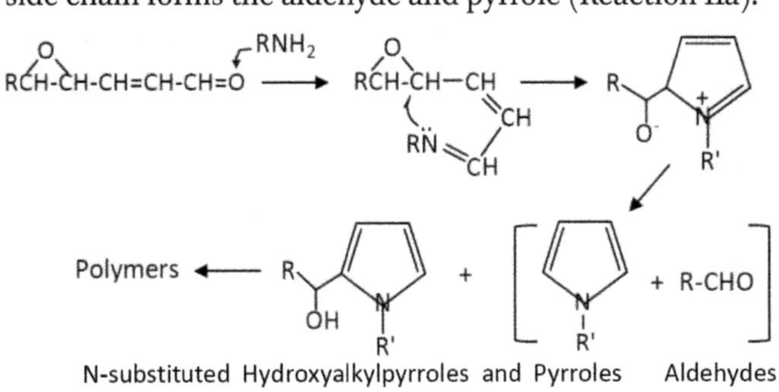

(11)

(11a)

b Alkyl pyrroles and alkyl furans are formed in polar organic solvents such as acetonitrile in addition to the compounds in Reaction 11a. Alkyl pyrroles form via N-addition to the first epoxide carbon (Reaction 12) (Zamora *et al.*, 2006; Zamora and Hidalgo, 2005). However, in competition with this, when the terminal carbonyl is released as a short-chain aldehyde before the amine can add, cyclization of the remaining aldehyde fragment incorporates the epoxide oxygen into an alkyl furan (Reaction 13). This reaction increases 10-fold at elevated temperatures.

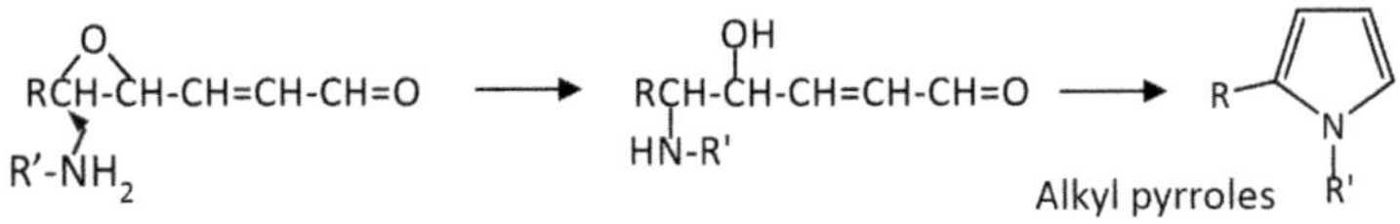

(12)

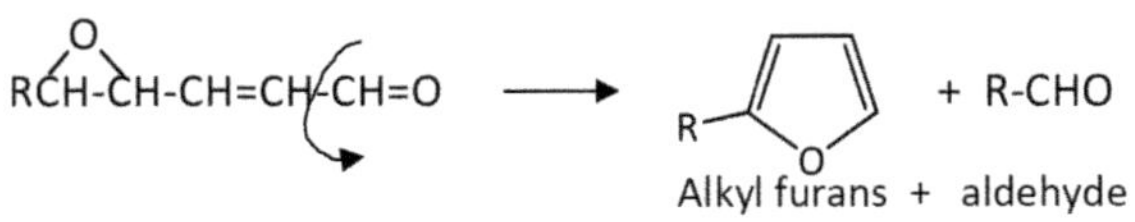

(13)

c) **Alkylpyridines** are formed under mildly acidic conditions. Amines form Schiff base adducts with the terminal carbonyl, the epoxide ring opens to form an alcohol, part of the amine side chain is eliminated as an aldehyde, and the remaining structure cyclizes (Reaction 14) (Zamora *et al.*, 2006; Zamora and Hidalgo, 2005).

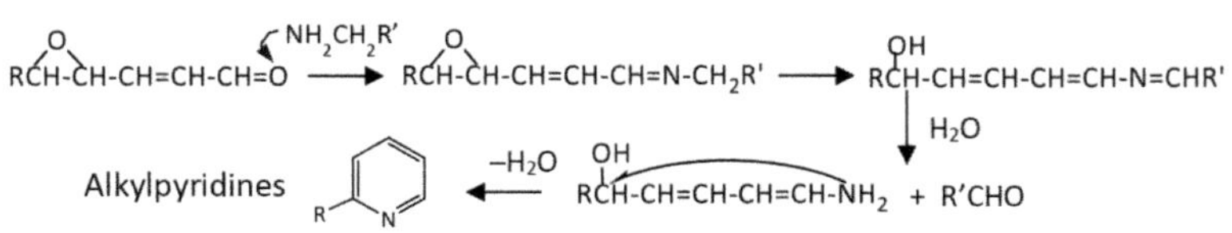

(14)

The epoxide-protein adducts just described have been demonstrated in model systems but finding them in proteins is analytically challenging first in detecting and identifying them, then in interpreting whether pyrroles and pyridines arise from epoxides or from aldehydes, which produce similar products. Nevertheless, the presence of epoxide adducts in proteins alters surface properties and hence functional properties of the proteins, they provide a source of off-flavors which remain to be characterized, and they destroy essential amino acids, reducing nutritional value, so it is important to begin tracking these co-oxidation products in real foods to understand conditions that control their formation and to develop procedures to intervene.

Additional epoxide-protein reaction details are available in (Schaich, 2008).

3.4 Carbonyls

Just as lipid oxidation increases in complexity as it progresses, so do lipid co-oxidations of proteins, in part because there are more different products available for reaction at higher levels, and as a result, there are also more competing alternative pathways for co-oxidizing proteins. In addition, as carbonyl concentrations increase, more varied combinations of aldehydes and protein groups into products become possible.

Lipid aldehydes of nearly every structure (Figure 6.3) react with nucleophilic groups on proteins to form adducts. Alkanals (saturated aldehydes) react with proteins, but slowly and only at terminal amines. Rate and extent of reactivity and damage increases with multiple carbonyls, unsaturation, and oxidation. Oxidized alkenals (4-hydroxy-2-alkenals and 4-oxo-2 alkenals) are the most reactive with proteins, and bifunctional aldehydes offer unique opportunities for crosslinking.

Lipid aldehydes generate three general structures of co-oxidation products with proteins.

Linear adducts. Schiff base condensations of amines with carbonyls, as in the well-known Maillard reaction, form the core of carbonyl reactions with proteins and are the only mechanism available to saturated carbonyls. Reactions between single carbonyls and amines give linear products – imines with primary amines (Reaction 15) and enamines with secondary amines (Reaction 16). The imine is fluorescent and provides an important marker for detecting lipid co-oxidation of proteins (Dillard and Tappel, 1973; 1984); the enamine is not fluorescent.

$$RCH_2C(H){=}O + R'\text{-}NH_2 \rightarrow H_3O^+ + \underset{\text{imine}}{RCH_2CH{=}N\text{-}R'} \leftrightarrow \underset{\text{enamine}}{RCH{=}CH\text{-}NH\text{-}R'} \quad (15)$$

$$CH(O)\text{-}CH_2\text{-}R + R_2\text{-}NH \leftrightarrow \underset{\text{enamine}}{R_2\text{-}NH\text{-}CH{=}CH\text{-}R} + H_3O^+ \quad (16)$$

Unsaturated carbonyls have a second competing pathway for carbonyl reactions with proteins – Michael addition which links protein nucleophilic N and S groups to the β-carbon of the double bond rather than to the carbonyl group, as shown for

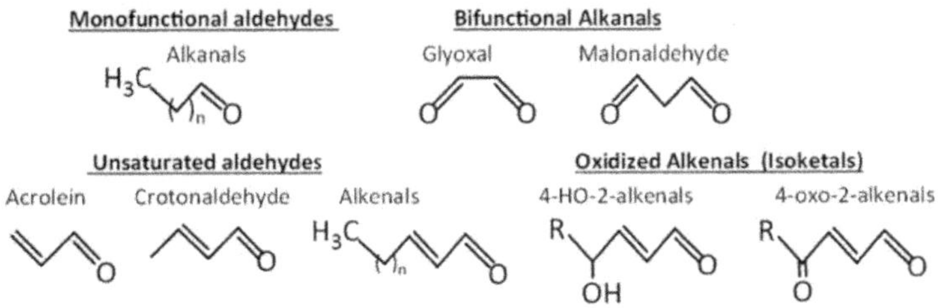

FIGURE 6.3 Structures of aldehydes that play important roles in oxidative degradation of proteins

cysteine and lysine residues in Reactions 17 and 18, respectively. Important in this interaction, the lipid carbonyl remains free and available for additional reactions, including Schiff base additions to a second protein to form crosslinks. These carbonyls are also detected in protein carbonyl assays.

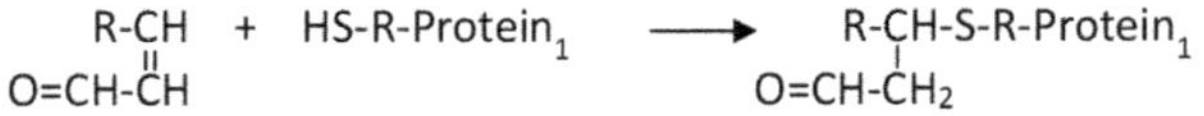

(17)

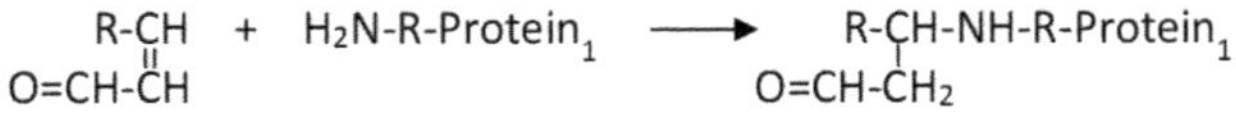

(18)

Linear products are the first formed at low lipid carbonyl levels and are also the first step in formation of more complex products. Not surprisingly, the main amino acids complexed are cysteine, lysine, and histidine (Esterbauer *et al.*, 1991), which contributes to loss of amino acid nutrition. Perhaps most recognized in foods, these are key reactions in browning, with colors and intensity varying with the amino acids involved and the concentration of aldehydes (Hidalgo and Zamora, 1993). Functionally just as important, though, the addition of aldehydes to protein surfaces alters surface properties and molecular interactions that control food behaviors such as gelling and foaming.

Cyclic adducts. Interaction of multiple aldehydes, especially if unsaturated, provides more options for combining Schiff base and Michael additions in different sequences to create families of products with the same core ring structure but ring adducts with different chain lengths and different polarity. These structural differences are responsible for the range of (off)flavors and protein functional alterations that accompany their formation. Formation of multiple product structures is one reason why there is no fixed pattern of protein damage to recognize in co-oxidations.

The simplest cyclic adduct example results from the reaction of saturated aldehydes with terminal amino acids with nucleophilic side chains, e.g. lysine. Saturated aldehydes are generally slow reactors with amines except when two amine groups or an amine plus a thiol group are available; then the carbonyl oxygen reacts rapidly with both groups to form cyclic products at the end of a peptide chain (Reactions 19, 20) with ring adduct size determined by the specific amino acid side chains (Gardner and Jursinic, 1981).

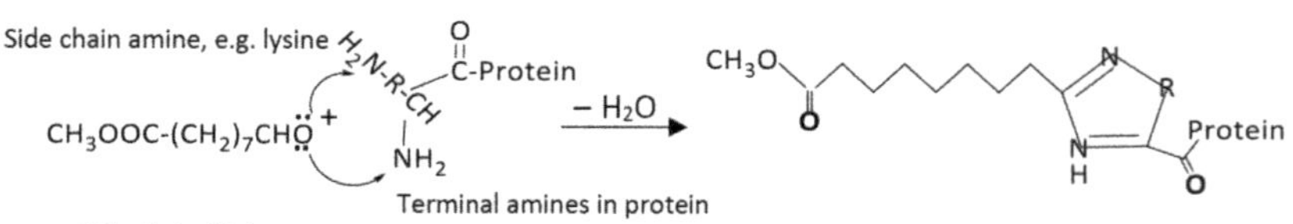

(19)

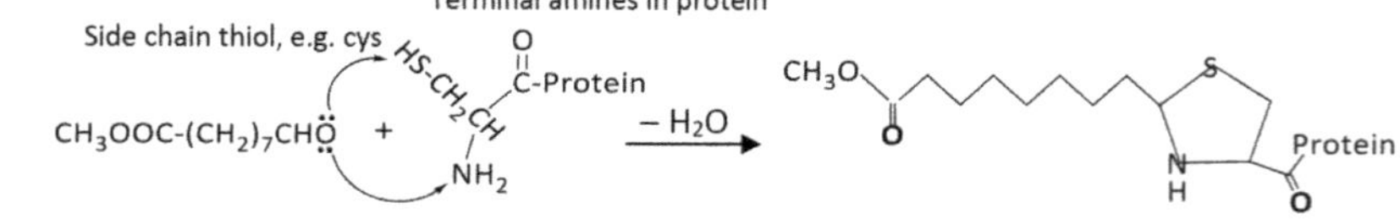

(20)

As lipid carbonyls increase in structural complexity and degree of excess over protein sites, reactions move to amino acid side chains and ring structures become the dominant product. Not surprisingly, the ε-amino groups on lysine side are major targets for reaction with lipid aldehydes. In contrast to the linear products formed at low aldehyde levels (Reactions 17 and 18), when lysine reacts with two molecules of unsaturated short-chain aldehydes such as acrolein in successive additions (one to a C=O, one to a C=C), the pyrrole derivative product shown in Reaction 21 is formed (Esterbauer *et al.*, 1991; Uchida *et al.*, 1998).

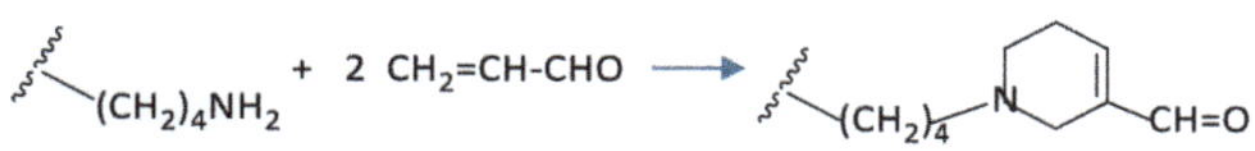

(21)

Reactions of three molecules of alkanals with lysine or other amino acids form alkylpyridines (Reaction 22), where R3 and R4 are new alkyl chains formed during reaction of the alkanals and P is the protein to which the amino acid is attached (Suyama & Adachi, 1979).

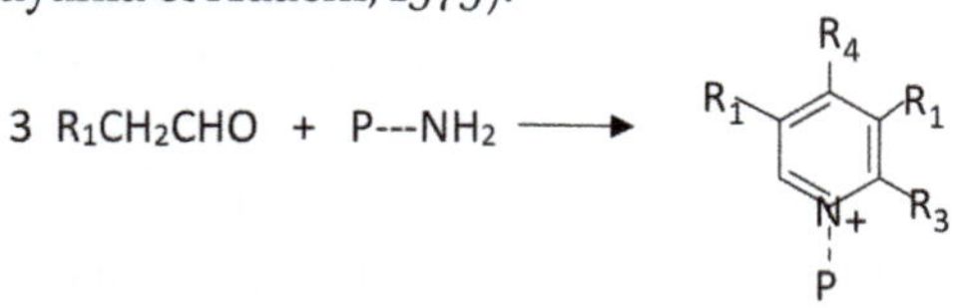

(22)

With longer unsaturated aldehydes, amines can add to either or both carbonyls and double bonds in sequence to generate still different ring structures. For example, when the amine from lysine first forms a Schiff base with the carbonyl of one crotonaldehyde, then the nitrogen also undergoes Michael addition to the double bond of a second crotonaldehyde, the final product is again an alkyl pyridine, ethylmethyl pyridinium lysine (EMP) (Reaction 23), but with different structure than (22) above (Ichihashi *et al.*, 2001).

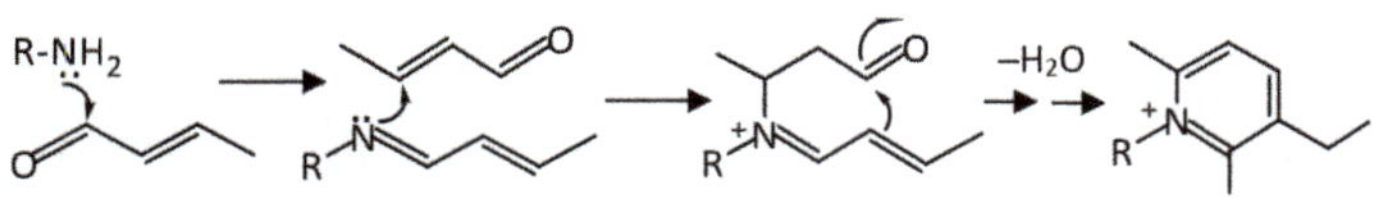

(23)

If lysine instead adds to the double bond in both crotonaldehydes, the final composite product is dehydropiperidino-lysine (FDP) (Reaction 24) (Ichihashi *et al.*, 2001).

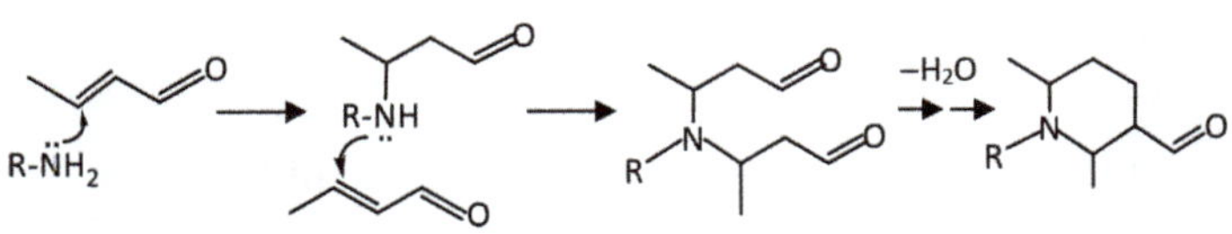

(24)

Reactions of difunctional aldehydes (two carbonyls) also add the possibility of polar ring adducts, as shown in the dihydropyridines formed when two molecules of malonaldehyde combine with one short chain aldehyde and an available amine such as in lysine in proteins (Reaction 25) (Freeman *et al.*, 2005; Kikugawa and Ido, 1984). A critical feature here is the two carbonyl ring adducts remain free to react, e.g. in browning or crosslinking.

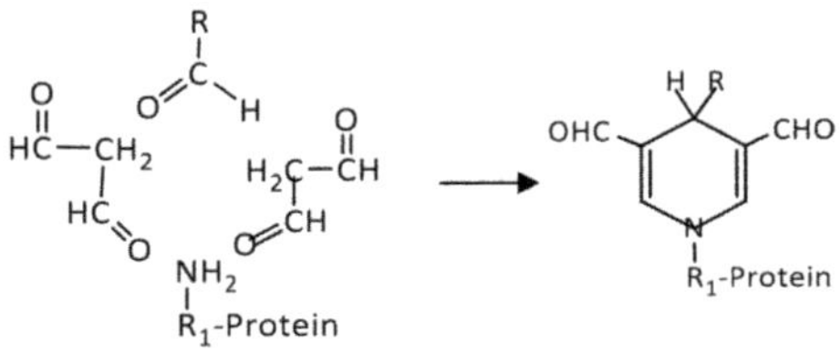

(25)

Regardless of the structure, the adducts shown above sit on the surfaces of proteins and have in common the following effects in food materials: increased browning potential, flavor and off-flavor formation, inhibition of digestive enzymes, interference with protease recognition of proteins, blockage of amino acid absorption, and modification of peptide associations important in protein functionalities such as gel and foam formation, and water binding.

Protein crosslinks, both intra- and inter-molecular. You can see in the examples shown above that often there are reactive carbonyl groups and double bonds remaining available in lipid carbonyl-protein amine complexes, both linear and cyclic. These provide important linkage points for dimerization and polymerization when proteins are in molar excess over aldehydes. Thus, the four major crosslinking mechanisms active with aldehydes are just extensions of initial carbonyl-amine interactions (Schaich, 2008).

1) **Schiff base condensation** of amines with both carbonyls of saturated bifunctional aldehydes provides bridges between protein chains (Reaction 26).

$$\mathbf{O{=}CHCH_2CH{=}O} + 2\ H_2NR\text{-Protein} \rightarrow \text{Protein-RN}{=}\mathbf{CH\text{-}CH{=}CH}\text{-NHR-Protein} + 2\ H_2\mathbf{O} \qquad (26)$$

This type of crosslink also forms between amines on one protein chain, P_1, and carbonyl oxidation sites on a second protein, P_2 (Reaction 27):

$$P_1\text{-lys-NH}_2 + P_2\text{-CHO} \rightarrow P_2\text{-CH=N-lys-}P_1 + H_2O \qquad (27)$$

2) **Michael addition** of thiols and amines (primarily lysine and histidine) to C-3 of α,β-unsaturated aldehydes and hydroxyl alkenals (Reaction 28) leave available aldehyde carbonyls for Schiff base addition to a second protein (Reaction 29):

(28)

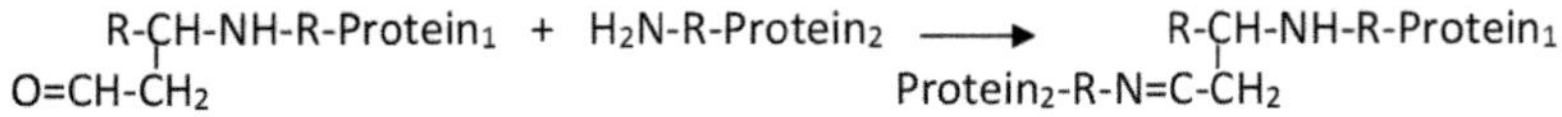

(29)

3) Pyrrole crosslinks form via sequential Schiff base and Michael additions under conditions of protein excess. One protein amine condenses with the carbonyl of a hydroxy- or keto-2-alkenal while a second protein amine undergoes Michael addition to the double bond; the aldehyde then cyclizes between the two protein nucleophiles, usually lysine, histidine, or cysteine (Reaction 30).

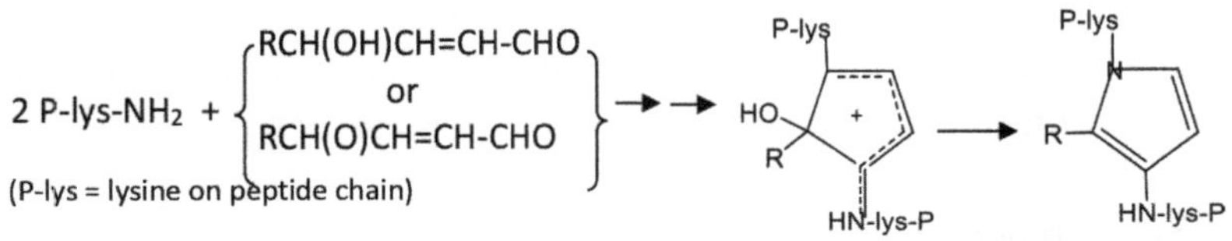

(30)

4) Pyridinium crosslinks form when pyridines with carbonyl adducts are generated under conditions of aldehyde excess, as in Reaction 25, then both available carbonyls form Schiff base linkages with protein amines (Reaction 31).

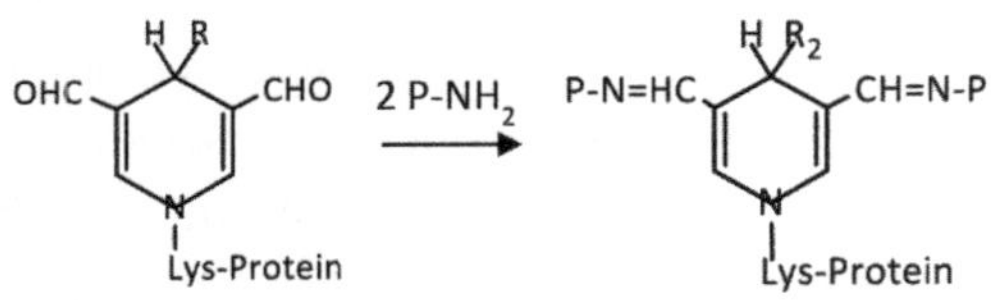

(31)

This distinctive diversity of crosslinks formed by lipid carbonyl reactions with proteins is a major reason for the variable texture changes occurring with co-oxidized proteins – they do not just get hard and stiff. In contrast, radical crosslinking tends to occur at positions farther apart and links proteins directly to each other, giving more rigid extended polymers. Be inserting molecules between multiple proteins and multiple sites on single proteins, carbonyl linkages generate more flexible polymers that tend to form aggregates rather than fixed structures.

A detailed review of the many types of products formed by lipid aldehyde reactions with amino acids and proteins is available in (Hidalgo and Zamora, 2016).

4 Analyses that document protein co-oxidation

The reactions outlined above show how different lipid oxidation products interact with proteins. The co-oxidation process has two effects: loss of lipid oxidation products to the analysis stream and damage to the protein. Thus, if only normally-measured lipid oxidation products are monitored in complex systems such as foods and co-oxidations are ignored, the true extent of oxidative degradation can be greatly underestimated.

Detailed protocols are beyond the scope of this book and routine protein analyses are available in too many molecular biology and biochemistry texts to reiterate here. Rather, analyzing co-oxidized proteins requires different thinking than with native proteins, even when the same assays are used. This section will focus on approaches that can provide information about the nature and extent of protein co-oxidation from lipids using instrumentation and equipment available in most research and quality control labs.

Further discussion and evaluation of the methods may be found in (Schaich, 2016). An introduction to advanced capabilities offered by HPLC in combination with mass spectrometry using omics approaches for tracking co-oxidation of proteins is available in (Ferré *et al.*, 2019; Kehm *et al.*, 2021; Raftery, 2014) for readers interested in exploring analyses beyond the scope of this chapter.

4.1 Preparation of proteins for analysis

Protein extraction. Procedures for all proteins cannot be given here because every protein has different structural organization and solubility characteristics that impart unique extraction requirements. Thus, general guidelines are offered, with focus on modifications required for oxidized proteins. One cardinal rule is that inability to extract proteins by normal methods merely reflects the protein modifications which are the target of investigation. Proteins that are not extracted cannot be analyzed. Therefore, be prepared to "think chemistry" and be creative in designing alternate extraction procedures that accommodate protein changes. Some typical modifications include the following:

Because aggregated or crosslinked proteins react poorly with most protein reagents, a denaturing solvent such as sodium dodecyl sulfate or dimethyl sulfoxide is often needed to extract co-oxidized proteins.

For proteins with reasonable content of sulfur amino acids, extraction both with and without a reducing agent such as 2-mercaptoethanol to disrupt disulfide bonds is required. This provides a more accurate picture of protein content, and an increase in solubility with reducing agent also reveals involvement of disulfide crosslinks.

In foods with higher fat content, protein yields may sometimes be improved by lipid extraction before protein solubilization. In processed foods where proteins become closely associated with gelatinized starches in the food matrix (e.g. baked or extruded animal foods), predigestion with mixtures of amylases may be necessary for accurate recovery of proteins (Strange & Schaich, 2000; Wang, 2000).

When proteins from novel sources are used in foods and feed, extraction methods need to be developed and optimized since few new proteins have been fully characterized, and they may not fit standard protein procedures.

When working with oxidized as compared to native proteins, keep in mind that pH extremes that are commonly used in protein extractions may decompose oxidation products and alter oxidation configurations, thus destroying important information about changes that occurred during oxidation.

Comparing levels and peptide patterns extracted by solvents that bind proteins by different mechanisms can often reveal types of protein interactions that are disrupted or created by co-oxidation, e.g. H bonding, charge interactions, van der Waals interactions.

Remember that oxidation never stops. (Baron *et al.*, 2007) found highly oxidized and polymerized proteins formed in fresh Rainbow trout just with normal handling for analyses. Thus, it is critical to work rapidly and protect protein and samples with inert gas (preferably argon) in headspace and solvents.

A detailed comparison of multiple protein extraction methods aimed for protection of bioactivity has been presented by (Franca-Oliveira *et al.*, 2021).

4.2 Analysis of key property changes

Solubility. Solubility reflects the availability of surface residues that associate with specific solvents, as well as the protein configuration of the residues, so like extraction, optimal solvent for maximum solubility varies with the protein or protein mixture. Solubility is determined by dissolving a specified weight of protein in the test solvent (known mg protein/ml solvent) and allowing time to hydrate and suspend, then the sample is centrifuged at 3000–5000 x *g* for 15–30 min to remove undissolved particles, and the supernatant is withdrawn for determination of protein content by reaction with a dye in solution, by nitrogen content, or by mass after drying. Solubility can be expressed on an absolute basis as mg protein/ml in the supernatant or on a relative basis as % protein=100*supernatant protein content/initial total protein content.

Solubility is supposed to reflect protein interactions with the solvent. Cross-linking, aggregation, change in protein surface hydrophobicity or hydrophilicity away from the solvent (e.g. by adduct formation or modification of amino acid side chains), or denaturation that increases accessibility of interior amino acid residues incompatible with the solvent all decrease physical solubility. Solubility increases when surface residues become more like the solvent, when peptides have undergone scission, and when denaturation increases accessibility of residues that interact with the solvent. However, an even more critical factor for co-oxidized proteins is whether amino acids involved in binding to detecting dyes remain available. When target amino acids are oxidized, they do not react with the detecting agents for the protein assays discussed below, so protein in solution appears to decrease chemically without corresponding change in physical mass. Not recognizing this issue has led to frequent misinterpretation of solubility data.

The following discussion outlines the most common assays for determining protein concentrations in solutions and includes limitations of applications of each method to oxidized proteins. For detailed protocols, see (Liu *et al.*, 2025; Olson and Markwell, 2007).

Lowry assay. Proteins first react with alkaline cupric sulfate in the presence of sodium potassium tartrate to form a tetradentate cuprous complex (four peptide bonds:one atom of copper). The Cu^{2+} in the complex then reduces phosphomolybdic-phosphotungstic acid (Folin-Ciocalteu phenol reagent) to generate a chromophore with intense blue color at 750 nm (Reaction 32) (Lowry et al., 1951).

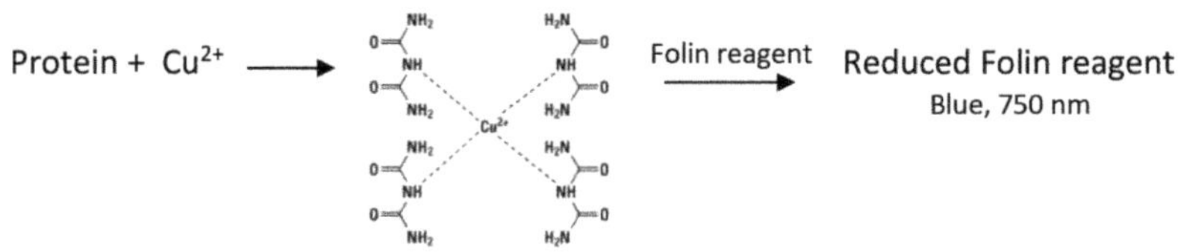

(32)

The Lowry assay has several disadvantages for oxidized proteins. The reaction is relatively slow with low sensitivity (5–100 μg) (Noble and Bailey, 2009). The Folin reaction itself is relatively constant from protein to protein (Dorsey et al., 1977), but the amount of copper bound varies with protein size and amino acid composition. Thus, color formation increases with peptide size and with content of tyrosine, tryptophan, cysteine, histidine and asparagine which contribute most of the reducing equivalents. Since these residues are readily attacked by oxidizing lipid radicals and products, this assay often exhibits very low reaction with oxidized proteins. High proline contents in the protein (e.g. collagen hydrolysates in muscle food products) also interfere with protein-copper complex formation and color development. Handling is awkward since Folin reagent must be added to each tube after precisely ten minutes reaction between protein and copper while simultaneously mixing each tube to avoid precipitation of the Folin reagent. This limits the number of samples that can be analyzed at the same time. The assay is subject to many detractions and interferences including reducing substances and free thiols, chelating agents, strong acids or strong bases, and detergents and potassium ions in the sample (Waterborg, 2009). However, the assay does tolerate up to 1% sodium dodecyl sulfate (SDS) in the sample, which is important considering the extensive use of SDS in extracting modified proteins.

Protocols for Lowry analyses are available in (Simonian and Smith, 2006). However, this traditional assay has been largely replaced with other assays that are more sensitive and less subject to interferences.

Bicinchoninic acid (BCA) assay. When proteins are mixed with cupric ions in alkaline solution, each copper molecule binds to four to six peptide bonds, cysteine or cystine, tyrosine or tryptophan residues, becoming reduced as a blue-colored complex is formed (Reaction 33a) (Smith *et al.*, 1985; Wiechelman *et al.*, 1988). With addition of bicinchoninic acid (BCA) which has stronger metal binding constants, the cuprous ions are released and transferred to the BCA to form intensely colored purple chelates (2 BCA:1 Cu^{1+}) that absorb at 562 nm (Smith et al., 1985) (Reaction 33b).

$$\text{Protein} + Cu^{2+} \rightarrow Cu^{1+}\text{-Protein (blue)} \qquad (33a)$$

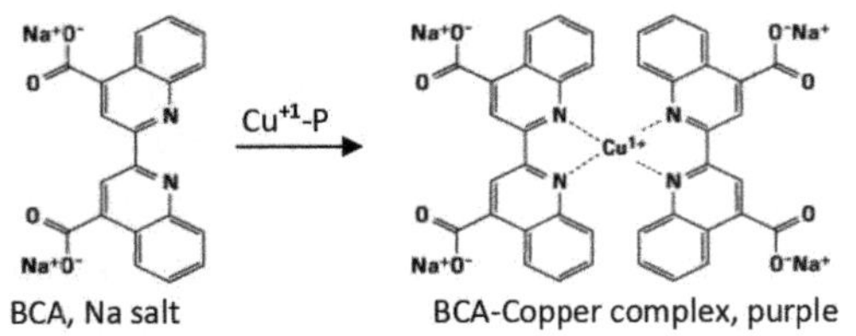

(33b)

The protein-copper-BCA mixtures are heated at 37 or 60 °C. Higher temperatures increase reaction response by partially denaturing the protein (Weichelman et al., 1988) but can only be used with proteins that do not aggregate at 60 °C. Behavior of damaged proteins often differs from that of their native forms, so both temperatures must be tested when using the BCA assay with oxidized or otherwise modified proteins. Also, damaged proteins may need longer incubation times than the standard 30 minutes for full reaction. Standard curves for quantitation are prepared from serial dilutions of bovine serum albumin (BSA) or a protein that closely matches the amino acid content (type and number) of the protein to be analyzed (Noble & Bailey, 2009).

Because oxidized and processed proteins often lose solubility, a solid-phase version of the BCA assay was developed in which the food material (e.g. animal feed) is suspended in the BCA reagent and incubated with sonication; the sample is centrifuged and the reacted BCA released into the supernatant is analyzed at 562 nm (Chan & Wasserman, 1993a). We have found that this modification works very well for native proteins but reports low protein concentrations when target cysteines or tryptophan are damaged.

As the most sensitive assay, BCA assay detects protein concentrations from 20 to 2,000 μg/mL in its standard protocol and 0.1–25 μg/mL in its micro version. It is compatible with up to 5% detergents so can be used with most sodium dodecyl sulfate extracts.

Unfortunately, the BCA assay has several drawbacks that limit its accuracy with oxidized proteins. First, the BCA reaction does not have a true end-point. Like Lowry analyses, the final color continues to develop over time, so reaction timing must be precisely controlled and starting times for multiple samples must be staggered to obtain accurate results. In addition, reducing substances, chelators, and strong acids or bases, and lipids must be removed from the sample to prevent non-specific formation of Cu^{+1} and erroneously high protein apparent concentrations. This can be accomplished by dialysis or protein precipitation in ice-cold acetone

(Gates, 1991) or deoxycholate and trichloroacetic acid (Brown et al., 1989), followed by washing, and by pre-extraction of lipids, respectively. Even more importantly, cysteine, cystine, tryptophan and tyrosine – the four amino acids that bind and reduce the copper in the BCA assay – are major targets for damage during processing and for reaction with oxidizing lipids. Consequently, proteins that are accurately quantitated in their native form are significantly underestimated by BCA when they are oxidized. This leads to overestimation of other chemical changes whose calculations are based on protein content, and it causes notable problems in applying consistent accurate quantities of protein to electrophoresis gels. Apparent loss of protein solubility by BCA must be verified by a method independent of amino acids, such as Kjeldahl, elemental, or mass analysis, as will be discussed below.

BCA assay protocols are available in (Olson, 2016; Olson and Markwell, 2007).

Bradford assay. The Bradford assay eliminates the copper interferences of the Lowry and BCA assays by binding Coomassie Brilliant Blue (CBB) G-250 dye directly to the protein (Bradford, 1976). Coomassie Blue G-250 binds to proteins via charge complexes between its sulfonic acid groups and positively-charged amino acids side-chains (arginine, histidine, lysine) on the protein (Tal et al., 1985; de Moreno et al., 1986) (Reaction 34).

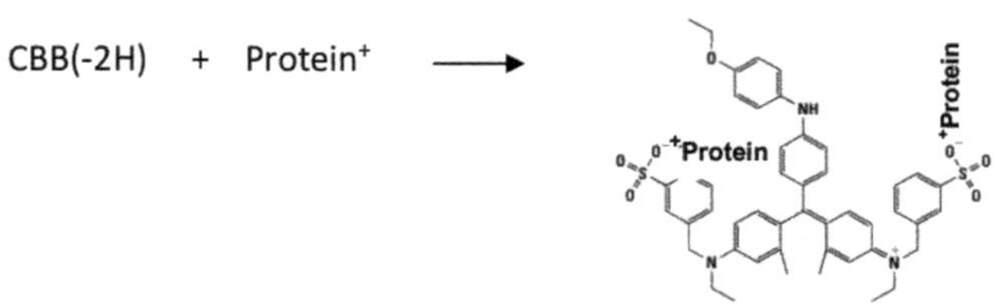

(34)

The complexes can then be quantitated by recording optical absorbance at 595 nm and determining protein concentration from standard curves prepared from a purified protein with amino acid composition comparable to those being analyzed. Purified bovine serum albumin or immunoglobulin G are common standards when protein compositions are unknown or pure proteins are not available.

As advantages, *t*he Bradford method is very sensitive, with lower detection limit of 0–50 μg in standard methods (Noble & Bailey, 2009) and nanograms in modified versions (Grintzalis et al., 2009). Requiring the smallest samples and shortest analysis times (Chutipongtanate et al., 2012), it tolerates normal levels of most salts, solvents, buffers, thiols, reducing substances and metal chelating agents encountered in protein samples (Noble & Bailey, 2009), and it is adaptable to both microplate and cuvette protocols.

However, the Bradford protein assay has two main disadvantages for oxidized proteins. It is inhibited by the presence of detergents, strong acids, or bases, so extracts containing these may need to be dialyzed and neutralized, or proteins must be isolated by precipitation as described for the BCA analysis. The linear response range is rather narrow, so samples should be tested with at least one dilution to make sure concentrations are below the plateau region.

Most importantly, the amino acid binding sites (arginine, histidine, and lysine) for the CBB dye are also major targets for reactions with oxidizing lipids. Loss of these binding sites reduces reactivity with CBB and apparent solubility in solution while also blocking staining on polyacrylamide gels (Schaich, 2014; Schaich, 2016). When CBB binding inhibition due to damaged lysine, arginine, and histidine is suspected in oxidized proteins (i.e. low reactivity is observed), apparent loss of protein solubility should be verified by a method independent of amino acids, such as Kjeldahl or elemental analysis.

Bradford assay protocols are available in (Kruger, 1994; Olson and Markwell, 2007).

Nitrogen analysis. One way to eliminate the inaccuracies caused by loss of dye-binding functional groups in amino acids when determining protein content is to analyze nitrogen content of original materials and solubility supernatants by Kjeldahl analysis or elemental analysis.

Kjeldahl analyses digest samples in concentrated sulfuric acid at very high temperature (~400 °C), under which conditions N in the sample is reduced to ammonia (NH_3) and complexed as ammonium sulfate $(NH_4)_2SO_4$. Addition of sodium hydroxide releases the ammonia, which is collected by distillation and titrated to quantitate the amount of released N. Kjeldahl analyses provide important validation of optical protein assays because they depend only on organic nitrogen content and are independent of structure as long as the nitrogen can be reduced and form ammonium sulfate. Hence, when Bradford or BCA methods indicate reduced solubility due only to loss of surface binding groups on oxidized proteins, Kjeldahl analyses may show little or no change from native proteins. However, nitrogen released as gaseous N_2, oxidized to various oxides, or incorporated into nitro or heterocyclic ring groups such as pyridines (Shirley and Becker, 1945), with triple bonds in azo groups or double bonds in any combination (e.g. -N=N-,-N=C-, -N=O, -N=S) are NOT detected in Kjeldahl analyses so oxidized proteins containing any of these structures will report lower N concentrations than their native counterparts. When this occurs, mass analysis is needed to determine whether apparent low solubility results from loss of dye-binding or Kjeldahl-active groups versus true loss of solubility.

Kjeldahl reactions and important controls are available in (Sáez-Plaza *et al.*, 2013).

Mass analysis. A simple approach to bypass dye-binding and Kjeldahl limitations in solubility analyses is to dialyze the supernatant collected as described at the beginning of the solubility section to remove non-protein contaminants, lyophilize the resulting solution, then weigh the mass of dissolved protein (Pacheco-Aguilar *et al.*, 2008). Solubility can then be expressed as:

$$\text{Solubility (\%)} = (\text{g Weight of soluble proteins}/\text{g Weight of total protein}) \times 100$$

This method is more chemically accurate but requires use of pure proteins in the analysis.

Which method is best? There is no best assay, only assays that are most appropriate for the particular protein mixtures being analyzed. The most important factor for co-oxidized proteins is protein binding sites – are they independent of oxidation? Other considerations that affect assay choice, such as handling, time, and instrumentation requirements are discussed in (Hayes, 2020; Moore et al., 2010; Sapan et al., 1999).

4.3 Sensory properties

Texture. Texture analyses measure mechanical properties of foods that are perceived by consumers, whether human or animal, and thus can connect observed molecular changes to food sensory characteristics and direct research to find causes of specific textures. Using a variety of probes, texture analyzers apply physical stress and strain as compression, shear, extension, bending, cutting, and piercing, then measure the movement or resistance of the food during application or after release of the force. Different combinations of probes and measurement conditions distinguish primary properties (reflect direct responses to force) such as hardness and cohesiveness from derivative secondary properties (reflect force over time) such as fracturability and chewiness (Table 6.2). Thus, texture analyses can be applied to all types of animal foods.

Texture analyses utilize two compression modes to model physical perceptions of consumers as they chew foods. One isolated compression measures the force required to push the material down and reflects forces experienced by consumers during their first impact ("bite") with the food material. In this mode, recording only peak force reports net hardness or stiffness, which can be useful in itself, but provides no clues as to textural elements contributing to the hardness. For example,

TABLE 6.2 Mechanical characteristics measured by texture profile analyses. Adapted from Szczesniak (1986).

Primary	Secondary	Common terminology
Hardness		soft, firm, hard
Viscosity		thin, viscous
Cohesiveness	Fracturability	crumbly, crunchy, brittle
	Chewiness	tender, chewy, tough
	Gumminess	short, mealy, pasty, gummy
Springiness		plastic, elastic
Adhesiveness		sticky, tacky, gooey

porous solids such as those found in animal kibbles and feeds are composed of molecular matrices and dispersed air spaces. The same hardness can be generated by a rigid matrix with large air spaces and a stiff but more flexible matrix with small air spaces. In such cases, greater differentiation can be gained by recording force over time to generate curves such as those shown in Figure 6.4 (top) that reveal yield stress points as well as force patterns. Narrow sharp peaks are generated by dense materials that are hard and rigid, while broader curves are presented by materials that may be firm but are more flexible. Single compression analyses have been used to relate hardness and disintegration of pet treats such as biscuits, rawhide, chews, and dental treats to digestibility (He *et al.*, 2020), and to determine mechanical properties of chicken feed supplemented with various levels of crude fiber concentrates (Urban *et al.*, 2024).

The "two-bite" mode applies two sequential compressions to model how food responds to and potentially recovers from "first-bite" forces, thus modifying the sensory experience in the "second bite". Force is applied, released, a wait time is added, then the force is applied and released again. Comparing force responses in the second versus first compression provides measures of secondary characteristics such as chewiness and gumminess, and the overall response profile can thus distinguish textural differences in greater detail than single compression analyses.

Two-bite texture profile analyses have been applied to evaluate texture differences of animal versus plant-based meats (Dunne *et al.*, 2025) and of pet food chunks made from soy protein isolate-wheat gluten meat analogues versus a standard poultry/beef/grain mixture versus cooked beef (Wehrmaker *et al.*, 2021). Particularly relevant to this chapter, two studies show that texture profiles can connect protein oxidation effects to chemistry. Lipid oxidation (malonaldehyde levels) and protein oxidation (loss of protein thiols) were associated with increased hardness in texture analyses in refrigerated grass carp muscle (Ramadhan *et al.*, 2025). When pork myofibrillar proteins were oxidized by hydrogen peroxide, texture profile analysis of their gels revealed that low levels of oxidation (1 mM H_2O_2) reduced hardness, springiness, gumminess and cohesiveness slightly, improving palatability (Xia *et al.*, 2019). These changes were accompanied by increases in protein carbonyls and some protein depolymerization, consistent with radical attack on the proteins. However, high oxidation markedly reduced the same texture values; these changes were accompanied by very high protein carbonyls, substantial loss of protein thiols, and extensive protein polymerization. The authors speculated that carbonyl crosslinking such as the reactions described in the previous section of this chapter were involved here.

Though few in number, these studies demonstrate that texture profile analyses can provide important indicators that properties are changing during reformulation, processing, or storage, even without knowing the chemistry. With routine analyses to establish a baseline of normal product characteristics, periodic texture analyses during processing and storage can flag potential problem batches, then chemical analyses can be focused to determine the molecular causes of the textures, e.g. crosslinking, scissions, protein subunit rearrangements, or intermolecular associations modified by the surface adducts described above. Many more such studies establishing connections between textures and protein oxidation states are needed. Chemometrics then can be applied to identify cut-off points when materials must be rejected.

Detailed information on texture analyzer designs, property definitions, examples of data profiles, and applications to several different types of foods relevant to animals, including wet foods, gels, and different forms of dry products are available in (Johnson, 2015, modified 2023; RheologyLab, 2025; Rosenthal and Chen, 2024; Smewing, 2013).

4.4 Functional properties

Enzymatic and In vitro digestibility. As outlined above, the extensive modifications in proteins co-oxidized by lipids create a strong likelihood that enzyme recognition and digestibility will be impaired, contributing to loss of nutritional value and failure to meet nutritional claims or standards. Thus, digestibility testing is commonly included in evaluation of foods and animal feed undergoing oxidation, evaluating

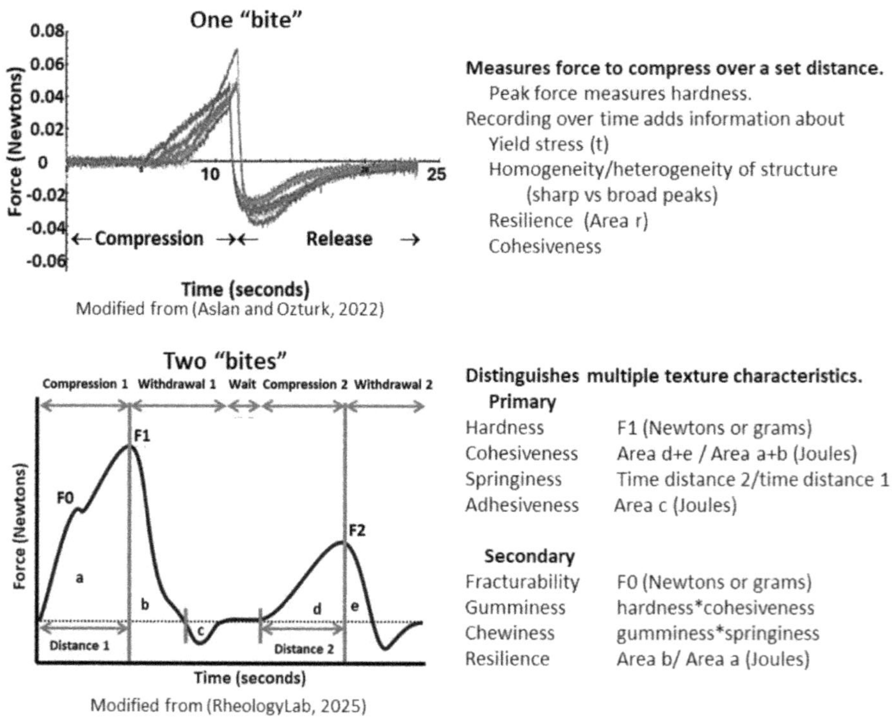

FIGURE 6.4 Graphical data patterns and textural characteristics derived in one and two "bite" compression modes of textural profile analyses

either simple inhibition of protease digestion or more complete reduction in digestibility in an *in vitro* model of physiological digestion, such as INFOGEST.

For initial investigations or quality control, inhibited trypsin, chymotrypsin, or pepsin (or a combination of the three) hydrolysis of test proteins is adequate to demonstrate functional modification of the protein that interferes with normal digestion, and it provides clues regarding amino acids damaged by the specificity of the enzyme inhibited – trypsin cleaves after lysine and arginine, chymotrypsin cleaves after phenylalanine, tryptophan, and tyrosine, and pepsin cleaves mainly after phenylalanine, leucine, and glutamic acid (Boisen, 2000). The method is simple and straightforward but requires isolated protein fractions. Protocols for each enzyme are available (Worthington, 2025).

If proteins must be tested in a full food or feed matrix, or if more extensive information about digestion interference by oxidized proteins is needed, several *in vitro* systems are available to model animal digestion. The most important of these, the INFOGEST model, was developed collaboratively by more than 45 countries to integrate and reconcile other methods and to provide protocols that could be followed with standard laboratory equipment and solutions (Brodkorb *et al.*, 2019). The model covers three stages of digestion with simulated salivary fluid, simulated gastric fluid, and simulated intestinal fluid, providing a means of identifying where specific protein modifications block digestion. Space limitations prevent full discussion of this new procedure. Readers are referred to (Brodkorb *et al.*, 2019) for explanation and protocols and (Egger *et al.*, 2019) for an application example.

Useful reviews of *in vitro* digestion methods are available in (Chernukha *et al.*, 2022; Santos-Sánchez *et al.*, 2024). Protocols have been outlined and methods have been compared and evaluated by (Franczyk, 2018).

***Water-holding capacity* (*WHC*).** As a critical function of proteins involved in cohesiveness, plasticity, and stability of foods, water binding involves hydrogen bonding or charge interactions of surface amino acids with water molecules. When any of these sites are modified by co-oxidation, water binding is altered and all associated functional properties such as gelling and foaming are also affected. If polar groups are blocked by lipid chains, the surface becomes less polar and water binding decreases. At the same time, if more polar groups are added to the protein surface, such as formation of various protein or adduct carbonyls, the surface becomes more polar and water-holding capacity increases. Thus, measuring water-holding capacity of food proteins first as an ingredient then after food processing and storage can report functional changes in the protein surface and provide useful clues to likely chemical and physical causes.

The methodology involves isolating and drying total protein or protein fractions from a food material, hydrating 0.02 g protein/g water overnight, centrifuging, then weighing the pellet before and after lyophilizing (Geerts *et al.*, 2018). The following quantities are then determined from the weights (Bühler *et al.*, 2020):

$$\text{Protein solubility} = (\text{mass}_{\text{dry powder}} - \text{mass}_{\text{dry pellet}}) \;/\; \text{mass}_{\text{dry powder}}$$

$$\text{WHC}_{\text{overall}} = (\text{mass}_{\text{wet pellet}} - \text{mass}_{\text{dry pellet}}) \;/\; \text{mass}_{\text{dry powder}}$$

$$\text{WHC}_{\text{insoluble}} = (\text{mass}_{\text{wet pellet}} - \text{mass}_{\text{dry pellet}}) \;/\; \text{mass}_{\text{dry pellet}}$$

Calculating $\text{WHC}_{\text{insoluble}}$ determines the contribution of solubility to WHC.

4.5 Protein surface modifications

Lipid radicals, hydroperoxides, aldehydes, and epoxides all react with amino acid side chains to modify protein surfaces in ways that interfere with normal protein function and degrade protein structure in foods. Surface modifications include oxidation of amines, alcohols, and thiols to oxides, carbonyls, and disulfides, respectively; deamination and decarboxylation; and complexation of lipid oxidation products through Michael addition and Schiff base condensations. Unfortunately, there is no way to test for all these modifications in a single assay; each modification must be analyzed independently. How many and which changes should be followed as markers of oxidation depends on the protein oxidized, the chemistry suspected to underlie observed macroscopic or functional modifications, and the loss of function or altered behavior that needs to be explained.

Hydrophobicity. Amino acids on the protein surface actively control interactions of each peptide with other molecules in creating food structures, so changes in the surface character have large impacts on protein functionality (e.g. foaming and gelling) and physical structures of foods. Measuring surface hydrophobicity, therefore, is one way to track protein denaturation and, more importantly, presumptive lipid addition to surface amino acid residues. In a simple form of the assay, extracted proteins of known concentration in buffer are reacted with aqueous bromophenol blue (BPB) which binds to hydrophobic sites on the proteins, the mixture is centrifuged to remove insoluble matter, then optical absorbances of the supernatants and blanks are measured. The buffer, pH, and wavelength monitored may vary with the protein analyzed. Tris buffer, pH 8, and 620 nm were used with bovine serum albumin (Bertsch *et al.*, 2003), while phosphate buffer, pH 6, and 595 nm were applied to oxidized porcine muscle fiber proteins (Chelh *et al.*, 2006). Results are expressed as either µg or mol BPB bound/unit protein.

Analysis protocols are available in (Zhang and Xiao, 2025).

Presence of fluorescent Schiff base adducts. The Reactions section above explained how Schiff base structures are important common products of lipid carbonyl additions to amino acids. Rapid presumptive evidence for Schiff base adducts having the requisite component -N=CH-CH=CH-NH- can be obtained by fluorescence spectra with emission scanned from 360–650 nm, excitation set at 350 nm. Under these conditions, the λ_{max} can distinguish linear (430 nm) from ring (470 nm) adducts (Chelh et al., 2007), as shown in Figure 6.5 for oxidized baked and fried tortilla chips versus ingredient cornmeal. Without disulfide reduction, fluorescent structures in both tortilla chips were cyclic, whereas after reduction, part of the fluorescent structures became linear. Either the rings contained S-S bonds or the mercaptoethanol reduced imine bonds in the rings of some products.

Fluorescence excitation/emission maxima and emission intensity vary with the protein components, the Schiff base structure, and the solvent. Shifts in λ_{max} for different lipid-protein complexes are compiled in (Schaich, 2016).

Cautionary note: careful sample handling is critical to avoid artifacts in fluorescence analyses. Contaminants in extracts catalyze secondary photoreactions, fluorescence quenching from molecular interactions occurs, especially in aqueous solvents, and fluorescence is sensitive to light. Hence, extracts should be prepared and analyzed as rapidly as possible and protected from light to avoid loss by decay and alteration of fluorescence.

Native gel electrophoresis. Native polyacrylamide gel electrophoresis (N-PAGE) provides one of the fastest ways to determine whether surface modifications have occurred on oxidized proteins. N-PAGE includes no denaturant so protein configurations are retained and migration is governed by molecular properties such as surface charge (the relative composition of NH_3^+ and COO^- groups on the peptide surface). When a current is applied to gels, acidic proteins move toward the cathode at the bottom and basic proteins distribute towards the anode at the top. Deamination, decarboxylation, or blocking of the charged amine and acid groups during co-oxidation alter the peptide migration patterns markedly. Comparison of bands visualized by Coomassie Blue versus silver staining can identify peptides in which basic amino acids vs acidic amino acids, respectively, have been modified and can diagnose deamination and decarboxylation (Ibadullah, 2013).

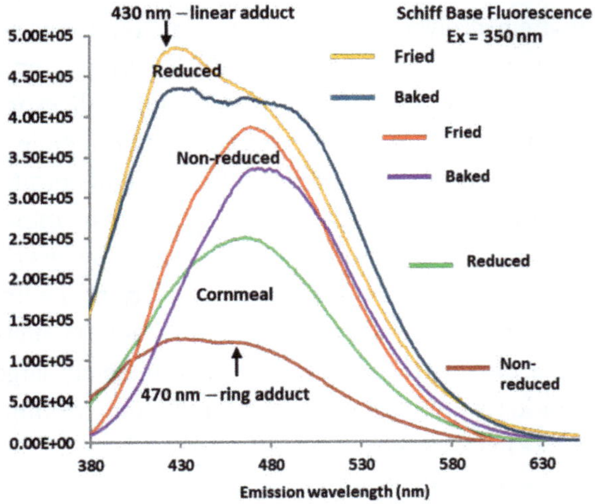

FIGURE 6.5 Fluorescence spectra of linear (430 nm) versus ring (470 nm) Schiff base adducts formed in cornmeal and baked vs fried tortilla chips. Adapted from Dong (2011), used with permission.

Native gels are particularly useful in distinguishing oxidized proteins that have been modified chemically but without molecular weight changes. For example, gliadins that are surface-modified in extrusion co-migrate in denaturing PAGE but can be completely separated in native gels (Partridge et al., 2003). N-PAGE also detects oxidative disruption of charge associations between quarternary structures, e.g. peanut arachins and conarachins (Ibadullah, 2013).

A few cautions must also be noted. Native gels generate considerable frictional heat, so apparatus cooling is mandatory to minimize protein denaturation and proteolysis and to prevent band blurring. This can be accomplished by using very cold water or ice slurries in the gel cooling jacket of most systems. Interpreting band patterns and tracking oxidation changes of a specific protein in a mixture is difficult due to mixed-mode migration. Proteins do not distribute by size on the gels, and peptide banding patterns are usually very different than in denaturing gels for the same proteins. Thus, pure controls must be run to identify positions of different proteins on the gels. Controls of individual unmodified proteins or extracts from native materials before processing should be run on each gel for comparison.

4.6 Analyses of key chemical changes

Amino acid content. Instrumentation and standard procedures for amino acid analyses are readily available so will not be discussed. Rather, this short overview highlights two critical impediments to amino acid analyses that are posed by co-oxidized proteins. First and foremost is that normal acid or alkaline hydrolysis procedures for releasing amino acids from proteins destroy amino acids that are oxidation targets, particularly tryptophan, cysteine, serine, threonine, asparagine, and glutamine (Hunt, 1985). Also, asparagine and glutamine are converted to aspartic acid and glutamic acid by deamination during acid hydrolysis, so normal procedures do not accurately report amino acid compositions of oxidized proteins. The alternative, enzymatic hydrolysis, runs under neutral conditions at moderate temperatures (30–60 °C) that minimize side reactions (Hou *et al.*, 2017), but it is a more complex process. Because of protein large sizes, complete digestion of proteins to amino acids rather than peptides requires mixtures of both endo- and exo- proteases that cleave amino acids in the middle and at the ends of protein chains, respectively, and sometimes di-peptidases as well (cleave two amino acid subunits). Digestion is slow and must be monitored periodically, e.g. by reversed phase HPLC, to determine completion. Proteases must be removed from the final digest before analysis, without loss or modification of amino acids. Potential methods for removing enzymes include ultracentrifugation (requires relatively large samples) and ultrafiltration (often loses amino acids by binding). Some examples of digestion protocols are available in (Riviere and Tempst, 2001).

Second, the various modifications in oxidized proteins inhibit or interfere with enzymes, making complete hydrolyses and accurate amino acid distributions difficult to obtain. Most proteases cleave only at specific amino acids; when altered target sites are not recognized, cleavage does not occur, and dimer or larger units rather than amino acids are released. Small peptides can be handled by LC-MS analyses but present problems in traditional amino acid analyzers. Fortunately, proteases with a range of amino acid specificities are available from plant, microbial, and insect sources, so with enough testing, proteases with specificity appropriate for different protein mixtures can be identified.

What information can amino acid analysis provide about oxidized proteins?

- Amino acid distributions reveal the degree of nutritional loss and amino acid imbalance that may cause physiological problems.
- Analyses reveal amino acids destroyed and converted to other amino acids, e.g. histidine → asparagine, aspartic acid, valine, ethylamine; asparagine → aspartic acid; glutamine → glutamic acid.
- Modified amino acids (even dimerized and with lipid adducts) can be separated from unmodified amino acids, detected by HPLC, and identified when interfaced with mass spectrometry. Obtaining product and adduct structures is extremely important in elucidating active co-oxidation mechanisms.

An excellent review of historical and contemporary amino acid analyses is available in (Fenaille *et al.*, 2004).

Free amines. Free amine contents of oxidized proteins decrease with adduct formation and deamination but increase with fragmentation, so this analysis is commonly included in the first battery of testing for oxidized proteins. Standard analyses have been based on reaction of free amines with detection reagents in solution, then quantitating the generated complexes by fluorescence or UV-VIS spectrophotometry.

A good starting point for new investigations is the familiar ninhydrin test (Moore and Stein, 1954) which has both uv-vis and fluorescence versions of the assays. In reactions, ninhydrin deaminates the amine and is reduced in the process (Reaction 35); the released ammonia then adds to one molecule of ninhydrin and one molecule of reduced ninhydrin to form colored condensation products which are violet-blue for most amino acids, yellow for hydroxyproline and proline, and brown for asparagine (Reactions 36) (Biocheminsider, 2021).

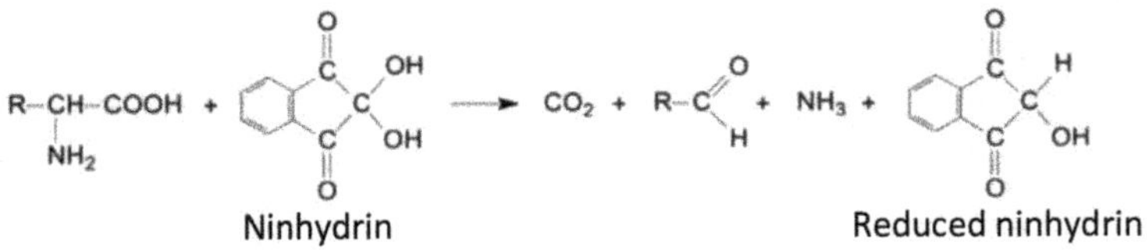

(35)

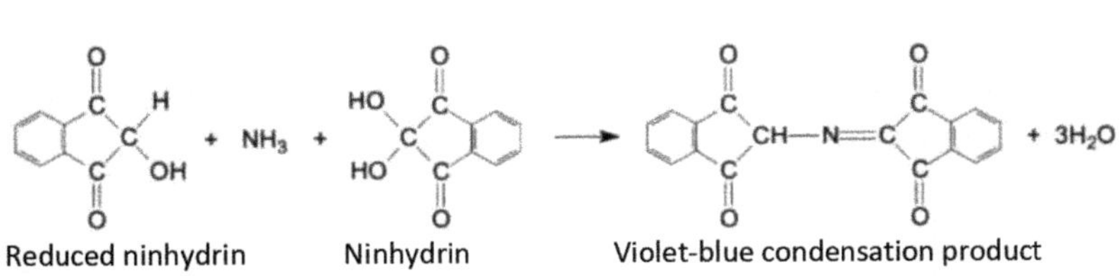

(36)

A simple version of this assay heats excess ninhydrin with buffered solutions of isolated protein(s) at 90 °C for 17 min, cools the mixture, reads the optical absorbance at 570 nm against a blank, then determines the concentration of free amines present by comparing the recorded absorbance to a standard curve prepared from pure amines. While a pH of 5.5 is generally recommended, the wide range of buffers, pH, and solvent modifiers cited in published applications make it clear that researchers routinely optimize reaction conditions for their specific proteins. Temperature and time of heating can also be adjusted to accommodate denaturation / aggregation properties of the test proteins. Protocols and considerations for reaction control are available in (Biocheminsider, 2021; Stauß *et al.*, 2024; Sun *et al.*, 2006)

Several factors limit accuracy and reproducibility of the ninhydrin assay for mixed protein systems such as found with animal and pet foods. Two major problems are unequal reactivity with the twenty amino acids present, so colors generated vary with the protein or protein mixtures, and ninhydrin reactions are extremely sensitive to reaction conditions so fresh standard curves must be prepared for each batch of material and each analysis. When protein mixtures contain notable levels of proline and hydroxyproline, as in protein by-products in pet foods, the assay absorbance must be read also at 430 nm where complexes with these amino acids absorb. The heating used to promote the reaction often alters test proteins, as is highly likely with oxidized proteins. Under such conditions, formation of the colored complexes can be accelerated by including hydrindantin, the reduced ninhydrin shown in Reaction 35, in the reaction mixture (Lamothe and McCormick, 1973). Finally, perhaps the greatest limitation is the low sensitivity: lower detection limits are only 10^{-4} to 10^{-5} M, which means larger samples and more replicates are required and substantial degradation must occur before the ninhydrin assay detects changes. Oxidation pathways thus may be missed in the mix of active changes caused by processing and other stresses.

Additional reagents have been developed to detect free amines faster, with less sample, and with greater sensitivity than ninhydrin. 2,4,6-Trinitrobenzenesulfonic acid (TNBS) reacts with free amino groups at pH 8.5 to 10 to form orange-colored complexes (Fields, 1972; Habeeb, 1966). After two hours incubation at room temperature, sodium dodecyl sulfate is added to fully solubilize the protein, the reaction is stopped with acid, and absorption at 340 nm is recorded. Although detection levels reportedly are double those of ninhydrin, the high pH used for the assay will hydrolyze some of the adducts formed in oxidized proteins, compromising accuracy of the assay for detecting loss of free amino groups during oxidation.

Two fluorescence assays for free amines are in use. Fluorescence increases sensitivity about 1000 times compared to uv-vis analyses, but the instrumentation is not universally available in labs, and close attention to contamination in glassware and reagents is required. There also are problems with interference from protein florescence from co-oxidized proteins.

O-Phthalaldehyde (or *o*-phthaldialdehyde) reacts with free amines at neutral pH in the presence of thiols to form a bifunctional product (Reaction 37) that can be detected at 340 nm in uv-vis assays (Church *et al.*, 1983) and at 455 nm with 340 nm excitation in fluorescence assays (Fisher *et al.*, 2001; Goodno *et al.*, 1981; Simons and Johnson, 1978).

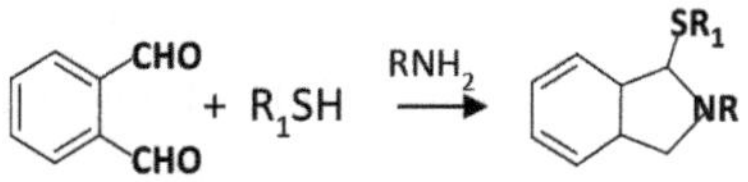

(37)

The reaction includes sodium dodecyl sulfate to denature the proteins and 2-mercaptoethanol to reduce S-S bonds; both ensure complete exposure of the free amine groups. With sample volumes of only 10 µl, reaction is complete in two minutes at room temperature and picomoles of free amines can be detected. By comparison, ninhydrin detected comparable levels of free amines in proteins but required 50–100 µl of the same protein solution and longer reaction times (Fisher *et al.*, 2001). The sensitivity of the fluorescence assay is a great advantage in early stages of protein oxidation. However, emissions are subject to fluorescence quenching and high backgrounds at high concentrations (>10^{-4} M) and will be complicated by overlapping Schiff base emissions in oxidized proteins. Under such circumstances, uv-vis versions may be a better alternative.

OPA protocols are available in (Held, 2006). A microplate version is described in (Robert-Peillard *et al.*, 2017).

Fluorescamine complexes free amines when reacted in neutral phosphate buffer at room temperature for 15 min to form a product that fluoresces at 475 nm emission with 375–390 nm excitation (Reaction 38) (Purcell *et al.*, 1976; Udenfriend *et al.*, 1972).

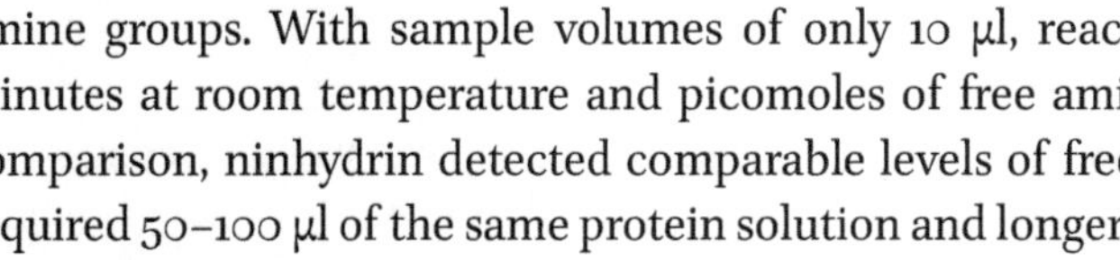

(38)

Picomoles of amine can be detected under optimal conditions, but the method has significant problems with high backgrounds from proteins and fluorescence quenching with high fluorescamine concentrations. Fluorescamine hydrolysis in aqueous solutions can be prevented by inclusion of an organic solvent such as ethanol in reactions, but how this affects solubility of co-oxidized proteins has not been ascertained. For oxidized proteins, both reduced solubility in neutral buffer and overlaps of protein Schiff base fluorescence impair sensitivity and accuracy.

A review of conditions affecting the fluorescein reaction are available in (Derayea and Samir, 2020). Protocols can be found in (ThermoFisher, 2008). Many commercial kits are available for this assay.

Cysteine/cystine content. Cysteine/cystine (SH/SS) contents and ratios are important indicators of normal protein configurations and, quite critically, of cysteine participation in protein co-oxidation via oxidation and adduct formation itself and by disulfide crosslinking to form polymers. SH/SS chemistry is among the first to show change. Thus, these two amino acids must be monitored regularly when following protein oxidation.

SH/SS determination is predominantly accomplished by reacting proteins in solution with Ellman's reagent, 5,5'-dithiobis-(2-nitrobenzoic acid) (DTNB) (Ellman, 1958; Ellman, 1959). The disulfide DTNB undergoes thiol-disulfide interchange with thiol groups on the proteins, PSH, to form a mixed disulfide and a 2-nitro-5-thiobenzoate anion (TNB^-) that converts to a yellow NB^{2-} dianion that can be detected optically at 412 nm (Reaction 39). One TNB is released for every thiol oxidized upon DTNB treatment.

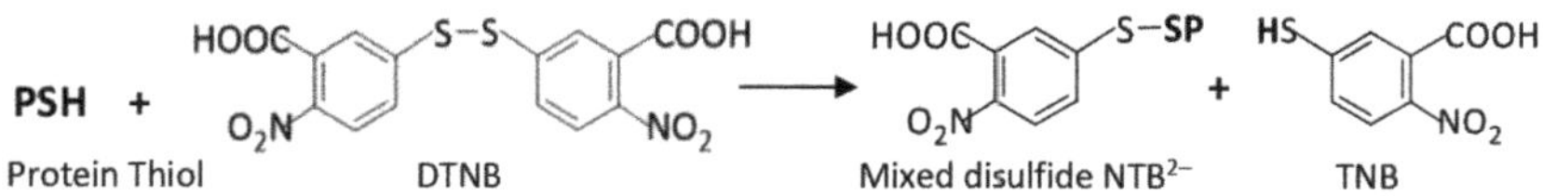

(39)

Thiol concentrations are then quantitated from Beer's Law using a published extinction coefficient if standard conditions are used or by comparison to a standard curve prepared from a pure thiol such as cysteine if reaction conditions have been modified. The extinction coefficient of the dianion varies with the solvent used (Schaich, 2016), so Ellman's original value of 13,600 $M^{-1}sec^{-1}$ should not be applied automatically. The lower detection limit is one thiol group per protein in 2 nanomoles of protein.

Disulfides can also be determined by DTNB in a reverse thiol-disulfide interchange reaction (Thannhauser et al., 1987). Disulfides are not very reactive, so to increase activity, S-S bonds in both proteins and DTNB are reduced with sodium sulfite at pH 9.5 to generate, respectively, protein thiyl anions, $P'S^-$ (Reaction 40) and the $NTSB^{2-}$ (disodium 2-nitro, 5-thiosulfobenzoate) anion plus NTB, nitrothiobenzoic acid. The protein thiyl anion then adds across the S-S bond of NTSB to give a protein mixed disulfide, P'S-SO_3, protein cysteinyl sulfite (Reaction 41) and the standard Ellmans' mixed disulfide which is detected at 412 nm as were the thiols and quantitated by comparison to a standard curve prepared from cystine.

$$PSSP' + SO_3^{2-} \leftrightarrow PSSO_3^- + P'S^- \qquad (40)$$

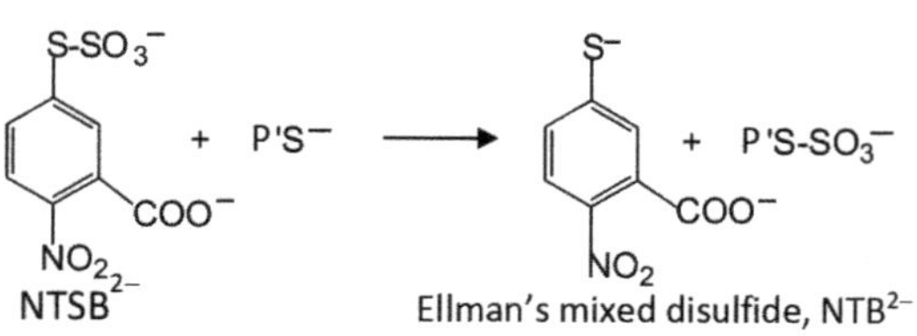

(41)

The thiol value measured is the total thiol content (both thiols and disulfides), so the free thiol content determined in unreduced proteins by the standard DTNB reaction must be subtracted from this total thiol content to calculate S-S content by difference. The assay is rapid, requiring 3–5 min for oligopeptides and 20 min for proteins, and detects as little as 10^{-8} mol of disulfide bonds.

Protein thiols are quite prone to oxidation during protein extraction, isolation, and storage. This can be avoided by using a solid-state version that combines the SH and SS assays just described to detect both thiol and disulfide groups in both soluble and particulate proteins without pre-extraction (Chan and Wasseran, 1993). Briefly, for free thiols, dried ground food samples (10–30 mg) are suspended in freshly prepared 8M urea, 10 mM DTNB, 3 mM ethylenediamine tetraacetic acid (EDTA), 1% SDS, and 0.2M Tris-HCl, pH 8 and allowed to react (required time determined by the material), then the samples are centrifuged to remove solids. The chromophore NTB^{2-} remains free in the supernatant, so absorbance at 412 nm is read for quantitation. Procedures for disulfides are basically the same as thiols except that Tris-HCl pH 9.5 and $NTSB^{2-}$ are substituted as solvent and reagent, respectively. The reaction is run without sulfite to determine disulfides and with sulfite reduction to determine total thiols. Disulfide content is calculated from the difference between total thiols after and before reduction of disulfide bonds with sulfite. Total cysteine is calculated as [-SH] + 2[-S-S-].

Two important cautions must be noted. Most critically, thiols are sensitive to oxidation and both TNB^- and $NTSB^{2-}$anions degrade in light (Walmsley et al., 1987), so accurate results require running reactions in dim light or dark under anoxic conditions, e.g. argon-sparged solutions and headspace maintained under argon or reaction run in a glove box. In addition, all reducing agents must be removed from extraction media by dialysis or by protein precipitation and resuspension before protein reaction with DTNB.

Further reaction considerations are available in (Schaich, 2016). Protocols are available in (Xiao *et al.*, 2025).

Sulfur oxides. Oxidized cysteines do not just form disulfide crosslinks and cause polymerization – they also oxidize to cysteine sulfoxides and sulfinic, sulfenic, and sulfonic acids which prevent disulfide crosslinking and increase protein surface polarity, opposite the effects of many lipid carbonyl and epoxide adducts. This makes sulfur oxides in proteins very important because, if formed, they interrupt productive as well as destructive disulfide crosslinking and alter functional and textural properties associated with them. Unfortunately, protein sulfur oxides have seldom been analyzed so little is yet known about the extent to which cysteine oxidizes to sulfur oxides (sulfoxide; sulfenic, sulfonic, and sulfinic acids; sulfones) in competition with or instead of recombining to form disulfide cystine, or the conditions that favor such transformations. The most straightforward technique for detecting sulfur oxides in proteins in solution or solid materials is by mid FT-IR (ChemTalk, 2025;

Schaich, 2016). The stretching vibrations listed for sulfur-oxy functions in proteins listed in Table 6.3 were used to detect sulfur oxides that prevented normal gluten development in aged flour (Resch, G. and Schaich, K.M., unpublished data, 2013). Additional methods and considerations for measuring oxidation states of sulfur in proteins are reviewed in (Rudyk and Eaton, 2014).

When mass spectrometry (MS) instrumentation is available, proteomics technology and MALDI-MS now make it possible to detect oxidation products of sulfur amino acids by the change in protein molecular weight (Raftery, 2014). ΔMass values for cysteine oxides, methionine sulfoxide, and methionine sulfone have been documented (Table 6.4).

Carbonyl groups in proteins. Carbonyls were one of the first protein oxidation products analyzed and are still considered to be the quintessential protein oxidation product, so straightforward accurate, sensitive assays are essential. The classical reaction of dinitrophenylhydrazine (DNPH) used to quantitate lipid carbon-

TABLE 6.3 Primary FT-IR absorption frequencies (cm^{-1}) of protein sulfur-oxy (S=O) groups. Excerpted from ChemTalk, 2025.

Frequency (cm^{-1})	Structure	Oxidation product
1380-1415	$R(S{=}O)_2(OH)_2$	sulfate
1380-1410	$R(S{=}O)_2Cl$	sulfonyl chloride
1342-1350	$R(S{=}O)_2OH$	sulfonic acid
1335-1372	$R(S{=}O)_2O^-$	sulfonate
1335-1370	$R(S{=}O)_2N(R^1)R^2$	sulfonamide
1300-1350	$R(S{=}O)_2R^1$	sulfone
1030-1070	$R(S{=}O)R^1$	sulfoxide

TABLE 6.4 Detection of oxidized sulfur amino acids by changes in mass ions in MS. Excerpted from Raftery, 2014.

Compound	Change	ΔMass	Residue mass
Cysteine	0	0	103.0092
Cys-disulphide	−2H	−2.014	100.9952
Cys-sulfinamide	+O(−2H)	+13.9809	116.9901
Cys-sulfenic acid	+O	+15.9949	119.0041
Cys-sulfinic acid	$+O_2$	+31.9898	134.9990
Cys-sulphonic acid	$+O_3$	+47.9847	150.9939
Met	none	0	131.0405
Met-sulphoxide	+O	+15.9949	147.0354
Met-sulphone	$+O_2$	+31.9898	163.0303

yls has also been applied to proteins in a variety of formats where the first step is always reaction of DNPH with proteins to form protein conjugated hydrazones (Reaction 42).

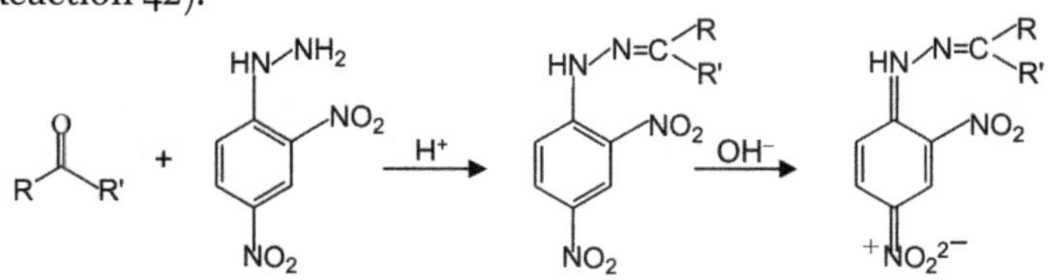

(42)

Protein carbonyl analyses by DNPH were initially developed as solution assays where oxidized proteins are reacted with DNPH in strong acid to form protein hydrazones that are detected optically at 265 or 370 nm and quantitated using $\varepsilon_{hydrazone} = 22{,}000\ M^{-1}cm^{-1}$ at 370 nm. Results are usually expressed as nmols DNP hydrazones per mg of protein (Levine *et al.*, 1990; Oliver *et al.*, 1987). The solution assay has been applied extensively in a wide range of protein systems, with solvents and conditions modified to accommodate different proteins. Nevertheless, it has numerous shortcomings that decrease reproducibility and sensitivity. The procedure is tedious and time-consuming with multiple reaction and centrifugation steps and very long reaction times. Conventional protein buffers such as TRIS that contain primary amines cannot be used in this assay because the amines react faster than DNPH with protein carbonyls, so they block hydrazone formation. Interfering substances (non-protein contaminants in the extracts before reaction and DNPH remaining unreacted after the reaction) must be removed and many blanks must be included for results to be valid. Acid conditions in the assay can modify or destroy target protein carbonyls, and they convert active DNPH to inactive $DNPH_2$. Thus, reaction pH should be no lower than 2.5 (Kandrac, 2018). Finally, a great detriment for quality control analyses is that the assay is difficult to handle with large numbers of samples, and the source of the carbonyls detected cannot be identified.

It is reasonable to expect that protein solubility will be less in animal food than in cells and tissues due to greater exposure and more extensive interactions with oxidizing lipids and processing stresses, so it is worth considering whether loss of solubility will interfere with detection of protein carbonyls. To address this question, three sequential steps were added to the traditional DNPH assay of (Levine *et al.*, 1990) to increase protein solubilization: dissolving in 5% sodium dodecyl sulfate (SDS), heating at 100 °C for 10 min, and ultrasonicating at 40 °C for 30 min (Soglia *et al.*, 2016). The increase in carbonyl levels detected after each treatment was attributed to increased protein unfolding and solubilization of the starting protein. Protein solubilization did increase, but heating and ultrasonicating also oxidize proteins and cause scissions that lead to carbonyl formation. Issues of protein solubilization and separating processing from lipid co-oxidation effects are critically important in analyses of oxidation in animal and pet food materials. Now that food oxidation is finally being recognized, it is time to focus research on these problems to provide assays that reflect true oxidation levels rather than handling.

For additional information about the solution DNPH assay and protocols, see (Colombo *et al.*, 2016; Levine *et al.*, 2000; Levine *et al.*, 1994) for biological applications and (Estévez, 2011) for meat systems.

In further attempts to improve the DNPH assay, decrease time and handling, increase sensitivity and specificity, and gain some information about which proteins carry the carbonyls, antibodies have been raised against DNP-labelled proteins and used as the basis for enzyme-linked immunosorbent assays (ELISA) and Western blot adaptations of the DNPH assays.

ELISA sets up reactions in layers in polystyrene microplates (typically 96 wells) in four variations of sequence. In indirect ELISA, most commonly with oxidized proteins, proteins that have bound DNPH at each carbonyl during reaction in solution are deposited in the plate wells, allowed to adsorb to the surface, then free protein is rinsed off. A primary anti-DNPH antibody is layered over the adherent protein where it binds to each bound DNPH, then a secondary overlaid antibody carrying an enzyme such as horseradish peroxidase (HRP) links to the first antibody to provide a marker reaction. Unreacted reagents are washed away between each step (Buss *et al.*, 1997). Finally, an agent is added to react with HRP and produce color for quantitation (see Figure 6.6). Protein carbonyls are quantitated from their absorbance at 265 or 379 nm read in a plate reader. This procedure greatly increases throughput while providing very low background and high specificity in results since only bound DNPH groups are detected and quantitated.

As drawbacks, a critical requirement for ELISA accuracy is complete initial binding of the test protein(s) to the wells, but this is very difficult to determine even with native proteins. Because there is little experience with ELISA of oxidized proteins where binding to plates varies and oxidation levels in foods should be much higher than proteins oxidized in tissues, alternate ELISA protocols must be tested and

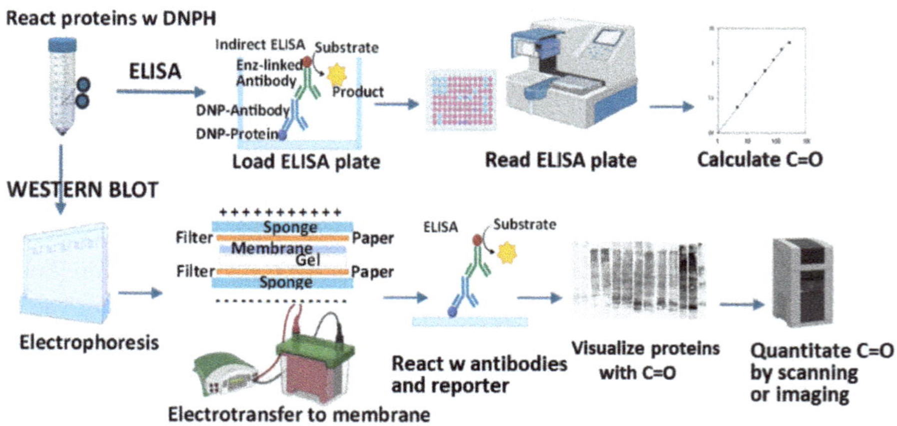

FIGURE 6.6 Analysis of protein carbonyls by reaction with DNPH and detection with DNPH-specific antibodies in ELISA and Western blot methods. Figure elements compiled from Novusbio.com, Microbenotes.com, and K.M. Schaich lab.

optimized for each new material analyzed. In addition, as with the solution assay, ELISA quantitates total carbonyls but can provide no information regarding location of the carbonyls, i.e. whether carbonyls are distributed among all proteins or are localized on specific peptides.

Reviews and protocols for ELISA versions of DNPH assays of protein carbonyls are available in (Ahirwar *et al.*, 2022; Bio-Rad, 2017; PierceBiotechnology, 2010).

Western blots with anti-DNPH antibodies fill this information gap by adding some extra steps to gain insights into the peptide distribution of carbonyls. First steps are the same as most other versions of the DNPH assay – extract and isolate the protein, then react with DNPH in solution. Next, labelled proteins are loaded onto two identical SDS-PAGE gels and separated by standard procedures. One gel is stained with an appropriate dye to visualize the total protein bands and the second gel is transferred via electroblotting onto PVDF (polyvinylidene difluoride) or nitrocellulose membranes in a Western blotting unit. The type of membrane used is determined by the binding characteristics of the proteins being analyzed – PVDF binds proteins by hydrophobic interactions (useful when lipid adducts are present), while nitrocellulose binds proteins by nitrogen dipole, H-bond, ionic, and hydrophobic interactions. The membrane is blocked then, as with ELISA, reacted in sequence with anti-DNPH antibodies, secondary antibodies, enzyme, and visualizing agents chosen for compatibility with available detectors. Protein carbonyl presence is verified by colored, fluorescent, or chemiluminescent peptide bands on the membranes (Figure 6.7D); carbonyls are quantitated from intensity of the bands, as determined by scanning the membrane with a densitometer, scanner or imager. Detection limits are generally less than one ng.

Western blot protocols can be found in (Colombo *et al.*, 2016; Mahmood and Yang, 2012; Sule *et al.*, 2023) and many vendor websites on the internet.

An excellent review has compared and evaluated multiple versions of DNPH procedures (Dalle-Donne et al., 2003).

Continued criticism of DNPH carbonyl assays, even ELISA and antibody versions, stimulated recent development of oxime blots where protein carbonyls are derivatized with aminooxy (oxime) reagents to generate oxime bonds with proteins that undergo hydrolysis less readily than hydrazone bonds with DNPH (Ladouce *et al.*, 2023) Oxime derivatizes protein carbonyls with a biotin-aminooxy probe in the presence of the *p*-phenylenediamine catalyst at a neutral pH, separates protein by PAGE with SDS, stains bands with Ponceau red, then transfers proteins to membranes in Western blots and reacts with antibodies to biotin-aminooxy probe. Carbonyl signals 60% stronger than with DNPH in oxidized proteins have been claimed, but the method has not been tested against proteins that are co-oxidized with lipids and have more hydrophobic surface properties. The use of alternate carbonyl probes with more selective antibodies offers some interesting opportunities for improving carbonyl detection and quantitation in complex food systems such as animal feeds, so exploration of such methods should be encouraged.

Tryptophan modification and environment. Intrinsic fluorescence measurements determine changes in both tryptophan content and its environment, so fluorescence is perhaps the best way to track whether tryptophan has been degraded or its environment has been modified in oxidized proteins. Native tryptophan in its normal protein environment emits at 340 nm, so the 340 nm peak intensity reflects the tryptophan content of analyzed proteins. This peak decreases when tryptophan is oxidized, modified, or adducted with lipids.

In standard procedures, extracted proteins are dissolved in phosphate buffer, pH 6, at concentration of 5 mg/ml, fluorescence spectra are scanned from 300 to at least 500 nm with excitation of 280 nm (Utrera *et al.*, 2012), emission peak wavelengths and intensity are recorded, and tryptophan content is calculated from a standard curve of 0.1 to 0.5 μM N-acetyl-L-tryptophan amide (NATA). Results are expressed as milligrams NATA equivalents per 100 g of sample.

Also diagnostic is the peak maximum wavelength, λ_{max}. If the molecular environment near the tryptophans becomes more polar, the emission λ_{max} shifts to higher wavelength; or if the molecular environment near the tryptophans becomes more hydrophobic, the emission λ_{max} shifts to lower wavelength (Hellmann and Schneider, 2019).

More polar environments arise when proteins have a more unwound or open structure, as could occur with fractionation or modification of amino acid side chains near the tryptophan; more hydrophobic environments develop when side chains near the tryptophan are deaminated or decarboxylated, when tryptophan residues are buried in the hydrophobic interior of the protein (Chakraborty et al., 2001), and when lipid adducts form near the tryptophan. In proteins with multiple tryptophan residues in different environments, the fluorescence spectra may have several peaks, each reporting a different environment with peak intensities revealing the number of tryptophan residues in that environment. Analytical protocols and guides to data interpretation are presented by (Hellmann and Schneider, 2019).

Cautionary note: tryptophan is damaged extensively by heat as well as oxidation. To distinguish loss from lipid co-oxidation versus processing, analyze spectra immediately after processing (thermal or other physical stress) and again after incubation or storage where oxidation occurs. The difference reflects tryptophan presumptively lost by co-oxidation.

Formation of specific amino acid products. AAS, AAA, and GGS. Looking for specific amino acid products is probably more relevant to basic research in protein co-oxidation than to quality control, sample preparation is lengthy and tedious, and HPLC systems and optimally also mass spectrometers are required for analyses. Thus, such detailed analyses are not the first chosen at the present time. Nevertheless, some brief method descriptions are included here because these products may become useful markers if they can be identified in foods as well as model systems, and they can verify pathways or processes (e.g. lipid oxidation vs metal-mediated oxidation vs heat) by which essential amino acids such as lysine are lost.

Three such markers are carbonyls produced by hydroxyl radical (HO·) from metal decomposition of hydrogen peroxide – aminoadipic semialdehyde (AAS) and aminoadipic acid (AAA) from lysine, and γ-glutamic semialdehyde (GGS) from arginine and proline (Akagawa *et al.*, 2006; Requena *et al.*, 2001). Given the catalytic activity of heme generation of oxyl radicals in meats, these marker products have been particularly useful in following protein oxidation during processing and storage in muscle foods (Estévez *et al.*, 2009; Estévez *et al.*, 2011). However, whether LOO· or LO· from lipid co-oxidation of proteins generate the same products needs to be verified. If so, these products can become sensitive co-oxidation markers in all food proteins; if not, they can distinguish damage from HO· radicals in processing from lipid oxyl radicals in co-oxidation.

Sample preparation is not trivial, requiring protein hydrolysis and carbonyl derivatization followed by chromatographic separation. Full method details and protocols are available for HPLC (Utrera *et al.*, 2011; Utrera *et al.*, 2012) and for GC (Requena *et al.*, 2003).

Crosslinking, scission, subunit reorganization. Texture hardening or softening is among the first property changes that consumers (human or animal) perceive in foods with co-oxidized proteins, so it is logical that electrophoresis ranks with solubility as the analyses most used as indicators of protein oxidation and modification. Normal protein patterns as well as increased molecular weights from crosslinking, decreased molecular weights from fragmentation, and shuffling of molecular weights by quarternary sub-unit reorganization (e.g. in soybean and seed proteins) can be visualized by polyacrylamide gel electrophoresis (PAGE), and results can help connect chemical changes to texture changes. For molecular weight determination, PAGE typically includes sodium dodecyl sulfate to denature the proteins and expose buried amino acids, then bind to proteins at a constant level of about 1.4 mg SDS/mg of protein. This creates a protein surface charge that is proportional to molecular weight, so when an electrical current is applied to the gel, protein components migrate through the gel by molecular weight, with smallest proteins moving to the bottom (+ electrode) and largest proteins remaining closer to the top (– electrode). Running parallel protein standards on the same plate allows calculation of molecular weights (Figure 6.7A).

SDS-PAGE thus makes it easy to determine which peptide fractions have been most modified (bands are blurred, lessened, change positions or are lost altogether) (Figure 6.7A), and it also reveals formation of new bands. Inclusion of 2-mercaptoethanol (2-MCE) in samples reduces disulfide bonds and releases the sub-units linked by them, so running identical samples without and with 2-MCE is an important technique to distinguish which peptides are S-S linked normally and which develop S-S bonds during oxidation (Figure 6.7A.)

Separating co-oxidized proteins with solubility limitations and a wide variety of modifications is a particular challenge in electrophoresis. Key points important specifically for electrophoresis of oxidized proteins will be highlighted below.

The fundamentals of gel electrophoresis and practical information are described in detail in (Schaich, 2016). Additional perspectives and protocols are available in (Hames, 1998; Hames and Rickwood, 1983; Hames and Rickwood, 1990; Lonza, 2015; Walker, 2009) so will not be covered here.

What acrylamide concentrations should be used? PAGE gels are generated with four components: base polymer acrylamide (CH_2=CH-CO-NH_2), crosslinker bis-acrylamide (CH=CH-CO-NH-CH_2-NH-CO-CH=CH_2), radical initiator ammonium persulfate, and accelerator TEMED (tetramethylethylenediamine). TEMED forms free radicals in the ammonium persulfate, which in turn initiates free radical chain reactions between acrylamide and bisacrylamide, and copolymerization of these two agents forms the gels. In routine analyses, total acrylamide levels (%T) levels typically range from 7 to 20 %. Table 6.5 shows the molecular weights generally separated by different %T levels. The acrylamide concentrations and acrylamide:bis proportions needed depends on the molecular weight distributions of proteins in the extracts and must be adjusted for mixtures of proteins in each analysis. When component peptides have a very wide molecular weight range or many peptide bands very similar in weight, gradient gels can effectively extend the working range, distribute protein bands more evenly, and improve separation of close bands. Gradient gels are particularly useful in separating co-oxidized proteins that have many surface and molecular weight modifications. Finding the optimum conditions for separating different native and modified proteins can require a lot of trial and error.

How are highly polymerized proteins handled? The maximum molecular weight limit that can enter polyacrylamide gels is 250 kDa. When protein polymers with even higher molecular weights are generated in foods by heat and shear stresses during processing and by co-oxidation during storage, separations must be shifted to agarose gels. This is seldom required, but presence of high polymers does require inclusion of stacking gels on top of resolving gels to focus protein groups and

TABLE 6.5 Separation ranges for proteins in polyacrylamide gels prepared with 5% crosslinker, Tris-Glycine buffer. Excerpted in part from Lonza, 2015.

Polyacrylamide	Concentration	Size separation range	Application
Isocratic	7.5%	50kDa - 200 kDa	General purpose
Isocratic	10%	25kDa - 200 kDa	General purpose
Isocratic	12%	20kDa - 100 kDa	Smaller proteins
Isocratic	15%	10kDa - 50 kDa	Smallest proteins
Gradient	4–20%	5 kDa - 200 kDa	Peptides → high polymer
Gradient	10–20%	5 kDa - 150 kDa	Peptides → polymer
Gradient	4–12%	25 kDa - 250 kDa	Small → high polymer
Gradient	8–16%	15 kDa - 200 kDa	Small → polymer

retain proteins that may be too large to enter the gel. When proteins are loaded directly onto PAGE gels, they diffuse slowly into the resolving gel at different rates, and this leads to blurring and reduced resolution of separated peptide bands. To obtain cleaner patterns and tighter peptide bands, a narrow stacking gel layer that has large pores (4–5% acrylamide) and is prepared in pH 6.8 buffer is added on top of the resolving gel to concentrate protein bands. Then electrophoresis is run in discontinuous mode, with Tris-HCl buffers pH 6.8 for stacking gels and pH 8.8 for resolving gels, and the reservoir buffer is Tris-glycine pH 8.3. When current is applied, chloride ions (−) from the two gels migrate in the front while glycine ions (+) from the reservoir buffer trail behind and proteins are stacked or focused between these two fronts in tight bands. Proteins remain stacked until glycine ions finally enter the resolving gel at higher pH, become deprotonated, and assume a net (−) charge. At this point, all molecules begin to migrate and separate according to molecular mass. Importantly for modified proteins, any polymers that are too large to enter the resolving gel remain in the stacking gel and can be detected there by staining (Figure 6.7B).

Staining issues. After electrophoresis, separated protein bands must be visualized by staining. Coomassie brilliant blue (CBB) and silver staining are routine, but specialty stains such as Ponceau red or amido black that have different binding sites may also be used to detect peptides with particular amino acids or lipid adducts (Steinberg, 2009). Choice of stains is a critical issue with co-oxidized proteins because the amino acid residues to which the stains bind are the same ones most modified by oxidation, deamination and decarboxylation, or adduct formation in co-oxidized proteins. As a result, less dye is bound and protein bands can fade or even disappear due to dye blockage as well as destruction (Figure 6.7A). Thus, band loss with co-oxidized proteins must be confirmed using at least two dyes with different binding specificity.

The most common two stains useful for deducing damage sites on proteins are CBB R-250 and silver nitrate silver. CBB binds to positively-charged arginine, histidine, and lysine by charge interactions and with aromatic amino acids and peptide regions by hydrophobic or van der Waals interactions. The reactions outlined in the previous section of this chapter show how strongly these amino acids are involved in lipid complexing, so CBB loses dye binding as proteins oxidize and band appearance weakens even when they are still present. In contrast, silver ions bind to negatively-charged sulfhydryl and carboxyl groups of aspartic and glutamic acids in proteins, where they are reduced to free metallic silver and deposit on the peptide backbones nearby (Rabilloud, 1990). Silver binding sites are less affected by protein co-oxidation, especially when 2-mercaptoethanol is present to reduce S-S bonds, and it is 100 times more sensitive, so many times when CBB sites are blocked and bands fade or are lost, silver stain will still be able to report presence of those peptide bands. Most of the time, proteins contain both (+) and (−) amino acids so exhibit comparable PAGE patterns with both dyes. However, when either basic or acidic amino acids dominate in proteins, as in peanut proteins, gels can look very different with the two dyes (Figure 6.7C). Thus, comparing pattern differences in parallel gels stained with CBB and silver can provide important clues to explain changes occurring in the proteins.

Two aspects of staining warrant further investigation with co-oxidized proteins. CBB is commonly used in solution form R-250 with electrophoresis but in colloidal form G-250 in the Bradford assay. CBB R-250 penetrates the entire gel matrix in addition to marking the proteins, so must be removed in destaining with acidic alcohol. The harshness of this destaining may remove some CBB bound weakly to oxidized proteins. Colloidal CBB G-250 has an extra two methyl groups that increase binding of the dye to proteins by van der Waals interactions and potentially replace lost charge associations with his, arg, and lys in oxidized proteins. Whether colloidal CBB offers advantages over soluble CBB and should be given preference for staining oxidized proteins needs to be investigated systematically.

Though quite familiar and comfortable for use with native, unmodified proteins, CBB and silver stains have demonstrated clearly that they have detriments when applied to co-oxidized proteins. To eliminate blockage of dye binding, other dyes with different binding modes need to be investigated for use with co-oxidized proteins.

Examples of staining protocols are available in (Arndt *et al.*, 2018) for CBB and in (Chevallet *et al.*, 2006; Kavran and Leahy, 2014) for silver.

Protein profiles. Mixtures of unmodified proteins run under proper loading concentrations, gel strength, and electrophoresis conditions will separate cleanly into bands with tight, crisp margins. However, co-oxidation introduces a wide variety of modifications to proteins randomly and non-homogeneously so that the resulting mixture now has major bands with discrete molecular weights plus some material with molecular weights distributed around main bands and throughout the gel. This makes smearing and overloaded bands commonplace with co-oxidized proteins (Figure 6.7B). To reduce this problem, investigators must experiment with lower protein loading, different protein buffers and pre-treatments, as well as different concentrations and types of dyes to improve appearance and focus of bands. For example, if bands fade but still smear and do not separate with lower protein loading, try a different acrylamide concentration and buffer.

Contaminant interference with PAGE. Salts, detergents, denaturants, traces of organic solvents, and carbohydrate contaminants in protein extracts all interfere with SDS-PAGE, changing migration and reducing band focusing. Thus, it is customary with native protein extracts and even more critical with oxidized proteins, to clean up the extracts before subjecting the proteins to PAGE. Several options are available depending on the nature of the contaminant, the amount of protein available, and the susceptibility of the test peptides to further oxidation (Evans *et al.*, 2009). Dialysis is commonly used but requires long times and large amounts of protein; dialysis slide cassettes are an option when sample volumes are limited.

Acid precipitation (e.g. trichloroacetic acid) is a standard method applied to native proteins, but it runs the risk of modifying oxidation sites in oxidized proteins. Precipitation in ice-cold acetone is effective, can be handled on microscale, does not damage modified proteins, and has the additional advantage of dissolving any unbound lipids that have been carried along in the extract. Protocols for acetone precipitation are available at (ThermoScientific, 2009).

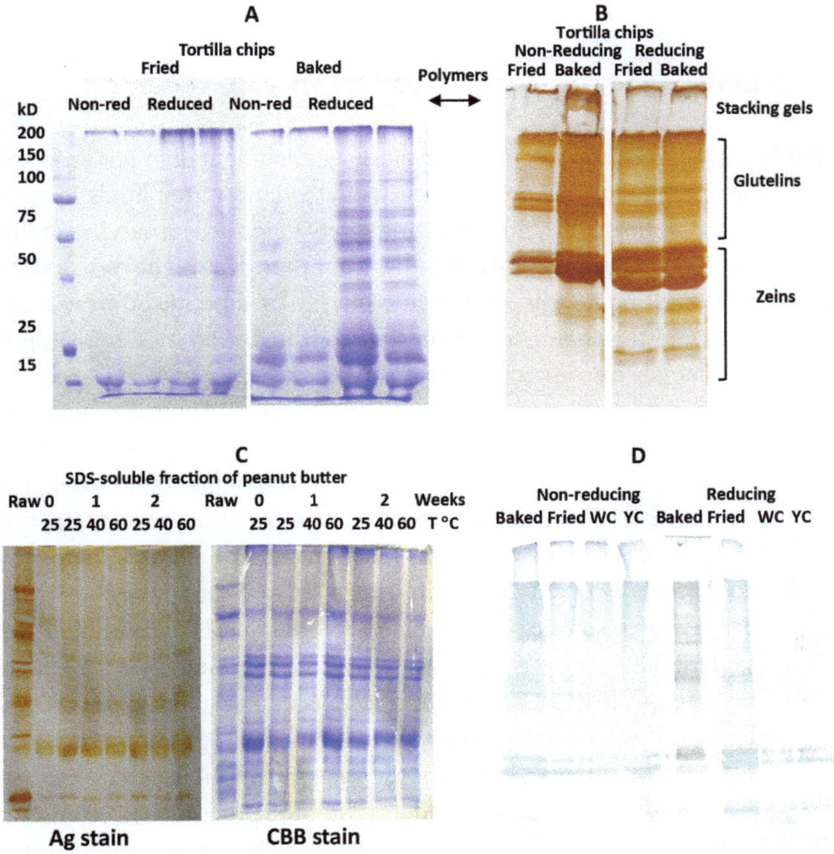

FIGURE 6.7 Polyacrylamide gel electrophoresis patterns and typical behaviors observed with co-oxidized proteins. SDS-PAGE gels showing a) peptide separations with protein standards for determining mol wts, fried chips show fading due to poor binding of Coomassie blue dye, lost bands due in part to crosslinking, mercaptoethanol reduction releases peptides crosslinked by S-S bonds, substantial smearing of bands; b) silver stain detects more proteins in both chips, condensing of bands causes overloaded bands after S-S reduction, high polymers are evident in stacking gel; c) silver and Coomassie blue stains differentially detect acidic and basic proteins in peanut butter, d) protein carbonyls in baked and fried tortilla chips detected by positive response to DNPH antibodies on Western blots; WC, white cornmeal; YC, yellow cornmeal. From Dong, 2011; Ibadullah, 2013; used with permission.

Handling issues. Because acrylamide crosslinking is greatly accelerated by oxygen, solutions must be carefully deaerated or the acrylamide solidifies before the gel can be cast. Keeping the acrylamide warm facilitates casting.

Toxicity. Always remember that acrylamide is a potent neurotoxin so must be handled in a hood with gloves, masks, and personal protective equipment, taking special care not to inhale any particles. Using commercially-prepared acrylamide solutions rather than making them from scratch limits contact exposure to dry acrylamide particles by touch or inhalation.

5 Summary and closing remarks

That lipids transfer oxidation to proteins has been documented in research for decades, yet the potential of this process to expand oxidation from lipids to systems and to compromise the quality, nutrition, and safety of food materials in significant ways has been largely ignored until recently. Even though protein co-oxidation is gaining more attention now, the sensory and physical symptoms of protein co-oxidation in foods and the chemistry responsible for the degradation remain poorly recognized and largely disconnected from each other.

To make these important connections and establish a solid foundation for studying and understanding lipid co-oxidation of proteins in both research and industrial settings, this chapter first reviewed the variety of changes that signal protein co-oxidation to validate the importance of protein co-oxidation in food quality degradation. Examples of the many reactions that cause these changes then showed how lipid and protein oxidation are closely integrated and cannot be considered in isolation when trying to solve oxidation problems in foods. Oxidizing lipids cause protein co-oxidation by nearly as many pathways as are present in lipid oxidation. In the process, protein reactions with lipid radicals and secondary products slow lipid oxidation while moving oxidation beyond them. Hence, when trying to stabilize feeds and food, both lipid and protein oxidation must be measured to accurately determine oxidation status of foods and interpret the processes involved.

To provide starting points for accomplishing such a Herculean task, this chapter also provided fundamental information about protein analyses that report important physical property changes as well as analyses that measure specific chemical changes. Both are needed to elucidate and quantitate how protein co-oxidation affects food properties and qualities. Medical research has amassed an extensive knowledge base of specific protein modifications mediated by various reactive oxygen species generated in cells. The same kind of detailed specific information about the hierarchy of degradation caused by different lipid oxidation products is now needed for foods to understand the 'what' and 'how' of flavor, odor, color, texture, and nutrition changes, and then to tailor food formulas, processing, and stabilization to control specific processes. For example, antioxidants are added to foods to stop lipid oxidation by scavenging radicals. If protein analyses find that lipid aldehydes are just as damaging as radicals (or more so), meta-polyphenols that bind carbonyls (Zamora *et al.*, 2017; Zamora and Hidalgo, 2018) may become vital new ingredients in animal feed. The protein chemical analyses described here may not be the most sophisticated available but were deliberately chosen because they can be handled with instrumentation available in most laboratories and can provide answers rapidly. Using these routinely will eventually develop a map of protein co-oxidation products and processes, along with their underlying chemistry and visible food characteristics, under different conditions. Results of these protein analyses with lipid oxidation analyses described in the previous chapter must then be coordinated to accurately assess the oxidation state of feed and foods and identify effective ways to control the process.

It is hoped that these two coordinated chapters provide knowledge and tools to encourage industry and academia alike to make coordinated lipid and protein analyses regular components of their research and efforts to improve stability, nutritional value, and quality of feed and foods.

Acknowledgements

This work was supported by the New Jersey Agricultural Experiment Station under Grant NJ10195.

References

Ahirwar, R., Bhattacharya, A. and Kumar, S., 2022. Unveiling the underpinnings of various non-conventional elisa variants: A review article. Expert Review of Molecular Diagnostics 22. 10.1080/14737159.2022.2117615

Akagawa, M., Sasaki, D., Ishii, Y., Kurota, Y., Yotsu-Yamashita, M., Uchida, K. and Suyama, K., 2006. New method for the quantitative determination of major carbonyls, α-aminoadipic and γ-glutamic semialdehydes: Investigation of the formation mechanism and chemical nature in vitro and in vivo. Chemical Research in Toxicology 19: 1059-1065. 10.1021/tx060026p

Arndt, C., Koristka, S., Feldmann, A., Bergmann, R. and Bachmann, M., 2018. Coomassie brilliant blue staining of polyacrylamide gels. In: B.T. Kurien and R.H. Scofield (Eds.), Protein gel detection and imaging: Methods and protocols. Springer New York, New York, NY, pp. 27-30. 10.1007/978-1-4939-8745-0_4

Baron, C.P., Kjaersgård, I.V., Jessen, F. and Jacobsen, C., 2007. Protein and lipid oxidation during frozen storage of rainbow trout (oncorhynchus mykiss). J Agric Food Chem 55: 8118-8125. 10.1021/jf070686f

Bertsch, M., Mayburd, A.L. and Kassner, R.J., 2003. The identification of hydrophobic sites on the surface of proteins using absorption difference spectroscopy of bromophenol blue. Analytical Biochemistry 313: 187-195. 10.1016/S0003-2697(02)00590-0

Bio-Rad, 2017. Elisa basics guide. Available at: https://www.bio-rad-antibodies.com/static/2017/an-introduction-to-elisa/elisa-basics-guide.pdf. Aug. 17, 2025.

Biocheminsider, 2021. Ninhydrin test: Principle, reagent preparation, procedure and result interpretation. Available at: https://biocheminsider.com/ninhydrin-test/. Aug. 16, 2025.

Boisen, S., 2000. In vitro digestibility methods: History and specific approaches. In: P.J. Moughan, M.W.A. Verstegen and M. Visser-Reynes (Eds.), Feed evaluation. Principles and practice. Wageningen Pers., The Netherlands, pp. 153-168.

Brodkorb, A., Egger, L., Alminger, M., Alvito, P., Assunção, R., Ballance, S., Bohn, T., Bourlieu-Lacanal, C., Boutrou, R., Carrière, F., Clemente, A., Corredig, M., Dupont, D., Dufour, C., Edwards, C., Golding, M., Karakaya, S., Kirkhus, B., Le Feunteun, S., Lesmes, U., Macierzanka, A., Mackie, A.R., Martins, C., Marze, S., McClements, D.J., Ménard, O., Minekus, M., Portmann, R., Santos, C.N., Souchon, I., Singh, R.P., Vegarud, G.E., Wickham, M.S.J., Weitschies, W. and Recio, I., 2019. Infogest static in vitro simulation of gastrointestinal food digestion. Nature Protocols 14: 991-1014. 10.1038/s41596-018-0119-1

Bühler, J., Dekkers, B., Bruins, M. and Goot, A.J., 2020. Modifying faba bean protein concentrate using dry heat to increase water holding capacity. Foods 9: 1077. 10.3390/foods9081077

Buss, H., Chan, T.P., Sluis, K.B., Domigan, N.M. and Winterbourn, C.C., 1997. Protein carbonyl measurement by a sensitive elisa method. Free Radical Biology and Medicine 23: 361-366. 10.1016/s0891-5849(97)00104-4

Capuano, E., Oliviero, T., Açar, Ö.Ç., Gökmen, V. and Fogliano, V., 2010. Lipid oxidation promotes acrylamide formation in fat-rich model systems. Food Research International 43: 1021-1026. 10.1016/j.foodres.2010.01.013

Castell, C.H., 1971. Metal-catalyzed lipid oxidation and changes of proteins in fish. Journal of the American Oil Chemists Society 48: 645-649. 10.1007/BF02638509

Chakraborty, S., Ittah, V., Bai, P., Luo, L., Haas, E. and Peng, Z., 2001. Structure and dynamics of the alpha-lactalbumin molten globule: Fluorescence studies using proteins containing a single tryptophan residue. Biochemistry 40: 7228-7238. 10.1021/bi010004w

Chan, K.-Y. and Wasseran, B.P., 1993. Direct colorimetric assay of free thiol groups and disulfide bonds in suspensions of solubilized and particulate cereal proteins. Cereal Chemistry 70: 22-26. No DOI available.

Chanadang, S., Koppel, K. and Aldrich, G., 2016. The impact of rendered protein meal oxidation level on shelf-life, sensory characteristics, and acceptability in extruded pet food. Animals 6: E44(41-17). 10.3390/ani6080044

Chelh, I., Gatellier, P. and Santé-Lhoutellier, V., 2006. Technical note: A simplified procedure for myofibril hydrophobicity determination. Meat Sci 74: 681-683. 10.1016/j.meatsci.2006.05.019

Chelh, I., Gatellier, P. and Santé-Lhoutellier, V., 2007. Characterisation of fluorescent Schiff bases formed during oxidation of pig myofibrils. Meat Sci 76: 210-215. 10.1016/j.meatsci.2006.10.028

ChemTalk, 2025. Table of IR spectroscopy values. Available at: https://chemistrytalk.org/ir-spectrum-values. Aug 17, 2025.

Chernukha, I., Meliashchenia, A., Kaltovich, I., Vasilevskaya, E., Aryzina, M., Smaliak, T., Senchenko, T. and Fedulova, L., 2022. Evolution of in vitro digestibility techniques: A systematic review. Theory and practice of meat processing 6: 300-310. 10.21323/2414-438X-2021-6-4-300-310

Chevallet, M., Luche, S. and Rabilloud, T., 2006. Silver staining of proteins in polyacrylamide gels. Nature Protocols 1: 1852-1858. 10.1038/nprot.2006.288

Church, F.C., Swaisgood, H.E., Porter, D.H. and Catignani, G.L., 1983. Spectrophotometric assay using o-phthaldialdehyde for determination of proteolysis in milk and isolated milk proteins. Journal of Dairy Science 66: 1219-1227. 10.3168/jds.S0022-0302(83)81926-2

Colombo, G., Clerici, M., Garavaglia, M.E., Giustarini, D., Rossi, R., Milzani, A. and Dalle-Donne, I., 2016. A step-by-step protocol for assaying protein carbonylation in biological samples. Journal of Chromatography B-Analytical Technologies in the Biomedical and Life Sciences 1019: 178-190. 10.1016/j.jchromb.2015.11.052

Davies, M.J., Fu, S. and Dean, R.T., 1995. Protein hydroperoxides can give rise to reactive free radicals. Biochemical Journal 305: 643-649. 10.1042/bj3050643

Derayea, S.M. and Samir, E., 2020. A review on the use of fluorescamine as versatile and convenient analytical probe. Microchemical Journal 156: 104835. 10.1016/j.microc.2020.104835

Desai, I.D. and Tappel, A.L., 1963. Damage to proteins by peroxidized lipids. Journal of Lipid Research 4: 204-207. 10.1016/S0022-2275(20)40348-7

Dietrich, R.B., Lincoln, L., Momen, S., Minkoff, B.B., Sussman, M.R. and Girard, A.L., 2025. Role of protein and lipid oxidation in hardening of high-protein bars during storage. Journal of Food Science 90: e17657. 10.1111/1750-3841.17663

Dillard, C.J. and Tappel, A.L., 1973. Fluorescent products from reaction of peroxidizing polyunsaturated fatty acids with phosphatidyl ethanolamine and phenylalanine. Lipids 8: 183-189. 10.1007/BF02544632

Dillard, C.J. and Tappel, A.L., 1984. Fluorescent damage products of lipid peroxidation. Methods in Enzymology 105: 337-341. 10.1016/S0076-6879(84)05044-8

Dong, Y., 2011. Thermal vs lipid co-oxidation of proteins in baked vs fried tortilla chips. M.S. Thesis, Food Science, Rutgers University, New Brunswick, NJ.

Dunne, R.A., Darwin, E.C., Perez Medina, V.A., Levenston, M.E., St Pierre, S.R. and Kuhl, E., 2025. Texture profile analysis and rheology of plant-based and animal meat. Food Res Int 205: 115876. 10.1016/j.foodres.2025.115876

Eckl, P.M. and Bresgen, N., 2017. Genotoxicity of lipid oxidation compounds. Free Radical Biology and Medicine 111: 244-252. 10.1016/j.freeradbiomed.2017.02.002

Eder, E., Scheckenbach, S., Deininger, C. and Hoffman, C., 1993. The possible role of α,β-unsaturated carbonyl compounds in mutagenesis and carcinogenesis. Toxicology Letters 67: 87-103. 10.1016/0378-4274(93)90048-3

Egger, L., Ménard, O., Baumann, C., Duerr, D., Schlegel, P., Stoll, P., Vergères, G., Dupont, D. and Portmann, R., 2019. Digestion of milk proteins: Comparing static and dynamic in vitro digestion systems with in vivo data. Food Research International 118: 32-39. 10.1016/j.foodres.2017.12.049

Esterbauer, H., Schaur, R.J. and Zollner, H., 1991. Chemistry and biochemistry of 4-hydroxynonenal, malonaldehyde, and related aldehydes. Free Radical Biology and Medicine 11: 81-128. 10.1016/0891-5849(91)90192-6

Estévez, M., 2011. Protein carbonyls in meat systems: A review. Meat Science 89: 259-279. 10.1016/j.meatsci.2011.04.025

Estévez, M., Ollilainen, V. and Heinonen, M., 2009. Analysis of protein oxidation markers α-aminoadipic and γ-glutamic semialdehydes in food proteins using liquid chromatography (LC)–electrospray ionization (ESI)–multistage tandem mass spectrometry (ms). Journal of Agricultural and Food Chemistry 57: 3901-3910. 10.1021/jf804017p

Estévez, M., Ventanas, S. and Heinonen, M., 2011. Formation of Strecker aldehydes between protein carbonyls – α-aminoadipic and γ-glutamic semialdehydes – and leucine and isoleucine. Food Chemistry 128: 1051-1057. 10.1016/j.foodchem.2011.04.012

Evans, D.R.H., Romero, J.K. and Westoby, M., 2009. Concentration of proteins and removal of solutes. Methods in Enzymology 463: 97-120. 10.1016/S0076-6879(09)63009-3

Farmer, L.J. and Mottram, D.S., 1990. Interaction of lipid in the Maillard reaction between cysteine and ribose: The effect of a triglyceride and three phospholipids on the volatile products. Journal of the Science of Food and Agriculture 53: 505-525. 10.1002/jsfa.2740530409

Farmer, L.J., Mottram, D.S. and Whitfield, F.B., 1989. Volatile compounds produced in Maillard reactions involving cysteine, ribose and phospholipid. Journal of the Science of Food and Agriculture 49: 347-368. 10.1002/jsfa.2740490311

Faustman, C., 2014. Myoglobin chemistry and modifications that influence (color and) color stability, Proceedings of the American Meat Science Association 67th Reciprocal Meat Conference, Madison, Wisconsin. Lulu.

Faustman, C., Liebler, D.C., McClure, T.D. and Sun, Q., 1999. α,β-unsaturated aldehydes accelerate oxymyoglobin oxidation. Journal of Agricultural and Food Chemistry 47: 3140-3144. 10.1021/jf990016c

Fenaille, F., Tabet, J.-C. and Guy, P.A., 2004. Study of peptides containing modified lysine residues by tandem mass spectrometry: Precursor ion scanning of hexanal-modified peptides. Rapid Communications in Mass Spectrometry 18: 67-76. 10.1002/rcm.1283

Ferré, S., González-Ruiz, V., Guillarme, D. and Rudaz, S., 2019. Analytical strategies for the determination of amino acids: Past, present and future trends. Journal of Chromatography B 1132: 121819. 10.1016/j.jchromb.2019.121819

Fields, R., 1972. The rapid determination of amino groups with TNBS. Methods in Enzymology 25: 464-468. 10.1016/s0076-6879(72)25042-x

Finley, J.W. and Lundin, R.E., 1980. Lipid hydroperoxide induced oxidation of cysteine in peptides. In: M.G. Simic and M. Karel (Eds.), Autoxidation in food and biological systems. Plenum, New York, pp. 223-236.

Fisher, G.H., Arias, I., Quesada, I., D'Aniello, S., Errico, F., Di Fiore, M.M. and D'Aniello, A., 2001. A fast and sensitive method for measuring picomole levels of total free amino acids in very small amounts of biological tissues. Amino Acids 20: 163-173. 10.1007/s007260170057

Franca-Oliveira, G., Fornari, T. and Hernández-Ledesma, B., 2021. A review on the extraction and processing of natural source-derived proteins through eco-innovative approaches. Processes 9: 1626. 10.3390/pr9091626

Franczyk, A.J., 2018. Evaluation of in vitro methodology to determine protein digestibility and quality of plant-based proteins. M.S. Thesis Food and Human Nutritional Sciences, University of Manitoba, Winnepeg, CA.

Freeman, T.L., Haver, A., Duryee, M.J., Tuma, D.J., Klassen, L.W., Hamel, F.G., White, R.L., Rennard, S.I. and Thiele, G.M., 2005. Aldehydes in cigarette smoke react with the lipid peroxidation product malonaldehyde to form fluorescent protein adducts on lysines. Chemical Research in Toxicology 18: 817-824. 10.1021/tx0500676

Fruebis, J., Parsasarathy, S. and Steinberg, D., 1992. Evidence for a concerted reaction between lipid hydroperoxides and polypeptides. Proceedings of the National Academy of Sciences of the U.S.A. 89: 10588-10592. 10.1073/pnas.89.22.10588

Gardner, H.W. and Jursinic, P.A., 1981. Degradation of linoleic acid hydroperoxides by a cysteine-$FeCl_3$ catalyst as a model for similar biochemical reactions. I. Study of oxygen requirement, catalyst and effect of pH. Biochimica et Biophysica Acta - Lipids and Lipid Metabolism 665: 100-112. 10.1016/0005-2760(81)90238-1

Gedeon, S., Montgomery, A., Gangapuram, M., Redda, K.K. and Ardley, T.W., 2023. The chemistry and pharmacology of tetrahydropyridines: Part 2. Available at: https://auctoresonline.org/article/the-chemistry-and-pharmacology-of-tetrahydropyridines-part-2. July 25, 2025.

Geerts, M.E.J., Dekkers, B.L., van der Padt, A. and van der Goot, A.J., 2018. Aqueous fractionation processes of soy protein for fibrous structure formation. Innovative Food Science & Emerging Technologies 45: 313-319. 10.1016/j.ifset.2017.12.002

Goodno, C.C., Swaisgood, H.E. and Catignani, G.L., 1981. A fluorimetric assay for available lysine in proteins. Analytical Biochemistry 115: 203-211. 10.1016/0003-2697(81)90547-9

Govindaraju, I., Sana, M., Chakraborty, I., Rahman, M.H., Biswas, R. and Mazumder, N., 2024. Dietary acrylamide: A detailed review on formation, detection, mitigation, and its health impacts. Foods 13. 10.3390/foods13040556

Habeeb, A.F.S.A., 1966. Determination of free amino groups in proteins by trinitrobenzenesulfonic acid. Analytical Biochemistry 14: 328-336. 10.1016/0003-2697(66)90275-2

Hames, B.D. (Ed.), 1998. Gel electrophoresis of proteins: A practical approach, Third edition. Oxford University Press, Oxford, UK. 10.1093/oso/9780199636402.001.0001

Hames, B.D. and Rickwood, D. (Eds.), 1983. Gel electrophoresis of proteins: A practical approach. First edition. Oxford University Press, Oxford, UK.

Hames, B.D. and Rickwood, D. (Eds.), 1990. Gel electrophoresis of proteins: A practical approach (second edition). Oxford University Press, Oxford, UK.

Hayes, M., 2020. Measuring protein content in food: An overview of methods. Foods 9. 10.3390/foods9101340

He, F., Holben, G. and De Godoy, M.R., 2020. Evaluation of selected categories of pet treats using in vitro assay and texture analysis. Translational Animal Science 4: 1023-1030. 10.1093/tas/txaa064

Held, P., 2006. Quantitation of total protein using OPA. Nature Methods | Application Notes. 10.1038/an1781

Hellmann, N. and Schneider, D., 2019. Hands on: Using tryptophan fluorescence spectroscopy to study protein structure. Methods in Molecular Biology 1958: 379-401. 10.1007/978-1-4939-9161-7_20

Hidalgo, F.J., Delgado, R.M. and Zamora, R., 2009. Degradation of asparagine to acrylamide by carbonyl-amine reactions initiated by alkadienals. Food Chemistry 116: 779-784. 10.1016/j.foodchem.2009.03.020

Hidalgo, F.J. and Zamora, R., 1993. Non-enzymatic browning and fluorescence development in a *(e)*-4,5-epoxy-*(e)*-2-heptenal/lysine model system. Journal of Food Science 58: 667-670. 10.1111/j.1365-2621.1993.tb04352.x

Hidalgo, F.J. and Zamora, R., 2000a. The role of lipids in nonenzymatic browning. Grasas y Aceites 51. 10.3989/gya.2000.v51.i1-2.405

Hidalgo, F.J. and Zamora, R., 2000b. Modification of bovine serum albumin structure following reaction with 4,5*(e)*-epoxy-2-*(e)*-heptenal. Chemical Research in Toxicology 13: 501-508. 10.1021/tx990205p

Hidalgo, F.J. and Zamora, R., 2008. Food anoxia and the formation of either flavor or toxic compounds by amino acid degradation initiated by oxidized lipids. Annals of the New York Academy of Sciences 1126: 25-29. 10.1196/annals.1433.020

Hidalgo, F.J. and Zamora, R., 2016. Amino acid degradations produced by lipid oxidation products. CRC Critical Reviews in Food Science and Nutrition 56: 1242-1252. 10.1080/10408398.2012.761173

Hou, Y., Wu, Z., Dai, Z., Wang, G. and Wu, G., 2017. Protein hydrolysates in animal nutrition: Industrial production, bioactive peptides, and functional significance. Journal of Animal Science and Biotechnology 8: 24. 10.1186/s40104-017-0153-9

Huang, Y., Hua, Y. and Qui, A., 2006. Soybean protein aggregation induced by lipoxygenase catalyzed linoleic acid oxidation. Food Research International 39: 240-249. 10.1016/j.foodres.2005.07.012

Hunt, S., 1985. Degradation of amino acids accompanying *in vitro* protein hydrolysis. In: G.C. Barrett (Ed.), Chemistry and biochemistry of the amino acids. Springer Netherlands, Dordrecht, pp. 376-398. 10.1007/978-94-009-4832-7_12

Ibadullah, W.Z.W., 2013. Lipid co-oxidation of proteins in peanut butter. Ph.D. Dissertation, Food Science, Rutgers University, New Brunswick, NJ.

Ichihashi, K., Osawa, T., Toyokuni, S. and Uchida, K., 2001. Endogenous formation of protein adducts with carcinogenic aldehydes. Journal of Biological Chemistry 276: 23903-23913.

Johnson, M., 2015, modified 2023. Overview of texture profile analysis. Available at: https://www.texturetechnologies.com/resources/texture-profile-analysis. Aug, 17, 2025.

Kanazawa, K., Ashida, H. and Natake, M., 1987. Autoxidizing process interaction of linoleic acid with casein. Journal of Food Science 52: 475-479. 10.1111/j.1365-2621.1987.tb06642.x

Kandrac, M., 2018. Factors affecting the 2,4-dinitrophenyl hydrazine reaction with lipid carbonyls. M.S. Thesis, Food Science, Rutgers University, New Brunswick, NJ.

Karel, M., Schaich, K.M. and Roy, R.B., 1975. Interaction of peroxidizing methyl linoleate with some proteins and amino acids. Journal of Agricultural and Food Chemistry 23: 159-163. 10.1021/jf60198a046

Kavran, J.M. and Leahy, D.J., 2014. Silver staining of SDS-polyacrylamide gel. Methods in Enzymology 541: 169-176. 10.1016/B978-0-12-420119-4.00014-8

Kehm, R., Baldensperger, T., Raupbach, J. and Höhn, A., 2021. Protein oxidation - formation mechanisms, detection and relevance as biomarkers in human diseases. Redox Biology 42: 101901. 10.1016/j.redox.2021.101901

Kikugawa, K. and Ido, Y., 1984. Studies on peroxidized lipids. V. Formation and characterization of 1,4-dihydropyridine-3,5-dicarbaldehydes as model of fluorescent components in lipofuschin. Lipids 19: 600-608. 10.1007/BF02534718

Kikugawa, K., Kato, T. and Hayasaka, A., 1991. Formation of dityrosine and other fluorescent amino acids by reaction of amino acids with lipid hydroperoxides. Lipids 26: 922-929. 10.1007/BF02535978

Krogull, M.K. and Fennema, O., 1987. Oxidation of tryptophan in the presence of oxidizing methyl linoleate. Journal of Agricultural and Food Chemistry 35: 66-70. 10.1021/jf00073a016

Kruger, N.J., 1994. The Bradford method for protein quantitation. Methods in Molecular Biology 32: 9-15. 10.1385/0-89603-268-x:9

Ladouce, R., Combes, G.F., Trajković, K., Drmić Hofman, I. and Merćep, M., 2023. Oxime blot: A novel method for reliable and sensitive detection of carbonylated proteins in diverse biological systems. Redox Biology 63: 102743. 10.1016/j.redox.2023.102743

Lamothe, P.J. and McCormick, P.G., 1973. Role of hydrindantin in the determination of amino acids using ninhydrin. Analytical Chemistry 45: 1906-1911. 10.1021/ac60333a021

Le Roux-Pullen, L. and Lessing, D., 2011. Should veterinarians consider acrylamide that potentially occurs in starch-rich foodstuffs as a neurotoxin in dogs? Journal of the South African Veterinary Association 82: 129-130. 10.4102/jsava.v82i2.47

Lea, C.H., 1957. Deteriorative reactions involving phospholipids and lipoproteins. Journal of the Science of Food and Agriculture 8: 1-14. 10.1002/jsfa.2740080102

Lederer, M.O., 1996. Reactivity of lysine moieties toward γ-hydroxy-α,β-unsaturated epoxides: A model study on protein-lipid oxidation product interaction. Journal of Agricultural and Food Chemistry 44: 2531-2537. 10.1021/jf950837r

Lederer, M.O., Schuler, A. and Ohmenhäuser, M., 1999. Reactivity of lysine moieties toward an epoxyhydroxylinoleic acid derivative: Aminolysis vs hydrolysis. Journal of Agricultural and Food Chemistry 47: 4611-4620. 10.1021/jf9903830

Levine, R.L., Garland, D., Oliver, C.N., Amici, A., Climent, I., Lenz, A.G., Ahn, B.W., Shaltiel, S. and Stadtman, E.R., 1990. Determination of carbonyl content in oxidatively modified proteins. Methods in Enzymology 186: 464-478. 10.1016/0076-6879(90)86141-H

Levine, R.L., Wehr, N., Williams, J.A., Stadtman, E.R. and Shacter, E., 2000. Determination of carbonyl groups in oxidized proteins. In: J.M. Walker and S.M. Keyse (Eds.), Stress response: Methods and protocols. Humana Press, Totowa, NJ, pp. 15-24. 10.1385/1-59259-054-3:15

Levine, R.L., Williams, J.A., Stadtman, E.P. and Shacter, E., 1994. Carbonyl assays for determination of oxidatively modified proteins. Methods in Enzymology 233: 346-357. 10.1016/S0076-6879(94)33040-90-9

Lewis, S.E. and Wills, E.D., 1962. Destruction of sulfhydryl groups of proteins and amino acids by peroxides of unsaturated fatty acids. Biochemical Pharmacology 11: 901-912. 10.1016/0006-2952(62)90139-9

Little, C. and O'Brien, P.J., 1967. Products of oxidation of a protein thiol group after reaction with various oxidizing agents. Archives of Biochemistry and Biophysics 122: 406-410. 10.1016/0003-9861(67)90212-3

Little, H. and O'Brien, P.J., 1968. The effectiveness of a lipid peroxide in oxidizing protein and non-protein thiols. Biochemical Journal 106: 419-423. 10.1042/bj1060419

Liu, C., Zhao, J., Gupta, S. and Carrillo, M., 2025. Quantification of crude and soluble protein content. In: Y. Li (Ed.), Plant-based proteins: Production, physicochemical, functional, and sensory properties. Springer US, New York, NY, pp. 107-122. 10.1007/978-1-0716-4272-6_9

Lonza, 2015. Section x. Protein separation in polyacrylamide gels. . Available at: https://www.lonzabio.jp/catalog/pdf/pd/PD065.pdf. Aug. 20, 2025.

Mahmood, T. and Yang, P.C., 2012. Western blot: Technique, theory, and trouble shooting. North American Journal of Medicine and Science 4: 429-434. 10.4103/1947-2714.100998

Matsushita, S., Kobayashi, M. and Nitta, Y., 1970. Inactivation of enzymes by linoleic acid hydroperoxides and linoleic acid. Agricultural and Biological Chemistry 34: 817-824. 10.1080/00021369.1970.10859695

Moore, J.C., DeVries, J.W., Lipp, M., Griffiths, J.C. and Abernethy, D.R., 2010. Total protein methods and their potential utility to reduce the risk of food protein adulteration. Comprehensive Reviews in Food Science and Food Safety 9: 330-357. 10.1111/j.1541-4337.2010.00114.x

Moore, S. and Stein, W.H., 1954. A modified ninhydrin reagent for the photometric determination of amino acids and related compounds. Journal of Biological Chemistry 211: 907-913. 10.1016/S0021-9258(18)71178-2

Morzel, M., Gatellier, P., Sayd, T., Renerre, M. and Laville, E., 2006. Chemical oxidation decreases proteolytic susceptibility of skeletal muscle myofibrillar proteins. Meat Science 73: 536-543. 10.1016/j.meatsci.2006.02.005

Nakamura, T. and Hama, Y., 1988. Prostaglandin-like substances, precursors of red pigment, formed during autoxidation of methyl arachidonate. Journal of Agricultural and Food Chemistry 36: 15-18. 10.1021/jf00079a004

Nakamura, T., Hama, Y., Tanaka, R., Taira, K. and Hatate, H., 1998. A new red coloration induced by the reaction of oxidized lipids with amino acids. Journal of Agricultural and Food Chemistry 46: 1316-1320. 10.1021/jf970798e

Nielsen, H.K., Finot, P.A. and Hurrell, R.F., 1985. Reactions of proteins with oxidizing lipids. 2. Influence on protein quality and on the bioavailability of lysine, methionine, cyst(e)ine and tryptophan as measured in rat assays. British Journal of Nutrition 53: 75-86. 10.1079/bjn19850012

Obanu, Z.A., Ledward, D.A. and Lawrie, R.A., 1980. Lipid protein interactions as agents of quality deterioration in intermediate moisture meats: An appraisal. Meat Science 4: 79-88. 10.1016/0309-1740(80)90033-9

Oliver, C.N., Ahn, B.W., Moerman, E.J., Goldstein, S. and Stadtman, E.R., 1987. Age-related changes in oxidized proteins. Journal of Biological Chemistry 262: 5488-5491. 10.1016/S0021-9258(18)45598-6

Olson, B.J., 2016. Assays for determination of protein concentration. Current Protocols in Pharmacology 73: A.3A.1-A.3A.32. 10.1002/cpph.3

Olson, B.J. and Markwell, J., 2007. Assays for determination of protein concentration. Current Protocols in Protein Science Chapter 3: Unit 3.4. 10.1002/0471140864.ps0304s48

Pacheco-Aguilar, R., Mazorra-Manzano, M.A. and Ramírez-Suárez, J.C., 2008. Functional properties of fish protein hydrolysates from pacific whiting (merluccius productus) muscle produced by a commercial protease. Food Chemistry 109: 782-789. 10.1016/j.foodchem.2008.01.047

PierceBiotechnology, 2010. Elisa technical guide and protocols. Available at: https://assets.thermofisher.com/TFS-Assets/LSG/Application-Notes/TR0065-ELISA-guide.pdf. Aug. 17, 2025.

Pokorný, J., Davídek, J., Tran, H.C., Valentová, H., Matějícek, J. and Dlasková, Z., 1988. Reactions of oxidized lipids with protein. Part 15. Mechanism of lipoprotein formation from interactions of oxidized ethyl linoleate with egg albumin. Nahrung 32: 343-350. 10.1002/food.19880320410

Purcell, J.M., Quimby, D.J. and Cavanaugh, 1976. New method for determination of free amino groups in intact pure proteins: Relationship to available lysine. Journal of the Association of the Official Analytical Chemists 59: 1251-1254.

Raftery, M.J., 2014. Determination of oxidative protein modifications using mass spectrometry. Redox Reports 19: 140-147. 10.1179/1351000214y.0000000089

Ramadhan, A.H., Yu, D., Hlaing, K.S.S., Jiang, Q., Xu, Y. and Xia, W., 2025. Effects of packaging and storage time on lipid and protein oxidation and modifications in texture characteristics of refrigerated grass carp (ctenopharyngodon idellus) fish muscles. International Journal of Food Science and Technology 60. 10.1093/ijfood/vvaf082

Requena, J.R., Chao, C.-C., Levine, R.L. and Stadtman, E.R., 2001. Glutamic and aminoadipic semialdehydes are the main carbonyl products of metal-catalyzed oxidation of proteins. Proceedings of the National Academy of Sciences of the U.S.A. 98: 69-74. 10.1073/pnas.98.1.69

Requena, J.R., Levine, R.L. and Stadtman, E.R., 2003. Recent advances in the analysis of oxidized proteins. Amino Acids 25: 221-226. 10.1007/s00726-003-0012-1

RheologyLab, 2025. Texture profile analysis of animal vs plant-based protein products. Available at: https://www.rheologylab.com/texture-profile-analysis-of-animal-vs-plant-based-protein-products/.

Riviere, L.R. and Tempst, P., 2001. Enzymatic digestion of proteins in solution. Current Protocols in Protein Science Chapter 11: Unit 11.11. 10.1002/0471140864.ps1101s00

Robert-Peillard, F., Barco, E.P., Ciulu, M., Demelas, C., Théraulaz, F., Boudenne, J.-L. and Coulomb, B., 2017. High throughput determination of ammonium and primary amine compounds in environmental and food samples. Microchemical Journal 133: 216-221. 10.1016/j.microc.2017.03.048

Rosenthal, A. and Chen, J., eds., 2024. Food texturology: Measurement and perception of food textural properties, second edition. Springer Nature Switzerland AG 1999, Cham, Switzerland. 10.1007/978-3-031-41900-3

Roubal, W.T. and Tappel, A.L., 1996. Damage to proteins, enzymes, and amino acids by peroxidizing lipids. Archives of Biochemistry and Biophysics 113: 5-8. 10.1016/0003-9861(66)90150-0

Roy, R.B. and Karel, M., 1973. Reaction products of histidine with autoxidized methyl linoleate. Journal of Food Science 38: 896-897. 10.1111/j.1365-2621.1973.tb02103.x

Rudyk, O. and Eaton, P., 2014. Biochemical methods for monitoring protein thiol redox states in biological systems. Redox Biology 2: 803-813. 10.1016/j.redox.2014.06.005

Sáez-Plaza, P., Michałowski, T., Navas, M.J., Asuero, A.G. and Wybraniec, S., 2013. An overview of the Kjeldahl method of nitrogen determination. Part I. Early history, chemistry of the procedure, and titrimetric finish. Critical Reviews in Analytical Chemistry 43: 178-223. 10.1080/10408347.2012.751786

Sante-Lhoutellier, V., Aubry, L. and Gatellier, P., 2007. Effect of oxidation on in vitro digestibility of skeletal muscle myofibrillar proteins. Journal of Agricultural and Food Chemistry 55: 5343-5348. 10.1021/jf070252k

Santos-Sánchez, G., Miralles, B., Brodkorb, A., Dupont, D., Egger, L. and Recio, I., 2024. Current advances for in vitro protein digestibility. Frontiers in Nutrition 11: 1404538. 10.3389/fnut.2024.1404538

Sapan, C.V., Lundblad, R.L. and Price, N.C., 1999. Colorimetric protein assay techniques. Biotechnology and Applied Biochemistry 29: 99-108. 10.1111/j.1470-8744.1999.tb00538.x

Schaich, K.M., 1980. Free radical initiation in proteins and amino acids by ionizing and ultraviolet radiations and lipid oxidation- part III: Free radical transfer from oxidizing lipids. CRC Critical Reviews in Food Science and Nutrition 13: 189-244. 10.1080/10408398009527290

Schaich, K.M., 2005. Lipid oxidation in fats and oils: Theoretical aspects. In: F. Shahidi (Ed.), Bailey's industrial fats and oils, 6th edition. John Wiley, New York, pp. 2681-2767. 10.1002/047167849X.bi0067

Schaich, K.M., 2008. Chapter 8. Co-oxidations of oxidizing lipids: Reactions with proteins. In: A. Kamal-Eldin and D. Min (Eds.), Lipid oxidation pathways, vol. 2. AOCS Press,, Champaign, IL, pp. 183-274. 10.1201/9781003040316

Schaich, K.M., 2014. Lipid co-oxidation of proteins: One size does not fit all. Inform 25: 134-139. 10.1016/B978-0-12-242352-9.00011-4

Schaich, K.M., 2016. Chapter 1 - analysis of lipid and protein oxidation in fats, oils, and foods. In: M. Hu and C. Jacobsen (Eds.), Oxidative stability and shelf life of foods containing oils and fats. AOCS Press, pp. 1-131. 10.1016/B978-1-63067-056-6.00001-X

Schaich, K.M., 2020. Toxicity of lipid oxidation products consumed in the diet, Bailey's industrial oil and fat products, pp. 1-88. 10.1002/047167849X.bi0116

Schaich, K.M., 2023. Epoxides: An underestimated lipid oxidation product. Free Radic Res: 1-48. 10.1080/10715762.2023.2277142

Schaich, K.M. and Karel, M., 1976. Free radical reactions of peroxidizing lipids with amino acids and proteins: An ESR study. Lipids 11: 392-400. 10.1007/BF02532846

Shirley, R.L. and Becker, W.w., 1945. Determination of nitrogen in pyridine ring-type compounds by Kjeldahl method. Industrial & Engineering Chemistry Analytical Edition 17: 437-438. 10.1021/i560143a011

Simonian, M.H. and Smith, J.A., 2006. Spectrophotometric and colorimetric determination of protein concentration. Current Protocols in Molecular Biology 76: 10.11.11-10.11A.19. 10.1002/0471142727.mb1001as76

Simons, S.S. and Johnson, D.F., 1978. Reaction of *o*-phthalaldehyde and thiols with primary amines: Fluorescence properties of 1-alkyl(and aryl)thio-2-alkylisoindoles. Analytical Biochemistry 90: 705-725. 10.1016/0003-2697(78)90163-X

Skowyra, M., Janiewicz, U., Salejda, A.M., Krasnowska, G. and Almajano, M.P., 2015. Effect of tara (caesalpinia spinosa) pod powder on the oxidation and colour stability of pork meat batter during chilled storage. Food Technology and Biotechnology 53: 419-427. 10.17113/ftb.53.04.15.3932

Smewing, J., 2013. Measuring textural properties. Available at:https://hosmed.fi/wp-content/uploads/measuring-textural-properties.pdf. Aug. 17, 2025.

Smith, D.M., Noormarji, S.H., Price, J.F., Bennink, M.R. and Herald, T.J., 1990. Effect of lipid oxidation on the functional and nutritional properties of washed chicken myofibrils stored at different water activities. Journal of Agricultural and Food Chemistry 38: 1307-1312. 10.1021/jf00096a003

Smith, P.K., Krohn, R.I., Hermanson, G.T., Mallia, A.K., Gartner, F.H., Provenzano, M.D., Fujimoto, E.K., Goeke, N.M., Olson, B.J. and Klenk, D.C., 1985. Measurement of protein using bicinchoninic acid. Analytical Biochemistry 150: 76-85. 10.1016/0003-2697(85)90442-7

Soglia, F., Petracci, M. and Ertbjerg, P., 2016. Novel DNPH-based method for determination of protein carbonylation in muscle and meat. Food Chemistry 197: 670-675. 10.1016/j.foodchem.2015.11.038

Stadtman, E.R. and Levine, R.L., 2003. Free radical-mediated oxidation of free amino acids and amino acid residues in proteins. Amino Acids 25: 207-218. 10.1007/s00726-003-0011-2

Stauß, A.C., Fuchs, C., Jansen, P., Repert, S., Alcock, K., Ludewig, S. and Rozhon, W., 2024. The ninhydrin reaction revisited: Optimisation and application for quantification of free amino acids. Molecules 29. 10.3390/molecules29143262

Steinberg, T.H., 2009. Protein gel staining methods: An introduction and overview. Methods in Enzymology 463: 541-563. 10.1016/s0076-6879(09)63031-7

Sugita, K., Yamamoto, J., Kaneshima, K., Sekimoto, M., Endo, O., Takagi, Y. and Kato-Yoshinaga, Y., 2021. Acrylamide in dog food. Fundamental Toxicological Sciences 8: 49-52. 10.2131/fts.8.49

Sule, R., Rivera, G. and Gomes, A.V., 2023. Western blotting (immunoblotting): History, theory, uses, protocol and problems. BioTechniques 75: 99-114. 10.2144/btn-2022-0034

Sun, S.-W., Lin, Y.-C., Weng, Y.-M. and Chen, M.-J., 2006. Efficiency improvements on ninhydrin method for amino acid quantification. Journal of Food Composition and Analysis 19: 112-117. 10.1016/j.jfca.2005.04.006

Swindler, J.M., 2013. Effect of enrichment-bleaching and low oxygen atmosphere storage on all-purpose wheat flour quality. M.S. Thesis, Department of Nutrition, Dietetics, and Food Science, Brigham Young University, Provo, UT.

Szczesniak, A.S., 1986. Sensory texture evaluation methodogy, Proceedings of the 39th American Meat Science Annual Meat Reciprocal Conference, Chicago. National Livestock and Meat Board, pp. 86-96.

ThermoFisher, 2008. Fluorescamine protein assay. Available at: https://assets.thermofisher.com/TFS-Assets/CAD/manuals/Fluorescamine-protocol.pdf. Aug. 15, 2025.

ThermoScientific, 2009. Acetone precipitation of proteins. Available at: https://documents.thermofisher.com/TFS-Assets/LSG/Application-Notes/TR0049-Acetone-precipitation.pdf. Aug, 17, 2025.

Uchida, K., Kanematsu, M., Sakai, K., Matsuda, T., Hattori, N., Mizuno, Y., Suzuki, D., Miyata, T., Noguchi, N., Niki, E. and Osawa, T., 1998. Protein-bound acrolein: Potential markers for oxidative stress. Proceedings of the National Academy of Sciences of the U.S.A. 95: 4882-4887. 10.1073/pnas.95.9.4882

Udenfriend, S., Stein, S., Böhlen, P., Dairman, W., Leimgruber, W. and Weigele, M., 1972. Fluorescamine: A reagent for assay of amino acids, peptides, proteins, and primary amines in the picomole range. Science 178: 871-872. 10.1126/science.178.4063.871

Urban, J., Michalczuk, M., Batorska, M., Marzec, A., Jaroszek, A. and Bień, D., 2024. Effect of crude fibre additives arbocel and vitacel on the physicochemical properties of granulated feed mixtures for broiler chickens. Animal Bioscience 37: 274-283. 10.5713/ab.23.0218

Utrera, M., Morcuende, D., Rodríguez-Carpena, J.-G. and Estévez, M., 2011. Fluorescent HPLC for the detection of specific protein oxidation carbonyls – α-aminoadipic and γ-glutamic semialdehydes – in meat systems. Meat Science 89: 500-506. 10.1016/j.meatsci.2011.05.017

Utrera, M., Rodríguez-Carpena, J.G., Morcuende, D. and Estévez, M., 2012. Formation of lysine-derived oxidation products and loss of tryptophan during processing of porcine patties with added avocado byproducts. Journal of Agricultural and Food Chemistry 60: 3917-3926. 10.1021/jf3001313

Vatansever, S., Chen, B. and Hall, C., 2024. Plant protein flavor chemistry fundamentals and techniques to mitigate undesirable flavors. Sustainable Food Proteins 2: 33-57. 10.1002/sfp2.1025

Vernin, G., 1982. Heterocyclic compounds in flavors and fragrances part III. Pyridine and derivatives. Perfumer and Flavorist 7: 23-35. 10.31579/2690-8794/135

Walker, J.M., 2009. SDS polyacrylamide gel electrophoresis of proteins. In: J.M. Walker (Ed.), The protein protocols handbook. Humana Press, Totowa, NJ, pp. 177-185. 10.1007/978-1-59745-198-7_21

Wan Ibadullah, W.Z., 2013. Lipid co-oxidation of proteins in peanut butter. Ph.D. Dissertation, Food Science, Rutgers University, New Brunswick, NJ.

Wehrmaker, A.M., Bosch, G. and van der Goot, A.J., 2021. Effect of sterilization and storage on a model meat analogue pet food. Animal Feed Science and Technology 271: 114737. 10.1016/j.anifeedsci.2020.114737

Wiechelman, K.J., Braun, R.D. and Fitzpatrick, J.D., 1988. Investigation of the bicinchoninic acid protein assay: Identification of the groups responsible for color formation. Analytical Biochemistry 175: 231-237. 10.1016/0003-2697(88)90383-1

Worthington, 2025. Worthington enzyme manual. Worthington Biochemical Corporation, Lakewood, NJ.

Xia, M., Chen, Y., Guo, J., Feng, X., Yin, X., Wang, L., Wu, W., Li, Z., Sun, W. and Ma, J., 2019. Effects of oxidative modification on textural properties and gel structure of pork myofibrillar proteins. Food Research International 121: 678-683. 10.1016/j.foodres.2018.12.037

Xiao, R., Hu, R., Kumar, N. and Li, Y., 2025. Free amino and sulfhydryl group content. In: Y. Li (Ed.), Plant-based proteins: Production, physicochemical, functional, and sensory properties. Springer US, New York, NY, pp. 123-132. 10.1007/978-1-0716-4272-6_10

Yong, S.H. and Karel, M., 1978. Reactions between peroxidizing lipids and histidyl residue analogues: Enhancement of lipid oxidation and browning by 4-methylimidazole. Lipids 13: 1-5. 10.1007/BF02533359

Yong, S.H. and Karel, M., 1979. Cleavage of the imidazole ring in histidyl residue analogs reacted with peroxidizing lipids. Journal of Food Science 22: 568-574. 10.1111/j.1365-2621.1979.tb03838.x

Yong, S.H., Lau, S., Hsieh, Y. and Karel, M., 1980. Degradation products of l-tryptophan reacted with peroxidizing methyl linoleate. In: M.G. Simic and M. Karel (Eds.), Autoxidation in foods and biological systems. Plenum Press, New York, pp. 237-247.

Zamora, R., Aguilar, I. and Hidalgo, F.J., 2017. Epoxyalkenal-trapping ability of phenolic compounds. Food Chemistry 237: 444-452. 10.1016/j.foodchem.2017.05.129

Zamora, R., Delgado, R.M. and Hidalgo, F.J., 2016. Chapter 15 - Use of nucleophilic compounds, and their combination, for acrylamide removal. In: V. Gökmen (Ed.), Acrylamide in food. Academic Press, London, UK, pp. 297-307. 10.1016/B978-0-12-802832-2.00015-2

Zamora, R., Gallardo, E. and Hidalgo, F.J., 2006. Amine degradation by 4,5-epoxy-2-decenal in model systems. Journal of Agricultural and Food Chemistry 54: 2398-2404. 10.1021/jf052757l

Zamora, R. and Hidalgo, F.J., 1994. Modification of lysine amino groups by the lipid peroxidation product 4,5(*e*)-epoxy-2(*e*)-heptenal. Lipids 29: 243-249.

Zamora, R. and Hidalgo, F.J., 1995. Linoleic acid oxidation in the presence of amino compounds produces pyrroles by carbonylamine reactions. Biochimica et Biophysica Acta 1258: 319-327.

Zamora, R. and Hidalgo, F.J., 2005. 2-alkylpyrrole formation from 4,5-epoxy-2-alkenals. Chemical Research in Toxicology 18: 342-348. 10.1021/tx049700y

Zamora, R. and Hidalgo, F.J., 2018. Carbonyl–phenol adducts: An alternative sink for reactive and potentially toxic lipid oxidation products. J Agric Food Chem 66: 1320-1324. 10.1021/acs.jafc.7b05360

Zhang, Y. and Xiao, Z., 2025. Surface hydrophobicity. In: Y. Li (Ed.), Plant-based proteins: Production, physicochemical, functional, and sensory properties. Springer US, New York, NY, pp. 279-286. 10.1007/978-1-0716-4272-6_21

PART 4

Applications in animal nutrition

CHAPTER 7

Insect-based ingredients in fish nutrition

C. Aragão[1,*], *A. Basto*[2], *S. Moutinho*[2] *and L.M.P. Valente*[2,3]

[1]*Centro de Ciências do Mar do Algarve (CCMAR/CIMAR LA), Universidade do Algarve, Campus de Gambelas, Ed. 7, 8005-139 Faro, Portugal*
[2]*Centro Interdisciplinar de Investigação Marinha e Ambiental (CIIMAR/CIMAR LA), Universidade do Porto, Av. General Norton de Matos, 4450-208 Matosinhos, Portugal*
[3]*Instituto de Ciências Biomédicas de Abel Salazar, Universidade do Porto, Rua de Jorge Viterbo Ferreira 228, 4050-313 Porto, Portugal*
**Correspondence: caragao@ualg.pt*

Abstract

The use of insects in aquafeeds has gained increasing attention following recent regulatory advancements, though legal frameworks vary significantly across countries. This chapter provides a comprehensive overview of insects as alternative ingredients in aquaculture diets, focusing on their nutritional and functional roles. Insect-derived products (primarily meals and oils, and more recently, hydrolysates) are promising sources of protein, energy, and bioactive compounds. The nutritional quality of insect-based ingredients is shaped by species, developmental stage, rearing substrate, and processing method. Fish performance outcomes mainly depend on ingredient type, processing technique, the nutrient source being replaced (e.g., fishmeal, fish oil, or plant-based ingredients), and inclusion level, as well as fish species-specific biology, physiology, and rearing conditions. Beyond nutrition, insect-based ingredients may enhance fish health through antioxidant and immunostimulatory effects, linked to compounds such as chitin, antimicrobial peptides, and short- and medium-chain fatty acids. However, these functional benefits are similarly shaped by factors such as insect species and developmental stage, rearing substrate, and processing methods, all of which influence the content and composition of bioactive compounds. To support broader adoption of insect-derived products, standardization of production and processing methods is essential to ensure consistent quality, safety, and efficacy. Additionally, regulatory flexibility, especially regarding permissible rearing substrates, improvements in scalability, and nutritional optimization will be key to unlocking the full potential of insects in aquaculture.

Keywords

alternative ingredients – functional ingredients – insect meal – insect oil – insects in aquafeeds

1 Introduction

Aquaculture is one of the fastest-growing food production sectors and plays an essential role in ensuring food security by supplying healthy fish products to the growing human population. Aquaculture-fed species represent approximately two-thirds of global aquaculture production, with feed representing the highest operational cost (FAO, 2024). Current feed formulations mainly rely on marine ingredients like fishmeal and oil, a variety of plant-based sources such as cereals, oilseeds, and pulses, and processed animal proteins (PAP) derived from terrestrial animal by-products through rendering (Glencross et al., 2024; Serra et al., 2024). However, these traditional ingredients face increasing price volatility and competition from other sectors. With the production of aquaculture-fed species projected to rise by 36 million tonnes by 2050, the demand for raw materials will also intensify, accelerating the search for alternative protein and lipid sources. Microbial ingredients (e.g., bacteria, yeast, fungi, and microalgae) and other terrestrial animal-derived protein sources (e.g., insects) are attracting growing global interest for aquafeed applications (FAO, 2024).

Over the past two decades, insect-derived ingredients have gained attention for inclusion in animal feeds due to their nutritional value, ability to convert waste and by-products into high-value biomass, minimal requirements for arable land and water, and high feed conversion efficiency. Indeed, insects are a natural part of the diet for many freshwater and marine fish species (Henry et al., 2015), making them a promising alternative source of protein and lipids for fish diets.

This chapter provides a comprehensive review of the use of insects as ingredients in aquafeeds, with a focus on their roles as sources of protein and energy, as well as their functional effects in fish. Before addressing these aspects, an overview of the regulatory framework, production numbers, and processing methods will be provided.

2 Regulatory framework for the use of insects in aquaculture

Recent regulatory advances have enabled the inclusion of insects into aquafeeds, and many commercial formulations already include insect-based ingredients. However, the legal frameworks governing their use vary widely across countries.

2.1 European Union

In a step towards promoting the use of insects in aquafeeds and reducing the industry's dependency on traditional protein sources, the European Union (EU) authorized the use of certain insect species in aquafeeds in 2017 (Commission Regulation (EU) 2017/893), prioritizing food and feed safety and enabling more opportunities for the sector (Meijer et al., 2025). From a legal perspective, insects fall into one of the following categories: 1) live insects; 2) PAP; and 3) derived products (fat, chitin, others).

The use of insects in animal feed, excluding ruminants, is authorized under Article 4 of Regulation (European Commission, EC) No 767/2009, which governs the marketing and use of feed. Insect-derived PAP are classified as Category 3 materials following Regulation (EC) No 1069/2009, meaning they must originate from animals fit for human consumption. These materials must be collected, handled, and processed by the methods detailed in Commission Regulation (EU) No 142/2011. Eight insect species have been authorized in the EU for processing into PAP, specifically for aquafeeds: black soldier fly (*Hermetia illucens* - HI), common housefly (*Musca domestica*), yellow mealworm (*Tenebrio molitor* - TM), lesser mealworm (*Alphitobius diaperinus*), house cricket (*Acheta domesticus*), banded cricket (*Gryllodes sigillatus*), field cricket (*Gryllus assimilis*), and more recently, the list was expanded to include the silkworm (*Bombyx mori*) (Commission Regulation (EU) 2017/893 and 2021/1925).

Insects intended for use in aquafeeds must be reared on approved rearing substrates, under Commission Regulation (EU) 2017/893, which prohibits the use of ruminant proteins, catering waste, meat-and-bone meal, and manure as insect feed. Permitted substrates include materials of vegetable origin and certain Category 3 animal by-products, such as egg- and milk-based products or fishmeal.

While insect protein is strictly regulated in the EU, there is currently no specific regulation dedicated to insect oil. The use of fats derived from insects is allowed for all animals, including ruminants. Processing requirements for both insect fats and hydrolysates are specified in Annex X of Commission Regulation (EC) No 142/2011.

2.2 Americas

In the United States of America (USA), the use of insects in animal feed is regulated by the Association of American Feed Control Officials (AAFCO) and the Food and Drug Administration (FDA) (Larouche et al., 2023). Unlike the EU, where insects fall under specific novel food regulations, in the USA, they are considered food and feed additives (Lotta, 2019). Since 2016, only HI has been approved for use in animal feed (both whole larvae and meal forms), and specifically for aquaculture purposes (Sogari et al., 2019). Canada has adopted a similar approach, authorizing the use of dried HI larvae meal in salmonid diets since 2017 (Lähteenmäki-Uutela et al., 2021; Sogari et al., 2019).

In 2024, Argentina's National Service for Agrifood Health and Quality (SENASA) added a new chapter, Resolution 1039/2024, to the *Reglamento de Inspección de Productos, Subproductos y Derivados de Origen Animal*, approved by Decree No 4.238 of July 19, 1968. This resolution regulates facilities that produce insect-based products and by-products intended for animal feed. It formally recognizes, for non-ruminant animals, the same insect species authorized in the EU for use in aquafeeds, except silkworm. The resolution establishes requirements for facility approval, including traceability systems, physical separation of production areas, mandatory thermal or freezing treatments during processing, and clear product labelling.

In Brazil, the *Ministério da Agricultura e Pecuária* (*MAPA*) approved HI meal and oil, TM meal, and superworm (*Zophobas morio*) meal as raw materials for use in animal feed (*Instrução Normativa* SDA/MAPA nº 110/2020, Art. 1º, Annex I). Similarly, in 2024, Chile's *Servicio Agrícola y Ganadero* (SAG) updated the list of authorized raw materials for animal feeds to include PAP and oil from HI (*Resolución Exenta 3860*).

2.3 Asia and Australia

Asia is a major player in the use of insects for both the food and feed industries, with a significant economic impact compared to other regions. Despite being the world's largest aquaculture producer, accounting for 87.9% of global production in 2022 (FAO, 2024), the region still faces some obstacles in incorporating insect-based ingredients into aquafeeds. The regulatory environment across Asian countries remains highly fragmented, with Singapore being the only country to have a comprehensive legal framework for the inclusion of insects in animal feed. Under the Singapore Food Agency, only approved low-risk insect species (like HI and TM), reared on safe substrates (excluding manure or decomposing organic waste) and processed under strict hygiene and safety standards, are permitted for use in aquafeeds. Furthermore, imported insect-derived products for aquaculture must meet additional requirements, including facility registration, prior approval, and compliance with limits on contaminants such as heavy metals, mycotoxins, and pathogenic microorganisms.

In contrast, other Asian countries such as Thailand, Malaysia, and China do not yet have specific legal frameworks governing the use of insects in aquafeeds. Thailand has established voluntary Good Agricultural Practices (GAP) for cricket farming, but these primarily target human consumption and export markets (Hanboonsong and Durst, 2020; Krongdang et al., 2023). Malaysia supports insect farming through its National Biotechnology Policy 2022–2030, though regulatory mechanisms specific to feed use are still under development. In China, there is no specific legislation regulating the use of insects in feed, and their use as feed additives is permitted as long as producers comply with the Measures for the Administration of Feeds and Feed Additives (van Huis and Oonincx, 2017). In contrast, in North Korea, the use of insects for animal feed is not yet regulated, while in South

Korea, their use is authorized but without specific legislation (Han et al., 2017; Sogari et al., 2019). According to the Asian Food and Feed Insect Association (AFFIA), the adoption of insects in aquafeeds is growing, yet their inclusion is limited by high production costs, limited scalability, and more restrictive regulations compared to other feed sectors, such as pet food.

In Australia, insects are permitted in aquafeeds across all states (DiGiacomo et al., 2019). However, strict regulations apply to insect rearing substrates, prohibiting the use of meat, manure, catering waste, or similar materials. Moreover, the use of raw insects (e.g., live or not heat-treated) is prohibited in feed formulations (DiGiacomo et al., 2019).

2.4 Africa

Insect production and use for animal feed in Africa have gained significant traction in recent years, but the sector remains in its early stages. Most production derives from small and medium-scale farms, while wild harvesting still plays a major role (Ssepuuya et al., 2017; Vernooij et al., 2019). Across the continent, approximately 2,300 insect farms exist, but fewer than 2% operate at a large scale. The farming of HI is the fastest-growing insect production sector, with an estimated annual production of 19,732 tonnes, about 9,780 of which are produced in Eastern Africa (Tanga and Kababu, 2023). More than 80% of farmed insects are directed toward livestock and aquaculture feed. Other commonly farmed species besides HI include the African palm weevil (*Rhynchophorus phoenicis*), silkworm, mealworms, crickets, and houseflies (Tanga and Kababu, 2023). Regulatory frameworks are also underdeveloped in many countries, creating challenges to standardization and commercialization (Tanga and Kababu, 2023).

Kenya introduced legislation in 2018 through the Kenya Bureau of Standards (KEBS) to regulate insect-based feed products. These standards focus on final product safety (covering contaminants such as heavy metals, microbes, and mycotoxins) rather than rearing substrates, unlike EU regulations. Insect meal in Kenya has proven cost-competitive with both fishmeal and soybean meal (Vernooij et al., 2019). Also, Uganda adopted the US 1712:2017 standard titled *"Dried Insect Products for Compounding Animal Feeds"* in June 2017, developed through the same regional initiative as Kenya (UNBS, 2022). In 2022, Rwanda, through the Rwanda Standards Board (RSB), established a formal regulatory framework setting national standards for the production and use of insects in feed. These standards were implemented alongside the country's first commercial HI production facility, supported by both government and international partners (ICIPE, 2022). Nevertheless, despite the lack of regulatory frameworks in most African countries and the prevalence of small-scale operations, insect farming in Africa is projected to generate crude protein worth up to USD 2.6 billion (US dollars) annually, holding significant promise in this sector (Tanga and Kababu, 2023; Verner et al., 2021).

3 Production numbers

Insect meal production is developing rapidly in China, Europe, North America, Australia, and Southeast Asian countries (Fantatto et al., 2024; Henry et al., 2015; Nogales-Mérida et al., 2019). The authorization of insects for aquafeeds in the EU, as well as in countries like the USA, Canada, and Singapore, has catalyzed innovation and investment.

In recent years, the European insect sector has grown exponentially, with an annual production of approximately 10,000 tonnes, including whole dried larvae and insect proteins and oils for use in feed for both farmed animals and pets. Prices range from €3,500 to €5,500 (euros) per tonne for insect meal, and from €500 to €1,500 per tonne for insect oil, depending on the type and quality of the ingredient (IPIFF, 2023). Despite this, the sector remains relatively small as it cannot yet meet the high inclusion levels needed for large-scale aquafeed applications. However, significant investments have been directed toward expanding production capacity in the EU. According to IPIFF (2023), the European insect-producing sector is projected to generate around 870,000 tonnes of insect feed ingredients by 2030, 40% of which is expected to be directed towards aquafeeds (de Jong and Nikolik, 2021). As the industry expands, it is estimated that the sale price of insect protein will drop from €3,500 to €1,500 per tonne (de Jong and Nikolik, 2021), increasing its competitiveness compared to other protein feed sources within the EU.

Globally, the insect protein market is expected to grow rapidly, with a projected compound annual growth rate of 32.5%. In Europe alone, cumulative investments in the sector have exceeded €1.5 billion (IPIFF, 2023). In North America, the insect protein market was valued at USD 59 billion in 2024 and is projected to reach USD 341 billion by 2031, with a compound annual growth rate of 24.4% (Data Bridge Market Research, 2024).

Despite this rapid growth, production volumes of insect meals remain low, limiting their widespread use in aquaculture and favoring high-value markets, such as human or pet foods. Whilst the insect oil represents a more affordable lipid source compared to fish oil, insect meal prices are still significantly higher than those of conventional feed ingredients like fishmeal and soybean meal (Gasco et al., 2023). To reduce production costs and improve competitiveness, advancements in infrastructure, automation, and the standardization of nutritional profiles are essential (Tran et al., 2022). Additionally, efforts should focus on expanding the list of approved substrates to include low-value and waste materials, thereby promoting circularity. This will require a thorough risk assessment to ensure their suitability for use in aquafeeds and ensure their safety for consumers.

4 Processing methods

Insect-derived ingredients for aquafeeds undergo various processing steps that affect their nutritional value, functional properties, and techno-functional characteristics. The nutritional quality of insects depends on several factors, including species, developmental stage, rearing substrate, and processing method applied before commercial use. For aquafeeds, most of these ingredients are obtained from larval stages. Typically, insect processing yields two main products: insect meal and oil. More recently, additional products such as insect hydrolysates (obtained through enzymatic hydrolysis) have gained interest due to their high digestibility and functional properties (Liceaga, 2019; Monteiro et al., 2025; Muñoz-Seijas et al., 2024).

Information on insect processing methods is still scarce, but can include fermentation, dehydration, heat treatment, high temperature drying, and high pressure (Meijer et al., 2025). These technological processes will determine the protein and fat content of the resulting meals and their bioavailability for fish. The proximate composition of insect meals is highly variable, even within the same species and developmental stage (Figure 7.1). Generally, full-fat TM meal tends to have higher crude protein contents than full-fat HI meal, while the latter typically contains more fat.

Defatting is an increasingly adopted processing step that can rely on various techniques, including the use of solvents, aqueous extraction, mechanical pressing, or supercritical CO_2 defatting (Gravel and Doyen, 2020), depending on insect species and intended applications. This process helps reduce variability in nutrient composition, increases the protein content, and reduces lipid levels in the meal, while

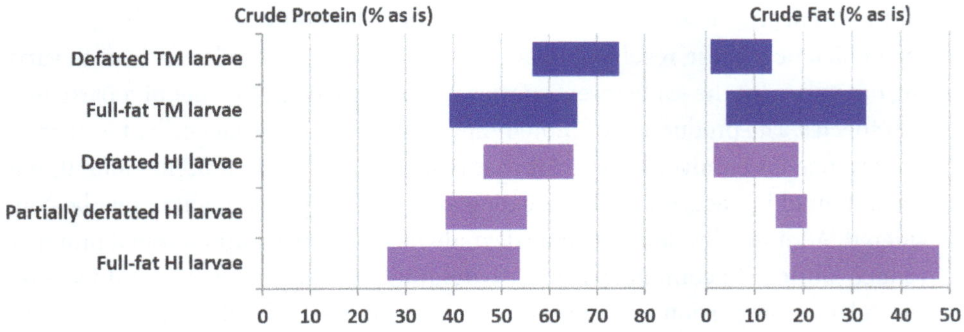

FIGURE 7.1 Ranges of crude protein and crude fat in insect meals from black soldier fly (*Hermetia illucens* – HI) and yellow mealworm (*Tenebrio molitor* – TM), according to processing method, as reported in published studies evaluating their use in fish diets (references for minimum and maximum values out of 157 studies analysed: Ardra et al., 2025; Chaklader et al., 2021a, b; Eggink et al., 2022; Fabrikov et al., 2021b; Fischer et al., 2022; Gebremichael et al., 2022; Hender et al., 2021; Lin et al., 2023; Mapanao et al., 2023; Mastoraki et al., 2025; Meesala et al., 2025; Melenchón et al., 2022; Rema et al., 2019; Sankian et al., 2018; Stejskal et al., 2020; Tippayadara et al., 2021; Wang et al., 2019; Yang et al., 2025).

generating insect oil as a valuable lipid-rich by-product (Cruz et al., 2025; Hossain et al., 2023). However, while defatted insect meals generally have higher protein and lower fat content, the proximate composition can vary so widely that defatted meals from the same species occasionally have protein and lipid levels comparable to those of full-fat meals (Figure 7.1).

Defatting can also alter the carbohydrate composition of insects, particularly their most abundant fraction, chitin. Chitin is a polysaccharide composed of glucosamine and N-acetyl-glucosamine, forming a major component of the insect exoskeleton. Its content in insect meals is highly variable, depending on insect species, developmental stage, and processing method. In aquaculture studies, reported chitin levels in TM larvae meals range from 3.2 to 6.5% in full-fat meals and from 4.2 to 13.6% in defatted meals (Coutinho et al., 2021; Ge et al., 2022; Melenchón et al., 2023; Willora et al., 2025). For HI larvae meals, chitin levels range from 1.8 to 11.5% in full-fat meals and from 2.7 to 7.7% in defatted meals (Huyben et al., 2019; Singha et al., 2025; Wang et al., 2019). Since fish species respond differently to dietary chitin (e.g., Karlsen et al., 2017), processing strategies that reduce or retain chitin may be desirable depending on the target species and the intended physiological or functional effects of the diet. However, further studies are required to define species-specific tolerance thresholders and optimal inclusion levels.

Defatting can improve the physical properties of feeds, as excessive lipid content in full-fat insect meals may impair the extrusion process and affect pellet buoyancy. For example, extrusion of pellets containing defatted HI larvae meal compared favorably to extruded pellets based on fishmeal (Edah and Owolabi, 2023). Other studies have reported that full-fat and de-chitinised insect meals reduced pellet hardness, expansion, and water stability in Atlantic salmon (*Salmo salar*) diets (Weththasinghe et al., 2022a). While these diets improved feed intake, growth rates, and cost-effectiveness compared to control or defatted diets, they also impaired feed efficiency. These results illustrate how processing methods can lead to varying outcomes for the same insect species when included in the diet of a particular fish species. The production of protein-dense meals with low fat content is of great importance for carnivorous species that require high dietary protein content, but the potential impacts on feed processing and physical pellet quality must be considered. Additionally, defatting may degrade heat-sensitive antimicrobial proteins, reduce lauric acid content (particularly relevant in HI larvae meal), and increase the relative proportion of chitin, affecting the functional value of insect meals (Chaklader et al., 2021a).

Following defatting, additional processes may be used to concentrate protein, for example, by adjusting pH and ionic strength to solubilize and recover proteins. Through key technological steps – defatting, protein solubilization and recovery, protein purification and drying – protein-rich ingredients can be produced, depending on the proximate composition of the raw material (Gravel and Doyen, 2020). These processes may further allow chitin separation (Liceaga, 2019).

Finally, drying is a critical processing method that can be applied either as an initial step to stabilize the insect meal or as a final step towards the obtention of protein concentrates. Common drying methods include oven-drying, spray-drying, or freeze-drying (Gravel and Doyen, 2020). The choice of a commercial processing technology needs to consider the impact on the techno-functional properties, safety, sensory (organoleptic), and nutritional quality of the final product.

Standardization of production and processing methods is necessary to reduce the nutritional variability and ensure consistent quality across insect meals. At an industrial scale, cost-effective and environmentally friendly processing conditions should be tackled to offer rich-protein insect meals able to support animal growth and welfare. As a nascent and still immature sector, the insect industry requires significant technological innovation to reduce production costs and enhance product competitiveness.

5 Insect meal as a protein source in fish diets

Most studies on novel aquafeed ingredients have relied on fish growth outcomes as the primary indicator of nutritional value. Since ingredient quality affects growth through a sequential cascade of physiological responses, this section evaluates the feasibility of insects as a protein source in aquafeeds by analyzing their effects on nutrient digestibility, growth performance, feed efficiency, and intermediary metabolism.

5.1 Nutrient digestibility

Proper formulation of aquafeeds to meet the nutritional needs of aquaculture species relies heavily on accurately determining nutrient bioavailability, which is the fraction of ingested nutrients that is absorbed and available for fish growth and metabolism. The bioavailability of nutrients is usually evaluated using *in vivo* methods to calculate the apparent digestibility coefficients (ADC) (Glencross et al., 2007). Aaqillah-Amr et al. (2024) conducted a meta-analysis of 32 studies published between 2013 and 2023, evaluating the nutrient digestibility in fish fed diets containing insect meal as a replacement for fishmeal or soybean meal. The analysis revealed that the average ADC for dry matter and crude lipid in fish fed insect meal-based diets was 74 and 89%, respectively, slightly lower than the 78 and 92% reported for insect meal-free diets. The reduced digestibility of insect meal-based diets has often been attributed to antinutritional factors, particularly chitin, but concrete evidence remains limited. According to the available studies on the effects of dietary chitin on nutrient digestibility, different fish species can tolerate chitin levels between 1 and 5% (Eggink et al., 2022, 2025; Karlsen et al., 2017; Pascon et al., 2024). However, further research is needed to define the maximum inclusion levels that do not compromise nutrient utilization across a broader range of fish species.

Processing techniques may also significantly influence the digestibility of insect meals in aquaculture species. Mastoraki et al. (2025) evaluated different processing methods on the digestibility of HI meal in European seabass (*Dicentrarchus labrax*), showing that defatted HI meal led to higher protein digestibility compared to full-fat meal, while enzymatic hydrolysis did not provide further improvements. Additional evidence for the benefits of defatting was also found for European seabass, where defatted TM meal outperformed the full-fat meal in terms of protein, fat, energy, phosphorus, and essential amino acids digestibility (Basto et al., 2020). Altogether, these studies underscore that defatting is a critical processing step that enhances nutrient digestibility in seabass and may also reduce antinutritional effects associated with poorly digested fats. However, this effect was not observed in Atlantic salmon (Weththasinghe et al., 2022a), likely reflecting the greater tolerance of salmonids to high dietary fat levels. Further investigation is therefore needed to clarify the impacts of defatting on nutrient digestibility, while also considering that, despite its apparent benefits, defatting may remove functional compounds with potential physiological value.

5.2 Growth performance and feed efficiency

The use of insect meal as a protein source in diets for both marine and freshwater fish species has received increasing attention in recent decades. Among the insect species evaluated for inclusion in aquafeed, nearly half of the research has focused on HI, while about one-quarter has investigated TM, reflecting advances in their large-scale production. Nonetheless, a meta-analysis performed by Tran et al. (2022) demonstrated that full-fat Oriental latrin fly (*Chrysomya megacephala*) and shea nut beetle (*Caryedon butyrospermi*) larvae, adults of black field cricket (*Gryllus bimaculatus*), and banded and variegated grasshoppers (*Oedaleus fuscovittatus* and *Zonocerus variegatus*, respectively) also show potential as a protein source for aquafeed. However, they are not currently approved for use in aquafeeds in certain geographies (e.g., Europe, USA).

Most studies on insect-based aquafeeds focus on HI meal, yet recommended inclusion levels and replacement rates vary considerably (Table 7.1). For example, in rainbow trout (*Oncorhynchus mykiss*) juveniles, diets with up to 40% HI meal (7–34% crude fat), replacing up to 50% of fishmeal at inclusion levels of 20–60%, did not impair growth or feed efficiency over six to 18 weeks (Biasato et al., 2022; Bordignon et al., 2022; Józefiak et al., 2019; Melenchón et al., 2021, 2022; Renna et al., 2017). However, complete replacement of fishmeal (at inclusion levels of 20–21%) with partially defatted HI meal (20% crude fat) has yielded contrasting results: no adverse effects on specific growth rate (SGR) and feed conversion ratio (FCR) at 32% HI inclusion (Biasato et al., 2022), but reduced growth performance at 26% inclusion (Dumas et al., 2018).

Similar to findings in rainbow trout, the effects of HI meals on African catfish

(*Clarias gariepinus*), Atlantic salmon, European seabass, gilthead seabream (*Sparus aurata*), Nile tilapia (*Oreochromis niloticus*), and other fish species are also variable (Belghit et al., 2019a; Fabrikov et al., 2020; Fawole et al., 2020; Hervé et al., 2025; Limbu et al., 2022; Moutinho et al., 2022; Tippayadara et al., 2021; Weththasinghe et al., 2021). For instance, in juvenile African catfish, dietary inclusion of 20% HI meal (15% crude fat), replacing 75% of fishmeal (28% in control diet), improved growth performance (Hervé et al., 2025), whereas inclusion of 17% HI meal (23% crude fat), replacing the same proportion of fishmeal, had no effect on weight gain (Fawole et al., 2020). Similarly, in European seabass juveniles, inclusion of HI meal – whether full-fat, partially defatted, or defatted, with fat contents from 6 to 20% and replacing up to 50% of fishmeal or 12% of a vegetable protein mix – did not adversely affect growth performance or feed efficiency over eight to 12 weeks (Abdel-Tawwab et al., 2020; Magalhães et al., 2017; Mastoraki et al., 2020; Randazzo et al., 2023; Zarantoniello et al., 2023). However, Reyes et al. (2020) reported that 18% HI meal (replacing half of the 36% fishmeal) impaired the growth performance and feed efficiency of European seabass.

The effects of TM meal inclusion in fish diets are similarly complex (Table 7.1). In rainbow trout, feeding trials lasting six to 22 weeks showed that including 5 to 20% full-fat TM meal (25–28% crude fat), replacing 15–100% of fishmeal in diets originally formulated with 20–37% fishmeal, had no significant impact on growth performance or feed efficiency (Chemello et al., 2020; Józefiak et al., 2019; Melenchón et al., 2021). Conversely, Rema et al. (2019) reported improved growth and feed utilization in trout fed diets containing 25% partially defatted TM meal (14% crude fat), fully replacing fishmeal (25% inclusion) over 13 weeks. In Nile tilapia, complete replacement of fishmeal with TM meal has also produced divergent outcomes. Growth performance and feed efficiency were impaired when fishmeal (10% in control diet) was fully substituted with TM meal (32% crude fat) over 60 days (Anany et al., 2025). By contrast, the use of a lower-fat TM meal (20% crude fat) to completely replace fishmeal in diets with a higher fishmeal content (20% inclusion) increased weight gain, despite a slight reduction in FCR after 12 weeks (Dernekbaşı and Karadayı, 2025). In European seabass, 50% inclusion of full-fat TM meal (24% crude fat) significantly reduced SGR after 13 weeks, despite no significant change in FCR (Gasco et al., 2016). In contrast, partially defatted TM meal (12% crude fat), included at levels replacing 80% of fishmeal (36% TM inclusion) or even all protein sources (60% TM inclusion), did not impair growth in seabass over 10 to 16 weeks; however, all diets contained 12–13% fish oil (Basto et al., 2021). Together, these findings underscore that insect meal efficacy depends not only on inclusion level, processing method, and fishmeal replacement rate, but also on their interactions. Species-specific biology, physiological traits, and rearing conditions further influence outcomes and must be considered in diet formulation.

TABLE 7.1 Summary of the results of the impact of insect meal dietary inclusion on fish growth and feed efficiency

Fish species	Insect meal (% crude fat)	Protein source replaced (% included in CTRL)	% insect meal inclusion (% substitution)	Trial duration (days)	Main outcomes	References
African catfitsh (*Clarias gariepinus*)	HI (14.8)	FM (28)	13.0 (50*), 19.5 (75*), 26.0 (100*)	12c	Increased growth, but no effects on feed efficiency (except at the 13% inclusion, which increased the feed conversion ratio).	Hervé et al. 2025
	HI (23)	FM (15)	5.7 (25*), 11.5 (50*), 17.2 (75*)	60	Reduced growth and feed efficiency (only at 11.5% inclusion).	Fawole et al. 2020
Atlantic salmon (*Salmo salar*)	HI (18)	FM (10)	4.9 (33.3), 9.8 (66.7), 14.8 (100)	114	No effects on growth and feed efficiency.	Belghit et al. 2019a
	HI (33.9)	FM (36.8)	5.6 (14.6), 10.9 (29.6)	46	No effects on growth performance and feed efficiency.	Fabrikov et al. 2020
	HI (32)	FM (25); soy protein concentrate (35.5); corn gluten (4); faba bean (1.85)	8.1 (6.25*), 16.1 (12.5*), 32.3 (25.0*)	49	Reduced growth (32.3% inclusion), but no effects on feed efficiency.	Weththasinghe et al. 2021
	TM (28.3)	FM (36.8)	5.1 (13.9), 10.7 (29.1)	46	No effects on growth performance and feed efficiency.	Fabrikov et al. 2020
European seabass (*Dicentrarchus labrax*)	HI reared on broilers diet (n.a.)	FM (35.9)	10.9 (29.5), 18.0 (49.9)	49	Reduced growth (10.9 and 18% inclusion) and feed efficiency (18% inclusion).	Reyes et al. 2020
	HI reared on fish by-products (n.a.)	FM (35.9)	18 (49.9)	49	Reduced growth and feed efficiency.	Reyes et al. 2020
	HI (5.6)	FM (32.4)	6.5 (15.1), 13.0 (30.0), 19.5 (45.1)	62	No effects on growth and feed efficiency.	Magalhães et al. 2017
	HI (6)	FM (65)	19.5 (30)	90	No effects on growth and feed efficiency.	Mastoraki et al. 2020
	HI (13)	FM (40)	8 (20)	90	No effects on growth and feed efficiency.	Zarantoniello et al. 2023
	HI (19.7)	FM (21)	7.3 (25.2), 10.1 (34.0), 14.8 (50.0)	56	No effects on growth and feed efficiency.	Abdel-Tawwab et al. 2020
	HI (20.8)	Vegetable protein mix (69)	8.1 (12.3)	126	No effects on growth and feed efficiency.	Randazzo et al. 2023
	TM reared on broilers diet (n.a.)	FM (35.9)	18 (49.9)	49	No effects on feed efficiency, but reduced growth.	Reyes et al. 2020

TABLE 7.1 Summary of the results of the impact of insect meal dietary inclusion on fish growth and feed efficiency *(cont.)*

Fish species	Insect meal (% crude fat)	Protein source replaced (% included in CTRL)	% insect meal inclusion (% substitution)	Trial duration (days)	Main outcomes	References
European seabass (*Dicentrarchus labrax*)	TM (23.6)	FM (70)	25 (50), 50 (71)	90	Reduced growth (50% inclusion), but without effects on feed efficiency.	Gasco et al. 2016
	TM (24)	FM (65)	19.5 (30)	90	No effects on growth, but reduced feed efficiency.	Mastoraki et al. 2020
	TM (12.1)	FM (45), soybean meal (10), soy protein concentrate (8), rapeseed meal (5)	18 (40% FM replacement), 36 (80% FM replacement), 60 (100% protein replacement)	70	No effects on growth, but increased feed efficiency (60% inclusion).	Basto et al. 2021
Gilthead seabream (*Sparus aurata*)	HI (33.9)	FM (36.8)	5.6 (14.6), 10.9 (29.6)	40	Reduced feed efficiency (10.9% inclusion), but no effects on growth.	Fabrikov et al. 2020
	HI (12.1)	FM (34.3), vegetable proteins (42.0)	15 (22.2; 48.2), 30 (60.3; 62.2), 45 (100; 74.8)	67	No effects on growth and feed efficiency.	Moutinho et al. 2022
	TM (28.3)	FM (36.8)	5.1 (13.9), 10.7 (29.1)	40	No effects on growth and feed efficiency.	Fabrikov et al. 2020
Nile tilapia (*Oreochromis niloticus*)	HI (19.4)	FM (42.4)	10.6 (25), 21.2 (50), 31.8 (75), 42.4 (100)	84	No effects on growth and feed efficiency (except at the 31.8% inclusion, which reduced the feed conversion ratio).	Limbu et al. 2022
	HI (n.a.)	FM (10)	1 (10), 2 (20), 4 (40), 6 (60), 8 (80), 10 (100)	84	No effects on growth and feed efficiency.	Tippayadara et al. 2021
	TM (32.2)	FM (10)	10 (100)	60	Reduced growth and feed efficiency.	Anany et al. 2025
	TM (20)	FM (20)	10 (50), 15 (75), 20 (100)	84	Improved growth (weight gain), with no change in specific growth rate (SGR). Feed efficiency improved at 15% inclusion, but was impaired at 10 and 20%.	Dernekbaşı and Karadayı 2025

TABLE 7.1 Summary of the results of the impact of insect meal dietary inclusion on fish growth and feed efficiency *(cont.)*

Fish species	Insect meal (% crude fat)	Protein source replaced (% included in CTRL)	% insect meal inclusion (% substitution)	Trial duration (days)	Main outcomes	References
Rainbow trout (*Oncorhynchus mykiss*)	HI (7.4)	FM (20)	6.2 (25), 12.4 (50)	76	Reduced specific growth rate in the first 26 days of feeding; no effects on feed efficiency.	Bordignon et al. 2022
	HI (18)	FM (60)	20 (25), 40 (50)	78	No effects on growth and feed efficiency.	Renna et al. 2017
	HI (20.3)	FM (20)	6.6 (25), 13.2 (50), 26.4 (100)	84	Reduced growth and feed efficiency.	Dumas et al. 2018
	HI (20.4)	FM (20.6)	8 (25), 16 (50), 32 (100)	123	No effects on growth performance and feed efficiency.	Biasato et al. 2022
	HI (25.6)	FM (35.9)	18 (49.8)	77	No effects on final body weight and daily feed intake but reduced specific growth rate and feed efficiency.	Melenchón et al. 2022
	HI (33.5)	FM (24)	20 (30.4)	71	No effects on growth and feed efficiency.	Józefiak et al. 2019
	HI (33.9)	FM (36.8)	5.6 (14.3), 10.9 (29.6)	46	No effects on growth and feed efficiency.	Melenchón et al. 2021
	TM (n.a.)	FM (20)	5 (25), 10 (50), 20 (100)	154	No effects on growth and feed efficiency.	Chemello et al. 2020
	TM (13.6)	FM (25)	5 (20), 7.5 (30), 15 (60), 25 (100)	90	Improved growth and feed efficiency.	Rema et al. 2019
	TM (25.3)	FM (20.6)	20 (41.8)	71	No effects on growth and feed efficiency.	Józefiak et al. 2019
	TM (27)	FM (35.9)	18 (49.8)	77	No effects on growth and feed efficiency.	Melenchón et al. 2022
	TM (28.3)	FM (36.8)	5.1 (14.8), 10.3 (29.6)	46	No effects on growth and feed efficiency.	Melenchón et al. 2021

*Substitution of protein content (%). CTRL: control diet; FM: fishmeal; HI: *Hermetia illucens* (black soldier fly); n.a.: not available; TM: *Tenebrio molitor* (yellowmeal worm).

5.3 Metabolism

Despite extensive research over the past decade, data on the metabolic effects of insect meal in fish remain limited. Most published studies use isoproteic, isolipidic, and isoenergetic diets, requiring adjustments to multiple dietary components when including insect meal. This complicates the attribution of observed metabolic changes specifically to the insect meal itself. In this context, several studies have used the activity of aspartate aminotransferase (AST) and alanine aminotransferase (ALT) as hepatic biomarkers, as elevated levels may indicate liver dysfunction or damage associated with metabolic disorders. Dietary inclusion of HI meal (full-fat or defatted) up to 20% and replacing 50% of fishmeal did not alter plasmatic or hepatic AST and ALT activities in several fish species, including rainbow trout, Atlantic salmon, European seabass, and gilthead seabream (Abdel-Tawwab et al., 2020; Fabrikov et al., 2021b; Mastoraki et al., 2020; Melenchón et al., 2021, 2022). Higher inclusion levels (up to 60%) or full fishmeal replacement also did not affect enzymatic activities in Atlantic salmon (Belghit et al., 2018, 2019a). Similarly, 20% TM meal fully replacing fishmeal did not affect AST or ALT activities in rainbow trout (Chemello et al., 2020), while in European seabass, 40% TM meal (100% fishmeal replacement) reduced hepatic ALT, suggesting a possible nutraceutical effect (Basto et al., 2023).

Many of the metabolic effects observed in fish fed insect meal appear to be linked to lipid metabolic pathways; however, the outcomes reported across studies are highly variable. For instance, Magalhães et al. (2017) reported reduced plasma cholesterol in European seabass fed a diet containing 20% defatted HI meal (6% crude fat), whereas Mastoraki et al. (2020) found no effect at a similar HI meal inclusion and fat content. In the same study, Mastoraki et al. (2020) reported reduced plasma cholesterol with 20% full-fat TM meal (24% crude fat), whereas Basto et al. (2021) observed increased levels with 18–60% inclusion of partially defatted TM meal (12% crude fat). Variations across studies may reflect differences in diet composition (e.g., inclusion levels and lipid sources) and in the lipid profiles of the ingredients used. Therefore, within the framework of precision nutrition in aquaculture, detailed characterization of feed ingredients and diets, particularly regarding lipid classes, such as triglycerides, cholesterol, and phospholipids, is increasingly needed.

6 Insect oil as a lipid source in fish diets

The effectiveness of insect oil as a dietary lipid source in aquafeeds depends on fish species, life stage, and formulation strategy. Up to this point, the number of studies on insect oil inclusion remains limited, especially compared to those on insect meals. Nevertheless, there is a growing body of research evaluating insect oils in aquafeeds, particularly those from HI, and their effects on fish growth performance, feed and nutrient utilization, digestibility, and product quality (Table 7.2). Studies have shown that insect oils can provide readily digestible energy and can partially or fully replace traditional oils, such as fish or vegetable oils, in diets for several fish species. However, their effectiveness ultimately depends on overall dietary formulation, particularly on meeting species-specific essential fatty acid requirements

(Turchini et al., 2009). This is particularly critical for marine carnivorous species, which require higher levels of n-3 long-chain polyunsaturated fatty acids (PUFA) (Sargent et al., 2002; Turchini et al., 2009). The low or absent levels of eicosapentaenoic acid (EPA; C20:5n−3) and docosahexaenoic acid (DHA; C22:6n−3) in insect oils constrain their ability to fully substitute fish oil, especially when fishmeal is also reduced or eliminated (Hossain et al., 2023). In such cases, combining insect oil with other lipid sources may help achieve a more balanced fatty acid profile.

TABLE 7.2 Key outcomes of insect oil inclusion in fish diets

Fish species	Insect oil	Oil being replaced	% insect oil inclusion (% substitution)	Main outcomes	Source
Atlantic salmon (*Salmo salar*)	HI oil	Rapeseed oil	4.8 (combined with HI meal) or 12.0 (100)	No effects on growth, feed efficiency, or fatty acid digestibility; reduced plasma and liver glucose	Belghit et al. 2019b
Barramundi (*Lates calcarifer*)	HI oil	Fish oil	1.05 (30)	No effects on growth or feed efficiency; no effect on plasma glucose or cholesterol; increased lauric acid in muscle	Hender et al. 2021
European catfish (*Silurus glanis*)	HI oil	Fish and rapeseed oil blend (50:50)	2.5 (50), 5.0 (100)	No effects on growth or feed efficiency; reduced dry matter and protein digestibility	Egessa et al. 2024
Gilthead seabream (*Sparus aurata*)	HI oil	Vegetable oil mix	4.0 (42), 7.9 (84), 9.5 (100)	Improved feed efficiency and energy digestibility (>7.9% inclusion) with no effects on growth; linear decrease in plasma lipids and triglycerides; increased n-6/n-3 ratio, while EPA+DHA remained stable	Moutinho et al. 2023, 2024b, 2025a
Jian carp (*Cyprinus carpio* var. Jian)	HI oil	Soybean oil	0.625 (25), 1.25 (50), 1.88 (75), 2.50 (100)	No effects on growth or feed efficiency; no effect on plasma metabolites; increased lauric acid, SFA, and DHA in muscle	Li et al. 2016
	Silkworm oil	Soybean oil	0.75 (25), 1.5 (50), 2.25 (75), 3.0 (100)	Up to 2.25% inclusion improved growth and feed utilization; increased muscle n-3 PUFA deposition and decreased n-6/n-3 ratio	Chen et al. 2017
Mirror carp (*Cyprinus carpio* var. *specularis*)	HI, TM, and silkworm oil	-	2.5 (-)	HI oil better lipid source than TM and silkworm oil, improving growth performance and feed efficiency	Xu et al. 2020
	HI oil (n-3 HUFA-enriched)	Soybean oil	0.625 (25), 1.25 (50), 1.88 (75), 2.50 (100)	>1.25% inclusion improved growth and feed efficiency; increased muscle EPA, DHA, and n-3 PUFA, while reducing n-6 PUFA >1.88% inclusion increased plasma protein and triglycerides and upregulated *cpt1* and *ppara* expression	Xu et al. 2021

TABLE 7.2 Key outcomes of insect oil inclusion in fish diets *(cont.)*

Fish species	Insect oil	Oil being replaced	% insect oil inclusion (% substitution)	Main outcomes	Source
Nile tilapia (*Oreochromis niloticus*)	HI oil	Soybean oil	1 (25), 2 (50), 3 (75), 4 (100)	1% inclusion improved growth and feed efficiency	Goda et al. 2024
Onychostoma macrolepis	HI oil	Fish oil	2.25 (25), 4.5 (50), 9.0 (100)	Up to 4.5% inclusion did not affect growth; reduced plasma triglycerides; increased *cpt1* and *ppara* expression	Gou et al. 2023
Rainbow trout (*Oncorhynchus mykiss*)	HI oil	Fish oil or soybean oil	16 (100)	No effects on growth or feed efficiency; plasma glucose decreased; EPA and DHA reduced with fish oil replacement; increased liver *fabp* and *fas* expression	Fawole et al. 2021
	HI oil	Fish oil	2.5 (25), 5.0 (50), 10.0 (100)	No effect on growth, feed efficiency, or lipid digestibility; plasma glucose increased at 2.5% inclusion	Dumas et al. 2018
Red hybrid tilapia (*Oreochromis* sp.)	HI oil	Fish oil	1.9 (25), 3.8 (50), 5.7 (75), 7.6 (100)	Reduced final body weight and weight gain; feed efficiency unchanged, reduced lipid digestibility (at 7.6% inclusion); increased plasma cholesterol	Abu Bakar et al. 2021
Striped catfish (*Pangasianodon hypophthalmus*)	HI oil	Fish oil	2.3 (100)	No effects on growth, feed efficiency, digestive enzymes activity, or plasma metabolites	Sudha et al. 2022
Totoaba (*Totoaba macdonaldi*)	HI oil	Fish oil	2 (30), 4 (60)	4% inclusion reduced final body weight and weight gain; increased muscle SFA and lauric acid, decreased EPA and n-3 PUFA	Maldonado-Othón et al. 2022

cpt1: carnitine palmitoyltransferase-1; EPA: eicosapentaenoic acid; DHA: docosahexaenoic acid; *fabp*: fatty acid binding protein; *fas*: fatty acid synthase; HI: *Hermetia illucens* (black soldier fly); HUFA: highly unsaturated fatty acids; *ppara*: peroxisome proliferator-activated receptor-α; PUFA: polyunsaturated fatty acids; SFA: saturated fatty acids; TM: *Tenebrio molitor* (yellowmeal worm).

6.1 Nutrient utilization and digestibility

Although the number of studies is still limited, the inclusion of insect oils has produced encouraging results in terms of nutrient digestibility and digestive physiology across fish species. In gilthead seabream, a Mediterranean carnivorous fish, the inclusion of 9.5% HI oil to replace a vegetable oil blend significantly improved the ADC of both protein and energy (Moutinho et al., 2023). This was accompanied by a linear increase in total protease activity, suggesting that HI oil may enhance digestive enzymes, particularly those involved in protein digestion. While lipase activity was not significantly affected, a positive trend was observed with increasing HI oil levels, suggesting also potential benefits for lipid digestion. Moreover, fatty acid digestibility followed the expected trend: saturated < monounsaturated < polyunsaturated fatty acids. However, the ADC of saturated fatty acids (SFA) improved significantly with higher HI oil inclusion (Moutinho et al., 2023). This was likely due to the high proportion of medium-chain fatty acids in HI oil, especially lauric acid (C12:0), which are more readily digestible than longer-chain SFA. Similar findings were reported in Atlantic salmon, where diets containing 4.8 or 12% HI oil from different sources resulted in high ADC (above 95%) for all fatty acid classes, including SFA and lauric acid (Belghit et al., 2019b). In contrast, in juvenile red hybrid tilapia (*Oreochromis* sp.), the dietary inclusion of 7.6% HI oil led to a significant reduction in lipid ADC compared to fish oil-based diets (Abu Bakar et al., 2021). Similarly, replacing a fish and rapeseed oil blend (50:50) with HI oil (5% inclusion) reduced dry matter and protein ADC in European catfish (*Silurus glanis*) (Egessa et al., 2024). In striped catfish (*Pangasianodon hypophthalmus*), 2.3% HI oil inclusion in replacement of fish oil had no significant effect on digestive enzyme activities, although diets' ADC were not reported (Sudha et al., 2022).

Overall, despite the limited number of publications, current research suggests that the inclusion of HI oil can have a positive effect on nutrient digestibility and digestive physiology in some fish species. However, responses appear to be species-specific and dependent on the inclusion level and overall dietary formulation.

6.2 Growth performance and feed efficiency

Most of the literature available on the effect of insect oil inclusion, mainly HI oil, is at the expense of vegetable oils, with outcomes varying according to inclusion level, replacement strategy, and species-specific responses. This is an important consideration, as the type of oil being replaced can significantly influence the fatty acid profile of the diet and ultimately affect fish performance and physiological responses.

In juvenile mirror carp (*Cyprinus carpio* var. *specularis*), a comparison of three insect oils – HI, TM, and silkworm – used as the sole lipid source at 2.5% inclusion demonstrated that HI oil was the most effective, improving both growth performance and feed efficiency compared to the other sources (Xu et al., 2020). A subsequent study showed that HI oil enriched in n-3 highly unsaturated fatty acids (HUFA) and low in lauric acid, when included at more than 1.25% (replacing 50% of soybean oil), significantly increased final body weight, weight gain, and feed efficiency in mirror carp (Xu et al., 2021). Additionally, total replacement of dietary soybean oil with 2.5% HI oil had no significant effect on growth performance or feed efficiency of juvenile Jian carp (*Cyprinus carpio* var. Jian), supporting the feasibility of HI oil as an alternative source to vegetable oils (Li et al., 2016). In Nile tilapia fry, while 1% HI oil (replacing 25% of soybean oil) improved growth performance, more than 3% inclusion (75% replacement) reduced final body weight (Goda et al., 2024). Positive outcomes have also been reported in gilthead seabream juveniles, where the inclusion of 9.5% HI oil (replacing a vegetable oil mix) had no adverse effect on final growth performance, while feed efficiency improved at inclusions above 7.9% (Moutinho et al., 2023).

Studies across several fish species evidenced variable outcomes when HI oil was used to replace fish oil in aquafeeds. In *Onychostoma macrolepis*, total replacement of fish oil with 9% HI oil reduced final body weight and weight gain, which was associated with a decrease in feed intake (Gou et al., 2023). Similarly, in juvenile red hybrid tilapia, HI oil included at 1.9%, replacing 25% of the fish oil fraction, compromised growth performance, with significant reductions in final body weight and weight gain (Abu Bakar et al., 2021). However, in juvenile totoaba (*Totoaba macdonaldi*), inclusion of 2% HI oil (replacing 30% of fish oil) had no detrimental effect on growth performance, although increasing the level to 4% significantly reduced both final body weight and weight gain (Maldonado-Othón et al., 2022). Likewise, in barramundi (*Lates calcarifer*), inclusion of 1.05% HI oil replacing 30% of fish oil (either alone or combined with 22% HI meal in diets containing 35–50% fishmeal) showed no impact on either growth performance or feed efficiency (Hender et al., 2021). Additionally, neutral or even positive outcomes were reported at higher replacement levels. Total replacement of fish oil (2.3%) or a fish and rapeseed oil blend (5%) with HI oil had no significant effect on the growth performance of striped catfish and European catfish (Egessa et al., 2024; Sudha et al., 2022). In Atlantic salmon, dietary inclusion of 12% HI oil (replacing rapeseed oil) or 4.8% HI oil in combination with HI meal did not affect growth performance or FCR (Belghit et al., 2019b). In rainbow trout, total replacement of fish oil with 10% HI oil showed no differences in growth performance (Dumas et al., 2018), and further inclusion up to 16%, replacing either fish or soybean oil, resulted in comparable growth performance (Fawole et al., 2021).

As previously mentioned, research on other insect oil sources other than HI oil remains limited. Nevertheless, a study by Chen et al. (2017) demonstrated that partial replacement of dietary soybean oil with silkworm oil (2.25% inclusion) led to improved growth performance and feed utilization in Jian carp juveniles.

In summary, insect oils, particularly HI oil, show great promise as a lipid source across various fish species, though the zootechnical outcomes depend on factors such as inclusion level, the type of oil being replaced, and species-specific requirements. While HI oil often performs well when replacing vegetable oils, its use in place of fish oil yields more variable results. Further research is needed to optimize its application and explore other insect oil sources.

6.3 Metabolism

The inclusion of HI oil in aquafeeds has shown varied effects on lipid and glucose metabolism across fish species, largely attributed to its distinctive fatty acid profile, particularly its high lauric acid content, which is rapidly oxidized for energy. In Atlantic salmon, dietary HI oil inclusion up to 12% significantly reduced plasma and liver glucose levels, likely due to enhanced fatty acid β-oxidation (Belghit et al., 2019b). Similarly, rainbow trout showed reduced plasma glucose concentrations when fish oil was replaced with 16% HI oil (Fawole et al., 2021). In gilthead seabream, vegetable oil replacement with HI oil led to a linear decrease in plasma total lipid and triglyceride content (Moutinho et al., 2025a). In contrast, no significant changes in plasma glucose or cholesterol levels were found in barramundi, Jian carp, and striped catfish, suggesting a neutral metabolic response to HI oil in these species (Hender et al., 2021; Li et al., 2016; Sudha et al., 2022).

Other species showed significant changes in lipid metabolism markers. For example, mirror carp fed diets containing 1.9% HI oil showed increased plasma protein, triglycerides, and low-density lipoprotein cholesterol compared to those fed diets based on soybean oil (Xu et al., 2021). This was accompanied by an upregulation of hepatic expression of peroxisome proliferator-activated receptor-α (*ppara*) and carnitine palmitoyltransferase-1 (*cpt1*), key genes involved in fatty acid uptake and β-oxidation, suggesting enhanced lipid catabolism. Similarly, Jian carp also demonstrated increased *ppara* expression with HI oil diets in two distinct studies (Li et al., 2016; Xu et al., 2020). In *O. macrolepis*, a reduction in triglyceride content was found when fish oil was completely replaced with 9% HI oil, accompanied by higher expression of *ppara* and *cpt1* in intraperitoneal fat, suggesting an increased capacity for fatty acid oxidation (Gou et al., 2023). Conversely, red hybrid tilapia showed increased total cholesterol levels when fish oil was substituted with HI oil (Abu Bakar et al., 2021), highlighting species-specific differences in cholesterol metabolism in response to this lipid source.

In conclusion, HI oil affects lipid and glucose metabolism in a species-specific manner, often promoting lipid catabolism, possibly due to its distinct fatty acid profile. However, responses vary, highlighting the need for further research to clarify its metabolic impacts across fish species.

7 Fillet quality in fish fed insect meals and oils

Fillet quality is a key parameter in aquaculture, as it influences consumer acceptance, marketability, and the economic value of the final product. Thus, evaluating the impact of insect meals and oils on fillet nutritional value is essential to ensure that the nutritional attributes expected by consumers and the industry are met. In fish, muscle fatty acid composition closely reflects that of the diet (Fabrikov et al., 2021a; Turchini et al., 2009). Consequently, dietary changes, such as the inclusion of insect-derived ingredients and the reduction of conventional ones, result in distinct fatty acid profiles, which can significantly influence the nutritional quality of the fillet.

One of the main nutritional drawbacks of insect meals and oils is their naturally low or absent content of n-3 HUFA, particularly EPA and DHA, well recognized for their health benefits. Therefore, aquafeed formulations, especially those for marine fish species, must be carefully balanced to meet n-3 HUFA requirements, sustain overall fish health, and ensure adequate deposition of EPA and DHA in edible tissues. Since fishmeal and oil are the main sources of these fatty acids in aquafeeds, their replacement with insect-derived ingredients must be carefully managed to ensure that the EPA and DHA content in fillets meets the recommendations set by health organizations for human health. Nevertheless, the strategic inclusion of n-3 HUFA-rich ingredients, such as oils from fish, algae, or genetically modified plants, can support higher inclusion levels of insect-derived products in aquafeeds without compromising fillet quality for human consumption.

Although the muscle fatty acid profile generally reflects that of the diet, species-specific responses have been observed. A meta-analysis conducted by Tran et al. (2025) suggested that HI meal inclusion generally increases SFA and n-6 PUFA while reducing EPA and DHA in the muscle of several fish species. In fact, in rainbow trout, inclusion of HI meal at 5.5-40% (14-100% fishmeal replacement) consistently increased muscle SFA and n-6 PUFA, such as linoleic acid (C18:2n-6), while reducing n-3 PUFA, particularly EPA and DHA, increasing the n-6/n-3 PUFA ratio, a key indicator of fillet fat quality for consumers (Fabrikov et al., 2021a; Mancini et al., 2018; Renna et al., 2017). With the ideal dietary n-6/n-3 PUFA ratio for human health ranging from 1:1 to 5:1, an increase in this fat quality biomarker is concerning, as higher ratios are often associated with activation of inflammatory pathways and a greater risk of metabolic disorders, cardiovascular disease, and other chronic diseases (Bishehkolaei and Pathak, 2024; Gonzalez-Becerra et al., 2023). Similarly, the full replacement of fish oil with HI oil at 16% of the diet (containing 24% fishmeal) decreased muscle EPA and DHA contents in this species (Fawole et al., 2021).

Similar trends have been reported in other species. In gilthead seabream, diets including 6-28% HI meal (replacing up to 75% of fishmeal) increased SFA, n-6 PUFA, and the n-6/n-3 PUFA ratio, while decreasing EPA and DHA levels (Oteri et al., 2022; Pulido et al., 2022). Likewise, when 2-4% HI oil was used to replace 30-60% fish oil (at 6.7% of the diet) in totoaba diets, dietary EPA and n-3 PUFA levels were reduced, resulting in lower muscle concentrations of these fatty acids (Maldonado-Othón et al., 2022). In contrast, in European seabass, moderate inclusion levels of HI larvae or pre-pupae meals (6.5-19.5% inclusion; 15–45% fishmeal replacement) resulted in minimal changes in muscle fatty acid composition, with the n-6/n-3 PUFA ratio remaining stable (Mastoraki et al., 2020; Moutinho et al., 2021). For barramundi, muscle levels of EPA, DHA, and the n-6/n-3 PUFA ratio were maintained when 1.05% HI oil replaced 30% of the fish oil (at 3.8% of the diet) (Hender et al., 2021). This lack of effect may be attributed not only to the low inclusion level of HI oil but also to the high fishmeal content (50%) of the diet, further highlighting the importance of the contribution of all dietary components for the fatty acid profile of the final feed (Hender et al., 2021).

In diets where insect oils replace vegetable oils rather than fish oils, the impacts on the fillet fatty acid profile also vary depending on species and dietary context. In gilthead seabream, substitution of a vegetable oil blend with 9.5% HI oil, while maintaining fish oil levels, increased the n-6/n-3 PUFA ratio, although total EPA and DHA levels remained stable across treatments (Moutinho et al., 2024b). In Jian carp, the gradual replacement of 2.5% soybean oil with HI oil (up to 100% substitution) improved the n-6/n-3 PUFA ratio and enhanced DHA retention, while EPA levels remained unchanged (Li et al., 2016). In mirror carp, the same inclusion and replacement levels but with HI oil enriched in n-3 HUFA (from larvae reared on media containing *Schizochytrium* residue) increased muscle EPA and DHA content and improved the n-6/n-3 PUFA ratio, without significantly affecting SFA levels (Xu et al., 2021).

Therefore, when evaluating the impact of insect meals and oils (or any novel ingredient) on fillet nutritional quality, it is essential to adopt an integrative approach that considers all dietary components, not just those being included or replaced. For example, fishmeal, although primarily a protein source, also contains a lipid fraction that supplies important nutrients, including EPA and DHA. Thus, when replacing fish or vegetable oils with insect-derived oils, the contribution of EPA and DHA from other ingredients, such as fishmeal, must be considered. The same reasoning applies when replacing fishmeal with insect meals: does the diet still include other sources of EPA and DHA? Was their inclusion sufficient to ensure that dietary EPA and DHA levels remained adequate, regardless of the inclusion level of the ingredient under study? These questions must be carefully considered when analyzing the nutritional implications of using novel ingredients, to ensure that not only fish growth and health, but also the nutritional quality of the fillet for human consumption are preserved.

Finally, it is essential to emphasize that evaluating the quality of fish products should encompass not only their nutritional value but also their organoleptic and sensory attributes, as these factors significantly influence consumer perception and acceptance. Sensory characteristics such as visual appearance, color, texture, odor, and flavor strongly influence purchasing behavior (Carlucci et al., 2015). Although studies on the sensory evaluation of insect-fed fish are still scarce, evidence from species such as Atlantic salmon, rainbow trout, or gilthead seabream indicates that partial replacement of fishmeal or fish oil with properly processed HI or TM meals does not compromise fillet sensory attributes and supports consumer acceptance comparable to conventional diets (Pulido et al., 2022; Shafique et al., 2021; Shaviklo, 2023). However, higher inclusion levels, processing methods (e.g., full-fat *vs.* defatted meals, hydrolysates), and species-specific responses may cause subtle changes in aroma or texture. Therefore, formulation optimization and validation through trained sensory panels are advised to ensure market acceptability. Given that these characteristics are influenced by dietary composition, further research is warranted when developing aquafeeds that incorporate novel ingredients, such as insect-based products.

8 Functional properties of insect-based ingredients

Insect-based ingredients such as insect meals, oils, and hydrolysates may enhance fish health through their antioxidant and immunostimulatory properties, largely due to bioactive compounds like chitin, antimicrobial peptides, and short- and medium-chain fatty acids (Mousavi et al., 2020). These compounds can influence two key physiological systems: the antioxidant defense, which protects against oxidative stress caused by an excess of reactive oxygen species (Hoseinifar et al., 2020), and the immune system, particularly the innate response, which defends against pathogens through mechanisms such as lysozyme activity and complement proteins (Rauta et al., 2012).

The immunostimulatory and antioxidant potential of insect-based ingredients has been widely studied, particularly for HI and TM meals used as alternative protein sources. These studies often involve high inclusion levels (>10%), yet excessive amounts, especially when replacing fishmeal, may negatively affect fish physiology. For instance, dietary inclusion of 25% full-fat HI meal (replacing 40% of fishmeal) reduced antioxidant capacity and increased intestinal inflammatory responses in largemouth bass (*Micropterus salmoides*) (Dong et al., 2024). In Siberian sturgeon (*Acipenser baerii*), 18.5 and 37.5% defatted HI meal (replacing up to 50% of fishmeal) led to increased oxidative stress (Caimi et al., 2020). In barramundi, up to 30% inclusion of common housefly meal did not impair growth, yet levels above 10% triggered oxidative stress and/or weakened innate immune responses (Lin and Mui, 2017).

By contrast, high inclusion levels may show beneficial effects on immune and antioxidant responses, but compromise fish performance. In olive flounder (*Paralichthys olivaceus*), for example, 39 and 52% inclusion of defatted TM meal (replacing up to 80% of fishmeal) improved immune and antioxidant responses, yet levels above 26% negatively affected growth, feed efficiency, and fillet nutritional value (Jeong et al., 2021b). Similarly, in Jian carp, defatted HI meal at 7.9 and 10.6% inclusion (up to 100% fishmeal replacement) boosted antioxidant status without impairing growth, but levels up to 5.3% were recommended to avoid stress and intestinal damage (Li et al., 2017).

Beyond inclusion levels, the functional effects of insect meals are also influenced by the insect species. For example, an increase in antioxidant enzyme activities and a reduction in lipid peroxidation were observed in rainbow trout fed a diet containing 18% full-fat TM meal, whereas no such effect was detected when an 18% full-fat HI meal was used (Melenchón et al., 2022). A previous study comparing the same full-fat insect meals in rainbow trout found that the inclusion of HI meal at 5.6% led to increased immune and antioxidant responses and decreased lipid peroxidation (Melenchón et al., 2021). Nonetheless, in the same study, the TM meal still elicited stronger functional effects than the HI meal, highlighting that both insect species and inclusion levels are important factors. However, the composition of the insect meal, particularly in HI, should not be overlooked, as it can be manipulated through rearing substrates.

Further differences in functional responses have been observed among mealworm species, with effects also varying depending on the fish species. In Atlantic salmon, diets with 10.7 or 21.7% defatted TM meal led to enhanced immune responses, whereas 12.6% full-fat lesser mealworm meal had no such effect (Habte-Tsion et al., 2024). In Nile tilapia, the inclusion of 15 or 30% of full-fat superworm meal (replacing soybean meal) improved innate immunity (Alves et al., 2021), while in gilthead seabream, immunomodulatory effects were only observed with inclusion levels up to 10% (replacing part of the 61% fishmeal from the control diet) (Henry et al., 2022). Similarly, 13.4% silkworm meal replacing dietary fishmeal enhanced immune responses in mirror carp (Wan et al., 2017), though lower levels (5.7–9.0%) had no effect or even reduced antioxidant capacity in Jian carp (Ji et al., 2015).

Few studies have analyzed the functional effects of cricket meals in fish diets. In catfish species (*C. gariepinus* and *Ictalurus punctatus*), 35–40% black field cricket meal fully replacing fishmeal enhanced health status and hepatic antioxidant capacity (Fan et al., 2023; Taufek et al., 2016), reduced intestinal inflammation (Fan et al., 2023), and increased innate immunity, leading to increased resistance against a bacterial challenge with *Aeromonas hydrophila* (Taufek et al., 2018). In olive flounder, the same cricket meal did not trigger an immune response, but levels up to 39% (replacing 60% of fishmeal) improved antioxidant enzyme activities, while 52% (80% fishmeal replacement) led to declines in both antioxidant response and growth performance (Jeong et al., 2021a). In snakehead (*Channa argus*), antioxidant capacity improved with up to 20% inclusion of cricket (*Gryllus testaceus*) meal (replacing up to 45% of fishmeal), but growth performance declined at levels above 20% (Liu et al., 2023).

The developmental stage of insects may also influence the functional properties of insect meals, as different stages vary in their proximate compositions and concentrations of bioactive compounds, such as chitin and fatty acids (Giannetto et al., 2020; Smets et al., 2020). For HI, studies have evaluated functional properties of insect meals derived from larvae (e.g., Chaklader et al., 2021a; Couto et al., 2022; Melenchón et al., 2021; Singha et al., 2025), prepupae (Cardinaletti et al., 2019; Moutinho et al., 2021, 2024a; Randazzo et al., 2023), and pupae (Randazzo et al., 2021) stages. However, most studies do not directly compare developmental stages, making it difficult to isolate their specific effects from other factors like processing method or diet composition. One study available suggests that HI larvae may perform better than prepupae in largemouth bass (Fischer et al., 2022). Thus, the role of insect developmental stage in shaping functional responses remains unclear and warrants further investigation.

Processing methods such as defatting and hydrolysis further influence insect meal functional effects. In a study with barramundi, fishmeal (72%) was completely replaced with a blend of poultry by-product and HI meal (70:30), either full-fat or defatted (Chaklader et al., 2021a). Fish fed the defatted meal showed reduced protection against oxidative stress and immune challenges, resulting in increased susceptibility to a bacterial infection caused by *Vibrio harveyi*. These results suggest that the full-fat HI meal provided superior protection, likely due to the presence of bioactive components such as lauric acid and antimicrobial peptides, which may be diminished during defatting. In gilthead seabream, full-fat superworm meal elicited stronger immunomodulatory effects than the defatted meal, leading the authors to suggest that the fat fraction may contain lipid-bound bioactive compounds lost during defatting (Henry et al., 2022). In contrast, Singha et al. (2025) reported that both full-fat and defatted HI meals enhanced gut health in rainbow trout compared to a soybean-based diet. However, only the full-fat meal treatment matched the performance observed in a fishmeal-based control. Therefore, the impact of defatting on the functional efficacy of insect meals warrants closer attention, especially considering the associated processing costs.

Enzymatic hydrolysates of insect meals are also being explored for their functional potential, though they are typically included at low levels (up to 10%) due to their high production costs. At this point, information is still scarce. In sea trout (*Salmo trutta m. trutta*), 10% inclusion of hydrolyzed TM or superworm meals (maintaining 15% dietary fishmeal) had no beneficial effects on growth or immune parameters (Mikołajczak et al., 2020). In mirror carp, 5% hydrolyzed silkworm meal (replacing 50% of fishmeal) improved antioxidant status but did not affect immune parameters (Xu et al., 2018). Although promising, the functional effects of insect hydrolysates remain to be confirmed.

In comparison with insect meals, studies evaluating the functional effects of insect oils in aquafeeds are still limited, and most available research has focused on HI oil. The available results suggest that replacing fish or vegetable oils with HI oil may enhance immune and antioxidant status in several fish species (Hender et al., 2021; Kumar et al., 2021; Moutinho et al., 2025b; Xu et al., 2020, 2021), likely due to the high lauric acid content of HI oil and the presence of antioxidant compounds such as tocopherols and phenolic compounds. Supporting this, a comparative study in mirror carp evaluated the inclusion of HI, TM, and silkworm oils as the sole dietary lipid source (at 2.5%, either individually or as 1:1:1 mixture). Only the diets containing HI oil led to reduced lipid peroxidation and inflammatory responses, along with increasing antioxidant capacity (Xu et al., 2020).

As with insect meals, the functional effects of insect oil inclusion are dose-dependent. In mirror carp, dietary inclusion of HI oil at 0.625% did not affect antioxidant and anti-inflammatory responses, whereas beneficial responses were observed at 1.25–2.5%, replacing up to 100% of soybean oil (Xu et al., 2021). Conversely, in Jian carp, low inclusion levels of silkworm oil (0.75 and 1.5%, replacing soybean oil) enhanced antioxidant enzyme activities, whereas higher levels (2.25 and 3%) showed no such effect (Chen et al., 2017). In gilthead seabream, 4.0–9.5% of HI oil (replacing 100% of a vegetable oil blend) enhanced hepatic antioxidant capacity and reduced intestinal lipid peroxidation (Moutinho et al., 2025b). Nevertheless, at the highest inclusion level, upregulation of both pro- and anti-inflammatory genes in muscle suggested a potential inflammatory response. Similarly, in rainbow trout, using HI oil as the sole lipid source (16%) led to an upregulation of pro-inflammatory cytokines, although the authors did not interpret this effect as detrimental (Kumar et al., 2021).

Interactions between insect protein and oil sources are also important. In barramundi, serum bactericidal activity increased when a partially defatted HI meal replaced 30% of the dietary fishmeal (maintaining nearly 20% of fishmeal). However, this immune benefit was not observed when the same diet also included the 30% replacement of fish oil with HI oil (Hender et al., 2021). Altogether, these findings underscore the multifactorial nature of functional responses in fish and highlight the increasing complexity of contemporary aquafeed formulations.

Functional properties may not always be evident under optimal conditions but may become more apparent when fish are exposed to pathogenic, environmental, or husbandry-related challenges. For instance, in barramundi fed poultry by-product meal diets supplemented with HI meal for six weeks, no significant effects were observed in skin barrier function or immune responses. However, following a challenge with *V. harveyi*, fish fed the HI-supplemented diets exhibited significantly higher survival, improved skin integrity, and enhanced immune responses (Chaklader et al., 2021b). Conversely, in rainbow trout, immunostimulatory effects of diets containing HI meal and oil were observed under low (4.9 kg/m^3) stocking density (biomass of fish kept in a given volume of water) but diminished under high (9.8 kg/m^3) stocking density. Nonetheless, even under high-density conditions, fish fed HI-based diets showed higher survival following a bacterial challenge with *Aeromonas salmonicida* (Cho et al., 2022). These findings underscore the need to test functional ingredients under challenging aquaculture conditions.

In addition to their immune and antioxidant effects, insect-based ingredients may also modulate the composition and function of the fish gut microbiota, a key player in nutrient digestion, immune modulation, and overall health. Reported microbiota responses to insect meals and oils are highly variable and species-specific, reflecting differences in insect species and developmental stages, inclusion levels, processing methods, and fish physiological traits (Antonopoulou et al., 2019; Drosdowech et al., 2024; Habte-Tsion et al., 2024; Huyben et al., 2019; Józefiak et al., 2023; Melenchón et al., 2023; Panteli et al., 2021; Rangel et al., 2022; Rawski et al., 2025; Weththasinghe et al., 2022b). The effects of insect meals on gut microbial communities have frequently been linked with chitin, which is indigestible for most fish species, functioning as an insoluble fibre with potential prebiotic activity. Interestingly, several studies report an enrichment of bacteria with chitinolytic capacity in fish fed insect meals, indicating a microbial adaptation to chitin-containing diets (Kalemi et al., 2025; Rawski et al., 2025; Weththasinghe et al., 2022b). Such shifts may contribute to improved handling of chitin and potentially support gut health. Given the complexity and context-dependency of these responses, we encourage readers to explore emerging research in this area, as understanding the interactions between insect ingredients and the gut microbiota could unlock further functional benefits in aquaculture nutrition.

In summary, insect-derived ingredients show promising functional benefits in aquaculture by enhancing immune responses and antioxidant capacity in fish. However, these effects depend on factors such as insect species, developmental stage, rearing substrate, and processing methods, all of which influence the composition of bioactive compounds. While most studies have focused on insect meals as alternative protein sources at high inclusion levels (>10%), lower levels (1–5%) targeting functional health benefits should be further explored. Given the high costs of insect-based ingredients, their targeted use as functional ingredients may offer a more feasible strategy. Future research should focus on optimizing inclusion strategies, understanding species-specific responses, and expanding the evaluation of immune and oxidative markers to fully capitalize on the potential of insect-derived functional ingredients in sustainable aquafeeds.

9 Future perspectives

Despite the growing interest and proven potential of insects as sustainable ingredients in aquafeeds, several challenges remain. Understanding these obstacles is essential to guiding future research, innovation, and policy development. Figure 7.2 provides a SWOT analysis of the use of insect-based ingredients in aquafeeds.

The insect production sector remains immature and requires technological innovation to improve efficiency and competitiveness. High infrastructure, labor, energy, and processing costs make insect meals more expensive than conventional protein sources, such as fishmeal and soybean meal. Scaling up is technically demanding and capital-intensive, with many producers still relying on small-scale or semi-automated systems (Biteau et al., 2024). Moreover, standardization of production and processing methods is crucial to ensure consistent quality, safety, and efficacy of insect meals for commercial use. Additionally, there is a growing recognition that insect farming practices must also account for insect welfare, with early-stage proposals for operational welfare indicators extending beyond survival and growth metrics (Johnson and Barrett, 2025).

In the EU, regulatory constraints, particularly those limiting the use of substrates, remain a major barrier to expanding insect farming. While rigorous monitoring and

Strengths	Weakness
– Require less land and water compared to conventional feed ingredients – Enable upcycling of organic waste, contributing to the circular economy – High species diversity allows for varied nutritional and functional profiles – Meals and oils generally exhibit good digestibility and palatability – High protein (25-75%) and essential amino acid contents – Favorable fatty acid profiles, especially when reared on tailored substrates	– Production sector remains immature and requires technological innovation – High production costs and limited scalability compared to fishmeal and soybean meal – Regulatory constraints in some regions, particularly regarding substrate sources – Nutritional variability due to differences in rearing substrate and processing methods – Limited data on nutritional composition across developmental stages – Chitin content may reduce digestibility in some fish species – Potential contamination risks (e.g., heavy metals, pathogens) depending on substrates
Opportunities	**Threats**
– Growing demand for alternative protein sources – Increasing regulatory support for the inclusion of insect-derived ingredients in aquafeeds – Potential for industrial symbiosis with agri-food sectors (e.g., brewery, vegetable waste valorization) – Rising interest in sustainability-labelled seafood products – Possibility to optimize insect strains through genetic and nutritional strategies	– Competition with well-established feed ingredients (e.g., soy protein concentrates, single-cell proteins) – Unknown long-term ecological effects of large-scale insect farming – Vulnerability to monoculture risks (e.g., disease outbreaks in insect colonies) – Limited comparative data with other alternative proteins (need for holistic life-cycle analysis) – Market consolidation may hinder innovation or access for smaller producers – Regulatory constraints may limit technological advances – Consumer skepticism and limited awareness of insect-based aquaculture feeds

FIGURE 7.2 Strategic overview of the strengths, weaknesses, opportunities, and threats (SWOT) associated with the use of insect-derived ingredients in aquafeeds.

risk assessment are crucial for ensuring food safety, expanding the list of authorized substrates is considered essential to lower operational costs and enable circular economy practices (Meijer et al., 2025; Niyonsaba et al., 2023). Insects offer enormous potential for upcycling agro-industrial by-products and converting low-value side streams into high-quality protein and lipid-rich biomass (Yadav et al., 2025). This approach not only enhances sustainability but can also improve the nutritional composition of insect meals and oils. However, the use of diverse substrates and by-products also introduces the risk of transferring allergens, pathogens, or contaminants (e.g., heavy metals), underscoring the need for careful safety monitoring (Tran et al., 2025).

Additionally, as scientific knowledge advances, the list of insect species authorized for use in aquafeeds should be regularly revised to reflect new evidence on their safety and nutritional value. Beyond regulatory revisions, selective breeding, and genetic improvement present promising avenues to enhance farming efficiency and product quality. By targeting key traits such as growth rate, nutrient composition, or disease resistance, insect strains can be optimized for aquafeed applications. Advances in genomics and gene editing may further accelerate this progress, though ethical and regulatory considerations must be carefully addressed.

Nutritional considerations are also key to determining the suitability of insect-based ingredients in aquafeeds. Insects are a valuable source of high-quality protein, lipids, and antioxidant compounds. However, standardization of production methods and comprehensive characterization of developmental stages are essential to reduce nutritional variability. The use of a wider range of rearing substrates presents an opportunity to tailor the nutritional profile of insects, particularly their lipid fraction and bioactive content (Lopes et al., 2025; Rodríguez-González et al., 2025). Importantly, insect-derived ingredients should increasingly be recognized as functional feed additives, supplying bioactive compounds that can enhance antioxidant defenses, immune responses, gut health, and stress resilience in fish. Among these compounds, chitin has received attention for its potential health benefits. However, some studies report adverse effects on growth performance and nutrient digestibility in certain fish species (Coutinho et al., 2021; Piccolo et al., 2017). As species-specific tolerance thresholds vary, permissible chitin levels should be determined on a case-by-case basis. Processing strategies that yield insect meals with varying chitin contents could broaden their applicability across fish species. Furthermore, while current evidence does not clearly indicate that defatting diminishes the functional benefits of insect meals, this aspect warrants further investigation, particularly given the costs associated with this processing step. In parallel, by-products from insect production, such as frass or exuviae, are gaining attention for their functional value in aquafeeds, with preliminary evidence suggesting immunomodulatory and gut health benefits (Rimoldi et al., 2023; Yildirim-Aksoy et al., 2020, 2023). Future research should aim to identify and optimize these functional properties.

In parallel with these efforts, broader environmental implications also warrant consideration. Although insects are widely recognized for their efficient resource use, requiring significantly less land and water than conventional feed crops, these environmental benefits should be weighed alongside other impact categories using a standardized full life-cycle analysis for better comparison across feed ingredients (Glencross et al., 2024). Moreover, the potential long-term ecological impacts of large-scale insect farming, including monoculture vulnerability and risks of biodiversity loss, should also be considered. Robust, holistic life-cycle analyses are needed to better benchmark their true environmental and nutritional value (Thévenot et al., 2018; Tran et al., 2022).

Ultimately, the wider adoption of insect-based aquafeeds will depend not only on regulatory flexibility, production scalability, and nutritional optimization but also on public perception and market acceptance. Consumer attitudes toward the use of insects in the diets of fish intended for human consumption are still evolving (Maulu et al., 2022). Improving communication and raising awareness about the sustainability and safety of insect-based feeds are crucial to consumer acceptance. Lifting barriers for the insect farming chain is and will continue to be needed, while maintaining a strong focus on safety and promoting circularity and sustainability goals.

Acknowledgements

This work was supported by Portuguese national funds from FCT - Foundation for Science and Technology through contracts UID/04326/2025, UID/PRR/04326/2025 and LA/P/0101/2020 (DOI:10.54499/LA/P/0101/2020) to CCMAR.

Note on the usage of AI tools

The authors of this chapter state that they used large language models (LLMs) strictly for English language correction and stylistic polishing of chapters, and not for content generation.

References

Aaqillah-Amr, M.A., Hidir, A., Ikhwanuddin, M., Fatan, N.A., Muhammad-Rahimi, R., Ning-Shahira, S., Siankhe, R.G.N., Vandenberg, G.W., Tanga, C.M. and Yossa, R., 2024. Effects of insect meals on fish digestibility, blood parameters, and economic performance: a meta-analysis. Journal of Insects as Food and Feed 11: 833–859. https://doi.org/10.1163/23524588-00001096

Abdel-Tawwab, M., Khalil, R.H., Metwally, A.A., Shakweer, M.S., Khallaf, M.A. and Abdel-Latif, H.M.R., 2020. Effects of black soldier fly (*Hermetia illucens* L.) larvae meal on growth performance, organs-somatic indices, body composition, and hemato-biochemical variables of European sea bass, *Dicentrarchus labrax*. Aquaculture 522: 735136. https://doi.org/10.1016/J.AQUACULTURE.2020.735136

Abu Bakar, N.-H., Abdul Razak, S., Mohd Taufek, N. and Alias, Z., 2021. Evaluation of black soldier fly (*Hermetia illucens*) prepupae oil as meal supplementation in diets for red hybrid tilapia (*Oreochromis* sp.). International Journal of Tropical Insect Science 41: 2093–2102. https://doi.org/10.1007/s42690-020-00398-z

Alves, A.P.D., Paulino, R.R., Pereira, R.T., da Costa, D. V and Rosa, P.V.E., 2021. Nile tilapia fed insect meal: growth and innate immune response in different times under lipopolysaccharide challenge. Aquaculture Research 52: 529–540. https://doi.org/10.1111/are.14911

Anany, E.M., Ibrahim, M.A., El-Razek, I.M.A., El-Nabawy, E.S.M., Amer, A.A., Zaineldin, A.I., Gewaily, M.S. and Dawood, M.A.O., 2025. Combined effects of yellow mealworm (*Tenebrio molitor*) and *Saccharomyces cerevisiae* on the growth performance, feed utilization intestinal health, and blood biomarkers of Nile tilapia (*Oreochromis niloticus*) fed fish meal-free diets. Probiotics and Antimicrobial Proteins 17: 1387–1398. https://doi.org/10.1007/S12602-023-10199-8

Antonopoulou, E., Nikouli, E., Piccolo, G., Gasco, L., Gai, F., Chatzifotis, S., Mente, E. and Kormas, K.A., 2019. Reshaping gut bacterial communities after dietary *Tenebrio molitor* larvae meal supplementation in three fish species. Aquaculture 503: 628–635. https://doi.org/10.1016/j.aquaculture.2018.12.013

Ardra, M., Pradhan, C., Das, S. and Vineetha, V.P., 2025. Effect of dietary incorporation of defatted yellow mealworm larvae as an alternative protein source on growth, digestive, antioxidant, metabolic enzyme activities, hepatic histomorphology and immunity in pangasius (*Pangasianodon hypophthalmus*). Animal Feed Science and Technology 319: 116177. https://doi.org/10.1016/J.ANIFEEDSCI.2024.116177

Basto, A., Matos, E. and Valente, L.M.P., 2020. Nutritional value of different insect larvae meals as protein sources for European sea bass (*Dicentrarchus labrax*) juveniles. Aquaculture 521: 735085. https://doi.org/10.1016/j.aquaculture.2020.735085

Basto, A., Calduch-Giner, J., Oliveira, B., Petit, L., Sá, T., Maia, M.R.G., Fonseca, S.C., Matos, E., Pérez-Sánchez, J. and Valente, L.M.P., 2021. The use of defatted *Tenebrio molitor* larvae meal as a main protein source is supported in European sea bass (*Dicentrarchus labrax*) by data on growth performance, lipid metabolism, and flesh quality. Frontiers in Physiology 12: 659567. https://doi.org/10.3389/fphys.2021.659567

Basto, A., Valente, L.M.P., Sousa, V., Conde-Sieira, M. and Soengas, J.L., 2023. Total fishmeal replacement by defatted *Tenebrio molitor* larvae meal induces alterations in intermediary metabolism of European sea bass (*Dicentrarchus labrax*). Journal of Animal Science 101: 1–11. https://doi.org/10.1093/JAS/SKAD040

Belghit, I., Liland, N.S., Waagbø, R., Biancarosa, I., Pelusio, N., Li, Y., Krogdahl, Å. and Lock, E.-J., 2018. Potential of insect-based diets for Atlantic salmon (*Salmo salar*). Aquaculture 491: 72–81. https://doi.org/10.1016/j.aquaculture.2018.03.016

Belghit, I., Liland, N.S., Gjesdal, P., Biancarosa, I., Menchetti, E., Li, Y., Waagbø, R., Krogdahl, Å. and Lock, E.-J., 2019a. Black soldier fly larvae meal can replace fish meal in diets of sea-water phase Atlantic salmon (*Salmo salar*). Aquaculture 503: 609–619. https://doi.org/10.1016/j.aquaculture.2018.12.032

Belghit, I., Waagbø, R., Lock, E.J. and Liland, N.S., 2019b. Insect-based diets high in lauric acid reduce liver lipids in freshwater Atlantic salmon. Aquaculture Nutrition 25: 343–357. https://doi.org/10.1111/ANU.12860

Biasato, I., Chemello, G., Oddon, S.B., Ferrocino, I., Corvaglia, M.R., Caimi, C., Resconi, A.,

Paul, A., van Spankeren, M., Capucchio, M.T., Colombino, E., Cocolin, L., Gai, F., Schiavone, A. and Gasco, L., 2022. *Hermetia illucens* meal inclusion in low-fishmeal diets for rainbow trout (*Oncorhynchus mykiss*): effects on the growth performance, nutrient digestibility coefficients, selected gut health traits, and health status indices. Animal Feed Science and Technology 290: 115341. https://doi.org/10.1016/J.ANIFEEDSCI.2022.115341

Bishehkolaei, M. and Pathak, Y., 2024. Influence of omega n-6/n-3 ratio on cardiovascular disease and nutritional interventions. Human Nutrition & Metabolism 37: 200275. https://doi.org/10.1016/J.HNM.2024.200275

Biteau, C., Bry-Chevalier, T., Crummett, D., Ryba, R. and Jules, M.S., 2024. Insect-based livestock feeds are unlikely to become economically viable in the near future. Food and Humanity 3: 100383. https://doi.org/10.1016/J.FOOHUM.2024.100383

Bordignon, F., Gasco, L., Birolo, M., Trocino, A., Caimi, C., Ballarin, C., Bortoletti, M., Nicoletto, C., Maucieri, C. and Xiccato, G., 2022. Performance and fillet traits of rainbow trout (*Oncorhynchus mykiss*) fed different levels of *Hermetia illucens* meal in a low-tech aquaponic system. Aquaculture 546: 737279. https://doi.org/10.1016/J.AQUACULTURE.2021.737279

Caimi, C., Gasco, L., Biasato, I., Malfatto, V., Varello, K., Prearo, M., Pastorino, P., Bona, M.C., Francese, D.R., Schiavone, A., Elia, A.C., Dörr, A.J.M. and Gai, F., 2020. Could dietary black soldier fly meal inclusion affect the liver and intestinal histological traits and the oxidative stress biomarkers of Siberian sturgeon (*Acipenser baerii*) juveniles? Animals 10: 155. https://doi.org/10.3390/ani10010155

Cardinaletti, G., Randazzo, B., Messina, M., Zarantoniello, M., Giorgini, E., Zimbelli, A., Bruni, L., Parisi, G., Olivotto, I. and Tulli, F., 2019. Effects of graded dietary inclusion level of full-fat *Hermetia illucens* prepupae meal in practical diets for rainbow trout (*Oncorhynchus mykiss*). Animals 9: 251. https://doi.org/10.3390/ani9050251

Carlucci, D., Nocella, G., De Devitiis, B., Viscecchia, R., Bimbo, F. and Nardone, G., 2015. Consumer purchasing behaviour towards fish and seafood products. Patterns and insights from a sample of international studies. Appetite 84: 212–227. https://doi.org/10.1016/J.APPET.2014.10.008

Chaklader, M.R., Howieson, J. and Fotedar, R., 2021a. Growth, hepatic health, mucosal barrier status and immunity of juvenile barramundi, *Lates calcarifer* fed poultry by-product meal supplemented with full-fat or defatted *Hermetia illucens* larval meal. Aquaculture 543: 737026. https://doi.org/10.1016/j.aquaculture.2021.737026

Chaklader, M.R., Howieson, J., Fotedar, R. and Siddik, M.A.B., 2021b. Supplementation of *Hermetia illucens* larvae in poultry by-product meal-based barramundi, *Lates calcarifer* diets improves adipocyte cell size, skin barrier functions, and immune responses. Frontiers in Nutrition 7: 320. https://doi.org/10.3389/fnut.2020.613158

Chemello, G., Renna, M., Caimi, C., Guerreiro, I., Oliva-Teles, A., Enes, P., Biasato, I., Schiavone, A., Gai, F. and Gasco, L., 2020. Partially defatted *Tenebrio molitor* larva meal in diets for grow-out rainbow trout, *Oncorhynchus mykiss* (Walbaum): effects on growth performance, diet digestibility and metabolic responses. Animals 10: 229. https://doi.org/10.3390/ani10020229

Chen, H., Tian, J., Wang, Y., Yang, K., Ji, H. and Li, J., 2017. Effects of dietary soybean oil replacement by silkworm, *Bombyx mori* L., chrysalis oil on growth performance, tissue fatty acid composition, and health status of juvenile Jian carp, *Cyprinus carpio* var. Jian. Journal of the World Aquaculture Society 48: 453–466. https://doi.org/10.1111/JWAS.12373

Cho, J.H., Bae, J. and Hwang, I.J., 2022. Effects of black soldier fly (*Hermetia illucens*) pre-pupae meal on the growth, stress, and immune responses of juvenile rainbow trout (*Oncorhynchus mykiss*) reared at different stocking densities. Aquaculture Reports 25: 101202. https://doi.org/10.1016/J.AQREP.2022.101202

Coutinho, F., Castro, C., Guerreiro, I., Rangel, F., Couto, A., Serra, C.R., Peres, H., Pousão-Ferreira, P., Rawski, M., Oliva-Teles, A. and Enes, P., 2021. Mealworm larvae meal in diets for meagre juveniles: growth, nutrient digestibility and digestive enzymes activity. Aquaculture 535: 736362. https://doi.org/10.1016/J.AQUACULTURE.2021.736362

Couto, A., Serra, C.R., Guerreiro, I., Coutinho, F., Castro, C., Rangel, F., Lavrador, A.S., Monteiro, M., Santos, R., Peres, H., Pousão-Ferreira, P., Gasco, L., Gai, F., Oliva-Teles, A. and Enes, P., 2022. Black soldier fly meal effects on meagre health condition: gut morphology, gut microbiota and humoral immune response. Journal of Insects as Food and Feed 8: 1281–1295. https://doi.org/10.3920/JIFF2021.0082

Cruz, V.A., Vicentini-Polette, C.M., Magalhaes, D.R. and de Oliveira, A.L., 2025. Extraction, characterization, and use of edible insect oil - A review. Food Chemistry 463: 141199. https://doi.org/10.1016/j.foodchem.2024.141199

Data Bridge Market Research, 2024. North America insect protein market report size, share, and trends analysis report – industry overview and forecast to 2032. Available at: https://www.databridgemarketresearch.com/reports/north-america-insect-protein-market?srsltid=AfmBOoqVhf8Bicp1-rKG70DHubV2AEl39tTIW0-gONh_035t3iIPMEP9.

de Jong, B. and Nikolik, G., 2021. No longer crawling: insect protein to come of age in the 2020s. Available at: https://insectfeed.nl/wp-content/uploads/2021/03/Rabobank_No-Longer-Crawling-Insect-Protein-to-Come-of-Age-in-the-2020s_Feb2021-1.pdf.

Dernekbaşı, S. and Karadayı, B.T., 2025. Effects of mealworm (*Tenebrio molitor*) larvae meal on growth, digestive and antioxidant parameters in Nile tilapia juvenile (*Oreochromis niloticus*). Fish Physiology and Biochemistry 51: 198. https://doi.org/10.1007/s10695-025-01610-y

DiGiacomo, K., Akit, H. and Leury, B.J., 2019. Insects: a novel animal-feed protein source for the Australian market. Animal Production Science 59: 2037. https://doi.org/10.1071/AN19301

Dong, W., Ran, X., He, G., Hu, W., Chen, Y., He, Y. and Lin, S., 2024. The effect of dietary full-fat *Hermetia illucens* larvae meal on growth performance and intestine physiology in largemouth bass (*Micropterus salmoides*). Animal Feed Science and Technology 317: 116089. https://doi.org/10.1016/J.ANIFEEDSCI.2024.116089

Drosdowech, S., Bezner, S., Daisley, B., Chiasson, M., Easton, A., Rooney, N. and Huyben, D., 2024. Influence of feeding black soldier fly (*Hermetia illucens*), cricket (*Gryllodes sigillatus*), and superworm (*Zophobas morio*) on the gut microbiota of rainbow trout (*Oncorhynchus mykiss*). Journal of Applied Microbiology 135: 295. https://doi.org/10.1093/JAMBIO/LXAE295

Dumas, A., Raggi, T., Barkhouse, J., Lewis, E. and Weltzien, E., 2018. The oil fraction and partially defatted meal of black soldier fly larvae (*Hermetia illucens*) affect differently growth performance, feed efficiency, nutrient deposition, blood glucose and lipid digestibility of rainbow trout (*Oncorhynchus mykiss*). Aquaculture 492: 24–34. https://doi.org/10.1016/j.aquaculture.2018.03.038

Edah, B. and Owolabi, O.D., 2023. Physical properties of defatted and extruded black soldier fly (*Hermetia illucens*) larvae-based aqua-feed using a twin-screw extruder. Discover Food 3: 1–10. https://doi.org/10.1007/S44187-023-00056-6

Egessa, R., Szűcs, A., Ardó, L., Biró, J., Lengyel-Kónya, É., Banjac, V., Stojkov, V., Lefler, K.K. and Sándor, Z.J., 2024. Dietary inclusion of insect oil: impact on growth, nutrient utilisation, lipid metabolism, antioxidant and immune-related responses in European catfish (*Silurus glanis* L). Aquaculture 592: 741213. https://doi.org/10.1016/J.AQUACULTURE.2024.741213

Eggink, K.M., Pedersen, P.B., Lund, I. and Dalsgaard, J., 2022. Chitin digestibility and intestinal exochitinase activity in Nile tilapia and rainbow trout fed different black soldier fly larvae meal size fractions. Aquaculture Research 53: 5536–5546. https://doi.org/10.1111/ARE.16035

Eggink, K.M., Aalto, S.L., Pedersen, P.B., Lund, I. and Dalsgaard, J., 2025. Effects of dietary chitin on nutrient digestibility, cholesterol metabolism, digestive enzyme activity, and gut microbiome in rainbow trout. Animal Feed Science and Technology 328: 116447. https://doi.org/10.1016/J.ANIFEEDSCI.2025.116447

Fabrikov, D., Sánchez-Muros, M.J., Barroso, F.G., Tomás-Almenar, C., Melenchón, F., Hidalgo, M.C., Morales, A.E., Rodriguez-Rodriguez, M. and Montes-Lopez, J., 2020. Comparative study of growth performance and amino acid catabolism in *Oncorhynchus mykiss*, *Tinca tinca* and *Sparus aurata* and the catabolic changes in response to insect meal inclusion in the diet. Aquaculture 529: 735731. https://doi.org/10.1016/j.aquaculture.2020.735731

Fabrikov, D., Barroso, F.G., Sánchez-Muros, M.J., Hidalgo, M.C., Cardenete, G., Tomás-Almenar, C., Melenchón, F. and Guil-Guerrero, J.L., 2021a. Effect of feeding with insect meal diet on the fatty acid compositions of sea bream (*Sparus aurata*), tench (*Tinca tinca*) and rainbow trout (*Oncorhynchus mykiss*) fillets. Aquaculture 545: 737170. https://doi.org/10.1016/J.AQUACULTURE.2021.737170

Fabrikov, D., Vargas-García, M. del C., Barroso, F.G., Sánchez-Muros, M.J., Cacua Ortíz, S.M., Morales, A.E., Cardenete, G., Tomás-Almenar, C. and Melenchón, F., 2021b. Effect on intermediary metabolism and digestive parameters of the high substitution of fishmeal with insect meal in *Sparus aurata* feed. Insects 12: 965. https://doi.org/10.3390/insects12110965

Fan, K., Liu, H., Pei, Z., Brown, P.B. and Huang, Y., 2023. A study of the potential effect of dietary fishmeal replacement with cricket meal (*Gryllus bimaculatus*) on growth performance, blood health, liver antioxidant activities, intestinal microbiota and immune-related gene expression of juvenile channel catfish. Animal Feed Science and Technology 295: 115542. https://doi.org/10.1016/J.ANIFEEDSCI.2022.115542

Fantatto, R.R., Mota, J., Ligeiro, C., Vieira, I., Guilgur, L.G., Santos, M. and Murta, D., 2024. Exploring sustainable alternatives in aquaculture feeding: The role of insects. Aquaculture Reports 37: 102228. https://doi.org/10.1016/J.AQREP.2024.102228

FAO, 2024. The state of world fisheries and aquaculture 2024 - blue transformation in action. FAO, Rome, Italy, 232 pp. https://doi.org/10.4060/CD0683EN

Fawole, F.J., Adeoye, A.A., Tiamiyu, L.O., Ajala, K.I., Obadara, S.O. and Ganiyu, I.O., 2020. Substituting fishmeal with *Hermetia illucens* in the diets of African catfish (*Clarias gariepinus*): effects on growth, nutrient utilization, haemato-physiological response, and oxidative stress biomarker. Aquaculture 518: 734849. https://doi.org/10.1016/J.AQUACULTURE.2019.734849

Fawole, F.J., Labh, S.N., Hossain, M.S., Overturf, K., Small, B.C., Welker, T.L., Hardy, R.W. and Kumar, V., 2021. Insect (black soldier fly larvae) oil as a potential substitute for fish or soy oil in the fish meal-based diet of juvenile rainbow trout (*Oncorhynchus mykiss*). Animal Nutrition 7: 1360. https://doi.org/10.1016/J.ANINU.2021.07.008

Fischer, H., Romano, N., Renukdas, N., Kumar, V. and Sinha, A.K., 2022. Comparing black soldier fly (*Hermetia illucens*) larvae versus prepupae in the diets of largemouth bass, *Micropterus salmoides*: effects on their growth, biochemical composition, histopathology, and gene expression. Aquaculture 546: 737323. https://doi.org/10.1016/J.AQUACULTURE.2021.737323

Gasco, L., Henry, M., Piccolo, G., Marono, S., Gai, F., Renna, M., Lussiana, C., Antonopoulou, E., Mola, P. and Chatzifotis, S., 2016. *Tenebrio molitor* meal in diets for European sea bass (*Dicentrarchus labrax* L.) juveniles: growth performance, whole body composition and in vivo apparent digestibility. Animal Feed Science and Technology 220: 34–45. https://doi.org/10.1016/j.anifeedsci.2016.07.003

Gasco, L., Renna, M., Oddon, S.B., Far, A.R., El Deen, S.N. and Veldkamp, T., 2023. Insect meals in a circular economy and applications in monogastric diets. Animal Frontiers 13: 81–90. https://doi.org/10.1093/AF/VFAD016

Ge, C., Cheng, H., Li, J., Wang, H., Ma, S., Qin, Y. and Xue, M., 2022. Effects of defatted yellow mealworm (*Tenebrio molitor*) on the feed qualities and the growth performance of largemouth bass (*Micropterus salmoides*). Journal of Insects as Food and Feed 8: 1265–1280. https://doi.org/10.3920/JIFF2021.0125

Gebremichael, A., Sándor, Z.J. and Kucska, B., 2022. Does dietary inclusion of defatted yellow mealworm (*Tenebrio molitor*) affect growth and body composition of juvenile common carp (*Cyprinus carpio*)? South African Journal of Animal Science 52: 444–451. https://doi.org/10.4314/SAJAS.V52I4.04

Giannetto, A., Oliva, S., Ceccon Lanes, C.F., de Araújo Pedron, F., Savastano, D., Baviera, C., Parrino, V., Lo Paro, G., Spanò, N.C., Cappello, T., Maisano, M., Mauceri, A. and Fasulo, S., 2020. *Hermetia illucens* (Diptera: Stratiomydae) larvae and prepupae: biomass production, fatty acid profile and expression of key genes involved in lipid metabolism. Journal of Biotechnology 307: 44–54. https://doi.org/10.1016/J.JBIOTEC.2019.10.015

Glencross, B.D., Booth, M. and Allan, G.L., 2007. A feed is only as good as its ingredients – a review of ingredient evaluation strategies for aquaculture feeds. Aquaculture Nutrition 13: 17–34. https://doi.org/10.1111/j.1365-2095.2007.00450.x

Glencross, B., Ling, X., Gatlin, D., Kaushik, S., Øverland, M., Newton, R. and Valente, L.M.P., 2024. A SWOT analysis of the use of marine, grain, terrestrial-animal and novel protein ingredients in aquaculture feeds. Reviews in Fisheries Science & Aquaculture 32: 396–434. https://doi.org/10.1080/23308249.2024.2315049

Goda, A.M.A., El-Haroun, E., Nazmi, H., Van Doan, H., Aboseif, A.M., Taha, M.K.S. and Abou Shabana, Nevine.M., 2024. Black soldier fly oil-based diets enriched in lauric acid enhance growth, hematological indices, and fatty acid profiles of Nile tilapia, *Oreochromis niloticus* fry. Aquaculture Reports 37: 102269. https://doi.org/10.1016/j.aqrep.2024.102269

Gonzalez-Becerra, K., Barron-Cabrera, E., Muñoz-Valle, J.F., Torres-Castillo, N., Rivera-Valdes, J.J., Rodriguez-Echevarria, R. and Martinez-Lopez, E., 2023. A balanced dietary ratio of n-6:n-3 polyunsaturated fatty acids exerts an effect on total fatty acid profile in RBCs and inflammatory markers in subjects with obesity. Healthcare 11: 2333. https://doi.org/10.3390/healthcare11162333

Gou, N., Wang, K., Jin, T. and Yang, B., 2023. First insights on the administration of insect oil (black soldier fly larvae) in the diet of juvenile *Onychostoma macrolepis*. Animals 13: 518. https://doi.org/10.3390/ani13030518

Gravel, A. and Doyen, A., 2020. The use of edible insect proteins in food: challenges and issues related to their functional properties. Innovative Food Science & Emerging Technologies 59: 102272. https://doi.org/10.1016/j.ifset.2019.102272

Habte-Tsion, H.M., Hawkyard, M., Sealey, W.M., Bradshaw, D., Meesala, K.M. and Bouchard, D.A., 2024. Effects of fishmeal substitution with mealworm meals (*Tenebrio molitor* and *Alphitobius diaperinus*) on the growth, physiobiochemical response, digesta microbiome, and immune genes expression of Atlantic salmon (*Salmo salar*). Aquaculture Nutrition 2024: 6618117. https://doi.org/10.1155/2024/6618117

Han, R., Shin, J.T., Kim, J., Choi, Y.S. and Kim, Y.W., 2017. An overview of the South Korean edible insect food industry: challenges and future pricing/promotion strategies. Entomological Research 47: 141–151. https://doi.org/10.1111/1748-5967.12230

Hanboonsong, A. and Durst, P., 2020. Guidance on sustainable cricket farming – A practical manual for farmers and inspectors. FAO, Bangkok, Thailand, 84 pp. https://doi.org/10.4060/CB2446EN

Hender, A., Siddik, M., Howieson, J. and Fotedar, R., 2021. Black soldier fly, *Hermetia illucens* as an alternative to fishmeal protein and fish oil: impact on growth, immune response, mucosal barrier status, and flesh quality of juvenile barramundi, *Lates calcarifer* (Bloch, 1790). Biology 10: 505. https://doi.org/10.3390/biology10060505

Henry, M., Gasco, L., Piccolo, G. and Fountoulaki, E., 2015. Review on the use of insects in the diet of farmed fish: past and future. Animal Feed Science and Technology 203: 1–22. https://doi.org/10.1016/j.anifeedsci.2015.03.001

Henry, M.A., Golomazou, E., Asimaki, A., Psofakis, P., Fountoulaki, E., Mente, E., Rumbos, C.I., Athanassiou, C.G. and Karapanagiotidis, I.T., 2022. Partial dietary fishmeal replacement with full-fat or defatted superworm (*Zophobas morio*) larvae meals modulates the innate immune system of gilthead seabream, *Sparus aurata*. Aquaculture Reports 27: 101347. https://doi.org/10.1016/J.AQREP.2022.101347

Hervé, M.K., Calice, M.D., Dzepe, D., Djouaka, R.F., Chia, S.Y., Efole, T. and Ndindeng, S.A., 2025. Black soldier fly (*Hermetia illucens*) larvae improve growth performance and flesh quality of African catfish (*Clarias gariepinus*). Discover Animals 2: 9. https://doi.org/10.1007/s44338-024-00045-8

Hoseinifar, S.H., Yousefi, S., Van Doan, H., Ashouri, G., Gioacchini, G., Maradonna, F. and Carnevali, O., 2020. Oxidative stress and antioxidant defense in fish: the implications of probiotic, prebiotic, and synbiotics. Reviews in Fisheries Science and Aquaculture 29: 198–217. https://doi.org/10.1080/23308249.2020.1795616

Hossain, M.S., Small, B.C. and Hardy, R., 2023. Insect lipid in fish nutrition: recent knowledge and future application in aquaculture. Reviews in Aquaculture 15: 1664–1685. https://doi.org/10.1111/RAQ.12810

Huyben, D., Vidaković, A., Werner Hallgren, S. and Langeland, M., 2019. High-throughput sequencing of gut microbiota in rainbow trout (*Oncorhynchus mykiss*) fed larval and pre-pupae stages of black soldier fly (*Hermetia illucens*). Aquaculture 500: 485–491. https://doi.org/10.1016/j.aquaculture.2018.10.034

ICIPE, 2022. Rwanda achieves landmarks in the insect-based food and feed sector; with the support of ICIPE and partners. Available at: https://www.icipe.org/news/rwanda-achieves-landmarks-insect-based-food-and-feed-sector-support-icipe-and-partners.

IPIFF, 2023. IPIFF perspectives on the evolution of the European insect sector towards 2030: current EU regulatory status, existing opportunities and prospects for development. Available at: https://ipiff.org/wp-content/uploads/2023/11/IPIFF-Brochure-1-1.pdf.

Jeong, S.M., Khosravi, S., Mauliasari, I.R., Lee, B.J., You, S.G. and Lee, S.M., 2021a. Nutritional evaluation of cricket, *Gryllus bimaculatus*, meal as fish meal substitute for olive flounder, *Paralichthys olivaceus*, juveniles. Journal of the World Aquaculture Society 52: 859–880. https://doi.org/10.1111/JWAS.12790

Jeong, S.M., Khosravi, S., Yoon, K.Y., Kim, K.W., Lee, B.J., Hur, S.W. and Lee, S.M., 2021b. Mealworm, *Tenebrio molitor*, as a feed ingredient for juvenile olive flounder, *Paralichthys olivaceus*. Aquaculture Reports 20: 100747. https://doi.org/10.1016/j.aqrep.2021.100747

Ji, H., Zhang, J.L., Huang, J.Q., Cheng, X.F. and Liu, C., 2015. Effect of replacement of dietary fish meal with silkworm pupae meal on growth performance, body composition, intestinal protease activity and health status in juvenile Jian carp (*Cyprinus carpio* var. Jian). Aquaculture Research 46: 1209–1221. https://doi.org/10.1111/ARE.12276

Johnson, M.G. and Barrett, M., 2025. Exploring correctness, usefulness, and feasibility of potential physiological operational welfare indicators for farmed insects to establish research priorities. Animal 19: 101501. https://doi.org/10.1016/J.ANIMAL.2025.101501

Józefiak, A., Nogales-Mérida, S., Mikołajczak, Z., Rawski, M., Kierończyk, B. and Mazurkiewicz, J., 2019. The utilization of full-fat insect meal in rainbow trout (*Oncorhynchus mykiss*) nutrition: the effects on growth performance, intestinal microbiota and gastrointestinal tract histomorphology. Annals of Animal Science 19: 747–765. https://doi.org/10.2478/aoas-2019-0020

Józefiak, A., Rawski, M., Kierończyk, B., Józefiak, D. and Mazurkiewicz, J., 2023. Effect of two insect meals on the gut commensal microbiome of healthy sea trout (*Salmo trutta* vr. *trutta*). BMC Veterinary Research 19: 1–12. https://doi.org/10.1186/S12917-023-03671-8

Kalemi, V., Rimoldi, S., Costa, R.S., Basto, A., Monteiro, M., Terova, G. and Valente, L.M.P., 2025. Replacing fishmeal with an insect meal blend: implications for intestinal microbiota in European seabass. Aquaculture Reports 43: 102939. https://doi.org/10.1016/J.AQREP.2025.102939

Karlsen, Ø., Amlund, H., Berg, A. and Olsen, R.E., 2017. The effect of dietary chitin on growth and nutrient digestibility in farmed Atlantic cod, Atlantic salmon and Atlantic halibut. Aquaculture Research 48: 123–133. https://doi.org/10.1111/are.12867

Krongdang, S., Phokasem, P., Venkatachalam, K. and Charoenphun, N., 2023. Edible insects in Thailand: an overview of status, properties, processing, and utilization in the food industry. Foods 12: 2162. https://doi.org/10.3390/foods12112162

Kumar, V., Fawole, F.J., Romano, N., Hossain, M.S., Labh, S.N., Overturf, K. and Small, B.C., 2021. Insect (black soldier fly, *Hermetia illucens*) meal supplementation prevents the soybean meal-induced intestinal enteritis in rainbow trout and health benefits of using insect oil. Fish & Shellfish Immunology 109: 116–124. https://doi.org/10.1016/j.fsi.2020.12.008

Lähteenmäki-Uutela, A., Marimuthu, S.B. and Meijer, N., 2021. Regulations on insects as food and feed: a global comparison. Journal of Insects as Food and Feed 7: 849–856. https://doi.org/10.3920/jiff2020.0066

Larouche, J., Campbell, B., Hénault-Ethier, L., Banks, I., Tomberlin, J., Preyer, C., Deschamps, M.H. and Vandenberg, G., 2023. The edible insect sector in Canada and the United States. Animal Frontiers 13: 16–25. https://doi.org/10.1093/af/vfad047

Li, S., Ji, H., Zhang, B., Tian, J., Zhou, J. and Yu, H., 2016. Influence of black soldier fly (*Hermetia illucens*) larvae oil on growth performance, body composition, tissue fatty acid composition and lipid deposition in juvenile Jian carp (*Cyprinus carpio* var. Jian). Aquaculture 465: 43–52. https://doi.org/10.1016/j.aquaculture.2016.08.020

Li, S., Ji, H., Zhang, B., Zhou, J. and Yu, H., 2017. Defatted black soldier fly (*Hermetia illucens*) larvae meal in diets for juvenile Jian carp (*Cyprinus carpio* var. Jian): growth performance, antioxidant enzyme activities, digestive enzyme activities, intestine and hepatopancreas histological structure. Aquaculture 477: 62–70. https://doi.org/10.1016/j.aquaculture.2017.04.015

Liceaga, A.M., 2019. Approaches for utilizing insect protein for human consumption: effect of enzymatic hydrolysis on protein quality and functionality. Annals of the Entomological Society of America 112: 529–532. https://doi.org/10.1093/aesa/saz010

Limbu, S.M., Shoko, A.P., Ulotu, E.E., Luvanga, S.A., Munyi, F.M., John, J.O. and Opiyo, M.A., 2022. Black soldier fly (*Hermetia illucens*, L.) larvae meal improves growth performance, feed efficiency and economic returns of Nile tilapia (*Oreochromis niloticus*, L.) fry. Aquaculture, Fish and Fisheries 2: 167–178. https://doi.org/10.1002/AFF2.48

Lin, Y.H. and Mui, J.J., 2017. Evaluation of dietary inclusion of housefly maggot (*Musca domestica*) meal on growth, fillet composition and physiological responses for barramundi, *Lates calcarifer*. Aquaculture Research 48: 2478–2485. https://doi.org/10.1111/are.13085

Lin, Y.H., Lee, Y.C., Chang, C.C. and Chen, Y.T., 2023. Evaluation of mealworm meal as an alternative to fish meal in the diet for giant grouper (*Epinephelus lanceolatus*) based on the protein level adjusted by the nitrogen content in chitin. Aquaculture Reports 33: 101828. https://doi.org/10.1016/J.AQREP.2023.101828

Liu, Y., Ding, X., Brown, P.B., Bai, Y., Liu, Z., Shen, J., Liu, H. and Huang, Y., 2023. The digestive enzyme activity, intestinal microbiota and immune-related genes expression of snakehead fish (*Channa argus*) juveniles affected by dietary cricket (*Gryllus testaceus*) meal. Animal Feed Science and Technology 304: 115721. https://doi.org/10.1016/J.ANIFEEDSCI.2023.115721

Lopes, I.G., Wiklicky, V. and Lalander, C., 2025. Bioconversion of aquaculture waste blended with vegetable by-products using *Hermetia illucens* larvae: Process parameters and larval quality. Aquaculture Reports 43: 102961. https://doi.org/10.1016/J.AQREP.2025.102961

Lotta, F., 2019. Insects as food: the legal framework. In: Sogari, G., Mora, C., Menozzi, D. (eds.) Edible insects in the food sector. Springer, Cham, Switzerland, p. 105–118. https://doi.org/10.1007/978-3-030-22522-3_8

Magalhães, R., Sánchez-López, A., Leal, R.S., Martínez-Llorens, S., Oliva-Teles, A. and Peres, H., 2017. Black soldier fly (*Hermetia illucens*) pre-pupae meal as a fish meal replacement in diets for European seabass (*Dicentrarchus labrax*). Aquaculture 476: 79–85. https://doi.org/10.1016/j.aquaculture.2017.04.021

Maldonado-Othón, C.A., De La Re-Vega, E., Perez-Velazquez, M. and González-Félix, M.L., 2022. Replacement of fish oil by camelina and black soldier fly larvae oils in diets for juvenile *Totoaba macdonaldi* and their effect on growth, fatty acid profile, and gene expression of pancreatic lipases. Aquaculture 552: 737985. https://doi.org/10.1016/j.aquaculture.2022.737985

Mancini, S., Medina, I., Iaconisi, V., Gai, F., Basto, A. and Parisi, G., 2018. Impact of black soldier fly larvae meal on the chemical and nutritional characteristics of rainbow trout fillets. Animal 12: 1672–1681. https://doi.org/10.1017/S1751731117003421

Mapanao, R., Jiwyam, W., Nithikulworawong, N. and Weeplian, T., 2023. Effects of black soldier fly (*Hermetia illucens*) larvae as a fish meal replacement on growth performance, feed utilisation, morphological characters and carcass composition of Thai climbing perch (*Anabas testudineus*). Journal of Applied Aquaculture 35: 1–15. https://doi.org/10.1080/10454438.2021.1923609

Mastoraki, M., Mollá Ferrándiz, P., Vardali, S.C., Kontodimas, D.C., Kotzamanis, Y.P., Gasco, L., Chatzifotis, S. and Antonopoulou, E., 2020. A comparative study on the effect of fish meal substitution with three different insect meals on growth, body composition and metabolism of European sea bass (*Dicentrarchus labrax* L.). Aquaculture 528: 735511. https://doi.org/10.1016/J.AQUACULTURE.2020.735511

Mastoraki, M., Paudel, K., Biasato, I., Bellezza Oddon, S., Caimi, C., Gasco, L. and Chatzifotis, S., 2025. Impact of processing technologies on insect meal digestibility in rainbow trout (*Oncorhynchus mykiss*) and European sea bass (*Dicentrarchus labrax*). Journal of Insects as Food and Feed 11: 49–60. https://doi.org/10.1163/23524588-00001425

Maulu, S., Langi, S., Hasimuna, O.J., Missinhoun, D., Munganga, B.P., Hampuwo, B.M., Gabriel, N.N., Elsabagh, M., Van Doan, H., Abdul Kari, Z. and Dawood, M.A.O., 2022. Recent advances in the utilization of insects as an ingredient in aquafeeds: a review. Animal Nutrition 11: 334–349. https://doi.org/10.1016/J.ANINU.2022.07.013

Meesala, K.M., Hong, J., Sealey, W.M., Popa, R., Bouchard, D.A. and Habte-Tsion, M., 2025. Effects of fishmeal substitution with defatted black soldier fly larvae and soy protein meals on the growth, physio-biochemical responses, and immune-related gene expression of Atlantic salmon (*Salmo salar*). Aquaculture 602: 742335. https://doi.org/10.1016/J.AQUACULTURE.2025.742335

Meijer, N., Safitri, R.A., Tao, W. and Hoek-Van den Hil, E.F., 2025. European Union legislation and regulatory framework for edible insect production – safety issues. Animal 19: 101468. https://doi.org/10.1016/j.animal.2025.101468

Melenchón, F., Larrán, A.M., de Mercado, E., Hidalgo, M.C., Cardenete, G., Barroso, F.G., Fabrikov, D., Lourenço, H.M., Pessoa, M.F. and Tomás-Almenar, C., 2021. Potential use of black soldier fly (*Hermetia illucens*) and mealworm (*Tenebrio molitor*) insect meals in diets for rainbow trout (*Oncorhynchus mykiss*). Aquaculture Nutrition 27: 491–505. https://doi.org/10.1111/ANU.13201

Melenchón, F., de Mercado, E., Pula, H.J., Cardenete, G., Barroso, F.G., Fabrikov, D., Lourenço, H.M., Pessoa, M.-F., Lagos, L., Weththasinghe, P., Cortés, M. and Tomás-Almenar, C., 2022. Fishmeal dietary replacement up to 50%: a comparative study of two insect meals for rainbow trout (*Oncorhynchus mykiss*). Animals 12: 179. https://doi.org/10.3390/ani12020179

Melenchón, F., Larrán, A.M., Hernández, M., Abad, D., Morales, A.E., Pula, H.J., Fabrikov, D., Sánchez-Muros, M.J., Galafat, A., Alarcón, F.J., Lourenço, H.M., Pessoa, M.-F. and Tomás-Almenar, C., 2023. Different diets based on yellow mealworm (*Tenebrio molitor*) — Part B: modification of the intestinal inflammatory response and the microbiota composition of rainbow trout (*Oncorhynchus mykiss*). Fishes 8: 284. https://doi.org/10.3390/fishes8060284

Mikołajczak, Z., Rawski, M., Mazurkiewicz, J., Kierończyk, B. and Józefiak, D., 2020. The effect of hydrolyzed insect meals in sea trout fingerling (*Salmo trutta* m. *trutta*) diets on growth performance, microbiota and biochemical blood parameters. Animals 10: 1031. https://doi.org/10.3390/ani10061031

Monteiro, M., Rodrigues-dos-Santos, L., Filipa-Silva, A., Marques, D.A., Pintado, M., Almeida, A. and Valente, L.M.P., 2025. Circular animal protein hydrolysates: a comparative approach of functional properties. Antioxidants 14: 782. https://doi.org/10.3390/antiox14070782

Mousavi, S., Zahedinezhad, S. and Loh, J.Y., 2020. A review on insect meals in aquaculture: the immunomodulatory and physiological effects. International Aquatic Research 12: 100–115. https://doi.org/10.22034/iar(20).2020.1897402.1033

Moutinho, S., Pedrosa, R., Magalhães, R., Oliva-Teles, A., Parisi, G. and Peres, H., 2021. Black soldier fly (*Hermetia illucens*) pre-pupae larvae meal in diets for European seabass (*Dicentrarchus labrax*) juveniles: effects on liver oxidative status and fillet quality traits during shelf-life. Aquaculture 533: 736080. https://doi.org/10.1016/j.aquaculture.2020.736080

Moutinho, S., Oliva-Teles, A., Martínez-Llorens, S., Monroig and Peres, H., 2022. Total fishmeal replacement by defatted *Hermetia illucens* larvae meal in diets for gilthead seabream (*Sparus aurata*) juveniles. Journal of Insects as Food and Feed 8: 1455–1468. https://doi.org/10.3920/JIFF2021.0195

Moutinho, S., Oliva-Teles, A., Pulido-Rodríguez, L., Magalhães, R., Monroig, Ó., Parisi, G. and Peres, H., 2023. Black soldier fly larvae oil as an alternative lipid source in diets for gilthead seabream (*Sparus aurata*) juveniles. Aquaculture 574: 739705. https://doi.org/10.1016/j.aquaculture.2023.739705

Moutinho, S., Oliva-Teles, A., Fontinha, F., Martins, N., Monroig, Ó. and Peres, H., 2024a. Black soldier fly larvae meal as a potential modulator of immune, inflammatory, and antioxidant status in gilthead seabream juveniles. Comparative Biochemistry and Physiology Part B: Biochemistry and Molecular Biology 271: 110951. https://doi.org/10.1016/J.CBPB.2024.110951

Moutinho, S., Oliva-Teles, A., Pulido-Rodríguez, L., Parisi, G., Magalhães, R., Monroig, Ó. and Peres, H., 2024b. Effects of black soldier fly (*Hermetia illucens*) larvae oil on fillet quality and nutritional traits of gilthead seabream. Aquaculture 579: 740219. https://doi.org/10.1016/j.aquaculture.2023.740219

Moutinho, S., Monroig, O., Peres, H., Villena-Rodriguez, A., Magalhaes, R., Pulido-Rodriguez, L., Parisi, G. and Oliva-Teles, A., 2025a. Effects of black soldier fly larvae oil on lipid metabolism, liver fatty acid composition, and plasma metabolite profiles in gilthead seabream juveniles. Comparative Biochemistry and Physiology Part B: Biochemistry and Molecular Biology 277: 111069. https://doi.org/10.1016/j.cbpb.2024.111069

Moutinho, S., Peres, H., Fontinha, F., Estevão-Rodrigues, T., Monroig, Ó., Magalhães, R., Pulido-Rodríguez, L., Parisi, G. and Oliva-Teles, A., 2025b. *Hermetia illucens* larvae oil as an alternative lipid source: effects on immune function, antioxidant activity, and inflammatory response in gilthead seabream juveniles. Comparative Biochemistry and Physiology Part B: Biochemistry and Molecular Biology 276: 111059. https://doi.org/10.1016/J.CBPB.2024.111059

Muñoz-Seijas, N., Fernandes, H., Outeiriño, D., Morán-Aguilar, M.G., Domínguez, J.M. and Salgado, J.M., 2024. Valorisation of *Tenebrio molitor* beetles: enzymes-based bioprocessing to obtain a highly digestible protein hydrolysate. Journal of Insects as Food and Feed 11: 101–113. https://doi.org/10.1163/23524588-00001054

Niyonsaba, H.H., Höhler, J., van der Fels-Klerx, H.J., Slijper, T., Alleweldt, F., Kara, S., Zanoli, R., Costa, A.I.A., Peters, M. and Meuwissen, M.P.M., 2023. Barriers, risks and risk management strategies in European insect supply chains. Journal of Insects as Food and Feed 9: 691–706. https://doi.org/10.3920/JIFF2022.0100

Nogales-Mérida, S., Gobbi, P., Józefiak, D., Mazurkiewicz, J., Dudek, K., Rawski, M., Kierończyk, B. and Józefiak, A., 2019. Insect meals in fish nutrition. Reviews in Aquaculture 11: 1080–1103. https://doi.org/10.1111/raq.12281

Oteri, M., Chiofalo, B., Maricchiolo, G., Toscano, G., Nalbone, L., Lo Presti, V. and Di Rosa, A.R., 2022. Black soldier fly larvae meal in the diet of gilthead sea bream: effect on chemical and microbiological quality of filets. Frontiers in Nutrition 9: 896552. https://doi.org/10.3389/FNUT.2022.896552

Panteli, N., Mastoraki, M., Lazarina, M., Chatzifotis, S., Mente, E., Kormas, K.Ar. and Antonopoulou, E., 2021. Configuration of gut microbiota structure and potential functionality in two Teleosts under the influence of dietary insect meals. Microorganisms 9: 699. https://doi.org/10.3390/microorganisms9040699

Pascon, G., Cardinaletti, G., Daniso, E., Bruni, L., Messina, M., Parisi, G. and Tulli, F., 2024. Effect of dietary chitin on growth performance, nutrient utilization, and metabolic response in rainbow trout (*Oncorhynchus mykiss*). Aquaculture Reports 37: 102244. https://doi.org/10.1016/j.aqrep.2024.102244

Piccolo, G., Iaconisi, V., Marono, S., Gasco, L., Loponte, R., Nizza, S., Bovera, F. and Parisi, G., 2017. Effect of *Tenebrio molitor* larvae meal on growth performance, in vivo nutrients digestibility, somatic and marketable indexes of gilthead sea bream (*Sparus aurata*). Animal Feed Science and Technology 226: 12–20. https://doi.org/10.1016/j.anifeedsci.2017.02.007

Pulido, L., Secci, G., Maricchiolo, G., Gasco, L., Gai, F., Serra, A., Conte, G. and Parisi, G., 2022. Effect of dietary black soldier fly larvae meal on fatty acid composition of lipids and sn-2 position of triglycerides of marketable size gilthead sea bream fillets. Aquaculture 546: 737351. https://doi.org/10.1016/J.AQUACULTURE.2021.737351

Randazzo, B., Zarantoniello, M., Gioacchini, G., Cardinaletti, G., Belloni, A., Giorgini, E., Faccenda, F., Cerri, R., Tibaldi, E. and Olivotto, I., 2021. Physiological response of rainbow trout (*Oncorhynchus mykiss*) to graded levels of *Hermetia illucens* or poultry by-product meals as single or combined substitute ingredients to dietary plant proteins. Aquaculture 538: 736550. https://doi.org/10.1016/j.aquaculture.2021.736550

Randazzo, B., Di Marco, P., Zarantoniello, M., Daniso, E., Cerri, R., Finoia, M.G., Capoccioni, F., Tibaldi, E., Olivotto, I. and Cardinaletti, G., 2023. Effects of supplementing a plant protein-rich diet with insect, crayfish or microalgae meals on gilthead sea bream (*Sparus aurata*) and European seabass (*Dicentrarchus labrax*) growth, physiological status and gut health. Aquaculture 575: 739811. https://doi.org/10.1016/J.AQUACULTURE.2023.739811

Rangel, F., Enes, P., Gasco, L., Gai, F., Hausmann, B., Berry, D., Oliva-Teles, A., Serra, C.R. and Pereira, F.C., 2022. Differential modulation of the European sea bass gut microbiota by distinct insect meals. Frontiers in Microbiology 13: 831034. https://doi.org/10.3389/FMICB.2022.831034

Rauta, P.R., Nayak, B. and Das, S., 2012. Immune system and immune responses in fish and their role in comparative immunity study: a model for higher organisms. Immunology Letters 148: 23–33. https://doi.org/10.1016/J.IMLET.2012.08.003

Rawski, M., Mazurkiewicz, J., Mikołajczak, Z., Kierończyk, B., Skrzypczak, P., Szymkowiak, P. and Józefiak, D., 2025. Black soldier fly meal as a gastrointestinal tract microbiota remodelling factor: a new natural and sustainable source of prebiotic substances for fish? Aquaculture Research 2025: 8852384. https://doi.org/10.1155/ARE/8852384

Rema, P., Saravanan, S., Armenjon, B., Motte, C. and Dias, J., 2019. Graded incorporation of defatted yellow mealworm (*Tenebrio molitor*) in rainbow trout (*Oncorhynchus mykiss*) diet improves growth performance and nutrient retention. Animals 9: 187. https://doi.org/10.3390/ANI9040187

Renna, M., Schiavone, A., Gai, F., Dabbou, S., Lussiana, C., Malfatto, V., Prearo, M., Capucchio, M.T., Biasato, I., Biasibetti, E., De Marco, M., Brugiapaglia, A., Zoccarato, I. and Gasco, L., 2017. Evaluation of the suitability of a partially defatted black soldier fly (*Hermetia illucens* L.) larvae meal as ingredient for rainbow trout (*Oncorhynchus mykiss* Walbaum) diets. Journal of Animal Science and Biotechnology 8: 1–13. https://doi.org/10.1186/S40104-017-0191-3

Reyes, M., Rodríguez, M., Montes, J., Barroso, F.G., Fabrikov, D., Morote, E. and Sánchez-Muros, M.J., 2020. Nutritional and growth effect of insect meal inclusion on seabass (*Dicentrarchuss labrax*) feeds. Fishes 5: 16. https://doi.org/10.3390/fishes5020016

Rimoldi, S., Ceccotti, C., Brambilla, F., Faccenda, F., Antonini, M. and Terova, G., 2023. Potential of shrimp waste meal and insect exuviae as sustainable sources of chitin for fish feeds. Aquaculture 567: 739256. https://doi.org/10.1016/J.AQUACULTURE.2023.739256

Rodríguez-González, E., da Cunha-Borges, V., Cantero-Bahillo, E., Fornari, T., García-Risco,

M.R. and Martin, D., 2025. Black soldier fly (*Hermetia illucens*) larvae accumulate bioactive compounds that modulate antioxidant activity when reared with bioactive agrifood by-products. Food Research International 219: 117013. https://doi.org/10.1016/J.FOODRES.2025.117013

Sankian, Z., Khosravi, S., Kim, Y.O. and Lee, S.M., 2018. Effects of dietary inclusion of yellow mealworm (*Tenebrio molitor*) meal on growth performance, feed utilization, body composition, plasma biochemical indices, selected immune parameters and antioxidant enzyme activities of mandarin fish (*Siniperca scherzeri*) juveniles. Aquaculture 496: 79–87. https://doi.org/10.1016/J.AQUACULTURE.2018.07.012

Sargent, J.R., Tocher, D.R. and Bell, J.G., 2002. The lipids. In: Halver, J.E., Hardy, R.W. (eds.) Fish nutrition. Academic Press, San Diego, USA, p. 181–257. https://doi.org/10.1016/B978-0-12-319652-1.X5000-9

Serra, V., Pastorelli, G., Tedesco, D.E.A., Turin, L. and Guerrini, A., 2024. Alternative protein sources in aquafeed: Current scenario and future perspectives. Veterinary and Animal Science 25: 100381. https://doi.org/10.1016/j.vas.2024.100381

Shafique, L., Abdel-Latif, H.M.R., Hassan, F., Alagawany, M., Naiel, M.A.E., Dawood, M.A.O., Yilmaz, S. and Liu, Q., 2021. The feasibility of using yellow mealworms (*Tenebrio molitor*): towards a sustainable aquafeed industry. Animals 11: 811. https://doi.org/10.3390/ani11030811

Shaviklo, A., 2023. Sensory and quality properties of aquaculture products influenced by insect-based and aquatic-origin diets: A review. Iranian Journal of Fisheries Sciences 22: 1228–1246. https://doi.org/10.22092/ijfs.2024.130761

Singha, K.P., Abanikannda, M.F., Ma, J., Romano, N., Koutsos, E., Adams, D. and Kumar, V., 2025. Complementing the high soybean meal diet with black soldier fly larvae meal as a functional feed ingredient to improve the performance, nutrient profile, and gut health of rainbow trout, *Oncorhynchus mykiss*. Aquaculture 603: 742405. https://doi.org/10.1016/J.AQUACULTURE.2025.742405

Smets, R., Verbinnen, B., Van De Voorde, I., Aerts, G., Claes, J. and Van Der Borght, M., 2020. Sequential extraction and characterisation of lipids, proteins, and chitin from black soldier fly (*Hermetia illucens*) larvae, prepupae, and pupae. Waste and Biomass Valorization 11: 6455–6466. https://doi.org/10.1007/S12649-019-00924-2

Sogari, G., Amato, M., Biasato, I., Chiesa, S. and Gasco, L., 2019. The potential role of insects as feed: a multi-perspective review. Animals 9: 119. https://doi.org/10.3390/ani9040119

Ssepuuya, G., Namulawa, V., Mbabazi, D., Mugerwa, S., Fuuna, P., Nampijja, Z., Ekesi, S., Fiaboe, K.K.M. and Nakimbugwe, D., 2017. Use of insects for fish and poultry compound feed in sub-Saharan Africa – a systematic review. Journal of Insects as Food and Feed 3: 289–302. https://doi.org/10.3920/JIFF2017.0007

Stejskal, V., Tran, H.Q., Prokesova, M., Gebauer, T., Giang, P.T., Gai, F. and Gasco, L., 2020. Partially defatted *Hermetia illucens* larva meal in diet of Eurasian perch (*Perca fluviatilis*) juveniles. Animals 10: 1876. https://doi.org/10.3390/ani10101876

Sudha, C., Ahilan, B., Felix, N., Uma, A., Chidambaram, P. and Prabu, E., 2022. Replacement of fish oil with black soldier fly larvae oil and vegetable oils: effects of growth, whole-body fatty acid profile, digestive enzyme activity, haemato-biochemical responses and muscle growth-related gene expression of juvenile striped catfish, *Pangasianodon hypophthalmus*. Aquaculture Research 53: 3097–3111. https://doi.org/10.1111/are.15823

Tanga, C.M. and Kababu, M.O., 2023. New insights into the emerging edible insect industry in Africa. Animal Frontiers 13: 26–40. https://doi.org/10.1093/AF/VFAD039

Taufek, N.M., Aspani, F., Muin, H., Raji, A.A., Razak, S.A. and Alias, Z., 2016. The effect of dietary cricket meal (*Gryllus bimaculatus*) on growth performance, antioxidant enzyme activities, and haematological response of African catfish (*Clarias gariepinus*). Fish Physiology and Biochemistry 42: 1143–1155. https://doi.org/10.1007/s10695-016-0204-8

Taufek, N.M., Simarani, K., Muin, H., Aspani, F., Raji, A.A., Alias, Z. and Razak, S.A., 2018. Inclusion of cricket (*Gryllus bimaculatus*) meal in African catfish (*Clarias gariepinus*) feed influences disease resistance. Journal of Fisheries 6: 623–631. https://doi.org/10.17017/J.FISH.23

Thévenot, A., Rivera, J.L., Wilfart, A., Maillard, F., Hassouna, M., Senga-Kiesse, T., Le Féon, S. and Aubin, J., 2018. Mealworm meal for animal feed: environmental assessment and sensitivity analysis to guide future prospects. Journal of Cleaner Production 170: 1260–1267. https://doi.org/10.1016/J.JCLEPRO.2017.09.054

Tippayadara, N., Dawood, M.A.O., Krutmuang, P., Hoseinifar, S.H., Doan, H. Van and Paolucci, M., 2021. Replacement of fish meal by black soldier fly (*Hermetia illucens*) larvae meal: effects on growth, haematology, and skin mucus immunity of Nile tilapia, *Oreochromis niloticus*. Animals 11: 193. https://doi.org/10.3390/ani11010193

Tran, H.Q., Nguyen, T.T., Prokešová, M., Gebauer, T., Doan, H. Van and Stejskal, V., 2022. Systematic review and meta-analysis of production performance of aquaculture species fed dietary insect meals. Reviews in Aquaculture 14: 1637–1655. https://doi.org/10.1111/RAQ.12666

Tran, H.Q., Nguyen, T.T., Øverland, M., Gasco, L., Valente, L.M.P. and Stejskal, V., 2025. Influence of dietary black soldier fly (*Hermetia illucens*) meal on the quality attributes of aquaculture fish products: a meta-analysis. Journal of Insects as Food and Feed 11: 2657-2673. https://doi.org/10.1163/23524588-bja10236

Turchini, G.M., Torstensen, B.E. and Ng, W.K., 2009. Fish oil replacement in finfish nutrition. Reviews in Aquaculture 1: 10–57. https://doi.org/10.1111/J.1753-5131.2008.01001.X

UNBS, 2022. UNBS executive director, Mr. David Livingstone Ebiru, Makerere University Chancellor, Professor Barnabas Nawangwe launch standard to facilitate commercialization of edible insects. Available at: https://www.unbs.go.ug/readmore-slider.php?sl=336&banner.

van Huis, A. and Oonincx, D.G.A.B., 2017. The environmental sustainability of insects as food and feed. A review. Agronomy for Sustainable Development 37: 43. https://doi.org/10.1007/s13593-017-0452-8

Verner, D., Roos, N., Halloran, A., Surabian, G., Tebaldi, E., Ashwill, M., Vellani, S. and Konishi, Y., 2021. Insect and hydroponic farming in Africa: the new circular food economy, The World Bank, Washington, DC, USA, 240 pp. https://doi.org/10.1596/978-1-4648-1766-3

Vernooij, A.G., Veldkamp, T. and Ndambi, A., 2019. Insects for Africa: developing business opportunities for insects in animal feed in Eastern Africa, Wageningen Livestock Research rapport, no. 1150. Wageningen Livestock Research, Wageningen, The Netherlands, 27 pp. https://doi.org/10.18174/470617

Wan, A.H.L., Snellgrove, D.L. and Davies, S.J., 2017. A comparison between marine and terrestrial invertebrate meals for mirror carp (*Cyprinus carpio*) diets: impact on growth, haematology and health. Aquaculture Research 48: 5004–5016. https://doi.org/10.1111/are.13318

Wang, G., Peng, K., Hu, J., Yi, C., Chen, X., Wu, H. and Huang, Y., 2019. Evaluation of defatted black soldier fly (*Hermetia illucens* L.) larvae meal as an alternative protein ingredient for juvenile Japanese seabass (*Lateolabrax japonicus*) diets. Aquaculture 507: 144–154. https://doi.org/10.1016/j.aquaculture.2019.04.023

Weththasinghe, P., Hansen, J., Nøkland, D., Lagos, L., Rawski, M. and Øverland, M., 2021. Full-fat black soldier fly larvae (*Hermetia illucens*) meal and paste in extruded diets for Atlantic salmon (*Salmo salar*): Effect on physical pellet quality, nutrient digestibility, nutrient utilization and growth performances. Aquaculture 530: 735785. https://doi.org/10.1016/J.AQUACULTURE.2020.735785

Weththasinghe, P., Hansen, J., Rawski, M., Józefiak, D., Ghimire, S. and Øverland, M., 2022a. Insects in Atlantic salmon (*Salmo salar*) diets – comparison between full-fat, defatted, and de-chitinised meals, and oil and exoskeleton fractions. Journal of Insects as Food and Feed 8: 1235–1248. https://doi.org/10.3920/JIFF2021.0094

Weththasinghe, P., Rocha, S.D.C., Øyås, O., Lagos, L., Hansen, J.Ø., Mydland, L.T. and Øverland, M., 2022b. Modulation of Atlantic salmon (*Salmo salar*) gut microbiota composition and predicted metabolic capacity by feeding diets with processed black soldier fly (*Hermetia illucens*) larvae meals and fractions. Animal Microbiome 4: 9. https://doi.org/10.1186/s42523-021-00161-w

Willora, F.P., Farris, N.W., Ghebre, E., Zatti, K., Bisa, S., Kiron, V., Verlhac-Trichet, V., Danielsen, M., Dalsgaard, T.K. and Sørensen, M., 2025. Full-fat black soldier fly larvae meal and yellow mealworm meal: Impact on feed protein quality, growth and nutrient utilization of Atlantic salmon (*Salmo salar*) post smolts. Aquaculture 595: 741648. https://doi.org/10.1016/J.AQUACULTURE.2024.741648

Xu, X., Ji, H., Yu, H. and Zhou, J., 2018. Influence of replacing fish meal with enzymatic hydrolysates of defatted silkworm pupa (*Bombyx mori* L.) on growth performance, body composition and non-specific immunity of juvenile mirror carp (*Cyprinus carpio* var. specularis). Aquaculture Research 49: 1480–1490. https://doi.org/10.1111/are.13603

Xu, X., Ji, H., Belghit, I. and Sun, J., 2020. Black soldier fly larvae as a better lipid source than yellow mealworm or silkworm oils for juvenile mirror carp (*Cyprinus carpio* var. specularis). Aquaculture 527: 735453. https://doi.org/10.1016/J.AQUACULTURE.2020.735453

Xu, X., Ji, H., Belghit, I., Liland, N.S., Wu, W. and Li, X., 2021. Effects of black soldier fly oil rich in n-3 HUFA on growth performance, metabolism and health response of juvenile mirror carp (*Cyprinus carpio* var. specularis). Aquaculture 533: 736144. https://doi.org/10.1016/J.AQUACULTURE.2020.736144

Yadav, N.K., Deepti, M., Patel, A.B., Kumar, P., Angom, J., Debbarma, S., Singh, S.K., Deb, S., Lal, J., Vaishnav, A., Das, R., Kashyap, S. and Meena, D.K., 2025. Dissecting insects as sustainable protein bioresource in fish feed for aquaculture sustainability. Discover Food 5: 47. https://doi.org/10.1007/S44187-025-00318-5

Yang, L., Cai, M., Zhong, L., Yin, Y., Xie, Y., Xie, S., Hu, Y. and Zhang, J., 2025. Yellow mealworm (*Tenebrio molitor*) meal in diets of grass carp (*Ctenopharyngodon idellus*): effects on growth performance, antioxidant capacity, immunity, intestinal morphology, and intestinal microbiota. Animal Nutrition 21: 70–83. https://doi.org/10.1016/J.ANINU.2025.01.001

Yildirim-Aksoy, M., Eljack, R., Schrimsher, C. and Beck, B.H., 2020. Use of dietary frass from black soldier fly larvae, *Hermetia illucens*, in hybrid tilapia (Nile x Mozambique, *Oreocromis niloticus* x *O. mozambique*) diets improves growth and resistance to bacterial diseases. Aquaculture Reports 17: 100373. https://doi.org/10.1016/j.aqrep.2020.100373

Yıldırım-Aksoy, M., Eljack, R., Aksoy, J. and Beck, B.H., 2023. Frass from black soldier fly larvae, *Hermetia illucens*, as a possible functional dietary ingredient in channel catfish feed. Fishes 8: 542. https://doi.org/10.3390/fishes8110542

Zarantoniello, M., de Oliveira, A.A., Sahin, T., Freddi, L., Torregiani, M., Tucciarone, I., Chemello, G., Cardinaletti, G., Gatto, E., Parisi, G., Bertolucci, C., Riolo, P., Nartea, A., Gioacchini, G. and Olivotto, I., 2023. Enhancing rearing of European seabass (*Dicentrarchus labrax*) in aquaponic systems: investigating the effects of enriched black soldier fly (*Hermetia illucens*) prepupae meal on fish welfare and quality traits. Animals 13: 1921. https://doi.org/10.3390/ani13121921

CHAPTER 8

Use of bacterial proteins from fermentation technologies in aquaculture nutrition

M. Longshaw[1*], *A. Chalaris*[2] *and M. Reshetiuk*[3]

[1]*Aseno Ltd, 52a St John Street, Ashbourne, Derbyshire, UK*
[2]*AC Aquaculture Consulting, Stirling, FK8 1 UZ, UK*
[3]*Calysta (Germany) GmbH, Carl-Friedrich-Gauß-Ring 5, 69124 Heidelberg, Germany*
**Correspondence: matt@aseno-partners.com*

Abstract

The use of cultured or single cell proteins as a partial or total replacement for conventional protein sources in aquaculture has yet to achieve full commercial success despite the large number of studies conducted. These studies have ranged from the relatively straightforward nutrition trials to determine inclusion level and digestibility, to more recent in-depth studies to determine mode of action and impacts on health and growth outcomes. This chapter focusses on the main bacteria grown using biomass fermentation techniques that utilize single carbon sources such as carbon dioxide, methane or methanol as the primary feedstock.

Keywords

single cell protein – digestibility – health – growth

1 Introduction

Without doubt the main source of protein in aquaculture is fish meal and efforts have been made to reduce its inclusion in feeds by trialing different protein sources including but not limited to, plants, insects, and single cell proteins. Despite their availability since the 1970s (Anon, 1978; Bewersdorff and Dostálek, 1971; Stringer, 1982; Tusé and and Miller, 1983), the production and incorporation of single cell proteins (SCPs) at commercial scale have yet to achieve success. These single cell or cultured proteins are derived from bacterial microbes grown using biomass fermentation that use single carbon sources such as methanol, methane, or carbon dioxide as a feedstock. Despite the lack of full commercial implementation, a large number of academic studies have assessed these microbial proteins for their potential as feed ingredients in aquaculture. Indeed, almost all of the SCPs tested have only been grown at lab- or pilot-scale due to the inherent difficulties in scaling up to full commercial production (Banks et al., 2022). However, despite this, the promise of partially or completely replacing various "traditional" protein sources with single cell proteins remains a viable option. These alternative protein sources have a high protein content that normally exceeds 60% and providing essential and non-essential amino acids that match traditional protein sources.

The main bacterial species tested include *Methylococcus capsulatus* (as FeedKind® from Calysta Inc., USA, PRO-DG from String Bio Pvt Ltd., India, and Uniprotein® from Unibio, Denmark), *Methylophilus methylotrophus* (=*Methylomonas clara*) (as Pruteen® from ICI Chemicals, UK and Probion® from Hoeschst, Germany), *Methylorubrum extorquens* (as KnipBio Meal® from KnipBio, USA), *Methylovorus menthalis* (Psomi®), *Clostridium autoethanogenum*, and a number of unidentified bacteria (Table 8.1).

The chapter does not include those proteins derived from bacteria that utilize waste side streams or are derived from sugars or those that are added at low level as biologically functional ingredients such as biotics (Salminen et al., 2021). Furthermore, it does not include precision fermentation where the use of gene editing and / or metabolic engineering are applied to produce specific strains of bacteria or to produce specific metabolites (Pereira et al., 2025). It focusses on palatability and feed intake, digestibility and excretion, sera biochemistry, antioxidant capacity, gut structure, disease, microbiota, body composition, production outcomes, technical properties of feed, costs and economics, and concludes with an assessment of limitations and the future direction of the cultured protein market in aquaculture. The number of aquaculture species tested is large, covering the most important cultivated species across the globe including shrimp, salmonids, cyprinids and a range of species in marine and freshwater environments (Table 8.2). Additionally, the studies have also considered the use of SCPs in unusual species such as mollusks, turtles, and sea cucumbers, reflecting the increasing importance of partially replacing fishmeal in the global aquaculture market.

2 Palatability and feed intake

There is a view that single cell proteins can be detrimental to palatability of feeds, with feed intake reduced as a result. However, few studies have explicitly addressed the underlying mechanisms for this apparent reduction and where data does exist, mitigation measures have been proposed. As with production outcomes dis-

TABLE 8.1 List of cultured proteins, their main feedstock and crude protein content as well as the hosts in which they have been tested. Where CP = Crude protein content, ND = No data

Protein source	Target species	Main substrate	References
Clostridium auto-ethanogenum (CP 85%)	*Acanthopagrus schlegelii*	Carbon Monoxide, Carbon Dioxide, Hydrogen ($CO/CO_2 + H_2$)	Chen et al. 2020
	Carassius gibelio		Yu et al. 2024
	Ctenopharyngodon idella		HongCheng et al. 2018; Liu et al. 2024; Xia et al. 2025; Xue et al. 2023; Yang et al. 2024
	Cyprinus carpio		X. Li et al. 2021
	Dicentrarchus labrax		Abdel-Tawwab et al. 2025
	Epinephelus fuscogut-tatus ♀ × *Epinephelus lanceolatus* ♂		B. Huang et al. 2023; Xu et al. 2023a; C.-Z. Zheng et al. 2023
	Haliotis discus hannai		Wu et al. 2024; Z. Wu et al. 2022
	Larimichthys crocea		Wang et al. 2023; Y. Wu et al. 2022; J. Zhang et al. 2022
	Litopenaeus vannamei		Cai et al. 2022; Chen et al. 2022; Y. Chen et al. 2022b; Jiang et al. 2021; Kuo et al. 2024; X. Li et al. 2022; Shi et al. 2025, 2023; Xu et al. 2023b; Yao et al. 2024; Yuan et al. 2024; C. Zheng et al. 2022a, 2022b
	Micropterus salmoides		Chen et al. 2025; Li et al. 2023; L. Li et al. 2022; Lu et al. 2021; Luo et al. 2023; Ma et al. 2022a; Shi et al. 2022; Yang et al. 2023, 2022a, 2022b; Zhu et al. 2022
	Oncorhynchus mykiss		H. Huang et al. 2023
	Oreochromis niloticus		Maulu et al. 2021a, 2021b
	Procambarus clarkii		Dai et al. 2023
	Scophthalmus maximus		Bai et al. 2024; Pan et al. 2025, 2024; Qi et al. 2024a, 2024b; J. Zheng et al. 2022
	Takifugu obscurus		Cui et al. 2022
Methylorubrum extorquens (KnipBio®) (CP 51%)	*Haemulon chrysargy-reum*	Methanol (CH_3OH)	Tlusty et al. 2017
	Litopenaeus vannamei		Qiu et al. 2018; Tlusty et al. 2017
	Oncorhynchus mykiss		Hardy et al. 2018
	Salmo salar		Salze and Tibbetts 2021; Tlusty et al. 2017

TABLE 8.1 List of cultured proteins, their main feedstock and crude protein content as well as the hosts in which they have been tested *(cont.)*

Protein source	Target species	Main substrate	References
Methylococcus capsulatus (FeedKind®, PRO-DG, UniProtein®) (CP 62-71%)	*Acanthopagrus schlegelii*	Methane (CH_4)	Xu et al. 2021
	Acipenser baerii		Safonov et al. 2025; Zenkovich et al. 2022
	Anguilla rostrata		Lu et al. 2023
	Coregonus nasus		Zenkovich et al. 2024
	Cyprinus carpio		Yu et al. 2022
	Dicentrarchus labrax		Moroni et al. 2024; Petereit et al. 2022; Vasilaki et al. 2023
	Epinephelus fuscoguttatus ♀ × *Epinephelus lanceolatus* ♂		Liang et al. 2022; Jiacheng Zhang et al. 2025; C.-Z. Zheng et al. 2023
	Haliotis laevigata × *Haliotis rubra*		Salini et al. 2025
	Hippoglossus hippoglossus		Aas et al. 2007
	Lateolabrax maculatus		Yu et al. 2023; Zhang et al. 2023
	Lates calcarifer		Samsing et al. 2024; Woolley et al. 2023
	Litopenaeus vannamei		Chen et al. 2022, 2021; Y. Chen et al. 2022a; Felix et al. 2023; Gisbert et al. 2025; Jintasataporn et al. 2021; X. Li et al. 2022; Nederlof et al. 2021, 2023
	Micropterus salmoides		Guo et al. 2022; Shi et al. 2022; Q. Zhang et al. 2022
	Oncorhynchus mykiss		Aas et al. 2006b; Øverland et al. 2006; Prakash et al. 2023; Rajesh et al. 2022; Ruiz et al. 2023; Vale Pereira et al. 2023
	Oreochromis niloticus		Chama et al. 2021; do Vale Pereira et al. 2024
	Pagrus major		Biswas et al. 2021
	Megalobrama amblycephala × *Culter alburnus*		Qiuxin et al. 2025
	Pelodiscus sinensis		Dai et al. 2024
	Salmo salar		Aas et al. 2006a; Berge et al. 2005; Glencross et al. 2023; Romarheim et al. 2013, 2011; Skrede et al. 1998; Storebakken et al. 2004, 1999, 1998
	Scophthalmus maximus		Hoerterer et al. 2022a, 2022b; J. Zheng et al. 2023; Zheng et al. 2024
	Seriola lalandi		Pilmer et al. 2025, 2022
	Seriola quinqueradiata		Biswas et al. 2020
	Sparus aurata		Carvalho et al. 2024, 2023; Fernandes et al. 2024; Piazzon et al. 2022; Tampou et al. 2024
	Trachinotus ovatus		Gao et al. 2023

TABLE 8.1 List of cultured proteins, their main feedstock and crude protein content as well as the hosts in which they have been tested *(cont.)*

Protein source	Target species	Main substrate	References
Methylophilus methylotrophus (Pruteen®, Probion®) (CP 71-82)	*Anguilla anguilla*	Methanol (CH_3OH)	Schmitz et al. 1984
	Coregonus schinzi		Dabrowski et al. 1984
	Cyprinus carpio		Atack et al. 1979
	Oncorhynchus mykiss		Kaushik and Luquet 1980; Matty and Smith 1978
	Saccostrea commercialis		Nell 1985; Nell and Wisely 1983
	Scophthalmus maximus		Voss 1988
Methylovorus menthalis (Psomi®) (CP 70-80)	*Salmo salar*	Methanol (CH_3OH)	Tibbetts et al. 2025b, 2025a
SCP (CP 73.5)	*Lutjanus malabaricus*	Methane (CH_4)	Ngoh et al. 2025
SCP (CP ND)	*Oncorhynchus mykiss*	ND	Buttle et al. 2024
SCP (Sasya®) (CP 64%)	*Oncorhynchus mykiss*	Carbon Dioxide (CO2)	Lee et al. 2020
SCP (CP ND)	*Oncorhynchus mykiss*	ND	Perera et al. 1995a, 1995b
SCP (CP 58.9)	*Stichopus monotuberculatus*	ND	Jiayin Zhang et al. 2025

TABLE 8.2 List of aquatic species in which single cell proteins have been tested. Where C = crustacean, E = echinoderm, M = mollusc, R = reptile, T = teleost

Target species	Protein source	References
Acanthopagrus schlegelii (T)	*Clostridium autoethanogenum*	Chen et al. 2020
	Methylococcus capsulatus	Xu et al. 2021
Acipenser baerii (T)	*Methylococcus capsulatus*	Safonov et al. 2025; Zenkovich et al. 2022
Anguilla anguilla (T)	*Methylophilus methylotrophus*	Schmitz et al. 1984
Anguilla rostrata (T)	*Methylococcus capsulatus*	Lu et al. 2023
Carassius gibelio (T)	*Clostridium autoethanogenum*	Yu et al. 2024
Coregonus nasus (T)	*Methylococcus capsulatus*	Zenkovich et al. 2024
Coregonus schinzi (T)	*Methylophilus methylotrophus*	Dabrowski et al. 1984
Ctenopharyngodon idella (T)	*Clostridium autoethanogenum*	HongCheng et al. 2018; Liu et al. 2024; Xia et al. 2025; Xue et al. 2023; Yang et al. 2024
Cyprinus carpio (T)	*Clostridium autoethanogenum*	X. Li et al. 2021
	Methylococcus capsulatus	Yu et al. 2022
	Methylophilus methylotrophus	Atack et al. 1979
Dicentrarchus labrax (T)	*Clostridium autoethanogenum*	Abdel-Tawwab et al. 2025
	Methylococcus capsulatus	Moroni et al. 2024; Petereit et al. 2022; Vasilaki et al. 2023

TABLE 8.2 List of aquatic species in which single cell proteins have been tested *(cont.)*

Target species	Protein source	References
Epinephelus fuscoguttatus ♀ × *Epinephelus lanceolatus* ♂ (T)	*Clostridium autoethanogenum*	E. Huang et al. 2023; Xu et al. 2023a; C.-Z. Zheng et al. 2023
	Methylococcus capsulatus	Liang et al. 2022; Jiacheng Zhang et al. 2025; C.-Z. Zheng et al. 2023
Haemulon chrysargyreum (T)	*Methylobacterium extorquens*	Tlusty et al. 2017
Haliotis discus hannai (M)	*Clostridium autoethanogenum*	Wu et al. 2024; Z. Wu et al. 2022
Haliotis laevigata × *Haliotis rubra* (M)	*Methylococcus capsulatus*	Salini et al. 2025
Hippoglossus hippoglossus (T)	*Methylococcus capsulatus*	Aas et al. 2007
Larimichthys crocea (T)	*Clostridium autoethanogenum*	Wang et al. 2023; Y. Wu et al. 2022; J. Zhang et al. 2022
Lateolabrax maculatus (T)	*Methylococcus capsulatus*	Yu et al. 2023; Zhang et al. 2023
Lates calcarifer (T)	*Methylococcus capsulatus*	Samsing et al. 2024; Woolley et al. 2023
Litopenaeus vannamei (C)	*Clostridium autoethanogenum*	Cai et al. 2022; Chen et al. 2022; Y. Chen et al. 2022b; Jiang et al. 2021; Kuo et al. 2024; X. Li et al. 2022; Shi et al. 2025, 2023; Xu et al. 2023b; Yao et al. 2024; Yuan et al. 2024; C. Zheng et al. 2022a, 2022b
	Methylobacterium extorquens	Qiu et al. 2018; Tlusty et al. 2017
	Methylococcus capsulatus	Chen et al. 2022, 2021; Y. Chen et al. 2022a; Felix et al. 2023; Gisbert et al. 2025; Jintasataporn et al. 2021; X. Li et al. 2022; Nederlof et al. 2021, 2023
Lutjanus malabaricus (T)	SCP	Ngoh et al. 2025
Megalobrama amblycephala × *Culter alburnus* (T)	*Clostridium autoethanogenum*	Qiuxin et al. 2025
Micropterus salmoides (T)	*Clostridium autoethanogenum*	Chen et al. 2025; Li et al. 2023; L. Li et al. 2022; Lu et al. 2021; Luo et al. 2023; Ma et al. 2022a; Shi et al. 2022; Yang et al. 2023, 2022a, 2022b; Zhu et al. 2022
	Methylococcus capsulatus	Guo et al. 2022; Shi et al. 2022; Q. Zhang et al. 2022
Oncorhynchus mykiss (T)	*Clostridium autoethanogenum*	H. Huang et al. 2023
	Methylobacterium extorquens	Hardy et al. 2018
	Methylococcus capsulatus	Aas et al. 2006b; Øverland et al. 2006; Prakash et al. 2023; Rajesh et al. 2022; Ruiz et al. 2023; Vale Pereira et al. 2023
	Methylophilus methylotrophus	Kaushik and Luquet 1980; Matty and Smith 1978
	Sasya®	Lee et al. 2020
	SCP	Buttle et al. 2024
	SCP	Papini et al. 2025
	SCP	Perera et al. 1995a, 1995b

TABLE 8.2 List of aquatic species in which single cell proteins have been tested *(cont.)*

Target species	Protein source	References
Oreochromis niloticus (T)	*Clostridium autoethanogenum*	Maulu et al. 2021a, 2021b
	Methylococcus capsulatus	Chama et al. 2021; do Vale Pereira et al. 2024
Pagrus major (T)	*Methylococcus capsulatus*	Biswas et al. 2021
Pelodiscus sinensis (R)	*Methylococcus capsulatus*	Dai et al. 2024
Procambarus clarkii (C)	*Clostridium autoethanogenum*	Dai et al. 2023
Saccostrea commercialis (M)	*Methylophilus methylotrophus*	Nell 1985; Nell and Wisely 1983
Salmo salar (T)	*Methylobacterium extorquens*	Salze and Tibbetts 2021; Tlusty et al. 2017
	Methylococcus capsulatus	Aas et al. 2006a; Berge et al. 2005; Glencross et al. 2023; Romarheim et al. 2013, 2011; Skrede et al. 1998; Storebakken et al. 2004, 1999, 1998
	Methylovorus menthalis	Tibbetts et al. 2025a, 2025b
Scophthalmus maximus (T)	*Clostridium autoethanogenum*	Bai et al. 2024; Pan et al. 2025, 2024; Qi et al. 2024a, 2024b; J. Zheng et al. 2022
	Methylococcus capsulatus	Hoerterer et al. 2022a, 2022b; J. Zheng et al. 2023; Zheng et al. 2024
	Methylophilus methylotrophus	Voss 1988
Seriola lalandi (T)	*Methylococcus capsulatus*	Pilmer et al. 2025, 2022
Seriola quinqueradiata (T)	*Methylococcus capsulatus*	Biswas et al. 2020
Sparus aurata (T)	*Methylococcus capsulatus*	Carvalho et al. 2024, 2023; Fernandes et al. 2024; Piazzon et al. 2022; Tampou et al. 2024
Stichopus monotuberculatus (E)	SCP	Jiayin Zhang et al. 2025
Takifugu obscurus (T)	*Clostridium autoethanogenum*	Cui et al. 2022
Trachinotus ovatus (T)	*Methylococcus capsulatus*	Gao et al. 2023

cussed later, there is often an optimal inclusion level for maximum feed intake and growth measures. In order to overcome some of the potential issues associated with reduced feed intake, some workers have utilized palatants and taste-masking approaches to improve palatability and feed intake. Addition of garlic powder and/or liquid tuna hydrolysate to diets containing up to 30% *M. capsulatus* in *Lates calcarifer* diets did not alter feed intake or growth outcomes and it was concluded that palatants were not required for this fish species (Woolley et al., 2023). Concerns with palatability of *M. capsulatus* in *Seriola quinqueradiata* diets were addressed by Biswas et al. (2020) who added enzyme-treated (hydrolysate) fishmeal to the diet. Although no significant differences were noted in growth outcomes in the different diet formulations, a lower protein retention efficiency was noted at the highest inclusion levels and in the fish given feeds containing hydrolysate. The authors suggested that this may have been due to the additional energy required to break down the increased bacterial protein content. Nell (1985) proposed that *M. methylotrophus* should be encapsulated to minimize fouling and bacterial contamination in oyster diets. These technologies could also be applied to single cell proteins to overcome palatability issues but comes with additional costs and the need to ensure appropriate digestion and uptake of these protein sources.

3 Digestibility

Elucidating the digestibility of a new ingredient is a crucial component of any nutrition study as it provides evidence of an animal's ability to digest and utilize these new ingredients whilst confirming that they do not interfere more broadly with any other ingredients in the formulation. Clearly, the best outcome for the inclusion of a new ingredient is to maintain or improve the digestibility scores of these components which typically include protein, lipid, energy, dry matter, and a number of important components such as copper, phosphorus, and calcium. Apparent digestibility coefficients (ADC) have been calculated previously for the major SCPs utilized in aquaculture feeds and although there are differences in values depending on the study design, ADC of protein is generally above 80% for all SCPs tested (Table 8.3).

The ADC of protein in *C. autoethanogenum* is typically higher than 85% with the SCP digestibility being assessed in *L. vannamei, M. salmoides, S. maximus, Takifugu obscurus* and *Epinephelus fuscoguttatus* ♀ × *Epinephelus lanceolatus* ♂ (J. Zheng et al., 2022; C.-Z. Zheng et al., 2023; X. Li et al., 2022; Shi et al., 2022; Zhu et al., 2022; Cui et al., 2022). In most cases, ADC's of amino acids, dry matter, crude protein and crude lipid increased with increasing replacement levels (J. Zheng et al., 2022; C.-Z. Zheng et al., 2023; X. Li et al., 2022; Shi et al., 2022), but decreased or plateaued above a certain inclusion level (Zhu et al., 2022; Cui et al., 2022).

Similar patterns of increasing then decreasing ADC values have been noted in fish fed diets with increasing *M. capsulatus* inclusion. Whilst ADCs of dry matter and crude lipid increased then decreased in *Acanthopagrus schlegelii* with increasing inclusion, no differences were noted in the ADC of crude protein (Xu et al., 2021). Similarly, ADCs increased then decreased in *O. mykiss* (Aas et al., 2006b; Øverland et al., 2006; Rajesh et al., 2022), in *Pagrus major* (Biswas et al., 2021), and in *S. maximus* (Hoerterer et al., 2022a) with increasing inclusion of *M. capsulatus*. Improvements in ADCs with increased *M. capsulatus* have been noted in *D. labrax* and *Sparus aurata* fed a mixture of SCPs (Vasilaki et al., 2023; Tampou et al., 2024), in *Haliotis laevigata* × *H. rubra* (Salini et al., 2025), in *L. vannamei* fed *M. capsulatus* subjected to different downstream processes (Nederlof et al., 2023), and in *M. salmoides* (Guo et al., 2022). In contrast, lower ADCs have been noted in a number of species with increased inclusion levels, irrespective of any downstream processes applied to the bacteria. This includes lower lipid digestibilities in *Lateolabrax mac-*

TABLE 8.3 Apparent digestibility coefficients (ADC) expressed as (%) for protein (Prot), total amino acids (AA), lipids (Lip), organic matter (OM), dry matter (DM), energy (E), carbohydrates (Carb), phosphorus (Phos), copper (Cu) and zinc (Zn) from the range of SCPs tested in aquaculture. Data are derived from references in the text.

Protein source	**ADC Prot**	**ADC AA**	**ADC Lip**	**ADC OM**	**ADC DM**	**ADC E**	**ADC Carb**	**ADC Phos**	**ADC Cu**	**ADC Zn**
Clostridium autoethanogenum	83.5–97.7	76.4–97.5	76.3–97.5	–	58.7–83.5	83.9	–	91.0	–	–
Methylococcus capsulatus	59.5–98.6	62.8–98.5	51.5–96.7	–	48.5–86.3	53.3–96.2	46.0–74.7	21.5–69.2	53.3–66.5	33.9–39.9
Methylophilus methylotrophus	89.5–96.2	–	–	47.0–90.0	51.1–53.7	44.5–49.4	–	–	–	–
Methylorubrum extorquens	42.6–92.0	69.6–99.3	62.6–94.0	–	-40.1–78.9	13.3–83.0	24.0–28.6	–	–	–
Methylovorus menthalis	92.4	85.5–96.3	–	–	77.5–85.3	84.7–87.9	–	–	–	–
SCP of Ngoh et al. (2025)	92.0	87.6–99.0	92.3	–	–	100	89.5	–	–	–
SCP of Buttle et al. (2024)	91.3–92.3	92.5–94.9	–	–	–	–	–	–	–	–
SCP of Lee et al. (2020)	79.8	–	92.9	72.0	59.9	73.5	–	–	–	–
SCP of Perera et al. (1995a, 1995b)	72.9–90.8	39.3–86.3	–	–	–	–	–	–	–	–

ulatus with increased *M. capsulatus* inclusion (Zhang et al., 2023), and lower ADCs in *S. salar* (Aas et al., 2006a; Romarheim et al., 2011; Skrede et al., 1998; Storebakken et al., 2004), in *Scophthalmus maximus* (J. Zheng et al., 2023) and in *Seriola quinqueradiata* that showed lower protein digestibility with increased SCP inclusion (Biswas et al., 2020). As noted by Glencross et al. (2023), inclusion level as well as acclimation time may affect the outcome of digestibility trials and so these data need to be treated with caution. Typical protein digestibility values for *M. capsulatus* exceed 85% as noted in a number of fish species including *Cyprinus carpio* (Yu et al., 2022), Hybrid grouper (C.-Z. Zheng et al., 2023), *Hippoglossus hippoglossus* (Aas et al., 2007), *M. salmoides* (Shi et al., 2022), *Oreochromis niloticus* (do Vale Pereira et al., 2024), *Seriola lalandi* (Pilmer et al., 2022) and salmonids (Prakash et al., 2023; Berge et al., 2005; Storebakken et al., 1998).

Digestibility of feeds incorporating *M. methylotrophus* have been tested in *A. anguilla, Cyprinus carpio* and *Oncorhynchus mykiss*. Although inclusion level typically decreased the digestibility coefficients, the effect was limited. Maximum ADC values were as follows: crude protein around 96%, organic matter 90% (but as low as 47% in *O. mykiss*), dry matter 53.7%, ash 22.4%, and energy 49.4% (Atack et al.,

1979; Kaushik and Luquet, 1980; Schmitz et al., 1984). Digestibility of *M. extorquens* has been assessed in *Litopenaeus vannamei* and *Salmo salar*. Maximum ADC values were 92% for crude protein, 94% for crude lipid, 74% in dry matter, 24% for carbohydrate, 83% for energy (Qiu et al., 2018; Salze and Tibbetts, 2021; Tlusty et al., 2017). ADCs for essential amino acids were typically greater than 85% in *S. salar* (Salze and Tibbetts, 2021; Tlusty et al., 2017). Similar ADC values were obtained in *Salmo salar* fed *Methylovorus menthalis* at 92.9% for crude protein, 77.5% for dry matter, 84.7% for energy and values for essential amino acids exceeding 90% (Tibbetts et al., 2025b).

Several unidentified SCPs have been assessed for digestibility in *Lutjanus malabaricus*, and *O. mykiss*. ADC values for the SCP tested in *L. malabaricus* were 100% for dry matter, 92% for crude protein, 92.4% for crude fats, and 89.5% for carbohydrates; ADC's for essential amino acids generally exceeded 90% (Ngoh et al., 2025). Improved ADCs were obtained in *O. mykiss* with crude protein ADCs exceeding 91% in two SCPs tested by Buttle et al. (2024); lower total amino acid digestibilities were observed but these did not affect growth outcomes. In contrast, ADC values for the SCP Sasya tested in the same animal were 59.9% for dry matter, 79.8% for crude protein, 92.9% for lipid, 72.0% for organic matter and 73.5% for energy (Lee et al., 2020).

The activity of digestive enzymes can give an indication of a fishes ability to utilize different feed sources, providing vital evidence of how well a new ingredient may be utilized and absorbed (Jiao et al., 2023). Arguably, a lack of difference in digestive enzymes between different dietary treatments would suggest that those diets would be well tolerated. Indeed, no differences in digestive enzymes were noted following the inclusion of *C. autoethanogenum* in *Dicentrarchus labrax* or *Acanthopagrus schlegelii* diets (Abdel-Tawwab et al., 2025; Chen et al., 2020), nor when *M. capsulatus* was included in *Haliotis laevigata* × *H. rubra* or *Litopenaeus vannamei* diets (Salini et al., 2025; Chen et al., 2022), suggesting that these SCPs were well tolerated. Increases in digestive enzyme activity have been reported and can be indicative of the availability of high-quality protein sources as well as lipopolysaccharides and biotics in the diet (Amenyogbe et al., 2024; Jiao et al., 2023; Wang et al., 2022). Improvements in the enzyme activities of *L. vannamei* and *M. salmoides* fed *C. autoethanogenum* (Cai et al., 2022; Chen et al., 2025; Luo et al., 2023; Yuan et al., 2024) and in *Acanthopagrus schlegelii, Lateolabrax maculatus* and *Sparus aurata* fed *M. capsulatus* (Tampou et al., 2024; Xu et al., 2021; Zhang et al., 2023) have been noted. In contrast, some SCP substitutions, do not elicit a positive response in enzyme activity, even in the same species where an increase has been reported. For example, reductions in enzyme activity have been reported in *M. salmoides,* and *O. mykiss* fed *C. autoethanogenum* (H. Huang et al., 2023; Yang et al., 2023), and in *Anguilla rostrata, L. maculatus, Scophthalmus maximus,* and *Stichopus monotuberculatus* fed *M. capsulatus* (Lu et al., 2023; Yu et al., 2023; Jiayin Zhang et al., 2025;

J. Zheng et al., 2023). However, these changes are not always clear cut and may be related to diet formulation, as well as the inclusion level of the cultured protein. For instance, an increase followed by a decrease in activities of trypsin in *L. crocea* and lipase in *M. salmoides* have been recorded in higher inclusion levels of *C. autoethanogenum*, despite other enzymes showing a linear relationship within the same study (Y. Wu et al., 2022; Zhu et al., 2022). It follows that careful consideration needs to be made of the formulations and the extremes of inclusion to fully understand the effects of SCP substitutions on aquaculture feeds.

4 Sera biochemistry

The effect of cultured proteins on overall health can be assessed by measuring a range of non-specific health markers in the sera as well as more specific immune measures. Broadly speaking, most blood parameters remain within normal limits and are not significantly different from control diets under experimental conditions. Elevated levels of alkaline phosphatase (AKP / ALP), alanine transaminase (ALT / GPT) and /or aspartate transaminase (AST / GOT) have been reported in *C. idella, D. labrax,* hybrid grouper, *L. vannamei, L. crocea, M. salmoides,* and *P. clarkii* given diets containing *C. autoethanogenum*, suggesting some level of hepatic damage caused by the SCP (Abdel-Tawwab et al., 2025; Dai et al., 2023; HongCheng et al., 2018; B. Huang et al., 2023; Jiang et al., 2021; Yang et al., 2024; Yuan et al., 2024; J. Zhang et al., 2022; Zhu et al., 2022). There are however, other studies that have shown the opposite with lower values for these measures of hepatic damage (M. Li et al., 2021; Ma et al., 2022a; Yang et al., 2023; J. Zheng et al., 2022). Variable results have also been obtained for immune related genes including lower expression of IL-1α, IL-8, TNFα in *M. salmoides* (Lu et al., 2021), but higher expression of C3, IFNγ, TGFβ, in *O. niloticus* (Maulu et al., 2021a). In contrast, no differences in IgM, IL-8, and IL-1α were reported in *C. idella* fed *C. autoethanogenum* (HongCheng et al., 2018).

In *M. capsulatus*-fed animals, values for ALP, ALT and AST are normally lower or not significantly different when compared to control diets (Chama et al., 2021; Chen et al., 2022; Dai et al., 2024; Guo et al., 2022; Lu et al., 2023; Pilmer et al., 2025; Woolley et al., 2023; Xu et al., 2021; Yu et al., 2022; Q. Zhang et al., 2022). For cytokines and similar, lower expression of NF-kB, TNFα, IL-8, IL-10, IL-16 and IFNy (Chama et al., 2021; Zhang et al., 2023) have been reported for *L. maculatus* and *O. niloticus* fed *M. capsulatus*. Additionally, higher expressions of TGFβ, IL-4 and IL-10 have been noted in *L. maculatus* given the same cultured protein (Zhang et al., 2023). No significant differences in blood parameters were noted in salmonids given Sasya or *M. menthalis* in test diets (Lee et al., 2020; Tibbetts et al., 2025a, 2025b).

5 Antioxidant capacity

Antioxidant capacity is a measure of an animals ability to prevent oxidative stress and to neutralize harmful free radicals. It is measured using a number of different tests including total antioxidant capacity (T-AOC) levels, and activity of superoxide dismutase (SOD), catalase (CAT), Glutathione peroxidase (GSH-Px), and malondialdehyde (MDA). In general, increases in T-AOC values are considered beneficial as they demonstrate the hosts ability to scavenge reactive oxygen species and in limiting oxidative damage; exceptionally increases in MDA values are indicative of oxidative stress. Whilst the addition of *C. autoethanogenum* led to reduced T-AOC levels in *L. vannamei*, *Megalobrama amblycephala* × *Culter alburnus* and *M. salmoides* (C. Zheng et al., 2022b; Qiuxin et al., 2025; Lu et al., 2021) and increased T-AOC in *M. salmoides* (Zhu et al., 2022), it had no effect on this measure in *Acanthopagrus schlegelii*, *Cyprinus carpio*, or *Larimichthys crocea* (Chen et al., 2020; M. Li et al., 2021; Wang et al., 2023). In addition, CAT activity increased in *Ctenopharyngodon idella* and *C. carpio* but was not significantly different between control and *C. autoethanogenum*-fed *A. schlegelii*, *Haliotis discus hannai*, *L. crocea*, *M. salmoides* or *Oreochromis niloticus* (Chen et al., 2020; Wu et al., 2024; Wang et al., 2023; Lu et al., 2021; Zhu et al., 2022; Chen et al., 2025; Maulu et al., 2021a). Generally, SOD activity increases with increased *C. autoethanogenum* inclusion (Yang et al., 2024; Shi et al., 2025; L. Li et al., 2022; Chen et al., 2025; Maulu et al., 2021a). In contrast, MDA levels either increase (Xia et al., 2025; Wu et al., 2024; Qiuxin et al., 2025; Lu et al., 2021) or decrease (Yang et al., 2024; M. Li et al., 2021; Shi et al., 2025; Zhu et al., 2022; Maulu et al., 2021a).

Addition of *Methylococcus capsulatus* to diets increased T-AOC levels in *C. carpio*, hybrid grouper, and *O. niloticus* (Yu et al., 2022; Liang et al., 2022; Chama et al., 2021) but decreased T-AOC levels in *S. maximus*, *T. ovatus*, and *Anguilla rostrata*, especially at high inclusion levels (J. Zheng et al., 2023; Gao et al., 2023; Lu et al., 2023). At high inclusion levels, SOD activity was decreased in *Pelodiscus sinensis*, *S. maximus*, and *A. rostrata* (Dai et al., 2024; J. Zheng et al., 2023; Lu et al., 2023), but generally SOD activity was increased when *M. capsulatus* was added (Liang et al., 2022; Gisbert et al., 2025; Guo et al., 2022; Gao et al., 2023). CAT levels are typically reduced in animals fed *M. capsulatus* (Dai et al., 2024; J. Zheng et al., 2023; Fernandes et al., 2024; Lu et al., 2023).

6 Gut structure / functionality

As the main organ responsible for digestion and absorption of nutrients, the gastrointestinal tract remains a strong focus of studies to understand the impact of feed ingredients on its functionality. Inclusion of *C. autoethanogenum* in the diet appears to have variable effects on gut structure, even within the same species and is thus likely to be related to inclusion level and feed formulation. Increases in intestinal fold height and goblet cell numbers have been noted in *C. idella* (Yang et al., 2024) and *M. salmoides* (Li et al., 2023; Luo et al., 2023), In contrast, increased inclusion led to decreased mucosal fold height, muscle thickness and a decreased number of goblet cells of hybrid grouper (B. Huang et al., 2023) and *M. salmoides* (H. Huang et al., 2023; L. Li et al., 2022; Yang et al., 2023, 2022a; Zhu et al., 2022), changes in villus height in *L. crocea* (Y. Wu et al., 2022; J. Zhang et al., 2022), a lower mucosal fold height in *S. maximus* (Bai et al., 2024) and reductions in villus height and damage to the brush borders as well as separation of the gut epithelium from the basement membrane in *L. vannamei* (Jiang et al., 2021; Xu et al., 2023b; C. Zheng et al., 2022a, 2022b). No histological differences were noted in the intestines of *C. carpio* (M. Li et al., 2021), of *D. labrax* (Abdel-Tawwab et al., 2025), of *L. vannamei* (Cai et al., 2022; Kuo et al., 2024), of *P. clarkii* (Dai et al., 2023), or of *M. salmoides* (Ma et al., 2022a) fed *C. autoethanogenum*.

Similarly, variable results have been noted in intestinal structure of animals fed *M. capsulatus*, with inclusion level greatly influencing the results obtained. Inclusion of *M. capsulatus* in the diets of *M. salmoides* led to an increase in intestinal wall thickness and villus height at lower inclusion levels and a significant decrease when inclusion levels were high (Q. Zhang et al., 2022). Similarly, a lower villus height and muscular thickness were noted in *A. rostrata* at high inclusion levels (Lu et al., 2023). Increases in microvillus density and villus height were noted at high inclusion levels in *A. schlegelii* (Xu et al., 2021) and hybrid grouper (Liang et al., 2022) whilst villus width was increased in *C. carpio* (Yu et al., 2022). Minor epithelial vacuolization of the midgut and pyloric caeca was noted in *H. hippoglossoides* by Aas et al. (2007); no other histological observations were noted. No negative histological observations were noted in *P. sinensis* (Dai et al., 2024), in *L. maculatus* (Yu et al., 2023; Zhang et al., 2023), in *L. calcarifer* (Woolley et al., 2023), in *L. vannamei* (Gisbert et al., 2025), in *M. salmoides* (Guo et al., 2022), in *O. mykiss* (Aas et al., 2006b; Rajesh et al., 2022; Ruiz et al., 2023), in *S. salar* (Berge et al., 2005; Romarheim et al., 2013, 2011; Storebakken et al., 2004), or in *S. lalandi* (Pilmer et al., 2025) fed *M. capsulatus*.

7 Microbiota

A recent innovation in assessing the wider biological impact of feed and feed ingredients on health and growth outcomes is the assessment of the gut microbiome. Alterations in microbiome diversity, richness, and composition have been noted, with studies focused on the inclusion of *C. autoethanogenum* and *M. capsulatus* (Chen et al., 2025; Dai et al., 2024, 2023; do Vale Pereira et al., 2024; Gisbert et al., 2025; Guo et al., 2022; Li et al., 2023; L. Li et al., 2022; Lu et al., 2023; Luo et al., 2023; Ma et al.,

2022a; Piazzon et al., 2022; Pilmer et al., 2025; Ruiz et al., 2023; Shi et al., 2025; Wu et al., 2024; Yang et al., 2024; Yu et al., 2023; Yuan et al., 2024; Jiacheng Zhang et al., 2025; J. Zhang et al., 2022; Zhang et al., 2023; C. Zheng et al., 2022a, 2022b; Zheng et al., 2024). Only one study to date appears to have been conducted on the effect of *M. extorquens* on gut microbiome in which no differences in the microbiome of *Haemulon chrysargyreum* were noted when fed diets containing up to 25% of the cultured protein for 41 days (Tlusty et al., 2017).

For *C. autoethanogenum* the majority of studies have focused on the role of the protein source on the microbiota of *L. vannamei* and *M. salmoides*. Overall, improvements in beneficial bacteria and decreases in the abundance of pathogenic bacteria are noted alongside a general increase in diversity measures in *L. vannamei* (Shi et al., 2025; Yuan et al., 2024; C. Zheng et al., 2022a, 2022b). Similarly, addition of *C. autoethanogenum* in the diets of *M. salmoides* led to increases in beneficial bacteria, with a notable increase in *Cetobacterium* in the gut (Chen et al., 2025; Li et al., 2023; L. Li et al., 2022; Luo et al., 2023; Ma et al., 2022a). Whilst no significant differences were noted in the microbiome of *C. idella, H. discus hannai*, or *L. crocea* when *C. autoethanogenum* was included in the diets (Wu et al., 2024; Yang et al., 2024; J. Zhang et al., 2022), there was a notable decrease in microbiome diversity in the intestines of *P. clarkii* attributed to its inclusion (Dai et al., 2023).

In contrast, the microbiome of a wider number of species have been assessed for *M. capsulatus*-fed animals. Compared with diets containing soybean meal, there is generally an increase in diversity and a reduction in pathogenic bacteria and the microbiome is implicated in the amelioration of soybean meal induced enteritis in *L. maculatus* (Zhang et al., 2023). Differences in diversity and richness have been noted as being non-significant in most studies, although there is a trend towards an improvement in these measures with increased inclusion levels of *M. capsulatus* (Gisbert et al., 2025; Guo et al., 2022; Lu et al., 2023; Jiacheng Zhang et al., 2025). Additionally, there is an increase in beneficial bacteria, particularly for genera such as *Cetobacterium* and *Lactobacillus* (Dai et al., 2024; do Vale Pereira et al., 2024; Pilmer et al., 2025; Yu et al., 2023; Zheng et al., 2024). A marked shift in the microbiome, with a notable increase in the pathogenic bacterium *Aeromonas schubertii* was noted in *L. calcarifer* fed *M. capsulatus* by Samsing et al. (2024). The microbial dysbiosis was attributed to the lower inclusion of fish meal in the diets and the concomitant outbreak of the bacterial infection. The understanding of the role of the microbiome in improving health and growth outcomes and in wider biological processes is in its infancy in fish, particularly when applied to feed ingredient studies. Future work should address the interactions between cultured proteins and the microbiome and widen the number of studies to new and novel protein sources and further assess the mycobiota as well as the more common microbiota in animal hosts.

8 Disease

One useful facet of single cell proteins in diets are their ability to confer additional health benefits such as resistance to disease challenges and to protect against dietary induced changes to gut function. Survival rates are typically equal or greater in *Litopenaeus vannamei* fed diets containing *Clostridium autoethanogenum* and *Methylococcus capsulatus* compared with fishmeal following challenge with *Vibrio parahaemolyticus* (Chen et al., 2022; Gisbert et al., 2025; Jintasataporn et al., 2021; Kuo et al., 2024; Shi et al., 2025; Yuan et al., 2024; C. Zheng et al., 2022b, 2022a). Similarly, survival of *Haliotis discus hannai* was markedly improved when challenged with *V. parahaemolyticus* following a 110 day feeding trial with *C. autoethanogenum* (Wu et al., 2024). Following a 56-day trial where *M. salmoides* were fed diets containing 19–38% *C. autoethanogenum*, 20 fish from each dietary groups were challenged by intraperitoneal injection with *Aeromonas hydrophila* (Chen et al., 2025). Survival rates were greatly improved in fish fed the SCP, particularly in the 19% *C. autoethanogenum* group. Rainbow trout challenged with *Aeromonas salmonicida* subsp. *salmonicida* after being fed *M. capsulatus* for 83 days showed higher survival compared to control groups, particularly at intermediate inclusion levels (Ruiz et al., 2023). Although no differences in infection-associated parameters of *Sparus aurata* fed a diet containing both *Corynebacterium glutamicum* and *M. capsulatus* were noted by Piazzon et al. (2022), the authors concluded that the presence of cultured proteins in the diets were not detrimental to overall health.

One side effect of the inclusion of soybean meal as an alternative protein in aquaculture diets is that for some fish species this can lead to a condition known as soybean meal induced enteritis (SBMIE). Alterations to gut morphology include cellular infiltration and destruction of the intestinal architecture. Whilst some of these changes are reversible, the short-term negative effects on growth and survival can strongly influence economic outcomes for farmers (Peng et al., 2020). Thus, the potential savings in feed costs derived from the inclusion of cheaper components such as soybean are offset by the detrimental effects of this ingredient on overall health and survival. Inclusion of *M. capsulatus* has been shown to alleviate SBMIE in salmon, turbot and spotted sea bass (Romarheim et al., 2013, 2011; Zhang et al., 2023; Zheng et al., 2024). Whole cell as well as cell fractions of *M. capsulatus* have been shown to prevent SBMIE and it appears that large, water insoluble components of the bacteria are responsible for the effects observed (Romarheim et al., 2013). More recently, it has been suggested that one of the underlying mechanisms for the protective nature of *M. capsulatus* against SMBIE is that it can modulate the expression on inflammatory factors in the intestine whilst promoting the growth of anti-inflammatory bacteria within the gut microbiome (Zhang et al., 2023). Furthermore, upregulation of tight junction proteins in the intestine following the inclusion of *M. capsulatus* in the diet appear to be responsible for further alleviation of the effects of soybean on intestinal health (Zheng et al., 2024).

9 Body composition

Ultimately, aquaculture products are being grown to provide a suitable food source that contains an appropriate balance of protein and fats with an acceptable flavor profile. The role of feed in helping to achieve this is critical with ingredients having an impact on these important aspects. No impact on body composition was noted in a number of fish following the inclusion of *Clostridium autoethanogenum* in test diets. These include *Ctenopharyngodon idella* by HongCheng et al. (2018), *Cyprinus carpio* by M. Li et al. (2021), *Dicentrarchus labrax* by Abdel-Tawwab et al. (2025), hybrid grouper by B. Huang et al. (2023) and Xu et al. (2023a), *Larimichthys crocea* by Wang et al. (2023), *Litopenaeus vannamei* by Kuo et al. (2024) and Yao et al. (2024), *Micropterus salmoides* by Yang et al. (2023) and Zhu et al. (2022), *Procambarus clarkii* by Dai et al. (2023), and *Takifugu obscurus* by Cui et al. (2022). However, although Kuo et al. (2024), Yao et al. (2024), Yang et al. (2023), Wang et al. (2023), and HongCheng et al. (2018) did not note any differences in body composition of *C. idella*, *M. salmoides*, *L. crocea* and *L. vannamei* with the inclusion of *C. autoethanogenum*, an increase in whole-body moisture content (Yang et al., 2024; Yao et al., 2022; J. Zhang et al., 2022) and crude protein (Xue et al., 2023; Luo et al., 2023; Yang et al., 2022a; Xia et al., 2025) has been noted in these animals. Decreases in whole-body crude protein have been recorded with the inclusion of *C. autoethanogenum* in the diets of *Megalobrama amblycephala* × *Culter alburnus* by Qiuxin et al. (2025), in *L. crocea* by Y. Wu et al. (2022), in *L. vannamei* by Shi et al. (2023), and in *Scophthalmus maximus* by Bai et al. (2024) and Qi et al. (2024a). Addition of *C. autoethanogenum* to diets increased whole-body crude protein content of *Haliotis discus hannai* (Z. Wu et al., 2022), and increased crude lipid content of *Scophthalmus maximus*, *M. salmoides*, *Oncorhynchus mykiss*, and *Oreochromis niloticus* (J. Zheng et al., 2022; Yang et al., 2022b; H. Huang et al., 2023; Maulu et al., 2021b). In contrast, inclusion of *C. autoethanogenum* decreased whole-body lipid content of *C. idella*, *L. crocea*, and *S. maximus* (Pan et al., 2024; Xia et al., 2025; J. Zhang et al., 2022).

As with *C. autoethanogenum*, inclusion of *M. capsulatus* in feeds appears to have different effects on body composition, likely dependent on feed formulation and inclusion level as well as the species of animal tested. For example, no differences in body composition were noted in *Cyprinus carpio* by Yu et al. (2022), in *Lates calcarifer* by Woolley et al. (2023), in *Hippoglossus hippoglossus* by Aas et al. (2007), in *Micropterus salmoides* by Q. Zhang et al. (2022), in *Oncorhynchus mykiss* by Aas et al. (2006b), in *Pelodiscus sinensis* by Dai et al. (2024), in *Salmo salar* by Aas et al. (2006a), in *Seriola quinqueradiata* by Biswas et al. (2020), in *Scophthalmus maximus* by Hoerterer et al. (2022a, 2022b), in *Sparus aurata* (Carvalho et al., 2023, 2024), and in *Litopenaeus vannamei* by Y. Chen et al. (2022a) and Felix et al. (2023). However, despite a lack of difference in some studies, others working with the same species have noted differences in body composition. For example, J. Zheng et al. (2023) noted increases in moisture, lipid and crude protein of *Scophthalmus maximus*, whilst Gisbert et al. (2025) noted decreases in monounsaturated fatty acids (MUFA) and polyunsaturated fatty acids (PUFA) in the bodies of *L. vannamei* (although body composition overall were similar). In contrast, crude protein and moisture levels were significantly higher in *L. vannamei* fed diets containing *M. capsulatus* (Chen et al., 2022). Reduced levels of body lipid and increased crude protein were noted by Ruiz et al. (2023) whilst lower crude protein and fat were noted by Vale Pereira et al. (2023) and Rajesh et al. (2022) in *O. mykiss*. Differences between the composition of viscera and the body of *S. salar* were noted by Berge et al. (2005) who showed that although significantly lower values for dry matter and crude fat occurred in the viscera, no differences were noted in body composition. Similarly, Vasilaki et al. (2023) noted an increase in whole-body fat content and a decrease in essential amino acids of *D. labrax* whilst Moroni et al. (2024) noted a decrease in whole-body fat and moisture content. In contrast, no differences in whole-body crude fat, moisture or ash of *D. labrax* fed diets containing up to 5% *Corynebacterium glutamicum*, 15% *Hermetia illucens* and 15% *M. capsulatus* were noted although whole-body crude protein content was increased (Petereit et al., 2022). Increases in crude fat content of whole fish were noted in *Trachinotus ovatus* (Gao et al., 2023), whilst decreases in whole-body fat content were noted in *Pagrus major* by Biswas et al. (2021) and in *Lateolabrax maculatus* (Zhang et al., 2023) who also showed a decreasing trend in whole-body crude protein alongside an increasing trend in body moisture.

Atack et al. (1979) tested *M. methylotrophus* alongside a number of protein sources, including yeast, algae, fishmeal and soybean in *C. carpio* diets and found few differences in the body composition of fish fed different diets. Similarly, no significant differences were apparent in *O. mykiss* fed *M. methylotrophus* although there was a slight increase in body lipid content at the higher inclusion (Kaushik and Luquet, 1980). *Saccostrea glomerata* (= *commercialis*) fed artificial diets containing *M. methylotrophus* had a higher glycogen content but a similar amino acid profile compared to control diets (Nell, 1985). A protein derived from an unidentified bacteria that uses carbon dioxide as its feedstock was included at up to 13.6% of *O. mykiss* diets (Lee et al., 2020). At the highest inclusion level whole body crude fat and gross energy were higher compared with fishmeal controls although C20:1n9 and C20:2n6 levels were lower. A second unidentified SCP was added to diets of *Stichopus monotuberculatus* by Jiayin Zhang et al. (2025), replacing up to 25% of the other protein sources. Crude protein content of the sea cucumbers increased at up to a 15% replacement whilst the fatty acids C16:0, C18:0, C20:5 n-3, C18:2 n-6 and C20:4 n-6 differed between groups. A single study testing the effect of *Methylorubrum extorquens* on body composition of *Haemulon chrysargyreum* showed that inclusion of the protein led to a lower whole-body protein content and a higher lipid content (Tlusty et al., 2017).

10 Production outcomes

One of the major outcomes measured in aquaculture production is the feed conversion and growth of an animal based on the diet provided with the aim to minimize the amount of feed given for the maximum growth whilst maintaining excellent flesh quality and low mortality. Typically, these measures include specific growth rate (SGR), body weight gain rate (WGR), feed efficiency / feed conversion efficiency (FE / FCE), and feed conversion ratio (FCR). In general, almost all nutrition studies involving cultured proteins include some element of these measurements. As a result, it is recommended that the reader consult the specific papers to obtain the details for each SCP and each species tested. However, it is clear that there is typically an optimum inclusion level that provides the best outcome; this is often not the maximum inclusion level tested.

11 Technical properties of feed

Despite the importance of the physical characteristics of feeds with regards to their behavior in storage, and when given to animals, few studies have considered the role of cultured proteins on pellet quality. As noted by Liu et al. (2023), that pellet quality is driven by the ingredients and by the physical characteristics of the extruder and variables such as temperature, pressure and extrusion speeds. Each of these factors need to be measured accurately and variations due to e.g. ingredients, need to be tested. Partially replacing fishmeal with *M. capsulatus* in a floating pellet, Liu et al. (2023) noted a number of significant changes in the physical characteristics of pellets. Peak-, trough- and breakdown-viscosities were higher in *M. capsulatus* feeds during the pasting phase whilst final viscosity, setback-viscosity and pasting temperature were lower. During extrusion, increased inclusion of *M. capsulatus* led to a significantly decreased bulk density and increased expansion ratio, improving pellet floatability. Pellet durability was improved, although inclusion level of *M. capsulatus*, moisture content and die temperature did not alter the overall hardness of the pellet. Interestingly, oil absorption and retention was improved in feeds containing the cultured protein, attributed to the pore microstructure within the pellet that was better able to absorb the oils, reducing oil loss when the pellet came into contact with water. A similar effect with *C. autoethanogenum* feeds was noted by Ma et al. (2022b). Overall, partially replacing fishmeal with *M. capsulatus* in a standard diet led to improvements in physical characteristics of the pellet, and lowered the water and energy consumption requirements during the production phase, thus reducing overall production costs.

Three studies examining the influence of *C. autoethanogenum* on pellet quality have been conducted. Loss of dry matter in pellets immersed for up to 48 hours was not significantly different across experimental diets where up to 100% of the fishmeal in a *Haliotis discus hannai* diet was replaced with *C. autoethanogenum* (Z. Wu et al., 2022). When fish meal was replaced with *C. autoethanogenum*, peak-and trough-viscosity was significantly increased (Ma et al., 2022b). Whilst breakdown viscosity increased when up to 75% of fishmeal was replaced, once fishmeal was fully replaced in the pellet, the breakdown viscosity decreased significantly. Additionally, pasting temperature and setback viscosity decreased with increased *C. autoethanogenum* inclusion, but no overall effect on final viscosity was noted, irrespective of inclusion level. Specific mechanical energy (SME) increased with higher *C. autoethanogenum* inclusion with the highest SME values noted in the 100% substitution group with low screw speed and low moisture content. Bulk density decreased up to 75% inclusion then increased once *C. autoethanogenum* replaced 100% of the fishmeal. As expected, expansion ratio showed the opposite trend. Improvements in oil absorption capacity were noted by the authors. In contrast to *M. capsulatus*, inclusion of *C. autoethanogenum* required a higher moisture content and thus a higher energy cost to dry the feed and an increased in SME during extrusion (Ma et al., 2022b). Finally, Cai et al. (2022) replaced 30% of the fish meal in a *L. vannamei* diet with either *Tenebrio molitor* meal, cottonseed protein concentrate, or *C. autoethanogenum*. For all three alternative protein sources, pellet hardness was significantly increased whilst dissolution loss and powder content in the cottonseed concentrate and *C. autoethanogenum* feeds were significantly decreased. Despite the importance of pellet quality on growth outcomes and profit margins throughout the supply chain, there remains a dearth of studies assessing the impact of protein sources on this parameter. It follows that studies examining the effect of microbial proteins and different feed types on final feed quality should be carried out, especially with untested new and novel protein sources and of the proteins they replace.

12 Costs and economics

The economic feasibility of bacterial proteins from fermentation is very important to their implementation in aquafeeds. According to FAO report from 2022 fishmeal has reached a production capacity of about 5 million tonnes annually and it plays a major role as the main ingredient of aquafeeds due to its high digestibility and balanced amino-acid profile. However, limited supply and rising demand from aquaculture will increase prices from approximately 1,600$ per tonne in 2022 to an estimated 2,545$ per tonne by 2030 (Future Market Insights, 2022), putting pressure on profitability and affordability of aquafeed, necessitating cost-effective alternatives (Miles and Chapman, 2024). Production costs of bacterial proteins vary depending on the process technology, feedstock used and microbial strain itself. First generation microbial proteins, like yeast-based or algal products, were highly expensive, reported in the range of 20–300$ per kilogram (Colombo et al., 2023). Recent technologies, like cultivation of methanotrophic bacteria on methane or biogas are showing much lower cost. El Abbadi et al. (2022) estimated that stranded methane-based SCP production can achieve price of 1.5–2$ per kilogram, which then can be further optimized as production scales and efficiency of the process improves. Shahzad et al. (2024) estimated that by using biogas and digestate streams, calculated production costs would be 1.54$ per kilogram, which places bacterial protein close to fishmeal on a cost and nutrient basis. Feed costs were significantly reduced when replacing fish meal with *C. autoethanogenum* in *S. maximus* and *D. labrax* diets (Abdel-Tawwab et al., 2025; Zheng et al., 2022). Usage of other feedstocks, such as molasses or agro-industrial byproducts, have also showed potential for low-cost microbial protein manufacturing and reduced production costs to around 1$ per kilogram of dry biomass (Lee and Kyun Kim, 2001). However, according to Glencross et al. (2020) when more energy intensive technologies are used, like those requiring expensive carbon sources, production costs in some cases can rise significantly, up to 5$ per kilogram.

In economic assessments of aquafeeds, bacterial ingredients are progressively evaluated by aquafeed manufacturers on a digestible nutrient cost basis. While soybean meal still represents the least expensive protein source (around 1$ per kilogram), bacterial proteins are already competitive with fishmeal and outperform it in terms of digestible protein as technological advances reduce production inputs (Ngoh et al., 2025). However, limitations in palatability and growth performance at higher inclusion rates reduce its current cost effectiveness comparative to plant-based proteins (Ngoh et al., 2025). To maximize economic returns, these performance constraints need to be addressed through bacterial strain selection and technology improvements.

Despite these challenges, bacterial proteins can replace up to 50% of fishmeal in aquafeeds without having negative impact on growth, which correspond to significant cost savings given current fishmeal market prices (Jannathulla et al., 2021). Furthermore, modular and decentralized bacterial protein facilities co-located with methane or biogas sources could reduce transportation and logistics costs while benefiting from economies of unit number, thus enhancing economic benefits (Zheng et al., 2024). Importantly, microbial protein costs are expected to stabilize or decrease with scale and innovation, while fishmeal remains unpredictable due to climate-driven fishery variability and geopolitical trade limitations. In summary, current evidence suggests that bacterial proteins production costs have transitioned from being excessively high to being within the same range of price as fishmeal. With ongoing optimization and economy of scale, bacterial proteins could not only provide cost savings in aquafeed formulations but also add economic value by stabilizing supply chains and reduce risks associated with fishmeal price volatility and limited quantity.

13 Limitations and future direction

The aquaculture industry is the fastest growing food production industry globally, driven by a rising demand for healthy seafood and its efficient use of resources compared to terrestrial farming. Bacterial proteins can contribute to the sustainable development of the aquaculture industry by providing a highly nutritious value alternative protein source, a stable quality raw material and, in many cases, a health boost. As with every new technology, the initial development steps are characterized by increased production costs, making commercial implementation challenging prior to scaling up the production and economies of scale come into action. While the industry focuses on optimizing the current sources of bacterial protein, other solutions will come into play, applying additional pressure for cost-competitive solutions. Comparable animal performance data using conventional aquafeeds together with a disproportionally higher production cost in bacterial protein sources do not allow significant and impactful application of the latter. The current infrastructure limits scaling up of any commercial production and significant investment will be required to have an impactful change to the industry. Given that current production of fermented bacterial proteins is still in its infancy, it is undeniable that production economics are also critical the implementation side of aquafeeds. Inclusion levels need to be optimized and application needs to be balanced to make financial sense to the aquafeed producers and with clear downstream performance benefits (e.g. fish growth, robustness, etc.).

A basic advantage of any farmed or cultivated raw material is the ability to provide a controlled production environment and stability between batches. However, various technical considerations need to be taken into account during extrusion, pelleting and storage. Starting with the fermentation process, the numerous variables (e.g. temperature, pH, substrate) need to be controlled under coherent quality standards. Processing at lower temperatures might be offering a higher yield, but it also raises spoilage risks. Moisture and water activity control pose a potential challenge (especially for the more hydrophilic products), unstable storage conditions could trigger microbial growth and risk safety. Furthermore, spores surviving the drying process can also allow growth of opportunistic bacteria. Evidences for the antioxidant capacity vary, and oxidative deterioration can still pose a significant risk. Given that bacterial proteins are still in their infancy in the aquafeed world, extensive quality control needs to be in place across the supply chain (i.e. production, transport, storage and use).

The regulatory framework needs to carefully safeguard the optimal use of the fermented bacterial proteins, and given a market pre-authorization, continuously updated to reflect new knowledge in this new field. Region-specific (e.g. use of GMO in the EU) or category-specific requirements (e.g. organic farming) need careful examination, together with a continuous market validation. Furthermore, the aquafeed market is taking advantage of the increasing demand of plant-based foods and advancements in fermentation technologies in other sectors, making the market acceptance easier compared to using other protein sources (e.g. insects). Nevertheless, the consumer's lack of understanding can cause an obstacle to market acceptance and wider social licensing.

All raw materials used in animal feeds, need to be safe, widely available and cost-effective. Aquafeeds are not an exception and this is where the industry needs to focus. Currently, the majority of bacterial protein production takes place in regions with high energy / high labor cost, and (in a global context) a relatively low aquaculture production (i.e. EU and USA). These constraints can confine commercial implementation. Strategic prioritization needs to focus on regions with the highest potential for financial, production and environmental growth (i.e. Asia and Africa), aiming impactful changes to the entire supply chain and the aquaculture industry as a whole. In that context, intensively farmed carnivorous species should also be prioritized, given the higher protein requirements. Aside from the financial roadblocks though, a technical roadmap is needed to overcome current scaling up challenges. For instance, the aquafeed industry is already facing drying capacity limitations and immense energy costs.

Practical considerations, such as oxidation levels, shelf-life, thermal tolerance (to survive the harsh extruding temperatures) and equilibrium moisture content, need to be extensively studied and matched with regional and species requirements. Fermentation can hydrolyze complex molecules, lower anti-nutritional factors and break down amino acids, aiming towards increased digestibility and offer better nutrient utilization. Long-term sustainability assessments which focus on environmental impacts and considers the entire supply chain need to be continuously assessed and develop universal metrics – comparable among species and regions. Consistent documentation is imperative to ensure adequate comparison between studies. Trial design, methods and interpretation should be based on universal methods, as currently industry reports and scientific studies use different metrics (% fish meal replaced vs % total diet replaced), often causing misleading conclusions.

Selective breeding for strains, targeting enhanced protein concentration and higher environmental resistance should be in line with tailor made solutions, based on species, environmental conditions and market priorities (yield, cost, quality, farming ease). Not everyone needs the same. Developing countries need affordable sources of protein, where more developed countries focus more on the sustainability accreditations and flesh quality parameters (e.g. ω-3 levels). The future direction needs to be part of a wider circular economy integration, out-of-silos development, precision nutrition and a clear industry buy-in. Together with adequate financial support, bacterial proteins will be a "resource for the future".

References

Aas, T.S., Grisdale-Helland, B., Terjesen, B.F., Helland, S.J., 2006a. Improved growth and nutrient utilisation in Atlantic salmon (*Salmo salar*) fed diets containing a bacterial protein meal. Aquaculture 259, 365–376. https://doi.org/10.1016/j.aquaculture.2006.05.032

Aas, T.S., Hatlen, B., Grisdale-Helland, B., Terjesen, B.F., Bakke-McKellep, A.M., Helland, S.J., 2006b. Effects of diets containing a bacterial protein meal on growth and feed utilisation in rainbow trout (*Oncorhynchus mykiss*). Aquaculture 261, 357–368. https://doi.org/10.1016/j.aquaculture.2006.07.033

Aas, T.S., Hatlen, B., Grisdale-Helland, B., Terjesen, B.F., Penn, M., Bakke-McKellep, A.M., Helland, S.J., 2007. Feed intake, growth and nutrient utilization in Atlantic halibut (*Hippoglossus hippoglossus*) fed diets containing a bacterial protein meal. Aquaculture Research 38, 351–360. https://doi.org/10.1111/j.1365-2109.2007.01672.x

Abdel-Tawwab, M., Khalil, R.H., Abo Selema, T.A.M., Abdelsalam, M., Abdelhakim, T.M.N., Sabry, E.A., Abd El-Ghaffar, H.A., Saad, A.H., 2025. Evaluating the inclusion of *Clostridium autoethanogenum* protein instead of fishmeal protein in diets for European seabass (*Dicentrarchus labrax*): Growth performance, digestive enzymes, health status, and tissues investigations. Animal Feed Science and Technology 324, 116318. https://doi.org/10.1016/j.anifeedsci.2025.116318

Amenyogbe, E., Droepenu, E.K., Ayisi, C.L., Boamah, G.A., Duker, R.Q., Abarike, E.D., Huang, J., 2024. Impact of probiotics, prebiotics, and synbiotics on digestive enzymes, oxidative stress, and antioxidant defense in fish farming: current insights and future perspectives. Front. Mar. Sci. 11. https://doi.org/10.3389/fmars.2024.1368436

Anon, 1978. Single-cell protein pilot unit starts up. C&EN May 15, 1978, 20.

Atack, T.H., Jauncey, K., Matty, A.J., 1979. The utilization of some single cell proteins by fingerling mirror carp (*Cyprinus carpio*). Aquaculture 18, 337–348. https://doi.org/10.1016/0044-8486(79)90037-1

Bai, N., Li, Q., Pan, S., Qi, Z., Deng, W., Gu, M., 2024. Effects of *Clostridium autoethanogenum* protein on growth performance, intestine and liver health of turbot *Scophthalmus maximus* during grow-out phase. Aquaculture Reports 36, 102076. https://doi.org/10.1016/j.aqrep.2024.102076

Banks, M., Johnson, R., Giver, L., Bryant, G., Guo, M., 2022. Industrial production of microbial protein products. Current Opinion in Biotechnology 75, 102707. https://doi.org/10.1016/j.copbio.2022.102707

Berge, G.M., Baeverfjord, G., Skrede, A., Storebakken, T., 2005. Bacterial protein grown on natural gas as protein source in diets for Atlantic salmon, *Salmo salar*, in saltwater. Aquaculture 244, 233–240. https://doi.org/10.1016/j.aquaculture.2004.11.017

Bewersdorff, M., Dostálek, M., 1971. The use of methane for production of bacterial protein. Biotechnology and Bioengineering 13, 49–62. https://doi.org/10.1002/bit.260130104

Biswas, A., Takakuwa, F., Yamada, S., Furukawa, A., Shiratori, M., Koshiishi, T., Tomokane, H., Tanaka, H., 2021. Partial replacement of fishmeal by a protein-rich product from methanotrophic bacteria in juvenile red sea bream (*Pagrus major*). Aquaculture Nutrition 27, 2726–2738. https://doi.org/10.1111/anu.13398

Biswas, A., Takakuwa, F., Yamada, S., Matsuda, A., Saville, R.M., LeBlanc, A., Silverman, J.A., Sato, N., Tanaka, H., 2020. Methanotroph (*Methylococcus capsulatus*, Bath) bacteria meal as an alternative protein source for Japanese yellowtail, *Seriola quinqueradiata*. Aquaculture 529, 735700. https://doi.org/10.1016/j.aquaculture.2020.735700

Buttle, L., Noorman, H., Roa Engel, C., Santigosa, E., 2024. Bridging the protein gap with single-cell protein use in aquafeeds. Front. Mar. Sci. 11, 1384083. https://doi.org/10.3389/fmars.2024.1384083

Cai, Y., Huang, H., Yao, W., Yang, H., Xue, M., Li, X., Leng, X., 2022. Effects of fish meal replacement by three protein sources on physical pellet quality and growth performance of Pacific white shrimp (*Litopenaeus vannamei*). Aquaculture Reports 25, 101210. https://doi.org/10.1016/j.aqrep.2022.101210

Carvalho, M., Ginés, R., Martín, I., Zamorano, M.J., Acosta, F., Fontanillas, R., Torrecillas, S., Montero, D., 2024. Genetic selection for high growth improves the efficiency of gilthead sea bream (*Sparus aurata*) in using novel diets with insect meal, single-cell protein and a DHA rich-microalgal oil. Aquaculture 578, 740034. https://doi.org/10.1016/j.aquaculture.2023.740034

Carvalho, M., Torrecillas, S., Montero, D., Sanmartín, A., Fontanillas, R., Farías, A., Moutou, K., Velásquez, J.H., Izquierdo, M., 2023. Insect and single-cell protein meals as replacers of fish meal in low fish meal and fish oil diets for gilthead sea bream (*Sparus aurata*) juveniles. Aquaculture 566, 739215. https://doi.org/10.1016/j.aquaculture.2022.739215

Chama, M.K.H., Liang, H., Huang, D., Ge, X., Ren, M., Zhang, L., Wu, L., Ke, J., 2021. Methanotroph (*Methylococcus capsulatus*, Bath) as an alternative protein source for genetically improved farmed tilapia (GIFT: *Oreochromis niloticus*) and its effect on antioxidants and immune response. Aquaculture Reports 21, 100872. https://doi.org/10.1016/j.aqrep.2021.100872

Chen, J., Huang, J., Pan, L., Song, F., Hu, Y., Zhang, S., Wang, Y., 2025. Composite single-cell protein as a sustainable fishmeal substitute: Enhancing growth and health in largemouth bass (*Micropterus salmoides*). Aquaculture Reports 42, 102810. https://doi.org/10.1016/j.aqrep.2025.102810

Chen, J., Wang, H., Yuan, H., Hu, N., Zou, F., Li, C., Shi, L., Tan, B., Zhang, S., 2022. Effects of dietary *Clostridium autoethanogenum* protein on the growth, disease resistance, intestinal digestion, immunity and microbiota structure of *Litopenaeus vannamei* reared at different water salinities. Front. Immunol. 13. https://doi.org/10.3389/fimmu.2022.1034994

Chen, K., Sagada, G., Xu, B., Liu, Y., Zheng, L., Tegomo, A.F., Yang, Y., Sun, Y., Wang, J., Longshaw, M., Shao, Q., 2022. Evaluation of methanotroph (*Methylococcus capsulatus*, Bath) bacteria meal as an alternative protein source for growth performance, digestive enzymes, and health status of Pacific white shrimp (*Litopenaeus vannamei*). Aquacult Int 30, 1693–1710. https://doi.org/10.1007/s10499-022-00869-9

Chen, Y., Chi, S., Zhang, S., Dong, X., Yang, Q., Liu, H., Tan, B., Xie, S., 2022a. Evaluation of Methanotroph (*Methylococcus capsulatus*, Bath) bacteria meal on body composition, lipid metabolism, protein synthesis and muscle metabolites of Pacific white shrimp (*Litopenaeus vannamei*). Aquaculture 547, 737517. https://doi.org/10.1016/j.aquaculture.2021.737517

Chen, Y., Chi, S., Zhang, S., Dong, X., Yang, Q., Liu, H., Zhang, W., Deng, J., Tan, B., Xie, S., 2021. Replacement of fish meal with Methanotroph (*Methylococcus capsulatus*, Bath) bacteria meal in the diets of Pacific white shrimp (*Litopenaeus vannamei*). Aquaculture 541, 736801. https://doi.org/10.1016/j.aquaculture.2021.736801

Chen, Y., Sagada, G., Xu, B., Chao, W., Zou, F., Ng, W.-K., Sun, Y., Wang, L., Zhong, Z., Shao, Q., 2020. Partial replacement of fishmeal with *Clostridium autoethanogenum* single-cell protein in the diet for juvenile black sea bream (*Acanthopagrus schlegelii*). Aquaculture Research 51, 1000–1011. https://doi.org/10.1111/are.14446

Chen, Y., Zheng, C., Zhang, X., Li, X., Yao, X., He, G., Cao, J., Chi, S., Zhang, S., Tan, B., Xie, S., 2022b. Evaluation of ratios of fish-to-soybean oils on growth, lipid and cholesterol metabolism and muscle metabolites of Pacific white shrimp (*Litopenaeus vannamei*) fed low fishmeal diets containing *Clostridium autoethanogenum* protein. Aquaculture Reports 27, 101417. https://doi.org/10.1016/j.aqrep.2022.101417

Colombo, S.M., Roy, K., Mraz, J., Wan, A.H.L., Davies, S.J., Tibbetts, S.M., Øverland, M., Francis, D.S., Rocker, M.M., Gasco, L., Spencer, E., Metian, M., Trushenski, J.T., Turchini, G.M., 2023. Towards achieving circularity and sustainability in feeds for farmed blue foods. Reviews in Aquaculture 15, 1115–1141. https://doi.org/10.1111/raq.12766

Cui, X., Ma, Q., Duan, M., Xu, H., Liang, M., Wei, Y., 2022. Effects of fishmeal replacement by *Clostridium autoethanogenum* protein on the growth, digestibility, serum free amino acid and gene expression related to protein metabolism of obscure pufferfish (*Takifugu obscurus*). Animal Feed Science and Technology 292, 115445. https://doi.org/10.1016/j.anifeedsci.2022.115445

Dabrowski, K., Charlon, N., Bergot, P., Kaushik, S., 1984. Rearing of coregonid (*Coregonus schinzi palea* cuv. et val.) larvae using dry and live food I. Preliminary data. Aquaculture 41, 11–20. https://doi.org/10.1016/0044-8486(84)90385-5

Dai, J., Chen, T., Guo, X., Dai, Z., He, Z., Hu, Y., 2023. Evaluation of fish meal replacement by *Clostridium autoethanogenum* protein in diets for juvenile red swamp crayfish (*Procambarus clarkii*). Aquaculture 570, 739379. https://doi.org/10.1016/j.aquaculture.2023.739379

Dai, J., Luo, H., Liu, Z., Hu, Y., 2024. Evaluation of fish meal replacement by *Methylcoccus capsulatus* protein in diets for juvenile Chinese softshell turtle (*Pelodiscus sinensis*). Aquaculture 740857. https://doi.org/10.1016/j.aquaculture.2024.740857

do Vale Pereira, G., Teixeira, C., Couto, J., Dias, J., Rema, P., Gonçalves, A.T., 2024. Dietary Protein Quality Affects the Interplay between Gut Microbiota and Host Performance in Nile Tilapia. Animals 14, 714. https://doi.org/10.3390/ani14050714

El Abbadi, S.H., Sherwin, E.D., Brandt, A.R., Luby, S.P., Criddle, C.S., 2022. Displacing fishmeal with protein derived from stranded methane. Nat Sustain 5, 47–56. https://doi.org/10.1038/s41893-021-00796-2

FAO, 2022. *The State of World Fisheries and Aquaculture 2022*. Rome: Food and Agriculture Organization of the United Nations.

Felix, N., Manikandan, K., Uma, A., Kaushik, S.J., 2023. Evaluation of single cell protein on the growth performance, digestibility and immune gene expression of Pacific white shrimp, *Penaeus vannamei*. Animal Feed Science and Technology 296, 115549. https://doi.org/10.1016/j.anifeedsci.2022.115549

Fernandes, A.M., Calduch-Giner, J.À., Pereira, G.V., Gonçalves, A.T., Dias, J., Johansen, J., Silva, T., Naya-Català, F., Piazzon, C., Sitjà-Bobadilla, A., Costas, B., Conceição, L.E.C., Fernandes, J.M.O., Pérez-Sánchez, J., 2024. Sustainable Fish Meal-Free Diets for Gilthead Sea Bream (*Sparus aurata*): Integrated Biomarker Response to Assess the Effects on Growth Performance, Lipid Metabolism, Antioxidant Defense and Immunological Status. Animals 14, 2166. https://doi.org/10.3390/ani14152166

Future Market Insights, 2022. *Fishmeal and Fish Oil Market Report 2022–2032*.

Gao W.-H., Peng S.-F., Liao C.-Y., Dong X.-H., Tan B.-P., Wang J., Chi S.-Y., 2023. Methanotroph (*Methylococcus capsulatus*, Bath) bacteria meal replacement of fish meal on growth, lipid deposition and antioxidant capacity of *Trachinotus ovatus*. Acta Hydrobiologica Sinica 47, 195–203. https://doi.org/10.7541/2023.2022.0245

Gisbert, E., Ruiz, A., Torrecillas, S., Cruz-Santana, Y., Bertomeu, E., Saromines, C., Melenchon, F., Ntokou, E., Pascual, I., Reinoso, S., Sarih, S., Solovyev, M., Furones, M., Thorringer, N.W., 2025. Methanotrophic bacterial meal as an alternative feed ingredient for whiteleg shrimp (*Litopenaeus vannamei*) diets. Aquaculture Reports 45, 103094.https://doi.org/10.1016/j.aqrep.2025.103094

Glencross, B.D., Huyben, D., Schrama, J.W., 2020. The Application of Single-Cell Ingredients in Aquaculture Feeds—A Review. Fishes 5, 22. https://doi.org/10.3390/fishes5030022

Glencross, B., Muñoz-Lopez, P., Matthew, C., MacKenzie, S., Powell, A., Longshaw, M., LeBlanc, A., 2023. Digestibility of bacterial protein by Atlantic salmon (*Salmo salar*) is affected by both inclusion level and acclimation time. Aquaculture 739137. https://doi.org/10.1016/j.aquaculture.2022.739137

Guo, B., He, X., Ge, C., Xue, M., Wang, Jia, Longshaw, M., Wang, Jie, Liang, X., 2022. A Natural Gas Fermentation Bacterial Meal (FeedKind®) as a Functional Alternative Ingredient for Fishmeal in Diet of Largemouth Bass, *Micropterus salmoides*. Antioxidants 11, 1479. https://doi.org/10.3390/antiox11081479

Hardy, R.W., Patro, B., Pujol-Baxley, C., Marx, C.J., Feinberg, L., 2018. Partial replacement of soybean meal with *Methylobacterium extorquens* single-cell protein in feeds for rainbow trout (*Oncorhynchus mykiss* Walbaum). Aquaculture Research 49, 2218–2224. https://doi.org/10.1111/are.13678

Hoerterer, C., Petereit, J., Lannig, G., Johansen, J., Conceição, L.E.C., Buck, B.H., 2022a. Effects of dietary plant and animal protein sources and replacement levels on growth and feed performance and nutritional status of market-sized turbot (*Scophthalmus maximus*) in RAS. Frontiers in Marine Science 9.

Hoerterer, C., Petereit, J., Lannig, G., Johansen, J., Pereira, G.V., Conceição, L.E.C., Pastres, R., Buck, B.H., 2022b. Sustainable fish feeds: potential of emerging protein sources in diets for juvenile turbot (*Scophthalmus maximus*) in RAS. Aquacult Int 30, 1481–1504. https://doi.org/10.1007/s10499-022-00859-x

HongCheng, W., HuanHuan, Y., XiaoMing, C., Wei, C., FangQi, Z., Pei, C., YinHua, Z., XiuFeng, W., XuFang, L., Min, X., 2018. Effects of soybean meal replaced by *Clostridium autoethanogenum* protein on growth performance, plasma biochemical indexes and hepatopancreas and intestinal histopathology of grass carp (*Ctenopharyngodon idellus*). Chinese Journal of Animal Nutrition 30, 4190–4201.

Huang, B., Shi, M., Pang, A., Tan, B., Xie, S., 2023. Effects of Fishmeal Replacement by *Clostridium Autoethanogenum* Protein Meal on Cholesterol Bile Acid Metabolism, Antioxidant Capacity, Hepatic and Intestinal Health of Pearl Gentian Grouper (*Epinephelus fuscoguttatus* ♀ × *Epinephelus lanceolatus* ♂). Animals 13, 1090. https://doi.org/10.3390/ani13061090

Huang, H., Li, X., Cao, K., Leng, X., 2023. Effects of Replacing Fishmeal with the Mixture of Cottonseed Protein Concentrate and *Clostridium autoethanogenum* Protein on the Growth, Nutrient Utilization, Serum Biochemical Indices, Intestinal and Hepatopancreas Histology of Rainbow Trout (*Oncorhynchus mykiss*). Animals 13, 817. https://doi.org/10.3390/ani13050817

Jannathulla, R., Sravanthi, O., Moomeen, S., Gopikrishna, G., Dayal, J.S., 2021. Microbial products in terms of isolates, whole-cell biomass, and live organisms as aquafeed ingredients: production, nutritional values, and market potential—a review. Aquacult Int 29, 623–650. https://doi.org/10.1007/s10499-021-00644-2

Jiang, X., Yao, W., Yang, H., Tan, S., Leng, X., Li, X., 2021. Dietary effects of *Clostridium autoethanogenum* protein substituting fish meal on growth, intestinal histology and immunity of Pacific white shrimp (*Litopenaeus vannamei*) based on transcriptome analysis. Fish & Shellfish Immunology 119, 635–644. https://doi.org/10.1016/j.fsi.2021.10.005

Jiao, F., Zhang, L., Limbu, S.M., Yin, H., Xie, Y., Yang, Z., Shang, Z., Kong, L., Rong, H., 2023. A comparison of digestive strategies for fishes with different feeding habits: Digestive enzyme activities, intestinal morphology, and gut microbiota. Ecol Evol 13, e10499. https://doi.org/10.1002/ece3.10499

Jintasataporn, O., Chumkam, S., Triwutanon, S., LeBlanc, A., Sawanboonchun, J., 2021. Effects of a Single Cell Protein (*Methylococcus capsulatus*, Bath) in Pacific White Shrimp (*Penaeus vannamei*) Diet on Growth Performance, Survival Rate and Resistance to *Vibrio parahaemolyticus*, the Causative Agent of Acute Hepatopancreatic Necrosis Disease. Frontiers in Marine Science 8, 1643. https://doi.org/10.3389/fmars.2021.764042

Kaushik, S.J., Luquet, P., 1980. Influence of bacterial protein incorporation and of sulphur amino acid supplementation to such diets on growth of rainbow trout, *Salmo gairdnerii* Richardson. Aquaculture 19, 163–175. https://doi.org/10.1016/0044-8486(80)90017-4

Kuo, Y.-C., Ho, T.H., Bharadwaj, A., Tran, H.T.Q., Chu, Y.-T., Wang, S.-H., Chen, T.-Y., Nan, F.-H., Lee, P.-T., 2024. Feasibility assessment of replacing fishmeal with *Clostridium autoethanogenum* protein in commercial whiteleg shrimp diets: Impacts on growth, muscle characteristics, and health. Animal Feed Science and Technology 309, 115916. https://doi.org/10.1016/j.anifeedsci.2024.115916

Lee, B.-K., Kyun Kim, J., 2001. Production of *Candida utilis* biomass on molasses in different culture types. Aquacultural Engineering 25, 111–124. https://doi.org/10.1016/S0144-8609(01)00075-9

Lee, S., Kumar, V., Cleveland, B., Romano, N., Vemuri, G.N., Yadav, A.K., Meiler, K., Hardy, R.W.,

2020. Fishmeal alternative from renewable CO2 for rainbow trout feed. Aquaculture Research 51, 4065–4074. https://doi.org/10.1111/are.14749

Li, L., Liu, X., Wang, Y., Huang, Y., Wang, C., 2022. Effects of alternate feeding between fish meal and novel protein diets on the intestinal health of juvenile largemouth bass (*Micropterus salmoides*). Aquaculture Reports 23, 101023. https://doi.org/10.1016/j.aqrep.2022.101023

Li, L., Wang, Y., Zhang, Z., Wang, C., 2023. Microbiomic and metabonomic analysis provide new insights into the enhanced intestinal health in large-size largemouth bass (*Micropterus salmoides*) when fed novel proteins: Novel proteins are promising foods for future aquaculture. Aquaculture 563, 739019. https://doi.org/10.1016/j.aquaculture.2022.739019

Li, M., Liang, H., Xie, J., Chao, W., Zou, F., Ge, X., Ren, M., 2021. Diet supplemented with a novel *Clostridium autoethanogenum* protein have a positive effect on the growth performance, antioxidant status and immunity in juvenile Jian carp (*Cyprinus carpio* var. Jian). Aquaculture Reports 19, 100572. https://doi.org/10.1016/j.aqrep.2020.100572

Li, X., Chen, Y., Zheng, C., Chi, S., Zhang, S., Tan, B., Xie, S., 2022. Evaluation of Six Novel Protein Sources on Apparent Digestibility in Pacific White Shrimp, *Litopenaeus vannamei*. Aquaculture Nutrition 2022, e8225273. https://doi.org/10.1155/2022/8225273

Li, X., Han, T., Zheng, S., Wu, G., 2021. Nutrition and Functions of Amino Acids in Aquatic Crustaceans. Adv Exp Med Biol 1285, 169–198. https://doi.org/10.1007/978-3-030-54462-1_9

Liang, W., Wang, J., Tan, B., Chi, S., 2022. Effects of methanotroph (*Methylococcus capsulatus*) bacteria meal replacement of fish meal on growth performance, serum biochemical indices and intestinal health of hybrid grouper (*Epinephelus fuscoguttatus* ♀ × *E. lanceolatus* ♂). Journal of Fisheries of China 46, 1969–1979. https://doi.org/10.11964/jfc.20220313407

Liu, G., Zhou, M., Mao, X., Gu, D., Chen, W., Long, X., Xie, S., Tan, Q., 2024. Evaluation of the appropriate *Clostridium autoethanogenum* protein level in grass carp (*Ctenopharyngodon idellus*) diets by growth performance, health status, and intestinal microbiota. Aquacult Int 32, 31–59. https://doi.org/10.1007/s10499-023-01163-y

Liu, R., Xing, S., Cheng, H., Liang, X., Wang, Jia, Wang, Jie, Longshaw, M., Hou, L., Xue, M., Wang, H., 2023. Effects of a novel bacteria meal (FeedKind®) on the physical properties of extrudates. Aquaculture Reports 33, 101744. https://doi.org/10.1016/j.aqrep.2023.101744

Lu, Q., Xi, L., Liu, Y., Gong, Y., Su, J., Han, D., Yang, Y., Jin, J., Liu, H., Zhu, X., Xie, S., 2021. Effects of Dietary Inclusion of *Clostridium autoethanogenum* Protein on the Growth Performance and Liver Health of Largemouth Bass (*Micropterus salmoides*). Front. Mar. Sci. 8. https://doi.org/10.3389/fmars.2021.764964

Lu, W., Yu, H., Liang, Y., Zhai, S., 2023. Evaluation of Methanotroph (*Methylococcus capsulatus*, Bath) Bacteria Protein as an Alternative to Fish Meal in the Diet of Juvenile American Eel (*Anguilla rostrata*). Animals 13, 681. https://doi.org/10.3390/ani13040681

Luo, K., Shi, Y., Yang, K., Xu, Q., Zhou, J., Gao, W., Wu, C., Wang, Z., 2023. Effects of dietary tributyrin supplementation in low fish meal diet containing high-*Clostridium autoethanogenum* protein on the growth performance, antioxidative capacity, lipid metabolism and intestinal microbiota of largemouth bass (*Micropterus salmoides*). Aquaculture Reports 32, 101722. https://doi.org/10.1016/j.aqrep.2023.101722

Ma, S., Liang, X., Chen, P., Wang, J., Gu, X., Qin, Y., Blecker, C., Xue, M., 2022a. A new single-cell protein from *Clostridium autoethanogenum* as a functional protein for largemouth bass (*Micropterus salmoides*). Animal Nutrition 10, 99–110. https://doi.org/10.1016/j.aninu.2022.04.005

Ma, S., Wang, H., Yang, J., Li, J., Xue, M., Cheng, H., Zou, F., Blecker, C., 2022b. Effects of *Clostridium autoethanogenum* protein inclusion levels and processing parameters on the physical properties of low-starch extruded floating feed. Aquaculture Reports 23, 101030. https://doi.org/10.1016/j.aqrep.2022.101030

Matty, A.J., Smith, P., 1978. Evaluation of a yeast, a bacterium and an alga as a protein source for rainbow trout: I. Effect of protein level on growth, gross conversion efficiency and protein conversion efficiency. Aquaculture 14, 235–246. https://doi.org/10.1016/0044-8486(78)90097-2

Maulu, S., Hualiang, L., Ke, J., Ren, M., Ge, X., Huang, D., Yu, H., 2021a. Dietary *Clostridium autoethanogenum* protein modulates intestinal absorption, antioxidant status, and immune response in GIFT (*Oreochromis niloticus*) juveniles. Aquaculture Research 52, 5787–5799. https://doi.org/10.1111/are.15454

Maulu, S., Liang, H., Ge, X., Yu, H., Huang, D., Ke, J., Ren, M., Mi, H., 2021b. Effect of dietary *Clostridium autoethanogenum* protein on growth, body composition, plasma parameters and hepatic genes expression related to growth and AMPK/TOR/PI3K signaling pathway of the genetically improved farmed tilapia (GIFT: *Oreochromis niloticus*) juveniles. Animal Feed Science and Technology 276, 114914. https://doi.org/10.1016/j.anifeedsci.2021.114914

Miles, R., Chapman, F., 2024. The Benefits of Fishmeal in Aquaculture Diets (No. FA122), IFAS Extension. University of Florida, Florida, USA.

Moroni, F., Carvalho, M., Di Rosa, A.R., Torrecillas, S., Fontanillas, R., Haffray, P., Allal, F., Bajek, A., Chiofalo, B., Terova, G., Montero, D., 2024. Genetic selection and novel feeds containing single cell protein as a substitute for fishmeal in European sea bass: Effects on growth, fatty acid profile and E-sensing analysis of fillets. Aquaculture Reports 35, 102021. https://doi.org/10.1016/j.aqrep.2024.102021

Nederlof, M., Harwinda, F., van Ginkel, R., Schrama, J., 2021. Effect of Different Types of Single Cell Protein on Feed Intake, Growth, Body Composition and Digestibility of Whiteleg Shrimp *Litopenaeus vannamei*. in: Aquaculture Europe 2021. Presented at the European Aquaculture Society Meetings, Funchal, Madeira, Spain.

Nederlof, M.A.J., Kaushik, S.J., Schrama, J.W., 2023. Effect of different types of bacterial single cell protein on feed intake, digestibility, growth and body composition of Pacific white shrimp (*Penaeus vannamei*). Aquaculture Reports 33, 101830. https://doi.org/10.1016/j.aqrep.2023.101830

Nell, J.A., 1985. Comparison of Some Single Cell Proteins in the Diet of the Sydney Rock Oyster (*Saccostrea commercialis*). The Progressive Fish-Culturist 47, 110–113. https://doi.org/10.1577/1548-8640(1985)47%253C110:COSSCP%253E2.0.CO;2

Nell, J.A., Wisely, B., 1983. Experimental feeding of Sydney rock oysters (*Saccostrea commercialis*). II. Protein supplementation of artificial diets for adult oysters. Aquaculture 32, 1–9. https://doi.org/10.1016/0044-8486(83)90264-8

Ngoh, S.Y., Shen, X., Nankervis, L., Chia, I., Hua, K., 2025. Evaluation of nutritional value and quality of aquafeed ingredients for Malabar snapper (*Lutjanus malabaricus*). Aquaculture 600, 742253. https://doi.org/10.1016/j.aquaculture.2025.742253

Øverland, M., Romarheim, O.H., Hovin, M., Storebakken, T., Skrede, A., 2006. Apparent total tract digestibility of unprocessed and extruded diets containing basic and autolyzed bacterial protein meal grown on natural gas in mink and rainbow trout. Animal Feed Science and Technology 129, 237–251. https://doi.org/10.1016/j.anifeedsci.2005.12.017

Pan, S., Bai, N., Qi, Z., Chen, C., Ma, D., Liang, Z., Gu, M., 2025. Integrated metabolomics and proteomics analysis to provide insights into muscle atrophy of turbot *Scophthalmus maximus* by dietary *Clostridium autoethanogenum* protein. Aquaculture Reports 40, 102603. https://doi.org/10.1016/j.aqrep.2024.102603

Pan, S., Gu, M., Chen, C., Ma, D., Liang, Z., Bai, N., 2024. Effects of combination of hydroxyproline, taurine and cholesterol on the growth and quality of turbot *Scophthalmus maximus* fed *Clostridium autoethanogenum* protein. Aquaculture Reports 39, 102411. https://doi.org/10.1016/j.aqrep.2024.102411

Papini, G., Opinion, A.G., Desmidt, J., Alloul, A., Vermeir, P., De Boeck, G., Vlaeminck, S.E., 2025. Successful high replacement of fishmeal in rainbow trout feed by consortia of aerobic heterotrophic bacteria results in equal growth, health, and muscle composition. Aquacult Int 33, 275. https://doi.org/10.1007/s10499-025-01955-4

Peng, K.-S., Wu, N., Cui, Z.-W., Zhang, X.-Y., Lu, X.-B., Wang, Z.-X., chen, D.-D., Zhang, Y.-A., 2020. Effect of the complete replacement of dietary fish meal by soybean meal on histopathology and immune response of the hindgut in grass carp (*Ctenopharyngodon idellus*). Veterinary Immunology and Immunopathology 221, 110009. https://doi.org/10.1016/j.vetimm.2020.110009

Pereira, A.A., Yaverino-Gutierrez, M.A., Monteiro, M.C., Souza, B.A., Bachheti, R.K., Chandel, A.K., 2025. Precision fermentation in the realm of microbial protein production: State-of-the-art and future insights. Food Research International 200, 115527. https://doi.org/10.1016/j.foodres.2024.115527

Perera, W.M., Carter, C.G., Houlihan, D.F., 1995a. Apparent absorption efficiencies of amino acids in rainbow trout, *Oncorhynchus mykiss* (Walbaum), fed diets containing bacterial single-cell protein. Aquaculture Nutrition 1, 95–103. https://doi.org/10.1111/j.1365-2095.1995.tb00024.x

Perera, W.M., Carter, C.G., Houlihan, D.F., 1995b. Feed consumption, growth and growth efficiency of rainbow trout (*Oncorhynchus mykiss* (Walbaum)) fed on diets containing a bacterial single-cell protein. Br J Nutr 73, 591–603. https://doi.org/10.1079/bjn19950061

Petereit, J., Hoerterer, C., Bischoff-Lang, A.A., Conceição, L.E.C., Pereira, G., Johansen, J., Pastres, R., Buck, B.H., 2022. Adult European Seabass (*Dicentrarchus labrax*) Perform Well on Alternative Circular-Economy-Driven Feed Formulations. Sustainability 14, 7279. https://doi.org/10.3390/su14127279

Piazzon, M.C., Naya-Català, F., Pereira, G.V., Estensoro, I., Del Pozo, R., Calduch-Giner, J.A., Nuez-Ortín, W.G., Palenzuela, O., Sitjà-Bobadilla, A., Dias, J., Conceição, L.E.C., Pérez-Sánchez, J., 2022. A novel fish meal-free diet formulation supports proper growth and does not impair intestinal parasite susceptibility in gilthead sea bream (*Sparus aurata*) with a reshape of gut microbiota and tissue-specific gene expression patterns. Aquaculture 558, 738362. https://doi.org/10.1016/j.aquaculture.2022.738362

Pilmer, L., Woolley, L., Lymbery, A., Dam, C., Elizur, A., Foysal, M.J., Partridge, G., 2025. Exploring single cell microbial protein as a sustainable fishmeal alternative in yellowtail kingfish (*Seriola lalandi*) diets: impacts on health and gut microbiome. J Animal Sci Biotechnol 16, 16. https://doi.org/10.1186/s40104-024-01146-w

Pilmer, L.W., Woolley, L.D., Lymbery, A.J., Salini, M., Partridge, G.J., 2022. Using dietary additives to improve palatability of diets containing single-cell protein from methanotrophic bacteria in yellowtail kingfish (*Seriola lalandi*) diets. Aquaculture Research 53, 5006–5017. https://doi.org/10.1111/are.15986

Prakash, S., Maas, R.M., Fransen, P.-M.M.M., Kokou, F., Schrama, J.W., Philip, A.J.P., 2023. Effect of feed ingredients on nutrient digestibility, waste production and physical characteristics of rainbow trout (*Oncorhynchus mykiss*) faeces. Aquaculture 574, 739621. https://doi.org/10.1016/j.aquaculture.2023.739621

Qi, Z., Bai, N., Li, Q., Pan, S., Gu, M., 2024a. Dietary fishmeal replacement by *Clostridium autoethanogenum* protein meal influences the nutritional and sensory quality of turbot (*Scophthalmus maximus*) via the TOR/AAR/AMPK pathways. Animal Nutrition 18, 84–95. https://doi.org/10.1016/j.aninu.2024.04.012

Qi, Z., Gu, M., Pan, S., Li, Q., Chen, C., Ma, D., Bai, N., 2024b. Effects of dietary replacement of fishmeal by defatted *Tenebrio molitor* meal, *Clostridium autoethanogenum* protein meal and *Chlorella vulgaris* meal on the freshness of turbot (*Scophthalmus maximus*) during chilled storage. Aquaculture Reports 38, 102296. https://doi.org/10.1016/j.aqrep.2024.102296

Qiu, X., Nguyen, L., Davis, D. a., 2018. Apparent digestibility of animal, plant and microbial ingredients for Pacific white shrimp *Litopenaeus vannamei*. Aquaculture Nutrition 24, 930–939. https://doi.org/10.1111/anu.12629

Qiuxin Y. a. N., Xianyong B.U., Yongtao L.I.U., Chuanwei Y. a. O., Yueru L.I., Kangsen M. a. I., Qinghui A.I., 2025. Effect of replacing soybean meal with *Clostridium autoethanogenum* protein on the growth performance, digestive capacity, antioxidant capacity, and protein metabolism of Hefang bream. scxb 49, 099608–14. https://doi.org/10.11964/jfc.20250515003

Rajesh, M., Kamalam, B.S., Sharma, P., Verma, V.C., Pandey, A., Dubey, M.K., Ciji, A., Akhtar, Md.S., Pandey, N., Sarma, D., Kaushik, S.J., 2022. Evaluation of a novel methanotroph bacteria meal grown on natural gas as fish meal substitute in rainbow trout, *Oncorhynchus mykiss*. Aquaculture Research 53, 2159–2174. https://doi.org/10.1111/are.15735

Romarheim, O.H., Landsverk, T., Mydland, L.T., Skrede, A., Øverland, M., 2013. Cell wall fractions from *Methylococcus capsulatus* prevent soybean meal-induced enteritis in Atlantic salmon (*Salmo salar*). Aquaculture 402–403, 13–18. https://doi.org/10.1016/j.aquaculture.2013.03.011

Romarheim, O.H., Øverland, M., Mydland, L.T., Skrede, A., Landsverk, T., 2011. Bacteria Grown on Natural Gas Prevent Soybean Meal-Induced Enteritis in Atlantic Salmon. The Journal of Nutrition 141, 124–130. https://doi.org/10.3945/jn.110.128900

Ruiz, A., Sanahuja, I., Thorringer, N.W., Lynegaard, J., Ntokou, E., Furones, D., Gisbert, E., 2023. Single cell protein from methanotrophic bacteria as an alternative healthy and functional protein source in aquafeeds, a holistic approach in rainbow trout (*Oncorhynchus mykiss*) juveniles. Aquaculture 576, 739861. https://doi.org/10.1016/j.aquaculture.2023.739861

Safonov, Y., Agapova, V., Randelin, D., Agapov, S., 2025. The influence of Gaprin on the dynamics of live weight, digestibility of nutrients in young sterlet. BIO Web Conf. 160, 01041. https://doi.org/10.1051/bioconf/202516001041

Salini, M.J., Wang, W., Mock, T.S., Siddik, M.A.B., Jago, M.K., Bews, K.L.F., Francis, D.S., 2025. Expanding the Ingredient Basket in Aquaculture: Growth Performance and Feed Utilization of Australian Hybrid Abalone (*Haliotis laevigata* × *H. rubra*) Fed Methanotrophic Single Cell Protein. Aquaculture Nutrition 2025, 7291857. https://doi.org/10.1155/anu/7291857

Salminen, S., Collado, M.C., Endo, A., Hill, C., Lebeer, S., Quigley, E.M.M., Sanders, M.E., Shamir, R., Swann, J.R., Szajewska, H., Vinderola, G., 2021. The International Scientific Association of Probiotics and Prebiotics (ISAPP) consensus statement on the definition and scope of postbiotics. Nat Rev Gastroenterol Hepatol 18, 649–667. https://doi.org/10.1038/s41575-021-00440-6

Salze, G.P., Tibbetts, S.M., 2021. Apparent digestibility coefficients of proximate nutrients and essential amino acids from a single-cell protein meal derived from *Methylobacterium extorquens* for pre-smolt Atlantic salmon (*Salmo salar* L.). Aquaculture Research 52, 6818–6823. https://doi.org/10.1111/are.15526

Samsing, F., Sullivan, R., Truong, H., Robenso, A., Sangster, C.R., Bannister, J., Longshaw, M., Becker, J.A., 2024. Replacement of fishmeal with a microbial single-cell protein induced enteropathy and poor growth outcomes in barramundi (*Lates calcarifer*) fry. Journal of Fish Diseases n/a, e13985. https://doi.org/10.1111/jfd.13985

Schmitz, O., Greuel, E., Preffer, E., 1984. Digestibility of crude protein and organic matter of potential sources of dietary protein for eels (*Anguilla anguilla*, L.). Aquaculture 41, 21–30. https://doi.org/10.1016/0044-8486(84)90386-7

Shahzad, H.M.A., Almomani, F., Shahzad, A., Mahmoud, K.A., Rasool, K., 2024. Challenges and opportunities in biogas conversion to microbial protein: A pathway for sustainable resource recovery from organic waste. Process Safety and Environmental Protection 185, 644–659. https://doi.org/10.1016/j.psep.2024.03.055

Shi, L., Yan, X., Yan, Z., Wang, H., Chen, J., Zhang, S., 2025. Combined microbiomic and transcriptomic analysis revealed that dietary *Clostridium autoethanogenum* protein could improve the disease resistance of *Litopenaeus vannamei* by regulating the oxidative phosphorylation. Aquaculture Reports 41, 102687. https://doi.org/10.1016/j.aqrep.2025.102687

Shi, M., Zheng, C., Sun, Y., Li, X., He, G., Cao, J., Tan, B., Xie, S., 2023. Effects of Dietary Chenodeoxycholic Acid Supplementation in a Low Fishmeal Diet Containing *Clostridium autoethanogenum* Protein on Growth, Lipid and Cholesterol Metabolism, and Hepatopancreas Health of *Litopenaeus vannamei*. Animals 13, 2109. https://doi.org/10.3390/ani13132109

Shi Y., Zhu S., Tan P., Zhang W., Zhou X., Gao W., Chi S., 2022. Digestibility of six non-grain protein sources in largemouth bass. ssswxb 46, 1187–1196. https://doi.org/10.7541/2022.2021.0350

Skrede, A., Berge, G.M., Storebakken, T., Herstad, O., Aarstad, K.G., Sundstøl, F., 1998. Digestibility of bacterial protein grown on natural gas in mink, pigs, chicken and Atlantic salmon. Animal Feed Science and Technology 76, 103–116. https://doi.org/10.1016/S0377-8401(98)00208-9

Storebakken, T., Baeverfjord, G., Skrede, A., Olli, J.J., Berge, G.M., 2004. Bacterial protein grown on natural gas in diets for Atlantic salmon, *Salmo salar*, in freshwater. Aquaculture 241, 413–425. https://doi.org/10.1016/j.aquaculture.2004.07.024

Storebakken, T., Kvien, I.S., Shearer, K.D., Grisdale-Helland, B., Helland, S.J., 1999. Estimation of gastrointestinal evacuation rate in Atlantic salmon (*Salmo salar*) using inert markers and collection of faeces by sieving: evacuation of diets with fish meal, soybean meal or bacterial meal. Aquaculture 172, 291–299. https://doi.org/10.1016/S0044-8486(98)00501-8

Storebakken, T., Kvien, I.S., Shearer, K.D., Grisdale-Helland, B., Helland, S.J., Berge, G.M., 1998. The apparent digestibility of diets containing fish meal, soybean meal or bacterial meal fed to Atlantic salmon (*Salmo salar*): evaluation of different faecal collection methods. Aquaculture 169, 195–210. https://doi.org/10.1016/S0044-8486(98)00379-2

Stringer, D.A., 1982. Industrial development and evaluation of new protein sources: Micro-organisms. Proceedings of the Nutrition Society 41, 289–300. https://doi.org/10.1079/PNS19820044

Tampou, A., Andreopoulou, S., Vasilaki, A., Nengas, I., Berillis, P., Gisbert, E., Karapanagiotidis, I.T., Antonopoulou, E., Mente, E., 2024. Growth performance of gilthead sea bream (*Sparus aurata*) fed a mixture of single cell ingredients for organic diets. Aquaculture Reports 36, 102105. https://doi.org/10.1016/j.aqrep.2024.102105

Tibbetts, S.M., Dumas, A., Thi Ha To, V.P., Tankovski, I., Chimsung, N., Patelakis, S.J.J., Colombo, S.M., Boates, T., Armenta, R.E., Stratton, B.O., 2025a. Single-cell protein meal produced from *Methylovorus menthalis* (Psomi™) in extruded aquafeeds for seawater phase Atlantic salmon (*Salmo salar* L.): Impacts on growth performance, nutrient utilization, fish health, and fillet quality. Aquaculture 743262. https://doi.org/10.1016/j.aquaculture.2025.743262

Tibbetts, S.M., Piercey, M.J., Patelakis, S.J.J., Stratton, B., 2025b. Single-cell protein (SCP) meals from *Methylovorus menthalis* as feed ingredients for freshwater Atlantic salmon (*Salmo salar* L.): Digestibility, growth performance, nutrient utilization, and fish health. Aquaculture 598, 741874. https://doi.org/10.1016/j.aquaculture.2024.741874

Tlusty, M., Rhyne, A., Szczebak, J.T., Bourque, B., Bowen, J.L., Burr, G., Marx, C.J., Feinberg, L., 2017. A transdisciplinary approach to the initial validation of a single cell protein as an alternative protein source for use in aquafeeds. PeerJ 5, e3170. https://doi.org/10.7717/peerj.3170

Tusé, D., and Miller, M.W., 1983. Single-cell protein: Current status and future prospects. C R C Critical Reviews in Food Science and Nutrition 19, 273–325. https://doi.org/10.1080/10408398409527379

Vale Pereira, G.D., Conceição, L.E.C., Soares, F., Petereit, J., Buck, B.H., Johansen, J., Dias, J., Faccenda, F., 2023. Alternative Feed Formulations Impact Growth Performance, Flesh Quality and Consumer Acceptance of Rainbow Trout (*Oncorhynchus mykiss*). Journal of Marine Science and Engineering 11. https://doi.org/10.3390/jmse11061135

Vasilaki, A., Mente, E., Fountoulaki, E., Henry, M., Nikoloudaki, C., Berillis, P., Kousoulaki, K., Nengas, I., 2023. Fishmeal, plant protein, and fish oil substitution with single-cell ingredients in organic feeds for European sea bass (*Dicentrarchus labrax*). Frontiers in Physiology 14.

Voss, B., 1988. Ersatz von Fischmehl in einer Ration für einjährige Steinbutte (*Scophthalmus maximus* L.) durch Bakterienprotein Probion und eine Mischung aus Geflügelabfall und hydrolysiertem Federmehl. Archiv fur Fischereiwissenschaft 38, 203–213.

Wang, S., Han, Z., Turchini, G.M., Wang, X., Fang, Z., Chen, N., Xie, R., Zhang, H., Li, S., 2022. Effects of Dietary Phospholipids on Growth Performance, Digestive Enzymes Activity and Intestinal Health of Largemouth Bass (*Micropterus salmoides*) Larvae. Front Immunol 12, 827946. https://doi.org/10.3389/fimmu.2021.827946

Wang, X., Wan, M., Wang, Z., Zhang, H., Zhu, S., Cao, X., Xu, N., Zheng, J., Bu, X., Xu, W., Mai, K., Ai, Q., 2023. Effects of Tributyrin Supplementation on Growth Performance, Intestinal Digestive Enzyme Activity, Antioxidant Capacity, and Inflammation-Related Gene Expression of Large Yellow Croaker (*Larimichthys crocea*) Fed with a High Level of Clostridium autoethanogenum Protein. Aquaculture Nutrition 2023, 2687734. https://doi.org/10.1155/2023/2687734

Woolley, L., Chaklader, M.R., Pilmer, L., Stephens, F., Wingate, C., Salini, M., Partridge, G., 2023. Gas to protein: Microbial single cell protein is an alternative to fishmeal in aquaculture. Science of The Total Environment 859, 160141. https://doi.org/10.1016/j.scitotenv.2022.160141

Wu, Y., Tian, S., Yuan, J., Zhang, Z., Zhou, H., Gao, W., Zhang, W., Mai, K., 2022. Effects of *Clostridium autoethanogenum* protein as substitute for dietary fishmeal on the growth, feed utilization, intestinal health and muscle quality of large yellow croaker *Larimichthys crocea*. Aquaculture 561, 738591. https://doi.org/10.1016/j.aquaculture.2022.738591

Wu, Z., Yu, X., Chen, P., Pan, M., Liu, J., Sahandi, J., Zhou, W., Mai, K., Zhang, W., 2024. Dietary *Clostridium autoethanogenum* protein has dose-dependent influence on the gut microbiota, immunity, inflammation and disease resistance of abalone *Haliotis discus hannai*. Fish & Shellfish Immunology 151, 109737. https://doi.org/10.1016/j.fsi.2024.109737

Wu, Z., Yu, X., Guo, J., Fu, Y., Guo, Y., Pan, M., Zhang, W., Mai, K., 2022. Replacement of dietary fish meal with *Clostridium autoethanogenum* protein on growth performance, digestion, mTOR pathways and muscle quality of abalone *Haliotis discus hannai*. Aquaculture 553, 738070. https://doi.org/10.1016/j.aquaculture.2022.738070

Xia, M., Hu, Z., Xue, R., Liu, R., Li, H., Ji, H., 2025. Dietary vitamin E further enhances the effectiveness of *Clostridium autoethanogenum* protein as an alternative source of soybean meal in grass carp (*Ctenopharygodon idellus*) diets based on growth performance, flesh quality and antioxidant status. Aquaculture 594, 741348. https://doi.org/10.1016/j.aquaculture.2024.741348

Xu, B., Liu, Y., Chen, K., Wang, L., Sagada, G., Tegomo, A.F., Yang, Y., Sun, Y., Zheng, L., Ullah, S., Shao, Q., 2021. Evaluation of Methanotroph (*Methylococcus capsulatus*, Bath) Bacteria Meal (FeedKind®) as an Alternative Protein Source for Juvenile Black Sea Bream, *Acanthopagrus schlegelii*. Frontiers in Marine Science 8, 1702. https://doi.org/10.3389/fmars.2021.778301

Xu, J., Huang, B., Chi, S., Zhang, S., Cao, J., Tan, B., Xie, S., 2023a. Replacement of Dietary Fishmeal with *Clostridium autoethanogenum* Protein on Lipidomics and Lipid Metabolism in Muscle of Pearl Gentian Grouper. Aquaculture Nutrition 2023, 6723677. https://doi.org/10.1155/2023/6723677

Xu, J., Zheng, C., Chi, S., Zhang, S., Cao, J., Tan, B., Xie, S., 2023b. *Clostridium autoethanogenum* protein substitution and bile acids addition altered intestinal health and transcriptome profiles of hepatopancreas in *Litopenaeus vannamei*. Aquaculture Reports 28, 101432. https://doi.org/10.1016/j.aqrep.2022.101432

Xue, R., Li, H., Liu, S., Hu, Z., Wu, Q., Ji, H., 2023. Substitution of soybean meal with *Clostridium autoethanogenum* protein in grass carp (*Ctenopharygodon idella*) diets: Effects on growth performance, feed utilization, muscle nutritional value and sensory characteristics. Animal Feed Science and Technology 295, 115547. https://doi.org/10.1016/j.anifeedsci.2022.115547

Yang, L., Cai, M., Xie, S., Hu, Y., Zhang, J., 2024. Evaluation of soybean meal replacement by *Clostridium autoethanogenum* protein in diets for juvenile grass carp (*Ctenopharyngodon idellus*). Aquaculture 592, 741218. https://doi.org/10.1016/j.aquaculture.2024.741218

Yang, P., Li, X., Song, B., He, M., Wu, C., Leng, X., 2023. The potential of *Clostridium autoethanogenum*, a new single cell protein, in substituting fish meal in the diet of largemouth bass (*Micropterus salmoides*): Growth, feed utilization and intestinal histology. Aquaculture and Fisheries 8, 67–75. https://doi.org/10.1016/j.aaf.2021.03.003

Yang, P., Li, X., Yao, W., Li, M., Wang, Y., Leng, X., 2022a. Dietary Effect of *Clostridium autoethanogenum* Protein on Growth, Intestinal Histology and Flesh Lipid Metabolism of Largemouth Bass (*Micropterus salmoides*) Based on Metabolomics. Metabolites 12, 1088. https://doi.org/10.3390/metabo12111088

Yang, P., Yao, W., Wang, Y., Li, M., Li, X., Leng, X., 2022b. Dietary effects of fish meal substitution with *Clostridium autoethanogenum* on flesh quality and metabolomics of largemouth bass (*Micropterus salmoides*). Aquaculture Reports 23, 101012. https://doi.org/10.1016/j.aqrep.2022.101012

Yao, W., Li, X., Zhang, X., Li, M., Wang, Y., He, H., Leng, X., 2024. The complete replacement of fish meal with *Clostridium autoethanogenum* protein in practical diet did not affect the growth, but reduced the flesh quality of Pacific white shrimp, *Litopenaeus vannamei*. Animal Feed Science and Technology 310, 115919. https://doi.org/10.1016/j.anifeedsci.2024.115919

Yao, W., Yang, P., Zhang, X., Xu, X., Zhang, C., Li, X., Leng, X., 2022. Effects of replacing dietary fish meal with *Clostridium autoethanogenum* protein on growth and flesh quality of Pacific white shrimp (*Litopenaeus vannamei*). Aquaculture 549, 737770. https://doi.org/10.1016/j.aquaculture.2021.737770

Yu, H., Liang, H., Longshaw, M., Wang, J., Ge, X., Ren, M., Zhang, L., 2022. Methanotroph (*Methylococcus capsulatus*, Bath) bacteria meal (FeedKind®) could effectively improve the growth, apparent digestibility coefficient, blood biochemical parameters, antioxidant indices of juvenile Jian carp (*Cyprinus carpio* var. Jian). Animal Feed Science and Technology 115293. https://doi.org/10.1016/j.anifeedsci.2022.115293

Yu, M., Li, X., Wang, J., Longshaw, M., Song, K., Wang, L., Zhang, C., Lu, K., 2023. Substituting fish meal with a bacteria protein (*Methylococcus capsulatus*, Bath) grown on natural gas: Effects on growth non-specific immunity and gut health of spotted seabass (*Lateolabrax maculatus*). Animal Feed Science and Technology 296, 115556. https://doi.org/10.1016/j.anifeedsci.2022.115556

Yu, Y., Wang, X., Jin, J., Han, D., Zhu, X., Liu, H., Zhang, Z., Yang, Y., Xie, S., 2024. Effects of the Replacement of Dietary Fishmeal by the Blend of *Tenebrio molitor* Meal, *Chlorella* Meal, *Clostridium autoethanogenum* Protein, and Cottonseed Protein Concentrate on Growth, Protein Utilization, and Intestinal Health of Gibel Carp (*Carassius gibelio*). Aquaculture Nutrition 2024, 5019899. https://doi.org/10.1155/2024/5019899

Yuan, H., Xie, M., Chen, J., Hu, N., Wang, H., Tan, B., Shi, L., Zhang, S., 2024. The role of dietary *Clostridium autoethanogenum* protein in the growth, disease resistance, intestinal health and transcriptome response of Pacific white shrimp under different stocking densities. Aquaculture 589, 740962. https://doi.org/10.1016/j.aquaculture.2024.740962

Zenkovich, P., Litvinenko, A., Zenkovich, R., Litvinenko, L., Korentovich, M., 2024. Using of dry biomass of methanotrophic bacteria for enrichment of starter artificial feeds when growing juveniles of broad whitefish *Coregonus nasus* (Pallas, 1776). BIO Web Conf. 118, 01011. https://doi.org/10.1051/bioconf/202411801011

Zenkovich, P.A., Korentovich, M.A., Shabalina, T.O., Litvinenko, A.I., Litvinenko, L.I., 2022. Experience of using dry biomass of methanotrophic bacteria in the enrichment of artificial feeds for Siberian sturgeon juveniles of the Ob population. IOP Conf. Ser.: Earth Environ. Sci. 949, 012136. https://doi.org/10.1088/1755-1315/949/1/012136

Zhang, Jiayin, Chen, N., Sha, Y., Fu, S., Li, C., Xu, Q., Gao, F., Wang, H., 2025. Single-cell protein as an alternative protein source in feed for juvenile sea cucumber. Aquaculture Reports 42, 102795. https://doi.org/10.1016/j.aqrep.2025.102795

Zhang, J., Dong, Y., Song, K., Wang, L., Li, X., Lu, K., Tan, B., Zhang, C., 2022. Substituting Fish Meal with a Bacteria Protein (*Clostridium autoethanogenum* Protein) Derived from Industrial-Scale Gas Fermentation: Effects on Growth and Gut Health of Juvenile Large Yellow Croakers (*Larimichthys crocea*). Fishes 7, 228. https://doi.org/10.3390/fishes7050228

Zhang, Jiacheng, Liang, W., Wang, J., Longshaw, M., Dong, X., Yang, Q., Liu, H., Deng, J., Tan, B., Chi, S., 2025. Methanotroph (*Methylococcus capsulatus*) bacteria meal, influences the intestinal microbiome, and improves intestinal barrier function and immunity in hybrid grouper (*Epinephelus fuscointestinetatus* ♀ × *E. lanceolatus* ♂). Animal Nutrition 22, 321–336. https://doi.org/10.1016/j.aninu.2025.03.020

Zhang, J., Yu, M., Wang, J., Longshaw, M., Song, K., Wang, L., Li, X., Zhang, C., Lu, K., 2023. Methanotroph (*Methylococcus capsulatus*, Bath) bacteria meal alleviates soybean meal-induced enteritis in spotted seabass (*Lateolabrax maculatus*) by modulating immune responses and the intestinal flora. Aquaculture 575, 739795. https://doi.org/10.1016/j.aquaculture.2023.739795

Zhang, Q., Liang, H., Longshaw, M., Wang, J., Ge, X., Zhu, J., Li, S., Ren, M., 2022. Effects of replacing fishmeal with methanotroph (*Methylococcus capsulatus*, Bath) bacteria meal (FeedKind®) on growth and intestinal health status of juvenile largemouth bass (*Micropterus salmoides*). Fish & Shellfish Immunology 122, 298–305. https://doi.org/10.1016/j.fsi.2022.02.008

Zheng, C., Cao, J., Chi, S., Dong, X., Yang, Q., Liu, H., Zhang, S., Xie, S., Tan, B., 2022a. Dietary phosphorus supplementation in the diet of Pacific white shrimp (*Litopenaeus vannamei*) alleviated the adverse impacts caused by high *Clostridium autoethanogenum* protein. Fish & Shellfish Immunology 131, 137–149. https://doi.org/10.1016/j.fsi.2022.10.005

Zheng, C., Gong, S., Cao, J., Dong, X., Chi, S., Yang, Q., Liu, H., Zhang, S., Xie, S., Tan, B., 2022b. Effects of Dietary Lipid Sources on Alleviating the Negative Impacts Induced by the Fishmeal Replacement With *Clostridium autoethanogenum* Protein in the Diet of Pacific White Shrimp (*Litopenaeus vannamei*). Front. Mar. Sci. 9. https://doi.org/10.3389/fmars.2022.879364

Zheng, C.-Z., Cao, J.-M., Dong, X.-H., Chi, S.-Y., Zhang, S., Yang, Q.-H., Liu, H.-Y., Deng, J.-M., Zhang, W., Tan, B.-P., Xie, S.-W., 2023. Apparent digestibility coefficients of seven protein sources for juvenile hybrid grouper (*Epinephelus fuscoguttatus* ♀ × *Epinephelus lanceolatus* ♂). Acta Hydrobiologica Sinica 47, 257–268. https://doi.org/10.7541/2023.2022.0172

Zheng, J., Zhang, W., Dan, Z., Cao, X., Cui, K., Zhu, S., Zhuang, Y., Mai, K., Ai, Q., 2023. Effects of fish meal replaced by methanotroph bacteria meal (*Methylococcus capsulatus*) on growth, body composition, antioxidant capacity, amino acids transporters and protein metabolism of turbot juveniles (*Scophthalmus maximus* L.). Aquaculture 562, 738782. https://doi.org/10.1016/j.aquaculture.2022.738782

Zheng, J., Zhang, W., Dan, Z., Cao, X., Gong, Y., Mai, K., Ai, Q., 2024. Dietary methanotroph bacteria meal alleviates soybean meal-induced enteritis by improving immune tolerance and intestinal flora profile of juvenile turbot (*Scophthalmus maximus* L.). Fish & Shellfish Immunology 148, 109463. https://doi.org/10.1016/j.fsi.2024.109463

Zheng, J., Zhang, W., Dan, Z., Zhuang, Y., Liu, Y., Mai, K., Ai, Q., 2022. Replacement of dietary fish meal with *Clostridium autoethanogenum* meal on growth performance, intestinal amino acids transporters, protein metabolism and hepatic lipid metabolism of juvenile turbot (*Scophthalmus maximus* L.). Front. Physiol. 13. https://doi.org/10.3389/fphys.2022.981750

Zheng, X., Liu, Q., Khademi, S., Khoshnevisan, B., Xu, M., Zhang, Y., Lou, Y., Liu, H., Duan, N., 2024. Methanotroph biotransformation for nutrient recovery: a review of current strategies and future opportunities. Biofuel Research Journal 11, 2065–2081. https://doi.org/10.18331/BRJ2024.11.2.2

Zhu, S., Gao, W., Wen, Z., Chi, S., Shi, Y., Hu, W., Tan, B., 2022. Partial substitution of fish meal by *Clostridium autoethanogenum* protein in the diets of juvenile largemouth bass (*Micropterus salmoides*). Aquaculture Reports 22, 100938. https://doi.org/10.1016/j.aqrep.2021.100938

CHAPTER 9

Microalgae and macroalgae as functional ingredients in fish nutrition

C. Aragão[1,*], *R. Teodósio*[1], *A.T. Gonçalves*[2,3] *and S. Engrola*[1]

[1]*Centro de Ciências do Mar do Algarve (CCMAR/CIMAR LA), Universidade do Algarve, Campus de Gambelas, Ed. 7, 8005-139 Faro, Portugal*
[2]*GreenCoLab - Associação Oceano Verde, Universidade do Algarve, Campus de Gambelas, Ed. 2, 8005-139 Faro, Portugal*
[3]*SPAROS Lda., Área Empresarial de Marim, Lote C, 8700-221 Olhão, Portugal*
**Correspondence: caragao@ualg.pt*

Abstract

Micro- and macroalgae have gained increasing attention as functional ingredients in aquafeeds due to their rich content of bioactive compounds and potential to improve fish health. In this context, they are not valued primarily as sources of dietary protein or energy, but as functional ingredients, typically used at low inclusion levels to deliver specific health benefits. This chapter reviews the functional roles of micro- and macroalgae in aquaculture, focusing on their effects on health, immune function, antioxidant status, and gut microbiota composition across different fish species. Microalgae consistently improve gut morphology, hematological parameters, and immune and antioxidant responses, particularly under stress or disease challenge. They also modulate gut microbiota composition, promoting microbial diversity and the abundance of beneficial taxa, although outcomes depend on species-, dose-, and processing-form. Macroalgae and their extracts stimulate immune function, but their effects on oxidative balance and microbiota are more variable. While some whole macroalgae show limited or inconsistent effects, processed forms and extracts, especially at low inclusion levels, tend to produce clearer immunological and microbiome-related benefits. Despite these promising results, the broader use of algae in fish feeds faces persistent challenges, including high production costs, scalability issues, and inconsistent biomass composition, especially in the case of macroalgae. Addressing these constraints will require standardized and cost-efficient production systems, and the application of precision nutrition principles to align algal inclusion with the specific nutritional requirements of the target fish species and their farming conditions. Overall, integration of algae-based ingredients into aquafeeds can enhance fish robustness, support immune and antioxidant defenses, and positively modulate the gut microbiota. With appropriate formulation and processing, micro- and macroalgae hold strong potential as key sustainable components of functional diets designed to improve fish health and resilience in aquaculture.

Keywords

antioxidant capacity – aquaculture – health status – immune responses – microbiota

1 Introduction

In recent years, both microalgae and macroalgae (also known as seaweeds) have attracted increased attention across multiple industries due to their rich biochemical composition and wide range of functional applications. In animal production, including aquaculture, algae have emerged as promising dietary components, either as partial replacements for conventional feed ingredients or as additives providing functional health benefits (Ansari et al., 2021; González-Meza et al., 2023; Valente et al., 2021; Vijayaram et al., 2024).

The use of functional feed additives in aquaculture has gained momentum as a sustainable strategy to improve stress resilience, enhance health status, and reduce dependence on chemotherapeutics. In this context, micro- and macroalgae are increasingly valued not primarily as sources of dietary protein or energy, but for their functional properties. At low inclusion levels, algae surpass their role as nutritional supplements and act as functional additives that provide benefits beyond basic nutrition by modulating physiological processes, including immune activity, oxidative balance, gut integrity, stress resilience, or disease resistance (González-Meza et al., 2023; Siddik et al., 2023, 2024b; Vijayaram et al., 2024; Wan et al., 2019). This functional-use approach is particularly relevant given the current economic and logistical constraints associated with large-scale algal biomass production, especially for microalgae, where high costs linked to production, harvesting, and downstream processing remain major barriers to extensive inclusion in fish feeds (Siddik et al., 2024b).

Microalgae commonly used in aquaculture include species from Chlorophyta (e.g., *Chlorella* spp., *Tetraselmis* spp.), Bacillariophyta (e.g., *Phaeodactylum* spp.), Haptophyta (e.g., *Isochrysis* spp.), and Ochrophyta (e.g., *Nannochloropsis* spp.). Although *Arthrospira platensis* (formerly *Spirulina platensis*) is taxonomically a cyanobacterium, it is often grouped with microalgae due to its similar applications and nutritional profile, a convention followed in this chapter. Microalgae are a source of high-quality proteins, lipids, polyunsaturated fatty acids (particularly omega-3 highly unsaturated fatty acids, such as eicosapentaenoic (EPA) and docosahexaenoic (DHA) acids), vitamins, and trace minerals. They also contain important bioactive compounds such as carotenoids (e.g., astaxanthin, lutein, and β-carotene), chlorophylls, phycobiliproteins (e.g., phycocyanin), as well as β-glucans and other polysaccharides, which have demonstrated antioxidant, immunostimulatory, and stress-mitigating effects that enhance immune function, gut health, and overall resilience in several fish species (Ansari et al., 2021; Siddik et al., 2024b). Beyond their functional properties, microalgae are already widely used as live feed in larviculture and as a source of pigments such as carotenoids used in salmonid production (Ansari et al., 2021; Siddik et al., 2024b).

Frequently used macroalgae in aquaculture are classified into three main groups based on their pigmentation and biochemical composition: green algae (Chlorophyta, e.g., *Ulva* spp.), brown algae (Phaeophyceae, e.g., *Sargassum* spp.), and red algae (Rhodophyta, e.g., *Gracilaria* spp.). They are incorporated into diets as whole biomass, extracts, or fermented products, with their composition varying by species, season, environmental conditions, and processing method. Macroalgae are a valuable source of proteins, polyunsaturated fatty acids, vitamins, and minerals. They are also a source of bioactive compounds, including sulphated polysaccharides (e.g., ulvans, fucoidans, and carrageenans), polyphenols (e.g., phlorotannins and phenolic acids), carotenoids (e.g., fucoxanthin, lutein, and β-carotene), and phycobiliproteins (e.g., phycoerythrin and phycocyanin). These compounds have demonstrated antioxidant, anti-inflammatory, antimicrobial, and immunomodulatory effects, supporting improved gut structure, immune function, and disease resistance in various fish species (González-Meza et al., 2023; Siddik et al., 2023).

Several studies have evaluated both micro- and macroalgae as alternatives to fishmeal, with some reviews summarizing their effects on fish growth performance, feed efficiency, or even health-related parameters (e.g., Ansari et al., 2021; Aragão et al., 2022; Siddik et al., 2024b; Valente et al., 2021; Vijayaram et al., 2024; Wan et al., 2019). This chapter focuses on the functional roles of micro- and macroalgae when included at low levels (<15%) in fish feeds. Emphasis is placed on their effects on health, immune function, oxidative status, and gut microbiota composition across different fish species.

2 Health status

Histological and hematological evaluations are key tools to assess the effects of micro- and macroalgae on overall health status. Histology reveals structural changes in tissues, particularly the intestine and liver, by measuring indicators such as villus height, goblet cell density, enterocyte integrity, and hepatocyte condition. Hematological parameters, including red blood cell (RBC) and white blood cell (WBC) counts, hemoglobin levels, and hematocrit, provide a systemic view of oxygen transport and immune function. Together with survival rates, these assessments offer valuable insights into how dietary algae influence fish responses to environmental and pathogenic stressors, supporting their potential role in functional diets.

2.1 Microalgae

The inclusion of microalgae in aquafeeds has shown substantial benefits for fish health, notably through improvements in gut architecture, hematological function, and disease resistance.

In beluga (*Huso huso*), dietary inclusion of *Arthrospira platensis* at 2.5–10% increased neutrophils and decreased lymphocytes at higher inclusion levels (Adel et al., 2016). Hemoglobin and hematocrit increased at 5 and 10%, while RBCs increased only at 10%. All inclusion levels improved fish survival following *Streptococcus iniae* infection, suggesting enhanced immunocompetence. In Nile tilapia (*Oreochromis niloticus*), dietary supplementation with *A. platensis* has generally shown beneficial or neutral effects on tissue structure and immune function. A 15% inclusion improved intestinal villus height and preserved organ integrity, despite mild liver congestion, and enhanced spleen lymphocyte populations and melanomacrophage center (MMC) formation, suggesting immune stimulation (El-Habashi et al., 2019). After *Aeromonas hydrophila* infection, both *A. platensis*- and *Chlorella vulgaris*-fed fish had less tissue damage and higher survival. Lower doses (0.5%) also increased MMC size and improved post-infection survival (Awad et al., 2022). Similarly, a 0.75% inclusion improved villus height and width, and reduced inter-villus spacing before and after *A. hydrophila* infection (Mohammed et al., 2025). Fermented *A. platensis* (0.05–0.8%) further supported gut structure by increasing villus height and goblet cells, while enhancing oxygen-carrying capacity without negatively affecting RBC size or immune cell levels (Sherif et al., 2024). These findings suggest improved nutrient absorption, gut health, and disease resistance, likely due to better oxygen transport and reduced inflammation.

Other microalgae, such as *Chlorella vulgaris* (El-Mashtoly et al., 2024), *Nannochloropsis gaditana* (Jorge et al., 2019), *Nannochloropsis oceanica* (Batista et al., 2020), and *Schizochytrium* sp. (de Souza et al., 2020), did not alter intestinal morphology in various fish species. Similarly, in golden pompano (*Trachinotus ovatus*), diets with 1–5% *Barranca yajiagengensis* had no impact on intestinal structure but mitigated liver inflammation (Zhao et al., 2022). *Haematococcus pluvialis* supplementation in the same species increased villus height and alleviated liver pathology (Zhao et al., 2021b), highlighting broader physiological benefits beyond gut morphology.

In juvenile gilthead seabream (*Sparus aurata*), dietary supplementation with microalgae has shown varied effects on intestinal and hematological health. Diets with 10% *Phaeodactylum tricornutum* or *Tetraselmis chui* increased villus height and intestinal diameter but also reduced goblet cells and intraepithelial leucocytes, widened enterocyte intracellular spaces, and, in the case of *T. chui*, caused enterocyte vacuolization and microvilli disruption, signs of mucosal impairment and intestinal stress (Cerezuela et al., 2012a). In early-juvenile seabream fed diets supplemented with *Tetraselmis suecica* or *Tisochrysis lutea* at 5 and 10%, intestinal morphology was also affected (Vizcaíno et al., 2016). *T. suecica* increased microvilli length at both inclusion levels, whereas fish fed *T. lutea* displayed values similar to the control. At 5% inclusion, *T. suecica* increased microvilli diameter and *T. lutea* decreased it, while at 10% both microalgae produced control-like values. In addition, fish fed 10% of either microalga exhibited increased enterocyte apical area, and all supplemented groups showed an overall enhancement of total enterocyte absorptive surface. In contrast, inclusion of 2.5–5% of whole or hydrolyzed *Nannochloropsis gaditana* caused no major histological changes, although 5% inclusion reduced lamina propria thickness (Sáez et al., 2022). Intestinal ultrastructural analysis revealed that whole biomass increased the diameter of microvilli, while the hydrolysates enhanced microvilli length and total absorptive surface, suggesting improved nutrient absorption. Similarly, 1% *P. tricornutum* (whole or broken cells) led to minimal hematological changes, except for increased hemoglobin and reduced WBCs and thrombocytes in the broken-cell group (Reis et al., 2021). Supplementation with 1% *P. tricornutum* extract rich in β-glucans had no effect on gut gene expression or blood markers under normal conditions but increased mean corpuscular volume (MCV) and mean corpuscular hemoglobin (MCH) following dextran sodium sulphate-induced intestinal inflammation, indicating immune activation (Teixeira et al., 2022). These findings suggest that algae processing can enhance the bioavailability of functional compounds, potentially improving their physiological effects.

In Nile tilapia, dietary supplementation with 1.2% *Schizochytrium* sp. did not alter hematological parameters except for reduced RBCs (de Souza et al., 2020). In contrast, 0.5–2% *Tetraselmis suecica* increased WBCs, RBCs, and hematocrit, while hemoglobin, MCV, and MCH increased only at a minimum of 1% inclusion (Abdel-Tawwab et al., 2024). Following an *Aeromonas sobria* challenge, all *T. suecica*-supplemented diets improved fish survival. Under the tested conditions, *T. suecica* demonstrated clear hematological and immunoprotective benefits, whereas *Schizochytrium* sp. had minimal physiological effects, emphasizing the importance of microalgal species, dosage, and experimental context in determining functional outcomes.

Low-dose microalgal extract blends have also yielded promising results. In rohu (*Labeo rohita*), diets containing a mixture of *Arthrospira platensis*, *Chlorella vulgaris*, and *Euglena viridis* (equal parts at 0.01–0.08%) enhanced hematological parameters and increased survival following *A. hydrophila* challenge, highlighting the functional potency of extract supplementation (Sattanathan et al., 2023).

Overall, the integration of microalgae into aquafeeds demonstrates significant potential to enhance fish health and disease resistance, even though its success depends on careful consideration of species, processing form, and dosage. Despite this variability, microalgae generally offer consistent benefits to gut morphology and blood parameters across a range of fish species.

2.2 Macroalgae

The incorporation of macroalgae into aquafeeds has shown promising effects on fish health, although results vary depending on algae species, inclusion levels, and experimental models.

Dietary supplementation with *Chaetomorpha aerea* has shown promising immunohematological and protective effects. In rohu, low inclusion levels (0.025–0.1%) increased hemoglobin, RBC, and WBC counts, suggesting better oxygen-carrying capacity and immune response (Sattanathan et al., 2020). Following *Aeromonas hydrophila* infection, fish fed 0.025 or 0.05% had much higher survival rates, suggesting early immunostimulation even at minimal doses. In a feeding trial with Mozambique tilapia (*Oreochromis mossambicus*) testing 0.1–1% inclusion, RBCs and hemoglobin increased at 0.2–1% and hematocrit was elevated at 0.2 and 0.5% (Sattanathan et al., 2024). The remaining parameters were unchanged, except for a decrease in MCH at 0.5% and neutrophil counts at 0.2–1%. These findings indicate improved erythropoietic activity and specific changes in innate immunity, without evidence of systemic hematological disruption. Importantly, all supplemented groups showed higher survival after *Edwardsiella tarda* infection, underscoring the immunoprotective potential of *C. aerea* even at low inclusion levels.

In European seabass (*Dicentrarchus labrax*), supplementation with *Gracilaria gracilis* produced variable effects on gut histology, depending on the dose and algae processing method. Diets with 8% *G. gracilis* had no impact on the anterior or posterior intestinal morphology (Batista et al., 2020). In contrast, lower inclusions (2.5 and 5%) of whole biomass increased goblet cell numbers and improved intestinal muscle layers and villus height in a dose-dependent manner (Passos et al., 2021b). An ethanolic extract (0.35%) also increased goblet cells and outer longitudinal thickness but reduced villus height and inner circular thickness. Hematological parameters were largely unchanged in all groups, except for decreased lymphocyte counts in fish fed the 5% whole biomass and the extract, suggesting mild immunomodulatory activity.

Similar trends were observed in gilthead seabream, where the dietary inclusion of whole *G. gracilis* (2.5 and 5%), ethanolic extract (0.15 and 0.35%), and a phycocolloid extract (0.2%) was tested (Passos et al., 2021a). Whole biomass induced moderate changes in intestinal cells without causing tissue damage, whereas the ethanolic extract (0.35%) had no effect on intestinal structure. In contrast, the phycocolloid extract caused minor changes in cellular organization. In fish fed whole biomass diets, hematological changes included increased MCV (both inclusion levels), hematocrit (2.5%), and thrombocyte counts, along with reduced lymphocyte counts (5%). The ethanolic extract at 0.15% increased thrombocytes without other changes. Following *Photobacterium damselae* subsp. *piscicida* challenge, only the group fed the 5% whole biomass diet showed improved survival, indicating a greater protective potential of the unprocessed macroalgal form.

Other *Gracilaria* species have shown variable outcomes across fish models. In golden pompano, dietary *G. lemaneiformis* had mild or no effects on intestinal histology (Xie and Niu, 2022). In contrast, a 10% inclusion of *G. vermiculophylla* reduced villus height and intestinal diameter in Nile tilapia and rainbow trout (*Oncorhynchus mykiss*; Araújo et al., 2016; Silva et al., 2015), while in gilthead seabream, a 5% thermally processed form improved survival after hypoxia without affecting blood parameters (Magnoni et al., 2017). The same study reported similar benefits with 5% processed *Ulva lactuca*. In contrast, a 10% *Ulva* sp. diet was unable to provoke histological changes in Nile tilapia, while 10% *Pyropia dioica* (formerly known as *Porphyra dioica*) increased villus height without altering goblet cells or muscle thickness (Silva et al., 2015). In golden pompano, 2% inclusion of *Pyropia haitanensis* (formerly known as *Porphyra haitanensis*) and *Saccharina japonica* (formerly known as *Laminaria japonica*) increased villus height, with *P. haitanensis* also enhancing intestinal muscle thickness, while goblet cell numbers remained unchanged (Xie and Niu, 2022). In largemouth bass (*Micropterus salmoides*), 5 and 9% inclusion of *S. japonica* improved both intestinal and hepatic conditions (Kong et al., 2024). In Persian sturgeon (*Acipenser persicus*), 0.5–1% *Gracilariopsis persica* (formerly known as *Gracilaria persica*) improved hematocrit, hemoglobin, RBCs, and WBCs, indicating enhanced oxygen transport and immune function (Adel et al., 2021). Similarly, in barramundi (*Lates calcarifer*), *Sargassum ilicifolium* at 3–9% increased RBCs at all levels, WBCs at 9%, and neutrophils at 6 and 9%, suggesting stimulated hematopoiesis and innate immune activation (Zeynali et al., 2020). Other parameters remained unchanged, except for a decrease in lymphocyte counts at 6 and 9%, possibly reflecting a shift toward a more innate-dominant immune profile. In contrast, in gilthead seabream, dietary inclusion of 5% *Chondrus crispus* or *Ulva lactuca* did not alter anterior intestinal structure (Guerreiro et al., 2019), while inclusion of 5–15% *Palmaria palmata* did not affect hematological parameters in Atlantic salmon (*Salmo salar*; Wan et al., 2016). Dietary inclusion of whole macroalgal biomass has shown species-specific and dose-dependent effects on fish health, particularly on gut morphology and hematology.

Processed macroalgal forms, such as enzymatic hydrolysates and extracts, appear effective at low doses. In golden pompano, diets supplemented with either 3% whole *Ulva prolifera* (formerly known as *Enteromorpha prolifera*) or its enzymatic hydrolysate at 0.1–0.3% increased villus height, suggesting enhanced nutrient absorption (Zhao et al., 2025). In common carp (*Cyprinus carpio*), *Padina australis* ethyl acetate extract (0.01–0.04%) selectively upregulated intestinal barrier genes, suggesting modulation of tight junction regulation, potentially improving gut integrity (Sheikhzadeh et al., 2022). Following *Aeromonas hydrophila* infection, all supplemented groups showed significantly increased survival. In another study with common carp, *Laurencia caspica* hydroethanolic extracts (0.5–2%) improved hematological parameters, elevating WBCs (1–2%), hemoglobin (1%), and neutrophils (2%), while RBCs and red cell indices remained unchanged (Khanzadeh et al., 2023). These changes indicate innate immune stimulation without altering systemic oxygen transport. All supplemented groups also showed improved survival against *Streptococcus agalactiae*, reinforcing the immunoprotective role of the extract. In olive flounder (*Paralichthys olivaceus*), *Pyropia yezoensis* water extract (0.5–2%) did not affect hemoglobin but improved survival against *Streptococcus iniae*, providing full protection at 1.5% inclusion (Choi et al., 2015). These findings highlight that processed macroalgal products, often effective at low doses, can enhance gut health, modulate immune responses, and improve disease resistance, making them promising candidates for targeted functional feed formulations.

Blended algal formulations, whether combining macroalgae, microalgae, or both, have shown potential to enhance fish immune responses, gut integrity, and survival against pathogenic challenges. In barramundi, equal parts of *Sargassum ilicifolium* and *Arthrospira platensis* (3–6%) led to changes in blood parameters, such as increased WBC counts and altered MCV and neutrophil levels, with all inclusion levels improving survival after *Vibrio harveyi* infection (Khanzadeh et al., 2025). In European seabass, a diet with 4% *Gracilaria gracilis* and 4% *Nannochloropsis oceanica* increased neutral goblet cells in the anterior intestine, pointing to improved mucosal defense (Batista et al., 2020). Meanwhile, in gilthead seabream, equal parts of *Chondrus crispus* and *Ulva lactuca* at 2.5% did not affect gut morphology (Guerreiro et al., 2019). Overall, these studies highlight the potential of algal blends to support fish health.

While responses to whole macroalgae vary, processed forms and extracts, especially at low doses, consistently enhance fish robustness, highlighting their potential for targeted use in functional aquafeeds.

3 Immune response

The immune system plays a crucial role in maintaining fish health, protecting the fish against pathogens. The fish immune system comprises both innate defenses (e.g., mucosal barriers, immune cells, lysozyme, complement system) and adaptive immunity (e.g., immunoglobulins). In this section, the potential benefits of micro- and macroalgae on the fish's immune system are revisited.

3.1 Microalgae

Given their abundance of bioactive compounds, microalgae are expected to stimulate immune responses in fish. A comprehensive review shows that most microalgae exert immunostimulatory effects in both freshwater and marine fish species, although these responses might be species-, dose-, and time-dependent (Table 9.1).

Different microalgae may elicit distinct immunostimulatory responses in the same fish species. In one of the few comparative studies available, Cerezuela et al. (2012b) evaluated diets containing 5 or 10% *Nannochloropsis gaditana*, *Phaeodactylum tricornutum*, or *Tetraselmis chui* in gilthead seabream (Table 9.1). Among these, *P. tricornutum* elicited the strongest immunostimulatory effect at both inclusion levels.

Regarding inclusion levels, immunostimulatory effects may only emerge above a certain threshold (Table 9.1). In grass carp (*Ctenopharyngodon idella*), 5 and 10% *Arthrospira platensis* improved skin immunity, but 1% did not (Faheem et al., 2022). Similarly, in golden pompano, dietary inclusion of 5% *Barranca yajiagengensis* enhanced immune parameters, whereas 1% had no effect (Zhao et al., 2022). However, higher doses do not always yield stronger responses. In gibel carp (*Carassius gibelio*), dietary levels of *Chlorella* sp. (0.4–2%) did not enhance immune responses in a dose-dependent manner, with the strongest effects observed at the lowest levels (Zhang et al., 2014).

Feeding duration may also impact immune responses. In rohu, bactericidal activity increased after 30, 60, and 90 days of feeding diets with 0.01–0.1% *Chlorella vulgaris* (Table 9.1). However, lysozyme activity varied over time: after 30 days, only the 0.1% group showed increased activity compared to the control; at 60 days, lysozyme activity was lower in all microalgae treatments; at 90 days, it was significantly elevated in all treated groups (Pradhan et al., 2023).

Time-dependent effects were also observed in European seabass fed diets with 1–5% *Arthrospira platensis* for 30 or 60 days (Table 9.1). At 30 days, plasma myeloperoxidase activity was unchanged, but increased at 60 days in the 2.5 and 5.0% groups. Conversely, respiratory burst activity was highest at 30 days in fish fed the highest inclusion levels, but this effect disappeared at 60 days. For lysozyme, the highest plasma activity at 30 days was observed in the 1% group, while at 60 days, the 5% group showed the highest levels (Guroy et al., 2022). Expression of immune-related and inflammatory response genes in head-kidney also revealed time-dependency. However, no significant changes occurred in the spleen, emphasizing the need for careful tissue selection in immune analyses.

In another study, gilthead seabream fed diets supplemented with 1% *Phaedactylum tricornutum* (whole or broken cells) showed increased alternative complement activity after two weeks, but not at 12 (Table 9.1). Conversely, bactericidal activity increased only at 12 weeks and only with broken cells (Reis et al., 2021). These findings, along with gene expression data, led the authors to recommend the use of broken rather than whole cells. Few studies have addressed the influence of microalgae processing on immune responses, but this subject deserves further attention.

In some cases, immune stimulation only becomes apparent after a challenge.

TABLE 9.1 Some examples of immunological effects observed in fish fed microalgae-supplemented diets

Microalgae species	Animal model species	Form of inclusion	Inclusion level (%)	Feeding period	Immunological status	References
Arthrospira platensis (formerly *Spirulina*)	European seabass (*Dicentrarchus labrax*)	Whole	0; 1; 2.5; 5	30 and 60 d	↑ RBA (except IL1); LYZ in IL1 ↔ MPO ↑ MPO (except IL1); LYZ in IL5 ↔ RBA	Guroy et al. 2022
	Grass carp (*Ctenopharyngodon idella*)	Whole	0; 1; 5; 10	90 d	↑ LYZ and AP (except IL1); POP and protease in IL5	Faheem et al. 2022
Barranca yajiagengensis	Golden pompano (*Trachinotus ovatus*)	Whole	0; 1; 5	42 d	↑ NO, TNOS, LYZ, and C4 in IL5	Zhao et al. 2022
Chlorella sp.	Gibel carp (*Carassius gibelio*)	Whole	0; 0.4; 0.8; 1.2; 1.6; 2.0	60 d	↑ Plasma IgM in IL0.4 and IgD in IL0.8 ↔ IgM and IgD in liver and kidney	Zhang et al. 2014
					↑ LYZ	Xu et al. 2014
Chlorella vulgaris	Rohu (*Labeo rohita*)	Whole	0; 0.01; 0.05; 0.1	30, 60, 90 d	↑ BA at 30, 60, and 90 d; LYZ in IL0.1 at 30 d and in all ILs at 90 d ↓ LYZ in IL0.01 & 0.05 at 30 d and in all ILs at 60 d	Pradhan et al. 2023
Nannochloropsis gaditana	Gilthead seabream (*Sparus aurata*)	Whole	0; 5; 10	2 and 4 w	↑ CH50 only at 4 w ↔ RBA, PA, and IgM	Cerezuela et al. 2012b
Phaeodactylum tricornutum	Gilthead seabream (*Sparus aurata*)	Whole	0; 5; 10	2 and 4 w	↑ CH50 and PA, RBA only at 2 w ↔ IgM	Cerezuela et al. 2012b
		Whole or broken cells	0; 1	2 and 12 w	↑ ACH50 only at 2 w; BA only for broken cells at 12 w	Reis et al. 2021
Schizochytrium sp.	Nile tilapia (*Oreochromis niloticus*)	Whole	0; 1; 2	105 d	↔ LYZ	de Souza et al. 2020
Tetraselmis chui	Gilthead seabream (*Sparus aurata*)	Whole	0; 5; 10	2 and 4 w	↑ CH50 only at 4 w in IL10 ↔ RBA, PA, and IgM	Cerezuela et al. 2012b

↑: significantly higher than the control (0% inclusion); ↔: without differing from control; ↓: significantly lower than the control; ACH50: alternative complement pathway activity; AP: antiprotease activity; BA: bactericidal activity; C4: complement 4; CH50: hemolytic complement activity; d: days; Ig: immunoglobulin; IL: microalgae inclusion level; LYZ: lysozyme activity; MPO: myeloperoxidase activity; NO: nitric oxide; PA: phagocytic activity; POP: peroxidase activity; RBA: respiratory burst activity; TNOS: total nitric oxide synthase activity; w: weeks.

In yellow catfish (*Pelteobagrus fulvidraco*), no differences in plasma complement 3 were observed after 40 days of feeding diets with 4% *Oedocladium* sp. or 5% *Tribonema utriculosum* (Chen et al., 2021b). However, after hypoxic stress (5 min of air exposure), complement 3 levels increased in microalgae-fed groups. In rohu, the immunomodulatory effects of a microalgae extract blend (equal parts of *Arthrospira platensis*, *Chlorella vulgaris*, and *Euglena viridis* at 0.01–0.8%) were more clearly defined after a bacterial challenge (Sattanathan et al., 2023). Serum immune responses during the feeding trial were time-dependent, but immune gene expression following *Aeromonas hydrophila* infection indicated better responses at 0.02–0.04%. These results highlight the importance of challenge tests in revealing immune benefits not evident under normal conditions.

3.2 Macroalgae

As with microalgae, macroalgae also exhibit immunostimulatory properties in fish, though responses vary depending on species, inclusion level, and duration of feeding (Table 9.2).

Direct comparisons of different macroalgae in the same fish species are rare but can reveal interesting findings. Xie and Niu (2022) evaluated the dietary inclusion of 2% *Gracilaria lemaneiformis*, *Pyropia haitanensis*, *Saccharina japonica*, or *Undaria pinnatifida* in golden pompano (Table 9.2). An immunostimulatory effect, reflected by upregulation of hepatic lysozyme expression, was observed only with *P. haitanensis*, while the other species caused downregulation.

The immunostimulatory effects of macroalgae often depend on the inclusion level. In common carp, diets with 0.2 or 0.4% *Padina australis* extract enhanced immune response, whereas inclusion of 0.1% had no effect (Sheikhzadeh et al., 2022; Table 9.2). However, higher inclusion levels are not always more effective. In barramundi fed diets containing equal parts of *Arthrospira platensis* and *Sargassum ilicifolium* (3–6%), immune responses were improved at 5% inclusion, which also led to the highest survival after a *Vibrio harveyi* challenge (Khanzadeh et al., 2025). However, at 6% inclusion, these parameters were reduced.

Immune responses to macroalgae supplementation can vary with feeding duration and tissue examined (Table 9.2). For instance, in Nile tilapia, after 25 days of feeding diets containing 0.5–2% *Laurencia caspica* extract, serum lysozyme activity was unchanged, but skin mucus lysozyme increased in all supplemented groups (Khanzadeh et al., 2023, 2024). On the same day, serum immunoglobulin M (IgM) and complement 3 levels were higher only at 1 and 2% inclusion. After 50 days, all supplemented groups showed increased serum lysozyme, IgM, and complement 3 levels. In rainbow trout, serum lysozyme activity increased only after 83 days of feeding diets supplemented with *Gracilariopsis persica*, *Hypnea flageliformis*, or *Sargassum boveanum*. However, expression of immune-related genes was more influenced by a shorter feeding period (47 days) and lower inclusion levels (Vazirzadeh et al., 2020).

TABLE 9.2 Main immunological outcomes in fish fed macroalgae-supplemented diets

Macroalgae species	Animal model species	Form of inclusion	Inclusion level (%)	Feeding period	Immunological status	References
Gracilaria gracilis	European seabass (*Dicentrarchus labrax*)	Whole	0; 2.5; 5	47 d	↔ LYZ and POP	Passos et al. 2021b
		Extract	0; 0.35			
	Gilthead seabream (*Sparus aurata*)	Whole	0; 2.5; 5	52 d	↔ LYZ in IL2.5; POP ↓ LYZ in IL5	Passos et al. 2021a
		Extract	0; 0.15; 0.35		↑ LYZ in IL 0.15 ↔ POP ↓ LYZ in IL 0.35	
Gracilaria lemaneiformis	Golden pompano (*Trachinotus ovatus*)	Whole	0; 2	8 w	↔ *c4* expression ↓ *lyz* expression	Xie and Niu 2022
Gracilariopsis persica	Rainbow trout (*Oncorhynchus mykiss*)	Whole	0; 5; 10	47 and 83 d	↑ LYZ in IL5 at 83 d	Vazirzadeh et al. 2020
Hypnea flageliformis	Rainbow trout (*Oncorhynchus mykiss*)	Whole	0; 5; 10	47 and 83 d	↑ LYZ at 83 d	Vazirzadeh et al. 2020
Laurencia caspica	Nile tilapia (*Oreochromis niloticus*)	Extract	0; 0.5; 1; 2	25 and 50 d	↑ LYZ (except IL0.5 at 50 d) and total Ig in skin mucus	Khanzadeh et al. 2024
					↑ Serum IgM and C3 at 25d, except IL0.5; LYZ, IgM, C3 at 50 d ↔ Serum LYZ at 25 d	Khanzadeh et al. 2023
	Rainbow trout (*Oncorhynchus mykiss*)	Extract	0; 0.5; 1; 1.5; 2	4 and	↑ LYZ in IL1.5 & 2; total Ig in IL0.5; ACH50	Kiadaliri et al. 2020
				8 w	↑ LYZ and total Ig (except IL0.5); ACH50	
Padina australis	Common carp (*Cyprinus carpio*)	Extract	0; 0.1; 0.2; 0.4	8 w	↑ LYZ, ACH50, and total Ig (except IL0.1)	Sheikhzadeh et al. 2022
Palmaria palmata	Atlantic salmon (*Salmo salar*)	Whole	0; 5; 10; 15	98 d	↔ LYZ and ACH50	Wan et al. 2016
Pyropia haitanensis	Golden pompano (*Trachinotus ovatus*)	Whole	0; 2	8 w	↑ *lyz* expression ↔ *c4* expression	Xie and Niu 2022
Saccharina japonica	Golden pompano (*Trachinotus ovatus*)	Whole	0; 2	8 w	↔ *c4* expression ↓ *lyz* expression	Xie and Niu 2022
Sargassum boveanum	Rainbow trout (*Oncorhynchus mykiss*)	Whole	0; 5; 10	47 and 83 d	↑ LYZ at 83 d	Vazirzadeh et al. 2020
Sargassum horneri	Blackhead seabream (*Acanthopagrus schlegelii*)	Whole	0; 3; 6; 9	8 w	↑ LYZ, C3, and C4 ↔ IgM	Shi et al. 2019
Undaria pinnatifida	Golden pompano (*Trachinotus ovatus*)	Whole	0; 2	8 w	↔ *c4* expression ↓ *lyz* expression	Xie and Niu 2022

↑: significantly higher than the control (0% inclusion); ↔: without differing from control; ↓: significantly lower than the control; ACH50: alternative complement pathway activity; c#: complement #; d: days; Ig: immunoglobulin; IL: macroalgae inclusion level; lyz: lysozyme; POP: peroxidase activity; w: weeks.

The processing form of macroalgae is also an important variable. Inclusion of whole-cell *Gracilaria gracilis* powder (2.5–5%) in gilthead seabream diets had negative or no immunostimulatory effects (Table 9.2), while low levels of the macroalgae extract (0.15%) increased plasma lysozyme activity (Passos et al., 2021a). However, effects were also species- and dose-dependent. A higher extract dose (0.35%) reduced lysozyme activity (Passos et al., 2021a), whereas neither the powder nor the extract elicited an immunostimulatory response in European seabass (Passos et al., 2021b).

Challenge tests are valuable for confirming immunological effects observed during feeding trials and for identifying effective dietary levels. For example, in rainbow trout, diets with 0.5–2.0% *Laurencia caspica* extract improved plasma immune responses after eight weeks of feeding, without a clear best inclusion level (Table 9.2). However, following a challenge with *Aeromonas hydrophila*, the highest survival and post-challenge immune gene expression was observed at 1.5% inclusion (Kiadaliri et al., 2020).

Environmental stressors may also reveal immune benefits. Blackhead seabream (*Acanthopagrus schlegelii*) fed diets containing 3–9% *Sargassum horneri* for eight weeks showed enhanced immune responses (Table 9.2). After a 24 h ammonia stress, fish fed the 9% diet showed reduced lysozyme activity and no changes in complement 3 or IgM levels compared to lower inclusion groups, indicating diminished protection (Shi et al., 2019). These findings were supported by lower serum cortisol and lactate levels in fish receiving the lower inclusion diets, indicating better stress resilience. These studies underscore the value of challenge and stress tests in validating the functional benefits of macroalgae. In fact, such tests should be more widely used to confirm immunomodulatory potential, particularly in cases where no benefits are observed under standard rearing conditions, as seen, for example (Table 9.2), with *Gracilaria gracilis* in European seabass (Passos et al., 2021b) or *Palmaria palmata* in Atlantic salmon (Wan et al., 2016).

4 Antioxidant capacity

Common aquaculture practices, including grading, handling, vaccination, and acute thermal changes, are known to trigger stress responses in fish, leading to increased generation of reactive oxygen species (ROS). While ROS are naturally produced as part of normal cellular metabolism, excessive ROS generation can overwhelm the fish's antioxidant defenses, resulting in oxidative stress. This, in turn, may cause lipid peroxidation, protein degradation, enzyme inactivation, and damage to cellular structures (Lennicke and Cochemé, 2021).

To counteract oxidative stress, fish possess an elaborate antioxidant defense system, comprising both enzymatic (e.g., superoxide dismutase – SOD, catalase – CAT, and glutathione peroxidase – GPx) and non-enzymatic (e.g., glutathione and ascorbic acid) antioxidants.

This section will explore the documented effects of incorporating micro- and macroalgae into aquafeeds, with a particular focus on their influence on redox homeostasis in fish, as summarized in Tables 9.3 and 9.4.

4.1 Microalgae

The dietary incorporation of microalgae has shown potential to influence the antioxidant status of fish, particularly when used as functional additives at inclusion levels typically below 15%. Based on current evidence, the effects of microalgae depend on the algae species, the level and form of inclusion, the fish species, and whether fish are subjected to physiological or pathogenic challenges.

In freshwater species such as common carp and rainbow trout, supplementation with *Chlorella vulgaris*, *Arthrospira platensis*, and *Haematococcus pluvialis* led to significant increases in some enzymatic antioxidants (Table 9.3), including SOD, CAT, and GPx, as well as reductions in malondialdehyde (MDA) content in blood or liver (El-Mashtoly et al., 2024; Sheikhzadeh et al., 2012; Teimouri et al., 2019). For instance, *C. vulgaris* at 5% improved SOD, CAT, and GPx activities and reduced MDA content in common carp (El-Mashtoly et al., 2024), while *A. platensis* at 5–15% decreased MDA and increased total antioxidant capacity (TAC) in gibel carp and rainbow trout (Cao et al., 2018b; Teimouri et al., 2019). Also, in rainbow trout, the inclusion of extracts of *Chlorella* sp. or *Schizochytrium* sp. at 0.5% increased the expression of antioxidant genes (Karataş, 2025).

In marine species, positive responses were also observed but varied more by species and inclusion level (Table 9.3). In European seabass, a 15% inclusion of *Chlorella* sp. or *Nannochloropsis* sp. was unable to change fish antioxidant responses, as well as lipid peroxidation, over 11 weeks (Castro et al., 2020). In turbot (*Scophthalmus maximus*), *Nannochloropsis* sp. at 5–10% increased serum TAC and reduced MDA across all inclusion levels (Qiao et al., 2019). In contrast, in Senegalese sole (*Solea senegalensis*), *Chlorella vulgaris*, *Nannochloropsis gaditana,* and *Phaeodactylum tricornutum* included at 3% for 20 days, did not significantly alter CAT, total glutathione, or MDA (Peixoto et al., 2021). The inclusion of defatted *Desmodesmus* sp. and *Nannochloropsis oceanica* at 10% in Atlantic salmon feeds only provoked very mild changes (Kiron et al., 2016; Sørensen et al., 2017).

Under challenge conditions, particularly bacterial infections (e.g., *Aeromonas hydrophila*) or other stressors, the inclusion of microalgae resulted in measurable protective effects (Table 9.3). In Nile tilapia, 15% inclusion of either *Arthrospira platensis* or *Chlorella vulgaris* increased SOD and CAT and reduced MDA levels in serum following bacterial challenge (El-Habashi et al., 2019). Similarly, *A. platensis* at 14% significantly increased SOD and reduced MDA in gibel carp after the bacterial challenge (Cao et al., 2018a). In Nile tilapia, dietary inclusion of *A. platensis* at 0.75% led to an upregulation of *cat* expression, while *gpx* expression remained unchanged prior to the challenge. Following the challenge, this pattern shifted, with *gpx* being upregulated and *cat* expression remaining stable (Mohammed et al., 2025).

TABLE 9.3 Impact of dietary microalgae as functional feed additives on redox status in aquaculture species

Microalgae species	Animal model species	Form of inclusion	Inclusion level (%)	Feeding period
Experiments without a final challenge				
Arthrospira platensis (formerly *Spirulina*)	Gibel carp* (*Carassius gibelio*)	Whole	0; 10; 15	60 d
	Rainbow trout* (*Oncorhynchus mykiss*)	Whole	0; 2.5; 5; 7.5; 10	10 w
Chlorella sp.	European seabass** (*Dicentrarchus labrax*)	Whole	0; 15	11 w
	Rainbow trout* (*Oncorhynchus mykiss*)	Extract	0; 0.5	60 d
Chlorella sorokiniana	Rainbow trout* (*Oncorhynchus mykiss*)	Whole	0; 5; 10	90 d
Chlorella vulgaris	Common carp* (*Cyprinus carpio*)	Whole	0; 5	12 w
	Senegalese sole** (*Solea senegalensis*)	Whole	0; 3	20 d
Desmodesmus sp.	Atlantic salmon** (*Salmo salar*)	Defatted	0; 10	10 w
Haematococcus pluvialis	Rainbow trout* (*Oncorhynchus mykiss*)	Whole	0; 0.1; 0.3; 1	30 d
Nannochloropsis sp.	European seabass** (*Dicentrarchus labrax*)	Whole	0; 15	11 w
	Turbot** (*Scophthalmus maximus*)	Whole	0; 2.5; 5; 7.5; 10	10 w
Nannochloropsis gaditana	Gilthead seabream** (*Sparus aurata*)	Whole	0, 0.5, 0.75; 1.5	37 d
	Senegalese sole** (*Solea senegalensis*)	Whole	0; 3	20 d
Nannochloropsis oceanica	Atlantic salmon** (*Salmo salar*)	Defatted	0; 10	12 w
Phaeodactylum tricornutum	Senegalese sole** (*Solea senegalensis*)	Whole	0; 3	20 d
Schizochytrium sp.	Rainbow trout* (*Oncorhynchus mykiss*)	Extract	0; 0.5	60 d
Chlorella sp. + *Schizochytrium* sp. (1:1)	Rainbow trout* (*Oncorhynchus mykiss*)	Extract	0; 0.5	60 d
Experiments with a challenge at the end				
Arthrospira platensis	Gibel carp* (*Carassius gibelio*)	Whole	0; 3; 7; 14	46 d
	Nile tilapia* (*Oreochromis niloticus*)	Whole	0; 0.75	60 d
	Nile tilapia* (*Oreochromis niloticus*)	Whole	0; 15	9 w
Chlorella vulgaris	Nile tilapia* (*Oreochromis niloticus*)	Whole	0; 15	9 w
Tetraselmis suecica	Nile tilapia* (*Oreochromis niloticus*)	Whole	0; 0.5; 1; 1.5; 2	12 w

*: freshwater species; **: marine species; ↑: significantly higher than the control (0% inclusion); ↔: without differing from control; ↓: significantly lower than the control; CAT: catalase; CuZnSod: copper-zinc-superoxide dismutase; d: days; G6PDH: glucose-6-phosphate dehydrogenase; GSH: reduced glutathione; GPx: glutathione peroxidase; GR: glutathione reductase; IL: microalgae inclusion level; MDA: malondialdehyde; MnSod: manganese superoxide dismutase; n.d.: not determined; SOD: superoxide dismutase; TAC: total antioxidant capacity; tGSH: total glutathione; w: weeks.

Antioxidant status	Challenge event (Y/N)	Post-challenge effects	References
↑ TAC in hepatopancreas ↔ SOD in plasma and hepatopancreas; MDA in hepatopancreas ↓ MDA in plasma	N	-	Cao et al. 2018b
↑ TAC in serum and liver, except IL2.5; *sod* and *cat* expression in liver in IL7.5 & 10 ↓ Serum and liver MDA, except IL2.5	N	-	Teimouri et al. 2019
↔ SOD, CAT, GR, and MDA ↓ G6PDH and GPx	N	-	Castro et al. 2020
↑ *sod*, *cat*, and *gpx* expression	N	-	Karataş 2025
↑ GPx and GSH in IL10; SOD and CAT ↓ MDA in IL10	N	-	Chen et al. 2021a
↑ SOD, CAT, and GPx ↓ MDA	N	-	El-Mashtoly et al. 2024
↔ CAT, tGSH, and MDA ↓ GSH	N	-	Peixoto et al. 2021
↔ SOD, CAT, TAC, and *sod* expression	N	-	Kiron et al. 2016
↑ TAC, except IL1 ↓ MDA	N	-	Sheikhzadeh et al. 2012
↔ SOD, CAT, G6PDH, GR, and MDA ↓ GPx	N	-	Castro et al. 2020
↑ Serum SOD in IL5 & 7.5; TAC, except IL2.5; GPx in IL5 ↑ Liver SOD, except IL10; TAC in IL5; GPx in IL5 & 7.5 ↓ Serum and liver MDA	N	-	Qiao et al. 2019
↔ *CuZnSod*, *MnSod*, and *cat* expression	N	-	Jorge et al. 2019
↔ CAT, GSH, tGSH, and MDA	N	-	Peixoto et al. 2021
↑ SOD ↔ TAC and CAT	N	-	Sørensen et al. 2017
↔ CAT, tGSH, and MDA ↓ GSH	N	-	Peixoto et al. 2021
↑ *sod*, *cat*, and *gpx* expression	N	-	Karataş 2025
↑ SOD, CAT, and GPx	N	-	Karataş 2025
↑ MDA, except IL14 ↔ SOD	Y Bacterial	↑ SOD in IL14 ↓ MDA in IL14	Cao et al. 2018a
↑ *cat* expression ↔ *gpx* expression	Y Bacterial	↑ *gpx* expression ↔ *cat* expression	Mohammed et al. 2025
↑ CAT and SOD ↓ MDA	Y Bacterial	↑ CAT and SOD ↓ MDA	El-Habashi et al. 2019
↑ CAT and SOD ↓ MDA	Y Bacterial	↑ CAT and SOD ↓ MDA	El-Habashi et al. 2019
↑ *sod*, *cat*, and *gpx* expression ↓ MDA	Y Bacterial	n.d.	Abdel-Tawwab et al. 2024

TABLE 9.4 Impact of dietary macroalgae as functional feed additives on redox status in aquaculture species

Macroalgae species	Animal model species	Form of inclusion	Inclusion level (%)	Feeding period	Antioxidant status	Challenge event (Y/N)	Post-challenge effects	References
Experiments without a final challenge								
Chondrus crispus	Gilthead seabream** (*Sparus aurata*)	Whole	0; 5	8 w	↑ MDA ↔ SOD, CAT, GPx, GSH, and G6PDH	N	-	Guerreiro et al. 2019
Gracilaria gracilis	European seabass** (*Dicentrarchus labrax*)	Extract	0; 0.35	47 d	↑ SOD ↔ CAT, GST, tGSH, and MDA	N	-	Passos et al. 2021b
	European seabass** (*Dicentrarchus labrax*)	Whole	0; 2.5; 5	47 d	↑ tGSH in IL2.5; SOD ↔ tGSH and MDA in IL5; CAT and GST ↓ MDA in IL2.5	N	-	Passos et al. 2021b
Porphyridium sp.	Golden pompano** (*Trachinotus ovatus*)	Whole	0; 1; 5	6 w	↑ CAT, GPx, and TAC in IL5; SOD and MDA	N	-	Zhao et al. 2021a
Saccharina japonica	Largemouth bass* (*Micropterus salmoides*)	Whole	0; 5; 9	8 w	↑ GPx in IL5 in liver ↔ MDA in IL5; SOD and CAT in serum ↔ CAT in IL5; GPx in IL9; SOD, TAC, and MDA in liver ↓ MDA in IL9 in serum	N	-	Kong et al. 2024
Sargassum aquifolium	Nile tilapia* (*Oreochromis niloticus*)	Whole	0; 5; 10	8 w	↑ SOD in IL10; CAT and GSH ↓ MDA	N	-	Negm et al. 2021
Ulva lactuca	Gilthead seabream** (*Sparus aurata*)	Whole	0; 5	8 w	↑ MDA ↔ SOD, CAT, GPx, GSH, and G6PDH	N	-	Guerreiro et al. 2019
Chondrus crispus + *Ulva lactuca* (1:1)	Gilthead seabream** (*Sparus aurata*)	Whole	0; 5	8 w	↑ MDA ↔ SOD, CAT, GPx, GSH, and G6PDH	N	-	Guerreiro et al. 2019

TABLE 9.4 Impact of dietary macroalgae as functional feed additives on redox status in aquaculture species *(cont.)*

Macroalgae species	Animal model species	Form of inclusion	Inclusion level (%)	Feeding period	Antioxidant status	Challenge event (Y/N)	Post-challenge effects	References
Experiments with a challenge at the end								
Gracilaria gracilis	Gilthead seabream** (*Sparus aurata*)	Extract	0; 0.15; 0.35	8 w	↑ MDA in IL0.15 ↔ CAT and MDA in IL0.35; GST ↓ CAT in IL0.15; SOD	Y Bacterial	n.d.	Passos et al. 2021a
	Gilthead seabream** (*Sparus aurata*)	Whole	0; 2.5; 5	8 w	↑ MDA in IL2.5 ↔ MDA in IL5; SOD, CAT, and GST	Y Bacterial	n.d.	Passos et al. 2021a
Gracilaria vermiculophylla	Gilthead seabream** (*Sparus aurata*)	Whole + temperature	0; 5	34 d	↔ CAT, GR, GPx, GST, and MDA	Y Hypoxia	↔ GR, GST, and MDA ↓ CAT and GPx	Magnoni et al. 2017
Laurencia caspica	Nile tilapia* (*Oreochromis niloticus*)	Extract	0; 0.5; 1; 2	50 d	↑ SOD and CAT	Y Bacterial	n.d.	Khanzadeh et al. 2023
Sargassum horneri	Blackhead seabream** (*Acanthopagrus schlegelii*)	Whole	0; 3; 6; 9	8 w	↑ SOD and TAC ↔ CAT ↓ MDA	Y Ammonia	↑ TAC, except IL3; SOD and CAT ↔ TAC in IL3; MDA	Shi et al. 2019
Ulva lactuca	Gilthead seabream** (*Sparus aurata*)	Whole + temperature	0; 5	34 d	↔ CAT, GR, GPx, GST, and MDA	Y Hypoxia	↔ CAT, GR, GPx, GST, and MDA	Magnoni et al. 2017

*: freshwater species; **: marine species; ↑: significantly higher than the control (0% inclusion); ↔: without differing from control; ↓: significantly lower than the control; CAT: catalase; d: days; G6PDH: glucose-6-phosphate dehydrogenase; GPx: glutathione peroxidase; GR: glutathione reductase; GSH: reduced glutathione; GST: glutathione S-transferase; IL: macroalgae inclusion level; MDA: malondialdehyde; n.d.: not determined; SOD: superoxide dismutase; TAC: total antioxidant capacity; tGSH: total glutathione; w: weeks.

Most studies do not apply a challenge event at the end of the experiment. In studies incorporating a challenge event, mortality is commonly used as the primary outcome, while the effects on the antioxidant defense system of fish are seldom evaluated. Freshwater species such as tilapia, carp, and trout responded very often to the inclusion of *Arthrospira platensis* and *Chlorella*, with consistent increases in enzymatic antioxidant activities and reductions in oxidative damage. In marine fish species, the beneficial effects of microalgae inclusion on redox responses were less consistent. These findings highlight the potential of microalgae as functional feed additives capable of reinforcing the antioxidant system and enhancing fish robustness, particularly under stress-prone conditions.

4.2 Macroalgae

Macroalgae are increasingly studied for their potential as functional additives in aquafeeds, particularly for modulating antioxidant defenses in both freshwater and marine fish. Their effects, however, vary according to species, form (whole *vs.* extract), inclusion level, and whether a challenge (e.g., bacterial, thermal, or hypoxia) is applied.

In studies without a final challenge, several macroalgae species (e.g., *Gracilaria gracilis*, *Porphyridium* sp., *Saccharina japonica*, and *Sargassum aquifolium*) led to some significant increases in antioxidant enzyme activities and some reduction of lipid peroxidation in both freshwater and marine fish species (Table 9.4).

In freshwater fish, Nile tilapia fed diets containing *Sargassum aquifolium* (5 and 10%) showed increased CAT and glutathione levels and reduced MDA at both inclusion levels, while SOD activity increased only at the higher inclusion level (Negm et al., 2021). Also, in Nile tilapia, an extract of *Laurencia caspica* enhanced SOD and CAT at 0.5–2% inclusions (Khanzadeh et al., 2023). Similarly, *Saccharina japonica* at 5% in largemouth bass increased GPx in the liver, while MDA in serum was reduced only at 9% inclusion (Kong et al., 2024; Table 9.4).

In marine species, gilthead seabream fed *Chondrus crispus*, *Ulva lactuca*, or their blend (1:1) at 5% presented no changes in SOD, CAT, GPx, glutathione, and glucose-6-phosphate dehydrogenase, and a significant increase in MDA levels (Guerreiro et al., 2019). In golden pompano, 5% inclusion of *Porphyridium* sp. increased SOD, CAT, GPx, and TAC (Zhao et al., 2021a). European seabass showed increased SOD after dietary inclusion of *Gracilaria gracilis* in whole form (2.5–5%), alongside a reduction in MDA at the lower level (Passos et al., 2021b). However, the extract form (0.35%) was only able to positively modify SOD levels (Table 9.4).

In fish exposed to stressors, macroalgae demonstrated an inconsistent antioxidant response (Table 9.4). In gilthead seabream challenged with *Photobacterium damselae* subsp. *piscicida*, dietary *Gracilaria gracilis*, before the challenge, both whole (2.5%) and extract (0.15%) forms, resulted in higher MDA, no changes in the antioxidant system, or even a decrease in SOD in the case of the extract (Passos et al., 2021a). Similarly, inclusion of *Gracilaria vermiculophylla* or *Ulva lactuca* at 5% for 34 days under hypoxic stress had very little effect on the antioxidant system of the fish (Magnoni et al., 2017). However, it is important to note the very comprehensive experimental design in this study. The authors have included three sampling periods: normoxia, hypoxia, and recovery (24 h after the previous sampling points). During the recovery period, MDA decreased with both macroalgae tested. These findings suggest that incorporating a recovery sampling point may help clarify the wide variability often observed in the responses. Conversely, blackhead seabream fed *Sargassum horneri* (3–9%) exhibited higher SOD and TAC levels and lower MDA concentrations prior to ammonia stress. Following the challenge, the enhanced redox response appeared to support detoxification processes, as reflected by increased SOD and CAT activities; however, MDA levels remained unchanged (Shi et al., 2019). Macroalgae inclusion appears to have a very limited impact on fish antioxidant responses, as only in a few studies was a clear positive effect observed.

5 Gut microbiota

The fish gut mucosa harbors a complex and dynamic microbial community that is essential for the host's well-being. This microbiota is critical for efficient nutrient digestion and metabolism (Sagaram et al., 2021; Zhang et al., 2022), the development and regulation of the immune system (Singh et al., 2025), and resistance to pathogens, ultimately supporting overall fish health and growth. Algae, as functional feed ingredients, possess a rich composition of bioactive compounds that will impact host health, particularly at the mucosal level. Their prebiotic potential is largely derived from unique indigestible polysaccharides (e.g., β-glucans, fucose-containing polymers, ulvans, and fucoidans), which are crucial for modulating the microbiota-host interface (Siddik et al., 2024a). Beyond these complex carbohydrates, other algal components like simple sugars (e.g., glucose, fructose) and polymeric extracellular substances can serve as energy sources for microbes (Oviedo-Olvera et al., 2025), while a range of secondary metabolites (e.g., polyphenols, phlorotannins, terpenes) can act as direct antimicrobial agents (Capillo et al., 2018; García-Beltrán et al., 2023; Pérez et al., 2016). Evidence strongly suggests that algae can be valuable modulators of the fish microbiome in aquafeeds (Aragão et al., 2022; Sagaram et al., 2021), but their effects are highly variable, depending on the algal and fish species, dietary inclusion rate, feeding duration, and the processing form of the algae (e.g., whole biomass or extract). A general overview of the most relevant impacts of algae as functional ingredients on the fish microbiota community is summarized in Figure 9.1.

5.1 Microalgae

The application of microalgae in aquafeeds has evolved from simple nutrient replacement to a more refined understanding of their role as functional ingredients that actively shape the gut environment. Microalgae from the genus *Chlorella* are the most extensively studied in aquaculture due to their known benefits (Zhang et al., 2022), but their effects on fish microbiome are not linear. A 2% inclusion of *C. vulgaris* in Nile tilapia diets increased alpha diversity (based on Chao1, Shannon, and Simpson indices), but only after 30 days of feeding (Huang et al., 2023). In Atlantic salmon, inclusion of 2–14% *C. vulgaris* modulated gut microbiota, specifically increasing the abundance of *Paenarthrobacter* and *Trichococcus*, known for metabolizing complex carbohydrates, while reducing lactic acid bacteria in digesta and *Floricoccus* in mucosa, both considered beneficial taxa (Mueller et al., 2024). The same inclusion level in Atlantic salmon also mitigated soy-induced deleterious effects, but without major impact on microbiome diversity (Grammes et al., 2013).

Similarly, studies with *Nannochloropsis* sp. showed variable results. An 8% inclusion of *N. oceanica* reduced gut microbial diversity (based on Shannon and Simpson indices) in European seabass, while beneficially increasing *Bacillus* and reducing *Vibrio* abundance (Ferreira et al., 2022). In gilthead seabream, lower inclusions (0.5–1.5%) of *N. gaditana* promoted a non-significant increase in community richness and diversity (Jorge et al., 2019), while a 5% inclusion of its hydrolyzed form did not affect the gut microbiota community (Cerezo-Ortega et al., 2021). These findings indicate that algae processing should be strategically implemented to achieve expected benefits.

Members of the genus *Arthrospira* have also been shown to beneficially reshape the fish microbial landscape. A 5% inclusion of *A. platensis* biomass increased gut microbial diversity in Sabah giant grouper (*Epinephelus fuscoguttatus* × *E. lanceolatus*), associated with higher growth (Man et al., 2020). In Nile tilapia, 5–15% inclusion promoted community richness (through Margalef index) and evenness, increasing the abundance of probiotic *Bacillus* and Actinomycetes (Al Mahrouqi et al., 2023). Also, in tilapia, 5% *A. maxima* reduced the relative abundance of *Aeromonas* and *Flavobacterium*, while increasing *Anaeromyxobacter*, *Deinococcus*, *Pseudorhodoplanes*, and *Kocuria*, all related to positive outcomes in fish performance (Rojas-Villalta et al., 2024). At higher inclusion levels (e.g., 15%), effects in European seabass were less promising, with microbiota modulation but reduced performance (Pérez-Pascual et al., 2020), although its positive outcome when provided at functional dosages is encouraging for future commercial applications.

Other microalgae species, such as *Phaeodactylum tricornutum*, have also been studied. In Senegalese sole, both its whole biomass and a polysaccharide-enriched extract reduced *Vibrio* abundance. However, while the biomass increased microbial diversity, the extract had the opposite effect (Carballo et al., 2019). In contrast, a 10% inclusion in gilthead seabream diets showed no negative effects on gut microbiome after four weeks but was associated with higher mortalities during a subsequent challenge (Cerezuela et al., 2012a). Other microalgae assessed as microbiome modulators include *Scenedesmus* sp. (Skalli et al., 2020), *Tetraselmis chui* (Cerezuela et al., 2012a), and *Schizochytrium* sp. (de Souza et al., 2020; Lyons et al., 2017). The latter has gained significant attention as a viable, sustainable, and functional alternative oil in aquafeeds. In rainbow trout, a 5% inclusion replacing fish and soybean oils increased diversity (Lyons et al., 2017). In Nile tilapia, inclusion in a blend with *Nannochloropsis oculata* and *Spirulina* sp. at equal parts increased the abundance of probiotic *Lactobacillus* and *Bacillus*, while reducing potentially pathogenic *Vibrio* and *Staphylococcus* (Ibrahim et al., 2022). Even at 1.2%, it prompted a beneficial shift in the tilapia microbiome, favoring Firmicutes over Fusobacteria and Proteobacteria (de Souza et al., 2020).

5.2 Macroalgae

Studies on the microbiome-modulatory capacity of macroalgae have been limited to *Ulva* sp., *Gracilaria* sp., and *Ascophylum* sp., all with noticeable encouraging results. Senegalese sole fed a diet supplemented with 5% *Ulva ohnoi* for 45 days had higher microbial diversity (Shannon index), and despite an observed increase in *Vibrio* abundance, no deleterious effects were reported (Cerezo et al., 2022; Tapia-Paniagua et al., 2019). Additionally, *Tenacibaculum* was reduced, representing a clear positive outcome (Tapia-Paniagua et al., 2019). However, 10% *U. rigida* biomass had no effect on Atlantic cod (*Gadus morhua*) after 12 weeks, whereas *A. nodosum* at the same level reduced growth and gut microbiota diversity, while increasing the abundance of *Psychromonas*, *Propionigenium*, and *Clostridium* (Keating et al., 2021). With divergent roles, this modulation may reflect a compensatory response to mitigate any deleterious effects of the algal inclusion in the diet (Keating et al., 2021). In European seabass, *G. gracilis* at 2.5 or 5% inclusion reduced *Photobacterium* abundance, with stronger effects observed using a 0.35% extract (Gonçalves et al., 2022). *Staphylococcus* abundance was regulated only by *G. gracilis* biomass, whereas *Sphingomonas* was increased by both the biomass and the extract, showing a positive outcome that depended on the form in which the algae were included (Gonçalves et al., 2022). Also in European seabass, an 8% inclusion of the same algae reduced diversity by lowering the abundance of *Opitutus* and *Rhizobium*, which was associated with lower nutrient digestibility, while increasing *Sulfitobacter* and *Methylobacterium*, potentiating competition with pathogens (Ferreira et al., 2022).

In contrast to these positive outcomes, a study on juvenile European seabass tested the viability of incorporating the invasive species *Rugulopteryx okamurae* into their diet (Fonseca et al., 2023). Over three months, fish were fed a control diet or one diet containing 5% of the macroalgae processed in different ways (crude, fermented, enzymatically treated, or both fermented and enzymatically treated). All algae-supplemented diets altered the intestinal microbiota, consistent with a dysbiosis frame, characterized by a sharp increase in potentially pathogenic Vibrionaceae and a decrease in beneficial families such as Flavobacteriaceae and Rhodobacteraceae. This effect was most pronounced in the group fed fermented algae,

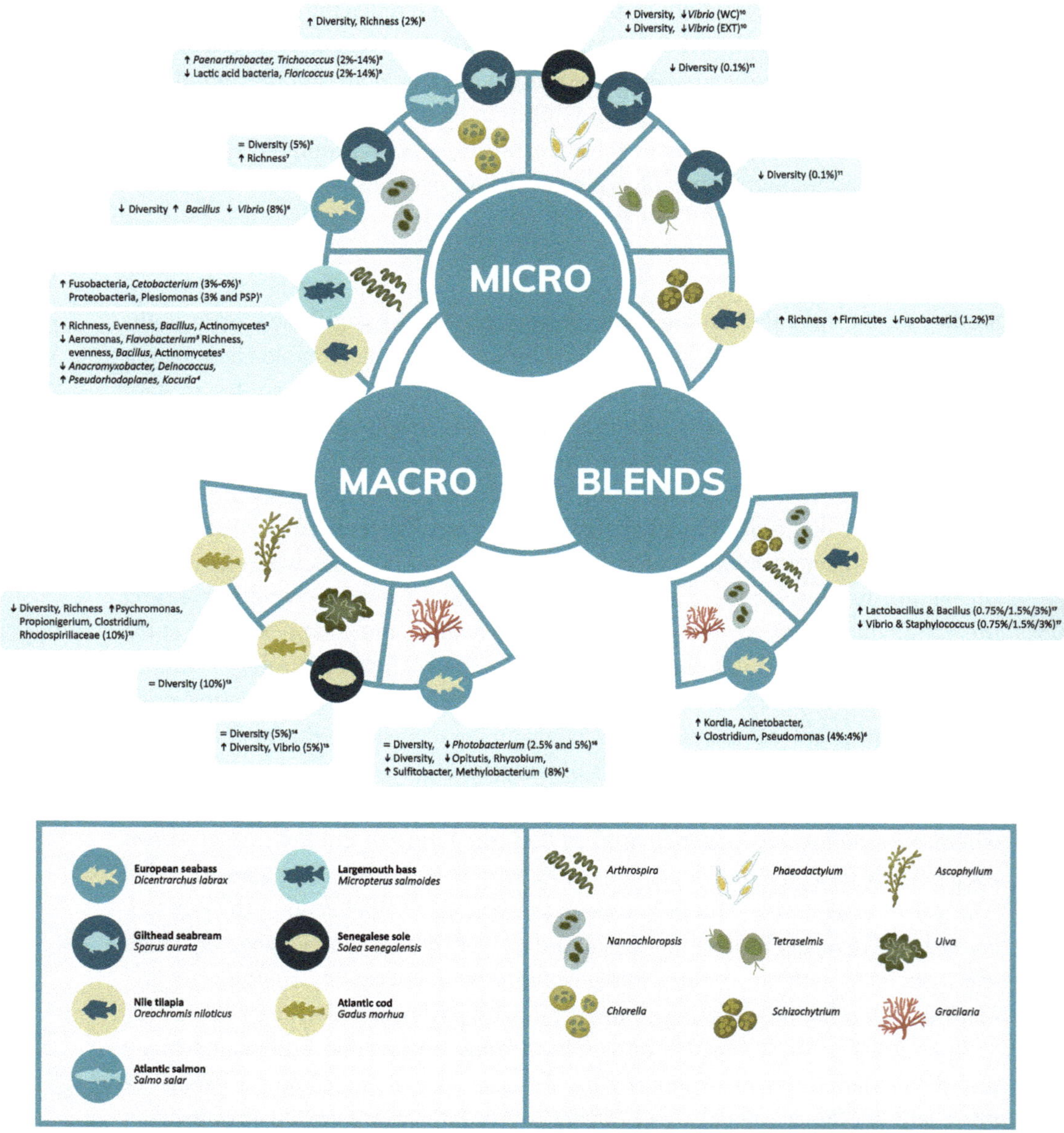

FIGURE 9.1 Summary of the most relevant intestinal microbiome modulations by the main aquaculture species fed with functional dosages of micro- and macroalgae or their blend. Only main features are shown, and taxonomical groups are presented until the Genus level. EXT: extract; PSP: polysaccharides; WC: whole cells. 1: Zhang et al. 2024; 2: Al Mahrouqi et al. 2023; 3: Rojas-Villalta et al. 2024; 4: Pérez-Pascual et al. 2020; 5: Cerezo-Ortega et al. 2021; 6: Ferreira et al. 2022; 7: Jorge et al. 2019; 8: Huang et al. 2023; 9: Mueller et al. 2024; 10: Carballo et al. 2019; 11: Cerezuela et al. 2012a; 12: de Souza et al. 2020; 13: Keating et al. 2021; 14: Cerezo et al. 2022; 15: Tapia-Paniagua et al. 2019; 16: Gonçalves et al. 2022; 17: Ibrahim et al. 2022

which rose Vibrionaceae abundance by 84%. These results suggest that while this invasive macroalgae could be a sustainable feed ingredient, effective pre-treatment is necessary to mitigate negative impacts on the gut microbial community. In contrast, invasive *Sargassum aquifolium* at 1 and 2% inclusion reduced gut aerobic and anaerobic bacterial counts in Nile tilapia under high stocking density, counteracting stress-related impacts (Negm et al., 2021) and representing a great opportunity to value a naturally growing macroalgae.

A promising strategy is the use of algal blends, which may boost the functionality of the ingredients as a mixture. In gilthead seabream, a 5% inclusion of a blend of *Scenedesmus almeriensis*, *Nannochloropsis gaditana*, and *Tisochrysis lutea* (25:15:5 ratio) promoted microbial diversity, although this effect was more evident at higher inclusion levels (15 and 25%; García-Márquez et al., 2023). In Nile tilapia, a blend of *Arthrospira* sp., *Schizochytrium* sp., and *Nannochloropsis oculata* increased probiotic genera and reduced potentially pathogenic *Vibrio* and *Staphylococcus* (Lyons et al., 2017). In European seabass, a blend of equal parts of *Gracilaria gracilis* and *Nannochloropsis oceanica* at 8% mitigated the loss of diversity observed when using micro- or macroalgae alone (Ferreira et al., 2022), indicating a synergistic effect.

6 Future perspectives and concluding remarks

Micro- and macroalgae are increasingly recognized as promising functional ingredients in aquafeeds, contributing to fish health and resilience through their antioxidant, immunomodulatory, and gut-supporting properties. While both groups share some biochemical traits, their effectiveness and consistency vary depending on the species, form of inclusion, and dosage. Microalgae generally provide more consistent benefits across a wider range of fish species, particularly improvements in gut morphology, hematological parameters, and oxidative status, with reliability strengthened when processed forms are used at appropriate inclusion levels. Macroalgae, although more variable in their whole-biomass form, demonstrate reproducible immunostimulatory effects when included as extracts or processed ingredients, even though their influence on antioxidant responses is comparatively limited. Algae also modulate gut microbiota, though outcomes are highly variable. While positive effects are frequently reported, null or negative effects may occur, especially at high inclusion levels. The complementary strengths of micro- and macroalgae, and emerging evidence for synergistic effects when different algal species are combined, highlight their potential as targeted functional components to enhance fish health and disease resistance.

To advance this field, encompassing the antioxidant, immunomodulatory, and gut microbiota effects of algae, future research must move beyond descriptive studies and adopt mechanistic, multi-omics approaches to elucidate the functional pathways involved. In parallel, optimizing biomass processing to improve compound bioavailability, clarifying the molecular mechanisms underlying immune and antioxidant modulation, and validating feeding trials under commercial conditions should be prioritized. Such an integrated framework, complemented by rigorous dose-response trials, will be essential to establish optimal inclusion levels and strengthen host resilience under environmental, nutritional, or pathogenic challenges.

Despite their functional potential, the widespread application of algae in aquaculture feeds faces several practical constraints. High production costs, limited scalability, and the lack of standardization in biomass quality and bioactive content remain major barriers to commercial adoption. Macroalgae are often harvested from the wild or cultivated in open systems, leading to variability in nutrient content, uncertainty in biomass yield, and potential contamination with harmful compounds such as heavy metals (Siddik et al., 2023). Although microalgae are often produced under more controlled conditions, intensive systems like photobioreactors involve high capital and operational costs (Siddik et al., 2024b). Moreover, batch-to-batch variability, arising from species-specific traits, seasonal fluctuations, and cultivation methods, poses a significant challenge to maintaining consistent product quality and bioactivity. These issues are particularly challenging for high-value algal strains with demonstrated antioxidant, immunostimulatory, or anti-inflammatory properties, many of which remain restricted to laboratory or pilot-scale production.

Overcoming these limitations will require the development of cost-effective, scalable, and standardized production systems. Promising strategies include advances in strain selection, controlled cultivation technologies, and biorefinery approaches that allow the extraction of multiple high-value compounds from a single biomass. This not only improves economic feasibility but also enables the targeted extraction and use of bioactive compounds from algae (such as fucoidan, laminarin, ulvan, phlorotannins, and carotenoids) as functional ingredients in aquafeeds.

Overall, both micro- and macroalgae hold significant promise as immunostimulants, antioxidants, and modulators of gut health, offering natural alternatives to synthetic additives and chemotherapeutics in aquaculture. Their successful integration in commercial feeds should be guided by precision nutrition principles to align specific algal ingredients and inclusion levels with the physiological needs and farming conditions of target species. Continued progress will depend on the integration of fundamental research, targeted feeding trials, and technological innovations to fully unlock the potential of algae in building a more resilient, efficient, and sustainable aquaculture sector.

Acknowledgements

The authors acknowledge Maria Eduarda Mendes for the design of Figure 9.1. This work was supported by Portuguese national funds from FCT - Foundation for Science and Technology through contracts UID/04326/2025, UID/PRR/04326/2025 and LA/P/0101/2020 (DOI:10.54499/LA/P/0101/2020) to CCMAR.

Note on the usage of AI tools

The authors of this chapter state that they used large language models (LLMs) strictly for English language correction and stylistic polishing of chapters, and not for content generation.

References

Abdel-Tawwab, M., Khalil, R.H., Abdel-Razek, N., Younis, N.A., Shady, S.H.H., Monier, M.N. and Abdel-Latif, H.M.R., 2024. Dietary effects of microalga *Tetraselmis suecica* on growth, antioxidant-immune activity, inflammation cytokines, and resistance of Nile tilapia fingerlings to *Aeromonas sobria* infection. Journal of Animal Physiology and Animal Nutrition 108: 511–526. https://doi.org/10.1111/jpn.13911

Adel, M., Yeganeh, S., Dadar, M., Sakai, M. and Dawood, M.A.O., 2016. Effects of dietary *Spirulina platensis* on growth performance, humoral and mucosal immune responses and disease resistance in juvenile great sturgeon (*Huso huso* Linnaeus, 1754). Fish and Shellfish Immunology 56: 436–444. https://doi.org/10.1016/j.fsi.2016.08.003

Adel, M., Omidi, A.H., Dawood, M.A.O., Karimi, B. and Shekarabi, S.P.H., 2021. Dietary *Gracilaria persica* mediated the growth performance, fillet colouration, and immune response of Persian sturgeon (*Acipenser persicus*). Aquaculture 530: 735950. https://doi.org/10.1016/j.aquaculture.2020.735950

Al Mahrouqi, H., Dobretsov, S. and Abdala-Daz, R.T., 2023. The effect of Spirulina (*Arthrospira platensis*) as a microbial dietary feed supplement for juvenile of Nile Tilapia (*Oreochromis niloticus*). Applied Environmental Biotechnology 8: 1–8. https://doi.org/10.26789/AEB.2023.01.001

Ansari, F.A., Guldhe, A., Gupta, S.K., Rawat, I. and Bux, F., 2021. Improving the feasibility of aquaculture feed by using microalgae. Environmental Science and Pollution Research 28: 43234–43257. https://doi.org/10.1007/s11356-021-14989-x

Aragão, C., Gonçalves, A.T., Costas, B., Azeredo, R., Xavier, M.J. and Engrola, S., 2022. Alternative proteins for fish diets: implications beyond growth. Animals 12: 1211. https://doi.org/10.3390/ani12091211

Araújo, M., Rema, P., Sousa-Pinto, I., Cunha, L.M., Peixoto, M.J., Pires, M.A., Seixas, F., Brotas, V., Beltrán, C. and Valente, L.M.P., 2016. Dietary inclusion of IMTA-cultivated *Gracilaria vermiculophylla* in rainbow trout (*Oncorhynchus mykiss*) diets: effects on growth, intestinal morphology, tissue pigmentation, and immunological response. Journal of Applied Phycology 28: 679–689. https://doi.org/10.1007/s10811-015-0591-8

Awad, L.Z., El-Mahallawy, H.S., Abdelnaeim, N.S., Mahmoud, M.M.A., Dessouki, A.A. and ElBanna, N.I., 2022. Role of dietary *Spirulina platensis* and betaine supplementation on growth, hematological, serum biochemical parameters, antioxidant status, immune responses, and disease resistance in Nile tilapia. Fish & Shellfish Immunology 126: 122–130. https://doi.org/10.1016/j.fsi.2022.05.040

Batista, S., Pereira, R., Oliveira, B., Baião, L.F., Jessen, F., Tulli, F., Messina, M., Silva, J.L., Abreu, H. and Valente, L.M.P., 2020. Exploring the potential of seaweed *Gracilaria gracilis* and microalga *Nannochloropsis oceanica*, single or blended, as natural dietary ingredients for European seabass *Dicentrarchus labrax*. Journal of Applied Phycology 32: 2041–2059. https://doi.org/10.1007/S10811-020-02118-Z

Cao, S., Zhang, P.Y., Zou, T., Fei, S.Z., Han, D., Jin, J.Y., Liu, H.K., Yang, Y.X., Zhu, X.M. and Xie, S.Q., 2018a. Replacement of fishmeal by spirulina *Arthrospira platensis* affects growth, immune related-gene expression in gibel carp (*Carassius auratus gibelio* var. CAS III), and its challenge against *Aeromonas hydrophila* infection. Fish & Shellfish Immunology 79: 265–273. https://doi.org/10.1016/j.fsi.2018.05.022

Cao, S.-P., Zou, T., Zhang, P.-Y., Han, D., Jin, J.-Y., Liu, H.-K., Yang, Y.-X., Zhu, X.-M. and Xie, S.-Q., 2018b. Effects of dietary fishmeal replacement with *Spirulina platensis* on the growth, feed utilization, digestion and physiological parameters in juvenile gibel carp (*Carassis auratus gibelio* var. CAS III). Aquaculture Research 49: 1320–1328. https://doi.org/10.1111/are.13590

Capillo, G., Savoca, S., Costa, R., Sanfilippo, M., Rizzo, C., Lo Giudice, A., Albergamo, A., Rando, R., Bartolomeo, G., Spanò, N. and Faggio, C., 2018. New insights into the culture method and antibacterial potential of *Gracilaria gracilis*. Marine Drugs 16: 492. https://doi.org/10.3390/md16120492

Carballo, C., Pinto, P.I.S., Mateus, A.P., Berbel, C., Guerreiro, C.C., Martinez-Blanch, J.F., Codoñer, F.M., Mantecon, L., Power, D.M. and Manchado, M., 2019. Yeast β-glucans and microalgal extracts modulate the immune response and gut microbiome in Senegalese sole (*Solea senegalensis*). Fish & Shellfish Immunology 92: 31–39. https://doi.org/10.1016/J.FSI.2019.05.044

Castro, C., Coutinho, F., Iglesias, P., Oliva-Teles, A. and Couto, A., 2020. *Chlorella* sp. and *Nannochloropsis* sp. inclusion in plant-based diets modulate the intestine and liver antioxidant mechanisms of European sea bass juveniles. Frontiers in Veterinary Science 7: 607575. https://doi.org/10.3389/FVETS.2020.607575

Cerezo, I.M., Fumanal, M., Tapia-Paniagua, S.T., Bautista, R., Anguís, V., Fernández-Díaz, C., Alarcón, F.J., Moriñigo, M.A. and Balebona, M.C., 2022. *Solea senegalensis* bacterial intestinal microbiota is affected by low dietary inclusion of *Ulva ohnoi*. Frontiers in Microbiology 12: 801744. https://doi.org/10.3389/FMICB.2021.801744

Cerezo-Ortega, I.M., Di Zeo-Sánchez, D.E., García-Márquez, J., Ruiz-Jarabo, I., Sáez-Casado, M.I., Balebona, M.C., Moriñigo, M.A. and Tapia-Paniagua, S.T., 2021. Microbiota composition and intestinal integrity remain unaltered after the inclusion of hydrolysed *Nannochloropsis gaditana* in *Sparus aurata* diet. Scientific Reports 11: 18779. https://doi.org/10.1038/s41598-021-98087-5

Cerezuela, R., Fumanal, M., Tapia-Paniagua, S.T., Meseguer, J., Moriñigo, M.A. and Esteban, M.A., 2012a. Histological alterations and microbial ecology of the intestine in gilthead seabream (*Sparus aurata* L.) fed dietary probiotics and microalgae. Cell and Tissue Research 350: 477–489. https://doi.org/10.1007/s00441-012-1495-4

Cerezuela, R., Guardiola, F.A., Meseguer, J. and Esteban, M.Á., 2012b. Enrichment of gilthead seabream (*Sparus aurata* L.) diet with microalgae: effects on the immune system. Fish Physiology and Biochemistry 38: 1729–1739. https://doi.org/10.1007/S10695-012-9670-9

Chen, W., Luo, L., Han, D., Long, F., Chi, Q. and Hu, Q., 2021a. Effect of dietary supplementation with *Chlorella sorokiniana* meal on the growth performance, antioxidant status, and immune response of rainbow trout (*Oncorhynchus mykiss*). Journal of Applied Phycology 33: 3113–3122. https://doi.org/10.1007/s10811-021-02541-w

Chen, W., Wang, Y., Han, D., Han, D., Zhu, X., Xie, S., Long, F., Jia, J. and Hu, Q., 2021b. Effects of dietary supplementation with filamentous microalgae (*Oedocladium* sp. or *Tribonema ultriculosum*) on growth performance, fillet fatty acid composition, skin pigmentation, and immune response of yellow catfish *Pelteobagrus fulvidraco*. Journal of the World Aquaculture Society 52: 1273–1289. https://doi.org/10.1111/JWAS.12839

Choi, Y.H., Lee, B.J. and Nam, T.J., 2015. Effect of dietary inclusion of *Pyropia yezoensis* extract on biochemical and immune responses of olive flounder *Paralichthys olivaceus*. Aquaculture 435: 347–353. https://doi.org/10.1016/j.aquaculture.2014.10.010

de Souza, F.P., de Lima, E.C.S., Urrea-Rojas, A.M., Suphoronski, S.A., Facimoto, C.T., da Silva Bezerra, J., de Oliveira, T.E.S., de Pádua Pereira, U., Di Santis, G.W., de Oliveira, C.A.L. and Lopera-Barrero, N.M., 2020. Effects of dietary supplementation with a microalga (*Schizochytrium* sp.) on the hemato-immunological, and intestinal histological parameters and gut microbiota of Nile tilapia in net cages. PLoS ONE 15: e0226977. https://doi.org/10.1371/JOURNAL.PONE.0226977

El-Habashi, N., Fadl, S.E., Farag, H.F., Gad, D.M., Elsadany, A.Y. and El Gohary, M.S., 2019. Effect of using *Spirulina* and *Chlorella* as feed additives for elevating immunity status of Nile tilapia experimentally infected with *Aeromonas hydrophila*. Aquaculture Research 50: 2769–2781. https://doi.org/10.1111/ARE.14229

El-Mashtoly, M., Magouz, F.I., Darwish, S., Amer, A.A., Zaineldin, A.I., Gewaily, M.S. and Dawood, M.A.O., 2024. Dietary *Chlorella vulgaris* and *Saccharomyces cerevisiae* enhanced the growth performance, blood biomarkers, and antioxidative capacity of common carp (*Cyprinus carpio*). Scientific African 26: e02407. https://doi.org/10.1016/J.SCIAF.2024.E02407

Faheem, M., Jamal, R., Nazeer, N., Khaliq, S., Hoseinifar, S.H., Van Doan, H. and Paolucci, M., 2022. Improving growth, digestive and antioxidant enzymes and immune response of juvenile grass carp (*Ctenopharyngodon idella*) by using dietary *Spirulina platensis*. Fishes 7: 237. https://doi.org/10.3390/fishes7050237

Ferreira, M., Abdelhafiz, Y., Abreu, H., Silva, J., Valente, L.M.P. and Kiron, V., 2022. *Gracilaria gracilis* and *Nannochloropsis oceanica*, singly or in combination, in diets alter the intestinal microbiota of European seabass (*Dicentrarchus labrax*). Frontiers in Marine Science 9: 1001942. https://doi.org/10.3389/FMARS.2022.1001942

Fonseca, F., Fuentes, J., Vizcaíno, A.J., Alarcón, F.J., Mancera, J.M., Martínez-Rodríguez, G. and Martos-Sitcha, J.A., 2023. From invasion to fish fodder: inclusion of the brown algae *Rugulopteryx okamurae* in aquafeeds for European sea bass *Dicentrarchus labrax* (L., 1758). Aquaculture 568: 739318. https://doi.org/10.1016/J.AQUACULTURE.2023.739318

García-Beltrán, J.M., Arizcun, M. and Chaves-Pozo, E., 2023. Antimicrobial peptides from photosynthetic marine organisms with potential application in aquaculture. Marine Drugs 21: 290. https://doi.org/10.3390/md21050290

García-Márquez, J., Rico, R.M., Acién, F.G., Mancera, J.M., Figueroa, F.L., Vizcaíno, A.J., Alarcón, F.J., Moriñigo, M.Á. and Abdala-Díaz, R.T., 2023. Dietary effects of a short-term administration of microalgae blend on growth performance, tissue fatty acids, and predominant intestinal microbiota in *Sparus aurata*. Microorganisms 11: 463. https://doi.org/10.3390/microorganisms11020463

Gonçalves, A.T., Simões, M., Costa, C., Passos, R. and Baptista, T., 2022. Modulatory effect of *Gracilaria gracilis* on European seabass gut microbiota community and its functionality. Scientific Reports 12: 1–16. https://doi.org/10.1038/S41598-022-17891-9

González-Meza, G.M., Elizondo-Luevano, J.H., Cuellar-Bermudez, S.P., Sosa-Hernández, J.E., Iqbal, H.M.N., Melchor-Martínez, E.M. and Parra-Saldívar, R., 2023. New perspective for macroalgae-based animal feeding in the context of challenging sustainable food production. Plants 12: 3609. https://doi.org/10.3390/plants12203609

Grammes, F., Reveco, F.E., Romarheim, O.H., Landsverk, T., Mydland, L.T. and Øverland, M., 2013. *Candida utilis* and *Chlorella vulgaris* counteract intestinal inflammation in Atlantic salmon (*Salmo salar* L.). PLoS ONE 8: e83213. https://doi.org/10.1371/journal.pone.0083213

Guerreiro, I., Magalhães, R., Coutinho, F., Couto, A., Sousa, S., Delerue-Matos, C., Domingues, V.F., Oliva-Teles, A. and Peres, H., 2019. Evaluation of the seaweeds *Chondrus crispus* and *Ulva lactuca* as functional ingredients in gilthead seabream (*Sparus aurata*). Journal of Applied Phycology 31: 2115–2124. https://doi.org/10.1007/S10811-018-1708-7

Guroy, B., Guroy, D., Bilen, S., Kenanoglu, O.N., Sahin, I., Terzi, E., Karadal, O. and Mantoglu, S., 2022. Effect of dietary Spirulina (*Arthrospira platensis*) on the growth performance, immune-related gene expression and resistance to *Vibrio anguillarum* in European seabass (*Dicentrarchus labrax*). Aquaculture Research 53: 2263–2274. https://doi.org/10.1111/are.15745

Huang, Z., Gao, J., Peng, C., Song, J., Xie, Z., Jia, J., Li, H., Zhao, S., Liang, Y. and Gong, B., 2023. The effect of the microalgae *Chlorella vulgaris* on the gut microbiota of juvenile Nile tilapia (*Oreochromis niloticus*) is feeding-time dependent. Microorganisms 11: 1002. https://doi.org/10.3390/microorganisms11041002

Ibrahim, D., Abd El-Hamid, M.I., Al-Zaban, M.I., ElHady, M., El-Azzouny, M.M., ElFeky, T.M., Al Sadik, G.M., Samy, O.M., Hamed, T.A., Albalwe, F.M., Alenezi, M.A. and Omar, A.E., 2022. Impacts of fortifying Nile tilapia (*Oreochromis niloticus*) diet with different strains of microalgae on its performance, fillet quality and disease resistance to *Aeromonas hydrophila* considering the interplay between antioxidant and inflammatory response. Antioxidants 11: 2181. https://doi.org/10.3390/antiox11112181

Jorge, S.S., Enes, P., Serra, C.R., Castro, C., Iglesias, P., Oliva Teles, A. and Couto, A., 2019. Short-term supplementation of gilthead seabream (*Sparus aurata*) diets with *Nannochloropsis gaditana* modulates intestinal microbiota without affecting intestinal morphology and function. Aquaculture Nutrition 25: 1388–1398. https://doi.org/10.1111/ANU.12959

Karataş, B., 2025. Effects of *Chlorella* sp. and *Schizochytrium* sp. extracts on growth indices, body composition, and gene expression profiles in rainbow trout (*Oncorhynchus mykiss*). Comparative Biochemistry and Physiology Part B: Biochemistry and Molecular Biology 276: 111047. https://doi.org/10.1016/j.cbpb.2024.111047

Keating, C., Bolton-Warberg, M., Hinchcliffe, J., Davies, R., Whelan, S., Wan, A.H.L., Fitzgerald, R.D., Davies, S.J., Ijaz, U.Z. and Smith, C.J., 2021. Temporal changes in the gut microbiota in farmed Atlantic cod (*Gadus morhua*) outweigh the response to diet supplementation with macroalgae. Animal Microbiome 3: 7. https://doi.org/10.1186/s42523-020-00065-1

Khanzadeh, M., Beikzadeh, B. and Hoseinifar, S.H., 2023. The effects of *Laurencia caspica* algae extract on hemato-immunological parameters, antioxidant defense, and resistance against *Streptococcus agalactiae* in Nile tilapia (*Oreochromis niloticus*). Aquaculture Nutrition 2023: 8882736. https://doi.org/10.1155/2023/8882736

Khanzadeh, M., Hoseinifar, S.H. and Beikzadeh, B., 2024. Investigating the effect of hydroalcoholic extract of red algae (*Laurencia caspica*) on growth performance, mucosal immunity, digestive enzyme activity and resistance to *Streptococcus iniae* and *Aeromonas hydrophila* in Nile tilapia (*Oreochromis niloticus*). Aquaculture Reports 35: 101984. https://doi.org/10.1016/J.AQREP.2024.101984

Khanzadeh, M., Hoseinifar, S.H., Zargari, A. and Van Doan, H., 2025. Effects of dietary supplements *Sargassum ilicifolium* and *Spirulina platensis* on growth parameters, immunity and gene expression in juvenile Asian seabass (*Lates calcarifer*). Journal of Agriculture and Food Research 19: 101689. https://doi.org/10.1016/J.JAFR.2025.101689

Kiadaliri, M., Firouzbakhsh, F. and Deldar, H., 2020. Effects of feeding with red algae (*Laurencia caspica*) hydroalcoholic extract on antioxidant defense, immune responses, and immune gene expression of kidney in rainbow trout (*Oncorhynchus mykiss*) infected with *Aeromonas hydrophila*. Aquaculture 526: 735361. https://doi.org/10.1016/J.AQUACULTURE.2020.735361

Kiron, V., Sorensen, M., Huntley, M., Vasanth, G.K., Gong, Y.Y., Dahle, D. and Palihawadana, A.M., 2016. Defatted biomass of the microalga, *Desmodesmus* sp., can replace fishmeal in the feeds for Atlantic salmon. Frontiers in Marine Science 3: 67. https://doi.org/10.3389/fmars.2016.00067

Kong, Q., Yang, Z., Su, Q., Yang, J., Xiao, G., Liu, D. and Tang, H., 2024. Effects of kelp meal (*Laminaria japonica*) on the growth, nutrition, and intestinal health of the largemouth bass (*Micropterus salmoides*). Aquaculture Reports 39: 102525. https://doi.org/10.1016/j.aqrep.2024.102525

Lennicke, C. and Cochemé, H.M., 2021. Redox metabolism: ROS as specific molecular regulators of cell signaling and function. Molecular Cell 81: 3691–3707. https://doi.org/10.1016/J.MOLCEL.2021.08.018

Lyons, P.P., Turnbull, J.F., Dawson, K.A. and Crumlish, M., 2017. Effects of low-level dietary microalgae supplementation on the distal intestinal microbiome of farmed rainbow trout *Oncorhynchus mykiss* (Walbaum). Aquaculture Research 48: 2438–2452. https://doi.org/10.1111/are.13080

Magnoni, L.J., Martos-Sitcha, J.A., Queiroz, A., Calduch-Giner, J.A., Gonçalves, J.F.M., Rocha, C.M.R., Abreu, H.T., Schrama, J.W., Ozorio, R.O.A. and Perez-Sanchez, J., 2017. Dietary supplementation of heat-treated *Gracilaria* and *Ulva* seaweeds enhanced acute hypoxia tolerance in gilthead sea bream (*Sparus aurata*). Biology Open 6: 897–908. https://doi.org/10.1242/BIO.024299/256652

Man, Y.B., Zhang, F., Ma, K.L., Mo, W.Y., Kwan, H.S., Chow, K.L., Man, K.Y., Tsang, Y.F., Li, W.C. and Wong, M.H., 2020. Growth and intestinal microbiota of Sabah giant grouper reared on food waste-based pellets supplemented with spirulina as a growth promoter and alternative protein source. Aquaculture Reports 18: 100553. https://doi.org/10.1016/j.aqrep.2020.100553

Mohammed, A.A.B., Saadd, A.H., Abdo, S.E., Fadl, S.E. and Hamouda, A.H., 2025. Immunomodulatory and growth promotors effects of rosemary (*Rosmarinus officinalis*) and/or *Spirulina* with respect to some gene expression on Nile tilapia (*Oreochromis niloticus*) in Aswan Governorate. Aquaculture 596: 741790. https://doi.org/10.1016/J.AQUACULTURE.2024.741790

Mueller, J., van Muilekom, D.R., Ehlers, J., Suhr, M., Hornburg, S.C., Bang, C., Wilkes, M., Schultheiß, T., Maser, E., Rebl, A., Goldammer, T., Seibel, H. and Schulz, C., 2024. Dietary *Chlorella vulgaris* supplementation modulates health, microbiota and the response to oxidative stress of Atlantic salmon. Scientific Reports 14: 1–19. https://doi.org/10.1038/S41598-024-72531-8

Negm, S.S., Ismael, N.E.M., Ahmed, A.I., Asely, A.M. El and Naiel, M.A.E., 2021. The efficiency of dietary *Sargassum aquifolium* on the performance, innate immune responses, antioxidant activity, and intestinal microbiota of Nile Tilapia (*Oreochromis niloticus*) raised at high stocking density. Journal of Applied Phycology 33: 4067–4082. https://doi.org/10.1007/s10811-021-02572-3

Oviedo-Olvera, M.V., Feregrino-Pérez, A.A., Nieto-Ramírez, M.I., Tovar-Ramírez, M.M., Aguirre-Becerra, H. and García-Trejo, J.F., 2025. Prebiotic emergent sources for aquaculture: microalgae and insects. Aquaculture and Fisheries 10: 19–26. https://doi.org/10.1016/J.AAF.2023.06.007

Passos, R., Correia, A.P., Ferreira, I., Pires, P., Pires, D., Gomes, E., do Carmo, B., Santos, P., Simões, M., Afonso, C. and Baptista, T., 2021a. Effect on health status and pathogen resistance of gilthead seabream (*Sparus aurata*) fed with diets supplemented with *Gracilaria gracilis*. Aquaculture 531: 735888. https://doi.org/10.1016/J.AQUACULTURE.2020.735888

Passos, R., Correia, A.P., Pires, D., Pires, P., Ferreira, I., Simões, M., do Carmo, B., Santos, P., Pombo, A., Afonso, C. and Baptista, T., 2021b. Potential use of macroalgae *Gracilaria gracilis* in diets for European seabass (*Dicentrarchus labrax*): health benefits from a sustainable source. Fish & Shellfish Immunology 119: 105–113. https://doi.org/10.1016/J.FSI.2021.09.033

Peixoto, D., Pinto, W., Goncalves, A.T., Machado, M., Reis, B., Silva, J., Navalho, J., Dias, J., Conceicao, L. and Costas, B., 2021. Microalgal biomasses have potential as ingredients in microdiets for Senegalese sole (*Solea senegalensis*) post-larvae. Journal of Applied Phycology 33: 2241–2250. https://doi.org/10.1007/s10811-021-02431-1

Pérez, M., Falqué, E. and Domínguez, H., 2016. Antimicrobial action of compounds from marine seaweed. Marine Drugs 14: 52. https://doi.org/10.3390/md14030052

Pérez-Pascual, D., Estellé, J., Dutto, G., Rodde, C., Bernardet, J.-F., Marchand, Y., Duchaud, E., Przybyla, C. and Ghigo, J.-M., 2020. Growth performance and adaptability of European sea bass (*Dicentrarchus labrax*) gut microbiota to alternative diets free of fish products. Microorganisms 8: 1346. https://doi.org/10.3390/microorganisms8091346

Pradhan, J., Sahu, S. and Das, B., 2023. Protective effects of *Chlorella vulgaris* supplemented diet on antibacterial activity and immune responses in rohu fingerlings, *Labeo rohita* (Hamilton), subjected to *Aeromonas hydrophila* infection. Life 13: 1028. https://doi.org/10.3390/life13041028

Qiao, H., Hu, D., Ma, J., Wang, X., Wu, H. and Wang, J., 2019. Feeding effects of the microalga *Nannochloropsis* sp. on juvenile turbot (*Scophthalmus maximus* L.). Algal Research 41: 101540. https://doi.org/10.1016/j.algal.2019.101540

Reis, B., Ramos-Pinto, L., Martos-Sitcha, J.A., Machado, M., Azeredo, R., Fernández-Boo, S., Engrola, S., Unamunzaga, C., Calduch-Giner, J., Conceição, L.E.C., Silva, T., Dias, J., Costas, B. and Pérez-Sánchez, J., 2021. Health status in gilthead seabream (*Sparus aurata*) juveniles fed diets devoid of fishmeal and supplemented with *Phaeodactylum tricornutum*. Journal of Applied Phycology 33: 979–996. https://doi.org/10.1007/s10811-021-02377-4

Rojas-Villalta, D., Gómez-Espinoza, O., Guillén-Watson, R., Murillo-Vega, F., Villalta-Romero, F., Vaquerano-Pineda, F., Chicas, M., Guerrero, M. and Núñez-Montero, K., 2024. Impact of *Arthrospira maxima* feed supplementation on gut microbiota and growth performance of tilapia fry (*Oreochromis niloticus*). Fishes 9: 374. https://doi.org/10.3390/FISHES9100374

Sáez, M.I., Galafat, A., Vizcaíno, A.J., Chaves-Pozo, E., Ayala, M.D., Arizcun, M., Alarcón, F.J., Suárez, M.D. and Martínez, T.F., 2022. Evaluation of *Nannochloropsis gaditana* raw and hydrolysed biomass at low inclusion level as dietary functional additive for gilthead seabream (*Sparus aurata*) juveniles. Aquaculture 556: 738288. https://doi.org/10.1016/j.aquaculture.2022.738288

Sagaram, U.S., Gaikwad, M.S., Nandru, R. and Dasgupta, S., 2021. Microalgae as feed ingredients: recent developments on their role in immunomodulation and gut microbiota of aquaculture species. FEMS Microbiology Letters 368: 71. https://doi.org/10.1093/FEMSLE/FNAB071

Sattanathan, G., Palanisamy, T., Padmapriya, S., Arumugam, V.A., Park, S., Kim, I.H. and Balasubramanian, B., 2020. Influences of dietary inclusion of algae *Chaetomorpha aerea* enhanced growth performance, immunity, haematological response and disease resistance of *Labeo rohita* challenged with *Aeromonas hydrophila*. Aquaculture Reports 17: 100353. https://doi.org/10.1016/J.AQREP.2020.100353

Sattanathan, G., Liu, W.-C., Padmapriya, S., Pushparaj, K., Sureshkumar, S., Lee, J.-W., Balasubramanian, B. and Kim, I.H., 2023. Effects of dietary blend of algae extract supplementation on growth, biochemical, haemato-immunological response, and immune gene expression in *Labeo rohita* with *Aeromonas hydrophila* post-challenges. Fishes 8: 7. https://doi.org/10.3390/fishes8010007

Sattanathan, G., Padmapriya, S., Kadaikunnan, S., Khaled, J.M., Malafaia, G. and Govindarajan, M., 2024. Marine macroalgae *Chaetomorpha aerea* as a dietary supplement: optimizing immunity and resistance to *Edwardsiella tarda* in tilapia (*Oreochromis mossambicus*). Fish & Shellfish Immunology 154: 109956. https://doi.org/10.1016/J.FSI.2024.109956

Sheikhzadeh, N., Tayefi-Nasrabadi, H., Khani Oushani, A. and Najafi Enferadi, M.H., 2012. Effects of *Haematococcus pluvialis* supplementation on antioxidant system and metabolism in rainbow trout (*Oncorhynchus mykiss*). Fish Physiology and Biochemistry 38: 413–419. https://doi.org/10.1007/s10695-011-9519-7

Sheikhzadeh, N., Ahmadifar, E., Soltani, M., Tayefi-Nasrabadi, H., Mousavi, S. and Naiel, M.A.E., 2022. Brown seaweed (*Padina australis*) extract can promote performance, innate immune responses, digestive enzyme activities, intestinal gene expression and resistance against *Aeromonas hydrophila* in common carp (*Cyprinus carpio*). Animals 12: 3389. https://doi.org/10.3390/ani12233389

Sherif, E.M., Abd El-Razek, I.M., El-Sharawy, M.E., Amer, A.A., Zaineldin, A.I., Gewaily, M.S., Ashry, A.M., Younis, N.A., Ahmed, H.A. and Dawood, M.A.O., 2024. Growth performance, antioxidative status, and immune response of Nile tilapia (*Oreochromis niloticus*) fed dietary fermented *Spirulina platensis*. Aquaculture Reports 39: 102324. https://doi.org/10.1016/j.aqrep.2024.102324

Shi, Q., Wen, X., Zhu, D., Aweya, J.J. and Li, S., 2019. Protective effects of *Sargassum horneri* against ammonia stress in juvenile black sea bream, *Acanthopagrus schlegelii*. Journal of Applied Phycology 31: 1445–1453. https://doi.org/10.1007/S10811-018-1637-5

Siddik, M.A.B., Francis, P., Rohani, M.F., Azam, M.S., Mock, T.S. and Francis, D.S., 2023. Seaweed and seaweed-based functional metabolites as potential modulators of growth, immune and antioxidant responses, and gut microbiota in fish. Antioxidants 12: 2066. https://doi.org/10.3390/antiox12122066

Siddik, M.A.B., Francis, P., Foysal, M.J. and Francis, D.S., 2024a. Dietary seaweed extract mitigates oxidative stress in Nile tilapia by modulating inflammatory response and gut microbiota. Frontiers in Immunology 15: 1471261. https://doi.org/10.3389/FIMMU.2024.1471261

Siddik, M.A.B., Sørensen, M., Islam, S.M.M., Saha, N., Rahman, M.A. and Francis, D.S., 2024b. Expanded utilisation of microalgae in global aquafeeds. Reviews in Aquaculture 16: 6–33. https://doi.org/10.1111/RAQ.12818

Silva, D.M., Valente, L.M.P., Sousa-Pinto, I., Pereira, R., Pires, M.A., Seixas, F. and Rema, P., 2015. Evaluation of IMTA-produced seaweeds (*Gracilaria*, *Porphyra*, and *Ulva*) as dietary ingredients in Nile tilapia, *Oreochromis niloticus* L., juveniles. Effects on growth performance and gut histology. Journal of Applied Phycology 27: 1671–1680. https://doi.org/10.1007/s10811-014-0453-9

Singh, B.K., Thakur, K., Kumari, H., Mahajan, D., Sharma, D., Sharma, A.K., Kumar, S., Singh, B., Pankaj, P.P. and Kumar, R., 2025. A review on comparative analysis of marine and freshwater fish gut microbiomes: insights into environmental impact on gut microbiota. FEMS Microbiology Ecology 101: 169. https://doi.org/10.1093/FEMSEC/FIAE169

Skalli, A., Firmino, J.P., Andree, K.B., Salomón, R., Estévez, A., Puig, P., Sabater-Martínez, M., Hechavarria, T. and Gisbert, E., 2020. The inclusion of the microalga *Scenedesmus* sp. in diets for rainbow trout, *Onchorhynchus mykiss*, juveniles. Animals 10: 1656. https://doi.org/10.3390/ani10091656

Sørensen, M., Gong, Y., Bjarnason, F., Vasanth, G.K., Dahle, D., Huntley, M. and Kiron, V., 2017.

Nannochloropsis oceania-derived defatted meal as an alternative to fishmeal in Atlantic salmon feeds. PLoS ONE 12: e0179907. https://doi.org/10.1371/journal.pone.0179907

Tapia-Paniagua, S.T., Fumanal, M., Anguís, V., Fernández-Díaz, C., Alarcón, F.J., Moriñigo, M.A. and Balebona, M.C., 2019. Modulation of intestinal microbiota in *Solea senegalensis* fed low dietary level of *Ulva ohnoi*. Frontiers in Microbiology 10: 171. https://doi.org/10.3389/fmicb.2019.00171

Teimouri, M., Yeganeh, S., Mianji, G.R., Najafi, M. and Mahjoub, S., 2019. The effect of *Spirulina platensis* meal on antioxidant gene expression, total antioxidant capacity, and lipid peroxidation of rainbow trout (*Oncorhynchus mykiss*). Fish Physiology and Biochemistry 45: 977–986. https://doi.org/10.1007/s10695-019-0608-3

Teixeira, C., Peixoto, D., Hinzmann, M., Santos, P., Ferreira, I., Pereira, G. V., Dias, J. and Costas, B., 2022. Dietary strategies to modulate the health condition and immune responses in gilthead seabream (*Sparus aurata*) juveniles following intestinal inflammation. Animals 12: 3019. https://doi.org/10.3390/ani12213019

Valente, L.M.P., Cabrita, A.R.J., Maia, M.R.G., Valente, I.M., Engrola, S., Fonseca, A.J.M., Ribeiro, D.M., Lordelo, M., Martins, C.F., Cunha, L.F., Almeida, A.M. and Freire, J.P.B., 2021. Microalgae as feed ingredients for livestock production and aquaculture. In: Galanakis, C.M. (ed.) Microalgae. Cultivation, recovery of compounds and applications. Academic Press, London, UK, p. 239–312. https://doi.org/10.1016/C2019-0-02811-7

Vazirzadeh, A., Marhamati, A., Rabiee, R. and Faggio, C., 2020. Immunomodulation, antioxidant enhancement and immune genes up-regulation in rainbow trout (*Oncorhynchus mykiss*) fed on seaweeds included diets. Fish & Shellfish Immunology 106: 852–858. https://doi.org/10.1016/J.FSI.2020.08.048

Vijayaram, S., Ringø, E., Ghafarifarsani, H., Hoseinifar, S.H., Ahani, S. and Chou, C.-C., 2024. Use of algae in aquaculture: a review. Fishes 9: 63. https://doi.org/10.3390/fishes9020063

Vizcaíno, A.J., Saéz, M.I., López, G., Arizcun, M., Abellán, E., Martínez, T.F., Cerón-García, M.C., Alarcón, F.J., 2016. *Tetraselmis suecia* and *Tisochrysis lutea* meal as dietary ingredients for gilthead sea bream (*Sparus aurata* L.) fry. Journal of Applied Phycology 28: 2843–2855. https://doi.org/10.1007/s10811-016-0845-0

Wan, A.H.L., Soler-Vila, A., O'Keeffe, D., Casburn, P., Fitzgerald, R. and Johnson, M.P., 2016. The inclusion of *Palmaria palmata* macroalgae in Atlantic salmon (*Salmo salar*) diets: effects on growth, haematology, immunity and liver function. Journal of Applied Phycology 28: 3091–3100. https://doi.org/10.1007/S10811-016-0821-8

Wan, A.H.L.L., Davies, S.J., Soler-Vila, A., Fitzgerald, R. and Johnson, M.P., 2019. Macroalgae as a sustainable aquafeed ingredient. Reviews in Aquaculture 11: 458–492. https://doi.org/10.1111/raq.12241

Xie, J. and Niu, J., 2022. Evaluation of four macro-algae on growth performance, anti-oxidant capacity and non-specific immunity in golden pompano (*Trachinotus ovatus*). Aquaculture 548: 737690. https://doi.org/10.1016/J.AQUACULTURE.2021.737690

Xu, W., Gao, Z., Qi, Z., Qiu, M., Peng, J.Q. and Shao, R., 2014. Effect of dietary *Chlorella* on the growth performance and physiological parameters of gibel carp, *Carassius auratus gibelio*.

Turkish Journal of Fisheries and Aquatic Sciences 14: 53–57. https://doi.org/10.4194/1303-2712-V14_1_07

Zeynali, M., Nafisi Bahabadi, M., Morshedi, V., Ghasemi, A. and Torfi Mozanzadeh, M., 2020. Replacement of dietary fishmeal with *Sargassum ilicifolium* meal on growth, innate immunity and immune gene mRNA transcript abundance in *Lates calcarifer* juveniles. Aquaculture Nutrition 26: 1657–1668. https://doi.org/10.1111/anu.13111

Zhang, Q., Qiu, M., Xu, W., Gao, Z., Shao, R. and Qi, Z., 2014. Effects of dietary administration of *Chlorella* on the immune status of gibel carp, *Carassius auratus gibelio*. Italian Journal of Animal Science 13: 3168. https://doi.org/10.4081/ijas.2014.3168

Zhang, Z., Xi, L., Liu, H., Jin, J., Yang, Y., Zhu, X., Han, D. and Xie, S., 2022. High replacement of fishmeal by *Chlorella* meal affects intestinal microbiota and the potential metabolic function in largemouth bass (*Micropterus salmoides*). Frontiers in Microbiology 13: 1016662. https://doi.org/10.3389/FMICB.2022.1016662

Zhang, W., Deng, Y., Yang, Z., Kong, Q., Liu, P., Liao, H. and Tang, H., 2024. Effects of partial replacement of fishmeal with Spirulina platensis powder and addition of Spirulina platensis polysaccharide on growth, nutrition, antioxidant capacity and gut microbiota of Micropterus salmoides. Aquaculture 586: 740802. https://doi.org/10.1016/j.aquaculture.2024.740802

Zhao, W., Fang, H.-H., Liu, Z.-Z., Chen, J.-M., Zhang, C.-W., Gao, B.-Y. and Niu, J., 2021a. Responses in growth performance, enzymatic activity, immune function and liver health after dietary supplementation of Porphyridium sp. in juvenile golden pompano (*Trachinotus ovatus*). Aquaculture Nutrition 27: 679–690. https://doi.org/10.1111/anu.13214

Zhao, W., Fang, H.H., Liu, Z.Z., Huang, M.Q., Su, M., Zhang, C.W., Gao, B.Y. and Niu, J., 2021b. A newly isolated strain of *Haematococcus pluvialis* JNU35 improves the growth, antioxidation, immunity and liver function of golden pompano (*Trachinotus ovatus*). Aquaculture Nutrition 27: 342–354. https://doi.org/10.1111/anu.13188

Zhao, W., Cui, X., Wang, Z.-Q., Yao, R., Chen, M.-D., Gao, B.-Y., Zhang, C.-W. and Niu, J., 2022. Effects of *Barranca yajiagengensis* powder in the diet of *Trachinotus ovatus* on the growth performance, antioxidant capacity, immunity and morphology of the liver and intestine. Antioxidants 11: 1220. https://doi.org/10.3390/antiox11071220

Zhao, W., Pu, H.Q., Wang, L.B., Dai, M.Q., Wang, P. and Niu, J., 2025. Modulatory effects of *Enteromorpha prolifera* powder and its enzymatic hydrolysate on growth performance, antioxidant capacity, and non-specific immunity of *Trachinotus ovatus*. Aquaculture 595: 741610. https://doi.org/10.1016/J.AQUACULTURE.2024.741610

CHAPTER 10

Microalgae and macroalgae as functional ingredients in shrimp nutrition

*A. Barreto[1], A.T. Gonçalves[1,2] and R.J.M. Rocha[1,3]**

[1]Riasearch Lda, Cais da Ribeira de Pardelhas nº 21, 3870-168 Murtosa, Portugal
[2]GreenCoLab - Associação Oceano Verde, Universidade do Algarve, Campus de Gambelas, Ed. 2, 8005-139 Faro, Portugal
[3]Centre for Environmental and Marine Studies (CESAM) and Department of Biology, University of Aveiro, 3810-193 Aveiro, Portugal
**Correspondence: ruimirandarocha@ua.pt*

Abstract

Both microalgae and macroalgae are increasingly valued as functional ingredients in shrimp aquaculture, combining balanced nutrition (proteins, essential fatty acids, vitamins, minerals) with diverse bioactive compounds such as carotenoids, phycobiliproteins, polysaccharides, β-glucans, and polyphenols. In the whiteleg shrimp (*Penaeus vannamei*), the world's leading farmed crustacean, dietary inclusion at low levels (<15%) consistently enhances immune competence (e.g., hemocyte count, lysozyme, phenoloxidase activity), improves gut health, supports oxidative balance, and increases resistance to bacterial and viral pathogens, while also conferring resilience against stressors such as salinity shifts, temperature fluctuations, hypoxia, and heavy metals. Outcomes vary according to ingredient form (whole biomass, extracts, fractions, fermented derivatives), algal group, dose–response dynamics, feeding duration, and feed processing technologies such as vacuum coating and low- or high-temperature extrusion. Algae further modulate the gut microbiota, acting as prebiotic substrates and quorum-sensing modulators that stabilize microbial communities, reduce *Vibrio* abundance, and enhance performance and survival. Future advances will require standardized methods for characterizing algal bioactives, reducing variability across species and processing, and integrating multi-omics tools to better link mechanisms with outcomes. In parallel, techno-economic and life-cycle assessments are essential to evaluate scalability and sustainability. In summary, beyond summarizing the outcomes, this chapter explores the underlying mechanisms, species-specific responses, and future directions for the strategic use of algal ingredients, positioning them as key contributors to the sustainability, resilience, and innovation of modern shrimp aquaculture.

Keywords

health status – immune responses – antioxidant capacity – microbiota – aquaculture

Introduction

Algae, encompassing both microalgae and macroalgae, have garnered increasing interest in aquaculture due to their rich nutrient composition and diverse bioactive compounds. In shrimp farming, their role extends beyond basic nutrition, as they are increasingly recognized as functional additives that enhance health, improve stress resilience, and optimize overall performance (Vijayaram et al., 2024). Beyond summarizing these outcomes, this chapter explores the underlying mechanisms, species-specific responses, and future directions for the strategic use of algal ingredients, positioning them as key contributors to the sustainability, resilience, and innovation of modern shrimp aquaculture.

While the term "crustaceans" includes a wide diversity of farmed species, this chapter focuses specifically on the Pacific whiteleg shrimp (*Penaeus vannamei*, formerly *Litopenaeus vannamei*), the leading farmed crustacean worldwide in both production volume and economic value. At the global level, *P. vannamei* was the most produced species in 2022, reaching 6.8 million tonnes (FAO, 2024), and has been the principal crustacean model in research on nutrition and functional feeds based on micro- and macroalgae. Its dominance in production and research, combined with well-established farming protocols and a documented responsiveness to dietary modulation, positions it as the most relevant model species for investigating the use of algal ingredients in crustacean aquaculture (Zhang et al., 2023; Vijayaram et al., 2024; Zhang et al., 2025).

In shrimp nutrition, algae are typically included at low dietary levels, serving as functional supplements rather than major replacements for protein or energy sources. This strategy is especially relevant in light of current constraints on large-scale algal biomass production (González-Meza et al., 2022; Siddik et al., 2024; Vijayaram et al., 2024).

Commonly used microalgae in shrimp culture include the genera *Chlorella*, *Tetraselmis*, *Phaeodactylum*, *Isochrysis*, and *Nannochloropsis*. In addition, *Arthrospira platensis* (commonly known as Spirulina), although taxonomically a cyanobacterium, is included here in the microalgae group due to its similar nutritional profile, functional roles, and the significant number of studies assessing its effects in *P. vannamei* (Ansari et al., 2021; Ahmad et al., 2022; Ma & Hu, 2024). These microalgae provide high-quality proteins, essential fatty acids (particularly EPA and DHA), vitamins, minerals, and bioactive compounds such as carotenoids, phycobiliproteins, β-glucans, and sulfated polysaccharides. Such components have been associated with enhanced immune function, improved gut health, mitigation of oxidative stress, and higher survival rates in *P. vannamei*. They are also indispensable in larviculture, serving both as live feed and as natural pigment sources that enhance shrimp coloration and market value (Zhang et al., 2023; Ma & Hu, 2024; Zhang et al., 2025).

From a phylogenetic perspective, macroalgae are distributed across three major groups: Chlorophyta (green algae), Rhodophyta (red algae), and Phaeophyceae (brown algae, within the Ochrophyta). They can be incorporated into shrimp diets as whole biomass, extracts, or fermented derivatives, providing proteins, polyunsaturated fatty acids, minerals, and vitamins, as well as bioactive compounds such as sulfated polysaccharides, polyphenols, and carotenoids. In *P. vannamei*, these components have demonstrated antioxidant, antimicrobial, anti-inflammatory, and immunomodulatory effects, frequently supporting improved gut structure and greater resistance to disease (Moreira et al., 2022; Zhang et al., 2023).

In *P. vannamei*, research has increasingly highlighted the functional potential of algae, particularly at low dietary inclusion levels (<15%), with robust evidence of positive effects on immune competence, gut integrity, oxidative balance, and disease resistance (Moreira et al., 2022; Zhang et al., 2023; Vijayaram et al., 2024).

1 Impacts on stress and immune response

This section provides an integrated overview of the mechanisms by which micro- and macroalgae modulate stress responses and immune function in whiteleg shrimp. Overall, the literature consistently reports enhancements across a range of cellular, humoral, and molecular immune markers. However, the magnitude and, in some cases, the direction of these responses is highly dependent on algal species, inclusion level, processing method, and experimental context.

1.1 Microalgae

Multiple studies have characterized the immunomodulatory effects of dietary microalgae in whiteleg shrimp. Namely, Pilotto et al. (2019) evaluated the dietary supplementation of *Arthrospira platensis* dry biomass at 0.6% compared to a control diet and observed significant increases in serum agglutination titers in whiteleg shrimp juveniles fed the microalgae-supplemented diet for 20 days. Agglutinins, particularly lectins, are involved in various immunological processes, including microbial recognition, agglutination, and exerting antibacterial and antiviral effects. Similarly, Novriadi et al. (2022) reported that the total hemocyte count (THC) and lysozyme activity in whiteleg shrimp juveniles were significantly enhanced with dietary inclusion of *A. platensis* at 0.2%, 0.4%, and 0.8%. Immunostimulatory effects were also evident at higher supplementation levels (3% and 6%), namely by significant increases in lysozyme activity, phagocytic activity, and clearance efficiency (Chen et al., 2016). Furthermore, enhanced innate immunity in whiteleg shrimp fed diets supplemented with *Chlorella vulgaris* (1.5, 2.0, 2.5, and 3.0 %), compared to a control diet, has been reported (Eissa et al., 2023). Improvements were observed in several immune parameters, including total hemocyte count, respiratory burst activity, lysozyme activity, phenoloxidase activity, and phagocytic activity indices. Yu et al. (2021) observed that the dietary inclusion of *Haematococcus pluvialis* powder (to achieve 35 and 70 mg carotenoids/kg of diet) significantly increased relative mRNA levels of immune-related genes toll-like receptors (*Toll*), *MyD88*, immune deficiency (*IMD*), crustin, phenoloxidase (*PO*) and lysozyme (*LYZ*) in the hepatopancreas of adult whiteleg shrimp after 15 days of feeding, which suggests their immune capacity was enhanced.

The immunomodulatory effects of dietary microalgae supplementation in diets for whiteleg shrimp are often dose-dependent, but the underlying responses are complex and varied. Whiteleg shrimp juveniles fed diets supplemented with an astaxanthin-rich *H. pluvialis* powder (0.05%, 0.10%, and 0.15%) for eight weeks exhibited a dose-dependent enhancement of immune responses compared to those fed a control diet (Eissa et al., 2025). Immune parameters increased linearly with inclusion level, as evidenced by elevated total hemocyte count (THC), respiratory burst activity, lysozyme activity, phenoloxidase activity, phagocytic activity, and upregulation of the *proPO* and *LYZ* genes. Similarly, Ozório et al. (2024) reported that whiteleg shrimp fed diets containing 2.0% crude extracts from the microalga *Porphyridium cruentum* for 30 days showed a significant increase in phenoloxidase activity, compared to both control group and shrimp fed diets supplemented with lower inclusion levels (0.5%, 1.0% and 1.5%). However, immune responses do not always increase linearly with higher supplementation levels, indicating a potential threshold or saturation effect. Hoffling et al. (2024) reported significant increases in THC in whiteleg shrimp fed diets supplemented with 1% *Aurantiochytrium* sp. meal over a 9-week period. In contrast, higher inclusion levels (2, 3 and 4%) did not differ statistically from the Control (no-supplementation). Conversely, intermediate concentrations may yield no significant effect or induce different immunological responses, indicating the complexity of dose-dependent interactions. For instance, Rocha et al. (2021) reported significant increases in immune parameters after a three-week feeding period: serum binder titer was elevated in shrimp fed diets containing 0.5% and 2% *Aurantiochytrium* sp. meal, while phenoloxidase activity was significantly higher in shrimp fed the 1% inclusion level, compared to both the control and the other inclusion levels (0.5% and 2%).

In addition to dosage, immunomodulatory effects can vary according to factors that influence the bioavailability and efficacy of bioactive compounds, such as the processing methods applied to microalgae prior to dietary inclusion, the feed technology used (e.g., inclusion as dry ingredient or through vacuum coating), and/or their combination with other dietary ingredients. Specifically, supplementation with *A. platensis* or with its free-lipid biomass (FL) led to enhanced innate immune responses in whiteleg shrimp post-larvae, namely increased lysozyme activity. However, FL supplementation (1%) significantly upregulated the expression of the gene prophenoloxidase (*PPO1*) compared to a control diet, suggesting enhanced immune function when the microalga was processed to remove lipid content (Ashour et al., 2024). In another study, Prabu et al. (2025) reported significant increases in the relative mRNA expression of *LYZ* in juvenile whiteleg shrimp fed diets supplemented with 1% *A. platensis* meal; 0.5% vacuum-coated *A. platensis*; and a combination of 0.5% *A. platensis* meal and 0.5% commercial probiotic, compared to a control diet and a 1% vacuum-coated *A. platensis* diet. Additionally, a significant upregulation of the *Toll* and *IMD* genes was observed in shrimp fed diets containing 1% *A. platensis* meal; 1% vacuum-coated *A. platensis;* and a combination of 0.5% *A. platensis* meal and 0.5% commercial probiotic. Furthermore, synergistic effects between dietary cell-free extracts of *Lactobacillus pentosus* and *A. platensis* were demonstrated by the up-regulation of immune-related gene expression in whiteleg shrimp following a feeding trial (Liu et al., 2022).

Microalgae supplementation can also modulate the immune responses of whiteleg shrimp after a pathogenic challenge. For example, the gene expression of *LGBP* (Lipopolysaccharide and β-1,3-Glucan Binding Protein) in hemocytes was significantly upregulated in shrimp fed diets containing *A. platensis* (6%), after being injected with a *Vibrio alginolyticus* suspension (Chen et al., 2016). Immunostimulatory effects were also observed in whiteleg shrimp fed diets supplemented with *P. cruentum* (0.5, 1.0, 1.5 and 2.0%) and challenged with *V. alginolyticus*, as evidenced by significantly higher THC levels than those fed a control diet. Additionally, shrimp fed the 2.0% diet, showed a significant increase in phenoloxidase activity compared to the control group (Ozório et al., 2024). Furthermore, Liu et al. (2022) reported that dietary supplementation with cell-free extracts of *L. pentosus* combined with *A. platensis* enhanced the expression of immune-related genes in whiteleg shrimp following a challenge test with *V. alginolyticus*. These findings suggest that this combination can improve disease resistance against *V. alginolyticus* by up-regulating immune gene expression. In a separate study, the gene expression analysis revealed significant upregulation of antibacterial-related genes (*alf1, alf2, crus, lyz, pen3a,* and *pen4*) and prophenoloxidase system-related genes (*sp, pexn, proPO I,* and *proPO II*) in the hepatopancreas of shrimp fed diets containing 3% of a 1:1 mixture of *Phaeodactylum tricornutum* and *Tetraselmis chui*, compared to those fed a control diet, following 8 hours of *V. parahaemolyticus* infection (Zhang et al., 2025). Similarly, supplementation with *H. pluvialis* (0.05%, 0.10%, and 0.15%) significantly increased THC, respiratory burst (RB) activity, lysozyme activity, PO activity, phagocytic activity, and upregulated the *proPO* and *LYZ* genes in shrimp challenged with *V. parahaemolyticus* (Eissa et al., 2025). Comparable enhancements in these innate immune parameters were also observed by Eissa et al., (2023) in shrimp fed diets supplemented with *C. vulgaris* at inclusion levels of 1.5%, 2.0%, 2.5%, and 3.0%, following infection with *V. parahaemolyticus.*

Moreover, microalgae supplementation can also play a role in reducing immune system impairments associated with environmental stress. For instance, *H. pluvialis* supplementation may help whiteleg shrimp acclimate to prolonged low-temperature conditions. Immunoregulatory effects were observed in shrimp fed a low fish meal diet supplemented with 0.03% *H. pluvialis* (0.03%) at lower rearing temperatures (20 °C), evidenced by significantly increased expression of genes associated with the Toll and IMD pathways, compared to shrimp fed a control diet (Lin et al., 2024).

1.2 Macroalgae

Macroalgae, like microalgae, have demonstrated ability to stimulate immune responses in whiteleg shrimp. Various species, including *Gracilaria verrucosa, Porphyra haitanensis, Ulva rigida,* and *Sargassum cristaefolium,* have been reported to enhance the immune function of whiteleg shrimp by improving cellular and humoral immune parameters such as THC, PO activity, phagocytic activity, lysozyme activity and/or differential hemocyte count (granular and hyaline cells (GCs and HCs)) (Akbary & Aminikhoei, 2018; Niu et al., 2018; Sudaryono et al., 2018; Jahromi et al., 2021; Elshopakey et al., 2024;). Additionally, immunomodulatory effects can also occur at the molecular level. In a comprehensive study, juvenile whiteleg shrimp fed diets containing *Gracilaria tenuistipitata* extracts for 16 weeks exhibited significantly upregulated expression of multiple immune- and stress-related genes. These included *LGBP*), *peroxinectin* (*PX*), *integrin β* (*IB*), *proPO I*, *proPO II*, *α-2-macroglobulin* (*α-2M*), *cytMnSOD*, *mtMnSOD*, *glutathione peroxidase* (*GPx*), *transglutaminase I* and *II* (*TG I*, *TG II*), *lysozyme*, and *heat shock protein 70* (*HSP70*). Also, shrimp showed significantly increased numbers of HCs, GCs, THC, PO activity, RB, and earlier hemolymph clotting time (Liu et al., 2019). Furthermore, a comparative study of different macroalgal species showed that dietary supplementation with *P. haitanensis, Undaria pinnatifida, Saccharina japonica,* or *Gracilaria lemaneiformis* significantly enhanced hepatopancreas immunity in whiteleg shrimp. This was evidenced by increased proPO activity, upregulation of *proPO* and *IκBα* gene expression, and down-regulation of *transforming growth factor β* (TGF-β) expression in the hepatopancreas after an eight-week feeding period (Niu et al., 2019). Although studies on the early developmental stages of whiteleg shrimp are relatively limited, Cheardkeattipol et al. (2017) reported that post-larvae (PL12) fed diets supplemented with *Macrocystis pyrifera* at 3% for 60 days exhibited significantly higher THC, phagocytosis, PO and bactericidal activities compared to these fed a control diet.

The immunomodulatory potential of macroalgae is often influenced by their inclusion level, suggesting a dose-dependent response. A linear dose-response was observed by Abdel-Razek et al. (2024), who reported that although total THC, phagocytic activity, and phagocytic index were significantly higher in shrimp fed diets supplemented with *Ulva* sp. extract at 0.05%, 0.1%, and 0.2% compared to the control group, serum total protein, acid phosphatase, alkaline phosphatase, lysozyme, and PO activities increased linearly with supplementation level, reaching their highest values at 0.2%. Similarly, the supplementation of polysaccharides extracted from *Sargassum dentifolium* in diets for whiteleg shrimp (0.1, 0.2 and 0.3%), enhanced the expression of immune-related genes (β-Glucan-binding protein (*β-Bgp*), *proPO*, *lysozyme*, and *Crustin*) in their muscle with a linear dose-response relationship, with expression levels peaking at the highest inclusion level (Abbas et al., 2023). Furthermore, PO, RB activities, and THC in whiteleg shrimp juveniles increased linearly with dietary inclusion levels of a fucoidan extracted from *Sargassum johnstonii* (Patel et al., 2025). Lee et al. (2020b) reported that whiteleg shrimp juveniles fed diet supplemented with *Sargassum horneri* hot-water extracts (0.25, 0.5 and 1.0%) for 4 weeks showed significantly higher THC, ROS production and phagocytic activity, compared to those fed a control diet, that increased linearly with inclusion level. Still, the gene expression analysis revealed that only shrimp fed diets containing 0.5% inclusion levels showed a significant modulation of immune-related genes, including *proPO I*, *proPO II*, *peroxinectin*, *α2macroglobulin*, *clotting protein*, *lysozyme*, *penaiedin2-4*, and *crustin*.

Still, clear dose-dependent patterns are not always easily established and can be transient. For instance, the dietary supplementation with hot water crude extracts of *Ulva intestinalis* at inclusion levels of 0.1%, 0.5%, and 1.0% resulted in variable expressions of several immune-related genes in the hemocytes and/or gills of whiteleg shrimp (*anti-lipopolysaccharide factor* (*ALF*), *proPO*, *TG*, *lysozyme*, *C-type lectin*, and *LGBP*). Despite all of them being upregulated compared to the control diet, no consistent relationship between inclusion levels and gene expression was observed (Klongklaew et al., 2021). Moreover, whiteleg shrimp fed diets supplemented with sulfated polysaccharides extracted from *Sargassum ilicifolium* (0.05%, 0.1%, 0.2%, and 0.4%) showed significantly higher relative expression of the *lysozyme* gene than those fed a control diet, with expression levels increasing linearly and peaking at the highest inclusion level, while the *proPO* peaked at 0.2% and *penaeidin 3-a* at the lowest level of inclusion (Pourazad et al., 2024). Lee et al. (2020) reported that the immune-enhancing effects in whiteleg shrimp fed diets supplemented with *Caulerpa racemosa* polysaccharides (0.2%, 0.6%, and 1.0%) peaked at 0.6% after 7–14 days; however, for most of the parameters analyzed, the effects were transient and diminished after 28 days.

The immunomodulatory effects of macroalgae may also be influenced by factors that affect the bioavailability and functionality of their bioactive compounds, including pre-inclusion processing methods such as extraction, physical or thermal processing, or fractioning. To illustrate this point, Lin et al. (2017) reported that whiteleg shrimp fed diets supplemented with heat-treated *S. cristaefolium* meal showed significantly higher phagocytic activity and clearance efficiency compared to those fed a diet supplemented with an untreated meal or a hot water extract of *S. cristaefolium*. In another study, supplementation with three types of alginates (acid, sodium or calcium), extracted from *Sargassum siliquosum*, stimulated the immune response in whiteleg shrimp by significantly increasing THC, PO, phagocytic activities and/or phagocytic index (Yudiati et al., 2016; Yudiati et al., 2019). However, these effects varied depending on the type of alginate, as well as the dose and duration of administration. Additionally, sodium alginate supplemented diets significantly up-regulated the relative expression of *lectin*, *proPO*, and *Toll* genes, while acid alginate had a more pronounced effect on the *LGBP* gene.

Macroalgae supplementation has also been shown to influence the immune responses of whiteleg shrimp following pathogen exposure. Niu et al. (2019) demonstrated that dietary inclusion of various macroalgal species, namely *P. haitanensis*, *U. pinnatifida*, *S. japonica*, and *G. lemaneiformis*, can enhance hepatopancreas immune function in whiteleg shrimp, following White Spot Syndrome Virus (WSSV) exposure. Supplementation with *G. verrucosa* or *G. birdiae* has also been shown to be able to promote immunestimulation in both juvenile and adult whiteleg shrimp, enhancing agglutinating capacity, THC, phagocytosis, PO and/or lysozyme activities following WSSV challenge (Cantelli et al., 2019; Elshopakey et al., 2024). Furthermore, whiteleg shrimp fed diets supplemented with polysaccharides from an *Ulva prolifera* cold water extract showed significantly higher hepatopancreas lysozyme and PO activities 24 to 96 h post challenge to *V. parahaemolyticus*, compared to those fed a control diet (Ge et al., 2019). Similarly, the supplementation of an *U. intestinalis* hot water crude extract significantly increased immune responses (phagocytic activity and clearance efficiency) against the same pathogen (Klongklaew et al., 2021).

The inclusion of macroalgae in whiteleg shrimp diets has also been associated with mitigating immune suppression caused by environmental stressors in whiteleg shrimp. As demonstrated by Chen et al. (2012), cellular (HCs, GCs, THC, RB) and humoral (PO activity and hemolymph protein) innate immune parameters of whiteleg shrimp fed diets containing a hot-water extract of *G. tenuistipitata* for 21 days and subsequently exposed to a 5 mg/L ammonia challenge returned to baseline values within 12–24 h post stress. In contrast, shrimp fed a control diet required 72–120 h for immune parameters to normalize. These findings suggest that the macroalgae extracts conferred a protective effect against ammonia-induced stress, as evidenced by the more rapid normalization of immune parameters.

In summary, both micro- and macroalgae have demonstrated strong immunomodulatory potential in whiteleg shrimp, enhancing a range of cellular, humoral, and molecular immune responses. These effects are influenced by algal species, inclusion level, processing methods, and environmental or pathogenic stressors. Overall, algae-based dietary supplements show promising potential for improving shrimp immunity and resilience under various conditions.

2 Impacts on antioxidant response

Dietary inclusion of algae has also been shown to modulate the antioxidant capacity of whiteleg shrimp. Across multiple studies, enhanced activity of key antioxidant enzymes, reduced lipid peroxidation, and dose-dependent response patterns are consistently reported. The following subsections examine the main factors influencing these antioxidant responses, including algal species, inclusion levels, and processing methods.

2.1 Microalgae

A vast body of literature highlights the antioxidant potential of several microalgae species in enhancing the health and oxidative status of whiteleg shrimp. For example, dietary supplementation with *H. pluvialis* or *A. platensis* in whiteleg shrimp diets has been shown to enhance antioxidant defense mechanisms by increasing superoxide dismutase (SOD) activity, catalase (CAT) activity, total antioxidant capacity (T-AOC), and/or glutathione S-transferase (GST), while simultaneously decreasing lipid peroxidation and/or malondialdehyde (MDA) levels (Yu et al., 2021; Huang et al., 2023; Said et al., 2023; Lin et al., 2024; Eissa et al., 2025). However, Fang et al. (2022) reported that whiteleg shrimp juveniles fed diets supplemented with *H. pluvialis* (0.05%) for 8 weeks had significantly lower levels of total SOD and glutathione peroxidase (GSH-Px) compared to those fed a control diet, as well as significantly lower mRNA expression levels of *SOD*, *GSH-Px*, and *catalase* (*CAT*) genes. Such reductions may indicate a lowered oxidative burden, reflecting an improved redox balance where less enzymatic activity is required to maintain homeostasis. This interpretation is further supported by the decreased MDA levels and reduced mRNA expression of the pro-inflammatory related gene *relish* observed in shrimp fed the microalgae-supplemented diets. Accordingly, Fang et al. (2022b) also reported that supplementation with *Klebsormidium* sp. (5%) in diets for whiteleg shrimp enhanced their antioxidant capacity by downregulating antioxidant parameters, including hepatopancreas SOD and GSH-Px, hemolymph SOD and MDA, and mRNA expression levels of antioxidant genes (*GSH-Px* and *CAT*). Furthermore, the authors state that dietary *Klebsormidium* sp. supplementation significantly improved hepatopancreas health by downregulating mRNA expression levels of pro-inflammatory related genes (*relish* and *rho*).

The antioxidant benefits of microalgae in shrimp diets can vary with dosage, making dose-response studies essential to better understand the complex relationships between inclusion levels and health outcomes. For instance, Ahmed et al. (2025) reported that whiteleg shrimp juveniles fed diets supplemented with *A. platensis* (0.2, 0.4, 0.6 and 0.8%) for 8 weeks exhibited significantly higher levels of GPx, SOD, and CAT activity and lower MDA levels, compared to shrimp fed a control diet. Notably, the two highest inclusion levels resulted in the highest levels of antioxidant enzymes and lowest lipid oxidation. Comparably, Xie et al. (2018) documented that the antioxidant-enhancing effects of dietary *H. pluvialis* supplementation in whiteleg shrimp diets reached a plateau at an inclusion level of 0.67%, as indicated by similar MDA levels in shrimp fed diets with higher inclusion rates. A similar threshold effect was observed when supplementing *C. vulgaris* hot water extracts in whiteleg shrimp diets. SOD activity peaked at a 10% inclusion level, whereas the 15% inclusion yielded activity comparable to that of the control group (Zhang et al., 2025). Interestingly, MDA levels were significantly lower at 15% inclusion compared to the control, while lower inclusion levels produced MDA values similar to the control diet.

Beyond dose-dependent effects, the form in which algae are included (e.g., whole biomass, fractions or extracts) also influences their antioxidant benefits, due to potential differences in bioavailability. As demonstrated by Ashour et al. (2024), supplementation of diets with whole *A. platensis* (2%) or with its free-lipid biomass (FL) at 3% improved the redox status of whiteleg shrimp PL, evidenced by increased SOD activity. Interestingly, FL supplementation at 1% significantly upregulated *peroxiredoxin* (*Prx*) gene expression compared to the control, whole *A. platensis*, and higher levels of FL. Meanwhile, MDA levels were significantly lower in shrimp fed diets supplemented with the whole *A. platensis* biomass.

Several studies have demonstrated that dietary supplementation with specific microalgae can enhance the antioxidant response in whiteleg shrimp under pathogenic stress. Zhang et al. (2025) reported that shrimp fed a diet containing 3% of a 1:1 mixture of *P. tricornutum* and *T. chui* exhibited significantly reduced MDA levels in the hepatopancreas 8 hours after being exposed to *V. parahaemolyticus*, alongside an upregulation of the antioxidant gene *catalase* (*cat*). Similarly, Félix et al. (2022) documented that juvenile shrimp fed diets supplemented with *Dunaliella* sp. (2%) every other day for 15 days displayed significantly higher SOD activity 48 hours after *V. parahaemolyticus* infection, compared to those fed a control diet. Additionally, Félix et al. (2019) reported a decrease in lipid oxidation in plasma triglycerides at 48-hour post inoculation in shrimp fed diets including *Dunaliella* sp.. In another study, Akbari et al. (2023) showed that shrimp post larvae (PL) fed diets supplemented with *Dunaliella salina* (0.05 and 0.1%) for 8 weeks and subsequently exposed to *Photobacterium damselae* had increased glutathione peroxidase (GPx) and SOD activities relative to the control (non-supplemented). Collectively, these findings highlight the capacity of microalgae to modulate the antioxidant defense system in whiteleg shrimp, potentially enhancing resilience against bacterial pathogens.

The inclusion of microalgae in whiteleg shrimp diets has also been linked to alleviating oxidative stress induced by environmental factors. Said et al. (2023) reported that juvenile whiteleg shrimp fed diets supplemented with selenium nanoparticles extracted from *A. platensis* (0.25, 0.50, and 1 mg/kg) for 56 days exhibited significantly enhanced antioxidant mechanisms, along with reduced oxidative damage levels in both hepatopancreas and muscle tissues following exposure to sublethal cadmium concentrations (0.2 mg/L) for 10 days, indicating improved resilience to oxidative stress. Likewise, Lin et al. (2024) suggested that *H. pluvialis* supplementation may assist whiteleg shrimp acclimate to prolonged low-temperature conditions and improve their antioxidant status, as evidenced by lower MDA levels and increased SOD activity in shrimp fed diets supplemented with the microalgae kept at 20 °C compared to those fed a control diet and kept at 30 °C for 8 weeks. Furthermore, *H. pluvialis* supplementation appears to alleviate oxidative stress induced by a low salinity challenge (28 to 5 ppt for 45 minutes) in whiteleg shrimp PL, as indicated by significantly lower MDA levels compared to controls (Xie et al., 2018).

2.2 Macroalgae

Building on the antioxidant benefits observed with microalgae, several macroalgae species have also been investigated for their capacity to reduce oxidative stress and enhance antioxidant defenses in whiteleg shrimp. Specifically, dietary supplementation with species such as *P. haitanensis*, *U. pinnatifida*, *S. japonica*, *G. lemaneiformis*, and *G. verrucosa* has been shown to significantly increase TAC, SOD, GPX and/or CAT activities, while concurrently decreasing MDA levels (Niu et al., 2019; Elshopakey et al., 2024).

As with microalgae, the antioxidant benefits of macroalgae in whiteleg shrimp diets also appear to be influenced by inclusion levels, highlighting the importance of dose-response studies to optimize health outcomes. For instance, Abbas et al. (2023) reported a linear increase in the expression of antioxidant genes *SOD* and *GPx* in whiteleg shrimp muscle with rising dietary supplementation of polysaccharides extracted from *S. dentifolium* (0.1, 0.2 and 0.3%). A similar pattern was documented by Abdel-Razek et al. (2024), who observed increasing SOD activity in whiteleg shrimp with rising inclusion levels of an ethanolic extract of *Ulva* sp. up to 0.2%. Moreover, whiteleg shrimp CAT and GPx activities increased, while MDA levels decreased linearly with dietary inclusion levels of *U. rigida* water-soluble polysaccharides extracts (0.05, 0.10 and 0.15%) (Akbary & Aminikhoei, 2018). Meanwhile, Pourazad et al. (2024) reported that supplementing diets with sulfated polysaccharides extracted from *S. ilicifolium* (at 0.05, 0.1, 0.2, and 0.4%) significantly upregulated *SOD* gene expression. However, increased GSH levels and reduced MDA concentrations were only observed at the highest inclusion level compared to the control. Antioxidant-promoting effects often plateau beyond certain thresholds and may exhibit variable patterns depending on the species and compounds involved. For example, whiteleg shrimp PL fed diets supplemented with *Iyengaria stellata* (0.05, 0.10 and 0.15%) for 8 weeks exhibited improved antioxidant capacity compared to those fed a control diet, namely by increased antioxidant enzymes activity and reduced MDA levels (Akbary et al., 2021). Notably, SOD, GPx and CAT activities peaked at 0.10% inclusion level and declined at both lower and higher concentrations.

Additionally, the antioxidant response to algal supplementation can be time-dependent, with some effects appearing transiently during specific periods of feeding. Illustratively, Lee et al. (2020) reported that superoxide anion (O_2^-) production was significantly higher in whiteleg shrimp fed diets supplemented with *C. racemosa* polysaccharides (0.2, 0.6 and 1%), after 1 day of feeding compared to the control. However, this effect persisted only during days 7 to 28 in shrimp fed the 0.6% inclusion level, while other groups showed values similar to the control. Accordingly, MDA levels in the hepatopancreas of shrimp fed the supplemented diets were significantly lower than in control shrimp from days 7 to 14, but effects lasted until day 28 only at the highest inclusion level. In another study, the dietary supplementation with S. horneri extracts increased whiteleg shrimp SOD activity after just 1 day of feeding, but levels returned to baseline by day 4, whereas GPx activity increased only after 28 days of feeding (Lee et al., 2020b).

The antioxidant efficacy of macroalgae in shrimp diets can also depend on how the algae are processed and incorporated, whether as whole biomass, fractionated compounds or extracted bioactive molecules. For example, dietary supplementation with heat-treated *S. cristaefolium* meal in whiteleg shrimp diets resulted in higher SOD activity and O_2^- production levels compared to shrimp fed unsupplemented diets (control), diets supplemented with untreated meal, or diets supplemented with a hot water extract of *S. cristaefolium* (Lin et al., 2017). Additionally, Yudiati et al. (2016) described that acid, sodium or calcium alginates extracted from *S. siliquosum* all enhanced whiteleg shrimp antioxidant enzymes activity levels. Specifically, these significantly elevated SOD activity levels compared to a control diet after 15 days of feeding. However, among the three types, sodium alginate supplementation resulted in the highest SOD activity observed.

In addition to enhancing antioxidant defenses under normal conditions, macroalgae supplementation has been shown to bolster shrimp antioxidant responses when exposed to pathogenic stress, potentially helping to mitigate oxidative damage during infections. For instance, dietary inclusion of species such as *P. haitanensis*, *U. pinnatifida*, *S. japonica*, *G. lemaneiformis*, and *G. verrucosa* significantly increased TAC, as well as activities of SOD, GPX and/or CAT activities, while simultaneously reducing MDA levels in the hepatopancreas or gills of whiteleg shrimp following WSSV exposure (Niu et al., 2019; Elshopakey et al., 2024).

In summary, the antioxidant effects of micro- and macroalgae in whiteleg shrimp are shaped by species-specific traits, dosage, time of administration, and inclusion method, as well as the shrimp's physiological state. By modulating key antioxidant enzymes and reducing oxidative damage, algae-derived ingredients can contribute to maintaining redox balance and enhancing overall health in standard or challenging rearing conditions.

3 Impacts on health and disease resistance

Improvements in immune function and oxidative balance associated with algal supplementation frequently translate into enhanced health status and disease resistance in whiteleg shrimp. Reported survival outcomes, however, vary according to pathogen type, stressor intensity, algal species, inclusion level, and feeding strategy, highlighting the context-dependent nature of these effects.

3.1 Microalgae

The enhancement of immune function and antioxidant mechanisms by microalgae are not isolated benefits but are closely intertwined processes that underpin improved overall health in whiteleg shrimp. Strengthening these physiological responses helps shrimp better withstand stressful conditions or pathogenic challenges, leading to enhanced health status and, ultimately, to higher survival rates. Microalgae species such as *Dunaliella* sp., *Tetraselmis suecica, T. chui, Chaetoceros muelleri,* and *P. tricornutum,* whether used individually or in combination, have been shown to enhance whiteleg shrimp resistance to virulent *V. parahaemolyticus,* leading to increased survival rates following pathogen exposure (Félix et al., 2017; Félix et al., 2022; González-Meza et al., 2022; Zhang et al., 2025). Similar resistance-enhancing effects against WSSV have also been observed in whiteleg shrimp fed diets supplemented with *A. platensis* or *Aurantiochytrium* sp. (Pilotto et al., 2019; Hoffling et al., 2024); and against *P. damselae* when fed diets supplemented with *D. salina* (Akbary et al., 2023). Additionally, *H. pluvialis* supplementation proved effective in mitigating *Fusarium solani* infection, reducing mortality rates from 65% (control diet) to a notable 35% (Eissa et al., 2025).

Furthermore, microalgae also play a crucial role in promoting localized health, particularly through their beneficial effects on the intestinal tract and hepatopancreas of whiteleg shrimp, organs paramount for immune responses, ultimately bolstering their resistance to pathogens. For instance, hepatopancreatic tissue from surviving shrimp fed T. *suecica, D. salina* and *C. muelleri* supplemented diets exhibited normal histological structure after pathogen exposure. In contrast, histopathological analysis of moribund shrimp from the control group revealed characteristic lesions of acute hepatopancreatic necrosis disease (AHPND), including extensive epithelial cell detachment from hepatopancreatic tubules and the presence of nodules in necrotic tissue (González-Meza et al., 2022). Similarly, Yao et al. (2022) reported that *Schizochytrium limacinum* supplementation (0.6%) in diets with reduced fish meal contents (15% vs 25% in the control diet) promotes hepatopancreas lipid metabolism, reduces apoptosis, promotes autophagy and improves intestinal health of whiteleg shrimp juveniles. Additionally, mitigating effects of *A. platensis* supplementation on histopathological alterations in the muscle and hepatopancreas of whiteleg shrimp caused by *V. parahaemolyticus* infection have been observed (Ahmed et al., 2025). Specifically, in muscle tissue, these effects included a reduction in the tearing of myofibrils, a decrease in spaces within the sarcomeres, and the restoration of a more normal muscle pattern. In the hepatopancreas, the supplementation helped reduce the sloughing of the lining of hepatopancreatic tubules, scattering of vacuoles, and distortion of hepatopancreatic cells. Moreover, *H. pluvialis* supplementation has been shown to induce positive histological changes, including increased secretory vesicles within B-cells in the hepatopancreas and larger muscle fibers of whiteleg shrimp (Eissa et al., 2025), and to also provide intestinal health benefits, by significantly increasing intestinal mucosal layer thickness and villi height in the mid-intestine of whiteleg shrimp (Fang et al., 2022).

Understanding these benefits further requires examining the dose-response effects, as the level of supplementation can influence outcomes. Evidence of this can be seen in the study by Ahmed et al. (2025), where whiteleg shrimp juveniles fed diets supplemented with *A. platensis* (0.2, 0.4, 0.6 and 0.8%) for 8 weeks were more resistant to *V. parahaemolyticus* infection than those fed a control diet. This was demonstrated by significantly lower cumulative mortality 15 days post-injection of the pathogen, with a clear dose-dependent effect observed as mortality progressively decreased with increasing supplementation, and the 0.8% inclusion leading to the lowest overall mortality reported. Similar linear dose-responses were observed by Chen et al. (2016), who documented that after a challenge test with *V. alginolyticus*, shrimp fed diets supplemented with dried *A. platensis* at 0.6% exhibited significantly higher survival rates than those fed the control diet or a 0.3% supplemented diet; and similarly, Eissa et al. (2023), found that shrimp fed a 0.30% supplemented diet with *C. vulgaris* exhibited the highest survival rates post-challenge with *V. parahaemolyticus* amongst all inclusion levels tested (0.15, 0.20, 0.25, and 0.30%). Conversely, Ozório et al. (2024) observed that a 1.0% inclusion level of crude extracts from the microalga *P. cruentum* in diets for whiteleg shrimp yielded the highest survival results after being challenged with *V. alginolyticus*, amongst all levels tested (0.5, 1.0, 1.5, and 2.0%).

The positive effects of microalgae are not confined to bolstering disease resistance; they also extend to fortifying whiteleg shrimp against diverse environmental stressors. Specifically, several studies have demonstrated a marked improvement in whiteleg shrimp survival during low salinity challenges when their feed was supplemented with *Sargassum* sp., *A. platensis*, or *H. pluvialis* (Xie et al., 2018; Yudiati et al., 2024). Beyond salinity, similar protective benefits have been observed against thermal shock. For example, supplementation with *Aurantiochytrium* sp. or *Nannochloropsis* sp. has been shown to improve survival rates under acute cold stress in whiteleg shrimp (Guimarães et al., 2021; Rocha et al., 2021). Furthermore, microalgae-derived compounds have also proven effective against heavy metal toxicity. Said et al. (2023), for instance, reported that histopathological alterations in the hepatopancreas of whiteleg shrimp juveniles, induced by sublethal cadmium exposure, were mitigated when these were fed diets supplemented with dietary selenium nanoparticles extracted from *A. platensis* at 0.25, 0.50, and 1 mg/kg inclusion levels.

3.2 Macroalgae

Macroalgae also represents a valuable resource to help shrimp more effectively combat adverse conditions and disease, leading to improved health outcomes and increased survival. A considerable amount of compelling evidence indicates that the dietary supplementation of several species of macroalgae is effective in improving whiteleg shrimp resistance to WSSV infection. Namely, research involving species such as *P. haitanensis*, *U. pinnatifida*, *Gracilaria birdiae*, *G. lemaneiformis*, or *G. verrucosa* has exhibited significant increases in survival of whiteleg shrimp post-challenge to WSSV, compared to control groups fed non-supplemented diets (Niu et al., 2018; Schleder et al., 2017; Cantelli et al., 2019; Niu et al., 2019; Elshopakey et al., 2024). Beyond viral challenges, macroalgae have also demonstrated efficacy against bacterial pathogens, notably *Vibrio* spp. For instance, whiteleg shrimp juveniles fed diets supplemented with *Ulva flexuosa* exhibited a survival rate twice as high as those fed a control diet, following a 24-hour challenge with *V. parahaemolyticus* (Gamboa-Barraza et al., 2021). Additionally, the supplementation with several bioactive compounds extracted from *Sargassum johnstonii*, *U. prolifera*, or *S. cristaefolium* has yielded similar results in *V. parahaemolyticus* challenge trials (Ge et al., 2019; Patel et al., 2025; Sudaryono et al., 2018). Similarly, *Ulva* sp., *G. tenuistipitata*, and *Kappaphycus alvarezii* have been linked to increased survival of whiteleg shrimp following dietary supplementation when exposed to pathogenic *V. harveyi* (Abdel-Razek et al., 2024; Liu et al., 2019; Suantika et al., 2018). Moreover, the effectiveness of supplementing diets with *I. stellata* in improving resistance against *P. damselae* has also been demonstrated by Akbary et al. (2020). Crucially, for macroalgae to optimally contribute to enhanced shrimp health and disease resistance, the form in which they are processed and incorporated into diets is a critical factor. For example, whiteleg shrimp fed diets supplemented with a heat-treated *S. cristaefolium* meal showed significantly higher survival rates after 168h of exposure to *V. alginolyticus* compared to those fed a control diet, or diets supplemented with *S. cristaefolium* hot water extracts or untreated meal (Lin et al., 2017).

Furthermore, macroalgae promote the health of key immune organs like the intestinal tract and hepatopancreas, enhancing overall health and the ability of whiteleg shrimp to resist pathogens. This was exemplified by Pratiwi et al. (2022), who reported that the dietary supplementation of *C. racemosa* extracts in diets for whiteleg shrimp for 15 days significantly improved the histopathological profile of the shrimp intestine compared to control-fed individuals. Additionally, clinical infection signs (pale hepatopancreas, empty intestines, reddened uropods and gills) were not observed in shrimp fed the supplemented diets, unlike those fed the control diet. Likewise, histopathological analyses revealed that the hepatopancreas of shrimp fed supplemented diets with *K. alvarezii* showed a decreased tubular epithelial cell lesion promoted by *V. harveyi* infection, compared to control individuals (Suantika et al., 2018).

While macroalgae clearly offers various health benefits, determining the optimal inclusion levels is critical, given that the efficacy of their dietary supplementation in whiteleg shrimp can be dose dependent. A linear dose-response pattern was described by Akbary & Aminikhoei (2018), who reported that whiteleg shrimp juveniles fed diets supplemented with *U. rigida* water-soluble polysaccharides extracts (0.05, 0.10, and 0.15%) for 8 weeks and then infected with *P. damselae* for 10 days exhibited significantly lower cumulative mortality, which progressively decreased with increasing inclusion levels, compared to those fed a control diet. However, resistance to *V. parahaemolyticus* enhancements through the supplementation of *M. pyrifera* appear to peak at 2%, as no further improvements in survival were observed for higher inclusion levels (Cherdkeattipol et al., 2017). Accordingly, Klongklaew et al. (2021) documented that an *U. intestinalis* hot water crude extract supplementation at 0.1% was effective in reducing whiteleg shrimp mortality until day 5 post infection with WSSV, while higher inclusion levels (0.5 and 1%) resulted in similar survival to a non-supplemented diet. Additionally, their findings indicated that the macroalgae supplementation at 0.1% yielded the lowest mortality rate (less than 10%) at day 10 post infection with yellow head virus, followed by 0.5% (around 50% mortality), while the 1% inclusion level and the control exhibited 100% mortality.

The beneficial impacts of macroalgae on whiteleg shrimp also extend to providing fortification against a spectrum of environmental challenges. To illustrate, whiteleg shrimp fed diets containing *S. cristaefolium* hot water extracts showed significantly higher survival rates than those fed a control diet, in a low dissolved oxygen stress tolerance test (Sudaryono et al., 2018). Beyond this, another study shows that the supplementation with *G. lemaneiformis* can improve whiteleg shrimp survival after a salinity stress for 5 hours (29 to 2.5 ppt), compared to when they are fed a control diet (Yu et al., 2016). Furthermore, Schleder et al. (2017) reported that whiteleg shrimp fed diets supplemented with *Sargassum filipendula* for 15 days showed significantly lower cumulative mortality after thermal shock (29 to 12.5 °C for 1 hour), compared to those fed a control diet.

Such findings suggest a promising strategy for enhancing the robustness of whiteleg shrimp in aquaculture, potentially improving their overall health and leading to reduced mortality rates in challenging rearing environments or during episodes of pathogenic outbreaks.

To provide a comprehensive synthesis, the effects of shrimp feeds enriched with micro- and macroalgae on stress and immune responses, antioxidant capacity, and overall health and disease resistance are presented in Table 10.1 and Table 10.2, respectively.

TABLE 10.1 Methods of inclusion and effects of microalgal incorporation in shrimp aquafeeds

Species	Product form	Effects
Arthrospira platensis	Biomass; free-lipid biomass; vacuum-coated; combined with probiotics or bacterial extracts	↑ THC, lysozyme, PO, RB, phagocytosis; ↑ immune gene expression; ↑ survival vs *Vibrio*, WSSV; ↑ antioxidant enzymes; ↓ MDA; improved histology; mitigates cold, salinity, metal stress
Aurantiochytrium sp.	Biomass	↑ THC; ↑ survival under WSSV & cold shock
Chaetoceros muelleri	Biomass	↑ disease resistance; improved hepatopancreas histology
Chlorella vulgaris	Biomass; hot-water extract	↑ THC, RB, lysozyme, PO, phagocytosis; ↑ survival vs *Vibrio*; antioxidant effects
Dunaliella salina	Biomass	↑ GPx, SOD; ↑ survival vs *Photobacterium*
Dunaliella sp.	Biomass; intermittent supplementation	↑ antioxidant enzymes; ↓ LPO; ↑ survival vs *Vibrio* and *Photobacterium*
Haematococcus pluvialis	Powder	↑ immune gene expression; ↑ THC, RB, PO; ↑ antioxidant enzymes; ↓ MDA; ↑ resilience vs *Vibrio*, fungi, cold & salinity stress
Klebsormidium sp.	Biomass	↓ oxidative stress; ↓ inflammatory gene expression
Nannochloropsis sp.	Biomass	↑ survival under cold shock
Phaeodactylum tricornutum + *Tetraselmis chui*	Biomass	↑ immune gene expression; ↓ MDA post-infection
Porphyridium cruentum	Crude extract	↑ PO; ↑ THC and PO post-*Vibrio* challenge; ↑ survival
Schizochytrium limacinum	Biomass	Improved hepatopancreas health; improved lipid metabolism, autophagy & intestinal structure
Tetraselmis chui	Biomass	↑ disease resistance vs *Vibrio*
Tetraselmis suecica	Biomass	↑ resistance to *Vibrio*; improved hepatopancreas morphology

↑: significantly higher than the control; ↓: significantly lower than the control; PO: Phenoloxidase; THC: Total Hemocyte Count; RB: Respiratory Burst; WSSV: White Spot Syndrome Virus; GPx: Glutathione Peroxidase; SOD: Superoxide Dismutase; LPO: Lipid Peroxidation.

TABLE 10.2 Methods of inclusion and effects of macroalgal incorporation in shrimp aquafeeds

Species	Product form	Effects
Caulerpa racemosa	Polysaccharides; extract	Improved intestinal health; ↑ immunity; ↓ mortality
Gracilaria birdiae	Biomass	↑ immunity; ↑ survival vs WSSV
Gracilaria lemaneiformis	Biomass	↑ proPO; ↓ TGF-β; ↑ antioxidant enzymes; ↑ survival vs WSSV; ↑ osmotic stress tolerance
Gracilaria spp.	Biomass; hot-water or crude extracts; polysaccharide-rich fractions	↑ THC, PO, RB, phagocytosis; ↑ immune/stress gene expression; ↑ antioxidant activity; improved tolerance to ammonia and osmotic stress; ↑ survival under WSSV and *Vibrio* challenge
Gracilaria tenuistipitata	Extract; hot-water extract	↑ immune/stress-related genes; ↑ THC, PO, RB; faster immune recovery
Gracilaria verrucosa	Biomass	↑ THC, PO, phagocytosis; ↑ survival *vs.* WSSV; ↑ antioxidant activity
Iyengaria stellata	Biomass	↑ antioxidant activity; ↓ MDA; ↑ resistance to *P. damselae*
Kappaphycus alvarezii	Biomass	Improved hepatopancreas histology (*Vibrio* challenge)
Macrocystis pyrifera	Biomass	↑ THC, phagocytosis, PO; ↑ bactericidal activity; improved disease resistance
Porphyra haitanensis	Biomass	↑ immune parameters; ↑ survival vs WSSV
Sargassum cristaefolium	Biomass (heat-treated or untreated); hot-water extract	Heat-treated strongly ↑ immunity, survival vs *Vibrio*; ↑ DO-stress tolerance; ↑ antioxidant response
Sargassum dentifolium	Polysaccharides	↑ immune and antioxidant gene expression
Sargassum filipendula	Biomass	↑ survival under thermal shock
Sargassum horneri	Hot-water extract	↑ THC, ROS, phagocytosis; immune gene modulation
Sargassum ilicifolium	Sulfated polysaccharides	↑ lysozyme and SOD gene expression; ↓ MDA
Sargassum johnstonii	Fucoidan extract	↑ THC, PO, RB; ↑ resistance vs *Vibrio*
Sargassum siliquosum	Alginates	↑ THC, PO, phagocytosis; sodium alginate strongly ↑ immune/antioxidant genes
Ulva flexuosa	Biomass	↑ survival after *Vibrio* challenge
Ulva intestinalis	Hot-water crude extract	↑ immune gene expression; ↑ survival
Ulva pinnatifida	Biomass	↑ immune gene expression; ↑ survival vs WSSV
Ulva prolifera	Cold-water extract	↑ lysozyme and PO post-*Vibrio* infection
Ulva rigida	Water-soluble polysaccharides extract	↑ THC, PO, lysozyme; ↑ SOD, GPx, CAT; ↓ MDA; ↑ survival
Ulva sp.	Ethanolic extract	↑ immune & antioxidant responses; ↑ *Vibrio* resistance

↑: significantly higher than the control; ↓: significantly lower than the control; ProPO: Prophenoloxidase; TGF-β: Transforming Growth Factor Beta; WSSV: White Spot Syndrome Virus; THC: Total Hemocyte Count; PO: Phenoloxidase; RB: Respiratory Burst; ROS: Reactive Oxygen Species; SOD: Superoxide Dismutase; GPx: Glutathione Peroxidase; CAT: Catalase; DMA: Malondialdehyde.

4 Impacts on gut microbiota

Mucosa associated microbial communities play a key role on the different functions of these systems, with particular attention given to the gut mucosa because of its central role in homeostasis and its multifunctional profile (Ringø et al., 2022). In shrimp, gut microbiota complexity is comparable to that of other animals. This microbial network acts as a facilitator of digestion and nutrient absorption through the secretion of extracellular enzymes (Tzuc et al., 2014); as promoters of immunity and resistance to diseases by regulating the proliferation of the opportunistic pathogens (Xiong et al., 2024; Holt et al., 2021; Zhou et al., 2024); as growth regulators and vitamin suppliers (Tzuc et al., 2014); and even as toxin degraders reducing physiological stress for the animal (Zeng et al., 2024). A healthy and well-balanced gut microbiota is therefore paramount for maintaining physiological functions of the gut and consequent animal health, and understanding its composition in shrimp, particularly in *P. vannamei*, has been a target over the last decade.

The gastro-intestinal (GI) associated microbiota in *P. vannamei* varies along the tract (Chaudhary et al., 2024), but studies have consistently identified Proteobacteria, Bacteroidetes and Actinobacteria as the most abundant phyla of healthy shrimp (revised in several studies, including Holt et al., 2021; Mao et al., 2023; Chaudhary et al., 2024; Li et al., 2024). However, communities in the different GI tract compartments differ quite largely, with a noticeable less diversity and richness in stomach and hepatopancreas compared with intestinal sections, each representing a unique ecological niche with a different community (Chaudhary et al., 2024). Stomach and hepatopancreas exhibit higher abundance of the dominant phyla, Proteobacteria, and seem to have a notable abundance of Spirochaetes and Fusobacteria when compared to intestines. These phyla harbor potential pathogenic organisms, and their lower abundance in the intestines stand as a good example to host autoregulation of the communities for the sake of health and performance (Chaudhary et al., 2024). However, shrimp microbiota presents a high plasticity and is altered by several factors, being the most prominent diet composition and surrounding environment (revised by Zhao et al., 2018; Rajeev et al., 2020). Therefore, modulation of GI tract microbiota in shrimp via nutrition is an avenue for health and performance in the production cycle. Among all ingredients that are currently tested, algae gained attention as a valuable one over the last 5 years. Both micro- or macroalgae have a compositional profile that goes beyond nutrition and modulates the microbiota community in the milieu, this being the rearing water or the GI tract (Peña-Rodriguez et al., 2024, Ringø et al., 2025).

4.1 Microalgae

Microalgae have been part of shrimp aquaculture for decades, being used to nutritionally enrich the rearing water in the early stages. This practice modulates the rearing water's microbiome composition, and it is one of the modes of action for the observed beneficial output in shrimp performance. Lian et al. (2024) inoculated shrimp rearing water with *Nannochloropsis oculata, Thalassiosira weissflogii* and their blend and identified different patterns of modulation according to the algae added, and *T. weissflogii* stabilized the rearing systems in a more effective manner. Although shrimp ingests the rearing water during feeding activities, higher focus has been given to algae in the feeds to promote gut microbiome and performance. In a 60 days feeding trial followed by a pathogenic challenge, shrimp were fed diets containing 1% or 3% of a blend of *Tetraselmis suecica, Dunaliella salina* and *Chaetoceros muelleri.* After exposure to *Vibrio parahaemolyticus*, no mortality was recorded in the algae-supplemented groups, whereas the control group presented 50% mortality (González-Meza et al., 2022). Moreover, shrimp fed the lowest inclusion (1%) had better performance with higher conversion ratio and weight gain.

This could have been related to the nutritional composition of microalgae, where a complete nutritional profile is complemented with a relevant pigment profile (e.g. astaxanthin, β-carotene, Phycocyanin) that stimulate antioxidant and immune response, but also complex polysaccharides and β-1,3-glucans that are known prebiotics (Sagaram et al., 2021). It becomes even more relevant that microalgae produce a variety of compounds with inhibitory action on the quorum sensing process (Kalia, 2018), but also antimicrobial compounds such as indol derivatives, beta-ionone, neophytadiene among others (Falaise et al., 2016), and this has been argued as one of the primary modes of action resulting in the microbiota modulation. Indeed, dietary supplementation with a natural astaxanthin extract from *Arthrospira platensis* (a cyanobacterium commonly classified within the microalgae group) for 90 days at inclusion levels ranging from 0.2% to 0.6% resulted in a marked reduction of total heterotrophic bacteria (THB) and total *Vibrio* counts (TVC) in both the rearing water and the intestinal microbiota (Mansour et al., 2022). Similarly, other microalgal species have demonstrated comparable antibacterial potential; for instance, Guimarães et al. (2021) tested *Nannochloropsis* at 0.5% up to 2% and fed for 2 weeks, resulting in less THB despite no differences on Vibrio abundance. Microalgae produce inhibitory compounds that overcount pathogens by affecting their activities (Molina-Cárdenas & Sánchez-Saavedra, 2017), which seems to be transversal to the available studies, but still require a dedicated effort to close the gaps in information and allow to understand microbiota modulation in shrimp via enrichment of aquafeeds with microalgae.

4.2 Macroalgae

Macroalgae have a unique composition of extremely valuable bioactive compounds with potential for microbiota modulation. Diverse polysaccharides found in macroalgae, including fucoidan, laminarin, alginate among others, act as prebiotics (Shannon et al., 2021). These are not readily digested in the gut and are fermented by the gut microbiota, producing short chain fatty acids (SCFA) that are central for gut cell energy supply, integrity, defense and other functions. Among all used species, members of genus *Ulva* have been highlighted, but other such as *Sargassum* genus members, considered invasive in certain regions, have also shown promising results for shrimp microbiota modulation. *Ulva clathrata* and *Ulva lactuca* when added to shrimp feed for 4 weeks changed the dominant phylum and reduced *Vibrio* abundance (Elizondo-González et al., 2018). Schleder et al. (2020) tested the effects of a blend of *Undaria pinnitifida* and *Sargassum filipendula* and observed that increasing levels of *U. pinnatifida* increased microbial richness and diversity, more evident with increasing dosages of *U. pinnatifida*. This was mainly due to the modulation of the Rhodobacteraceae, Flavobacteriaceae and Cellulomadaceae families, indicating a regulation of potentially pathogenic microorganisms. In another study, the application of biorefinery strategies to concentrate or purify bioactive compounds resulted in positive outcomes regarding deleterious microorganisms. Abbas et al., (2023) found that inclusions from 0.1% to 0.3% of a polysaccharide extract of *Sargassum dentifolium* reduced the THB in both shrimp rearing water and intestine, which was concomitant with a reduction in TVC. Other species have been investigated as well, and Niu et al (2019) compared the effects of four macroalgae species (*Porphyra haitanensis*, *Undaria pinnatifida*, *Saccharina japonica*, *Gracilaria lemaneiformis*) at 2% inclusion in diets provided for 56 days followed by a challenge. The inclusion of different macroalgae increased overall microbial richness and diversity. All species, except *Saccharina*, enhanced the relative abundance of Bacteroidetes at the expense of Proteobacteria. In particular, *Gracilaria* reduced the abundance of potentially pathogenic families such as Flavobacteriaceae and Vibrionaceae, while all macroalgae promoted the proliferation of beneficial taxa, including members of the Bacillaceae family, indicating a positive modulation of the gut-associated microbiota. González-Colunga et al (2019) explored the metabolites of *Ecklonia arborea* in shrimp and concluded that the phlorotannin's could be used as additives in shrimp diets and could work as antibiotic agent against *V. parahaemolyticus*. The polysaccharides derived from the green algae *Caulerpa lentilifera* were used to supplement *P. vannamei* diets from 0.1% up to 0.4% for 5 weeks and the lowest dosage increased Bacteroides abundance in the intestine (Tong et al 2024). The middle dosage increased diversity and the abundance of TM7 and Firmicutes. According to the functions associated with these microbial groups, results suggest a possible higher carbohydrate absorption by shrimp due to TM7, and an improved nutrient processing and energy conversion due to *Firmicutes*.

Despite the known importance of the gut microbiota in shrimp health, and the evident positive impact of algae in the GI tract microbial community, it has been only in the last 5 years that research community has been pressured to test, assess and understand associated dynamics. From the available studies, the beneficial role of micro- and macroalgae biomass and extracts is a common denominator, but the precise functions of the modulated microbial groups remain largely unknown, limiting further optimizations for industrial applications. The intricate mechanisms by which the gut microbiota communicates with the host, and how these signals for an inflammatory response and metabolic pathways, remain largely unresolved. Nevertheless, from an applied perspective, the inclusion of algae in shrimp feeds emerges as a tool to regulate pathogenic organisms, particularly members of the Vibrionaceae family. Given that Vibrionaceae are among the main causes of mortality and performance losses in shrimp farming, algae-based functional ingredients represent a promising strategy for disease control via microbiota modulation. Further studies are needed to clarify species-specific effects and to develop cost-effective solutions that can be implemented at industrial scale in the near future.

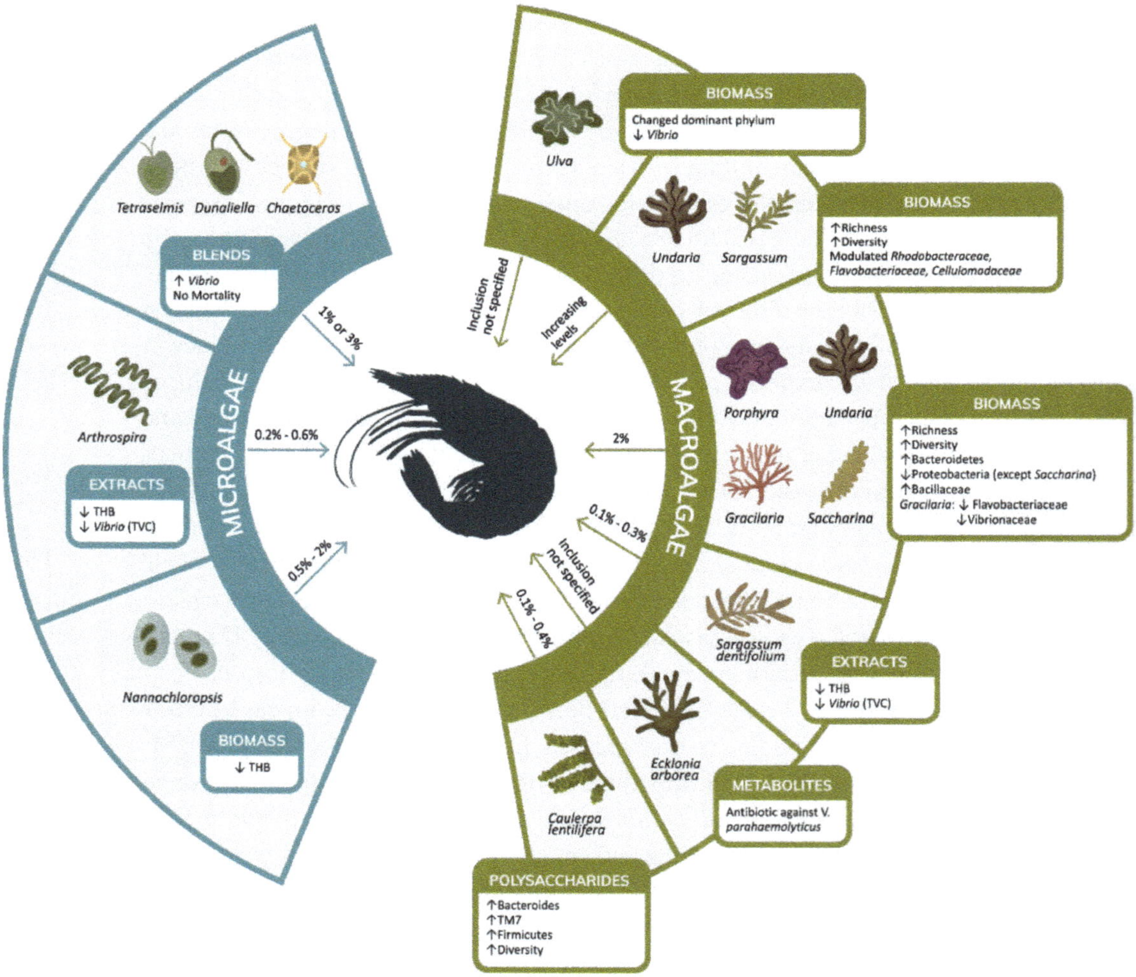

FIGURE 10.1 Summary of the effects of different micro- and macroalgae on the intestinal microbiota of *Penaeus vannamei*. Abbreviations: THB, total heterotrophic bacteria; TVC, total *Vibrio* counts. The authors acknowledge Maria Eduarda Mendes for the design of this figure.

A general overview of the most relevant effects of algae as functional ingredients on the shrimp microbiota is presented in Figure 10.1.

5 Final considerations

Despite the recognized nutritional value of algae, particularly regarding protein content, its application as a bulk ingredient in shrimp feeds remains limited. The primary barriers include high production costs, difficulties with scalability, and significant variation in nutritional quality depending on the species and culture conditions. Furthermore, issues with palatability and digestibility, especially regarding macroalgae containing complex polysaccharides and phenolic compounds, require processing to improve nutrient availability, which further increases commercial costs. Consequently, the incorporation of micro- and macroalgae into *Penaeus vannamei* diets represents a shift from basic nutritional supplementation to their use as strategic functional ingredients.

Current evidence highlights that these algal biomasses and derivatives consistently improve immune competence, antioxidant capacity, and gut integrity, while increasing resistance against critical pathogens like *Vibrio* spp. and WSSV. Moreover, algal supplementation provides a physiological buffer against environmental stressors, such as temperature fluctuations, salinity changes, and heavy metal exposure, thereby supporting production resilience. However, several complexities and knowledge gaps must still be addressed. The physiological response to algae is not always linear; while some parameters improve with higher inclusion levels, others plateau or even decline at high concentrations. This underscores the need to establish precise, species-specific optimal dosages that prioritize health outcomes. Additionally, the efficacy of bioactive compounds is heavily influenced by the product form (e.g., biomass, extracts, fermented derivatives) and processing methods (e.g., drying, extrusion), and these factors must be optimized to maximize bioavailability. To reduce variability across studies, standardized methodologies for assessing health benefits are also required. Future research integrating multi-omics with physiological assessments will help link cellular mechanisms to phenotypic outcomes, advancing the field of shrimp's precision functional nutrition. Finally, transitioning from research to industrial application requires techno-economic and life-cycle assessments to determine scalability and guide the industry toward cost-effective, sustainable solutions.

References

Abbas, E.M., Al-Souti, A.S., Sharawy, Z.Z., El-Haroun, E. and Ashour, M., 2023. Impact of dietary administration of seaweed polysaccharide on growth, microbial abundance, and growth and immune-related genes expression of the pacific whiteleg shrimp (*Litopenaeus vannamei*). Life 13(2): 344. https://doi.org/10.3390/life13020344

Abdel-Razek, N., Khalil, R.H., Afifi, A.A.M., Alkhuriji, A.F. and Metwally, D.M., 2024. Nutritional innovation using green seaweed (*Ulva* sp.) and garlic powder extracts for white-leg shrimp (*Litopenaeus vannamei*) challenged by *Vibrio harveyi*. Veterinary Medicine and Science 10(6): e70052. https://doi.org/10.1002/vms3.70052

Ahmad, A., W. Hassan, S., & Banat, F., 2022. An overview of microalgae biomass as a sustainable aquaculture feed ingredient: food security and circular economy. Bioengineered, 13(4), 9521-9547. https://doi.org/10.1080/21655979.2022.2061148

Ahmed, R.A., Jastaniah, S.D., Alaidaroos, B.A., Shafi, M.E., El-Haroun, E., Abd El-Aziz, Y.M., Abd El Megeed, O.H., AL-Qurashi, M.M., Bahshwan, S.M.A., Munir, M.B., Abdul Kari, Z., Mathew, R.T., Eissa, M.E.H., Eissa, E.-S.H. and Elfeky, A., 2025. Effects of dietary *Spirulina platensis* supplementation on growth performance, whole body composition, antioxidant activity, histological alterations, and resistance to *Vibrio parahaemolyticus* in Pacific white shrimp, *Litopenaeus vannamei*. Aquaculture Reports 38: 102606. https://doi.org/10.1016/j.aqrep.2024.102606

Akbari, P., Gholamhosseini, A., Ali, M., Aminikhoei, Z., Rezaei Tavabe, K., & Samadi Kuchaksaraei, B., 2023. Growth, Fatty Acid Composition, Antioxidant Activity and Resistance of *Litopenaeus vannamei* Fed With *Dunaliella salina*. Iranian Journal of Science, 47(1), 35-45. https://doi.org/10.1007/s40995-022-01396-1

Akbari, P., Gholamhosseini, A., Ali, M., Rahimi, M., and Yousefi, M., 2023. Growth, Fatty Acid Composition, Antioxidant Activity and Resistance of *Litopenaeus vannamei* Fed With *Dunaliella salina*. Iranian Journal of Science 47(1): 35–45. https://doi.org/10.1007/s40995-022-01396-1

Akbary, P. and Aminikhoei, Z., 2018. Effect of polysaccharides extracts of algae *Ulva rigida* on growth, antioxidant, immune response and resistance of shrimp, *Litopenaeus vannamei* against *Photobacterium damselae*. Aquaculture Research 54(6): 1885–1894. https://doi.org/10.1111/are.13710

Akbary, P., Gholamhosseini, A., Ali, M., Aminikhoei, Z., Tavabe, K.R., and Kuchaksaraei, B.S., 2021. Growth Yield, Fatty Acid Profile and Antioxidant Status of *Litopenaeus vannamei* Fed *Iyengaria stellata* Supplemented Diet. Iranian Journal of Science and Technology, Transactions A: Science 45(1): 111–119. https://doi.org/10.1007/s40995-020-01004-0

Ansari, F. A., Guldhe, A., Gupta, S. K., Rawat, I., & Bux, F., 2021. Improving the feasibility of aquaculture feed by using microalgae. Environmental Science and Pollution Research, 28(32), 43234-43257. https://doi.org/10.1007/s11356-021-14989-x

Ashour, M., Mabrouk, M.M., Mansour, A.I.A., Abdelhamid, A.F., AbdEl Kader, M.F., Elokaby, M.A., El-Nawsany, M.M., Abdelwarith, A.A., Younis, E.M., Davies, S.J., El-Haroun, E., Naiel,

M.A.E., 2024. Impact of dietary administration of *Arthrospira platensis* free-lipid biomass on growth performance, body composition, redox status, immune responses, and some related genes of Pacific whiteleg shrimp, *Litopenaeus vannamei*. PLOS ONE 20(1): e0300748. https://doi.org/10.1371/journal.pone.0300748

Cantelli, L., Goncalves, P., Guertler, C., Kayser, M., Pilotto, M.R., Barracco, M.A., and Perazzolo, L.M., 2019. Dietary supplementation with sulfated polysaccharides from *Gracilaria birdiae* promotes a delayed immunostimulation in marine shrimp challenged by the white spot syndrome virus. Aquaculture International 27(2): 349–367. https://doi.org/10.1007/s10499-018-0328-1

Chaudhary, D. K., Kim, S.-E., Park, H.-J. & Kim, K.-H., 2024. Unveiling the Bacterial Community across the Stomach, Hepatopancreas, Anterior Intestine, and Posterior Intestine of Pacific Whiteleg Shrimp. Journal of Microbiology and Biotechnology 34(6): 1260-1269. https://doi.org/10.4014/jmb.2403.03039

Chen, Y.Y., Chen, J.C., Tayag, C.M., Li, H.F., Putra, D.F., Kuo, Y.H., Bai, J.C., and Chang, Y.H., 2016. Spirulina elicits the activation of innate immunity and increases resistance against *Vibrio alginolyticus* in shrimp. Fish & Shellfish Immunology 55: 690–698. https://doi.org/10.1016/j.fsi.2016.06.042

Chen, Y.-Y., Sim, S.S., Chiew, S.L., Yeh, S.-T., Liou, C.-H., and Chen, J.-C., 2012. Dietary administration of a *Gracilaria tenuistipitata* extract produces protective immunity of white shrimp *Litopenaeus vannamei* in response to ammonia stress. Aquaculture 388–391: 164–170. http://dx.doi.org/10.1016/j.aquaculture.2012.09.031

Cherdkeattipol, K., Keetanon, A., Chuchird, N., Wudtisin, I., Frisbie, M.L., and Tacon, A.G., 2017. Effect of Giant Bladder Kelp (*Macrocystis pyrifera*) Feed Additive on Growth, Survival and Immune Responses of Pacific White Shrimp (*Litopenaeus vannamei*) Injected with *Vibrio parahaemolyticus*. Journal of Fisheries and Environment 41(3): 37–47. Retrieved from https://li01.tci-thaijo.org/index.php/JFE/article/view/88097

Eissa, E.H., Hendam, B.M., Dighiesh, H.S., Abd Elnabi, H.E., Sakr, S.E., Kabary, H., Abdel Rahman, A.N., Eissa, M.E.H., and Ahmed, N.H., 2025. The benefits of astaxanthin-rich microalgal powder on growth, health, and disease resistance against *Fusarium solani* in Pacific white shrimp. Fish & Shellfish Immunology 138: 110059. https://doi.org/10.1016/j.fsi.2024.110059

Eissa, E.-S.H., Aljarari, R.M., Elfeky, A., Abd El-Aziz, Y.M., Munir, M.B., Jastaniah, S.D., Alaidaroos, B.A., Shafi, M.E., Abd El-Hamed, N.N.B., AL-Farga, A., Dighiesh, H.S., Okon, E.M., Abd El-Hack, M.E., Ezzo, O.H., Eissa, M.E.H. and ElBanna, N.I., 2023. Protective effects of *Chlorella vulgaris* as a feed additive on growth performance, immunity, histopathology, and disease resistance against *Vibrio parahaemolyticus* in the Pacific white shrimp. Aquaculture International 32(3): 2821–2840. https://doi.org/10.1007/s10499-023-01298-y

Elizondo-González R, Quiroz-Guzmán E, Escobedo-Fregoso C, Magallón-Servín P, Peña-Rodríguez A. 2018. Use of seaweed Ulva lactuca for water bioremediation and as feed additive for white shrimp *Litopenaeus vannamei*. PeerJ 6:e4459 https://doi.org/10.7717/peerj.4459

Elshopakey, G.E., Abdelwarith, A.A., Younis, E.M., Davies, S.J., and Elbahnaswy, S., 2024. Alleviating effects of *Gracilaria verrucosa* supplement on non-specific immunity, antioxidant capacity and immune-related genes of pacific white shrimp (*Litopenaeus vannamei*) provoked with white spot syndrome virus. BMC Veterinary Research 20: 155. https://doi.org/10.1186/s12917-024-04304-4

Falaise, C., François, C., Travers, M.-A., Morga, B., Haure, J., Tremblay, R., Turcotte, F., Pasetto, P., Gastineau, R., Hardivillier, Y., Leignel, V., & Mouget, J.-L., 2016. Antimicrobial Compounds from Eukaryotic Microalgae against Human Pathogens and Diseases in Aquaculture. Marine Drugs, 14(9), 159. https://doi.org/10.3390/md14090159

Fang, H., Zhuang, Z., Huang, L., Niu, J. and Zhao, W., 2022. A Newly Isolated Strain of *Haematococcus pluvialis* GXU-A23 Improves the Growth Performance, Antioxidant and Anti-Inflammatory Status, Metabolic Capacity and Mid-intestine Morphology of Juvenile *Litopenaeus vannamei*. Frontiers in Physiology 13: 882091. https://doi.org/10.3389/fphys.2022.882091

Fang, H., Zhuang, Z., Huang, L., Zhao, W. and Niu, J., 2022b. Dietary *Klebsormidium* sp. supplementation improves growth performance, antioxidant and anti-inflammatory status, metabolism, and mid-intestine morphology of *Litopenaeus vannamei*. Frontiers in Nutrition 9: 857351. https://doi.org/10.3389/fnut.2022.857351

FAO (2024). The state of world fisheries and aquaculture 2024 blue transformation in action. Available at: https://openknowledge.fao.org/items/06690fd0-d133-424c-9673-1849e414543d

Félix, D.M., Córdova, Á.I.C., Elías, J.A.L., Córdova, L.R.M., Preciado, G.F., Jacinto, E.C., González, A.L., Cano, F.M. and Aldaz, N.H., 2019. Dosage and frequency effects of the microalgae *Dunaliella* sp. on the diet of *Litopenaeus vannamei* challenged with *Vibrio parahaemolyticus*. Journal of Invertebrate Pathology 160: 31–38. https://doi.org/10.1016/j.jip.2018.12.010

Félix, D.M., Jacinto, E.C., Córdova, Á.I.C., Elías, J.A.L., Córdova, L.R.M., González, A.L. and Soto, S.D.L., 2022. Physiological and antioxidant response of *Litopenaeus vannamei* against *Vibrio parahaemolyticus* infection after feeding supplemented diets containing *Dunaliella* sp. flour and β-glucans. Journal of Invertebrate Pathology 187: 107702. https://doi.org/10.1016/j.jip.2021.107702

Félix, M.D., Elías, J.A.L., Córdova, Á.I.C., Córdova, L.R.M., González, A.L., Jacinto, E.C., Aldaz, N.H., Mendoza, F.C. and Zazueta, M.G.B., 2017. Survival of *Litopenaeus vannemei* shrimp fed on diets supplemented with *Dunaliella* sp. is improved after challenges by *Vibrio parahaemolyticus*. Journal of Invertebrate Pathology 148: 118–123. https://doi.org/10.1016/j.jip.2017.06.003

Gamboa-Barraza, L.G., Bolan-Mejía, M.C., Osuna-Ruiz, I., Martínez-Rodríguez, I.E., Calvario-Martínez, O., and Morales-Covarrubias, M.S., 2021. Evaluation of aqueous extracts of *Gracilaria vermiculophylla* and *Ulva flexuosa* as treatment in challenged *Penaeus vannamei* with *Vibrio parahaemolyticus*. Aquaculture Research 49(5): 739-749. https://doi.org/10.3856/vol49-issue5-fulltext-2724

Ge, H., Ni, Q., Chen, Z., Li, J., and Zhao, F., 2019. Effects of short period feeding polysaccharides from marine macroalga, *Ulva prolifera* on growth and resistance of *Litopenaeus vannamei* against *Vibrio parahaemolyticus* infection. Journal of Applied Phycology 31: 2085-2092. https://doi.org/10.1007/s10811-018-1663-3

González-Colunga, D., Antunes-Ricardo, M., Gutiérrez-Uribe, J. A., & Cruz-Suárez, L. E., 2019.

Bioactivity-guided identification of anti-AHPND (acute hepatopancreatic necrosis disease) metabolites of *Ecklonia arborea*. Journal of Applied Phycology, 31(5), 3189–3199. https://doi.org/10.1007/s10811-019-01818-5

González-Meza, G., Herrera-Castro, K., Casillas-Hernández, R., Rentería-Mexía, A., Gortáres-Moroyoqui, P., Gil-Núñez, J., Ibarra-Gámez, J., & Ulloa-Mercado, G., 2022. Dietary supplementation effect of three microalgae on *Penaeus vannamei* growth, biochemical composition, and resistance to *Vibrio parahaemolyticus* (AHPND). Latin American Journal of Aquatic Research, 50(1), 88-98. http://dx.doi.org/10.3856/vol50-issue1-fulltext-2773

Guimarães, A.M., Guertler, C., Pereira, G.V., Coelho, J.R., Rezende, P.C., Nóbrega, R.O. and Vieira, F.N., 2021. *Nannochloropsis* spp. as feed additive for the Pacific white shrimp: Effect on midgut microbiology, thermal shock resistance and immunology. Animals 11(1): 150. https://doi.org/10.3390/ani11010150

Hoffling, F.B., Marquezi, A.S., Pinheiro, I., Simon, C., Rombenso, A.N., Seiffert, W.Q., Vieira, F.N. and Schleder, D.D., 2024. *Aurantiochytrium* sp. meal as feed additive for Pacific white shrimp reared under low temperature and challenged by WSSV in association with thermal stress. Fishes 9(3): 108. https://doi.org/10.3390/fishes9030108

Holt, C. C., Bass, D., Stentiford, G. D. and van der Giezen, M., 2021. Understanding the role of the shrimp gut microbiome in health and disease. Journal of Invertebrate Pathology 186: 107387. https://doi.org/10.1016/j.jip.2020.107387

Huang, F., Xue, D., Li, Z., Dong, S., Jiang, F., Hu, Z., and Zheng, Y., 2023. Evaluation of dietary *Haematococcus pluvialis* in pilot-scale farming of *Litopenaeus vannamei*: Effects on production performance, physiological health, bacterial communities, and eukaryotic plankton. Aquaculture 562: 738858. https://doi.org/10.1016/j.aquaculture.2022.738858

Jahromi, S.T., Pourmozaffar, S., Jahanbakhshi, A., Rameshi, H., Gozari, M., Khodadadi, M., Sohrabipour, J., Behzadi, S., Bazrkar, N., Nahavandi, R., Zahedi, M.R., and Moezzi, M., 2021. Effect of different levels of dietary *Sargassum cristaefolium* on growth performance, hematological parameters, histological structure of hepatopancreas and intestinal microbiota of *Litopenaeus vannamei*. Aquaculture 536: 736130. https://doi.org/10.1016/j.aquaculture.2020.736130

Kalia, V.C. (Ed.). 2018. Quorum sensing and its biotechnological applications. Springer, Berlin

Klongklaew, N., Praiboon, J., Tamtin, M., and Srisapoome, P., 2021. Chemical composition of a hot water crude extract (HWCE) from *Ulva intestinalis* and its potential effects on growth performance, immune responses, and resistance to white spot syndrome virus and yellowhead virus in Pacific white shrimp (*Litopenaeus vannamei*). Fish & Shellfish Immunology 112: 8–22. https://doi.org/10.1016/j.fsi.2021.02.004

Lee, P.-T., Lee, M.-C., Ho, Y.-T., Liao, Z.-H., Huang, H.-T., Ho, Y.-S., and An, F.-H., 2020. Polysaccharides from the green alga *Caulerpa racemosa* (Agardh, 1873) improve the immune response and antioxidant status in the white shrimp *Litopenaeus vannamei* (Boone, 1931) (Dendrobranchiata, Penaeidae). Crustaceana 93(6): 611–632. https://doi.org/10.1163/15685403-bja10045

Lee, P.-T., Tran, H.T.Q., Huang, H.-T., Nan, F.-H., and Lee, M.-C., 2020. *Sargassum horneri* extracts stimulate innate immunity, enhance growth performance, and upregulate immune genes in the white shrimp *Litopenaeus vannamei*. Fish & Shellfish Immunology 102: 276–285. https://doi.org/10.1016/j.fsi.2020.04.049

Li, M., Ghonimy, A., Chen, D. Q., Li, J. T., He, Y. Y., López Greco, L. S., ... & Chang, Z. Q., 2024. Profile of the gut microbiota of Pacific white shrimp under industrial indoor farming system. Applied Microbiology and Biotechnology, 108(1), 225. https://doi.org/10.1007/s00253-024-13046-0

Lian, C., Xiang, J., Cai, H., Ke, J., Ni, H., Zhu, J., ... & Yang, W., 2024. Microalgae inoculation significantly shapes the structure, alters the assembly process, and enhances the stability of bacterial communities in shrimp-rearing water. Biology, 13(1), 54. https://doi.org/10.3390/biology13010054

Lin, S., Chen, M., Chen, X., Li, Y., Liu, Y., Zhang, P., Hou, X., Tan, B. and Niu, J., 2024. Supplemental effects of *Haematococcus pluvialis* in a low-fish meal diet for *Litopenaeus vannamei* at varying temperatures: growth performance, innate immunity and gut bacterial community. Frontiers in Immunology 15: 1501753. https://doi.org/10.3389/fimmu.2024.1501753

Lin, Y.-H., Su, Y.-C., and Cheng, W., 2017. Simple heat processing of brown seaweed *Sargassum cristaefolium* supplementation in diet can improve growth, immune responses and survival to *Vibrio alginolyticus* of white shrimp, *Litopenaeus vannamei*. Journal of Marine Science and Technology 25(2): 242–248. https://doi.org/10.6119/JMST-016-1129-1

Liu, L., Cai, X., Ai, Y., Li, J., Long, H., Ren, W., Huang, A., Zhang, X., and Xie, Z., 2022. Effects of *Lactobacillus pentosus* combined with *Arthrospira platensis* on the growth performance, immune response, and intestinal microbiota of *Litopenaeus vannamei*. Fish & Shellfish Immunology 120: 345–352. https://doi.org/10.1016/j.fsi.2021.12.005

Liu, P.-C., Lin, P.-W., Huang, C.-L., Hsu, C.-H., and Chen, J.-C., 2019. Long-term administration of diets containing *Gracilaria tenuistipitata* extract induce the expression of immune-related genes and increase the immune response and resistance against *Vibrio harveyi* in white shrimp *Litopenaeus vannamei*. Gene Reports 15: 100378. https://doi.org/10.1016/j.genrep.2019.100378

Ma, M., & Hu, Q., 2024. Microalgae as feed sources and feed additives for sustainable aquaculture: Prospects and challenges. Reviews in Aquaculture, 16(2), 818-835. https://doi.org/10.1111/raq.12869

Mansour A.T., Ashour M., Abbas E.M., Alsaqufi A.S., Kelany M.S., El-Sawy M.A. and Sharawy Z.Z., 2022. Growth performance, immune-related and antioxidant genes expression, and gut bacterial abundance of Pacific white leg shrimp, *Litopenaeus vannamei*, dietary supplemented with natural astaxanthin. Front. Physiol. 13:874172. https://doi.org/10.3389/fphys.2022.874172

Mao, J., Lu, J., Chen, J., & Xiong, J., 2023. Consistent features of the gut microbiota in response to diverse shrimp *Litopenaeus vannamei* diseases: A meta-analysis. Fish and Fisheries, 24(6), 1103–1117. https://doi.org/10.1111/faf.12787

Molina-Cárdenas, C.A.; Sánchez-Saavedra, M.P., 2017. Inhibitory effect of benthic diatom species on three aquaculture pathogenic vibrios; Algal Research; 27:131-139; https://doi.org/10.1016/j.algal.2017.09.004

Moreira, A., Cruz, S., Marques, R., & Cartaxana, P., 2022. The underexplored potential of green macroalgae in aquaculture. Reviews in Aquaculture, 14(1), 5-26. https://doi.org/10.1111/raq.12580

Niu, J., Xie, J.-J., Guo, T.-Y., Fang, H.-H., Zhang, Y.-M., Liao, S.-Y., Xie, S.-W., Liu, Y.-J., and Tian, L.-X., 2019. Comparison and evaluation of four species of macro-algaes as dietary ingredients in *Litopenaeus vannamei* under normal rearing and WSSV challenge conditions: Effect on growth, immune response, and intestinal microbiota. Frontiers in Physiology 9: 1880. https://doi.org/10.3389/fphys.2018.01880

Niu, J., Xie, S.-W., Fang, H.-H., Xie, J.-J., Guo, T.-Y., Zhang, Y.-M., Liu, Z.-L., Liao, S.-Y., He, J.-Y., Tian, L.-X., and Liu, Y.-J., 2018. Dietary values of macroalgae *Porphyra haitanensis* in *Litopenaeus vannamei* under normal rearing and WSSV challenge conditions: Effect on growth, immune response and intestinal microbiota. Fish & Shellfish Immunology 81: 135–149. https://doi.org/10.1016/j.fsi.2018.06.010

Novriadi, R., Irawan, F., Malahayati, S., Khotimah, N., Bosman, O., Tanaka, B. and Nugroho, J.E., 2022. Supplemental effects of microalgae *Spirulina Arthrospira* platensis to the plant-based diet for Pacific white shrimp *Litopenaeus vannamei*. Indonesian Aquaculture Journal 17(2): 139–145. https://doi.org/10.15578/iaj.17.2.2022.139-145

Ozório, R.Á., Lopes, R.G., Vieira, F.d.N., Bolívar-Ramírez, N.C., Oliveira, C.Y.B.d., Barracco, M.A.A.M., Owatari, M.S., Fracalossi, D.M. and Derner, R.B., 2024. Crude polysaccharide extract from the microalga *Porphyridium cruentum* improved nonspecific immune responses and resistance in *Penaeus vannamei* exposed to *Vibrio alginolyticus*. Aquaculture Journal 4(3): 104–113. https://doi.org/10.3390/aquacj4030008

Patel, J.K., Parmar, P., Alwinpeter, M., Bajaniya, V.C., Chavda, V.M., Parmar, H., Nanjiyani, R., Rathod, V.J., and Khoraba, M.P., 2025. Effect of fucoidan extracted from the brown seaweed, *Sargassum johnstonii* on growth, survival, immune response and disease resistance to *Vibrio parahaemolyticus* in *Penaeus vannamei* Boone, 1931. International Journal of Agriculture and Food Science 7(1): 258–267. https://doi.org/10.33545/2664844X.2025.v7.i1d.266

Peña-Rodríguez, A., Omont, A., Quiroz-Guzmán, E. et al. Microbiome changes of an integrated aquaculture system of shrimp *Litopenaeus vannamei* and seaweed Ulva lactuca with different water exchanges. Aquacult Int 32, 1955–1973 (2024). https://doi.org/10.1007/s10499-023-01253-x

Pilotto, M.R., Milanez, S., Moreira, R.T., Rosa, R.D. and Perazzolo, L.M., 2019. Potential immunomodulatory and protective effects of the *Arthrospira*-based dietary supplement on shrimp intestinal immune defenses. Fish & Shellfish Immunology 88: 47–52. https://doi.org/10.1016/j.fsi.2019.02.062

Pourazad, M., Bagheri, D., Sotoudeh, E., and Ghasemi, A., 2024. Effects of dietary sulfated polysaccharides extracted from brown macroalgae *Sargassum ilicifolium* on growth indices, antioxidant status, and immune genes expression in white shrimp (*Litopenaeus vannamei*). Aquaculture Reports 39: 102387. https://doi.org/10.1016/j.aqrep.2024.102387

Prabu, E., Felix, N., Sivasankar, S., Thirumalai, A., Uma, A., Manikandan, K., Sathishkumar, G. and Thiruvasagam, T., 2025. Effects of dietary algae meal supplementation on growth, whole-body composition, histological and immune responses of Pacific white shrimp, *Penaeus vannamei*. Annals of Animal Science. https://doi.org/10.2478/aoas-2025-0007

Pratiwi, A.F., Satyantini, W.H., Mahasri, G., Mukti, A.T., and Isnansetyo, A., 2022. The administration of *Caulerpa racemosa* extract in feed of whiteleg shrimp (*Litopenaeus vannamei*) after infected by *Vibrio parahaemolyticus*. Journal of Aquaculture and Fish Health 11(2): 1–10. https://doi.org/10.20473/jafh.v11i2.29722

Rajeev, R., Adithya, K.K., Kiran, G.S. and Selvin, J., 2020. Healthy microbiome: a key to successful and sustainable shrimp aquaculture. Rev. Aquacult., 13: 238-258. https://doi.org/10.1111/raq.12471

Ringø E, Ashour M, Ahmed S, et al., 2025. Microalgae and seaweeds as feed additives for aquatic animals: effects on growth, immunity, and disease resistance. Algae - Science and Applications. IntechOpen. Available at: http://dx.doi.org/10.5772/intechopen.1009224

Ringø, E., Harikrishnan, R., Soltani, M., & Ghosh, K., 2 022. The effect of gut microbiota and probiotics on metabolism in fish and shrimp. Animals 12(21): 3016. https://doi.org/10.3390/ani12213016

Rocha, J.S., Soares, M., Guimarães, A.M., Schleder, D.D., Seiffert, W.Q. and Vieira, F.D., 2021. Addition of *Aurantiochytrium* sp. meal in the diet affects immunity and thermal shock resistance of the Pacific white shrimp. Boletim do Instituto de Pesca 47: e631. https://doi.org/10.20950/1678-2305/bip.2021.47.e631

Sagaram, U. S., Gaikwad, M. S., Nandru, R., & Dasgupta, S., 2021. Microalgae as feed ingredients: recent developments on their role in immunomodulation and gut microbiota of aquaculture species. FEMS Microbiology Letters, 368(11), fnab071. https://doi.org/10.1093/femsle/fnab071

Said, R.M., Nassar, S.E., Alaidaroos, B.A., Jastaniah, S.D., Dighiesh, H.S., Eissa, E.-S.H., Al-Farga, A., Kari, Z.A., Téllez-Isaías, G. and Attia, M.S., 2023. Impacts of dietary selenium nanoparticles from *Spirulina platensis* on growth performance, physio-biochemical components and alleviating effect against cadmium toxicity in Pacific white shrimp *Litopenaeus vannamei*. Catalysts 13(11): 1389. https://doi.org/10.3390/catal13111389

Schleder, D. D., Blank, M., Peruch, L. G. B., Poli, M. A., Gonçalves, P., Rosa, K. V., Fracalossi, D. M., Vieira, F. do N., Andreatta, E. R., & Hayashi, L., 2020. Impact of combinations of brown seaweeds on shrimp gut microbiota and response to thermal shock and white spot disease. Aquaculture, 519, 734779. https://doi.org/10.1016/j.aquaculture.2019.734779

Schleder, D.D., Rosa, J.R., Guimarães, A.M., Ramlov, F., Maraschin, M., Seiffert, W.Q., Vieira, F.N., Hayashi, L., and Andreatta, E.R., 2017. Brown seaweeds as feed additive for white-leg shrimp: effects on thermal stress resistance, midgut microbiology, and immunology. Journal of Applied Phycology 29: 2471–2477. https://doi.org/10.1007/s10811-017-1129-z

Shannon, E., Conlon, M., & Hayes, M., 2021. Seaweed components as potential modulators of the gut microbiota. Marine Drugs, 19(7), 358. https://doi.org/10.3390/md19070358

Siddik, M. A., Sørensen, M., Islam, S. M., Saha, N., Rahman, M. A., & Francis, D. S., 2024. Expanded utilisation of microalgae in global aquafeeds. Reviews in Aquaculture, 16(1), 6-33. https://doi.org/10.1111/raq.12818

Suantika, G., Situmorang, M.L., Khakim, A., Wibowo, I., Aditiawati, P., Suryanarayan, S., Nori, S.S., Kumar, S., and Putri, F., 2018. Effect of red seaweed *Kappaphycus alvarezii* on growth, survival, and disease resistance of Pacific white shrimp *Litopenaeus vannamei* against *Vibrio harveyi* in the nursery phase. Journal of Aquaculture Research & Development 9(2). https://doi.org/10.4172/2155-9546.1000523

Sudaryono, A., Chilmawati, D., and Susilowati, T., 2018. Oral administration of hot-water extract of tropical brown seaweed, *Sargassum cristaefolium*, to enhance immune response, stress tolerance, and resistance of white shrimp, *Litopenaeus vannamei*, to *Vibrio parahaemolyticus*. Journal of World Aquaculture Society 49(5): 877–888. https://doi.org/10.1111/jwas.12527

Tong, Y., Li, Q., Yang, C., Peng, M., Chen, X., Liu, Q., Huang, Y., Feng, P., Zeng, D., Wang, F., Hu, T., Liu, Q., & Zhao, Y., 2024. Enhancing shrimp growth and immunity with green algal *Caulerpa lentillifera* polysaccharides through gut microbiota regulation. Algal Research, 82, 103627. https://doi.org/10.1016/j.algal.2024.103627

Tzuc, J. T., Rendíz Escalante, D., Rojas Herrera, R., Gaxiola Cortés, G., & Arena Ortiz, M. L., 2014. Microbiota from *Litopenaeus vannamei*: digestive tract microbial community of Pacific white shrimp (*Litopenaeus vannamei*). SpringerPlus 3: 280. https://doi.org/10.1186/2193-1801-3-280

Vijayaram, S., Ringø, E., Ghafarifarsani, H., Hoseinifar, S. H., Ahani, S., & Chou, C. C., 2024. Use of algae in aquaculture: a review. Fishes, 9(2), 63. https://doi.org/10.3390/fishes9020063

Xie, S., Fang, W., Wei, D., Liu, Y., Yin, P., Niu, J. and Tian, L., 2018. Dietary supplementation of *Haematococcus pluvialis* improved the immune capacity and low salinity tolerance ability of post-larval white shrimp, *Litopenaeus vannamei*. Fish & Shellfish Immunology 80: 452–457. https://doi.org/10.1016/j.fsi.2018.06.039

Xiong, J. B., Sha, H. N., & Chen, J., 2024. Updated roles of the gut microbiota in exploring shrimp etiology, polymicrobial pathogens, and disease incidence. Zoological Research 45(4): 910-923. https://doi.org/10.24272/j.issn.2095-8137.2024.158

Yao, X., Lin, Y., Shi, M., Chen, L., Qu, K., Liu, Y., Tan, B. and Xie, S., 2022. Effect of *Schizochytrium limacinum* supplementation to a low fish-meal diet on growth performance, lipid metabolism, apoptosis, autophagy and intestinal histology of *Litopenaeus vannamei*. Frontiers in Marine Science 9: 1090235. https://doi.org/10.3389/fmars.2022.1090235

Yu, W., Zhang, M., Wang, B., Xu, R., Pang, T. and Liu, J., 2021. Dietary *Haematococcus pluvialis* powder supplementation affect carotenoid content, astaxanthin isomer, antioxidant capacity and immune-related gene expression in Pacific white shrimp, *Litopenaeus vannamei*. Aquaculture Research 00: 1-12. https://doi.org/10.1111/are.15090

Yu, Y.-Y., Chen, W.-D., Liu, Y.-J., Niu, J., Chen, M., and Tian, L.-X., 2016. Effect of different dietary levels of *Gracilaria lemaneiformis* dry powder on growth performance, haematological parameters and intestinal structure of juvenile Pacific white shrimp (*Litopenaeus vannamei*). Aquaculture 450: 356–362. https://doi.org/10.1016/j.aquaculture.2015.07.037

Yudiati, E., Isnansetyo, A., & Handayani, C. R., 2016. Innate immune-stimulating and immune genes up-regulating activities of three types of alginate from *Sargassum siliquosum* in Pacific white shrimp, *Litopenaeus vannamei*. Fish & shellfish immunology, 54, 46-53.

Yudiati, E., Isnansetyo, A., Murwantoko, Triyanto, & Handayani, C. R., 2019. Alginate from *Sargassum siliquosum* simultaneously stimulates innate immunity, upregulates immune genes, and enhances resistance of Pacific white shrimp (*Litopenaeus vannamei*) against white spot syndrome virus (WSSV). Marine Biotechnology, 21(4), 503-514. https://doi.org/10.1016/j.fsi.2016.03.022

Yudiati, E., Kurniawan, P. N. R., Putri, A. N., Al Falah, I. H., Azhar, N., Sedjati, S., ... & Santosa, G. W., 2024. Benefits of Alginate and Spirulina Supplementation to the Growth and Resistance Against Salinity Exposure in *Litopenaues vannamei* Post-Larvae. Ribarstvo, Croatian Journal of Fisheries, 82(2), 65-72. https://doi.org/10.2478/cjf-2024-0008

Zeng, S., He, J. & Huang, Z., 2024. The intestine microbiota of shrimp and its impact on cultivation. Applied Microbiology and Biotechnology 108: 362. https://doi.org/10.1007/s00253-024-13213-3

Zhang, L., Liao, K., Shi, P., Guo, J., Xie, F., & Xu, J., 2025. Dietary inclusion of microalgae meal for Pacific white shrimp (*Litopenaeus vannamei*): effects on growth performance, flesh quality, and immunity. Animal Feed Science and Technology, 320, 116205. https://doi.org/10.1016/j.anifeedsci.2024.116205

Zhang, Z., Shi, X., Wu, Z., Wu, W., Zhao, Q., & Li, E., 2023. Macroalgae improve the growth and physiological health of white shrimp (*Litopenaeus vannamei*). Aquaculture Nutrition, 2023(1), 8829291. https://doi.org/10.1155/2023/8829291

Zhao, Y., Duan, C., Zhang, X. X., Chen, H., Ren, H., Yin, Y., & Ye, L., 2018. Insights into the gut microbiota of freshwater shrimp and its associations with the surrounding microbiota and environmental factors. Journal of microbiology and biotechnology, 28(6), 946-956. https://doi.org/10.4014/jmb.1710.09070

Zhou, K., Liu, P.-P., Yao, J.-Y., Vasta, G. R., et al., 2024. Shrimp intestinal microbiota homeostasis: dynamic interplay between the microbiota and host immunity. Reviews in Aquaculture 17(2): 12986. https://doi.org/10.1111/raq.12986

CHAPTER 11

The multifaceted value of camelina oil as a novel aquafeed ingredient

*S.K. Das and L. Nankervis**

Centre for Sustainable Tropical Fisheries and Aquaculture (CSTFA), College of Science and Engineering, James Cook University, Australia
**Correspondence: Leo.nankervis@jcu.edu.au*

Abstract

The global aquaculture industry has been seeking sustainable alternatives to fish oil, driven by its limited sustainable availability and escalating cost. Camelina oil (CO), derived from *Camelina sativa*, has garnered notable interest as a possible plant-based alternative due to its richness in alpha-linolenic acid and gamma-tocopherol. Additionally, CO provides significant nutritional benefits and stability against oxidation, making it an excellent choice for aquafeeds. This chapter explores the biochemical composition, digestibility, and physiological effects of CO in aquaculture diets for different species. Freshwater fish such as rainbow trout and Nile tilapia, benefit markedly from dietary CO supplementation due to their ability to desaturate and elongate alpha-linolenic acid (ALA) into long-chain ω-3 fatty acids like eicosapentaenoic acid (EPA, 20:5n-3) and docosahexaenoic acid (DHA, 22:6n-3). Marine species have lower activities of the necessary enzymes for this process, and the fatty acid profile benefits of CO are restricted to direct supply of ALA or total ω-3 levels in diets. In addition to its nutritional value, growing camelina brings environmental benefits, such as minimal water and fertilizer needs, suitability for less productive lands, and lower greenhouse gas emissions than close alternatives such as canola. Overall, CO is not a total replacement for fish oil, but forms part of the potential future basket of expandable, eco-friendly oils that allow some degree of decreased reliance on marine-sourced lipids in aquaculture nutrition.

Keywords

omega-3 fatty acids – plant-based lipids – sustainable aquafeeds – antioxidant stability – fish health

1 Introduction

Camelina (*Camelina sativa*) is an oil seed plant in the family Brassicaceae. Its agricultural origins date back as far as 4000 BC (Berti et al. 2016), however modern agriculture has favored its higher yielding mustard seed and canola/rapeseed relatives. Despite its lower yields, camelina has received renewed interest in recent years due to its adaptability to a range of agricultural environments, with its tolerance to cold and water stress particularly advantageous traits (Berti et al. 2016; Hosseini Sanehkoori et al. 2021). Cold-climate oil seed growth promotes production of polyunsaturated fatty acids and camelina grown in cold environments contains particularly high levels of the ω-3 a-linolenic acid (ALA) (Righini et al. 2019; Clemente et al. 2025). While this often makes oils prone to oxidation, camelina further contains high gamma-tocopherol (vitamin E) content (Clemente et al. 2025) which imparts antioxidant protection. Camelina therefore presents a unique value-proposition as a relatively stable source of plant-based ω-3 fatty acids.

1.1 Aquafeeds and oil supply

Oils in aquafeed contribute to the dietary supply of energy, essential fatty acids, and other non-triglyceride lipid nutrients. Fish oil traditionally comprised the primary source of lipids in aquafeeds, which in turn consume the majority of the world's fish oil supply, with 74% of global fish oil production estimated to be consumed in aquaculture in 2021 (IFFO estimates in FAO 2024). Fish oil production reached its peak in 1994, with a reasonably static annual supply since this time necessitating development of alternatives to facilitate the expansion of aquafeed production (FAO 2024). The energy supply from fish oil can be derived from many sources of lipids, and global aquafeed production has increasingly utilized a range of plant and animal-based lipid sources for this purpose. There has thus been a shift over the last 30 years away from marine oils to a predominance of terrestrial-derived oil sources.

The marine food chain has a unique fatty acid profile. Relatively little bioconversion of fatty acid classes occurs between trophic levels for non-ruminant animals, meaning that the fatty acid profile of consumers largely reflects their food. Animals are capable to lipogenesis of palmitate from other energy sources; however, the specificity of enzymes involved in elongation and desaturation of the carbon chain limits the resultant fatty acids. Specific limitations in fatty acid synthesis occur due to a lack of fatty acid desaturation at the third and sixth carbon from the methyl end, also known as the omega (ω) end, of the fatty acid chain. The lack of these ω-3 and ω-6 desaturase enzymes in animal systems means that the functions of these fatty acids must be satisfied by a nutritional supply. The physiological need for ω-3 and ω-6 fatty acids for normal growth and development makes their requirements nutritionally 'essential' or 'indispensable'. In the marine food chain, these fatty acids are derived from algae, where the ω-3 fatty acid group predominate. In contrast, lipids of the modern terrestrial agricultural supply chain contain a higher proportion of fatty acids in the ω-6 group.

The search for cost-effective and scalable sources of digestible lipids have driven aquafeed producers towards oils from terrestrial plants and byproducts from terrestrial animal processing. This shift away from fish oil has therefore changed the fatty acid profile of aquafeeds from ω-3 dominated towards more ω-6 dominated fatty acid profiles.

1.2 Omega 3 and omega 6 fatty acids

Fatty acids in ω-3 and ω-6 classes are polyunsaturated, long-chain fatty acids, and as such have a low melting point and are well digested by fish. They have similar roles in maintaining fluidity when incorporated into phospholipids of cell membranes and are involved in cell signaling processes (Innes and Calder 2018). The specific roles of these fatty acid classes in cell signaling can however be opposed. Fatty acids in the ω-6 class form precursors to proinflammatory leukotrienes and prostaglandins, initiating cellular response to injury and disease, while ω-3 fatty acids form precursors to the equivalent anti-inflammatory signaling mechanisms (Simopoulos 2002; Innes and Calder 2018). While inflammation is a normal part of damage and disease mitigation, excessive inflammation can cause significant tissue damage, and the de-escalation of this process is critical for healing after damage occurs. These pro- and anti-inflammatory signaling pathways utilize the same enzyme pathways, meaning that an excess of either ω-3 or ω-6 fatty acid classes is likely to unbalance inflammatory processes. Therefore, nutritionists have focused on the balance of ω-3 and ω-6 fatty acids to support effective immune function and recovery after injury.

The fatty acid profile of fish feed is firstly important to meet the essential fatty acid requirements of fish, secondly to control inflammatory signaling balance and ultimately for the health benefits to the human consumer.

1.3 Other lipid-soluble components of oils

1.3.1 Potential for the residual meal

While oilseeds are largely grown for their oil yield, extracted oil accounts for 30–50% of the seed dry matter, with the remainder sold as a secondary meal product (Bouallegue et al. 2020). Camelina seeds contain naturally low glucosinolate and erucic acid content (Clemente et al. 2025), which are antinutritional elements prevalent in unselected rapeseed and mustard strains. They further contain polyphenolic compounds that contribute to antioxidant processes within production animals (Kurasiak-Popowska and Stuper-Szablewska 2020).

2 Nutritional composition of camelina oil

2.1 Fatty acid composition and antioxidant profile

CO is particularly rich in α-linolenic acid (ALA), comprising 28–40% of its total fatty acids, a level much higher than in soybean, corn, or canola (Garcia-Mendoza et al. 2021; Hixson et al. 2014b). This makes camelina a unique plant oil source for aquafeeds, providing a favorable ω-3: ω-6 ratio that supports fish health and inflammation balance (Innes & Calder, 2018). In addition, CO contains moderate levels of linoleic acid (LA, 18:2n-6), oleic acid (18:1n-9), and gondoic acid (20:1n-9), which may contribute to membrane fluidity and energy metabolism. CO also contains high γ-tocopherol (700–900 mg/kg), an antioxidant that protects the oil from oxidation during storage and feed processing, helping preserve the nutritional quality of aquafeeds (Table 11.1).

CO also demonstrates higher oxidative stability than fish oil but is less stable than sunflower, corn, sesame, olive, and rapeseed oils under accelerated storage at 65 °C (Eidhin et al. 2003). This relative stability is mainly influenced by its high α-linolenic acid content, which increases susceptibility to oxidation compared with more saturated oils such as palm oil. Despite this, CO retains sufficient stability for aquafeed use, particularly when protected from light and stored appropriately. Its favorable biochemical composition further highlights its potential for sustainable and nutritionally appropriate inclusion in aquafeed formulations.

2.2 Metabolic relevance and application in aquafeeds

Although CO lacks eicosapentaenoic acid (EPA, 20:5n-3) and docosahexaenoic acid (DHA, 22:6n-3), it is a valuable source of ALA, the metabolic precursor to these long-chain omega-3 PUFAs. Freshwater species such as rainbow trout (*Oncorhynchus mykiss*) (Hixson et al. 2014a) and carp (*Cyprinus carpio*) possess enzymatic pathways (Δ6-desaturase, elongase, and Δ5-desaturase) that enable partial conversion of ALA into EPA and DHA (Figure 11.1) enabling them to utilize CO more effectively. In contrast, most marine fish exhibit limited capacity for this conversion resulting in reduced EPA and DHA levels in tissues when fish oil is replaced with CO (Glencross et al. 2024; Morais et al. 2012; Huyben et al. 2020). Thus, the nutritional role of CO depends strongly on species-specific biosynthetic capacity.

3 Camelina oil in aquafeeds: effects on fish growth, physiology, and health outcomes

3.1 Growth performance and nutritional utilization

While CO shows promise in effectively supplying lipid energy and essential ω-3 fatty acids in aquafeed formulation, its effects need to be placed in the context in

TABLE 11.1 Fatty acid and nutrient composition of camelina and other vegetable oils

Component	Camelina oil	Linseed oil	Soybean oil	Corn oil	Palm oil	Rapeseed oil	References
α-Linolenic Acid (18:3 n-3)	28–40%	45–55%	7–8%	<1%	<1%	8–10%	Günç Ergönül and Aksoylu Özbek, 2020
Linoleic Acid (18:2 n-6)	13–20%	15–18%	~50%	~60%	~10%	~20%	Günç Ergönül and Aksoylu Özbek, 2020
Oleic Acid (18:1 n-9)	15–19%	~20%	~24%	~28%	~40%	~60%	Günç Ergönül and Aksoylu Özbek, 2020
Gondoic Acid (20:1 n-9)	10–16%	–	–	–	–	1–2%	Günç Ergönül and Aksoylu Özbek, 2020
Saturated Fatty Acids (SFA)	7–9%	~9–10%	~14–16%	~13%	~50%	~7%	Günç Ergönül and Aksoylu Özbek, 2020
EPA (20:5 n-3)	0% (wild type)	–	–	–	–	–	Hixson et al., 2014b
DHA (22:6 n-3)	0% (wild type)	–	–	–	–	–	Hixson et al., 2014b
Total PUFA	~50-60%	~70%	~55%	~60%	~10%	30%	Garcia-Mendoza et al., 2021
Total MUFA	~25-30%	~20%	~23%	~28%	~40%	~60%	Garcia-Mendoza et al., 2021
Tocopherols (mg/kg)	700–1250	500–800	~800	~600	~500	~600	Günç Ergönül and Aksoylu Özbek, 2020

Note. Information adapted from: Günç Ergönül and Aksoylu Özbek (2020); Hixson et al. (2014b); Garcia-Mendoza et al. (2021). Values represent typical composition ranges for cold-pressed or refined oils reported in the cited literature. EPA and DHA values refer to wild-type Camelina oil.

which it is tested. Several studies have demonstrated that CO can effectively replace fish oil in aquafeeds without impairing growth or nutrient utilization, especially in freshwater species with active LC-PUFA biosynthetic pathways. For example, Atlantic salmon and rainbow trout achieved comparable weight gain and feed conversion ratios (FCR) when dietary fish oil was fully replaced with CO (Hixson et al. 2014b; Lu et al. 2020). However, this is only the case when any EPA and DHA specific requirements are met by the residual oil in the fish meal component of the diet. Salmonid fish have some ability to elongate and desaturate approximately 25% of their long chain PUFA from ALA (Hixson et al. 2014b; Hixson et al. 2014a), however this occurs most effectively when EPA and DHA content of the diet is low. Therefore,

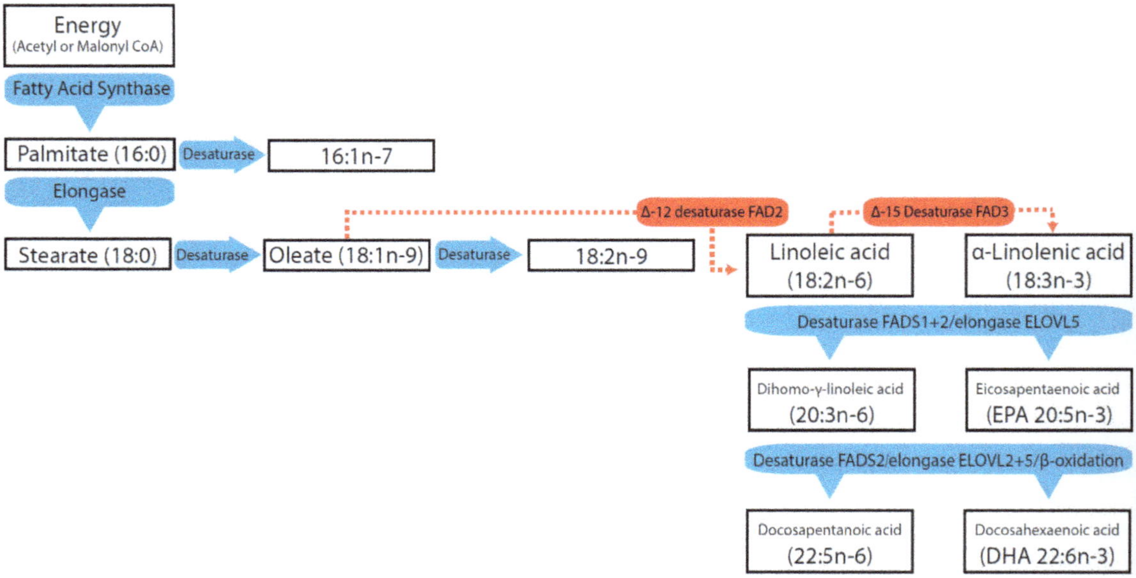

FIGURE 11.1 Fatty acid biosynthesis pathway in fish. Pathway is simplified for clarity. Blue symbols indicate pathways present in fish, while red color indicates pathways present in plants but absent in fish, necessitating dietary supply. Adapted from Bell and Tocher, 2009; Tocher, 2010

the degree to which ALA from CO can supplement EPA and DHA from fish oil in these feeds is highly dependent on the EPA and DHA targets in the edible portion of the fish being grown. Likewise, CO can enrich long chain PUFA in tilapia muscle when dietary levels are insufficient and present a potential non-fish oil pathway to increase health benefits to the human consumer without affecting production rates or efficiency (Toyes-Vargas et al. 2020).

In contrast, Atlantic cod catabolize the majority of the ALA from CO when it fully replaces fish oil, resulting in impaired growth and reduced tissue EPA and DHA levels (Hixson et al. 2014b). These findings exemplify the species-specific nature of responses, with freshwater fish generally more capable of utilizing ALA from CO compared to marine species (as previously discussed in detail in Section 2.2)

3.2 Physiological and health parameters

Physiological assessments in camelina-fed fish generally show minimal or no adverse effects. The liver indices, intestinal morphology, and hepatosomatic index (HSI) are generally unaffected at moderate inclusion levels of CO. However, adding more than 15% of CO to salmon diets has been linked to slight intestinal inflammation, suggesting this as a practical safe limit (Ye et al. 2016). Similar villi shortening have been observed in salmon fed other plant oils, likely reflecting reduced EPA and DHA availability for intestinal proliferation (Moldal et al. 2014).

The γ-tocopherol content of CO contributes to the antioxidant capacity of feeds and fish consumers, inhibiting peroxidation which prolongs shelf life of feed and fish products, as well as reducing oxidative stress in cells (Eidhin et al. 2003). While antioxidant effects are not unique to camelina, its relatively high γ-tocopherol concentration enhances these protective benefits compared to many other plant oils. Immune-related outcomes have also been reported, for example cytokine gene expression upregulation has been demonstrated in red seabream fed diets containing CO (Mzengereza et al. 2022). However, whether these immune responses are specific to CO remains to be clarified.

Overall, the nutritional value of CO has been validated across multiple freshwater and marine species. Findings from controlled feeding trials are summarized in Table 11.2 providing insights into species-specific growth, physiological, and health outcomes following CO and meal inclusion.

4 Industrial, economic, and sustainability applications of camelina oil

CO offers a distinctive array of nutritional, industrial, and environmental benefits. Camelina functions as a lipid source in aquafeeds and offers a viable and scalable

TABLE 11.2 Effects of camelina oil, alone and in combination with camelina meal on fish growth, physiology, and nutritional outcomes

Fish species	Environment	Oil/meal type	Inclusion level	Key outcomes	References
Rainbow trout (*Oncorhynchus mykiss*)	Fresh water	Camelina oil	100% lipid replacement	No difference in growth or tissue fatty acid composition	Lu et al. 2020
Rainbow trout (*Oncorhynchus mykiss*)	Fresh water	Camelina oil	100% lipid replacement	Stable growth, enhanced LC-PUFA biosynthesis	Hixson et al. 2014a
Nile tilapia (*Oreochromis niloticus*)	Fresh water	Camelina oil	100% lipid replacement	No impact on growth, survival, or tissue fatty acid profile	Toyes-Vargas et al. 2020
Nile tilapia (*Oreochromis niloticus*)	Fresh water	Camelina meal + oil	~25% meal + 5% oil	Stable performance, growth and muscle lipid composition	Toyes-Vargas et al. 2020
Atlantic salmon (*Salmo salar*)	Marine	Camelina oil	100% lipid replacement	Stable growth, muscle FA profile, enhanced LC-PUFA biosynthesis	Hixson et al. 2014b
Atlantic Cod (*Gadus morhua*)	Marine	Camelina oil	100% lipid replacement	Reduced growth, reduced EPA/DHA content	Hixson et al. 2014b
Gilthead seabream (*Sparus aurata*)	Marine	Camelina + chia oil	100% lipid replacement	Upregulation of immune genes, limited ALA to EPA/DHA conversion	Toyes-Vargas et al. 2020; Ofori-Mensah et al. 2020
Atlantic Salmon (*Salmo salar*)	Marine	Transgenic camelina Oil	60% oil + 20% meal	Improved tissue LC-PUFA retention, stable growth	Betancor et al. 2015

solution for sustainable agriculture and feed processing. As a hardy, short-cycle crop with low input demands, camelina can be cultivated on marginal lands with minimal water and fertilizer use, supporting climate-resilient farming (Berti et al. 2011). Its suitability for crop rotation and low competition with food crops further enhance its appeal in sustainable feed ingredient supply chains.

4.1 Environmental and economic advantages

The extraction of CO using cold pressing or solvents yields oil rich in α-linolenic acid (ALA) and γ-tocopherol, along with a protein-rich meal. This co-product can serve as animal feed or fertilizer, thus enhancing the crop's overall value. Life cycle assessment studies indicate that CO generates approximately 100–150 kg of CO_2 per tonne, far lower than fish oil, which produces 600–900 kg of CO_2 per tonne (Li and Mupondwa 2014; Mupondwa et al. 2016). Additional sustainability benefits include reduced water usage and smaller ecological footprint compared to marine derived fish oil.

While CO is lower yielding than the related canola or rapeseed crops, its ability to grow in colder and dryer climates may enable its cultivation in areas that are unsuitable for these higher yielding crops. Its viability is a combination of yield, production efficiency, and value proposition as an ingredient. With recent market prices estimated at USD 1,200 to 1,600 per ton, CO compares favorably to fish oil priced at USD 1,800 to 2,500 per ton (Jafari et al., 2022; USDA, 2020). Sustainably sourced fish oil is limited and global production unlikely to increase, likely enhancing the future value of alternative sources of ω-3 fatty acids.

Feeding trial evidence further supports its economic potential: Hixson et al. (2014b) demonstrated that substituting up to 60% of fish oil with CO in salmonid diets had no negative impact on economic feed conversion rations or gross profit margins. Similarly, Wei et al. (2019) reported that high-oil residue camelina meal could be included at levels up to 40 g/kg in trout and salmon diets without impairing growth or health outcomes. Together, these findings affirm that CO can reduce feed costs while maintaining production efficiency and product quality.

4.2 Innovation and future integration

Recent advancements in plant biotechnology have produced transgenic camelina lines that can synthesize EPA and DHA, providing a sustainable source of LC-PUFAs previously obtained solely from marine fish oil or marine algae. Feeding trials with European sea bass and Atlantic salmon demonstrated that these transgenic oils achieve performance and tissue enrichment results comparable to traditional fish oil (Betancor et al. 2021; Betancor et al. 2015).

Beyond aquafeeds, camelina contributes to a circular bioeconomy. Its oil and co-products are used in the production of biofuels, bioplastics, and specialty industrial oils, diversifying revenue streams and enhancing supply chain resilience.

Such versatility reinforces camelina's market value while supporting environmentally responsible aquaculture and eco-certified markets (Stamenković et al. 2023; Arshad et al. 2022).

The entire camelina production cycle, encompassing cultivation, oil extraction, and aquafeed integration, illustrates the crop's adaptability and minimal environmental impact (Figure 11.2). A comparative analysis of the sustainability attributes and economic indicators of CO in relation to fish oil is presented in Table 11.3.

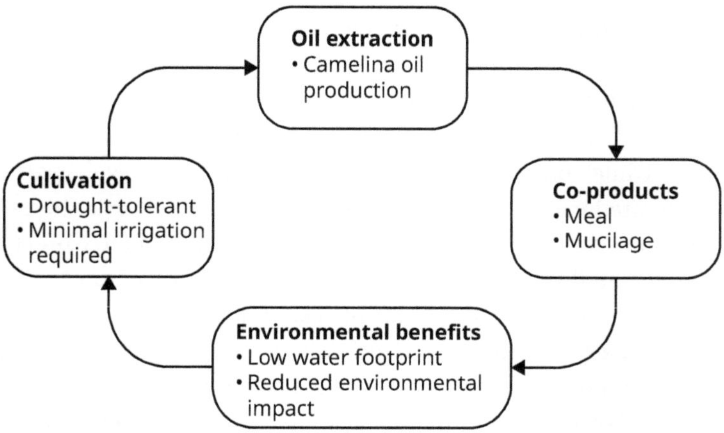

FIGURE 11.2 Diagram illustrating camelina's agronomic cycle, oil extraction, co-product streams (meal, mucilage), and aquafeed application. Adapted from Berti et al. 2011; Waraich et al. 2013; Mupondwa et al. 2016; Eidhin et al. 2003

TABLE 11.3 Sustainability and economic comparison of camelina oil and fish oil

Parameter	Camelina oil	Fish oil	References
LC-PUFA content (EPA+DHA)	0% native, 15–20% (transgenic camelina)	20–30% (natural)	Betancor et al. 2021
CO_2 emissions (kg/ton oil)	100–150	600–900	Mupondwa et al. 2016
Water use & ecological footprint	Low (minimal irrigation; marginal land suitability)	Moderate-High (linked to fisheries and processing impacts)	Li and Mupondwa 2014; Avadí and Fréon 2013
Estimated market price (USD/ton)	~ $1,200–1,600	~$1,800–2,500	Jafari et al. 2022
Co-product utility	High (meal, biodiesel, mucilage, fertilizer, bioplastics)	Moderate (fishmeal only)	Waraich et al. 2013;Arshad et al. 2022; Stamenković et al. 2023
Oxidative stability	High (γ-tocopherol preserves PUFA)	Moderate (prone to rancidity)	Hixson et al. 2014c

5 Effect on fish quality: nutritional composition, texture, flavour, and consumer acceptance

Replacing FO with CO in aquafeeds can affect multiple aspects of fish quality, including nutritional composition, oxidative stability, texture, flavor, and consumer acceptance. While CO provides a favorable omega-3-to-omega-6 ratio and is rich in α-linolenic acid (ALA), it lacks long-chain omega-3 polyunsaturated fatty acids (LC-PUFAs), EPA, and DHA, which are critical for human nutrition and desirable in aquaculture products. These limitations are particularly important when evaluating fish nutritional quality and consumer acceptance.

5.1 Nutritional composition of fillets

Replacement of FO with CO into fish diets significantly increases ALA concentrations in fillets, while concurrently decreasing EPA and DHA levels, particularly in species with limited capacity to synthesize LC-PUFA. Nile tilapia (Toyes-Vargas et al. 2020), rainbow trout (Lu et al. 2020), and Atlantic salmon (Hixson et al. 2014a) have all demonstrated this pattern although the magnitude of change varies with species, life stage, and feeding duration. The resulting enzymatic conversion of fatty acids via desaturases and elongases clearly differs between species and life stages within species. Salmonids exhibit moderate conversion of dietary ALA to EPA and DHA, whereas Atlantic cod and tilapia have much lower efficiency (Hixson et al. 2014b; Xue et al. 2015). Consequently, salmonid fillets typically retain higher LC-PUFA concentrations when fed CO-based diets. Selective breeding for genes such as elovl2 and elovl5 has been suggested as a mechanism to improve ALA conversion and preserve nutritional quality. These species-specific differences in lipid metabolism and LC-PUFA biosynthesis are fully detailed in Section 2.2. A summary of key fillet quality responses to camelina oil inclusion across species is presented in Table 11.4.

5.2 Fillet texture, oxidative stability

The relatively high PUFA content in CO could make fillets more susceptible to oxidation; however, camelina's naturally high γ-tocopherol along with phenolic compounds and flavonoids, helps prevent lipid peroxidation. This renders fillets from fish fed CO-enriched feeds more stable compared to those containing other PUFA-rich oils such as flaxseed oil (Sydor et al. 2022).

In terms of texture, most studies indicate no adverse effects when CO replaces up to about 40–50% of the added fish oil fraction (Betancor et al. 2015). At higher substitution levels, however, outcomes become more variable. Some studies such as (Betancor et al. 2016) reported no major changes in fillet texture even when more than half of the added fish oil was replaced, whereas others noted slight softening of fillets, likely linked to alterations in muscle lipid class composition and water-holding capacity (Jafari et al. 2022; Hixson et al. 2014b). In commercial aquafeeds, lipid inclusion typically ranges from 10–20% of the diet (equivalent to ~ 25–30 % of dietary energy as lipids) (Salem et al. 2023). Within this lipid fraction, fish oil is only one of the several raw materials, and thus it is more precise to express CO substitution as a percentage of the added fish oil fraction, rather than as a percentage of total dietary lipid. At the research level, much higher substitution rates of this fish oil fraction have been tested. For example in Atlantic cod, CO successfully replaced 40% or even 80% of the fish oil component of the diet without compromising growth (Booman and Rise 2013).

Together, these findings suggest that CO can replace a substantial proportion of fish oil in aquafeeds without negatively affecting fillet texture or oxidative stability under practical feeding scenarios. At very high substitution levels, some variability has been observed, underscoring the importance of considering both the proportion of added fish oil replaced and the absolute amount of lipid included in the diet.

5.3 Flavor and consumer acceptance

Sensory evaluations suggest that substitution of FO with CO at levels up to 40%–50% of the added fish oil fraction does not significantly alter fillet flavor or consumer acceptance (Table 11.4; Hussin et al. 2019; Betancor et al. 2015). At higher substitution levels, however, mild off-flavors have been reported in certain species (Booman et al. 2014). Detailed sensory evaluations show this can include the development of a slight "plant oil" or "bitter" note in species like Atlantic salmon or cod when CO substitution exceeds the 50% threshold, highlighting the importance of balancing substitution rates with consumer panel outcomes and market expectations (Betancor et al. 2015).

Recent research indicates that fish oil replacement with CO can lead to subtle changes in fillet fatty acid composition and flavor, but these do not necessarily reduce consumer acceptance. In fact, sensory panel studies have shown that moderate inclusion of CO maintains acceptable taste and texture profiles. For example, Betancor et al. (2015) reported no significant differences in overall consumer preference or acceptability for steamed rainbow trout fillets fed diets containing 50% or 100% fish oil replacement with CO. Similarly, Hussin et al. (2019) found that Nile tilapia fillets from CO based diets exhibited only minor flavor and texture differences, with no negative impact on consumer preference or willingness to purchase. These findings provide empirical support for the sustainability narrative by confirming that sensory quality remains acceptable at moderate substitution levels.

Moreover, consumer perception of sustainability can further enhance acceptance. When panel participants were informed that the fish were raised on sustainable CO based diets, overall acceptance scores increased significantly (Hussin et al. 2019). This suggests that clear communication of environmental and nutritional benefits can enhance marketability, even if slight sensory variations occur. Consequently, retailers and processors are encouraged to integrate sustainability and health messaging to strengthen consumer confidence in CO fed fish products.

TABLE 11.4 Summary of key fillet quality changes in fish fed camelina oil

Parameter	Effect of camelina oil	Fish species	Inclusion level (fish oil substitution)	References
ALA Content ↑	Significant increase in fillets	Rainbow trout *(O. mykiss),* Nile tilapia *(O. niloticus),* Atlantic salmon *(S. salar)*	40–100%	Lu et al. 2020; Xue et al. 2015
EPA & DHA ↓	Marked reduction	Atlantic salmon *(S. salar)* Nile tilapia *(O. niloticus)*	40–100%	Betancor et al. 2015; Xue et al. 2015
Flavor changes	Mild off-flavor at high CO	Atlantic salmon *(S. salar)*	>50%	Betancor et al. 2015
Texture	Generally unaffected	Nile tilapia *(O. niloticus),* Rainbow trout *(O. mykiss)*	<50%	Betancor et al. 2015; Lu et al. 2020
Oxidative stability	Improved vs flax oil	Multiple species	Up to 50%	Sydor et al. 2022

Note: ↑ = increase; ↓ = decrease. ALA: Alpha-linolenic acid (18:3n-3); EPA: Eicosapentaenoic acid (20:5n-3); DHA: Docosahexaenoic acid (22:6n-3).

6 Digestibility and lipid metabolism in fish

CO is gaining prominence as a sustainable lipid source in aquafeeds due to its beneficial fatty acid profile and generally high digestibility across many species. Beyond growth outcomes and fillet quality (Section 3 and Section 5), understanding digestibility and lipid metabolism is essential to evaluate how fish utilize CO and how it can be integrated into sustainable diets.

6.1 Digestibility across fish species

Several studies confirm that CO is highly digestible in aquaculture species (Table 11.5). For example, rainbow trout fed CO-based diets showed lipid digestibility exceeding 90%, comparable to or slightly lower than traditional fish oil diets (Fraser et al. 2017; Lu et al. 2020). Similar digestibility values have been reported in Nile tilapia and common carp, with no significant impairment in lipid assimilation compared to fish oil diets (Chen et al. 2018).

Digestive and metabolic efficiency, however, varies by species due to differences in fatty acid desaturation and elongation pathways. Freshwater fish such as tilapia and carp possess robust enzymatic activity, particularly Δ 6 desaturase (fads2)

and elongase (elovl5) that supports conversion of ALA into EPA and DHA (Chen et al. 2018) while marine carnivores such as salmon, seabream, and cod have lower enzymatic capacity (Tocher et al. 2024; Venegas-Calerón and Napier 2023). In gilthead seabream, for example, digestibility estimates are more difficult to specify due to limitations in experimental design, but fatty acid balance and gene expression analyses indicate altered metabolism and reduced EPA/DHA retention (Ofori-Mensah et al. 2020). These differences, outlined in Section 2.2, explain why digestibility outcomes do not always translate into equivalent retention of long-chain PUFAs. As illustrated in Figure 11.1, fish possess the core biosynthetic pathway for fatty acid desaturation and elongation. However, species-specific differences in enzymatic activity are more clearly represented in Table 11.5, which contrasts freshwater species (higher fads2/elovl5 activity and omega-3 retention) with marine species (lower conversion efficiency).

6.2 Lipid metabolism, gene regulation

Beyond digestion, replacement of FO with CO influences lipid metabolic pathways, gene expression, and tissue lipid composition. In Atlantic salmon, dietary CO inclusion has been associated with hepatic upregulation of desaturase (fads2) and elongase (elovl5) gene expression (Betancor et al. 2016; Xue et al. 2015), reflecting an adaptive response to maintain LC-PUFA homeostasis. However, these adjustments do not fully offset the reduced EPA and DHA deposition in tissues associated with the reduced dietary supply of these fatty acids (Hixson et al. 2014a).

In addition, dietary CO has been shown to modify the fatty acid profile of tissue phospholipids in liver and muscle, rather than changing overall lipid class distribution. These shifts in phospholipid composition may influence membrane fluidity, oxidative balance, and energy metabolism, underscoring the need to align dietary formulations with species-specific metabolic capacity Betancor et al. (2021).

Taken together, these interspecies metabolic responses highlight a key point: while digestibility of CO is consistently high across species, the efficiency of ALA conversion and long-chain PUFA retention is highly variable. Table 11.5 provides a comparative overview, showing that freshwater species such as tilapia and carp retain more EPA and DHA due to higher desaturase and elongase activity, whereas marine species such as salmon and seabream exhibit lower conversion efficiency despite high digestibility.

6.3 Strategic formulation and nutritional implications

Feed formulation must take interspecies metabolic differences into account. In freshwater species with high desaturase and elongase activity (e.g., tilapia and carp), higher levels of CO are generally well utilized. In marine species, blending CO with DHA-rich oils or using transgenic sources may help maintain tissue EPA/DHA levels. These strategies complement the performance outcomes already discussed in Section 3 and the product quality outcomes covered in Section 5.

Advances in nutrigenomics and selective breeding offer further solutions. Selective breeding or genome editing of genes such as fads2, elovl2, and elovl5 may enhance LC-PUFA biosynthesis and improve the compatibility of sustainable oils like camelina in aquafeeds (Venegas-Calerón and Napier 2023). These approaches link nutritional biochemistry with future-oriented aquaculture sustainability.

7 Antinutritional considerations in camelina oil

While some antinutritional properties have been associated with whole camelina seed or the residual camelina meal after oil extraction, refined CO is essentially devoid of antinutritional compounds. The primary antinutritional factors typically associated with camelina seed such as glucosinolates, sinapine, phytic acid, trypsin inhibitor activity and condensed tannins are polar molecules which do not strongly associate with the non-polar CO (Pozzo et al. 2023). Antinutritional factors are therefore concentrated in the residual meal fraction after oil extraction (Zubr 2003; Eidhin et al. 2003; Pozzo et al. 2023; Zago et al. 2015).

TABLE 11.5 Digestibility and lipid metabolism response to camelina oil in different fish species

Fish species	Environment	Digestibility (%)	ALA to EPA conversion	EPA/DHA retention (tissue)	References
Rainbow Trout (*O. mykiss*)	Fresh Water	>90%	Moderate	Retention declines with CO diet	Fraser et al. 2017; Lu et al. 2020
Nile Tilapia (*O. niloticus*)	Fresh water	>90%	High	Better retention	Chen et al. 2018
Common Carp (*C. carpio*)	Fresh water	>90%	Moderate	Maintained at moderate CO	Tocher, 2010
Atlantic Salmon (*S. salar*)	Marine	>90%	Low	Limited, despite fads2↑	Hixson et al. 2014b; Xue et al. 2015
Gilthead Seabream (*S. aurata*)	Marine	Not clearly specified; difficult to determine due to study design	Low	Reduced retention (not composition); changes in phospholipid fatty acid profiles	Monroig et al. 2013 Ofori-Mensah et al. 2020

Note: Table values reflect both apparent lipid digestibility and subsequent tissue retention of LC-PUFAs where data are available. Note: ↑ = upregulation; CO = Camelina Oil; ALA = alpha-linolenic acid; EPA = Eicosapentaenoic acid; DHA = Docosahexaenoic acid.

Because CO is commonly evaluated alongside its co-product meal in aquafeed research, some studies report physiological effects that are attributable to meal inclusion rather than to the refined oil itself. While sinapine content is considered to be low in camelina meal, other antinutrients are present in variable concentrations according to genetics and growing conditions (Russo and Reggiani 2017; Pozzo et al. 2023; Jiang et al. 2016). This likely contributes to the large variation in responses reported in fish experiments with camelina meals, which have resulted in recommendations ranging from 4–20% of the feed (Mzengereza et al. 2022; Koskela et al. 2021; Wei et al. 2019; Ye et al. 2016). Excessive camelina meal inclusion results in growth impairment as well as diverse physiological impacts including intestinal alterations (Ye et al. 2016; Wei et al. 2019), impaired immune and growth regulation, digestive enzyme activity and altered antioxidant status (Mzengereza et al. 2021). This indicates diverse and variable antinutrients in camelina meal that require further species-specific elucidation, contrasting with the lack of negative impact associated with dietary CO inclusion (Table 11.2).

For the purposes of this chapter, which focuses specifically on CO, antinutritional constrains are considered negligible and are discussed only to highlight the distinction between oil and meal fractions.

8 Regulatory status and feed formulation challenges

The regulatory acceptance of CO for aquafeeds has progressed in recent years, driven by increased interest in sustainable feed alternatives and the growing body of research on camelina's nutritional and safety profiles. This section outlines the current status of regulatory approvals and highlights key challenges and opportunities in formulating camelina-based feeds for aquaculture.

8.1 Regulatory status and safety approvals

Regulatory agencies such as the European Food Safety Authority (EFSA) and the U.S. Food and Drug Administration (FDA) have issued guidance and approvals regarding the inclusion of CO in aquafeeds. EFSA has established safety thresholds for undesirable substances in animal feeds to ensure compliance with feed safety standards (EFSA, 2008). In parallel, the FDA has granted Generally Recognized As Safe (GRAS) status to CO for use in animal feeds, including aquaculture, under defined inclusion limits (FDA, 2016). These regulatory endorsements reflect institutional confidence in camelina's safety and open pathways for its broader adoption.

8.2 Formulation challenges and technological solutions

Despite regulatory approval, several nutritional and processing challenges must be addressed for optimal inclusion of CO in commercial aquafeeds. One primary limitation is its fatty acid profile CO is rich in alpha-linolenic acid (ALA; 18:3n-3) but contains negligible amounts of long-chain omega-3 fatty acids such as EPA and DHA. Marine fish, which have limited enzymatic capacity to elongate ALA into EPA and DHA, may experience reduced tissue LC-PUFA content when fed CO based diets. To overcome this, feed manufacturers may employ 'finishing diets' that include marine oils or algal-based EPA/DHA supplements to restore fillet nutritional quality prior to harvest (Betancor et al. 2021).

Another concern is oxidative stability. While more stable than FO, the high PUFA content in CO poses a risk of lipid peroxidation, leading to rancidity and reduced feed palatability. Further supplementing tocopherols or adding alternative natural antioxidants such as rosemary extract can further improve oxidative stability in storage, transport and feed manufacture as reviewed for PUFA-rich fish oils by Shahidi and Zhong (2010). More conventional synthetic antioxidants (e.g., BHT, BHA) are also widely used in the feed industry, through their long-term viability may be limited by regulatory restriction and consumer preference for clean label ingredients. Given that camelina is already relatively expensive compared with other oils, reliance on costly additive strategies could constrain its competitiveness. Gentle processing methods such as cold-pressing and low-temperature extrusion further help maintain the oil's bioactive and nutritional qualities.

Overall, CO offers considerable promise as a sustainable, cost-effective alternative to marine-derived oils in aquaculture. Its success depends on continued regulatory alignment, strategic feed formulation, and technological innovation. Addressing these factors will be key to unlocking CO's full potential and advancing sustainable aquafeed production.

9 Future perspectives and research gaps

CO has shown considerable promise as a sustainable alternative to fish oil in aquaculture. However, to fully unlock its potential, future research must take a systems-level approach, addressing species-specific nutritional responses, feed formulation innovations, and environmental scalability. This section outlines key priorities and interdisciplinary directions that will shape the next frontier of CO application in aquaculture.

9.1 Nutritional physiology and species-specific considerations

A critical area for future research lies in understanding interspecies variability in LC-PUFA biosynthesis. Freshwater species such as rainbow trout and carp exhibit some capacity to convert alpha-linolenic acid (ALA, 18:3n-3) to EPA and DHA, whereas most marine species lack adequate enzymatic activity in key pathways regulated by FADS and ELOVL gene families (Glencross et al. 2024; Monroig and Kabeya 2018). Genomic, transcriptomic, and enzymatic studies can guide selective breeding programs to enhance this capacity and optimize species-specific feed formulations.

Additionally, there is a need to investigate CO's effects during early developmental stages and broodstock nutrition. Parameters such as reproductive performance, embryonic development, larval health, and transgenerational epigenetic modifications require evaluation to determine life-stage-specific safety thresholds (Glencross et al. 2024; Izquierdo et al. 2001). Long-term and multigenerational studies are needed to ensure the safe integration of CO across the aquaculture production cycle.

9.2 Innovations in feed formulation and camelina crop engineering

Feed formulation strategies must evolve to overcome CO's limitations in LC-PUFA content. Blending CO with alternative lipid sources such as algal oil, insect lipids, or shrimp waste oil offers a promising avenue for balancing essential fatty acid profiles while reducing environmental impact (Azcuy et al. 2025; Glencross et al. 2024). Precision nutrition strategies that tailor lipid composition to species and developmental stages can further enhance feed efficiency and fish health.

Simultaneously, biotechnological advances in camelina crop improvement particularly transgenic and CRISPR-based modifications offer transformative potential. GM camelina lines engineered to produce EPA and DHA have already shown success (Betancor et al. 2016; Petrie et al. 2014). However, limitations with social license exist, as not all market accept GM ingredients in the aquaculture supply chain, with stricter resistance in regions such as Australia and Europe. Broader adoption will therefore depend not only on biosafety evaluations, and long-term feeding trials, but also on consumer acceptance, regulatory harmonization and clear sustainability narratives to support their commercial deployment.

9.3 Sustainability, scalability, and future research directions

Evaluating the broader environmental and economic sustainability of CO adoption requires robust life cycle assessments (LCA), techno-economic modeling, and regional feasibility analyses. Studies suggest camelina cultivation and oil extraction result in lower greenhouse gas emissions, land use, and water consumption compared to fish oil (IFFO 2025; Li and Mupondwa 2014).

Moreover, camelina's compatibility with regenerative agricultural systems and marginal land cultivation aligns with global goals for climate-adaptive food production. Future research should assess camelina's role in circular bioeconomy frameworks and investigate market integration, supply chain resilience, and policy incentives to support its large-scale adoption.

10 Conclusion

CO represents a highly promising feed ingredient with wide-reaching implications for sustainable aquaculture. It offers a unique combination of nutritional, economic, and environmental benefits, particularly due to its high ALA content and natural antioxidants such as γ-tocopherol, which support fish health, preserve feed quality, and reduce reliance on marine-derived lipids. CO can be utilized without affecting growth, feed efficiency, or tissue quality in numerous freshwater species capable of converting ALA into EPA and DHA. In marine species, where this conversion is limited, strategies such as blending CO with other lipid sources or employing transgenic camelina enriched with EPA and DHA may provide additional value in some markets.

From a sustainability perspective, camelina demonstrates several agronomic advantages. It can be cultivated on marginal land with minimal input and integrates well into crop rotation systems. Its production requires less water and results in a lower environmental footprint compared to fish oil. Nonetheless, careful processing and regulatory oversight remain important to ensure safe and consistent use in commercial aquafeeds.

Future research should focus on species specific responses in LC-PUFA biosynthesis, continued optimization of feed formulation strategies, and biotechnological crop improvements. Regulatory frameworks and consumer acceptance will also play key roles in CO's broader adoption. Overall, CO offers a strategic and sustainable pathway toward resilient, and nutritionally adequate aquaculture systems.

References

Arshad M, Mohanty AK, Van Acker R, Riddle R, Todd J, Khalil H, Misra M (2022) Valorization of camelina oil to biobased materials and biofuels for new industrial uses: a review. RSC Adv 12 (42):27230-27245. https://doi.org/10.1039/d2ra03253h

Avadí A, Fréon P (2013) Life cycle assessment of fisheries: A review for fisheries scientists and managers. Fisheries Research 143:21-38. https://doi.org/10.1016/j.fishres.2013.01.006

Azcuy RL, Hernández AJ, Márquez L, Casaretto ME, Morales GA, Dantagnan P (2025) Effects of replacing fish oil with shrimp waste, canola, and camelina oils on growth and fillet lipid quality in juvenile rainbow trout fed low-fishmeal diets. Aquaculture Reports 43:102956. https://doi.org/10.1016/j.aqrep.2025.102956

Bell MV, Tocher DR (2009) Biosynthesis of polyunsaturated fatty acids in aquatic ecosystems: general pathways and new directions. In: Lipids in Aquatic Ecosystems. pp 211-236. https://doi.org/10.1007/978-0-387-89366-2_9

Berti M, Gesch R, Eynck C, Anderson J, Cermak S (2016) Camelina uses, genetics, genomics, production, and management. Industrial Crops and Products 94:690-710. https://doi.org/10.1016/j.indcrop.2016.09.034

Berti M, Wilckens R, Fischer S, Solis A, Johnson B (2011) Seeding date influence on camelina seed yield, yield components, and oil content in Chile. Industrial Crops and Products 34 (2):1358-1365. https://doi.org/10.1016/j.indcrop.2010.12.008

Betancor MB, MacEwan A, Sprague M, Gong X, Montero D, Han L, Napier JA, Norambuena F, Izquierdo M, Tocher DR (2021) Oil from transgenic *Camelina sativa* as a source of EPA and DHA in feed for European sea bass (*Dicentrarchus labrax* L.). Aquaculture 530:735759. https://doi.org/10.1016/j.aquaculture.2020.735759

Betancor MB, Sprague M, Montero D, Usher S, Sayanova O, Campbell PJ, Napier JA, Caballero MJ, Izquierdo M, Tocher DR (2016) Replacement of Marine Fish Oil with de novo Omega-3 Oils from Transgenic *Camelina sativa* in Feeds for Gilthead Sea Bream (*Sparus aurata* L.). Lipids 51 (10):1171-1191. https://doi.org/10.1007/s11745-016-4191-4

Betancor MB, Sprague M, Usher S, Sayanova O, Campbell PJ, Napier JA, Tocher DR (2015) A nutritionally-enhanced oil from transgenic *Camelina sativa* effectively replaces fish oil as a source of eicosapentaenoic acid for fish. Sci Rep 5:8104. https://doi.org/10.1038/srep08104

Booman M, Rise ML (2013) Evaluation of the impact of camelina oil-containing diets on the expression of genes involved in the innate anti-viral immune response in Atlantic cod. Fish & Shellfish Immunology 34 (6). https://doi.org/10.1016/j.fsi.2013.03.018

Booman M, Xu Q, Rise ML (2014) Evaluation of the impact of camelina oil-containing diets on the expression of genes involved in the innate anti-viral immune response in Atlantic cod (*Gadus morhua*). Fish Shellfish Immunol 41 (1):52-63. https://doi.org/10.1016/j.fsi.2014.05.017

Bouallegue K, Allaf T, Younes R, Téllez-Pérez C, Besombes C, Allaf K (2020) Pressure, temperature and processing time in enhancing Camelina sativa oil extraction by Instant Controlled Pressure-Drop (DIC) texturing pre-treatment. Grasas y Aceites 71:365. https://doi.org/10.3989/gya.0337191

Chen C, Guan W, Xie Q, Chen G, He X, Zhang H, Guo W, Chen F, Tan Y, Pan Q (2018) n-3 essential fatty acids in Nile tilapia, *Oreochromis niloticus*: Bioconverting LNA to DHA is relatively efficient and the LC-PUFA biosynthetic pathway is substrate limited in juvenile fish. Aquaculture 495:513-522. https://doi.org/10.1016/j.aquaculture.2018.06.023

Clemente C, Rossi A, Angelini LG, Villalba RG, Moreno DA, Tomás-Barberán FA, Tavarini S (2025) Effect of environmental conditions on seed yield and metabolomic profile of camelina (*Camelina sativa* (L.) Crantz) through on farm multilocation trials. Journal of Agriculture and Food Research 21:101814. https://doi.org/10.1016/j.jafr.2025.101814

Eidhin DN, Burke J, Lynch B, O'Beirne D (2003) Effects of Dietary Supplementation with Camelina Oil on Porcine Blood Lipids. Journal of Food Science 68 (2):671-679. https://doi.org/10.1111/j.1365-2621.2003.tb05730.x

Fraser JM, Collins SA, Chen Z, Tibbetts SM, Lall SP, Anderson DM (2017) Effects of dietary Camelina sativa products on digestible nutrient compositions for rainbow trout (*Oncorhynchus mykiss*). Aquaculture Nutrition 23 (5):973-982. https://doi.org/10.1111/anu.12465

Garcia-Mendoza MdP, Espinosa-Pardo FA, Savoire R, Harscoat-Schiavo C, Cansell M, Subra-Paternault P (2021) Improvement of the oxidative stability of camelina oil by enrichment with phospholipid-quercetin formulations. Food Chemistry 341:128234. https://doi.org/10.1016/j.foodchem.2020.128234

Glencross BD, Bachis E, Betancor MB, Calder P, Liland N, Newton R, Ruyter B (2024) Omega-3 Futures in Aquaculture: Exploring the Supply and Demands for Long-Chain Omega-3 Essential Fatty Acids by Aquaculture Species. Reviews in Fisheries Science & Aquaculture 33 (2):167-216. https://doi.org/10.1080/23308249.2024.2388563

Günç Ergönül P, Aksoylu Özbek Z (2020) Cold pressed camelina (*Camelina sativa* L.) seed oil. In: Cold Pressed Oils. pp 255-266. https://doi.org/10.1016/b978-0-12-818188-1.00021-9

Hixson SM, Parrish CC, Anderson DM (2014a) Changes in tissue lipid and fatty acid composition of farmed rainbow trout in response to dietary camelina oil as a replacement of fish oil. Lipids 49 (1):97-111

Hixson SM, Parrish CC, Anderson DM (2014b) Full substitution of fish oil with camelina (*Camelina sativa*) oil, with partial substitution of fish meal with camelina meal, in diets for farmed Atlantic salmon (*Salmo salar*) and its effect on tissue lipids and sensory quality. Food Chem 157:51-61. https://doi.org/10.1016/j.foodchem.2014.02.026

Hixson SM, Parrish CC, Anderson DM (2014c) Use of camelina oil to replace fish oil in diets for farmed salmonids and Atlantic cod. Aquaculture 431:44-52. https://doi.org/10.1016/j.aquaculture.2014.04.042

Hosseini Sanehkoori F, Pirdashti H, Bakhshandeh E (2021) Quantifying water stress and temperature effects on camelina (*Camelina sativa* L.) seed germination. Environmental and Experimental Botany 186:104450. https://doi.org/10.1016/j.envexpbot.2021.104450

Hussin SN, Azlan A, Khoo HE, Abdul Kadir NAA, Razman MR (2019) Comparison of fat composition and chemical properties of fat extracts between fish fillets of selected warm-water and cold-water fish. Bioscience Journal 35 (6):1968-1978. https://doi.org/10.14393/BJ-v35n6a2019-42782

Huyben D, Rimoldi S, Ceccotti C, Montero D, Betancor M, Iannini F, Terova G (2020) Effect of dietary oil from *Camelina sativa* on the growth performance, fillet fatty acid profile and gut microbiome of gilthead Sea bream (*Sparus aurata*). PeerJ 8:e10430. https://doi.org/10.7717/peerj.10430

IFFO (2025) The Marine Ingredients Organisation: Global fish oil production and trends. https://wwwiffocom/global-fishmeal-and-fish-oil-production-year-year

Innes JK, Calder PC (2018) Omega-6 fatty acids and inflammation. Prostaglandins, Leukotrienes and Essential Fatty Acids 132:41-48. https://doi.org/10.1016/j.plefa.2018.03.004

Izquierdo MS, Fernández-Palacios H, Tacon AGJ (2001) Effect of broodstock nutrition on reproductive performance of fish. Aquaculture 197 (1):25-42. https://doi.org/10.1016/S0044-8486(01)00581-6

Jafari S, Ebrahimi M, Assatarakul K, Jafari SM (2022) Plant Oils Rich in Essential Fatty Acids. In: Handbook of Food Bioactive Ingredients. pp 1-24. https://doi.org/10.1007/978-3-030-81404-5_21-1

Jiang Y, Li J, Caldwell CD (2016) Glucosinolate Content of Camelina Genotypes as Affected by Applied Nitrogen and Sulphur. Crop Science 56 (6):3250-3262. https://doi.org/10.2135/cropsci2016.01.0018

Koskela J, Leskinen H, Mattila P, Airaksinen S, Rinne M, Pihlava JM, Pihlanto A (2021) The effect of gradual addition of camelina seeds in the diet of rainbow trout (*Oncorhynchus mykiss*) on growth, feed efficiency and meat quality. Aquaculture Research 52 (10):4681-4692. https://doi.org/10.1111/are.15302

Kurasiak-Popowska D, and Stuper-Szablewska K (2020) The phytochemical quality of Camelina sativa seed and oil. Acta Agriculturae Scandinavica, Section B — Soil & Plant Science 70 (1):39-47. https://doi.org/10.1080/09064710.2019.1665706

Li X, Mupondwa E (2014) Life cycle assessment of camelina oil derived biodiesel and jet fuel in the Canadian Prairies. Sci Total Environ 481:17-26. https://doi.org/10.1016/j.scitotenv.2014.02.003

Lu J, Tibbetts SM, Lall SP, Anderson DM (2020) Use of dietary oil, solvent-extracted meal and protein concentrate from *Camelina sativa* for rainbow trout, *Oncorhynchus mykiss*, at the early fry stage. Aquaculture 524. https://doi.org/10.1016/j.aquaculture.2020.735252

Moldal T, Løkka G, Wiik-Nielsen J, Austbø L, Torstensen BE, Rosenlund G, Dale OB, Kaldhusdal M, Koppang EO (2014) Substitution of dietary fish oil with plant oils is associated with shortened mid intestinal folds in Atlantic salmon (Salmo salar). BMC veterinary research 10 (1):60

Monroig Ó, Kabeya N (2018) Desaturases and elongases involved in polyunsaturated fatty acid biosynthesis in aquatic invertebrates: a comprehensive review. Fisheries Science 84 (6):911-928. https://doi.org/10.1007/s12562-018-1254-x

Monroig O, Tocher DR, Navarro JC (2013) Biosynthesis of polyunsaturated fatty acids in marine invertebrates: recent advances in molecular mechanisms. Mar Drugs 11 (10):3998-4018. https://doi.org/10.3390/md11103998

Morais S, Edvardsen RB, Tocher DR, Bell JG (2012) Transcriptomic analyses of intestinal gene expression of juvenile Atlantic cod (*Gadus morhua*) fed diets with Camelina oil as replacement for fish oil. Comp Biochem Physiol B Biochem Mol Biol 161 (3):283-293. https://doi.org/10.1016/j.cbpb.2011.12.004

Mupondwa E, Li X, Tabil L, Falk K, Gugel R (2016) Technoeconomic analysis of camelina oil extraction as feedstock for biojet fuel in the Canadian Prairies. Biomass and Bioenergy 95:221-234. https://doi.org/10.1016/j.biombioe.2016.10.014

Mzengereza K, Ishikawa M, Koshio S, Shadrack RS, Zhang Y, Dossou S, Kotani T, Shahin SA, Zaineldin AI, Waqalevu V, Dawood MAO, Hassan AM, Al-Sharif MM, El Basuini MF (2022) Responses of growth, blood health, pro-inflammatory cytokines genes, intestine and liver histology in Red Seabream (*Pagrus major*) to camelina meal. Aquaculture Reports 24. https://doi.org/10.1016/j.aqrep.2022.101175

Mzengereza K, Ishikawa M, Koshio S, Yokoyama S, Yukun Z, Shadrack RS, Seo S, Kotani T, Dossou S, Basuini MFE, Dawood MAO (2021) Growth Performance, Growth-Related Genes, Digestibility, Digestive Enzyme Activity, Immune and Stress Responses of de novo Camelina Meal in Diets of Red Seabream (*Pagrus major*). Animals 11 (11):3118

Ofori-Mensah S, Yıldız M, Arslan M, Eldem V (2020) Fish oil replacement with different vegetable oils in gilthead seabream, *Sparus aurata* diets: Effects on fatty acid metabolism based on whole-body fatty acid balance method and genes expression. Aquaculture 529. https://doi.org/10.1016/j.aquaculture.2020.735609

Petrie JR, Shrestha P, Belide S, Kennedy Y, Lester G, Liu Q, Divi UK, Mulder RJ, Mansour MP, Nichols PD, Singh SP (2014) Metabolic Engineering *Camelina sativa* with Fish Oil-Like Levels of DHA. PLOS ONE 9 (1):e85061. https://doi.org/10.1371/journal.pone.0085061

Pozzo S, Piergiovanni AR, Ponzoni E, Brambilla IM, Galasso I (2023) Evaluation of nutritional and antinutritional compounds in a collection of *Camelina sativa* varieties. Journal of Crop Improvement 37 (6):934-952. https://doi.org/10.1080/15427528.2022.2163017

Righini D, Zanetti F, Martínez-Force E, Mandrioli M, Toschi TG, Monti A (2019) Shifting sowing of camelina from spring to autumn enhances the oil quality for bio-based applications in response to temperature and seed carbon stock. Industrial Crops and Products 137:66-73. https://doi.org/10.1016/j.indcrop.2019.05.009

Russo R, Reggiani R (2017) Glucosinolates and Sinapine in Camelina Meal. Food and Nutrition Sciences 08 (12):1063-1073. https://doi.org/10.4236/fns.2017.812078

Salem MOA, Tastan Y, Bilen S, Terzi E, Sonmez AY (2023) Dietary flaxseed (*Linum usitatissimum*) oil supplementation affects growth, oxidative stress, immune response, and diseases resistance in rainbow trout (*Oncorhynchus mykiss*). Fish Shellfish Immunol 138:108798. https://doi.org/10.1016/j.fsi.2023.108798

Shahidi F, Zhong Y (2010) Lipid oxidation and improving the oxidative stability. Chem Soc Rev 39 (11):4067-4079. https://doi.org/10.1039/b922183m

Simopoulos AP (2002) The importance of the ratio of omega-6/omega-3 essential fatty acids. Biomedicine & Pharmacotherapy 56 (8):365-379. https://doi.org/10.1016/S0753-3322(02)00253-6

Stamenković OS, Gautam K, Singla-Pareek SL, Dhankher OP, Djalović IG, Kostić MD, Mitrović PM, Pareek A, Veljković VB (2023) Biodiesel production from camelina oil: Present status and future perspectives. Food and Energy Security 12 (1):e340. https://doi.org/10.1002/fes3.340

Sydor M, Kurasiak-Popowska D, Stuper-Szablewska K, Rogoziński T (2022) *Camelina sativa*. Status quo and future perspectives. Industrial Crops and Products 187. https://doi.org/10.1016/j.indcrop.2022.115531

Tocher DR (2010) Fatty acid requirements in ontogeny of marine and freshwater fish. Aquaculture Research 41 (5):717-732. https://doi.org/10.1111/j.1365-2109.2008.02150.x

Tocher DR, Sprague M, Han L, Sayanova O, Norambuena F, Napier JA, Betancor MB (2024) Inclusion of oil from transgenic *Camelina sativa* in feed effectively supplies EPA and DHA to Atlantic salmon (*Salmo salar*) grown to market size in seawater pens. Food Chem 456:139414. https://doi.org/10.1016/j.foodchem.2024.139414

Toyes-Vargas EA, Parrish CC, Viana MT, Carreón-Palau L, Magallón-Servín P, Magallón-Barajas FJ (2020) Corrigendum to "Replacement of fish oil with camelina (*Camelina sativa*) oil in diets for juvenile tilapia (var. GIFT *Oreochromis niloticus*) and its effect on growth, feed utilization and muscle lipid composition" [Aquaculture, Volume 523, 30 June 2020, 735177]. Aquaculture 522. https://doi.org/10.1016/j.aquaculture.2020.735235

Venegas-Calerón M, Napier JA (2023) New alternative sources of omega-3 fish oil. Adv Food Nutr Res 105:343-398. https://doi.org/10.1016/bs.afnr.2023.01.001

Waraich EA, Ahmed Z, Ahmad R, Ashraf MY, Saifullah, Naeem SMS, Rengel Z (2013) *Camelina sativa*, a climate proof crop, has high nutritive value and multiple-uses: a review. Australian Journal of Crop Science 7 (10):1551-1559

Wei M, Anderson DM, Zhang Z, Colombo SM (2019) High-oil residue camelina meal, a viable source of protein at low levels in diets for juvenile salmonids. Aquaculture Nutrition 26 (2):558-567. https://doi.org/10.1111/anu.13017

Xue X, Hixson SM, Hori TS, Booman M, Parrish CC, Anderson DM, Rise ML (2015) Atlantic salmon (*Salmo salar*) liver transcriptome response to diets containing Camelina sativa products. Comp Biochem Physiol Part D Genomics Proteomics 14:1-15. https://doi.org/10.1016/j.cbd.2015.01.005

Ye CL, Anderson DM, Lall SP (2016) The effects of camelina oil and solvent extracted camelina meal on the growth, carcass composition and hindgut histology of Atlantic salmon (*Salmo salar*) parr in freshwater. Aquaculture 450:397-404. https://doi.org/10.1016/j.aquaculture.2015.08.019

Zago E, Lecomte J, Barouh N, Aouf C, Carré P, Fine F, Villeneuve P (2015) Influence of rapeseed meal treatments on its total phenolic content and composition in sinapine, sinapic acid and canolol. Industrial Crops and Products 76:1061-1070. https://doi.org/10.1016/j.indcrop.2015.08.022

Zubr J (2003) Qualitative variation of *Camelina sativa* seed from different locations. Industrial Crops and Products 17 (3):161-169. https://doi.org/10.1016/s0926-6690(02)00091-2

CHAPTER 12

Current status and future horizons in sustainable aquafeed certification and circular production practices

*M.D. Cyrus, R.C. Neil, K. Condon and L. Nankervis**

Centre for Sustainable Tropical Fisheries and Aquaculture (CSTFA), College of Science and Engineering, James Cook University, Australia
**Correspondence: Leo.nankervis@jcu.edu.au*

Abstract

Progress in the certification sector has not kept pace with the speed of innovation in alternative aquafeed components, particularly when incorporating the impact of including agricultural waste streams. Assessing the sustainability of the diverse range of agricultural waste into aquafeed ingredients has necessitated a review of assessment matrices to adequately capture the impact of aquafeeds across a large range of supply chains. Moving away from relatively simplified matrices relating to fish meal usage, current assessments in sustainable practices are moving towards Life Cycle Analysis. Life Cycle Analysis captures the transfer of nutrients and energy within a system to assess the long-term sustainability of the process. The application of the LCA tool however risks overestimating the impact of aquafeed ingredients by attributing the impact of the entire production line of agricultural products into life cycle footprint of aquafeed production. Ironically, the process of LCA inadvertently leads to poor assessment outcomes that contradict the benefits of circularity in efficiently reducing and utilising waste streams to productive food outputs or improved ecological performance. Consequently, the traditional LCA tool is not ideal, and further developments such as current efforts towards economic allocation of environmental burden of co-products show promise.

Although aquaculture has turned to LCA as a measuring tool, the potential to utilise the tool to develop rather than measure more sustainable practices has rarely been recognised. Developing circularity and integrated processes like IMTA present opportunities to utilise nutrient cycling to substantially improve production both through contributions in feed and waste usage as well energy retention. However, IMTA is a complex system, with a large technical barrier to entry and success. To date there has been little government support to motivate businesses to integrate their production systems. Certification could become the driver in the next evolutionary step towards sustainable practice in aquaculture. This chapter outlines the contribution and evolution of certification programs to drive aquafeed innovation and presents the prospect that a similar drive in innovation could be achieved with certification programs that recognise the sustainability in harnessing nutrient cycling via circularity in aquaculture.

1 Background and context

As global demand for seafood increases, aquaculture continues to emerge as a pivotal solution to meet the needs of the future while simultaneously relieving pressure on wild fisheries. However, the growth of the aquaculture sector must occur in tandem with environmental and social responsibility. Sustainability refers to the capacity of aquaculture systems to meet current production needs without compromising the health of natural ecosystems or the ability of future generations to meet their own needs (Garlock et al., 2024). Sustainable practice includes minimizing resource depletion, maintaining ecosystem services, and ensuring economic and social equity. Sustainable aquaculture certification programs have evolved as both the impetus and illustration of sustainable practices in the aquaculture sector. Aquafeeds are the predominant mechanism for nutrient flow in aquaculture, and as such carry the major burden of environmental impact and economic cost through the aquaculture value chain. Aquafeeds have therefore emerged as a focal point of measuring and certifying sustainable production in aquaculture.

2 Certification of sustainable aquafeeds

Certification programs to demonstrate sustainable aquaculture first emerged in the 1990s through the work of Organics International to create a framework for aquaculture to ensure seafood quality without creating ecological damage (Saha, 2022). As the aquaculture sector evolved, multiple entities have created sustainability certifications, and standards have become more complex to incorporate a range of matrices to demonstrate economic, environmental and socially acceptable practice.

Certification programs have delivered a mechanism to drive change towards sustainability within the aquaculture sector by delivering clearly defined criteria that promote business innovation whilst also providing a point-of-sale mechanism to enable the conscientious consumer to provide an economic return to certified businesses. The programs vary considerably and have evolved into complex standards that contain elements beyond core criteria of food safety, efficient use of resource, animal welfare and habitat destruction.

Interestingly, in contrast to terrestrial livestock production, certification programs are rarely mandated in the country of production but rather compliance in sustainable practice has been driven by access to import markets. With an estimated global value of $192 Billion USD large import consumers in the EU (32%) and USA (17%) have particularly driven the need for demonstrated sustainable practice in global seafood production (FAO, 2024). Critical to the integrity of these programs is the third-party certification system, where the quality assurance auditing is performed independently of both the certified business or certifying body. However, uptake of the programs has been limited with the combined two largest certification programs, ASC (Aquaculture Stewardship Council) and BAP (Best Aquaculture Practices), accounting for approximately 3% of global aquaculture production (Naylor et al., 2021).

To complement certification programs which target the farmer and supply chains, the sustainable certification environment also includes ratings programs aimed at guiding the consumer towards purchasing of sustainably rated options in seafood. Ratings programs do not require subscription by producers and with an estimated 53% of global aquaculture rated by US Seafood Watch (Naylor et al., 2021) fill a significant gap to drive sustainable practice through the ethical consumer. In addition to aquaculture production several certifications have emerged within the aquafeed ingredient sector. Aquaculture certification and ratings environment include aquafeed assessment programs at the level of:

a. aquaculture producers (certification)
b. aquafeed ingredient producers (certification)
c. retail consumer (ratings)

2.1 Scale of Impact of certification on global aquaculture practice

Naylor et al. (2021) estimated 30–50 voluntary certification, labelling and rating schemes have been introduced with the intent to address sustainable aquaculture production. Until recently, the programs acted without collaboration or harmonisation which undermined both the efficacy of the certification program and presented a disjointed message to farmers, stakeholders, and consumers.

Five certification or ratings programs including the Aquaculture Stewardship council (ASC), Fair Trade USA, Marine Stewardship Council (MSC), Monterey Bay Aquarium Seafood Watch Program and the Sustainable Fisheries Partnership Foundation (SFP) recently formed the Certification and Ratings Collaboration (C&RC). The C&RC website www.certificationandratings.org hosts the Sustainable Seafood Data Tool (SSDT) (https://certificationandratings.org/datatool/) that summarises assessment data from each of their member platforms. In addition, organisations within the Global Seafood Alliance's Best Aquaculture Practices (BAP) program and Qingdao Marine Conservation Society are captured and the SSDT provides a significant resource for consolidation of data on environmental performance and human rights abuses and risks in seafood production. Latest results from the SSDT indicate approximately 52.9% of worldwide seafood production has been certified or rates for environmentally sustainable practice but within the seafood sector, when aquatic plants are excluded only 5.05% (4,965,478 mt) of aquaculture is certified by a third party certification program and a further 35% (34,468,531 mt) rated as green choices by ratings organisations (33,052,917 mt) (Seafood Certification & Ratings Collaboration 2025; www.certificationandratings.org accessed 24/4/2026).

3 Evolution of aquafeed certification: adoption to alternative ingredients

In recognition of the major contribution played by aquafeed on the three pillars of sustainable practice (economic, social and environmental), contemporary certification programs have extended criteria from production systems to also evaluate feed mills and aquafeed ingredients. At its most basic, the mass balance between feed consumed and fish growth is represented by feed conversion ratio (FCR). While FCR describes the biological transformation of feed mass to fish growth, in practice there are losses of stock associated with escapes or mortality, and feed is not entirely eaten. FCR is therefore often specified as biological (bFCR) and economic FCR (eFCR) to better describe feeding efficiency in aquaculture practice.

FCR or bFCR = Feed intake (g) / animal growth (g)

eFCR = Feed offered (kg) / biomass gain (kg)

Where biomass gain is the entire weight gain of the population for the given period.

Early recognition of the dependence of aquaculture on limited resources were initially primed towards use of wild fishery catches to produce fish meal and fish oil (Tacon & Metian, 2008). Using average yield of fish meal (22.5%) and oil (5%) from pelagic fisheries and applying an economic feed conversion ratio (eFCR) a Fish in: Fish out ratio was used to describe the degree to which wild fish was converted into more valuable aquaculture products. Thus, fish meal and oil were combined in a single value, representing the dependence of the aquaculture activity on reduction fisheries; those that primarily catch fish for processing into fish meal and oil rather than for more direct uses. In this context, the terms 'wild', 'reduction' and 'forage' are used interchangeably.

$$FIFO = eFCR\ x\ \frac{\%\,fish\ meal\ inclusion + \%\,fish\ oil\ inclusion}{Fish\ meal\ yield\ (22.5\%) + Fish\ oil\ yield\ (5\%)}$$

FIFO values less than one describe net production of fish, meaning that the aquaculture practice produces more fish than consumed in its feed. Efficiency of resource use into aquafeeds has improved in the first two decades of this century, with eFCR reported to have reached an average of 0.71 across the global aquaculture industry, with an average FIFO of 0.27 (Glencross et al., 2024). These values are skewed by the predominance of cyprinids in global aquaculture volume (FAO, 2024), however all species groupings are now reported to have FIFO less than one, apart from turtles, frogs and eels (Glencross et al., 2024). While production of these net fish consumers generally converts lower value forage fish into higher value products, the concept of value can be very subjective. Sustainability credentials are challenged by ascribing a higher value based on financial gain compared to social benefits of directly consuming forage fish.

Further challenges with FIFO arise through the calculation of yield from reduction fisheries. Even though they are derived from the same fisheries, fish meal and oil differ in their importance across aquafeeds, with salmon farming notably using similar inclusion of fish meal and oil despite the disparity of yields from source fisheries (Aas et al., 2022). A single calculation may not adequately describe dependency on the limiting components of the fishery effectively, leading to the development of separate forage fish dependency ratio calculations for fish meal and fish oil (Tacon et al., 2022).

Over time, the proportion of fish meal and oil has increasingly come from trimmings, with FAO (2024) estimates at 34.5 and 53.4% for trimmings contribution to global fish meal and oil supply, respectively, in 2022. FIFO was originally used to highlight the dependence of aquaculture on reduction fisheries, and trimmings were excluded from the calculation to support the reuse of these by-products. The literal interpretation of FIFO to include all fish inputs compared to all fish outputs may lead to inconsistency in allocation of trimmings. Therefore the Forage Fish Dependency Ratio (FFDR) was developed to intentionally articulate a reduction fisheries resource utilisation efficiency, while also dividing the calculations for meal ($FFDR_{FM}$) and oil ($FFDR_{FO}$). The FFDR also facilitates specific targets to be set for oil and meal utilisation efficiencies depending on the state-of-the-art nutrition science for different industries.

$$FFDR_{FM} = \%\ \text{reduction fish meal} / \text{fish meal yield } (24\%) \times \text{eFCR}$$

$$FFDR_{FO} = \%\ \text{reduction fish oil} / \text{fish oil yield } (5\%) \times \text{eFCR}$$

Further refinements on apportioning dependency on wild fish stocks have been suggested based on nutrient dependency (Ytrestøyl et al., 2015) or cost (Kok et al., 2020). Both methods recognise allocation based on motivation and value of the ingredient. In particular, economic Fish In: Fish Out (eFIFO) has been applied to capture the value of the fish input compared to the aquaculture product. For example, the eFIFO recognises fish by-products, which would typically enter the waste disposal stream, have a higher value in aquafeed than direct consumption of the by-product (Glencross et al., 2024).

Both the Aquaculture Stewardship Council (ASC) and Best Aquaculture Practices (BAP) contain standards for Feed Mills in addition to efficiency matrices within species specific aquaculture standards (grow out and hatchery) that assess sustainability of feed usage. Many criteria within Feed Mill certifications of both BAP and ASC relate to food safety and hygiene standards. However, both programs also require demonstrated traceability of feed ingredients that ensures both product providence along with assurance that employment at the ingredient source is neither child labour nor modern slavery conditions.

BAP Farm standard (3.1) for certification requires efficient use of fishmeal and fish oil for all production systems. The standard applies to non-salmonid fish or salmonids farmed outside of traditional sea cage systems, with the latter production system covered separately under the BAP Salmon Farm Standard (Issue 3.0 due for implementation 2026). The farm standard specifies that the farm shall obtain feed from BAP feed mills or feed mills that comply with feed mill standards regarding species of fishery origins and responsibly sourced soy ingredients. Farms must comply with Fish In: Fish Out (FIFO) and Forage Fish Dependency Ratios (FFDR). FIFO limits have been set based on demonstrated efficient industry examples, differentiated between species where data is available; whiteleg (1.0) and black tiger shrimp (1.2), tilapia (0.5), pangasius (0.3) and channel catfish (0.3); rainbow trout (not raised in sea cages, 1.2) and Atlantic salmon (raised in recirculation systems only, 1.4). In addition, the BAP Salmon Farm Standards (V2.4, 2023) require FIFO of 1.5 or less for salmonid fish production in marine sea cage aquaculture. Species for which reliably efficient industry data is not yet available shall not exceed the FIFO value of 4 where byproducts are not included in the calculation or 5 if fish processing by products are included. In addition, the BAP Feed Mill Standard (Issue 3.3) also currently dictates that 75% of fish meal and oil must come from certified fisheries (e.g. Marin Trust, MSC) or recognised fishery improvement programs (FIPs), 50% of soy products and 100% of palm oil also must come from certified sources and feed additives must be listed in current EU Register of feed additives or equivalent register. This standard has a clear aim for all fish, soy and palm ingredients to be certified, with progressively staged targets to achieve 100% certification. The feed mill must also certify that the feed does not contain aquatic protein from the same species as being farmed, however hydrolysates <10,000 Daltons of the same species are permissible.

In 2023 the ASC established a Feed Standard (V1.1 effective 1st November 2025). Environmental risk management in the ASC Feed standard recognises the risk of harvesting raw materials contributing to marine resource depletion through overfishing, illegal-and-unregulated-fishing, deforestation and land conversion associated with agricultural practices to produce non-marine ingredients. Social risks with aquafeed production in the ASC standard also prevent the use of child and forced labour in agriculture and fisheries and the management of adverse impacts of feed mills on surrounding communities (smell and industrial discharges). ASC focusses on improvement, setting higher standards for implementation of certified ingredient supply chains over time. This is not dissimilar to BAP's approach, since BAP have recently increased from 50% of marine ingredient certification to 75% and have targets stretching to 100% by 2030. The staged approach enables the supply chain to adapt over time, driving improvement without leaving the supply chain with a material shortage that would impede certified aquaculture production.

3.1 Marine ingredients certification

The traceability and providence of a range of marine by-products used in aquafeed created logistical difficulties to demonstrate sustainable practice in aquafeed standards. Sustainable practice in marine sourced components can occur through the Marine Stewardship Council (MSC) which certifies fisheries however the MSC program does not certify the supply chain of the ingredients beyond the harvest. Consequently, supplier groups have created specialised Aquafeed ingredient certifications that support the use of marine by-products including programs devised by The International Fishmeal and Fish Oil Organisation (IFFO) (https://www.iffo.com) leading to the development of Marin Trust (https://www.marin-trust.com). Approximately 45% of marine ingredients are certified as Marin Trust compliant materials (https://www.marin-trust.com/impacts/global-impact). In addition, in areas where the certification standard cannot be met, Marin Trust has a Fisheries Improver Project that can be deployed towards reaching sustainable practice. The SSDT also contains a marine ingredients subsector which indicates only a small proportion of marine ingredients are ranked as certified (currently 6.5%; MSC or BAP certified), best choice (4.9%; Seafood Watch Green) or good alternative choices (5.8% Seafood Watch Yellow) with respect to environmental assessment due to the low uptake of MSC and BAP certifications worldwide.

3.2 Alternative ingredients

Assessment and certification of the range of ingredients beyond marine sources in aquafeed is difficult to compile as each sector has varying degrees of maturity towards demonstrating sustainable practice. The US Soybean Meal Association refers to the third party certification of the soy sustainability assurance protocol (American Soybean Association (ASA), 2025). The degree of detail within certifications varies depending on the end user application of the ingredient. Similarly, several certification collaborations have been formed by stakeholders of the palm oil industry such as the EU palm oil alliance and Roundtable on Sustainable Palm Oil (RSPO) which contains organisations that have entered a voluntary agreement to advance and promote sustainable palm oil production (https://palmoilalliance.eu/).

The multitude of stakeholders within the aquafeed sector creates a complex network of data collection to meet the needs of certification of the aquaculture product. Global Good Agricultural Practices (Global G.A.P.) has recognised the challenges that alternative practice poses to certification and has adopted a life cycle analysis approach for assessment of sustainable practice.

3.3 Life cycle analysis

Life cycle analysis (LCA) recognises that however well intentioned, the adoption of ingredients derived from plants, single cell culture and animals proposed to replace fish meal and fish oil, can result in burden shifting rather than improved environmental/economic or social outcomes. For example, the assessed environmental benefit of the use of soybean meal to replace fish meal may be reduced if habitat destruction and use of water to grow the soy crop is considered in the assessment matrix over improvements to the FIFO ratio (Clawson et al., 2025).

Life Cycle Analysis (LCA) has been proposed as a robust matrix to ensure legitimate improved environmental outcomes is achieved with feed ingredient purchasing decisions (e.g. Maiolo et al., 2020). The LCA approach evaluates the impacts of aquafeed ingredients from "cradle to grave" which includes all stages of production of the ingredient from cropping to processing and distribution, consumption and waste disposal. In the case of many alternative ingredients, LCA is somewhat controversial as the approach creates a matrix that is in direct conflict with the intent of the LCA. In the case of agricultural or livestock by-product to create aquafeed ingredients, the assessment of environmental contributions in the production of the ingredient incorporates the entire product supply chain history rather than starting at the point of "by-product" generation. The extended supply chain history of circular products therefore expands the LCA footprint, making reuse of byproducts appear less environmentally sustainable than linear supply chain alternatives. In contrast, removing any allocation of by-products to LCA calculation impacts the scope of the analysis and ignores the increasing economic return that occurs as by-products transition to co-products. A promising approach to solve this issue comes in the allocation of LCA impact based on economics rather than mass. Where ingredients are traditionally assigned burden based on mass, environmental impacts are equally divided between primary and secondary products. However, in a similar manner to described for eFIFO above, economic allocation to LCA divides impact based on the drivers of production, and likely better represents practical environmental impacts. This approach has recently been applied to poultry by-products, demonstrating an order of magnitude difference in LCA measures for rendered animal materials, demonstrating lower impacts than fish meal and oil (Campos et al., 2020).

Feed clearly has the largest impact on sustainable production of animal products. Regardless of the approach taken to assess sustainability, the impact of feeds is largely dictated by the amount of the feed used per unit production of aquaculture product. The ingredients themselves have importance, however optimised sustainable production requires efficient use of nutrients as they are upscaled to more highly valued forms. Therefore efficient feed practices and optimising feeds towards species-specific nutrient requirements is a critical need for aquaculture sustainability. Similarly, retaining nutrients in the value chain and maximising production per nutrient input are further pillars of aquaculture sustainability.

3.4 Is circularity in nutrient cycling the next step toward sustainable aquaculture?

In aquaculture, circularity largely involves the reuse of nutrients and organic matter, valorisation of by-products, and development of feed and farming systems that align with natural cycles (Cunningham et al., 2022). When properly implemented, circular approaches can enhance sustainability by improving resource efficiency, reducing environmental impacts, and creating additional value streams, often in the form of alternative crops or co-products (Chary et al., 2024). When circular economy practices are embedded into operational models of well-managed aquaculture systems, the synergistic nature of sustainability can simultaneously improve environmental, economic, and social performance (Garlock et al., 2024). The ability to implement circularity varies depending on the aquaculture production system and associated nutrient streams.

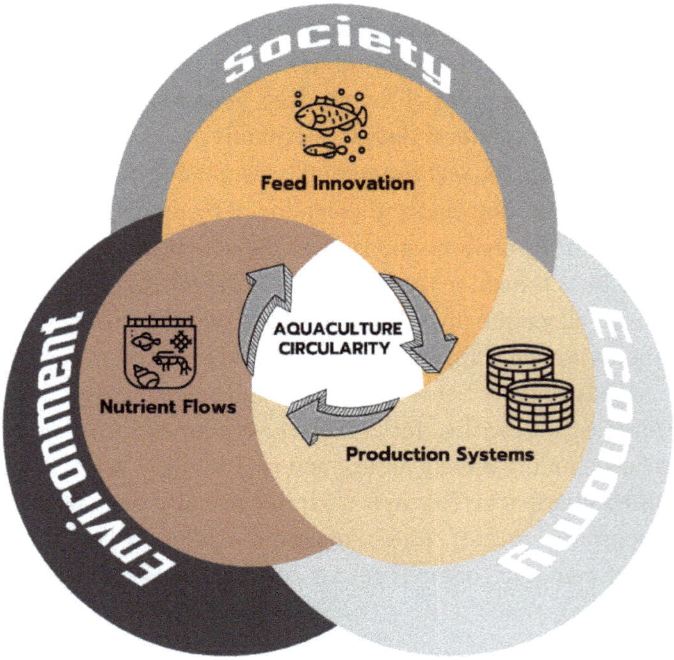

FIGURE 12.1 Interaction of the 3 pillars of sustainability (Environment, Society and Economy) with 3 key focus areas required for aquaculture circularity (feed innovation, production systems, and nutrient flows)

The concept of circularity builds on closing resource loops, transforming linear production models that follow a 'take-make-dispose' trajectory into regenerative systems where waste is minimized, and outputs are reused. The concept of circularity is relevant to aquafeed because circularity models involve the conversion of waste into aquafeed ingredients however, circularity potential can be added to both ends of the production chain to optimise the retention and value of nutrients.

4 Aquaculture production systems and nutrient flow

Aquaculture production systems vary widely in complexity, intensity, and environmental footprint. These systems generally fall into categories such as extensive pond systems, open-water cages, flow-through raceways, and Recirculating Aquaculture Systems (RAS), each interacting with their environment differently and presenting trade-offs in energy, land, and water use. However, in general standard monoculture systems retain less than 45% of nitrogen and 40% of phosphorus from feed inputs in the harvested biomass (Boyd, 2003; Dalbem Barbosa et al., 2024; Summerfelt & Penne, 2007), with the remainder entering the receiving environment if not managed adequately.

Extensive pond systems, which often rely on natural productivity (e.g., plankton, detritus), typically require little or no commercial feed. However, they can still generate nutrient-rich sediments and discharges that may affect nearby ecosystems, particularly when not integrated with crop or livestock production. Feed addition and energy use increase with the shift from extensive to semi-intensive farming practices, broadening the impact due to a more complex supply chain. However, efficiency through increased management and increased yield in semi-intensive systems can offset additional feed burden when expressed per unit production (Wilfart et al., 2013). Open cage systems typically have low infrastructure and energy needs but release nutrients and organic matter into surrounding waters, contributing to local eutrophication and benthic degradation if management or the farm environment are not conducive to nutrient dispersion. RAS, by contrast, excel in waste capture and water reuse (recycling up to 99% of water), but their high energy demands (especially for pumping, filtration, temperature control) can elevate greenhouse gas emissions particularly when powered by non-renewable energy. In general, though, intensification improves both bFCR and eFCR, outweighing other impacts on LCA metrics and improving sustainability measures (Ghamkhar et al., 2021).

All these systems, do however, ultimately result in the release of nutrient waste in the form of uneaten feed, faeces, and metabolic by-products into the surrounding environment be it at varying levels. Improving nutrient retention is therefore consistently identified in LCAs as a critical pathway for reducing aquaculture's environmental footprint and improving sustainability (Bohnes et al., 2019). Traditionally viewed as a liability, nutrient waste is increasingly being reconsidered as a potential resource within Circular Economy frameworks. One approach that aims to improve aquaculture circularity by addressing feed innovation, production systems, and nutrient flows simultaneously is Integrated Multi-Trophic Aquaculture (IMTA). IMTA is a sustainable method that enhances the production of aquatic organisms from multiple functional groups by leveraging nutrient flows within the same system. This is achieved through transforming waste into resources for reuse and recycling. Through this process, organic waste from higher-trophic species, such as fish, abalone or prawns, becomes a nutrient source for lower-trophic extractive species like bivalves (which filter particulates) and macroalgae (which absorb dissolved nutrients). Macroalgae such as *Ulva*, *Gracilaria*, and *Cladophora* are highly efficient at assimilating dissolved nitrogen and phosphorus from aquaculture effluents (Hughes et al., 2025; Morais et al., 2020; Moreira et al., 2022; Prazukin et al., 2024). Nederlof et al. (2022) estimate that well-balanced IMTA systems can retain up to 75% of nutrients from feed inputs, depending on species choice, water flow, and stocking ratios. When cultivated within IMTA systems, algae can transform otherwise polluting nutrients into valuable biomass often of improved quality compared to both wild and monocultures sources.

Importantly, macroalgae is currently establishing a competitive position within the global feed additive market, which was valued at $38.86 billion in 2022 and is expected to grow at a CAGR of 3.9% through to 2030 (World Bank, 2023). Trends indicate that the macroalgae based segment alone could reach a value of $1.12 billion by 2030. This growth is driven largely by increasing recognition of their nutritional advantages for animal health and the rising demand for sustainable, natural feed solutions. Numerous studies highlight the potential of macroalgae as supplementary and functional feed additives/ingredients in animal feed. Feeding fresh *Ulva* for instance can partially replace formulated feeds in abalone and sea urchin feeding regimes, enhancing growth and consumption while positively influencing health, physiology, pigmentation and gut microbiome to support improved metabolic and immune function (Brand, 2023; Cyrus et al., 2015). Similarly, *Gracilaria* and *Ulva* have shown promise in aquatic feeds at inclusion levels up to 20%, without compromising growth performance (Cyrus et al., 2013; Hughes et al., 2025; Wan et al., 2019). Wan et al. (2019) reviewed >50 finfish feeding trials with macroalgae and reported positive or no negative effects with inclusion levels between 5–30% of diet. In these trials macroalgae was often used to replace other dietary ingredients like of fishmeal and/or soy which could help to reduce feed costs and improve overall sustainability. Optimum inclusion was often species specific but inclusions of 15% and bellow did not impair growth in many formulations, except in those fed to carnivorous fish, which had lower limits of between 5–10% (Wan et al., 2019). Given the success of macroalgae inclusion of up to 20% as a supplementary feed ingredient in formulations to produce the 3rd largest aquaculture finfish species group worldwide, Cichlids (6,628,000 tonnes in 2022, FAO (2024)), the potential for its development in the aquafeed industry is very positive. El-Tawil (2010) demonstrated improved FCR's with up to 20% inclusion in diets fed to red tilapia (*Oreochromis* sp.). Carvalho et al. (2025) tested 0, 5, 10, 15% *Ulva lactuca* meal, inclusion produced in an integrated system, on juvenile tilapia (*Oreochromis niloticus*). Results showed no differences in growth performance or carcass composition among treatments. However, 5–10% inclusion improved antioxidant capacity and stress resilience relative to higher inclusions. *Cladophora* spp., contain up to 60% carbohydrate, 25% protein, and demonstrated digestibility and growth performance benefits in fish and livestock diets (Prazukin et al., 2024).

Macroalgae ingredients also offer bioactive benefits, such as antioxidant and antimicrobial properties, due to their high content of sulphated polysaccharides, phenolics, antimicrobial peptides (AMP's) and fatty acids (García-Vaquero, 2018; Morais et al., 2020; Silveira, 2022). Their incorporation supports organism health and resilience, which is particularly relevant in the context of reducing antibiotic use in production systems. AMPs in particular, have gained attention as a valuable feed additive due to their diverse biological functions and strong potential to limit the emergence of resistant microbes (Silveira, 2022). The extent to which macroalgae will be included in aquafeeds will however vary on a species by species bases and the purpose of these additions may therefore also be species dependent (eg: protein source vs antioxidant). This will be most evident when producing species who naturally consume algal or more generalist diets.

Even though the importance of algal feed additives is being recognised the development of seaweed-based feed additives is often constrained by several technical and logistical challenges. One of which is a reliable and traceable supply chain. To ensure access to large, continuous volumes of identified macroalgal products, downstream users have now started to investigate the potential of vertical integration in their production facilities, for which IMTA systems are ideally suited. When macroalgae are cultivated on farm (e.g., IMTA systems), their use as feed ingredients achieves dual circularity both from a nutrient recycling and feed generating perspective while simultaneously producing an improved quality product. For instance, *Ulva lactuca* grown in IMTA conditions has demonstrated protein contents as high as 33% dry weight and essential amino acid content suitable for inclusion in aquafeeds (García-Vaquero, 2018; Moreira et al., 2022). This localized on farm production and use also eliminate the need for transportation and reduces external resource usage. This type of site-specific circularity therefore creates value-added feedback loops. Moreover, localized production circumvents dependency on fluctuating global commodity prices. However, it is important to understand that these systems require careful species selection and nutrient balancing, as noted by Nederlof et al. (2022). The biomass production and nutrient uptake capacity of extractive species must be matched to the nutrient outputs of the fed species. Overstocking seaweed in nutrient-poor effluent can limit growth and reduce quality, while understocking may result in incomplete nutrient assimilation and increase environmental loading.

5 Could certification drive the next generation of sustainable aquaculture?

Despite growing awareness, Kusumowardani and Tjahjono (2020) found that the practical implementation of circular economy principles in aquafeed manufacturing remains low. As most companies do not yet incorporate circularity indicators into their design or procurement processes. A key barrier is the absence of standardized metrics and incentives. The benefits of circularity are not hypothetical and have been demonstrated through early adopters, for example those in the South African Abalone aquaculture sector. However, certification programs have not been established to accurately prescribe sustainable practice in circularity.

6 Case Study: commercial IMTA of Abalone & Ulva (South Africa)

South Africa presents one of the most established commercial applications of land-based Integrated Multi-Trophic Aquaculture (IMTA), where *Haliotis midae* (South African abalone) are integrated with the green seaweed *Ulva lacinulata*. This model highlights the core principles of circular economy in aquaculture by linking feed innovation, production system efficiency, and nutrient reuse in a closed-loop framework (Figure 12.3).

The IMTA approach employed on several South African farms involves the capture and treatment of effluent-rich water from abalone tanks using *Ulva* spp cultivated in adjacent paddle-raceway systems. The effluent, rich in dissolved nitrogen (ammonia, nitrate, nitrite), phosphate, and other organic matter, supports *Ulva* growth. This seaweed is then either reintroduced directly as a feed supplement or dried and processed into *Ulva* meal for formulated feeds. This process transforms a waste stream into a valuable resource, creating an internal nutrient loop that reduces environmental discharge and enhances input efficiency (Bolton, 2006; Bolton et al., 2016; Checa et al., 2024; Macey et al., 2024; Robertson-Andersson et al., 2008). Inclusion of up to 6% IMTA *Ulva* meal can also effectively replace equivalent levels of fish and soy meal and has been shown to enhance growth rates and feed utilisation in sub-adult (65–75 g) *H. midae* (Falade, 2023).

The economic and ecological incentives for this model are considerable. Cultivating *Ulva* on-site using effluent nutrients reduces reliance on wild-harvested seaweeds (e.g., kelp) and formulated feeds, thereby decreasing dependence on fishmeal and other resource-intensive inputs. Checa et al. (2024) demonstrated this using data from Buffeljags Abalone Farm (Viking Aquaculture), which produces 450 tonnes of abalone and 600 tonnes of *Ulva lacinulata* annually. Feed regimes at this farm incorporating IMTA-produced *Ulva* showed improved feed conversion ratios (FCRs) of 4.01 compared to 6.16 in monoculture systems. Uneaten feed fractions de-

FIGURE 12.2 An example of commercially operating Integrated Multitrophic Aquaculture (IMTA) in South Africa on Buffeljags Abalone Farm, owned by Viking Aquaculture. (A) Aerial photograph showing the entire farm and modular units; (B) Photograph of the module Abalone-*Ulva* Units (Photos: ASTRAL Project https://www.astral-project.eu)

creased from 1.14% to 0.71%, and feed linearity defined as the proportion of ingredients not sourced from valorisation dropped from 100% in monoculture systems to 71.42% in the abalone-*Ulva* IMTA system. Feed cost savings from *Ulva* supplementation were estimated at approximately 20% (Pers. comm. to Mark Cyrus; Macey et al. (2024) Macey et al., 2025). Furthermore, producing supplementary feed inputs on-site supports localized production cycles and enhances resilience to price fluctuations and supply chain disruptions (Chary et al., 2024; Checa et al., 2024).

From a system design perspective, South African IMTA farms exhibit high water recirculation levels (approximately 50%), significantly reducing water use and effluent output (Checa et al., 2024; Neori et al., 2004). *Ulva* serves both as a biofilter and a co-product removing up to 85% of total ammonia nitrogen while generating biomass for feed and potential secondary markets (e.g., nutraceuticals, fertilizers). The modular structure of these systems allows for spatial separation of species while maintaining functional connectivity via shared water and nutrient flows. Increased on-site recirculation also reduces the IMTA system's energy demand compared to monoculture systems, from 10.35 kWh to 6.80 kWh per ton of biomass harvested (Checa et al., 2024).

The integration of *Ulva* into abalone farming demonstrates how Circular Economy principles can be effectively implemented into aquaculture enterprises. Thereby contributing to aquafeed production through integrating waste recovery, on-site resource production, and ecological design demonstrating that circular aquaculture can deliver both economic and environmental benefits when appropriately implemented.

Challenges to its adoption currently still include spatial design, market value of co-products, and lack of integrated licensing (e.g., seaweed and shellfish may require separate permits). Nonetheless, the growing market for seaweed (food, feed, bioplastics, carbon credits) is improving IMTA viability.

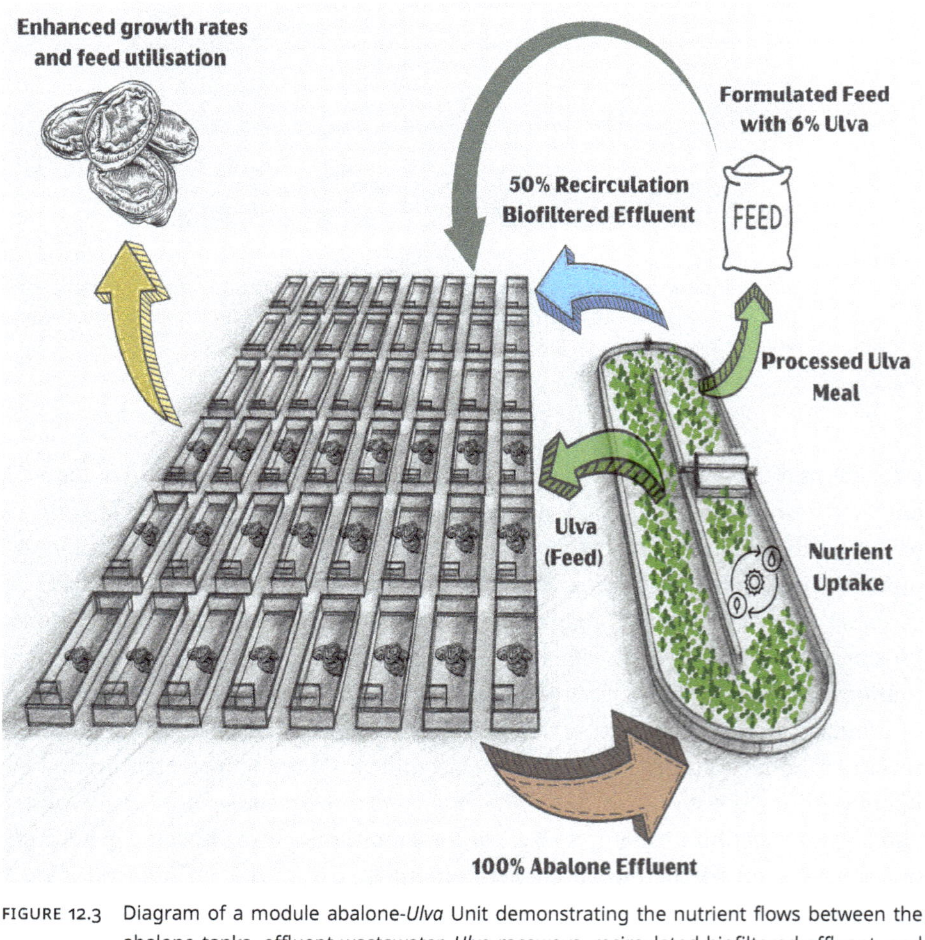

FIGURE 12.3 Diagram of a module abalone-*Ulva* Unit demonstrating the nutrient flows between the abalone tanks, effluent wastewater, *Ulva* raceways, recirculated biofiltered effluent, and co-products used for value adding in onsite feeding

7 Summary

The future of sustainable aquaculture lies in integrating innovations across feed development, production system design, and nutrient management. These focus areas, while distinct, are inherently interlinked and must be approached holistically to realize the full potential of circularity and sustainability.

Circularity in aquaculture particularly regarding aquafeeds is often framed as a system of closed loops or cycles. However, the progressive flow of materials, energy, and value through different stages of the aquaculture system is more accurately illustrated as a vertical spiral. Multiple modes of circularity can operate simultaneously at various levels, each contributing uniquely to overall sustainability. We therefore propose an adaptation of the circularity model; (SPINE™ Sustainable Production of Integrated Network Efficiency) which represents a dynamic framework for understanding and implementing circularity in aquaculture systems. SPINE captures how sustainability is advanced through interconnected pathways that integrate resources, emphasize practical application, and strengthen linkages across production scales from external resource loops to market-driven impacts. Unlike closed-loop models, the vertical spine concept reflects the layered, adaptive nature of circularity, enabling upward and downward feedback that accelerates transformation in aquafeeds, farm systems, and supply chains and provides the backbone to accelerate sustainable practice in aquaculture.

At the core of circular integration is the recognition that aquafeed formulation determines not just input efficiency, but also the scale and nature of waste outputs. High-quality, low-impact feeds reduce the demand on wild fish stocks, minimize land and water use, and influence the digestibility and nutrient retention of cultured species. Improved feed conversion ratios (FCRs), achievable through advanced nutrition, genetics and animal health, reduce the quantity of feed required per unit of biomass, thus decreasing environmental loading.

Nutrient flow and energy retention assessed in LCA ultimately tracks the fate of feed-derived nitrogen, phosphorus, and organic carbon thus becomes a focal point for both sustainability and circularity. The low nutrient retention efficiencies in conventional monoculture systems highlight the need for smarter waste management. Integrated Multi-Trophic Aquaculture (IMTA) provides a compelling model, where waste from one species becomes a resource for another. This not only mitigates environmental impacts but also creates diversified income streams and ecosystem resilience. The concept of circularity therefore amplifies sustainability by embedding feedback loops into the production cycle.

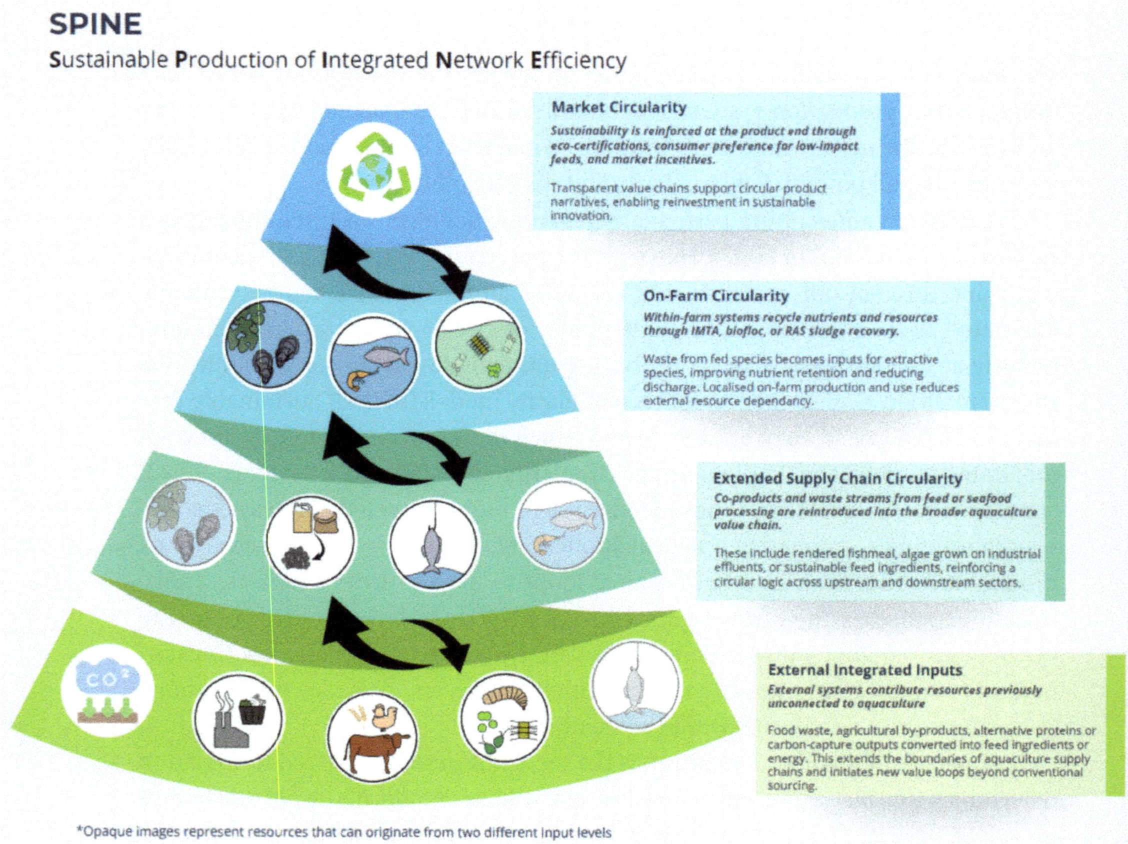

FIGURE 12.4 Visual representation of the SPINE (Sustainable Production of Integrated Network Efficiency) model, depicting how circularity in aquaculture and aquafeeds advances through an interconnected framework of integrated resources use, enabling upward and downward feedback, encompassing (1) External Integrated Inputs, (2) Extended Supply Chain Circularity, (3) On-Farm Circularity and (4) Market Circularity. *Opaque images represent resources that can originate from two different model levels.

References

Aas, T. S., Åsgård, T., & Ytrestøyl, T. (2022). Utilization of feed resources in the production of Atlantic salmon (*Salmo salar*) in Norway: An update for 2020. *Aquaculture Reports*, *26*, 101316. https://doi.org/10.1016/j.aqrep.2022.101316

American Soybean Association (ASA). (2025). *Soy Sustainability Assurance Protocol*. https://soygrowers.com/education-resources/grower-education/sustainability-assurance-protocol/

Bohnes, F. A., Hauschild, M. Z., Schlundt, J., & Laurent, A. (2019). Life cycle assessments of aquaculture systems: a critical review of reported findings with recommendations for policy and system development. *Reviews in Aquaculture*, *11*(4), 1061-1079. https://doi.org/10.1111/raq.12280

Bolton, J. J. (2006). Do we have the vision to integrate our marine aquaculture? *African Journal of Science*, *102*(11-12), 507-508.

Bolton, J. J., Cyrus, M. D., Brand, M. J., Joubert, M., & Macey, B. M. (2016). Why grow *Ulva*? Its potential role in the future of aquaculture. *Perspectives in Phycology*, *3*(3), 113-120. https://doi.org/10.1127/pip/2016/0058

Boyd, C. E. (2003). Bottom Soil and Water Quality Management in Shrimp Ponds. *Journal of Applied Aquaculture*, *13*(1-2), 11-33. https://doi.org/10.1300/J028v13n01_02

Brand, M. (2023). *Ulva as a functional feed: a practical investigation into the effects of Ulva lacinulata on the growth, consumption, health and gut microbiota of the farmed abalone Haliotis midae* [PhD thesis, University of Cape Town]. http://hdl.handle.net/11427/39286

Campos, I., Pinheiro Valente, L. M., Matos, E., Marques, P., & Freire, F. (2020). Life-cycle assessment of animal feed ingredients: Poultry fat, poultry by-product meal and hydrolyzed feather meal. *Journal of Cleaner Production*, *252*, 119845. https://doi.org/10.1016/j.jclepro.2019.119845

Carvalho, A., Müller, L., Rosas, V., Tesser, M. B., Ventura-Lima, J., Turan, G., Pias, M., & Poersch, L. H. (2025). Evaluation of the Inclusion of the Seaweed Ulva lactuca Produced in an Integrated System with Biofloc in the Diet of Juvenile Tilapia Oreochromis niloticus. *Applied Sciences*, *15*(12). https://doi.org/10.3390/app15126410

Chary, K., van Riel, A. J., Muscat, A., Wilfart, A., Harchaoui, S., Verdegem, M., Filgueira, R., Troell, M., Henriksson, P. J. G., de Boer, I. J. M., & Wiegertjes, G. F. (2024). Transforming sustainable aquaculture by applying circularity principles. *Reviews in Aquaculture*, *16*(2), 656-673. https://doi.org/10.1111/raq.12860

Checa, D., Macey, B. M., Bolton, J. J., Brink-Hull, M., O'Donohoe, P., Cardozo, A., Poersch, L. H., & Sánchez, I. (2024). Circularity Assessment in Aquaculture: The Case of Integrated Multi-Trophic Aquaculture (IMTA) Systems. *Fishes*, *9*(5). https://doi.org/10.3390/fishes9050165

Clawson, G., Blanchard, J. L., Cormery, M., Fulton, E. A., Halpern, B. S., Hamilton, H. A., O'Hara, C. C., & Cottrell, R. S. (2025). Continued transitions from fish meal and oil in aquafeeds require close attention to habitat impact trade-offs. *Cell Reports Sustainability*, 100457. https://doi.org/10.1016/j.crsus.2025.100457

Cunningham, R., Barclay, K., Jacobs, B., Sharpe, S., & McClean, N. (2022). *Circular economy opportunities for fisheries and aquaculture in Australia*. https://doi.org/10.13140/RG.2.2.25222.65606

Cyrus, M., Bolton, J., De Wet, L., & Macey, B. (2013). The development of a formulated feed containing *Ulva* (Chlorophyta) to promote rapid growth and enhanced production of high quality roe in the sea urchin *Tripneustes gratilla* (Linnaeus). *Aquaculture Research*, *45*(1), 159-176. https://doi.org/10.1111/j.1365-2109.2012.03219.x

Cyrus, M. D., Bolton, J. J., & Macey, B. M. (2015). The role of the green seaweed *Ulva* as a dietary supplement for full life-cycle grow-out of *Tripneustes gratilla*. *Aquaculture*, *446*, 187-197. https://doi.org/10.1016/j.aquaculture.2015.05.002

Dalbem Barbosa, A. P., Kosten, S., Muniz, C. C., & Oliveira-Junior, E. S. (2024). From Feed to Fish—Nutrients' Fate in Aquaculture Systems. *Applied Sciences*, *14*(14). https://doi.org/10.3390/app14146056

El-Tawil, N. E. (2010). Effects of Green Seaweeds (*Ulva* sp.) as Feed Supplements in Red Tilapia (*Oreochromis* Sp.) Diet on Growth Performance, Feed Utilization And Body Composition. *Journal of the Arabian Aquaculture Society*, *5*(2), 179-194.

Falade, A. E. (2023). *Optimising Integrated Multitrophic Aquaculture (IMTA) on a South African Abalone Farm* [PhD thesis, Rhodes University].

FAO. (2024). *The State of World Fisheries and Aquaculture 2024 – Blue Transformation in action*. https://doi.org/10.4060/cd0683en

García-Vaquero, M. (2018). Seaweed Proteins and Applications in Animal Feed. In *Novel Proteins for Food, Pharmaceuticals and Agriculture* (pp. 139-161). https://doi.org/10.1002/9781119385332.ch7

Garlock, T. M., Asche, F., Anderson, J. L., Eggert, H., Anderson, T. M., Che, B., Chávez, C. A., Chu, J., Chukwuone, N., Dey, M. M., Fitzsimmons, K., Flores, J., Guillen, J., Kumar, G., Liu, L., Llorente, I., Nguyen, L., Nielsen, R., Pincinato, R. B. M., . . . Tveteras, R. (2024). Environmental, economic, and social sustainability in aquaculture: the aquaculture performance indicators. *Nature Communications*, *15*(1), 5274. https://doi.org/10.1038/s41467-024-49556-8

Ghamkhar, R., Boxman, S. E., Main, K. L., Zhang, Q., Trotz, M. A., & Hicks, A. (2021). Life cycle assessment of aquaculture systems: Does burden shifting occur with an increase in production intensity? *Aquacultural Engineering*, *92*, 102130. https://doi.org/10.1016/j.aquaeng.2020.102130

Glencross, B. D., Bachis, E., Robb, D., & Newton, R. (2024). The Evolution of Sustainability Metrics for the Marine Ingredient Sector: Moving Towards Holistic Assessments of Aquaculture Feed. *Reviews in Fisheries Science & Aquaculture*, *32*(4), 545-561. https://doi.org/10.1080/23308249.2024.2337426

Hughes, A. D., Twigg, G. C., Msuya, F. E., Padmakumar, K. P., & Tocher, D. R. (2025). The use of macroalgae in feeds for finfish aquaculture [Review]. *Frontiers in Aquaculture*, *4*, 1570842. https://doi.org/10.3389/faquc.2025.1570842

Kok, B., Malcorps, W., Tlusty, M. F., Eltholth, M. M., Auchterlonie, N. A., Little, D. C., Harmsen, R., Newton, R. W., & Davies, S. J. (2020). Fish as feed: Using economic allocation to quantify the Fish In : Fish Out ratio of major fed aquaculture species. *Aquaculture*, *528*, 735474. https://doi.org/10.1016/j.aquaculture.2020.735474

Kusumowardani, N., & Tjahjono, B. (2020). Circular economy adoption in the aquafeed manufacturing industry. *Procedia CIRP*, *90*, 43-48. https://doi.org/10.1016/j.procir.2020.01.088

Macey, B. M., Brink-Hull, M., Bolton, J. J., Cyrus, M. D., De Vos, B., Loubser, N., Fourie, S., & Loubser, M. (2024). *Land based, pump ashore IMTA production manual*. EU H2020 project 'ASTRAL'. https://www.astral-project.eu/training-education-materials

Maiolo, S., Parisi, G., Biondi, N., Lunelli, F., Tibaldi, E., & Pastres, R. (2020). Fishmeal partial substitution within aquafeed formulations: life cycle assessment of four alternative protein sources. *The International Journal of Life Cycle Assessment*, *25*(8), 1455-1471. https://doi.org/10.1007/s11367-020-01759-z

Morais, T., Inácio, A., Coutinho, T., Ministro, M., Cotas, J., Pereira, L., & Bahcevandziev, K. (2020). Seaweed Potential in the Animal Feed: A Review. *Journal of Marine Science and Engineering*, *8*(8). https://doi.org/10.3390/jmse8080559

Moreira, A., Cruz, S., Marques, R., & Cartaxana, P. (2022). The underexplored potential of

CHAPTER 13

Are insects sensory-friendly as pet food ingredients? Approaches to increase acceptability

V. Pinilla, M. Sepúlveda, V. Villanueva, B. Padilla, S. Castro and C. Valenzuela

Facultad de Ciencias Veterinarias y Pecuarias, Universidad de Chile

Abstract

The growing global demand for food, driven by both population growth and an increase in pet ownership, poses significant challenges for sustainability. Pet food production, traditionally based on animal-derived ingredients, places significant pressure on natural resources and the environment. In this context, there is a need to develop alternative protein sources that are sustainable, efficient, and of high nutritional quality. Insects have been proposed by the Food and Agriculture Organization of the United Nations (FAO) as a viable option due to their high protein content (35–60% on a dry basis), balanced essential amino acid profile, high digestibility (77–98%), and complementary contribution of lipids, vitamins, minerals, and dietary fiber. In addition, they offer notable environmental advantages, such as greater feed conversion efficiency, lower water and land requirements, and the possibility of breeding them on substrates derived from organic waste, thus contributing to the circular economy. In the pet food industry, insects have become increasingly important, with the use of meal and fats obtained mainly from black soldier fly larvae, mealworms, crickets, and silkworms. Acceptance by pet owners has increased, especially when insect-derived products are processed and supported by veterinarians. Recent studies indicate that dogs and cats tolerate diets containing between 5% and 37.5% insect protein well, showing palatability comparable to traditional diets. While the chitin present in insects may slightly reduce protein digestibility, it could also provide benefits to intestinal health. Finally, emerging technologies, such as encapsulation, fermentation, hydrolysis, and protein concentration, seek to optimize the sensory, nutritional, and functional properties of these ingredients, promoting their integration into pet foods. Overall, insects represent a promising source of sustainable protein, with the potential to transform animal nutrition and reduce environmental impact.

green macroalgae in aquaculture. *Reviews in Aquaculture*, *14*(1), 5-26. https://doi.org/10.1111/raq.12580

Naylor, R. L., Hardy, R. W., Buschmann, A. H., Bush, S. R., Cao, L., Klinger, D. H., Little, D. C., Lubchenco, J., Shumway, S. E., & Troell, M. (2021). A 20-year retrospective review of global aquaculture. *Nature*, *591*(7851), 551-563. https://doi.org/10.1038/s41586-021-03308-6

Nederlof, M. A. J., Verdegem, M. C. J., Smaal, A. C., & Jansen, H. M. (2022). Nutrient retention efficiencies in integrated multi-trophic aquaculture. *Reviews in Aquaculture*, *14*(3), 1194-1212. https://doi.org/10.1111/raq.12645

Neori, A., Chopin, T., Troell, M., Buschmann, A. H., Kraemer, G. P., Halling, C., Shpigel, M., & Yarish, C. (2004). Integrated aquaculture: rationale, evolution and state of the art emphasizing seaweed biofiltration in modern mariculture. *Aquaculture*, *231*(1), 361-391. https://doi.org/10.1016/j.aquaculture.2003.11.015

Prazukin, A. V., Anufriieva, E. V., & Shadrin, N. V. (2024). Biomass of *Cladophora* (Chlorophyta, Cladophorales) is a promising resource for agriculture with high benefits for economics and the environment. *Aquaculture International*, *32*(3), 3637-3673. https://doi.org/10.1007/s10499-023-01342-x

Robertson-Andersson, D. V., Potgieter, M., Hansen, J., Bolton, J. J., Troell, M., Anderson, R. J., Halling, C., & Probyn, T. (2008). Integrated seaweed cultivation on an abalone farm in South Africa. *Journal of Applied Phycology*, *20*(5), 579-595. https://doi.org/10.1007/s10811-007-9239-7

Saha, C. K. (2022). Emergence and evolution of aquaculture sustainability certification schemes. *Marine Policy*, *143*, 105196. https://doi.org/10.1016/j.marpol.2022.105196

Silveira, R. F. (2022). *Antimicrobial peptides as additives to beef cattle nutrition* [PhD thesis, São Paulo State University]. http://hdl.handle.net/11449/217068

Summerfelt, R., & Penne, C. (2007). Nutrient Retention by Fish in a Multispecies Recirculating Aquaculture Facility. *International Journal of Recirculating Aquaculture*, *8*. https://doi.org/10.21061/ijra.v8i1.1417

Tacon, A. G. J., & Metian, M. (2008). Global overview on the use of fish meal and fish oil in industrially compounded aquafeeds: Trends and future prospects. *Aquaculture*, *285*(1-4), 146-158. https://doi.org/10.1016/j.aquaculture.2008.08.015

Tacon, A. G. J., Metian, M., & McNevin, A. A. (2022). Future Feeds: Suggested Guidelines for Sustainable Development. *Reviews in Fisheries Science & Aquaculture*, *30*(2), 135-142. https://doi.org/10.1080/23308249.2020.1860474

Wan, A. H. L., Davies, S. J., Soler-Vila, A., Fitzgerald, R., & Johnson, M. P. (2019). Macroalgae as a sustainable aquafeed ingredient. *Reviews in Aquaculture*, *11*(3), 458-492. https://doi.org/10.1111/raq.12241

Wilfart, A., Prudhomme, J., Blancheton, J. P., & Aubin, J. (2013). LCA and emergy accounting of aquaculture systems: towards ecological intensification. *J Environ Manage*, *121*, 96-109. https://doi.org/10.1016/j.jenvman.2013.01.031

World Bank. (2023). Global Seaweed New and Emerging Markets Report.

Ytrestøyl, T., Aas, T. S., & Åsgård, T. (2015). Utilisation of feed resources in production of Atlantic salmon (*Salmo salar*) in Norway. *Aquaculture*, *448*, 365-374. https://doi.org/10.1016/j.aquaculture.2015.06.023

Keywords

insect – feed – acceptability – dog – cat

1 Introduction

By 2050, one of the greatest challenges facing humanity will be feeding a projected population of 9.6 billion, which will require a 56% increase in global food production (FAO, 2009; Tomiyama et al., 2020). Simultaneously, pet ownership has risen sharply worldwide. According to the World Organization for Animal Health (OMSA), there are an estimated 900–970 million dogs and 600–750 million cats globally (Hosie and Hofmann-Lehmann, 2022; AVMA, 2022; GfK, 2016). The pet food industry has increasingly relied on animal-based ingredients, which are environmentally unsustainable (Alexander et al., 2020; Valdés et al., 2022). When the demand for such ingredients from both humans and companion animals is considered together, the environmental impact becomes even more concerning. In this context, there is an urgent need to develop novel food ingredients that are nutritionally equivalent to animal-derived products but significantly more sustainable (OECD/FAO, 2021).

Among the available alternatives are improved protein ingredients derived from plant-based raw materials such as soybeans, chickpeas, lentils, common beans, peas, fava beans, and others (Freeman et al., 2018; Øverland et al., 2009; Reilly et al., 2020). However, these plant-based proteins are generally of lower nutritional quality compared to animal-derived proteins, due to differences in essential amino acid profiles and reduced digestibility (Emkani et al., 2022; Rumpold and Schlüter, 2013). Other alternatives include protein sources derived from microalgae, seaweed, yeasts, fungi, microorganisms, and reprocessed animal or marine by-products (Brain-Isasi et al., 2021; Dale and Valenzuela, 2016; Fasolin et al., 2019; Luigia, 2019; Mosna et al., 2021). Despite their potential, these sources are limited by low and highly variable production yields, as well as high production costs. Processing steps such as fermentation (Thirunathan and Manickavasagan, 2019) or enzymatic hydrolysis (Marciniak et al., 2018; Tavano, 2013) applied to conventional plant proteins can modestly improve digestibility but rarely alter their essential amino-acid profiles. (Dodd et al., 2018; Kanakubo et al., 2015).

In the urgent search for alternative protein sources for both human and animal nutrition, the Food and Agriculture Organization of the United Nations (FAO) has proposed the use of insects (FAO, 2012). This recommendation is primarily based on insects' excellent nutritional properties and their low environmental footprint, which make them a highly sustainable food ingredient (Van Huis, 2013; Van Huis and Oonincx, 2017). Nutritionally, insects offer a high crude protein content, ranging from 35% to 60% on a dry matter basis (DM) (Melgar-Lalanne et al., 2019), along with a favorable essential amino acid profile (Cappelli et al., 2020; Kępińska-Pacelik and Biel, 2022) and high digestibility levels (77–98%), comparable to certain animal-derived proteins, like whey protein concentrate (Oghenesuvwe and Paul, 2019; Do et al., 2020; Oibiokpa et al., 2018). Additionally, they contain significant lipid levels (13–33% DM), and are rich in vitamins, minerals (Hawkey et al., 2021), and dietary fiber, up to 10%, mainly in the form of chitin (Hawkey et al., 2021). However, the nutritional composition of insects can vary considerably depending on species, developmental stage, feeding substrate, and processing methods (Van Huis and Oonincx, 2017).

From a sustainability perspective, insect farming presents substantial advantages over conventional livestock production. Insects have significantly higher reproductive rates and lower demands for feed, space, and water. For example, many species produce between 500 and 900 eggs and reach adulthood in just 25–45 days (Singh and Kumari, 2019). Their feed conversion efficiency is also notably high (Oonincx et al., 2010). Black soldier fly larvae (BSFL), for instance, have a feed conversion ratio (FCR) of 2, meaning they require only 2 kg of feed to produce 1 kg of edible biomass – compared to FCRs of 2.3 for chickens, 4.0 for pigs, and 8.8 for cattle (Oonincx et al., 2015). Moreover, insects can be raised in vertical systems using stacked plastic containers – so-called "mini-farms" – allowing for highly space-efficient production. To produce 1 kg of edible protein, mealworms (*Tenebrio molitor*) require only 10% of the land area needed to produce the same amount of beef (Baiano, 2020). Water usage is also dramatically reduced: producing 1 kg of protein from mealworms requires approximately 23,000 liters of water, compared to 34,000 liters for chicken, 57,000 liters for pork, and 112,000 liters for beef (Miglietta et al., 2015). Additionally, insects can be fed on organic waste streams that would otherwise contribute to environmental pollution, positioning them as a prime example of circular economy innovation (Ramos-Elorduy et al., 2002).

In recent decades, growing efforts have been directed toward incorporating insects as food ingredients for human consumption (Van Huis and Rumpold, 2023). However, this remains a significant challenge, as insects often elicit aversion among populations unfamiliar with entomophagy (Higa et al., 2021; La Barbera et al., 2018; Pozharlieva et al., 2023). A more promising niche lies in animal feed, where the inclusion of insects has yielded encouraging results, particularly in aquaculture (Gasco et al., 2016; Magalhães et al., 2017; Singh et al., 2024). Their application in livestock diets has also been explored; however, the high cost of insect-derived meals and oils remains a major constraint, limiting their widespread use as substitutes for conventional animal- or plant-based ingredients in species such as pigs, laying hens, and broiler chickens (Barragan-Fonseca et al., 2017; Crosbie et al., 2021; Gasco et al., 2023; Veldkamp and Vernooij, 2021). In contrast, the use of insects in pet food, particularly for dogs and cats is gaining attraction. This trend is driven by pet owners' increasing acceptance of novel ingredients (Baptista da Silva et al., 2024; Siddiqui et al., 2023; Villanueva et al., 2023), the favorable palatability observed especially in dogs (Kierończyk, 2018; Paßlack and Zentek, 2018; Wei et al., 2024), and growing evidence indicating that insect-based ingredients are not only safe for pets (Do et al., 2022; Freel et al., 2021; Jarett et al., 2019; Mouithys-Mickalad et al., 2020; Valdés et al., 2022), but also offer nutritional benefits, due to the high protein content of insects similar to soybean meal (40% to 45%) and meat meal (40% to 50%). Insect amino acids have high digestibility (76% to 98%) as they are similar to animal proteins and are rich in essential amino acids (Abd El-Wahab et al., 2021; Kierończyk et al., 2022; Valdés et al., 2022).

2 Insect meal and fat: methods of production and extraction

To produce insect meal, larvae are first separated from the rearing substrate using sieving or flotation systems. They are then washed and blanched at 90–100 °C for 30–60 seconds to inactivate enzymes and reduce microbial load. Following this, the larvae are dried using various technologies, such as solar drying or hot air drying. Once dried, oils or fats can be extracted and the dried larvae milled to produce a fine powder or meal, which can be further sieved to obtain a more homogeneous particle size distribution (Queiroz et al., 2023).

Several methods for extracting insect fat have been reported in the literature. One of the most commonly used techniques is mechanical pressing (Son et al., 2019), which may involve cold or hot pressing through a hydraulic or screw press. This method is simple, safe (as it avoids chemical use), cost-effective, and helps preserve thermolabile bioactive compounds in the case of cold pressing. Alternatively, fat can be extracted using organic solvents such as hexane, ethanol, or isopropanol. These solvents dissolve the lipids, which are then recovered through evaporation (Almeyda et al., 2022; Srisuksai et al., 2024). Solvent extraction offers higher efficiency and greater fat yield than mechanical methods; however, it poses several disadvantages, including the need for careful handling due to potential solvent toxicity, the high cost of specialized equipment, and environmental concerns related to solvent disposal (Kim et al., 2021; Bitwell et al., 2023).

A more advanced but less commonly used method is supercritical CO_2 extraction, which employs carbon dioxide under high pressure and temperature to extract lipids (Fornari et al., 2023; Cruz et al., 2023). This technique is environmentally friendly, leaves no toxic residues, and preserves heat-sensitive compounds. Nonetheless, it requires costly equipment and involves a high degree of technical complexity (Yıldırım et al, 2024; Nozari and Kander, 2025).

Insect meal and fat are the most commonly utilized insect-derived ingredients in pet food formulations. Insect meals are primarily incorporated as a protein source, replacing conventional animal- or plant-based protein ingredients, while insect fat is used mainly as an energy source (Valdés et al., 2022). However, fat can also provide nutritional benefits due to its fatty acid profile. Some insects are rich in lauric acid, others in oleic acid, whereas some species contain abundant omega-6 or omega-3 polyunsaturated fatty acids (PUFA), so insect oil is a potentially healthy and functional oil (Huang et al., 2025). These ingredients are predominantly obtained from four insect species: the black soldier fly (*Hermetia illucens*), the mealworm (*Tenebrio molitor*), the house cricket (*Acheta domesticus*) and the silkworm (*Bombyx mori*). Table 13.1 summarizes the proximate chemical analysis of these ingredients.

TABLE 13.1 Proximate chemical analysis of common insect species used in pet food

Parameters (%)	**BSFL**	**Mealworm**	**Cricket**	**Silkworm**
Moisture	5.9-9.7[1-5]	3.3-5.3[1-5]	6.9-10.7[1-5]	48,7-69,84[13-15]
Crude protein	35.0-42.4[2,4-9]	49.5-58.8[2,4-9]	42.0-70.0[2,4,6-9]	36,46-69,84[13-15]
Crude Fat	19.4-33.5[2,4-11]	25.2-38.1[2,4-11]	11.2-29.1[2,4-11]	8,09-30,85[13-15]
Crude fiber	5.3-9.7[4,6-10]	4.6-8.5[4,6-10]	3.7-9.3[4,6-10]	5,95-6,38[13,14]
Ash	7.1-21.5[4,6,7,9]	2.8-4.5[4,6,7,9]	4.7-6.6[4,6,7,9]	5,79-11,11[13-15]
Non-nitrogenous extract	15.8[4,9,10,12]	13.7-15.3[4,9,10,12]	5.6-22.0[4,9,10,12]	-
Amino acids				
Histidine	3.3-4.7[17]	1.47[16]	1.72[18]	1.36[19]
Isoleucine	1.6-2.6[17]	1.43[16]	2.9[18]	2.03[19]
Leucine	2.9-3.7[17]	2.75[16]	3.8[18]	3.42[19]
Lysine	4.1-4.7[17]	1.74[16]	3.22[18]	3.81[19]
Methionine	6.1-7.9[17]	0.81[16]	0.98[18]	-
Phenylalanine	1.9-4.6[17]	1.92[16]	2.38[18]	2.2[19]
Threonine	-	1.62[16]	1.65[18]	2.13[19]
Valine	1.2-9.3[17]	2.08[16]	4.5[18]	2.59[19]
Fatty acids				
C14:0	7.3[20]	4.37[20]	0.96[20]	0.16[20]
C16:0	15.74[20]	18.27[20]	24.55[20]	19.27[20]
C17:0	0.14[20]	0.18[20]	0.13[20]	-
C18:0	4.06[20]	4.21[20]	8.55[20]	10.62[20]
C20:0	0.13[20]	0.14[20]	0.40[20]	0.48[20]
C22:0	0.01[20]	0.02[20]	0.02[20]	0.65[20]

BSFL: black soldier fly larvae. References: Khanal et al.,2023[1], Medrano, 2019[2], Mitchaothai et al., 2022[3], Reátegui et al.,2020[4], Rocha et al., 2021[5], Kępińska-Pacelik et al., 2022[6], Kierończyk et al.,2018[7], Montowska et al., 2019[8], Rawski et al., 2020[9], Mustafa et al., 2025[10], Jajić, et al., 2020[11], Kulma et al., 2019[12], Rumpold and Schluter, 2013[13], Kim et al., 2016[14], Kuntadi et al., 2018[15]. Oliveira et al., 2024[16], Shumo et al., 2019[17], Farkas et al., 2025[18], Anootthato et al., 2018[19], Huang et al., 202520, [16,18]Expressed in (g/100g), [17]Expressed in (mg/g),[19]Expressed in (mg/100mg), [20]Expressed in (g/100 g fatty acid).

3 Pet owners' willingness to feed insects to their pets

Table 13.2 summarizes several studies that have investigated pet owners' willingness to feed their animals with insect-based ingredients or foods. The findings indicate a growing level of acceptance, although this remains influenced by various factors, including the species of the pet, the presentation and form of the product, the type of insect used, cultural attitudes, and the general lack of information regarding the benefits and safety of insect-based ingredients. Overall, attitudes tend to be favorable, particularly when insects are processed into meal or incorporated into pet food or treats in a way that renders them unrecognizable (Bae Seong-moon et al., 2020; Higa et al., 2021; Triggs et al., 2025; Villanueva et al., 2023). Acceptance is especially high among consumers with previous exposure to insect-based products or those without aversion to entomophagy. Among pet owners who expressed reluctance or refusal to use insect-based products, the most commonly cited reasons included feelings of disgust, limited knowledge about insect farming and food safety, and ethical concerns related to insect use as a form of animal exploitation (Villanueva et al., 2023).

Emotions play a significant role in the acceptance of insects as food ingredients. In a study by Fernandes da Silva et al. (2023), two distinct consumer groups were identified: (1) a flexible group, which associated positive emotions with feeding insects to their pets, and (2) an inflexible group, more resistant to the idea, though potentially open to reconsideration when provided with reliable and educational information. One of the key factors contributing to a more favorable perception of insect-based diets was veterinary endorsement. Many pet owners expressed greater confidence in these products when their use was recommended by a trusted animal health professional. The availability of clear, detailed information on the safety, nutritional value, and benefits of insect-based ingredients is therefore essential for building trust and promoting their wider adoption (Baptista da Silva et al., 2024; Higa et al., 2021).

In conclusion, the acceptance of insect-based pet foods among pet owners is generally favorable, particularly when their sustainability and nutritional benefits are taken into account (Baptista da Silva et al., 2024). To support wider adoption, future efforts should prioritize addressing consumer concerns, optimizing production and safety standards, and enhancing the palatability of insect-derived ingredients. Raising awareness and providing education, both for pet owners and veterinary professionals, regarding the nutritional value and safety of insect-based diets is essential to building trust and promoting their use as a viable alternative protein source in companion animal nutrition.

TABLE 13.2 Survey-based studies on pet owners' willingness to feed their animals insect-based ingredients or foods

Species	Insect	Format	N	Main Findings
Cats and dogs	*Tenebrio molitor*	Diet	25	Insect-based diets were well accepted by pets and their owners, and were more trusted when recommended by veterinarians[1]
Cats and dogs	Various insects	Diet, snacks	300	Over half were aware of insect-based products, and 48.5% showed interest in their nutritional and hypoallergenic benefits. Rejection was mainly due to disgust[2]
Dogs	Crickets, Black soldier fly larvae, Ants, Mealworms	Whole insects, insect meal, snacks	214	Respondents preferred dog food with cricket or BSFL meal over other insects and were more willing to accept insect meal or insect-based snacks than whole insects[3]
Cats and dogs	General insects	Diet	400	Between 41–54% of pet owners would consider buying insect-based pet food[4]
Cats and dogs	General insects	Diet	385	Emotions influence acceptance of insect consumption by pets. People who initially disagreed showed improved willingness after receiving information[5]
Cats	*Hermetia illucens, Tenebrio molitor, Acheta domesticus*	Whole insects, insect meal, and snacks	1684	Sixty-four percent of cat owners expressed willingness to feed their pets insect-based ingredients, showing a preference for insect-based snacks first, followed by meals, and lastly whole insects. The main reasons for rejection included feelings of disgust, lack of familiarity with insect-based foods, and attachment to traditional feeding practices[6]
Cats and dogs	General insects	Diet	435	Acceptance of insect-based food was favorable, highlighting sustainability and health; cat owners were more receptive[7]
Dogs	Crickets, Black soldier fly larvae, Mealworms	Diet	201	Participants were receptive to insect-based dog food, preferring crickets and mealworms[8]

N: number of people surveyed. BSFL: black soldier fly larvae. References: Leriche et al., 2016[1], Bae Seong-moon et al., 2020[2], Higa et al., 2021[3], Duijnisveld, 2022[4], Fernandes da Silva et al., 2023[5], Villanueva et al., 2023[6], Baptista da Silva et al., 2024[7], Triggs et al., 2025[8].

4 Acceptability of insects by pets

Appearance, aroma, texture, and flavor are sensory characteristics that are important in determining the acceptability of pet foods (Koppel, 2014). Palatability refers to the ability of a food, based on its sensory attributes, to be accepted and consumed by pets. It is a critical aspect in pet food formulation and is influenced by the sensory factors mentioned above (Le Guillas et al., 2024; Calderón et al., 2024). In pets such as dogs and cats, insect meal and fat are most commonly used as ingredients incorporated in varying amounts into the formulation of extruded feeds (kibbles) and solid treats, as illustrated in Figure 13.1.

In terms of acceptance, various studies show that insects are well accepted by dogs and cats, with inclusion rates ranging from 5% to 37.5% in the diet (Table 13.3). Regarding studies in dogs, Freel et al. (2021) observed that both BSFL meal and fat were highly accepted by dogs, suggesting good palatability of these ingredients when included at rates of 5–20% for meal and 1–5% for fat. The BSFL meal was included as an ingredient in a complete diet, the fat was used as external application. Kierończyk et al. (2018) evaluated various insects as olfactory attractants for dogs, and all proved to be appealing. Interestingly, differences in preference were observed between males and females, with males showing a greater preference for *Tenebrio molitor* and females favoring *Shelfordella lateralis*. They used dried insect species. Feng et al. (2020) analyzed the palatability and preference for food sprayed with *Tenebrio molitor* as an olfactory attractant compared to chicken liver hydrolysate, a widely used and preferred palatant for pets. The insect attractant showed lower preference than the hydrolysate, but palatability could potentially be enhanced by adding key volatile compounds. The key volatile compounds af-

FIGURE 13.1 Flow diagram illustrating the transformation of Black Soldier Fly larvae (BSFL) into meal and fat, and their incorporation into pet food products. Images provided by F4F Company (Chile, https://f4f.cl/) and Circular Pet (Chile, https://www.circular.pet)

fecting palatability in this study were 2-methyl-3-furyl disulfide, indole, methional, 2-(methyl thio) phenol, gamma-butyrolactone, furfuryl alcohol, acetic acid, and isovaleraldehyde, which could improve palatability of *L. edodes* and *T. molitor* attractant. Ahmed and Inal (2025) reported that diets containing *Locusta migratoria* meal for dogs have high acceptability, comparable to traditional diets formulated with chicken meal, the most commonly used animal-based ingredient in dog foods. Wei et al. (2024) incorporated BSFL at inclusion levels of 5–15% in dog diets, which positively influenced initial sniffing, first bite, and overall food intake, especially at the 15% inclusion level. Lee et al. (2021) showed that Silkworm pupae meal was highly palatable to dogs, with optimal acceptance observed at 20% inclusion and no adverse effects on fecal quality or blood parameters.

Regarding studies in cats, Paßlack and Zentek (2018) evaluated BSFL meal inclusion at levels of 22% and 35%. The results indicated good diet tolerance; however, three out of ten cats rejected the diet containing 35% insect meal. In the group fed with 22% insect meal, one cat vomited and subsequently refused the food, while two cats showed reduced intake. Do et al. (2022) reported that diets containing 5–10% BSFL were preferred over a poultry-based control diet, with higher consumption and being the first choice. Li et al. (2024) examined the palatability of a protein hydrolysate from *Schizochytrium* in cats, obtaining positive results in palatability tests, initial sniffing, first bite, and food intake with BSFL inclusion levels ranging from 5% to 15%. Bosch et al. (2024) incorporated 37.5% BSFL meal into extruded cat food, concluding that the diet was well accepted by cats.

5 Digestibility of insects by pets

Studies in dogs have shown that the apparent total tract digestibility of macronutrients and amino acids can be maintained similar to the digestibility levels of control diets, and not present significant differences, or have a higher digestibility, when fed with insect meal and fat, being a promising alternative source of dietary protein for this species (Lei et al., 2019; Kilburn et al., 2020; Freel et al., 2021; Penazzi et al., 2021; Lisenko et al., 2023).

In cats, studies have shown diverse results, with some studies indicating good digestibility (86–89%) (Hu et al., 2020; Bosch et al., 2024), while others consider it moderate (73– 77%) (Paßlack and Zentek, 2018). Therefore, more studies are needed to evaluate insect-based pet foods. Recent studies have evaluated the digestibility of cat diets containing BSFL reared on different organic substrates, demonstrating that the larval diet is an influencing factor (Kim et al., 2025; Jeon et al., 2025). Moreover, the nutritional value of BSFL as an animal feed ingredient is directly influenced by the substrates on which they are reared. A meta-analysis concluded that all types of substrates influence the performance and nutrients of BSFL. Animal feed substrate showed a more effective substrate among the substrates. The

TABLE 13.3 Studies conducted on preference, palatability, and acceptability tests of insect-based products in pets

Species	Insect	Format	Inclusion	Main Findings
Cats	BSFL	Meal	22 and 35%	Acceptability test. Good tolerance and general acceptance. However, a few cats had low food intake or rejected the diet[1]
Dogs	BSFL	Meal and fat	5-10% meal 2.5-5% fat	Acceptability test. Highly accepted by dogs[2]
Dogs	*Tenebrio molitor, Shelfordella lateralis, Hermetia illucens, Gryllodes sigillatus*	Whole	100%	Olfactory attractants test. All studied insects were attractive to dogs[3]
Dogs	*Tenebrio molitor*	Hydrolysate	Sprayed on pellets	Preference and palatability test. Lower preference than chicken liver[4]
Dogs	*Locusta migratoria*	Meal	140 g/kg	Palatability test. No significant differences in diet intake (chicken *versus* locust meal diets)[5]
Dogs	BSFL	Hydrolysate	5, 10, and 15%	Palatability test. Positively affected first sniff, first bite, and food intake, especially at 15%, suggesting good palatability[6]
Cats	BSFL	Meal, fat, whole insect	10% meal, 1.5% fat, 5% whole insect	Palatability test. Diets were palatable and did not negatively affect fecal or serum characteristics, though nutrient digestibility may be slightly lower in adult cats[7]
Cats	BSFL	Hydrolysate	5, 10, and 15%	Palatability test. First sniff, first bite, and food consumption significantly increased compared to control diet[8]
Cats	BSFL	Meal	37.5%	Intake test. Well accepted by cats[9]
Dogs	BSF	Hydrolysate	3%	Palatability and digestibility test. Diets supplemented with BSF hydrolysates are well-accepted by dogs, but effects depend on the enzymatic hydrolysis performed[10]

References: Paßlack and Zentek, 2018[1], Freel et al., 2021[2], Kieronczyk et al., 2018[3], Feng et al., 2020[4], Ahmed and Inal, 2025[5], Wei et al., 2024[6], Do et al., 2022[7], Li et al., 2024[8], Bosch et al., 2024[9], Leal et al., 2024[10]

presence of poly-unsaturated fatty acid are recommended as substrate considering their positive effect on high polyunsaturated fatty acid in BSFL (Fitriana et al., 2022). In another study, fat composition of the BSFL was found to be affected by the diet and the population, demonstrating the possibility to modify the lipid profile, adapting it to technological and food needs (Tognocchi et al., 2024).

The reported digestibility levels vary depending on the insect species, for example, in *Hermetia illucens*, lipid digestibility in cats values between 93% and 96% have been reported, however, apparent protein digestibility was between 73% and 77% (Paßlack and Zentek, 2018). For *Nauphoeta cinerea*, *Gromphadorhina portentosa* and *Zophobas morio* larvae the total tract apparent digestibility of DM in cats was 86.5%-88.1%, organic matter 88.9%-90.6%, fat 90.1%-92.3% and crude protein 86.3%-89.4% (Hu et al., 2020). For *Gryllodes sigillatus* the apparent digestibility of nutrients in dogs was: DM 84% to 89%; crude protein 88%; fat 95%-96%; fiber 44%-61% and gross energy 88%-92% (Kilburn et al., 2020).

A drawback in the digestibility of insects is the presence of chitin, a structural component present in insects, fungi, and crustaceans that may increase the content of indigestible carbohydrates (Jacuńska et al., 2024). Dogs, being primarily carnivores, lack the enzymes required for the breakdown of components such as chitin in their digestive tract. The active enzyme (chitinase) that breaks down chitin is primarily present in omnivorous animals. This indicates that dogs may encounter difficulties in digesting it effectively (Tabata et al., 2018; Gałęcki et al., 2025). High chitin levels may slightly reduce protein digestibility by encapsulating some nutrients, but lower levels may have positive effects, such as stimulating beneficial gut microbiota (acting as a fermentable fiber) and improving stool quality (Gałęcki et al., 2025). However, ingestion of insects by dogs is also not an uncommon event and chitin is usually well-tolerated (Tabata et al., 2018).

The same difficulties with chitin occur in cats; a reduction in nutrient digestibility has been described in cats fed BSFL meal, which could be partly related to the chitin content of said meal, which can negatively affect nutrient digestibility (Do et al., 2022).

6 Technologies to enhance the acceptance of insects as food ingredients

Positive sensory experiences play a fundamental role in the acceptance of new foods. In this regard, the sensory properties of insects are one of the main obstacles to their widespread adoption as food ingredients for human consumption (La Barbera et al., 2021). Their flavors and aromas are highly variable and are determined by the presence of pheromones (Skotnicka et al., 2021), such as the ketones 2-heptanone and 2-nonanone found in house cricket oil (Tzompa-Sosa et al., 2019), as well as terpenes like limonene (Lee et al., 2021; Ssepuuya et al., 2020; Tzompa-Sosa et al., 2019). Additionally, during the thermal processing of insects into flour, compounds are formed through Maillard reactions (Khatun et al., 2021; Kiatbenjakul et al., 2015) and lipid oxidation, including alcohols, aldehydes, ketones, esters, organic acids, and hydrocarbons (Lee et al., 2021a; Ribeiro et al., 2019; Ssepuuya et al., 2020).

The high lipid content of insects, particularly polyunsaturated fatty acids, makes them prone to oxidation, a process that is intensified by high temperatures. Lipid oxidation produces odors and flavors commonly described as "rancid and unpleasant" (Larouche et al., 2019; Singh et al., 2020), generating substances such as hexanal (herbal aroma), 1-octen-3-ol (fishy flavor), and 2-pentylfuran (buttery aroma) (Lee et al., 2021a; Ssepuuya et al., 2020). In general, insect flours have been characterized by strong and unpleasant odors, such as mushroom, algae, fish, and earthy notes (Dion-Poulin et al., 2021; Grossmann et al., 2021; Sogari et al., 2018). In a study by Sogari et al. (2018) with Italian consumers, texture and appearance were more decisive factors than flavor in the rejection of insect-based foods. Whole insects may have a crunchy texture (due to the exoskeleton, as in adult crickets) or a soft one (in larvae with lower chitin content). Insect flour has been negatively perceived due to its grainy and sandy texture, a result of its coarse particle size (1.2–1.4 mm) and heterogeneous particle distribution (de Oliveira et al., 2017; Purschke et al., 2018; Dion-Poulin et al., 2021; Sogari et al., 2018). Color is another limiting factor. It is generally determined by the presence of melanin (Kim et al., 2016), giving raw insects a gray/brown/black hue, which is perceived as unattractive. Upon thermal processing, these colors usually darken to brown shades (Kouřimská & Adámková, 2016; Ssepuuya et al., 2020). These color changes can result from: i) phenolic oxidation reactions (enzymatic browning) and protein denaturation during drying (Ravi et al., 2020); ii) iron–polyphenol complex formation during grinding (Janssen et al., 2017, 2019); iii) Maillard reactions (non-enzymatic browning) during thermal processing (David-Birman et al., 2018); and iv) pigment degradation or leaching (Seladi & Mabelebele, 2021; Azzollini et al., 2016).

Furthermore, incorporating insect-based ingredients into foods can impair their technological properties, such as solubility, water and fat absorption capacity, and foaming, emulsifying, and gelling properties (Biró et al., 2019; Kim et al., 2022; Haber et al., 2019). In this context, improving the techno-functional properties of insect ingredients is key to expanding their use in the food industry. Therefore, efforts to incorporate insects into the diet should focus on the development of new ingredients that are more appealing from a sensory perspective, while also exhibiting optimal technological performance and maintaining or enhancing their nutritional qualities (Tan et al., 2016).

In this context, over the past two decades, research has increasingly focused on making insect meals and oil more appealing from a sensory standpoint. As shown in Table 13.4, the technologies with the greatest potential to improve the acceptance of insect-derived ingredients are those that have successfully reduced unpleasant odors, enhanced flavor and texture, and even generated functional compounds. The most commonly used processes include: hydrolysis, encapsulation (including emulsification), fermentation, and protein concentration. All of these technologies have primarily been applied to increase consumer acceptance of insects for human consumption. Few studies have focused on pet food applications, but research on these new products in pets is expected to grow in the near future. Regarding pet nutrition, the main technology used is extrusion, a widely adopted process for manufacturing pet food. Insect meal and fat are mixed with other dietary ingredients during extrusion (Azzollini et al., 2018; Smetana et al., 2018), as shown in Table 13.3.

TABLE 13.4 Technologies applied to insect-based products to improve their sensory characteristics

Insect	Format	Technology	Product	Improvements
Housefly larvae	Flour	Encapsulation	Micropowder	Lighter color and only 4 volatiles found out of 22 present in the flour[1]
Tenebrio molitor, BSFL, Silkworm	Whole insect	Encapsulation and hydrolysis	Liquid palatant for animals	Improved preference and palatability compared to control insects without applied technology[2]
Tenebrio molitor	Protein	Hydrolysis	Peptides	Obtained active peptides as functional ingredients[3]
Gryllodes sigillatus	Whole insect	Hydrolysis	Hydrolysate	Increased bioactivity and reduced allergenicity[4]
Silkworm	Pupae	Protein concentration	Protein powder	High-protein powder with good solubility, foaming, and emulsifying capacity[5]
Tenebrio molitor	Paste	Fermentation	Fermented paste	Increased glutamic and aspartic acid content[6]
Tenebrio molitor	Dried larva	Extrusion	Extruded snack	Improved protein digestibility[7]
Alphitobius diaperinus	Protein concentration	Extrusion	Meat analogue	Fiber-like texture similar to meat in meat substitutes[8]
Housefly larvae	Meal	Encapsulation	Alginate beads ('fly caviar')	Encapsulation reduced volatile compounds[9]
Cricket and *Tenebrio molitor*	Protein powder	Hydrolysis	Protein hydrolysate	Improved flavoring potential of insect proteins[10]
Gryllus bimaculatus	Flour	Solvent extraction	Defatted powder	Increased protein concentration and reduced antinutritional compounds[11]
Acheta domesticus	Flour	Extrusion	Extruded pellet	Adding insect flour significantly reduced bulk density and durability of the pellets[12]
Tenebrio molitor	Oil	Oleogelation with waxes	Oleogel	Improved oxidative stability and rheological properties[13]

Sánchez et al., 2021[1]; Chen et al., 2022[2]; Rivero Pino et al., 2020[3]; Hall et al., 2018[4]; Yao y Dai 2019[5]; Borremans et al., 2019[6]; Azzollini et al., 2018[7]; Smetana et al., 2018[8]; Sanchez et al., 2022[9]; Grossman et al., 2021[10]; Damasceno et al., 2023[11]; Wójtowicz et al., 2023[12]; Kim and Oh 2022[13].

7 Future prospects

Despite advances in the use of insects as an alternative source of protein in pet food, there are still significant gaps in knowledge that need to be addressed. More studies are needed to evaluate the long-term effects of these proteins on health, digestibility, and nutrient bioavailability, as well as their impact on the composition and function of the gut microbiota. There is also an opportunity to innovate in the development of functional formulations, incorporating bioactive compounds derived from insects that can provide additional health benefits for dogs and cats. Another promising line of research is personalized nutrition for pets, aimed at adapting the composition of diets according to age, physiological condition, activity level, or presence of pathologies. In addition, the integration of emerging technologies, such as encapsulation, fermentation, and protein engineering could optimize the palatability, stability, and nutritional profile of products. These prospects pave the way for the development of more sustainable, functional foods tailored to the specific needs of pets, helping to transform the market and reduce the industry's environmental impact.

Note on the usage of AI tools

The authors of this chapter state that they used large language models (LLMs) strictly for English language correction and stylistic polishing of chapters, and not for content generation.

References

Abd El-Wahab, A., Meyer, L., Kölln, M., Chuppava, B., Wilke, V., Visscher, C. and Kamphues, J., 2021. Insect Larvae Meal (Hermetia illucens) as a Sustainable Protein Source of Canine Food and Its Impacts on Nutrient Digestibility and Fecal Quality. Animals, 11(9): 2525. https://doi.org/10.3390/ani11092525

Ahmed, I. and İnal, F., 2025. Replacement of chicken meal with full-fat locust meal in Golden Retriever dog food and its influence on palatability, digestibility, fecal quality, and immunity. Animal Feed Science and Technology, 320(9): 116191. https://doi.org/10.1016/j.anifeedsci.2024.116191

Alexander, P., Berri, A., Moran, D., Reay, D. and Rounsevell, M. D. A., 2020. The global environmental paw print of pet food. Global Environmental Change, 65(9): 102153. https://doi.org/10.1016/j.gloenvcha.2020.102153

Almeida, C., Murta, D., Nunes, R., Baby, A. R., Fernandes, Â., Barros, L., Rijo, P. and Rosado, C., 2022. Characterization of lipid extracts from the Hermetia illucens larvae and their bioactivities for potential use as pharmaceutical and cosmetic ingredients. Heliyon, 8(5): e09455. https://doi.org/10.1016/j.heliyon.2022.e09455

American Veterinary Medical Association (AVMA), 2022. 2022 AVMA pet ownership and demographics sourcebook. American Veterinary Medical Association. Available at: https://ebusiness.avma.org/ProductCatalog/product.aspx?ID=2050

Anootthato, S., Therdthai, N., & Ritthiruangdej, P. 2019. Characterization of protein hydrolysate from silkworm pupae (Bombyx mori). Journal of Food Processing and Preservation, 43(8): 1–15. https://doi.org/10.1111/jfpp.14021

Azzollini, D., Derossi, A. and Severini, C., 2016. Understanding the drying kinetic and hygroscopic behaviour of larvae of yellow mealworm (Tenebrio molitor) and the effects on their quality. Journal of Insects as Food and Feed, 2(4): 233–244. https://doi.org/10.3920/JIFF2016.0001

Bae, S.-M., Lee, S., Kim, J. and Hwang, Y., 2020. Analysis of Consumer Receptivity to Pet Food Containing Edible Insects in South Korea. Korean Journal of Applied Entomology, 59(2): 139–143. https://doi.org/10.5656/KSAE.2020.02.1.055

Baiano, A., 2020. Edible insects: An overview on nutritional characteristics, safety, farming, production technologies, regulatory framework, and socio-economic and ethical implications. Trends in Food Science & Technology, 100(4): 35–50. https://doi.org/10.1016/j.tifs.2020.03.040

Baptista da Silva, C., Santos Neto, E. L. dos, Broeckx, B. J. G., Ruiz-Suárez, N., Eeckhout, M. and Hesta, M., 2024. Assessing pet owners' willingness to embrace insects in pet food. Journal of Animal Physiology and Animal Nutrition, 108(6): 1833–1843. https://doi.org/10.1111/jpn.14019

Barragan-Fonseca, K. B., Dicke, M. and van Loon, J. J. A., 2017. Nutritional value of the black soldier fly (Hermetia illucens L.) and its suitability as animal feed – a review. Journal of Insects as Food and Feed, 3(2): 105–120. https://doi.org/10.3920/JIFF2016.0055

Biró, B., Fodor, R., Szedljak, I., Pásztor-Huszár, K. and Gere, A., 2019. Buckwheat-pasta enriched with silkworm powder: Technological analysis and sensory evaluation. LWT, 116(2): 108542. https://doi.org/10.1016/j.lwt.2019.108542

Bitwell, C., Indra, S. Sen, Luke, C., & Kakoma, M. K. 2023. A review of modern and conventional extraction techniques and their applications for extracting phytochemicals from plants. Scientific African, 19: e01585. https://doi.org/10.1016/j.sciaf.2023.e01585

Borremans, A., Sam, C., Dries, V., Ruben, S., Christel, V., Mik, V., Bart, L. and Leen, V., 2019. Comparison of Six Commercial Meat Starter Cultures for the Fermentation of Yellow Mealworm (Tenebrio molitor) Paste. Microorganisms, 7(11): 540. https://doi.org/10.3390/microorganisms7110540

Bosch, G., Loureiro, B. A., Schokker, D., Kar, S. K., Paul, A. and Sluczanowski, N., 2024. Black soldier fly larvae meal in an extruded food: effects on nutritional quality and health parameters in healthy adult cats. Journal of Insects as Food and Feed, 10(9): 1595–1606. https://doi.org/10.1163/23524588-00001093

Brain-Isasi, S., Carú, C. and Lienqueo, M. E., 2021. Valorization of the green seaweed Ulva rigida for production of fungal biomass protein using a hypercellulolytic terrestrial fungus. Algal Research, 59(11): 102457. https://doi.org/10.1016/j.algal.2021.102457

Calderón, N., White, B. L. and Seo, H., 2024. Measuring palatability of pet food products: Sensory components, evaluations, challenges, and opportunities. Journal of Food Science, 89(12): 8175–8196. https://doi.org/10.1111/1750-3841.17511

Cappelli, A., Cini, E., Lorini, C., Oliva, N. and Bonaccorsi, G., 2020. Insects as food: A review on risks assessments of Tenebrionidae and Gryllidae in relation to a first machines and plants development. Food Control, 108(12): 106877. https://doi.org/10.1016/j.foodcont.2019.106877

Chen, J.; Zhang, Z.; Liu, Y.; Zhang, Y., 2022. Preparation Method of Pet Food Palatability Enhancer Taking Insects as Main Materials, Palatability Enhancer and Application. CN PATENT CN115088786.

Crosbie, M., Zhu, C., Karrow, N. A. and Huber, L.-A., 2021. The effects of partially replacing animal protein sources with full fat black soldier fly larvae meal (Hermetia illucens) in nursery diets on growth performance, gut morphology, and immune response of pigs. Translational Animal Science, 5(2): 106877. https://doi.org/10.1093/tas/txab057

Cruz, V. A., Ferreira, N. J., Cornelio-Santiago, H. P., Santos, G. M. T. and Oliveira, A. L., 2023. Oil extraction from black soldier fly (Hermetia illucens L.) larvae meal by dynamic and intermittent processes of supercritical CO2 – Global yield, oil characterization, and solvent consumption. The Journal of Supercritical Fluids, 195: 105861. https://doi.org/10.1016/j.supflu.2023.105861

Dale, N. and Valenzuela, C., 2016. Nutritional properties of dried salmon silage for broiler feeding. Animal Science Journal, 87(6): 791–795. https://doi.org/10.1111/asj.12480

Damasceno, B. C., Nakajima, M., Taarji, N., Kobayashi, I., Ichikawa, S. and Neves, M. A., 2023. Improvements in Visual Aspects and Chemical, Techno-Functional and Rheological Characteristics of Cricket Powder (Gryllus bimaculatus) by Solvent Treatment for Food Utilization. Foods, 12(7): 1422. https://doi.org/10.3390/foods12071422

David-Birman, T., Raften, G. and Lesmes, U., 2018. Effects of thermal treatments on the colloidal properties, antioxidant capacity and in-vitro proteolytic degradation of cricket flour. Food Hydrocolloids, 79(7): 48–54. https://doi.org/10.1016/j.foodhyd.2017.11.044

de Oliveira, L. M., da Silva Lucas, A. J., Cadaval, C. L. and Mellado, M. S., 2017. Bread enriched with flour from cinereous cockroach (*Nauphoeta cinerea*). Innovative Food Science & Emerging Technologies, 44: 30–35. https://doi.org/10.1016/j.ifset.2017.08.015

Dion-Poulin, A., Turcotte, M., Lee-Blouin, S., Perreault, V., Provencher, V., Doyen, A. and Turgeon, S. L., 2021. Acceptability of insect ingredients by innovative student chefs: An exploratory study. International Journal of Gastronomy and Food Science, 24(7): 100362. https://doi.org/10.1016/j.ijgfs.2021.100362

Do, S., Koutsos, E. A., McComb, A., Phungviwatnikul, T., de Godoy, M. R. C. and Swanson, K. S., 2022. Palatability and apparent total tract macronutrient digestibility of retorted black soldier fly larvae-containing diets and their effects on the fecal characteristics of cats consuming them. Journal of Animal Science, 100(4): 100362. https://doi.org/10.1093/jas/skac068

Do, S., Koutsos, L., Utterback, P. L., Parsons, C. M., de Godoy, M. R. C. and Swanson, K. S., 2020. Nutrient and AA digestibility of black soldier fly larvae differing in age using the precision-fed cecectomized rooster assay. Journal of Animal Science, 98(1): 100362. https://doi.org/10.1093/jas/skz363

Dodd, S. A. S., Adolphe, J. L. and Verbrugghe, A., 2018. Plant-based diets for dogs. Journal of the American Veterinary Medical Association, 253(11): 1425–1432. https://doi.org/10.2460/javma.253.11.1425

Duijnisveld, K., 2022. Insect based pet food. Available at: https://libstore.ugent.be/fulltxt/RUG01/003/118/108/RUG01-003118108_2023_0001_AC.pdf

Emkani, M., Oliete, B. and Saurel, R., 2022. Effect of Lactic Acid Fermentation on Legume Protein Properties, a Review. Fermentation, 8(6): 244. https://doi.org/10.3390/fermentation8060244

Farkas, V. I., Máté, M., Takács, K., & Jánosi, A. 2025. The House Cricket (Acheta domesticus Linnaeus) in Food Industry: Farming, Technological Challenges, and Sustainability Considerations. Applied Sciences, 15(17): 9494. https://doi.org/10.3390/app15179494

Fasolin, L. H., Pereira, R. N., Pinheiro, A. C., Martins, J. T., Andrade, C. C. P., Ramos, O. L. and Vicente, A. A., 2019. Emergent food proteins – Towards sustainability, health and innovation. Food Research International, 125(6): 108586. https://doi.org/10.1016/j.foodres.2019.108586

Feng, T., Hu, Z., Tong, Y., Yao, L., Zhuang, H., Zhu, X., Song, S. and Lu, J., 2020. Preparation and evaluation of mushroom (Lentinus edodes) and mealworm (Tenebrio molitor) as dog food attractant. Heliyon, 6(10): e05302. https://doi.org/10.1016/j.heliyon.2020.e05302

Fernandes da Silva, C. B., Tavares Filho, E. R., Pimentel, T. C., Pagani, M. M., Mársico, E. T., Cruz, A. G. and Esmerino, E. A., 2023. The impact of information on the perception, emotional profile, and insights of Brazilian pet owners with different degrees of entomophobia. Food Quality and Preference, 110(10): 104967. https://doi.org/10.1016/j.foodqual.2023.104967

Fitriana, E. L., Laconi, E. B., Astuti, D. A., & Jayanegara, A. 2022. Effects of various organic substrates on growth performance and nutrient composition of black soldier fly larvae: A meta-analysis. Bioresource Technology Reports, 18: 101061. https://doi.org/10.1016/j.biteb.2022.101061

Food and Agriculture Organization of the United Nations (FAO), 2009. Feeding the world in 2050. World Summit on Food Security. Available at: https://openknowledge.fao.org/server/api/core/bitstreams/2a91baaf-c2e2-4a86-a327-f75c244108fe/content

Food and Agriculture Organization of the United Nations (FAO), 2012. Assessing the Potential of Insects as Food and Feed in Assuring Food Security. Technical Consultation Meeting 23–25 January, Rome, Italy. Available at: https://www.fao.org/4/an233e/an233e00.pdf

Fornari, T., Vázquez, L., Villanueva-Bermejo, D., Hurtado-Ribeira, R., Martín Hernández, D. and Martin, D., 2023. Effect of Moisture and Oil Content in the Supercritical CO2 Defatting of Hermetia illucens Larvae. Foods, 12(3): 490. https://doi.org/10.3390/foods12030490

Freel, T. A., McComb, A. and Koutsos, E. A., 2021. Digestibility and safety of dry black soldier fly larvae meal and black soldier fly larvae oil in dogs. Journal of Animal Science, 99(3): 104967. https://doi.org/10.1093/jas/skab047

Freeman, L. M., Stern, J. A., Fries, R., Adin, D. B. and Rush, J. E., 2018. Diet-associated dilated cardiomyopathy in dogs: what do we know? Journal of the American Veterinary Medical Association, 253(11): 1390–1394. https://doi.org/10.2460/javma.253.11.1390

Gałęcki, R., Pszczółkowski, B. and Zielonka, Ł., 2025. Experiences in Formulating Insect-Based

Feeds: Selected Physicochemical Properties of Dog Food Containing Yellow Mealworm Meal. Animals, 15(14): 2087. https://doi.org/10.3390/ani15142087

Gasco, L., Henry, M., Piccolo, G., Marono, S., Gai, F., Renna, M., Lussiana, C., Antonopoulou, E., Mola, P. and Chatzifotis, S., 2016. Tenebrio molitor meal in diets for European sea bass (Dicentrarchus labrax L.) juveniles: Growth performance, whole body composition and in vivo apparent digestibility. Animal Feed Science and Technology, 220(11): 34–45. https://doi.org/10.1016/j.anifeedsci.2016.07.003

Gasco, L., Oddon, S. B., Vandenberg, G. W., Veldkamp, T. and Biasato, I., 2023. Factors affecting the decision-making process of using insect-based products in animal feed formulations. Journal of Insects as Food and Feed, 220(11): 1–12. https://doi.org/10.3920/JIFF2022.0164

Grossmann, K. K., Merz, M., Appel, D., De Araujo, M. M. and Fischer, L., 2021. New insights into the flavoring potential of cricket (Acheta domesticus) and mealworm (Tenebrio molitor) protein hydrolysates and their Maillard products. Food Chemistry, 364(11): 130336. https://doi.org/10.1016/j.foodchem.2021.130336

Growth from Knowledge (GFK), 2016. Global pet ownership survey. Available at: https://cdn2.hubspot.net/hubfs/2405078/cmspdfs/fileadmin/user_upload/country_one_pager/nl/documents/global-gfk-survey_pet-ownership_2016.pdf.

Haber, M., Mishyna, M., Martinez, J. J. I. and Benjamin, O., 2019. The influence of grasshopper (Schistocerca gregaria) powder enrichment on bread nutritional and sensorial properties. LWT, 115(11): 108395. https://doi.org/10.1016/j.lwt.2019.108395

Hall, F., Johnson, P. E. and Liceaga, A., 2018. Effect of enzymatic hydrolysis on bioactive properties and allergenicity of cricket (Gryllodes sigillatus) protein. Food Chemistry, 262(11): 39–47. https://doi.org/10.1016/j.foodchem.2018.04.058

Hawkey, K. J., Lopez-Viso, C., Brameld, J. M., Parr, T. and Salter, A. M., 2021. Insects: A Potential Source of Protein and Other Nutrients for Feed and Food. Annual Review of Animal Biosciences, 9(1): 333–354. https://doi.org/10.1146/annurev-animal-021419-083930

Higa, J. E., Ruby, M. B. and Rozin, P., 2021. Americans' acceptance of black soldier fly larvae as food for themselves, their dogs, and farmed animals. Food Quality and Preference, 90(1): 104119. https://doi.org/10.1016/j.foodqual.2020.104119

Hosie, M. J. and Hofmann-Lehmann, R., 2022. Special Issue: Viral Infections in Companion Animals. Viruses, 14(2): 320. https://doi.org/10.3390/v14020320

Hu, Y., He, F., Mangian, H., Lambrakis, L., Saad, F. M. and de Godoy, M. R. C., 2020. PSVI-26 Insect meals as novel protein sources in wet pet foods for adult cats. Journal of Animal Science, 98(Supplement_4): 315–315. https://doi.org/10.1093/jas/skaa278.561

Huang, G., Zhang, Y., Liu, F., Xiao, J., & Huang, D. 2025. Fatty acid profile of insect oil and regulation mechanism as nutritious and functional oil: An integrative review. Journal of Food Composition and Analysis, 137: 106809. https://doi.org/10.1016/j.jfca.2024.106809

Jajic, I., Popovic, A., Urosevic, M., Krstovic, S., Petrovic, M., Guljas, D. and Samardzic, M., 2020. Fatty and amino acid profile of mealworm larvae (Tenebrio molitor L.). Biotechnology in Animal Husbandry, 36(2): 167–180. https://doi.org/10.2298/BAH2002167J

Jacuńska, W., Biel, W. and Zych, K., 2024. Evaluation of the Nutritional Value of Insect-Based

Complete Pet Foods. In Applied Sciences (Switzerland) (Vol. 14, Issue 22), 19 pp. https://doi.org/10.3390/app142210258

Janssen, R. H., Vincken, J.-P., van den Broek, L. A. M., Fogliano, V. and Lakemond, C. M. M., 2017. Nitrogen-to-Protein Conversion Factors for Three Edible Insects: *Tenebrio molitor*, *Alphitobius diaperinus*, and *Hermetia illucens*. Journal of Agricultural and Food Chemistry, 65(11): 2275–2278. https://doi.org/10.1021/acs.jafc.7b00471

Janssen, R. H., Canelli, G., Sanders, M. G., Bakx, E. J., Lakemond, C. M. M., Fogliano, V. and Vincken, J.-P., 2019. Iron-polyphenol complexes cause blackening upon grinding Hermetia illucens (black soldier fly) larvae. Scientific Reports, 9(1): 2967. https://doi.org/10.1038/s41598-019-38923-x

Jarett, J. K., Carlson, A., Rossoni Serao, M., Strickland, J., Serfilippi, L. and Ganz, H. H., 2019. Diets with and without edible cricket support a similar level of diversity in the gut microbiome of dogs. PeerJ, 7(1): e7661. https://doi.org/10.7717/peerj.7661

Jeon, K., Lee, J., Song, M., Park, K., Chang, S., Song, D., Kim, H., Yang, J. and Cho, J., 2025. Alternative nutritional evaluation of black soldier fly with various substrates in cat diets and its effects on fecal microbiota. Journal of Animal Science and Technology, 67(4): 922–943. https://doi.org/10.5187/jast.2025.e58

Kanakubo, K., Fascetti, A. J. and Larsen, J. A., 2015. Assessment of protein and amino acid concentrations and labeling adequacy of commercial vegetarian diets formulated for dogs and cats. Journal of the American Veterinary Medical Association, 247(4): 385–392. https://doi.org/10.2460/javma.247.4.385

Kępińska-Pacelik, J. and Biel, W., 2022. Insects in Pet Food Industry—Hope or Threat? Animals, 12(12): 1515. https://doi.org/10.3390/ani12121515

Khanal, P., Pandey, D., Næss, G., Cabrita, A. R. J., Fonseca, A. J. M., Maia, M. R. G., Timilsina, B., Veldkamp, T., Sapkota, R. and Overrein, H., 2023. Yellow mealworms (Tenebrio molitor) as an alternative animal feed source: A comprehensive characterization of nutritional values and the larval gut microbiome. Journal of Cleaner Production, 389(12): 136104. https://doi.org/10.1016/j.jclepro.2023.136104

Khatun, H., Claes, J., Smets, R., De Winne, A., Akhtaruzzaman, M. and Van Der Borght, M., 2021. Characterization of freeze-dried, oven-dried and blanched house crickets (Acheta domesticus) and Jamaican field crickets (Gryllus assimilis) by means of their physicochemical properties and volatile compounds. European Food Research and Technology, 247(5): 1291–1305. https://doi.org/10.1007/s00217-021-03709-x

Kiatbenjakul, P., Intarapichet, K.-O. and Cadwallader, K. R., 2015. Characterization of potent odorants in male giant water bug (Lethocerus indicus Lep. and Serv.), an important edible insect of Southeast Asia. Food Chemistry, 168(5): 639–647. https://doi.org/10.1016/j.foodchem.2014.07.108

Kierończyk, B., Rawski, M., Pawełczyk, P., Różyńska, J., Golusik, J., Mikołajczak, Z. and Józefiak, D., 2018. Do insects smell attractive to dogs? A comparison of dog reactions to insects and commercial feed aromas – a preliminary study. Annals of Animal Science, 18(3): 795–800. https://doi.org/10.2478/aoas-2018-0012

Kierończyk, B., Rawski, M., Mikołajczak, Z., Homska, N., Jankowski, J., Ognik, K., Józefiak, A., Mazurkiewicz, J. and Józefiak, D., 2022. Available for millions of years but discovered through the last decade: Insects as a source of nutrients and energy in animal diets. Animal Nutrition, 11(3): 60–79. https://doi.org/10.1016/j.aninu.2022.06.015

Kilburn, L. R., Carlson, A. T., Lewis, E. and Serao, M. C. R., 2020. Cricket (Gryllodes sigillatus) meal fed to healthy adult dogs does not affect general health and minimally impacts apparent total tract digestibility. Journal of Animal Science, 98(3): 60–79. https://doi.org/10.1093/jas/skaa083

Kim, H. W., Setyabrata, D., Lee, Y. J., Jones, O. G. and Kim, Y. H. B., 2016. Pre-treated mealworm larvae and silkworm pupae as a novel protein ingredient in emulsion sausages. Innovative Food Science and Emerging Technologies, 38: 116–123. https://doi.org/10.1016/j.ifset.2016.09.023

Kim, T.-K., Yong, H. I., Kim, Y.-B., Jung, S., Kim, H.-W., & Choi, Y.-S. 2021. Effects of organic solvent on functional properties of defatted proteins extracted from Protaetia brevitarsis larvae. Food Chemistry, 336: 127679. https://doi.org/10.1016/j.foodchem.2020.127679

Kim, D. and Oh, I. 2022. The Characteristic of Insect Oil for a Potential Component of Oleogel and Its Application as a Solid Fat Replacer in Cookies. Gels, 8(6): 355. https://doi.org/10.3390/gels8060355

Kim, T.-K., Cha, J. Y., Yong, H. I., Jang, H. W., Jung, S. and Choi, Y.-S., 2022. Application of Edible Insects as Novel Protein Sources and Strategies for Improving Their Processing. Food Science of Animal Resources, 42(3): 372–388. https://doi.org/10.5851/kosfa.2022.e10

Koppel, K., 2014. Sensory analysis of pet foods. Journal of the Science of Food and Agriculture, 94(11): 2148–2153. https://doi.org/10.1002/jsfa.6597

Kouřimská, L. and Adámková, A., 2016. Nutritional and sensory quality of edible insects. NFS Journal, 4(11): 22–26. https://doi.org/10.1016/j.nfs.2016.07.001

Kulma, M., Kouřimská, L., Plachý, V., Božik, M., Adámková, A. and Vrabec, V., 2019. Effect of sex on the nutritional value of house cricket, Acheta domestica L. Food Chemistry, 272(11): 267–272. https://doi.org/10.1016/j.foodchem.2018.08.049

Kuntadi, K., Adalina, Y. and Maharani, K. E., 2018. NUTRITIONAL COMPOSITIONS OF SIX EDIBLE INSECTS IN JAVA. Indonesian Journal of Forestry Research, 5(1): 57–68. https://doi.org/10.59465/ijfr.2019.5.1.57-68

La Barbera, F., Verneau, F., Amato, M. and Grunert, K., 2018. Understanding Westerners' disgust for the eating of insects: The role of food neophobia and implicit associations. Food Quality and Preference, 64(11): 120–125. https://doi.org/10.1016/j.foodqual.2017.10.002

La Barbera, F., Verneau, F., Amato, M., Grunert, K. G. and Schnettler, B., 2021. Acceptance of insect-based food in Chile: Evidence from a survey using the entomophagy attitude questionnaire (EAQ). Food Quality and Preference, 93(11): 104269. https://doi.org/10.1016/j.foodqual.2021.104269

Larouche, J., Deschamps, M.-H., Saucier, L., Lebeuf, Y., Doyen, A. and Vandenberg, G. W., 2019. Effects of Killing Methods on Lipid Oxidation, Colour and Microbial Load of Black Soldier Fly (Hermetia illucens) Larvae. Animals, 9(4): 182. https://doi.org/10.3390/ani9040182

Le Guillas, G., Vanacker, P., Salles, C. and Labouré, H. 2024. Insights to Study, Understand and Manage Extruded Dry Pet Food Palatability. Animals, 14(7): 1095. https://doi.org/10.3390/ani14071095

Leal, D., Barroso, C., Fontes-Sousa, A.P., Borges, S., Maia, M.R.G., Almeida, A., Pintado, M., Fonseca, A.J.M, Cabrita, A.R.J. 2024. Influence of dietary inclusion of black soldier fly hydrolysates on palatability, digestibility and cardiac biomarkers of adult Beagle dogs. Congress Proceedings. 28th Congress of the European Society of Veterinary and Comparative Nutrition.

Lee, H. E., Kim, J., Kim, Y., Bang, W. Y., Yang, J., Lee, S.-J. and Jung, Y. H., 2021. Identification and improvement of volatile profiles of Allomyrina dichotoma larvae by fermentation with lactic acid bacteria. Food Bioscience, 43(7): 101257. https://doi.org/10.1016/j.fbio.2021.101257

Lei, X. J., Kim, T. H., Park, J. H. and Kim, I. H., 2019. Evaluation of Supplementation of Defatted Black Soldier Fly (Hermetia illucens) Larvae Meal in Beagle Dogs. Annals of Animal Science, 19(3): 767–777. https://doi.org/10.2478/aoas-2019-0021

Leriche, I., Chala, V., Ereau, C., 2016. Pet owners' perception of insects as a protein source for cats and dogs. Available at: https://vet-it.virbac.com/files/live/sites/virbac-b2b-it/files/hpm-physio-diet/Nutrition-resources/poster%20CAT%20HYPOALL%20OWNERS%20SEVC%202017.pdf

Li, Y., Liu, M., Wei, Y., Li, L., Ma, D., Weng, Y., Wang, H. and Xu, X., 2024. Influence of a Mixture of Protein Hydrolysate from Black Soldier Fly Larvae and Schizochytrium on Palatability, Plasma Biochemistry, and Antioxidative and Anti-Inflammatory Capacity in Cat Diets. Animals, 14(5): 751. https://doi.org/10.3390/ani14050751

Lisenko, K. G., Godoy, M. R. C., Oliveira, K. R. B., Oliveira, M. R. D., Silva, T. V., Fontes, T. V., Lacerda, R. F., Ferreira, L. G., Gonçalves, T. M., Zangeronimo, M. G., da Costa, D. V. and Saad, F. M. O. B., 2023. Digestibility of insect meals for dogs and their effects on blood parameters, faecal characteristics, volatile fatty acids, and gut microbiota. Journal of Insects as Food and Feed, 9(7): 907–918. https://doi.org/10.3920/JIFF2022.0114

Luigia, D. S., 2019. New protein Sources: Novel Foods. In Encyclopedia of Food Security and Sustainability (Vol. 9, Issue 7, pp. 276–279). Elsevier. https://doi.org/10.1016/B978-0-08-100596-5.22129-X

Magalhães, R., Sánchez-López, A., Leal, R. S., Martínez-Llorens, S., Oliva-Teles, A. and Peres, H., 2017. Black soldier fly (Hermetia illucens) pre-pupae meal as a fish meal replacement in diets for European seabass (Dicentrarchus labrax). Aquaculture, 476(7): 79–85. https://doi.org/10.1016/j.aquaculture.2017.04.021

Marciniak, A., Suwal, S., Naderi, N., Pouliot, Y. and Doyen, A., 2018. Enhancing enzymatic hydrolysis of food proteins and production of bioactive peptides using high hydrostatic pressure technology. Trends in Food Science & Technology, 80(7): 187–198. https://doi.org/10.1016/j.tifs.2018.08.013

Medrano, L., 2019. Larvas de gusano de harina (Tenebrio molitor) como alternativa proteica en la alimentación animal. Universidad Nacional Abierta y a Distancia. Bogotá, Colombia: 1-44.

Melgar-Lalanne, G., Hernández-Álvarez, A. and Salinas-Castro, A., 2019. Edible Insects Processing: Traditional and Innovative Technologies. Comprehensive Reviews in Food Science and Food Safety, 18(4): 1166–1191. https://doi.org/10.1111/1541-4337.12463

Miglietta, P., De Leo, F., Ruberti, M. and Massari, S., 2015. Mealworms for Food: A Water Footprint Perspective. Water, 7(11): 6190–6203. https://doi.org/10.3390/w7116190

Mitchaothai, J., Grabowski, N. T., Lertpatarakomol, R., Trairatapiwan, T., Chhay, T., Keo, S. and Lukkananukool, A., 2022. Production Performance and Nutrient Conversion Efficiency of Field Cricket (Gryllus bimaculatus) in Mass-Rearing Conditions. Animals, 12(17): 2263. https://doi.org/10.3390/ani12172263

Montowska, M., Kowalczewski, P. Ł., Rybicka, I. and Fornal, E., 2019. Nutritional value, protein and peptide composition of edible cricket powders. Food Chemistry, 289(17): 130–138. https://doi.org/10.1016/j.foodchem.2019.03.062

Mosna, D., Bottani, E., Vignali, G. and Montanari, R., 2021. Environmental benefits of pet food obtained as a result of the valorisation of meat fraction derived from packaged food waste. Waste Management, 125(17): 132–144. https://doi.org/10.1016/j.wasman.2021.02.035

Mouithys-Mickalad, A., Schmitt, E., Dalim, M., Franck, T., Tome, N. M., van Spankeren, M., Serteyn, D. and Paul, A., 2020. Black Soldier Fly (Hermetia illucens) Larvae Protein Derivatives: Potential to Promote Animal Health. Animals, 10(6): 941. https://doi.org/10.3390/ani10060941

Mustafa, F., Sajjad, A., Sajjad, M., Ali, M., Bashir, H. S., Abbas, M. G., Binyameen, M. and Mozūratis, R., 2025. Comparative evaluation of Acheta domesticus and Hermetia illucens as alternative protein sources for the growth, health, and meat quality of the broiler. Frontiers in Animal Science, 6(6): 244. https://doi.org/10.3389/fanim.2025.1531761

Nozari, B., & Kander, R. 2025. Supercritical CO2 technology for biomass extraction: Review. Industrial Crops and Products, 233: 121348. https://doi.org/10.1016/j.indcrop.2025.121348

Oghenesuvwe, E. E. and Paul, C., 2019. Edible Insects bio-actives as anti-oxidants: Current status and perspectives. Journal of Complementary Medicine Research, 10(2): 89-102. https://doi.org/10.5455/jcmr.20190130100319

Oibiokpa, F. I., Akanya, H. O., Jigam, A. A., Saidu, A. N. and Egwim, E. C., 2018. Protein quality of four indigenous edible insect species in Nigeria. Food Science and Human Wellness, 7(2): 175–183. https://doi.org/10.1016/j.fshw.2018.05.003

Oliveira, L. A., Pereira, S. M. S., Dias, K. A., Paes, S. da S., Grancieri, M., Jimenez, L. G. S., de Carvalho, C. W. P., de Oliveira, E. E., Martino, H. S. D., & Della Lucia, C. M. 2024. Nutritional content, amino acid profile, and protein properties of edible insects (Tenebrio molitor and Gryllus assimilis) powders at different stages of development. Journal of Food Composition and Analysis, 125: 105804. https://doi.org/10.1016/j.jfca.2023.105804

Oonincx, D. G. A. B., van Itterbeeck, J., Heetkamp, M. J. W., van den Brand, H., van Loon, J. J. A. and van Huis, A., 2010. An Exploration on Greenhouse Gas and Ammonia Production by Insect Species Suitable for Animal or Human Consumption. PLoS ONE, 5(12): e14445. https://doi.org/10.1371/journal.pone.0014445

Organisation for Economic Co-operation and Development/ Food and Agriculture Organization of the United Nations (OECD/FAO), 2021. OECD-FAO Agricultural Outlook 2021-2030, OECD Publishing, Paris. https://doi.org/10.1787/19428846-en.

Øverland, M., Sørensen, M., Storebakken, T., Penn, M., Krogdahl, Å. and Skrede, A., 2009. Pea protein concentrate substituting fish meal or soybean meal in diets for Atlantic salmon (Salmo salar)—Effect on growth performance, nutrient digestibility, carcass composition, gut health, and physical feed quality. Aquaculture, 288(3–4): 305–311. https://doi.org/10.1016/j.aquaculture.2008.12.012

Paßlack, N. and Zentek, J., 2018. Akzeptanz, Verträglichkeit und scheinbare Nährstoffverdaulichkeit von Alleinfuttermitteln auf Basis von Hermetia-illucens-Larvenmehl bei Katzen. Tierärztliche Praxis Ausgabe K: Kleintiere / Heimtiere, 46(04): 213–221. https://doi.org/10.15654/TPK-180372

Penazzi, L., Schiavone, A., Russo, N., Nery, J., Valle, E., Madrid, J., Martinez, S., Hernandez, F., Pagani, E., Ala, U. and Prola, L., 2021. In vivo and in vitro Digestibility of an Extruded Complete Dog Food Containing Black Soldier Fly (Hermetia illucens) Larvae Meal as Protein Source. Frontiers in Veterinary Science, 8(04): 213–221. https://doi.org/10.3389/fvets.2021.653411

Pozharliev, R., De Angelis, M., Rossi, D., Bagozzi, R. and Amatulli, C., 2023. I might try it: Marketing actions to reduce consumer disgust toward insect-based food. Journal of Retailing, 99(1): 149–167. https://doi.org/10.1016/j.jretai.2022.12.003

Purschke, B., Brüggen, H., Scheibelberger, R. and Jäger, H., 2018. Effect of pre-treatment and drying method on physico-chemical properties and dry fractionation behaviour of mealworm larvae (Tenebrio molitor L.). European Food Research and Technology, 244(2): 269–280. https://doi.org/10.1007/s00217-017-2953-8

Queiroz, L. S., Nogueira Silva, N. F., Jessen, F., Mohammadifar, M. A., Stephani, R., Fernandes de Carvalho, A., Perrone, Í. T. and Casanova, F., 2023. Edible insect as an alternative protein source: a review on the chemistry and functionalities of proteins under different processing methods. Heliyon, 9(4): e14831. https://doi.org/10.1016/j.heliyon.2023.e14831

Ramos-Elorduy, J., González, E. A., Hernández, A. R. and Pino, J. M., 2002. Use of Tenebrio molitor (Coleoptera: Tenebrionidae) to Recycle Organic Wastes and as Feed for Broiler Chickens. Journal of Economic Entomology, 95(1): 214–220. https://doi.org/10.1603/0022-0493-95.1.214

Ravi, H. K., Degrou, A., Costil, J., Trespeuch, C., Chemat, F. and Vian, M. A., 2020. Effect of devitalization techniques on the lipid, protein, antioxidant, and chitin fractions of black soldier fly (Hermetia illucens) larvae. European Food Research and Technology, 246(12): 2549–2568. https://doi.org/10.1007/s00217-020-03596-8

Rawski, M., Mazurkiewicz, J., Kierończyk, B. and Józefiak, D., 2020. Black Soldier Fly Full-Fat Larvae Meal as an Alternative to Fish Meal and Fish Oil in Siberian Sturgeon Nutrition: The Effects on Physical Properties of the Feed, Animal Growth Performance, and Feed Acceptance and Utilization. Animals, 10(11): 2119. https://doi.org/10.3390/ani10112119

Reátegui, J., Barriga, X., Obando, A., Moscoso, G., Manrique, P. and Salazar, I. 2020. Hermetia illucens larva (Diptera: Stratiomyidae) meal as a protein ingredient for partial replacement of soybean meal in the feed of Cavia porcellus (Guinea pig): effect on the consumption, weight gain and feed conversion. Scientia Agropecuaria, 11(4): 513–519. https://doi.org/10.17268/sci.agropecu.2020.04.06

Reilly, L. M., von Schaumburg, P. C., Hoke, J. M., Davenport, G. M., Utterback, P. L., Parsons, C. M. and de Godoy, M. R. C., 2020. Macronutrient composition, true metabolizable energy and amino acid digestibility, and indispensable amino acid scoring of pulse ingredients for use in canine and feline diets. Journal of Animal Science, 98(6): 513–519. https://doi.org/10.1093/jas/skaa149

Rivero Pino, F., Pérez Gálvez, R., Espejo Carpio, F. J. and Guadix, E. M., 2020. Evaluation of Tenebrio molitor protein as a source of peptides for modulating physiological processes. Food & Function, 11(5): 4376–4386. https://doi.org/10.1039/D0FO00734J

da Costa Rocha, C., de Andrade, J.C. and de Oliveira, D., 2021. Perspective on integrated biorefinery for valorization of biomass from the edible insect Tenebrio molitor. Trends in Food Science & Technology, 116(5): 480–491. https://doi.org/10.1016/j.tifs.2021.07.012

Rumpold, B. A. and Schlüter, O. K., 2013. Nutritional composition and safety aspects of edible insects. Molecular Nutrition & Food Research, 57(5): 802–823. https://doi.org/10.1002/mnfr.201200735

Sánchez, M., Gómez, C., Avendaño, C., Harmsen, I., Ortiz, D., Ceballos, R., Villamizar-Sarmiento, M. G., Oyarzun-Ampuero, F., Wacyk, J. and Valenzuela, C., 2021. House fly (Musca domestica) larvae meal as an ingredient with high nutritional value: Microencapsulation and improvement of organoleptic characteristics. Food Research International, 145(5): 110423. https://doi.org/10.1016/j.foodres.2021.110423

Sánchez, M., Villamizar-Sarmiento, M. G., Harmsen, I., Valdés, F., Villanueva, V., Ceballos, R., Wacyk, J., Oyarzun-Ampuero, F. and Valenzuela, C., 2022. Encapsulation of house fly larvae (Musca domestica) meal by ionic gelation as a strategy to develop a novel nutritive food ingredient with improved aroma and appearance. LWT, 163(5): 113597. https://doi.org/10.1016/j.lwt.2022.113597

Selaledi, L. and Mabelebele, M., 2021. The Influence of Drying Methods on the Chemical Composition and Body Color of Yellow Mealworm (Tenebrio molitor L.). Insects, 12(4): 333. https://doi.org/10.3390/insects12040333

Shumo, M., Osuga, I. M., Khamis, F. M., Tanga, C. M., Fiaboe, K. K. M., Subramanian, S., Ekesi, S., van Huis, A., & Borgemeister, C. 2019. The nutritive value of black soldier fly larvae reared on common organic waste streams in Kenya. Scientific Reports, 9(1): 10110. https://doi.org/10.1038/s41598-019-46603-z

Siddiqui, S. A., Brunner, T. A., Tamm, I., van der Raad, P., Patekar, G., Alim Bahmid, N., Aarts, K. and Paul, A., 2023. Insect-based dog and cat food: A short investigative review on market, claims and consumer perception. Journal of Asia-Pacific Entomology, 26(1): 102020. https://doi.org/10.1016/j.aspen.2022.102020

Singh, A. and Kumari, K., 2019. An inclusive approach for organic waste treatment and valorisation using Black Soldier Fly larvae: A review. Journal of Environmental Management, 251(1): 109569. https://doi.org/10.1016/j.jenvman.2019.109569

Singh, Y., Cullere, M., Kovitvadhi, A., Chundang, P. and Dalle Zotte, A., 2020. Effect of different killing methods on physicochemical traits, nutritional characteristics, in vitro human digestibility and oxidative stability during storage of the house cricket (Acheta domesticus L.). Innovative Food Science & Emerging Technologies, 65(1): 102444. https://doi.org/10.1016/j.ifset.2020.102444

Singh, S. K., Pawar, L., Thomas, A. J., Debbarma, R., Biswas, P., Ningombam, A., Devi, A. G., Waikhom, G., Patel, A. B., Meena, D. K. and Chakraborty, G., 2023. The current state of research and potential applications of insects for resource recovery and aquaculture feed. Environmental Science and Pollution Research, 31(53): 62264–62282. https://doi.org/10.1007/s11356-023-29068-6

Skotnicka, M., Karwowska, K., Kłobukowski, F., Borkowska, A. and Pieszko, M., 2021. Possibilities of the Development of Edible Insect-Based Foods in Europe. Foods, 10(4): 766. https://doi.org/10.3390/foods10040766

Smetana, S., Ashtari Larki, N., Pernutz, C., Franke, K., Bindrich, U., Toepfl, S. and Heinz, V., 2018. Structure design of insect-based meat analogs with high-moisture extrusion. Journal of Food Engineering, 229(4): 83–85. https://doi.org/10.1016/j.jfoodeng.2017.06.035

Sogari, G., Menozzi, D. and Mora, C., 2018. Sensory-liking Expectations and Perceptions of Processed and Unprocessed Insect Products. International Journal on Food System Dynamics, 9(4): 314–320. https://doi.org/10.18461/ijfsd.v9i4.942

Son, Y.-J., Lee, J.-C., Hwang, I.-K., Nho, C. W. and Kim, S.-H., 2019. Physicochemical properties of mealworm (Tenebrio molitor) powders manufactured by different industrial processes. LWT, 116: 108514. https://doi.org/10.1016/j.lwt.2019.108514

Srisuksai, K., Limudomporn, P., Kovitvadhi, U., Thongsuwan, K., Imaram, W., Lertchaiyongphanit, R., Sareepoch, T., Kovitvadhi, A. and Fungfuang, W., 2024. Physicochemical properties and fatty acid profile of oil extracted from black soldier fly larvae (Hermetia illucens). Veterinary World, 9(4): 518–526. https://doi.org/10.14202/vetworld.2024.518-526

Ssepuuya, G., Nakimbugwe, D., De Winne, A., Smets, R., Claes, J. and Van Der Borght, M., 2020. Effect of heat processing on the nutrient composition, colour, and volatile odour compounds of the long-horned grasshopper Ruspolia differens serville. Food Research International, 129(4): 108831. https://doi.org/10.1016/j.foodres.2019.108831

Tabata, E., Kashimura, A., Kikuchi, A., Masuda, H., Miyahara, R., Hiruma, Y., Wakita, S., Ohno, M., Sakaguchi, M., Sugahara, Y., Matoska, V., Bauer, P. O. and Oyama, F., 2018. Chitin digestibility is dependent on feeding behaviors, which determine acidic chitinase mRNA levels in mammalian and poultry stomachs. Scientific Reports, 8(1): 1461. https://doi.org/10.1038/s41598-018-19940-8

Tan, H. S. G., Fischer, A. R. H., van Trijp, H. C. M. and Stieger, M., 2016. Tasty but nasty? Exploring the role of sensory-liking and food appropriateness in the willingness to eat unusual novel foods like insects. Food Quality and Preference, 48(4): 293–302. https://doi.org/10.1016/j.foodqual.2015.11.001

Tavano, O. L., 2013. Protein hydrolysis using proteases: An important tool for food biotechnology. Journal of Molecular Catalysis B: Enzymatic, 90(4): 1–11. https://doi.org/10.1016/j.molcatb.2013.01.011

Thirunathan, P. and Manickavasagan, A., 2019. Processing methods for reducing alpha-galactosides in pulses. Critical Reviews in Food Science and Nutrition, 59(20): 3334–3348. https://doi.org/10.1080/10408398.2018.1490886

Tognocchi, M., Abenaim, L., Adamaki-Sotiraki, C., Athanassiou, G. C., Rumbos, I. C., Mele, M., Conti, B., & Conte, G. 2024. Effect of different diet composition on the fat profile of two different black soldier fly larvae populations. Animal, 18(7): 101205. https://doi.org/10.1016/j.animal.2024.101205

Tomiyama, J.-M., Takagi, D. and Kantar, M. B., 2020. The effect of acute and chronic food shortage on human population equilibrium in a subsistence setting. Agriculture & Food Security, 9(1): 6. https://doi.org/10.1186/s40066-020-00261-x

Triggs, A., Bless, I., Danner, L., Saarela, M. and Wilkinson, K., 2025. Australian Dog Owners' Acceptance of Insect-Based Pet Food. Insects, 16(3): 290. https://doi.org/10.3390/insects16030290

Tzompa-Sosa, D. A., Yi, L., van Valenberg, H. J. F. and Lakemond, C. M. M., 2019. Four insect oils as food ingredient: physical and chemical characterisation of insect oils obtained by an aqueous oil extraction. Journal of Insects as Food and Feed, 5(4): 279–292. https://doi.org/10.3920/JIFF2018.0020

Valdés, F., Villanueva, V., Durán, E., Campos, F., Avendaño, C., Sánchez, M., Domingoz-Araujo, C. and Valenzuela, C., 2022. Insects as Feed for Companion and Exotic Pets: A Current Trend. Animals, 12(11): 1450. https://doi.org/10.3390/ani12111450

van Huis, A., 2013. Potential of Insects as Food and Feed in Assuring Food Security. Annual Review of Entomology, 58(1): 563–583. https://doi.org/10.1146/annurev-ento-120811-153704

van Huis, A. and Oonincx, D. G. A. B., 2017. The environmental sustainability of insects as food and feed. A review. Agronomy for Sustainable Development, 37(5): 43. https://doi.org/10.1007/s13593-017-0452-8

van Huis, A. and Rumpold, B., 2023. Strategies to convince consumers to eat insects? A review. Food Quality and Preference, 110(5): 104927. https://doi.org/10.1016/j.foodqual.2023.104927

Veldkamp, T. and Vernooij, A. G., 2021. Use of insect products in pig diets. Journal of Insects as Food and Feed, 7(5): 781–793. https://doi.org/10.3920/JIFF2020.0091

Villanueva, V., Valdés, F., Zavala, S., Yáñez, J. M. and Valenzuela, C., 2023. Perception of cat owners on the use of insects as feed ingredients for cats. Journal of Insects as Food and Feed, 9(11): 1431–1443. https://doi.org/10.1163/23524588-20220106

Wei, Y., Xue, L., Ma, D., Weng, Y., Liu, M., Li, L., Dai, Z., Zhao, Z., Wang, H. and Xu, X., 2024. The Effect of Dietary Protein Hydrolysate from Black Soldier Fly Larvae and Schizochytrium on Palatability, Nutrient Metabolites and Health Status in Beagle Dogs. Metabolites, 14(3): 165. https://doi.org/10.3390/metabo14030165

Wójtowicz, A., Combrzyński, M., Biernacka, B., Oniszczuk, T., Mitrus, M., Różyło, R., Gancarz, M. and Oniszczuk, A, 2023. Application of Edible Insect Flour as a Novel Ingredient in Fortified Snack Pellets: Processing Aspects and Physical Characteristics. Processes, 11(9): 2561. https://doi.org/10.3390/pr11092561

Yao, C., Dai, P., 2019. Preparation method of silkworm pupa protein powder. CN Patent CN109170118.

Yıldırım, M., Erşatır, M., Poyraz, S., Amangeldinova, M., Kudrina, N. O., & Terletskaya, N. V. 2024. Green Extraction of Plant Materials Using Supercritical CO2: Insights into Methods, Analysis, and Bioactivity. Plants, 13(16): 2295. https://doi.org/10.3390/plants13162295

CHAPTER 14

Use of bacterial proteins from fermentation technologies in pet nutrition

A.P. Wesker[1], R. Alexa[2], M. Tesla[3], Z. Salimi[3], A.M.M de Matos[4], L. Giver[5] and M. Longshaw[1]*

[1]Aseno Ltd, 52a St John Street, Ashbourne, Derbyshire. DE6 1GH, UK
[2]Pet food Hydrocolloids Consultant, 36137 Großenlüder, Germany
[3]Passion4food AS, Skurvebakkane 32, 4331 Ålgård, Norway
[4]Microharvest Unipessoal, Lda., Avenida Infante Dom Henrique 143, Unit N. 04, 1950-406 Lisbon, Portugal
[5]Calysta Inc., 135 West 25th Ave 1027, San Mateo, CA 94403, USA
**Correspondence: alex@aseno-partners.com*

Abstract

Production of microbial proteins by biomass fermentation has been established for a long time, but the inclusion of these complementary proteins in pet foods remains in its infancy. This is despite a reduction in the availability of animal proteins, and the societal shift towards more sustainable and cruelty-free protein sources. To date, three main sources of microbial proteins derived from biomass fermentation have been evaluated in dogs and cats. These include *Methylophilus methylotrophus* (a methylotrophic bacterium), *Methylococcus capsulatus* (a methanotrophic bacterium), and microbial protein produced via the MicroHarvest fermentation platform, using natural bacterial strains from a variety of genera including *Bacillus* spp, *Vibrio* spp., and *Geobacillus* spp. This chapter discusses the different methods used to produce microbial biomass and the subsequent safety to workers, the nutritional profile of these bacteria, their palatability, digestibility and safety in cats and dogs, as well as observed health and immune responses. Additionally, the technical functionality of these proteins in wet and dry pet foods, along with consumer perceptions of these cultured proteins are examined. The chapter concludes with a summary of current limitations on the use of these proteins in pet food and provides a view on potential additional studies required to understand their mode of action and to unlock their full potential as sustainable complementary protein sources.

Keywords

single cell protein – cultured protein – microbial protein – cat – dog

1 Introduction

This chapter covers the production and application of bacteria obtained via biomass fermentation to produce proteins that are explicitly used to replace or complement existing plant and animal proteins in pet food. Peer-reviewed studies on the application of microbial proteins in pet food are limited and so, where appropriate, additional data from unpublished early-phase studies are included. It does not cover bacteria that are added as a biologically functional feed ingredient where the primary purpose is where it acts as a biotic. To date only three microbial proteins derived from fermentation have been tested in cats and dogs, including *Methylococcus capsulatus* (FeedKind® Pet from Calysta Inc., USA and fermented protein from String Bio Pvt Ltd., India), *Methylophilus methylotrophus* (as *Methylomonas clara* (Probion®) from Hoeschst, Germany) and a microbial protein marketed as MPX Care® (MicroHarvest GmbH). Notably, the bacterial groups used for the production of MPX Care® are included in the EFSA Qualified Presumption of Safety (QPS) list, and can come from a variety of genera such as *Bacillus* spp., *Vibrio* spp. and/or *Geobacillus* spp. The protein in this case is produced through biomass fermentation using agricultural side streams as substrates, offering a scalable and sustainable solution of high-quality protein. FeedKind Pet cultured protein contains a consortia of bacteria including *M. capsulatus* which comprises >90% of the total dried biomass, alongside *Aneurinibacillus danicus*, *Brevibacillus agri* and *Cupriavidus cauae* and is grown on methane in a loop fermenter (Banks et al., 2022). For brevity, it will be referred to as *Methylococcus capsulatus.* Publications on Probion applications in pet food originally referred to the bacteria as *Methylomonas clara* (characterized by Hohnloser et al. (1978)). Subsequently, this was considered a junior synonym of *Methylophilus methylotrophus* by Eisen (1995), Lloyd and Sharp (1993) and Schoch (2020).

The three microbial proteins tested for potential as pet food ingredients are all produced via dedicated fermentation processes. The biomass resulting from the growth of the microbial strains is the main product of the fermentation (the biomass is not a side-stream or byproduct from production of a different product), and the whole cells are the final product (there is no purification/enrichment of a specific protein as there is in a precision fermentation process). The cells from these fermentations are generally inactivated, usually by a heating step, unlike live cell products that are made for probiotic products. The fermentation processes for these three microbial proteins differ substantially. As defined above they use different microbial strains, which impacts the composition of the final product. The inputs to the fermentation are different to optimize growth of these strains. For the *M. capsulatus* product, the process is a gas-fed fermentation where methane and oxygen are the key inputs into a specialized loop reactor designed to maximize gas utilization, with capacity to produce up to 20,000 tonnes per year being possible. The *Methylophilus methylotrophus* process, on the other hand, used methanol as its feedstock, grown in a stirred tank fermentation system, with unique optimization to avoid high concentrations of methanol. Finally, the microbial protein that could come from the variety of genus *Bacillus* spp., *Vibrio* spp., and *Geobacillus* spp., is produced through biomass fermentation using agricultural side streams as substrates in a stirred-tank system.

All these microbial products have sustainability advantages over the products they displace in pet food applications. Land and water use required for production of the microbial proteins is significantly lower than that required for plant products and for human grade animal ingredients that are used in super premium categories of pet food (Banks et al., 2022). Use of rendered animal proteins or by-products can offset some of these concerns. While all the fermentation processes will produce carbon dioxide (CO_2) as the cells grow, the total CO_2 footprint of the final products are lower than meat products, depending on the allocation approach used in life cycle assessments (LCA). Although rendered ingredients might have a lower LCA, as the most environmental burdens are attributed to the human fraction, when impacts are expressed per kilogram of digestible protein, microbial proteins show lower land and water use, particularly when produced using renewable energy sources or optimized fermentation processes. In addition, while not yet implemented at commercial scale, there are attractive opportunities to capture and use the CO_2 produced from these fermentations in different ways, including converting back into methane or methanol and recycling back into the process. These modifications aim to reduce the carbon footprint of these products to net zero or net negative, further increasing their appeal in pet food applications. Additional advantages of microbial proteins include reduced time of production, improved shelf-life stability, consistency of quality and reduced risk of contaminants (such as pesticides, antibiotics, pathogens).

Agrifood systems account for about 31% of total anthropogenic greenhouse gas emissions (GHGs), of which animal production is a major contributor. Further, 50% of habitable land and 70% of freshwater withdrawals are used for agriculture (FAO, 2021; Poore and Nemecek, 2018). The global need for protein continues to grow as a result of both population growth and expected increase in meat consumption with growth varying per continent (Le Mouël and Forslund, 2017). The global population has grown from an estimated 2.5 billion in 1950, to 8 billion in 2022, and projected to reach 9.7 billion in 2050 and 10.4 billion in 2100, from which point figures are expected to stabilize (UN WPP, 2022). In 2018, global meat commodity volumes totaled 333 million tons (Smith et al., 2022). Meat is a dense source of nutrients and a disproportionally large contributor to global availability of nutrients (Smith et al., 2022). A review reports livestock products are estimated at 22–25% of global average food diet in 2025 (Le Mouël and Forslund, 2017) with average livestock product consumption per capita predicted to be at 48–49 kg/year by 2050, totaling 475 million tons (Alexandratos & Bruinsma, 2012; Rosegrant et al., 2013). The increase in meat demand will mostly be seen in Sub-Saharan Africa and South-Asia (Le Mouël and Forslund, 2017).

Pet food diets often have a prominent inclusion of ingredients of animal origin, which can contribute to their environmental impact (Pedrinelli et al., 2022). Traditionally, pets were fed table scraps from the human food table in what are nowadays considered established markets for commercial pet food, including most of Europe. Pet food became commercially produced about a century ago, starting off in the USA with wet food in cans. Currently, table scraps play a more significant role in feeding pets in emerging markets than in established markets. Recent research covering Australia, Canada, New Zealand, the UK and the USA found most pets (79% dogs, 90% cats) were offered commercial diets, although only 13% dogs, 32% cats were offered conventional diets exclusively and with pets offered home-made and/or raw diets also (64% dogs, 46% cats, and 66% dogs, 53% cats, respectively) (Dodd et al., 2020). The transition to commercially produced pet food allowed for specifically the by-products from the human food industry to be used for pet food. Nowadays, most commercial pet food still heavily relies on by-products. Commonly used animal proteins in commercial pet food products include chicken, beef, fish (including salmon and whitefish), lamb and pork.

A recent study evaluated the environmental impact of pet food diets by assessing carbon-dioxide equivalent (CO_2eq), land and stress-weighted water use, plus sulfur dioxide equivalent and phosphate emissions (Pedrinelli et al., 2022). These measures give an indication of global warming potential, competition of land area and freshwater availability for humans, plus acidification and eutrophication of the environment, respectively. The 938 diets ranged across dry, wet and homemade formats in Brazil, home to the second-largest population of dogs and third-largest of cats globally. Impact was expressed per unit of energy (per 1000 kcal of dietary metabolizable energy) in order to standardize comparison based on amount consumed by the pet. Wet diets for dogs represented a significantly greater impact on the environment than the other diets for all parameters. Dry diets for dogs scored lower against the other diets for all parameters. Homemade diets were intermediary between wet and dry diets for most parameters, apart from cat diets where sulfur dioxide equivalent and water use were similar between dry and commercial homemade diets. Through extrapolation of CO_2eq emission alone, the authors indicate that dry diets for dogs would be 12.4% of the emission of a Brazilian citizen, while wet diets would be 97.8%. At a population level this would be 2.9% for dry diets and 24.6% for wet diets. Diet composition and ingredient choice therefore greatly influences the environmental impact of pet diets (Pedrinelli et al., 2022). The potential sustainability impact warrants consideration of alternative protein sources for pet food diets.

Industries are familiar with adapting to the rise in global protein requirement. For example, selective breeding programs or genetic modifications of plant and animal species have resulted in higher yields. Also, alternative growing substrates have provided faster or greater return on investment. Furthermore, developments in technological compounds have resulted in fewer illnesses and deaths of agricultural crops and livestock, and increased gain from nutrients, altogether resulting in greater efficiency and less waste. A big difference in recent decades is the ease of access to mass communication and end-consumers can give their opinion on developments more publicly.

Aside from the sustainability aspects of alternative protein sources, the protein needs to meet various aspects for pets and pet food manufacturers, including processing, palatability, nutrition, safety and health. These have become established over the years for traditional protein sources, but the use of any alternative protein is subjected to scrutiny for at least these aspects. In traditional commercial pet food diets based on poultry, lamb or pork, the first limiting amino acid is often methionine, with lysine second. Alternative proteins such as insect can be very low in taurine, which is a specific nutritional requirement for cats more so than for dogs. Black soldier fly (*Hermetia illucens*) and yellow mealworm (*Tenebrio molitor*) are the most frequented candidates for use in pet food. Taurine levels are around 240 mg per kilogram dry matter for both, but is about ten times that level in poultry meal (Kouřimská et al., 2023) and therefore requires different levels of added taurine in the diet to meet requirements. Although bacteria do not produce taurine, alternative proteins derived from biomass fermentation typically have a balanced amino acid profile, with essential and non-essential amino acids being available (Table 14.1). The use of any alternative protein is subjected to scrutiny based on a range of factors compared to more traditional protein sources; this chapter provides an overview of current knowledge in dogs and cats.

2 Technical functionality

It is important that any new or innovative ingredient does not unduly influence final product cost, visual appearance, palatability, digestibility, or production, nor impact shelf life and storage of the finished product. As such, producers extensively test new ingredients in different recipes and different matrices to find the best inclusion level that balances these various factors. Øverland et al. (2007) assessed the effect of partially replacing fish meal with whole cell and / or homogenates of

TABLE 14.1 Amino acid profiles of different protein sources derived from biomass fermentation, expressed as g/100g dry matter for *M. capsulatus* and MPX, and as proportion of total amino acids in *M. methylotrophus*

Amino acid	*Methylophilus methylotrophus*[1]	*Methylococcus capsulatus*[2]	MPX[3]
Histidine	2.65	1.34	1.04
Isoleucine	5.48	2.65	2.79
Leucine	8.88	4.61	4.68
Lysine	5.68	3.41	3.61
Methionine	2.72	1.61	1.12
Phenylalanine	4.47	2.49	2.43
Threonine	6.14	2.62	2.39
Tryptophan	1.00	0.77	-
Valine	7.41	3.46	3.52
Alanine	8.34	4.27	4.17
Arginine	5.56	3.89	2.90
Aspartic acid	10.39	5.21	5.31
Cystine	0.50	0.37	0.40
Glutamic acid	12.50	6.51	7.77
Glycine	7.20	3.06	3.17
Proline	3.83	2.49	1.84
Serine	4.84	2.07	2.04
Tyrosine	3.89	1.60	0.63

[1]Values converted from molar percentages given in Teller et al. (1985). [2]Longshaw et al. (2025). [3]Graham et al. (2026).

M. capsulatus in dog kibbles. Additionally, the authors assessed the effect of mild and moderate processing on a range of outcomes. Mild processing included a lower feed rate and lower temperature in the conditioning phase, and a lower screw speed and lower temperature during the extruder phase compared with the moderately processed feed (Øverland et al., 2007). Neither whole cell, bacterial homogenate nor inclusion level influenced fat leakage, durability or breaking force of kibbles. In contrast, mild processing increased extruded kibble density and decreased durability compared with the moderate processing. Inclusion of either bacterial form generally led to shorter kibbles with increased expansion, and a reduction in the percentage of dust produced as well as reduced breaking force. The authors concluded that the technical properties of dog kibble depended on processing conditions, which would need to be optimized for each formulation.

In an unpublished Calysta study (data on file, Calysta, 2023a), isonitrogenous diets were produced where low ash poultry meal in a control kibble was partially

TABLE 14.2 Recipe for feeds incorporating various protein sources to assess their technical functionality

Ingredient (%)	Control poultry meal	*Mc* meal	BSFL meal	MDM
Poultry meal	43	35	34.89	34.61
Potato flakes	28.62	28.14	26.91	29.32
Pea starch meal	15	15	15	15
Poultry fat	8.53	9.01	8.15	8.58
Palatant	3	3	3	3
Mineral premix	0.7	0.7	0.7	0.7
Pea fiber	0.5	0.5	0.5	0.5
Pure inulin	0.5	0.5	0.5	0.5
Taurine	0.1	0.1	0.1	0.1
Methylococcus capsulatus	0	8	0	0
BSFL	0	0	10.2	0
MDM	0	0	0	7.65
Antioxidant	0.05	0.05	0.05	0.05
Proximate values				
Moisture	8	8	8	8
Crude protein	35.11	35.21	35	35
Crude fat	15	15	15	15
Crude fiber	1.69	1.76	2.54	1.70
Ash	8.11	7.73	7.62	7.61

replaced with either *M. capsulatus* (*Mc* meal), insect meal (Black Soldier Fly Larvae (BSFL)), or chicken mechanically deboned meat (MDM) derived from dried chicken (Table 14.2).

In brief, ingredients were weighed and mixed prior to being added to a Coperion ZSK30 extruder with a 6.9 mm die via a high precision twin screw feeder. Samples were taken once the material achieved a bulk density of 440 g/L. Following extrusion, kibbles were dried and coated in warm poultry fat followed by a digest coating under vacuum. Kibbles were tested for microbial contamination, moisture, pH, color, viscosity, expansion, bulk density, kibble size, water activity, texture and Doris durability. Production of kibble containing *M. capsulatus* required less water (7.5 kg/h) than the control or MDM groups (8.4 kg/h), a lower die temperature (123 °C vs. 130 °C) and lower energy consumption compared with other (78.09 kW/t in *M. capsulatus* vs. 82.39–84.4 kW/t in other feeds). The pH of the kibble was lower in the control, MDM and *M. capsulatus* kibbles (5.6–5.7) compared with the insect meal kibble (6.1). Moisture content and water activity was lowest in the *M. capsulatus* kibble at 8.04% and 0.62 respectively, compared with the other kibbles (8.58% and 0.66 in the control). Hardness, crunchiness, and elasticity were lower in the *M. capsulatus* group, which was reflected in the lower Doris durability score of 90.06 compared with 98.05 in the control group. Inclusion of alternative protein sources typically made the kibble lighter, more red and less yellow in color.

Blood plasma and chicken liver are used in wet pet food production to provide additional strength to chunks due to their emulsion, water retention and gelling properties. In an unpublished report, Calysta (data on file, Calysta, 2024) assessed the effect of partially or wholly replacing 1% blood plasma and/or 18% chicken liver in a model meat recipe (Table 14.3) on a range of texture measurements. Ingredients, including ground meat products, were blended into an emulsion, transferred to individual pouches, vacuum sealed and steam cooked for 10 minutes at 100 °C. Following cooling, the emulsions were cut into 1 cm^3 chunks and analyzed for hardness, adhesiveness, elasticity, and cohesiveness using a Ta.XTPlusC Texture Analyzer from Stable Microsystems.

Although replacing 100% of the blood plasma did not affect elasticity, or cohesiveness of the chunks, replacing 50% and 100% of the plasma reduced the adhesiveness, whilst replacing 100% of the plasma reduced the hardness of the chunk suggesting that *M. capsulatus* can replace at least 50% of plasma in a chunk recipe. In contrast, replacing 50% or 100% of liver protein with *M. capsulatus* in the recipe

TABLE 14.3 Ingredients in basal recipe used to assess impact of replacing blood plasma with cultured protein

Description	**% of Recipe**
Poultry muscle tissue (cut pieces)	36
Animal-derived by-products	16
Chicken liver	18
Blood plasma (porcine, spray-dried)	1
Wheat flour (Type 405)	6.75
Pea protein	1
Tetrapotassium pyrophosphate ($K_4P_2O_7$)	0.8
Potassium chloride (KCl)	0.12
Sodium chloride (NaCl)	0.15
Potassium sorbate	0.02
Vegetable oil	2
Water	18.16
Sum (w/w)	**100**

led to an increase in elasticity for both groups, an increase in cohesiveness and a decrease in hardness when replacing it at 100%, but had no effect on adhesiveness. Approaches to mitigate these changes in chunk texture with increasing *M. capsulatus* inclusion could include the addition of vegetable proteins or hydrocolloids. Interestingly, when *M. capsulatus* was added to a matrix containing hydrocolloids such as xanthan and/or guar gum, clear phase separation occurred after sterilization. This phase separation can be avoided by mixing *M. capsulatus* with oil prior to making a gravy. That preliminary study was subsequently refined to include an additional number of reference proteins that were tested at either 1:2 or 1:5 protein to water ratio. Proteins were mixed with water and then heat treated at either 75 °C or 121 °C under pressure, allowed to cool overnight then assessed for phase separation and hardness at ambient temperature. In the 1:2 or the 1:5 mixtures, *M. capsulatus* does not form a gel at either temperature unlike plasma and egg protein. Additionally, in the 1:5 mixture pea, soy and *M. capsulatus* mix with water whilst yeast and BSFL meals drop out of solution (Figure 14.1).

In an unpublished Calysta report (data on file, Calysta, 2025a), proteins including *M. capsulatus*, plasma, whole egg, BSFL, pea protein isolate, soy protein isolate and yeast were compared for their technical properties. Proteins, water and oil were mixed in the ratios 1:3:3, 1:5:1, 1:5:7, 1:7:7, or 1:10:10, to form an emulsion. Subsequently, these mixtures were heated to 75 °C then either examined after cooling or further heated to 121 °C for 5 minutes in a high-pressure canner. Texture of the samples

FIGURE 14.1 Results of reference protein mixtures tested at a 1:5 protein to water ratio and heated to 121 °C to assess their water binding capabilities as denoted by the apparent homogeneity of the mixtures. Colour differences between mixtures containing protein sources are apparent. Where PI = protein isolate, FeedKind = *Methylococcus capsulatus* cultured protein, and BSFL – Black soldier fly larvae.

were analyzed in a Brookfield DVE viscometer whilst gel strength was measured using a CTE Texture Analyzer. When viscosity was assessed in the 1:3:3, 1:7:7 and 1:10:10 mixtures, plasma formed highly viscous pastes or gels with centipoise (cPs) values exceeding 20,000 in both heat-treated groups. In contrast, viscosity measures for *M. capsulatus* were <8600 cPs in the 1:3:3 mixture, <1200 cPs in the 1:7:7 mixture and <960 cPs in the 1:10:10 mixture. In the 1:5:1 mixtures heated to 121 °C, cPs values were as follows for the different protein sources: *M. capsulatus* (400), Plasma (>20,000), whole egg (>20,000), BSFL (580), pea protein isolate (820), soy protein isolate (3480) and yeast (120); plasma and whole egg formed gels after heat treatment. Stable emulsions were formed in *M. capsulatus*, pea and soy protein isolates in a 1:5:7 mixture at room temperature, at 75 °C and at 121 °C whilst yeast showed some minor separation. Plasma, whole egg and BSFL were not able to form homogeneous emulsions, irrespective of processing temperature (Figure 14.2). Emulsion values in samples heated to 121 °C were 6860 cPs for *M. capsulatus*, 3820 cPs for pea protein isolate, 18240 cPs for soy protein isolate and 7420 cPs for yeast. The data suggest that increasing the fat content increased the viscosity and emulsion properties of mixtures containing *M. capsulatus* supporting its inclusion in gravies and other liquid phases of feeds.

In a separate study (data on file, Calysta, 2023b), *M. capsulatus*, produced with three different particle sizes, was included at 5% in a wet dog food recipe, added to cans, sealed and sterilized for 30 minutes at 121 °C. Loaves were then assessed for a number of parameters including overall appearance, hardness, elasticity, ad-

FIGURE 14.2 Results of reference protein mixtures tested at a 1:5:1 protein to water to oil ratio and heated to 121 °C to assess their emulsion forming capabilities. Differences in emulsion formation and colour are apparent. Where FK = *Methylococcus capsulatus* cultured protein, BSFL = Black soldier fly larvae, and PI = protein isolate.

hesiveness, and weight loss after sterilization. Inclusion of *M. capsulatus* reduced hardness (611 g in control, 335–499 g in test loaves), elasticity (5.5 mm in control, 4.2–5.3 mm in test loaves), adhesiveness (-218 g/mm in control, -320 to -483 g/mm in test loaves) and weight loss of the loaves (8.02% in control, 4.16–6.01% in test loaves) suggesting that the inclusion of *M. capsulatus* in loaves can influence its overall quality.

Additional studies that incorporate microbial proteins in commercially relevant feed types, including treats and snacks, kibble, loaves, pâtés and chunks in gravy should be conducted to ensure that the most nutritionally appropriate formulations are developed with good technical functionality. Microbial proteins have now progressed from pilot testing to commercial application in pet foods, confirming their practical feasibility while highlighting the need for continued independent research.

3 Palatability / feed intake

One of the most important aspects of any feed formulation, apart from ensuring it is safe and nutritious to eat, is to make sure that it is readily accepted. Measuring palatability and feed intake are of paramount importance, especially for pets where food choice is often determined by the pet owners, rather than by the animal selecting food in the wild. From a commercial perspective, palatability tests typically take the form of a consumption test with animals either given a choice of two feeds (preference test) or a monadic, single feed (acceptance test) (Aldrich and Koppel, 2015; Watson et al., 2023). Whilst both have validity, preference tests tend to provide more robust data, enabling ranking of different ingredients (Tobie et al., 2015). Measurement of feed intake provides a useful metric for understanding if there are aversive components of the diet that might limit feed intake. Using Arctic foxes as a proxy for dogs, Skrede and Ahlstrøm (2002) showed that increasing dietary levels of up to 12% *M. capsulatus* had no effect on feed intake. In contrast, feed intake in mink declined with increased *M. capsulatus* inclusion in the short term, but increasing towards the end of the study, perhaps as the animals acclimated to the diets (Ahlstrøm et al., 2006). Feed consumption in dogs fed diets containing 10% *M. capsulatus* were comparable to control diets over a 3-week study (Babu et al., 2025); similarly feed intake was unaffected when dogs were fed diets with up to 8% *M. capsulatus* over a six-month period (Longshaw et al., 2025). Even with a two-week adaptation period, a significant 30% decrease in feed intake was recorded in dogs fed diets containing up to 9.2% *M. methylotrophus* (Giesecke et al., 1982).

More recently, de Matos & Cruz (2025) conducted preference tests in both cats and dogs, using diets in which 10% of the protein was provided by microbial protein from genus such as *Bacillus* spp., *Vibrio* spp. and/or *Geobacillus* spp. In the test formulation; 3.5% fish meal and 6.5% chicken meal were replaced by microbial protein, while the final analytical composition was maintained comparable between test and control diets. In cats (n=30), first approach was to the diet with the inclusion of microbial protein in 67% of the cases, with first consumed also favouring this diet (65%). Total food intake was significantly higher for the microbial protein diet (72% vs. 28%, $p<0.05$). In dogs (n=30), first approach (63%) and first consumed (70%) also favoured the diet with microbial protein, and overall intake was significantly greater (68% vs. 32%, $p<0.05$). At the end of the trial, pet owners reported a clear perception that animals consumed more of the diets containing microbial protein. This study provides the first direct evidence of high palatability of microbial protein in companion animals, contrasting with earlier studies in non-companion species and strengthening the case for microbial protein as a viable ingredient in pet nutrition.

4 Digestibility

To date, the only *in vivo* studies on digestibility of microbial SCPs in pets have been conducted on *M. capsulatus*. Early studies focused on digestibility in mink and in foxes as a proxy for dogs (Ahlstrøm et al., 2006; Ahlstrøm and Skrede, 1998; Øverland et al., 2007, 2006; Schøyen et al., 2005; Skrede et al., 2009). More recent work has focused on the use of a cecectomized rooster model to predict amino acid digestibility and availability in cats and dogs (Oba et al., 2023), with confirmation of protein digestibility values being completed in dogs (Longshaw et al., 2025) and cats (data on file, Calysta, 2023c).

Ahlstrøm and Skrede (1998) suggested that foxes can be used as a proxy for dogs in digestibility trials as the figures for dogs and foxes obtained in their study were comparable. Furthermore, they contend that the digestive system of foxes and dogs were functionally and anatomically similar. Whilst mink had lower digestibility scores compared with dogs and foxes, there was a higher correlation with dog data suggesting that mink was an appropriate proxy for dogs as long as an appropriate regression analysis was used. At typical inclusion rates of 5–10%, protein digestibility in mink fed whole cell *M. capsulatus* is above 80% (Ahlstrøm et al., 2006), whilst in blue foxes this was greater than 92% (Skrede and Ahlstrøm, 2002). In beagles fed diets containing up to 8% *M. capsulatus*, the fecal digestibility of protein was >80% (Longshaw et al., 2025). When *M. capsulatus* was grown on methanol or was hydrolyzed or autolyzed, digestibility of protein was increased (Mydland et al., 2008; Øverland et al., 2006; Skrede et al., 2009). Crude protein digestibility in cats is markedly lower (around 69.8% in a diet containing 8% *M. capsulatus*), but compares favorably with other protein sources including insect meal, MDM and poultry meal that had protein digestibility values of 68, 68.6, and 65.5% respectively (data on file, Calysta, 2023c). In the same study, fecal digestibility values for dry matter, fat, energy, and copper in *M. capsulatus* were 80.65, 92.6, 86.4, and 40.9% respectively. In contrast, protein digestibility of *M. capsulatus* in foxes was marginally lower compared with soybean meal and meat meal (Vhile et al., 2005). Fecal digestibility values for dry matter, fat, energy and copper in dogs fed kibble containing 8% *M. capsulatus* are 78.56, 91.44, 82.35 and 20.11% respectively (Longshaw et al., 2025). Using a cecectomized rooster model, (Oba et al., 2023) demonstrated that digestibility of essential amino acids was above 85% for *M. capsulatus* (Figure 14.3). They showed that, using Association of American Feed Control Officials (AAFCO) guide-

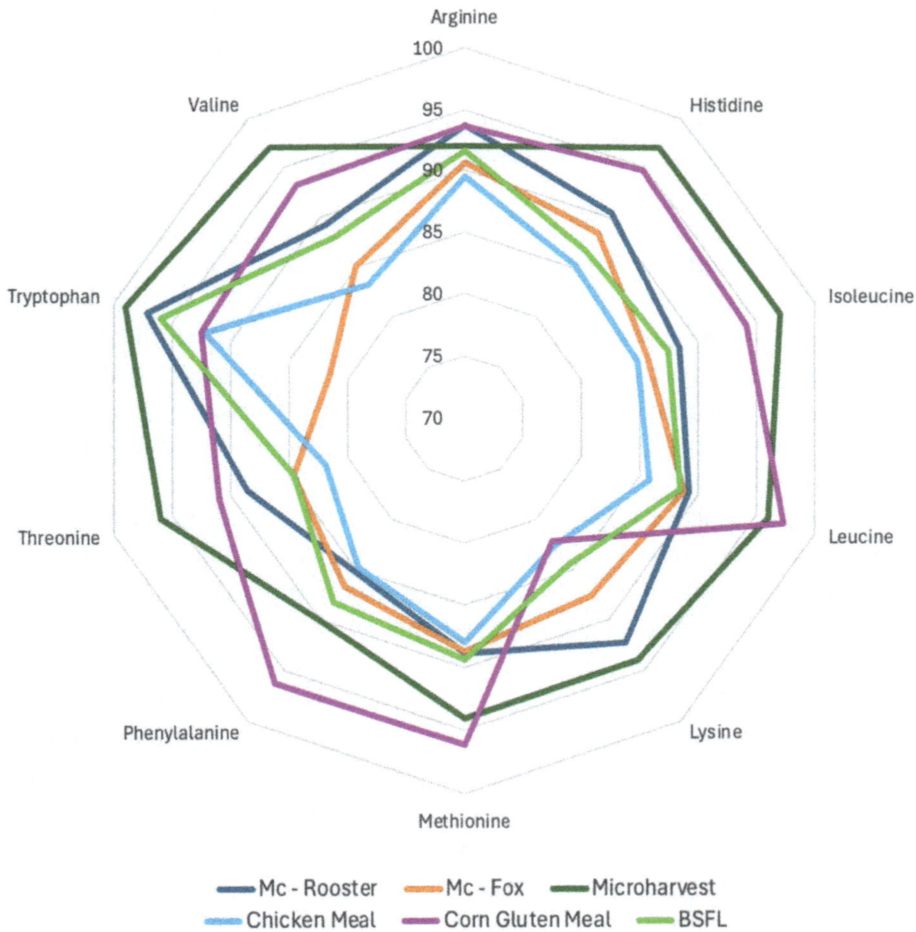

FIGURE 14.3 Digestibility of essential amino acids in *Methylococcus capsulatus* in a cecectomised rooster model (Mc - Rooster), in foxes (Mc - Fox), in microbial protein from Microharvest (Microharvest), in chicken meal, corn gluten meal and black soldier fly larvae (BSFL). Data from Microharvest derived from Graham et al. (2026), Mc - Fox data from Vhile et al. (2005), remaining data from Oba et al. (2023). Scale showing percentage digestibility.

lines, that all DIAAS-like values for the cultured protein were greater than 100 and determined that it could be used as the sole protein source in adult cats and dogs, recognizing of course the absence of taurine in the bacteria. In contrast, essential amino acid total tract digestibility values found in foxes were generally lower (Vhile et al., 2005).

In addition to these target animal-based and rooster models, more recent studies have applied the standardized INFOGEST *in vitro* digestion model, which is widely recognized for its robustness and reproducibility across protein sources. Using this approach, Graham et al. (2026) reported that the protein that comes from genus such as *Bacillus* spp., *Vibrio* spp. and/or *Geobacillus* spp. achieved a mean essential amino acid digestibility of 91.6%, comparable to casein and pea protein, and higher than fish meal and black soldier fly larvae protein. The DIAAS-like score of 0.96 further confirms the nutritional adequacy of this protein as a high-quality protein source for companion animals. These findings illustrate the scientific progress from proxy animal models to validated *in vitro* methodologies, providing greater confidence in predicting amino acid availability for dogs and cats.

5 Health and microbiome

In addition to acting as a protein replacement in feeds, one aspect that has been generally overlooked in mammalian systems, is the potential for microbial proteins to provide additional health benefits such as improved microbiome, changes in immune responses and enhanced disease resistance. Whilst these ideas have been tested to a greater or lesser extent in other animal models such as fish and shellfish (Abdel-Tawwab et al., 2025; Behera et al., 2024; Pilmer et al., 2025; Salini et al., 2025; Zhang et al., 2023), there have been fewer studies in pets, possibly due to the relatively limited examples of cultured proteins incorporated into pet foods. Focus on early phase studies such as palatability, digestibility and safety of these microbial proteins has rightly meant that more detailed studies such as health have not yet been completed. No specific disease challenges in dogs or cats fed microbial proteins have been conducted to date. However, inclusion of 10% *M. methylotrophus* in diets of mink exposed to Infectious Mink Plasmacytosis (IMP) led to reductions in growth and higher mortalities (Almlid et al., 1979). It was promulgated that the increased pathogenicity of the viral infection may have been exacerbated by endotoxins produced by the bacterial meal. Changes in the fecal microbiome of dogs fed up to 10% *M. capsulatus* were described by Babu et al. (2025) and Longshaw et al. (2025). Whilst there were limited changes in the bacterial diversity of the bacterial community, there were some changes noted in the abundance of multiple genera. Interestingly, despite differences in numbers of dogs tested, inclusion levels, and study duration, some similarities in microbial diversity were recorded.

The dominant phylum in both studies was Firmicutes with changes in the microbiome happening in both the short- and long-term study. Longshaw et al. (2025) showed that the changes that occurred were not fixed, with the general pattern of microbiome diversity and abundance reverting back to baseline values once the cultured protein was removed from the dog's diet. In both studies, decreases in Actinobacteriota, and *Bifidobacterium*, and increases in Bacteroidota, *Bacteroides* and *Fusobacterium* were noted in dogs fed *M. capsulatus* (Babu et al., 2025; Longshaw et al., 2025). The role of the fecal microbiome in supporting various physiological processes, including weight control, purine digestion, gut health and in the production of short chain fatty acids were considered by Longshaw et al. (2025). As reported below, cultured microbial proteins are generally safe as a pet food ingredient, but further work to understand their potential role as health promoters in pets needs to be conducted with urgency.

6 Target animal safety

Target animal safety studies are designed to demonstrate safety of a product at a maximum inclusion level and, for feed ingredients, typically include both long-term feeding studies as well as studies that assess genotoxicity and potential reproductive effects (Buchanan et al., 2011). Hazards associated with single cell proteins include high nucleic acid content / purines, potential allergens, heavy metals and toxicity (Das et al., 2023), but the use of multi-species bacterial consortia and high-quality feedstocks can reduce the risk of unwanted contamination and endotoxin production in biomass fermentation (Hadi and Brightwell, 2021).

Two preliminary studies on the safety of dietary *M. capsulatus* for cats have been conducted (VKM, 2006). A one-month trial was completed where cats were fed either a control diet or a diet containing 2.5% *M. capsulatus* with an assessment of body weight maintenance and fecal quality being measured. No differences in these measures were noted in either group. A second study reported in VKM (2006) concerned the feeding of adult cats a diet containing 0, 5, 10 or 20% *M. capsulatus* for eight weeks. A lower body weight gain was noted in females fed 20% *M. capsulatus* after one month, but this was not considered as treatment related according to the report authors. Additionally, there was an increase in the absolute and relative weight of the lymph nodes in the same group. No other treatment-related changes were observed. Histopathological assessment of mesenteric lymph nodes from rats, mini-pigs, pigs, foxes and cats fed *M. capsulatus* did not show any pathological changes in the lymph nodes, suggesting that the increased weight of the mesenteric lymph nodes in high doses was not problematic.

More recently, Longshaw et al. (2025) described a safety study conducted on adult beagle dogs that were fed kibble diets containing up to 8% *M. capsulatus* for six months, followed by a two-month wash out / recovery period. No adverse effects were noted in body condition score, sera chemistry, hematology or urinalysis of the dogs at the end of six-month feed trial. However, a transient and reversible increase in white blood cell counts were noted in the 6% and 8% groups in the first two months. These values had returned to normal by the next sample point. The authors concluded that *M. capsulatus* was a safe feed ingredient for dogs. A three-week study by Babu et al. (2025) concluded that a 10% inclusion of *M. capsulatus* was well tolerated by dogs with no digestive problems such as poor fecal quality or notable side effects. No detrimental shifts in microbiome diversity or richness have been noted in dogs fed *M. capsulatus* (Babu et al., 2025; Longshaw et al., 2025).

One major concern associated with the use of microbial proteins is the potential increase in purines derived from the fast growth of these protein sources (Das et al., 2023). Digestion of feed-derived DNA and RNA leads to the production of pyrimidines and purines, with purines degrading to uric acid and allantoin (Defarges et al., 2020). Excess allantoin is associated with the production of purine uroliths or urinary bladder stones (Osborne et al., 2010a, 2010b, 2009). Increased levels of dietary *M. capsulatus* led to increased urinary excretion of allantoin in pigs and mink (Ahlstrøm et al., 2006; Hellwing et al., 2007b), rather than the more problematic uric acid. The absence of urogenic effects in dogs associated with *M. capsulatus* was also noted following a six-month feed trial which was partly attributed to the relatively low levels of purines in the bacteria (2.5% of biomass), but also due to the potential actions of the gut microbiome in purine metabolism (Longshaw et al., 2025). In contrast, Giesecke et al. (1982) suggested that the microbiome of Dalmatian dogs lacked the microbial activity to degrade dietary purines and that any metabolized purines are largely excreted as uric acid. Although higher levels of purines were reported in *M. methylotrophus* fed to Dalmatian dogs, only around 50% was bioavailable with the metabolized purines being excreted as allantoin and urate in equal amounts (Gaebler et al., 1981; Giesecke et al., 1982; Giesecke and Tiemeyer, 1982). The potential detrimental effects of high purines from *M. methylotrophus* were not reported by these authors.

7 Human safety

Production of microbial proteins follows a similar pattern with the bacteria being grown in a liquid medium within sealed bioreactors. Once the appropriate concentration of the cultured protein is obtained, the resultant liquid is heat-killed then concentrated by methods such as centrifugation followed by drying to a powder. The main risk of occupational exposure to these dusts occurs after the drying phase unless appropriate mitigation is put in place. Industrial exposure to dusts are known to increase the risk of unfavorable health effects including sensitization, allergies, and acute or chronic respiratory infections (Stobnicka and Górny, 2015). Regular health and environmental monitoring of workplaces to ensure air quality remains within legislated exposure limits can help to minimize the risk of exposure and harm (Saleiro et al., 2019). Typical respiratory symptoms following occupational exposure to dusts include coughing fits, tightness of the chest, breathlessness and dyspnea (Saleiro et al., 2019). These symptoms, alongside more influenza-like symptoms such as headaches and aching limbs, have been reported for workers exposed to dusts arising from the production of *M. methylotrophus* and *M. capsulatus* (Ekenvall et al., 1983; Mayes, 1982; Skogstad et al., 2005), possibly driven by exposure to endotoxins present in the cultured proteins (Skogstad et al., 2009, 2005). These effects were transitory and generally self-resolving within 24–48 hours after exposure. Circulating antibodies in the sera of exposed workers and reductions in lung function have been reported (Mayes, 1982; Sikkeland et al., 2009; Skogstad et al., 2009, 2005). Similar results have been obtained across the food industry where exposure to dusts is an occupational hazard (Liebers et al., 2020). Implementation of appropriate monitoring and personal protective equipment coupled with improvements in downstream processing have almost entirely eliminated the issues associated with production. In particular, methods such as improvements in dewatering to produce a flocculated suspension that produced non-respirable proteins (Ekenvall et al., 1983), pelletizing of the bacterial product (Mayes, 1982), and the inclusion of 1% soya oil to minimize dusting followed by pelletization (Hellwing et al., 2007a) have reduced the issues reported.

8 Public perception

With shifting consumer perceptions, the inclusion of alternative protein sources in pet foods is seen as more acceptable. A number of surveys have been conducted to determine the public perception of single cell proteins in the human food chain and of insect proteins in the pet food space (Baptista da Silva et al., 2024; Benmeridja et al., 2025; Cedeno et al., 2025; Chezan et al., 2022; Pinney and Costa-Font, 2024; Siddiqui et al., 2023, 2022). However, no data have been published on consumer perceptions of the potential to include microbial proteins in pet food, reflecting the limited number of microbial sources available. However, two unpublished studies are available – one examining customer attitudes to *M. capsulatus* (data on file, Calysta, 2025b) in the USA and the other towards the microbial protein microbial protein from genus such as *Bacillus* spp., *Vibrio* spp. and/or *Geobacillus* spp., in Germany and the UK (EDERA Lab. Consulting, 2024). Both surveys were conducted online and questioned over 1000 respondents for UK and Germany and 500 for the USA. For the USA the preferred common name was cultured protein, whilst in Europe the name microbial protein was preferred. In Europe, 78.4% said they would be willing to buy complete dog foods containing microbial protein which dropped to 53% in the USA; 43% of respondents in the USA expected it to cost more than current protein sources, perhaps partly explaining this lower result. Beyond willingness to buy, the European survey also highlighted key drivers of acceptance, including perceptions of sustainability, health benefits and naturalness, while main barriers related to lack of familiarity with the ingredient, concerns about safety, and price expectations. Younger consumers and those with higher environmental awareness were significantly more open to microbial proteins. Importantly, when compared with insect protein, microbial protein was perceived as a cleaner and more natural option. These findings suggest that framing microbial protein within narratives of circularity and planetary health could further enhance consumer acceptance. Whilst there is a need for further work in this space, it is clear that there is an appetite for microbial proteins in pet foods, providing support for their inclusion.

9 Future direction

The global pet food industry is estimated to be worth greater than $100 billion annually. Despite an increase in cat and dog ownership and an increased desire to move towards sustainable diets that minimize environmental harm, the transition to include microbial proteins in feeds has not matched this trend, with limited numbers of commercial products available compared with mainstream products. Although microbial proteins have been tested in feeds since the 1970's (Anon, 1978; Bewersdorff and Dostálek, 1971; Braude, 1976; Liven and Staveland, 1976), only two of these are at or near commercial implementation in the industry. *Methylophilus methylotrophus* showed promise as a feed ingredient but a combination of poor health outcomes in dogs and competition from other more traditional protein sources effectively destroyed the business case for the product. Application of new and novel microbial proteins in pet foods is still in its infancy despite the apparently large numbers of potential microbial proteins being tested in terrestrial and aquatic food animals. There is therefore a need to conduct trials that assess palatability, safety, digestibility and related health outcomes in cats and dogs with these new protein sources to provide a wider choice for the industry. It is likely that each new source will impart different technical and biological functional properties to feeds and pets that could be leveraged by industry.

The majority of research and application of microbial proteins in pet foods has focused on dogs, with a much smaller number of studies considering their potential in cat foods. However, the pet industry encompasses a wider variety of animals such as invertebrates, reptiles, birds, ornamental fish and a range of small mammals such as rodents (e.g. hamsters, mice, rats, guinea pigs), and rabbits. Although there have been extensive studies on aquaculture species and poultry which have applicability to pets, there are no publications where microbial protein has been tested in other pet species, with one exception. Fatufe et al. (2010) reported on the inclusion of up to 6.8% *Streptomyces* sp. in the diets of farmed rabbits and showed that growth outcomes were favorable. Therefore, it follows that studies should be conducted that test established and newer microbial proteins in a wider variety of pet species to determine their applicability. Development of age-, stage-, species- and condition-specific diets that incorporate microbial proteins have the potential to revolutionize the industry. Additionally, the inclusion of microbial proteins in different feed types such as snacks, treats and wet and dry foods with concomitant understanding of appropriate inclusion levels will need to be conducted. Coupled with the development of new diets there should be studies that assess the mode of action. Studies to date have provided tantalizing evidence that microbial proteins have a role in palatability enhancement, improving gut health, influencing the microbiome and affecting weight gain in dogs. Use of -omics and other molecular tools to elucidate how microbial proteins exert this influence should be conducted both from an academic perspective but also to provide supporting data for the commercial sector.

More broadly, upscaling from idealistic lab scale through to pilot scale and then up to industrial scale is perhaps the biggest stumbling block to commercial success for microbial proteins. Although lab scale production is feasible and generally cost-effective, once production shifts to pilot and commercial scale, costs can become prohibitively high. Support from investors, government and industry can help to mitigate some of these costs, allowing the microbial protein space to develop to its full potential. Improvements in downstream processing that minimize dusting, reduce potential toxicants, and advance technical functionality including the production of functional fractions and lysates will further develop the industry. At the same time, recent advances are beginning to generate more robust evidence for microbial proteins in pet foods. Alongside earlier work on *M. capsulatus* and *M. methylotrophus*, which provided initial insights into digestibility and feed intake, newer studies have applied standardized *in vitro* digestibility models, companion animal palatability trials, and consumer perception surveys. Together, these findings support the feasibility of microbial proteins as safe, nutritious and desirable ingredients in pet diets. In parallel, regulatory recognition – including the listing of relevant microbial groups on the EFSA Qualified Presumption of Safety (QPS) list and inclusion in the EU Catalogue of Feed Materials – provides a strong foundation for their safe commercialization and wider adoption in the industry.

Conflict of interest statement

The authors declare that some data and studies cited in this chapter were generated within company research programs at Calysta Ltd. and MicroHarvest GmbH.

Authors affiliated with these companies contributed technical expertise and experimental data but did not influence the interpretation of results or the editorial review. All efforts were made to present information in an objective and scientifically balanced manner.

References

Abdel-Tawwab, M., Khalil, R.H., Abo Selema, T.A.M., Abdelsalam, M., Abdelhakim, T.M.N., Sabry, E.A., Abd El-Ghaffar, H.A., Saad, A.H., 2025. Evaluating the inclusion of *Clostridium autoethanogenum* protein instead of fishmeal protein in diets for European seabass (*Dicentrarchus labrax*): Growth performance, digestive enzymes, health status, and tissues investigations. Animal Feed Science and Technology 324, 116318. https://doi.org/10.1016/j.anifeedsci.2025.116318

Ahlstrøm, Ø., Skrede, A., 1998. Comparative Nutrient Digestibility in Dogs, Blue Foxes, Mink and Rats. The Journal of Nutrition 128, 2676S-2677S. https://doi.org/10.1093/jn/128.12.2676S

Ahlstrøm, Ø., Tauson, A.-H., Hellwing, A.L.F., Mydland, L.T., Skrede, A., 2006. Growth performance, nitrogen balance and urinary purine derivatives in growing-furring mink (*Mustela vison*) fed bacterial protein produced from natural gas. J. Anim. Feed Sci. 15, 491–504. https://doi.org/10.22358/jafs/66919/2006

Aldrich, G.C., Koppel, K., 2015. Pet Food Palatability Evaluation: A Review of Standard Assay Techniques and Interpretation of Results with a Primary Focus on Limitations. Animals 5, 43–55. https://doi.org/10.3390/ani5010043

Almlid, T., Frøysedal, K., Nordstoga, K., 1979. Accelerated Development of Infectious Mink Plasmacytosis (IMP) in Animals Fed Single-cell Protein Produced by *Pseudomonas methylotropha* (Pruteen). Acta Agriculturae Scandinavica 29, 363–368. https://doi.org/10.1080/00015127909435249

Anon, 1978. Single-cell protein pilot unit starts up. C&EN May 15, 1978, 20.

Babu, R., Padmanabhan, S., Ganesan, R., Subbian, E., Van, T.T.H., Eri, R., 2025. Sustainable, greenhouse gas derived fermented protein in canine diets—a pilot study. Front. Vet. Sci. 11. https://doi.org/10.3389/fvets.2024.1477182

Banks, M., Johnson, R., Giver, L., Bryant, G., Guo, M., 2022. Industrial production of microbial protein products. Current Opinion in Biotechnology 75, 102707. https://doi.org/10.1016/j.copbio.2022.102707

Baptista da Silva, C., Santos Neto, E.L. dos, Broeckx, B.J.G., Ruiz-Suárez, N., Eeckhout, M., Hesta, M., 2024. Assessing pet owners' willingness to embrace insects in pet food. Journal of Animal Physiology and Animal Nutrition 108, 1833–1843. https://doi.org/10.1111/jpn.14019

Behera, G., Kawade, S., Reddy, B., Sedyaaw, P., 2024. Single cell protein in aquaculture: a comprehensive review. International Journal of Development Research Vol. 14, 65379–65383.

Benmeridja, N., Deltomme, B., Sakarika, M., Rini, L., De Steur, H., Gellynck, X., 2025. Feeding the future: consumer willingness to try bacterial protein, a comparative study with fungi, algae and cultured meat. BFJ. https://doi.org/10.1108/BFJ-10-2024-1095

Bewersdorff, M., Dostálek, M., 1971. The use of methane for production of bacterial protein. Biotechnology and Bioengineering 13, 49–62. https://doi.org/10.1002/bit.260130104

Braude, R., 1976. New sources of protein for pigs. Proceedings of the Nutrition Society 35, 93–101. https://doi.org/10.1079/PNS19760015

Buchanan, R.L., Baker, R.C., Charlton, A.J., Riviere, J.E., Standaert, R., 2011. Pet food safety: a shared concern. British Journal of Nutrition 106, S78–S84. https://doi.org/10.1017/S0007114511005034

Cedeno, F.R.P., Olubiyo, O.J., Ferreira, S., 2025. From microbial proteins to cultivated meat for alternative meat-like products: a review on sustainable fermentation approaches. Journal of Biological Engineering 19, 44. https://doi.org/10.1186/s13036-025-00509-9

Chezan, D., Flannery, O., Patel, A., 2022. Factors affecting consumer attitudes to fungi-based protein: A pilot study. Appetite 175, 106043. https://doi.org/10.1016/j.appet.2022.106043

Das, R.S., Dong, G., Tiwari, B.K., Garcia-Vaquero, M., 2023. Chapter 22 - Food safety concerns of alternative proteins and regulatory guidelines for their commercialization in the human food market, in: Tiwari, B.K., Healy, L.E. (Eds.), Future Proteins. Academic Press, pp. 469–508. https://doi.org/10.1016/B978-0-323-91739-1.00022-2

Defarges, A., Evason, M., Dunn, M., Berent, A., 2020. Urolithiasis in Small Animals, in: Clinical Small Animal Internal Medicine. John Wiley & Sons, Ltd, pp. 1123–1156. https://doi.org/10.1002/9781119501237.ch123

de Matos, A.M.M., & Cruz, A.L.B. (2025). Microbial protein as a novel ingredient in pet food: a sensory acceptance study in dogs and cats. Research Square. https://doi.org/10.21203/rs.3.rs-6706331/v1

EDERA Lab Consulting (2024). *Consumer perceptions of microbial protein in pet food in Germany and the UK*. Unpublished market research report, commissioned by MicroHarvest GmbH.

Eisen, J.A., 1995. The RecA Protein as a Model Molecule for Molecular Systematic Studies of Bacteria: Comparison of Trees of RecAs and 16S rRNAs from the Same Species. J Mol Evol 41, 1105–1123. https://doi.org/10.1007/BF00173192

Ekenvall, L., Dölling, B., Göthe, C.J., Ebbinghaus, L., Stedingk, L.V. von, Wasserman, J., 1983. Single cell protein as an occupational hazard. Occupational and Environmental Medicine 40, 212–215. https://doi.org/10.1136/oem.40.2.212

Fatufe, A.A., Matanmi, I.O., Alalade, A.O., 2010. Performance and carcass characteristics of growing rabbits fed bacterial protein meal. African Journal of Food, Agriculture, Nutrition and Development 10. https://doi.org/10.4314/ajfand.v10i4.55332

Gaebler, S., Maier, J., Tiemeyer, W., Giesecke, D., 1981. Über den Einfluß von Diäten mit definiertem Nucleinsäuren- und Puringehalt auf den Harnsäurespiegel im Blutplasma von Dalmatiner-Hunden. Zentralblatt für Veterinärmedizin Reihe A 28, 494–503. https://doi.org/10.1111/j.1439-0442.1981.tb01217.x

Giesecke, D., Gaebler, S., Tiemeyer, W., 1982. Purine Availability and Metabolism in Dogs Fed Single-Cell Protein or RNA. The Journal of Nutrition 112, 1822–1826. https://doi.org/10.1093/jn/112.10.1822

Giesecke, D., Tiemeyer, W., 1982. Availability and metabolism of purines of single-cell proteins in monogastric animals. Proceedings of the Nutrition Society 41, 319–327. https://doi.org/10.1079/PNS19820047

Graham, R., de Matos, A.M., Parr, T., Brameld, J.M., 2026. Evaluating quality of novel alternative protein sources (microbial and insect proteins) using the INFOGEST *in vitro* digestion model. Food Chemistry 501, 147468. https://doi.org/10.1016/j.foodchem.2025.147468

Hadi, J., Brightwell, G., 2021. Safety of Alternative Proteins: Technological, Environmental and Regulatory Aspects of Cultured Meat, Plant-Based Meat, Insect Protein and Single-Cell Protein. Foods 10, 1226. https://doi.org/10.3390/foods10061226

Hellwing, A.L.F., Tauson, A.-H., Kjos, N.P., Skrede, A., 2007a. Bacterial protein meal in diets for growing pigs: effects on protein and energy metabolism. Animal 1, 45–54. https://doi.org/10.1017/S1751731107283879

Hellwing, A.L.F., Tauson, A.-H., Skrede, A., 2007b. Excretion of purine base derivatives after intake of bacterial protein meal in pigs. Livestock Science, 10th International Symposium on Digestive Physiology in Pigs, Denmark 2006, Part 2 109, 70–72. https://doi.org/10.1016/j.livsci.2007.01.070

Hohnloser, W., Lingens, F., Präve, P., 1978. Characterization of a new methylotrophic strain, *Methylomonas clara*. European J. Appl. Microbiol. Biotechnol. 6, 167–179. https://doi.org/10.1007/BF00504428

Liebers, V., Brüning, T., Raulf, M., 2020. Occupational endotoxin exposure and health effects. Arch Toxicol 94, 3629–3644. https://doi.org/10.1007/s00204-020-02905-0

Liven, E., Staveland, K., 1976. Single Cell Protein in the Diets of Pigs and Chickens. Acta Vet Scand 17, 441–450. https://doi.org/10.1186/BF03547899

Lloyd, A.T., Sharp, P.M., 1993. Evolution of the recA gene and the molecular phylogeny of bacteria. J Mol Evol 37, 399–407. https://doi.org/10.1007/BF00178869

Longshaw, M., Quest, B., Miller, W., Oba, P.M., Swanson, O.R., Swanson, K.S., Miller, K., 2025. The safety of FeedKind Pet® (*Methylococcus capsulatus*, Bath) as a cultured protein source in the diet of adult dogs and its effect on feed digestibility, fecal microbiome, and health status. Animals 15 (13), 1975. https://doi.org/10.3390/ani15131975

Matos, A.M.M. de, Cruz, A.L.B., 2025. Microbial Protein as a Novel Ingredient in Pet Food A Sensory Acceptance Study in Dogs and Cats. https://doi.org/10.21203/rs.3.rs-6706331/v1

Mayes, R.W., 1982. Lack of allergic reactions in workers exposed to Pruteen (bacterial single-cell protein). Occupational and Environmental Medicine 39, 183–186. https://doi.org/10.1136/oem.39.2.183

Motta, A., 2024. Microbial Protein: Pet Owners perceptions and Pet Acceptance.

Mydland, L.T., Frøyland, J.R.K., Skrede, A., 2008. Composition of individual nucleobases in diets containing different products from bacterial biomass grown on natural gas, and digestibility in mink (*Mustela vison*). Journal of Animal Physiology and Animal Nutrition 92, 1–8. https://doi.org/10.1111/j.1439-0396.2007.00674.x

Oba, P.M., Utterback, P.L., Longshaw, M., Parsons, C.M., Swanson, K.S., 2023. Comparing the standardized amino acid digestibility of an alternative protein source with commercially available protein-based ingredients using the precision-fed cecectomized rooster assay. Journal of Animal Science 101, skad236. https://doi.org/10.1093/jas/skad236

Osborne, C.A., Bartges, J.W., Lulich, J.P., Albasan, H., Weiss, C., 2010a. Canine Purine Urolithiasis: Causes, Detection, Management and Prevention, in: Small Animal Clinical Nutrition.

Osborne, C.A., Lulich, J.P., Forrester, D., Albasan, H., 2009. Paradigm Changes in the Role of Nutrition for the Management of Canine and Feline Urolithiasis. Veterinary Clinics of North America: Small Animal Practice, Changing Paradigms in Diagnosis and Treatment of Urolithiasis 39, 127–141. https://doi.org/10.1016/j.cvsm.2008.10.001

Osborne, C.A., Lulich, J.P., Ulrich, L.K., 2010b. Canine Urolithiasis: Definitions, Pathophysiology and Clinical Manifestations.

Øverland, M., Romarheim, O.H., Ahlstrøm, Ø., Storebakken, T., Skrede, A., 2007. Technical quality of dog food and salmon feed containing different bacterial protein sources and processed by different extrusion conditions. Animal Feed Science and Technology 134, 124–139. https://doi.org/10.1016/j.anifeedsci.2006.05.014

Øverland, M., Romarheim, O.H., Hovin, M., Storebakken, T., Skrede, A., 2006. Apparent total tract digestibility of unprocessed and extruded diets containing basic and autolyzed bacterial protein meal grown on natural gas in mink and rainbow trout. Animal Feed Science and Technology 129, 237–251. https://doi.org/10.1016/j.anifeedsci.2005.12.017

Pilmer, L., Woolley, L., Lymbery, A., Dam, C., Elizur, A., Foysal, M.J., Partridge, G., 2025. Exploring single cell microbial protein as a sustainable fishmeal alternative in yellowtail kingfish (*Seriola lalandi*) diets: impacts on health and gut microbiome. J Animal Sci Biotechnol 16, 16. https://doi.org/10.1186/s40104-024-01146-w

Pinney, J., Costa-Font, M., 2024. A Model for Consumer Acceptance of Insect-Based Dog Foods among Adult UK Dog Owners. Animals 14, 1021. https://doi.org/10.3390/ani14071021

Saleiro, S., Rocha, L., Bento, J., Antunes, L., Costa, J.T. da, 2019. Occupational exposure to dust: an underestimated health risk? J. bras. pneumol. 45, e20170396. https://doi.org/10.1590/1806-3713/e20170396

Salini, M.J., Wang, W., Mock, T.S., Siddik, M.A.B., Jago, M.K., Bews, K.L.F., Francis, D.S., 2025. Expanding the Ingredient Basket in Aquaculture: Growth Performance and Feed Utilization of Australian Hybrid Abalone (*Haliotis laevigata* × *H. rubra*) Fed Methanotrophic Single Cell Protein. Aquaculture Nutrition 2025, 7291857. https://doi.org/10.1155/anu/7291857

Schoch, C., 2020. *Methylophilus methylotrophus* [WWW Document]. NCBI Taxonomy: a comprehensive update on curation, resources and tools. URL https://www.ncbi.nlm.nih.gov/Taxonomy/Browser/wwwtax.cgi?lvl=0&id=17 (accessed 6.5.25).

Schøyen, H.F., Frøyland, J.R.K., Sahlström, S., Knutsen, S.H., Skrede, A., 2005. Effects of autolysis and hydrolysis of bacterial protein meal grown on natural gas on chemical characterization and amino acid digestibility. Aquaculture, Fish Nutrition and Feeding 248, 27–33. https://doi.org/10.1016/j.aquaculture.2005.04.017

Siddiqui, S.A., Alvi, T., Sameen, A., Khan, S., Blinov, A.V., Nagdalian, A.A., Mehdizadeh, M., Adli, D.N., Onwezen, M., 2022. Consumer Acceptance of Alternative Proteins: A Systematic Review of Current Alternative Protein Sources and Interventions Adapted to Increase Their Acceptability. Sustainability 14, 15370. https://doi.org/10.3390/su142215370

Siddiqui, S.A., Brunner, T.A., Tamm, I., van der Raad, P., Patekar, G., Alim Bahmid, N., Aarts, K., Paul, A., 2023. Insect-based dog and cat food: A short investigative review on market, claims and consumer perception. Journal of Asia-Pacific Entomology 26, 102020. https://doi.org/10.1016/j.aspen.2022.102020

Sikkeland, L.I.B., Eduard ,W., Stangeland ,A. M., Thorgersen ,E. B., Haug ,T., Aukrust ,P., Halvorsen ,B., Mollnes ,T. E., and Kongerud, J., 2009. Occupational exposure to bacterial single cell protein induces inflammation in lung and blood. Inhalation Toxicology 21, 674–681. https://doi.org/10.1080/08958370802491425

Skogstad, M., Eduard, W., Holme, J., Qvenild, T., Einarsdottir, E., 2005. Short term changes in lung function, leukocytosis in blood, and lachrymal fluid among bacterial single cell protein workers after an episode with high exposure to endotoxins. Occup Environ Med 62, 576–580. https://doi.org/10.1136/oem.2004.019273

Skogstad, M., Sikkeland, L.I.B., Øvstebø, R., Bente, K., Haug, F., Kierulf, P., Heldal, K.K., Kongerud, J., 2009. Does endotoxin exposure affect lung function and induce systemic inflammation in workers handling bacterial single cell protein? SJWEH suppl 7, 36–40.

Skrede, A., Ahlstrøm, Ø., 2002. Bacterial Protein Produced on Natural Gas: A New Potential Feed Ingredient for Dogs Evaluated Using the Blue Fox as a Model. The Journal of Nutrition 132, 1668S-1669S. https://doi.org/10.1093/jn/132.6.1668S

Skrede, A., Mydland, L.T., Øverland, M., 2009. Effects of growth substrate and partial removal of nucleic acids in the production of bacterial protein meal on amino acid profile and digestibility in mink. J. Anim. Feed Sci. 18, 689–698. https://doi.org/10.22358/jafs/66442/2009

Stobnicka, A., and Górny, R.L., 2015. Exposure to flour dust in the occupational environment. International Journal of Occupational Safety and Ergonomics 21, 241–249. https://doi.org/10.1080/10803548.2015.1081764

Teller, E., Godeau, J.-M., van Nevel, C.J., Demeyer, D.I., 1985. Comparative in vitro and in vivo evaluation of protein degradability in the rumen using soybean meal and Pruteen. Zeitschrift für Tierphysiologie Tierernährung und Futtermittelkunde 54, 121–130. https://doi.org/10.1111/j.1439-0396.1985.tb01523.x

Thibault, G., Mazen, R., 2022. Extracts from fast growing microbes. WO 2022/268842 A1.

Tobie, C., Péron, F., Larose, C., 2015. Assessing Food Preferences in Dogs and Cats: A Review of the Current Methods. Animals (Basel) 5, 126–137. https://doi.org/10.3390/ani5010126

Vhile, S.G., Skrede, A., Ahlstrøm, Ø., Szymeczko, R., Hove, K., 2005. Ileal and total tract nutrient digestibility in blue foxes (*Alopex lagopus*) fed extruded diets containing different protein sources. Archives of Animal Nutrition 59, 61–72. https://doi.org/10.1080/17450390512331342395

VKM, 2006. Opinion on the safety of BioProtein® by the Scientific Panel on Animal Feed of the Norwegian Scientific Committee for Food Safety. Revised version (No. 2006:43). Norwegian Scientific Committee for Food Safety.

Watson, P.E., Thomas, D.G., Bermingham, E.N., Schreurs, N.M., Parker, M.E., 2023. Drivers of Palatability for Cats and Dogs—What It Means for Pet Food Development. Animals 13, 1134. https://doi.org/10.3390/ani13071134

Zhang, J., Yu, M., Wang, J., Longshaw, M., Song, K., Wang, L., Li, X., Zhang, C., Lu, K., 2023. Methanotroph (*Methylococcus capsulatus*, Bath) bacteria meal alleviates soybean meal-induced enteritis in spotted seabass (*Lateolabrax maculatus*) by modulating immune responses and the intestinal flora. Aquaculture 575, 739795. https://doi.org/10.1016/j.aquaculture.2023.739795

CHAPTER 15

Emerging and underutilized aquatic ingredients in canine nutrition

A.R.J. Cabrita and A.J.M. Fonseca

REQUIMTE, LAQV, ICBAS, School of Medicine and Biomedical Sciences, University of Porto, Rua Jorge Viterbo Ferreira 228, 4050-313 Porto, Portugal
**Correspondence: arcabrita@icbas.up.pt; ajfonseca@icbas.up.pt*

Abstract

The rising global pet population, coupled with the increasing humanization of companion animals, has intensified concerns regarding food quality, safety, and the health implications of diet. Pet owners now demand foods that not only meet nutritional requirements but also support overall well-being and align with environmentally sustainable practices. Meeting this demand requires novel ingredients capable of enhancing feed efficiency while delivering both nutritional and functional benefits, with minimal environmental, economic, and societal impact. Aquatic plants and animals represent promising but mainly underexplored resources in canine nutrition. They are rich in nutrients and bioactive compounds with potential to promote health and sustainability. This review examines emerging aquatic ingredients with potential for dog food, focusing on macroalgae, duckweed, mussels, squid, krill, and shrimp. While these resources are already used in the feed industry at a variable extent, their use in dog food remains limited, and their nutritional composition, digestibility, and functional properties are not yet fully characterized. However, their inclusion could support canine health while it has potential to contribute to the development of more resource-efficient and ethically responsible food systems. Current evidence suggests that these ingredients may improve diet digestibility, modulate gut microbiota, and provide compounds with a wide range of biological activities. However, *in vivo* studies with dogs remain scarce, and robust data on palatability, long-term safety, and health outcomes are lacking. Future research should address these gaps by conducting controlled feeding trials in dogs, assessing both physiological and functional effects, while also considering market feasibility and consumer acceptance. Advancing knowledge in these areas will be critical for integrating aquatic-derived ingredients into canine diets and fostering innovation that supports both animal health and the sustainable growth of the pet food industry.

Keywords

dog nutrition – health effects – novel ingredients – sustainability

1 Introduction

According to the latest report from the European Pet Food Industry Federation (FEDIAF, 2024), the dog population in Europe was approximately 106 million in 2022, with 25% of households owning at least one dog. In 2024, the number of dogs in the United States was estimated at around 90 million, with roughly half of all households owning a pet dog (AVMA, 2024). In Asia, China has the largest population of pet dogs, with approximately 58 million urban dogs in 2024 (Statista, 2025). Drivers of the increasing pet population trend include sociodemographic changes such as household size, generational shifts in attitudes toward pets, growing environmental concerns, and the impact of the COVID-19 pandemic. These figures also highlight the importance of the human-dog relationship in modern society, emphasizing that when this relationship is successful, both members benefit mutually, promoting the well-being and health of the dyad.

Additionally, the growing humanization of companion animals has heightened owners' concerns about food quality, safety, and the impact of food on their pets' health and well-being. In fact, owners are increasingly looking for complete foods that meet their pets' nutritional needs and have functional effects, particularly in disease prevention, promoting well-being, and ensuring healthy longevity (Euromonitor, 2022). Consequently, production of premium compound foods has increased. To meet this demand and promote the sustainability of the sector, we must urgently identify ingredients and supplements that increase food efficiency and provide nutritional, health, and welfare benefits that are environmentally, economically, and socially sustainable. A vast array of aquatic plants and animals can serve as sources of nutrients and bioactive compounds that have significant potential for use in dog food. Incorporating these sources into the pet food industry could make it more sustainable by reducing the environmental impact of land-based food production. These new or underutilized ingredients can also benefit dogs with food sensitivities and create more resource-efficient and ethical food sources. This addresses pet owners' concerns about sustainability and the availability of health-conscious options, while also allowing for innovation and differentiation in the market.

In this context, microalgae have been proposed as a means of achieving sustainable food production due to their higher production efficiency compared to terrestrial plants. The literature on *in vivo* studies examining the biological effects of various microalgae species, including *Arthrospira platensis*, *Chlorella* sp., *Euglena gracilis*, *Haematococcus pluvialis*, *Nannochloropsis oceanica*, *Tetradesmus obliquus*, and *Schizochytrium* sp. was recently reviewed (Cabrita *et al.*, 2025). The results indicated health benefits for dogs and support the growing market for complete foods or treats containing microalgal biomass or extracts. Similarly, fish oil and fish protein hydrolysates have long been used in dog food, and their health benefits have been well-documented (Ribeiro *et al.*, 2024; Rodrigues *et al.*, 2024). From a circular economy and zero waste perspective, obtaining these food resources from by-products and co-products of the fish industry and incorporating them into pet food contributes to the sustainability of the fisheries, aquaculture, and pet food sectors, as well as to animal health (Cabrita *et al.*, 2024a; Cabrita *et al.*, 2024b), ultimately promoting the health of aquatic ecosystems, an essential component of the One Health concept."

This chapter aims to review the available information on emerging aquatic food sources that are underexplored in canine nutrition and not yet well characterized concerning their nutritional and functional potential. Within this framework, we highlight macroalgae, duckweed, mussels, squid, krill, and shrimp as promising new ingredients for dog food that are already used in the compound feed industry at a variable extent. Selected amino acid profiles, fatty acid profiles, and proximate composition for these materials can be found in Part 2 of this volume. These values are intended for orientation and to facilitate comparison with conventional protein and lipid sources, but should be interpreted with caution, as the composition of novel aquatic ingredients may vary considerably depending on species, origin, cultivation conditions, and processing methods. To the best of our knowledge, no comprehensive or officially standardized nutrient tables, such as those available from INRAE or FAO for conventional feed ingredients, exist for these materials. For each source, we provide an overview of its chemical composition and bioactive compounds. Then, we discuss the limited *in vivo* data available, as well as current trends, future directions, and key concerns. We hope this chapter encourages creative thinking and the development of innovative solutions.

2 Macroalgae

Macroalgae are eukaryotic, photosynthetic organisms found in marine and freshwater environments. They are classified into three main groups based on pigmentation: green (Chlorophyta), red (Rhodophyta), and brown (Phaeophyta). Green algae derive their color from chlorophyll; red algae derive theirs from phycoerythrin and phycocyanin, which mask other pigments such as chlorophyll and carotenoids; and brown algae derive theirs from fucoxanthin, which conceals chlorophyll and other carotenoids (O'Sullivan *et al.*, 2010). The cultivation of macroalgae demands substantially fewer resources, such as freshwater and arable land, compared with conventional feed crops (O'Sullivan *et al.*, 2010). Furthermore, macroalgae contribute to mitigating nutrient pollution and eutrophication by assimilating excess nitrogen and phosphorus from wastewater and aquaculture effluents (Marinho-Soriano et al., 2009). Because of their sustainability and nutritional value, macroalgae and their extracts have gained attention as alternative ingredients in human and livestock diets, potentially reducing reliance on land-based resources (Cabrita *et al.*, 2019; Makkar *et al.*, 2016; Wells *et al.*, 2017).

2.1 Chemical composition and bioactive compounds

Macroalgae possess a diverse nutritional profile comprising carbohydrates, proteins, lipids, and essential minerals such as iodine, iron, calcium, and magnesium. They are also a source of water-soluble vitamins, including the B-complex and vitamin C, as well as fat-soluble vitamins A, D, E, and K (Cabrita *et al.*, 2016; Kumar *et al.*, 2008). Although macroalgae contain only 1–5% lipids, they are primarily polyunsaturated fatty acids (PUFA), including eicosapentaenoic acid (EPA, 20:5 n3) and arachidonic acid (ARA, 20:4 n6). These fatty acids are particularly prevalent in brown and red algae (Ganesan *et al.*, 2019). Macroalgae contain unique undigestible polysaccharides (e.g., alginate, laminarins and fucoidans in brown algae; carrageenans and agar in red algae; ulvans in green algae) that can act as prebiotics, supporting the growth of beneficial gut bacteria and improving nutrient digestion and absorption in mammals (O'Sullivan *et al.*, 2010). Several *in vivo* studies in monogastric farm animals, including pigs, poultry, and fish, have reported beneficial effects on intestinal microbiota composition, immune system response, and overall health when using intact brown macroalgae or its extracts (Okolie *et al.*, 2017; Øverland *et al.*, 2019). Conversely, the evidence remains limited in companion animals.

The chemical composition of macroalgae depends on several factors, including species, habitat, harvest season, and environmental conditions such as light exposure, temperature, salinity, and nutrient availability (Marsham *et al.*, 2007). Generally, green and red algae are considered to have higher nutritional value than brown algae. Although brown algae have lower protein content (typically 5–15%, dry matter, DM, basis) than red and green algae, they have a broader and more diverse repertoire of secondary metabolites of commercial significance (Mišurcová, 2011). Macroalgae present high-levels of nonprotein nitrogen, reduced the adequate conversion factors of nitrogen to protein of 5.38, 4.92 and 5.13 for brown, red and green macroalgae, respectively (Morais *et al.*, 2020). Red macroalgae are recognized for having higher-quality protein, with higher concentrations of methionine and lysine, than brown and green macroalgae (Angell *et al.*, 2016). Furthermore, macroalgae contain high levels of glutamic acid in both free and protein-bound forms that contributes to the umami flavor (Øverland *et al.*, 2019).

One of the most remarkable features of macroalgae is their abundance in bioactive compounds. These include sulphated polysaccharides, PUFA, polyphenols, phlorotannin compounds, carotenoids, and phycobiliproteins, peptides, and bioactive amino acids (AA). These compounds are associated with a wide range of biological activities, including antioxidant, antimicrobial, antiviral, anti-inflammatory, immunomodulatory, prebiotic, and anticancer properties (Chojnacka *et al.*, 2012; Gomez-Zavaglia *et al.*, 2019; Healy *et al.*, 2023).

2.2 *In vivo* experimental evidence

Despite the availability of commercial diets and treats for dogs containing macroalgae and the recognized potential of macroalgae-based foods as sustainable and nutrient-rich components of animal feed, the scarcity of robust scientific studies has hindered their broader adoption in companion animal nutrition. *Ascophyllum nodosum*, abundantly distributed throughout the northwest coast of Europe and the northeastern coast of North America, is the most widely researched brown, intertidal macroalgae. Several studies have highlighted the abundance of bioactive compounds in *A. nodosum* with functional properties, namely antioxidant, anti-inflammatory, immunostimulant, antibacterial, anticancer, prebiotic, and antidiabetic properties (Dutot *et al.*, 2012; Ramnani *et al.*, 2012; Riou *et al.*, 1996; Turner *et al.*, 2002; Zhang *et al.*, 2007). The functional properties of *A. nodosum* are primarily attributed to the prebiotic effects of its complex polysaccharides, complemented by other bioactive constituents such as phlorotannins and antioxidants, being also a source of vitamins and chelated microminerals, known to be more bioavailable for the animal than inorganic sources (Evans and Critchley, 2014; MacArtain *et al.*, 2007). Although *A. nodosum* is widely used in companion animal products, particularly due to its beneficial effects on oral health (Gawor and Jank, 2023; Gawor *et al.*, 2018; Gawor *et al.*, 2021), few studies have attempted to evaluate its influence on other parameters, being most published studies conducted in livestock species (Morais *et al.*, 2020).

Isidori *et al.* (2021) performed a clinical trial to evaluate the effects of dietary supplementation of *A. nodosum* and/or probiotic microorganism *Bacillus subtilis* C-3102, in dogs diagnosed with canine chronic inflammatory enteropathy. Authors did not find effects on clinical signs, besides increased numbers of beneficial carbohydrate-fermenting bacteria, suggesting a potentially positive effect on gut health. Table 15.1 presents results from *in vivo* studies focused on evaluating effects of macroalgae inclusion in dogs' diets. Regarding effects on diet palatability, Isidori *et al.* (2019; Table 15.1) investigated the effect of adding *A. nodosum* at low (0.3%) or high (1.0%) concentrations to a control extruded food and reported no differences in first choice, but food consumption and intake ratio were lower with diet containing 1.0% *A. nodosum*, when compared to the control diet. As no effects were observed in first choice, these results suggest that the negative effects on palatability are primarily related to gustatory factors rather than to olfactory stimuli, contradicting the belief that the pungent smell of *A. nodosum* is responsible for diminished food consumption.

Pinna *et al.* (2021) investigated the dietary effects of supplementing dog diets with *A. nodosum*, *Undaria pinnatifida*, *Saccharina japonica* (brown macroalgae), and *Palmaria palmata* (red macroalgae), and reported no significant effects on the measured parameters. Given that total polysaccharide content typically ranges from 10% to 75% in brown macroalgae and from 10% to 59% in red algae (Sardari and Nordberg Karlsson, 2018), the lack of observable effects is difficult to explain. The authors suggested that the quantity of indigestible polysaccharides reaching the hindgut may have been insufficient to induce measurable modulation on the fecal parameters and further noted that only a limited number of key taxa of the canine microbiome was assessed. Overall, the study demonstrated that intact macroalgae supplementation at 1.5% of diet was well tolerated and did not compromise diet digestibility, but it failed to produce detectable changes in fecal microbiota and immunological outcomes.

Recently, Zhang *et al.* (2025) found that dietary supplementation with a commercial macroalgae extract (2%) improved the loose stools of dogs, increased serum antioxidase activities, immunoglobulin (Ig)A, IgM, and interleukin (IL)-10 levels and lowered serum malondialdehyde concentrations, inflammatory markers (IL-1β, C-reactive protein, and tumor necrosis factor-α) and intestinal barrier markers (d-lactate, lipopolysaccharide, intestinal fatty acid-binding protein, and diamine oxidase). Moreover, macroalgae extract increased short-chain fatty acids (SCFA)-producing gut microbiota abundance, particularly *Bifidobacterium* and *Holdemanella*, and fecal propionate concentrations. Overall, the results suggest that dietary supplementation with the studied macroalgae extract enhances antioxidant and prebiotic effects, suppresses the activation of nuclear factor kappa-B (NF-κB) and thereby attenuates inflammatory responses, helps regulate the balance between proinflammatory and anti-inflammatory pathways, and strengthens both the immune response and gut barrier function in dogs, potentially supporting intestinal health. Because the macroalgae extract contains multiple constituents, the specific mechanisms of individual components and their potential interactions warrant further investigation.

The effect of dietary supplementation with two tropical red macroalgae, *Kappaphycus alvarezii* and *Gracilaria salicornia* at three different levels (0, 1.5 and 3%) was studied by Srinivas *et al.* (2024). Conversely to the findings of Isidori *et al.* (2019), supplementation with the tropical red macroalgae studied had no adverse effect on food acceptability. These contradictory results may be explained by the higher phlorotannin concentrations typically found in brown macroalgae, which reduce palatability (Pavia and Brock, 2000), whereas tropical red macroalgae do not contain such compounds. Dietary supplementation with *K. alvarezii* and *G. salicornia* also did not affect BW, food intake, and nutrient digestibility. However, diet with 3% of *G. salicornia* decreased fecal ammonia production, improved counts of *Bifidobacterium* and *Lactobacillus*, and increased lactic acid production, suggesting that the fermentation shifted from proteolytic to saccharolytic type. This could be promoted by the sulphated polysaccharides of red macroalgae that being resistant to gastrointestinal enzymes, become available as a substrate for the bacteria in the colon, increasing saccharolytic bacteria that use ammonia for their growth. The mean fecal *Bifidobacterium* population was higher, while the mean clostridial population was lower in dogs fed a diet with 3% *G. salicornia*. This negative correlation suggests that the reduction in clostridia counts may be partially attributed to increased SCFA production by saccharolytic bacteria, as it is well established that probiotic bacteria can suppress pathogenic species through competitive inhibition, SCFA production, and the secretion of anti-bacterial peptides (Torok *et al.*, 2008). Furthermore, higher concentrations of reduced glutathione and lower levels of lipid peroxides were observed in the high-supplementation diets. A tendency toward improved cell-mediated immunity was also evident with 3% *G. salicornia* inclusion. Overall, supplementation with either macroalga at 1.5% had no significant effect on most of the parameters examined, apart from fecal SCFA and lactate concentrations, whereas both macroalgae at 3% inclusion demonstrated beneficial effects on gut health, antioxidant capacity, and immune function. Notably, supplementation with *G. salicornia* at 3% was more effective than *K. alvarezii* at the same level, particularly in enhancing gut health, antioxidant status and immunity. This superior effect is likely due to the higher phenolic content and more electronegative functional groups in *G. salicornia* which confer greater free radical scavenging capability (Ganesan *et al.*, 2008).

Following a different approach, Mota *et al.* (2024) evaluated the effects of dietary supplementation with a commercial algal blend, composed of two macroalgae (*Ulva rigida* and *Fucus vesiculosus*) and one microalga (*Chlorella vulgaris*) in healthy adult Beagle dogs. Although supplementation did not affect first approach or taste, dogs showed a preference for the control diet over the diet containing 1.5% algal blend, likely due to sensory attributes such as the powdery texture of the supplement, which was added on top of the kibble rather than incorporated during formulation. Compared with the control diet, algal blend-supplemented diets improved the digestibility of most nutrients and energy. Based on the structural polysaccharide composition of the blend (ulvans, cellulose, mannose and galactose in *U. Lactuca*; cellulose, alginates, and fucans in *F. vesiculosus*, and algaenan, mannose, and chitin-like polysaccharides of *C. vulgaris*), authors hypothesized that more pronounced effects might be achieved if the blend was pretreated with chemical, mechanical, or enzymatic techniques to disrupt algal cell walls and enhance nutrient digestibility and bioaccessibility. Fecal metabolite analyses suggested that algal blend supplementation promoted carbohydrate fermentation, while proteolytic fermentation remained unchanged. However, no effects on fecal microbiota diversity and abundance were observed. Overall, while the findings indicate that algal blend represents a promising alternative dietary supplement for dogs, further research is required to elucidate their potential health-promoting effects.

TABLE 15.1 *In vivo* studies focused on evaluating effects of macroalga inclusion in dog diets

Animals	Experimental design	Experimental diets	Effects	References
11 large/medium-sized (6 females and 5 males; 25.6 ± 7.5 kg BW), mixed breed, <8 years, naïve, client-owned dogs	3 two-bowl tests: CTR vs. AN-low, CTR vs. AN-high and AN-low vs. AN-high	Extruded dog food with 0% *Ascophyllum nodosum* (CTR), and two experimental diets with *A. nodosum* at 0.3% (AN-low) and 1.0% (AN-high) in replacement of wheat middlings	Lower food consumption and intake ratio with AN-high vs. CTR; No differences on first choice	Isidori *et al.* 2019
10 healthy adults (8 females and 2 males; 22.8 ± 6.1 kg BW), different breed, 1 and 6 years, client-owned dogs	5 × 5 replicated Latin square; 28-d feeding periods and 7-d wash-out between periods	CTR diet; CTR + *A. nodosum*; CTR + *Undaria pinnatifida*; CTR + *Saccharina japonica*; CTR + *Palmaria palmata*; macroalgae added to food at a daily dose of 1.5%	No effects on fecal DM content, ammonia percentage and concentrations, SCFA, BCFA, biogenic amines and IgA, nutrient digestibility and fecal microbial populations	Pinna *et al.* 2021
Two bowl-test: 12 healthy adult Beagle dogs (6 females and 6 males; 3.1±0.05 years; 13.3±1.65 kg BW) Digestibility trials: 6 adult dogs (3 females and 3 males; 2.9±0.02 years; 15.0±2.04 kg BW) randomly selected from the initial group of 12 dogs	Two-bowl test: CTR vs 1.5% algae blend; 10-d trial to determine CTR diet digestibility; Replicated 3x3 Latin square with 10-d periods	CTR diet without algal blend; Three levels of commercial algal blend (Algaessence® food; *Ulva rígida, Fucus vesiculosus* and *Chlorella vulgaris*) inclusion (0.5, 1.0 and 1.5%) in replacement of the CTR diet	No effect on first approach and taste, but lower IR with algae blend diet; No effect on food intake, number of defecations, fecal score, and fecal DM content; Organic matter, fiber, and energy digestibility increased with 1.5% inclusion compared with 0.5 %; 1.5% inclusion promoted fecal total SCFA and acetate; 0.5% inclusion decreased fecal propionate and increased fecal production; Compared with CTR, algal blend promoted most nutrients and energy digestibility, diet ME content, and fecal quality, while reducing fecal butyrate; Fecal microbiota diversity and abundance were mostly unaffected.	Mota *et al.* 2024
30 adult Golden Retrievers (15 male, 15 female; 2.73±1.17 years; 30.34±2.66 kg BW). 10 dogs with fecal scores between 2 and 3; 20 with loose stools (4-6; 7-point scale)	28-d feeding trial; 10 replicates per group	Healthy dogs fed a CTR diet; dogs with loose stools fed a CTR diet (LC); dogs with loose stools fed a CTR diet supplemented on-top with 2% macroalgae extract (LS); Macroalgae extract: 12.4% laminarin, 9.9% fucoidan, 10.2% mannitol, 0.9% protein, and 65.1% ash, with 12.7% potassium, 13.8% sodium, 16.8% chloride, 2.3% calcium, 0.7% phosphorus, and 1.8% magnesium (Shidai Marine Biotechnology Co., Ltd., Weihai, China)	LS improved loose stools, increased serum antioxidase activities and lowered serum malondialdehyde concentrations compared to the LC group. Higher serum levels of IgA, IgM, and IL-10 and lower serum levels of IL-1β, C-reactive protein, tumor necrosis factor-α, d-lactate, lipopolysaccharide, intestinal fatty acid-binding protein, and diamine oxidase with LS than LC; LS increased SCFA-producing gut microbiota abundance and fecal propionate concentrations	Zhang *et al.* 2025

TABLE 15.1 *In vivo* studies focused on evaluating effects of macroalga inclusion in dog diets *(cont.)*

Animals	Experimental design	Experimental diets	Effects	References
5 adult Labrador dogs (6 years, 15.36 ± 2.63 kg BW)	5 × 5 Latin square; 21-d periods and 7-d wash-out between periods	CTR; CTR + *Kappaphycus alvarezii* at 1.5% (diet DM, KA_L); CTR + *K. alvarezii* at 3% (diet DM, KA_H); CTR + *Gracilaria salicornia* at 1.5% (diet DM, GS_L); and CTR + *G. salicornia* at 3% (diet DM, GS_H).	No effect on food intake, BW, digestibility of DM and nutrients, fecal score, daily mean defecation frequency, fecal pH, individual and total BCFA, mean fecal *Coliform* counts; Fecal ammonia concentration lower in GS_L and GS_H groups than CTR, KA_L, and KA_H groups; Higher fecal lactate concentration in GS_L and GS_H groups; Lactate concentration higher in GS_H group than GS_L group; Acetate concentration higher in KA_H and GS_H groups; Propionate higher in all macroalga-supplemented groups; Butyrate, total SCFA, and total SCFA higher in KA_H, GS_L and GS_H; Fecal count of *Bifidobacterium* higher and of *Clostridium* lower in GS_H group; Concentration of reduced glutathione higher and lipid peroxides lower in KA_H and GS_H groups; Tendency of improved cell-mediated immunity observed in the GS_H group	Srinivas *et al.* 2024

Abbreviations: BCFA, branched-chain fatty acids; BW, body weight, CTR, control diet; DM, dry matter; Ig, immunoglobulin; IL, interleukin; IR, intake ratio; ME, metabolizable energy; SCFA, short-chain fatty acids.

2.3 Current trends, future directions, and key concerns

Despite scientific evidence supporting the beneficial role of macroalgae and their extracts as health-promoting livestock feed, studies with dogs are limited. Given the strong influence of environmental and biological factors on macroalgae composition, it is essential to assess the biochemical and nutritional profiles of different species cultivated under varying conditions. A comprehensive understanding of optimal growth sites, harvesting periods, and processing methods is critical to producing macroalgae with desirable attributes such as high nutritional value and improved digestibility. Moreover, to ensure the products' viability in the market and their year-round availability, large quantities of macroalgal biomass are required; thus, economically viable, large-scale production is essential. Approaches such as large-scale macroalgae cultivation through aquaculture or the valorization of invasive species align with circular economy principles and contribute to mitigating greenhouse gas emissions, treating agro-industrial effluents, and lowering the overall carbon footprint (Weerakkody *et al.*, 2023).

Notably, strategies for extracting proteins from macroalgae to produce protein concentrates and bioactive peptides are being increasingly explored, driven by the growing demand for alternative protein sources alongside the declining availability of land and water resources (Gordalina *et al.*, 2021; Kazir *et al.*, 2019). Furthermore, extracts of bioactive compounds from macroalgae can stimulate the immune system, improve health, and enhance antioxidant and antimicrobial properties. However, because the extraction of bioactive compounds demands extensive processing, specialized equipment, and the use of solvents, the development of innovative technologies – such as hydrothermal and microwave-assisted techniques, along with green extraction methods – should be prioritized (González-Meza *et al.*, 2023).

Future research should focus on longitudinal designs, multiomics tools, and component purification to better understand the effectiveness of dogs' ability to use macroalgae, particularly the ability of canine gut microbiota to ferment algal polysaccharides, and for a more comprehensive exploration of the specific mechanisms and assessment of the long-term impacts on animal health. Additionally, investigating the long-term effects is relevant because it has been suggested that only regular consumption of macroalgae can exert selective pressure on intestinal microbiota, favoring bacteria that ferment algal polysaccharides (Bhattacharya *et al.*, 2015). Finally, while these organisms have great potential to serve as alternative sources of protein, energy, and bioactive compounds in human and animal diets, concerns about their capacity to bioaccumulate toxic substances from the environment warrant careful evaluation prior to widespread use (Zokm *et al.*, 2022).

3 Duckweed

Duckweed is a ubiquitous group of aquatic plants, distributed across tropical and temperate regions and found in diverse freshwater environments such as ponds, lakes, and pools (Tippery and Les, 2020). This group comprises 35 species and two natural hybrids distributed among five genera: *Spirodela* (2 species), *Landoltia* (1 species), *Lemna* (11 species), *Wolffiella* (10 species), and *Wolffia* (11 species; Braglia *et al.*, 2024; Sree and Appenroth, 2020). Duckweeds predominantly reproduce asexually through vegetative propagation, allowing for rapid spreading on any surface on which they grow (Ziegler *et al.*, 2015). As a result, their productivity is exceptionally high, with reported DM yields ranging from 10 to over 100 tons per hectare per year, surpassing most land-based crops (Cheng *et al.*, 2002; Fourounjian *et al.*, 2020). *Lemna* species, commonly referred to as water lentils, are of particular importance within this group. One of the most widely studied species is *Lemna minor*, which is characterized by fast growth, broad environmental tolerance, and flexible biochemical composition, including protein and starch content that varies according to abiotic and nutrient conditions (Sosa *et al.*, 2024). Due to their rapid growth rate, ability to grow in wastewater, and abundance in natural ecosystems, duckweeds have been proposed as sustainable alternatives to conventional feed ingredients that do not compete with valuable agricultural land (Goopy and Murray, 2003). However, the nutritional quality and safety of duckweed biomass depend heavily on growth conditions, which influence nutrient levels and the presence of bioactive compounds or pollutants (Sosa et al., 2024).

3.1 Chemical composition and bioactive compounds

Dry matter content typically ranges from 4 to 8% of fresh weight, and ash from 5% to 25% DM, reflecting the ability to absorb and concentrate minerals from water (Appenroth *et al.*, 2017; Hu *et al.*, 2022; Pagliuso *et al.*, 2022). Calcium, potassium, magnesium, phosphorus, iron, manganese, and zinc are the most prominent elements (Muller *et al.*, 2025). Dietary fiber ranges from 5% to 60% DM, with cellulose reported to be between 20% and 40% in duckweed cell walls, hemicelluloses from 3.5% to 35%, pectin from 20% to 54%, and lignin lower than 10% of the cell wall (Hu *et al.*, 2022; Pagliuso *et al.*, 2022; Sowinski *et al.*, 2019). Starch, which is mainly stored in the chloroplasts of the fronds, has been reported to range from 4% to over 70% DM, and fat from 0.7% to 14% of the whole plant biomass (Muller *et al.*, 2025; Pagliuso *et al.*, 2022). Approximately 33% of duckweeds are composed of saturated fatty acids, mainly palmitic acid, whereas PUFA represent between 48% and 71% of the total fatty acids, with the most prevalent being n-3 alpha-linolenic acid (ALA, 18:3, around 37.1% to 42.8%) and n-6 linoleic acids (LA, 18:2, 16.1% to 25.0%; Muller *et al.*, 2025). Remarkably, the n-6 to n-3 ratio is typically below 1, and the phytosterol content is relatively high (ca. 5% of total fat), which might confer significant health benefits, namely cholesterol-lowering properties (Appenroth *et al.*, 2017; Muller *et al.*, 2025).

Duckweed is notable for its high protein content, which typically ranges between 15% and 45% DM, being richer in lysine than wheat and providing higher levels of methionine and cysteine than soybeans, two AA often limiting in cereal- and legume-based diets (Iqbal *et al.*, 2006; Muller *et al.*, 2025; Pagliuso *et al.*, 2022). Duckweed proteins can be broadly divided into soluble and insoluble fractions, each constituting approximately 50% of the total protein content (Muller and Bazinet, 2025; Nieuwland *et al.*, 2021). The soluble fraction includes cytoplasmic proteins and enzymes, with ribulose-1,5-bisphosphate carboxylase/oxygenase (RubisCO), known for its high solubility, nutritional quality, functional properties, and digestibility, being the dominant protein (Nieuwland *et al.*, 2021; Pearce and Brunke, 2023). The insoluble fraction is composed mainly of membrane-associated proteins involved in structural integrity, photosynthesis, and ATP synthesis (Muller and Bazinet, 2025; Nieuwland *et al.*, 2021). Beyond their nutritional value, duckweed proteins comprise a source of bioactive peptides with antimicrobial, antihypertensive, antidiabetic, anticancer, and antioxidant effects (Muller *et al.*, 2025). Additionally, as recently reviewed by Ofoedu *et al.* (2025), duckweeds are rich in other bioactive compounds such as α-tocopherol, vitamin E, phytosteorls, polyphenols, flavonoids and carotenoids (e.g., zeaxanthin, lutein), which have known antioxidant, anti-inflammatory, and anti-cancer properties.

Despite its favorable nutritional profile, duckweed also contains secondary metabolites that may act as anti-nutritional factors, including oxalic acid, phytic acid, and tannins (Hu *et al.*, 2022; Pagliuso *et al.*, 2022). Oxalic acid, commonly found in *L. minor*, can bind with calcium to form calcium oxalate, thereby reducing calcium bioavailability (Franceschi, 1987). In contrast, in *Wolffia* species, oxalate does not occur as calcium oxalate, reducing the associated health risks (Appenroth *et al.*, 2018). Phytic acid can chelate proteins and minerals such as iron, zinc, phosphorus, and calcium, lowering their bioavailability, while tannins can impart astringency and bitterness, potentially reducing iron bioavailability (Pereira *et al.*, 2021; Petroski and Minich, 2020). Although these compounds are often considered detrimental, recent reviews stress that may also confer health benefits, including antioxidant, cardioprotective, and prebiotic effects (Petroski and Minich, 2020). The greater concern for duckweed consumption lies not in its intrinsic metabolites, but in contamination risks from pathogens (e.g., *Escherichia coli, Clostridium botulinum*) or from environmental pollutants such as heavy metals, phenols, or pesticides, which can be readily absorbed during growth and processing (van der Spiegel *et al.*, 2013).

3.2 *In vivo* experimental evidence

The potential use of duckweed as animal feed has been recognized for decades, and has been frequently reviewed for aquaculture, poultry, pigs and cattle (Goopy and Murray, 2003; Islam, 2002; Minich and Michael, 2024; Takács *et al.*, 2025). However, little research has been conducted on companion animals. Table 15.2 summarizes the two *in vivo* studies we are aware of that evaluate the effects of including duckweed in dog diets. Brown *et al.* (2013) evaluated *in vivo* the effects of dietary inclusion of *Landoltia punctata* at 10%, 20% and 30% in replacement of whole corn and maize gluten, and observed a linear decrease in DM, gross energy and crude protein (CP) digestibility, most probably attributed to the increasing content of insoluble non-starch polysaccharides with increasing level of duckweed inclusion. The reduction of magnesium availability might be explained by the presence of phytic acid in duckweed, whereas the increased potassium of the diets with increasing level of duckweed might have contributed to the reduced digestibility of this nutrient.

The addition of phytase significantly enhanced CP digestibility in diets containing 30% duckweed. This improvement can be attributed to the enzyme's ability to release phytate-bound protein. Duckweed inclusion had no significant effect on diet palatability. However, a significant increase in fecal consistency score was observed with duckweed supplementation, likely due to differences in dietary fiber content.

More recently, Lin *et al.* (2024) evaluated the effects of including duckweed protein (*Lemna*; MCSelect; Parabel; Vero Beach, FL) as a partial replacement for pea protein in dog diets at 0%, 5%, and 10%. Diets containing 5% and 10% duckweed protein showed reduced palatability, with dogs preferring the control diet. However, no adverse effects on nutrient digestibility were observed at either inclusion level. Stools from dogs fed duckweed diets were firmer, although a distinct green coloration was consistently reported, which may be perceived negatively by dog owners. To improve the suitability of duckweed as a protein source in canine nutrition, approaches such as decolorization or pigment removal warrant further exploration. The contrasting findings compared to Brown *et al.* (2013) may be attributed to the specific type of duckweed used. *Lemna* species are known to contain elevated levels of oxalic acid and phenolic compounds, which can impart an unpleasant odor and taste, potentially explaining the reduced palatability observed in dogs.

3.3 Current trends, future directions, and key concerns

Although duckweed holds strong potential as a sustainable food source, research and commercialization are still at an early stage, particularly regarding its application in canine nutrition. Limited attention has been given to how preservation methods (e.g., freezing, drying, powdering) influence nutritional and functional stability. Beyond compositional analyses, *in vivo* studies with dogs are needed to assess effects on intake, feed efficiency, and health parameters. As an attractive sustainable source of protein and bioactive peptides, future research should focus on protein extraction, characterization, and incorporation into food matrices, since the high fiber content and other leaf-derived compounds in duckweed can limit protein functionality. Investigating the effects of duckweed hydrolysates is especially valuable. In this context, it is important to clarify whether duckweed hydrolysates can be used directly as functional ingredients, or whether duckweed should instead be regarded as a reservoir of novel peptide sequences for targeted applications. Furthermore, the unknown stability of these peptides during compound feed manufacturing and throughout animal digestion, highlights the need for further studies on duckweed protein hydrolysis and peptide production. It is worth noting that while mineral absorption capacity is advantageous as a source of nutrients or a phytoremediation method, concerns arise when it is cultivated for human or animal consumption due to the potential accumulation of heavy metals, such as lead, cadmium, and mercury (van der Spiegel *et al.*, 2013). Therefore, it is essential to implement robust contaminant control plans, especially since heavy metal contamination in animal feed is strictly regulated due to its potential impact on animal health and food chain safety.

4 Mussels

A total of 12 mussel species (mainly from the *Mytilus* and *Perna* genera) are farmed in 44 countries worldwide. China, Chile, Spain, Thailand, and New Zealand are among the leading producers of mussel (Avdelas *et al.*, 2021; FAO, 2020). New Zealand's mussel aquaculture industry is based entirely on the endemic green-lipped or Greenshell™ mussel (*Perna canaliculus*), which has long been an important food source for the indigenous Māori people (Miller *et al.*, 2023; Skelton *et al.*, 2022). This

TABLE 15.2 *In vivo* studies focused on evaluating effects of duckweed inclusion in dog diets

Animals	Experimental design	Experimental diets	Effects	References
18 adult beagles (13 males and 5 females; 11.9 ± 1.9 kg BW).	10-d feeding study conducted with a completely randomized design; 6 dogs per diet	Complete diets with 0% (CTR), 5%, and 10% duckweed protein (*Lemna*; MC Select; Parabel; Vero Beach, FL, USA)	No effects on BW, feed intake, DM, OM, CP, fat and energy digestibility, daily fecal output and fecal DM content; Firmer stool consistency in dog fed duckweed; Stools displayed a green coloration	Lin *et al.* 2024
30 adult beagles	Two-bowl tests: CTR diet vs. 5% duckweed protein diet; and CTR vs. 10% duckweed protein diet	Complete diets with 0% (CTR), 5%, and 10% duckweed protein (*Lemna*; MC Select; Parabel; Vero Beach, FL, USA)	Negative impact on diet palatability (first choice, daily consumption and intake ratio).	Lin *et al.* 2024
30 privately owned dogs (15 labradors and 15 beagles; 7 males and 23 females; 1–12 years)	12-d randomized block design; 3 replicates of 10 dogs; each breed equally represented in each dietary treatment and within each replicate; 6 dogs per diet.	Diet without duckweed (CTR); CTR plus 10%, 20% or 30% duckweed and the CTR plus 30% duckweed and phytase	Duckweed decreased DM, GE, CP, Mg, Na, K, S digestibility, and molar proportion of propionic acid, increased fecal output and molar proportion of acetate and butyrate and promoted firmer stools; Addition of phytase increased CP digestibility; No effects on daily DM intake, fecal pH, fecal DM content, total concentrations of SCFA.	Brown *et al.* 2013
	Two-bowl test: CTR vs. 30% duckweed diet.		No effects on amount consumed or percentage intake; tendency to select CTR	

Abbreviations: BW, body weight; CP, crude protein; CTR, control; DM, dry matter; GE, gross energy; OM, organic matter; SCFA, short-chain fatty acids.

mussel is distinguished by its characteristic green shell margin and larger size compared to other commercial mussels such as the blue mussel (*Mytilus galloprovincialis*). These mussels are farmed for food as well as for nutraceutical products, such as oil extracts. These are marketed internationally in diverse forms, including fresh and processed foods, nutraceutical extracts, and powders (Coulson *et al.*, 2015). Moreover, green-lipped mussels aquaculture is considered highly sustainable. As filter feeders, mussels obtain nutrients directly from the water column. This eliminates the need for external feed inputs and minimizes environmental impacts. Mussels also provide ecosystem services, including nutrient removal and improved water quality, and the creation of habitats for other marine organisms (Naylor *et al.*, 2009; Stenton-Dozey and Broekhuizen, 2019). Undersized mussels that are not suitable for human consumption, as well as those cultivated to reduce nitrogen and phosphorus pollution near urban areas or finfish farms, can be processed into mussel meal for use in feed formulations (Lindahl, 2013). From the perspective of a circular economy and zero waste, byproducts such as extracellular fluid, shell, meat, and byssus threads – the latter two of which comprise around 60% of protein – can be used as food resources (Naik and Hayes, 2019).

4.1 Chemical composition and bioactive compounds

Fresh whole green-lipped mussels comprise approximately 12–14% protein, 3–6% carbohydrates, 2–3% ash, 1.6–2.2% fat and 76–82% moisture, whereas dried whole meat is composed of 36–67% protein, 10–25% carbohydrates, 2–25% ash, and 6–12% fat (Miller *et al.*, 2023). Green-lipped mussels also contain significant amounts of minerals, such as iron, zinc, magnesium, calcium and iodine (Coulson *et al.*, 2015). The chemical composition depends, among others, on the harvest season, aquaculture location, mussel diet, and the procedure used to produce the powder (Miller and Tian, 2018). Although the green-lipped mussel has low fat and carbohydrate content and a high protein content, its non-protein components are of the most value due to their bioactivity. They contain a diverse array of bioactive compounds, including lipids such as EPA, docosahexaenoic acid (DHA, 22:6), sphingolipids, phytosterols, diacylglycerols, diterpenes, and sesquiterpenes; saponins; and antioxidants such as carotenoids, xanthophylls, and anthocyanins; glycosaminoglycans, namely chondroitin sulphate with its high glucosamine content (Hielm-Björkman *et al.*, 2009; Miller *et al.*, 2023; Miller *et al.*, 2014). Most of these compounds are likely biosynthesized in marine algae and then accumulate in the shellfish that feed on them.

4.2 *In vivo* experimental evidence

In recent years, green-lipped mussels have gained prominence as a functional ingredient in pet food, as consumption has been shown to alleviate chronic orthopedic pain in dogs (Bui and Bierer, 2001; Cave *et al.*, 2024; Hielm-Björkman *et al.*,

2009). However, being green-lipped mussels a complex natural product, with glucosamine and chondroitin sulfate, n-3 PUFA (e.g., EPA and DHA), other lipids, including pro-resolving lipid mediators, and bioactive peptides, all these chemical compounds could contribute to the overall physiological effect (Mehler *et al.*, 2016; Monteiro *et al.*, 2023; Rialland *et al.*, 2013). Indeed, these nutrients may act synergistically to reduce inflammation, limit further cartilage degeneration, and support the regeneration of damaged joint cartilage and synovial fluid (Eason *et al.*, 2018).

4.3 Current trends, future directions, and key concerns

To the best of our knowledge, no study has evaluated the *in vivo* effects of green-lipped mussel consumption beyond alleviating osteoarthritis symptoms. However, the mussel's abundance of bioactive compounds makes it an interesting functional food. Besides its content of glycosaminoglycans, such as glucosamine and chondroitin sulfate, the lipid fraction of green-lipped mussels is particularly rich in DHA and EPA, which are known to provide several health benefits for dogs (Carlisle *et al.*, 2024; Vendramini *et al.*, 2025). Additionally, green-lipped mussels contain other unique anti-inflammatory n-3 PUFA, such as 5,9,12,15-octodecatetraenoic acid (OTA, C18:4), 5,9,12,16-nonadecatertraenoic acid (C19:4), 7,11,14,17-eicosatetraenoic acid (ETA, C20:4) and 5,9,12,15,18-heneicosapentaenoic acid (C21:5), which are not reported in other marine sources (Treschow *et al.*, 2007), and plasmalogens (Miller *et al.*, 2023). With the increased market demand for n-3 marine oil supplements, green-lipped mussels have the advantage of presenting an n-3 long-chain PUFA content higher than certain fish oils such as those obtained from sardines, anchovies, and cod liver. Previous studies reported a concentration of PUFA in green-lipped mussel lipids ranging from 19 to 49.1%, with 6–12% DHA, and 8–24% EPA (Mclean and Bulling, 2005; Miller *et al.*, 2011). Another critical aspect of green-lipped mussel oil is the bioavailability of its fatty acids, which increases due to the glycerol's molecular position within triglycerides (Treschow *et al.*, 2007). Along with their high content of n-3 PUFA, mussels are also a sustainable source of high-quality protein, phytosterols, and other key micronutrients such as vitamin B12 (Yaghubi *et al.*, 2021), with potential to be used in pet food. Future research should also focus on other species, such as the freshwater mussel *Sinanodonta woodiana*. This species is one of the most invasive and has spread throughout Europe over the last two decades via fish shipments from Asia. It has been shown to have excellent properties for inclusion in pet food, including solubility, emulsifying capacity, gel-forming ability, and high taurine content (Konieczny *et al.*, 2022). However, any valorization approach must be implemented within a well-defined framework of ecological monitoring, legal restrictions on live handling and movement, and controlled processing practices, in order to mitigate the risk of unintended spread of invasive populations (Douda *et al.*, 2025).

5 Marine crustacean and cephalopod

The present section addresses crustaceans and cephalopods collectively as prospective ingredients or dietary supplements for canine nutrition. Although these marine resources exhibit considerable nutritional and functional potential, the current body of research involving dogs is scarce, underscoring the necessity for further systematic investigation.

5.1 Squid

The mantle, head and tentacles of squids are mainly used for human consumption whereas squid waste (around 50% of the whole weight) usually including viscera, but also head, tentacles, fin, skin and pen, as well as whole squids that do not meet quality standards for human consumption, are processed into squid meal (Hertrampf and Piedad-Pascual, 2000b). Squid meal used for animal feeding commonly comes from the giant squid (*Dosidicus gigas*), and is often used as feed ingredient in aquaculture diets, particularly in shrimp diets (Córdova-Murueta and García-Carreño, 2002; Ennayer *et al.*, 2025). By-products of the squid industry can be developed into protein hydrolysates or used to extract bioactive peptides with anti-hypertensive, antioxidant, and anticancer properties (Alemán *et al.*, 2011). The chemical composition of squid meal depends on the species, growth stage, feeding environment, parts of the squid used, and the processing techniques (Singh *et al.*, 2022). Overall, squid meal stands out for its high CP content, which ranges from 68% to 81% (Córdova-Murueta and García-Carreño, 2002; Guilherme-Fernandes *et al.*, 2024a). This content is higher than that of the animal and vegetable protein sources commonly used in dog food, such as poultry by-product meal, meat and bone meal, and corn gluten meal (NRC, 2006). Squid meal also contains high levels of lysine (10% of total protein), and glutamic acid (15% of total protein; Carranco-Jáuregui *et al.*, 2021), with the methionine content (3% DM) being higher than that found in several common protein sources (poultry by-product meal, meat and bone meal, soybean meal). Therefore, the AA profile of squid meal makes this by-product interesting for the pet food industry to overcome the dietary deficiency of AA rather than employing synthetic AA (e.g., methionine) supplements. Despite having a low fat content (around 3% DM; Guilherme-Fernandes *et al.*, 2024a), PUFA group is reported to be the most abundant (ca. 51%), with appreciable amounts of DHA (29.5–30.8%) and EPA (15.9–17.2 %; Aubourg *et al.*, 2021), which have known benefits for dog health.

As far as we know, Guilherme-Fernandes *et al.* (2024a) study was the only one conducted to evaluate the effects of increasing dietary levels of squid meal on palatability, digestibility, fecal characteristics and metabolites, and microbiota in adult healthy Beagle dogs. First approach and taste were unaffected by the inclusion of 15% squid meal in replacement of the control diet, but dogs showed a preference for the control diet despite the high levels of glutamic and aspartic acids, which induce the umami taste highly attractive to dogs (Tanase *et al.*, 2022). We suggested that the observed effect might be explained by adding squid meal on top, rather than including it in the kibble. The inclusion levels of squid meal (5%, 10% and 15%) did not affect fecal output or characteristics, but CP digestibility was highest with 15% inclusion. Compared to the control, digestibility of DM, most nutrients and energy was higher with squid meal inclusion, the major difference observed being for CP. Moreover, squid meal supplementation reduced fecal butyrate concentration without altering fecal microbiota composition. These findings indicate that, beyond its high potential as a novel, highly digestible protein source for canine diets, squid meal may also be suitable for incorporation into therapeutic diets aimed at managing adverse food reactions.

5.2 Krill

Krill are small crustaceans belonging to the *Euphausiacea* family. Antarctic krill (*Euphausia superba*) are widely distributed around Antarctica, and account for the largest commercial krill fishery worldwide (FAO, 2020). Catches are concentrated in the Southern Ocean, and Antarctic krill must be processed immediately upon capture to mitigate the nutrient loss due to their highly active endogenous enzymes (Haard and Simpson, 2000); with krill meal being the most predominant product processed onboard. The manufacturing process of krill meal, which is primarily used as animal feed, affects the content of its bioactive compounds, such as astaxanthin (Yang *et al.*, 2024). Whole dried krill is mainly used in the form of a high-protein krill meal. Krill meal has a high CP content, often exceeding 56% DM (Hertrampf and Piedad-Pascual, 2000a; Tou *et al.*, 2007), and has a balanced AA profile, that is rich in alanine (8.2%-9.6% of total AA), glycine (8.2%-9.3% of total AA), lysine (7.8%-8.3% of total AA), aspartic acid (9.9%-11.5% of total AA), and glutamic acid (12.0%-13.2% of total AA; Nicol *et al.*, 2000; Saether and Mohr, 1987). Additionally, 25% of krill meal is typically krill oil, which is a valuable source of n-3 PUFA, such as EPA (14% oil) and DHA (6.5 g/100 g oil; Burri *et al.*, 2016). Krill lipids are primarily in the phospholipid form and consist mainly of phosphatidylcholine, which is believed to be more bioavailable than the triglycerides (Köhler *et al.*, 2015). This increases the tissue integration of n-3 PUFA. Bioactive antioxidant compounds, including choline, selenium, astaxanthin, and vitamin E, are also present in krill meal (Tou *et al.*, 2007).

Studies on krill meal in dogs are limited and focus mainly on its n-3 PUFA profile. This is because plant-based oils (e.g., flaxseed, canola, walnut, soybean) provide ALA, which confers health benefits only indirectly through its conversion to the long chain n-3 PUFA. However, this conversion is inefficient in mammals due to the rate-limiting activity of the Δ6-desaturase enzyme (Calder, 2013). To address this limitation, direct dietary inclusion of long-chain n-3 PUFA is recommended.

These are commonly found in marine sources such as fatty fish (e.g., menhaden, anchovies, herring and mackerel) crustaceans, and algae. However, since krill lipids are primarily in the phospholipid form, which is believed to be more bioavailable than the triglycerides found in fish oil, a higher tissue incorporation of EPA and DHA is expected with the former. Several studies have used the n-3 index, which expresses the content of EPA + DHA in red blood cell membranes as a percentage of all fatty acids in red blood cells, to validate this hypothesis. The n-3 index is strongly correlated with overall n-3 tissue status (Fenton *et al.*, 2016) and is used in humans as a biomarker of long-term n-3 PUFA intake, as well as a potential risk indicator for coronary heart disease (Harris and Von Schacky, 2004).

Burri *et al.* (2018) demonstrated that 8% krill meal supplementation led to a significantly higher n-3 index and reduced exercise-induced inflammation. They also observed a tendency for reduced muscle damage after a long-distance sled dog competition. Additionally, Burri *et al.* (2020) assessed the effectiveness of incorporating equal amounts of n-3 PUFA from krill meal phospholipids and fish oil triglycerides into the diet of dogs to enhance their red blood cell n-3 PUFA profile. The authors found that krill meal supplementation increased EPA/DHA levels and reduced the production of ARA-derived pro-inflammatory mediators, shifting the immune state toward anti-inflammation. Similar results were obtained by Lindqvist *et al.* (2023) after four weeks of feeding a diet containing 3% krill meal, compared to diets containing fish oil or flaxseed. The available results show that adding krill meal to the diet improves the n-3 index and markers for inflammation and muscle damage, but the optimal level of the n-3 index for canines has not yet been determined. Moreover, it should also be mentioned that work in healthy animals with excess EPA and DHA is scarce, deserving further research toward incorporating ingredients with high EPA and DHA into dogs' diets.

5.3 Shrimp

In 2022, global production of farmed marine shrimp for human consumption reached 5.5 million tons, with *Litopenaeus vannamei* accounting for over half of the total output (FAO, 2023). Processing shrimp for human consumption generates substantial waste because non-edible body parts such as heads, shells, and tails account for nearly half of the animal's weight (Roy *et al.*, 2023). While some of these by-products are used in animal feed, most remain unexploited. This underutilization results to the loss of valuable bioactive components and increases environmental pollution (Nirmal *et al.*, 2020). Overall, shrimp waste contains valuable bioactive components including protein, peptides, chitin, chitosan, pigments, enzymes, lipids, minerals and vitamins (Cahú *et al.*, 2012; Gómez-Estaca *et al.*, 2019; Nair *et al.*, 2017; Senphan and Benjakul, 2012; Senphan *et al.*, 2014). The chemical composition of shrimp waste depends on its sources and the processing conditions. The CP content ranges from 20 to 50%, being the AA profile of the dried shrimp shells of higher value than that of shrimp heads (Ibrahim *et al.*, 1999). Shrimp waste is also an important source of n-3 PUFA, such as EPA and DHA (Senphan and Benjakul, 2012). More than 70% of the protein in shrimp waste can be recovered as a hydrolysate with several bioactive properties, including anti-hypertensive, anti-cancer, antimicrobial, and antioxidant activities, when the waste is subjected to hydrolysis with proteolytic enzymes (Djellouli *et al.*, 2020; Gildberg *et al.*, 2011; Guilherme-Fernandes *et al.*, 2024a; Kannan *et al.*, 2011; Leal *et al.*, 2025).

Several studies have reported the health and performance benefits of shrimp meals and hydrolysates across a range of species, particularly fish (Nirmal *et al.*, 2020). To our knowledge, however, no studies have examined the impact of shrimp meal on dogs, while the effects of shrimp hydrolysate have only recently been investigated by our group. In a short-term study, Guilherme-Fernandes *et al.* (2024a) reported potential benefits of shrimp hydrolysate for dogs, attributable to its methionine content, antioxidant activity, and high digestibility. In that study, dietary inclusion of shrimp hydrolysate at 5%, 10%, and 15% (replacing the control diet) did not alter the parameters measured; however, compared to the control diet, it increased total and individual SCFA concentrations, with the exception of butyrate. Significant shifts were also observed in the abundances of *Oscillosperaceae* (UCG-005), *Firmicutes*, and *Lactobacillus*. Collectively, these findings suggest that shrimp hydrolysate may function as a prebiotic, supporting gut health.

Following these promising preliminary results, the long-term effects of replacing wheat gluten with 5% shrimp hydrolysate in an extruded complete dog food were recently evaluated with respect to palatability, intake, digestibility, fecal characteristics, metabolites and microbiota, oral volatile sulfur compounds, coat quality, hematological parameters, serum chemistry, and both innate and adaptive immune function (Guilherme-Fernandes *et al.*, 2025; Guilherme-Fernandes *et al.*, 2024b). No differences were observed in first approach, first taste, or intake ratio. The inclusion of shrimp hydrolysate did not affect intake, nutrient and energy digestibility or oral volatile sulfur compounds. However, it increased fecal output and decreased fecal butyrate concentration. Both diets promoted high coat quality. Furthermore, the inclusion of shrimp hydrolysate increased neutrophils, platelets, and white blood cells, while decreasing eosinophils and having no effect on Ig E levels, suggesting a role in modulating immune function and potentially reducing inflammation or allergic responses. In addition, shrimp hydrolysate decreased blood glucose levels, probably associated with the decreased production of superoxide. Serum IL-8 levels tended to be higher with shrimp hydrolysate inclusion, suggesting an anti-inflammatory effect. Notably, the inclusion of shrimp hydrolysate promoted the proliferation of CD4+ T cells stimulated with LipL32. This indicates a potential positive effect on the generation of memory CD4+ T cells, which suggests possible modulation of immune parameters in dogs. According to the results of the previous short-term study, the inclusion of shrimp hydrolysate increased the abundance of the *Oscillo-*

spiraceae genus but decreased the abundance of *Sellimonas*. The presence of chitin in the shrimp hydrolysate might contribute to the effects on fecal metabolites and microbiota. However, in this study, the soluble and insoluble fiber contents were similar among diets, and the chitin content is suggested to be residual due to the process of obtaining of shrimp hydrolysate.

Though limited, current evidence suggests that shrimp hydrolysate may support canine health. Reported benefits include immunomodulatory effects, potential use in hypoallergenic diets, and improved metabolic outcomes, such as reduced blood glucose and superoxide production, which are particularly relevant to diabetic dogs. Enhanced CD4+ T-cell proliferation in response to LipL32 suggests possible benefits for vaccination responses. Moreover, favorable shifts in gut microbiota composition, along with high digestibility, support its application as a prebiotic and for promoting gastrointestinal health.

5.4 Current trends, future directions, and key concerns

Squid, krill, and shrimp by-products are widely used in aquaculture and livestock feed, but their use in canine nutrition is limited. Few studies have evaluated their effects on dogs. These marine-derived ingredients have several beneficial properties, such as high-quality protein, favorable fatty acid profiles, and bioactive compounds that may have biological activities. Their high digestibility and nutrient density make them ideal candidates for hypoallergenic and functional diets, especially for dogs with allergies or metabolic disorders. Preliminary findings with shrimp hydrolysate suggest beneficial effects on immune function and metabolic markers. However, long-term feeding trials are lacking for all three by-products. Future research should address this gap by evaluating not only basic nutritional parameters, but also immune, metabolic, and gastrointestinal outcomes. Intervention trials in dogs with chronic conditions, such as diabetes or food allergies, are especially needed to explore targeted applications. The responsible integration of these by-products into canine diets should consider nutritional adequacy, safety, environmental performance, and regulatory compliance. From a sustainability standpoint, using them supports circular economy goals by valorizing waste, improving resource efficiency, and reducing dependence on terrestrial protein sources. However, their environmental impact varies and is influenced by factors such as fishery status, processing methods, and transportation. As noted by the FAO (2020) energy-intensive post-harvest stages, such as drying and distribution, can substantially increase overall environmental burdens. Therefore, ingredient-specific life cycle assessments are essential to accurately assessing their sustainability.

6 Concluding remarks

The available literature demonstrates that the chemical composition and bioactive compound content vary considerably among the aquatic food resources reviewed, resulting in different potential applications as dietary ingredients or supplements. Indeed, some sources show promise as highly digestible protein providers with favorable AA profiles, whereas others are particularly rich in n-3 PUFA. In addition, many contain bioactive constituents with potential health benefits for companion animals. Nevertheless, this review underscores the limited number of *in vivo* studies – including long-term trials – needed to clarify their effects and to optimize inclusion levels for achieving desired outcomes. For their practical application, ensuring consistent market availability at competitive costs and evaluating pet owners' acceptance of these alternative ingredients are key parameters. Looking ahead, their integration into pet food formulations may represent a viable strategy to enhance the sustainability of the sector, by reducing carbon footprint and minimizing competition with human food sources, while simultaneously supporting animal health and welfare. This strategy can provide additional environmental benefits by using invasive species as food sources, contributing to the restoration of aquatic habitats through their control, and simultaneously reducing pressure on finite natural resources.

Note on the usage of AI tools

The authors of this chapter state that they used large language models (LLMs) strictly for English language correction and stylistic polishing of chapters, and not for content generation.

References

Alemán, A., Pérez-Santín, E., Bordenave-Juchereau, S., Arnaudin, I., Gómez-Guillén, M.C. and Montero, P., 2011. Squid gelatin hydrolysates with antihypertensive, anticancer and antioxidant activity. Food Research International 44: 1044-1051. https://doi.org/10.1016/j.foodres.2011.03.010

Angell, A.R., Angell, S.F., de Nys, R. and Paul, N.A., 2016. Seaweed as a protein source for monogastric livestock. Trends in Food Science & Technology 54: 74-84. https://doi.org/10.1016/j.tifs.2016.05.014

Appenroth, K.-J., Sree, K.S., Bog, M., Ecker, J., Seeliger, C., Böhm, V., Lorkowski, S., Sommer, K., Vetter, W. and Tolzin-Banasch, K., 2018. Nutritional value of the duckweed species of the genus Wolffia (*Lemnaceae*) as human food. Frontiers in Chemistry 6: 483. https://doi.org/10.3389/fchem.2018.00483

Appenroth, K.-J., Sree, K.S., Böhm, V., Hammann, S., Vetter, W., Leiterer, M. and Jahreis, G., 2017. Nutritional value of duckweeds (*Lemnaceae*) as human food. Food Chemistry 217: 266-273. https://doi.org/10.3389/fchem.2018.00483

Aubourg, S.P., Trigo, M., Prego, R., Cobelo-García, A. and Medina, I., 2021. Nutritional and healthy value of chemical constituents obtained from Patagonian squid (*Doryteuthis gahi*) by-products captured at different seasons. Foods 10: 2144. https://doi.org/10.3390/foods10092144

Avdelas, L., Avdic-Mravlje, E., Borges Marques, A.C., Cano, S., Capelle, J.J., Carvalho, N., Cozzolino, M., Dennis, J., Ellis, T. and Fernández Polanco, J.M., 2021. The decline of mussel aquaculture in the European Union: causes, economic impacts and opportunities. Reviews in Aquaculture 13: 91-118. https://doi.org/10.1111/raq.12465

AVMA, 2024. U.S. pet ownership statistics, American Veterinary Medical Association.

Bhattacharya, T., Ghosh, T.S. and Mande, S.S., 2015. Global profiling of carbohydrate active enzymes in human gut microbiome. PLoS One 10: e0142038. https://doi.org/10.1371/journal.pone.0142038

Braglia, L., Ceschin, S., Iannelli, M., Bog, M., Fabriani, M., Frugis, G., Gavazzi, F., Gianì, S., Mariani, F. and Muzzi, M., 2024. Characterization of the cryptic interspecific hybrid Lemna× mediterranea by an integrated approach provides new insights into duckweed diversity. Journal of Experimental Botany 75: 3092-3110. https://doi.org/10.1093/jxb/erae059

Brown, W.Y., Choct, M. and Pluske, J.R., 2013. Duckweed (*Landoltia punctata*) in dog diets decreases digestibility but improves stool consistency. Animal Production Science 53: 1188-1194. https://doi.org/10.1071/AN13198

Bui, L.M. and Bierer, R.L., 2001. Influence of green lipped mussels (*Perna canaliculus*) in alleviating signs of arthritis in dogs. Journal of Veterinary Pharmacology and Therapeutics 2: 101-111. https://doi.org/19753702

Burri, L., Heggen, K. and Storsve, A.B., 2020. Higher omega-3 index after dietary inclusion of omega-3 phospholipids versus omega-3 triglycerides in Alaskan Huskies. Veterinary World 13: 1167-1173. https://doi.org/10.14202/vetworld.2020.1167-1173

Burri, L., Hoem, N., Monakhova, Y.B. and Diehl, B.W., 2016. Fingerprinting krill oil by 31P, 1H and 13C NMR spectroscopies. Journal of the American Oil Chemists' Society 93: 1037-1049. https://doi.org/10.1007/s11746-016-2836-3

Burri, L., Wyse, C., Gray, S.R., Harris, W.S. and Lazzerini, K., 2018. Effects of dietary supplementation with krill meal on serum pro-inflammatory markers after the Iditarod sled dog race. Research in Veterinary Science 121: 18-22. https://doi.org/10.1016/j.rvsc.2018.10.002

Cabrita, A.R.J., Barroso, C., Fontes-Sousa, A.P., Correia, A., Teixeira, L., Maia, M.R.G., Vilanova, M., Yergaliyev, T., Camarinha-Silva, A. and Fonseca, A.J.M., 2024a. Assessing functional properties of diet protein hydrolysate and oil from fish waste on canine immune parameters, cardiac biomarkers and fecal microbiota. Frontiers in Veterinary Science 11. https://doi.org/10.3389/fvets.2024.1449141

Cabrita, A.R.J., Maia, M.R.G., Alves, A.P., Aires, T., Rosa, A., Almeida, A., Martins, R. and Fonseca, A.J.M., 2024b. Protein hydrolysate and oil from fish waste reveal potential as dog food ingredients. Frontiers in Veterinary Science 11. https://doi.org/10.3389/fvets.2024.1372023

Cabrita, A.R.J., Maia, M.R.G. and Fonseca, A.J.M., 2025. Chapter 33 - Performance of microalgae-based food on pet health. In: A. Pérez-Gálvez, E. Jacob-Lopes, L. Queiroz Zepka and M. Roca (Eds.), Microalgae and One Health. Academic Press, pp. 521-536. https://doi.org/10.1016/B978-0-443-22080-7.00018-1

Cabrita, A.R.J., Maia, M.R.G., Oliveira, H.M., Sousa-Pinto, I., Almeida, A.A., Pinto, E. and Fonseca, A.J.M., 2016. Tracing seaweeds as mineral sources for farm-animals. Journal of Applied Phycology 28: 3135-3150. https://doi.org/10.1007/s10811-016-0839-y

Cabrita, A.R.J., Valente, I.M., Oliveira, H.M., Fonseca, A.J.M. and Maia, M.R.G., 2019. Effects of feeding with seaweeds on ruminal fermentation and methane production. In: L. Pereira, Bahcevandziev, K., Joshi, N.H. (Ed.), Seaweeds as Plant Fertilizer, Agricultural Biostimulants and Animal Fodder. CRC Press, pp. 187-210. https://doi.org/10.1201/9780429487156-10

Cahú, T.B., Santos, S.D., Mendes, A., Córdula, C.R., Chavante, S.F., Carvalho Jr, L.B., Nader, H.B. and Bezerra, R.S., 2012. Recovery of protein, chitin, carotenoids and glycosaminoglycans from Pacific white shrimp (*Litopenaeus vannamei*) processing waste. Process Biochemistry 47: 570-577. https://doi.org/10.1016/j.procbio.2011.12.012

Calder, P.C., 2013. Omega-3 polyunsaturated fatty acids and inflammatory processes: nutrition or pharmacology? British Journal of Clinical Pharmacology 75: 645-662. https://doi.org/10.1111/j.1365-2125.2012.04374.x

Carlisle, C., Metzger, B.T., Tintle, N.L., Polley, K., Jackson, K.H., Le Brun-Blashka, S., Griffiths, J. and Harris, W.S., 2024. The effects of omega-3 supplementation on the omega-3 index and quality of life and pain scores in dogs. Animals (Basel) 14. https://doi.org/10.3390/ani14213108

Carranco-Jáuregui, M., Fuente-Martínez, B., Ramírez-Poblano, M., Calvo-Carrillo, M. and Ávila-González, E., 2021. Giant squid meal (*Dosidicus gigas*) inclusión, as a source of protein in laying hens' diet. Abanico Veterinario 10: 1-12. https://doi.org/10.21929/abavet2020.14

Cave, N., Bolton, S. and Cogger, N., 2024. A pilot study to detect the effects of a green-lipped mussel (*Perna canaliculus*) nutraceutical on working farm dogs with musculoskeletal abnormalities using accelerometry. New Zealand Veterinary Journal 72: 323-333. https://doi.org/10.1080/00480169.2024.2379966

Cheng, J., Landesman, L., Bergmann, B., Classen, J.J., Howard, J. and Yamamoto, Y., 2002. Nutrient removal from swine lagoon liquid by *Lemna minor* 8627. Transactions of the ASAE 45: 1003. https://doi.org/10.13031/2013.9953

Chojnacka, K., Saeid, A., Witkowska, Z. and Tuhy, L., 2012. Biologically active compounds in seaweed extracts-the prospects for the application, The Open Conference Proceedings Journal. Bentham Science Publishers Ltd., pp. 20-28.

Córdova-Murueta, J.H. and García-Carreño, F.L., 2002. Nutritive value of squid and hydrolyzed protein supplement in shrimp feed. Aquaculture 210: 371-384. https://doi.org/10.1016/S0044-8486(02)00011-X

Coulson, S., Palacios, T. and Vitetta, L., 2015. *Perna canaliculus* (Green-Lipped Mussel): bioactive components and therapeutic evaluation for chronic health conditions. Progress in Drug Research: 91-132. https://doi.org/10.1007/978-3-0348-0927-6_3

Djellouli, M., López-Caballero, M.E., Arancibia, M.Y., Karam, N. and Martínez-Alvarez, O., 2020. Antioxidant and antimicrobial enhancement by reaction of protein hydrolysates derived from shrimp by-products with glucosamine. Waste and Biomass Valorization 11: 2491-2505. https://doi.org/10.1007/s12649-019-00607-y

Douda, K., Zieritz, A., Vodáková, B., Urbańska, M., Bolotov, I.N., Marková, J., Froufe, E., Bogan, A.E. and Lopes-Lima, M., 2025. Review of the globally invasive freshwater mussels in the genus *Sinanodonta Modell*, 1945. Hydrobiologia 852: 1243-1273. https://doi.org/10.1007/s10750-023-05457-3

Dutot, M., Fagon, R., Hemon, M. and Rat, P., 2012. Antioxidant, anti-inflammatory, and anti-senescence activities of a phlorotannin-rich natural extract from brown seaweed *Ascophyllum nodosum*. Applied Biochemistry and Biotechnology 167: 2234-2240. https://doi.org/10.1007/s12010-012-9761-1

Eason, C.T., Adams, S.L., Puddick, J., Romanazzi, D., Miller, M.R., King, N., Johns, S., Forbes-Blom, E., Hessian, P.A., Stamp, L.K. and Packer, M.A., 2018. Greenshell™ Mussels: a review of veterinary trials and future research directions. Veterinary Sciences 5. https://doi.org/10.3390/vetsci5020036

Ennayer, I., Chadli, H., Chebbaki, K., Chiaar, A., Bouhdid, S., Nhhala, H. and Chairi, H., 2025. Utilization of squid (*Loligo vulgaris*) co-products meal as fishmeal replacer in the diet of European seabass (*Dicentrarchus labrax*) farmed in Morocco. Egyptian Journal of Aquatic Research 51: 117-124. https://doi.org/10.1016/j.ejar.2024.12.001

Euromonitor, 2022. Top five trends in pet care.

Evans, F. and Critchley, A., 2014. Seaweeds for animal production use. Journal of Applied Phycology 26: 891-899. https://doi.org/10.1007/s10811-013-0162-9

FAO, 2020. The state of world fisheries and aquaculture 2020. Sustainability in action, Rome. https://openknowledge.fao.org/handle/20.500.14283/ca9229en

FAO, 2023. GLOBEFISH Highlights – International markets for fisheries and aquaculture products – Second issue 2023, with January–December 2022 Statistics. GLOBEFISH Highlights, No. 2–2023. Rome. https://doi.org/10.4060/cc7781en.

FEDIAF, 2024. Annual Report 2024.

Fenton, J.I., Gurzell, E.A., Davidson, E.A. and Harris, W.S., 2016. Red blood cell PUFAs reflect the phospholipid PUFA composition of major organs. Prostaglandins, Leukotrienes and Essential Fatty Acids 112: 12-23. https://doi.org/10.1016/j.plefa.2016.06.004

Fourounjian, P., Fakhoorian, T. and Cao, X.H., 2020. Importance of duckweeds in basic research and their industrial applications, The Duckweed Genomes. Springer, pp. 1-17. https://doi.org/10.1007/978-3-030-11045-1_1

Franceschi, V., 1987. Oxalic acid metabolism and calcium oxalate formation in *Lemna minor* L. Plant, Cell & Environment 10: 397-406. https://doi.org/10.1111/1365-3040.ep11603639

Ganesan, A.R., Tiwari, U. and Rajauria, G., 2019. Seaweed nutraceuticals and their therapeutic role in disease prevention. Food Science and Human Wellness 8: 252-263. https://doi.org/10.1016/j.fshw.2019.08.001

Ganesan, P., Kumar, C.S. and Bhaskar, N., 2008. Antioxidant properties of methanol extract and its solvent fractions obtained from selected Indian red seaweeds. Bioresource Technology 99: 2717-2723. https://doi.org/10.1016/j.biortech.2007.07.005

Gawor, J. and Jank, M., 2023. *Ascophyllum nodosum* as a nutrient supporting oral health in dogs and cats: a review. Polish Journal of Veterinary Science 26: 511-520. https://doi.org/10.24425/pjvs.2023.145053

Gawor, J., Jank, M., Jodkowska, K., Klim, E. and Svensson, U.K., 2018. Effects of edible treats containing *Ascophyllum nodosum* on the oral health of dogs: a double-blind, randomized, placebo-controlled single-center study. Frontiers in Veterinary Science 5: 168. https://doi.org/10.3389/fvets.2018.00168

Gawor, J.P., Wilczak, J., Svensson, U.K. and Jank, M., 2021. Influence of dietary supplementation with a powder containing A.N. ProDen™ (*Ascophyllum nodosum*) algae on dog saliva metabolome. Frontiers in Veterinary Science Volume 8 - 2021. https://doi.org/10.3389/fvets.2021.681951

Gildberg, A., Arnesen, J.A., Sæther, B.-S., Rauø, J. and Stenberg, E., 2011. Angiotensin I-converting enzyme inhibitory activity in a hydrolysate of proteins from Northern shrimp (*Pandalus borealis*) and identification of two novel inhibitory tri-peptides. Process Biochemistry 46: 2205-2209. https://doi.org/10.1016/j.procbio.2011.08.003

Gómez-Estaca, J., Alemán, A., López-Caballero, M.E., Baccan, G.C., Montero, P. and Gómez-Guillén, M.C., 2019. Bioaccessibility and antimicrobial properties of a shrimp demineralization extract blended with chitosan as wrapping material in ready-to-eat raw salmon. Food Chemistry 276: 342-349. https://doi.org/10.1016/j.foodchem.2018.10.031

Gomez-Zavaglia, A., Prieto Lage, M.A., Jimenez-Lopez, C., Mejuto, J.C. and Simal-Gandara, J., 2019. The potential of seaweeds as a source of functional ingredients of prebiotic and antioxidant value. Antioxidants 8: 406. https://doi.org/10.3390/antiox8090406

González-Meza, G.M., Elizondo-Luevano, J.H., Cuellar-Bermudez, S.P., Sosa-Hernández, J.E., Iqbal, H.M.N., Melchor-Martínez, E.M. and Parra-Saldívar, R., 2023. New perspective for macroalgae-based animal feeding in the context of challenging sustainable food production. Plants 12: 3609. https://doi.org/10.3390/plants12203609

Goopy, J. and Murray, P., 2003. A review on the role of duckweed in nutrient reclamation and as a source of animal feed. Asian-Australasian Journal of Animal Sciences 16: 297-305. https://doi.org/10.5713/ajas.2003.297

Gordalina, M., Pinheiro, H.M., Mateus, M., da Fonseca, M.M.R. and Cesário, M.T., 2021. Macroalgae as protein sources - a review on protein bioactivity, extraction, purification and characterization. Applied Sciences 11: 7969. https://doi.org/10.3390/app11177969

Guilherme-Fernandes, J., Aires, T., Fonseca, A.J.M., Yergaliyev, T., Camarinha-Silva, A., Lima, S.A.C., Maia, M.R.G. and Cabrita, A.R.J., 2024a. Squid meal and shrimp hydrolysate as novel protein sources for dog food. Frontiers in Veterinary Science 11. https://doi.org/10.3389/fvets.2024.1360939

Guilherme-Fernandes, J., Barroso, C., Correia, A., Aires, T., Yergaliyev, T., Camarinha-Silva, A., Vilanova, M., Fonseca, A.J.M., Lima, S.A.C., Maia, M.R.G. and Cabrita, A.R.J., 2025. Unraveling the role of shrimp hydrolysate as a food supplement in the immune function and fecal microbiota of beagle dogs. Scientific Reports 15: 25510. https://doi.org/10.1038/s41598-025-09942-8

Guilherme-Fernandes, J., Fonseca, A.J.M., Aires, T., Lima, S.A.C., Maia, M.R.G. and Cabrita, A.R.J., 2024b. Unveiling the effects of shrimp hydrolysate as a dietary supplement in healthy adult Beagle dogs. Journal of Animal Science 3:102:skae280. https://doi.org/10.1093/jas/skae280

Haard, N.F. and Simpson, B.K., 2000. Seafood enzymes: utilization and influence on postharvest seafood quality. CRC Press. https://doi.org/10.1201/9781482289916

Harris, W.S. and Von Schacky, C., 2004. The omega-3 index: a new risk factor for death from coronary heart disease? Preventive medicine 39: 212-220. https://doi.org/10.1016/j.ypmed.2004.02.030

Healy, L.E., Zhu, X., Pojić, M., Sullivan, C., Tiwari, U., Curtin, J. and Tiwari, B.K., 2023. Biomolecules from macroalgae—nutritional profile and bioactives for novel food product development. Biomolecules 13: 386. https://doi.org/10.3390/biom13020386

Hertrampf, J.W. and Piedad-Pascual, F., 2000a. Krill meal. In: J.W. Hertrampf and F. Piedad-Pascual (Eds.), Handbook on Ingredients for Aquaculture Feeds. Springer Netherlands, Dordrecht, pp. 221-228. https://doi.org/10.1007/978-94-011-4018-8

Hertrampf, J.W. and Piedad-Pascual, F., 2000b. Squid meal. In: J.W. Hertrampf and F. Piedad-Pascual (Eds.), Handbook on Ingredients for Aquaculture Feeds. Springer Netherlands, Dordrecht, pp. 403-410. https://doi.org/10.1007/978-94-011-4018-8_43

Hielm-Björkman, A., Tulamo, R.M., Salonen, H. and Raekallio, M., 2009. Evaluating complementary therapies for canine osteoarthritis Part I: Green-lipped mussel (*Perna canaliculus*). Evidence-Based Complementary and Alternative Medicine 6: 365-373. https://doi.org/10.1093/ecam/nem136

Hu, Z., Fang, Y., Yi, Z., Tian, X., Li, J., Jin, Y., He, K., Liu, P., Du, A., Huang, Y. and Zhao, H., 2022. Determining the nutritional value and antioxidant capacity of duckweed (*Wolffia arrhiza*) under artificial conditions. LWT 153: 112477. https://doi.org/10.1016/j.lwt.2021.112477

Ibrahim, H.M., Salama, M.F. and El-Banna, H.A., 1999. Shrimp's waste: Chemical composition, nutritional value and utilization. Food/Nahrung 43: 418-423. https://doi.org/10.1002/(SICI)1521-3803(19991201)43:6

Iqbal, A., Khalil, I.A., Ateeq, N. and Khan, M.S., 2006. Nutritional quality of important food legumes. Food Chemistry 97: 331-335. https://doi.org/10.1016/j.foodchem.2005.05.011

Isidori, M., Rueca, F., Massacci, F.R., Diaferia, M., Giontella, A., Caldin, M., Furlanello, T., Corbee, R.J., Mannucci, G., Pezzotti, G. and Trabalza-Marinucci, M., 2021. The use of *Ascophyllum nodosum* and *Bacillus subtilis* C-3102 in the management of canine chronic inflammatory enteropathy: a pilot study. Animals 11: 3417. https://doi.org/10.3390/ani11123417

Isidori, M., Rueca, F. and Trabalza-Marinucci, M., 2019. Palatability of extruded dog diets supplemented with *Ascophyllum nodosum* L. (Fucaceae, Phaeophyceae). Journal of Applied Phycology 31: 3275-3281. https://doi.org/10.1007/s10811-019-01799-5

Islam, K., 2002. Feasibility of duckweed as poultry feed-A review. The Indian Journal of Animal Sciences 72. https://epubs.icar.org.in/index.php/IJAnS/article/view/37651

Kannan, A., Hettiarachchy, N.S., Marshall, M., Raghavan, S. and Kristinsson, H., 2011. Shrimp shell peptide hydrolysates inhibit human cancer cell proliferation. Journal of the Science of Food and Agriculture 91: 1920-1924. https://doi.org/10.1002/jsfa.4464

Kazir, M., Abuhassira, Y., Robin, A., Nahor, O., Luo, J., Israel, A., Golberg, A. and Livney, Y.D., 2019. Extraction of proteins from two marine macroalgae, *Ulva* sp. and *Gracilaria* sp., for food application, and evaluating digestibility, amino acid composition and antioxidant properties of the protein concentrates. Food Hydrocolloids 87: 194-203. https://doi.org/10.1016/j.foodhyd.2018.07.047

Köhler, A., Sarkkinen, E., Tapola, N., Niskanen, T. and Bruheim, I., 2015. Bioavailability of fatty acids from krill oil, krill meal and fish oil in healthy subjects–a randomized, single-dose, cross-over trial. Lipids in Health and Disease 14: 19. https://doi.org/10.1186/s12944-015-0015-4

Konieczny, P., Andrzejewski, W., Yang, T., Urbańska, M., Stangierski, J., Tomczyk, Ł. and Mikołajczak, B., 2022. Selected quality attributes of freshwater mussel powder as a promising ingredient for pet food. Animals 12: 90. https://doi.org/10.3390/ani12010090

Kumar, C.S., Ganesan, P., Suresh, P. and Bhaskar, N., 2008. Seaweeds as a source of nutritionally beneficial compounds-a review. Journal of Food Science and Technology 45: 1-13.

Leal, D., Borges, S., Almeida, A., Pintado, M., Fonseca, A.J.M. and Cabrita, A.R.J., 2025. Protein hydrolysates from *Tenebrio molitor* and *Hermetia illucens* as novel food sources for dogs. Translational Animal Science 19. https://doi.org/10.1093/tas/txaf106

Lin, C.Y., Kerr, K.R., Panasevich, M.R., Daristotle, L. and Frantz, N.Z., 2024. Duckweed protein as an alternative plant-based protein source for dog and cat dry diets. Journal of Animal Science 102. https://doi.org/10.1093/jas/skae244

Lindahl, O., 2013. Mussel meal production based on mussels from the Baltic Sea, Finnish Game and Fisheries Research Institute, Helsinki.

Lindqvist, H., Dominguez, T., Dragøy, R., Ding, Y. and Burri, L., 2023. Comparison of fish, krill and flaxseed as omega-3 sources to increase the omega-3 index in dogs. Veterinary Sciences 10: 162. https://doi.org/10.3390/vetsci10020162

MacArtain, P., Gill, C.I., Brooks, M., Campbell, R. and Rowland, I.R., 2007. Nutritional value of edible seaweeds. Nutrition Reviews 65: 535-543. https://doi.org/10.1301/nr.2007.dec.535-543

Makkar, H.P., Tran, G., Heuzé, V., Giger-Reverdin, S., Lessire, M., Lebas, F. and Ankers, P., 2016. Seaweeds for livestock diets: A review. Animal Feed Science and Technology 212: 1-17. https://doi.org/10.1016/j.anifeedsci.2015.09.018

Marsham, S., Scott, G.W. and Tobin, M.L., 2007. Comparison of nutritive chemistry of a range of temperate seaweeds. Food Chemistry 100: 1331-1336. https://doi.org/10.1016/j.foodchem.2005.11.029

Mclean, C.H. and Bulling, K.R., 2005. Differences in lipid profile of New Zealand marine species over four seasons. Journal of Food Lipids 12: 313-326. https://doi.org/10.1111/j.1745-4522.2005.00026.x

Mehler, S.J., May, L.R., King, C., Harris, W.S. and Shah, Z., 2016. A prospective, randomized, double blind, placebo-controlled evaluation of the effects of eicosapentaenoic acid and docosahexaenoic acid on the clinical signs and erythrocyte membrane polyunsaturated fatty acid concentrations in dogs with osteoarthritis. Prostaglandins, Leukotrienes and Essential Fatty Acids 109: 1-7. https://doi.org/10.1016/j.plefa.2016.03.015

Miller, M.R., Abshirini, M., Wolber, F.M., Tuterangiwhiu, T.R. and Kruger, M.C., 2023. Greenshell mussel products: a comprehensive review of sustainability, traditional use, and efficacy. Sustainability 15: 3912. https://doi.org/10.3390/su15053912

Miller, M.R., Pearce, L. and Bettjeman, B.I., 2014. Detailed distribution of lipids in Greenshell™ Mussel (*Perna canaliculus*). Nutrients 6: 1454-1474. https://doi.org/10.3390/nu6041454

Miller, M.R., Perry, N.B., Burgess, E.J. and Marshall, S.N., 2011. Regiospecific analyses of triacylglycerols of hoki (*Macruronus novaezelandiae*) and Greenshell™ mussel (*Perna canaliculus*). Journal of the American Oil Chemists' Society 88: 509-516. https://doi.org/10.1007/s11746-010-1690-y

Miller, M.R. and Tian, H., 2018. Changes in proximate composition, lipid class and fatty acid profile in Greenshell™ mussels (*Perna canaliculus*) over an annual cycle. Aquaculture Research 49: 1153-1165. https://doi.org/10.1111/are.13565

Minich, J.J. and Michael, T.P., 2024. A review of using duckweed (*Lemnaceae*) in fish feeds. Reviews in Aquaculture 16: 1212-1228. https://doi.org/10.1111/raq.12892

Mišurcová, L., 2011. Chemical composition of seaweeds. Handbook of Marine Macroalgae: Biotechnology and Applied Phycology: 171-192. https://doi.org/10.1002/9781119977087

Monteiro, B.P., Lascelles, B.D.X., Murrell, J., Robertson, S., Steagall, P.V.M. and Wright, B., 2023. 2022 WSAVA guidelines for the recognition, assessment and treatment of pain. Journal of Small Animal Practice 64: 177-254. https://doi.org/10.1111/jsap.13566

Morais, T., Inácio, A., Coutinho, T., Ministro, M., Cotas, J., Pereira, L. and Bahcevandziev, K., 2020. Seaweed potential in the animal feed: a review. Journal of Marine Science and Engineering 8: 559. https://doi.org/10.3390/jmse8080559

Mota, C.S.C., Cabrita, A.R.J., Yergaliyev, T., Camarinha-Silva, A., Almeida, A., Abreu, H., Silva, J., Fonseca, A.J.M. and Maia, M.R.G., 2024. Macroalgae and microalga blend in dogs' food: Effects on palatability, digestibility, and fecal metabolites and microbiota. Algal Research 84: 103775. https://doi.org/10.1016/j.algal.2024.103775

Muller, T. and Bazinet, L., 2025. Water lentil (duckweed) protein purification by chemical (HCl) and electrochemical (Electrodialysis with bipolar membranes) acidification: Composition, structure and functional properties vs commercial protein isolates. Food Chemistry: 144901. https://doi.org/10.1016/j.foodchem.2025.144901

Muller, T., Cournoyer, A. and Bazinet, L., 2025. Emerging potentials of duckweed (*Lemnaceae*): From composition to protein uses in food and nutraceuticals – A review. Food Research International 219: 116777. https://doi.org/10.1016/j.foodres.2025.116777

Naik, A.S. and Hayes, M., 2019. Bioprocessing of mussel by-products for value added ingredients. Trends in Food Science & Technology 92: 111-121. https://doi.org/10.1016/j.tifs.2019.08.013

Nair, S., Gagnon, J., Pelletier, C., Tchoukanova, N., Zhang, J., Ewart, H.S., Ewart, K.V., Jiao, G. and Wang, Y., 2017. Shrimp oil extracted from the shrimp processing waste reduces the development of insulin resistance and metabolic phenotypes in diet-induced obese rats. Applied Physiology, Nutrition, and Metabolism 42: 841-849. https://doi.org/10.1139/apnm-2016-0644

Naylor, R.L., Hardy, R.W., Bureau, D.P., Chiu, A., Elliott, M., Farrell, A.P., Forster, I., Gatlin, D.M., Goldburg, R.J. and Hua, K., 2009. Feeding aquaculture in an era of finite resources. Proceedings of the National Academy of Sciences 106: 15103-15110. https://doi.org/10.1073/pnas.0905235106

Nicol, S., Forster, I. and Spence, J., 2000. Products Derived from Krill, Krill, pp. 262-283. https://doi.org/10.1002/9780470999493.ch10

Nieuwland, M., Geerdink, P., Engelen-Smit, N.P., Van Der Meer, I.M., America, A.H., Mes, J.J., Kootstra, A.M.J., Henket, J.T. and Mulder, W.J., 2021. Isolation and gelling properties of duckweed protein concentrate. ACS Food Science & Technology 1: 908-916. https://doi.org/10.1021/acsfoodscitech.1c00009

Nirmal, N.P., Santivarangkna, C., Rajput, M.S. and Benjakul, S., 2020. Trends in shrimp processing waste utilization: An industrial prospective. Trends in Food Science & Technology 103: 20-35. https://doi.org/10.1016/j.tifs.2020.07.001

NRC, 2006. Nutrient Requirements for dogs and cat. The National Academies Press, Washington, DC, USA.

O'Sullivan, L., Murphy, B., McLoughlin, P., Duggan, P., Lawlor, P.G., Hughes, H. and Gardiner, G.E., 2010. Prebiotics from marine macroalgae for human and animal health applications. Marine Drugs 8: 2038-2064. https://doi.org/10.3390/md8072038

Ofoedu, C.E., Bozkurt, H. and Mortimer, J.C., 2025. Towards sustainable food security: Exploring the potential of duckweed (*Lemnaceae*) in diversifying food systems. Trends in Food Science & Technology 161: 105073. https://doi.org/10.1016/j.tifs.2025.105073

Okolie, C.L., CK Rajendran, S.R., Udenigwe, C.C., Aryee, A.N. and Mason, B., 2017. Prospects of brown seaweed polysaccharides (BSP) as prebiotics and potential immunomodulators. Journal of Food Biochemistry 41: e12392. https://doi.org/10.1111/jfbc.12392

Øverland, M., Mydland, L.T. and Skrede, A., 2019. Marine macroalgae as sources of protein and bioactive compounds in feed for monogastric animals. Journal of the Science of Food and Agriculture 99: 13-24. https://doi.org/10.1002/jsfa.9143

Pagliuso, D., Grandis, A., Fortirer, J.S., Camargo, P., Floh, E.I.S. and Buckeridge, M.S., 2022. Duckweeds as promising food feedstocks globally. Agronomy 12: 796. https://doi.org/10.3390/agronomy12040796

Pavia, H. and Brock, E., 2000. Extrinsic factors influencing phlorotannin production in the brown alga *Ascophyllum nodosum*. Marine Ecology Progress Series 193: 285-294. https://doi.org/10.3354/meps193285

Pearce, F.G. and Brunke, J.E., 2023. Is now the time for a Rubiscuit or Ruburger? Increased interest in Rubisco as a food protein. Journal of Experimental Botany 74: 627-637. https://doi.org/10.1093/jxb/erac414

Pereira, A.M., Maia, M.R.G., Fonseca, A.J.M. and Cabrita, A.R.J., 2021. Zinc in dog nutrition, health and disease: a review. Animals 11: 978. https://doi.org/10.3390/ani11040978

Petroski, W. and Minich, D.M., 2020. Is there such a thing as "anti-nutrients"? A narrative review of perceived problematic plant compounds. Nutrients 12: 2929. https://doi.org/10.3390/nu12102929

Pinna, C., Vecchiato, C.G., Grandi, M., Stefanelli, C., Zannoni, A. and Biagi, G., 2021. Seaweed supplementation failed to affect fecal microbiota and metabolome as well as fecal iga and apparent nutrient digestibility in adult dogs. Animals 11. https://doi.org/10.3390/ani11082234

Ramnani, P., Chitarrari, R., Tuohy, K., Grant, J., Hotchkiss, S., Philp, K., Campbell, R., Gill, C. and Rowland, I., 2012. In vitro fermentation and prebiotic potential of novel low molecular weight polysaccharides derived from agar and alginate seaweeds. Anaerobe 18: 1-6. https://doi.org/10.1016/j.anaerobe.2011.08.003

Rialland, P., Bichot, S., Lussier, B., Moreau, M., Beaudry, F., del Castillo, J.R., Gauvin, D. and Troncy, E., 2013. Effect of a diet enriched with green-lipped mussel on pain behavior and functioning in dogs with clinical osteoarthritis. Canadian Journal of Veterinary Research 77: 66-74. PMID: 23814358

Ribeiro, T.B., Maia, M.R.G., Fonseca, A.J.M., Marques, B., Caleja, C., Rosa, A., Martins, R., Almeida, A., Mota, M.J., Aires, T., Rocha, C.M.R., Teixeira, J.A., Cabrita, A.R.J., Barros, L. and Pintado, M., 2024. Sustainable pet nutrition: a comprehensive review of fish protein hydrolysates in functional pet food formulations. Food Reviews International 41: 1321-1359. https://doi.org/10.1080/87559129.2024.2430660

Riou, D., Colliec-Jouault, S., Pinczon du Sel, D., Bosch, S., Siavoshian, S., Le Bert, V., Tomasoni, C., Sinquin, C., Durand, P. and Roussakis, C., 1996. Antitumor and antiproliferative effects of a fucan extracted from *Ascophyllum nodosum* against a non-small-cell bronchopulmonary carcinoma line. Anticancer Research 16: 1213-1218. PMID: 8702239

Rodrigues, M., Rosa, A., Almeida, A., Martins, R., Ribeiro, T., Pintado, M., Gonçalves, R.F.S., Pinheiro, A.C., Fonseca, A.J.M., Maia, M.R.G., Cabrita, A.R.J., Barros, L. and Caleja, C., 2024. Omega-3 fatty acids from fish by-products: Innovative extraction and application in food and feed. Food and Bioproducts Processing 145: 32-41. https://doi.org/10.1016/j.fbp.2024.02.007

Roy, V.C., Islam, M.R., Sadia, S., Yeasmin, M., Park, J.-S., Lee, H.-J. and Chun, B.-S., 2023. Trash to treasure: An up-to-date understanding of the valorization of seafood by-products, targeting the major bioactive compounds. Marine Drugs 21: 485. https://doi.org/10.3390/md21090485

Saether, O. and Mohr, V., 1987. Chemical composition of North Atlantic krill. Comparative Biochemistry and Physiology Part B: Comparative Biochemistry 88: 157-164. https://doi.org/10.1016/0305-0491(87)90095-2

Sardari, R.R. and Nordberg Karlsson, E., 2018. Marine poly-and oligosaccharides as prebiotics. Journal of Agricultural and Food Chemistry 66: 11544-11549. https://doi.org/10.1021/acs.jafc.8b04418

Senphan, T. and Benjakul, S., 2012. Compositions and yield of lipids extracted from hepatopancreas of Pacific white shrimp (*Litopenaeus vannamei*) as affected by prior autolysis. Food Chemistry 134: 829-835. https://doi.org/10.1016/j.foodchem.2012.02.188

Senphan, T., Benjakul, S. and Kishimura, H., 2014. Characteristics and antioxidative activity of carotenoprotein from shells of Pacific white shrimp extracted using hepatopancreas proteases. Food Bioscience 5: 54-63. https://doi.org/10.1016/j.fbio.2013.11.004

Singh, A., Mittal, A. and Benjakul, S., 2022. Full utilization of squid meat and its processing by-products: revisit. Food Reviews International 38: 455-479. https://doi.org/10.1080/87559129.2020.1734611

Skelton, B.M., South, P.M. and Jeffs, A.G., 2022. Inefficiency of conversion of seed into market-ready mussels in New Zealand's Greenshell™ mussel (*Perna canaliculus*) industry. Aquaculture 560: 738584. https://doi.org/10.1016/j.aquaculture.2022.738584

Sosa, D., Alves, F.M., Prieto, M.A., Pedrosa, M.C., Heleno, S.A., Barros, L., Feliciano, M. and Carocho, M., 2024. *Lemna minor*: unlocking the value of this duckweed for the food and feed industry. Foods 13. https://doi.org/10.3390/foods13101435

Sowinski, E.E., Gilbert, S., Lam, E. and Carpita, N.C., 2019. Linkage structure of cell-wall polysaccharides from three duckweed species. Carbohydrate Polymers 223: 115119. https://doi.org/10.1016/j.carbpol.2019.115119

Sree, K. and Appenroth, K.-J., 2020. Worldwide genetic resources of duckweed: stock collections, The Duckweed Genomes. Springer, pp. 39-46. https://doi.org/10.1007/978-3-030-11045-1_3

Srinivas, K., Das, A., Reddy, P.B. and Verma, A., 2024. Supplementation of tropical red seaweeds improved gut health indices, antioxidant status and immunity in adult dogs. Journal of Applied Phycology 36: 2183-2198. https://doi.org/10.1007/s10811-024-03229-7

Statista, 2025. Number of domestic cats and dogs in urban China 2020-2024.

Stenton-Dozey, J. and Broekhuizen, N., 2019. Provision of ecological and ecosystem services by mussel farming in the Marlborough sounds: a literature review in context of the state of the environment pre- and post-mussel farming; National Institute of Water & Atmospheric Research Ltd. 2019020CH (project MFI19201). In: N.I.o.W.A.R. Ltd. (Ed.), Auckland, New Zealand. https://www.marinefarming.co.nz/site_files/24792/upload_files/Fullreport_28.07.2021update.pdf?dl=1

Takács, K., Végh, R., Mednyánszky, Z., Haddad, J., Allaf, K., Du, M., Chen, K., Kan, J., Cai, T., Molnár, P., Bársony, P., Maczó, A., Zalán, Z. and Dalmadi, I., 2025. New insights into duckweed as an alternative source of food and feed: key components and potential technological solutions to increase their digestibility and bioaccessibility. Applied Sciences 15: 884. https://doi.org/10.3390/app15020884

Tanase, R., Senda, R., Matsunaga, Y. and Narukawa, M., 2022. Taste characteristics of various amino acid derivatives. Journal of Nutritional Science and Vitaminology 68: 475-480. https://doi.org/10.3177/jnsv.68.475

Tippery, N.P. and Les, D.H., 2020. Tiny plants with enormous potential: phylogeny and evolution of duckweeds, The Duckweed Genomes. Springer, pp. 19-38. https://doi.org/10.1007/978-3-030-11045-1_2

Torok, V.A., Ophel-Keller, K., Loo, M. and Hughes, R.J., 2008. Application of methods for identifying broiler chicken gut bacterial species linked with increased energy metabolism. Applied and Environmental Microbiology 74: 783-791. https://doi.org/10.1128/AEM.01384-07

Tou, J.C., Jaczynski, J. and Chen, Y.-C., 2007. Krill for human consumption: nutritional value and potential health benefits. Nutrition Reviews 65: 63-77. https://doi.org/10.1111/j.1753-4887.2007.tb00283.x

Treschow, A., Hodges, L., Wright, P., Wynne, P., Kalafatis, N. and Macrides, T., 2007. Novel anti-inflammatory ω-3 PUFAs from the New Zealand green-lipped mussel, *Perna canaliculus*. Comparative Biochemistry and Physiology Part B: Biochemistry and Molecular Biology 147: 645-656. https://doi.org/10.1016/j.cbpb.2007.04.004

Turner, J., Dritz, S., Higgins, J. and Minton, J., 2002. Effects of *Ascophyllum nodosum* extract on growth performance and immune function of young pigs challenged with Salmonella typhimurium. Journal of Animal Science 80: 1947-1953. https://doi.org/10.2527/2002.8071947x

van der Spiegel, M., Noordam, M.Y. and van der Fels-Klerx, H.J., 2013. Safety of novel protein sources (insects, microalgae, seaweed, duckweed, and rapeseed) and legislative aspects for their application in food and feed production. Comprehensive Reviews in Food Science and Food Safety 12: 662-678. https://doi.org/10.1111/1541-4337.12032

Vendramini, T.H.A., Marchi, P.H., Olivindo, R.F.G., Pedrinelli, V., Amaral, A.R., Miranda, M.S.d., Príncipe, L.A., Cesar, C.G.L., Zafalon, R.V.A., Perini, M.P., Lima, L.O.C., Balieiro, J.C.d.C. and Brunetto, M.A., 2025. Exploring the efficacy and optimal dosages of omega-3 supplementation for companion animals. Nutrition and Research Reviews: 1-16. https://doi.org/10.1017/S0954422425100115

Weerakkody, W.S., Ling, K.H., Hsieh, H.-H., Abedneko, V.G., Shyu, J.-F., Lee, T.-M., Shih, Y.-Y., Ranatunga, R., Santschi, P.H. and Hung, C.-C., 2023. Carbon capture by macroalgae *Sarcodia suae* using aquaculture wastewater and solar energy for cooling in subtropical regions. Science of The Total Environment 855: 158850. https://doi.org/10.1016/j.scitotenv.2022.158850

Wells, M.L., Potin, P., Craigie, J.S., Raven, J.A., Merchant, S.S., Helliwell, K.E., Smith, A.G., Camire, M.E. and Brawley, S.H., 2017. Algae as nutritional and functional food sources: revisiting our understanding. Journal of Applied Phycology 29: 949-982. https://doi.org/10.1007/s10811-016-0974-5

Yaghubi, E., Carboni, S., Snipe, R.M.J., Shaw, C.S., Fyfe, J.J., Smith, C.M., Kaur, G., Tan, S.-Y. and Hamilton, D.L., 2021. Farmed mussels: a nutritive protein source, rich in omega-3 fatty acids, with a low environmental footprint. Nutrients 13: 1124. https://doi.org/10.3390/nu13041124

Yang, L., Zheng, Y., Zhang, S., Wang, X. and Guo, Q., 2024. Effects of boiling, centrifuging and drying methods on bioactive compounds in Antarctic krill meal. International Journal of Food Science and Technology 59: 6409-6420. https://doi.org/10.1111/ijfs.17386

Zhang, J., Tiller, C., Shen, J., Wang, C., Girouard, G.S., Dennis, D., Barrow, C.J., Miao, M. and Ewart, H.S., 2007. Antidiabetic properties of polysaccharide-and polyphenolic-enriched fractions from the brown seaweed Ascophyllum nodosum. Canadian Journal of Physiology and Pharmacology 85: 1116-1123. https://doi.org/10.1139/Y07-105

Zhang, M., Mo, R., Qu, Y., Li, M., Wang, H., Liu, P., Liu, T. and Wu, Y., 2025. Effects of supplemental seaweed extract on antioxidant properties, loose stools, gut microbiota, and its metabolite composition in adult dogs. Journal of Animal Science 103. https://doi.org/10.1093/jas/skaf149

Ziegler, P., Adelmann, K., Zimmer, S., Schmidt, C. and Appenroth, K.J., 2015. Relative in vitro growth rates of duckweeds (*Lemnaceae*)–the most rapidly growing higher plants. Plant Biology 17: 33-41. https://doi.org/10.1111/plb.12184

Zokm, G.M.E., Ismail, M.M. and Okbah, M.A.E., 2022. Seaweed as bioindicators of organic micropollutants polycyclic aromatic hydrocarbons (PAHs) and organochlorine pesticides (OCPs). Environmental Science and Pollution Research 29: 34738-34748. https://doi.org/10.1007/s11356-022-18634-z

CHAPTER 16

Pulse proteins for pet dogs and cats

A. Rankovic[1]*, A.K. Shoveller*[1] *and G. Bosch*[2*]

[1]*Department of Animal Biosciences, University of Guelph, 50 Stone Rd E, Guelph, ON N1G 2W1, Canada*
[2]*Animal Nutrition Group, Wageningen University & Research, De Elst 1, 6708 WD, Wageningen, The Netherlands*
**Correspondence: guido.bosch@wur.nl*

Abstract

We review the suitability of pulse ingredients (peas, faba beans, common beans, chickpea, lentils, cowpeas, and lupins) for use in dog and cat food and compare these to poultry by-product meal (PBM), focusing on palatability, nutritional quality, non-nutrient compounds, and environmental impact. Across acceptance and preference studies with varying experimental design, pulses incorporated into extruded or homemade foods were generally well consumed by dogs and cats. In canine palatability studies, some pulse-rich foods matched or outperformed grain-based controls, although macronutrient content should be considered. Pulses provide higher protein and lower carbohydrate than grains, with lower ash than PBM, but higher phytate. Pulse proteins are typically rich in Lys and Arg, but lower in Met, Gly and Pro relative to animal proteins. Total tract protein and starch digestibility of pulse-containing extruded diets is high, yet ileal amino acid digestibility varies by pulse species and processing methods employed, indicating the need for the use of complementary proteins or supplemental amino acids, especially the sulfur amino acids. Starch in pulses have higher amylose, resulting in slower digestion rates. Pulse fibers are mostly insoluble and modestly fermentable, though the oligosaccharides are rapidly fermentable. Pulses contain bioactive compounds whose activities are variably reduced by dehulling, cooking, or extrusion and residual activity and batch variability monitoring is warranted. Life-cycle assessment data indicate lower global warming potential per kg for some pulses in contrast to PBM, though rankings change when normalized to crude protein. Overall, pulses are viable ingredients that contribute protein and carbohydrate in pet foods when diets are balanced for amino acids, processed appropriately, and evaluated for digestibility. However, there is still a paucity of data in dogs and cats with regards to ingredient suitability, nutrient requirements and formulation level environmental assessments that warrant further research.

Keywords

palatability – nutritional quality – bioactive compounds – environmental impact

1 Introduction

Pet dog and cat nutrition has evolved rapidly over recent decades, driven by increased pet ownership, the humanization of pets, and growing consumer interest in health and sustainability. Traditional pet food formulations have relied heavily on rendered animal meals and cereal grains as primary protein and carbohydrate sources. However, concerns about environmental sustainability, supply chain stability, and ingredient diversification have intensified industry interest in alternative plant-based ingredients including pulses. Pulses are edible seeds of plants in the legume family. Selective breeding of species yielded hundreds of cultivars or varieties within pulse species (Table 16.1). Pulses represent a category of alternative ingredients that can serve as sources of both protein and carbohydrate in canine and feline diets. In human diets, these crops offer several potential advantages, including relatively high protein content, established safety profiles from human food applications, and a lower environmental footprint compared to animal-derived proteins such as meat. Despite growing commercial interest in pulse ingredients, their impact on palatability and nutritional adequacy in pet foods must be carefully evaluated. The objective of this chapter is to provide background information on pet foods as well as the current state of knowledge regarding a selection of pulses as ingredients in dog and cat foods. Specifically, this review examines the palatability, nutritional quality, including digestibility and amino acid profiles, non-nutrient bioactive compounds, and environmental impact of pulses in pet food formula-

TABLE 16.1 Scientific names of selected pulses and vernacular names of cultivars*

Scientific name	**Vernacular names of cultivars**
Pisum sativum L.	Pea, field pea, garden pea
Vicia faba L.	Broad bean, faba bean, horse bean
Phaseolus vulgaris L.	Common bean, kidney bean, navy bean, pinto bean
Cicer arietinum L.	Chickpea, garbanzo, Egyptian pea
Lens culinaris L.	Lentil, bing dou
Vigna unguiculata (L.) Walp	Cow pea, black eye pea
Several *Lupinus* L. species	Lupine

*Names are sourced from the taxonomic Tropicos (2025) database.

tions, with poultry by-product meal (PBM) serving as a reference protein source commonly used in both dog and cat dry extruded foods. Though pulse cultivars may differ, data in tables are presented on the species level, allowing comparisons among pulse species. To improve readability, the first vernacular names of common cultivars mentioned in Table 16.1 are mentioned in the tables instead of the scientific name of the species.

2 Pet foods

The global pet food industry has experienced continual growth over recent decades, driven by increasing pet ownership and the humanization of companion animals. In the United States alone, it is estimated that 68 million households own at least one dog and 49 million own at least one cat (APPA, 2025). Additionally, dogs and cats are increasingly considered family members, leading consumers to expect high-quality, nutritious, and health-focused products for their pets (Simonensen et al., 2014; Schleicher et al., 2019; Vinassa et al., 2020). Increased humanization has encouraged the expansion of premium and functional pet foods that often mirror human nutrition trends. Although rendered meats and cereal grains remain the most common traditionally used ingredients in the pet food industry, consumers are becoming more concerned with sustainability, ingredient transparency, and the ethical sourcing of ingredients, driving further innovation and diversification of ingredient use (Dodd et al., 2019; Vinassa et al., 2020; Fantechi et al., 2024).

Pet owners today can choose from a wide range of products formulated to support specific life stages, breeds, sizes, and health conditions. In 2024, global pet food sales were approximately $151 billion USD (Statista, 2025). Of this, dry extruded (kibble) and wet retorted (canned) foods remain the most common formats globally. The popularity of dry extruded food is driven by its convenience, extended shelf life, storage ease, and cost-effectiveness (Zicker, 2008), while wet foods are often preferred for palatability, particularly in cats and older animals. Formats such as freeze-dried, fresh-cooked, and raw represent a smaller segment of the global pet food industry, but interest in feeding these alternative formats is growing among consumers (Dodd et al., 2020). Globally, the dog food market exceeds cat food in size, reflecting greater dog ownership and larger portion requirements. Although both dog and cat food segments continue to grow steadily, the cat food category has been growing at a faster pace (Euromonitor International, 2024), likely driven by increasing cat ownership worldwide and a preference for more premium formats.

Consumer purchasing decisions are influenced by multiple factors, including price, convenience, packaging design, brand reputation, and marketing claims (Simonensen et al., 2014; Schleicher et al., 2019; Vinassa et al., 2020). However, pet owners consistently rank perceived health and nutritional quality as the most important factor when choosing a product (Schleicher et al., 2019; Banton et al., 2021; Rombach and Dean, 2021; Prata, 2022), informed primarily by ingredient lists and marketing claims (Vinassa et al., 2020; Banton et al., 2021; Rombach and Dean, 2021). Ongoing consumer satisfaction is also strongly tied to pet outcomes such as stool quality, coat condition, and perceived enjoyment of the diet (Vinassa et al., 2020). In particular, stool consistency and appearance were important factors for owners when purchasing their pet foods. This may reflect owners' preference for ease of waste management, as well as the tendency to view stool quality as a practical indicator of digestive health (Bosch and Swanson, 2021).

Palatability is a critical factor for pet food manufacturers to consider, not only for ensuring animal acceptance of the food but also because feeding interactions reinforce the human–animal bond (Bradshaw and Cook, 1996; Kienzle et al., 1998). Owners derive satisfaction from observing their pets enjoy their food, making palatability a central consideration in product development. From a technical standpoint, successful pet food manufacturing requires the production of a safe, palatable, and nutritionally adequate product in an efficient and cost-effective manner. Thermal processing techniques such as extrusion for dry foods and retorting for wet foods are therefore widely used to ensure microbial safety, extend shelf life, and achieve desirable texture, shape, digestibility, and palatability (Hendriks et al., 1999; Tran et al., 2008).

Nutritional adequacy depends on the inclusion of digestible and bioavailable nutrients that meet species- and life stage-specific requirements. Historically, standardized ileal digestibility assays were used to evaluate nutrient absorption in dogs, similar to methods applied in production animal nutrition. However, due to animal welfare concerns, these procedures are no longer applied in companion animal research. Instead, apparent total tract (fecal) digestibility studies are typically used. While this approach is practical and non-invasive, it is recognized to yield less precise estimates of nutrient absorption (Hendriks et al., 2012). In vitro assays and alternative animal models, such as the precision-fed cecectomized rooster assay, are also used to provide complementary insights into digestibility and nutrient availability (Johnson et al., 1998; Smeets-Peeters et al., 1999).

To ensure that diets meet the nutritional needs of dogs and cats, formulation is guided by established guidelines where the National Research Council (NRC) of the United States provides the scientific foundation by publishing recommended nutrient allowances for different life stages, based on evidence gathered from academic and industry research (NRC, 2006). Building on these recommendations, regulatory and trade bodies translate these recommendations and build on them to produce practical guidance for pet food formulators and manufacturers. In Europe, the European Pet Food Industry Federation (FEDIAF) publishes nutrient recommendations for dogs and cats that reflect NRC data, along with more recent research findings (FEDIAF, 2025). In the United States, the Association of American Feed Control Officials (AAFCO) similarly publishes nutrient guidelines and protocols that draw on NRC recommendations and other available scientific data (AAFCO, 2024). Pet foods marketed as "complete and balanced" may only make this claim if they meet or exceed the nutrient requirements set by AAFCO in the USA or FEDIAF in Europe.

Sustainability and innovation are emerging key drivers of growth in the pet food industry. Rising environmental concerns have accelerated interest in alternative protein sources such as pulses, insects, algae, and upcycled by-products, alongside the adoption of more sustainable packaging solutions (Conway and Saker, 2018; Acuff et al., 2021; Bosch and Swanson, 2021; Klinmalai et al., 2025). At the same time, technology and e-commerce have reshaped distribution and marketing channels, allowing for greater product personalization and closer engagement between brands and consumers. Despite these opportunities, manufacturers must balance consumer expectations for transparency and natural ingredients with the need for safe, digestible, palatable and nutritionally adequate products. Given the need for diverse and sustainable ingredients within the pet food supply chain, understanding the functional properties and limitations of pulse ingredients is critical for successful pet food formulation. The current knowledge on the effects of pulses on palatability, as well as their nutritional quality and environmental impact properties, are therefore discussed below.

3 Palatability

Palatability is a critical consideration in the formulation of commercial pet food, as diets must meet an animal's nutrient requirements and be readily consumed by the animal. Palatability can be assessed through acceptance tests, in which a single food is offered alone, or preference tests, where two foods are compared to each other (Aldrich and Koppel, 2015). Currently, available data suggest that pulse ingredients are well accepted by both dogs and cats when included as a component of an extruded or homemade diet. To our knowledge, only one study has offered a pulse ingredient on its own to dogs. In this study, Rankovic et al. (2020) fed dogs 25 g of green lentils (weighed dry prior to cooking), which all dogs voluntarily consumed.

In studies where pulse ingredients were incorporated into extruded foods, reported food acceptance has generally been high. Pulses, including faba beans, peas, lentils, chickpeas, and garbanzo beans, have been successfully included at levels up to 64% in canine extruded or homemade diets and up to 67% in feline extruded diets, with animals readily consuming the foods and reported food intake not differing from foods containing lower levels of pulses or other protein or starch sources at similar inclusion levels (de Oliveira et al., 2008; Adolphe et al., 2015; Alvarenga et al., 2020; Reilly et al., 2021a; Reis et al., 2021; Bakke et al., 2022; Pattanaik and Kore, 2022;

Singh et al., 2023; Kaelle et al., 2024; Penazzi et al., 2024; Curso-Almeida et al., 2025). Minor differences in intake have been reported in some feline studies, including a reduced daily energy intake of approximately 27 kcal (7%) and 22 kcal (6%) for cats fed extruded diets containing 7.5% and 30% raw garbanzo beans, respectively, compared with a 0% garbanzo control (30% rice) (Reilly et al., 2021a). Additionally, Curso-Almeida et al. (2025) reported refusals and a significant reduction in food intake (numerical values not reported) when cats were fed an extruded food with 30% unfermented pea starch, as compared to diets containing 30% fermented pea starch or 30% corn starch.

A small number of studies have compared pulse-containing foods in dogs using preference test methods. Alvarenga et al. (2020) evaluated dehulled faba bean inclusion at 0% (control), 10%, 20%, and 30% in extruded canine foods and reported that while total food intake did not differ across treatments, dogs preferred the control food (45% rice and 10% corn gluten meal) over the 10% and 30% faba bean foods, with the 20% faba food showing similar preference to the control. In contrast, Penazzi et al. (2024) evaluated the palatability of red lentil pasta by-product using five extruded foods using preference testing in dogs, where first choice referred to the first product consumed and "preferred" to the food consumed in greater quantity. The basal control food (CO) contained 58% rice, while three test foods substituted rice with red lentil pasta by-product at 33, 66, or 100% of the rice inclusion level (LP33, LP66 and LP100), although increasing inclusion of red lentil pasta was associated with increasing dietary protein and may have confounded the results. The authors reported that the dogs showed a preference for the LP100 food (containing 58% red lentil pasta) over the other foods, both in terms of first choice and intake. Similarly, Kaelle et al. (2024) reported that dogs preferred extruded foods containing chickpea (~64% of the formulation) or pea (~57%) compared with a corn-based control food (~43% corn). Preference was assessed using six paired two-bowl tests, with each test food offered alongside the corn food once daily for two consecutive days. Food was provided at 30% above maintenance requirements to ensure leftovers, and bowl positions were alternated on the second day. Palatability was determined based on intake ratio and first choice. Authors reported that the dogs showed a significant preference for the chickpea and pea foods over corn in terms of intake ratio, whereas a difference in first choice was observed only for the chickpea food. Overall, these findings suggest that pulse ingredients can perform as well as, or in some cases better than, traditional starch and protein sources in canine preference trials. To the authors' knowledge, no published studies have conducted preference testing in cats with pulse-based foods. While acceptance studies indicate that cats generally consume pulse-containing foods readily, additional work is needed to determine whether pulses influence preference.

4 Nutritional quality

4.1 Chemical composition

As with all ingredients, there is variation in chemical composition among the selected pulses, but there can also be quite some variability within the pulse due to variety in cultivars and growing conditions (Hall et al., 2017). We retrieved data on food and feed composition from large databases (Government of Canada, 2024; INRAE-CIRAD-AFZ, 2025) and present averages as well as ranges for a selection of constituents (Table 16.2). These databases also include more detailed information on energy contents and other constituents like vitamins, minerals, trace elements and fatty acids, which are not reported here but are important when formulating complete foods for dogs and cats. As pulses are rich in proteins and carbohydrates, i.e. starches and fibers, specific nutritional properties of these components are described in more detail.

Average crude protein levels (all in as is basis) vary from 18.6% in chickpeas to 33.5% in lupines, whereas ranges can vary from ~3% up to ~12% from the average values (Table 16.2). Generally, pulses contain per gram less indispensable amino acids relative to animal-derived proteins (Young and Pellett, 1994), and there are differences in the amino acid composition of the proteins (see below). Next to protein, starch is a major constituent of pulses with average levels ranging from 36.2% for common beans to 44.7% for peas. The exception is lupines, which have an average starch content of 6.2% and are somewhat richer in sugars (6.0%) and more fibrous (18.9% neutral detergent fiber) than the other selected pulses. The ash contents in pulses are on average between 3 and 4%, which is considerably lower than the average 10% for PBM. The levels of phosphorus (P) are higher than the calcium (Ca) levels, whereas this is the opposite in PBM. Based on the averages in Table 16.2, Ca to P ratios are 1.8 for PBM, 0.7 for lupines and 0.3 to 0.4 for the other pulses. Part of P is stored in the form of phytic acid, rendering it less bioavailable and capable of binding other nutrients, which ranges between 33% of total P in common beans and 73% in cowpeas.

4.1.1 Amino acids

To allow comparison of the amino acid composition of the proteins in pulses and PBM, contents of amino acids are expressed on a g/16 g N basis. Methionine is commonly one of the first limiting indispensable amino acids for dogs and cats fed animal- and pulse-based diets, whereas grains are first limiting in lysine for humans and likely dogs (Marinangeli et al., 2024). The proteins of cow peas contain on average the most methionine (1.5 g/16 g N), whereas those in broad beans, lentils and lupines (0.7 to 0.8 g/16 g N) contain less than half of that in PBM (1.7 g/16 g N). Cysteine levels in pulse proteins, however, are roughly similar to that in PBM protein, with lupine protein being richer. Although most pulses are lower in tryptophan compared to cereals (Young and Pellett, 1994), tryptophan contents are similar or higher than in the proteins of PBM. Pulse proteins are also lower in glycine and proline than the proteins in PBM, which is likely related to their use in collagen for various types of tissues in animals (Li and Wu, 2018). Except for lupines, pulse proteins are relatively rich in lysine. Moreover, proteins in peas, broad beans, chickpeas and lupines are richer in arginine than animal-based proteins. Relative to PBM proteins, proteins of the common bean, chickpea and cow peas are rich in phenylalanine whereas common bean and lupine proteins are rich in tyrosine.

4.1.2 Starch

Structurally, starch is composed of two main glucose polymers: amylose and amylopectin. Amylose is a mostly linear polymer of α-1,4-linked glucose units, containing less than <0.1% α-1,6 linkages (Ball et al., 1996), while amylopectin is a highly branched polymer containing both α-1,4 and α-1,6 linkages (Ball et al., 1996). The relative proportion of amylose to amylopectin has strong implications for physicochemical and nutritional properties, including gelatinization, enzymatic susceptibility, and retrogradation (Srichuwong and Jane, 2007; Schirmer et al., 2013). Pulse starches are distinctive in their elevated amylose content (30–45% or higher), compared to 15–30% in cereals and tubers (Hoover, 2001; Wani et al., 2016; Bajaj et al., 2018). For example, pea starch averages 35–40% amylose, lentil and common bean starches may exceed 40%, while chickpea and faba bean starches are typically lower (30–35%) (Hoover and Ratnayake, 2002; Singh et al., 2004; Hoover et al., 2010). Substantial intra-species variation has also been reported, with amylose levels in field peas reported between 21 and 58% depending on the strain and environment (Singh et al., 2010). Pulse starches are less susceptible to enzymatic digestion in their raw form because their higher amylose content and compact granule structure limit enzymatic hydrolysis. This reduced susceptibility reflects a slower rate of digestion, not a reduction in overall digestibility once the starch is gelatinized during processing.

4.1.3 Fiber

Pulses can be rich sources of dietary fiber, containing on average 11 to 28% total dietary fiber on a dry matter basis (DMB) (Table 16.2), with cultivar and processing method being factors that can impact these levels (Tosh and Yada, 2010). Pulses contain more complex fibers that are insoluble but also oligosaccharides that are soluble, although the majority is insoluble fiber (Chen et al., 2016). The insoluble fiber content across pulse species ranges from 10 to 28% DMB (Tosh and Yada, 2010; Chen et al., 2016), which roughly aligns with ranges in values for neutral detergent fiber contents in Table 16.2. However, between pulse species, fiber content and composition show variability. Lentils and peas often contain a higher percentage of soluble fiber, whereas chickpeas and beans are often richer in insoluble fiber (Tosh and Yada, 2010).

The majority of fiber is found within the hull (Dalgetty and Baik, 2003), which represents approximately 10% the weight of the pulse seed. Specifically, the hull contains large amounts of insoluble cellulose, along with some hemicellulose, pectins, and oligosaccharides (Weightman et al., 1994). As a result, whole seeds contain greater amounts of insoluble fiber, and therefore total dietary fiber, as compared to their dehulled counterparts (Dalgetty and Baik, 2003). Soluble fiber is more abundant in the cotyledon, and its content remains relatively consistent in raw vs. dehulled seeds (Dalgetty and Baik, 2003). Pulses also contain oligosaccharides, specifically a-galactosides, raffinose, stachyose and verbascose, which typically account for 2–10% DMB, depending on the pulse species (Guillon and Champ, 2002).

TABLE 16.2 Average (and ranges) in chemical composition (% as is) of poultry by-product meal and selected pulses*

Constituent[a]	Poultry meal	Pea	Broad bean	Common bean	Chickpea	Lentil	Cow pea	Lupine
DM	93.6 (86.0-99.0)	87.2 (79.2-94.2)	86.6 (79.5-92.5)	88.9 (82.9-95.3)	87.3 (83.3-93.0)	88.4 (85.8-92.1)	89.7 (80.7-97.3)	88.1 (77.1-94.6)
CP	63.2 (60.0-70.0)	20.3 (14.5-25.4)	25.5 (19.7-30.1)	22.1 (17.0-25.7)	18.6 (15.8-23.9)	23.4 (20.3-29.0)	22.4 (19.1-25.1)	33.5 (24.8-46.1)
EE	21.7 (9.4-31.7)	1.2 (0.5-2.3)	1.3 (0.7-2.0)	1.4 (0.8-2.7)	5.5 (4.2-7.1)	1.4 (0.5-4.4)	1.4 (0.6-2.9)	8.5 (3.7-11.9)
Starch	0	44.7 (35.5-50.1)	38.3 (31.7-42.9)	36.2 (20.1-42.0)	43.8 (39.8-48.1)	40.1 (35.3-46.7)	42.2 (36.4-49.6)	6.2 (0.0-14.6)
Sugars	0	4.2 (2.8-5.5)	3.0 (2.2-5.0)	4.3 (2.6-5.1)	5.2 (2.3-7.6)	4.3 (-)	4.2 (0.3-6.3)	6.0 (3.6-8.3)
Crude fiber	0.9 (0.3-2.0)	5.6 (3.6-8.1)	7.8 (5.0-10.6)	4.6 (2.6-7.1)	3.4 (2.0-5.0)	4.3 (2.6-6.8)	4.2 (1.2-5.6)	8.5 (3.7-11.9)
NDF	4.8	12.7 (8.2-20.0)	13.4 (10.2-20.8)	11.8 (-)	9.9 (6.9-15.6)	12.6 (10.9-25.3)	12.9 (8.7-19.3)	18.9 (14.5-27.4)
ADF	3.1	6.4 (4.0-8.8)	9.7 (7.0-12.2)	6.6 (4.6-8.8)	3.9 (2.9-6.2)	4.8 (3.3-5.6)	5.6 (3.6-6.9)	14.1 (8.8-21.1)
Lignin	0.5	0.5 (0.0-1.8)	0.8 (0.1-2.3)	0.2 (0.1-0.3)	0.2 (0.0-0.7)	1.4 (1.0-1.8)	0.4 (0.3-0.6)	0.9 (0.3-4.4)
TDF	-	-	14.9	15.5-24.9	10.0	10.7	-	18.9
Ash	10.0 (2.6-26.3)	3.1 (2.3-4.2)	3.3 (2.5-4.2)	4.1 (3.4-8.4)	2.9 (2.2-3.8)	3.1 (2.4-9.3)	3.9 (2.9-6.9)	3.5 (2.2-6.2)
Ca	2.2 (0.5-10.7)	1.0 (0.1-3.0)	1.3 (0.5-2.6)	1.8 (1.1-4.2)	1.2 (0.3-1.8)	1.3 (0.6-4.9)	1.1 (0.3-2.5)	2.7 (0.9-6.0)
P	1.2 (0.1-4.6)	3.8 (2.5-5.2)	4.8 (3.6-6.0)	4.5 (1.9-5.6)	4.1 (2.6-5.3)	3.9 (2.9-5.8)	3.3 (0.2-4.9)	3.8 (2.7-5.1)
Phytate-P, %	-	45	60	33	49	64	73	61

*Data for for poultry meal and pulses are sourced from the INRAE-CIRAD-AFZ (2025) feed and Government of Canada (2024) food composition databases.

[a]DM, dry matter; CP, crude protein; EE, ether extract; NDF, neutral detergent fiber; ADF, acid detergent fiber; TDF, total dietary fiber. Phytate-P as % of total P.

TABLE 16.3 Average and ranges in amino acid (g/16 g N) composition of poultry by-product meal and selected pulses*

Amino acid	Poultry meal	Pea	Broad bean	Common bean	Chickpea	Lentil	Cow pea	Lupine
Lysine	4.8 (3.0-6.4)	7.3 (6.4-8.0)	6.5 (5.7-6.8)	6.6 (4.0-8.4)	6.8 (5.8-7.5)	6.1 (4.3-7.0)	6.5 (5.2-7.1)	4.9 (3.9-5.7)
Threonine	4.0 (2.6-4.3)	3.8 (3.3-4.4)	3.6 (3.0-4.0)	4.3 (3.2-5.9)	3.6 (2.8-4.2)	3.5 (2.5-4.2)	3.8 (3.0-5.3)	3.7 (3.0-4.5)
Methionine	1.7 (1.1-2.1)	1.0 (0.7-1.5)	0.7 (0.5-0.9)	1.3 (0.3-2.5)	1.2 (0.7-1.7)	0.8 (0.4-1.1)	1.5 (1.2-1.6)	0.8 (0.4-1.2)
Cysteine	1.3 (0.6-3.1)	1.4 (1.1-1.9)	1.3 (1.0-1.6)	1.0 (0.8-1.2)	1.2 (0.7-1.9)	1.2 (0.9-1.5)	1.1 (0.8-1.4)	1.7 (1.0-2.3)
Tryptophan	0.7 (0.5-1.0)	0.9 (0.7-1.2)	0.8 (0.7-1.0)	1.1 (0.9-2.1)	0.9 (0.6-1.0)	0.7 (0.5-0.8)	1.1 (0.9-1.3)	0.7 (0.5-1.0)
Isoleucine	4.0 (2.4-4.2)	4.2 (3.4-4.9)	4.1 (3.4-4.5)	4.4 (2.8-5.9)	3.9 (2.6-4.7)	3.9 (3.1-4.7)	4.0 (2.8-5.2)	4.6 (3.1-5.0)
Valine	5.2 (3.0-6.1)	4.7 (3.8-5.5)	4.5 (3.8-5.5)	5.1 (3.4-6.7)	4.0 (2.9-5.0)	4.2 (3.8-4.8)	4.7 (3.4-5.5)	4.3 (3.2-5.0)
Leucine	7.1 (4.5-7.3)	7.1 (6.3-8.0)	7.6 (6.4-7.9)	8.0 (5.5-10.0)	7.3 (6.4-8.1)	6.8 (5.6-7.4)	7.4 (5.8-11.3)	7.3 (5.7-8.2)
Phenylalanine	4.0 (2.4-4.2)	4.7 (4.1-5.4)	4.2 (3.6-4.4)	5.4 (3.6-7.1)	5.5 (4.8-6.1)	4.4 (3.6-5.5)	5.5 (4.4-6.4)	3.9 (3.1-4.7)
Tyrosine	2.7 (2.1-3.3)	3.1 (2.3-4.2)	3.2 (1.6-3.4)	3.7 (2.7-4.4)	2.3 (1.4-4.6)	2.7 (2.3-3.2)	3.0 (2.6-3.6)	4.7 (2.9-5.0)
Histidine	1.9 (1.4-2.7)	2.5 (2.1-3.0)	2.5 (2.3-2.9)	2.6 (2.1-3.4)	2.5 (2.0-3.1)	2.3 (1.0-3.3)	3.3 (2.4-5.1)	2.2 (1.9-3.5)
Arginine	6.7 (4.4-8.3)	8.5 (6.7-10.9)	9.5 (7.6-10.5)	6.5 (4.9-9.2)	8.3 (6.6-11.3)	7.1 (3.9-8.8)	6.7 (5.0-8.7)	10.8 (7.7-12.5)
Alanine	5.6 (4.2-6.7)	4.4 (3.6-5.4)	4.1 (3.3-4.6)	4.2 (3.0-5.3)	3.8 (3.0-5.1)	3.8 (2.4-4.3)	4.2 (3.4-5.1)	3.4 (2.6-4.2)
Aspartic acid	7.4 (4.3-7.9)	11.7 (9.1-13.3)	12.0 (-)	11.7 (5.0-14.7)	11.2 (10.2-13.2)	10.9 (9.2-12.4)	10.4 (9.2-12.7)	10.6 (8.4-11.5)
Glutamic acid	11.9 (8.5-12.9)	16.3 (14.7-20.0)	17.6 (14.7-18.8)	15.7 (9.7-24.4)	17.8 (14.6-21.0)	14.9 (13.8-16.3)	15.8 (14.1-18.7)	20.7 (17.1-24.2)
Glycine	8.9 (6.0-11.5)	4.4 (3.7-5.0)	4.3 (3.7-4.6)	4.0 (3.1-5.6)	3.5 (3.2-4.3)	3.8 (3.3-4.4)	3.9 (3.1-4.8)	4.0 (3.2-4.6)
Serine	4.4 (2.6-7.2)	4.7 (3.8-5.5)	5.1 (4.2-5.1)	6.2 (4.4-7.2)	4.9 (4.6-5.9)	4.2 (2.9-6.0)	4.9 (3.8-5.6)	5.3 (4.0-6.1)
Proline	6.7 (5.7-7.9)	4.2 (3.5-5.8)	4.1 (3.7-4.5)	3.8 (2.1-5.1)	4.2 (3.4-5.8)	3.9 (2.6-4.4)	4.6 (3.8-5.7)	4.2 (3.1-4.5)

*Data for poultry meal and pulses are sourced from the INRAE-CIRAD-AFZ (2025) feed composition database.

4.1.4 Non-nutrient bioactive compounds

Pulse are equipped with an arsenal of bioactive compounds functional as protection against animals, microorganisms and viruses, making the seeds unpalatable (e.g. saponins, tannins), anti-nutritive (e.g. digestive enzyme inhibitors, phytic acid) or toxic (e.g. some lectins, alkaloids) (Enneking and Wink, 2000). These compounds have been named antinutritional factors or antinutrients, given their potential negative effects on nutrient availability and health in animals, but these compounds can also have positive effects on health (see e.g. Champ, 2002) (Table 16.4). There is a large body of studies on the potential health effects of isolated compounds using in vitro models (e.g. cells) or in vivo models (e.g. mice) (see examples in Campos-Vega et al., 2010). As pulses typically contain multiple compounds, it is of interest to also look into potential antagonistic or synergistic effects on health. Given that dogs and cats are carnivorous species and typically consume only limited amounts of plant material containing these compounds in nature (Plantinga et al., 2011; Bosch et al., 2015), their potential species-specific metabolic idiosyncrasies may make results from other animal models not be directly applicable. Therefore, longer-term studies are needed to assess the health effects of pulse-based diets rich in non-nutrient bioactive compounds in companion animals. To what extent such effects are to be expected depends obviously on the formulation with choices in pulse product type(s) as well as product inclusion rates.

Types and amounts of bioactive compounds in selected pulses are shown in Table 16.5 with the exception of lupins, which can have relatively high alkaloid levels and no digestive enzyme inhibitors, lectins or polyphenolic compounds (De Lange et al., 2000). There are also alkaloid-free varieties of lupins, which allow their use

TABLE 16.4 Potential beneficial and detrimental effects of non-nutrient bioactive compounds in pulses (Champ, 2002)

Compound	Beneficial	Detrimental
Protease inhibitors	Anticarcinogenic (?)	↑ Carcinogenesis (?), growth inhibition
α-Amylase inhibitors	Therapeutic in diabetes (?)	↓ Starch digestibility
Phytic acid	Hypocholesterolaemic effect (?), anticarcinogenic (?)	↓ Mineral bioavailability
Oxalates		↓ Mineral bioavailability
Flavonoids, isoflavones	↓ Risk of hormone-dependent cancer (?)	Infertility syndrome
Condensed tannins		Astringent taste
Lectins	Therapeutic in obesity (??),↓ tumor growth (??)	Growth inhibition, ↓ nutrient absorption
Saponins		Bitter taste

as ingredients in foods and feeds. The types and amounts of the compounds can vary considerably as indicated in the ranges. Such variability in compounds relates to the use of different varieties (genetics), cultivation conditions and post-harvest processing conditions but also to differences in analytical methods used for quantification. The ranges shown in Table 16.5 therefore provide only a rough idea of the potential levels of the compounds in the pulses and it is advised to check the compound levels with the supplier when acquiring pulse-based ingredients for pet food applications. Inhibitors of digestive enzymes (trypsins, chymotrypsins, α-amylases) can be relatively high in kidney beans. Phytic acid levels can be quite similar among the selected pulses, with faba beans having levels up to 3.2 g/100 g DMB (Adamidou et al., 2011). Tannin levels can be considerably higher in kidney beans than in the other pulses, with levels up to 6.7 g/100 g DMB (Khattab and Arntfield, 2009), although there are also studies that reported low levels (e.g. 0.4 g/100 g DMB in Alonso et al., 2000). Lectin levels also vary considerably among pulses with values up to 500 HU/mg DMB in lentils reported by Pedrosa et al. (2012).

TABLE 16.5 Ranges in non-nutrient bioactive compounds in selected pulses[a]

Compound[b]	Pea	Broad bean	Common bean	Chickpea	Lentil	Cow pea
Trypsin inhibitor	1.1-7.8	4.5-6.7	3.1-33.7	1.0-19.0	4.0-8.4	1.6-2.4
Chymotrypsin inhibitor	2.7-4.9	1.1-3.6	3.9-24.5	11.8-13.6	3.5-4.9	
α-Amylase inhibitor	0-17	19	248-1370			
Phytic acid	0.3-1.3	0.5-3.2	0.2-1.9	0.4-2.1	0.2-2.3	0.3-1.5
Oxalates	0.7		0.1-9.6	0.1-7.0	0.1-1.6	0.0-0.8
Polyphenolic compounds	0.0-0.1	0.4-2.8	0.2			0.1-0.3
Tannins	0.0-3.1	0.0-2.1	0.0-6.7	0.0-0.5	0.0-1.0	2.3-2.6
Lectins	5.1-6.2	49.3	74.5	6.2	500	
Saponins	0.1-1.1	0.0-3.6	0.2	0.1-6.0	0.4-0.5	

a Data are sourced from studies and literature reviews on pea (Zdunczyk et al., 1997; Alonso et al., 1998; Hernández-Infante et al., 1998; Wang and Daun, 2004; Khattab and Arntfield, 2009; Campos-Vega et al., 2010; Adamidou et al., 2011; Shi et al., 2017), broad bean (Makkar et al., 1997; Alonso et al., 2000; Campos-Vega et al., 2010; Adamidou et al., 2011; Luo and Xie, 2013; Shi et al., 2017), common bean (Hernández-Infante et al., 1998; Alonso et al., 2000; Quinteros et al., 2003; Judprasong et al., 2006; Khattab and Arntfield, 2009; Campos-Vega et al., 2010; Wang et al., 2010; Shi et al., 2017), chickpea (Hernández-Infante et al., 1998; El-Adawy, 2002; Campos-Vega et al., 2010; Adamidou et al., 2011; Pedrosa et al., 2012; Shi et al., 2017), lentil (Hernández-Infante et al., 1998; Quinteros et al., 2003; Wang and Daun, 2006; Wang et al., 2009; Campos-Vega et al., 2010; Shi et al., 2017), cow pea (Judprasong et al., 2006; Khattab and Arntfield, 2009; Gonçalves et al., 2016).

b Trypsin inhibitor in IU/mg, chymotrypsin inhibitor in IU/mg, α-amylase inhibitor in IU/g, phytic acid in g/100 g, tannins in g/100 g, polyphenolic compounds in g/100 g, oxalates in g/100 g, lectins in HU/mg, and saponins in g/100 g. All units per unit of dry matter.

The negative consequences of the consumption of the non-nutrient bioactive compounds has led to a considerable research interest on the elimination or reduction of these consequences (see e.g. Melcion and Van der Poel, 1993; Enneking and Wink, 2000; Patterson, 2017). Processing of the pulses (e.g. dehulling, soaking, germination) or processes applied when making complete foods (e.g. cooking, extrusion) can impact the bioactivity of the compounds. The effects of processing of pulses on, for example, protein digestibility can vary from biologically non-significant to large (13%) improvement when measured in vitro (reviewed by Cargo-Froom et al., 2020). Tannins and other phenolic compounds as well as oxalates mainly present in the seed coat can be removed by dehulling (Champ, 2002; Patterson et al., 2017). With other non-nutrient bioactive compounds mainly present in the cotyledon, dehulling increases the levels of these compounds per unit of product. Some of these compounds are heat-sensitive and activity can be reduced via heat processing treatments with variation in effects depending on compound types as well as pulse.

Patterson et al. (2017) reviewed the literature on the impact of extrusion and (pressure-)cooking on the presence of compounds in ground pulses. The phytate reduction by extrusion was limited with reduction levels of 5.9–13% in peas, 16–25.5% in kidney beans, and 26.7% in faba beans. Effects of cooking (boiling) or pressure-cooking can be higher with ranges of 5.3–33% in peas, 0–66% in kidney beans, 24.1–39 in faba beans, 0–41% in chickpeas and 5.7–41.6% in lentils. Tannins of peas and kidney beans seem to be less heat stable during extrusion (75–91.7% reduction) than faba beans and chickpeas (7–54.4%) though with (pressure-)cooking these differences are less clear with reductions of tannins up to 76 and 93% in faba beans. Extrusion can reduce phenolic compounds between 23.7 and 54% and (pressure-)cooking between 58 and 76.1%. The trypsin inhibitor activity can be greatly reduced by extrusion in peas (~95%) and in faba beans (99%) but the chymotrypsin inhibitor activity is less reduced in peas (~65%) and faba beans (53%). In kidney beans, stability of trypsin inhibitor was larger (69.4–93.6% reduction) than chymotrypsin inhibitor (96.2–100%). Extrusion largely to completely eliminate α-amylase inhibitor and hemagglutinating activities in peas, kidney beans and faba beans. Alkaloids in lupins are heat-stable (Schryvers et al., 2023) also under extrusion conditions (Chamone et al., 2023). Studies on the stability and presence of the non-nutrient bioactive compounds in pet foods are limited. Dog foods containing 45% green lentils or 44% garbonzo beans had trypsin inhibitor levels of 0.7 and 4.4 IU/mg DMB before processing, which were reduced to 0.5 and 1.2 IU/mg DMB after extrusion and to 0 IU/mg DMB in the final dog foods (Reilly et al., 2021b). In another study, Reilly et al. (2021a) evaluated graded inclusion levels of raw garbonzo beans (0, 7.5, 15, 30%) or cooked garbonzo beans (30%) in extruded foods for cats and reported trypsin inhibitor levels of 0, 0, 0.4, 2.0 and 0.3 IU/mg DMB before processing and levels of 0.7, 0, 0.3, 0 and 0 in the final cat food products, respectfully.

The above studies on the stability of the various non-nutrient bioactive compounds in pulses illustrate that both extrusion and (pressure-)cooking may not completely remove or inactivate these compounds. It remains therefore relevant to assess the nutritional quality and potential health effects of dry extruded and moist retorted pet foods containing ingredients rich in these compounds.

4.2 Digestibility and fermentability

The nutritional properties of pulses not only depend on the chemical composition (Tables 16.2 and 16.3) but also on how these pulse constituents are handled by the digestive tract of dogs and cats, potentially in interaction with non-nutrient bioactive compounds like polyphenols, phytic acid or oxalate. There is little information on the bioavailability of minerals in pulses in dogs and cats as well as how non-nutrient bioactive compounds in pulses or processed pulse-containing pet foods might impact the mineral bioavailability. It is clear that this aspect warrants more research, especially considering the stability of some of these compounds during processing (see above). The digestibility of proteins and, more specifically, amino acids as well as starches in raw and processed pulses has been studied in dog and cats. Moreover, pulse fiber fermentation by canine and feline large intestinal microbiota has also been investigated. The insights from these studies are summarized below.

4.2.1 Nitrogen and amino acids

Ultimately, the combination of nutrient content and subsequent processing will predict the total tract apparent digestibility of protein (more specifically nitrogen) and the ileal digestibility and metabolic availability of the amino acids. While commercial standards for pet food represent nutrient density on a gross basis, the overall impact of pet foods that combine these ingredients should consider ileal digestibility of the amino acids. With respect to digestibility: adding 25% cooked navy beans yielded an apparent total tract digestibility (ATTD) of 78%, which did not differ from a control diet (Forster et al., 2012). Diets using garbanzo beans (43%) or green lentils (44%) as the main protein sources produced >80% ATTD for crude protein (Reilly et al., 2021b). Stepwise inclusion of faba beans at 0, 10, 20, and 30% maintained ATTD at ~90% across levels (Alvarenga et al., 2020). Still, ATTD is a crude indicator of protein/AA digestion as compared to using ileal digestibility and more specifically, standardized ileal digestibility which accounts for endogenous losses, but not ingredient specific losses. Because of ethical and logistical considerations in companion-animal research, dogs are not used to quantify ileal digestibility of AA and instead, the data from animal models are widely used to estimate ileal AA digestibility and support protein-quality calculations. Templeman and Shoveller (2022) used data from the pig model and the precision-fed cecectomized rooster assay. Validity of these models is unknown with only one validation study for the precision-fed cecectomized rooster assay (Johnson et al., 1998). The pig and rooster models may yield different standardized digestibility values, not only because of the differences in species-specific digestive physiology but also because the ileal-cannulated pigs are growing and have higher feed intakes to support growth than the adult cecectomized roosters fed to maintain body weight. We present data from studies using both models not with the intention to compare these models but to allow for understanding differences in pulse ileal AA digestibility within these models. Regardless, for comparison among ingredients, the pig model offers a plethora of data in contrast to the cecectomized rooster.

Table 16.6 compiles standardized ileal indispensable AA (IAA) digestibility values for pulse ingredients using the ileal cannulated pig model and cecectomized rooster assay. Broadly, results show >80% ileal digestibility for most IAAs, with Met emerging as the limiting amino acid when pulses are used in dog and cat foods (Reilly et al., 2020b; Reilly et al., 2020c; Oba et al., 2023; Banton et al., 2025); however, chickpeas consistently had <70% ileal digestibility for the IAA in (Banton et al., 2025). While understanding the digestibility of ingredients can provide data to support the replacement of one ingredient with another of similar IAA content and ileal digestibility, when ingredients are mixed and extruded to produce a food, the IAA ileal digestibility values are not additive (Banton et al., 2025) and currently the ileal digestibility of the IAA in the final product cannot be predicted with a high level of accuracy. Consequently, while pulses can be valuable sources of protein and IAAs, formulations that rely on them as primary proteins should be balanced with crystalline AA(s) and/or complementary ingredients to meet requirements, especially for Met and, at times, Trp (see also below). Furthermore, IAA digestibility of final product should be measured until sufficient data exists to understand nutrient interactions of different ingredients during cooking in order to model final product IAA bioavailability.

Protein quality can be further characterized by the profile of digestible IAAs relative to species minimal requirements (Batterham, 1992). Templeman and Shoveller (2022) estimated DIAAS-like values for numerous ingredients, including pulses, for use in dog diets, anchoring calculations to NRC (2006) minimal requirements whereas recommended allowances and regulatory amino-acid guidance from AAFCO (2024) and FEDIAF (2018) could be used to assess AA scores without correction for digestibility, although the validity of this approach remains unvalidated. Using standardized ileal AA digestibility, a DIAAS≥1.00 indicates that the digestible content of a given IAA meets or exceeds the target requirement (Templeman and Shoveller, 2022). Aside from one case for whole field pea evaluated against FEDIAF's late-growth recommendations, where Trp was limiting, Met was the limiting amino acid for a range of pulse ingredients (Templeman and Shoveller, 2022).

We calculated the DIAAS-like scores and determined the first limiting IAA using the standardized ileal IAA digestibility coefficients in the INRAE-CIRAD-AFZ (2025) feed composition database and the adult maintenance minimal IAA requirements for dogs (NRC, 2006) (in line with FAO, 2014). In contrast to PBM where the DIAAS-like score is 0.36 and Trp is the first limiting amino acid, all the pulses we evaluated were first limiting in Met or the total sulfur amino acids (Met + Cys). While the protein quality is generally lower in beans, chickpeas (0.33), common bean (0.30) and cow peas (0.34) are close to relatively similar to PBM and broad beans have the poorest DIAAS-scores (0.17). The other pulses had intermediate scores being 0.25 for peas, 0.21 for lentils, and 0.24 for lupins. It is critical to remember that DIAAS-scores speaks to the ability of an ingredient to meet the amino acid requirements of the species of interest and we combine ingredients when formulating complete and balanced diets for dogs and cats. As such, ingredients should complement the amino acid content, resulting in improved DIAAS-scores for complete diets. The petfood industry can also add supplemental Met, Trp, Thr, and Lys to balance these amino acids and improve protein quality.

While digestibility accounts for the majority of variation in the metabolic availability of AA, it does not measure the cellular bioavailability of AA. As such, the indicator AA oxidation technique was first adapted to measure AA metabolic availability using the slope ratio method in growing pigs (Moehn et al., 2005) and more recently adapted for use in dogs (Crosbie et al., 2024) although a suitable reference protein remains unknown (Crosbie et al., 2025). Interestingly, Moehn et al. (2005) validated the procedure by comparing the availability of Lys from raw peas, heat-treated peas, and heat-treated peas with added Lys showing that added Lys resulted in the greatest improvement in Lys bioavailability in contrast to equivalent Lys from raw and heated peas. It is notable that the reference patterns of AA requirements among species greatly differs and so where lysine may be first limiting in pulses for swine, Met is believed to be first limiting in pulses for dogs and cats as the first limiting AA is a function of the ability of the diet to meet the AA requirements of the animal. Crosbie et al. (2025) observed that the metabolic availability of Met was 31% greater in dogs fed peas in contrast to chicken meal; however, subsequent studies using the same diet and methodology design, but different batches of peas and chicken meal found the metabolic availability of Met was 66% and 51% of chicken meal when measured two days and five-days after adaptation, respectively. The measures of metabolic availability emphasize batch to batch differences in ingredients and the variability introduced in processing. Overall, in vitro species-specific digestibility assay development is needed to reduce the use of animals in research and because total tract nitrogen digestibility does not account for differences in individual AA digestibility.

In addition to the IAA digestibility and metabolic availability measures of ingredients and/or diets, to assess protein quality a standard reference protein or IAA reference pattern is needed. For example, the FAO recommends age-specific reference patterns, 0.5 to 3 yrs, older children and adults (3+ yrs) be used. Similarly, regulatory guidance for both dog and cat food includes IAA requirements for growth/

reproduction and adult maintenance; however, these commercial recommendations have a correction for digestibility. In contrast, the minimal IAA requirements for growth/reproduction and adult maintenance presented in NRC, represent the physiological requirements for an amino acid and is the reference pattern that should be used along with IAA ileal digestibility of the ingredient or diet of interest to generate DIAAS-like scores. However, while there are book values for the IAA requirements of dogs and cats, there are a dearth of estimates of the IAA of dogs and cats and with most empirical estimates using growth or nitrogen balance as the primary outcome. More recent research using the more sensitive indicator amino acid oxidation technique has repeatedly measured IAA requirements greater than the recommended allowance in NRC and the regulatory recommendations from AAFCO and FEDIAF (Mansilla et al., 2018; Mansilla et al., 2020a; Mansilla et al., 2020b; Sutherland et al., 2020). More research into determining the IAA requirements of dogs is needed focusing on breed differences, the effect of age on IAA needs, and with different diet formats and formulations. In terms of considerations for cats, the IAA recommendations are based off of even less data with two estimates of IAA requirements using the indicator amino acid oxidation technique have been completed and these also suggested that the Phe (Lambie et al., 2024) and the Met requirement (Pezzali et al., 2024) are greater than the current recommendations. Indeed, greater work defining the IAA of cats is needed.

TABLE 16.6 Average standardized ileal amino acid digestibility coefficients of poultry by-product meal and selected raw and pre-cooked pulses in pigs and cecectomized roosters*

Amino acid	Poultry meal	Pea		Broad bean		Common bean	Chickpea		Lentil		Cow pea	Lupine
	Pig	*Pitg*	*CR*	*Pig*	*CR*	*Pig*	*Pig*	*CR*	*Pig*	*CR*	*Pig*	*Pig*
Lysine	80	83	89	85	87	85	94	79	82	88	82	87
Threonine	77	76	89	80	87	76	77	79	81	89	81	83
Methionine	82	80	85	80	84	87	89	75	86	73	86	86
Tryptophan	70	73	91	70	88	82	83	75	82	88	82	77
Isoleucine	83	79	90	82	88	72	82	77	84	88	84	88
Valine	80	77	90	79	87	65	81	77	83	86	83	84
Leucine	82	80	91	84	89	71	86	78	86	86	86	88
Phenylalanine	83	80	91	83	91	74	89	79	86	90	86	90
Histidine	68	84	90	85	87	66	87	80	86	81	85	92
Arginine	87	89	94	89	93	94	96	82	90	92	90	94

*Data based on pigs are sourced from the INRAE-CIRAD-AFZ (2025) feed composition database and on cecectomized roosters (CR) from multiple studies on pea (Reilly et al., 2020a; b; Oba et al., 2023), broad bean (Reilly et al., 2020a; b; Banton et al., 2025), chickpea (Reilly et al., 2020a; Banton et al., 2025) and lentil (Reilly et al., 2020a).

4.2.2 Starch

Starch digestibility is often classified into three fractions: rapidly digestible starch (RDS), slowly digestible starch, and resistant starch (RS) (Englyst et al., 1992). Slowly digestible starch is completely digested in the small intestine, but at a slower rate, relative to RDS. Resistant starch, particularly RS2 (native ungelatinized granules) and RS3 (retrograded starch formed after cooking/cooling), escapes small intestinal digestion and undergoes colonic fermentation, yielding straight-chain fatty acids (SCFA) such as acetate, propionate, and butyrate (Englyst et al., 1992; Annison and Topping, 1994). Functionally, RS is similar to dietary fiber. Cooked pulses generally contain ~4% RS (de Almeida Costa et al., 2006; Brummer et al., 2015; Fabbri et al., 2016). In contrast, commercial extruded canine and feline diets have been reported to contain less than 1.6% RS relative to total starch, regardless of whether they contained pulses, cereal grains or tubers, reflecting the gelatinization that occurs during extrusion (Alvarenga and Aldrich, 2020; Rankovic et al., 2020).

To the authors' knowledge, the digestibility of raw pulse starches has not been tested in vivo in dogs or cats. However, in vitro assays simulating digestion in dogs demonstrated that ground legumes (beans, lentils and peas) had substantially lower ileal digestible starch (IDS; 21%) and total digestible starch (TDS; 31%) than ground cereal grains (60% and 67%, respectively) (Bednar et al., 2001). However, thermal processing, such as retort or extrusion, induces starch gelatinization (Cheftel, 1986; Lin et al., 1997; Whistler et al., 2012), thereby improving the digestibility of pulse starches to levels more comparable with those of cereal starches. In dogs and cats, the total tract apparent digestibility of starch was evaluated in vivo using extruded diets formulated with cassava flour, brewer's rice, corn, sorghum, peas, or lentils. Although the quantity of RS was not analyzed, starch digestibility was high across all diets (>93% for cats and >98% for dogs) (Carciofi et al., 2008; de-Oliveira et al., 2008). However, in both studies, the pea- and lentil-based diets had the lowest values numerically for apparent starch digestibility.

The high amylose content in pulse starches contributes to a slower rate of digestion when compared to that in cereal and tuber starches (see above). The digestive properties of dietary starch impacts the postprandial blood glucose and insulin response. Feeding cats and dogs food with pulses as main starch sources results in a slower rise in and lower peaks of blood glucose (Carciofi et al., 2008; de-Oliveira et al., 2008; Adolphe et al., 2012; Quilliam et al., 2021). As such, pulse starch may serve as an alternative to starch from cereal grains for dogs and cats that require special foods supporting better glucose control.

4.2.3 Fiber

Pulses contain complex insoluble dietary fibers as well as more simple oligosaccharides that are soluble (Chen et al., 2016). The dietary fiber in pulses resists the digestive processes in the upper digestive tract and are potential substrates for fermentation by the microbiota, mainly residing in the large intestine. Moreover, fibers may act as bulking agents, binding water that subsequently impacts fecal consistency. The fermentability of fibers depend on their chemical nature and the enzymatic capacity of the microbiota to hydrolyze these specific chemical structures. Insoluble fibers are generally more difficult to ferment within the colon of monogastric animals, including dogs and cats, but the soluble fibers can be readily fermented. In vivo models used to predict the small intestinal fermentability of pea fiber and pea hulls in dogs have observed less SCFA production and lower organic matter disappearance with said pulse fibers, as compared to fermentable fiber sources, such as beet pulp or soybean oligosaccharides (Bosch et al., 2008; Donadelli et al., 2019; Kim and Aldrich, 2022). Beet pulp fiber is largely composed of pectin, which is considered highly soluble and fermentable. Similarly, in vitro canine models reported that pea fiber resulted in less SCFA production in vitro as apple pomace (Holt et al., 2023), which similarly contains significant amounts of pectin (Guillon et al., 1995; Lebet et al., 1998). However, pea fiber had greater production of butyrate (Holt et al., 2023). In an earlier study, pea hulls were reported to have lower fermentability as compared to apple pomace, carrot pomace, tomato pomace and flaxseed, as measured by organic matter disappearance using an in vitro canine model (Swanson et al., 2001). Additionally, when the fermentability of whole ground legumes was assessed as compared to ground cereals, similarly using an in vitro canine model, the legumes on average resulted in greater SCFA production, however, this was attributed to the fermentation of not only fiber but also the starch components of the ground ingredients (Bednar et al., 2001). More recently, Traughber et al. (2020) incubated filter residues of pre-digested pulses and reported lower to similar SCFA production in 12 h of incubation as for beet pulp. Moreover, clear differences in SCFA profiles were noted among the pulse substrates with, for example, garbanzo beans yielding considerably lower amounts of acetate and propionate but higher amounts of butyrate. As the latter study applied filtration for preparing the substrates, soluble fibers would have been lost and not accounted for in the assessment of the fermentability. Feeding dogs and cats foods containing too high amounts of rapidly fermentable fibers could result in loose feces, and therefore the oligosaccharide levels and their fermentability require attention, as well as feeding studies assessing the impact of pulse inclusion on fecal consistency.

5 Environmental impact

The environmental footprints of ingredients and complete pet foods is becoming increasingly important. The footprints of pulses have been quantified in various studies, but differences in methodologies hamper direct comparisons. The Global Feed LCA Institute provides a life cycle analysis database with impact values for an-

TABLE 16.7 Global warming potential and land use of poultry by-product meal and selected pulses*

Impact category[a]	Poultry meal	Pea	Broad bean	Lupine
GWP, g CO_2-eq/kg	778	517	453	1010
GWP, g CO_2-eq/kg CP	1231	2547	1776	3015
Land use, m^2/kg	1	4	3	7
Land use, m^2/kg CP	2	22	13	22

*Data retrieved from Global Feed LCA Institute FeedPrint International (Global Feed LCA Institute, 2021).
[a]GWP, global warming potential; CP, crude protein with assumed contents (on as is basis, Table 16.2) of 63.2% for poultry meal, 20.3% for field peas, 25.5% for faba beans and 33.5% for lupins.

imal feed ingredients with data for several pulses as well as PBM (Table 16.7). Based on economic allocation and per kg of product, field peas and faba beans have lower global warming potential values than PBM whereas lupins have higher values. Land use is higher for these pulses than for PBM. The footprint can also be expressed per kg of crude protein. Due to the differences in crude protein contents (Table 16.1), the relative differences between these different protein sources also shift. PBM has the lowest global warming potential and land use followed by faba beans. Peas and lupins have the highest impact values. The standardization of footprints can also be done at the level of the first limiting amino acid (Table 16.6), which would again impact the differences between the ingredients. Moreover, it should be noted that pulses in a formulation (recipe) do not only contribute to protein but also to other components like starch. This means, for example, that less starch from rice or corn is needed in the formulation, which would then impact the overall environmental footprint of the food. These examples illustrate the complexity of comparing pet food ingredients for environmental footprints and that comparisons should be done at the level of the formulation of the pet food with set nutrient targets.

6 Concluding remarks

With increasing constraints on our global food systems, it is critical that the ingredients available to the pet food industry are diversified. Further to the diversification of ingredients for the pet food industry, the environmental impact of the ingredients used also needs to be a primary consideration and warrants further research. In addition to ingredient considerations, further research to replicate and build consensus around the accuracy of dog and cat nutrient requirements would help move pet foods towards precision nutrition. Overall, this would minimize nutrient waste and maximize animal health and well-being. Pulse species and their cultivars can vary greatly in their properties, with potential beneficial but also detrimental effects on pet health, which clearly warrants more and long-term studies. The inclusion of pulses in pet foods is relatively new, and we are learning about their suitability when used in diets for dogs and cats. Nutritional guidelines for commercial pet foods have been developed based on data from research and practice, with the vast majority of foods containing animal-based protein sources. Including pulses as protein sources also brings in non-nutrient bioactive compounds that are not foreseen in the nutritional guidelines. It is therefore of interest to develop a better understanding on the availability of, in particular, minerals when pulses are used as protein sources, which might result in adapted guidelines for pulse-based pet foods.

References

AAFCO, 2024. Official publication of the Association of American Feed Control Officials. AAFCO Incorporated, Atlanta, GA, USA.

Acuff, H.L., Dainton, A.M., Dhakal, J., Kiprotich, S. and Aldrich, G., 2021. Sustainability and pet food: is there a role for veterinarians? Veterinary Clinics of North America - Small Animal Practice 51: 563–581. https://doi.org/10.1016/j.cvsm.2021.01.010

Adamidou, S., Nengas, I., Grigorakis, K., Nikolopoulou, D. and Jauncey, K., 2011. Chemical composition and antinutritional factors of field peas (*Pisum sativum*), chickpeas (*Cicer arietinum*), and faba beans (*Vicia faba*) as affected by extrusion preconditioning and drying temperatures. Cereal Chemistry 88: 80–86. https://doi.org/10.1094/CCHEM-05-10-0077

Adolphe, J.L., Drew, M.D., Huang, Q., Silver, T.I. and Weber, L.P., 2012. Postprandial impairment of flow-mediated dilation and elevated methylglyoxal after simple but not complex carbohydrate consumption in dogs. Nutrition Research 32: 278–284. https://doi.org/10.1016/j.nutres.2012.03.002

Adolphe, J.L., Drew, M.D., Silver, T.I., Fouhse, J., Childs, H. and Weber, L.P., 2015. Effect of an extruded pea or rice diet on postprandial insulin and cardiovascular responses in dogs. Journal of Animal Physiology and Animal Nutrition 99: 767–776. https://doi.org/10.1111/jpn.12275

Aldrich, G.C. and Koppel, K., 2015. Pet food palatability evaluation: A review of standard assay techniques and interpretation of results with a primary focus on limitations. Animals 5: 43–55. https://doi.org/10.3390/ani5010043

Alonso, R., Aguirre, A. and Marzo, F., 2000. Effects of extrusion and traditional processing methods on antinutrients and in vitro digestibility of protein and starch in faba and kidney beans. Food Chemistry 68: 159–165. https://doi.org/10.1016/s0308-8146(99)00169-7

Alonso, R., Orúe, E. and Marzo, F., 1998. Effects of extrusion and conventional processing methods on protein and antinutritional factor contents in pea seeds. Food Chemistry 63: 505–512. https://doi.org/10.1016/S0308-8146(98)00037-5

Alvarenga, I.C. and Aldrich, C.G., 2020. Starch characterization of commercial extruded dry pet foods. Translational Animal Science 4: 1017–1022. https://doi.org/10.1093/tas/txaa018

Alvarenga, I.C., Holt, D. and Aldrich, C.G., 2020. Evaluation of faba beans as an ingredient in dog diets: apparent total tract digestibility of extruded diets with graded levels of dehulled faba beans (*Vicia faba* L.) by dogs. Journal of Animal Science 98: skaa085. https://doi.org/10.1093/jas/skaa085

Annison, G. and Topping, D.L., 1994. Nutritional role of resistant starch: chemical structure vs physiological function. Annual Review of Nutrition 14: 297–320. https://doi.org/10.1146/annurev.nu.14.070194.001501

APPA, 2025. The American Pet Products Association (APPA) releases 2025 State of the Industry Report, America Pet Products Association, Inc.

Bajaj, R., Singh, N., Kaur, A. and Inouchi, N., 2018. Structural, morphological, functional and digestibility properties of starches from cereals, tubers and legumes: a comparative study. Journal of Food Science and Technology 55: 3799–3808. https://doi.org/10.1007/s13197-018-3342-4

Bakke, A.M., Wood, J., Salt, C., Allaway, D., Gilham, M., Kuhlman, G., Bierer, T., Butterwick, R. and O'Flynn, C., 2022. Responses in randomised groups of healthy, adult Labrador retrievers fed grain-free diets with high legume inclusion for 30 days display commonalities with dogs with suspected dilated cardiomyopathy. BMC Veterinary Research 18: 157. https://doi.org/10.1186/s12917-022-03264-x

Ball, S., Guan, H.P., James, M., Myers, A., Keeling, P., Mouille, G., Buléon, A., Colonna, P. and Preiss, J., 1996. From glycogen to amylopectin: a model for the biogenesis of the plant starch granule. Cell 86: 349–352. https://doi.org/10.1016/s0092-8674(00)80107-5

Banton, S., Baynham, A., Pezzali, J.G., van Massow, M. and Shoveller, A.K., 2021. Grains on the brain: A survey of dog owner purchasing habits related to grain-free dry dog foods. PLoS ONE 16: e0250806. https://doi.org/10.1371/journal.pone.0250806

Banton, S., Templeman, J.R., Oba, P.M., Swanson, K.S., Saunders-Blades, J. and Shoveller, A.K., 2025. Standardized ileal amino acid digestibility estimates are not additive in extruded canine diets with an increasing inclusion of whole pulse flours. Journal of Animal Science 103: skaf305. https://doi.org/doi.org/10.1093/jas/skaf305

Batterham, E.S., 1992. Availability and utilization of amino acids for growing pigs. Nutrition Research Reviews 5 https://doi.org/10.1079/NRR19920004

Bednar, G.E., Patil, A.R., Murray, S.M., Grieshop, C.M., Merchen, N.R. and Fahey, G.C., Jr., 2001. Starch and fiber fractions in selected food and feed ingredients affect their small intestinal digestibility and fermentability and their large bowel fermentability in vitro in a canine model. Journal of Nutrition 131: 276–286. https://doi.org/10.1093/jn/131.2.276

Bosch, G., Hagen-Plantinga, E.A. and Hendriks, W.H., 2015. Dietary nutrient profiles of wild wolves: insights for optimal dog nutrition? British Journal of Nutrition 113: S40–54. https://doi.org/10.1017/S0007114514002311

Bosch, G., Pellikaan, W.F., Rutten, P.G.P., van der Poel, A.F.B., Verstegen, M.W.A. and Hendriks, W.H., 2008. Comparative in vitro fermentation activity in the canine distal gastrointestinal tract and fermentation kinetics of fiber sources. Journal of Animal Science 86: 2979–2989. https://doi.org/10.2527/jas.2007-0819

Bosch, G. and Swanson, K.S., 2021. Effect of using insects as feed on animals: pet dogs and cats. Journal of Insects as Food and Feed 7: 795–805. https://doi.org/10.3920/JIFF2020.0084

Bradshaw, J.W.S. and Cook, S.E., 1996. Patterns of pet cat behaviour at feeding occasions. Applied Animal Behaviour Science 47: 61–74. https://doi.org/10.1016/0168-1591(95)01011-4

Brummer, Y., Kaviani, M. and Tosh, S.M., 2015. Structural and functional characteristics of dietary fibre in beans, lentils, peas and chickpeas. Food Research International 67: 117–125. https://doi.org/10.1016/j.foodres.2014.11.009

Campos-Vega, R., Loarca-Piña, G. and Oomah, B.D., 2010. Minor components of pulses and their potential impact on human health. Food Research International 43: 461–482. https://doi.org/10.1016/j.foodres.2009.09.004

Carciofi, A.C., Takakura, F.S., de-Oliveira, L.D., Teshima, E., Jeremias, J.T., Brunetto, M.A. and Prada, F., 2008. Effects of six carbohydrate sources on dog diet digestibility and post-prandial glucose and insulin response. Journal of Animal Physiology and Animal Nutrition 92: 326–336. https://doi.org/10.1111/j.1439-0396.2007.00794.x

Cargo-Froom, C., Shoveller, A.K., Marinangeli, C.P.F. and Columbus, D.A., 2020. Methods for processing pulses to optimize nutritional functionality and maximize amino acid availability in foods and feeds. Cereal Foods World 65 https://doi.org/10.1094/CFW-65-6-0068

Chamone, M.E.R., Ascheri, J.L.R., Vargas-Solórzano, J.W., Stephan, M.P. and Carvalho, C.W.P., 2023. Chemical characterization of white lupin (*Lupinus albus*) flour treated by extrusion cooking and aqueous debittering processes. Plant Foods for Human Nutrition 78: 292–298. https://doi.org/10.1007/s11130-023-01050-0

Champ, M.M.J., 2002. Non-nutrient bioactive substances of pulses. British Journal of Nutrition 88: S307–319. https://doi.org/10.1079/BJN2002721

Cheftel, J.C., 1986. Nutritional effects of extrusion-cooking. Food Chemistry 20: 263–283. https://doi.org/10.1016/0308-8146(86)90096-8

Chen, Y., McGee, R., Vandemark, G., Brick, M. and Thompson, H.J., 2016. Dietary fiber analysis of four pulses using AOAC 2011.25: implications for human health. Nutrients 8: 829. https://doi.org/10.3390/nu8120829

Conway, D.M. and Saker, K.E., 2018. Consumer attitude toward the environmental sustainability of grain-free pet foods. Frontiers in Veterinary Science 24: 170. https://doi.org/10.3389/fvets.2018.00170

Crosbie, M., Templeman, J.R., Pezzali, J.G., Courtney-Martin, G., Levesque, C.L., Hancock, L., Buff, P.R., Columbus, D.A., Verbrugghe, A. and Shoveller, A.K., 2024. Chicken meal is not an appropriate reference protein for estimating protein quality of ingredients used in extruded diets intended for dogs. Journal of Animal Science 102: skae265. https://doi.org/10.1093/jas/skae265

Crosbie, M., Templeman, J.R., Pezzali, J.G., Courtney-Martin, G., Levesque, C.L., Hancock, L., Buff, P.R., Columbus, D.A., Verbrugghe, A. and Shoveller, A.K., 2025. Semi-synthetic reference diets containing crystalline amino acids at 120% of recommendations for adult dogs are not appropriate for estimating protein quality of ingredients used in extruded diets. Journal of Animal Science: skaf306. https://doi.org/10.1093/jas/skaf306

Curso-Almeida, P., Subramaniam, M., Costa, M.D., Adolphe, J.L., Drew, M.D., Loewen, M.E. and Weber, L.P., 2025. Effects of feeding *Candida utilis*-fermented pea starch on overall, metabolic and intestinal health of dogs and cats. Frontiers in Veterinary Science 12: 1542484. https://doi.org/10.3389/fvets.2025.1542484

Dalgetty, D.D. and Baik, B.K., 2003. Isolation and characterization of cotyledon fibers from peas, lentils, and chickpeas. Cereal Chemistry 80: 310–315. https://doi.org/10.1094/CCHEM.2003.80.3.310

de-Oliveira, L.D., Carciofi, A.C., Oliveira, M.C., Vasconcellos, R.S., Bazolli, R.S., Pereira, G.T. and Prada, F., 2008. Effects of six carbohydrate sources on diet digestibility and postprandial glucose and insulin responses in cats. Journal of Animal Science 86: 2237–2246. https://doi.org/10.2527/jas.2007-0354

de Almeida Costa, G.E., da Silva Queiroz-Monici, K., Reis, S.M. and de Oliveira, A.C., 2006. Chemical composition, dietary fibre and resistant starch contents of raw and cooked pea, common bean, chickpea and lentil legumes. Food Chemistry 94: 327–330. https://doi.org/10.1016/j.foodchem.2004.11.020

De Lange, C.F.M., Nyachoti, C.M. and Verstegen, M.W.A., 2000. The significance of antinutritional factors in feedstuffs for monogastric animals, in: Moughan, P.J., Verstegen, M.W.A. and Visser-Reyneveld, M.I. (Eds.), Feed evaluation: principles and practice, Wageningen Pers, Wageningen, The Netherlands, pp. 169–188.

de Oliveira, L.D., Carciofi, A.C., Oliveira, M.C., Vasconcellos, R.S., Bazolli, R.S., Pereira, G.T. and Prada, F., 2008. Effects of six carbohydrate sources on diet digestibility and postprandial glucose and insulin responses in cats. Journal of Animal Science 86: 2237–2246. https://doi.org/10.2527/jas.2007-0354

Dodd, S., Cave, N., Abood, S., Shoveller, A.K., Adolphe, J. and Verbrugghe, A., 2020. An observational study of pet feeding practices and how these have changed between 2008 and 2018. Veterinary Record 186: 643. https://doi.org/10.1136/vr.105828

Dodd, S.A., Cave, N.J., Adolphe, J.L., Shoveller, A.K. and Verbrugghe, A., 2019. Plant-based (vegan) diets for pets: A survey of pet owner attitudes and feeding practices. PLoS ONE 14: e0210806. https://doi.org/10.1371/journal.pone.0210806

Donadelli, R.A., Titgemeyer, E.C. and Aldrich, C.G., 2019. Organic matter disappearance and production of short-and branched-chain fatty acids from selected fiber sources used in pet foods by a canine in vitro fermentation model. Journal of Animal Science 97: 4532–4539. https://doi.org/10.1093/jas/skz302

El-Adawy, T.A., 2002. Nutritional composition and antinutritional factors of chickpeas (*Cicer arietinum* L.) undergoing different cooking methods and germination. Plant Foods for Human Nutrition 57: 83–97. https://doi.org/10.1023/a:1013189620528

Englyst, H.N., Kingman, S.M. and Cummings, J.H., 1992. Classification and measurement of nutritionally important starch fractions. European Journal of Clinical Nutrition 46: S33–50.

Enneking, D. and Wink, M., 2000. Towards the elimination of anti-nutritional factors in grain legumes, in: Knight, R. (Ed.), Linking research and marketing opportunities for pulses in the 21st century, Springer Netherlands, Dordrecht, The Netherlands, pp. 671–683.

Euromonitor International, 2024. Pet humanisation and premium products drive global pet care sales up by 5.9% to USD 197.6 billion, Euromonitor International.

Fabbri, A.D., Schacht, R.W. and Crosby, G.A., 2016. Evaluation of resistant starch content of cooked black beans, pinto beans, and chickpeas. NFS Journal 3: 8–12. https://doi.org/10.1016/j.nfs.2016.02.002

Fantechi, T., Califano, G., Caracciolo, F. and Contini, C., 2024. Puppy power: How neophobia, attitude towards sustainability, and animal empathy affect the demand for insect-based pet food. Food Research International 177: 113879. https://doi.org/10.1016/j.foodres.2023.113879

FAO, 2014. Research approaches and methods for evaluating the protein quality of human foods, Food and Agriculture Organization of the United Nations, Rome, Italy, p. 59.

FEDIAF, 2018. Nutritional guidelines for complete and complementary pet food for cats and dogs, European Pet Food Industry Federation, Bruxelles, Belgium.

FEDIAF, 2022. Nutritional guidelines for complete and complementary pet food for cats and dogs, European Pet Food Industry Federation, Bruxelles, Belgium.

FEDIAF, 2025. Nutritional guidelines for complete and complementary pet food for cats and dogs, European Pet Food Industry Federation, Bruxelles, Belgium.

Forster, G.M., Hill, D., Gregory, G., Weishaar, K.M., Lana, S., Bauer, J.E. and Ryan, E.P., 2012. Effects of cooked navy bean powder on apparent total tract nutrient digestibility and safety in healthy adult dogs. Journal of Animal Science 90: 2631–2638. https://doi.org/10.2527/jas.2011-4324

Global Feed LCA Institute, 2021. GFLI FeedPrint International version 2020.52.

Gonçalves, A., Goufo, P., Barros, A., Domínguez-Perles, R., Trindade, H., Rosa, E.A.S., Ferreira, L. and Rodrigues, M., 2016. Cowpea (*Vigna unguiculata* L.Walp), a renewed multipurpose crop for a more sustainable agri-food system: nutritional advantages and constraints. Journal of the Science of Food and Agriculture 96: 2941–2951. https://doi.org/10.1002/jsfa.7644

Government of Canada, 2024. Canadian Nutrient File (CNF), Government of Canada.

Guillon, F. and Champ, M.J., 2002. Carbohydrate fractions of legumes: uses in human nutrition and potential for health. British Journal of Nutrition 88: S293–306. https://doi.org/10.1079/BJN2002720

Guillon, F., Renard, C.M., Hospers, J., Thibault, J.F. and Barry, J.L., 1995. Characterisation of residual fibres from fermentation of pea and apple fibres by human faecal bacteria. Journal of the Science of Food and Agriculture 68: 521–529. https://doi.org/10.1002/jsfa.2740680417

Hall, C., Hillen, C. and Garden Robinson, J., 2017. Composition, nutritional value, and health benefits of pulses. Cereal Chemistry 94: 11–31. https://doi.org/10.1094/CCHEM-03-16-0069-FI

Hendriks, W.H., Emmens, M.M.A., Trass, B. and Pluske, J.R., 1999. Heat processing changes the protein quality of canned cat foods as measured with a rat bioassay. Journal of Animal Science 77: 669–676. https://doi.org/10.2527/1999.773669x

Hendriks, W.H., van Baal, J. and Bosch, G., 2012. Ileal and faecal protein digestibility measurement in monogastric animals and humans: a comparative species view. British Journal of Nutrition 108: S247–257. https://doi.org/10.1017/S0007114512002395

Hernández-Infante, M., Sousa, V., Montalvo, I. and Tena, E., 1998. Impact of microwave heating on hemagglutinins, trypsin inhibitors and protein quality of selected legume seeds. Plant Foods for Human Nutrition 52: 199–208. https://doi.org/10.1023/a:1008033610737

Holt, D.A., Alvarenga, I.C., Donadelli, R.A. and Aldrich, C.G., 2023. Substrate degradation and postbiotic analysis of alternative fiber ingredients fermented using an in vitro canine fecal inoculum model. Journal of Animal Science 101: skad078. https://doi.org/10.1093/jas/skad078

Hoover, R., 2001. Composition, molecular structure, and physicochemical properties of tuber and root starches: a review. Carbohydrate Polymers 45: 253–267. https://doi.org/10.1016/S0144-8617(00)00260-5

Hoover, R., Hughes, T., Chung, H.J. and Liu, Q., 2010. Composition, molecular structure, properties, and modification of pulse starches: A review. Food Research International 43: 399–413. https://doi.org/10.1016/j.foodres.2009.09.001

Hoover, R. and Ratnayake, W.S., 2002. Starch characteristics of black bean, chick pea, lentil, navy bean and pinto bean cultivars grown in Canada. Food Chemistry 78: 489–498. https://doi.org/10.1016/S0308-8146(02)00163-2

INRAE-CIRAD-AFZ, 2025. INRAE-CIRAD-AFZ Feed tables, INRAE-CIRAD-AFZ.

Johnson, M.L., Parsons, C.M., Fahey, G.C., Jr., Merchen, N.R. and Aldrich, C.G., 1998. Effects of species raw material source, ash content, and processing temperature on amino acid digestibility of animal by-product meals by cecectomized roosters and ileally cannulated dogs. Journal of Animal Science 76: 1112–1122. https://doi.org/10.2527/1998.7641112x

Judprasong, K., Charoenkiatkul, S., Sungpuag, P., Vasanachitt, K. and Nakjamanong, Y., 2006. Total and soluble oxalate contents in Thai vegetables, cereal grains and legume seeds and their changes after cooking. Journal of Food Composition and Analysis 19: 340–347. https://doi.org/10.1016/j.jfca.2005.04.002

Kaelle, G.C.B., Bastos, T.S., dos Santos de Souza, R.B., Fernandes, E.L., Santos, L.N.A., de Oliveira, S.G. and Félix, A.P., 2024. Starch sources and their influence on extrusion parameters, kibble characteristics and palatability of dog diets. Italian Journal of Animal Science 23: 388–396. https://doi.org/10.1080/1828051X.2024.2313084

Khattab, R.Y. and Arntfield, S.D., 2009. Nutritional quality of legume seeds as affected by some physical treatments 2. Antinutritional factors. LWT - Food Science and Technology 42: 1113–1118. https://doi.org/10.1016/j.lwt.2009.02.004

Kienzle, E., Bergler, R. and Mandernach, A., 1998. A comparison of the feeding behavior and the human–animal relationship in owners of normal and obese dogs. Journal of Nutrition 128: S2779–2782. https://doi.org/10.1093/jn/128.12.2779S

Kim, H.S.S. and Aldrich, C.G., 2022. Organic matter disappearance and production of short- and branched-chain fatty acids from whole soybeans and selected fiber sources used in pet foods by a canine in vitro fermentation model. Journal of Animal Science 100: 51–52. https://doi.org/10.1093/jas/skac247.101

Klinmalai, P., Kamonpatana, P., Thongpech, A., Sodsai, J., Promhuad, K., Srisa, A., Laorenza, Y., Kovitvadhi, A., Areerat, S., Seubsai, A. and Sablani, S.S., 2025. Comprehensive review of alternative proteins in pet food: Research publications, patents, and product trends in plant, aquatic, insect, and cell-based sources. Foods 14: 2640. https://doi.org/10.3390/foods14152640

Lambie, J.L., Pezzali, J.G., Richards, T.L., Ellis, J.L., Verbrugghe, A. and Shoveller, A.K., 2024. Phenylalanine requirements using the direct amino acid oxidation technique, and the effects of dietary phenylalanine on food intake, gastric emptying, and macronutrient oxidation in adult cats. Journal of Animal Science 102: skae009. https://doi.org/10.1093/jas/skae009

Lebet, V., Arrigoni, E. and Amadò, R., 1998. Measurement of fermentation products and substrate disappearance during incubation of dietary fibre sources with human faecal flora. LWT - Food Science and Technology 31: 473–479. https://doi.org/10.1006/fstl.1998.0401

Li, P. and Wu, G., 2018. Roles of dietary glycine, proline, and hydroxyproline in collagen synthesis and animal growth. Amino Acids 50: 29–38. https://doi.org/10.1007/s00726-017-2490-6

Lin, S., Hsieh, F. and Huff, H.E., 1997. Effects of lipids and processing conditions on degree of starch gelatinization of extruded dry pet food. LWT - Food Science and Technology 30: 754–761. https://doi.org/http://dx.doi.org/10.1006/fstl.1997.0271

Luo, Y.W. and Xie, W.H., 2013. Effect of different processing methods on certain antinutritional factors and protein digestibility in green and white faba bean (*Vicia faba* L.). Journal of Food 11: 43–49. https://doi.org/10.1080/19476337.2012.681705

Makkar, H.P.S., Becker, K., Abel, H. and Pawelzik, E., 1997. Nutrient contents, rumen protein degradability and antinutritional factors in some colour- and white-flowering cultivars of *Vicia faba* beans. Journal of the Science of Food and Agriculture 75: 511–520. https://doi.org/10.1002/(SICI)1097-0010(199712)75:4<511::AID-JSFA907>3.0.CO;2-M

Mansilla, W.D., Fortener, L., Templeman, J.R. and Shoveller, A.K., 2020a. Adult dogs of different breed sizes have similar threonine requirements as determined by the indicator amino acid oxidation technique. Journal of Animal Science 98: skaa066. https://doi.org/10.1093/jas/skaa066

Mansilla, W.D., Gorman, A., Fortener, L. and Shoveller, A.K., 2018. Dietary phenylalanine requirements are similar in small, medium, and large breed dogs using the indicator amino acid oxidation technique. Journal of Animal Science 96: 3112–3120. https://doi.org/10.1093/jas/sky208

Mansilla, W.D., Templeman, J.R., Fortener, L. and Shoveller, A.K., 2020b. Minimum dietary methionine requirements in miniature Dachshunds, Beagle, and Labrador Retriever adult dogs using the indicator amino acid oxidation technique. Journal of Animal Science 98: skaa324. https://doi.org/10.1093/jas/skaa324

Marinangeli, C.P.F., Nosworthy, M.G. and Shoveller, A.K., 2024. Cereal proteins in the human diet: Reflecting on their contributions to daily protein intake. Journal of Cereal Science 117: 103908. https://doi.org/10.1016/j.jcs.2024.103908

Melcion, J.-P. and Van der Poel, A.F.B., 1993. Process technology and antinutritional factors: principles, adequacy and process optimization, in: Van der Poel, A.F.B., Huisman, J. and Saini, H.S. (Eds.), Recent advances of research in antinutritional factors in legume seeds, Wageningen Academic Press, Wageningen, the Netherlands, pp. 419–434.

Moehn, S., Bertolo, R.F.P., Pencharz, P.B. and Ball, R.O., 2005. Development of the indicator amino acid oxidation technique to determine the availability of amino acids from dietary protein in pigs. Journal of Nutrition 135: 2866–2870. https://doi.org/10.1093/jn/135.12.2866

NRC, 2006. Nutrient requirements of dogs and cats. National Academies Press, Washington, D.C., U.S.

Oba, P.M., Utterback, P.L., Longshaw, M., Parsons, C.M. and Swanson, K.S., 2023. Comparing the standardized amino acid digestibility of an alternative protein source with commercially available protein-based ingredients using the precision-fed cecectomized rooster assay. Journal of Animal Science 101: skad236. https://doi.org/10.1093/jas/skad236

Pattanaik, A.K. and Kore, K.B., 2022. Partial replacement of soybean meal with red gram and lentil as vegetable proteins in limited ingredients homemade diets for pet dogs. Legume Science 4: e119. https://doi.org/10.1002/leg3.119

Patterson, C.A., 2017. Effect of processing on antinutrient compounds in pulses. Cereal Chemistry 94: 2–10. https://doi.org/10.1094/CCHEM-05-16-0144-FI

Patterson, C.A., Curran, J. and Der, T., 2017. Effect of processing on antinutrient compounds in pulses. Cereal Chemistry 94: 2–10. https://doi.org/10.1094/CCHEM-05-16-0144-FI

Pedrosa, M.M., Cuadrado, C., Burbano, C., Allaf, K., Haddad, J., Gelencsér, E., Takács, K., Guillamón, E. and Muzquiz, M., 2012. Effect of instant controlled pressure drop on the oligosaccharides, inositol phosphates, trypsin inhibitors and lectins contents of different legumes. Food Chemistry 131: 862–868. https://doi.org/10.1016/j.foodchem.2011.09.061

Penazzi, L., Freire, T.G., Theodoro, S.D., Frias, J.L., Ala, U., Carciofi, A.C. and Prola, L., 2024. Lentils pasta by-product in a complete extruded diet for dogs and its effect on extrusion, digestibility, and carbohydrate metabolism. Frontiers in Veterinary Science 11: 1429218. https://doi.org/10.3389/fvets.2024.1429218

Pezzali, J.G., Lambie, J.L., Verbrugghe, A. and Shoveller, A.K., 2024. Minimum methionine requirement in adult cats as determined by the indicator amino acid oxidation. Journal of Animal Science 102: skad411. https://doi.org/10.1093/jas/skad411

Plantinga, E.A., Bosch, G. and Hendriks, W.H., 2011. Estimation of the dietary nutrient profile of free-roaming feral cats: possible implications for nutrition of domestic cats. British Journal of Nutrition 106: S35–S48. https://doi.org/10.1017/S0007114511002285

Prata, J.C., 2022. Survey of pet owner attitudes on diet choices and feeding practices for their pets in Portugal. Animals 12: 2775. https://doi.org/10.3390/ani12202775

Quilliam, C., Ren, Y., Ai, Y. and Weber, L.P., 2021. The effects of 7 days of feeding pulse-based diets on digestibility, glycemic response and taurine levels in domestic dogs. Frontiers in Veterinary Science 8: 654223. https://doi.org/10.3389/fvets.2021.654223

Quinteros, A., Farré, R. and Lagarda, M.J., 2003. Effect of cooking on oxalate content of pulses using an enzymatic procedure. International Journal of Food Sciences and Nutrition 54: 373–377. https://doi.org/10.1080/09637480310001595270

Rankovic, A., Adolphe, J.L., Ramdath, D.D., Shoveller, A.K. and Verbrugghe, A., 2020. Glycemic response in nonracing sled dogs fed single starch ingredients and commercial extruded dog foods with different carbohydrate sources. Journal of Animal Science 98: skaa241. https://doi.org/10.1093/jas/skaa241

Reilly, L.M., He, F., Rodriguez-Zas, S.L., Southey, B.R., Hoke, J.M., Davenport, G.M. and de Godoy, M.R.C., 2021a. Effects of graded inclusion levels of raw garbanzo beans on apparent total tract digestibility, fecal quality, and fecal fermentative end-products and microbiota in extruded feline diets. Journal of Animal Science 99: 1–17. https://doi.org/10.1093/jas/skab297

Reilly, L.M., He, F., Rodriguez-Zas, S.L., Southey, B.R., Hoke, J.M., Davenport, G.M. and de Godoy, M.R.C., 2021b. Use of legumes and yeast as novel dietary protein sources in extruded canine diets. Frontiers in Veterinary Science 8: 667642. https://doi.org/10.3389/fvets.2021.667642

Reilly, L.M., Von Schaumburg, P.C., Hoke, J.M., Davenport, G.M., Utterback, P.L., Parsons, C.M. and de Godoy, M.R.C., 2020a. Macronutrient composition, true metabolizable energy and amino acid digestibility, and indispensable amino acid scoring of pulse ingredients for use in canine and feline diets. Journal of Animal Science 98: skaa149. https://doi.org/10.1093/jas/skaa149

Reilly, L.M., von Schaumburg, P.C., Hoke, J.M., Davenport, G.M., Utterback, P.L., Parsons, C.M. and de Godoy, M.R.C., 2020b. Use of precision-fed cecectomized rooster assay and digestible indispensable amino acid scores to characterize plant-and yeast-concentrated proteins for inclusion in canine and feline diets. Translational Animal Science 4: txaa133. https://doi.org/10.1093/tas/txaa133

Reilly, L.M., von Shaumburg, P.C., Hoke, J.M., Davenport, G.M., Utterback, P.L. and de Godoy, M.R.C., 2020c. Use of precision-fed cecectomized rooster assay and digestible indispensable amino acid scores to characterize plant- and yeast-concentrated proteins for inclusion in canine and feline diets. Translational Animal Science 4: txaa133. https://doi.org/10.1093/tas/txaa133

Reis, L.G., Morris, T., Quilliam, C., Rodrigues, L.A., Loewen, M.E. and Weber, L.P., 2021. The effect of fermentation of high-or low-tannin fava bean on glucose tolerance, body weight, cardiovascular function, and blood parameters in dogs after 7 days of feeding: comparison with commercial diets with normal vs. high protein. Frontiers in Veterinary Science 8: 653771. https://doi.org/10.3389/fvets.2021.653771

Rombach, M. and Dean, D.L., 2021. It keeps the good boy healthy from nose to tail: Understanding pet food attribute preferences of US consumers. Animals 11: 3301. https://doi.org/10.3390/ani11113301

Schirmer, M., Höchstötter, A., Jekle, M., Arendt, E. and Becker, T., 2013. Physicochemical and morphological characterization of different starches with variable amylose/amylopectin ratio. Food Hydrocolloids 32: 52–63. https://doi.org/10.1016/j.foodhyd.2012.11.032

Schleicher, M., Cash, S.B. and Freeman, L.M., 2019. Determinants of pet food purchasing decisions. Canadian Veterinary Journal 60: 644–650.

Schryvers, S., Arinzechukwu, C., Miserez, B., Eeckhout, M. and Jacxsens, L., 2023. The fate of quinolizidine alkaloids during the processing of lupins (*Lupinus* spp.) for human consumption. Food Chemistry 429: 136847. https://doi.org/10.1016/j.foodchem.2023.136847

Shi, L., Mu, K., Arntfield, S.D. and Nickerson, M.T., 2017. Changes in levels of enzyme inhibitors during soaking and cooking for pulses available in Canada. Journal of Food Science and Technology 54: 1014–1022. https://doi.org/10.1007/s13197-017-2519-6

Simonensen, J.E., Fasenko, G.M. and Lillywhite, J.M., 2014. The value-added dog food market: Do dog owners prefer natural or organic dog foods? Journal of Agricultural Science 6: 86–97.

Singh, N., Kaur, N., Rana, J.C. and Sharma, S.K., 2010. Diversity in seed and flour properties in field pea (*Pisum sativum*) germplasm. Food Chemistry 122: 518–525. https://doi.org/10.1016/j.foodchem.2010.02.064

Singh, N., Sandhu, K.S. and Kaur, M., 2004. Characterization of starches separated from Indian chickpea (*Cicer arietinum* L.) cultivars. Journal of Food Engineering 63: 441–449. https://doi.org/10.1016/j.jfoodeng.2003.09.003

Singh, P., Banton, S., Raheb, S., Templeman, J.R., Saunders-Blades, J., Kostiuk, D., Kelly, J., Marinangeli, C.P., Verbrugghe, A., Verton-Shaw, S. and Shoveller, A.K., 2023. The pulse of it: dietary inclusion of up to 45% whole pulse ingredients with chicken meal and pea starch in a complete and balanced diet does not affect cardiac function, fasted sulfur amino acid status, or other gross measures of health in adult dogs. Journal of Nutrition 153: 1461–1475. https://doi.org/10.1016/j.tjnut.2023.03.027

Smeets-Peeters, M.J.E., Minekus, M., Havenaar, R., Schaafsma, G. and Verstegen, M.W.A., 1999. Description of a dynamic *in vitro* model of the dog gastrointestinal tract and an evaluation of various transit times for protein and calcium. ATLA Alternatives to Laboratory Animals 27: 935–949. https://doi.org/10.1177/026119299902700608

Srichuwong, S. and Jane, J., 2007. Physicochemical properties of starch affected by molecular composition and structures: a review. Food Science and Biotechnology 16: 663–674.

Statista, 2025. Pet food market worldwide – Topic overview, Statista, Inc.

Sutherland, K.A.K., Mansilla, W.D., Fortener, L. and Shoveller, A.K., 2020. Lysine requirements in small, medium and large breed adult dogs using the indicator amino acid oxidation technique. Translational Animal Science 4: txaa082. https://doi.org/10.1093/tas/txaa082

Swanson, K.S., Grieshop, C.M., Clapper, G.M., Shields Jr, R.G., Belay, T., Merchen, N.R. and Fahey Jr, G.C., 2001. Fruit and vegetable fiber fermentation by gut microflora from canines. Journal of Animal Science 79: 919–926. https://doi.org/10.2527/2001.794919x

Templeman, J.R. and Shoveller, A.K., 2022. Digestible indispensable amino acid scores of animal and plant ingredients potentially used in dog diet formulation: how this protein quality metric is affected by ingredient characteristics and reference amino acid profile. Journal of Animal Science 100: 1–21. https://doi.org/10.1093/jas/skac279

Tosh, S.M. and Yada, S., 2010. Dietary fibres in pulse seeds and fractions: Characterization, functional attributes, and applications. Food Research International 43: 450–460. https://doi.org/10.1016/j.foodres.2009.09.005

Tran, Q.D., Hendriks, W.H. and Van der Poel, A.F.B., 2008. Effects of extrusion processing on nutrients in dry pet food. Journal of the Science of Food and Agriculture 88: 1487–1493 https://doi.org/10.1002/jsfa.3247

Traughber, Z.T., He, F., Hoke, J.M., Davenport, G.M. and de Godoy, M.R.C., 2020. Chemical composition and in vitro fermentation characteristics of legumes using canine fecal inoculum. Journal of Animal Science 4: 1–10. https://doi.org/10.1093/tas/txaa200

Tropicos, 2025. Tropicos database (v4.0.0), Missouri Botanical Garden, St Louis, MO.

Vinassa, M., Vergnano, D., Valle, E., Giribaldi, M., Nery, J., Prola, L., Bergero, D. and Schiavone, A., 2020. Profiling Italian cat and dog owners' perceptions of pet food quality traits. BMC Veterinary Research 16: 131. https://doi.org/10.1186/s12917-020-02357-9

Wang, N. and Daun, J.K., 2004. Effect of variety and crude protein content on nutrients and certain antinutrients in field peas (*Pisum sativum*). Journal of the Science of Food and Agriculture 84: 1021–1029. https://doi.org/10.1002/jsfa.1742

Wang, N. and Daun, J.K., 2006. Effects of variety and crude protein content on nutrients and anti-nutrients in lentils (*Lens culinaris*). Food Chemistry 95: 493–502. https://doi.org/10.1016/j.foodchem.2005.02.001

Wang, N., Hatcher, D.W., Toews, R. and Gawalko, E.J., 2009. Influence of cooking and dehulling on nutritional composition of several varieties of lentils (*Lens culinaris*). LWT - Food Science and Technology 42: 842–848. https://doi.org/10.1016/j.lwt.2008.10.007

Wang, N., Hatcher, D.W., Tyler, R.T., Toews, R. and Gawalko, E.J., 2010. Effect of cooking on the composition of beans (*Phaseolus vulgaris* L.) and chickpeas (*Cicer arietinum* L.). Food Research International 43: 589–594. https://doi.org/10.1016/j.foodres.2009.07.012

Wani, I.A., Sogi, D.S., Hamdani, A.M., Gani, A., Bhat, N.A. and Shah, A., 2016. Isolation, composition, and physicochemical properties of starch from legumes: A review. Starch/Stärke 68: 834–845. https://doi.org/10.1002/star.201600007

Weightman, R.M., Renard, C.M. and Thibautl, J.F., 1994. Structure and properties of the polysaccharides from pea hulls. Part 1: Chemical extraction and fractionation of the polysaccharides. Carbohydrate Polymers 24: 139–148. https://doi.org/10.1016/0144-8617(94)90024-8

Whistler, R.L., BeMiller, J.N. and Paschall, E.F. (Eds.), 2012. Starch: chemistry and technology. Academic Press.

Young, V.R. and Pellett, P.L., 1994. Plant proteins in relation to human protein and amino acid nutrition. American Journal of Clinical Nutrition 59: S1203–1212. https://doi.org/10.1093/ajcn/59.5.1203S

Zdunczyk, Z., Godycka, I. and Amarowicz, R., 1997. Chemical composition and content of antinutritional factors in Polish cultivars of peas. Plant Foods for Human Nutrition 50 https://doi.org/10.1007/BF02436041

Zicker, S.C., 2008. Evaluating pet foods: how confident are you when you recommend a commercial pet food? Topics in Companion Animal Medicine 23: 121–126. https://doi.org/10.1053/j.tcam.2008.04.003

CHAPTER 17

Mushrooms in pet food: potential, applications and current limitations

F.D. Henne[1] *and F. Zorn*[2]

[1]*Pet Food Ingredient Specialist, Applications and Palatability, Verona, Italy*
Correspondence: florian.d.henne@science4climate.org
[2]*Head of New Business, saturn petcare gmbh, Bremen, Germany*
Correspondence: fzorn@saturn-petcare.de

Abstract

The projected growth of the global population toward 10 billion by the end of the 21st century is intensifying demand for high-quality proteins and increasing competition among human food, livestock, aquaculture, and companion animal nutrition, thereby driving the search for sustainable alternatives. Within this context, fungi, a kingdom comprising filamentous fungi, yeasts and macrofungi (mushroom-forming fungi), are gaining increasing interest as a source of sustainable proteins. These organisms efficiently convert agro-industrial by-products into nutritionally valuable biomass with comparatively modest environmental inputs.

While yeasts and filamentous fungi are already established in pet nutrition, this chapter will give an overview of edible macrofungi (mushrooms) and their relevance for companion animal diets. It examines macrofungal production systems, emphasizing their distinction from classical microbial fermentation and their reliance on lignocellulosic solid-substrate cultivation aligned with circular bioeconomy principles. Essential amino acid profiles of mushrooms suggest potential to meet requirements when formulated appropriately in pet food diets. Beyond nutrition, mushrooms provide technofunctional benefits relevant to pet food processing, including emulsification and water-binding properties, and they provide bioactive compounds such as β-glucan, chitin, ergosterol, and fungal immunomodulatory proteins.

Sustainability assessments position mushroom production as resource-efficient relative to conventional livestock proteins, particularly through agricultural by-product valorization and co-product reuse such as spent mushroom substrate. Safety and regulatory frameworks in the EU and the US indicate that mushroom-derived ingredients can be considered safe for use in pet food when established edible species are used and production is conducted under controlled conditions, with appropriate contaminant monitoring and compositional standardization. Emerging research and early commercial applications suggest functional health benefits and growing industry interest, although inclusion levels and protein replacement strategies remain in developmental stages. Collectively, mushrooms can represent a multifunctional, sustainable, and nutritionally relevant ingredient class for innovation in companion animal nutrition.

Keywords

novel protein – edible macrofungi – functional ingredients – circular economy – waste valorization – lignocellulosic substrates

1 Introduction

The continuous growth in human population (expected to reach 10 billion by the end of the 21st century) requires an increase in nutrients such as high-quality proteins and essential fatty acids (FAO, 2020; UN, 2022). It is expected that this demand cannot fully be satisfied from traditional plant and animal sources such as soy, beef, pork, poultry and fish because of their relatively high land and feed requirements. In addition, competition for land and protein resources between human food, livestock, companion animals and aquaculture puts additional pressure on the availability and overall distribution of the resources (Poore and Nemecek, 2018; Tilman et al., 2011). That global challenge led to increased interest in novel, sustainable protein sources. Fungi are emerging as a promising alternative, especially in pet food. They can convert diverse, low-cost substrates into protein-rich biomass with comparatively modest resource inputs (Amara and El-Baky, 2023). This chapter aims to highlight mushrooms, a specific group of organisms within the fungi kingdom, and their use as novel protein source in pet food.

The global pet food market is increasingly shaped by consumer demand for sustainability, transparency, and ethical sourcing (Fletcher, 2025). Unlike livestock and aquaculture feeds, which are primarily cost-driven and formulated around bulk protein and energy, pet foods can economically accommodate premium, function-rich ingredients. Mushrooms fit this niche well: they are available from established food-grade supply chains, provide both nutritional and bioactive components, and can be positioned for health attributes that are highly valued by pet owners. Among these alternatives, fungal proteins have gained attention for their nutritional value, functional properties, and ecological efficiency (Amara and El-Baky, 2023; Wang B. et al., 2023; Scholtmeijer et al., 2023).

Fungi constitute a diverse kingdom of eukaryotic organisms encompassing molds (filamentous fungi), mushrooms (macrofungi), and yeasts. Yeasts, such as *Saccharomyces* spp. in inactive or dried forms, have been established ingredients in pet nutrition for decades. Similarly, filamentous fungi are well-characterized sources of high-quality protein and dietary fiber. In comparison, although evidence of mushroom consumption dates back nearly 20,000 years, edible macrofungi (mushrooms) represent a distinct and comparatively less-explored source of protein for pet foods (Li et al., 2021). They offer unique technical functionality, bioactive compounds, and sustainable cultivation possibilities (Ionescu et al., 2025; Niego et al., 2021; Singh A. et al., 2025). This chapter therefore focuses exclusively on mushroom-derived proteins, highlighting their nutritional quality, digestibility, functional properties, and potential application in companion animal diets. Unlike yeast and filamentous fungi, which are commonly produced via submerged fermentation (SmF) or classical solid-state fermentation (SSF), edible macrofungi are predominantly cultivated on lignocellulosic substrates under solid-substrate conditions for fruiting-body production. While this process may be considered a form of SSF from a microbial fermentation perspective, it is traditionally described as mushroom cultivation, reflecting its distinct biological and technological characteristics (Chang and Wasser, 2017; Sánchez, 2010; Valverde et al., 2015).

This chapter reviews fungal proteins for pet food applications by examining production technologies, nutrient quality and digestibility, technical functionality and biofunctional attributes, sustainability and resource considerations, and safety aspects. The aim is to provide pet food formulators and nutrition scientists with a clear, structured overview of macrofungal-based proteins, highlighting opportunities, limitations, and research gaps specific to companion-animal diets. To date, peer-reviewed canine and feline feeding trials evaluating mushroom-derived proteins as partial replacements for conventional primary protein sources in complete diets remain scarce, whereas most available evidence relates to functional supplementation or palatability applications. Therefore, this chapter clearly distinguishes between (i) functional supplementation, (ii) palatability/processing uses, and (iii) true protein replacement strategies, highlighting gaps and practical implications for formulation. The fungal species referenced throughout this chapter include a range of edible macrofungi (mushrooms) and selected microfungi used in fermentation processes. For clarity, Table 17.1 summarizes the scientific names and commonly used English names of the fungal taxa mentioned.

2 Biotechnological production of macrofungi proteins

Edible macrofungi (mushrooms) differ fundamentally from yeasts and filamentous microfungi in their growth form and cultivation methods. In food and feed applications, mushrooms are obtained primarily through controlled cultivation of fungal

TABLE 17.1 Scientific and commonly used English names of fungal species mentioned throughout the chapter

Scientific name	Common name	Context in chapter
Agaricus bisporus	Button mushroom, champignon, portobello	Nutritional composition and amino acid comparisons
Agaricus macrosporus	Large button mushroom	Digestibility example
Auricularia polytricha	Wood ear mushroom	Nutrient composition
Boletus edulis	Porcini, king bolete	Fatty acid and nutrient composition
Cantharellus cibarius	Chanterelle	Nutritional composition
Flammulina velutipes	Enoki, velvet foot	Protein isolate functional properties
Lentinula edodes	Shiitake	Protein isolate functionality
Pleurotus ostreatus	Oyster mushroom	Technofunctional properties and sustainability example
Pleurotus eryngii	King oyster mushroom	Amino acid composition
Pleurotus djamor	Pink oyster mushroom	Amino acid composition
Pleurotus sajor-caju	Grey oyster mushroom	Fatty acid composition
Pleurotus citrinopileatus	Golden oyster mushroom	Nutritional composition
Tricholoma terreum	Grey knight mushroom	Digestibility example
Terfezia claveryi	Desert truffle	Digestibility example
Saccharomyces cerevisiae	Baker's yeast, brewer's yeast	Fermentation example referenced in reviewer discussion
Fusarium spp.	Fusarium molds	Mycoprotein production
Rhizopus spp.	Rhizopus molds	Fermentation-based protein production

fruiting bodies, rather than through microbial fermentation in bioreactors (Montebello et al. 2025; Wang B. et al., 2023). The edible portion typically corresponds to the fruiting body, while the vegetative mycelium develops within a solid substrate.

Global edible mushroom production is dominated by a small number of cultivated species, including shiitake (*Lentinula edodes*), wood ear (*Auricularia* spp.), oyster mushrooms (*Pleurotus* spp.), button mushroom (*Agaricus bisporus*), and velvet foot (*Flammulina velutipes*), which together represent the majority of worldwide production volumes (Singh M. et al., 2020). Commercial mushroom production relies almost exclusively on solid lignocellulosic substrates, such as straw, sawdust, wood chips, corn cobs, or agricultural residues (Miskiewicz et al., 2025). These substrates provide both structural support and nutrients, allowing macrofungi to utilize their enzymatic capacity to degrade cellulose, hemicellulose, and lignin (de Figueiredo Trindade et al., 2025). As a result, mushroom cultivation is strongly linked to waste valorization and circular bioeconomy concepts (Kumla et al., 2020). Unlike submerged fermentation (SmF), which is characteristic of yeast and filamentous fungal protein production, mushroom cultivation occurs under solid-state conditions that more closely resemble natural growth environments. The process typically involves substrate preparation (moistening and sterilization or pasteurization), inoculation with mushroom spawn (mycelium-colonized grain), mycelial colonization, and induction of fruiting through controlled environmental triggers such as temperature, humidity, light, and CO_2 concentration (Miles and Chang, 2004; Valverde et al., 2015).

Recent research has also explored harvest of mycelial biomass from macrofungi cultivated on lignocellulosic substrates under solid-state fermentation conditions as a novel protein source. Producing mycelium instead of fruiting bodies can shorten production cycles, reduce energy inputs, and allow greater flexibility in strain selection. However, such approaches remain less established than traditional mushroom cultivation and are still emerging in food and feed applications (Scholtmeijer et al., 2023).

For pet food applications, mushroom cultivation offers several potential advantages. Fruiting bodies provide a complex matrix of protein, dietary fiber (notably β-glucan and chitin), minerals, and bioactive compounds, while the solid-substrate cultivation systems enable the efficient valorization of agricultural side streams (straw, sawdust, and other lignocellulosic residues). These characteristics make mushroom production particularly attractive within sustainability-oriented feed strategies (Scholtmeijer et al., 2023; Valverde et al., 2015). However, compared with fermentation-based microbial protein production, mushroom cultivation generally results in lower protein concentrations and longer production cycles, which must be considered when evaluating scalability and cost-effectiveness (Liu et al., 2025; Valverde et al., 2015).

Importantly, the biological and technological distinction between fermentation-based fungal protein production and mushroom cultivation must be considered when assessing nutritional value, sustainability, and regulatory classification. Fermentation-based systems rely on the controlled growth of yeasts or filamentous fungi in submerged or solid-state fermentations to produce single-cell protein or mycoprotein, often yielding protein contents exceeding 30–50% of dry matter under optimized conditions (Finnigan, 2011; Paz Cedeño et al., 2025). In contrast, edible mushrooms are produced through the formation of macroscopic fruiting bodies by basidiomycetous fungi grown on solid lignocellulosic substrates, resulting in a structurally and compositionally distinct form of biomass.

Moreover, protein estimation in mushroom-based ingredients may be influenced by non-protein nitrogen originating from chitin and other nitrogen-containing compounds, necessitating careful consideration of nitrogen-to-protein conversion factors when comparing protein levels across different fungal protein sources (Scholtmeijer et al., 2023). Table 17.2 summarizes the main distinguishing characteristics between fermentation-based fungal protein production (yeasts and filamentous fungi) and mushroom cultivation, highlighting key differences in biology, production technology, protein concentration, and suitability for pet food applications.

TABLE 17.2 Comparison of fermentation-based fungal protein production and mushroom cultivation

Aspect	Fermentation (SmF / SSF)	Mushroom cultivation
Typical organisms	Yeasts (*Saccharomyces* spp.), filamentous fungi (*Fusarium*, *Rhizopus*)	Basidiomycetous macrofungi (*Pleurotus*, *Agaricus*, *Lentinula*)
Biological form harvested	Microbial biomass (cells or mycelium)	Fruiting bodies (edible mushrooms)
Main production route	Submerged fermentation (SmF) or solid-state fermentation (SSF)	Solid-substrate cultivation (SSF-like)
Growth environment	Liquid media (SmF) or moist solids (SSF)	Moist lignocellulosic substrates
Typical substrates	Sugars, starch hydrolysates, industrial by-products	Straw, sawdust, wood chips, agricultural residues
Primary product	Protein-rich biomass (SCP, mycoprotein)	Whole mushroom tissue
Protein concentration	High (40–60% dry matter)	Moderate (≈10–35% dry matter)
Production time	Hours to days	Weeks
Process control	Highly controlled bioreactors	Environmental control (humidity, temperature, CO_2, light)
Relevance to pet food	Protein concentrates, palatants, functional ingredients	Whole-food ingredient, fiber-rich protein source

These technological and biological differences also influence the compositional properties of mushroom-derived ingredients, including their protein concentration, amino acid profile, and digestibility which are examined in greater detail in the following section.

3 Nutrient composition and digestibility of mushroom proteins

Fungal proteins derived from yeasts, microfungi, and macrofungi are increasingly discussed as novel protein sources in animal nutrition due to their generally balanced amino acid profiles, functional properties and good digestibility reported for several fermentation-derived fungal products (Khan et al., 2024; Montebello et al., 2025; Wang R. et al., 2023). However, edible mushrooms (macrofungi) differ substantially from these fermentation-derived fungal proteins in terms of composition, structure, and production systems. Mushroom fruiting body biomass typically contains 8–37% protein on a dry matter basis, depending on species, cultivation substrate, growth conditions, and post-harvest processing (Bach et al., 2017). Thus, mushrooms typically contain lower protein concentrations compared to 40–60% protein on a dry matter basis from yeast and filamentous fungi (microfungi) but higher levels of structural polysaccharides such as chitin and β-glucans, as well as other bioactive compounds (Drzewiecka et al., 2024; Flores et al., 2024; Noreen et al., 2025; Pérez-Bassart et al., 2023; Singh A. et al., 2025; Zhao et al., 2023).

Consequently, while mushrooms can contribute nutritionally valuable amino acids and functional components, their use in companion animal diets should therefore be considered as a complementary protein ingredient rather than a direct replacement. Their lower protein concentrations compared with high-protein meals from traditional sources (approximately 50–90% crude protein e.g., typical animal by-product meals at the lower end and blood meal at the upper end) would require substantially higher inclusion levels to achieve equivalent dietary protein contributions.

Table 17.3 summarizes the nutritional profiles of selected edible mushrooms, including crude protein, crude fat, crude fiber, nitrogen-free extract (NFE), and ash contents on a dry matter basis. Given that more than 2,500 edible mushroom species have been described (Li et al., 2021), the table focuses on the most commonly

TABLE 17.3 Nutritional profiles of common edible mushrooms and traditional animal meals (% per dry weight)

Species	Crude protein	Crude fat	Nitrogen-free extract	Crude fiber	Ash
Macrofungi (mushrooms)					
Agaricus bisporus[1,2,3]	23.9 – 34.8	1.4 – 8.0	51.0 – 62.5	2.9 – 29.4	7.7 – 12.0
Auricularia polytricha[3]	8.5	5.2	5.9	3.5	5.2
Boletus edulis[2,3]	21.9 - 29.7	2.6 - 3.1	59.2 - 59.7	8.0 – 9.9	6.4 - 7.5
Cantharellus cibarius[2,3]	19.9 - 26.3	2.2 - 9.1	18.0 – 43.5	12.4 - 13.6	13.2 - 25.8
Flammulina velutipes[1,3]	17.6 – 19.0	1.9 – 2.0	70.5 – 73.1	3.7 – 32.1	7.4 – 8.5
Lentinus edodes[1,2,3]	13.4 – 20.7	1.3 – 8.0	59.5 – 78.0	3.8 – 41.9	3.7 – 8.8
Pleurotus ostreatus[1,3]	10.5 – 37.0	0.6 – 2.2	53.3 – 81.8	7.5 – 46.6	5.8 – 9.8
Pleurotus sajor-caju[3]	26.6	2.0	50.7	13.3	6.5
Pleurotus citrinopileatus[2]	37.6	2.2	36.3	7.0	7.9
Meat and bone meal[4]	24.5 – 55.6	3.9 – 14.4	-	1.8 – 9.6	30.0 – 33.6
Blood meal[4]*	89.7	0.8	-	-	2.8
Poultry by-product meal[5]	66.5	13.6	-	-	12.6
White fish meal[6,7,8]	53.0 – 71.9	4.0 – 10.0	-	<0.2	5.3 – 25.0
Salmon (viscera) meal[9]	74.0	11.0	-	-	8.5
Soybean meal[4,7,10]	40.0 – 54.9	1.1 – 3.5	-	3.0 – 7.0	5.4 – 7.0

Data from: [1]Bach et al., 2017; [2]Dimopoulou et al., 2022; [3]Rahi and Malik, 2016; [4]Sauvant et al., 2004; [5]Yu et al., 2023; [6]alaska pollock: FAO, 1997; [7]anchovy: Cho and Kim, 2011; [8]tuna, sardine, and herring: Lizárraga-Hernández et al., 2024; [9]Bayrakli and Yildiz, 2024; [10]Banaszkiewicz, 2011; *dried blood of slaughtered, warm-blooded animals; methods used: crude protein via AOAC No.920.87 (Kjeldahl); crude fat via soxhlet extraction; Nitrogen-free extract (NFE) calculated by difference; crude fiber (soluble and insoluble) via AOAC No.991.43 (enzymatic digestion); ash via AOAC No.923.03 (calcination at 550°C for 5h-12h); for more details, please open the corresponding original literature.

cultivated species, as well as representative wild species such as boletus (*Boletus edulis*) and chanterelle (*Cantharellus cibarius*). Reported nutrient values vary substantially depending on cultivation conditions, developmental stage at harvest, and postharvest processing, which must be considered when interpreting compositional data and applying them in pet food formulation. For comparison, Table 17.3 also includes values from traditional protein sources such as poultry, fish and soybean meal among others. Nitrogen-free extracts are not reported for these meals because carbohydrate concentrations in these ingredients are generally negligible.

4 Protein content and amino acid profile

While yeast and filamentous fungal biomasses produced via fermentation typically yield around 40% crude protein or more on a dry matter basis, edible mushroom fruiting bodies typically contain approximately 8%–37% crude protein (Table 17.3). However, crude protein values in mushrooms are commonly calculated from total nitrogen using conventional nitrogen-to-protein conversion factors (i.e. 6.25). This approach can overestimate nutritionally available protein (i.e., nitrogen present in amino acids) because mushrooms contain substantial amounts of non-protein nitrogen, notably from chitin and other nitrogenous cell wall components. Several studies therefore propose lower nitrogen-to-protein conversion factors for mushrooms, typically around 4.4 instead of the conventional factor of 6.25. Applying the adjusted factor may reduce estimated protein values by approximately 30%, depending on species and analytical assumptions (Mariotti et al., 2008).

True nitrogen digestibility (TND), historically determined by feeding trials and fecal nitrogen excretion in rats, provides a preliminary estimate of the nutritionally available protein fraction in mushrooms. Reported TND values for edible mushrooms vary considerably depending on species and processing conditions (e.g., *Tricholoma terreum* 52.6%, *Terfezia claveryi* 61.4%, *Pleurotus ostreatus* 73.4%, and *Agaricus macrosporus* 80.5%; Dabbour and Takruri, 2002). However, digestibility values based on fecal nitrogen must be interpreted cautiously, because part of the nitrogen that escapes digestion in the small intestine can be metabolized by microbial fermentation in the large intestine. The absorption of microbial nitrogenous metabolites may therefore lead to an overestimation of true nitrogen digestibility when assessed at the fecal level. An alternative and often more precise approach is the quantification of the ileal digestibility of individual amino acids, where the sum of amino acids approximates the true protein fraction and avoids overestimation caused by non-protein nitrogen compounds such as chitin.

Processing strategies such as fermentation, enzymatic pretreatment, or mechanical disruption have been shown to improve protein accessibility and digestibility in mushroom-derived ingredients. For example, enzymatic or fermentation-based

TABLE 17.4 Range of essential amino acid (EAA) profiles of selected edible mushrooms (g/100 g dry weight; n = species count)

Amino acid	Black Sea region (Turkey; n = 11)[1]	Yunnan province (China, n = 13)[2]	Global review (n = 10)[3]	Paraná state (Brazil, n = 9)[4]	Kastamonu province (Turkey, n = 5)[5]
Arginine	0.6 – 2.1	0.01 – 0.7	-	1.4 – 2.6	0.5 – 1.5
Histidine	0.3 – 2.5	0.03 – 0.4	-	0.4 – 1.1	0.2 – 0.5
Isoleucine	0.6 – 1.4	0.03 – 0.7	-	0.6 – 1.3	0.4 – 0.6
Leucine	0.9 – 1.8	0.03 – 1.2	1.1 – 16.3	0.9 – 2.4	0.6 – 1.1
Lysine	0.5 – 1.4	0.01 – 0.5	0.7 – 9.9	0.9 – 1.9	0.6 – 1.0
Methionine	0.1 – 0.9	0.01 – 0.2	0.3 – 3.5	0.3 – 0.6	0.06 – 1.1
Met + Cystine	0.2 – 1.3	0.07 – 1.1	-	0.9 – 1.8	0.2 – 2.0
Phenylalanine	0.4 – 1.0	0.02 – 0.8	0.7 – 7.0	0.7 – 1.4	0.5 – 0.7
Phe + Tyrosine	0.6 – 1.6	0.07 – 1.6	-	1.4 – 2.8	0.9 – 1.3
Threonine	0.5 – 1.3	0.03 – 0.7	1.0 – 9.5	0.8 – 1.7	0.4 – 0.8
Tryptophan	0.1 – 0.5	0.02 – 0.2	0.1 – 2.0	0.2 – 0.6	-
Valine	0.4 – 1.0	0.1 – 1.0	0.9 – 8.9	0.9 – 1.9	0.7 – 1.1

Data include mushrooms from different global regions; ranges illustrate natural variability; Sources: [1]Ayaz et al., 2011; [2]Sun L. et al., 2017; [3]Rahi and Malik, 2016; [4]Bach et al., 2017; [5]Bakir et al., 2018.

treatments increased true nitrogen digestibility by approximately 10–25%, depending on species and processing conditions (Wang D. et al., 2024). While most studies report digestibility at the nitrogen level, these interventions are also expected to enhance amino acid availability, although further ileal studies are needed to confirm this effect in companion animals.

The AA composition of edible mushrooms generally includes all essential amino acids (EAA's), though their relative proportions vary among species. Table 17.4 summarizes ranges of EAA's in edible mushrooms reported across multiple reviews in different geographic regions such as Brazil, China, and Turkey. Values are expressed per 100 g dry matter because several of the underlying studies did not report protein contents required for conversion to AA concentrations per unit of protein. Therefore, this table primarily illustrates the presence and overall magnitude of EAA's across mushroom species and geographic regions rather than allowing direct comparison of protein quality. For detailed AA levels, a closer look into the original literature is suggested.

However, for nutritional comparison with conventional protein ingredients used in pet food formulations, Table 17.5 presents essential amino acid profiles of selected mushroom species (*Agaricus bisporus*, *Flammulina velutipes*, *Lentinula edo-*

TABLE 17.5 Essential amino acid (EAA) composition of selected edible mushrooms (g/100 g crude protein)

Amino acid	*Agaricus bisporus* (Champignon / Portobello)	*Flammulina velutipes* (Velvet foot)	*Lentiuns edodes* (Shiitake)	*Pleurotus djamor* (Pink oyster)	*Pleurotus eryngii* (King oyster)	*Pleurotus ostreatus* (white / black oyster)
Arginine	1.41 / 1.76	1.71	1.38	1.64	1.95	1.41 / 2.58
Histidine	0.69 / 0.77	0.58	0.52	0.59	0.41	0.69 / 1.05
Isoleucine	1.26 / 1.21	0.77	0.88	0.88	0.61	0.85 / 1.27
Leucine	1.87 / 2.02	1.1	1.28	1.46	0.87	1.41 / 2.09
Lysine	1.52 / 1.69	1.43	1.14	1.22	0.87	1.23 / 1.74
Methionine	0.59 / 0.62	0.34	0.37	0.45	0.29	0.37 / 0.62
Met + Cystine	1.49 / 1.44	1.22	1.58	1.48	0.92	1.32 / 1.75
Phenylalanine	1.21 / 1.32	1.02	0.91	0.94	0.73	1.05 / 1.41
Phe + Tyrosine	2.22 / 1.35	2.0	1.72	2.29	1.36	1.97 / 2.74
Threonine	1.50 / 1.50	1.01	1.11	1.14	0.84	1.29 / 1.73
Tryptophan	0.50 / 0.56	0.26	0.30	0.30	0.19	0.24 / 0.32
Valine	1.58 / 1.58	1.08	1.20	1.21	0.87	1.36 / 1.85

Original data are reported in g/100 g dry matter. For better comparability, data were converted into g/100 g protein using reported protein content data: *A. bisporus*: 27.0% (champignon) and 29.8% (portobello), *F. velutipes*: 19.0%, *L. edodes*: 18,9%, *P. djamor*: 22.5%, *P. eryngii*: 16.5%, *P. ostreatus*: 22.5% (white oyster) and 37.0% (black oyster); source: Bach et al., 2017.

TABLE 17.6 Essential amino acid (EAA) composition of selected conventional protein ingredients (g/100 g crude protein)

Amino acid	Soybean meal[1]	Meat and bone meal (pork)[2]	Poultry meal[3]	Fish meal[1]
Arginine	3.5	3.6	5.6	3.8
Histidine	1.3	1.3	2.4	1.6
Isoleucine	2.2	2.0	4.2	2.5
Leucine	3.5	3.8	5.7	4.5
Lysine	2.9	3.3	6.4	4.6
Methionine	0.7	1.0	2.4	1.6
Met + Cystine	1.4	1.6	3.4	2.2
Phenylalanine	2.4	2.1	4.3	2.4
Phe + Tyrosine	3.9	3.5	8.1	4.3
Threonine	1.8	1.9	3.7	2.6
Tryptophan	0.6	0.5	-	0.6
Valine	2.3	2.7	4.5	3.0

Source: [1]Sauvant et al., 2004; [2]Yu et al., 2023; [3]Bolarinwa et al., 2012.

des, *Pleurotus eryngii*, and *Pleurotus ostreatus*), expressed per 100 g protein. These species represent global cultivation importance and commercial availability as they account for a substantial proportion of global mushroom production and are therefore the most relevant candidates for potential use in pet food applications. Furthermore, Table 17.6 presents EAA of traditional protein sources (soybean meal, pork meat and bone meal, poultry meal, and fish meal). This approach allows a more meaningful evaluation of amino acid balance relative to other dietary protein sources. In comparison to the traditional protein sources most essential amino acids in mushrooms occur at lower concentrations (Tables 17.5 and 17.6). In particular, sulfur-containing amino acids such as methionine (and methionine + cystine) as well as tryptophan are present at relatively low levels and may represent limiting amino acids in formulations relying heavily on mushroom-derived proteins. In contrast, amino acids such as arginine, threonine, and valine occur in moderate proportions relative to the other mushroom amino acids, although still below the concentrations typically observed in protein meals from conventional sources.

To assess the nutritional adequacy of mushroom-derived proteins relative to companion animal requirements, EAA ratios were calculated for *Agaricus bisporus*, soybean meal, and poultry by-product meal using the minimum amino acid requirements reported by the National Research Council (NRC, 2006; Table 17.7). For cats, values represent either minimum requirements (MR) or adequate intake (AI). Dog and cat values for each AA are shown in the same cell of the respective AA, cat values shown in parentheses. NRC values originally expressed as g/1000 kcal metabolizable energy (ME) were converted to g/100 g crude protein according to Equation 1 using the NRC reference protein intakes of 20 g crude protein per 1000 kcal ME for adult dogs and 40 g crude protein per 1000 kcal ME for adult cats.

$$\text{AA requirement (g/100 g protein)} = \frac{\text{AA requirement(g/100 g dry matter)}}{\text{crude protein per 1000 kcal ME (g)}} \qquad (1)$$

AA ratios (%) were subsequently calculated according to Equation 2 to illustrate the extent to which the evaluated protein sources meet species-specific amino acid requirements.

$$\text{AA ratio} = \frac{\text{AA content (g/100 g protein)}}{\text{AA requirement (g/100 g protein)}} \qquad (2)$$

Ratios below 1 indicate that the respective requirement is not met by the individual protein source when considered in isolation, whereas ratios equal to or greater than 1 indicate that the requirement is satisfied. The resulting ratios shown in Table 17.7 indicate that proteins from *Agaricus bisporus* remain below the minimum amino acid requirements for adult dogs for essential amino acids. For cats, several amino acids approach the requirement levels, particularly lysine and methionine + cystine, but remain limiting overall compared with conventional protein sources.

TABLE 17.7 Essential amino acid requirements for adult dogs and cats (cats value in parenthesis), expressed as g/100 g crude protein and calculated amino acid ratios for *Agaricus bisporus* (Champignon), soybean meal and poultry by-product meal according to Equation 2

Amino acid	**Requirement (g/100g protein)**[1]	**AA ratio for Agaricus bisporus**[2]	**AA ratio for soybean meal**[3]	**AA ratio for poultry by-product meal**[4]
Arginine	3.50 (4.83*)	0.40 (0.29)	1.00 (0.73)	1.60 (1.16)
Histidine	1.85 (1.63*)	0.37 (0.42)	0.70 (0.80)	1.30 (1.47)
Isoleucine	3.75 (2.70*)	0.34 (0.47)	0.59 (0.81)	1.12 (1.56)
Leucine	6.75 (6.38*)	0.28 (0.29)	0.52 (0.55)	0.84 (0.89)
Lysine	3.50 (1.70)	0.43 (0.89)	0.83 (1.71)	1.83 (3.76)
Methionine	3.25 (0.85)	0.18 (0.69)	0.22 (0.82)	0.74 (2.82)
Met + Cystine	6.50 (1.70)	0.23 (0.88)	0.22 (0.82)	0.52 (2.00)
Phenylalanine	4.50 (2.50*)	0.27 (0.48)	0.53 (0.96)	0.96 (1.72)
Phe + Tyrosine	7.40 (9.58*)	0.30 (0.23)	0.53 (0.41)	1.10 (0.85)
Threonine	4.25 (3.25*)	0.35 (0.46)	0.42 (0.55)	0.87 (1.14)
Tryptophan	1.40 (0.83*)	0.36 (0.60)	0.43 (0.72)	- (-)
Valine	4.90 (3.20*)	0.32 (0.49)	0.47 (0.72)	0.92 (1.41)

Source: [1]NRC, 2006; [2]Bach et al., 2017; [3]Sauvant et al., 2004; [4]Bolarinwa et al., 2012. NRC amino acid requirements originally expressed as g/1000 kcal metabolizable energy were converted according to Equation 1 to g/100 g crude protein assuming reference protein intakes (adult dog: 20 g crude protein/1000 kcal ME, adult cat: 40 g crude protein/1000 kcal ME). For cats adequate intake (*) replace minimum requirement where data were not available. Amino acid ratios (%) were calculated according to Equation 2.

These results suggest that mushroom proteins can contribute to essential amino acids but are more appropriately considered complementary protein ingredients within balanced companion animal diets.

5 Lipids, carbohydrates, and fiber fractions

Although mushrooms are not lipid-rich compared with animal tissues (approximately 1–5% of dry matter), the lipid fraction is usually dominated by unsaturated fatty acids (UFAs) such as linoleic and oleic acids. Alpha-linolenic acid (ALA, C18:3ω-3) occurs at lower concentrations, and long-chain polyunsaturated fatty acids (LCPUFAs) such as arachidonic acid (ARA, C20:4ω-6), eicosapentaenoic acid (EPA, C20:5ω-3) and docosahexaenoic acid (DHA, C20:6ω-3) are generally absent or present only in trace amounts in macrofungi (Table 17.8). Table 17.8 summarizes from multiple reviews ranges of fatty acid (FAs)s in edible mushrooms globally or from specific geographic regions such as Korea and Turkey. Values are expressed

TABLE 17.8 Fatty acid profiles of selected edible mushroom species (% of total fatty acids)

Fatty acid	Global review (n = 55)[1]	Global review (n = 19)[2]	Turkey (n = 16)[3]	Korea (n = 15)[4*]
Myristic (C14:0)	not given	0.0–2.70	not given	1.0–17.0
Palmitic (C16:0)	not given	0.0–21.3	0.0–23.3	44.5–557.8
Palmitoleic (C16:1)	not given	0.0–12.9	not given	2.62–118.7
Stearic (C18:0)	not given	0.0–25.33	0.0–13.6	13.8–247.6
Oleic (C18:1)	0.0–58.8	0.0–58.4	4.2–48.7	7.7–257.0
Linoleic (C18:2ω-6) LA	0.0–76.4	3.0–75.6	26.5–67.6	326.2–1385.2
α-Linolenic (C18:3ω-3) ALA	0.0–20.3	0.0–13.0	0.0–19.0	0.6–507.1
Arachidic (C20:0)	not given	0.0–3.16	not given	0.0–35.6
Eicosenic (C20:1)	not given	0.0–0.8	not given	not given
ΣSFA	not given	0.0–38.3	0.0–38.3	94.1–884.7
ΣMUFA	0.0–58.8	0.0–66.5	0.0–48.9	25.8–320.7
ΣPUFA	0.0–76.4	0.0–76.1	26.5–67.6	336.1–1394.8

Sum (Σ) SFA: saturated fatty acid, MUFA: monounsaturated fatty acid, PUFA: polyunsaturated fatty acid; * values in mg/100 g of dry weight – not directly comparable; sources: [1]Sande et al. 2019; [2]Tan et al., 2024; [3]Bulam et al., 2021; [4]Saini et al., 2021; because only selected FAs are presented, the sum of SFA + MUFA + PUFA ≠ 100.

per % of total fatty acids except the study about mushrooms from Korea, where, because of missing information about the fat content in dry matter, data are presented in mg/100 g dry weight. This table primarily illustrates the presence and overall magnitude of fatty acids across diverse mushroom species and geographic regions rather than allowing direct comparison of lipid quality. For detailed fatty acid levels, a closer look into the original literature is suggested.

However, Table 17.9 shows that the fat in specific mushroom species (*Agaricus bisporus*, *Boletus edulis*, *Lentinula edodes*, *Pleurotus ostreatus*, and *Pleurotus sajor-caju*) contain relatively high proportions of mono- and polyunsaturated fatty acids (MUFA and PUFA) as a percentage of total lipids, with oleic acid (OA, C18:1) and linoleic acid (LA, C18:2ω-6) often exceeding 50% of total fatty acids (Sande et al., 2019). Consequently, while mushrooms can support baseline ω-6 fatty acid supply, they cannot replace animal or marine lipid sources required to meet feline ARA needs or to deliver physiologically relevant levels of marine ω-3 LC-PUFAs during growth and reproduction.

Mushrooms are a non-animal source of vitamin D, containing ergosterol, which can be converted to vitamin D_2 upon UV exposure (Boro et al., 2025; Phillips et al., 2011). However, dogs and cats rely primarily on dietary vitamin D_3 (cholecalciferol) to meet their metabolic requirements, and the capacity to utilize ergosterol-derived vitamin D_2 has not been clearly demonstrated in companion animals.

TABLE 17.9 Essential fatty acid (EFA) composition of selected edible mushrooms (% of total fatty acids)

Fatty acid	*Agaricus bisporus* (n = 4)	*Boletus edulis* (n = 7)	*Lentinula edodes*	*Pleurotus ostreatus* (n = 8)	*Pleurotus sajor-caju* (n = 4)
Oleic (C18:1) OA	6.1–55.0	0.0–31.1	3.5	4.8–27.0	16.4–41.7
Linoleic (C18:2ω-6) LA	3.3–67.3	0.0–75.6	75.8	4.3–76.5	0.0–53.8
α-Linolenic (C18:3ω-3) ALA	1.5–4.6	0.0–1.7	0.3	0.0–6.6	0.0–11.7
ΣMUFA	6.1–55.0	0.0–31.1	3.5	4.8–27.0	16.4–41.7
ΣPUFA	4.8–71.9	0.0–77.1	76.1	4.3–83.1	0.0–65.5

Sum (Σ) MUFA: monounsaturated fatty acid, PUFA: polyunsaturated fatty acid; Sources: Sande et al., 2019; because only selected FAs are presented, the sum of MUFA + PUFA ≠ 100.

Carbohydrates represent a major fraction of mushroom dry matter, accounting for approximately 20–40%, and include both structural and storage polysaccharides such as β-glucan, chitin, glycogen, trehalose, and mannitol. β-glucan and chitin function as dietary fibers and have attracted particular interest due to their potential effects on gut health.

6 Technical functionality and bioactive properties of mushroom proteins in pet nutrition

Beyond their nutritional value, mushroom proteins offer technical functionality and bioactive properties that can improve product quality, animal health, and feed performance. Technical functionality refers to the properties of an ingredient that affect how it behaves during processing and in the final product, such as water binding, texture formation, emulsification, or stability. The technical functionality of fungal proteins such as water-holding capacity, gelation, emulsification, and flavor contribution, varies substantially among molds, macrofungi (mushrooms), and yeasts. These technical functionality properties are particularly relevant in wet and semi-moist pet foods, where protein functionality plays a key role in stabilizing fat–water emulsions, retaining moisture, and determining the texture and cohesiveness of chunks, pâté, or gravy systems during thermal processing and storage (Narsipur et al., 2024; Rodríguez et al., 2023; Wang Z. et al., 2025). In such high-moisture formulations, proteins and binders contribute to water retention, gelling, and emulsion stability, which are essential for maintaining product structure and preventing separation of fat and aqueous phases during retort sterilization and shelf life (Rivera et al., 2000; Vasconcellos et al., 2023). Mushroom proteins, particularly from oyster mushrooms (*Pleurotus* spp.), have demonstrated notable techno-functional properties, including water-holding capacities of approximately 6.5 g/g, oil-holding capacities of 8.6 g/g, emulsifying activity indices of around 2.1 m^2/g, and foaming capacities exceeding 50 %. These effects are likely due to enzymatic hydrolysis, which exposes functional groups and alters protein structure (Khapudang and Siddiqui, 2026). In another study, mushroom protein-isolates from *Lentinula edodes*, *Flammulina velutipes*, and *Pleurotus* spp. exhibit enhanced emulsifying and foaming properties, which were associated with a higher proportion of β-sheet structures. Beta-sheets are extended protein secondary structures formed by inter-strand hydrogen bonding between polypeptide chains, and in its context, they may influence the technofunctional properties of mushroom protein during processing (Qian et al., 2025). These properties can enhance stability and interaction in complex matrices due to the surface activity and interfacial properties of mushroom polysaccharides, supporting their use in emulsified wet foods alongside traditional protein sources (Zhang S. et al., 2023).

Bioactive compounds are molecules that exert physiological effects beyond basic nutrition, such as modulating gut health, immune function, metabolism, and skin and coat condition (Ma et al., 2018). These properties stem from the unique cell wall structure (chitin, β-glucan, and mannans; Bach et al., 2017), polysaccharide–protein interactions (β-glucan-protein-complexes and fungal immunomodulatory proteins (FIPs); Singh A. et al., 2025), and secondary metabolites found in different macrofungi. Beta-glucans are major constituents of fungal cell walls, including mushrooms and yeast. Yeast-derived β-(1,3)/(1,6)-glucans have been shown to modulate innate and adaptive immune responses in dogs (Fries-Craft et al., 2023; Paris et al., 2020). While mushrooms contain β-glucans with similar backbone structures, evidence for their immunomodulatory effects in dogs is currently limited and may depend on extraction, solubility, and branching patterns. Chitin and chitosan fractions further aid in gut barrier function and may have mild anti-inflammatory and hypocholesterolemic effects (Edo et al., 2025). Secondary metabolites such as phenolic compounds show antioxidant activity, terpenoids are reported to have anti-inflammatory and antimicrobial effects, and sterols, more specifically ergosterol, are precursors of vitamin D2 (El-Ramady et al., 2022; Kumar et al., 2021; Moniruzzaman et al., 2025; Phillips et al., 2011; Sun X. et al., 2024).

Collectively, these features support the emerging concept of "functional fungal proteins", ingredients that provide both high nutritional value and measurable physiological benefits for companion animals. From a formulation perspective, mushroom-derived ingredients can be seen as multifunctional agents rather than simple protein sources. Their inclusion in functional pet foods may support immune and digestive health via natural β-glucans and mannan-oligosaccharides (MOS), while also supplying bioactive compounds that benefit dermatological and gastrointestinal health.

It should be noted that most evidence on the biofunctional and health-promoting properties of edible macrofungi comes from in vitro studies, rodent models, and human clinical and nutritional research. Although many mechanistic pathways like immunomodulation, antioxidant activity, gut microbiota modulation, and metabolic regulation are largely conserved among mammals, direct extrapolation to dogs and cats should be approached cautiously. Variations in digestive physiology, nutrient metabolism, effective dosages, and long-term safety parameters highlight the need for targeted research in companion animals. Thus, the functional interpretations in this chapter should be considered translationally informed rather than fully species-validated, emphasizing both the innovation potential and existing research gaps in fungal applications for pet nutrition.

7 Sustainability and resource efficiency

Environmental sustainability of protein ingredients can be evaluated using multiple life-cycle assessment (LCA) indicators, including land use, water use, energy demand, and greenhouse gas emissions. However, results for land and water footprints often vary substantially between studies due to differences in system boundaries, allocation methods, geographic production conditions, and processing technologies. For aquatic resources such as fishmeal, land use is largely associated with fuel production or feed ingredients rather than the fishing activity itself, while water consumption depends strongly on processing systems and regional water availability. Consequently, direct comparison of land and water footprints across heterogeneous production systems is challenging. In many comparative assessments of food and feed proteins, global warming potential (GWP) or greenhouse gas (GHG) emissions are used as primary indicators because they are generally reported more consistently across LCA studies and allows a clearer comparison of relative environmental impacts. Fungal protein production, spanning filamentous microfungi and mushroom-forming macrofungi, is usually associated with high resource-efficacy compared with conventional livestock-based proteins (Zhang Z. et al., 2024; Table 17.10). However, in many cases this comparison reflects fungal protein production being evaluated against meat production for human consumption where animals are raised and slaughtered specifically for meat. In contrast, animal proteins used in pet food are typically generated through rendering processes that utilize by-products from livestock slaughter. From this perspective, the life cycle assessments for pet food protein ingredient should begin with the collection of slaughterhouse by-products, placing the rendering industry within the circular bioeconomy framework where waste streams are upcycled into nutritionally valuable ingredients under relatively efficient conditions. A similar situation exists for soybean meal in the feed industry, which is typically considered a co-product of soybean oil production. Consequently, soybean meal also represents a valuable component of circular bioeconomic systems through the utilization of agricultural co-products.

Macrofungal systems, though more resource-intensive than yeast and filamentous fungi due to solid-state growth and longer cycles, convert lignocellulosic residues such as straw, sawdust, or coffee grounds into protein-rich biomass with minimal inputs of water, arable land, and energy (Zhang Z. et al., 2024). Mushroom cultivation also exemplifies circular bioeconomy principles by converting low-value organic residues into protein-rich biomass with a relatively low ecological footprint, addressing both the nutritional and environmental challenges of modern food and pet food production (Amara and El-Bakey, 2023; Boro et al., 2025; Majumder et al., 2024; Scholtmeijer et al., 2023). This bioconversion aligns with European Union waste-reduction targets and FAO circular economy guidelines (EU Commission, 2025). However, one of the unique sustainability advantages of macrofungal production lies in co-product utilization. Spent mushroom substrate (SMS) typically amounts to 4–5 kg per kg of harvested mushrooms and contains residual mycelium and nutrients. It can be revalorized as animal feed to replace up to 20% of conventional ruminant or monogastric feed without adverse effects on performance, or as a growth substrate for other fungi, closing nutrient loops and reducing organic waste (Leong et al., 2022; Martín et al., 2023; Mbambalala et al., 2026; Ravlikovsky et al., 2024).

The metrics presented in Table 17.10 briefly illustrate global warming potential (GWP) as an environmental indicator for selected protein ingredients relevant for pet food production (poultry by-product meal, fish meal and soy meal) and compares them to GWP data of selected but not representative filamentous microfungi (mycoprotein) production and mushroom fruiting bodies from a mushroom farm. Insect meal is included into the table as another representative of a novel protein source. GWP values represent literature ranges normalized per kg protein where possible. The GWP usually summarizes the energy demand for the generation of electricity and steam for cooking (poultry and fish rendering), frying (insect) and drying processes, the production of fuel for transportation and delivery purposes as well as production steps such as cultivation, cleaning and packaging just to name a few. Energy demand represents a major contribution to GWP values and consideration of renewable energy sources compared to fossil ones or biodiesel compared to conventional fuel can have a huge impact on the outcome of a LCA. Therefore, comparisons in this table should be interpreted cautiously and are most appropriate when evaluating individual parameters reported under comparable methodological conditions.

The LCA of the poultry by-product meal was conducted in two different rendering units in Portugal (Campos et al., 2020), the LCA of the fish meal was done in three Peruvian fish meal plants (Fréon et al., 2017) and for the soybean meal several production regions in the USA (Adom et al., 2012) were considered. The LCAs of insect meals were conducted in Italy (black soldier fly/*Hermetia illucens*; Salomone et al., 2017) and The Netherlands (mealworm/*Tenebrio molitor*; Oonincx and de Boer, 2012) and several LCAs from mycoprotein as representative for filamentous micro-

TABLE 17.10 Reported environmental performance indicators of selected protein-rich production systems

Protein source	Protein content (% dry matter)	GWP (kg CO_2 eq./kg protein)	Production time
Poultry by-product meal[1]	60–70	1.1 (rendering)	weeks (broiler cycle)
Fish meal[2]	60–70	4.0–15.0	weeks–months (fish harvest)
Soybean meal[3]	30–45	1–1.5[#]	4–6 months (crop growing season)
Insect meal (black soldier fly[4], mealworm[5])	40–60	2.0–5.0	1–2 weeks
Filamentous fungi[6]	40–60	4.5–6.8[‡]	2–5 days
Mushroom fruiting body[7]	10–40	7.5–30*	weeks

Data from: [1]Campos et al., 2020; [2]Fréon et al., 2017; [3]Adom et al., 2012; [4]Salomone et al., 2017; [5]Oonincx and de Boer, 2012; [6]Shahid et al., 2024; [7]Dorr et al., 2021; [#]reported as average 0.46 g CO_2 eq./kg dry feed converted into per kg protein using 30–45% protein content; [‡]reported as average 2.7 kg CO_2 eq./kg product converted into per kg protein using 40–60% protein content; *reported as 3 kg CO_2 eq./kg fresh mushroom converted into per kg protein using 10–40% protein content. GWP values originate from published life cycle assessment studies and may differ in methodological assumptions, including allocation procedures used to distribute environmental impacts among co-products.

fungi (Shahid et al., 2024). In case of an oyster mushroom (*Pleurotus ostreatus*; Dorr et al. 2021) farm in France supplying regional markets, the calculated GWP per kg protein appears highest in this comparison, primarily due to the lower protein concentration of mushrooms relative to the other protein sources.

8 Safety, quality control, and regulatory considerations of mushroom-derived ingredients

The industrial cultivation and use of edible macrofungi as food and feed ingredients are generally regarded as safe when established species and controlled production systems are employed. However, their incorporation into pet food requires careful consideration of regulatory compliance, quality control, and toxicological aspects. In the European Union, mushroom-derived ingredients are regulated under general food and feed safety legislation, and those that are novel in terms of species, processing, or intended use may require authorization under the Novel Food Regulation (EU) 2015/2283 (EU Parliament, 2015). Safety assessments conducted by the European Food Safety Authority (EFSA) typically evaluate taxonomic identity, production process, compositional consistency, nutritional impact, potential allergenicity, and toxicological endpoints. EFSA assessments have demonstrated that mushroom-derived ingredients, including vitamin D_2-enriched mushroom powders produced via UV treatment, are safe under defined conditions of use, provided that compositional specifications, contaminant limits, and microbiological quality criteria are met (EFSA, 2021). In the United States, edible mushrooms and mushroom mycelium are considered suitable for food and feed use when produced on acceptable substrates and labeled accurately, with oversight by the FDA and state-level feed control authorities often referencing AAFCO ingredient definitions for pet food applications.

From a quality-control perspective, industrial mushroom production must adhere to good agricultural and manufacturing practices, including substrate control, hygiene management, and traceability of raw materials. Because mushrooms can accumulate heavy metals or other environmental contaminants from their growth substrates, routine monitoring for elements such as cadmium, lead, and arsenic is essential. Microbiological safety is also critical, particularly for dried or powdered mushroom ingredients, requiring verification of the absence of pathogens such as *Salmonella* spp. and *Listeria monocytogenes*. In addition, compositional standardization, including protein content, fiber fractions (β-glucan and chitin), moisture, and bioactive compounds is increasingly important for consistent nutritional performance in pet food formulations. Analytical methods commonly applied include nitrogen analysis with adjusted conversion factors, enzymatic or chromatographic quantification of polysaccharides, and spectroscopic or chromatographic techniques for sterols and phenolic compounds.

Toxicological considerations for mushroom-based ingredients primarily relate to species authentication, contaminant control, and processing-related changes rather than intrinsic toxicity, as edible macrofungi have a long history of safe consumption. Unlike some filamentous fungi, cultivated mushrooms are not known to produce mycotoxins under commercial cultivation; nevertheless, cross-contamination risks and substrate quality must be managed within hazard analysis and preventive control systems. At present, neither EFSA nor the FDA has established specific maximum inclusion levels for edible mushrooms in pet food, and their use is evaluated within the broader framework of ingredient safety, nutritional adequacy, and labeling accuracy. Consequently, mushroom-derived proteins are best positioned as complementary ingredients that combine nutritional, technofunctional, and bioactive properties, provided that standardized production, robust quality control, and regulatory compliance are ensured throughout the value chain.

While mushroom proteins are generally safe, species-specific sensitivities must be considered. Certain wild mushrooms, edible for humans, have caused adverse effects in cats due to metabolic differences. In a case report, a 3-month-old female domestic shorthair ate several raw mushrooms of the species *Boletus edulis*, *Boletus aereus*, and *Amanita caesarea* (Liatis et al., 2018). Even if these mushrooms are edible for humans, the cat experienced vomiting, hyperventilation, horizontal head oscillation, and limb muscle tremors within the first 6 hours after consumption. After two days, the cat was admitted with more severe symptoms and remained hospitalized for 5 days. The kitten recovered completely a week later following treatment with intravenous fluids and vitamin E and was still healthy 8 months later.

In vivo studies in companion animals, such as dogs, indicate good tolerance and no adverse effects on health, serum biochemistry, or macronutrient digestibility when fed diets supplemented with spent mushroom substrate (SMS; Kayser et al., 2024). Collectively, these findings support mushroom proteins, when properly manufactured, as a nutritionally safe and environmentally sustainable alternative to conventional animal proteins in pet food formulations, with additional benefits from by-product valorization and circular economy integration.

9 Recent development and application in pet food

While dedicated clinical studies on mushroom protein inclusion in complete pet food formulations are limited, several reports and industry implementations show growing interest and practical use. Protein from shiitake (*Lentinula edodes*) mushroom was evaluated as a palatant in dog food formulations, highlighting its potential as a novel protein and flavor enhancer in commercial diets (Feng et al., 2020). Initially, a palatant made from shiitake mushroom as well as another made from mealworm (*Tenebrio molitor*) did not show higher palatability to dogs compared to the industry benchmark made from chicken liver. However, because of their intensive HS-SPME/GC-MS and SDE/GC-MS (Headspace-Solid Phase Microextraction Gas Chromatography-Mass Spectrometry and Simultaneous Distillation/Gas Chromatography-Mass Spectrometry) analysis of all three palatants, key flavors and volatile compounds affecting palatability such as 2-methyl-3-furyl disulfide, indole, methional, 2(methyl thio)phenol (among others) were identified. Enriching the mushroom and mealworm palatants with these specific aroma molecules significantly improved their palatability to dogs.

A very recent reprint of a veterinary study has reported cognitive benefits from a medicinal mushroom supplement blend made of Lion's Mane (*Hericium erinaceus*), Reishi (*Ganoderma lucidum*) and *Cordyceps* sp. in senior dogs (Ehrenzweig and Easler, 2025). In this proof-of-concept clinical trial, a mushroom-derived nutraceutical was evaluated in client-owned dogs diagnosed with Canine Cognitive Decline (CCD). CCD is an age-related neurodegenerative condition in dogs and presents clinical signs closely resembling those of human Alzheimer's disease. Of the 22 dogs that participated in the study, 13 dogs (57%) showed improvements in cognitive scores, 5 dogs (23%) remained stable (a favorable outcome in a progressive condition) and 4 dogs (18%) declined. "The greatest benefit was seen in dogs with moderate CCD, representing the optimal therapeutic window" (Ehrenzweig and Easler, 2025). The mushroom supplement was well tolerated, easily administered and palatable. No supplement-related adverse events were reported.

The industry is currently utilizing mushrooms such as lion's mane as functional ingredients by incorporating them into pet products for gut health and wellness applications or as separate supplements. Commercial extracts (e.g., β-glucan and triterpene blends) are already marketed to support immune and cardiovascular health in companion animals, reflecting early adoption of mushroom-derived compounds in the pet food sector. A growing but still limited body of peer-reviewed research supports the inclusion of mushroom ingredients in companion animal nutrition. For example, Shiitake mushroom powder supplementation increased antioxidative enzyme activity and modulated lipid metabolism in dogs in a controlled trial (Kusaba and Arai, 2024). Healthy dogs were supplemented for 4 weeks with Shiitake powder (800 mg/kg body weight) and showed a significant increase in plasma superoxide dismutase (SOD) activity and elevated leukocyte SIRT1 mRNA expression, while plasma total cholesterol decreased. SOD is a marker of antioxidative capacity, therefore direct evidence supporting antioxidative enzyme activity and modulated lipid metabolism in dogs. No negative effects, such as increase in body weight or body condition score, were observed during and after the supplementation.

Another study aimed to evaluate the effects of dietary inclusion of Reishi on peripheral blood mononuclear cells (PBMC; T cells, B cells, monocytes, and natural killers), vaccine response, nutrient digestibility, fecal fermentative end-products, and skin and coat quality of adult dogs (Kayser et al., 2024). Forty adult beagles were used in a completely randomized study and were fed over 28 days a commercial dry extruded diet plus Reishi supplement topping. The supplementation was well accepted and no detrimental effects on macronutrient apparent total tract digestibility were observed. The study results indicate immunomodulatory effects when provided at a daily dose of 15 mg/kg body weight.

There are multiple commercial mushroom-based supplements developed for companion animals. Examples include mushroom bioactives such as β-glucans and triterpenes, targeting respiratory and cardiovascular, immune, cognitive, and neurological, as well as gut health and microbiome systems. These products illustrate how medicinal mushroom derivatives are already being marketed in the pet industry, mainly based on general evidence for bioactive compounds but more and more clinical validation is ongoing. This highlights the need for more targeted trials on digestibility, health outcomes, and species-specific effects to fully establish therapeutic claims and optimal inclusion levels in pet foods. Another research gap is the fact that direct inclusion of mushroom-derived proteins to fully or partially replace animal proteins in complete diets remains in early stages of development. Industry adoption is growing but not yet mainstream for complete mushroom-derived protein inclusion.

10 Conclusion and future perspectives

Edible macrofungi represent a distinct and increasingly relevant category of novel proteins for companion animal nutrition. Unlike fermentation-derived fungal biomasses, mushrooms are produced through solid-substrate cultivation systems that can utilize lignocellulosic by-products, aligning their production with circular bioeconomy principles. Although their crude protein concentrations are generally lower than those of yeast or filamentous fungal proteins, mushroom-derived ingredients provide moderate amounts of amino acid and contain structural carbohydrates such as β-glucans and chitin as well as other fungal metabolites that may contribute functional properties.

From a formulation perspective, mushrooms may serve as multifunctional ingredients that can influence texture, water-binding capacity, and product structure in certain pet food matrices. Components such as β-glucans and chitin have been associated with immunomodulatory and gastrointestinal effects in other species and experimental systems. However, evidence for specific health benefits in dogs and cats remain limited and requires further investigation. Similarly, available digestibility data are scarce and indicate that fiber fractions, including chitin, may influence apparent nitrogen digestibility, highlighting the need for additional studies on amino acid digestibility and processing strategies.

Mushroom cultivation is often discussed within sustainability frameworks due to its ability to utilize agricultural residues as growth substrates and to generate co-products such as spent mushroom substrate that can be further valorized. When edible and well-characterized species are used and production is conducted under controlled conditions, existing safety assessments and regulatory frameworks suggest that mushroom-derived materials can be suitable for use in pet food formulations. Nevertheless, controlled feeding studies in companion animals remain limited, and further research is required to determine optimal inclusion levels, nutritional contribution, and long-term health outcomes.

Overall, the available literature suggests that edible macrofungi represent a promising but still relatively underexplored ingredient class in companion animal nutrition. While their protein concentrations are generally lower than those of conventional protein ingredients, their composition, cultivation systems, and potential functional components indicate opportunities for their use as complementary ingredients in pet food formulations.

In conclusion, mushroom-derived ingredients should currently be viewed less as primary protein replacements and more as complementary components that may contribute nutritional diversity and functional properties in pet food formulations. Their integration into pet foods may support broader diversification of ingredient sources while further research clarifies their nutritional value, digestibility, and potential health effects in dogs and cats.

11 Key takeaways

The following points summarize the main conclusions of this chapter.
- Edible macrofungi can convert lignocellulosic residues into nutritionally valuable biomass, supporting sustainable ingredient production.
- Mushroom-derived proteins contain essential amino acids and can complement conventional proteins in pet food formulations.
- Functional polysaccharides (β-glucans, chitin) and bioactives may contribute to immune and gastrointestinal functions, as shown for other animal species.
- Some of their technofunctional properties such as water-binding and emulsification have the potential to influence pet food production processes.
- Further research is needed to determine optimal inclusion levels, digestibility, long-term health effects, and protein replacement strategies in companion animals

References

Adom, F.; Maes, A.; Workman, C.; Clayton-Nierderman, Z.; Thoma, G.; Shonnard, D. 2012. Regional carbon footprint analysis of dairy feeds for milk production in the USA. *International Journal of Life Cycle Assessment (Int. J. Life Cycle Assess.)* 17:520-534. https://doi.org/10.1007/s11367-012-0386-y

Amara, A.A.; El-Baky, N.A. 2023. Fungi as a source of edible proteins and animal feed. *Journal of Fungi*, 9:73. https://doi.org/10.3390/jof9010073

Ayaz, F.A.; Chuang, L.T.; Torun, H.; Colak, A.; Sesli, E.; Presley, J.; Smith, B.R.; Glew, R.H. 2011. Fatty acid and amino acid composition of selected wild-edible mushrooms consumed in Turkey. *International Journal of Food Sciences and Nutrition (Int. J. Food Sci. Nutr.)*, 62: 328-335. https://doi.org/10.3109/09637486.2010.533160

Bach, F.; Helm, C.V.; Bellettini, M.B.; Maciel, G.M.; Haminiuk, C.W.I. 2017. Edible mushrooms: a potential source of essential amino acids, glucans and minerals. *International Journal of Food Science and Technology (Int. J. Food Sci. Technol.)* 52:2382-2392. https://doi.org/10.1111/ijfs.13522.

Bakir, T. kan; Boufars, M.; Karadeniz, M.; Ünal, S. 2018. Amino acid composition and antioxidant properties of five edible mushroom species from Kastamonu, Turkey. *African Journal of Traditional, Complementary and Alternative Medicines (Afr. J. Tradit. Complement. Altern. Med.)* 15:80-87. https://doi.org/10.21010/ajtcam.v15i2.10.

Banaszkiewicz, T. 2011. Nutritional value of soybean meal. *Soybean and Nutrition*, 12:120. https://doi.org/10.5772/23306

Bayrakli, B.; Yildiz, S. 2024. Comparative Analysis of Nutritional Values of Fishmeals Produced From Whole Anchovy and Sprat and Farmed Salmon Viscera in the Black Sea Region. *Acta Natura Et Scientia (Act. Nat. Sci.)* 5:150-159. https://doi.org/10.61326/actanatsci.v5i2.299

Bolarinwa, O. A.; Olukosi, O. A.; Adeola, O. 2012. Metabilozable energy value of porcine meat and bone meal for broiler chickens. *Canadian Journal of Animal Science (Can. J. Anim. Sci.)* 92: 73-78. https://doi.org/10.4141/cjas2011-103

Boro, S.; Kambhampati, V.; Das, S.; Saikia, D. 2025. Edible mushrooms as meat analogues: A comprehensive review of nutritional, therapeutic, and market potential. *Food Research International*, 214:116632. https://doi.org/10.1016/j.foodres.2025.116632

Bulam, S.; Üstün, N.; Pekşen, A. 2021. A review on fatty acid profiles of edible wild mushrooms from Turkey. *European Food Science and Engineering (Eur. Food Sci. Eng.)* 2:46-51. https://izlik.org/JA98DG47MG

Campos, I.; Pinheiro Valente, L.M.; Matos, E.; Marques, P.; Freire, F. 2020. Life-cycle assessment of animal feed ingredients: Poultry fat, poultry by-product meal and hydrolyzed feather meal. *Journal of Cleaner Production (J. Clean. Prod.)* 252:119845. https://doi.org/10.1016/j.jclepro.2019.119845

Chang S.T.; Wasser, S.P. 2017. The Cultivation and Environmental Impact of Mushrooms. *Oxford Research Encyclopedia of Environmental Science*. https://doi.org/10.1093/acrefore/9780199389414.013.231

Cho, J. H.; Kim, I. H. 2011. Fish meal – nutritive value. *Journal of Animal Physiology and Animal Nutrition (J. Anim. Physiol. Anim. Nutr.*) 95: 685-692. https://doi.org/10.1111/j.1439-0396.2010.01109.x

Dabbour, I.R.; Takruri, H.R. 2002. Protein Digestibility using Corrected Amino Acid Score method (PDCAAS) of four types of mushroom grown in Jordan. *Plant Foods for Human Nutrition (Plant Foods Hum. Nutr.)* 57:13-24. https://doi.org/10.1023/a:1013110707567

de Figueiredo Trindade, A.R.; de Brito Hilario, I.; Gimenes da Rocha, E.A.; da Rosa Borges dos Santos, L.A.; Giatti Marques de Souza, C.; Proença Dantas, M.; Roldão Ferreira, B.M.; Carvalho Gomes Corrêa, R.; Yamaguchi, N.U.; Bracht, A.; Peralta, R.M. 2025. Sustainable Production of Alternative Proteins from Basidiomycetes: Valorization of Mycelial and Fruiting Body Biomass. *Processes* 13:3746. https://doi.org/10.3390/pr13113746.

Dimopoulou, M.; Kolonas, A.; Mourtakos, S.; Androutsos, O.; Gortzi, O. 2022. Nutritional Composition and Biological Properties of Sixteen Edible Mushroom Species. *Applied Sciences (Appl. Sci.)* 12:8074. https://doi.org/10.3390/app12168074.

Dorr, E.; Koegler, M.; Gabrielle, B.; Aubry, C. 2021. Life cycle assessment of a circular, urban mushroom farm. *Journal of Cleaner Production (J. Clean. Prod.)* 288:125668. https://doi.org/10.1016/j.jclepro.2020.125668

Drzewiecka, B.; Wessely-Szponder, J.; Świeca, M.; Espinal, P.; Fusté, E.; Fernández-De La Cruz, E. 2024. Bioactive Peptides and Other Immunomodulators of Mushroom Origin. *Biomedicines* 12:1483. https://doi.org/10.3390/biomedicines12071483.

Edo, G.I.; Mafe, A.N.; Ali, A.B.M.; Akpoghelie, P.O.; Yousif, E.; Apameio, J.I.; Isoje, E.F.; Igbuku, U.A.; Garba, Y.; Essaghah, A.E.A.; Ahmed, D.S.; Umar, H.; Ozsahin, D.U. 2025 Chitosan and its derivatives: A novel approach to gut microbiota modulation and immune system enhancement. *International Journal of Biological Macromolecules (Int. J. Biol. Macromol.)* 289:138633. https://doi.org/10.1016/j.ijbiomac.2024.138633

(EFSA) EFSA Scientific Opinion. 2021. Safety of Vitamin D 2 mushroom powder (*Agaricus bispo-*

rus) as a Novel food pursuant to Regulation (EU) 2015/2283. *EFSA Journal* 19:6516. https://doi.org/10.2903/j.efsa.2021.6516

Ehrenzweig, J.; Easler, C. 2025. Prospective Evaluation of a Mushroom-Derived Nutraceutical in Dogs with Cognitive Dysfunction Syndrome: A Decentralized Proof-of-Concept Trial. Veterinary Health Research Centers (VHRC). Available at: https://www.researchgate.net/publication/395594468_Prospective_Evaluation_of_a_Mushroom_Derived_Nutraceutical_in_Dogs_with_Cognitive_Decline

El-Ramady, H.; Abdalla, N.; Badgar, K.; Llanaj, X.; Törős, G.; Hajdú, P.; Eid, Y.; Prokisch, J. 2022. Edible Mushrooms for Sustainable and Healthy Human Food: Nutritional and Medicinal Attributes. *Sustainability* 14-4941. https://doi.org/10.3390/su14094941.

(EU Commission) European Commission. 2025. A Strategic Framework for a Competitive and Sustainable EU Bioeconomy. Communication from the Commission to the European Parliament. Available at: https://eur-lex.europa.eu/legal-content/EN/TXT/?uri=celex-%3A52025DC0960

(EU Parliament) November 2015. Regulation (EU) 2015/2283 of the European Parliament and of the Council on novel Foods, amending Regulation (EU) No 1169/2011 of the European Parliament and of the Council and repealing Regulation (EC) No 258/97 of the European Parliament and of the Council and Commission Regulation (EC) No 1852/2001. Available at: https://eur-lex.europa.eu/eli/reg/2015/2283/oj/eng

(FAO) Food and Agriculture Organization of the United Nations. 1997. Feed ingredients and fertilizers for farmed aquatic animals – Sources and composition. FAO Fisheries Technical Paper No. 375/2. Available at: https://www.fao.org/4/i1142e/i1142e04.pdf

(FAO) Food and Agriculture Organization of the United States. 2020. The State of Food Security and Nutrition in the World 2020. Available at: https://openknowledge.fao.org/server/api/core/bitstreams/e612e779-ec47-44c2-a3e0-499569c3422d/content.

Feng, T.; Hu, Z.; Tong, Y.; Yao, L.; Zhuang, H.; Zhu, X.; Song, S.; Lu, J. 2020. Preparation and evaluation of mushroom (*Lentinus edodes*) and mealworm (*Tenebrio molitor*) as dog food attractant. *Heliyon* e05302. https://doi.org/10.1016/j.heliyon.2020.e05302

Finnigan, T.J.A., 2011. Mycoprotein: Origins, production and properties. In: *Handbook of Food Proteins, Woodhead Publishing Series in Food Science, Technology and Nutrition*, p. 335–352. https://doi.org/10.1533/9780857093639.335

Fletcher, A. October 29 2025. What do pet owners want from pet food products in 2026? [Interview]. *Food Ingredients (Fi) Global Insights*. https://insights.figlobal.com/sustainability/what-do-pet-owners-want-from-pet-food-products-in-2026-interview. Accessed Jan 2026.

Flores, G.A.; Cusumano, G.; Venanzoni, R.; Angelini, P. 2024. The Glucans Mushrooms: Molecules of Significant Biological and Medicinal Value. *Polysaccharides* 5, 212–224. https://doi.org/10.3390/polysaccharides5030016

Fréon, P.; Durand, H.; Avadí, A.; Huaranca, S.; Orozco Moreyra, R. 2017. Life cycle assessment of three Peruvian fishmeal plants: Toward a cleaner production. *Journal of Cleaner Production (J. Clean. Prod.)* 145:50-63. https://doi.org/10.1016/j.jclepro.2017.01.036

Fries-Craft, K.; Kilburn-Kappeler, L. R.; Aldrich, C. G.; Bobeck, E. A. 2023. Dietary yeast beta

1,3/1,6 glucan supplemented to adult Labrador Retrievers alters peripheral blood immune cell responses to vaccination challenge without affecting protective immunity. *Journal of Animal Science (J. Anim. Sci.)* 101:1-9. https://doi.org/10.1093/jas/skad029

Ionescu, M.; Dincă, M.-N.; Ferdeș, M.; Zăbavă, B.-S.; Paraschiv, G.; Moiceanu, G. 2025. Proteins from Edible Mushrooms: Nutritional Role and Contribution to Well-Being. *Foods* 14:3201. https://doi.org/10.3390/foods14183201

Kayser, E.; Castaneda, P.L.; Soto-Diaz, K.; Steelman, A.J.; Murphy, A.; Spindola, M.; He, F.; de Godoy, M.R.C. 2024. Functional properties of *Ganoderma lucidum* supplementation in canine nutrition. *Journal of Animal Science (J. Anim. Sci.)* 102:skae051. https://doi.org/10.1093/jas/skae051

Khan, R.; Brishti, F.H.; Arulrajah, B.; Goh, Y.M.; Abd Rahim, M.H.; Karim, R.; Hajar-Azhari, S.; Sam, K.K.; Anwar, F. Saari, N. 2024. Mycoprotein as a meat substitute: production, functional properties, and current challenges-a review. *International Journal of Food Science and Technology (Int. J. Food Technol.)* 59:522-544. https://doi.org/10.1111/ijfs.16791.

Khapudang, R.; Siddiqui, S. 2026. Structural, thermal and functional properties of protein isolates from oyster mushroom (Pleurotus ostreatus) isolated by alkaline iso-electric, ultra-sonication and enzymatic extraction techniques. *Food Physics* 3:100072. https://doi.org/10.1016/j.foodp.2025.100072

Kumar, K.; Mehra, R.; Guiné, R.P.F.; Lima, M.J.; Kumar, N.; Kaushik, R.; Ahmed, N.; Yadav, A.N.; Kumar, H. 2021. Edible Mushrooms: A Comprehensive Review on Bioactive Compounds with Health Benefits and Processing Aspects. *Foods* 10:2996. https://doi.org/10.3390/foods10122996

Kumla, J.; Suwannarach, N.; Sujarit, K.; Penkhrue, W.; Kakumyan, P.; Jatuwong, K.; Vadthanarat, S.; Lumyong, S. 2020. Cultivation of Mushrooms and Their Lignocellulolytic Enzyme Production Through the Utilization of Agro-Industrial Waste. *Molecules* 25:2811. https://doi.org/10.3390/molecules25122811.

Kusaba, A.; Arai, T. 2024. Shiitake mushroom powder supplementation increase antioxidative activity in dogs. *Frontiers in Veterinary Science (Front. Vet. Sci.)* 11:1355560. https://doi.org/10.3389/fvets.2024.1355560

Leong, Y.K.; Ma, T.W.; Chang, J.S.; Yang, F.C. 2022. Recent advances and future directions on the valorization of spent mushroom substrate (SMS): A review. *Bioresource Technology* 344:126157. https://doi.org/10.1016/j.biortech.2021.126157

Li, H.; Tian, Z.; Menolli Jr, N.; Ye, L.; Karunarathna, S. C.; Perez-Moreno, J.; Rahman, M. M.; Rashid, Md. H.; Phengsintham, P.; Rizal, L.; Kasuya, T.; Lim, Y. W.; Dutta, A. K.; Khalid, A. N.; Huyen, L. T.; Parungao Balolong, M.; Baruah, G.; Madawala, S.; Thongklang, N.; Hyde, K. D.; Kirk, P. M.; Xu, J.; Sheng, J.; Boa, E.; Mortimer, P. E. 2021. Reviewing the world's edible mushroom species: A new evidence-based classification system. *Comprehensive Reviews in Food Science and Food Safety (Compr. Rev. Food Sci. Food Saf.)* 20:1982-2014. https://doi.org/10.1111/1541-4337.12708.

Liatis, T.; Adamama-Moraitou, K.; Pardali, D.; Kavarnos, I.; Bates, N.; Rallis, T. 2018. Exposure of a cat to human-edible mushrooms: Were they toxic? *Hellenic Journal of Companion Ani-*

mal Medicine, 7:50–54. Available at: https://hjcam.hcavs.gr/images/vol7iss2/izs_v7i2-exposure-cat-human-edible-en.pdf

Liu, P.; Zhang, Z.; Wu, D.; Li, W.; Chen, W.; Yang, Y. 2025. The prospect of mushroom as an alternative protein: From acquisition routes to nutritional quality, biological activity, application and beyond. *Food Chemistry (Food Chem.)*, 469:142600. https://doi.org/10.1016/j.foodchem.2024.142600

Lizárraga-Hernández, F. D.; Soberanes-Yepiz, M. L.; Cortés-Jacinto, E.; Rangel-Dávalos, C.; Cientuegos-Martínez, K. 2024. Comparative analysis of the chemical quality of fishmeal produced on the Northwest coast of Mexico. *Agro Productividad*, 17: 177-184. https://doi.org/10.32854/agrop.v17i4.2870

Ma, G.; Yang, W.; Zhao, L.; Pei, F.; Fang, D.; Hu, Q. 2018. A critical review on the health promoting effects of mushrooms nutraceuticals. *Food Science and Human Wellness* 7:125133. https://doi.org/10.1016/j.fshw.2018.05.002.

Majumder, R.; Kiatur, S.; Saha, A.; Hossain, S. 2024. Mycoprotein: production and nutritional aspects: a review. *Sustainable Food Technology (Sustain. Food Technol.)* 2:81-91. https://doi.org/10.1039/d3fb00169e

Mariotti, F.; Tomé, D.; Mirand, P.P. 2008. Converting Nitrogen into Protein – Beyond 6.35 and Jones' Factors. *Critical Reviews in Food Science and Nutrition (Crit. Rev. Food Sci. Nutr.)* 48:177-184. https://10.1080/10408390701279749.

Martín, C.; Zervakis, G.I.; Xiong, S.; Koutrotsiosc, G.; Strætkvern, K.O. 2023. Spent substrate from mushroom cultivation: exploitation potential toward various applications and value-added products. *Bioengineered* 14:2252138. https://doi.org/10.1080/21655979.2023.2252138

Mbambalala, L.; Mwanda, L.; Cembi, S.K.; Mnukwa, M.L.; Mhlongo, L.C. 2026. Potential of spent mushroom substrate as a sustainable alternative feed resource in ruminant nutrition and its limitations. *Discover Animals* 3:10. https://doi.org/10.1007/s44338-025-00148-w

Miles, P.G.; Chang, S.-T. 2004. Mushrooms: Cultivation, Nutritional Value, Medicinal Effect, and Environmental Impact. 2nd Edition, CRC Press (Taylor & Francis). ISBN: 9780849310430 / 9781135515775

Miskiewicz, K.; Gendaszewska, D.; Sieczynska, K.; Gutarowska, B. 2025. Agri-food wastes as substrates for oyster mushroom (Pleurotus ostreatus) cultivation and their agricultural potential. *Scientific Reports* 15:42617. https://doi.org/10.1038/s41598-025-26843-y.

Moniruzzaman, M.; Tuong, D. T. C.; Chin, S.; Anjana, S.; Karthikeyan, A.; Min, T.; Kim, I.-J. 2025. A review on pharmacological insights of edible and medicinal mushroom based βglucans. *Applied Biological Chemistry (Appl. Biol. Chem.)* 68:41. https://doi.org/10.1186/s13765-025-01006-9.

Montebello, R; Derossi, A.; Caporizzi, R.; Taherzadeh, M.J.; Rousta, N.; Severini, C. 2025. Pooling scientific information on the nutritional, sensory, and technological properties of mycoprotein to support its role in creating a more sustainable food system. *Food Bioscience (Food Biosci)* 69:106882. https://doi.org/10.1016/j.fbio.2025.106882

Narsipur, S.; Kew, B.; Ferreira, C.; El-Gendy, R.; Sarkar, A. 2024. Emulsion stabilised by yeast proteins and biomass: a mini review. *Current Option in Food Science* 57:101167. https://doi.org/10.1016/j.cofs.2024.101167

Niego, A.G.; Rapior, S.; Thongklang, N.; Raspé, O.; Jaidee, W.; Lumyong, S.; Hyde, K.D. 2021. Macrofungi as a Nutraceutical Source: Promising Bioactive Compounds and Market Value. *Journal of Fungi (J. Fungi)* 7:397. https://doi.org/10.3390/jof7050397.

Noreen, S.; Sultan, H.; Hashmi, B.; Aja, P.M.; Atoki, A.V. 2025. Mushroom marvels: understanding their role in human health. *Frontiers in Nutrition (Front. Nutr.)* 12:1654911. https://doi.org/10.3389/fnut.2025.1654911

(NRC) National Research Council. 2006. Nutrient Requirements of Dogs and Cats. Washington, DC. National Academies Press. https://nap.nationalacademies.org/catalog/10668/nutrient-requirements-of-dogs-and-cats

Oonincx, D.G.A.B.; de Boer, I.J.M. 2012. Environmental Impact of the Production of Mealworms as a Protein Source for Humans – A Life Cycle Assessment. *PloS ONE* 7:e51145. https://doi.org/10.1371/journal.pone.0051145

Paris, S.; Chapat, L.; Pasin, M.; Lambiel, M.; Sharrock, T. E.; Shukla, R.; Sigoillot-Claude, C.; Bonnet, J.-M.; Poulet, H.; Freyburger, L.; De Luca, K.; 2020. β-Glucan-Induced Trained Immunity in Dogs. *Frontiers in Immunology (Front. Immunol.)* 11:566893. https://doi.org/10.3389/fimmu.2020.566893

Paz Cedeño, F.R.; Olubiyo, O.J.; Ferreira, S. 2025. From microbial proteins to cultivated meat for alternative meatlike products: a review on sustainable fermentation approaches. *Journal of Biological Engineering (J. Biol. Eng.)* 19:44. https://doi.org/10.1186/s13036-025-00509-9

Pérez-Bassart, Z.; Fabra, M.J.; Martínez-Abad, A.; López-Rubio, A. 2023. Compositional differences of β-glucan-rich extracts from three relevant mushrooms obtained through a sequential extraction protocol. *Food Chemistry (Food Chem.)* 402:134207. https://doi.org/10.1016/j.foodchem.2022.134207

Phillips, K. M; Ruggio, D. M.; Horst, R. L.; Minor, B.; Simon, R. R.; Feeney, M. J.; Byrdwell, W. C.; Haytowitz, D. B. 2011. Vitamin D and Sterol Composition of 10 Types of Mushrooms from Retail Suppliers in the United States. *Journal of Agricultural and Food Chemistry (J. Agric. Food Chem.)* 59:7841-7853. https://doi.org/10.1021/jf104246z.

Poore, J.; Nemecek, T. 2018. Reducing food's environmental impacts through producers and consumers. *Science* 360: 987-992. https://doi.org/10.1126/science.aaq0216.

Rahi, D. K.; Malik, D. 2016. Diversity of Mushrooms and Their Metabolites of Neutraceutical and Therapeutic Significance. *Journal of Mycology (J. Mycol.)* 7654123. http://dx.doi.org/10.1155/2016/7654123

Qian, Z.; Yang, M.; Chen, X.; Zhan, Q.; Zhong, L.; Hu, Q.; Zhao, L. 2025. Evaluation of physicochemical and functional properties of protein isolates from 5 edible mushroom species. *Food Science and Human Wellness* 14:9250043. https://doi.org/10.26599/FSHW.2024.9250043

Ravlikovsky, A.; Pinheiro, M.N.C.; Dinca, L.; Crisan, V.; Symochko, L. 2024. Valorization of Spent Mushroom Substrate: Establishing the Foundation for Waste-Free Production. *Recycling* 9:44. https://doi.org/10.3390/recycling9030044

Rivera, J.A.; Sebranek, J.G.; Rust, R.E. 2000. Functional properties of meat by-products and mechanically separated chicken (MSC) in a high-moisture model petfood system. *Meat Science* 55:61-66. https://doi.org/10.1016/S0309-1740(99)00126-6

Rodríguez, M.-C.; Cámara, C.; Morera, S.; Saborido, N.; Polo, J. 2023. Technological Benefits Associated with the Use of Spray-Dried Animal Plasma in Fish-Based Chunks for Canned Pet Food. *Animals* 13:3460. https://doi.org/10.3390/ani13223460

Saini, R.K.; Rauf, A.; Khalil, A.A.; Kod, E.-Y.; Keuma, Y.-S.; Anware, S.; Alamrie, A.; Rengasamy, K.R.R. 2021. Edible mushrooms show significant differences in sterols and fatty acid compositions. *South African Journal of Botany (S. Afr. J. Bot.)* 141:344-356. https://doi.org/10.1016/j.sajb.2021.05.022

Salomone, R.; Saija, G.; Mondello, G.; Giannetto, A.; Fasulo, S.; Savastano, D. 2017. Environmental impact of food waste bioconversion by insects: Application of Life Cycle Assessment to process using Hermetia illucens. *Journal of Cleaner Production (J. Clean. Prod.)* 140:890-905. https://doi.org/10.1016/j.jclepro.2016.06.154

Sánchez, C. 2010. Cultivation of Pleurotus ostreatus and other edible mushrooms. *Applied Microbiology and Biotechnology (Appl. Microbiol. Biotechnol.)* 85:1321-1337. https://doi.org/10.1007/s00253-009-2343-7.

Sande, D.; de Oliveira, G.P.; Moura, M.A.F.E; Martins, B.d.A.; Nogueira Silva Lima, M.T.; Takahashi, J.A. 2019. Edible mushrooms as a ubiquitous source of essential fatty acids. Food Research International 125:108524. https://doi.org/10.1016/j.foodres.2019.108524

Sauvant, D.; Perez, J.-M. and Tran G. 2004. Tables of Composition and Nutritional Values of Feed Animals. Wageningen Academic Publishers, Wageningen, the Netherlands, 192-284. https://doi.org/10.3920/978-90-8686-668-7. (English transation of the French original work)

Scholtmeijer, K.; van den Broek, L. A. M.; Fischer, A. R. H.; van Peer, A. 2023. Potential protein production from lignocellulosic materials using edible mushroom forming fungi. *Journal of Agricultural and Food Chemistry (J. Agric. Food Chem.)* 71:4450–4457. https://doi.org/10.1021/acs.jafc.2c08828

Shahid, M.; Shah, P.; Mach, K.; Rodgers-Hunt, B.; Finnigan, T.; Frost, G.; Neal, B.; Hadjikakou, M. 2024. The environmental impact of mycoprotein-based meat alternatives compared to plant-based meat alternatives: A systematic review. *Future Foods* 10:100410. https://doi.org/10.1016/j.fufo.2024.100410

Singh, A.; Saini, R. K.; Kumar, A.; Chawla, P.; Kaushik, R. 2025. Mushrooms as Nutritional Powerhouses: A Review of Their Bioactive Compounds, Health Benefits, and Value-Added Products. *Foods* 14:741. https://doi.org/10.3390/foods14050741.

Singh, M.; Kamal, S.; Sharma, V.P. 2020. Status and trends in world mushroom production-III World production of different mushroom species in 21st century. *Mushroom Research* 29:75-111. https://doi.org/10.36036/MR.29.2.2020.113703

Sun, L.; Liu, Q.; Bao, C.; Fan, J. 2017. Comparison of Free Total Amino Acid Compositions and Their Functional Classifications in 13 Wild Edible Mushrooms. *Molecules*, 22:350-359. https://doi.org/10.3390/molecules22030350

Sun, X.; Shi, Y.; Shi, D.; Tu, Y.; Liu, L. 2024. Biological Activities of Secondary Metabolites from the Edible-Medicinal Macrofungi. *Journal of Fungi (J. Fungi)* 10:144. https://doi.org/10.3390/jof10020144

Tan, R.Y.; Ilham, Z.; Wan-Mohtar, W.A.A.Q.I.; Halim-Lim, S.A.; Usuldin, S.R.A.; Ahmad, R.; Adlim,

M. 2024. Mushroom oils: A review of their production, composition, and potential applications. *Heliyon* 10:e31594. https://doi.org/10.1016/j.heliyon.2024.e31594.

Tilman, D.; Blazer, C.; Hill, J.; Bedfort, B. L. 2011. Global food demand and the sustainable intensification of agriculture. *Proceedings of the National Academy of Sciences of the United States of America (PNAS)* 108:20260-20264. https://doi.org/10.1073/pnas.1116437108

(UN) United Nations Department of Economic and Social Affairs, Population Division. 2022. World Population Prospects 2022: Summary of Results. United Nations. Available at: https://population.un.org/wpp/. Accessed January 2026.

Valverde, M. E.; Hernández-Pérez, T.; Paredes-López, O. 2015. Edible mushrooms: Improving human health and promoting quality life. *International Journal of Microbiology (Int. J. Microbiol.)* 376387. https://doi.org/10.1155/2015/376387

Vasconcellos, R.S.; Henríquez, L.B.F.; Lourenço, P.d.S. 2023. Spray-Dried Animal Plasma as a Multifaceted Ingredient in Pet Food. Animals 13:1773. https://doi.org/10.3390/ani13111773

Wang, B; Shi, Y.; Lu, H.; Chen, Q. 2023. A critical review of fungal proteins: Emerging preparation technology, active efficacy and food application. *Trends in Food Science & Technology (Trends Food Sci. Technol.)* 141:104178. https://doi.org/10.1016/j.tifs.2023.104178.

Wang, D.; Zhang, M.; Wan, J.; Liu, H.; Wang, Y.; Yang, R.; Wu, Y.; Bao, D.; Chen, H.; Zou, G.; Zhao, Y. 2024. Enhancing Digestibility and Intestinal Peptide Release of *Pleurotus eryngii* Protein: An Enzymatic Approach. *Fungi* 10:890. https://doi.org/10.3390/jof10120890.

Wang, R.; Sar, T.; Mahboubi, A.; Fristedt, R.; Taherzadeh, M.J.; Undeland, I. 2023. In vitro protein digestibility of edible filamentous fungi compared to common food protein sources. *Food Bioscience* 54:102862. https://doi.org/10.1016/j.fbio.2023.102862

Wang, Z.; Xu, X.; Liu, Z.; Xu, T.; Ma, L.; Song, F. 2025. High-pressure homogenization treatment on yeast protein: Effect on structure and emulsifying properties. *Food Research International (Food Res. Int.)* 213:116550. https://doi.org/10.1016/j.foodres.2025.116550

Yu, H.; Li, M.; Yu, L.; Ma, X.; Wang, S.; Yuan, Z.; Li, L. 2023. Partial Replacement of Fishmeal with Poultry By-Product Meal in Diets for Coho Salmon (*Oncorhynchus kisutch*) Post-Smolts. *Animals*, 13: 2789 https://doi.org/10.3390/ani13172789

Zhang, S.; Wang, K.; Qin, Y.; Zhu, S.; Gao, Q.; Liu, D. 2023. The synthesis, biological activities and applications of protein–polysaccharide conjugates in food system: a review. *Food Quality and Safety (Food Qual. Saf.)* 7:1–17. https://doi.org/10.1093/fqsafe/fyad006

Zhang, Z.; Liu, H.; Jin, J.; Zhu, X.; Han, D.; Xie, S. 2024. Toward a low-carbon footprint: Current status and prospects for aquaculture. *Water Biology and Security* 100290. https://doi.org/10.1016/j.watbs.2024.100290

Zhao, J.; Hu, Y.; Qian, C.; Hussain, M.; Liu, S.; Zhang, A.; He, R.; Sun, P. 2023. The Interaction between Mushroom Polysaccharides and GutMicrobiota and Their Effect on Human Health: A Review. *Biology* 12:122. https://doi.org/10.3390/biology12010122

CHAPTER 18

Insect-based ingredients in livestock feed

B. Palumbo and A. Dalle Zotte

Department of Animal Medicine, Production and Health (MAPS), University of Padova, Agripolis, Viale dell'Università 16, 35020 Legnaro, Padova, Italy
Correspondence: bianca.palumbo@unipd.it; antonella.dallezotte@unipd.it

Abstract

This chapter systematically evaluates three commercially viable insect species for livestock nutrition: black soldier fly (*Hermetia illucens*, BSF), yellow mealworm (*Tenebrio molitor*, YMW), and mulberry silkworms (*Bombyx mori*, MSW). While BSF and YMW are widely produced for feed due to their scalability and bioconversion efficiency, MSW is included because its pupae are readily available as a sericulture by-product. These species exhibit favorable biological characteristics, such as rapid growth rates and nutrient-dense profiles rich in immunomodulatory bioactive compounds; consequently, they offer significant potential for industrial scalability. In poultry production systems, dietary BSF incorporation at 10–12% inclusion rates sustains or enhances growth performance, feed conversion efficiency, and intestinal functionality while inducing beneficial shifts in cecal microbiota composition. Comparable responses are observed in laying hens, with improvements in egg quality indices and yolk pigmentation. YMW demonstrates similar efficacy, in terms of performance and feed efficiency at inclusion thresholds below 7.5%, while the MSW maintain adequate performance at moderate levels (4–10%), though antinutritional factors limit higher inclusion rates. In swine diets, the inclusion of 4–12% BSF larvae meal maintains productive performance while enhancing gastrointestinal health, with its lipid fraction also functioning as an efficient energy substrate. The inclusion of YMW at levels up to 10% supports growth dynamics and augments apparent ileal protein digestibility with beneficial microbiota modulation. In rabbits and ruminants, despite relevant EU regulatory limitations, insect lipids and protein meals can effectively replace conventional feedstuffs while maintaining performance. Moreover, they offer potentiality in mitigating methane production in ruminants. From an environmental sustainability perspective, insect-based ingredients demonstrate potential advantages in greenhouse gas mitigation and land-use efficiency. Nevertheless, life cycle assessments indicate that benefits exhibit substantial context-dependency, being maximized when production systems utilize low-value agro-industrial by-products as rearing substrates. Contemporary constraints encompass economic competitiveness, regulatory frameworks, compositional variability, and stakeholder acceptance. Future advancement necessitates technological innovation, process optimization, standardization, and regulatory harmonization. Overall, insect-derived ingredients constitute promising alternative feed resources whose sustainability credentials require validation within specific production contexts.

Keywords

alternative protein sources – terrestrial livestock nutrition – animal performance – gut health – feed sustainability

1 Introduction

Projections indicate that the global human population will exceed 10 billion by 2050, with a concomitant sharp increase in the consumption of animal protein, particularly from poultry and pork (United Nations, 2019). This trend is expected to substantially increase the global demand for protein feed ingredients, further exacerbating environmental and economic pressures on feed production systems. Consequently, the rapid expansion of the livestock sector has intensified pressure on conventional feed resources such as soybean meal (SBM) and fish meal (FM). These ingredients are associated with significant environmental concerns, including deforestation, biodiversity loss, overfishing, greenhouse gas emissions, and competition with human food supply (van Huis and Oonincx, 2017; Voora et al., 2020). In this context, the search for sustainable, efficient, and resilient alternative protein sources for animal feed has become a strategic priority for both researchers and industry stakeholders, to address climate change and promote the circularity of food systems (Van Zanten et al., 2023).

Insects have emerged as a promising solution to address these challenges, owing to their high nutritional value, efficient feed conversion ratio (FCR), the ability to be reared on a wide range of organic substrates, including agro-industrial by-products and the high environmental sustainability potential in terms of emissions, land and water uses (Gasco et al., 2021; Oonincx et al., 2010; van Huis et al., 2024). Exploiting insects in animal nutrition is not a novel concept; in fact, insects constitute a natural component of the diet of different farmed species, particularly poultry and pigs (Bovera et al., 2016; van Huis et al., 2024). However, recent advances in insect farming technologies, coupled with increasing regulatory recognition, have renewed scientific and commercial interest in their large-scale application as feed ingredients.

Among the various insect species investigated for feed applications, the black soldier fly (*Hermetia illucens*, BSF) is currently the most widely produced. Mealworms, particularly the yellow mealworm (*Tenebrio molitor*, YMW), have also attracted considerable attention (van Huis and Gasco, 2023). In fact, these species combine favorable biological traits, such as rapid growth rates, high reproductive capacity, and adaptability to large-scale rearing systems. In parallel, mulberry silkworm (MSW) has emerged as a relevant and complementary insect species, mainly due to the large availability of MSW pupae as a by-product of the sericulture industry. Although several insect species are currently authorized for feed use in the European Union (EU), this chapter will also cover, in addition to BSF and YMW, the MSW, due to its unique position within the insect production landscape. Unlike many other species primarily reared specifically for feed or food applications, MSW pupae are widely available as a large-scale by-product of the sericulture industry. This characteristic provides an opportunity to valorize existing agro-industrial residue, contributing to circular bioeconomy strategies and improving the sustainability of insect-based feed production systems. Despite differences in production systems, BSF, YMW and MSW share a rich and well-balanced nutritional profile (Kröncke et al., 2019; Makkar et al., 2014; van Huis and Oonincx, 2017). In addition to their high protein and lipid content, BSF and YMW provide bioactive compounds, including antimicrobial peptides and chitin, which have been associated with positive effects on animal health. These compounds may enhance immune responses, contribute to protection against pathogenic contamination, and improve overall gut health (Elahi et al., 2022; Swiatkiewicz et al., 2015).

Consequently, insect-derived meals from BSF, YMW and MSW are increasingly regarded as viable and sustainable alternatives to conventional protein sources in livestock diets. Their relevance in current research and practical applications is supported by their nutritional value and increasing regulatory acceptance, although the underlying drivers differ among species, with BSF and YMW mainly associated with scalable production systems and efficient bioconversion, and MSW primarily linked to the availability of pupae as a sericulture by-product.

Insect-based products may be incorporated into animal diets in three main forms: live insects, processed animal proteins (PAPs), fats and other insect-derived products. In the EU, insects can be legally used live as feed materials for several animal species, except for ruminants and rabbits (SCoPAFF, 2024). Alternatively, insects may be processed into PAPs. To date, eight insect species are authorized in the production of PAPs in the EU: the BSF, the common housefly (*Musca domestica*), YMW, the lesser mealworm (*Alphitobius diaperinus*), the house cricket (*Acheta domesticus*), the banded cricket (*Gryllodes sigillatus*), the field cricket (*Gryllus assimilis*), and the MSW (EFSA, 2015).

Under current EU legislation, PAPs, including those derived from insects, are not permitted in feed for ruminants or rabbits. This restriction is rooted in the general feed ban introduced to prevent the risk of transmissible spongiform encephalopathies (TSEs), such as bovine spongiform encephalopathy. Accordingly, the EU TSE Regulation (Regulation (EC) No 999/2001) continues to prohibit the use of all PAPs in ruminant feed, without specific exemptions for insect-derived proteins. Conversely, insect PAPs are currently authorized for use in feed for non-ruminant species, including aquaculture animals, poultry, and swine, but excluding rabbits. This authorization is governed by specific EU regulations, notably the Regulation (EU) 2017/893 and its subsequent amendments. Regarding insect fats, no restrictions currently apply concerning the animal species to which these ingredients may be fed.

Despite their potential, the large-scale adoption of insect-based feeds is still challenged by factors such as variability in nutrient composition depending on rearing substrates, production costs, regulatory constraints, and consumer acceptance. Addressing and optimizing these aspects is essential to enable the widespread and economically viable use of insects as feed ingredients.

This chapter aims to provide an overview of the role of insects in animal nutrition, with a specific focus on BSF, YMW and MSW. It will address the inclusion of these insect species in the diets of poultry, swine, rabbits and ruminants, examining their effects on animal performance and metabolism, functional properties, nutrient digestibility, growth performance, and feed efficiency (FE), as well as the impact of insect-based diets on the quality of animal-derived products (meat and eggs).

To facilitate comparisons across insect species, livestock categories, and production stages, consolidated Tables are provided at the end of this chapter. These Tables synthesize nutritional profiles, optimal inclusion levels, and the resulting effects on animal performance, health, and product quality.

2 Application in livestock nutrition

2.1 Poultry nutrition

2.1.1 BSF in broiler nutrition

In poultry, the use of BSF larvae and derived products has been extensively investigated across different species, production systems and forms of administration, including whole larvae, meal and oil/fat, generally yielding positive outcomes in terms of growth performance, welfare and health of birds (Table 18.1). Moderate replacement of SBM with BSF meal (generally up to 10–12% inclusion level or 25% substitution) consistently showed to maintain growth performance, feed intake (FI) and FCR, with several studies reporting results comparable to those obtained with conventional diets, particularly during the starter phase (Baderuddin et al., 2024; Cullere et al., 2016; Facey et al., 2023; Fruci et al., 2023; Gariglio et al., 2025; Rytlewski et al., 2025; Saidani et al., 2025). Comparable outcomes have also been reported at substantially higher substitution rates, with BSF meal replacing up to 75% of SBM without adverse effects on overall growth performance, carcass traits, gut morphology or health status in broilers (Machado et al., 2023). In some cases, BSF inclusion has been associated with improved growth parameters, including increased average daily gain (ADG), body weight gain (BWG) and final body weight (BW), without detrimental effects on feed intake (FI) or nutrient utilization. This was observed in broilers, slow-growing chickens and turkeys fed live, dehydrated or defatted BSF products (Dörper et al., 2024a; Fiorilla et al., 2024a; Zampiga et al., 2025).

The productive response to BSF appears to be both dose- and form-dependent. While larvae included at low levels (3–5%) or moderate proportions of voluntary FI generally support normal growth performance and nutrient conversion efficiency, higher inclusion levels or excessive substitution rates may impair FI, energy intake or protein utilization efficiency (Facey et al., 2023; Fruci et al., 2023; La Mantia et al., 2024; Seyedalmoosavi et al., 2022). Conversely, the use of live or unprocessed larvae, particularly when provided as environmental enrichment, has frequently resulted in improved BWG, reduced FCR and higher final BW, even in the presence of reduced total dry matter (DM) intake, thus suggesting enhanced nutrient utilization (Dörper et al., 2024a; Fiorilla et al., 2024a; Ipema et al., 2022).

Beyond productive performance, BSF inclusion has been consistently associated with beneficial effects on animal health, welfare and gut functionality. Several studies have reported improved litter quality, reduced incidence and severity of footpad dermatitis and hock burns, and increased expression of active and exploratory behaviors, particularly when dried or live larvae were scattered in the pen rather than incorporated into the feed (Bongiorno et al., 2022; Dörper et al., 2024a; Ipema et al., 2022). Health-related parameters, including blood metabolites, immune organ weights and indicators of stress or immune activation, were generally not adversely affected by BSF inclusion, with some studies reporting enhanced immune organ development or reduced markers of hepatic stress (Bongiorno et al., 2022; Fiorilla et al., 2024a). The replacement of conventional lipid sources with BSF fat has been shown to improve FCR and lipid digestibility in broilers and turkeys, with enhanced apparent ileal digestibility of ether extract and increased pancreatic lipase activity when BSF fat was used as a partial or total substitute for soybean oil (Aslam et al., 2025; Kierończyk et al., 2022). In addition, BSF fat supplementation has been associated with improved immune responses, reduced pro-inflammatory markers and favorable modulation of thyroid hormone status (Aslam et al., 2025; Kierończyk et al., 2022).

Moreover, BSF-based diets have been shown to modulate gut health and microbiota composition, reducing potentially harmful bacterial populations such as Enterobacteriaceae and increasing beneficial taxa, including *Lactobacillus* spp. and *Faecalibacterium*, alongside favorable changes in short-chain fatty acids (SCFA) production and gut metabolomic profiles (Fiorilla et al., 2024a; Saidani et al., 2025; Zampiga et al., 2025). These effects likely contribute to the observed improvements in growth performance and FE, supporting the role of BSF ingredients not only as nutrient sources but also as functional feed components.

A larger body of literature has focused on the effects of BSF inclusion on poultry meat quality (Table 18.2). Overall, the available evidence indicates that the dietary use of BSF meal, larvae, or derived fats does not compromise the main physicochemical or sensory traits of poultry meat. Across chickens, quails, ducks, turkeys, pheasants and game birds, parameters such as pH, color, water holding capacity (WHC), tenderness, cooking loss and overall sensory acceptability generally remain comparable to those observed in birds fed conventional soybean-based diets, even when substantial replacement levels are applied (Cullere et al., 2016; Cullere et al., 2018; Cullere et al., 2019a; Fiorilla et al., 2024b; Gariglio et al., 2021; Pieterse et al., 2019; Rytlewski et al., 2025; Schiavone et al., 2017). Minor variations in meat color, particularly increased yellowness, lighter appearance, or slight reductions in pH, have occasionally been reported following BSF inclusion; however, these changes consistently fall within commercially acceptable ranges and do not translate into negative sensory perceptions (Altmann et al., 2020; Cullere et al., 2016; Popova et al., 2020; Secci et al., 2018).

A recurrent and well documented effect of BSF-based diets is the modification of the fatty acid (FA) profile of poultry meat, largely reflecting the lipid composition of BSF ingredients. An increase in saturated fatty acids (SFA), mainly lauric and myristic acids, often occurs at the expense of polyunsaturated fatty acids (PUFA) when BSF meals or fats replace SBM or oil (Cullere et al., 2018; Cullere et al., 2019a; Kierończyk et al., 2023; Kim et al., 2021; Popova et al., 2020; Schiavone et al., 2019). Similar shifts have been reported in quails, ducks, partridges and Muscovy ducks: BSF inclusion increased SFA proportions while leaving cholesterol content and most technological traits unaffected (Mbhele et al., 2019; Gariglio et al., 2021; Gariglio et al., 2025; Secci et al., 2018). Despite these lipid profile changes, no consistent detrimental effects on meat quality indices relevant to consumer acceptance have been reported, and in several studies BSF inclusion was associated with improved oxidative stability or reduced lipid oxidation during refrigerated storage (Choi et al., 2021; Cullere et al., 2018; Kim et al., 2020).

Additional studies have explored more specific compositional and functional aspects, such as volatile compounds, bioactive peptides, amino acid profiles and antioxidant capacity, showing that insect-based diets may induce moderate biochemical or metabolic changes without impairing overall meat quality or sensory characteristics (de Souza Vilela et al., 2021; Gkarane et al., 2020a,b). In particular, BSF-based diets showed to modify the volatile profile of meat and the concentration of endogenous bioactive compounds, mainly reflecting changes in lipid composition, while maintaining acceptable flavor-related characteristics (Gkarane et al., 2020a,b). Evidence from chickens, quails and turkeys also confirms that BSF meal or fat inclusion at moderate levels maintains meat quality while inducing only limited alterations in metabolomic profiles, FA composition or antioxidant status (Cullere et al., 2019b; Kim et al., 2020; Zampiga et al., 2025).

Moreover, it has been suggested that targeted dietary strategies, such as modulation of insect rearing substrates or the combination of BSF-based ingredients with PUFA-rich feedstuffs, may help mitigating SFA accumulation and optimize the nutritional quality of meat obtained from insect-fed poultry (Cullere et al., 2019b). Taken together, these findings support the suitability of BSF as a sustainable feed ingredient for poultry, ensuring meat quality while contributing to the diversification of protein and lipid sources in poultry nutrition.

2.1.2 BSF in laying chicken and quail nutrition

In addition to broilers, dietary inclusion of BSF has been extensively investigated in laying hens, showing generally positive and dose-dependent responses across different inclusion levels, ingredient forms and production phases. Partial or total replacement of SBM or FM with BSF meal, either full-fat or defatted, typically maintained laying performance, FI and FCR, with several studies reporting comparable or slightly improved egg production, egg weight or FCR at inclusion levels ranging from approximately 3 to 12% (Attivi et al., 2023; Lalev et al., 2023b; Nassar et al., 2023; Petkov et al., 2022; Veldkamp et al., 2025; Zhao et al., 2022) (Table 18.3). In some cases, BSF inclusion resulted in improved FE associated with reduced FI, and in increased BW development, particularly at moderate to high substitution rates of SBM or FM (Attivi et al., 2023; Petkov et al., 2022; Veldkamp et al., 2025; Zhao et al., 2022). Similar neutral or positive responses have been observed when BSF larvae were incorporated within alternative protein-based diets or combined with locally sourced plant ingredients, supporting its suitability as part of diversified and regionally adapted feeding strategies (Montalbán et al., 2024).

Long-term studies further demonstrated that BSF meal can be safely used throughout the rearing and laying phases, with positive effects on BW development, reproductive traits and circulating reproductive hormones at higher substitution rates of FM, without adverse effects on laying rate or FCR (Attivi et al., 2023).

Egg quality parameters were generally preserved across studies, with several reports describing improvements in yolk color, albumen height and Haugh unit, while shell characteristics were largely unaffected, apart from occasional reductions in shell thickness at complete SBM replacement (Dörper et al., 2024; Mancini et al., 2025; Navasero et al., 2022). Dietary BSF inclusion consistently modified egg FA profiles, typically increasing SFA and monounsaturated fatty acids (MUFA) at the expense of PUFA, although birds appeared to partially regulate the transfer of medium-chain FA such as lauric and myristic acids into eggs (Chatzidimitriou et al., 2022; Kim et al., 2022; Lokaewmanee et al., 2023).

Beyond productivity, BSF-based diets influenced gut microbiota composition, increasing microbial diversity and the abundance of potentially beneficial bacterial genera, and inducing structural shifts in microbial communities associated with gut functionality and metabolic responses (Huang et al., 2024; Ndotono et al., 2022). The provision of live or dried BSF larvae also acted as an effective form of environmental enrichment, reducing fearfulness, feather pecking and physiological stress indicators, while maintaining laying performance and integument status under both experimental and commercial farming conditions (Dabbou et al., 2025; Tahamtani et al., 2025).

In laying quails, dietary inclusion of BSF meal, whole larvae or BSF-derived fats likewise supported satisfactory egg production and FE, with optimal responses often observed at moderate inclusion levels (10–20%). Defatted BSF meal improved egg shape index, shell characteristics and yolk color, while maintaining sensory quality, and enhancing oxidative stability during storage at intermediate inclusion rates (Dalle Zotte et al., 2019). Whole BSF larvae positively affected FCR, antioxidant capacity, and immune responses, as well as egg amino acid and FA profiles, with 20% inclusion level emerging as a favorable compromise between productive performance and metabolic responses (Liu et al., 2025). The use of BSF fat or its calcium salts further demonstrated improvements in FCR, egg weight, yolk and albumen protein content, along with reductions in blood and egg cholesterol, accompanied by modulation of genes involved in lipid metabolism (Armaghoza et al., 2025).

Overall, the available evidence indicates that BSF-based ingredients can be effectively incorporated into the diets of laying hens and quails, contributing to sustainable egg production while maintaining productivity, egg quality, gut health and animal welfare.

2.1.3 YMW in broiler nutrition

Similarly to BSF, the inclusion of YMW in poultry diets has attracted considerable research interest, revealing a generally dose-dependent response across species, production systems and productive traits. In relation to growth performance, the available evidence consistently indicates that low to moderate inclusion levels are well tolerated and may even be beneficial, particularly during early growth phases (Table 18.4). In broiler chickens, dietary YMW meal at inclusion levels around 2.5–5% showed to support BWG and FCR without detrimental effects on FI or mortality, with performance improvements being especially evident during the starter phase at the lowest inclusion rates (Elahi et al., 2020; Sedgh-Gooya et al., 2022). Low levels (around 3%) have also been associated with improved metabolic profiles and enhanced growth at slaughter age, including lower serum triglyceride and serum aspartate aminotransferase activities, together with shifts in caecal metabolic pathways involved in lipid and bile acid metabolism (Jiang et al., 2024). When YMW is used within balanced partial replacement strategies, growth performance and FCR appear comparable to, or in some cases better than, conventional SBM-based diets, as shown in grower diets where YMW supported final BW and improved FCR relative to other alternative insect-based strategies (Pietras et al., 2021). It is noteworthy that complete replacement of SBM with YMW meal did not impair growth performance or carcass traits in broilers, while also leading to an enhanced FCR over the entire growing period. However, this dietary modification resulted in changes in intestinal development and lower ileal digestibility coefficients for DM, organic matter and crude protein compared with SBM-based diets (Bovera et al., 2016). Conversely, when YMW meal is included at levels generally ≥7.5–10%, a progressive reduction in FE and growth performance has been observed, especially when YMW meal was used as a major protein source or at inclusion levels of 7.5-15%, with adverse effects extending to carcass weight and breast yield at the highest levels (Biasato et al., 2017; Petkov et al., 2024), highlighting the importance of inclusion level optimization. These responses appear to be partly species-specific, as YMW inclusion had neutral effects in free-range chickens at moderate levels (7.5%) (Biasato et al., 2016). Differently, in quails, YMW inclusion up to 3% was associated with improved BW, FCR and carcass yield (Zadeh et al., 2019).

Regarding health status, most studies converge in indicating that YMW meal inclusion at low to moderate levels does not negatively affect hematological, serum biochemical or intestinal morphometric parameters, and gut morphology and histopathological features generally remain unaffected, supporting the overall safety of YMW-based diets in poultry (Biasato et al., 2016, 2017). Nevertheless, targeted investigations suggest that higher inclusion levels (10–15%) may influence intestinal homeostasis, as reflected by shifts in cecal microbiota structure and altered mucin dynamics, without necessarily translating into overt pathological outcomes (Biasato et al., 2019a). Moreover, when YMW-derived lipid sources are used to replace conventional fats, improvements in lipid-related traits have been reported, including reduced liver size, hepatic triglycerides and total cholesterol, together with modifications of tissue FA profiles (Benzertiha et al., 2019). Under challenging environmental conditions, such as heat stress, YMW-based diets have also shown potential to support growth and reduce oxidative stress markers suggesting a role in enhancing physiological resilience (Stelios et al., 2024). In addition, the provision of live YMW larvae at limited amounts has been proposed as an effective form of environmental enrichment, improving FCR and modulating immune organ development (e.g., higher spleen relative weight) without impairing health status (Bellezza Oddon et al., 2021).

Concerning meat quality, YMW inclusion generally preserves the main physicochemical traits of poultry meat while inducing moderate changes in lipid composition. In broilers and free-range chickens, YMW meal inclusion at moderate levels (7.5%) does not affect pH, color or proximate composition, while simultaneously enhancing the content of oleic and α-linolenic acids and lowering atherogenic and thrombogenic indices (Dabbou et al., 2020a). Comparable trends were observed when YMW meal fully replaced SBM, as most chemical and physical properties of breast meat remained unaffected, although slight increases in breast pH and cooking losses were reported (Bovera et al., 2016). In slow-growing chickens, partial replacement of SBM with YMW meal (50–75%) improved meat tenderness and meat moisture, whereas complete substitution appeared to compromise certain quality parameters (Nieto et al., 2024). Higher inclusion levels have occasionally been linked to less favorable sensory attributes, including increased cooking losses and reduced juiciness and tenderness, particularly when YMW was used at elevated rates or in combination with other insect meals (Shaviklo et al., 2021; Biasato et al., 2025). In game birds such as Barbary partridge, the partial replacement of dietary protein with YMW meal (25–50%) increased carcass weight without negatively affecting meat pH, color or shear force, while inducing favorable modifications in FA profiles and maintaining cholesterol content comparable to conventional diets (Secci et al., 2018). In quails, dietary YMW up to 3% positively influenced meat color, water retention, carcass yield and jejunal morphology (Zadeh et al., 2019), while in Muscovy ducks, supplementation with live YMW larvae did not negatively impact slaughter performance or overall meat quality, causing only minor variations in proximate composition and FA profile (Gariglio et al., 2025).

Altogether, these findings indicate that YMW can serve as an effective alternative protein and lipid source in poultry diets, if inclusion levels are carefully managed to optimize growth performance, gut health and meat quality.

2.1.4 YMW in laying chicken and quail nutrition

In laying poultry, the inclusion of YMW has been explored using different formulations (meal, whole larvae and live larvae) and inclusion levels, with overall evidence indicating that low to moderate dietary incorporation can be achieved without compromising laying performance while positively modulating egg quality traits (Table 18.5). In laying hens, dietary inclusion of YMW meal up to 3–5% is well tolerated and does not impair laying rate, FI or egg weight, while exerting beneficial effects on egg quality. Literature results show that, at inclusion levels between 1 and 3%, YMW meal supplementation does not modify egg production but consistently improves yolk color and enriches the yolk FA profile, increasing the proportion of unsaturated FA, including linoleic and α-linolenic acids, together with a higher total PUFA content (Ko et al., 2020). Slightly higher inclusion levels (2.5–5%) have been associated with improvements in egg production, egg mass and FCR, particularly at the lower inclusion rate, without affecting FI, egg weight, BWG or blood biochemical parameters (Sedgh-Gooya et al., 2021). Under alternative production systems, the partial replacement of SBM and vegetable oil with whole YMW meal at approximately 8% maintained egg physicochemical quality and sensory attributes, with only marginal reductions in Haugh unit, and no detectable differences in consumer perception, supporting the suitability of YMW as a sustainable ingredient in free-range laying hen nutrition (Remón et al., 2025).

In laying quails, the response to YMW inclusion appears more variable and strongly influenced by dietary level and replacement strategy compared to laying hens. YMW meal can successfully replace FM up to 3% without adverse effects on FE, hematological traits or immune-related parameters, while reducing serum cholesterol concentrations at the highest inclusion levels (Shariat Zadeh et al., 2020). Conversely, the progressive replacement of soybean protein with YMW meal at increasing levels induces a linear reduction in laying rate and egg mass, despite concomitant increases in egg weight, and yolk and albumen proportion, indicating a less favorable productive balance at high inclusion rates (Secci et al., 2021). Accordingly, lower inclusion levels (around 1.4–2.8%) emerge as the most suitable compromise, supporting intestinal morphology, mucous cell activity and short-chain FA production, thereby contributing to improved gut health without penalizing productive performance (Addeo et al., 2022). Additional findings confirms that YMW larvae or YMW-derived products can be incorporated into quail diets up to 12%, maintaining egg production while reducing yolk lipid and cholesterol content (Nuraini et al., 2022, 2023). More recently, the inclusion of YMW meal as part of symbiotic formulations has highlighted its functional potential, promoting beneficial gut microbiota without affecting overall laying performance (Mohammadi et al., 2025). Live YMW larvae supplementation, used primarily as nutritional enrichment at approximately 10% of expected daily FI, stimulated feeding activity and supported nutrient digestibility while maintaining egg physicochemical quality and storage stability, although without clear improvements in laying performance and with minor modifications in sensory attributes (Dalle Zotte et al., 2024a).

Overall, the available evidence indicates that YMW can be effectively included in diets for laying poultry, provided that inclusion levels are carefully optimized to balance laying performance, egg quality and gut health outcomes, with low to moderate inclusion levels offering the most consistent benefits across production systems.

2.1.5 MSW in poultry nutrition

Alongside BSF and YMW, MSW-derived ingredients (full-fat or defatted pupae meal) are increasingly explored as alternative protein and lipid sources in poultry nutrition, with available evidence indicating a generally favorable response when inclusion levels are carefully calibrated. In broiler chickens, MSW meal can effectively replace conventional protein sources, supporting growth performance and FE at low inclusion level (Table 18.6). Incremental replacement of FM with MSW meal improves FI, BWG and protein utilization up to intermediate inclusion rates (approximately 4–10%), whereas excessive inclusion tends to compromise performance efficiency, likely due to reduced digestibility and palatability (Emadi et al., 2015; Kongsup et al., 2022). Comparable outcomes are reported when MSW pupae meal partially replaces SBM and oil, with moderate substitution levels (around 25%) supporting optimal growth, carcass traits and breast yield, while higher substitution levels require caution (Miah et al., 2020). When used within balanced replacement strategies, including combinations with plant proteins such as lupine, MSW pupae meal maintains growth performance and FCR comparable to conventional soybean-based diets, without impairing slaughter traits or meat sensory quality (Pietras et al., 2021).

Beyond growth performance, MSW-based diets generally maintain health status and metabolic homeostasis. Blood biochemical and immune parameters remain unaffected when MSW pupae meal is included at moderate levels (7–10%) in chickens and laying hens, supporting its safety as a dietary ingredient (Lalev et al., 2022a; Lalev et al., 2023b). Improvements in the net utilization of dietary energy and protein at the carcass level further highlight the nutritional value of MSW pupae meal, particularly at inclusion levels of 10% (Lalev et al., 2023a). In laying birds, partial replacement of SBM with MSW pupae meal (7%) does not adversely affect egg pH or major physical traits, while positively modulating yolk FA composition by increasing the proportion of unsaturated FA (Lalev et al., 2022b). In addition, fermented MSW pupae show promising results, improving FE, immune organ development, and meat quality when used as partial (30%) or total substitutes for FM, suggesting that processing strategies may mitigate digestive and antinutritional constraints associated with raw MSW products (Zhao et al., 2023).

MSW inclusion generally maintains meat physicochemical traits while positively modulating nutritional composition. Both defatted and full-fat MSW meals increase the proportion of *n-3* PUFAs and improve the *n-6/n-3* ratio in chicken meat, without adversely affecting pH, oxidative stability or overall sensory attributes when inclusion levels and feeding phases are appropriately managed (Dalle Zotte et al., 2024b; Miah et al., 2020; Zsedely et al., 2023). Feeding phase plays a relevant role, as inclusion during the grower–finisher phases appears more effective in improving meat nutritional quality than early dietary exposure (Dalle Zotte et al., 2024b; Zsedely et al., 2023). In quails, MSW pupae meal supports carcass yield and productive performance when replacing conventional animal protein sources, while in laying quails full-fat MSW pupae meal up to 12% sustains egg production and edible egg fraction, albeit with a deterioration of FCR and some alterations in eggshell and albumen traits at higher inclusion levels, highlighting the need for careful dose optimization (Gokulakrishnaa et al., 2025; Singh et al., 2023). Excessive inclusion levels may reduce nutrient digestibility and feed preference due to chitin content and the presence of specific bioactive compounds (see below), emphasizing the importance of technological processing or dietary adjustments (Dalle Zotte et al., 2021; Gokulakrishnaa et al., 2025).

Overall, current evidence supports MSW-derived meals as viable and sustainable ingredients for poultry diets, capable of partially replacing traditional protein sources while maintaining health status and enhancing product quality. However, production performances are MSW dose-dependent. In fact, it should be noted that silkworm pupae contain 1-deoxynojirimycin (1-DNJ), an iminosugar inhibitor of α-glucosidase and β-glucosidase enzymes, thereby limiting glucose absorption and reducing carbohydrates digestibility. This antinutritional factor may negatively affect growth performance, particularly at higher inclusion levels (Dalle Zotte et al., 2024b; Marchetti et al., 2022). Nevertheless, responses remain strongly dose-dependent and influenced by species, processing method and feeding phase, underlining the importance of optimized inclusion strategies to fully exploit the nutritional potential of MSW-based ingredients.

2.2 Swine nutrition

2.2.1 BSF in swine nutrition

Within swine nutrition, BSF-derived ingredients are increasingly regarded as viable alternatives to conventional protein and energy sources, supported by a substantial body of evidence across different production stages (Table 18.7). In weaned piglets, dietary inclusion of BSF meal in either partially defatted or full-fat form, typically at levels between 4 and 12%, is generally well tolerated and does not negatively affect growth performance, FCR, nutrients digestibility, blood profile or intestinal morphology (Biasato et al., 2019b; Phaengphairee et al., 2023; Spranghers et al., 2018; Yordanova et al., 2025). The response to BSF inclusion in swine mirrors observations in poultry, showing a largely dose-dependent pattern. Moderate replacement of SBM (approximately 20–25%) has been associated with improvements in ADG, FI, nutrient digestibility, digestive enzyme activity and intestinal morphology, along with beneficial shifts in gut microbiota, whereas higher substitution rates (≥75–100%) tend to impair performance and alter microbial profiles, likely due to increased dietary chitin levels and the limited chitin-degrading capacity of piglets (Liu et al., 2023). Consistently, increasing inclusion of full-fat BSF meal up to 19% may slightly reduce protein digestibility and ADG, while preserving overall gut integrity and showing only minor changes in colonic microbiota, suggesting that relatively high inclusion levels may still be feasible when diets are carefully balanced (Håkenåsen et al., 2021). From a protein utilization standpoint, defatted BSF meal shows good standardized ileal digestibility of crude protein and amino acids, albeit lower than FM but superior to adult BSF, confirming its suitability as a protein source for growing pigs (Kim et al., 2023). In piglet nursery diets, partial replacement of animal protein sources with full-fat BSF meal (up to 50%) maintains overall ADG, FI, gut morphology and immune indices, with no deleterious effects relative to appropriate controls (Crosbie et al., 2021). Likewise, full-fat and defatted BSF meals show high standardized ileal digestibility for most amino acids, with full-fat forms providing markedly higher digestible, metabolizable and predicted net energy than defatted forms, highlighting their dual function as both protein and energy sources (Crosbie et al., 2020).

Similar outcomes are reported when BSF is used as a FM substitute in weaning pigs through defatted or hydrolyzed preparations, with overall maintenance of growth performance and, depending on processing, improvements in digestibility or reductions in fecal *Escherichia coli* counts (Chang et al., 2024; Lee et al., 2024). Defatted BSF larvae can replace multiple protein sources up to 30% without compromising ADG, FI, FCR, fecal score or jejunal morphology, while inducing selective shifts in intestinal microbiota (Song et al., 2024). Additionally, dietary supplementation of young pigs with defatted BSF meal has been associated with improved survival, BWG and FCR, accompanied by changes in blood protein profiles, immune indices, Ca-P metabolism and markers of energy metabolism (Razanova et al., 2025). BSF may also provide a valuable source of calcium and phosphorus, as these minerals exhibit high apparent total tract digestibility in pigs, similar to that observed from inorganic sources, potentially contributing substantially to mineral supply (Grundmann, et al., 2024).

The energy suitability of BSF is further supported by studies on BSF-derived lipids. Replacement of conventional lipid sources such as corn oil with BSF fat at inclusion levels up to 6% maintains or improves growth performance and FE in nursery pigs, supporting the high energy value of this ingredient, although modest increases in serum cholesterol have been reported while remaining within physiological ranges (van Heugten et al., 2022). Similarly, inclusion of BSF fat blends as alternatives to soy and palm kernel oil has been shown to preserve post-weaning performance and microbiome development, reinforcing BSF fat as a valid alternative energy source in nursery feeding strategies (Buffoni et al., 2025). Energy characterization studies further confirm that insect fat from BSF larvae provides high digestible energy values comparable to conventional vegetable oils, supporting its incorporation as an alternative energy ingredient in pig diets (Yordanova et al., 2024).

BSF has also attracted interest for its potential functional and health-promoting properties. Overall, dietary inclusion of BSF larvae is associated with improved indicators of gut health and immune status, including favorable intestinal morphology and modulation of gut microbiota, together with signs of reduced inflammatory tone (Boontiam et al., 2022; Kar et al., 2021; Phaengphairee et al., 2023). However, under challenging post-weaning conditions, partial replacement of SBM with BSF meal in low-complexity diets maintains growth performance but does not consistently match the protective effects of highly digestible diets supplemented with animal proteins and zinc oxide in mitigating digestive disturbances (Christensen et al., 2023). Comparable functional effects have also been reported when BSF meal is included in sow and pre-weaning piglet diets. Moderate inclusion levels are associated with improved indicators of gut environment, such as enhanced fecal fermentation patterns and reduced ammonia production, without negative effects on piglet growth, whereas higher inclusion rates may require careful formulation due to chitin-related constraints (Kawasaki et al., 2023; Zhao et al., 2024). These functional responses are partly attributed to the lauric acid-rich lipid fraction of BSF, which exhibits antimicrobial properties and contributes to a moderate modulation of gut microbiota without compromising performance (Spranghers et al., 2018). Evidence further suggests that the functional value of BSF may be enhanced through appropriate dietary strategies. For instance, the combination of full-fat BSF larvae with antioxidant compounds improved pork fat oxidative stability and shelf life without affecting growth, while excessive inclusion in the absence of antioxidant support may negatively affect some hematological traits (Szczepanik et al., 2022).

Most research on dietary BSF inclusion in swine has focused on weaning and growing phases, with relatively few studies investigating its effects on finishing pigs and pork quality. Specifically, the partial or complete replacement of FM with BSF meal ensured growth performance and carcass traits and, in some cases, resulted in heavier carcasses and improved meat quality attributes, including increased intramuscular fat content and marbling scores (Chia et al., 2021; Yu et al., 2019; Zhu et al., 2022). Furthermore, pork quality is generally preserved when SBM is replaced with partly defatted BSF meal, and the increased lauric acid concentrations in backfat may serve as a dietary biomarker of BSF inclusion (Altmann et al., 2019).

BSF larvae can also be administered in live form as edible environmental enrichment. Live BSF larvae provision stimulates exploratory behavior, reduces harmful social interactions and oral manipulation of pen mates, and decreases neophobia in weaned pigs, while maintaining overall growth performance and net energy intake (Ipema et al., 2021a,b).

Taken together, the available evidence supports the effective inclusion of BSF larvae, meal, fat, and derived products in swine diets, contributing not only to sustainable nutrition and efficient growth, but also to improved gut health, resilience and animal welfare, provided that inclusion levels and dietary formulation are carefully optimized.

2.2.2 YMW in swine nutrition

Following the growing body of evidence on BSF-based ingredients in swine nutrition, YMW meal has likewise attracted increasing interest as an alternative protein source, particularly during the critical post-weaning and growing phases (Table 18.8). Across studies, YMW is generally included at levels ranging from approximately 1.5% to 10%. From a growth performance perspective, literature overall indicates that YMW supplementation is well tolerated and capable of sustaining productive outcomes at the tested inclusion levels. Linear increases in BW, ADG and FI during the early post-weaning period have been reported with increasing YMW meal inclusion, together with improved nitrogen retention and enhanced digestibility of DM and crude protein (Jin et al., 2016). Moreover, partial replacement of FM with YMW meal (50%) has been shown to improve ADG and FCR ratio without impairing immune responsiveness (Ko et al., 2020).

Not all studies, however, reported clear performance advantages compared to conventional feed ingredients. Complete replacement of FM with dried YMW larvae at low inclusion levels (1–2%) resulted in growth performance and nitrogen digestibility comparable to, but not exceeding, FM-based diets in weaned pigs, indicating that YMW does not necessarily confer additional benefits under all feeding conditions (Ao et al., 2020). In growing-finishing pigs, YMW inclusion at 5–10% does not negatively affect growth performance, with metabolomic and transcriptomic analyses revealing only minor alterations in intermediary metabolism, thereby supporting the suitability of YMW as a substitute for conventional protein sources (Meyer et al., 2020). More recent studies further confirm that dried or hydrolyzed YMW meals, even at inclusion levels approaching 10%, do not compromise growth, nutrient digestibility or fecal consistency across different production stages (Kolawole et al., 2025; Pereira et al., 2025). Consistently, Ferri et al. (2025) reported that replacing fermented soy protein concentrate with 5% YMW meal in growing pigs did not affect growth curves or FI, confirming neutral effects on productive performance under isoenergy and isonitrogenous conditions.

Beyond growth performance, YMW meal also exhibits functional properties related to nutrient utilization, gut health and metabolic stability. Several studies report improved digestibility of essential amino acids, including lysine, methionine and threonine, when YMW replaces traditional animal-derived protein sources (Cho et al., 2019; Yoo et al., 2019). At the intestinal level, YMW supplementation modulates gut microbiota composition, favoring beneficial bacterial populations while reducing potentially harmful taxa, without impairing overall digestive efficiency (Ferri et al., 2025; Zacharis et al., 2023). Hematological and metabolic indicators generally remain within physiological ranges, with stable immunoglobulin levels and unchanged inflammatory markers, supporting the overall safety of YMW based diets (Ao et al., 2020; Ferri et al., 2025). Moreover, YMW inclusion does not induce oxidative stress in liver or skeletal muscle tissues and may even enhance antioxidant enzyme activity at the muscular level (Ringseis et al., 2021).

With respect to meat quality, YMW-based diets generally preserve physicochemical traits while inducing moderate and potentially beneficial modifications in lipid composition. YMW inclusion has been associated with improvements in FA profiles and increased total phenolic content of pork, without affecting meat color or proximate composition, particularly when insects are reared on enriched substrates (Zacharis et al., 2023). Furthermore, when combined with additional bioactive compounds such as chitosan, YMW supplementation enhances meat oxidative stability and improves FA composition, suggesting added functional value beyond its role as a protein source (Zacharis et al., 2024).

In summary, these findings demonstrate that YMW meal is a technically viable and versatile ingredient for swine nutrition. At the inclusion levels evaluated, it maintains growth performance and physiological homeostasis while providing distinct functional benefits and improving the sustainability profile of the production cycle.

2.3 Rabbit nutrition

2.3.1 BSF, YMW and MSW in rabbit nutrition

In rabbit nutrition, the use of insect-derived ingredients remains less extensively investigated than in poultry and swine, as the rabbit represents a geographically and economically marginal meat production species. A key limiting factor is the current EU regulatory framework (Commission Regulation EU 2021/1372), which authorizes insect meals for aquaculture, poultry, and swine, but does not allow it for rabbits, thereby strongly limiting research funding opportunities and commercial applications. From an ecological perspective, wild rabbits may ingest insects incidentally or opportunistically, particularly under conditions of nutritional stress or limited forage availability (Dearing et al., 2005). This behavioral evidence, together with the urgent need for sustainable feed alternatives, has motivated experimental research into insect-based feeds for meat rabbits (Table 18.9).

Regarding BSF, most studies have focused on the use of BSF fat as a replacement for conventional vegetable lipid sources. Partial or total substitution of soybean or sunflower oil with BSF fat did not impair growth performance, carcass traits, nutrient digestibility, blood parameters or gut morphology in growing rabbits, despite a slight reduction in total tract apparent digestibility compared with linseed-based diets (Dalle Zotte et al., 2018; Gasco et al., 2019a; Martins et al., 2018). From a meat quality perspective, and consistent with observations in other livestock species, BSF fat increased the proportion of SFA, particularly lauric and myristic acids, while reducing PUFA compared with linseed or YMW oils, although rabbit meat from BSF-fed animals was generally less susceptible to lipid oxidation (Dalle Zotte et al., 2018; Gasco et al., 2019b). When BSF meal was used as the main dietary protein source (10% inclusion level), growth performance and meat quality were maintained or improved compared with SBM-based diets, supporting the feasibility of BSF as an alternative protein ingredient in rabbit feeding (Ogbon et al., 2025).

Evidence regarding YMW indicates that both meal and fat can be included in rabbit diets as alternative protein and lipid sources, respectively, with responses largely dependent on inclusion level and dietary balance. Partial replacement of SBM with YMW meal at moderate inclusion levels (around 4%) has been shown to improve final BW and daily weight gain (DWG), without negatively affecting FE or meat quality, while also supporting favorable changes in amino acid composition and maintaining overall carcass traits (Kowalska et al., 2020; Strychalski et al., 2021). In these studies, YMW inclusion was associated with improved apparent digestibility of ether extract and selected fiber fractions, alongside adaptive changes in cecal microbial enzyme activity, indicating an active utilization of chitin by the intestinal microbiota.

When YMW meal was used as the main dietary protein source, rabbits showed a transient reduction in DWG and FI during the early fattening phase; however, no detrimental effects were observed on final BW, FCR or coefficients of total tract apparent digestibility over the entire fattening period (Volek et al., 2021). Importantly, YMW-based diets significantly reduced urinary and total nitrogen excretion and increased nitrogen retention, suggesting an efficient utilization of dietary protein. Comparable findings were reported when YMW meal was compared with SBM and cricket meal, with no adverse effects on growth performance or nitrogen output, despite slightly lower digestibility coefficients relative to SBM (Volek et al., 2023).

From a lipid perspective, YMW oil proved to be a suitable substitute for conventional vegetable oils. Partial or total replacement of soybean oil with YMW oil did not affect growth performance, morbidity, mortality, nutrient digestibility, gut morphology or blood parameters in growing rabbits (Gasco et al., 2019a). Moreover,

YMW oil exerted positive functional effects on cecal fermentation and microbiota composition, increasing volatile FA (VFA) concentrations and enriching bacterial taxa associated with gut health (Dabbou et al., 2020b). In terms of meat quality, YMW oil maintains carcass characteristics and sensory acceptance while leading to a more favorable FA profile compared with BSF fat thanks to the higher proportions of PUFAs (Gasco et al., 2019b).

Studies on MSW-derived ingredients include both MSW pupae meal and MSW oil. Low to moderate inclusion levels of MSW pupae meal (up to 4–5%) generally improved growth performance, DWG and FE, without negatively affecting carcass traits or meat quality (Kowalska et al., 2020; Strychalski et al., 2021). However, higher inclusion levels (≥10%) were associated with reduced growth performance, nutrient digestibility and nitrogen retention, likely due to 1-DNJ and chitin-related constraints, suggesting that careful dose optimization is required (Gugołek et al., 2019, 2021). In contrast, MSW oil consistently emerged as a promising alternative lipid source, fully replacing sunflower oil without adverse effects on growth performance, carcass traits, digestibility or sensory quality of rabbit meat, while markedly improving the FA profile by increasing *n-3* FA and reducing the *n-6/n-3* ratio (Cullere et al., 2022, 2024; Dalle Zotte et al., 2022).

Several comparative and combined approaches further support the suitability of insect ingredients in rabbit nutrition. Diets including MSW pupae meal or YMW meal at moderate levels enhanced growth performance, carcass meat content and selected aspects of meat nutritional quality, while complete replacement of SBM with combinations of insect meals and white lupine seeds maintained productive performance, albeit with some alterations in intestinal short-chain FA profiles (Kowalska et al., 2025). Collectively, these findings indicate that BSF, YMW and MSW ingredients can be successfully included in rabbit diets, particularly at low to moderate inclusion levels, contributing to sustainable feeding strategies without compromising animal performance, gut health or product quality.

2.4 Ruminant nutrition

2.4.1 BSF, YMW and SW in ruminant nutrition

As in rabbit nutrition, the use of insect-derived ingredients in ruminant feeding is currently not authorized under EU legislation (Commission Regulation EU 2021/1372), which permits the use of PAPs-including the above-mentioned insect species-, only for aquaculture, poultry, and swine. Despite this regulatory constraint, research in ruminants has proceeded over the past 5 years, mainly focusing on rumen microbial activity and derived physiological and productive consequences. This research activity has been sustained by the substantial economic importance of ruminant farming and the prospect of future regulatory amendments. Ruminants differ fundamentally from monogastric in their reliance on a complex rumen microbial ecosystem, which strongly influences nutrient availability, fermentation patterns and overall animal performance (Tardiolo et al., 2025). Consequently, the inclusion of insects in ruminant diets must consider not only their nutritional value, but also their effects on ruminal degradation, microbial activity and fermentation end-products. In this context, insect meals and fats have attracted growing interest as alternative feed ingredients capable of supplying high-quality protein and energy, while potentially modulating rumen fermentation and methane emissions, with responses strongly dependent on the form of the tested ingredient (meal, oil/ fat, or functional fractions) and inclusion level (Table 18.10).

For BSF, *in vivo* studies in dairy cows and growing ruminants demonstrated a good potential to partially replace conventional protein and fat sources without compromising animal performance. In lactating dairy cows, the complete replacement of SBM or combined at a 50:50 ratio with defatted BSF meal, maintained milk production, while fiber digestibility was improved. Furthermore, FE and nitrogen use efficiency were maintained, despite a slight reduction in DM and crude protein intake at full replacement was observed (Braamhaar et al., 2025). The lipid fraction of BSF has also shown considerable potential. Replacing hydrogenated palm fat with BSF fat at approximately 3% of the concentrate, increased milk yield and FE and was associated with a reduction in oxidative stress markers, overall indicating an improved cows' metabolic status (Rastello et al., 2025). Similarly, supplementation with BSF fat at daily doses ranging from 10 to 100 g per head enhanced ruminal fermentation and VFA production and resulted in modest improvements in milk yield and milk fat content, without impairing digestion or rumen function (Nekrasov et al., 2022).

The suitability of BSF meal has also been explored in beef cattle and growing ruminants fed low-quality forages. When supplied at isonitrogenous levels, BSF meal consistently stimulated FI and total digestible organic matter intake compared with standard diets, showing responses broadly comparable to those obtained with conventional protein supplements such as cottonseed meal or soybean-based concentrates (Carrasco and Drewery, 2024; Fukuda et al., 2022; Maggit et al., 2025). In parallel, BSF-derived lipids have been tested in young ruminants: incorporation of BSF oil into calf milk replacer fat blends, at levels up to 40% of the fat fraction, did not affect growth performance or FE and, during early life, reduced the probability of abnormal fecal events in some treatments, supporting its practical feasibility as a milk replacer fat ingredient (Quigley et al., 2025). Although small shifts in specific microbial populations were observed, overall rumen fermentation patterns and digestibility were largely unaffected, suggesting good ruminal tolerance.

Similar conclusions were drawn in fattening lambs, where BSF meal effectively replaced SBM without compromising growth performance, ruminal fermentation or stress-related biochemical indicators (Odeon et al., 2025). Evidence from *in vitro* studies further supports the use of BSF, indicating that replacing 20–40% of SBM in the concentrate can enhance DM and fiber digestibility while reducing ammonia production, although excessive inclusion levels may alter gas production due to the combined effects of lipids and chitin (Kahraman et al., 2023).

Additional insights come from studies focusing on rumen fermentation and methane mitigation. In Thai indigenous steers, supplementation with BSF fat at 1–4% DM reduced FI at the highest inclusion level but improved ether extract digestibility; notably, protozoal populations declined and estimated methane production decreased progressively with increasing BSF fat inclusion, with 2% BSF fat emerging as a particularly balanced level (Prachumch et al., 2025). Mechanistic evidence from *in vitro* studies aligns with these observations, indicating that BSF meal, especially that characterized by higher fat and chitin-associated fractions, tend to produce less methane than SBM, although they may also show lower *in vitro* organic matter digestibility depending on insect composition and processing (Jayanegara et al., 2017; Renna et al., 2022).

YMW meal is characterized by relatively low ruminal nitrogen degradability and high intestinal digestibility of rumen-undegraded protein, making it a particularly attractive source of metabolizable protein for ruminant diets. Multiple evaluation approaches -*in vitro*, *in situ* and regression techniques have confirmed substantial rumen YMW bypass and efficient post-ruminal digestion of the undegraded fraction (Toral et al., 2022). In fattening lambs, the inclusion of YMW meal at around 6% DM resulted in DM intake comparable to FM, improved total-tract digestibility compared with SBM, and a favorable nitrogen balance, while maintaining normal rumen and intestinal morphology (Robles-Jiménez et al., 2025). However, ADG tended to be lower than that observed with conventional protein sources, suggesting that YMW meal may be more effectively used as a partial rather than a complete replacement (Robles-Jiménez et al., 2025). In forage-based beef systems, whole dried YMW larvae supplied at isonitrogenous levels also enhanced forage utilization and, in some cases, elicited a stronger intake response than BSF meal (Carrasco and Drewery, 2024). In addition, several *in vitro* studies indicate that YMW inclusion may contribute to reduced methane emissions compared with SBM, although nutrients digestibility can decline when inclusion levels become excessive, largely due to the high lipid content (Jayanegara et al., 2017; Renna et al., 2022).

Information on MSW-derived ingredients in ruminant nutrition is currently more limited and largely based on *in vitro* approaches. Nevertheless, available studies suggest that MSW pupae meal can be used as a viable alternative protein source in cows. When included at approximately 20–25% of dietary DM to replace conventional protein sources, MSW pupae meal showed digestibility comparable to meat and bone meal and was associated with lower total gas and methane production, without major alterations in ruminal degradability (Purnamia et al., 2026). Consistently, partial substitution of SBM (25%) with MSW pupae meal did not impair nutrient digestibility and was accompanied by a reduction in methane output (Ahmed et al., 2021). Dose response effects have also been highlighted: *in vitro* evaluations indicate that MSW-derived ingredients can be included up to about 20% of the diet without adverse effects on rumen fermentation, whereas higher inclusion levels reduce gas production, VFA synthesis and nutrient digestibility, likely due to the high lipid content of the insect material (Ahmed and Nishida, 2023). Beyond whole meals, MSW-derived chitosan has attracted attention for its functional properties. Included at around 6% of substrate DM, chitosan fractions markedly reduced methane production and shifted rumen fermentation toward higher propionate proportions, although these effects were accompanied by a reduction in DM digestibility, highlighting the need for careful dose optimization (Sagala et al., 2024).

In conclusion, insect-derived ingredients represent promising and sustainable alternatives for use in ruminant nutrition. When appropriately formulated and managed, they can maintain or enhance animal performance, improve nitrogen efficiency, modulate rumen function, and contribute to environmental sustainability goals. Future research should focus on long-term feeding trials, detailed characterization of rumen microbial responses, evaluation of product quality (dairy and meat), assessment of methane mitigation potential, and economic and regulatory pathways to facilitate commercial adoption.

3 Sustainablity aspects

As already discussed in the introduction, beyond their favorable nutritional profile and high production efficiency, insects are attracting increasing interest due to their potential contribution to environmental sustainability. Insect production systems are often associated with low environmental impacts, particularly in terms of greenhouse gas (GHG) emissions, land use, and water consumption (van Huis and Oonincx, 2017). This aspect is especially relevant in a context where the environmental footprint of the livestock sector is under growing scrutiny. Livestock production, particularly ruminant species (cattle, sheep, goats), is considered one of the major contributors to global GHG emissions, considering both direct emissions and indirect sources, such as those associated with feed production (Campbell et al., 2017; Steinfeld et al., 2006). Moreover, most agricultural land is currently devoted to livestock production, either as pasture or for feed crop cultivation (Steinfeld et al., 2006; Voora et al., 2020), and the expected increase in future demand for animal products is likely to exacerbate land-use pressure. In this context, soybean production has been identified as a key driver of land-use change and deforestation, posing a significant threat to biodiversity (van Huis and Oonincx, 2017; Voora et al., 2020). Similarly, FM production is widely recognized as a major source of pressure on marine ecosystems due to its reliance on wild-caught biomass, contributing to stock depletion, bycatch of non-target and juvenile species, and the persistence of illegal, unreported, and unregulated fishing practices (Olsen and Hasan, 2012). For these reasons, insects have increasingly been proposed as a promising alternative to conventional feed ingredients. However, a few critical aspects must be carefully considered to determine whether insects can represent a key solution to the aforementioned challenges.

From a life cycle assessment (LCA) perspective, insect-based proteins can be compared with conventional protein sources such as meat, milk, FM, and SBM, as demonstrated for species including *Tenebrio molitor, Hermetia illucens, Acheta domesticus*, and *Musca domestica* (Halloran et al., 2017; Miglietta et al., 2015; Oonincx and de Boer, 2012). One of the main limitations highlighted by LCA studies is the relatively high energy demand required to maintain optimal rearing temperatures (van Huis and Oonincx, 2017). In addition, a substantial share of total energy consumption is associated with post-harvest processing, especially drying, which is necessary to obtain a stable and storable insect meal (van Zanten et al., 2015). Improving energy efficiency is therefore considered a key priority for enhancing the overall sustainability of insect production systems.

The environmental performance of insect-based feed systems is strongly influenced by the type of substrate used for rearing and by the efficiency of the production process (van Huis and Oonincx, 2017). Across livestock systems, feed production represents the main contributor to land use, energy demand, and GHG emissions, and insects may offer advantages when they efficiently convert feed into biomass. In particular, the use of low-value agri-food by-products as rearing substrates can significantly reduce environmental burdens (van Huis, 2022). This approach represents a clear advantage in terms of resource valorization, as it enables the transformation of organic waste into valuable products rather than its disposal (Fowley and Nansen, 2020). Conversely, the use of conventional protein and lipid-rich feed ingredients in insect diets largely offsets these potential sustainability benefits.

A life cycle inventory (LCI) analysis published by Smetana et al. (2016) confirms the strong influence of substrate choice and processing intensity on the environmental performance of insect-based feed ingredients. To produce one functional unit (1 kg of product plus 0.9 kg of fat), insect protein powders require highly variable amounts of raw materials. While waste-based substrates reduce dependence on conventional agricultural inputs, they are associated with lower insect biomass and protein yields, resulting in increased resource use during the rearing stage and additional requirements for biowaste treatment. In contrast, high-quality diets enhance insect productivity but are linked to higher upstream environmental impacts due to their agricultural origin. Energy demand is also strongly affected by processing intensity, with dried insect protein powders requiring substantially more heat than high-moisture intermediates, mainly due to drying operations.

These findings indicate that insects cannot be considered inherently more sustainable than conventional protein sources such as SBM or FM; instead, their environmental performance is highly context dependent.

Although the use of waste and by-product-based substrates is often proposed as a key advantage of insect production systems, only a limited number of such scenarios achieve a lower environmental impact than conventional feed benchmarks. In particular insect production may outperform SBM and FM when based on nutritionally suitable low-value by-products, such as distillers' dried grains, or waste streams that provide environmental credits through waste reduction. Conversely, substrates associated with low biomass yields or high processing and energy requirements can offset these advantages, leading to environmental impacts similar to or greater than those of conventional protein sources.

Therefore, the sustainability advantage of insect-based feeds over SBM and FM cannot be generalized and must be demonstrated on a case-by-case basis through detailed LCA. In this context, the assessment explicitly accounted for multiple interacting variables, including substrate type, insect biomass and protein yields, energy demand for rearing and processing, waste treatment requirements, co-product allocation, functional unit definition, and impact assessment methodology, highlighting how each of these factors contributes to the observed variability in environmental performance (Smetana et al., 2016).

When BSF-based protein is compared with conventional protein sources under defined production scenarios, clear trade-offs emerge across different impact categories. The energy demand associated with the production of one ton of protein from BSF larvae is lower than that reported for FM, but higher than that of SBM, reflecting both the biological efficiency of insects and the energy requirements of rearing and processing systems (Salomone et al., 2017). However, when global warming potential is evaluated in combination with land use, BSF-based protein shows lower GHG emissions than both FM and SBM, primarily due to reduced pressure on land resources and the avoidance of land-use change associated with soybean cultivation (Salomone et al., 2017; van Zanten et al., 2015). In terms of land occupation, BSF protein production requires more land than FM, which is largely decoupled from terrestrial land use, but substantially less land than SBM. These results confirm that the relative sustainability of BSF protein depends on the impact category considered and reinforce the need for multi-criteria LCA approaches rather than single-metric comparisons.

Overall, insects represent a promising, but not universally sustainable, alternative to conventional protein sources used in livestock feed. Their potential environmental advantages over SBM and FM are highly dependent on production context, particularly substrate selection, energy efficiency, processing intensity, and system scale. While insect-based feed ingredients can contribute to reduced GHG and land use under optimized conditions, inappropriate substrate choices or energy-intensive processing may offset these benefits. Consequently, sustainability claims related to insect-based feeds must be supported by case-specific LCA rather than generalized assumptions. Identifying production pathways that balance biological efficiency, resource valorization, and environmental performance will be essential for the successful integration of insects into sustainable livestock feed systems.

4 Challenges and future perspectives

Despite the growing interest in insect-based feeds, several challenges still limit their widespread adoption in livestock production systems. Current EU regulations permit the use of insect-derived PAPs for omnivorous monogastric species (poultry, pigs) and aquaculture, while prohibiting their use in ruminants and non-omnivorous monogastric species, such as the rabbit. These legislative frameworks constrain both the scope of research and the practical implementation of insect proteins in the livestock sector. Furthermore, the large-scale deployment of insect-based feed ingredients is influenced by economic competitiveness, safety assurance, processing efficiency and biological variability, as well as to the acceptance of insect-based feeds by stakeholders and consumers. Addressing these interconnected aspects will be essential to fully exploit the potential of insects as sustainable feed ingredient.

Economic factors play a key role in shaping the future of the insect feed sector. At present, insect-based feed ingredients generally remain more expensive than conventional protein sources, mainly due to high production costs associated with labor, energy demand and limited automation (Baiano et al., 2020; Cadinu et al., 2020). In this context, one significant challenge for the large-scale implementation of insect-based feeds lies in the identification of low-cost, non-competitive rearing substrates capable of ensuring a consistent and standardized final product. The use of organic waste materials and agro-industrial by-products as rearing substrates represents a promising strategy to improve sustainability and reduce production costs. However, their variable nutritional composition, regulatory constraints, and limited availability often restrict their practical application at an industrial scale (van Huis, 2022).

Moreover, the choice of the feeding substrate plays a pivotal role in determining the nutritional quality of insects. Insect nutrient profiles are highly variable and depend on several factors, including insect species, developmental stage, substrate type, and processing methods (Oonincx and Finke, 2021; Payne et al., 2016). This variability, particularly in protein and lipid content and quality, as well as in other components such as chitin, complicates the production of standardized feed ingredients (Meyer-Rochow et al., 2021). Consequently, precise diet formulation is challenging, and the comparability of experimental results across studies is often limited.

Improving process efficiency through technological innovation, greater mechanization and optimized production workflows represents a critical step towards enhancing the competitiveness of insect-derived products relative to conventional feed ingredients (Kröncke et al., 2020). Farm productivity in insect rearing systems is shaped by a complex interplay of factors that extend beyond the choice of feeding substrates. Rearing conditions and dietary strategies directly influence insect physiological status, growth dynamics and final biomass quality (Kazemi and Valizadeh, 2025). Advancing knowledge on these interactions and translating it into optimized production practices will be essential to support the long-term development and sustainability of insect farming (Cadinu et al., 2020; van Huis, 2022). In parallel, increasing attention is being paid to the biological aspects of insect production. Research has highlighted substantial variability among strains within the same species in terms of growth rate, body composition, reproductive performance and resistance to pathogens (de Miranda et al., 2021; Zhou et al., 2013). Given the short life cycles of insects, targeted genetic selection offers a realistic opportunity to improve productivity and resilience, thereby supporting the long-term sustainability of insect farming for feed purposes.

In addition to economic aspects, ensuring the safety of insect-derived feed ingredients remains a fundamental requirement. Potential concerns include the presence of pathogenic microorganisms, chemical contaminants and allergens, which may originate from rearing substrates, farming conditions or processing steps (Braide et al., 2011; Eilenberg et al., 2015; Palumbo et al., 2025). While most viruses naturally associated with insects are species-specific and not harmful to vertebrates, they may cause significant production losses and therefore represent a practical challenge for insect producers (Leuschner et al., 2010). Overall, these risks can be effectively managed through appropriate substrate selection, strict hygiene measures and suitable processing technologies, highlighting the importance of robust quality control systems along the entire production chain (ANSES, 2015; EFSA, 2015).

Beyond technical, economic and safety-related considerations, the successful integration of insect-based feeds into livestock production systems also depends on their acceptance by downstream stakeholders and consumers. Even when safety and nutritional adequacy are ensured, perceptions related to feed origin, production practices and product quality may influence the willingness to adopt insect-derived ingredients, particularly for food-producing animals. Compared with direct human consumption, the use of insects as feed ingredients tends to be more readily accepted by consumers (Kostecka et al., 2017; Mancuso et al., 2016). Surveys conducted in different countries consistently show higher acceptance of insect-fed fish and poultry, whereas lower acceptance is observed for pigs and, especially, cattle (Ankamah-Yeboah et al., 2018; de Faria Domingues et al., 2020; Mancuso et al., 2016; Verbeke et al., 2015). These differences appear to be influenced by cultural, ethical and safety-related perceptions, indicating that consumer attitudes, although indirect, may still affect the market development of insect-based feeds.

Looking ahead, the role of insects in livestock nutrition will largely depend on the ability to integrate technological innovation, regulatory evolution and biological optimization into economically viable production systems. Continued progress in production efficiency, standardization and safety management, combined with clearer communication along the feed and food chains, will be key to building confidence among producers and consumers. In this sense, insects should be viewed not as a universal solution, but as feed ingredients whose value and sustainability depend on specific production contexts and well-defined conditions of use.

Tables

TABLE 18.1 Impact of dietary black soldier fly (*Hermetia illucens*) inclusion on broiler growth performances and health status

Variables	Avian species	Insect form	Substitution level range (%)	Inclusion level range (%)	Impact	References
Growth performance, feed conversion ratio (FCR), health, welfare, litter quality	chicken	meal, oil, dried, live		8	P<0.05	Ipema et al. 2022
Growth performance, FCR, slaughter yield	chicken	meal, oil, live		3.49-10	P<0.05	Dörper et al. 2024a
Growth and gut health	chicken	meal	25–75		NS	Machado et al. 2023
Feed intake (FI), FCR	chicken	meal		3-5	P<0.05	Attia et al. 2023
Growth performance, FRC, mortality, gut organs	chicken	meal	12.5-100		P<0.05	Facey et al. 2023
Growth performance, excreta metabolome	chicken	meal	12.5-100		P<0.05	Fruci et al. 2023
Growth performance, FCR, slaughter yield	chicken	meal		6-12	P<0.05	Baderuddin et al. 2024
Slaughter weight (SW)	chicken	meal	50-100	12-25.6	P<0.05	La Mantia et al. 2024
Growth performance, FCR	chicken	meal		5-10	NS	Saidani et al. 2025
Gut health	chicken	meal		5-10	P<0.05	Saidani et al. 2025
Growth performance, FCR	chicken	whole		10-30[a]	P<0.05	Seyedalmoosavi et al. 2022
SW, propionic and lactic acid of excreta	chicken	live		5[a]	P<0.05	Fiorilla et al. 2024a
SW, slaughter yield	chicken	live		10[a]	P<0.05	Bongiorno et al. 2022
FCR, slaughter yield, gut health, immune markers	chicken	live		5	NS	Fiorilla et al. 2024a
Growth performance, FCR	chicken	fat		3-9	P<0.05	Kierończyk et al. 2023
FCR	chicken	fat	100		P<0.05	Aslam et al. 2025
Daily weight gain (DWG), FI		fat	100		NS	Aslam et al. 2025
Growth performance, FCR	turkey	meal	-	5	P<0.05	Zampiga et al. 2025
Growth performance	turkey	fat	50-100		NS	Kierończyk et al. 2022
Nutrients' digestibility, gut health, immune markers	turkey	fat	50-100		P<0.05	Kierończyk et al. 2022
Growth performance, slaughter performance	Muscovy-duck	live		5[a]	NS	Gariglio et al. 2025
Growth performance, FCR, health, feed choice, slaughter yield	quail	meal		10–15	NS	Cullere et al. 2016
Ether extract digestibility	quail	meal		10–15	P<0.05	Cullere et al. 2016
Growth performance	pheasant	meal	50-100	12.5-28		Rytlewski et al. 2025
	pheasant	whole	50-100	12.5-28		Rytlewski et al. 2025

[a] of the expected average daily feed intake.

TABLE 18.2 Impact of dietary black soldier fly (*Hermetia illucens*) inclusion on broiler meat quality

Variables	Avian species	Insect form	Substitution level range (%)	Inclusion level range (%)	Impact	References
pH, cooking loss[1], shear force[1]	chicken	meal	50–75	0.5–1.0	NS	Choi et al. 2021
				5–15	NS	[1]Altmann et al. 2020
					NS	Schiavone et al. 2019
Color[1,2]; TBARS[a], DPPH radical scavenging	chicken	meal	25–50	0.5–1.0	P<0.05	Choi et al. 2021
				5–15	P<0.05	[1]Kim et al. 2021
					P<0.05	[2]Schiavone et al. 2019
pH[1,2]; color[1,2]; lipid oxidation, cooking loss, shear force	chicken	meal	50–75	5–20	NS	de Souza Vilela et al. 2021
				5	P<0.05	[1]Popova et al. 2020
					P<0.05	[2]Altmann et al. 2020
Amino acid (AA), fatty acid (FA) profiles	chicken	meal		5–20	P<0.05	de Souza Vilela et al. 2021
FA profile	chicken	meal	25–50		NS	Kim et al. 2021
				5	P<0.05	Popova et al. 2020
			50–75		P<0.05	Altmann et al. 2020
				5–15	P<0.05	Schiavone et al. 2019
Mineral profile	chicken	meal	25–50	–	P<0.05	Kim et al. 2021
Bioactive peptides, volatile profile	chicken	meal	100	–	P<0.05	Gkarane et al. 2020a; 2020b
Proximate composition	chicken	meal		5	P<0.05	Popova et al. 2020
			50–75		NS	Altmann et al. 2020
				5–15	P<0.05	Schiavone et al. 2019
Sensory evaluation	chicken	meal	50–75	–	NS	Altmann et al. 2020
pH, color, water holding capacity (WHC), proximate composition, AA, FA and mineral profiles, sensory evaluation	chicken	meal		5–15	NS	Pieterse et al. 2019
FA profile		whole dry/live		5[b]	NS	Fiorilla et al. 2024a
pH, color, thaw loss, shear force, proximate composition, lipid oxidation	chicken	whole dry/live		5[b]	NS	Fiorilla et al. 2024b
FA profile	chicken	live		5[b]	P<0.05	Fiorilla et al. 2024b
FA profile	chicken	fat	100		P<0.05	Kim et al. 2020
			50–100		P<0.05	Cullere et al. 2019a
			50–100		P<0.05	Schiavone et al. 2017
FA profile, WHC, color	chicken	fat		3–9	P<0.05	Kierończyk et al. 2023

TABLE 18.2 Impact of dietary black soldier fly (*Hermetia illucens*) inclusion on broiler meat quality (*cont.*)

Variables	Avian species	Insect form	Substitution level range (%)	Inclusion level range (%)	Impact	References
pH[1], thaw loss, proximate composition[1], TBARS	chicken	fat	50–100		NS	Cullere et al. 2019a
					NS	Schiavone et al. 2017
Cholesterol	chicken	fat	50–100		P<0.05 / NS	Cullere et al. 2019a
Sensory evaluation	chicken	fat	50–100		NS	Cullere et al. 2019a
pH, color, total moisture loss, shear force, heme iron, shelf life	quail	meal		10	NS	Cullere et al. 2019b
Proximate composition, cholesterol, AA, FA profiles, sensory evaluation	quail	meal		10	P<0.05	Cullere et al. 2019b
pH, color, WHC, shear force	quail	meal	25–100		NS	Mbhele et al. 2019
Proximate composition, sensory evaluation, cholesterol, TBARS	quail	meal		10–15	NS	Cullere et al. 2018
Mineral, FA, AA profiles	quail	meal		10–15	P<0.05	Cullere et al. 2018
pH, cooking loss	quail	meal		10–15	P<0.05	Cullere et al. 2016
Color, shear force, AA profile	quail	meal		10–15	NS	Cullere et al. 2016
pH, drip loss cooking loss, shear force, proximate composition	turkey	meal		5	NS	Zampiga et al. 2025
Color (yellowness)	turkey	meal		5	P<0.05	Zampiga et al. 2025
pH, shear force, cook loss, proximate composition	Barbary partridge	meal	25–50		NS	Secci et al. 2018
Color, FA profile	Barbary partridge	meal	25–50		P<0.05	Secci et al. 2018
pH, color, proximate composition, TBARS	Muscovy duck	meal		3–9	NS	Gariglio et al. 2021
FA profile, mineral profile	Muscovy duck	meal		3–9	P<0.05	Gariglio et al. 2021
pH, color, WHC, shear force, proximate composition	Muscovy duck	live		5[c]	NS	Gariglio et al. 2025
FA profile	Muscovy duck	live		5[c]	P<0.05	Gariglio et al. 2025
FA profile	pheasant	meal, whole	50-100	12.5-28	P<0.05	Rytlewski et al. 2025
Sensory analysis	pheasant	meal, whole	50-100	12.5-28	NS	Rytlewski et al. 2025

[a]TBARS = thiobarbituric acid reactive substances; [b]of the expected daily DM intake; [c]of the expected average daily feed intake.
Source: Dalle Zotte, 2021, modified.

TABLE 18.3 Impact of dietary black soldier fly (*Hermetia illucens*) inclusion on laying chicken and quail

Variables	Avian species	Insect form	Substitution level range (%)	Inclusion level range (%)	Impact	References
Gut health	chicken (pullet)	meal	25–100		P<0.05	Ndotono et al. 2022
Feed conversion ratio (FCR), egg production	chicken	meal		5	P<0.05	Petkov et al. 2022
Fatty acid (FA) profile	chicken	meal		12-24	P<0.05	Chatzidimitriou et al. 2022
Production performance, egg quality	chicken	meal		1.5-3	NS	Zhao et al. 2022
Egg weight, yolk weight	chicken	meal		1.5-3	P<0.05	Zhao et al. 2022
Production performance, egg weight, sensory analysis	chicken	meal		1.5-3	NS	Navasero et al. 2022
Nutrient and energy digestibility	chicken	meal		1.5-3	P<0.05	Navasero et al. 2022
Production performance, egg quality, physiological status, economic efficiency	chicken	meal		3-12	P<0.05	Nassar et al. 2023
Production performance, physiological and immunological status	chicken	meal		7	NS	Lalev et al. 2023b
Production performance, egg weight, reproductive physiology	chicken	meal		4-8	P<0.05	Attivi et al. 2023
Production performance, shell thickness	chicken	meal	50-100	7-13	P<0.05	Mancini et al. 2025
Production performance, egg quality, sensory analysis	chicken	meal		5-10	NS	Veldkamp et al. 2025
Feed efficiency, egg yolk color	chicken	meal		5-10	P<0.05	Veldkamp et al. 2025
Egg weight, yolk color	chicken	meal + fat		4.1-8.1 + 0.8-1.6	P<0.05	Dörper et al. 2024
Production performance, egg quality	chicken	whole dry		5	NS	Montalbán et al. 2024

TABLE 18.3 Impact of dietary black soldier fly (*Hermetia illucens*) inclusion on laying chicken and quail *(cont.)*

Variables	Avian species	Insect form	Substitution level range (%)	Inclusion level range (%)	Impact	References
Welfare, fearfulness, feather pecking	chicken	whole dry		1[a]	P<0.05	Tahamtani et al. 2025
Water consumption, yolk color	chicken	live		5-10[a]	P<0.05	Dörper et al. 2024
Production performance, egg weight	chicken	live		1-3	NS	Lokaewmanee et al. 2023
Yolk color, FA profile	chicken	live		1-3	P<0.05	Lokaewmanee et al. 2023
Gut microbiota	chicken	live		10-20[b] - *ad-lib.*	P<0.05	Huang et al. 2024
Behavior	chicken	live		10-20[b] - *ad-lib.*	NS	Huang et al. 2024
Production performance, health status	chicken	live		15-30[c]	NS	Dabbou et al. 2025
Corticosterone levels	chicken	live		15-30[c]	P<0.05	Dabbou et al. 2025
Production performance, egg quality	chicken	fat	50-100		NS	Kim et al. 2022
Yolk color, FA profile	chicken	fat	50-100		P<0.05	Kim et al. 2022
Production performance, egg weight, yolk cholesterol content, sensory analysis	quail	meal		10-15	NS	Dalle Zotte et al. 2019
Eggshell thickness, yolk color, proximate composition, FA profile, shelf life	quail	meal		10-15	P<0.05	Dalle Zotte et al. 2019
Production performance, egg quality, protein digestibility, health status	quail	whole unprocessed		10-30	P<0.05	Liu et al. 2025
Microbiome, metabolome	quail	whole unprocessed		10-30	NS	Liu et al. 2025
Production performance, egg quality, cholesterol, gene expression lipid metabolism	quail	fat		1-2	P<0.05	Armaghaza et al. 2025

[a] One g/hen/day; [b] of the expected average daily DM intake: [c] of the expected daily feed intake (as fed basis)

TABLE 18.4 Impact of dietary yellow mealworm (YMW, *Tenebrio molitor*) inclusion on broiler growth performance, health status and meat quality

Variables	Avian species	Insect form	Substitution level range (%)	Inclusion level range (%)	Impact	References
Growth performance, hematological and serum parameters, gut health	chicken	meal		7.5	NS	Biasato et al. 2016
Growth performance, feed conversion ratio (FCR), carcass traits, cecal microbiota*, intestinal mucin*	chicken	meal		5-15	P<0.05	Biasato et al. 2017, 2019a*
Growth performance, carcass traits, meat physicochemical traits	chicken	meal	100		NS	Bovera et al. 2016
FCR, nutrients digestibility	chicken	meal	100		P<0.05	Bovera et al. 2016
Meat pH, color, drip loss, proximate composition, fatty acid (FA) profile	chicken	meal		7.5	NS	Dabbou et al. 2020a
Meat pH, color, water holding capacity (WHC), shear force	chicken	meal		2-8	NS	Elahi et al. 2020
Meat proximate composition, sensory evaluation	chicken	meal		17	NS	Pietras et al. 2021
Meat pH, color, WHC, proximate composition, TVBN[a], sensory evaluation	chicken	meal		1-3	P<0.05	Shaviklo et al. 2021
Growth performance, gut health	chicken	meal		2.5-5	NS	Sedgh-Gooya et al. 2022
Growth performance, FCR, carcass weight, breast yield, meat physical traits	chicken	meal		2.5-10	P<0.05	Petkov et al. 2024
Carcass traits, meat physicochemical traits	chicken	meal	50-100		NS	Nieto et al. 2024
FA profile	chicken	meal	50-100		P<0.05	Nieto et al. 2024

TABLE 18.4 Impact of dietary yellow mealworm (YMW, *Tenebrio molitor*) inclusion on broiler growth performance, health status and meat quality *(cont.)*

Variables	**Avian species**	**Insect form**	**Substitution level range (%)**	**Inclusion level range (%)**	**Impact**	**References**
Growth performance, serum biochemical parameters	chicken	meal		3-9	NS	Jang et al. 2024
Oxidative stress, resilience under heat stress	chicken	meal		5	P<0.05	Stelios et al. 2024
FCR, carcass weight, FA profile, meat quality, sensory traits	chicken	meal		5-10	P<0.05	Biasato et al. 2025
Growth performance, slaughter performance, hematological and serum parameters, gut health	chicken	live		5[d]	NS	Bellezza Oddon et al. 2021
FCR	chicken	live		5[b]	P<0.05	Bellezza Oddon et al. 2021
Growth performance, nutrients digestibility	chicken	fat		5	NS	Benzertiha et al. 2019
Blood biochemical parameters, meat FA profile	chicken	fat		5	P<0.05	Benzertiha et al. 2019
Growth performance, FCR, carcass weight*, carcass yield*, FA classes*	Muscovy duck	live		5[b]	NS	Gariglio et al. 2023, 2025*
Welfare, meat composition*	Muscovy duck	live		5[b]	P<0.05	Gariglio et al. 2023, 2025*
pH, shear force, cook loss, proximate composition, cholesterol	Barbary partridge	meal	25-50		NS	Secci et al. 2018
Color, FA profile	Barbary partridge	meal	25-50		P<0.05	Secci et al. 2018
WHC	quail	meal		7.5-30	NS	Zadeh et al. 2019
Color	quail	meal		7.5-30	P<0.05	Zadeh et al. 2019

[a] TVBN = Total volatile basic nitrogen; [b] of the expected average daily feed intake

TABLE 18.5 Impact of dietary yellow mealworm (YMW, *Tenebrio molitor* inclusion on laying chicken and quail

Variables	Avian species	Insect form	Substitution level range (%	Inclusion level range (%	Impact	References
Yolk color, fatty acid (FA) profile, histidine	chicken[a]	meal		1-3	P<0.05	Ko et al. 2020
Egg production, egg mass, feed conversion ratio (FCR)	chicken	meal		2.5-5	P<0.05	Sedgh-Gooya et al. 2021
Egg quality, blood parameters	chicken	meal		2.5-5	NS	Sedgh-Gooya et al. 2021
Egg quality, sensory traits	chicken	meal		8	NS	Remón et al. 2025
Health, FCR	quail[b]	meal	25-100	0.75-3		Shariat Zadeh et al. 2020
Serum cholesterol	quail	meal	25-100	0.75-3	P<0.05	Shariat Zadeh et al. 2020
Production performance, egg mass, FCR, nutrients digestibility, yolk FA profile	quail	meal		1.4-5.6	P<0.05	Secci et al. 2021
Gut health, cecal short chain FA	quail	meal		1.4-5.6	P<0.05	Addeo et al. 2022
Production performance, egg mass, FCR	quail	meal		3-12	NS	Nuraini et al. 2022
Egg yolk fat & cholesterol	quail	meal		3-12	P<0.05	Nuraini et al. 2023
Production performance, egg mass, FCR, blood metabolites	quail	meal		1	NS	Sosan et al. 2025
Egg weight, gut health	quail	meal		1	P<0.05	Sosan et al. 2025
Production performance, gut health, egg quality	quail	meal		1	NS	Mohammadi et al. 2025
Egg weight, yolk weight, cecal *L. plantarum*	quail	meal		1	P<0.05	Mohammadi et al. 2025
Nutrients and energy digestibility, chitin digestibility, sensory traits	quail	live		10[c]-100	P<0.05	Dalle Zotte et al. 2024a
Production performance, egg quality, egg storage stability	quail	live		10[c]-100	NS	Dalle Zotte et al. 2024a

[a] *Gallus gallus domesticus*; [b] *Coturnix japonica*; [c] of the expected daily feed intake.

TABLE 18.6 Impact of dietary mulberry silkworm (MSW; *Bombyx mori*) inclusion on poultry growth performances, health status and meat quality

Variables	Avian species/ type	Insect form	Substitution level range (%)	Inclusion level range (%)	Impact	References
Growth performance	chicken[a]/ broiler	meal		4-16	P<0.05	Emadi et al. 2015
Growth performance, feed conversion ratio (FCR), meat proximate composition	chicken/ broiler	meal	25-50	7-14	NS	Miah et al. 2020
Fatty acid (FA) profile	chicken/ broiler	meal	25-50	7-14	P<0.05	Miah et al. 2020
Growth performance, slaughter weight, amino acid profile, C18:3n-3	chicken/ broiler	meal		10	P<0.05	Lalev et al. 2022a
Growth performance, health status, carcass traits, meat quality, skin yellowness	chicken/ broiler	meal[b]		5-15	P<0.05	Kongsup et al. 2022
FCR, meat shear force, immune organs	chicken/ broiler	meal[c]	30-100	0.45-3	P<0.05	Zhao et al. 2023
Growth performance, FCR, carcass traits, rheological meat quality	chicken/ broiler	meal		4	NS	Zsedely et al. 2023
Meat proximate composition	chicken/ broiler	meal		4	P<0.05	Zsedely et al. 2023
Meat FA profile, sensory evaluation	chicken/ broiler	meal		4	P<0.05	Dalle Zotte et al. 2024b
Net utilization of energy and protein	chicken/ broiler	meal		10	P<0.05	Lalev et al. 2023a
Egg yolk color, FA profile	chicken/ hen	pupae		7	P<0.05	Lalev et al. 2022b
Production performance, FCR, physiological and immunological status	chicken/ hen	meal		7	NS	Lalev et al. 2023b
Nutrient and energy digestibility, feed choice, fecal microbiome	quail[d]/ broiler	meal		12.5	P<0.05	Dalle Zotte et al. 2021
Production performance, FCR, egg physical quality	quail /hen			4-12	P<0.05	Singh et al. 2023
Growth performance, FCR	quail / broiler	meal[c]	25-100	7.5-30	P<0.05	Gokulakrishnaa et al. 2025

[a] *Gallus gallus domestics;* [b] *Samia ricini;* [c] Fermented pupae; [d] *Coturnix japonica.*

TABLE 18.7 Impact of dietary black soldier fly (BSF; *Hermetia illucens*) inclusion on swine growth performances, health status and meat quality

Variables	Feeding length	Insect form	Substitution level range (%)	Inclusion level range (%)	Impact	References
Milk proximate composition, piglets' growth performance, feed conversion ratio (FCR)	lactating sow/ piglets	meal		2.3-4.6 /12-25	NS	Zhao et al. 2024
Piglets fecal free amino acids (AA), ammonia, and SCFA[a]	lactating sow/ piglets	meal		2.3-4.6 /12-25	P<0.05	Zhao et al. 2024
Growth performance	35 days (35-70 days of age)	meal		2.5-5	NS	Szczepanik et al. 2021
Blood parameters, carcass backfat	35 days (35-70 days of age)	meal		2.5-5	P<0.05	Szczepanik et al. 2021
Growth performance, FCR, gut health	28 days (nursery pigs)	meal		10-30	NS	Song et al. 2024
Growth performance, fecal digestibility	15 days (weanling)	meal		4-8	NS	Spranghers et al. 2018
Growth performance	14 days (weanling)	meal		30	P<0.05	Christensen et al. 2023
Fecal *E. coli* CFU, fecal consistency	14 days (weanling)	meal		30	NS	Christensen et al. 2023
Nutrient digestibility, growth performance	28 days (28-56 days of age)	meal	25-100	4.5-18.3	P<0.05	Liu et al. 2023
Growth performance, FCR, nutrient digestibility, blood and serum parameters, gut health	61 days (growing)	meal		5-10	NS	Biasato et al. 2019b
Growth performance, gut health, blood parameters	42 days (weanling)	meal		4-25	NS	Kawasaki et al. 2023
Ileal digestible AA	7 days (weanling)	meal	36.5-50		NS	Crosbie et al. 2020
Growth performance, gut health, immune response	38 days	meal	25-50	0.5-14.8	NS	Crosbie et al. 2021
Growth performance, gut health		meal		4.8-19.1	NS	Håkenåsen et al. 2021
Growth performance, nutrient digestibility, microbiota		meal		4.8-19.1	P<0.05	Håkenåsen et al. 2021
Gut health, functional properties	21 days	meal		30	P<0.05	Kar et al. 2021
Growth performance, nutrient digestibility, gut health	28 days	meal		6-12	P<0.05	Boontiam et al. 2022
Growth performance, nutrient digestibility, gut health, anti-oxidative capacity	28 days	meal		12	P<0.05	Phaengphairee et al. 2023
Ca and P digestibility	14 days	meal		16.6-21.9	P<0.05	Grundmann et al. 2024
AA digestibility, blood parameters	42 days	meal	100	5	NS	Lee et al. 2024

TABLE 18.7 Impact of dietary black soldier fly (BSF; *Hermetia illucens*) inclusion on swine growth performances, health status and meat quality *(cont.)*

Variables	Feeding length	Insect form	Substitution level range (%)	Inclusion level range (%)	Impact	References
Growth performance, FCR	42 days (weanling)	meal	100	5	P<0.05	Lee et al. 2024
Growth performance, FCR, fecal microbiota	42 days (weanling)	meal		3	NS	Chang et al. 2024
Nutrients' digestibility	42 days (weanling)	meal		3	P<0.05	Chang et al. 2024
Piglets' growth performance, blood parameters	70 days	meal		2	P<0.05	Razanova et al. 2025
Growth performance	38 days (119-157 days of age)	meal		3-12	NS	Yordanova et al. 2025
Oxidative stress	38 days (119-157 days of age)	meal		3-12	P<0.05	Yordanova et al. 2025
Growth performance, carcass weight, proximate composition, mineral profile	98 days	meal	25-100	6-14	P<0.05	Chia et al. 2021
pH, color, drip loss, shear force	46 days	meal		4-8	NS	Yu et al. 2019
Growth performance, FCR, marbling score, IMF[b], IMP[c], mRNA expression lipid metabolism & MyHC-IIa[d], fatty acid (FA) profile	46 days	meal		4-8	P<0.05	Yu et al. 2019
pH, color, cooking loss, shear force, proximate composition, TBARS[e]	25-110 kg LW	meal	50-100	8.2-12.3	NS	Altman et al. 2019
Sensory evaluation	25-110 kg LW	meal	50-100	8.2-12.3	P<0.05	Altman et al. 2019
Growth performance, meat quality	113 days (28-141 days of age)	meal		4-8	P<0.05	Zhu et al. 2022
Amino acid digestibility	2 days	meal / adults		59[e]	P<0.05	Kim et al. 2023
Growth performance, FCR, serum cholesterol	40 days (weanling)	fat		2-6	P<0.05	van Heugten et al. 2022
Growth performance, faecal microbiome	39 days (weanling)	fat		1.5-3	NS	Buffoni et al. 2025
Growth performance	11 days	live		75-150[f]	NS	Ipema et al. 2021a
Environmental enrichment	7 days (28-35 days of age)	live		Feed choice	P<0.05	Ipema et al. 2021b

[a]SCFA: short chain fatty acids; [b]IMF: intramuscular fat; [c]IMP: inosine monophosphate; [d]MyHC-IIa: myosin heavy-chain; [e]TBARS: Thiobarbituric acid reactive substances; [e]g/piglet/day; [f]percentage dietary crude protein.

TABLE 18.8 Impact of dietary yellow mealworm (YMW; *Tenebrio molitor*) inclusion on swine growth performances, health status and meat quality

Variables	Feeding length	Insect form	Substitution level range (%)	Inclusion level range (%)	Impact	References
Growth performance, nutrient digestibility, blood profiles	35 days (weanling)	meal		1.5-6	P<0.05	Jin et al. 2016
Nutrient and amino acids (AA) digestibility	3 days (weanling)	meal		10	P<0.05	Yoo et al. 2019
Nutrient and AA digestibility	2 days (weanling)	meal		10	P<0.05	Cho et al. 2019
Growth performance, blood parameters	28 days (weanling)	meal	50-100		P<0.05	Ko et al. 2020
Growth performance, nutrient digestibility	35 days (weanling)	meal		1-2	P<0.05	Ao et al. 2020
Blood profiles, fecal noxious gas emission	35 days (weanling)	meal		1-2	NS	Ao et al. 2020
Antioxidant status, oxidative stress response	35 days (35-70 days of age)	meal		5-10	NS	Ringseis et al. 2020
Growth performance, carcass yield, meat proximate composition, meat pH and L*a*b* color	42 days (34-76 days of age)	meal		10	NS	Zacharis et al. 2023
Gut microbiota, blood cholesterol, meat fatty acid (FA) profile	42 days (34-76 days of age)	meal		10	P<0.05	Zacharis et al. 2023
Growth performance, fecal microbiota, red blood cells, meat antioxidants, oxidative stability, FA profile, L*a*b* color	42 days (34-76 days of age)	meal		5-10	P<0.05	Zacharis et al. 2024
Growth performance, nutrient digestibility, health status	28 days (80-108 days of age)	meal		5	NS	Ferri et al. 2025
Gut microbiota	28 days (80-108 days of age)	meal		5	P<0.05	Ferri et al. 2025
Metabolic effects	28 days (35-70 days of age)	meal		5-10	NS	Meyer et al. 2020
Growth performance, nutrient digestibility, fecal score	35 days (weanling)	meal[a]		10	NS	Kolawole et al. 2025
Growth performance, hemoglobin content	32 days (weanling)	meal		6[b]	NS	Pereira et al. 2025

[a]dried or hydrolyzed mealworm; [b]with or without Fe or Mn enrichments

TABLE 18.9 Impact of insects' dietary inclusion on rabbit growth performance, health status, and meat quality

Variable	Insect species[a]	Insect form	Substitution level range (%)	Inclusion level range (%)	Impact	References
Meat color, TBARS[b], fatty acid (FA) profile	BSF	fat	100	3-6	P<0.05	Dalle Zotte et al. 2018
Mortality, growth performance, slaughter traits	BSF	fat	100	3-6	NS	Martins et al. 2018
Nutrients' digestibility	BSF	fat	100	3-6	P<0.05	Martins et al. 2018
Carcass yield, meat proximate composition	BSF	meal		10	NS	Ogbon et al. 2025
Feed choice, growth performance, feed conversion ratio (FCR), slaughter weight, meat Zn content	BSF	meal		10	P<0.05	Ogbon et al. 2025
Growth performance, nutrient digestibility, gut health, blood parameters	BSF/YMW	fat	50-100	0.75-1.5	NS	Gasco et al. 2019a
Meat pH, color, WHC[c], shear force, proximate composition, sensory evaluation	BSF/YMW	fat	50-100		NS	Gasco et al. 2019b
Meat FA profile, TBARS	BSF/YMW	fat	50-100	0.75-1.5	P<0.05	Gasco et al. 2019b
Cecal fermentation, gut microbiota	BSF/YMW		100		P<0.05	Dabbou et al. 2020b
Growth performance, nutrients digestibility	YMW	meal		3	NS	Volek et al. 2021
Growth performance, nitrogen balance	YMW	meal	100	3.9	NS	Volek et al. 2023
Organic matter and energy digestibility	YMW	meal	100	3.9	P<0.05	Volek et al. 2023
Gut volatile FA, short chain FA	YMW	meal		4	P<0.05	Strychalski et al. 2021
Meat proximate composition	YMW	meal		4	NS	Kowalska et al. 2020
Meat FA profile, amino acid (AA) profile	YMW	meal		4	P<0.05	Kowalska et al. 2020
Growth performance, nutrient digestibility, gut physiology	YMW/MSW	meal		4	P<0.05	Strychalski et al. 2021
Growth performance, carcass traits, meat cholesterol content	YMW/MSW	meal		5	NS	Kowalska et al. 2025
Gut short chain FA, meat proximate composition, FA profile, AA profile	YMW/MSW	meal		5	P<0.05	Kowalska et al. 2025
Meat pH, WHC, proximate composition, sensory evaluation, TBARS[c]	MSW	fat	100	1.30	NS	Cullere et al. 2022
Growth performance, carcass traits, nutrient digestibility	MSW	fat	100	1.30	NS	Cullere et al. 2024
Meat FA profile	MSW	fat	100	1.30	P<0.05	Dalle Zotte et al. 2022
Meat proximate composition	MSW	meal		4	NS	Kowalska et al. 2020
Meat FA profile, AA profile	MSW	meal		4	P<0.05	Kowalska et al. 2020
Meat proximate composition	MSW	meal	50-100	5-10	NS	Gugołek et al. 2019
Meat FA profile	MSW	meal	50-100	5-10	P<0.05	Gugołek et al. 2019
Gut physiology, nutrient digestibility, blood parameters	MSW	meal	50-100	5-10	P<0.05	Gugołek et al. 2021

[a] BSF: black soldier fly (*Hermetia illucens*), YMW: yellow mealworm (*Tenebrio molitor*), MSW: mulberry silkworm (*Bombyx mori*);
[b] TBARS: Thiobarbituric acid reactive substances; [c] WHC: water holding capacity; * Source: Dalle Zotte (2021), modified.

TABLE 18.10 Impact of insects' dietary inclusion on ruminants' live performance, health status, and derived food quality

Variable	Insect species[a]	Insect form	Substitution level range (%)	Inclusion level range (%)	Impact	References
Ruminal fermentation	BSF	meal	100		NS	Fukuda et al. 2022
Digestible energy intake	BSF	meal	100		P<0.05	Fukuda et al. 2022
Digestibility, *in vitro* fermentation	BSF	meal	20-40	3.2-6.4	P<0.05	Kahraman et al. 2023
Milk yield, composition	BSF	meal	50-100	4.5-8.9[b]	NS	Braamhaar et al. 2025
Fibre fractions digestibility, N intake	BSF	meal	50-100	4.5-8.9[b]	P<0.05	Braamhaar et al. 2025
Lamb growth performance, ruminal health, stress markers	BSF	meal		10	NS	Odeon et al. 2025
Milk yield	BSF	fat		10-100 g/head/day	NS	Nekrasov et al. 2022
Milk SCFA[c]	BSF	fat		10-100 g/head/day	P<0.05	Nekrasov et al. 2022
Nutrient digestibility, volatile fatty acids (VFA), methane emission, protozoal population	BSF	fat		1-4	P<0.05	Prachumchai et al. 2025
Feed palatability, nutrient digestibility, milk composition	BSF	fat		3	NS	Rastello et al. 2025
Feed conversion ratio (FCR), milk production, antioxidant status	BSF	fat		3	P<0.05	Rastello et al. 2025
Calves' growth, FCR	BSF[c]	fat	20-40		NS	Quigley et al. 2025
In vitro rumen fermentation: nutrient digestibility, methane emission	BSF/YMW	meal			P<0.05	Jayanegara et al. 2017

TABLE 18.10 Impact of insects' dietary inclusion on ruminants' live performance, health status, and derived food quality *(cont.)*

Variable	Insect species[a]	Insect form	Substitution level range (%)	Inclusion level range (%)	Impact	References
In vitro rumen[e] fermentation: VFA, ammonia content, methane emission	BSF/YMW	meal			P<0.05	Renna et al. 2022
Nutrients' digestibility	BSF/YMW	meal		100 mg N/kg BW	NS	Carrasco & Drewery 2024
Forage intake	BSF/YMW	meal		100 mg N/kg BW	P<0.05	Carrasco & Drewery 2024
Forage utilization, rumen microbial community	BSF/FRASS	meal		100 mg N/kg BW	P<0.05	Maggitt et al. 2025
Ruminal[e] degradation, intestinal digestibility	YMW	meal			P<0.05	Toral et al. 2022
In vitro rumen[e] fermentation, gut health	YMW	meal			NS	Robles-Jimenez et al. 2025
Lamb growth performance, digestibility, nitrogen balance	YMW	meal		6[d]	P<0.05	Robles-Jimenez et al. 2025
In vitro rumen fermentation profile, nutrient digestibility	MSW	meal	25	10	NS	Ahmed et al. 2021
Methane emission	MSW	meal	25	10	P<0.05	Ahmed et al. 2021
In vitro rumen fermentation: nutrient digestibility, gas production, VFA	MSW	meal		10-40	P<0.05	Ahmed & Nishida 2023
Rumen methane emission, fermentation, microbial population	MSW	chitosan		6	P<0.05	Sagala et al. 2024
In vitro rumen fermentation: methane emission, VFA	MSW	meal		20	P<0.05	Purnamia et al. 2026

[a]BSF: black soldier fly (*Hermetia illucens*), YMW: yellow mealworm (*Tenebrio molitor*), MSW: mulberry silkworm (*Bombyx mori*); [b]Dry matter basis; [c]SCFA: short chain fatty acids; [d]In calf milk replacer; [e]ovine.

References

Addeo, N.F., Randazzo, B., Olivotto, I., Messina, M., Tulli, F., Vozzo, S., Attia, Y.A., Mahayri, T.M., Iannaccone, F., Asiry, K.A., Moniello, G., Bovera, F., 2022. Low inclusion levels of *Tenebrio molitor* larvae meal in laying Japanese quail (*Coturnix japonica*, Gould, 1837) diet improve the intestinal morphometry, enzymatic activity and caecal short chain fatty acids profile. Research in Veterinary Science 149: 51-59. https://doi.org/10.1016/j.rvsc.2022.06.007.

Ahmed, E., Fukuma, N., Hanada, M., Nishida, T., 2021. Insects as novel ruminant feed and a potential mitigation strategy for methane emissions. Animals 11: 2648. https://doi.org/10.3390/ani11092648.

Ahmed, E., Nishida, T., 2023. Optimal inclusion levels of cricket and silkworm as alternative ruminant feed: a study on their impacts on rumen fermentation and gas production. Sustainability 15: 1415. https://doi.org/10.3390/su15021415.

Altmann, B.A., Neumann, C., Rothstein, S., Liebert, F. Mörlein, D., 2019. Do dietary soy alternatives lead to pork quality improvements or drawbacks? A look into micro-alga and insect protein in swine diets. Meat Science 153: 26-34. https://doi.org/10.1016/j.meatsci.2019.03.001.

Altmann, B.A., Wigger, R., Ciulu, M., Mörlein, D., 2020. The effect of insect or microalga alternative protein feeds on broiler meat quality. Journal of the Science of Food and Agriculture 100: 4292-4302. https://doi.org/10.1002/jsfa.10473.

Ankamah-Yeboah, I., Jacobsen, J.B., Olsen, S.B., 2018. Innovating out of the fishmeal trap: The role of insect-based fish feed in consumers' preferences for fish attributes. British Food Journal 120: 2395-2410. https://doi.org/10.1108/BFJ-11-2017-0604.

ANSES (French Agency for Food, Environmental and Occupational Health and Safety). 2015. Opinio non the use of insects as food and feed and the review of scientific knowledge on the health risks related to the consumption of insects.

Ao, X., Yoo, J.S., Wu, Z.L., Kim, I.H., 2020. Can dried mealworm (*Tenebrio molitor*) larvae replace fish meal in weaned pigs? Livestock Science 239: 104103. https://doi.org/10.1016/j.livsci.2020.104103.

Armaghoza, H., Nugroho, S., Sungkono, S., Widodo, S.L., Saputra, B.C., Aprinato, M.A., Anas, M., 2025. Evaluation of black soldier fly larvae oil (*Hermetia illucens* L.) calcium salt as an alternative fat source for laying quail diets. Animal Bioscience 38: 1497-1507. https://doi.org/10.5713/ab.24.0289.

Aslam, M.R., Kierończyk, B., Rawski, M., Szymkowiak, P., Stuper-Szablewska, K., Kołodziejski, P., Mikuła, R., Dankowiakowska, A., Józefiak , D., 2025. Insect fat influences broiler performance, meat quality, and the cecal microbiota similarly to plant oils rather than animal fats. Scientific Reports 15: 18086. https://doi.org/10.1038/s41598-025-02820-3.

Attia, Y.A., Bovera, F., Asiry, K.A., Alqurashi, S., Alrefaei, M.S., 2023). Fish and black soldier fly meals as partial replacements for soybean meal can affect sustainability of productive performance, blood constituents, gut microbiota, and nutrient excretion of broiler chickens. Animals 13: 2759. https://doi.org/10.3390/ani13172759.

Attivi, K., Agboka, K., Mlaga, K.G., Dassidi, N., Oke, O.E., Karou, S.D., Onagbesan, O., Hai, L.,

Tona, K., 2023. Productive performance, follicular development, and reproductive hormone profiles of laying hens fed with black soldier fly (*Hermetia illucens*) larvae meal as a substitute for fish meal. European Poultry Science 87: 1-17. https://doi.org/10.1399/eps.2023.368.

Baderuddin, S.H., David, L.S., Wester, T.J., Morel, P.C.H., 2024. Influence of different levels of black soldier fly larvae meal on growth performance and carcass quality of broiler chickens. Livestock Science 290: 105588. https://doi.org/10.1016/j.livsci.2024.105588.

Baiano, A., 2020. Edible insects: An overview on nutritional characteristics, safety, farming, production technologies, regulatory framework, and socio-economic and ethical implications. Trends in Food Science & Technology 100: 35-50. https://doi.org/10.1016/j.tifs.2020.03.040.

Bellezza Oddon, S., Biasato, I., Imarisio, A, Pipan, M., Dekleva, D., Colombino, E., Capucchio, M.T., Meneguz, M., Bergagna, S., Berbero, R., Gariglio, M., Dabbou, S., Fiorilla, E., Gasco, L, Schiavone, A., 2021. Black soldier fly and yellow mealworm live larvae for broiler chickens: Effects on bird performance and health status. Journal of Animal Physiology and Animal Nutrition 105:10-18. https://doi.org/10.1111/jpn.13567.

Benzertiha, A., Kierończyk, B., Rawski, M., Kołodziejski, P., Bryszak, M., Józefiak, D., 2019. Insect oil as an alternative to palm oil and poultry fat in broiler chicken nutrition. Animals 9: 116. https://doi.org/10.3390/ani9030116.

Biasato, I., De Marco, M., Rotolo, L., Renna, M., Lussiana, C., Dabbou, S., Capucchio, M.T., Biasibetti, E., Costa, P., Gai, F., Pozzo, L., Dezzutto, D., Bergagna, S., Martínez, S., Tarantola, M., Gasco, L., Schiavone, A., 2016. Effects of dietary *Tenebrio molitor* meal inclusion in free-range chickens. Journal of Animal Physiology and Animal Nutrition 100: 1104-1112. https://doi.org/10.1111/jpn.12487.

Biasato, I., Ferrocino, I., Grego, E., Dabbou, S., Gai, F., Gasco, L., Cocolin, L., Capucchio, M. T., Schiavone, A., 2019a. Gut microbiota and mucin composition in female broiler chickens fed diets including yellow mealworm (*Tenebrio molitor*, L.). Animals 9: 213. https://doi.org/10.3390/ani9050213.

Biasato, I., Gariglio, M., Bongiorno, V., Fiorilla, E., Cappone, E.E., Bellezza Oddon, S., Loiotine, Z., Caimi, C., Mendez Rondo, K.J., Renna, M., Lussiana, C., Brugiapaglia, A., Hernández, F., Schiavone, A., Gasco, L., 2025. Can a mixture of *Hermetia illucens* and *Tenebrio molitor* meals be feasible to feed broiler chickens? A focus on bird productive performance, nutrient digestibility, and meat quality. Poultry Science 104: 105150. https://doi.org/10.1016/j.psj.2025.105150.

Biasato, I., Gasco, L., De Marco, M., Renna, M., Rotolo, L., Dabbou, S., Capucchio, M.T., Biasibetti, E., Tarantola, M., Bianchi, C., Cavallarin, L., Gai, F., Pozzo, L., Dezzutto, D., Bergagna, S., Schiavone, A., 2017. Effects of yellow mealworm larvae (*Tenebrio molitor*) inclusion in diets for female broiler chickens: implications for animal health and gut histology. Animal Feed Science and Technology 234: 253-263. https://doi.org/10.1016/j.anifeedsci.2017.09.014.

Biasato, I., Renna, M., Gai, F., Dabbou, S., Meneguz, M., Perona, G., Martinez, S., Cristina, A., Lajusticia, B., Bergagna, S., Sardi, L., Capucchio, M.T., Bressan, E., Dama, A., Schiavone, A., Gasco, L., 2019b. Partially defatted black soldier fly larva meal inclusion in piglet diets: effects on the growth performance, nutrient digestibility, blood profile, gut morphology and histological features. Journal of Animal Science and Biotechnology 10: 12. https://doi.org/10.1186/s40104-019-0325-x.

Bongiorno, V., Gariglio, M., Zambotto, V., Cappone, E.E., Biasato, I., Renna, M., Forte, C., Coudron, C., Bergagna, S., Gai, F., Schiavone, A., 2022. Black soldier fly larvae used for environmental enrichment purposes: Can they affect the growth, slaughter performance, and blood chemistry of medium-growing chickens? Frontiers in Veterinary Science 9: 1064017. https://doi.org/10.3389/fvets.2022.1064017.

Boontiam, W., Phaengphairee, P., Hong, J., Kim, Y.Y., 2022. Full-fatted *Hermetia illucens* larva as a protein alternative: effects on weaning pig growth performance, gut health, and antioxidant status under poor sanitary conditions. Journal of Applied Animal Research 50: 732-739. https://doi.org/10.1080/09712119.2022.2147185.

Bovera, F., Loponte, R., Marono, S., Piccolo, G., Parisi, G., Iaconisi, V., Gasco, L., and Nizza, A. 2016. Use of *Tenebrio molitor* larvae meal as protein source in broiler diet: Effect on growth performance, nutrient digestibility, and carcass and meat traits. Journal of Animal Science 94: 639-647. https://doi.org/10.2527/jas.2015-9201

Bovera, F., Loponte, R., Marono, S., Piccolo, G., Parisi, G., Iaconisi, V., Gasco, L., Nizza, A., 2016. Use of *Tenebrio molitor* larvae meal as protein source in broiler diet: Effect on growth performance, nutrient digestibility, and carcass and meat traits. Journal of Animal Science 94: 639-647. https://doi.org/10.2527/jas.2015-9201.

Braamhaar, D.J.M., Pellikaan, W.F., List, D., Korir, D., Tanga, C.M., Oosting, S.J., 2025. Defatted black soldier fly larvae meal as a substitute of soybean meal in dairy cow diets. Animal 19: 101476. https://doi.org/10.1016/j.animal.2025.101476.

Braide, W., Oranusi, S., Udegbunamm L.I., Oguoma, O., Akobondu, C., Nwaoguikpe, R.N., 2011. Microbiological quality of an edible caterpillar of an emperor moth, *Bunaea alcinoe*. Journal of Ecology and the Natural Environment 3: 176-180. http://www.academicjournals.org/jene.

Buffoni, M., Viveen, M., Arjan, J., De Visser, G.M., Mughini-Gras, L., Willems, R.J.L., Schürch, A.C., Kers, J.G., 2025. Effect of black soldier fly (*Hermetia illucens*) larvae oil feed inclusion on growth performance and fecal microbiome development in post-weaning pigs. Livestock Science: 301, 105819. https://doi.org/10.1016/j.livsci.2025.105819.

Cadinu, L.A., Barra, P., Torre, F., Delogu, F., Madau, F.A., 2020. Insect rearing: potential, challenges, and circularity. Sustainability 12: 4567. https://doi.org/10.3390/su12114567.

Campbell, B.M., Beare, D.J., Bennett, E.M., Hall-Spencer, J.M., Ingram, J.S.I., Jaramillo, F., Ortiz, R., Ramankutty, N., Sayer, J.A., Shindell, D., 2017. Agriculture production as a major driver of the Earth system exceeding planetary boundaries. Ecology and Society 22: 8. https://doi.org/10.5751/ES-09595-220408.

Carrasco, M.N., Drewery, M.L., 2024. Mealworm larvae and black soldier fly larvae as novel protein supplements for cattle consuming low-quality forage. Translational Animal Science 8: txae122. https://doi.org/10.1093/tas/txae122.

Chang, S.Y., Kim, K.H., Lee, B.K., Lee, J.H., Oh, H.J., An, J.W., Dong, C.S., Hyun, A.C., Se, H.P., Kyeong, H.J., Ji, Y.C., Cho, J.H., 2024.Defatted or hydrolyzed black soldier fly larvae have sufficient potential as an alternative to fishmeal for weaned pigs. Animal Feed Science and Technology 315: 116003. https://doi.org/10.1016/j.anifeedsci.2024.116003.

Chatzidimitriou, E., Davis, H., Maurer, V., Leiber, F., Leifert, C., Stergiadis, S., Butler, G. 2022. Egg fatty acid profiles and potential health risk from defatted insect meal in laying hens' diets. Journal of Insects as Food and Feed 8: 1085-1096. https://doi.org/10.3920/JIFF2022.0027.

Chia, S.Y., Tanga, C.M., Osuga, I.M., Alaru, A.O., Mwangi, D.M., Githinji, M., Dubois, T., Ekesi, S., van Loon, J.J.A., Dicke, M., 2021. Black soldier fly larval meal in feed enhances growth performance, carcass yield and meat quality of finishing pigs. Journal of Insects as Food and Feed 7: 433-448. https://doi.org/10.3920/JIFF2020.0072.

Cho, K.H., Kang, S.W., Yoo, J.S., Song, D.K., Chung, Y.H., Kwon, G.T., Kim, Y.Y., 2019. Effects of mealworm (*Tenebrio molitor*) larvae hydrolysate on nutrient ileal digestibility in growing pigs compared to those of defatted mealworm larvae meal, fermented poultry by-product, and hydrolyzed fish soluble. Asian-Australasian Journal of Animal Sciences 33: 490-500. https://doi.org/10.5713/ajas.19.0793.

Choi, S.U., Choi, I.H., Chung, T.H., 2021. Investigation of breast meat traits of broilers fed different amounts of *Hermetia illucens* and *Protaetia brevitarsis seulensis* powder. Entomological Research 51: 343-348. https://doi.org/10.1111/1748-5967.12504.

Christensen, B., Zhu, C., Niazy, M., McCullough, T., Ricker, N., Huber, L., 2023. Partial replacement of soybean meal with full-fat black soldier fly larvae meal in plant-based nursery diets did not influence fecal *Escherichia coli* colony forming units or improve fecal consistency when pigs were weaned into non-disinfected pens. Translational Animal Science 7: txad121. https://doi.org/10.1093/tas/txad121.

Commission Regulation (EU) 2021/1372 of 17 August 2021 amending Annex IV to Regulation (EC) No 999/2001 of the European Parliament and of the Council as regards the prohibition to feed non-ruminant farmed animals, other than fur animals, with protein derived from animals. https://eur-lex.europa.eu/eli/reg/2021/1372/oj/eng.

Crosbie, M., Zhu, C., Karrow, N.A., Huber, L.A., 2021. The effects of partially replacing animal protein sources with full fat black soldier fly larvae meal (*Hermetia illucens*) in nursery diets on growth performance, gut morphology, and immune response of pigs. Translational Animal Science 5: txab057. https://doi.org/10.1093/tas/txab057.

Crosbie, M., Zhu, C., Shoveller, K., Huber, L.A., 2020. Standardized ileal digestible amino acids and net energy contents in full fat and defatted black soldier fly larvae meal (*Hermetia illucens*) fed to growing pigs. Translational Animal Science 4: 1-10. https://doi.org/10.1093/tas/txaa104.

Cullere, M., Schiavone, A., Dabbou, S., Gasco, L., Dalle Zotte, A., 2019a. Meat quality and sensory traits of finisher broiler chickens fed with black soldier fly (*Hermetia illucens* L.) larvae fat as alternative fat source. Animals 9: 140. https://doi.org/10.3390/ani9040140.

Cullere, M., Singh, Y., Gerencsér, Zs., Matics, Zs., Cappellozza, S., Dalle Zotte, A., 2022. Silkworm (*Bombyx mori* L.) oil in growing rabbit nutrition: effects on meat physicochemical traits, sensory profile and shelf-life. Journal of Insects as Food and Feed 8: 733-741. https://doi.org/10.3920/JIFF2021.0084.

Cullere, M., Szendrő, Zs., Matics, Zs., Gerencsér, Zs., Singh, Y., Dalle Zotte, A., 2024. Effect of silkworm (*Bombyx mori* L.) pupae oil dietary inclusion on growth performance, digestibility and carcase traits of growing rabbits. Italian Journal of Animal Science 23: 751-757. https://doi.org/10.1080/1828051X.2024.2344851.

Cullere, M., Tasoniero, G., Giaccone, V., Acuti, G., Marangon, A., Dalle Zotte, A., 2018. Black soldier fly as dietary protein source for broiler quails: Meat proximate composition, fatty acid and amino acid profile, oxidative status and sensory traits. Animal 12: 640-647. https://doi.org/10.1017/S1751731117001860.

Cullere, M., Tasoniero, G., Giaccone, V., Miotti-Scapin, R., Claeys, E., De Smet, S., Dalle Zotte, A., 2016. Black soldier fly as dietary protein source for broiler quails: apparent digestibility, excreta microbial load, feed choice, performance, carcass and meat traits. Animal 10: 1923-1930. https://doi.org/10.1017/S1751731116001270.

Cullere, M., Woods, M.J., van Emmenes, L., Pieterse, E., Hoffman, L.C., Dalle Zotte, A., 2019b. *Hermetia illucens* larvae reared on different substrates in broiler quail diets: Effect on physicochemical and sensory quality of the quail meat. Animals 9: 925. https://doi.org/10.3390/ani9080525.

Dabbou, S., Cattaneo, A., Cattapan, F., Malinverni, R., Macchi, E., Manenti, I., Todesco, S., Poloni, E., Rosà, R., Santori, D., Meneguz, M., Pertot, I., Sezzi, E., 2025. Impact of live black soldier fly larvae supplementation on laying hen performance, stress levels and excreta microbiota. Poultry Science 104: 105418. https://doi.org/10.1016/j.psj.2025.105418.

Dabbou, S., Ferrocino, I., Gasco, L., Schiavone, A., Trocino, A., Xiccato, G., Barroeta, A.C., Maione, S., Soglia, D., Biasato, I., Cocolin, L., Gai, F., Nucera, D.M., 2020b. Antimicrobial effects of black soldier fly and yellow mealworm fats and their impact on gut microbiota of growing rabbits. Animals 10: 1292. https://doi.org/10.3390/ani10081292.

Dabbou, S., Gasco, L., Lussiana, C., Brugiapaglia, A., Biasato, I., Renna, M., Cavallarin, L., Gai, F., Schiavone, F., 2020a. Yellow mealworm (*Tenebrio molitor* L.) larvae inclusion in diets for free-range chickens: Effects on meat quality and fatty acid profile. Renewable Agriculture and Food Systems 35: 571-578. https://doi.org/10.1017/S1742170519000206.

Dalle Zotte, A. 2021. Do insects as feed ingredient affect meat quality? Theory and practice of meat processing. 6: 200-209. https://doi.org/10.21323/2414-438X-2021-6-3-200-209

Dalle Zotte, A., Cullere, M., Martins, C., Alves, S.P., Freire, J.P.B., Falcão-e-Cunha, L., Bessa, R.J., 2018. Incorporation of black soldier fly (*Hermetia illucens* L.) larvae fat or extruded linseed in diets of growing rabbits and their effects on meat quality traits including detailed fatty acid composition. Meat Science 146: 50-58. https://doi.org/10.1016/j.meatsci.2018.08.002.

Dalle Zotte, A., Singh, Y., Gerencsér, Zs., Matics, Zs., Szendrő, Zs., Cappellozza, S., Cullere, M., 2022. Feeding silkworm (*Bombyx mori* L.) oil to growing rabbits improves the fatty acid composition of meat, liver and perirenal fat. Meat Science 193: 108944. https://doi.org/10.1016/j.meatsci.2022.108944.

Dalle Zotte, A., Singh, Y., Michiels, J., Cullere, M., 2019. Black soldier fly (*Hermetia illucens*) as dietary source for laying quails: live performance, and egg physico-chemical quality, sensory profile and storage stability. Animals 9: 115. https://doi.org/10.3390/ani9030115.

Dalle Zotte, A., Singh, Y., Palumbo, B., Contiero, B., Cullere, M., 2024a. Live yellow mealworm (*Tenebrio molitor*) larvae: a promising nutritional enrichment for laying quails. Poultry Science 103: 103759. https://doi.org/10.1016/j.psj.2024.103759

Dalle Zotte, A., Singh, Y., Squartini, A., Stevanato, P., Cappellozza, S., Kovitvadhi, A., Subaneg, S., Bertelli, D., Cullere, M., 2021. Effect of a dietary inclusion of full-fat or defatted silkworm pupa meal on the nutrient digestibility and faecal microbiome of fattening quails. Animal 15: 100112. https://doi.org/10.1016/j.animal.2020.100112.

Dalle Zotte, A., Singh, Y., Zsedely, E., Contiero, B., Palumbo, B., Cullere, M. (2024b). Dietary inclusion of defatted silkworm (*Bombyx mori* L.) pupa meal in broiler chickens: phase feeding effects on nutritional and sensory meat quality. Poultry Science: 103: 103812. https://doi.org/10.1016/j.psj.2024.103812.

de Faria Domingues, C.H.F., Borges, J.A.R., Ruviaro, C.F., Guidolin, D.C.F., Carrijo-Mauad, J., 2020. Understanding the factors influencing consumer willingness to accept the use of insects to feed poultry, cattle, pigs and fish in Brazil. PLoS ONE 15: e0224059. https://doi.org/10.1371/journal.pone.0224059.

de Miranda, J.R., Granberg, F., Onorati, P., Jansson, A., Berggren, Å., 2021. Virus prospecting in crickets discovery and strain divergence of a novel filovirus in wild and cultivated *Acheta domesticus*. Viruses 13: 364. https://doi.org/10.3390/v13030364.

de Souza Vilela, J., Alvarenga, T.I.R.C., Andrew, N.R., McPhee, M., Kolakshyapati, M., Hopkins, D.L., Ruhnke, I., Eyres, G.T., 2021. Technological quality, amino acid and fatty acid profile of broiler meat enhanced by dietary inclusion of black soldier fly larvae. Foods 10: 297. https://doi.org/10.3390/foods10020297.

Dearing, M.D., Foley, W.G., McLean, S., 2005. The Influence of plant secondary metabolites on the nutritional ecology of herbivorous terrestrial vertebrates. Annual Reviews 36: 169-189. https://doi.org/10.1146/annurev.ecolsys.36.102003.152617.

Dörper, A., Berman, H.M., Gort, G., Harn, J. van, Dicke, M., Veldkamp, T., 2024a. Effects of different black soldier fly larvae products on slow-growing broiler performance and carcass characteristics. Poultry Science 103. https://doi.org/10.1016/j.psj.2024.103481.

EFSA Scientific Committee, 2015. Risk profile related to production and consumption of insects as food and feed. EFSA journal 13: 4257. https://doi.org/10.2903/j.efsa.2015.4257

Eilenberg, J., Vlak, J.M., Nielsen-LeRoux, C., Cappellozza, S., Jensen, A.B., 2015. Diseases in insects produced for food and feed. Journal of Insects as Food and Feed 1: 87-102. https://doi.org/10.3920/JIFF2014.0022.

Elahi, U., Wang, J., Ma, Y., Wu, S., Wu, J., Qi, G., Zhang, H., 2020. Evaluation of yellow mealworm meal as a protein feedstuff in the diet of broiler chicks. Animals 10: 224. https://doi.org/10.3390/ani10020224.

Elahi, U., Xu, C.C., Wang, J., Lin, J., Wu, S.G., Zhang, H.J. and Qi, G.H., 2022. Insect meal as a feed ingredient for poultry. Animal Bioscience 35: 332-346. https://doi.org/10.5713/ab.21.0435

Emadi, M., Vafa, T., Moghadam, S., 2015. Effect of silkworm pupae meal as protein supplement on performance of broiler chickens. Journal of Pure and Applied Microbiology 9: 161-165.

Facey, H., Kithama, M., Mohammadigheisar, M., Huber, L.A., Shoveller, A.K., Kiarie, E.G., 2023.

Complete replacement of soybean meal with black soldier fly larvae meal in feeding program for broiler chickens from placement through to 49 days of age reduced growth performance and altered organs morphology. Poultry Science 102. https://doi.org/10.1016/j.psj.2022.102293.

Ferri, I., Dell'Anno, M., Quiese, A., Castiglioni, B., Cremonesi, P., Biscarini, F., Canala, B., Santoru, M., Colombini, A., Ruffo, G., Baldi, A., Rossi, L., 2025. Microbiota modulation by the inclusion of *Tenebrio molitor* larvae as alternative to fermented soy protein concentrate in growing pigs diet. Veterinary Research Communications 49: 26. https://doi.org/10.1007/s11259-024-10588-6.

Fiorilla, E., Ferrocino, I., Gariglio, M., Gai, F., Zambotto, V., Ozella, L., Franciosa, I., Giribaldi, M., Antoniazzi, S., Raspa, F., Cappone, E.E., Fabrikov, D., Pozzo, S., Bongiorno, V., Calissano, A., Ippolito, D., Bergagna, S., Srikanthithasan, K., Forte, C., Capucchio, M.T., Schiavone, A., 2024a. Black soldier fly larvae: a one health approach to investigate gut, and organ health and meat quality response in slow-growing chickens. BMC Veterinary Research 20: 580. https://doi.org/10.1186/s12917-024-04394-0.

Fiorilla, E., Gariglio, M., Gai, F., Zambotto, V., Bongiorno, V., Cappone, E.E., Rødbotten, R., Koga, S., Rieder, A., Tengstrand, E., Pozzo, S., Daniele, G.M., Cianciabella, M., Predieri, S., Forte, C., Schiavone, A., 2024b. Breaking down barriers: live or dehydrated dietary whole black soldier fly larvae supplementation in slow growing chickens preserve meat quality and sensory traits. Poultry Science 103: 104120. https://doi.org/10.1016/j.psj.2024.104120.

Fowley, T.M., Nansen, C., 2020. Insect-based bioconversion: Value from food waste. In: Närvänen, E., Mesiranta, N., Mattila, M., Heikkinen, A. (Eds.), Food Waste Management. Palgrave Macmillan, Cham, Switzerland, pp. 321-346.

Fruci, M., Kithama, M., Kiarie, E.G., Shao, S., Liu, H., Topp, E., Diarra, M.S., 2023. Effects of partial or complete replacement of soybean meal with commercial black soldier fly larvae (*Hermetia illucens*) meal on growth performance, cecal short chain fatty acids, and excreta metabolome of broiler chickens. Poultry Science 102: 102463. https://doi.org/10.1016/j.psj.2022.102463.

Fukuda, E.P., Cox, J.R., Wickersham, T.A., Drewery, M.L., 2022. Evaluation of Black Soldier Fly larvae (*Hermetia illucens*) as a protein supplement for beef steers consuming low quality forage. Translational Animal Science 6: 1-6. https://doi.org/10.1093/tas/txac018.

Gariglio, M., Dabbou, S., Gai, F., Trocino, A., Xiccato, G., Holodova, M., Gresakova, L., Nery, J., Bellezza Oddon, S., Biasato, I., Gasco, L., Schiavone, A., 2021. Black soldier fly larva in Muscovy duck diets: effects on duck growth, carcass property, and meat quality. Poultry Science 100: 101303. https://doi.org/10.1016/j.psj.2021.101303.

Gariglio, M., Dabbou, S., Renna, M., Biasato, I., Bellezza Oddon, S., Meneguz, M., Miazzo, R.D., Bergagna, S., Colombino, E., Macchi, E., Schiavone, A., 2023. Does the provision of live black soldier fly and yellow mealworm larvae improve Muscovy duck welfare? Journal of Animal Science and Biotechnology 14: 153. https://doi.org/10.1186/s40104-023-00949-7.

Gariglio, M., Fiorilla, E., Trocino, A., Bordignon, F., Brugiapaglia, A., Gai, F., Bianchi, C., Zambotto, V., Schiavone, A., 2025. Effects of live black soldier fly and yellow mealworm larvae supplementation on slaughter performance and meat composition of Muscovy ducks. Scientific Reports 15: 13597. https://doi.org/10.1038/s41598-025-97903-6.

Gasco, L., Dabbou, S., Gai, F., Brugiapaglia, A., Schiavone, A., Birolo, M., Xiccato, G., Trocino, A., 2019b. Quality and consumer acceptance of meat from rabbits fed diets in which soybean oil is replaced with black soldier fly and yellow mealworm fats. Animals 9: 629. https://doi.org/10.3390/ani9090629.

Gasco, L., Dabbou, S., Trocino, A., Xiccato, G., Capucchio, M.T., Biasato, I., Dezzutto, D., Birolo, M., Meneguz, M., Schiavone, A., Gai, F., 2019. Effect of dietary supplementation with insect fats on growth performance, digestive efficiency and health of rabbits. Journal of Animal Science and Biotechnology 10: 4. https://doi.org/10.1186/s40104-018-0309-2.

Gasco, L., Józefiak, A. and Henry, M., 2021. Beyond the protein concept: health aspects of using edible insects on animals. Journal of Insects as Food and Feed 7: 715-741. https://doi.org/10.3920/JIFF2020.0077

Gkarane, V., Ciulu, M., Altmann, B. A., Schmitt, A. O., Mörlein, D., 2020b. The effect of algae or insect supplementation as alternative protein sources on the volatile profile of chicken meat. Foods 9: 1235. https://doi.org/10.3390/foods9091235.

Gkarane, V., Ciulu, M., Altmann, B.A., Mörlein, D., 2020a. Effect of alternative protein feeds on the content of selected endogenous bioactive and flavour-related compounds in chicken breast meat. Foods 9: 392. https://doi.org/10.3390/foods9040392.

Gokulakrishnaa, R.K., Selvamuthukumaran, T., Kothandaraman, S., 2025. Eri silkworm pupae meal as a fish meal replacement in Japanese quail starter diet. Journal of Animal and Plant Sciences 35: 273-284. https://doi.org/10.36899/JAPS.2025.1.0022.

Grundmann, S.M., Herrero-Encinas, J., Eder, K., 2024. Apparent total tract digestibility of calcium and phosphorus from partially defatted *Hermetia illucens* larvae meal in pigs. Journal of Insects as Food and Feed 10(11), 1783-1789. https://doi.org/10.1163/23524588-00001134

Gugołek, A., Kowalska, D., Strychalski, J., Ognik, K., Juśkiewicz, J., 2021. The effect of dietary supplementation with silkworm pupae meal on gastrointestinal function, nitrogen retention and blood biochemical parameters in rabbits. BMC Veterinary Reserch 17: 204. https://doi.org/10.1186/s12917-021-02906-w.

Gugołek, A., Strychalski, J., Kowalska, D., 2019. Growth performance and meat composition of rabbits fed diets supplemented with silkworm pupae meal. Spanish Journal of Agricultural Research 17: e0607. https://doi.org/10.5424/sjar/2019173-14882.

Håkenåsen, I.M., Grepperud, G.H., Hansen, J.Ø., Øverland, M., Ånestad, R.M., Mydland, L.T., 2021. Full-fat insect meal in pelleted diets for weaned piglets: Effects on growth performance, nutrient digestibility, gastrointestinal function, and microbiota. Animal Feed Science and Technology 281: 115086. https://doi.org/10.1016/j.anifeedsci.2021.115086.

Halloran, A., Hanboonsong, Y., Roos, N., Bruun, S. 2017. Life cycle assessment of cricket farming in north-eastern Thailand. Journal of Cleaner Production 156: 83-94. https://doi.org/10.1016/j.jclepro.2017.04.017.

Huang, C., Hernandez, C.E., Wall, H., Tahamtani, F.M., Ivarsson, E., Sun, L.,2024. Live black soldier fly (*Hermetia illucens*) larvae in feed for laying hens: effects on hen gut microbiota and behavior. Poultry Science 103: 103429. https://doi.org/10.1016/j.psj.2024.103429.

Ipema, A.F., Bokkers, E.A.M., Gerrits, W.J.J., Kemp, B., Bolhuis, J.E., 2022. Provision of black soldier fly larvae (*Hermetia illucens*) in different ways benefits broiler welfare and performance, with largest effects of scattering live larvae. Physiology & Behavior 257: 113999. https://doi.org/10.1016/j.physbeh.2022.113999.

Ipema, A.F., Bokkers, E.A.M., Gerrits, W.J.J., Kemp, B., Bolhuis, J.E., 2021a. Providing live black soldier fly larvae (*Hermetia illucens*) improves welfare while maintaining performance of piglets post-weaning. Scientific Reports 11: 7371. https://doi.org/10.1038/s41598-021-86765-3.

Ipema, A.F., Gerrits, W.J.J., Bokkers, E.A.M., Kemp, B., Bolhuis, J.E., 2021b. Live black soldier fly larvae (*Hermetia illucens*) provisioning is a promising environmental enrichment for pigs as indicated by feed- and enrichment-preference tests. Applied Animal Behaviour Science 244: 105481. https://doi.org/10.1016/j.applanim.2021.105481.

Jayanegara, A., Yantina, N., Novandri, B., Laconi, E.B., Nahrowi, N., Ridla, M., 2017. Evaluation of some insects as potential feed ingredients for ruminants: chemical composition, *in vitro* rumen fermentation and methane emissions. Journal of the Indonesian Tropical Animal Agriculture 42: 247-254. https://doi.org/10.14710/jitaa.42.4.247-254.

Jiang, T., Zhao, X., Li, H., Zhang, L., Tang, B., Ding, Y., Wu, Y., 2024. Effects of yellow mealworm (*Tenebrio molitor*) larvae meal on the growth performance, serum biochemical parameters and caecal metabolome in broiler chickens. Italian Journal of Animal Science 23: 813-823. https://doi.org/10.1080/1828051X.2024.2350230.

Jin, X.H., Heo, P.S., Hong, J.S., Kim, N.J., Kim, Y.Y., 2016. Supplementation of dried mealworm (*Tenebrio molitor* larva) on growth performance, nutrient digestibility and blood profiles in weaning pigs. Asian-Australas Journal of Animal Sciences 29: 979-986. https://doi.org/10.5713/ajas.15.0535.

Kahraman, O., Gülşen, N., İnal, F., Alataş, M.S., İnanç, Z.S., Ahmed, İ., ¸Sişman, D., Küçük, A.E., 2023. Comparative analysis of *in vitro* fermentation parameters in total mixed rations of dairy cows with varied levels of defatted black soldier fly larvae (*Hermetia illucens*) as a substitute for soybean meal. Fermentation 9: 652. https://doi.org/10.3390/fermentation9070652.

Kar, S.K., Schokker, D., Harms, A.C., Kruijt, L., Smits, M.A., Jansman, A.J.M., 2021. Local intestinal microbiota response and systemic effects of feeding black soldier fly larvae to replace soybean meal in growing pigs. Scientific Reports 11: 15088. https://doi.org/10.1038/s41598-021-94604-8.

Kawasaki, K., Zhao, J., Takao, N., Sato, M., Ban, T., Tamamaki, K., Kagami, M., Yano, K., 2023. Sustenance trial to analyze the effects of black soldier fly larvae meal on the reproductive efficiency of sows and the hematological properties of suckling and weaning piglets. Animals 13: 3410. https://doi.org/10.3390/ani13213410.

Kazemi, M., Valizadeh, R., 2025. Overcoming feed scarcity: how insects can transform sustainable animal nutrition. Cogent Food & Agriculture 11. https://doi.org/10.1080/23311932.2025.2566440.

Kierończyk, B., Rawski, M., Mikołajczak, Z., Szymkowiak, P., Stuper-Szablewska, K., Józefiak, D., 2023. Black soldier fly larva fat in broiler chicken diets affects breast meat quality. Animals 13: 1137. https://doi.org/10.3390/ani13071137.

Kierończyk, B., Sypniewski, J., Mikołajczak, Z., Rawski, M., Pruszyńska-Oszmałek, E., Sassek, M., Kołodziejski, P. Józefiak, D., 2022. Replacement of soybean oil with cold-extracted fat from *Hermetia illucens* in young turkey diets: Effects on performance, nutrient digestibility, selected organ measurements, meat and liver tissue traits, intestinal microbiota modulation, and physiological and immunological status. Animal Feed Science and Technology 286: 115210. https://doi.org/10.1016/j.anifeedsci.2022.115210.

Kim, B., Kim, H.R., Lee, S., Baek, Y.C., Jeong, J.Y., Bang, H.T., Kim, Y.B., 2021. Effects of dietary inclusion level of microwave-dried and press-defatted black soldier fly (*Hermetia illucens*) larvae meal on carcass traits and meat quality in broilers. Animals 11: 665. https://doi.org/10.3390/ani11030665.

Kim, B., Kim, M., Jeong, J.Y., Kim, H.R., Ji, S.Y., Jung, H., Park, S.H., 2022. Black soldier fly (*Hermetia illucens*) larvae oil as an alternative fat ingredient to soybean oil in laying hen diets. Anim Bioscience 35: 1408-1417. https://doi.org/10.5713/ab.22.0052.

Kim, J., Ji, S.Y., Kim, B.G., 2023. Ileal digestibility of amino acids in defatted black soldier fly larva meal and adult black soldier fly fed to pigs. Journal of Insects as Food and Feed 9: 1345-1352. https://doi.org/10.3920/JIFF2023.0013.

Kim, Y.B., Kim, D.H., Jeong, S.B., Lee, J.W., Kim, T.H., Lee, H.G., Kim, Y.Y., 2020. Black soldier fly larvae oil as an alternative fat source in broiler nutrition. Poultry Science 99: 3133–3143. https://doi.org/10.1016/j.psj.2020.01.018.

Ko, H., Kim, Y., Kim, J., 2020. The produced mealworm meal through organic wastes as a sustainable protein source for weanling pigs. Journal of Animal Science and Technology 62: 365-373. https://doi.org/10.5187/jast.2020.62.3.365.

Ko, H.S., Choi, Y.H., Khun, S., Cho, E.S., Kim, Y.Y., Pi, J.S., Park, K.H., Kim, J.D., Lee, S.H., Kim, J.S., 2020. Laying performance, egg quality, haematological traits, and faecal noxious gas emission of laying hens fed with *Tenebrio molitor* meal. European Poultry Science 84: 1-13. https://doi.org/10.1399/eps.2020.307.

Kolawole, U.K., Lee, K.J., Kim, I.H., 2025. Dietary inclusion effects of dried mealworm, hydrolyzed mealworm, fermented poultry by-product, and hydrolyzed fish soluble protein on weaning pigs' performance, fecal score, and blood profiles. Animals 15: 1507. https://doi.org/10.3390/ani15111507.

Kongsup, P., Lertjirakul, S., Chotimanothum, B., Chundang, P., Kovitvadhi, A., 2022. Effects of eri silkworm (*Samia ricini*) pupae inclusion in broiler diets on growth performances, health, carcass characteristics and meat quality. Anim Bioscience 35: 711-720. https://doi.org/10.5713/ab.21.0323.

Kostecka, J., Konieczna, K., Cunha, L.M., 2017. Evaluation of insect-based food acceptance by representatives of polish consumers in the context of natural resources processing retardation. Journal of Ecological Engineering 18: 166-174. https://doi.org/10.12911/22998993/68301.

Kowalska, D., Gugołek, A., Strychalski, J., 2020). Evaluation of slaughter parameters and meat quality of rabbits fed diets with silkworm pupae and mealworm larvae meals. Annals of Animal Science 20: 551-564. https://doi.org/10.2478/aoas-2019-0080.

Kröncke, N., Baur, A., Böschen, V., Demtröder, S., Benning, R., Delgado, A., 2020. Automation of insect mass rearing and processing technologies of mealworms (*Tenebrio molitor*). In African edible insects as alternative source of food, oil, protein and bioactive components. Springer International Publishing, pp. 123-139.

Kröncke, N., Grebenteuch, S., Keil, C., Demtröder, S., Kroh, L., Thünemann, A.F., Benning, R. and Haase, H. 2019. Effect of different drying methods on nutrient quality of the yellow mealworm (*Tenebrio molitor* L.). Insects 10: 84. https://doi.org/10.3390/insects10040084

La Mantia, M.C., Calì, M., Petrocchi Jasinski, L., Contò, M., Meo Zilio, D., Renzi, G., Guarino Amato, M., 2024. Black soldier meal in feed could adversely affect organic broiler meat quality when used for the total or half replacement of diet proteins. Poultry 3: 66-84. https://doi.org/10.3390/poultry3020007.

Lalev, M., Hristakieva, P., Mincheva, N., Oblakova, M. & Ivanova, I., 2022a. Insect meal as alternative protein ingredient in broiler feed. Bulgarian Journal of Agricultural Science 28: 743-751. https://www.agrojournal.org/28/04-22.pdf.

Lalev, M., Mincheva, N., Hristakieva, P., Ivanova, I., Oblakova, M., 2023b. Effect of inclusion insect meals in feed on performance, blood biochemical and immune parameters of layer hens. Bulgarian Journal of Agricultural Science 29: 938–945.

Lalev, M., Penkov, D., Hristakieva, P., Oblakova, M., Mincheva, N., Ivanova, I., 2023a. Influence of the inclusion of insect meals on the net utilization of energy and protein in broiler chickens. Bulgarian Journal of Agricultural Science 29: 507-513.

Lalev, M., Vlahova-Vangelova, D.B., Kolev, N.D., 2022b. Quality of eggs after feed supplements with silkworm pupae (*Bombyx mori*) and black fly larvae (*Hermetia illucens*). BIO Web of Conferences 45: 01011. https://doi.org/10.1051/bioconf/20224501011.

Lee, J., Park, Y., Song, D., Chang, S., Cho, J., 2024. Effects of defatted and hydrolyzed black soldier fly larvae meal as an alternative fish meal in weaning pigs. Animals 14: 1692. https://doi.org/10.3390/ani14111692.

Leuschner, R.G.K., Robinson, T.P., Hugas, M., Cocconcelli, P.S., Richard-Forget, F., Klein, G., Licht, T.R., Nguyen-The, C., Querol, A., Richardson, M., Suarez, J.E., Thrane, U., Vlak, J.M., von Wright, A., 2010. Qualified presumption of safety (QPS): a generic risk assessment approach for biological agents notified to the European Food Safety Authority (EFSA). Trends in Food Science and Technology 21: 425-435. https://doi.org/10.1016/j.tifs.2010.07.003.

Liu, K., Zhang, G., Li, Y., Jiao, M., Guo, J., Shi, H., Ji, X., Zhang, W., Quan, K., Xia, W., 2025. Effects of feeding unprocessed whole black soldier fly (*Hermetia illucens*) larvae on performance, biochemical profile, health status, egg quality, microbiome and metabolome patterns of quails. Poultry Science 104: 105374. https://doi.org/10.1016/j.psj.2025.105374.

Liu, S., Wang, J., Li, L., Duan, Y., Zhang, X., Wang, T., Zang, J. Piao, X., Ma, Y., Li, D., 2023. Endogenous chitinase might lead to differences in growth performance and intestinal health of piglets fed different levels of black soldier fly larva meal. Animal Nutrition 14: 411-424. https://doi.org/10.1016/j.aninu.2023.05.008.

Lokaewmanee, K., Suttibak, B., Sukthanapirat, R., Sriyoha, R., Chansakhatana, N., Baotong, S., Trithailen, U., 2023. Laying hen performance, feed economy, egg quality and yolk fatty acid profiles from laying hens fed live black soldier fly larvae. Czech Journal of Animal Science 68: 169-177. https://doi.org/10.17221/174/2022-CJAS.

Machado, M., Catarino, J., Almeida, J., Lopes, I., Alvarado, A., Matos, R., Ribeiro, T., Faisca, P., Moreira, O., Murta, D., 2023. Effects of soybean meal replacement by *Hermetia illucens* larvae meal on growth performance, meat quality and gastrointestinal health in broilers. Journal of Insects as Food and Feed, 9: 569-582. https://brill.com/view/journals/jiff/9/5/article-p569_4.xml?rskey=Q3cyok&result=1

Maggitt, S., Cox, J., Dobson, K., McCann, J., Wickersham, T.A., Drewery, M.L., 2025. Intake, digestion, and rumen microbial impacts of black soldier fly larvae and frass provided as protein supplements to cattle consuming forage. Journal of Animal Science 103: skaf077. https://doi.org/10.1093/jas/skaf077.

Makkar, H.P., Tran, G., Heuzé, V. and Ankers, P. 2014. State of the-art on use of insects as animal feed. Animal. Feed Science and Technology 197: 1-33. https://doi.org/10.1016/j.anifeedsci.2014.07.008

Mancini, S., Vitali, A., Meneguz, M., Volterrani, C., Battisti, S., Cucci, S., Carvelli, A., Santori, D., Grossi, G., Batistini, E., Lacetera, N., Ferrarini. N., Sezzi, E., 2025. Full-fat *Hermetia illucens* larvae meal as a sustainable soybean meal substitute in laying hen diets: effects on performance, egg quality, and shelf-life. Italian Journal of Animal Science 24: 1586-1595. https://doi.org/10.1080/1828051X.2025.2539251.

Mancuso, T., Pippinato, L., Gasco, L., 2016. The European insects sector and its role in the provision of green proteins in feed supply. Quality – Access to success 20: 374-381.

Marchetti, L., Truzzi, E., Frosi, I., Papetti, A., Cappellozza, S., Saviane, A., Pellati, F., Bertelli, D., 2022. *In vitro* bioactivity evaluation of mulberry leaf extracts as nutraceuticals for the management of diabetes mellitus. Food and Function 13: 4344–4359. https://doi.org/10.1039/D2FO00114D.

Martins, C., Cullere, M., Dalle Zotte, A., Cardoso, C., Alves, S.P., Bessa, R.J.B. Freire, J.P.B., Falcao-e-Cunha L., 2018. Incorporation of two levels of black soldier fly (*Hermetia illucens* L.) larvae fat or extruded linseed in diets of growing rabbits: effects on growth performance and diet digestibility. Czech Journal of Animal Science 63: 356-362. https://doi.org/10.17221/22/2018-CJAS.

Mbhele, F.G.T., Mnisi, C.M., Mlambo, V., 2019. A nutritional evaluation of insect meal as a sustainable protein source for jumbo quails: Physiological and meat quality responses. Sustainability 11: 6592. https://doi.org/10.3390/su11236592.

Meyer, S., Gessner, D.K., Braune, M.S., Friedhoff, T., Most, E., Höring, M., Liebisch, G., Zorn, H., Eder, K., Ringseis, R., 2020. Comprehensive evaluation of the metabolic effects of insect meal from *Tenebrio molitor* L. in growing pigs by transcriptomics, metabolomics and lipidomics. Journal of Animal Science and Biotechnology 11: 20. https://doi.org/10.1186/s40104-020-0425-7.

Meyer-Rochow, V.B., Gahukar, R.T., Gosh, S., Jung, C., 2021. Chemical composition, nutrient quality and acceptability of edible insects are affected by species, developmental stage, gender, diet, and processing method. Foods 10: 1036. https://doi.org/10.3390/foods10051036.

Miah, M.Y., Singh, Y., Cullere, M., Tenti, S. Dalle Zotte, A., 2020. Effect of dietary supplementation with full-fat silkworm (*Bombyx mori* L.) chrysalis meal on growth performance and meat quality of Rhode Island Red x Fayoumi crossbred chickens. Italian Journal of Animal Science 19: 447-456. https://doi.org/10.1080/1828051X.2020.1752119.

Miglietta, P.P., De Leo, F., Ruberti, M., Massari, S., 2015. Mealworms for food: a water footprint perspective. Water 7: 6190-6203. https://doi.org/10.3390/w7116190.

Mohammadi, S., Torki, M., Darbemamieh, M., 2025. Effects of mealworm-based synbiotics on productive performance, egg quality, blood metabolites, and gut health in laying Japanese quails. Food Science and Nutrition 13: e71281. https://doi.org/10.1002/fsn3.71281.

Montalbán, A., Madrid, J., Hernández, F., Schiavone, A., Ruiz, E., Sánchez, C.J., Ayala, L., Fiorilla, E., Martínez-Miró, S., 2024. The influence of alternative diets and whole dry black soldier fly larvae (*Hermetia illucens*) on the production performance, blood status, and egg quality of laying hens. Animals 14: 2550. https://doi.org/10.3390/ani14172550.

Nassar, F.S., Alsahlawi, A.M., Abbas, A.O., Alaqil, A.A., Kamel, N.N., Abdelwahab, A.M., 2023. Impact of dietary inclusion of black soldier fly larvae (*Hermetia illucens*) as a replacement for soybean-corn ingredients on egg production, physiological status, and economic efficiency of laying hens. Advances in Animal and Veterinary Sciences 11: 295-304. https://dx.doi.org/10.17582/journal.aavs/2023/11.2.295.304.

Navasero, J.M.M., Angeles, A.A., Adinova, C.B., Merca, F.E., 2022. Effect of dietary black soldier fly, *Hermetia illucens* (Linnaeus) larvae meal and poultry meal on production performance, egg quality, and nutrient digestibility in post-peak chicken layers. Journal of the International Society for Southeast Asian Agricultural Sciences 28: 93-106. EID: 2-s2.0-85132175055 (abstract).

Ndotono, E.W., Khamis, F.M., Bargul, J.L., Tanga, C.M.,2022. Gut microbiota shift in layer pullets fed on black soldier fly larvae-based feeds towards enhancing healthy gut microbial community. Scientific Reports 12: 16714. https://doi.org/10.1038/s41598-022-20736-0.

Nekrasov, R.V., Ivanov, G.A., Chabaev, M.G., Zelenchenkova, A.A., Bogolyubova, N.V., Nikanova, D.A., Sermyagin, A.A., Bibikov, S.O., Shapovalov, S.O., 2022. Effect of Black Soldier Fly (*Hermetia illucens* L.) Fat on Health and Productivity Performance of Dairy Cows. Animals 12: 2118. https://doi.org/10.3390/ani12162118.

Nieto, J., Plaza, J., Hernández-Jiménez, M., Revilla, I., Palacios, C., 2024. Substitution of soybean meal for yellow mealworm meal in the diet of slow-growing chickens provides comparable carcass traits and meat quality. British Poultry Science 65: 730-739. https://doi.org/10.1080/00071668.2024.2369671.

Nuraini, N., Nur, Y.S., Djulardi, A., Amizar, R., Sari, Y.C., 2022. The effect of *Tenebrio molitor* caterpillar in the diet on production performance of laying quail. Advances in Animal and Veterinary Sciences 10: 2090-2099. https://dx.doi.org/10.17582/journal.aavs/2022/10.10.2090.2099.

Nuraini, N., Nur, Y.S., Djulardi, A., Amizar, R., Sari, Y.C., 2023. The effect of dietary supplementation of Hong Kong caterpillar (*Tenebrio molitor*) on quail egg quality. Journal of World's Poultry Science 13: 253-260. https://dx.doi.org/10.36380/jwpr.2023.28.

Odeon, M.M., Fernández-Arhex, V., García - Falabella, B., Villar, L., Caballero, V., Hara, S., Cancino, A.K., Villagra, E.S., 2025. Innovative protein sources for sustainable sheep farming:

Assessing black soldier fly larvae meal in merino lamb diet. Small Ruminant Research 250: 107556. https://doi.org/10.1016/j.smallrumres.2025.107556.

Ogbon, E.A., Balè, A., Santos, C., Seïdou, L.M., Dzepe, D., Behanzin, J.G., Djouaka, R., 2025. *Azolla* spp and *Hermetia illucens* meals as main protein sources for rabbit nutrition: impact on feed quality, growth performance, and meat quality. Current Developments in Nutrition 9: 107595. https://doi.org/10.1016/j.cdnut.2025.107595.

Olsen, R.L., Hasan, M.R., 2012. A Limited Supply of Fishmeal: Impact on Future Increases in Global Aquaculture Production. Trends in Food Science & Technology 27: 120-128. https://doi.org/10.1016/j.tifs.2012.06.003.

Oonincx, D.G.A.B., De Boer, I.J.M., 2012. Environmental impact of the production of mealworms as a protein source for humans – a life cycle assessment. PLoS One 7: e51145. https://doi.org/10.1371/journal.pone.0051145.

Oonincx, D.G.A.B., Finke, M.D., 2021. Nutritional value of insects and ways to manipulate their composition. Journal of Insects as Food and Feed 7: 639-659. https://doi.org/10.3920/JIFF2020.0050.

Oonincx, D.G.A.B., van Itterbeeck, J., Heetkamp, M.J.W., van den Brand, H., van Loon, J.J.A. and van Huis, A., 2010. An exploration on greenhouse gas and ammonia production by insect species suitable for animal or human consumption. PLoS One 5: 14445. https://doi.org/10.1371/journal.pone.0014445

Palumbo, B., Cullere, M., Singh, Y., Pontalti, E., Bacci, A., Contiero, B., Dalle Zotte, A., 2025. Effect of killing and drying methods on physicochemical traits, microbial count and sensory properties of *Tenebrio molitor* larvae. Journal of Insects as Food and Feed in press. https://doi.org/10.1163/23524588-bja10308.

Payne, C.L.R., Scarborough, P., Rayner, M., Nonaka, K., 2016. A systematic review of nutrient composition data available for twelve commercially available edible insects, and comparison with reference values. Trends in Food Science & Technology 47: 69-77. https://doi.org/10.1016/j.tifs.2015.10.012.

Pereira, J.C., da Silva, A., Rossiti, B., da Silva, E.R., da Silva, L., Evangelista, M., dos Santos Ruiz, U., Tse, M., 2025. Insect meal (*Tenebrio molitor*) has high nutrient digestibility for newly weaned piglets. Animal Science Journal 96: e70036. https://doi.org/10.1111/asj.70036.

Petkov, E., Ignatova, M., Popova, T., 2022. Layers' performance and egg hatchability as affected by the dietary inclusion of two meals of black soldier fly (*Hermetia illucens*). Journal of Insects as Food and Feed 8: 1077-1084. https://doi.org/10.3920/JIFF2021.0206.

Petkov, E., Popova, T., Dimov, K., Vlahova-Vangelova, D., Balev, D., Kolev, N., Dragoev, S., 2024. Low-fat *Tenebrio molitor* meal as a component in the broiler diet: growth performance and carcass composition. Insects 15: 979. https://doi.org/10.3390/insects15120979.

Phaengphairee, P., Boontiam, W., Wealleans, A., Hong, J., Kim, Y.Y., 2023. Dietary supplementation with full-fat *Hermetia illucens* larvae and multi-probiotics, as a substitute for antibiotics, improves the growth performance, gut health, and antioxidative capacity of weaned pigs. BMC Veterinary Research 19: 7. https://doi.org/10.1186/s12917-022-03550-8.

Pieterse, E., Erasmus, S.W., Uushona, T., Hoffman, L.C., 2019. Black soldier fly (*Hermetia illu-*

cens) pre-pupae meal as a dietary protein source for broiler production ensures a tasty chicken with standard meat quality for every pot. Journal of the Science of Food and Agriculture 99: 893-903. https://doi.org/10.1002/jsfa.9261.

Pietras, M., Orczewska-Dudek, S., Szczurek, W. Pieszka, M., 2021. Effect of dietary lupine seeds (*Lupinus luteus* L.) and different insect larvae meals as protein sources in broiler chicken diet on growth performance, carcass, and meat quality. Livestock Science 250: 104537. https://doi.org/10.1016/j.livsci.2021.104537.

Popova, T., Petkov, E., Ignatova, M., 2020. Effect of black soldier fly (*Hermetia illucens*) meals on the meat quality in broilers. Agricultural and Food Science 29: 177-188. https://doi.org/10.23986/afsci.88098.

Prachumchai, R., Suntara, C., Kanakai, N., Cherdthong, A., 2025. Inclusion of black soldier fly larval oil in ruminant diets influences feed consumption, nutritional digestibility, ruminal characteristics, and methane estimation in Thai-indigenous steers. Journal of Animal Physiology and Animal Nutrition 109: 812-820. https://doi.org/10.1111/jpn.14101.

Purnamia, S.A.A., Ridla, M., Nahrowi, N., Sarwono, K.A., Utami, S., Kahono, S., Ridwan, R., Fidriyanto, R., Fitri, A., Rohmatussolihat, Firmansyah, D., 2025. Utilization of Silkworm (*Bombyx mori*) pupae meal and Javanese bird grasshopper (*Valanga nigricornis*) meal as alternative protein sources to replace meat and bone meal in ruminant feed: An *in vitro* study. Veterinary Integrative Sciences 24: 1-12 in press. https://doi.org/10.12982/VIS.2026.018.

Quigley, J.D., Zontini, A., Schroeder, G.F., Roman-Garcia, Y., Houbiers, L., Bach, A., 2025. Nutritional value of black soldier fly larvae oil in calf milk replacers. Journal of Dairy Science 108: 2481-2488. https://doi.org/10.3168/jds.2024-25666.

Rastello, L., Gasco, L., Coppa, M., Gerbelle, M., Colombini, S., Battelli, M., Badino, P., Vernetti-Prot, L., Toral, P.G., Brugiapaglia, A., Gardini, G., Malfatto, V., Constant, I., Galli, A., Trespeuch, C., Renna, M., 2025. *Hermetia illucens* oil vs. hydrogenated palm fat in dairy cow nutrition: effects on digestive parameters, oxidative stress, and milk production performance. Journal of Animal Science and Biotechnology 16: 90. https://doi.org/10.1186/s40104-025-01222-9.

Razanova, O.P., Beznosyuk, A.M., Holubenko, T.L., Skoromna, O.I., Tkachenko, T.Y., Farionik, T.V., Solomon, A.M., 2025. Hematological blood profile of pigs fed with protein powder extracted from the black soldier fly (*Hermetia illucens*). Regulatory Mechanisms in Biosystems 16: e25026. https://doi.org/10.15421/0225026.

Regulation (EC) No 999/2001 of the European Parliament and of the Council of 22 May 2001 laying down rules for the prevention, control and eradication of certain transmissible spongiform encephalopathies. Official Journal of the European Union L147: 1-40. http://data.europa.eu/eli/reg/2001/999/oj

Regulation (EU) 2017/893 of 24 May 2017 amending Annexes I and IV to Regulation (EC) No 999/2001 of the European Parliament and of the Council and Annexes X, XIV and XV to Commission Regulation (EU) No 142/2011 as regards the provisions on processed animal protein (Text with EEA relevance.). Official Journal of the European Union L138: 92-116. http://data.europa.eu/eli/reg/2017/893/oj

Remón, S., Fondevila, G., Fondevila, M., 2025. Inclusion of *Tenebrio molitor* meal in diets for free-range hens does not affect egg sensory quality. Journal of Insects as Food and Feed in press. https://doi.org/10.1163/23524588-bja10346.

Renna, M., Coppa, M., Lussiana, C., Le Morvan, A., Gasco, L., Maxin, G., 2022. Full-fat insect meals in ruminant nutrition: *in vitro* rumen fermentation characteristics and lipid biohydrogenation. Journal of Animal Science and Biotechnology 13: 138. https://doi.org/10.1186/s40104-022-00792-2.

Ringseis, R., Peter, L., Gessner, D. K., Meyer, S., Most, E., Eder, K., 2021. Effect of *Tenebrio molitor* larvae meal on the antioxidant status and stress response pathways in tissues of growing pigs. Archives of Animal Nutrition 75: 237-250. https://doi.org/10.1080/1745039X.2021.1950106.

Robles-Jimenez, L.E., Angeles, S., Ramirez-Perez, A.H., Fuente, B., Velazquez-Ordoñez, V., Cardoso-Gutierrez, E., Renna, M., Rastello, L., Capucchio, M.T., Hassan, T., Gasco, L., Pino-Moreno, J.M., Ghavipanje, N., Dominguez-Vara, I.A., Gonzalez-Ronquillo, M., 2025. *In vitro* and *in vivo* investigations on the use of yellow mealworm (*Tenebrio molitor*) as a novel protein feed ingredient for fattening lambs. Animal Feed Science and Technology 320: 116224. https://doi.org/10.1016/j.anifeedsci.2025.116224.

Rytlewski, G., Flis, M., Grela, E.R., 2025. Effect of meal and whole larvae of black soldier fly (*Hermetia illucens*) on the performance, blood lipid profile, slaughter characteristics, sensory properties and fatty acid composition of pheasant (*Phasianus colchicus* L.) muscles. Animals 15: 3215. https://doi.org/10.3390/ani15213215.

Sagala, Y.G., Andadari, L., Handayani, T.H., Sholikin, MM., Fitri, A., Fidriyanto, R., Rohmatussolihat, R., Ridwan, R., Astuti, W.D., Widyastuti, Y., Fassah,. DM., Wijayanti, I., Sarwono, K.A., 2024. The effect of silkworms (*Bombyx mori*) chitosan on rumen fermentation, methanogenesis, and microbial population in vitro. Veterinary World 17: 1216-1226. https://doi.org/10.14202/vetworld.2024.1216-1226.

Saidani, M., Dabbou, S., Larbi, M.B., Slimen, I.B., Fraihi, W., Arbi, T., Amraoui, M., M'Hamdi, N., 2025. Effect of black soldier fly (*Hermetia illucens* L.) larvae meal on growth performance, carcass characteristics, meat quality, and cecal microbiota in broiler chickens. Frontiers in Animal Science 6: 1531773. https://doi.org/10.3389/fanim.2025.1531773.

Salomone, R., Saija, G., Mondello, G., Giannetto, A., Fasulo, S., Savastano, D., 2017. Environmental impact of food waste bioconversion by insects: Application of Life Cycle Assessment to process using *Hermetia illucens*. Journal of Cleaner Production 140: 890-905. https://doi.org/10.1016/j.jclepro.2016.06.154.

Schiavone, A., Cullere, M., De Marco, M., Meneguz, M., Biasato, I., Bergagna, S., Dezzutto, D., Gai, F., Dalle Zotte, A., Gasco, L., 2017. Partial or total replacement of soybean oil by black soldier fly larvae (*Hermetia illucens* L.) fat in broiler diets. Italian Journal of Animal Science 16: 93-100. https://doi.org/10.1080/1828051X.2016.1249968.

Schiavone, A., Dabbou, S., Petracci, M., Zampiga, M., Sirri, F., Biasato, I., Gai, F., Gasco, L., 2019. Black soldier fly defatted meal as a dietary protein source for broiler chickens: Effects on carcass traits, breast meat quality and safety. Animal 13: 2397-2405. https://doi.org/10.1017/S1751731119000685.

SCoPAFF (Standing Committee on Plants, Animals, Food and Feed Section Animal Nutrition), 2024. Sante.g.3(2024)2112308: Summary Report 15-16 February 2024. Available online at: https://food.ec.europa.eu/document/download/48254f32-f70b-4758-b385-db7edf45627b_en?filename=reg-com_ani-nutrit_20240215_sum.pdf.

Secci, G., Addeo, N.F., Pulido Rodriguez, L.F., Bovera, F., Moniello, G., Parisi, G., 2021. *In vivo* performances, ileal digestibility, and physicochemical characterization of raw and boiled eggs as affected by *Tenebrio molitor* larvae meal at low inclusion rate in laying quail (*Coturnix japonica*) diet. Poultry Science 100: 101487. https://doi.org/10.1016/j.psj.2021.101487.

Secci, G., Moniello, G., Gasco, L., Bovera, F., Parisi, G., 2018. Barbary partridge meat quality as affected by *Hermetia illucens* and *Tenebrio molitor* larva meals in feeds. Food Research International 112: 291-298. https://doi.org/10.1016/j.foodres.2018.06.045.

Sedgh-Gooya, S., Torki, M., Darbemamieh, M., Khamisabadi, H., Abdolmohamadi, A., 2022. Growth performance and intestinal morphometric features of broiler chickens fed on dietary inclusion of yellow mealworm (*Tenebrio molitor*) larvae powder. Veterinary Medicine and Science 8: 2050-2058. https://doi.org/10.1002/vms3.881.

Sedgh-Gooya, S., Torki, M., Darbemamieh, M., Khamisabadi, H., Abdolmohamadi, A., 2021. Effect of dietary inclusion of yellow mealworm (*Tenebrio molitor*) larvae meal on productive performance, egg quality indices and blood parameters of laying hens. Animal Production Science 61: 1365-1372. https://doi.org/10.1071/AN20102.

Seyedalmoosavi, M.M., Mielenz, M., Görs, S., Wolf, P., Daş, G., Metges, C.C., 2022. Effects of increasing levels of whole Black Soldier Fly (*Hermetia illucens*) larvae in broiler rations on acceptance, nutrient and energy intakes and utilization, and growth performance of broilers. Poultry Science 101. https://doi.org/10.1016/j.psj.2022.102202.

Shariat Zadeh, Z., Kheiri, F., Faghani, M., 2020. Productive performance, egg-related indices, blood profiles, and interferon-Ɣ gene expression of laying Japanese quails fed on *Tenebrio molitor* larva meal as a replacement for fish meal. Italian Journal of Animal Science 19: 274-281. https://doi.org/10.1080/1828051X.2020.1722970.

Shaviklo, A.R., Alizadeh-Ghamsari, A.H., Hosseini, S.A., 2021. Sensory attributes and meat quality of broiler chickens fed with mealworm (*Tenebrio molitor*). Journal of Food Science and Technology 58: 4587-4597. https://doi.org/10.1007/s13197-020-04946-w

Singh, Y., Cullere, M., Bertelli, D., Segato, S., Franzo, G., Frangipane di Regalbono, A., Catellani, P., Taccioli, C., Cappellozza, S., Dalle Zotte, A., 2023. Potential of full-fat silkworm-based diets for laying quails: performance and egg physical quality. Animals 13: 1510. https://doi.org/10.3390/ani13091510.

Smetana, S., Palanisamy, M., Mathys, A., Heinz, V., 2016. Sustainability of insect use for feed and food: Life Cycle Assessment perspective. Journal of Cleaner Production 137: 741-751. https://doi.org/10.1016/j.jclepro.2016.07.148.

Song, Y.S., Ha, D.U., Park, K., Kim, B.G., 2024. Dietary full-fat or defatted black soldier fly larvae can replace protein sources with no detrimental effect on growth performance or intestinal health of nursery pigs. Journal of Animal Science 102: skae333. https://doi.org/10.1093/jas/skae333.

Spranghers, T., Michiels, J., Vrancx, J., Ovyn, A., Eeckhout, M., De Clercq, P., De Smet, S., 2018. Gut antimicrobial effects and nutritional value of black soldier fly (*Hermetia illucens* L.) prepupae for weaned piglets. Animal Feed Science and Technology 235: 33-42. https://doi.org/10.1016/j.anifeedsci.2017.08.012.

Steinfeld, H., Gerber, P., Wassenaar, T., Castel, V., Rosales, M., De Haan, C., 2006. Livestock's long shadow: environmental issues and options. Food and Agriculture Organization of the United Nations (FAO), Rome, Italy.

Stelios, V., Ilias, G., Ioannis, P., Athanassiou, C., Papadopoulos, E., Fortomaris, P., 2024. Effect of three different insect larvae on growth performance and antioxidant activity of thigh, breast, and liver tissues of chickens reared under mild heat stress. Tropical Animal Health and Production 56: 80. https://doi.org/10.1007/s11250-024-03923-1.

Strychalski, J., Juśkiewicz, J., Kowalska, D., Gugołek, A., 2021. Performance indicators and gastrointestinal response of rabbits to dietary soybean meal replacement with silkworm pupae and mealworm larvae meals. Archives of Animal Nutrition 75: 294-310. https://doi.org/10.1080/1745039X.2021.1962171.

Swiatkiewicz, S., Swiatkiewicz, M., Arczewska-Wlosek, A., and Jozefiak, D., 2015. Chitosan and its oligosaccharide derivatives (chito-oligosaccharides) as feed supplements in poultry and swine nutrition. Journal of Animal Physiology and Animal Nutrition 99: 1-12. https://doi.org/10.1111/jpn.12222

Szczepanik K., Furgał-Dierżuk I., Gala Ł., Świątkiewicz M., 2022. Effects of *Hermetia illucens* larvae meal and astaxanthin as feed additives on health and production indices in weaned pigs. Animals 13: 163. https://doi.org/10.3390/ani13010163.

Tahamtani, F.M., Kittelsen, K., Vasdal, G., 2025. Dried black soldier fly larvae (*Hermetia illucens*) as environmental enrichment for laying hens – a full-scale commercial study. Journal of Insects as Food and Feed 11: 2617-2631. https://doi.org/10.1163/23524588-bja10220.

Tardiolo, G., La Fauci, D., Riggio, V., Daghio, M., Di Salvo, E., Zumbo, A., Sutera, A.M., 2025. Gut microbiota of ruminants and monogastric livestock: an overview. Animals 15: 758. https://doi.org/10.3390/ani15050758.

Toral, P.G., Hervás, G., González-Rosales, M.G., Mendoza, A.G., Robles-Jimenez, L.E., Frutos, P., 2022. Insects as alternative feed for ruminants: comparison of protein evaluation methods. Journal of Animal Science and Biotechnology 13: 21. https://doi.org/10.1186/s40104-021-00671-2.

United Nations, Department of Economic and Social Affairs, Population Division, 2019. World Population Prospects 2019: Highlights. United Nations, New York.

van Heugten, E., Martinez, G., McComb, A., Koutsos, L., 2022. Improvements in performance of nursery pigs provided with supplemental oil derived from black soldier fly (*Hermetia illucens*) larvae. Animals 12: 3251. https://doi.org/10.3390/ani12233251.

van Huis, A. and Gasco, L., 2023. Insects as feed for livestock production. Science 379: 139. https://doi.org/10.1126/science.adc9165

van Huis, A. and Oonincx, D.G.A.B., 2017. The environmental sustainability of insects as food and feed. A review. Agronomy for Sustainable Development 37: 43. https://doi.org/10.1007/s13593-017-0452-8

van Huis, A., 2022. Edible insects: Challenges and prospects. Entomological Research 52: 161-177. https://doi.org/10.1111/1748-5967.12582.

van Huis, A., Rumbos, C.I, Oonincx, D.G.A.B., Rojo, S., Kovitvadhi, A. and Gasco, L. 2024. From traditional to industrial use of insects as feed: a review. Journal of Insects as Food and Feed 11: 115-137. https://doi.org/10.1163/23524588-00001232

van Zanten, H.H.E., Mollenhorst, H., Oonincx, D.G.A.B., Bikker, P., Meeburg, B.G., de Boer, I.J.M., 2015. From environmental nuisance to environmental opportunity: housefly larvae convert waste to livestock feed. Journal of Cleaner Production 102: 362-369. https://doi.org/10.1016/j.jclepro.2015.04.106.

van Zanten, H.H.E., Simon, W., Van Selm, B., Wacker, J., Maindl, T.I., Frehner, A., Hijbeek, R., van Ittersum, M.K. and Herrero, M., 2023. Circularity in Europe strengthens the sustainability of the global food system. NatuRE Food 4: 320-330. https://doi.org/10.1038/s43016-023-00734-9

Veberke, W., 2015. Profiling consumers who are ready to adopt insects as a meat substitute in a Western society. Food Quality and Preference 39: 147-155. https://doi.org/10.1016/j.foodqual.2014.07.008

Veldkamp, T., Rezaei Far, A., Caimi, C., Gasco, L., Lima, R.C., Cunha, L.M., 2025. Partial or complete replacement of soybean meal with black soldier fly larvae meal improves feed efficiency in laying hens between 22 to 30 weeks of age. Journal of Insects as Food and Feed 11: 73-88. https://doi.org/10.1163/23524588-00001228.

Volek, Z., Adámková, A., Zita, L., Adámek, M., Plachý, V., Mlček, J., Marounek, M., 2021. The effects of the dietary replacement of soybean meal with yellow mealworm larvae (*Tenebrio molitor*) on the growth, nutrient digestibility and nitrogen output of fattening rabbits. Animal Feed Science and Technology 280: 115048. https://doi.org/10.1016/j.anifeedsci.2021.115048.

Volek, Z., Zita, L., Adámková, A., Adámek, M., Mlček, J., Plachý, V., 2023. Dietary inclusion of crickets (*Acheta domesticus*) and yellow mealworm meal (*Tenebrio molitor*) in comparison with soybean meal: effect on the growth, total tract apparent digestibility, and nitrogen balance of fattening rabbits. Animals 13: 1637. https://doi.org/10.3390/ani13101637.

Voora, V., Larrea, C. and Bermudez, S., 2020. Global Market Report: Soybeans. The International Institute for Sustainable Development.

Yoo, J.S., Cho, K.H., Hong, J.S., Jang, H.S., Chung, Y.H., Kwon, G.T., Shin., DG., Kim, Y.Y., 2019. Nutrient ileal digestibility evaluation of dried mealworm (*Tenebrio molitor*) larvae compared to three animal protein by-products in growing pigs. Asian-Australas Journal of Animal Sciences 32: 387-394. https://doi.org/10.5713/ajas.18.0647.

Yordanova, G., Nedeva, R.D., Apostolov, A.P., Mansbridge, S.C., Whiting, I.M., Mackenzie, A.M., Nikolova, G.D., Karamalakova, Y.D., Pirgozliev, V.R., 2025. Partial replacement of soyabean meal with defatted black soldier fly (*Hermetia illucens* L.) larvae meal influences blood biochemistry and modulate oxidative stress, but not growth performance of pigs. Animals 15: 1077. https://doi.org/10.3390/ani15081077.

Yu, M., Li, Z., Chen, W., Rong, T., Wang, G., Li, J., Ma, X., 2019. Use of *Hermetia illucens* larvae as a dietary protein source: effects on growth performance, carcass traits, and meat quality in finishing pigs. Meat Science 158: 107837. https://doi.org/10.1016/j.meatsci.2019.05.008.

Zacharis, C., Bonos, E., Giannenas, I., Skoufos, I., Tzora, A., Voidarou, C., Tsinas, A., Fotou, K., Papadopoulos, G., Mitsagga, C., Athanassiou, C., Antonopoulou, E., Grigoriadou, K., 2023. Utilization of *Tenebrio molitor* larvae reared with different substrates as feed ingredients in growing pigs. Veterinary Sciences 10: 393. https://doi.org/10.3390/vetsci10060393.

Zacharis, C., Bonos, E., Voidarou, C., Magklaras, G., Fotou, K., Giannenas, I., Giavasis, I. Mitsagga, C., Athanassiou, C., Antonopoulou, E., Grigoriadou, K., Tzora, A., Skoufos, I., 2024. Combined dietary supplementation of *Tenebrio molitor* larvae and chitosan in growing pigs: a pilot study. Veterinary Sciences 11: 73. https://doi.org/10.3390/vetsci11020073.

Zadeh, Z.S., Kheiri, F., Faghani, M., 2019. Use of yellow mealworm (*Tenebrio molitor*) as a protein source on growth performance, carcass traits, meat quality and intestinal morphology of Japanese quails (*Coturnix japonica*). Veterinary and Animal Science 8: 100066. https://doi.org/10.1016/j.vas.2019.100066.

Zampiga, M., De Cesare, A., Laghi, L., Soglia, F., Indio, V., Trespeuch, C., Benini, E., Dayan, J., Petracci, M., Sirri, F., 2025. Growth performance, meat quality, cecal microbiota and metabolomics profile of turkeys fed diets containing black soldier fly (*Hermetia illucens*) meal. Scientific Reports 15: 22040. https://doi.org/10.1038/s41598-025-05624-7.

Zhao, J., Kawasaki, K., Miyawaki, H., Hirayasu, H., Izumo, A., Iwase, S., Kasai, K., 2022. Egg quality and laying performance of Julia laying hens fed with black soldier fly (*Hermetia illucens*) larvae meal as a long-term substitute for fish meal. Poultry Science 101: 101986. https://doi.org/10.1016/j.psj.2022.101986.

Zhao, J., Sato, M., Takao, N., Ban, T., Tamamaki, K., Kagami, M., Yano, K., Kawasaki, K., 2024. Defatted black soldier fly larvae meal in lactating sow and pre-weaning piglet diets: impacts on growth performance, fecal microbiota, and metabolic pathways. Journal of Insects as Food and Feed 10: 1791-1809. https://doi.org/10.1163/23524588-00001120.

Zhao, X., Kuang, Z., Yang, J., Yang, M.B.R., 2023. Effect of fermented silkworm (*Bombyx mori*) pupae on growth performance, immune index and meat quality in lingnan yellow broiler chicken. Pakistan Journal of Zoology 55: 1891-1898. https://dx.doi.org/10.17582/journal.pjz/20220217030205.

Zhou, F., Tomberlin, J.K., Zheng, L., Yu, Z., Zhang, J., 2013. Developmental and waste reduction plasticity of three black soldier fly strains (Diptera: Stratiomyidae) raised on different livestock manures. Journal of Medical Entomology 50: 1224-1230. https://doi.org/10.1603/me13021.

Zhu, M., Liu, M., Yuan, B., Jin, X., Zhang, X., Xie, G., Wang, Z., Lv, Y., Wang, W., Huang, Y., 2022. Growth Performance and Meat Quality of Growing Pigs Fed with Black Soldier Fly (*Hermetia illucens*) Larvae as Alternative Protein Source. Processes 10: 1498. https://doi.org/10.3390/pr10081498.

Zsedely, E., Cullere, M., Takacs, G., Herman, Z., Szalai, K., Singh, Y., Dalle Zotte, A., 2023. Dietary inclusion of defatted silkworm (*Bombyx mori* L.) pupa meal for broiler chickens at different ages: growth performance, carcass and meat quality traits. Animals 13: 119. https://doi.org/10.3390/ani13010119.

CHAPTER 19

Microbial proteins from biomass fermentation used in livestock rearing

M. Longshaw[1]* *and M. Reshetiuk*[2]

[1]*Aseno Ltd, 52a St John Street, Ashbourne, Derbyshire. DE6 1GH, UK*
[2]*Calysta Germany GmbH, Carl-Friedrich-Gauß-Ring 5, 69124 Heidelberg, Germany*
**Correspondence: matt@aseno-partners.com*

Abstract

Microbial proteins from bacteria grown predominately in methanol or methane have been assessed as potential protein sources in livestock feeds for several years. However, despite a number of promising avenues, most have not achieved commercial success, partly due to a lack of uptake from the wider industry and partly because of the inherent technological and regulatory hurdles in upscaling from lab to full scale production. However, despite these constraints, several single cell proteins (SCPs) have been added to livestock feeds with a strong focus on pigs and poultry. To a lesser extent, proteins in cattle and turkey feeds have been partially replaced on an experimental basis with these SCPs. Typical studies have focused on growth and production outcomes in livestock although more recent studies have addressed the potential of SCPs to influence health outcomes. This chapter evaluates the different studies completed to date, including providing data on positive and negative outcomes associated with the inclusion of SCPs in feeds. It concludes with a view on future needs for cultured proteins from fermentation.

Keywords

single cell protein – microbiome – growth – health – pigs – poultry

1 Introduction

A major challenge for the livestock industry is to replace conventional proteins in feeds with complementary protein sources whilst balancing the biological requirements of the animal, ensuring commercial viability, achieving both equal or better growth, and favourable environmental outcomes. A number of so-called alternative proteins have been tested, mainly to replace animal proteins and / or soybean. One potential source of new proteins are bacteria grown using biomass fermentation. This typically involves the use of microbes grown at an industrial scale using carbon sources such as methane or methanol as a feedstock. Although several potential microbial sources have been assessed at a lab scale, few have been tested in livestock. Part of the reason for this has been the inherent difficulties involved in upscaling from pilot to full commercial scale (Banks et al., 2022). Indeed, as noted below, only data from five major bacterial strains have been tested as alternative proteins in the livestock space.

This chapter covers the production and application of bacteria obtained via biomass fermentation to produce proteins that are explicitly used to replace or complement existing plant and animal proteins in livestock feed where livestock are defined as poultry, cattle, and pigs. No specific studies on inclusion of microbial proteins as a protein source in sheep appear to have been published. Furthermore, whilst mink and foxes could be considered as livestock as they are reared for fur, they are considered in the pet food chapter (Chapter 14) as the purpose of those studies was predominantly to provide proxy data for cats and dogs. This chapter does not cover bacteria that are added as a biologically functional feed ingredient where the primary purpose is where it acts as a biotic (Salminen et al., 2021a), nor does it include precision fermentation where modern tools such as gene editing and metabolic engineering are utilised to produce specific strains and products (Pereira et al., 2025). Where appropriate the nomenclature of the various bacteria have been updated to reflect the latest taxonomy of the bacteria used as a protein source. The majority of studies have been completed in poultry and in pigs, with a smaller number of trials in cattle and other poultry species. In addition, most studies have assessed the effects of *Methylococcus capsulatus* (as FeedKind® from Calysta Inc., USA, PRO-DG from String Bio Pvt Ltd., India, and Uniprotein® from Unibio, Denmark) (crude protein (CP) = 62–71%), *Methylophilus methylotrophus* (as Pruteen® from ICI Chemicals, UK and as *Methylomonas clara* (Probion®) from Hoeschst, Germany) (CP = 71–82%), and *Methylomonas methanolis* (CP = 78%) from Mitsubishi Gas Chemical Inc., Japan. *Methylococcus capsulatus* cultured protein contains a consortia of bacteria including *M. capsulatus* which comprises >90% of the total dried biomass, alongside *Aneurinibacillus danicus, Brevibacillus agri* and *Cupriavidus cauae* and is grown on methane in a loop fermenter (Banks et al., 2022). For brevity, it will be referred to as *Methylococcus capsulatus.* Other species examined as potential microbial protein sources include *Clostridium autoethanogenum* (CP = 85%), *Cupriavidus necator* (=*Alcaligenes eutrophus*) (CP = 80–90%), and *Methylomicrobium album* (=*Methylomonas albus*). Despite their potential as protein sources, *Corynebacterium ammoniagenes, Corynebacterium glutamicum* (=*Bacillus lactofermentum*) (as Eurolysine bacterial protein), and *Clostridium acetobutylicum* are excluded from the chapter as they are by-products of the fermentation of sugars and are typically incorporated as biotics, rather than as protein sources. Further information on their potential as feed ingredients in pigs and poultry can be found in a number of publications (An et al., 2018; Braude and Foot, 1942; Cheng et al., 2021; Näsi, 1982a, 1982b; Näsi and Huida, 1982; Wang et al., 2013; Zhang et al., 2013; Zhao et al., 2012).

2 Palatability / feed intake

One of the main concerns with any new ingredient is the need to ensure that they are palatable and readily accepted by the animal. Reductions in feed intake can directly impact on growth, health and production either through reduced energy consumption or by limiting essential nutrients required to thrive. Furthermore,

TABLE 19.1 Main characteristics of bacteria used as microbial protein sources (SCP) in animal nutrition. Source: Adapted from the cited references. Where AOC = Antioxidant capacity, FA = Fatty Acids, FCR = Food Conversion Ratio, NA = nucleic acids, N = Nitrogen

Bacterium / Commercial product	Metabolic type	Main substrate	Crude protein (% DM)	Livestock species tested	Main outcomes / observations	Key references
Methylococcus capsulatus	Methanotrophic	Methane (CH_4)	62–71	Poultry, pigs, calves	High digestibility (~75–85%); improved FCR ≤6%; modulates gut microbiota; NA ≈5–10%	Øverland et al., 2001; Skrede et al., 1998; Hellwing et al., 2007a, b; Hombegowda et al., 2021
Methylophilus methylotrophus	Methylotrophic	Methanol (CH_3OH)	71–82	Poultry, pigs, cattle	Good growth ≤10%; arginine limiting; digestibility ~75–90%	D'Mello et al., 1976; Braude et al., 1977; Plavnik et al., 1981
Methylomonas methanolis	Methylotrophic	Methanol (CH_3OH)	≈78	Poultry (hens, ducks, turkeys)	Feed intake ↓ >10%; no effect on laying ≤11%	Yoshida et al., 1981, 1988; Richter et al., 1989
Cupriavidus necator (=*Alcaligenes eutrophus*)	Hydrogenotrophic / chemolithotrophic	$H_2 + CO_2$	80–90	Poultry, pigs	Digestibility 85–87%; good N retention ≤10%; possible FA profile benefits	Greife et al., 1980, 1984
Clostridium autoethanogenum	Acetogenic	$CO/CO_2 + H_2$ (gas fermentation)	85	Poultry (broilers)	Improved growth, FCR, AOC and microbiota; activates mTOR pathway	Chen et al., 2024; Shan et al., 2025; Wu et al., 2022

conditioned aversion of unpalatable feeds may limit the use of such ingredients if there is an initial negative response to these components (Duncan and Young, 2002; Sarabian et al., 2023; Villalba et al., 2019). It is known that some SCPs can be unpalatable at high inclusion levels in fish which can be overcome by the addition of palatants such as hydrolysates (Montoya-Martínez et al., 2018; Pilmer et al., 2022; Siddik et al., 2024). However, the use of palatants to increase SCP intake may exacerbate negative effects associated with these protein sources and, as far as the authors are aware, palatants and taste-masking technologies have not been applied to SCPs in terrestrial animal feeds.

Improved or equal feed intake in broilers fed up to 5% *C. autoethanogenum* over a 42-day period have been reported (Chen et al., 2024; Wu et al., 2022). A slight decrease in feed intake was noted in hens fed either *C. necator* at 12.99% or *M. methanolis* at 10.96% for over two years but this was not significant and had no impact on growth outcomes for these animals (Yoshida, 1988; Yoshida et al., 1981). In contrast, reduced feed intakes were noted in hens, in Muscovy ducks and in turkeys fed between 7.5 and 15% *M. methanolis* which led the authors to suggest that lower inclusion levels would be need if the SCP were to be used as a protein source (Richter et al., 1991, 1989, 1988). Inclusion up to 6% *M. capsulatus* in broiler diets for up to 30 days had no effect on feed intake (Hellwing et al., 2003, 2006). Reduced feed intake at higher inclusion levels of 7.5–10% of *M. capsulatus* were attributed to the formation of clumps of microbial protein around the beaks of broilers that restricted the birds ability to consume the feed (Hombegowda et al., 2021; Skrede et al., 2003). Production of appropriate diets that ensure correct assimilation of these protein sources may alleviate this issue and help to determine if the reduced feed intake is because of overall feed quality or due to palatability (Plavnik et al., 1981; Whittemore et al., 1978b). The reduced feed intake and feed efficiency noted in broilers fed autolysed *M. capsulatus* at 8–12% of the diet was attributed to poor palatability associated with the greater proportion of free amino acids available as a result of treatment (Øverland et al., 2010). Additionally, pellets made with autolysed *M. capsulatus* were harder with a stronger binding capacity compared with control and whole cell *M. capsulatus* feeds which may have influenced feed intake (Øverland et al., 2010). The reduced feed intake of whole cell *M. capsulatus* diets in the same study did not affect growth rate and led to an improved feed conversion efficiency suggesting that high inclusion is possible. No effect on feed intake due to autolysed or whole cell *M. capsulatus* was apparent at a lower inclusion of 6% (Schøyen et al., 2007a, 2007b). Conversely, reduced feed intake of feeds containing as little as 3.65% of *M. methylotrophus* have been noted which were attributed to lower levels of arginine in these diets or to palatability due to the microbial protein (Plavnik et al., 1981). Feed intake and weight gain were markedly reduced in chicks fed up to 20% spray-dried *M. methylotrophus* but improvements were noted in these parameters when the chicks were fed diets containing 9.2% flash-dried granules of the same microbial protein (D'Mello and Acamovic, 1976). The reduction in feed intake may partially be explained by the powdery nature of the product in the spray-dried form compared with the granules although this effect was not noted in the study of D'Mello and Acamovic (1976) or Waldroup and Payne (1974) where reduced feed intake was apparent at higher inclusion levels. Supplementation with various amino acids does not improve feed intake or feed conversion efficiency in chicks fed 15% *M. methylotrophus* (D'Mello, 1979a). In contrast, turkeys fed diets containing 20% *M. methylotrophus* and supplemented with amino acids showed similar feed intake to control diets (D'Mello, 1979b).

In a twelve-week study where adult cattle given silage supplemented with either soybean meal, groundnut cake or *M. methylotrophus*, feed intake was lower in the *M. methylotrophus* diets compared with control but higher than the groundnut cake diet (Castle and Watson, 1979). At the same time, water intake was reduced in the *M. methylotrophus* group. One cow occasionally left some feed containing *M. methylotrophus* suggesting that there may have been some individual dietary preferences and that these may need to be considered in subsequent feed formulations. When *M. methylotrophus* was used as a milk replacer at 20% of the total diet in weaning calves, intake and associated growth parameters were adversely affected, leading to the early cessation of the trial; lower inclusion levels were well tolerated and feed intake was unaffected (Hinks, 1977).

Feed intake in pigs during the weaning, growing and finishing period is unaffected by the inclusion of *M. capsulatus* at up to 12% (Hedemann et al., 2023; Øverland et al., 2004, 2001; Rønn et al., 2022). Pigs fed diets containing 28.6% *M. methylotrophus* showed a reduced appetite with concomitant diarrhoea and growth in weeks 2–4 of a study by Farstad et al. (1980), but recovered and returned to full appetite afterwards. It is unclear if the reduced appetite was due to the animals feeling unwell although diarrhoea has been noted in pigs fed *M. methylotrophus* without impacting feed intake (Hanssen and Farstad, 1980). Variations in feed intake are noted, perhaps related to overall feed composition. For example, when *M. methylotrophus* was included at 13.5% of the diet there was a reduction in feed intake of young pigs although lower inclusion levels had no effect (Schneider and Bronsch, 1977); in contrast no effect on feed intake was noted when piglets were given diets containing 20% *M. methylotrophus* (Whittemore et al., 1976).

3 Digestibility

Digestibility is a useful comparative measure of how well a food source is broken down by the action of digestive enzymes and other components in the gastrointestinal tract. A further refinement of this measure would be to consider bioavailability – a measure of how well the breakdown products are absorbed by the body and subsequently utilised in metabolic processes. Unfortunately, no measures of bioavailability of microbial proteins in livestock appear to have been conducted.

Clostridium autoethanogenum digestibility in the diets of Arbor Acres broiler chicks aged 1–42 days old was tested and compared with that of soybean meal (Chen et al., 2024). Apparent ileal digestibility of protein was 82.04% whilst the standardized ileal digestibility was 93.81%. Apparent and standardized amino acid digestibility showed similar values to each other and were generally above 85% with the exception of histidine and cysteine which were 78% and 73% respectively. Similar values of between 84.9% and 87.3% for nitrogen-digestibility were obtained in broilers fed diets containing up to 20% *C. necator* (as *Alcaligenes eutrophus*) (Greife et al., 1980). Higher inclusion levels of the microbial protein led to a reduction in digestibility values. In 1-year old Leghorn roosters fed diets containing 33% *M. capsulatus* as the sole protein sources, apparent total tract digestibility score for nitrogen was 80.5% (Skrede et al., 1998). Amino acid digestibility scores in the same study were >72% for most amino acids with the exception of cystine which had a digestibility score of 44.7%. Minor differences in ileal amino acid digestibility were noted in diets containing *M. capsulatus* to partially replace either fish meal or soybean meal in 32-day old broiler chickens (Schøyen et al., 2007b). Increasing *M. capsulatus* inclusion levels generally increased the amino acid digestibility scores although this increase was minor. Total tract amino acid digestibilities were unaffected by duration of feeding when broilers chicks were fed diets containing 6% whole cell or autolysed *M. capsulatus* for up to 5 weeks (Schøyen et al., 2007a). Although limited differences in total tract or ileal digestibility were noted for whole cell or autolysed *M. capsulatus*, the authors speculated that dietary whole cell *M. capsulatus* led to a net synthesis of amino acids in the hind gut of broilers, possibly as a result of a higher supply of fermentable nitrogen compounds compared with autolysed *M. capsulatus* or soybean meal. Digestibility of two bacteria – *Aerobactor* (sic) sp. and *M. methanolis* (as *Pseudomonas* sp.) – grown on methanol were assessed in colostomized laying hens by Yamazaki et al. (1977). Protein digestibilities were 84.1% and 85% respectively whilst amino acid digestibilities ranged from 74.3% for tryptophan to 91.9% for methionine in the *M. methanolis* bacterial meal.

Limited differences in nitrogen digestibility values for *M. methylotrophus* compared with other protein sources such as urea and soybean meal were noted in lactating cows (Teller and Godeau, 1986). Digestibility of nitrogen in test diets, which contained around 12% *M. methylotrophus*, was 58.4%. In comparison, inclusion of *M. methylotrophus* in a milk replacer diet for veal calves, markedly improved the apparent digestibility coefficient of nitrogen to 93.7% which compared favourably to a standard milk replacer diet (Van Weerden and Huisman, 1977; VanderWal, 1976). However, values for total ileal nitrogen digestibility of *M. methylotrophus* in veal calves were lower at around 71–78% in a later study with digestibility decreasing with increased levels of the microbial protein (Sedgman et al., 1985). Total tract digestibility of crude protein in dairy cattle fed diets supplemented with 2500g of live *M. capsulatus* were estimated to be 68–70% with lower values recorded in the small and large intestines (Maigaard et al., 2024).

Ileal digestibility of nitrogen in weaning piglets fed up to 15% *M. capsulatus* was between 75% and 77%, marginally lower than control diets at 78%; no effect of duration of feeding was noted in these piglets (Hellwing et al., 2007a). Higher values for ileal digestibility at 78% and 85.4% for total tract digestibility of nitrogen were recorded in pigs fed diets containing 23% *M. capsulatus* as the only protein source (Skrede et al., 1998). Whilst iIeal digestibility of amino acids ranged from 54.9% for cystine, to 88% for arginine in these pigs, the majority of total tract and ileal amino acid digestibilities were >75% (Skrede et al., 1998). In contrast, the apparent ileal digestibility of protein in weaner pigs fed diets with the same inclusion level of *M. capsulatus* was lower at 68.4% in comparison to other protein sources such as potato protein, fishmeal and soy protein concentrate (Rønn et al., 2022). Essential amino acid digestibility scores were also lower compared to these other protein sources (Rønn et al., 2022) and to those reported by Skrede et al. (1998). However, standardised ileal digestibility values for crude protein were significantly higher at 80.7% and although standardised ileal digestibility scores for essential amino acids were higher compared with apparent ileal digestibility scores, they were still lower than the other protein sources tested (Rønn et al., 2022). Apparent nitrogen digestibility of *M. methylotrophus* fed at an inclusion level of either 3% or 7% to pigs was 73.9% and 77.5% respectively (Braude et al., 1977). Amino acid digestibilities were similar to those of white fishmeal and were typically >75% with the highest apparent digestibility score of 91% being recorded for glutamic acid. Higher apparent and true digestibility values for amino acids and nitrogen have been noted in pigs fed *M. methylotrophus* (D'Mello et al., 1976; Whittemore et al., 1976; Whittemore and Moffat, 1976). Finally, nitrogen digestibilities of an unnamed SCP fed to young pigs were 81% and 83% when fed at 12 or 24% of the diet respectively (Zimmerman and Tegbe, 1977).

TABLE 19.2 Digestibility, inclusion levels and nucleic acid content of bacterial single-cell proteins used in animal nutrition. Where AA = amino acid, DM = dry matter.

Bacterial SCP	Protein / AA digestibility (%)	Safe inclusion level in diets (%)	Nucleic acids (% DM)
Methylococcus capsulatus	75–85 / 70–88	≤10 (broilers, pigs)	5–10
Methylophilus methylotrophus	73–90 / 75–91	≤15 (broilers)	16–20
Clostridium autoethanogenum	82–94 / >85	≤5 (broilers)	<5 (est.)

Source: Compiled from Chen et al. (2024), Øverland et al. (2001), Braude et al. (1977), Skrede et al. (1998), Hellwing et al. (2007), Greife et al. (1980), Yamazaki et al. (1977), Teller and Godeau (1986) and Braude et al. (1977).

4 Nitrogen balance

Nitrogen is a critical component of agriculture, utilised in crops as well as animal production (Hristov et al., 2019). One important facet of animal production involves the determination and efficiency of nitrogen utilisation derived from feeds. Nitrogen not utilised in growth or other production parameters such as milk, is excreted in the faeces and urine potentially contributing to pollution (Hristov et al., 2019). Methods for assessing nitrogen metabolism are discussed in Hristov et al. (2019). Broadly, these techniques involve the calculation of known quantities of nitrogen digestibility in a given timeframe followed by an assessment of nitrogen concentration in milk, faeces and urine, taking into account any production of protein by gut microbes as well as recycling of nitrogen from the blood. When replacing more than 30% of soybean meal in a poultry diet, *C. necator* (as 10% of total diet) had a slightly lower nitrogen digestibility although nitrogen retention was unaffected (Greife et al., 1980). At higher inclusion levels nitrogen digestibility decreased and urinary excretion of nitrogen increased. Similarly, nitrogen retention was unaffected in poultry fed up to 6% *M. capsulatus* (Hellwing et al., 2006). In pigs, faecal nitrogen excretion increases significantly but renal nitrogen excretion remains unchanged when they are fed *C. necator* (Greife et al., 1984). Further work in other livestock species with new and novel microbial proteins are required to fully understand the role of these protein sources on nitrogen turnover and their impact on growth as well as the potential for mitigating or contributing to pollution from farmed animals.

5 Growth and production

Broilers fed up to 5% *C. autoethanogenum* for 42 days had lower feed conversion ratios (FCR), and increased final body weight, as well as increased average daily weight gains (Chen et al., 2024; Shan et al., 2025, 2024; Wu et al., 2022). Egg production was delayed and the numbers of eggs produced was reduced in hens fed an unnamed bacterial protein at either 7.5 or 15% inclusion (Richter et al., 1989). No effect on sperm quality was apparent although sperm volume was reduced and sperm concentration increased in cocks fed bacterial protein. Live weight gain was reduced at the higher inclusion level and the authors concluded that a maximum inclusion of 5% was appropriate for the methanol-grown bacteria (Richter et al., 1988). Inclusion of up to 6% *M. capsulatus* in poultry diets supports a normal weight gain with some studies suggesting that higher inclusion levels of 8% and 10% reduced weight gain (Hellwing et al., 2003; Schøyen et al., 2007b; Skrede et al., 2003). One explanation for the reduced growth rates at the higher inclusions is due to reduced feed intake as a result of clumping of bacterial meal around the beaks (Hombegowda et al., 2021; Skrede et al., 2003; Waldroup and Payne, 1974). Whilst

Hombegowda et al. (2021) suggested that improved FCR values were obtained at 2.5% and 5% inclusions of *M. capsulatus*, FCR improvements were still apparent at up to 10% inclusion compared to soybean meal. Addition of 6% autolysed *M. capsulatus* to broiler diets did not improved growth outcomes (Schøyen et al., 2007a). Chicks fed diets containing either 10% *M. album*, maize-soybean, ground maize, or whole egg showed best growth parameters when fed maize-soybean (D'Mello, 1973). A slower growth rate and a marginally better FCR was noted in the *M. album* group compared with the control but these were not significant. In contrast, weight gain and FCR values were poorer in the ground maize and while egg groups. No impact on egg production, egg weight, viability, fertility or body weights were noted in hens fed 11% *M. methanolis* for up to 4 years (Yoshida et al., 1981). However, sexual maturity was delayed by 5 days compared with controls in a longer-term study and egg weight was lower in the *M. methanolis* group (Yoshida, 1988). Reduced body weight gain and higher FCR values were noted in turkeys and Muscovy ducks fed either 7.5 or 15% *M. methanolis* (Richter et al., 1991). The authors recommended that high inclusion levels were not appropriate for turkeys or ducks and proposed that additional studies to determine a lower inclusion level would be necessary (Richter et al., 1991). Reductions in weight gain rate of chicks with increased inclusion of *M. methylotrophus*, particularly above 10% inclusion were noted, mostly driven by reduced feed intake (Farstad, 1977; Plavnik et al., 1981). Marked improvements in FCR and weight gain are apparent at lower inclusions of 3% *M. methylotrophus* in chick diets (Mak and Oh, 1983). Arginine supplementation of diets containing *M. methylotrophus* can improve growth outcomes in chicks, suggesting that arginine is the first limiting amino acid in this bacterial protein (D'Mello, 1979a). Supplementation by L- or DL- forms of methionine and tryptophan in *M. methylotrophus* broiler chick diets raised growth rates and dry matter utilisation compared with non-supplemented chicks (D'Mello, 1987). In turkeys, the first limiting amino acid in *M. methylotrophus* diets is methionine (D'Mello, 1979b). Unlike chickens, turkeys appear to tolerate inclusions of up to 20% with no impact on growth or FCR. Downstream processing by flash-drying or spray-drying *M. methylotrophus* did not improve weight gain, feed intake or feed efficiency in chicks fed up to 29% of the bacterial protein (D'Mello and Acamovic, 1976). No effect was noted in egg production or feed efficiency In laying hens fed diets containing up to 13% *M. methylotrophus* for up to a year, although egg weight and shell stability worsened in a linear fashion (Vogt et al., 1983; Whittemore et al., 1978a).

When calves were given *M. capsulatus* as a milk replacer at inclusion levels of between 4% and 24% to replace skimmed milk, weight gain, nitrogen retention and FCR was reduced above 12%, possibly due to nutritional imbalance (The Scientific Committee for Animal Nutrition and the Scientific Committee for Food, 1995). Additionally, at higher inclusion levels faecal quality was diminished and had an aberrant grey green colour. No effect on feed intake, milk yield or milk quality were noted in lactating cattle when 2500g per day of live *M. capsulatus* was added to diets of lactating cattle (Maigaard et al., 2024). Similarly, milk yield and quality was unaffected by inclusion of 12% to 18% *M. methylotrophus* in diets given to lactating cattle (Castle and Watson, 1979; Teller and Godeau, 1986). Veal calves given a milk replacer containing 20% *M. methylotrophus* suffered with diarrhoea, loss of appetite, apathy and weight loss within 14 days (Hinks, 1977). Up to 10% inclusion, no such effects were seen and indeed, growth was unaffected (Hinks, 1977; Roth and Kirchgessner, 1976).

Daily weight gain and feed conversion efficiency were improved in pigs fed up to 14% *C. necator* to replace soybean meal (Greife et al., 1984). Whilst average daily growth was decreased by the inclusion of 6% or 12% *M. capsulatus* in growing-finishing pigs during the growing phase, no overall effect on growth performance was seen in the finisher or the overall period (Øverland et al., 2004). Average daily growth was reduced in piglets fed 15% *M. capsulatus*, with no effect on FCR or average daily feed intake in the same study. However, at a lower inclusion of 4–12%, average daily feed intake and average daily growth were improved in weanling pigs (Øverland et al., 2001). Early post-weaning piglets given diets containing up to 10% *M. capsulatus* in the first 15 days had a higher FCR although feed intake was unaffected; during the subsequent grower phase no effect on performance was noted, even at the higher inclusion levels (Hedemann et al., 2023). In a separate study, inclusion of 10% *M. capsulatus* in diets of pigs 14 days post-weaning had no effect on feed intake but did reduce weight gain and thus reduced feed efficiency (Rønn et al., 2022). No significant difference in growth rate or FCR compared to fishmeal were apparent when a lower inclusion of between 3.3 and 7.7% *M. methylotrophus* was added to diets of growing pigs (Braude et al., 1977). In the early phase of the same study, pigs given *M. methylotrophus* showed a lower final weight, lower daily gain and a higher FCR associated with a lower appetite and diarrhoea. Addition of up to 25% *M. methylotrophus* in finishing pigs did not affect growth rate or FCR suggesting that higher inclusion levels are better tolerated by older pigs (Hanssen and Farstad, 1980). When given as a milk replacer, 100g/l of *M. methylotrophus* had no effect on performance of 2–28 day old piglets with higher levels of 150–200 g/l reducing live weight gain and increasing FCR values (Newport and Keal, 1980). Pigs reared on 15% *M. methylotrophus*, followed by 11% during the grower phase and then at 8% during a pregnancy and lactating phase were assessed for the effect on growth and breeding success (Waterworth and Heath, 1981). No detrimental effects were seen on growth parameters, reproductive performance, litter size and weights, or milk production. Reduced feed intake, weight gain, and poorer FCR values were noted in pigs fed an unnamed single cell protein by (Zimmerman and Tegbe, 1977).

6 Meat quality / body composition

Aside from good feed conversion and improvements in growth parameters, perhaps the most important measure in production animals are those related to body (meat) composition / quality and flavour, which can include taste, smell, and appearance (Spence, 2013). Significant improvements in eviscerated yield, as well as protein content and yield of the breast meat of broilers fed diets containing up to 4% *C. autoethanogenum* have been noted (Shan et al., 2024; Wu et al., 2022). Differences in carcass quality have also been noted in broilers fed *M. capsulatus* which appear to be dependant on inclusion levels (Hombegowda et al., 2021; Øverland et al., 2010; Schøyen et al., 2007b). One major consideration for poultry meat quality is its fatty acid composition which can influence storage duration and sensory parameters. Whilst Skrede et al. (2003) noted a reduction in abdominal fat in broilers fed diets containing 6% or 9% *M. capsulatus*, lower inclusion levels had no effect on this measure (Hellwing et al., 2006; Skrede et al., 2003). Increases in saturated fatty acids (SFA), particularly C14:0, C16:0, and C20:0, and in monounsaturated fatty acids (MUFA), alongside a decrease in polyunsaturated fatty acids (PUFA), occur in the thigh meat of broilers fed up to 12% *M. capsulatus* (Øverland et al., 2011). No marked differences were noted in volatile oxidation or in sensory attributes such as odour intensity or rancidity of thigh meat (Øverland et al., 2011; Øverland and Skrede, 2012; Schøyen et al., 2007b; Skrede et al., 2003). In a study by Yoshida et al. (1982), sensory qualities of eggs and soup made from hens fed *M. methanolis* were compared with those of hens fed yeast. In a further refinement of the study, eggs hatched from adults fed the protein for 2 months were hatched and the offspring reared for 4 years on the same bacterial and yeast proteins. Subsequently, the offspring were assessed for sensory qualities. Significantly more than half of the panellists preferred the soup derived from chickens fed the bacterial protein but showed no preference for egg flavour. Whilst no differences were noted in carcass quality of Muscovy ducks or turkeys fed up to 15% *M. methanolis*, reduced growth performance at either 7.5 or 15% inclusion precluded its use at these high levels (Richter et al., 1991). Significant increases in carcass dry matter and lipid alongside reductions in nitrogen content of chicks fed diets containing 19–29% *M. methylotrophus* have been noted (D'Mello and Acamovic, 1976). However, reductions in weight gain in these animals at higher inclusion levels and marked reductions in feed intake and FCR values were noted in the 29% inclusion group.

Pigs fed up to 12% *M. capsulatus* were assessed for backfat thickness, firmness, colour, odour and flavour by Øverland et al. (2001). Reduced carcass quality as measured by increased backfat thickness, alongside decreases in carcass lean percentage and meat area in cutlets with increasing cultured protein were noted (Øverland et al., 2004). The authors suggested that lower lysine levels in the diets and higher daily energy intake in *M. capsulatus*-fed pigs may have been responsible for this (Øverland et al., 2005). However, improved firmness, colour, odour and flavour of backfat were noted and attributed to the fatty acid composition of the bacteria. Whilst no overall differences in SFA's are noted, increases in C18:0 and C20:0 alongside decreases in C24:0, increases in the MUFA's C14:1, C16:1, and C18:1 and decreases in the PUFA's in the backfat were reported (Øverland et al., 2005). The oxidative stability of the meat and backfat following storage was attributed to the inclusion of *M. capsulatus* in the pig diets. Furthermore, reductions in rancid odours and taste were noted in pigs fed *M capsulatus* compared with control diets possibly due to the increased levels of C16:1 fatty acids and the reduced PUFA levels (Øverland et al., 2005). No differences in carcass or meat quality of pigs fed either control or *M. methylotrophus* diets were noted by Braude and Rhodes (1977) or Hanssen and Farstad (1980). As with *M. capsulatus*, changes in fatty acid composition of the backfat are noted, with no impact on organoleptic or storage qualities (Hanssen and Farstad, 1980). Limited differences in the flavour of ham derived from pigs fed *M. methanolis* compared with control diets was noted by Yoshida et al. (1982).

7 Health outcomes

Microbial proteins have been shown to elicit health benefits when added to feeds, in particular influencing immune responses, haematology, antioxidant capacity (AOC) and the microbiome. Dried bacterial cells incorporated into feeds that elicit health benefits are referred to as postbiotics, irrespective of whether they impart any beneficial benefits whilst alive (Aguilar-Toalá et al., 2021; Salminen et al., 2021b, 2021a). This area of research is increasingly being applied to commercially relevant SCPs, moving these microbes from simple protein replacements to functional ingredients providing additional health benefits to animals. Specific target animal safety studies with increasing inclusion levels to determine an appropriate upper inclusion level have rarely been conducted for microbial proteins in livestock. Whilst many authors explicitly state that no clinical health problems have been noted in specific studies, a number of studies have reported on pathological conditions associated with these protein sources, as discussed below alongside specific safety studies.

Addition of *C. autoethanogenum* at up to 5% of the total diet of broilers improved the AOC and had a minor influence on the intestinal microbiome (Chen et al., 2024; Shan et al., 2025; Wu et al., 2022). In particular, the positive shifts in microbiome led to improvements in SCFA and bile acid production as well an upregulation of beneficial bacteria within the gut. No differences in serum biochemical indicators are apparent although there is an increase in haemoglobin and haematocrit, as well as serum IgA and IgG (Chen et al., 2024; Shan et al., 2024). The increases in haemoglobin and haematocrit can be associated with improved growth outcomes as well as improved AOC whilst the increased immunoglobulin production was considered by the authors to provide protection against bacterial and viral infections and to improve the intestinal mucosal barrier. Furthermore, metabolomic analysis of broilers showed that *C. autoethanogenum* enhanced a number of metabolic pathways including those involved in the citric acid cycle and amino acid metabolism (Shan et al., 2025). Shan et al. (2025) noted an increase in blood hydroxyproline with increased inclusion of *C. autoethanogenum*; higher hydroxyproline levels in blood are correlated with scoliosis in chickens. Whilst the authors did not record any incidences of the condition in their study, they did caution against high inclusion levels of *C. autoethanogenum* to minimise the risk. An inflammatory response was noted in the liver of hens fed *C. autoethanogenum* and although the authors suggested that this was not feed related, the responses were more marked and occurred earlier in the treatment groups (Wu et al., 2022). Improvements in villi length and decreased intestinal inflammation of broilers fed *M. capsulatus* were attributed to changes in the microbiome and their associated compounds influencing gut health (Hombegowda et al., 2021). Additionally, improvements in AOC have been noted in broilers fed *M. capsulatus* (Øverland et al., 2011). An early study on the effect of *M. methylotrophus* on broilers showed that whilst there was a dose-dependant elevated generalised immune response in hens fed the SCP, there was no specific immune response to the bacterial protein (Farstad, 1977). Additionally, no significant differences in bacterial loads in the intestines were noted in these hens (Farstad, 1977). Twenty-five percent of turkeys fed 15% *M. methylotrophus* died and at postmortem showed evidence of "turkey hepatosis". The aetiology of this condition is not clear from the paper although the authors did suggest that dietary stress due to an imbalance of sodium, potassium and chlorine may have exacerbated the condition (D'Mello, 1979b). The relationship with the liver cirrhosis reported in two chicks fed 12.4 and 24.8% *M. methylotrophus* by Farstad (1977) is not clear. The authors suggested that the condition may have been due to the presence of endotoxins in the bacterial biomass. No other pathologies were reported. Mortalities associated with enlarged fatty livers with focal haemorrhaging were reported in layer hens by Whittemore et al. (1978a). The absence of these lesions in broilers and in co-housed hens in the same study fed control diets alongside similar findings in the other studies likely implicate *M. methylotrophus* as the causative agent. However, Whittemore et al. (1978a) were reluctant to attribute the mortality in their study to the SCP and recommended additional studies be conducted.

Villous height, mucous cell count and lymphocyte infiltration in the small intestine increased in pigs fed up to 6% *M. capsulatus* and led to an overall improvement in intestinal health (Hedemann et al., 2023). The absence of weight increase in the liver and of the intestine was evidence of a lack of toxicological effects associated with *M. capsulatus* in these pigs. The absence of significant changes in haematological and sera chemistry measures in pigs fed *M. capsulatus* supports the view that the cultured protein does not affect health outcomes (Hellwing et al., 2007b). Likewise, there were no significant differences in blood parameters of pigs fed *M. methylotrophus* indicating a lack of effect on these animals according to Farstad et al. (1980). However, significant increases in leucocytes, blood urea, and sera alanine transaminase of pigs fed diets containing 24.7% *M. methylotrophus* were recorded by Hanssen and Farstad (1980). No pathological changes were reported by the authors. Farstad et al. (1980) reported reduced appetite and diarrhoea of pigs fed *M. methylotrophus* in the first month of feeding. At the end of the study, one anaemic, emaciated pig was reported. Histological examination of renal tissues from pigs in this group revealed enlarged glomeruli, inflammation, and fibrosis as well as formation of hyaline casts in the glomeruli.

One important safety concern associated with novel feed ingredients are the presence of variable amounts of DNA and RNA that are metabolised into nucleotides, consisting mainly of purines and pyrimidines. Nucleotides can also be produced endogenously, potentially increasing their availability within the animal. Notably higher concentrations are reported in fast growing animals including single-celled organisms but these nucleotides are also reported in a wide range of feed sources including animals and plants (Clifford and Story, 1976; Mani et al., 2023; Sabolová et al., 2023, 2021). They are an essential dietary component for animals being involved in a range of metabolic processes, including the *de novo* synthesis of nucleic acids (Ding et al., 2021), but excess levels of purines in particular have been associated with disease in animals. Purines can be of concern due to the production of urinary bladder stones or uroliths following the metabolic degradation of purines to uric acid (Longshaw et al., 2025). Uric acid can be further metabolised to the less harmful allantoin, reducing the risk of urogenic effects reported. Alongside DNAase in the intestine (Liven and Staveland, 1976), it has been promulgated that the gut microbiome may have a role to play in the degradation of nucleotides (Zhen et al., 2022). It is likely that dietary nucleic acids are all catabolised, and breakdown products excreted via the urine (Braude et al., 1977). Nucleic acid content of *M. methylotrophus* is estimated to be around 16–20% of the bacterial mass (Braude et al., 1977; Liven and Staveland, 1976), whilst levels of purines, consisting of adenine and guanine, in *M. capsulatus* are around 2.5% of dried product; total nucleic acid is around 5% (Longshaw et al., 2025). Some estimates put the nucleic acid content for *M. capsulatus* as high as 9.5% (Skrede et al., 1998). Hellwing et al. (2007c) showed that increased inclusion of *M. capsulatus* in pig diets significantly increased allantoin and uric acid excretion. However, there was a significant decrease in plasma uric acid concentrations with increased inclusion, suggesting that the SCP did not exert any urogenic effect in pigs (Hellwing et al., 2007b). Urinary excretion of allantoin in pigs increased with increasing *M. capsulatus* inclusion, accounting for up to 90% of the excreted purine base derivatives (Hellwing et al., 2007d); renal function, as measured by creatinine excretion, was unaffected by the inclusion of *M. capsulatus*. Similarly, allantoin excretion increased markedly in pigs fed *C. necator* and *M. methylotrophus* with no discernible change in sera creatinine levels, confirming their safety in pig diets (Braude et al., 1977; D'Mello et al., 1976; Greife et al., 1984). In contrast to *M. capsulatus*, inclusion of *M. methylotrophus* as a milk replacer for suckling pigs led to an increase in uric acid and allantoin concentrations in the blood plasma and urine (Newport and Keal, 1980).

Overall, the number of studies on the health benefits and safety assessment of microbial proteins in livestock have disproportionately focussed on *M. methylotrophus* and *M. capsulatus*, perhaps reflecting their importance as a protein source in animal feeds and their commercial readiness.

8 Human safety

Detailed discussion on human safety aspects of *M. methylotrophus* and *M. capsulatus* are covered in the Biomass / Petfood chapter (Chapter 4.8). In brief, transient flu-like symptoms have been reported in workers exposed to dusts containing these microbial proteins (Mayes, 1982; Sikkeland et al., 2009, 2008; Sikkeland, 2007; Skogstad et al., 2009, 2005). Recovery within 24–48 hours typically occurred once workers were no longer exposed to the dust and mitigation measures included the pelletization and addition of oil to the dried powder to reduce dusting effects (Ekenvall et al., 1983; Hellwing et al., 2007a; Mayes, 1982).

9 Sustainability

Sustainability is a general concern in livestock production because of its significant contribution to global greenhouse gas emissions, land degradation, biodiversity loss, and nutrient pollution. Traditional raw materials used in feed, such as soybean meal and fishmeal, have a heavy environmental footprint with some soybean cultivation associated with deforestation, biodiversity loss, heavy pesticide use, and high water and land usage, while fishmeal production depends on limited wild fisheries (Glencross et al., 2024; Song et al., 2021). Thus, the search for alternative protein sources such as bacterial proteins made by fermentation can offer a promising sustainable feed option.

Bacterial proteins have important advantages compared to traditional crop-based raw materials for feed as they do not require arable land or excessive freshwater, and production takes place in closed bioreactors, reducing their planetary impact (Buttle et al., 2024). When powered by renewable energy and carbon captured from industrial sources, bacterial protein production can have a very low or even negative carbon footprint (Marcellin et al., 2022; Matassa et al., 2016). Another advantage of bacterial proteins is alignment with circular bioeconomy principles. Some of the fermentation technologies can turn waste streams, such as agricultural byproducts, food waste, or biogas, into high-quality microbial biomass (Pereira et al., 2022), helping to recycle nutrients that otherwise would be lost to the environment. Studies suggest that fermentation systems could recycle up to 18.5 million tonnes of nitrogen and 6.5 million tonnes of phosphorous annually, reducing the need for synthetic fertilizers and reducing eutrophication risks (Pereira et al., 2022). Facilities that produce bacterial proteins can be located on non-arable land and operate independently of seasonal and geographical limitations (Buttle et al., 2024; Khoshnevisan et al., 2022).

However, there are still challenges as fermentation processes use a lot of energy, and when this energy comes from fossil-based sources, the environmental benefits are reduced (Banks et al., 2022; Khoshnevisan et al., 2020). Switching bacterial protein production to renewable energy inputs, including feedstocks and energy sources, is thus critical. Promising new technologies, such as hydrogen-oxidizing bacteria, that use renewable hydrogen from water electrolysis, show some of the largest sustainability potential for lowering emissions (Li et al., 2025). While these technologies are still in early stages, they highlight the long-term potential of bacterial proteins to contribute to carbon-neutral livestock production systems. In summary, using bacterial proteins in livestock feeds could help reduce the industry's most urgent sustainability issues by reducing GHG emissions, minimizing land and water use, closing nutrient cycles, and thus lowering biodiversity impacts leading to a lower-carbon, resource-efficient model for livestock farming while still meeting global protein demands.

10 Future direction

The paucity of bacterial proteins derived from biomass fermentation that have been tested in livestock feeds may be reflective of the cost and effort required to effectively test the safety and applicability of these protein sources. Indeed, despite early promises for Pruteen as a protein replacement, to date, only biomass obtained from the fermentation of *M. capsulatus* have achieved commercial scale up as a feed ingredient. However, despite this, the inclusion of *M. capsulatus* as a complementary protein in feeds has yet to achieve full commercial success. It follows that efforts by investors, businesses, researchers and governments to support the incorporation of SCPs into animals feeds need to be increased. In particular, the identification and testing of other biomass-derived proteins needs to be conducted in order to address concerns associated with environmental sustainability of the feed industry, and to increase the number of protein sources available. Studies should therefore be conducted to determine appropriate inclusion levels, their safety to the target animals and to assess possible biological benefits that may arise from their inclusion. Tantalizing evidence from more recent papers have shown that these cultured proteins have the potential to influence biological and health outcomes such as immune responses, improved growth outcomes and influencing the microbiome. However, for livestock feeds, these studies are in their infancy and would need to be applied more widely. It is clear from the studies to date that both subtle and broad effects have been observed in different breeds and in animals at different ages and stages and with various inclusion levels. Therefore, studies to account for these biological variations need to be considered. Finally, clarity on potential routes to full commercial implementation for all new SCPs as well as strong economic arguments for their inclusion needs to be undertaken to support the industry and provide a clear rationale for their mainstream uptake in feeds.

References

Aguilar-Toalá, J.E., Arioli, S., Behare, P., Belzer, C., Berni Canani, R., Chatel, J.-M., D'Auria, E., de Freitas, M.Q., Elinav, E., Esmerino, E.A., García, H.S., da Cruz, A.G., González-Córdova, A.F., Guglielmetti, S., de Toledo Guimarães, J., Hernández-Mendoza, A., Langella, P., Liceaga, A.M., Magnani, M., Martin, R., Mohamad Lal, M.T., Mora, D., Moradi, M., Morelli, L., Mosca, F., Nazzaro, F., Pimentel, T.C., Ran, C., Ranadheera, C.S., Rescigno, M., Salas, A., Sant'Ana, A.S., Sivieri, K., Sokol, H., Taverniti, V., Vallejo-Cordoba, B., Zelenka, J., Zhou, Z., 2021. Postbiotics - when simplification fails to clarify. Nat Rev Gastroenterol Hepatol 18, 825–826. https://doi.org/10.1038/s41575-021-00521-6

An, B.-K., Choi, Y.-I., Kang, C.-W., Lee, K.-W., 2018. Effects of dietary *Corynebacterium ammoniagenes*-derivedsingle cell protein on growth performance, blood and tibia bonecharacteristics, and meat quality of broiler chickens. J. Anim. Feed Sci. 27, 140–147. https://doi.org/10.22358/jafs/91966/2018

Banks, M., Johnson, R., Giver, L., Bryant, G., Guo, M., 2022. Industrial production of microbial protein products. Current Opinion in Biotechnology 75, 102707. https://doi.org/10.1016/j.copbio.2022.102707

Braude, R., Foot, A.S., 1942. Pig-feeding experiment using dried *Clostridium* residue. The Journal of Agricultural Science 32, 324–329. https://doi.org/10.1017/S0021859600048103

Braude, R., Hosking, Z.D., Mitchell, K.G., Plonka, S., Sambrook, I.E., 1977. Pruteen, a new source of protein for growing pigs. I. Metabolic experiment: Utilization of nitrogen. Livestock Production Science 4, 79–89. https://doi.org/10.1016/0301-6226(77)90022-7

Braude, R., Rhodes, D.N., 1977. Pruteen, a new source of protein for growing pigs. II. Feeding trial: Growth rate, feed utilization and carcass and meat quality. Livestock Production Science 4, 91–100. https://doi.org/10.1016/0301-6226(77)90023-9

Buttle, L., Noorman, H., Roa Engel, C., Santigosa, E., 2024. Bridging the protein gap with single-cell protein use in aquafeeds. Front. Mar. Sci. 11, 1384083. https://doi.org/10.3389/fmars.2024.1384083

Castle, M.E., Watson, J.N., 1979. Silage and milk production: a comparison between soya, groundnut and single-cell protein as silage supplements. Grass and Forage Science 34, 101–106. https://doi.org/10.1111/j.1365-2494.1979.tb01453.x

Chen, X., Zheng, A., Shoaib, A.P., Chen, Z., Qiu, K., Wang, Z., Chang, W., Cai, H., Liu, G., 2024. Evaluation of *Clostridium autoethanogenum* protein as a new protein source for broiler chickens in replacement of soybean meal. Anim Biosci 37, 1236–1245. https://doi.org/10.5713/ab.23.0419

Cheng, Y.-C., Duarte, M.E., Kim, S.W., 2021. Nutritional and functional values of lysed *Corynebacterium glutamicum* cell mass for intestinal health and growth of nursery pigs. Journal of Animal Science 99, skab331. https://doi.org/10.1093/jas/skab331

Clifford, A.J., Story, D.L., 1976. Levels of Purines in Foods and Their Metabolic Effects in Rats. The Journal of Nutrition 106, 435–442. https://doi.org/10.1093/jn/106.3.435

Ding, T., Song, G., Liu, X., Xu, M., Li, Y., 2021. Nucleotides as optimal candidates for essential nutrients in living organisms: A review. Journal of Functional Foods 82, 104498. https://doi.org/10.1016/j.jff.2021.104498

D'Mello, J.P.F., 1987. Utilisation of L- and DL-amino acids by broiler chicks fed diets containing Pruteen. Animal Feed Science and Technology 17, 1–13. https://doi.org/10.1016/0377-8401(87)90048-4

D'Mello, J.P.F., 1979a. Amino acid supplementation of methanol-grown dried microbial cells in diets for young chicks. Animal Feed Science and Technology 4, 63–69. https://doi.org/10.1016/0377-8401(79)90024-5

D'Mello, J.P.F., 1979b. Limiting amino acids in dried microbial cells given to young turkeys. Animal Feed Science and Technology 4, 209–220. https://doi.org/10.1016/0377-8401(79)90014-2

D'Mello, J.P.F., 1973. The use of methane-utilising bacteria as a source of protein for young chicks 1. British Poultry Science 14, 291–301. https://doi.org/10.1080/00071667308416031

D'Mello, J.P.F., Acamovic, T., 1976. Evaluation of methanol-grown bacteria as a source of protein and energy for young chicks. British Poultry Science 17, 393–401. https://doi.org/10.1080/00071667608416293

D'Mello, J.P.F., Peers, D.G., Whittemore, C.T., 1976. Utilization of dried microbial cells grown on methanol in a semi-purified diet for growing pigs. British Journal of Nutrition 36, 403–410. https://doi.org/10.1079/BJN19760095

Duncan, A.J., Young, S.A., 2002. Can goats learn about foods through conditioned food aversions and preferences when multiple food options are simultaneously available?. J Anim Sci 80, 2091–2098. https://doi.org/10.1093/ansci/80.8.2091

Ekenvall, L., Dölling, B., Göthe, C.J., Ebbinghaus, L., Stedingk, L.V. von, Wasserman, J., 1983. Single cell protein as an occupational hazard. Occupational and Environmental Medicine 40, 212–215. https://doi.org/10.1136/oem.40.2.212

Farstad, L., 1977. Biological Effects of Feeding Single Cell Protein Produced by *Pseudomonas methylotropha* from Methanol to Chickens. Acta Agriculturae Scandinavica 27, 129–137. https://doi.org/10.1080/00015127709435121

Farstad, L., Flatlandsmo, K., Almlid, T., Koppang, N., 1980. Health and Performance in Growing Pigs Fed Two Sources of Single Cell Protein in Comparison with Soybean. https://doi.org/10.1080/00015128009435689

Glencross, B., Ling, X., Gatlin, D., Kaushik, S., Øverland, M., Newton, R., Valente, L.M.P., 2024. A SWOT Analysis of the Use of Marine, Grain, Terrestrial-Animal and Novel Protein Ingredients in Aquaculture Feeds. Reviews in Fisheries Science & Aquaculture 32, 396–434. https://doi.org/10.1080/23308249.2024.2315049

Greife, H.A., Molnar ,S., Bos ,Teuntje, Gussmann ,M., Günther, K.-D., 1984. N-Stoffwechsel wachsender Schweine bei Austausch von Sojaextraktionsschrot durch einen bakteriellen Proteinträger (*Alcaligenes eutrophus*). Archiv für Tierernährung 34, 179–190. https://doi.org/10.1080/17450398409424670

Greife, H.A., Molnar, S., Gunther, K.-D., 1980. Nitrogen metabolism in broiler chickens consuming the bacterial strain *Alcaligenes eutrophus*. Animal Feed Science and Technology 5, 241–253. https://doi.org/10.1016/0377-8401(80)90034-6

Hanssen, J.T., Farstad, L., 1980. Effects of Feeding Large Amounts of "Pruteen" and "Toprina" on Some Biological Parameters in Growing Finishing Pigs. Acta Agriculturae Scandinavica 30, 74–80. https://doi.org/10.1080/00015128009435698

Hedemann, M.S., Rønn, M., van der Heide, M.E., Julegaard, I.K., Nielsen, M.O., 2023. Dietary inclusion of methanotrophic microbial cell-derived protein in the early postweaning period sustains growth performance and intestinal health of weaner piglets. animal 17, 100798. https://doi.org/10.1016/j.animal.2023.100798

Hellwing, A.L.F., Tauson, A.-H., Chwalibog, A., 2003. Energy and nitrogen balance in slaughter chickens fed bacterial protein. In: Progress in research on energy and protein metabolism, Brill. https://doi.org/10.3920/9789086865208_029

Hellwing, A.L.F., Tauson, A.-H., Kjos, N.P., Skrede, A., 2007a. Bacterial protein meal in diets for growing pigs: effects on protein and energy metabolism. Animal 1, 45–54. https://doi.org/10.1017/S1751731107283879

Hellwing, A.L.F., Tauson, A.-H., Skrede, A., 2007b. Blood parameters in growing pigs fed increasing levels of bacterial protein meal. Acta Veterinaria Scandinavica 49, 33. https://doi.org/10.1186/1751-0147-49-33

Hellwing, A.L.F., Tauson, A.-H., Skrede, A., 2007c. Excretion of purine base derivatives after intake of bacterial protein meal in pigs. Livestock Science, 10th International Symposium on Digestive Physiology in Pigs, Denmark 2006, Part 2 109, 70–72. https://doi.org/10.1016/j.livsci.2007.01.070

Hellwing, A.L.F., Tauson, A.-H., Skrede, A., 2006. Effect of bacterial protein meal on protein and energy metabolism in growing chickens. Archives of Animal Nutrition 60, 365–381. https://doi.org/10.1080/17450390600884351

Hellwing, A.L.F., Tauson, A.-H., Skrede, A., Kjos, N.P., Ahlstrøm, Ø., 2007d. Bacterial protein meal in diets for pigs and minks: Comparative studies on protein turnover rate and urinary excretion of purine base derivatives. Archives of Animal Nutrition 61, 425–443. https://doi.org/10.1080/17450390701565248

Hinks, C.E., 1977. The replacement of skim milk by dried microbial cells and whey in milk diets for calves. Animal Feed Science and Technology 2, 85–92. https://doi.org/10.1016/0377-8401(77)90044-X

Hombegowda, G.P., Suresh, B.N., Shivakumar, M.C., Ravikumar, P., Girish, B.C., Rudrappa, S.M., Indresh, H.C., 2021. Growth performance, carcass traits and gut health of broiler chickens fed diets incorporated with single cell protein. Anim Biosci 34, 1951–1962. https://doi.org/10.5713/ab.20.0844

Hristov, A.N., Bannink, A., Crompton, L.A., Huhtanen, P., Kreuzer, M., McGee, M., Nozière, P., Reynolds, C.K., Bayat, A.R., Yáñez-Ruiz, D.R., Dijkstra, J., Kebreab, E., Schwarm, A., Shingfield, K.J., Yu, Z., 2019. *Invited review:* Nitrogen in ruminant nutrition: A review of measurement techniques. Journal of Dairy Science 102, 5811–5852. https://doi.org/10.3168/jds.2018-15829

Khoshnevisan, B., He, L., Xu, M., Valverde-Pérez, B., Sillman, J., Mitraka, G.-C., Kougias, P.G., Zhang, Y., Yan, S., Ji, L., Carbajales-Dale, M., Elyasi, S.N., Marami, H., Tsapekos, P., Liu, H., Angelidaki, I., 2022. From renewable energy to sustainable protein sources: Advancement, challenges, and future roadmaps. Renewable and Sustainable Energy Reviews 157, 112041. https://doi.org/10.1016/j.rser.2021.112041

Khoshnevisan, B., Tabatabaei, M., Tsapekos, P., Rafiee, S., Aghbashlo, M., Lindeneg, S., Angelidaki, I., 2020. Environmental life cycle assessment of different biorefinery platforms valorizing municipal solid waste to bioenergy, microbial protein, lactic and succinic acid. Renewable and Sustainable Energy Reviews 117, 109493. https://doi.org/10.1016/j.rser.2019.109493

Li, R., Jiang, Y., Luo, K., Chen, B., Wang, H., Fan, X., Pan, J., Litti, Y.V., Fu, S., Guo, R., 2025. Hydrogen-oxidizing bacteria: A promising contributor to environmental sustainability and resource recycling. Green Carbon. https://doi.org/10.1016/j.greenca.2025.02.004

Liven, E., Staveland, K., 1976. Single Cell Protein in the Diets of Pigs and Chickens. Acta Vet Scand 17, 441–450. https://doi.org/10.1186/BF03547899

Longshaw, M., Quest, B., Miller, W., Oba, P.M., Swanson, O.R., Swanson, K.S., Miller, K., 2025. The Safety of FeedKind Pet® (*Methylococcus capsulatus*, Bath) as a Cultured Protein Source in the Diet of Adult Dogs and Its Effect on Feed Digestibility, Fecal Microbiome, and Health Status. Animals (Basel) 15, 1975. https://doi.org/10.3390/ani15131975

Maigaard, M., Weisbjerg, M.R., Nielsen, M.O., Hellwing, A.L.F., Lund, P., 2024. Effects of dietary nitrate, fumaric acid, and methanotrophic bacteria supplementation on rumino-intestinal nutrient metabolism and enteric gas exchange in dairy cows. Livestock Science 289, 105572. https://doi.org/10.1016/j.livsci.2024.105572

Mak, T., Oh, B., 1983. The use of "Pruteen" in broiler diets. Pertanika 6, 111–115.

Mani, K., Vitenberg, T., Ben-Mordechai, L., Schweitzer, R., Opatovsky, I., 2023. Comparative untargeted metabolic analysis of natural- and laboratory-reared larvae of black soldier fly, *Hermetia illucens* (L.) (Diptera: Stratiomyidae). Comparative Biochemistry and Physiology Part B: Biochemistry and Molecular Biology 266, 110851. https://doi.org/10.1016/j.cbpb.2023.110851

Marcellin, E., Angenent, L.T., Nielsen, L.K., Molitor, B., 2022. Recycling carbon for sustainable protein production using gas fermentation. Current Opinion in Biotechnology 76, 102723. https://doi.org/10.1016/j.copbio.2022.102723

Matassa, S., Boon, N., Pikaar, I., Verstraete, W., 2016. Microbial protein: future sustainable food supply route with low environmental footprint. Microb Biotechnol 9, 568–575. https://doi.org/10.1111/1751-7915.12369

Mayes, R.W., 1982. Lack of allergic reactions in workers exposed to Pruteen (bacterial single-cell protein). Occupational and Environmental Medicine 39, 183–186. https://doi.org/10.1136/oem.39.2.183

Montoya-Martínez, C., Nolasco-Soria, H., Vega-Villasante, F., Carrillo-Farnés, O., Álvarez-González, A., Civera-Cerecedo, R., 2018. Attractability and palatability of ingredients in longarm river prawn *Macrobrachium tenellum* feed. Latin American Journal of Aquatic Research 46, 615–620. https://doi.org/10.3856/vol46-issue3-fulltext-17

Näsi, M., 1982a. Performance of laying hens on diets containing Eurolysine bacterial protein or Pekilo protein. AFSci 54, 271–278. https://doi.org/10.23986/afsci.72106

Näsi, M., 1982b. Nutritive value of Eurolysine bacterial protein and Pekilo protein for growing pigs. AFSci 54, 263–269. https://doi.org/10.23986/afsci.72105

Näsi, M., Huida, L., 1982. Digestibility of amino acids in pig diets containing Eurolysine bacterial protein or Pekilo protein, with special reference to a gas chromatographic method used in amino acid determination. AFSci 54, 279–285. https://doi.org/10.23986/afsci.72107

Newport, M.J., Keal, H.D., 1980. Artificial rearing of pigs: 10. Effect of replacing dried skim-milk by a single-cell protein (Pruteen) on performance and digestion of protein. British Journal of Nutrition 44, 161–170. https://doi.org/10.1079/BJN19800023

Øverland, M., Borge, G.I., Vogt, G., Schøyen, H.F., Skrede, A., 2011. Oxidative stability and sensory quality of meat from broiler chickens fed a bacterial meal produced on natural gas. Poultry Science 90, 201–210. https://doi.org/10.3382/ps.2010-00784

Øverland, M., Kjos, N.P., Olsen, E., Skrede, A., 2005. Changes in fatty acid composition and improved sensory quality of backfat and meat of pigs fed bacterial protein meal. Meat Science 71, 719–729. https://doi.org/10.1016/j.meatsci.2005.05.017

Øverland, M., Petter Kjos, N., Skrede, A., 2004. Effect of bacterial protein meal grown on natural gas on growth performance and carcass traits of pigs. Italian Journal of Animal Science 3, 323–336. https://doi.org/10.4081/ijas.2004.323

Øverland, M., Schøyen, H.F., Skrede, A., 2010. Growth performance and carcase quality in broiler chickens fed on bacterial protein grown on natural gas. Br Poult Sci 51, 686–695. https://doi.org/10.1080/00071668.2010.522556

Øverland, M., Skrede, A., 2012. Fatty acid composition, oxidative stability and sensory quality of meat from broiler chicken fed autolysate from bacteria grown on natural gas. Journal of Animal Physiology and Animal Nutrition 96, 739–746. https://doi.org/10.1111/j.1439-0396.2011.01202.x

Øverland, M., Skrede, A., Matre, T., 2001. Bacterial protein grown on natural gas as feed for pigs. Acta Agriculturae Scandinavica, Section A — Animal Science 51, 97–106. https://doi.org/10.1080/090647001750193422

Pereira, A.A., Yaverino-Gutierrez, M.A., Monteiro, M.C., Souza, B.A., Bachheti, R.K., Chandel, A.K., 2025. Precision fermentation in the realm of microbial protein production: State-of-the-art and future insights. Food Research International 200, 115527. https://doi.org/10.1016/j.foodres.2024.115527

Pereira, A.G., Fraga-Corral, M., Garcia-Oliveira, P., Otero, P., Soria-Lopez, A., Cassani, L., Cao, H., Xiao, J., Prieto, M.A., Simal-Gandara, J., 2022. Single-Cell Proteins Obtained by Circular Economy Intended as a Feed Ingredient in Aquaculture. Foods 11, 2831. https://doi.org/10.3390/foods11182831

Pilmer, L.W., Woolley, L.D., Lymbery, A.J., Salini, M., Partridge, G.J., 2022. Using dietary additives to improve palatability of diets containing single-cell protein from methanotrophic bacteria in yellowtail kingfish (*Seriola lalandi*) diets. Aquaculture Research 53, 5006–5017. https://doi.org/10.1111/are.15986

Plavnik, I., Bornstein ,S., Hurwitz, S., 1981. Evaluation of methanol-grown bacteria and hydrocarbon-grown yeast as sources of protein for poultry: Studies with young chicks. British Poultry Science 22, 123–140. https://doi.org/10.1080/00071688108447871

Richter, G., Richter ,H., Bentz, H., 1991. Untersuchungen zur Einsatzwürdigkeit von Methanol-Bakterienmasse bei Mastputen und Mastcairina. Archiv für Tierernährung 41, 195–202. https://doi.org/10.1080/17450399109428460

Richter, G., Richter, H., Bentz, H., 1989. Einsatz von Methanol-Bakterienmasse im Multigenerationsversuch am Huhn. Archiv für Tierernaehrung 39, 131–139. https://doi.org/10.1080/17450398909432956

Richter, G., Richter ,H., Bentz ,H., Hennig, A., 1988. Einsatz von Methanol-Bakterienmasse im Multigenerationsversuch am Huhn. Archiv für Tierernaehrung 38, 921–928. https://doi.org/10.1080/17450398809430919

Rønn, M., Thorsteinsson, M., Johannsen, J.C., Nørgaard, J.V., Julegaard, I.K., Nielsen, M.O., 2022. Evaluation of nutritional quality for weaner piglets of a new methanotrophic microbial cell-derived protein feed. Animal Feed Science and Technology 294, 115498. https://doi.org/10.1016/j.anifeedsci.2022.115498

Roth, F.X., Kirchgessner, M., 1976. Methanol-fermentation protein in veal calf nutrition. Animal Feed Science and Technology 1, 33–44. https://doi.org/10.1016/0377-8401(76)90005-5

Sabolová, M., Kulma, M., Kouřimská, L., 2021. Sex-dependent differences in purine and uric acid contents of selected edible insects. Journal of Food Composition and Analysis 96, 103746. https://doi.org/10.1016/j.jfca.2020.103746

Sabolová, M., Kulma, M., Škvorová, P., Veselá, K., Kurečka, M., Kouřimská, L., 2023. Investigating purine and uric acid contents of various development stages of artificially reared edible insects. https://doi.org/10.3920/JIFF2022.0011

Salminen, S., Collado, M.C., Endo, A., Hill, C., Lebeer, S., Quigley, E.M.M., Sanders, M.E., Shamir, R., Swann, J.R., Szajewska, H., Vinderola, G., 2021a. The International Scientific Association of Probiotics and Prebiotics (ISAPP) consensus statement on the definition and scope of postbiotics. Nat Rev Gastroenterol Hepatol 18, 649–667. https://doi.org/10.1038/s41575-021-00440-6

Salminen, S., Collado, M.C., Endo, A., Hill, C., Lebeer, S., Quigley, E.M.M., Sanders, M.E., Shamir, R., Swann, J.R., Szajewska, H., Vinderola, G., 2021b. Reply to: Postbiotics - when simplification fails to clarify. Nat Rev Gastroenterol Hepatol 18, 827–828. https://doi.org/10.1038/s41575-021-00522-5

Sarabian, C., Wilkinson, A., Sigaud, M., Kano, F., Tobajas, J., Darmaillacq, A.-S., Kalema-Zikusoka, G., Plotnik, J.M., MacIntosh, A.J.J., 2023. Disgust in animals and the application of disease avoidance to wildlife management and conservation. Journal of Animal Ecology 92, 1489–1508. https://doi.org/10.1111/1365-2656.13903

Schneider, D., Bronsch, K., 1977. Methanolbakterien (*Methylomonas methylotropha*) als Proteinquelle für die Ernährung von Ferkeln. Zeitschrift für Tierphysiologie Tierernährung und Futtermittelkunde 39, 313–325. https://doi.org/10.1111/j.1439-0396.1977.tb00285.x

Schøyen, H.F., Hetland, H., Rouvinen-Watt, K., Skrede, A., 2007a. Growth Performance and Ileal and Total Tract Amino Acid Digestibility in Broiler Chickens Fed Diets Containing Bacterial Protein Produced on Natural Gas. Poultry Science 86, 87–93. https://doi.org/10.1093/ps/86.1.87

Schøyen, H.F., Svihus, B., Storebakken, T., Skrede, A., 2007b. Bacterial protein meal produced on natural gas replacing soybean meal or fish meal in broiler chicken diets. Archives of Animal Nutrition 61, 276–291. https://doi.org/10.1080/17450390701431953

Sedgman, C.A., Roy, J.H.B., Thomas, J., Stobo, I.J.F., Ganderton, P., 1985. Digestion, absorption and utilization of single-cell protein by the preruminant calf: The true digestibility of milk and bacterial protein and the apparent digestibility and utilization of their constituent amino acids. British Journal of Nutrition 54, 219–244. https://doi.org/10.1079/BJN19850107

Shan, C., Liu, Y., Liu, S., Li, C., Ma, C., Yu, H., Li, J., Jiang, G., Tian, J., 2025. Exploring the Mechanism of *Clostridium autoethanogenum* Protein for Broiler Growth Based on Gut Microbiota and Serum Metabolomics. Biology 14, 29. https://doi.org/10.3390/biology14010029

Shan, C., Liu, Y., Ma, C., Li, C., Liu, Q., Liu, S., Jiang, G., Tian, J., 2024. Dietary supplementation with *Clostridium autoethanogenum* protein improves growth performance and promotes muscle protein synthesis by activating the mTOR signaling pathway of the broiler. Front. Vet. Sci. 11. https://doi.org/10.3389/fvets.2024.1389738

Siddik, M.A.B., Julien, B.B., Islam, S.M.M., Francis, D.S., 2024. Fermentation in aquafeed processing: Achieving sustainability in feeds for global aquaculture production. Reviews in Aquaculture 16, 1244–1265. https://doi.org/10.1111/raq.12894

Sikkeland, L.I.B., Eduard, W., Stangeland, A. M., Thorgersen, E. B., Haug, T., Aukrust, P., Halvorsen, B., Mollnes, T. E., and Kongerud, J., 2009. Occupational exposure to bacterial single cell protein induces inflammation in lung and blood. Inhalation Toxicology 21, 674–681. https://doi.org/10.1080/08958370802491425

Sikkeland, L.I.B., Skogstad, M., Øvstebø, R., Brusletto, B., Haug, K.B.F., Kongerud, J., Eduard, W., Kierulf, P., 2008. Circulating lipopolysaccharides in the blood from "bioprotein" production workers. Occupational and Environmental Medicine 65, 211–214. https://doi.org/10.1136/oem.2007.032938

Sikkeland, L.I.B.T., E. B; Haug, T; Mollnes, T.E., 2007. Complement activation and cytokine response by BioProtein, a bacterial single cell protein. Clinical and experimental immunology 148, 146–152. https://doi.org/10.1111/j.1365-2249.2007.03339.x

Skogstad, M., Eduard, W., Holme, J., Qvenild, T., Einarsdottir, E., 2005. Short term changes in lung function, leukocytosis in blood, and lachrymal fluid among bacterial single cell protein workers after an episode with high exposure to endotoxins. Occup Environ Med 62, 576–580. https://doi.org/10.1136/oem.2004.019273

Skogstad, M., Sikkeland, L.I.B., Øvstebø, R., Bente, K., Haug, F., Kierulf, P., Heldal, K.K., Kongerud, J., 2009. Does endotoxin exposure affect lung function and induce systemic inflammation in workers handling bacterial single cell protein? SJWEH suppl 7, 36–40.

Skrede, A., Berge, G.M., Storebakken, T., Herstad, O., Aarstad, K.G., Sundstøl, F., 1998. Digestibility of bacterial protein grown on natural gas in mink, pigs, chicken and Atlantic salmon. Animal Feed Science and Technology 76, 103–116. https://doi.org/10.1016/S0377-8401(98)00208-9

Skrede, A., Faaland Schøyen, H., Svihus, B., Storebakken, T., 2003. The effect of bacterial protein grown on natural gas on growth performance and sensory quality of broiler chickens. Can. J. Anim. Sci. 83, 229–237. https://doi.org/10.4141/A02-047

Song, X.-P., Hansen, M.C., Potapov, P., Adusei, B., Pickering, J., Adami, M., Lima, A., Zalles, V., Stehman, S.V., Di Bella, C.M., Conde, M.C., Copati, E.J., Fernandes, L.B., Hernandez-Serna, A., Jantz, S.M., Pickens, A.H., Turubanova, S., Tyukavina, A., 2021. Massive soybean expansion in South America since 2000 and implications for conservation. Nat Sustain 4, 784–792. https://doi.org/10.1038/s41893-021-00729-z

Spence, C., 2013. Multisensory flavour perception. Current Biology 23, R365–R369. https://doi.org/10.1016/j.cub.2013.01.028

Teller, E., Godeau, J.-M., 1986. Evaluation of the nutritive value of single-cell protein (Pruteen) for lactating dairy cows. The Journal of Agricultural Science 106, 593–599. https://doi.org/10.1017/S0021859600063462

The Scientific Committee for Animal Nutrition and the Scientific Committee for Food, 1995. First report of the Scientific Committee for Animal Nutrition on Question 69 by the Commission on the use of protein products of fermentation from natural gas obtained by culture of *Methylococcus capsulatus* (Bath), *Alcaligenes acidovorans*, *Bacillus brevis*.

Van Weerden, E.J., Huisman, J., 1977. Digestibility of protein and amino acids of a fermentation single-cell protein for veal calves. Animal Feed Science and Technology 2, 377–383. https://doi.org/10.1016/0377-8401(77)90010-4

VanderWal, P., 1976. Experience with SCP in animal feeding in Europe. PAG Bulletin 6, 7–18.

Villalba, J., Spackman, C., Lobón, S., 2019. The interplay between exposure and preference for unpalatable foods by lambs. Applied Animal Behaviour Science 212, 44–51. https://doi.org/10.1016/j.applanim.2019.01.011

Vogt, V.H., Harnisch, S., Krieg, R., Löliger, H.Ch., Rauch, H.W., 1983. Bakterien - Bioprotein im Legehennenfutter. European Poultry Science 47, 25–33. https://doi.org/10.1016/S0003-9098(25)02667-0

Waldroup, P.W., Payne, J.R., 1974. Feeding Value of Methanol-Derived Single Cell Protein for Broiler Chicks. Poultry Science 53, 1039–1042. https://doi.org/10.3382/ps.0531039

Wang, J.P., Kim, J.D., Kim, J.E., Kim, I.H., 2013. Amino acid digestibility of single cell protein from *Corynebacterium ammoniagenes* in growing pigs. Animal Feed Science and Technology 180, 111–114. https://doi.org/10.1016/j.anifeedsci.2012.12.006

Waterworth, D.G., Heath, M.E., 1981. Pruteen in the diet of breeding pigs: Reproductive performance. Animal Feed Science and Technology 6, 297–307. https://doi.org/10.1016/0377-8401(81)90008-0

Whittemore, C.T., Moffat, I.W., 1976. The digestibility of dried microbial cells grown on methanol in diets for growing pigs. The Journal of Agricultural Science 86, 407–410. https://doi.org/10.1017/S0021859600054885

Whittemore, C.T., Moffat, I.W., Taylor, A.G., 1976. Evaluation by digestibility, growth and slaughter of microbial cells as a source of protein for young pigs. Journal of the Science of Food and Agriculture 27, 1163–1170. https://doi.org/10.1002/jsfa.2740271213

Whittemore, C., Gilchrist Shirlaw, D., McDonald, D., Taylor, A., 1978a. Performance of broilers and layers given diets containing dried microbial cells (Pruteen). British Poultry Science 19, 283–287.

Whittemore, C.T., Tullis, J.B., Hastie, S.W., 1978b. Efficiency of use of nitrogen from dried microbial cells after a period of N deprivation in growing pigs. British Journal of Nutrition 39, 193–200. https://doi.org/10.1079/BJN19780025

Wu, Y., Wang, J., Jia, M., Huang, S., Cao, Y., Yao, T., Li, J., Yang, Y., Gu, X., 2022. *Clostridium autoethanogenum* protein inclusion in the diet for broiler: Enhancement of growth performance, lipid metabolism, and gut microbiota. Front. Vet. Sci. 9. https://doi.org/10.3389/fvets.2022.1028792

Yamazaki, M., Ando, M., Kubota, D., 1977. Studies on the Digestibility of Single Cell Protein Grown on Various Nutrient Substrates for Colostomized Laying Hens. Japanese Poultry Science 14, 232–235. https://doi.org/10.2141/jpsa.14.232

Yoshida, M., 1988. Effect of Feeding Yeast and Bacteria Grown on Methanol and Ethanol on the Performance of Breeding Hens. Japanese poultry science 25, 1–14. https://doi.org/10.2141/jpsa.25.1

Yoshida, M., Hagano, D., Koba, K., Iwamoto, T., Matsushima, M., Bansho, H., Iino, M., Kato, T., 1981. Comparison of Three Feeding Experiments with Hens to Check Unknown or Unexpected Factor in a Novel Feed Ingredient I. Lifetime Feeding Experiment. Jpn. poult. sci. 18, 205–216. https://doi.org/10.2141/jpsa.18.205

Yoshida, M., Hagano, D., Koba, K., Iwamoto, T., Matsushima, M., Seima, T., Bansho, H., Kotaki, M., Kato, T., Iesaka, H., 1982. Comparison of flavor of eggs, chicken and pork produced by feeding yeast and bacteria cultured in methanol. Jap. J. Poult. Sci. 19, 76–82. https://doi.org/10.2141/jpsa.19.76

Zhang, H.Y., Piao, X.S., Li, P., Yi, J.Q., Zhang, Q., Li, Q.Y., Liu, J.D., Wang, G.Q., 2013. Effects of Single Cell Protein Replacing Fish Meal in Diet on Growth Performance, Nutrient Digestibility and Intestinal Morphology in Weaned Pigs. Asian Australas. J. Anim. Sci 26, 1320–1328. https://doi.org/10.5713/ajas.2013.13200

Zhao, Y., Yu, B., Mao, X., He, J., Huang, Z., Mao, Q., Chen, D., 2012. Effect of dietary bacterial lysine by-product meal supplementation on growth performance and excretion of purine base derivatives in growing-finishing pigs. Livestock Science 149, 18–24. https://doi.org/10.1016/j.livsci.2012.06.017

Zhen, Y., Chen, Y., Ge, L., Wei, W., Wang, Y., Hu, L., Loor, J.J., Wang, M., Yin, J., 2022. The Short-Day Cycle Induces Intestinal Epithelial Purine Metabolism Imbalance and Hepatic Disfunctions in Antibiotic-Mediated Gut Microbiota Perturbation Mice. International Journal of Molecular Sciences 23, 6008. https://doi.org/10.3390/ijms23116008

Zimmerman, D.R., Tegbe, S.B., 1977. Evaluation of a Bacterial SCP for Young Pigs and Rats. Journal of Animal Science 45, 469–475. https://doi.org/10.2527/jas1977.453469x

CHAPTER 20

Algae as sustainable source of ingredients for livestock feed

E.M. Melchor-Martínez[1], A.G. Reyes-Alvarado[2], O.J. Portillo-Castillo[1], V.M. Rivas-Galindo[1] and R. Parra-Saldívar[3]*

[1]Facultad de Medicina, Universidad Autonoma de Nuevo León. Av. Madero y Gonzalitos s/n, Colonia Mitras Centro, Monterrey Nuevo León, 64460, Mexico
[2]Centro de Investigaciones Biológicas del Noroeste (CIBNOR), Programa de Agricultura en Zonas Áridas,Calle IPN # 195, La Paz, 23096, Baja California Sur, Mexico;
[3]Magan Centre of Applied Mycology (MCAM), Cranfield University, MK43 0AL, UK
**Correspondence: elda.melchormr@uanl.edu.mx*

Abstract

Livestock faces significant challenges stemming from the global population growth, scarcity of renewable resources that ensure food availability, and emerging policies worldwide that aim to reduce greenhouse gas emissions, agroindustrial waste, and water consumption. Algae investigation is increasingly recognized, and offer a promising and sustainable alternative to conventional livestock ingredients. They are rich in proteins, lipids and bioactive compounds that can address nutritional and animal health needs for livestock production. Algae organisms exhibit advantages over other organisms, like they are widely distributed in diverse ecosystems, having rapid growth rates, and that do not altering the arable land. In addition, its incorporation in feeding leads to benefits for meat production, enhances sensorial appeal and longer shelf life. Although, there are gaps in the utilization of algae in animal feed, including high costs production, variable composition, and challenges on the replication of the studies *in vivo* and optimization of the sequences on a large-scale, their unique composition in metabolites and easy incorporation of whole algae biomass or special extracts make them a highly sustainable source of ingredients that allows to reduce ingredients that can be used in human nutrition enhancing global food security.

Keywords

livestock feeding– microalgae–macroalgae– sustainable ingredients

1 Introduction

The growing global demand for animal-based foods, with meat production expected to reach 498 million tons by 2050, has intensified the pressure on conventional feed resources such as soybean meal, maize, and fishmeal. These ingredients, although nutritionally efficient, are associated with extensive environmental costs. Soybean cultivation, for example, is a major driver of deforestation in South America, while fishmeal production continues to strain marine ecosystems. In addition to ecological concerns, feed remains the most expensive component of livestock production, accounting for up to 80% of total production costs (Abo Bark et al., 2024; Kossmann et al., 2025). Livestock products are considered food rich in essential amino acids for human development and nutrition. In recent years, concerns about feed-food competition and environmental impacts of the global livestock sector are increasing. Approaches to explore alternative protein feed ingredients is mandatory to transform the traditional livestock production into a more sustainable and resilient livestock industry, making more efficient the use of land, water and energy resources. Considering that livestock is a piece of the future food security, policies and regulations on this matter are on discussion by the governmental authorities and non-governmental organizations. The future adoption of novel and alternative ingredients in livestock feed is expected leading to reduce the negative impacts associated with proteins crops (Pexas et al., 2023).

Alternative ingredients can be obtained from sustainable sources which are being explored to meet the controversial needs along with challenges. The inclusion of algae in animal diets has gained attention to improve the growth performance, meat and milk quality, health and reduce the environmental impact of methane emissions in the ruminant production (Figure 20.1).

The whole algae biomass or special extracts, fresh or dried material can be added to enhance animal feed formulations, particularly is common adding to aquatic organisms feeding (Ahmed et al., 2023). Algae metabolites are used as fortifiers in basal animals feed, considering the main source of ingredients, traditional sources like soybean proteins, fishmeal or fish oil. It is known that algae could provide antinutritional ingredients and higher rates of inclusion interferes with nutrients utilization and triggering disadvantages on growth performance. Therefore, studies showing effects of inclusion levels of algae on growth performance, improvement of health, and livestock production should be encouraging to demonstrate the influence of species, their composition, and administration time.

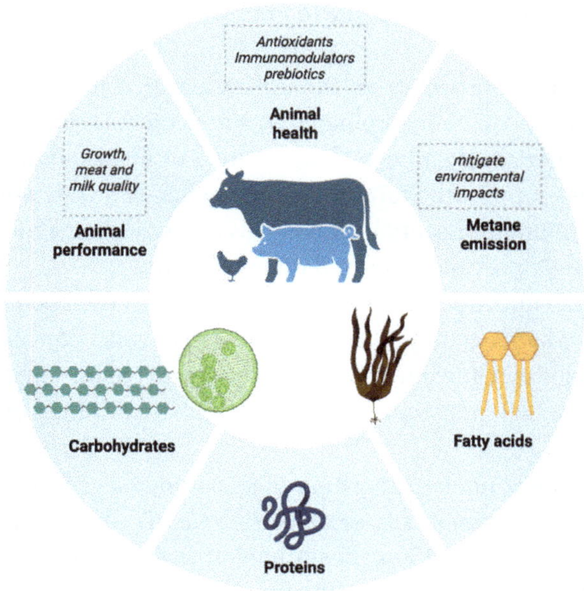

FIGURE 20.1 Algae as novel source of sustainable ingredients for livestock feed

2 Nutritional contributions of algae

Sustainable efforts are underway to face challenges in the livestock industry; algae are novel feed ingredients that possess elements capable of meeting the requirements of modern livestock systems and may replace or partially substitute the conventional sources in animals feeding. Algae are classified into two categories: microalgae, which are small unicellular organisms and macroalgae or seaweed, which are multicellular and can be referred as large size. The inclusion of micro- and macroalgae in animal diets has gained attention for its potencial enhance growth performance, improve meat and milk quality, and contribute health-promoting compounds. As feed additives, they help mitigate adverse environmental impacts due to their potential to reduce the methane emissions (Ahmed et al., 2023).

2.1 Macroalgae

Macroalgae are grouped in three categories, brown (Phaeophyceae), red (Rhodophyceae), and green (Chlorophyceae). These organisms contain high-quality nutrients in their biomass composition, including, proteins, carbohydrates, lipids, vitamins and other micronutrients such as minerals. Several bioactive components have been identified in macroalgae such as unsaturated fatty acids, carotenoids and functional polysaccharides that are responsible for additional properties like antiviral, antimicrobial, prebiotic, antioxidant and immunomodulatory effect. Including macroalgae in diets can contribute to improve growth conditions and enhance animals health.

There are several approaches on the utilization of macroalgae as ingredients in animal feed. They can be incorporated as whole biomass or special extracts, as fresh or dried material. Regardless of the method of inclusion of algae nutrients, numerous studies support the fortification of animal feed with whole macroalgae biomass, which contains essential amino acids and other metabolites relevant for growth, immune support and improved protein quality.

Noteworthy, the nutritional value of algae depends on their composition, which can be influenced by geography distribution, season of collection and specie among others. This variability, concentrates external efforts to chemically characterize the biomass to ensure high nutritional profiles and enhance animal nutrition (Min et al., 2021).

Although the nutritional composition of some microalgae species, such as *Euglena gracilis*, *Schizochytrium* sp., or *Chlorella vulgaris*, shows a favorable profile of essential amino acids and long-chain polyunsaturated fatty acids, their large-scale application as direct replacements for conventional protein or lipid sources remains limited. The current use of both micro- and macroalgae in livestock feed is primarily functional rather than nutritional, aiming to improve oxidative stability, gut health, immune response, and product quality rather than to serve as full substitutes for soybean meal or fish oil.

For a detailed comparison of amino acid and fatty acid profiles of these and other novel ingredients against conventional feed proteins and oils, readers are referred to the Nutritional Tables section at the beginning of this book.

2.1.1 Impact on methane reduction emissions

Bioactive secondary metabolites synthetized by macroalgae have the capability to improve animal health and exhibit anti-methanogenic properties in livestock. Studies on the effects of brown, green, and red macroalgae on methane production in ruminants have demonstrated significant action on methane reduction without compromising animal performance. For example, the inclusion of seaweeds from the genus *Asparagopsis* at levels below 1% in the feed for cattle and sheep have inhibited methanogenesis by up to 98%. Red algae produce halogenated methane analogue compounds which inhibit key steps in the methanogenesis. Bromoform is considered the most important compound for the anti-methanogenic effect in *Asparagopsis*. However, some concerns have been raised on bromoform potential carcinogenesis and ozone-depleting action. In addition, the total effectiveness depends on the bioactive content and the final formulation of the basal feed (Glasson et al., 2022).

Brown seaweeds from *Ascophyllum* genus have shown a less substantial impact on enteric methane production, even at high inclusion levels. Seaweeds like *Lami-*

naria provide omega-3, omega-6 and polyunsaturated fatty-acids which add benefits to ruminant reproduction, improving conception rates and reducing pregnancy losses. They also reduce oxidative stress and the incidence of ketosis. Phlorotannins are the main compounds identified as contributors of the anti-methanogenic activity of brown algae (Vijn et al., 2020).

The main green seaweeds species explored for their anti-methanogenic potential include *Ulva*, *Cladophora*, and *Caulerpa* spp. Studies on *Ulva* sp. in ruminant feed found that inclusion levels of 4% dry matter (DM) reduced methane emissions compared with the control diet, while levels of 25% DM yielded a 55% reduction relative to a meadow hay-based control diet. Other species like *Clandophora* were evaluated *in vitro* at inclusion levels of 20% w/w of the total basal feed, the authors observed 66% reduction in methane in comparison with the control. Green macroalgae have been reported as a source of saponins. Evaluations on anti-methanogenic action of saponins concluded that high levels of these compounds reduce methane emissions and that the saponins concentrations depend on the species (McGurrin et al., 2023).

2.1.2 Impact on animal performance

Macroalgae are rich in several macro-nutrients and micronutrients including a lipid fraction (e.g., 0.5 to 1.0 % of DM in brown algae), proteins (0.6 to 1.6 % of DM in brown algae and 24.1 to 27.1 % DM in red algae), ash content (ranging from 10 to 50% DM), minerals like bromine (up to 816–972 mg/kg, w/dw in brown algae) and iodine (50–5000 ppm, w/dw), vitamins (B1, B2, B3, C, E and A – with brown and green algae meeting livestock vitamin requirements), carotenoids, and carbohydrates (with wide variability among species). However, these contents may vary between species and factors based on macroalgal cultivation conditions (Costa et al., 2021).

Research on the use of macroalgae as feed supplements for livestock highlights its effectiveness in improving productivity and providing health benefits to animals growing on these specialised diets.

Potential benefits of macroalgae inclusion in pork production have been noted since early 2000s, in response to the increasing global meat demand. Red, green and brown macroalgae offer nutraceuticals for pigs which improve their immune system and provide prebiotic components to support gastrointestinal health. Polysaccharides such as laminarin, fucoidan, ulvans and carrageenan are known to modulate cytokine responses and provide anti-inflammatory properties (González-Meza et al., 2023).

However, other studies have reported inconsistent results on pig growth depending on the algae species used. *Ascophyllum nodosum* extracts have shown an increase on the average daily gain and final body weight when was evaluated on weaned pigs, whereas, brown algae, reduced these parameters in finisher pigs

(Turner et al., 2002). Regarding meat quality, the use of brown seaweed (*Macrocystis pyrifera*) in pigs diets did not significantly affect overall quality, except for increased redness intensity and, at high inclusion levels, around 4%, increased ash content and reduced lipids levels (Jerez-Timaure et al., 2021).

Seaweed has also been incorporated into ruminant diets as a feed supplement or ingredient to improve growth performance. For example, the whole matrix of *Porphyra* sp. has been used to replace soybean meal in sheep diets. No significant differences in average daily weight gain were observed between algae-based and soybean-based diets, supporting the use of *Porphyra* sp. as a viable protein source for ruminants (Lind et al., 2020). In contrast, green seaweed has limited reports on their use and growth performance in ruminants. *Ulva Lactuca* was included at 20% of the diet on male lambs, without negative effects on palatability. It showed low protein degradability (40%) and moderate digestibility (60%), values comparable to cereal grains parameters (Morais, et al., 2020). Other studies have investigated, the effect of a seaweed mixture (9% *Laminaria digitata* + 91% *Ascophyllum nodosum*) in dairy cows diets at different inclusion levels. The authors found that the seaweed supplementation did not change the milk yield or composition, except for a slight reduction in protein and casein content (Newton et al., 2021).

Even though, there are more than 12, 000 species of seaweed in the world, only a limited number can be considered for inclusion in animal feeding. Several challenges must be addressed to solve this underutilization. For example, the potential heavy metal contamination by their adsorptive capacities, animal health risks associated with bromoform accumulation in seaweeds, potential environmental impact of bromoform that can destroy ozone via chemical reactions. Additionally, the cost of meat and milk products from ruminants fed macroalgae-supplemented diets could increase by up to 20%. Therefore, sustainable strategies for cultivation and processing must be developed to support a viable supply chain (Liu et al., 2025).

2.2 Microalgae

Microalgae are divided into four different groups: eukariotic diatoms, (*Bacillariophyceae*), green algae (*Chlorophyceae*), golden algae (*Chrysophyceae*), and blue-green algae (*Cyanophyceae*). They are photosynthetic organisms with a great potential of CO_2 capture and can be found in both marine and freshwater ecosystems (Martins et al., 2021). Microalgae are a promising source of proteins (60%–70% dry weight), essential amino acids, fatty acids, carotenoids, chlorophylls, vitamins (4% dry weight), and minerals. Due to this rich composition, their biomass can be valorized in a circular economy strategy to provide nutritional components and benefits for livestock feed (Lopez-Pacheco et al. 2025).

2.2.1 Impact on animal performance

Euglena gracilis (EG), is a freshwater microalga known by its high protein and fat content. Its effect has been evaluated *in vitro* to determine its impact on methane production. The findings demonstrated that an inclusion up to 10% of *Euglena* in ruminant diets has the potential to mitigate methane emissions, owing to its rich content of amino acids and fatty acids (Aemiro et al., 2016).

Spirulina and *Chlorella* are well-known for their high protein and high fatty acids content and have been successfully tested in feeding trials with pigs, ruminants and rabbits. The dietary incorporation of 5% *Chlorella vulgaris* in finishing pigs was assessed, showing increased levels of pigments and polyunsaturated fatty acids (PUFAs), without affecting growth performance or meat quality (Coelho et al., 2020). Other researchers have focused efforts on using microalgae to modulate rumen microbiota. *Spirulina* is a sustainable microalga with antioxidant, immunomodulatory, and microbial-modulating properties. Its use has been investigated to improve animal productivity and immune function. High supplementation level (15 g of Spirulina/day) was assessed in ewes, suggesting a slight modulation of microbial populations, particularly those related to cellulolytic bacteria (Christodoulou et al. 2023).

Several authors have reported the inclusion of *Schizochytrium* spp. in livestock diets to enhance nutritional attributes of meat due to its high DHA content and rapid growth under heterotrophic conditions. Dietary supplementation of male cattle with *Schizochytrium* sp. resulted in elevated levels of DHA and EPA, as well as an increased total antioxidant capacity and glutathione peroxidase activity in meat compared to the control. Additionally, twenty-four volatile compounds were identified – mainly aldehydes, alcohols and ketones – which contributed to improve the sensory scores of beef (Xu et al., 2021).

Interestingly, both whole microalgal biomass and extracted fractions can offer similar functionality in animal feeding. For example, post-lipid extraction biomass of *Scenedesmus* sp. biomass significantly reduced methane emissions by approximately 50% compared to the control diet (47% reduction) and those containing whole cells (51% reduction). This finding, support the concept of algae as sustainable resource within a biorefinery framework, where multiple high-value products can be extracted from different microalgal fractions (M. Tibbetts et al. 2017).

Despite the unique nutritional characteristics of the microalgae and their potential as feed additives or supplements for ruminants, several challenges need to be faced. One major issue is the high cost of microalgae cultivation compared with conventional feed sources. This is why autotrophic culture methods, despite their low productivity, are still widely used. Moreover, harvesting is energy-intensive and expensive, and the large-scale cultivation has not yet achieve viable methods to ensure microalgae consistent growth in quantity and quality. Additional challenges include the limited number of microalgal species approved for animal feed and reduced intake rate due to unfavorable flavor. Therefore, further research is essential to address these limitations and improve the adoption of microalgae as feed additives or supplements in livestock nutrition (Zhu et al., 2024).

3 Future perspective

Considering the issues related to human nutrition, the increasing demand for animal-based food, and the challenges associated with a green environmental footprint. The global trend is clear; the current food systems need to be reshaped for sustainable systems. Exploration of novel ingredients for animal feeding will leave edible grain available for human nutrition. However, global heating will affect livestock productivity to the point of being unsustainable and limited. Therefore, the valorization of algae biomass seems also a solution to obtain nutrients for human nutrition and manage circular economy models facing the zero waste challenges.

References

Abo Bakr, S., Khidr, R. E., Farrag, B., & El-Awady, R. (2024). Agricultural residues and their utilization as alternative feed resources in Egypt. *Egyptian Journal of Animal Production*. https://doi.org/10.21608/ejap.2024.282457.1081

Aemiro, A., Watanabe, S., Suzuki, K., Hanada, M., Umetsu, K., & Nishida, T. (2016). Effects of Euglena (Euglena gracilis) supplemented to diet (forage: concentrate ratios of 60: 40) on the basic ruminal fermentation and methane emissions in in vitro condition. *Animal Feed Science and Technology*, 212, 129-135.

Ahmed, E., Suzuki, K., & Nishida, T. (2023). Micro-and macro-algae combination as a novel alternative ruminant feed with methane-mitigation potential. *Animals*, 13(5), 796.

Christodoulou, C., Mavrommatis, A., Loukovitis, D., Symeon, G., Dotas, V., Kotsampasi, B., & Tsiplakou, E. (2023). Effect of Spirulina dietary supplementation in modifying the rumen microbiota of ewes. *Animals*, 13(4), 740.

Coelho, D., Pestana, J., Almeida, J. M., Alfaia, C. M., Fontes, C. M., Moreira, O., & Prates, J. A. (2020). A high dietary incorporation level of Chlorella vulgaris improves the nutritional value of pork fat without impairing the performance of finishing pigs. *Animals*, 10(12), 2384.

Costa, M., Cardoso, C., Afonso, C., Bandarra, N. M., & Prates, J. A. (2021). Current knowledge and future perspectives of the use of seaweeds for livestock production and meat quality: A systematic review. *Journal of Animal Physiology and Animal Nutrition*, *105*(6), 1075-1102.

Glasson, C. R., Kinley, R. D., de Nys, R., King, N., Adams, S. L., Packer, M. A., ... & Magnusson, M. (2022). Benefits and risks of including the bromoform containing seaweed Asparagopsis in feed for the reduction of methane production from ruminants. *Algal Research*, 64, 102673.

González-Meza, G. M., Elizondo-Luevano, J. H., Cuellar-Bermudez, S. P., Sosa-Hernández, J.

E., Iqbal, H. M., Melchor-Martínez, E. M., & Parra-Saldívar, R. (2023). New perspective for macroalgae-based animal feeding in the context of challenging sustainable food production. *Plants*, *12*(20), 3609.

Jerez-Timaure, N., Sánchez-Hidalgo, M., Pulido, R., & Mendoza, J. (2021). Effect of dietary brown seaweed (Macrocystis pyrifera) additive on meat quality and nutrient composition of fattening pigs. *Foods*, 10(8), 1720.

Kossmann, H., Moess, T., & Breunig, P. (2025). The climate impact and land use of cultivated meat: Evaluating agricultural feedstock production. *PloS one*, *20*(1), e0316480. https://doi.org/10.1371/journal.pone.0316480

Lind, V., Weisbjerg, M. R., Jørgensen, G. M., Fernandez-Yepes, J. E., Arbesú, L., & Molina-Alcaide, E. (2020). Ruminal fermentation, growth rate and methane production in sheep fed diets including white clover, soybean meal or Porphyra sp. *Animals*, 10(1), 79.

Liu, Y., Zhou, M., Diao, Q., Ma, T., & Tu, Y. (2025). Seaweed as a feed additive to mitigate enteric methane emissions in ruminants: Opportunities and challenges. *Journal of Integrative Agriculture*, 24(4), 1327-1341.

López-Pacheco, I. Y., González-Meza, G. M., González-González, R. B., Parra-Saldívar, R., & Melchor-Martínez, E. M. (2025). Algae bioremediation of swine and domestic wastewater promotes a reduction of coliforms and antibiotic-resistant bacteria. *Environmental Pollution*, 126294.

M. Tibbetts, S., & H. Fredeen, A. (2017). Nutritional Evaluation of Whole and Lipid-Extracted Biomass of the Microalga Scenedesmus sp. AMDD for Animal Feeds: Simulated Ruminal Fermentation and In Vitro Monogastric Digestibility. Current Biotechnology, 6(3), 264-272.

Martins, C. F., Ribeiro, D. M., Costa, M., Coelho, D., Alfaia, C. M., Lordelo, M., ... & Prates, J. A. (2021). Using microalgae as a sustainable feed resource to enhance quality and nutritional value of pork and poultry meat. *Foods,* 10(12), 2933.

McGurrin, A., Maguire, J., Tiwari, B. K., & Garcia-Vaquero, M. (2023). Anti-methanogenic potential of seaweeds and seaweed-derived compounds in ruminant feed: current perspectives, risks and future prospects. *Journal of animal science and biotechnology*, 14(1), 145.

Min, B. R., Parker, D., Brauer, D., Waldrip, H., Lockard, C., Hales, K., ... & Augyte, S. (2021). The role of seaweed as a potential dietary supplementation for enteric methane mitigation in ruminants: Challenges and opportunities. *Animal Nutrition*, 7(4), 1371-1387.

Morais, T., Inácio, A., Coutinho, T., Ministro, M., Cotas, J., Pereira, L., & Bahcevandziev, K. (2020). Seaweed potential in the animal feed: A review. *Journal of Marine Science and Engineering*, 8(8), 559.

Newton, E. E., Pétursdóttir, Á. H., Ríkharðsson, G., Beaumal, C., Desnica, N., Giannakopoulou, K., ... & Stergiadis, S. (2021). Effect of dietary seaweed supplementation in cows on milk macrominerals, trace elements and heavy metal concentrations. *Foods,* 10(7), 1526.

Pexas, G., Doherty, B., & Kyriazakis, I. (2023). The future of protein sources in livestock feeds: implications for sustainability and food safety. *Frontiers in Sustainable Food Systems*, 7, 1188467. https://doi.org/10.3389/fsufs.2023.1188467

Turner, J. L., Dritz, S. S., Higgins, J. J., & Minton, J. E. (2002). Effects of Ascophyllum nodosum extract on growth performance and immune function of young pigs challenged with Salmonella typhimurium. *Journal of Animal Science*, 80, 1947–1953.

Vijn, S., Compart, D. P., Dutta, N., Foukis, A., Hess, M., Hristov, A. N., ... & Kurt, T. D. (2020). Key considerations for the use of seaweed to reduce enteric methane emissions from cattle. *Frontiers in Veterinary Science*, 7, 597430.

Xu, C., Zhang, S., Sun, B., Xie, P., Liu, X., Chang, L., ... & Zhang, S. (2021). Dietary supplementation with microalgae (Schizochytrium sp.) improves the antioxidant status, fatty acids profiles and volatile compounds of beef. *Animals*, 11(12), 3517.

Zhu, M., Singer, S. D., Guan, L. L., & Chen, G. (2024). Emerging microalgal feed additives for ruminant production and sustainability. *Advanced Biotechnology*, 2(2), 17.

CHAPTER 21

Utilization of agricultural waste for enhancing livestock nutrition

E.M. Melchor-Martínez[1], A.G. Reyes-Alvarado[2], O.J. Portillo-Castillo[1], V.M. Rivas-Galindo[1] and R. Parra-Saldívar[3]*

[1]Facultad de Medicina, Universidad Autonoma de Nuevo León. Av. Madero y Gonzalitos s/n, Colonia Mitras Centro, Monterrey Nuevo León, 64460, Mexico
[2]Centro de Investigaciones Biológicas del Noroeste (CIBNOR), Programa de Agricultura en Zonas Áridas,Calle IPN # 195, La Paz, 23096, Baja California Sur, Mexico
[3]Magan Centre of Applied Mycology (MCAM), Cranfield University, MK43 0AL, UK
**Correspondence: elda.melchormr@uanl.edu.mx*

Abstract

Several models of agricultural waste valorization have been explored to obtain animal feed ingredients due to their massive production and high impact on greenhouse gas emissions. Common residues can include fruit and vegetable byproducts, cereal processing waste, olive and grape waste, and brewer's spent grain, among others. These materials typically contain moderate levels of protein, high fiber content, and compounds with diverse biological activities. The features and composition of waste make them an alternative for the extraction of ingredients that can be incorporated into animal feed diets. Mostly, waste requires microbial transformation or bioprocessing such as solid-state fermentation or enzymatic treatment to make nutrients more accessible. In addition, extraction technologies including microwave-assisted extraction and supercritical fluids have enhanced the recovery of nutrients in high-quality products, promoting the transformation of diets. The integration of agricultural waste into livestock diets has demonstrated tangible benefits on ruminates and poultry with satisfactory metrics of production on growth, digestibility, milk production, and health. However, waste utilization remains challenges to be addressed, such as the presence of heavy metals, mycotoxins, and pathogens that can be removed under a continued refinement of bioprocessing. Therefore, waste will continue to be an inspiration to investigate novel ingredients with an impact on sustainability and still be discussed in matters of economic and environmental aspects.

Keywords

livestock – agricultural waste – bioprocessing-ruminants

1 Introduction

Considering environmental pressures, the valorization of agricultural waste as livestock feed has gained considerable interest. Crop residues, fruit and vegetable by-products, and agro-industrial discards are no longer seen merely as waste but as renewable and regionally available resources. Egypt alone generates approximately 27.7 million tons of agricultural residues annually, much of which could be incorporated into livestock diets following processing through drying, chemical treatment, or fermentation (Abo Bakr et al., 2024).

This trend is mirrored in other regions. In Colombia's La Guajira department, annual biomass waste estimates range between 292,000 and 522,000 tons, and efforts are underway to repurpose these materials for both energy and nutritional applications in rural communities (Rodríguez Romero et al.,2025). Similarly, in southern Italy, dairy farms in the Sorrento Peninsula have successfully integrated olive pruning residues and mill leaves into the diets of lactating cows. This inclusion has been linked to both the increment of feed intake and the improvement of milk fat composition, particularly in terms of healthier fatty acid profiles (Masucci et al., 2025).

Nutritionally, agricultural residues are characterized by their fiber content and often contain modest levels of protein, minerals, and bioactive compounds such as polyphenols and flavonoids (Figure 21.1). However, their raw form often contains indigestible components and anti-nutritional factors, such as lignin or tannins. Modern bioprocessing techniques have played an important role in overcoming these barriers. For instance, microwave-assisted extraction has been shown to significantly increase the yield of flavonoids and phenolic compounds from hop leaves, materials typically discarded in brewing industries, indicating their potential value in functional feed applications (Carbone et al., 2025). Olive leaf pruning waste, when subjected to pressurized liquid and supercritical fluid extraction, yielded compounds such as luteolin and maslinic acid, both of which exhibit potent antioxidant activity and contribute to animal health when included in feed formulations (Mushtaq et al., 2025).

Fermentation, particularly solid-state fermentation using fungi or lactic acid bacteria, is a widely adopted method for improving the digestibility and bioavailability of nutrients in lignocellulosic waste. Through microbial action, structural carbohydrates are broken down and anti-nutritional factors are reduced. This change of composition enhances the energy value of the feed, and at the same time, enriches it with microbial proteins, probiotics, enzymes, and organic acids that contribute to gut health and feed efficiency in livestock (Yasmeen and Ahmad, 2024).

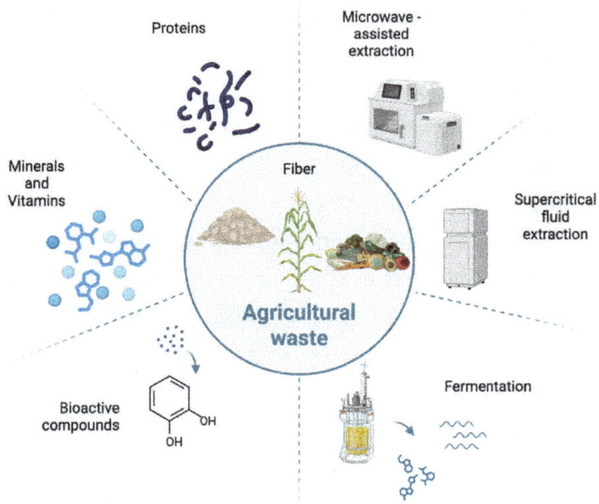

FIGURE 21.1 Valorization of agricultural waste for high-value compounds extraction

Nonetheless, the use of agricultural waste in livestock feed is challenging. Nutrient composition can vary widely between feedstocks, seasons, and regions. There are also concerns about contamination from mycotoxins, pesticide residues, or heavy metals. Addressing these issues requires standardized protocols for quality assurance, as well as integrated assessments that combine chemical analysis with *in vivo* animal performance trials.

As the global livestock sector shifts toward more sustainable practices, the nutritional evaluation of agricultural waste will play a central role in shaping future feed strategies. These residues, once properly processed and evaluated, offer a viable path toward reducing environmental impact, lowering production costs, and improving regional feed self-sufficiency.

2 Agricultural waste: nutritional composition and variability

Agricultural waste streams exhibit remarkable diversity in both their nutrient composition and their biological value as livestock feed ingredients. This variability arises from differences in botanical origin, post-harvest handling, and regional agroecological conditions. Residues like cottonseed meal, olive pomace, and red pepper processing waste vary substantially in their dry matter (DM), protein, fat, fiber, and ash content. For instance, cottonseed meal shows the highest levels of crude protein (up to 15.9%) and fat (15.9%), while olive pomace contains the most structural fiber, with Acid Detergent Fiber (ADF) and neutral detergent fiber (NDF) values exceeding 70% and 55%, respectively (Ücok and Yang, 2025).

These compositional differences are critical in determining the role of each residue in feed formulation. High-fiber wastes like rice straw and wheat bran, although low in protein, serve as effective roughage sources in ruminant diets. On the other hand, fruit and vegetable residues offer moderate protein, high moisture, and valuable bioactive compounds such as polyphenols and antioxidants, though their rapid perishability often requires immediate processing (Ab Razak et al., 2024).

Among the most extensively studied agro-industrial residues is grape pomace, a by-product of winemaking rich in fiber, phenolic compounds, and unsaturated fatty acids. Its nutritional value is highly variable depending on the grape cultivar and origin. Typical compositions include 10–13% crude protein, 45–55% NDF, and 7–12% fat on a dry matter basis. When included in livestock diets at controlled levels (5–10%), grape pomace improves the antioxidant status of animals, enhances meat oxidative stability, and enriches fatty acid profiles – especially omega-3s. Studies on pigs, for instance, demonstrated improved intramuscular fat content and a more favorable omega-6 to omega-3 ratio without affecting growth performance when 5–10% dehydrated grape pomace was added to their diets (da Silveira Almeida et al., 2024).

Similarly, tomato pomace and mixed fruit pomaces also exhibit valuable nutritional traits. These contain unfermentable sugars, lycopene, tannins, and antioxidants that contribute to improved gut health, immunity, and reproductive performance in ruminants and poultry (Lu et al., 2022).

Another promising group of residues includes wheat middlings and bran, which are rich in crude protein (ranging from 15–17%), fiber, B vitamins, and minerals. These are by-products of the wheat milling industry and are highly digestible in monogastric animals, particularly pigs and poultry, due to their favorable fiber and energy balance. In livestock nutrition, wheat middlings are typically used as concentrate supplements. A recent compositional study confirmed the presence of 7.2% crude protein, 6.0% fiber, and appreciable levels of minerals such as calcium (52.7 mg/100g) and magnesium (65.3 mg/100g), reinforcing their value as nutrient-rich feed ingredients (Remi and Esther, 2025).

Brewer's spent grain (BSG), the most abundant by-product of beer production, is similarly rich in protein (25–34%) and fiber (35–60%), and it also provides essential amino acids and micronutrients. Its inclusion in ruminant diets, particularly in dairy cattle and sheep, has consistently been shown to increase total dry matter intake and milk yield while maintaining or improving feed conversion efficiency. In a 2025 study, BSG inclusion at 10–15% dry matter in Simmental crossbred cattle diets enhanced nutrient digestibility, favorably influenced blood metabolites, and improved fecal microbiota composition, suggesting improved gastrointestinal health and feed efficiency. Together, wheat by-products and BSG represent effective, low-cost alternatives to conventional cereal grains and protein meals, contributing to sustainable feed systems without compromising animal productivity or health (Fan et al., 2025).

On the other hand, beyond cereal by-products, novel plant residues such as those from *Agave lechuguilla* are emerging as promising candidates in sustainable feed strategies. Native to arid regions of Mexico, this plant produces substantial amounts of fibrous residue, locally known as "guishe", during Tampico fiber extraction. These residues, typically discarded, are rich in secondary metabolites including saponins and flavonoids, compounds recognized for their antioxidant, anti-inflammatory, and lipid-regulating properties (Morreeuw et al., 2021).

3 Impact on animal performance

The integration of agricultural waste into livestock diets has demonstrated tangible and species-specific benefits, particularly in ruminants and poultry. One of the most consistent outcomes is a reduction in feed costs, estimated at 10–30%, when agro-industrial by-products partially substitute conventional feed components (Kazemi, 2025). These savings stem from the significantly lower cost of residues like brewer's spent grain, fruit pomace, and vegetable peels compared to commercial ingredients such as soybean meal or maize, while maintaining satisfactory production metrics like weight gain and milk yield (Khairi et al., 2025).

In dairy cattle, for instance, the inclusion of pasteurized waste milk as a liquid feed instead of commercial milk replacer significantly enhanced weight gain, feed efficiency, and health indicators, while lowering incidences of diarrhea and pneumonia (Ma et al., 2022).

Similarly, the inclusion of by-products such as olive pomace, tomato waste, and spent mushroom substrate in ruminant diets has been consistently associated with improved feed intake, nutrient digestibility, and favorable rumen fermentation dynamics. For instance, olive pomace, traditionally limited by high fiber and low crude protein content, can be nutritionally enhanced through fungal bioprocessing. A 2024 study demonstrated that treatment with oyster mushrooms significantly reduced neutral detergent fiber while increasing crude protein, resulting in improved gas production, fermentation efficiency, and *in vitro* organic matter degradability during rumen incubation, making it more suitable for ruminant nutrition (Mnisi et al., 2024).

In the case of spent mushroom substrate (SMS), reuse in cattle feeding systems has shown promise as a low-cost roughage substitute. A 2025 investigation into shiitake SMS indicated potential reuse through nutritional recycling, enhancing sustainability while contributing fiber and residual nutrients to the feed system (Junior et al., 2025).

Furthermore, a 2025 study evaluating olive pomace, red pepper waste, and cottonseed meal confirmed their diverse fiber and protein profiles, supporting their strategic use in ruminant diets. Olive pomace, with its high ADF and NDF values (over 70% and 55%, respectively), contributed to optimal ruminal fermentation when incorporated into forage-based systems (Ücok and Yang, 2025).

Moreover, non-traditional protein sources like fermented rapeseed meal and vitamin B12 fermentation waste (VFW) have been studied. Fermented rapeseed meal included up to 10% DM has proven effective in replacing soybean meal in broiler diets, without impairing growth performance, nutrient digestibility or immune responses (Shaikh et al., 2024). The inclusion of (VFW) in lamb diets, replacing soybean meal at levels up to 40%, maintained growth rate and nutrient digestibility while enhancing serum immunoglobulin M and antioxidant status. Moderate VFW inclusion also modified the rumen microbiota, increasing beneficial genera such as Prevotella and Selenomonas and improving immune responsiveness without negative effects on performance (Zhang et al., 2025).

In parallel, black soldier fly larvae (BSFL) grown on food and agricultural residues continue to demonstrate superior bioconversion efficiencies. A recent study recorded net growth efficiencies between 0.33 and 0.56, showcasing their dual potential as waste recyclers and sustainable protein sources in livestock diets (Nielsen et al., 2025).

These results collectively emphasize the dual advantage of incorporating agricultural wastes into livestock feeding systems; they reduce reliance on costly conventional feeds, and also deliver measurable enhancements in animal productivity and physiological health. Whether through improved rumen fermentation in cattle, or efficient bioconversion via insect larvae, the repurposing of agro-residues supports both economic viability and environmental stewardship. Continued refinement of bioprocessing techniques, such as fermentation, enzymatic treatment, and probiotic inoculation, will be essential to standardize performance outcomes and broaden the safe, effective integration of these materials across diverse species and production systems.

Recent advances in bioconversion technologies have unlocked the potential of agricultural waste to serve as feedstock for high-protein feed ingredients through microbial and enzymatic processing. These strategies aim not only to upcycle waste but also to generate nutrient-dense, cost-effective alternatives to conventional feed components.

A striking example of this technology is the upcycling of agricultural cabbage waste using genetically engineered *Escherichia coli* strains. Researchers engineered strains to express the crtB and crtE genes for phytoene synthesis, integrated the mevalonate pathway to boost carbon flux, and introduced poly(3-hydroxybutyrate) biosynthetic pathways to enhance storage and photostability. This system achieved a production of 9.51 mg/L phytoene with a threefold increase in photostability compared to base strains (Jeong et al., 2025).

Similarly, recent advances in fungal bioprocessing have explored *Rhizopus oligosporus* and *Aspergillus oryzae* grown on wheat milling by-products and potato peels, yielding SCP with protein contents ranging between 35–60% dry matter, and improved digestibility for poultry and pigs (Furlan et al., 2024).

While the integration of agricultural waste into animal nutrition has gained considerable scientific and commercial interest, several knowledge gaps and operational challenges must be addressed to realize its potential fully. Most notably, variability in nutrient composition, driven by botanical source, climate, and post-harvest handling, requires a shift toward dynamic feed formulation systems (Singh and Jana, 2024). In parallel, the safety of agricultural residues remains a persistent concern, particularly in terms of contamination by mycotoxins, pesticides, heavy metals, and microbial pathogens. Although bioprocessing techniques such as fermentation and extrusion can reduce many of these contaminants, systematic risk assessments and harmonized quality control protocols are urgently needed to establish regulatory confidence across international markets (Adelere et al., 2025).

Lastly, broader multi-actor collaboration is essential to scale innovations beyond experimental or local use. Partnerships across academia, government, and industry will be critical to invest in infrastructure for waste collection, processing, and distribution, as well as in animal trials that assess long-term health, productivity, and product quality.

4 Future perspective

The valorization of agricultural and agro-industrial by-products represents one of the most tangible pathways toward circular livestock production. This topic, including the concept of *low opportunity cost biomass* (LOCB) and its implications for sustainability assessment, is further explored in detail in the chapter by Siddique and Pelletier in the Sustainability section of this book.

References

Ab Razak, A. F., Mohamed, H. N., Abdullah, M. S., Sulaiman, M. S., Mohd Basri, S. N., Ting, U. H., & Razak, A. A. (2024). Waste agriculture material in animal feed trend evaluation: a bibliometric analysis. Journal of Tourism, Hospitality and Environment Management, 9(38), 242–258 https://doi.org/10.35631/JTHEM.938017

Abo Bakr, S., Khidr, R. E., Farrag, B., & El-Awady, R. (2024). Agricultural residues and their utilization as alternative feed resources in Egypt. *Egyptian Journal of Animal Production*. https://doi.org/10.21608/ejap.2024.282457.1081

Adelere, I. A., Aboyeji, D. O., Auta, H. S., Barasarathi, J., & Ahmed, A. (2025). Mycotoxins and Global Food Safety. In Biotoxins in Food. *CRC Press*. Boca Raton, FL. pp. 96-134.

Carbone, K. (2025). Evaluation of drying methods and green extraction techniques to enhance the recovery of bioactive compounds from hop leaves: A sustainable approach for the valorisation of agricultural by-products. *Foods*, 14(1), 55.

da Silveira Almeida, B. C., Ludke, M. D. C. M. M., Bertol, T. M., Ludke, J. V., Bernardi, D. M., Cunha Jr, A., & Coldebella, A. (2024). Growth performance, meat quality, and lipid oxidation in pigs' fed diets containing grape pomace. *Applied Biosciences*, 3(3), 378-391.

Fan, Z., He, S., Lin, Q., Lin, S., Zhu, L., Yang, R., ... & Guo, P. (2025). Brewer's Grains on Growth Performance, Nutrient Digestibility, Blood Metabolites, and Fecal Microbiota in Simmental Crossbred Cattle Finished in Feedlot. *Agriculture*, 15(9), 977. https://doi.org/10.3390/agriculture15090977

Furlan, O., Subtil de Oliveira, N., Caldas de Paula, R., Takaki Rosa, R., Michelotto, P. V., Wéber, H., Bianchini, L. F., & Ribeiro Rosa, E. A. (2024). Pilot scale production of high-content mycoprotein using Rhizopus microsporus var. oligosporus by submerged fermentation and agro-industrial by-products. *Bioresource Technology*, 404, 131515.

Jeong, S. W., Kang, C. K., & Choi, Y. J. (2018). Metabolic engineering of Deinococcus radiodurans for the production of phytoene. *Journal of Microbiology and Biotechnology*, 28(10), 1691-1699.

Junior, W. G. V., Caitano, C. E. C., da Silva Alves, L., Teixeira, P. A. G., Noble, R., Pardo, J. E., & Zied, D. C. (2025). From waste to resource: Sustainable reuse of spent shiitake mushroom substrate in subsequent production cycles. *International Biodeterioration & Biodegradation*, 200, 106034.

Kazemi M. Recycling Agricultural Waste: Sustainable Solutions for Enhancing Livestock Nutrition. Vet Med Sci. 2025 May;11(3):e70321. https://doi.org/10.1002/vms3.70321

Khairi, F., Maharani, C. P., & Aditya, F. A. (2025). Sustainable livestock management: strategies for enhancing productivity while reducing environmental impact. *The Journal of Academic Science*, *2*(1), 382-391. https://doi.org/10.59613/efybvy75

Lu, S., Chen, S., Li, H., Paengkoum, S., Taethaisong, N., Meethip, W., ... & Paengkoum, P. (2022). Sustainable valorization of tomato pomace (Lycopersicon esculentum) in animal nutrition: a review. *Animals*, *12*(23), 3294.

Ma, Y., Khan, M. Z., Xiao, J., Alugongo, G. M., Chen, X., Li, S., ... & Cao, Z. (2022). An overview of waste milk feeding effect on growth performance, metabolism, antioxidant status and immunity of dairy calves. *Frontiers in veterinary science*, 9, 898295. https://doi.org/10.3389/fvets.2022.898295

Masucci, F., Romano, R., Proto, V., & Varricchio, M. (2025). Circular economy on a small scale: The sustainable use of olive tree biomass residues as feed for lactating cows in the Sorrento Peninsula. *Sustainability*, 17(2), 831.

Mnisi, C. M., Kunene, S. I., Soko, N. N., Egbu, C. F., & Mlambo, V. (2024). Oyster mushroom bioprocessing enhances the nutritional value of olive pomace for ruminants. *South African Journal of Animal Science*, 54(2), 226-235.

Morreeuw, Z. P., Castillo-Quiroz, D., Ríos-González, L. J., Martínez-Rincón, R., Estrada, N., Melchor-Martínez, E. M., ... & Reyes, A. G. (2021). High-throughput profiling of flavonoid abundance in Agave lechuguilla residue-valorizing, an under-explored Mexican plant. *Plants*, 10(4), 695.

Mushtaq, H., Ishtiaq, M., & Azeem, M. (2025). Recovery of bioactive constituents from olive leaf pruning waste: A comparison of green extraction techniques. *Foods*, 14(3), 614.

Nielsen, F. K., Hansen, R. J., Muurmann, A. T., Bahrndorff, S., & Eriksen, N. T. (2025). Metabolic Performance of Mealworms and Black Soldier Fly Larvae Reared on Food and Agricultural Waste and By-Products. *Animals*, 15(2), 233.

Remi, V. O., & Esther, O. M. (2025). Characterization of phytochemical and nutritional composition of wheat flour (*Triticum aestivum*). GSC Advanced Research and Reviews. eISSSN:2582-4597.

Rodríguez Romero, T. E., Cabello Eras, J. J., Sagastume Gutierrez, A., Mendoza Fandiño, J. M., & Rueda Bayona, J. G. (2025). The Energy Potential of Agricultural Biomass Residues for Household Use in Rural Areas in the Department La Guajira (Colombia). Sustainability (2071-1050), 17(3).

Shaikh, J. M. (2024). Comparative Study of Rapeseed Mealand fermented Rapeseed Meal Replacing Soybean Meal in Diet on Performance of Broilers (Doctoral dissertation, MAFSU, Nagpur).

Singh, B., & Jana, A. K. (2024). Forecast of agri-residues generation from rice, wheat and oilseed crops in India using machine learning techniques: Exploring strategies for sustainable smart management. *Environmental Research*, 245, 117993.

Üçok, S., & Yang, X. (2025). Chemical composition and methane production potential of agricultural residues: Olive pomace, cottonseed meal and red pepper processing waste. Tekirdağ Ziraat Fakültesi Dergisi, 22(1), 195-204.

Yasmeen, R., & Ahmad, F. (2024). Microbial fermented agricultural waste-based broiler feed: A sustainable alternative to conventional feed. *World's Poultry Science Journal*, 80(1), 25–40.

Zhang, Xinlong and Xie, X.L and Xin, Hangshu and Wang, Qi and Lv, Tuoyu and Liu, Jincheng and Lin, Xu and Hou, Yu and Jiao, Peixin and Yang, Wenzhu, Impact of Partially Substitution of Vitamin B12 Fermentation Waste for Soybean Meal in the Diet on Growth Performance, Nutrient Digestibility, Immune Responses, Ruminal Fermentation and Microbiota of Finishing Lambs. http://dx.doi.org/10.2139/ssrn.5071929

CHAPTER 22

Camelina sativa in livestock feeding: opportunities and challenges

E. Murcia García

UCAM Catholic University of Murcia, Campus de Guadalupe, Avda. de Los Jerónimos, s/n, 30107, Murcia, Spain
Correspondence: emurcia@ucam.edu

Abstract

The search for alternative feed ingredients is intensifying as livestock production faces increasing challenges related to sustainability, dependence on conventional limited protein sources, and consumer demand for healthier animal products. *Camelina sativa* has re-emerged as promising candidate due to its interesting agronomic profile and distinctive nutritional composition, characterized by high oil content rich in α-linolenic acid and a moderate protein fraction. Oil extraction generates several co-products – camelina oil, cake, meal, and hulls – each with specific compositional traits that offer opportunities for inclusion in animal diets. Alongside valuable nutrients and functional bioactives such as tocopherols and phenolic compounds, camelina also contains antinutritional factors, notably glucosinolates, phytate, and fiber, which can limit dietary use. To overcome these constraints, a variety of approaches are being explored. Current evidence suggests that camelina can contribute to protein and energy supply, enhance the fatty acid composition of animal-derived foods, and support the development of more sustainable feeding strategies. In this chapter, it is provided a review of the nutritional composition, antinutritional factors, and feeding applications of camelina and its co-products, providing an integrated perspective on opportunities, challenges, and future directions for their use in livestock nutrition.

Keywords

ruminants – monogastrics –antinutrients – protein – fatty acids

1 Introduction

The search for sustainable feed ingredients that are efficiently utilized by animals has intensified in recent years, driven by environmental concerns, consumer demand for healthier animal products, and limited global protein resources. Among emerging alternatives, *Camelina sativa* has re-emerged as a promising oilseed crop that may contribute to diversifying feed resources while supporting the transition toward more sustainable and resilient livestock systems (Matthäus and Zubr, 2000; Riaz et al., 2022).

Camelina, also known as false flax or gold of pleasure, belongs to the Brassicaceae family (Cullere et al., 2023; Sydor et al., 2022). It is native to Northern Europe and Central Asia and was historically cultivated for its oil before being displaced by higher-yielding crops. In recent decades, interest in camelina has resurged due to its agronomic advantages, as it combines a short growing cycle, low input requirements (fertilizers, pesticides, and water), and tolerance to poor soils, making it suitable for cultivation in marginal environments. These features make camelina suitable for cultivation in regions where conventional oilseeds such as soybean, sunflower, or canola are less competitive (Delver and Smith, 2024; Veljković et al., 2022). Such adaptability has positioned camelina as a strategic candidate for integrating biofuel production with feed applications, as the crop can be cultivated in crop rotation systems or in less productive lands without directly competing with food crops (Halmemies-Beauchet-Filleau et al., 2018; Riaz et al., 2022).

From a compositional perspective, *Camelina sativa* stands out for its balanced content of oil and protein, which together define its dual value as an energy- and nutrient-rich crop. The oil fraction is rich in polyunsaturated fatty acids (PUFAs), especially in α-linolenic acid (ALA), conferring a distinctive lipid profile compared with conventional oilseeds such as soybean or sunflower (Hurtaud and Peyraud, 2007; Quezada and Cherian, 2012). In addition, camelina contains natural antioxidants, including tocopherols and phytosterols, which contribute to the stability of its oil and may offer health-promoting effects (Zubr and Matthaus, 2002). These compositional traits not only support animal health and performance but also enhance the nutritional quality of animal-derived products such as milk, meat, and eggs (Cullere et al., 2023; Ryhänen et al., 2007; Smit and Beltranena, 2017).

Following oil extraction, camelina yields several co-products of nutritional interest – mainly camelina meal, camelina cake, and hulls – that differ in their composition and applications (Riaz et al., 2022). These by-products are valued for their protein content and the presence of bioactive compounds, making them suitable ingredients of sustainable livestock feeding systems (Kurasiak-Popowska et al., 2019). Nevertheless, some challenges remain that limit the broader use of camelina. Among them is the presence of antinutritional factors (ANFs), including glucosinolates, which can affect nutrient utilization and feed intake, particularly in monogastric animals (Egorova, 2023). Despite these limitations, camelina aligns closely with the global search for renewable, sustainable protein sources that can alleviate dependence on feedstuffs such as soybean meal and fishmeal (Delver and Smith, 2024).

This chapter provides an updated and comprehensive review of the nutritional composition of camelina seeds and co-products, with particular emphasis on ANFs and their mitigation. It also summarizes current evidence from feeding trials in various livestock species, highlighting both opportunities and challenges, and concludes with future perspectives for the inclusion of *Camelina sativa* as a sustainable feed ingredient.

2 Chemical composition and nutritional value

The nutritional profile of camelina varies depending on the plant part used (whole seed, oil, cake/meal), the processing method (cold-pressing, solvent extraction), and the specific cultivar. *Camelina sativa* seeds underpin its value as both an oilseed crop and a source of protein-rich co-products for animal feeding. Camelina seeds typically contain 35–40% oil and 20–25% protein, levels comparable to those of canola and flax (Matthäus and Zubr, 2000; Riaz et al., 2022). The oil fraction is dominated by polyunsaturated fatty acids, particularly α-linolenic acid, which accounts for up to 35–40% of total fatty acids. Although camelina has a lower oil yield than canola, it provides a substantially higher proportion of α-linolenic acid, making it a distinctive source of n-3 fatty acids among oilseeds (Matthäus and Zubr, 2000). Moreover, camelina oil exhibits a beneficial fatty acid composition, characterized by a balanced n-6/n-3 ratio- that is, an optimal proportion between linoleic and α-linolenic acids- which supports healthy lipid metabolism and contributes to the nutritional quality of animal products, together with significant levels of antioxidant tocopherols, especially when cold-pressed (Piravi-vanak et al., 2022; Rabiej-Kozioł et al., 2023; Rakita et al., 2024).

Beyond lipids, camelina seeds contain a protein fraction characterized by a balanced amino acid profile, with arginine and glutamic acid as predominant components and lysine and methionine at levels comparable to soybean meal (F. N. Almeida et al., 2013; Chen et al., 2010; Pekel et al., 2015; Ryhänen et al., 2007; Thacker and Widyaratne, 2012). Camelina also supplies minerals such as phosphorus, potassium, and magnesium, although the presence of phytate may reduce their bioavailability (Blades and Zubr, 2010).

Cake and expeller meal typically contain high crude protein (CP) levels – around 30–45% of DM – depending on processing: solvent-extracted meals often report 30–35% CP, while expeller cake may reach ~38.8% CP with residual oil (Delver and Smith, 2024). This difference is attributed to the less intensive extraction process of expeller pressing, which retains more residual oil and results in lower protein denaturation or loss during processing compared to solvent extraction. These co-products also retain significant amounts of residual oil rich in polyunsaturated fatty acids (PUFAs). For instance, expeller meal can contain 10–20% residual oil, of which ~30% is α-linolenic acid (Juodka et al., 2022; Zubr, 2003).

From a nutritional perspective, the main constraints of camelina relate to its antinutritional factors. Glucosinolate concentrations in seeds can range between 15 and 35 mmol/kg DM, depending on genotype and growing conditions, and may impair feed intake or thyroid function in monogastric animals when inclusion levels are excessive (Egorova, 2023). Similarly, phytate reduces mineral availability, and the relatively high fiber content can limit digestibility in some species (Lanzoni et al., 2024; Luise et al., n.d.). These aspects underline the need for careful formulation and processing strategies when using camelina-based ingredients in livestock diets.

Camelina oil is the primary product obtained from camelina seeds and represents the main rationale for its cultivation, due to its favorable traits and nutritional profile. The average yield of oil from seeds is about 40% on DM basis making it competitive with other additional oilseeds such as canola or flax, though slightly lower in absolute yield compared to a rapeseed (Riaz et al., 2022).

2.1 Camelina oil

Camelina oil is characterized by a high degree of unsaturation, comprising more than 50% polyunsaturated, around 30% monounsaturated, and about 10% saturated fatty acids. Its fatty acid composition is largely influenced by the cultivar as well as the environmental conditions cultivation (Decker et al., 2025). The oil is particularly valued for its distinctive fatty acid profile: α-linolenic acid often constitutes 30–40% of total fatty acids, with linoleic acid contributing 15–20% and oleic acid around 15–17%. (Belayneh et al., 2015; Hajiazizi et al., 2024; Kahindi et al., 2014). This confers a balanced n-6/n-3 ratio, often below 1, which is markedly more advantageous than that of soybean, sunflower, or corn oils (Delver and Smith, 2024; Rakita et al., 2024).

In addition, camelina oil is a rich source of natural antioxidants, notably tocopherols and phytosterols. Total tocopherol content can reach 700–750 mg/kg oil, dominated by y-tocopherol (Aziza et al., 2010; Hajiazizi et al., 2024; Quezada and Cherian, 2012). Although α-tocopherol is the most biologically active form in animals – being preferentially incorporated into tissues due to its high affinity for α-tocopherol transfer protein – γ-tocopherol exerts stronger anti-inflammatory activity and is more effective in scavenging reactive nitrogen species. Therefore, the predominance of γ-tocopherol in camelina oil may provide complementary antioxidant protection and contribute to its functional value in animal nutrition (Atkinson et al., 2008). Recent analyses have also identified phenolic compounds in camelina oil, adding to its functional bioactivity and potential contributions to animal health (Kurasiak-Popowska et al., 2019; Riaz et al., 2022; Zając et al., 2020).

From a feed perspective, dietary supplementation with camelina oil has shown promise for enriching animal-derived products with n-3 fatty acids. Studies in dairy cows have demonstrated improvements in the fatty acid profile of milk without impairing yield when moderate levels of camelina oil are included in the ration (Riaz et al., 2022a). However, in ruminants, a substantial proportion of dietary unsaturated fatty acids is subjected to ruminal biohydrogenation, where microorganisms convert PUFAs – such as α-linolenic acid – into more saturated fatty acids (e.g., stearic acid). This process reduces the direct transfer of n-3 fatty acids to milk and meat, meaning that only partial enrichment is typically achievable unless rumen-protection strategies or specific feeding approaches are used (Jenkins et al., 2008). Similarly, poultry and swine feeding trials have reported enhanced deposition of n-3 fatty acids in eggs and meat, respectively, reflecting the efficient transfer of α-linolenic acid to edible tissues (Aziza et al., 2010; Luise et al., 2024).

Nevertheless, the high degree of unsaturation also makes camelina oil highly susceptible to oxidative degradation. Unless adequately stabilized (via antioxidants or careful storage), rancidity and loss of functional properties can occur rapidly (Fröhlich et al., 2012). For this reason, while camelina oil is a valuable feed ingredient, its use in animal nutrition requires attention to processing, storage, and inclusion levels to ensure product stability and animal health.

2.2 Camelina cake

Camelina cake is obtained after mechanical pressing of the seeds to extract oil and constitutes both an energy- and protein-rich feed ingredient (Delver and Smith,

TABLE 22.1 Comparative fatty acid profile of camelina and its co-products (% total fatty acids)*

% Fatty acid	Camelina seed	Camelina meal	Camelina cake	Camelina oil
Palmitic acid (C16:0)	5.1–10.3	9.1–9.2	7.19–7.46	5.2–7.0
Palmitoleic (cis-9 C16:1)	0.1–0.9	0.3–0.5	-	0.1
Stearic (C18:0)	2.2–2.9	2.3–2.9	-	2.2–3.1
Oleic (C18:1)	11.9–19.9	17.7–21.7	-	15.1–18.7
Linoleic (C18:2)	13.5–20.9	24.4–28.8	-	16.0–19.6
Linolenic (C18:3 n-3)	28.6–41.3	24.2–46.3	-	28.0–38.1
Arachidic (C20:0)	1.2–1.8	1.2	-	1.2–2.3
Eicosenoic (C20:1)	13.3–25.4	10.1–13.3	-	11.6–15.1
Erucic (C22:1 n-9)	1.6–4.2	0.8	2.8–3.3	1.6–4.2
Total SFA	9.0–13.1	9.7–9.9	-	10.2–11.3
Total MUFA	31.0–37.7	33.5–33.9	-	31.6–34.6
PUFA	51.8–57.4		-	55.2

*Data extracted from multiple sources (Abramovič and Abram, 2005; Budin et al., 1995; Hurtaud and Peyraud, 2007a; Quezada and Cherian, 2012; Sarramone et al., 2020)

2024; Matthäus and Zubr, 2000). The residual lipid fraction maintains the characteristic fatty acid profile of camelina oil, dominated by α-linolenic acid, followed by linoleic and oleic acids. This confers potential to enrich animal-derived products with n-3 fatty acids when included in diets (Smit and Beltranena, 2017). In addition to its macronutrient value, camelina cake contains notable levels of bioactive compounds such as tocopherols and phenolic compounds, contributing antioxidant activity (Lanzoni et al., 2024). However, as with other Brassicaceae, it also carries antinutritional factors, primarly glucosinolates (Colombini et al., 2014; Matthäus and Zubr, 2000; Russo et al., 2017, 2012; Thacker and Widyaratne, 2012; Woyengo et al., 2017). At high inclusion levels, these compounds can impair feed intake, growth performance, and metabolic function, mainly in monogastric animals. For instance, studies have demonstrated that increasing the inclusion of camelina press cake in pig diets can negatively affect diet digestibility and growth performance (Zhu et al., 2021). Fiber content is also substantial, especially when hulls are not removed, which may limit digestibility. Juodka et al. 2022 reported that crude fiber in camelina by-products ranges from 9.7–17.4%, which is higher than in soybean meal and comparable to rapeseed meal. The presence of non-starch polysaccharides in camelina cake can increase intestinal digesta viscosity, reducing nutrient digestibility (Juodka et al., 2022). In general, the inclusion of camelina cake in animal diets has been evaluated in several studies across different livestock species as a partial substitute for conventional protein sources and showing controversial results in growth performance, feed intake, and product-derived quality in a dose-dependent manner (Halmemies-Beauchet-Filleau et al., 2011; Hurtaud and Peyraud, 2007; Riaz et al., 2022). Additionally, Yalçin et al. 2025 found that inclusion of camelina cake in broiler diets decreased apparent digestibility of DM, CP, and gross energy, which was attributed to the increased fiber content in the diet (Yalçın et al., 2025).

2.3 Camelina meal

Camelina meal is obtained after solvent extraction of oil from camelina seeds, which allows higher oil recovery compared to mechanical pressing. Unlike camelina cake, which retains 6–12% residual oil, solvent-extracted meal contains very low residual fat (<6%) and a higher concentration of CP, usually ranging between 35–40% of DM (Cerisuelo et al., 2023; Piergiovanni et al., 2025; Riaz et al., 2022). This makes camelina meal the most protein-dense co-product of camelina processing and the one most frequently evaluated in feeding trials.

The amino acid profile of camelina meal closely resembles that of other oilseed meals, as summarized in Table 22.2. Arginine, glutamic acid, and leucine are among the most abundant amino acids, while lysine and methionine are present at moderate levels (Aziza et al., 2013; Colombini et al., 2014; Kasiga et al., 2020; Woyengo et al., 2016). Lysine content is slightly lower than in soybean meal, while methionine (combined with cystine) is present at levels similar to rapeseed meal and higher than those in soybean meal (F N Almeida et al., 2013; Pekel et al., 2015; Thacker and Widyaratne, 2012). From a nutritional standpoint, this positions camelina meal as a potentially valuable protein source in livestock feeding, especially when combined with complementary protein sources to balance essential amino acids.

Solvent extraction significantly reduces the residual oil content in camelina meal compared to camelina cake, resulting in lower gross energy values (Piergiovanni et al., 2025). Although detailed micronutrient data in solvent-extracted meal are limited, minerals such as phosphorus, potassium, and magnesium are present at appreciable levels (Aziza et al., 2013; Lawrence et al., 2016; Paula et al., 2019).

Like other Brassicaceae species, camelina also contains certain ANFs, mainly glucosinolates, which may affect nutrient availability and digestibility. However, the occurrence, variability, and impact of these compounds are discussed in detail in the next section of this chapter.

Feeding trials with camelina meal have revealed both opportunities and challenges that vary depending on the animal species and the level of inclusion in the diet. In ruminants, diets containing camelina meal generally show good tolerance and may lead to improvements in the fatty acid profile of animal-derived products. (Delver and Smith, 2024; Hurtaud and Peyraud, 2007; Mutsvangwa et al., 2022). In poultry, however, higher inclusion levels have been associated with reduced growth performance and feed efficiency, largely due to glucosinolates and fiber (Hajiazizi et al., 2023; Pekel et al., 2015; Ryhänen et al., 2007; Thacker and Widyaratne, 2012). In pigs, camelina meal has been used as a partial replacement for soybean meal with moderate success, although tolerance appears to decrease when higher proportions are used (Luise et al., n.d.). These aspects, together with species-specific responses, are further analyzed in the next section of this chapter "Applications of camelina in livestock feeding".

2.4 Camelina hulls

Camelina hulls are the fibrous outer seed coats obtained as a by-product during dehulling or oil extraction. Their nutritional value is limited, as they contain less than 10% CP, while being rich in structural carbohydrates, lignin, and mucilage. This composition confers a high fiber content but low digestibility, particularly in monogastric animals (Salas, Castillejos & Ferret, 2017). Although their direct use as a feed ingredient is marginal, camelina hulls may provide a source of dietary fiber in ruminant diets, where their inclusion in small proportions does not appear to compromise performance. However, their relevance is modest compared with other fibrous by-products such as soybean hulls or beet pulp (Jolazadeh et al., 2025). Beyond the feed sector, hulls may also represent a raw material for bioenergy, composting, or bioproduct development within circular economy frameworks (Boardman et al., 2025).

TABLE 22.2 Comparison of camelina meal with other conventional oilseed meals*

Component (%)	Camelina meal	Canola meal	Soybean meal	Rapeseed meal
Dry matter	92.2	91.4	90.2	87.7
Organic matter	93.9	92.2	92.7	-
Crude protein	41.9	39.4–40.1	49.6–54.9	34.5
NDF	33.4	28.5	10.3–18.8	31.8
ADF	23.8	19.4–27.6	6.2–19.5	21.6
Ether extract	7.03	4.56	1.1–1.92	4.11
Ash	5.98	7.69	6.9	8.4
Minerals, % DM				
P	0.82	1.11	0.73	1.1
Ca	0.31	0.89	0.7	0.8
Amino acids, % CP				
Arginine	4.13	5.9	-	5.58
Alanine	2.81	4.43	4.26	4.16
Glutamic acid	7.6	22.7	14.99	15.6
Histidine	1.72	2.52	2.55	2.13
Lysine	2.27	4.87	5.91	4.8
Methionine and cystine	1.64	1.77	1.51	1.2
Leucine	3.24	6.39	7.52	6.49
Isoleucine	2.17	3.53	3.89	3.76
Serine	2.81	4.13	5.44	4.11
Threonine	1.59	3.87	4.07	4.32
Tryptophan	-	1.35	-	1.14

*Data extracted from multiple sources (Juodka et al., 2022; Riaz et al., 2022; Thacker and Widyaratne, 2012; Zubr, 2003); NDF: Neutral Detergent Fiber; ADF: Acid Detergent Fiber.

2.5 Antinutritional factors in camelina and mitigation strategies

In addition to its valuable nutrients, camelina contains several antinutritional factors (ANFs) that may constrain its use in animal feeding. The most relevant are glucosinolates, phytate, tannins, trypsin inhibitors, and fiber/lignin fractions (Egorova, n.d.; Ghidoli et al., 2023).

Glucosinolates (GSL) are the main antinutritional compounds in camelina seeds and co-products. Reported concentrations range from 12 to 36 mmol/kg DM, with glucocamelinin as the predominant form (Alexander et al., 2008; Russo et al., 2012). Upon hydrolysis by the plant's own enzyme myrosinase, glucosinolates yield biologically active products such as isothiocyanates, nitriles, and thiocyanates, which are known to impair thyroid function and feed palatability in sensitive species (Tripathi and Mishra, 2007).

Other ANFs include phytate, which reduces the bioavailability of essential minerals such as phosphorus, zinc, and iron (Colombini et al., 2014; Matthäus and Zubr, 2000; Russo et al., 2012); condensed tannins, which can lower protein digestibility by precipitating proteins and inhibiting proteolytic enzymes (Juodka et al., 2022; Russo et al., 2012); and trypsin inhibitors, which can reduce amino acid utilization in monogastric animals despite being present at relatively low levels in camelina meal (Delver and Smith, 2024). Camelina hulls, although not widely used, contain substantial amounts of lignin and insoluble fibre, which further limit digestibility in monogastric species (Lanzoni et al., 2024).

To reduce the impact of ANFs in camelina cake/meal, several mitigation strategies have been investigated. Thermal processing, such as toasting or extrusion, effectively reduces glucosinolate levels and their hydrolysis products, while also inactivating enzymes like myrosinase and protease inhibitors (Egorova, n.d.). Solvent extraction, commonly applied to produce camelina meal, not only concentrates protein but also reduces oil-soluble compounds and certain glucosinolate derivatives. Dehulling lowers fiber and lignin, thereby improving protein digestibility. In addition, enzyme supplementation, particularly with phytase and carbohydrases, has proven effective in enhancing phosphorus release from phytate and improving energy and nutrient availability in monogastric animals fed camelina meal. Copper supplementation, for instance with copper sulfate, can mitigate glucosinolate-associated performance issues in poultry diets (Pekel et al., 2015a). Finally, plant breeding efforts are underway to develop camelina varieties with lower glucosinolate content, which could represent a long-term strategy to enhance its feed value (Hölzl et al., 2023; Piergiovanni et al., 2025).

TABLE 22.3 Antinutritional factors of camelina meal, cake and expeller*

ANFs	Camelina meal	Camelina cake	Camelina expellers
GSL (mmol kg^{-1})	19.9–24.3	36.3	36.3
Phytic acid (g kg^{-1})	12–24.8	-	-
Condensed tannins (g kg^{-1})	1.81–2.11	2	1.9
Sinapine	2.19–3.27	-	-

*Data extracted from multiple sources (Colombini et al., 2014; Kahindi et al., 2014; Kiarie et al., 2016; Lawrence et al., 2016)

3 Applications of camelina in livestock feeding

The potential of *Camelina sativa* and its co-products as alternative feed ingredients has been evaluated across a wide range of livestock species. Most studies have focused on the inclusion of camelina meal, cake or seed in diets for ruminants, pigs, and poultry. The following subsections summarize the available experimental evidence for each major livestock group, highlighting nutritional responses, product quality outcomes, and constraints identified in feeding trials.

3.1 Ruminants: effects on performance and milk/meat quality

3.1.1 Dairy cows

Camelina by-products (mainly cake and meal) have been widely tested in dairy cows as partial replacements for conventional protein sources such as soybean meal or rapeseed meal. Across multiple studies, the most consistent finding is an enhancement of milk fatty acid composition, notably increased α-linolenic acid (ALA), vaccenic acid, and rumenic acid- while milk yield is generally maintained. These fatty acids are of particular nutritional relevance: ALA acts as a metabolic precursor of long-chain omega-3 polyunsaturated fatty acids (EPA and DHA) and contributes to improved lipid metabolism and anti-inflammatory activity (Bauman and Lock, 2006); vaccenic acid serves as a substrate for the endogenous synthesis of rumenic acid (conjugated linoleic acid, CLA) in the mammary gland (Shingfield and Griinari, 2007); and rumenic acid itself has been associated with beneficial effects on immune modulation, antioxidant status, and potential anticarcinogenic properties (Destaillats et al., 2005).

In contrast, the proportion of saturated fatty acids typically declines (Halmemies-Beauchet-Filleau et al., 2011; Hurtaud and Peyraud, 2007; Mutsvangwa et al., 2022; Sarramone et al., 2020; Toma et al., 2015) . These changes are nutritionally desirable from a human health perspective, since they improve the n-6/n-3 ratio and raise bioactive lipid intermediates linked with cardiometabolic benefits (Ventto et al., 2017).

Mechanistically, these responses are attributed to the high unsaturation of camelina lipids, which partially inhibit ruminal biohydrogenation and shift fermentation towards propionate at the expense of acetate (Paula et al., 2019). While this may reduce methane production, it also explains the risk of milk fat depression (MFD) when high lipid loads are supplied (Kokić et al., 2024; Ventto et al., 2017). Trials using >10 % DM camelina by-products or >2 kg/day of whole seed have reported reductions in milk fat content, without necessarily affecting milk yield (Hurtaud and Peyraud, 2007). This phenomenon is consistent with the well-described "biohydrogenation theory" of MFD, which proposes that intermediates formed during the incomplete ruminal hydrogenation of unsaturated fatty acids inhibit the expression of key lipogenic enzymes in the mammary gland, thereby reducing de novo milk fat synthesis (Shingfield and Griinari, 2007).

From a methane mitigation perspective, inclusion of camelina co-products has been associated with lower methane yield per unit of milk. Experimental studies – including in vitro fermentations and continuous culture systems – demonstrated that PUFAs from camelina reduce methanogenesis by acting as hydrogen sinks and shifting ruminal fermentation toward propionate, although total volatile fatty acids may decrease (Riaz et al., 2022). However, reductions in neutral detergent fiber digestibility have also been observed, highlighting a trade-off between environmental benefits and nutritional risks at higher inclusion levels (Riaz et al., 2022; Wang et al., 2017).

3.1.2 Beef cattle and sheep

In growing cattle and sheep, camelina meal has been evaluated as a protein-rich substitute for conventional meals like soybean or canola. In dairy heifers, inclusion levels up to 10% of diet DM maintained growth performance and carcass weight comparable to diets containing soybean or canola meals (Lawrence et al., 2016). However, neutral-detergent fiber digestibility may decline modestly. In lambs, supplementation with camelina meal has been shown to increase n-3 PUFA deposition and improve the n-6/n-3 ratio in muscle (Gómez-Cortés et al., 2018). However, some studies note a reduction in fiber digestibility at higher inclusion levels – likely due to glucosinolates and fiber – and while sheep seem reasonably tolerant, the data remain limited compared to dairy cow studies (Ponnampalam et al., 2024).

Taken together, current evidence indicates that camelina by-products can be incorporated into ruminant diets at moderate inclusion rates that maximize nutritional benefits while minimizing potential disturbances to rumen fermentation. In dairy cows, camelina cake or meal is generally well tolerated at 5–10% of diet DM, supporting neutral milk yield and improving the milk fatty acid profile; however, inclusion above 10% DM or supplying >2 kg/day of whole seed increases the risk of milk fat depression due to incomplete ruminal biohydrogenation. Beef cattle typically maintain growth performance with inclusion levels up to 10% DM, provided that total dietary fat remains within recommended limits for rumen function. Sheep appear somewhat more tolerant and may utilize camelina meal at up to ~12% DM, although reductions in fiber digestibility can occur at higher levels, especially in forage-rich diets. Across all ruminant species, gradual dietary adaptation, maintaining adequate effective fiber, and keeping total dietary fat below 6–7% of DM help ensure safe and efficient use of camelina co-products in practical feeding systems.

3.2 Swine: growth performance and gut health

Camelina co-products have been tested in swine mainly as partial substitutes for soybean meal, with research focusing on growth performance, nutrient digestibility, and health outcomes. Results highlight both opportunities and constraints, particularly in young pigs where digestive physiology is highly sensitive to fiber and glucosinolates.

3.2.1 Weaned pigs

A recent study by Luise et al. (2024) (Luise et al., 2024) investigated the effects of including camelina cake at 4, 8, and 12% of diet DM in weaned pigs for 28 days. The trial demonstrated a linear reduction in average daily gain with increasing camelina inclusion, most pronounced during the first week post-weaning. Feed intake was also depressed in the initial phase, while feed conversion ratio worsened modestly across treatments. Diets were formulated to be isoenergetic, and amino acid concentrations were adjusted to meet weaned pig requirements; however, full amino acids balancing was not achieved because camelina cake has lower standardized ileal digestibility of essential amino acids – particularly lysine and methionine – compared with soybean meal. This mismatch likely contributed to the decline in performance at higher inclusion levels.

Despite these performance penalties, physiological parameters remained largely unchanged. Jejunal morphology and thyroxine (T4) levels were not altered, suggesting no direct impairment of intestinal health or thyroid function at the tested levels. Notably, pigs fed camelina cake exhibited a higher alpha-diversity of the gut microbiota by day 28 together with shifts in bacterial taxa, pointing to possible gut-stabilizing effects. Additionally, liver weight increased in dietary level, consistent with enhanced metabolism of glucosinolates.

In conclusion, camelina cake can be used in weaned pig diets at ≤4% inclusion, where performance impacts are minimal and potential gut health benefits may emerge. Higher inclusions (>8–10%) are not advisable without further processing (e.g., heat treatment to reduce glucosinolates) or amino acid balancing, as camelina proteins show lower standardized ileal digestibility of lysine and methionine compared to soybean meal.

3.2.2 Growing-finishing pigs

Research in grower-finisher pigs is more limited, but available trials indicate that camelina meal and cake can replace part of soybean meal without major performance losses included at ≤8–10% of diet DM (Delver and Smith, 2024). At higher levels, reductions in feed intake and growth are frequently observed, attributed to the high fiber content and antinutritional factors.

Digestibility studies confirm that solvent-extracted camelina meal offers higher protein content and lower residual oil than expeller cake, but both contain appreciable levels of glucosinolates and phytic acid (Cerisuelo et al., 2023). Consequently, nutrient availability - particularly of lysine – is lower than in soybean meal, which remains the benchmark protein source. Processing strategies such us toasting, enzyme supplementation (e.g., phytase), or partial dehulling have been suggested to improve nutritional value (Delver and Smith, 2024).

3.3 Poultry: growth, egg and meat enrichment with n-3 fatty acids

Poultry feeding studies with camelina co products (meal, cake, oil, and seed) have primarily focused on their ability to enrich poultry products with n-3 fatty acids. The results consistently show improvements in the fatty acid profile of meat and eggs, but performance outcomes are more variable and depend on the inclusion level, the specific product used, and the presence of ANFs, especially glucosinolates (Juodka et al., 2022).

3.3.1 Broilers

Several studies have tested camelina meal and cake in broiler diets. Inclusions of 5–10% meal or cake are generally well tolerated, maintaining body weight gain (BWG) and feed conversion ratio (FCR) while enriching meat lipids (Hajiazizi et al., 2024). For example, Aziza et al. (2010) (Aziza et al., 2014) reported no adverse effects on growth when 5–10% camelina meal was included, while breast and thigh meat showed significantly higher n-3 PUFA. By contrast, Thacker and Widyaratne (Thacker and Widyaratne, 2012) observed a linear reduction in BWG and feed intake when camelina meal was included at 3–15%, with the strongest effects above 10%. Pekel et al. (2015) (Pekel et al., 2015) also found that diets containing 10–20% meal impaired BWG, feed intake, and efficiency during the grower phase.

On the other hand, whole seed at 5% of diet DM generally had neutral effects, but at 10% BWG reductions have been reported (Aziza et al., 2010; Hajiazizi et al., 2024).

The negative effects at higher inclusions are attributed to the presence of glucosinolates and fiber, which reduce feed palatability and may impair thyroid function (Hajiazizi et al., 2024c; Tripathi and Mishra, 2007). Interestingly, supplementation with copper (~150 mg/kg) alleviated growth depression in broilers fed 10% camelina meal, suggesting that mineral interactions can mitigate the ANF effects (Pekel et al., 2015).

In addition, camelina increases n-3 PUFA in breast and thigh meat. In some cases, total n-3 PUFA were doubled at ~10% inclusion, and the n-6/n-3 ratio decreased significantly (Hajiazizi et al., 2024; Orczewska-Dudek and Pietras, 2019). This demonstrates the potential of camelina as a functional ingredient to produce poultry meat with improved lipid quality.

In summary, for broilers, camelina meal or cake can be included at 5–10% of diet DM to enrich meat with n-3 PUFA while maintaining growth. Higher levels (>10%) are not recommended without mitigation of ANFs (processing, supplementation), as they consistently depress performance.

3.3.2 Laying hens

Camelina is highly effective at enriching egg yolks with n-3 PUFA. Inclusion of 5–10% meal or cake increases yolk α-linolenic acid and long-chain n-3 PUFA up to eightfold compared with control diets (Cherian et al., 2009).

Most studies report no negative effects on laying rate, feed intake, or egg weight at moderate inclusion levels (≤10%). For example, Pilgeram et al. (2007) found that hens fed camelina meal up to 15% maintained production. However, Cherian et al. (2009) observed that at >15% meal, egg production and yolk weight declined, and yolk total lipid content was reduced.

As in broilers, glucosinolates and fiber are the main limitations, especially at high inclusion. Reviews emphasize the variability in GSL concentration across camelina products, ranging from ~13–36 µmol/g in seeds to ~15–45 µmol/g in cake (Sydor et al., 2022). Since poultry tolerances for GSL are not yet well defined, conservative inclusion levels are recommended.

3.4 Genetically modified camelina

Recent advances in plant biotechnology have turned *Camelina sativa* into a promising crop for genetic improvement aimed at producing feed ingredients with en-

TABLE 22.4 Summary of feeding trials with camelina co-products in different species

Animal species	Product	Inclusion level	Main effects	Comments
Dairy cows	Meal, cake	Up to ~10% DM	↑ n-3 FA in milk, ↑ CLA; neutral yield; risk of milk-fat depression at high fat	Excess lipid load → trans-FA & MFD risk
Beef cattle/ sheep	Meal, cake	Up to ~10–15% DM	Neutral growth; ↑ meat FA profile; intake issues >15% DM	Fiber & glucosinolates at high levels
Pigs	Meal, cake	4–12% (weaned pigs); grow-finish lower tested	≤4–8% tolerated; ↑ gut diversity; ↓ growth/intake at higher doses	Glucosinolates & fiber limit; lysine lower vs. SBM
Poultry (broilers)	Meal, cake	5–15% (meal/ cake); 2–4%	≤10% enriches meat n-3 without impairing growth; >10% reduces FI, BW, FCR	Sensitive to glucosinolates; copper may mitigate
Poultry (layers)	Meal, cake, oil	5–15% (meal/ cake); 2–4% oil	5–10% ↑ n-3 in yolk (up to 8-fold); >15% ↓ egg production, yolk weight	High inclusions reduce egg output; careful with ANFs

*Data extracted from multiple sources (Almeida et al., 2013; Aziza et al., 2013; Cerisuelo et al., 2023; Hajiazizi et al., 2024; Hilbrands et al., 2021; Kahindi et al., 2014; Riaz et al., 2022; Ryhänen et al., 2007)

hanced nutritional and functional properties. Through metabolic engineering, researchers have developed genetically modified (GM) and genome-edited camelina lines capable of modifying the composition of seed oil and protein to improve their value in animal nutrition (Yuan and Li, n.d.). Some of these developments focus on increasing the content of essential fatty acids and modifying the profile of PUFAs, resulting in oils richer in omega-3 compounds that can positively influence lipid metabolism, immune function, and the fatty acid composition of animal-derived products (Lu and Kang, 2008). Other research efforts have targeted the enhancement of amino acid composition, particularly lysine, or the reduction of antinutritional compounds such as GSL and phytic acid, which currently limit the inclusion of camelina co-products in monogastric diets (Blume et al., 2024).

Although these advances are promising, experimental data in livestock species remain limited. Further controlled studies are required to assess feed digestibility, animal performance, nutrient transfer to milk, meat, and eggs, and the long-term effects of modified camelina products on productivity and product quality . Evaluating the safety of these novel traits is important, since metabolic or compositional changes resulting from genetic modification may alter secondary metabolites that affect palatability, digestibility, or feed chemistry (Jayakody et al., 2024; Kawall, 2021).

The regulatory landscape of GM camelina varies across regions. In the United States, genome-edited plants that no longer contain exogenous DNA sequences from *Agrobacterium tumefaciens* T-DNA, but retain only the intended modification, are not regulated by the U.S. Department of Agriculture's Animal and Plant Health Inspection Service (USDA–APHIS) as potential plant health risks. In 2018, Yield10 Bioscience, an agricultural biotechnology company, submitted to the authorities a genome-edited *Camelina sativa* line in which all six copies of a gene involved in oil biosynthesis and lipid turnover were modified through single-base editing. In 2021, field evaluations conducted across multiple geographical regions in the United States assessed the agronomic performance of this line, confirming genotype stability and a 5% increase in seed oil content, expressed as a percentage of seed weight, compared to control plants (Yield10 Biosciences, 2022).

In the European Union, however, organisms developed through genome editing are not exempt from specific regulations that continue to determine which legal framework applies to such crops. In Italy, genetically modified crops and new breeding techniques are considered potential tools to address the challenges posed by climate change. In June 2022, during a session at the European Parliament, the Italian Minister of Agriculture called for the liberalisation of new plant breeding techniques by "untying them from the GMO rules of the 1999 Directive."

Currently, within the EU, plants developed through genome editing can be protected by patents, provided that the origin of the mutation is disclosed, even if the same genetic modification could theoretically occur naturally (European Parliament, 2022).

In general, several approaches can be used to improve *Camelina sativa*, keeping in mind that intraspecific and interspecific crosses can generate new varieties. Conversely, through recurrent backcrossing, mutagenesis, transgenesis, or genome editing, it is possible to obtain essentially derived varieties that are regulated under EU legislation (Kiewiet, 2006).

In summary, GM camelina represent an important opportunity to enhance the nutritional quality and sustainability of livestock feeding. Their ability to improve the balance of beneficial fatty acids, increase essential amino acid content, and reduce ANFs could make camelina co-products more versatile and valuable in ruminant and monogastric diets. However, large-scale adoption will depend on the generation of solid experimental evidence in farm animals and on clear, harmonized regulatory frameworks that ensure feed safety, environmental protection, and consumer acceptance.

4 Conclusions

In this chapter, the nutritional potential and feeding applications of *Camelina sativa* and its co-products have been reviewed. Camelina represents a valuable oilseed crop with distinctive features, particularly its high α-linolenic acid content and moderate protein fraction, which make it attractive as an alternative feed ingredient. Its co-products (oil, cake, meal) can contribute to protein and energy supply and enrich animal-derived product with n-3 fatty acids.

Nevertheless, the presence of antinutritional factors imposes inclusion limits that are more restrictive in pigs and poultry than in ruminants. Processing and dietary strategies can mitigate these constraints, but further refinement is required.

Future research should focus on optimizing processing technologies to reduce antinutritional compounds, establishing safe inclusion levels under commercial conditions, and evaluating long-term impacts on animal performance, product quality, and environmental outcomes. In addition, there is a clear need for standardized characterization of camelina varieties and expanded trials in monogastrics, together with integrative studies on environmental and human health impacts. These knowledge gaps define new lines of research that will be crucial to fully unlock camelina's role in sustainable animal production systems.

References

Abramovič, H., Abram, V., 2005. Physico-Chemical Properties, Composition and Oxidative Stability of *Camelina sativa* Oil. Food Technol Biotechnol 43, 63–70.

Alexander, J., Auðunsson, G.A., Benford, D., Cockburn, A., Cravedi, J.-P., Dogliotti, E., Domenico, A. Di, Luisa Férnandez-Cruz, M., Fürst, P., Fink-Gremmels, J., Corrado, L., Galli, P., Grandjean, J., Gzyl, G., Heinemeyer, N., Johansson, A., Mutti, J., Schlatter, R., Van Leeuwen, C. Van, Peteghem, P.V., 2008. Glucosinolates as undesirable substances in animal feed - Scientific Opinion of the Panel on Contaminants in the Food Chain. EFSA Journal 6, 590. https://doi.org/10.2903/J.EFSA.2008.590

Almeida, F. N., Htoo, J.K., Thomson, J., Stein, H.H., 2013. Amino acid digestibility in camelina products fed to growing pigs. Can J Anim Sci 93, 335–343. https://doi.org/10.4141/CJAS2012-134

Aziza, A.E., Awadin, W.F., Quezada, N., Cherian, G., 2014. Gastrointestinal morphology, fatty acid profile, and production performance of broiler chickens fed camelina meal or fish oil. European Journal of Lipid Science and Technology 116, 1727–1733. https://doi.org/10.1002/EJLT.201400019

Aziza, A.E., Panda, A.K., Quezada, N., Cherian, G., 2013. Nutrient digestibility, egg quality, and fatty acid composition of brown laying hens fed camelina or flaxseed meal. Journal of Applied Poultry Research 22, 832–841. https://doi.org/10.3382/JAPR.2013-00735

Aziza, A.E., Quezada, N., Cherian, G., 2010. Antioxidative effect of dietary Camelina meal in fresh, stored, or cooked broiler chicken meat. Poult Sci 89, 2711–2718. https://doi.org/10.3382/PS.2009-00548,

Aziza, A.E., Quezada, N., Cherian, G., 2010. Feeding *Camelina sativa* meal to meat-type chickens: Effect on production performance and tissue fatty acid composition. Journal of Applied Poultry Research 19, 157–168. https://doi.org/10.3382/JAPR.2009-00100

Bauman, D.E., Lock, A.L., 2006. Conjugated Linoleic Acid: Biosynthesis and Nutritional Significance. Advanced Dairy Chemistry 2, 93–136. https://doi.org/10.1007/0-387-28813-9_3

Belayneh, H.D., Wehling, R.L., Cahoon, E., Ciftci, O.N., 2015. Extraction of omega-3-rich oil from *Camelina sativa* seed using supercritical carbon dioxide. J Supercrit Fluids 104, 153–159. https://doi.org/10.1016/J.SUPFLU.2015.06.002

Yield10 Bioscience Inc. 2022. Yield10 Bioscience announces an update on the Camelina line E3902 development program for producing low-carbon feedstock oil for renewable diesel. Accessed 20 November 2022. Available at: https://www.globenewswire.com/en/news-release/2022/03/03/2396309/34378/en/Yield10-Bioscience-Announces-an-Update-on-the-Camelina-Line-E3902-Development-Program-for-Producing-Low-carbon-Feedstock-Oil-for-Renewable-Diesel.html

Blades, M., Zubr, J., 2010. Carbohydrates, vitamins and minerals of *Camelina sativa* seed. Nutr Food Sci 40, 523–531. https://doi.org/10.1108/00346651011077036

Blume, R.Y., Hotsuliak, V.Y., Nazarenus, T.J., Cahoon, E.B., Blume, Y.B., 2024. Genome-wide identification and diversity of FAD2, FAD3 and FAE1 genes in terms of biotechnological importance in *Camelina* species. BMC Biotechnol 24, 1–21. https://doi.org/10.1186/S12896-024-00936-4

Boardman, K., Sun, X., Yao, D., Chen, C., van Lierop, L., Hu, B., 2025. Increasing the Nutritional Value of Camelina Meal via Trametes versicolor Solid-State Fermentation with Various Co-Substrates. Fermentation 2025, Vol. 11, Page 77 11, 77. https://doi.org/10.3390/FERMENTATION11020077

Budin, J.T., Breene, W.M., Putnam, D.H., 1995. Some compositional properties of camelina (*Camelina sativa* L. Crantz) seeds and oils. J Am Oil Chem Soc 72, 309–315. https://doi.org/10.1007/BF02541088

Cerisuelo, A., Ferrer, P., Angel Gómez, E., Woyengo, T.A., Stein, H.H., Martínez, M., Luís Cano, J., Piquer, O., 2023. Nutritional value of Spanish *Camelina sativa* co-products for pigs. Anim Feed Sci Technol 301, 115665. https://doi.org/10.1016/j.anifeedsci.2023.115665

Chen, C.C., Shih, Y.C., Chiou, P.W.S., Yu, B., 2010. Evaluating nutritional quality of single stage- and two stage-fermented soybean meal. Asian-Australas J Anim Sci 23, 598–606. https://doi.org/10.5713/AJAS.2010.90341

Cherian, G., Campbell, A., Parker, T., 2009. Egg quality and lipid composition of eggs from hens fed *Camelina sativa*. Journal of Applied Poultry Research 18, 143–150. https://doi.org/10.3382/JAPR.2008-00070

Colombini, S., Broderick, G.A., Galasso, I., Martinelli, T., Rapetti, L., Russo, R., Reggiani, R., 2014. Evaluation of *Camelina sativa* (L.) Crantz meal as an alternative protein source in ruminant rations. J Sci Food Agric 94, 736–743. https://doi.org/10.1002/JSFA.6408

Cullere, M., Singh, Y., Pellattiero, E., Berzuini, S., Galasso, I., Clemente, C., Dalle Zotte, A., 2023. Effect of the dietary inclusion of *Camelina sativa* cake into quail diet on live performance, carcass traits, and meat quality. Poult Sci 102, 102650. https://doi.org/10.1016/J.PSJ.2023.102650

Decker, S., Craine, W., Paulitz, T., Chen, C., Lu, C., 2025. Genomic Analysis of the Natural Variation of Fatty Acid Composition in Seed Oils of *Camelina sativa*. Biology (Basel) 14, 1199. https://doi.org/10.3390/BIOLOGY14091199/S1

Delver, J.J., Smith, Z.K., 2024. Opportunities for Camelina Meal as a Livestock Feed Ingredient. Agriculture (Switzerland) 14. https://doi.org/10.3390/AGRICULTURE14010116

Destaillats, F., Japiot, C., Chouinard, P.Y., Arul, J., Angers, P., 2005. Short Communication: Rearrangement of Rumenic Acid in Ruminant Fats: A Marker of Thermal Treatment. J Dairy Sci 88, 1631–1635. https://doi.org/10.3168/JDS.S0022-0302(05)72834-4

Egorova, T.A., n.d. Glucosinolates in rape and camelina: composition, concentrations, toxicity and anti-nutritive effects in poultry, methods of neutralization — a mini-review. https://doi.org/10.15389/agrobiology.2023.6.1021eng

Fröhlich, A., O'Dea, G., Hackett, R., O'Beirne, D., Ni Eidhin, D., Burke, J., 2012. Stabilization of camelina oil with synthetic and natural antioxidants. JAOCS, Journal of the American Oil Chemists' Society 89, 837–847. https://doi.org/10.1007/S11746-011-1973-Y/METRICS

Ghidoli, M., Ponzoni, E., Araniti, F., Miglio, D., Pilu, R., 2023. Genetic Improvement of *Camelina sativa* (L.) Crantz: Opportunities and Challenges. Plants 2023, Vol. 12, Page 570 12, 570. https://doi.org/10.3390/PLANTS12030570

Gómez-Cortés, P., Galisteo, O.O., Ramírez, C.A., Blanco, F.P., De La Fuente, M.A., Sánchez, N.N., Marín, A.L.M., 2018. Intramuscular fatty acid profile of feedlot lambs fed concentrates with alternative ingredients. Anim Prod Sci 59, 914–920. https://doi.org/10.1071/AN17885

Hajiazizi, F., Sadeghi, A., Ibrahim, S., 2024. *Camelina sativa* (L. Crantz) products; an alternative feed ingredient for poultry diets with its nutritional and physiological consequences. Trop Anim Health Prod 56. https://doi.org/10.1007/S11250-024-03904-4,

Hajiazizi, F., Sadeghi, A., Karimi, A., 2023. The effects of dietary inclusion of *Camelina sativa* on performance and ascites incidence in broilers subjected to cold exposure. Vet Med Sci 9, 1711–1725. https://doi.org/10.1002/VMS3.1185,

Halmemies-Beauchet-Filleau, A., Kokkonen, T., Lampi, A.M., Toivonen, V., Shingfield, K.J., Vanhatalo, A., 2011. Effect of plant oils and camelina expeller on milk fatty acid composition in lactating cows fed diets based on red clover silage. J Dairy Sci 94, 4413–4430. https://doi.org/10.3168/jds.2010-3885

Halmemies-Beauchet-Filleau, A., Rinne, M., Lamminen, M., Mapato, C., Ampapon, T., Wanapat, M., Vanhatalo, A., 2018. Review: Alternative and novel feeds for ruminants: nutritive value, product quality and environmental aspects. Animal 12, s295–s309. https://doi.org/10.1017/S1751731118002252

Hilbrands, A.M., Johnston, L.J., Cox, R.B., Forcella, F., Gesch, R., Li, Y.Z., 2021. Effects of increasing dietary inclusion of camelina cake on growth performance of growing-finishing pigs. Transl Anim Sci 5, txab140. https://doi.org/10.1093/TAS/TXAB140

Hölzl, G., Rezaeva, B.R., Kumlehn, J., Dörmann, P., 2023. Ablation of glucosinolate accumulation in the oil crop *Camelina sativa* by targeted mutagenesis of genes encoding the transporters GTR1 and GTR2 and regulators of biosynthesis MYB28 and MYB29. Plant Biotechnol J 21, 189–201. https://doi.org/10.1111/PBI.13936

Hurtaud, C., Peyraud, J.L., 2007a. Effects of feeding camelina (seeds or meal) on milk fatty acid composition and butter spreadability. J Dairy Sci 90, 5134–5145. https://doi.org/10.3168/jds.2007-0031

Jayakody, T.B., Zarka, D., Cho, K.H., Jensen, J., Sikora, S., Buell, C.R., Douches, D.S., Nadakuduti, S.S., 2024. Genome-wide evaluation of gene editing outcomes using CRISPR/Cas9 in seed propagated *Camelina sativa* and vegetatively propagated Solanum tuberosum. Front Plant Sci 15, 1496861. https://doi.org/10.3389/FPLS.2024.1496861/BIBTEX

Jolazadeh, A., Azizi, A., Sharifi, A., Memarzadeh, F., Fallali, N., Imani-Rad, E., 2025. Evaluation of camelina seed pods as a novel feed ingredient for ruminants: nutritional value, fermentation characteristics and nutrient digestibility. Vet Anim Sci 29, 100477. https://doi.org/10.1016/J.VAS.2025.100477

Juodka, R., Nainienė, R., Juškienė, V., Juška, R., Leikus, R., Kadžienė, G., Stankevičienė, D., 2022. Camelina (*Camelina sativa* (L.) Crantz) as Feedstuffs in Meat Type Poultry Diet: A Source of Protein and n-3 Fatty Acids. Animals (Basel) 12, 295. https://doi.org/10.3390/ANI12030295

Kahindi, R.K., Woyengo, T.A., Thacker, P.A., Nyachoti, C.M., 2014. Energy and amino acid digestibility of camelina cake fed to growing pigs. Anim Feed Sci Technol 193, 93–101. https://doi.org/10.1016/J.ANIFEEDSCI.2014.03.012

Kasiga, T., Karki, B., Croat, J., Kaur, J., Gibbons, W.R., Muthukumarappan, K., Brown, M.L., 2020. Process effects on carinata *Brassica carinata* and camelina *Camelina sativa* seed meal compositions and diet palatability in Rainbow Trout *Oncorhynchus mykiss*. Anim Feed Sci Technol 267, 114578. https://doi.org/10.1016/J.ANIFEEDSCI.2020.114578

Kawall, K., 2021. Genome-edited *Camelina sativa* with a unique fatty acid content and its potential impact on ecosystems. Environ Sci Eur 33, 1–12. https://doi.org/10.1186/S12302-021-00482-2/METRICS

Kiarie, E., Walsh, M.C., He, L., Velayudhan, D.E., Yin, Y.L., Nyachoti, C.M., 2016. Phytase improved digestible protein, phosphorous, and energy contents in camelina expellers fed to growing pigs. J Anim Sci 94, 215–218. https://doi.org/10.2527/JAS.2015-9735

Kokić, B., Rakita, S., Vujetić, J., 2024. Impact of Using Oilseed Industry Byproducts Rich in Linoleic and Alpha-Linolenic Acid in Ruminant Nutrition on Milk Production and Milk Fatty Acid Profile. Animals 14, 539. https://doi.org/10.3390/ANI14040539/S1

Kurasiak-Popowska, D., Rynska, B., Stuper-Szablewska, K., 2019. Analysis of Distribution of Selected Bioactive Compounds in *Camelina sativa* from Seeds to Pomace and Oil. Agronomy 2019, Vol. 9, Page 168 9, 168. https://doi.org/10.3390/AGRONOMY9040168

Lanzoni, D., Grassi Scalvini, F., Petrosillo, E., Nonnis, S., Tedeschi, G., Savoini, G., Buccioni, A., Invernizzi, G., Baldi, A., Giromini, C., 2024a. Antioxidant capacity and peptidomic analysis of in vitro digested *Camelina sativa* L. Crantz and Cynara cardunculus co-products. Sci Rep 14, 1–14. https://doi.org/10.1038/S41598-024-64989-3

Lawrence, R.D., Anderson, J.L., Clapper, J.A., 2016a. Evaluation of camelina meal as a feedstuff for growing dairy heifers. J Dairy Sci 99, 6215–6228. https://doi.org/10.3168/jds.2016-10876

Lu, C., Kang, J., 2008. Generation of transgenic plants of a potential oilseed crop *Camelina sativa* by Agrobacterium-mediated transformation. Plant Cell Rep 27, 273–278. https://doi.org/10.1007/S00299-007-0454-0

Luise, D., Correa, F., Cestonaro, G., Sattin, E., Conte, G., Mele, M., Archetti, I., Virdis, S., Negrini, C., Galasso, I., Stefanelli, C., Mazzoni, M., Nataloni, L., Trevisi, P., Costanzo, E., 2024. Effect of different doses of camelina cake inclusion as a substitute of dietary soyabean meal on growth performance and gut health of weaned pigs. British Journal of Nutrition 131, 1962–1974. https://doi.org/10.1017/S0007114524000722,

Matthäus, B., Zubr, J., 2000. Variability of specific components in *Camelina sativa* oilseed cakes. Ind Crops Prod 12, 9–18. https://doi.org/10.1016/S0926-6690(99)00040-0

Mutsvangwa, T., Abeysekara, S., Thompson, S., 2022. Effects of feeding graded levels of camelina expeller meal up to 10% of dietary dry matter on production and milk fatty acid profiles in dairy cows. Can J Anim Sci 103, 101–106. https://doi.org/10.1139/CJAS-2022-0027

Orczewska-Dudek, S., Pietras, M., 2019. The Effect of Dietary *Camelina sativa* Oil or Cake in the Diets of Broiler Chickens on Growth Performance, Fatty Acid Profile, and Sensory Quality of Meat. Animals 2019, Vol. 9, Page 734 9, 734. https://doi.org/10.3390/ANI9100734

Paula, E.M., da Silva, L.G., Brandao, V.L.N., Dai, X., Faciola, A.P., 2019. Feeding Canola, Camelina, and Carinata Meals to Ruminants. Animals 2019, Vol. 9, Page 704 9, 704. https://doi.org/10.3390/ANI9100704

Pekel, A.Y., Kim, J.I., Chapple, C., Adeola, O., 2015. Nutritional characteristics of camelina meal for 3-week-old broiler chickens. Poult Sci 94, 371–378. https://doi.org/10.3382/PS/PEU066

Piergiovanni, A.R., Alberghini, B., Zanetti, F., Ponzoni, E., Brambilla, I.M., Galasso, I., 2025a. The effect of year and variety on the nutritional value of *Camelina sativa* meal. European Food Research and Technology 251, 841–849. https://doi.org/10.1007/S00217-025-04672-7

Piravi-vanak, Z., Azadmard-Damirchi, S., Kahrizi, D., Mooraki, N., Ercisli, S., Savage, G.P., Rostami Ahmadvandi, H., Martinez, F., 2022. Physicochemical properties of oil extracted from camelina (*Camelina sativa*) seeds as a new source of vegetable oil in different regions of Iran. J Mol Liq 345, 117043. https://doi.org/10.1016/J.MOLLIQ.2021.117043

Ponnampalam, E.N., Kearns, M., Kiani, A., Santhiravel, S., Vahmani, P., Prache, S., Monahan, F.J., Mapiye, C., 2024. Enrichment of ruminant meats with health enhancing fatty acids and antioxidants: feed-based effects on nutritional value and human health aspects – invited review. Frontiers in Animal Science 5, 1329346. https://doi.org/10.3389/FANIM.2024.1329346

Quezada, N., Cherian, G., 2012. Lipid characterization and antioxidant status of the seeds and meals of *Camelina sativa* and flax. European Journal of Lipid Science and Technology 114, 974–982. https://doi.org/10.1002/EJLT.201100298

Rabiej-Kozioł, D., Momot-Ruppert, M., Stawicka, B., Szydłowska-Czerniak, A., 2023. Health Benefits, Antioxidant Activity, and Sensory Attributes of Selected Cold-Pressed Oils. Molecules 2023, Vol. 28, Page 5484 28, 5484. https://doi.org/10.3390/MOLECULES28145484

Rakita, S., Spasevski, N., Savić, I., Savić Gajić, I., Lazarević, J., Dragojlović, D., Đuragić, O., 2024. Comparative Evaluation of Camelina Seed Oils Obtained by Cold-Pressing and Solvent Extraction. Foods 2024, Vol. 13, Page 3605 13, 3605. https://doi.org/10.3390/FOODS13223605

Riaz, R., Ahmed, I., Sizmaz, O., Ahsan, U., 2022. Use of *Camelina sativa* and By-Products in Diets for Dairy Cows: A Review. Animals 2022, Vol. 12, Page 1082 12, 1082. https://doi.org/10.3390/ANI12091082

Russo, R., Reggiani, R., Russo, R., Reggiani, R., 2017. Glucosinolates and Sinapine in Camelina Meal. Food Nutr Sci 8, 1063–1073. https://doi.org/10.4236/FNS.2017.812078

Russo, R., Reggiani, R., Russo, R., Reggiani, R., 2012a. Antinutritive Compounds in Twelve *Camelina sativa* Genotypes. Am J Plant Sci 3, 1408–1412. https://doi.org/10.4236/AJPS.2012.310170

Ryhänen, E.L., Perttilä, S., Tupasela, T., Valaja, J., Eriksson, C., Larkka, K., 2007. Effect of *Camelina sativa* expeller cake on performance and meat quality of broilers. J Sci Food Agric 87, 1489–1494. https://doi.org/10.1002/JSFA.2864

Sarramone, J.P., Gervais, R., Benchaar, C., Chouinard, P.Y., 2020. Lactation performance and milk fatty acid composition of lactating dairy cows fed *Camelina sativa* seeds or expeller. Anim Feed Sci Technol 270, 114697. https://doi.org/10.1016/J.ANIFEEDSCI.2020.114697

Shingfield, K.J., Griinari, J.M., 2007. Role of biohydrogenation intermediates in milk fat depression. European Journal of Lipid Science and Technology 109, 799–816. https://doi.org/10.1002/EJLT.200700026

Smit, M.N., Beltranena, E., 2017. Effects of feeding camelina cake to weaned pigs on safety, growth performance, and fatty acid composition of pork. J Anim Sci 95, 2496–2508. https://doi.org/10.2527/JAS.2016.1265

Sydor, M., Kurasiak-Popowska, D., Stuper-Szablewska, K., Rogoziński, T., 2022a. *Camelina sativa*. Status quo and future perspectives. Ind Crops Prod 187, 115531. https://doi.org/10.1016/J.INDCROP.2022.115531

Thacker, P., Widyaratne, G., 2012. Effects of expeller pressed camelina meal and/or canola meal on digestibility, performance and fatty acid composition of broiler chickens fed wheat-soybean meal-based diets. Arch Anim Nutr 66, 402–415. https://doi.org/10.1080/1745039X.2012.710082

Toma, S., Dragomir, C., Habeanu, M., Ropota, M., Cismileanu, A., Grosu, H., 2015. Effects of partial or total replacement of sunflower meal with camelina meal on dairy cows' milk fatty acids profile. Archiva Zootechnica 18, 85–94.

Tripathi, M.K., Mishra, A.S., 2007. Glucosinolates in animal nutrition: A review. Anim Feed Sci Technol 132, 1–27. https://doi.org/10.1016/J.ANIFEEDSCI.2006.03.003

Veljković, V.B., Kostić, M.D., Stamenković, O.S., 2022. Camelina seed harvesting, storing, pretreating, and processing to recover oil: A review. Ind Crops Prod 178, 114539. https://doi.org/10.1016/J.INDCROP.2022.114539

Ventto, L., Leskinen, H., Kairenius, P., Stefański, T., Bayat, A.R., Vilkki, J., Shingfield, K.J., 2017. Diet-induced milk fat depression is associated with alterations in ruminal biohydrogenation pathways and formation of novel fatty acid intermediates in lactating cows. British Journal of Nutrition 117, 364–376. https://doi.org/10.1017/S0007114517000010,

Wang, S., Kreuzer, M., Braun, U., Schwarm, A., 2017. Effect of unconventional oilseeds (safflower, poppy, hemp, camelina) on in vitro ruminal methane production and fermentation. J Sci Food Agric 97, 3864–3870. https://doi.org/10.1002/JSFA.8260,

Woyengo, T.A., Beltranena, E., Zijlstra, R.T., 2017. Effect of anti-nutritional factors of oilseed co-products on feed intake of pigs and poultry. Anim Feed Sci Technol 233, 76–86. https://doi.org/10.1016/J.ANIFEEDSCI.2016.05.006

Woyengo, T.A., Patterson, R., Slominski, B.A., Beltranena, E., Zijlstra, R.T., 2016. Nutritive value of cold-pressed camelina cake with or without supplementation of multi-enzyme in broiler chickens. Poult Sci 95, 2314–2321. https://doi.org/10.3382/PS/PEW098

Yalçın, Sakine, Ramay, M.S., Yalçınkaya, H., Kardoğan, Ö., Erkurt, A., Kılınç, B., Gebeş, E.S., Bundur, A., Onbaşılar, E.E., Yalçın, Suzan, Subaşı, İ., Şafak, C.E., Kocadaş, E., 2025. Evaluation of high levels of solvent extracted *Camelina sativa* meal in diets on performance, blood indices, cecal microorganisms, and nutrient digestibility in broilers. Trop Anim Health Prod 57, 1–13. https://doi.org/10.1007/S11250-025-04339-1

Yuan, L., Li, R., n.d. Metabolic Engineering a Model Oilseed *Camelina sativa* for the Sustainable Production of High-Value Designed Oils. https://doi.org/10.3389/fpls.2020.00011

Zając, M., Kiczorowska, B., Samolińska, W., Klebaniuk, R., 2020. Inclusion of Camelina, Flax, and Sunflower Seeds in the Diets for Broiler Chickens: Apparent Digestibility of Nutrients, Growth Performance, Health Status, and Carcass and Meat Quality Traits. Animals 2020, Vol. 10, Page 321 10, 321. https://doi.org/10.3390/ANI10020321

Zhu, Y., Cox, R., Johnston, L.J., Reese, C., Forcella, F., Gesch, R.W., Li, Y.Z., 2021. Effects of increasing inclusion of camelina press cake in diets fed to growing-finishing pigs on pork quality. Applied Animal Science 37, 357–366. https://doi.org/10.15232/AAS.2021-02161

Zubr, J., 2003. Dietary fatty acids and amino acids of *Camelina sativa* seed. J Food Qual 26, 451–462. https://doi.org/10.1111/J.1745-4557.2003.TB00260.X

Zubr, J., Matthaus, B., 2002. Effects of growth conditions on fatty acids and tocopherols in *Camelina sativa* oil. Ind Crops Prod 15, 155–162. https://doi.org/10.1016/S0926-6690(01)00106-6

PART 5

Regulatory framework

CHAPTER 23

Use of innovative feed in the EU: legal enablers and restrictions

A. Bouxin

European Feed Manufacturers' Federation (FEFAC), Rue de la Loi, 223-1040, Brussels, Belgium
Correspondence: abouxin@fefac.eu

Abstract

The contribution of the livestock sector to efficient circular bioeconomy and sustainable food systems is gaining recognition as we (re-)discover the symbiotic relationship between crop and livestock farming and the high biological value that animal products bring to sustainable diets. Enhancing circularity in food systems requires minimising input of finite resources (fertiliser, fossil fuels, water and land), encouraging regenerative practices and stimulating reuse and recycling of residual streams in a way to add highest value to unavoidable food system residues.

The livestock sector is under pressure to increase its circularity while at the same time upscale its practices to minimise its environment footprint, contribute better to food security and secure animal health and welfare. In addition, a high priority is put on reducing EU dependency upon Third Countries. Feed plays a pivotal role in addressing many of these challenges.

This paper outlines the main features of EU feed legislation and its interlinks with waste and animal by-products legislation. It identifies current restrictions on access to certain resources and proposes legislative and non-legislative initiatives to foster greater feed circularity.

Keywords

waste – regulation – animal-by-products – prohibitions – safety – innovation

1 Introduction

The Von der Leyen Commission I has prioritized sustainability through the Green Deal, aiming to transition towards more sustainable practices, particularly in environmental and social domains. Recent global challenges, such as the COVID-19 pandemic, the Russian aggression of Ukraine and rising protectionism, have shifted the policy priorities of the Von der Leyen Commission II towards economic resilience in the Competitiveness Compass (EU Commission, 2025), building upon from the Draghi report (Draghi, 2024) and the concept of Open Strategic Autonomy aiming among others at more food security and sovereignty. As far as the EU food systems are concerned, this new focus complements existing environmental goals, requiring the EU farming sector to adopt more environmentally friendly practices while remaining profitable and resilient.

Over the past decade, the livestock sector has faced criticism for its environmental impact, particularly regarding feed production (e.g., soy imports linked to deforestation, marine resource depletion) and consumption (e.g., greenhouse gas emissions, nitrates, phosphates). However, recent years have seen growing recognition of the sector's essential role in healthy diets and sustainable food systems (FAO, 2023; Peyraud et al., 2020). Livestock farming and aquaculture contribute significantly to food system circularity by producing food from non-edible biomass, closing nutrient loops via manure, and providing non-food goods (FAO, 2024; Van Zanten et al., 2019).

Still, there is a margin of improvement in terms of mitigation of the environment footprint of livestock production by innovative feeding practices involving techniques to reduce livestock emissions and also to extend the range of nutrient resources not used directly by consumers (FEFAC, 2023) and for increased circularity of the livestock sector (FAO, 2024).

In the EU, there is a long tradition of collecting co-products of food processing as well as non-feed/food industries such as bioenergy or oleochemical industry (FEFAC, 2019, 2022). Although the majority of the proteins consumed by farm animals come from unprocessed crops (forages and grains), balancing the diets to meet animals' requirements in amino acids is possible thanks to many co-products from the food processing industry with medium to high levels of proteins like oilseed meals or maize gluten feed. The vast majority of what is available as possible circular resources for feed is actually used in the EU so that the potential for further circularity is basically composed of:

- Underused existing feed resources for example from marine resources (e.g. krill, zooplankton) or co-products from the fermentation industry, or former foodstuffs;
- Innovative feed ingredients (insects, single cell proteins, grass protein concentrates, micro-algae) and technologies (biorefinery, fermentation);

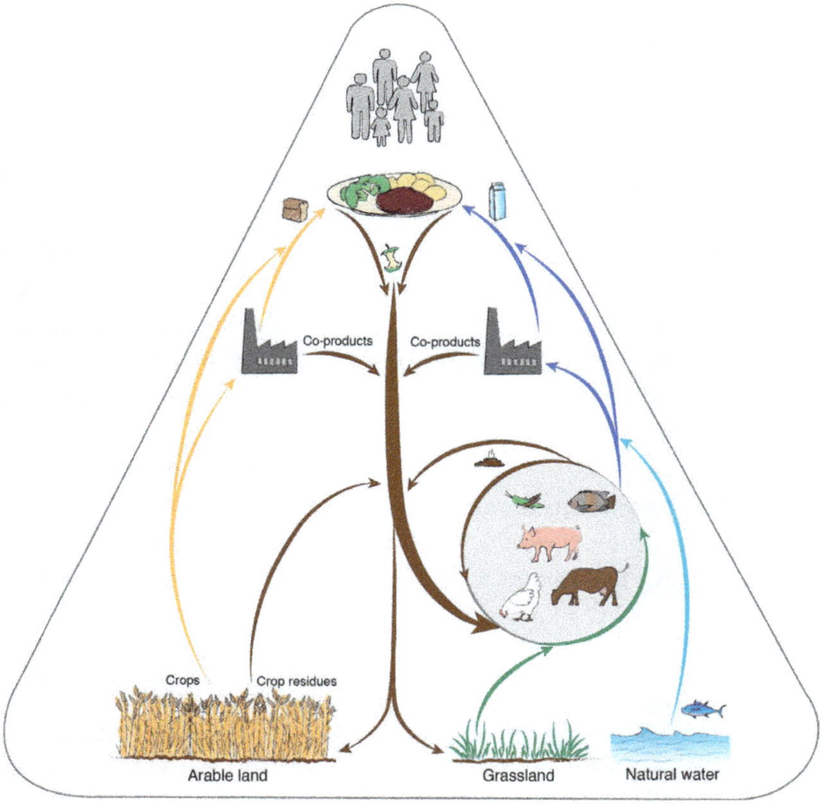

FIGURE 23.1 The biophysical concept of circularity (Zanten et al., 2019)

- Feed materials non-compliant with legal standards for contaminants or pesticides;
- Prohibited materials (waste, certain animal by-products).

In this paper, we intend to provide to the 'newcomers' in the feed chain an overview of the current EU legislation governing the feed chain, with a main focus on feed safety, identify the current restrictions and enablers and, as appropriate, identify possible legal pathways towards more sustainable feeding practices and protein autonomy.

1 Understanding the legal requirements for placers on the feed market and the 'feed safety culture'

The EU feed legislation was among the first to be harmonised at European level. The first EU Directive on feed additives dates back to 1970 and the first Directive on the marketing of feed from 1973. In the wake of the food and feed safety crisis in the 90es, the feed and food legislations have been completely revised in the direction of more consistency, more harmonisation, and more responsibility put on operators. The review process was finalised as far as feed is concerned in 2019 (with the medicated feed Regulation (EU) 2019/4 – not addressed in this article) and is composed now exclusively of regulations directly applicable in the same way in all EU Member States, exception being the EU Directive on undesirable substances (see Figure 23.2). In short, the feed legislation is based on two regulations governing the safety of feed and its marketing. It is complemented by specific requirements related to maximum levels of contaminants[1] and an authorisation procedure for certain types of feed (feed additives and feed materials produced from genetically modified organisms).

1.1 The framework legislation: the General Food Law

The foundation of the food and the feed legislation since 2002 is the General Food Law Regulation (EC) No 178/2002[2]). It defines the respective responsibilities and

1 There are also maximum residues limits for pesticides in feed, regulated by Regulation (EC) No 396/2005. https://eur-lex.europa.eu/eli/reg/2005/396/oj/eng

2 Regulation (EC) No 178/2002 of the European Parliament and of the Council of 28 January 2002 laying down the general principles and requirements of food law. https://eur-lex.europa.eu/legal-content/EN/TXT/?uri=CELEX%3A32002R0178

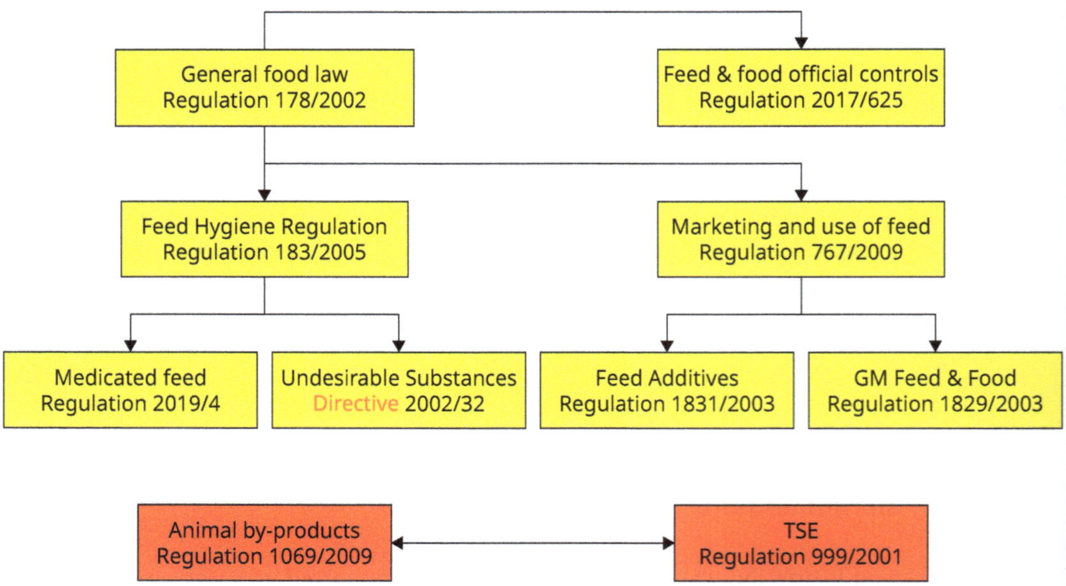

FIGURE 23.2 Overview of the structure of the main texts ruling the sourcing, manufacturing and delivery of feed to animals in the EU in 2025

obligations of the various operators of the food and feed chain and authorities, which can be summarised by the following four principles: safety, responsibility, traceability, transparency:

- Safety: only products that are deemed as safe can be placed on the market. A feed is deemed unsafe when it has an adverse effect on human or animal health or it makes the food derived from food producing animals unsafe for human consumption. The safety of a product is established among others by reference to existing legal safety standards for contaminants and / or conditions of authorization for regulated products (i.e. feed additives or genetically modified feed as far as feed is concerned), further to a risk assessment performed by the European Food Safety Authority (EFSA). However, compliance with legal standards may not be sufficient, in particular in case of presence of a non-regulated contaminant at EU level, in which case national standards apply when they exist. Otherwise, safety is established on the basis of existing scientific information. This is especially relevant for new feed resources with limited scientific information available on their risk profile. Indeed, unlike in the food sector, there is no 'Novel Feed' Regulation, meaning that placers on the market of new feed materials hold the full responsibility for assuring the safety of the product they place on the market and cannot rely on the backing of EFSA.
- Responsibility: the operator is the first responsible for the safety of the feed/food it places on the market. Additional requirements to fulfill this principle are set in secondary legislation. The scope is wide and includes any operator who produces, manufactures, processes, stores, transports or distributes feed/food.
- Traceability: operators must have in place a system ensuring traceability one step back – one step forward.
- Transparency: operators who are aware that a product (feed/food) they import, produce, process or distribute does not meet the safety requirements shall notify it to the authorities. The Member State who receives the notification from the operator shall notify the incident to the EU Rapid Alert System for Food and Feed (RASFF) in case the incident would concern other EU Member States.

Guidance on the implementation of the General Food Law was issued by the EU Commission to facilitate a harmonized enforcement of these principles by Member States and operators[3].

3 Guidance on the implementation of articles 11, 12, 14, 17, 18, 19 and 20 of Regulation (EC) No 178/2002 on General Food Law. 2010. https://food.ec.europa.eu/document/download/4cea21a6-ec96-49a7-a3a0-e1747c2c595d_en?filename=gfl_req_implementation-guidance_en.pdf

1.2 The secondary legislation for feed safety: the Feed Hygiene Regulation

The legislator laid down minimum requirements to be fulfilled by feed business operators to comply with their responsibilities in terms of feed safety via the Feed Hygiene Regulation (EC) No 183/2005[4]. The main requirements are as follows:

- All feed business operators shall be registered by national authorities based on an application from operators; this is meant mostly for traceability purpose: feed business operators and farmers shall indeed only source and use feed from registered and/or approved (see below) establishments.
- For some operations presenting a higher risk, an approval is required, which means that the license to operate is delivered after inspection by competent authorities. This concerns operations involving certain feed additives, premixtures and compound feedingstuffs, as well as i) the processing of crude vegetable oil except those under the scope of food legislation, ii) oleochemical manufacturing of fatty acids, iii) manufacturing of biodiesel and iv) fat blending. National authorities may also decide that certain operators located in their country are subject to approval. Producers of innovative feed materials whose risk profile is uncertain (new type of biomass or new process) may be regarded as deserving an approval.
- Minimum requirements are set regarding good agriculture practice for the primary production and use of feed. Minimum requirements for good hygiene practice are also set for other operators, including prerequisites regarding the design of establishments, pest controls, light, cleaning, waste management, etc…) as well as an obligation to establish a feed safety management system involving procedures for traceability, record and sample keeping, staff training, quality controls of products, management of complaints and recall, etc. Furthermore, for operations other than primary production, the feed safety management systems must integrate HACCP principles. Representative European organisations of the feed chain are encouraged to draft guides to good hygiene practice to assist operators for the implementation of these feed safety management systems. A procedure for the development and evaluation of the sector guides is specified in the Regulation and the EU Commission publishes a list of sector guides they have endorsed[5]. Although the use of these guides to good practice remains voluntary, it is strongly recommended to follow them. A number of third party-audited feed safety assurance schemes have been developed based on these guides. Although participation in these assurance schemes is not mandatory by law, they

---

4 Regulation (EC) No 183/2005 of the European Parliament and of the Council of 12 January 2005 laying down requirements for feed hygiene. https://eur-lex.europa.eu/legal-content/EN/TXT/?uri=CELEX%3A32005R0183

5 Guides to Good Practice (https://food.ec.europa.eu/food-safety/animal-feed/feed-hygiene/guides-good-practice_en)

are widely implemented in many EU Member States and worldwide and illustrate the importance of the 'Feed Safety Culture' in the EU (Bouxin, 2023).
- A very important provision requires that 'The labelling of the products shall clearly indicate whether they are intended for feed or other purposes. If a certain batch of a product is declared not intended for feed use, this declaration shall not be subsequently altered by an operator at a later stage of the chain'. This provision introduced in 2015 further to an incident involving the use of technical oil as feed, has a considerable impact on the interface with other non-feed/food legislations.
- Additional requirements for the manufacturing of feed materials from animal origin are laid down in the Animal By-Products (ABP) Regulation (Regulation No 396/2009[6]) and its implementing rules (Regulation (EU) No 142/2011[7]), as well as in the Transmissible Spongiform Encephalopathies (TSE) Regulation (EC) No 999/2001[8] (article 7 and Annex IV related to the so-called 'feed ban').

The EU Commission published guidance on the implementation of the Feed Hygiene Regulation[9].

1.3 The secondary legislation for placing on the market and labelling

The requirements for the placing on the market and labelling of feed are laid down in Regulation (EC) No 767/2009[10]. The scope is limited to feed materials and compound feed, whereas the requirements for the placing on the market and labelling of feed additives and premixtures are laid down in Regulation (EC) No 1831/2003[11].

The main requirements are as follows:
- Feed placed on the market must be sound, genuine, unadulterated, fit for purpose and of merchantable quality.

---

6 Regulation (EC) No 1069/2009 of the European Parliament and of the Council of 21 October 2009 laying down health rules as regards animal by-products. https://eur-lex.europa.eu/eli/reg/2009/1069/oj/eng

7 Commission Regulation (EU) No 142/2011 of 25 February 2011 implementing Regulation (EC) No 1069/2009 of the European Parliament and of the Council. https://eur-lex.europa.eu/eli/reg/2011/142/oj/eng

8 Regulation (EC) No 999/2001 of the European Parliament and of the Council of 22 May 2001 laying down rules for the prevention, control and eradication of certain transmissible spongiform encephalopathies. https://eur-lex.europa.eu/eli/reg/2001/999/oj/eng

9 Guidance document on the implementation of certain provisions of Regulation (EC) No 183/2005 laying down requirements for feed hygiene. 2019. https://eur-lex.europa.eu/legal-content/EN/TXT/PDF/?uri=CELEX:52019XC0705(01)

10 Regulation (EC) No 767/2009 of the European Parliament and of the Council of 13 July 2009 on the placing on the market and use of feed. https://eur-lex.europa.eu/eli/reg/2009/767/oj/eng

11 Regulation (EC) No 1831/2003 of the European Parliament and of the Council of 22 September 2003 on additives for use in animal nutrition. https://eur-lex.europa.eu/eli/reg/2003/1831/oj/eng

- The notion of labelling includes any communication via any medium referring to or accompanying the feed, such as packaging, container, notice, label, document, ring, collar or the Internet, including for advertising purposes. This a broader definition than for food and means that claims made on internet regarding feed materials are also included. This may be relatively important for new feed materials for which the ability to communicate to customers on the characteristics and function of the product may be a key parameter for successful market uptake.
- The regulation provides also definitions of the different types of feed falling under the scope in particular feed materials and compound feed. This is also important for new feed ingredients, considering that the legal status (feed material vs. feed additive) is not always easy to determinate and there is no pre-marketing authorisation requirement for (most) feed materials, unlike feed additives. The EU Commission provided guidance on the determination of the status of a product in Recommendation 2011/25[12] and FEFANA developed a substance classification tool[13] available to operators.
- Feed materials must be either listed on the non-exhaustive catalogue of feed materials Regulation (EC) 68/2013[14] or notified by the person placing for the first time the product on the EU market to the register of feed materials. More guidance to operators regarding the notification obligation is provided on the Online Feed Materials Register[15]. An innovative feed material being new is not expected to be listed on the current version of the EU Catalogue of feed materials and therefore any placer on the market should notify it to the register of feed materials, unless it would have been already registered by another placer on the market. It must be underlined that, whereas the presence of a product in the EU Catalogue of feed materials guarantees officially its feed material status, this is not the case when listed on the register of feed materials since all information provided on the register is under the sole responsibility of the notifier.
- The Regulation displays the labelling requirements, which may be split in 3 categories:
  - Information related to the identity/traceability of the consignment: the feed material should be named in a manner that should not mislead the user as to its real nature; the identity of the supplier/manufacturer and the lot identification number shall be provided as well;

12 Commission Recommendation 2011/25 of 14 January 2011 establishing guidelines for the distinction between feed materials, feed additives, biocidal products and veterinary medicinal products. https://eur-lex.europa.eu/legal-content/EN/TXT/?uri=CELEX%3A32011H0025&qid=1775492266151

13 Substances classification. FEFANA. https://fefana.org/tools/substances-classification/

14 Commission Regulation (EU) No 68/2013 of 16 January 2013 on the Catalogue of feed materials. https://eur-lex.europa.eu/eli/reg/2013/68/oj/eng

15 Register of feed materials. https://feedmaterialsregister.eu/

  - Information related to the analytical constituents (as specified in the EU Catalogue of feed materials for those listed herewith or required in Annex V of the Regulation for the others) and the presence of feed additives when needed in accordance with Annex VII of the Regulation.
  - Information on instructions for use where needed or any specific restriction of use.
- Beyond mandatory labelling particulars, the regulation makes it possible to provide additional voluntary information as well as making claims, provided they are substantiated.
- The Regulation establishes also purity criteria for feed materials: feed materials shall be free from chemical impurities resulting from their manufacturing process and from processing aids, unless a specific maximum content is fixed in the Catalogue of feed materials. This requirement is meant to encourage operators whose main activity is not feed production to preserve the integrity of the co-products they generate.

2 Additional safety-related legal requirements and restrictions to the placing on the market and use of feed

2.1 Prohibited materials

Although there is no positive list of feed materials, some materials are subject to prohibition or restrictions of use. Regulation (EC) No 767/2009 provides in its Annex III a list of materials whose use as feed is restricted or prohibited. Among the listed materials, we can find manure, packaging materials or solid urban waste and waste waters.

Some products of animal origin are also subject to prohibition. The ABP Regulation (EC) No 1069/2009 establishes in particular that:

- Only animal by-products from so-called Category 3 materials (i.e. fit for use for food but not consumed by human) may be used for feed purposes;
- The feeding of farmed animals other than fur animals with catering waste or feed materials containing or derived from catering waste is prohibited;
- Terrestrial animals from a species may not be fed with processed animal proteins derived from the bodies or parts of bodies of animals of the same species; farmed fish may not be fed with processed animal protein derived from the bodies or parts of bodies of farmed fish of the same species.

There is no endpoint to these prohibitions, i.e. that the material remains prohibited, irrespective of any further processing/treatment. Regulation (EC) 142/2011 still exempts compound feed containing such material from certain provisions.

Finally, Regulation (EC) 999/2001 on TSE establishes additional restrictions (so-called feed ban provisions (EU Commission, 2021) (see Figure 23.3).

Situation as of 7 Sept. 2021	Feed for farmed animals other than fur animals					Feed for pets and fur animals
	Ruminants	Non-ruminants (except fish)			Fish	
		Pigs	Poultry	Others		
• Ruminant PAP, including ruminant blood meal						
• Blood products from ruminants						
• Gelatine and collagen from ruminants		2021	2021	2021	2021	
• Hydrolysed proteins <u>other than those</u> derived from non-ruminants or from ruminant hides and skins						
• Pig PAP			2021			
• Poultry PAP		2021			2013	
• Other non-ruminant PAP, including non-ruminant blood meal but excluding fishmeal						
• Insect PAP		2021	2021		2017	
• Fishmeal						
• Blood products from non-ruminants						
• Di and tricalcium phosphate of animal origin						
• Animal proteins other than those mentioned elsewhere in the table						
• Hydrolysed proteins from non-ruminants or from ruminant hides and skins						
• Gelatine and collagen from non-ruminants						
• Egg, egg products, milk, milk products, colostrum						

FIGURE 23.3 Species specific prohibitions for use of certain products of animal origin other than fat

2.2 Materials subject to authorisation

The placing on the market of a feed additive requires an authorisation as provided for in Regulation (EC) No 1831/2003, based on an assessment by EFSA regarding not only the safety of the product but also its efficacy in relation to defined feed additives functions. As indicated earlier, the correct determination of the legal status of a feed ingredient is essential for new products meant for feed use considering the huge differences in terms of requirements for their placing on the market as feed additive vs. feed material.

Feed materials produced from genetically modified micro-organisms undergo also an authorisation procedure as provided for in Regulation (EC) 1829/2003[16]. The latter applies in particular whatever the nature of subsequent processing stages (in all theory, phosphorous recovered from the incineration of a GM derived product such as soybean meal may not be usable if the GM soybean is not authorised for feed and food use in the EU).

16 Regulation (EC) No 1829/2003 of the European Parliament and of the Council of 22 September 2003 on genetically modified food and feed. https://eur-lex.europa.eu/eli/reg/2003/1829/oj/eng

2.3 Presence of contaminants

Contaminants in feed are of 3 different natures: chemical, physical and microbiological.

Physical contaminants are not specifically regulated. It is expected via the Regulation on feed hygiene, that the presence of physical hazards is prevented or they are removed (e.g. via magnets for metallic hazards).

The main microbiological contaminants of relevance for animal and human health and the safety of products of animal origin are viruses, *Enterobacteriaceae*, especially *Salmonella* spp. and prions. There is no specific requirement as regards viruses (it is expected that feed is free from active viruses, which is the case most of the time, when good manufacturing / trading practices are applied). For *Enterobacteriaceae* and *Salmonella*, microbiological criteria apply but only for products of animal origin (animal by-products). The microbiological risk with prion concerns exclusively products of animal origin and is ruled by the ABP and the TSE legislation.

For chemical contaminants, standards are set in the EU Regulation (EC) No 395/2005[17] on Maximum Residue Levels (MRL) for Pesticides and the EU Directive on Undesirable Substances and Products (Directive 2002/32/EC[18]). For the latter, the maximum limits are set for feed materials based on the As Low As Reasonably Achievable (ALARA) principle, usually following the '95 percentile' approach, i.e. in a way that 95% of the monitoring data collected by EFSA do not exceed the maximum limit.

Maximum limits may be set for all feed materials or some, depending on the nature of the contaminant and its specificity. This means that, for new feed materials / additives, maximum limits for certain contaminants may exist already. It is therefore important, before considering any marketing, to check the compliance of the product with existing limits and, where necessary, to request a modification of the maximum limit, if justified and possible without compromising feed safety.

Regulation (EC) No 396/2005 and Directive 2002/32/EC have in common to prohibit the mixing with other consignments of the same feed material to reduce the average contamination to a level below the maximum permitted limit (so-called 'ban on dilution'). Directive 2002/32/EC however authorises detoxification by means of approved treatments, subject to assessment by EFSA (Regulation (EU) 2015/786[19]).

17 Regulation (EC) No 396/2005 of the European Parliament and of the Council of 23 February 2005 on maximum residue levels of pesticides in or on food and feed. https://eur-lex.europa.eu/eli/reg/2005/396/oj/eng

18 Directive 2002/32/EC of the European Parliament and of the Council of 7 May 2002 on undesirable substances in animal feed. https://eur-lex.europa.eu/legal-content/EN/TXT/?uri=CELEX%3A32002L0032&qid=1775492950520

19 Commission Regulation (EU) 2015/786 of 19 May 2015 defining acceptability criteria for detoxification processes applied to products intended for animal feed. https://eur-lex.europa.eu/eli/reg/2015/786/oj/eng

3 Navigating in the troubled waters of the boundaries between different legal frameworks (feed, food, TSE, ABP and waste legislations)

3.1 General considerations

Establishing the lawfulness of a product for feed use requires not only a good understanding of the feed legislation itself but also diving into the troubled waters of the boundaries with other legal frameworks. Before even considering whether a product is a feed additive or a feed material, a preliminary question concerns the boundaries of the feed chain, i.e. when does a product start being a feed? A number of materials are common to different sectors/destinations and for some of them, feed is a minor outlet. This is especially important since the Feed Hygiene Regulation imposes to a feed business operator to buy feed from establishments registered or approved as feed business operators. The EU Commission provided some guidance on this in its notice 2019/C 225/01[20] as regards the starts of the feed chain, listing criteria such as whether the operator has the intention to distribute his product for feed use, the precedence of another legal status according to another EU legislation (i.e. waste, food, by-product, whether from animal origin or not, etc.), the labelling (whenever the legal status of the product triggers specific labelling requirements) or the legal status of the operator himself (e.g. whether approved as animal-by-products processor). However, there subsists always a grey zone.

Waste vs. by-product: the Waste Hierarchy is a backbone of the Waste Framework Directive 2008/98/EC[21], which encourages the minimisation of waste (see Figure 23.4). This conditions not only the ability or not to use a material (here as feed) but also introduces a hierarchy among different sectors for the access to the material. The Directive excludes from its scope substances destined for use as feed materials, as illustrated by the Joint Research Center in its Food & Waste Use Hierarchy (De Laurentiis et al., 2024) (see Figure 23.5), placing feed destination in the section of the pyramid dealing with resources usages below food destination and above the legal status of waste.

However, the question remains whether a material qualifying as waste, being reprocessed / recycled can pretend to a 'end-of-waste' status and be used as feed. The Belgian authorities in their circular to operators (FASFC, 2021), stress that, since feed materials are excluded from the scope of the Waste Framework Directive, any material labelled as waste is not intended to be used as feed materials and cannot therefore be used as feed, considering that the Feed Hygiene Regulation orders that,

20 Commission notice 2019/C 225/01- Guidance document on the implementation of certain provisions of Regulation (EC) No 183/2005 laying down requirements for feed hygiene. https://eur-lex.europa.eu/legal-content/EN/TXT/PDF/?uri=CELEX:52019XC0705(01):52019XC0705(01)

21 Directive 2008/98/EC of the European Parliament and of the Council of 19 November 2008 on waste. https://eur-lex.europa.eu/eli/dir/2008/98/oj/eng

FIGURE 23.4 Waste hierarchy (EU Commission)

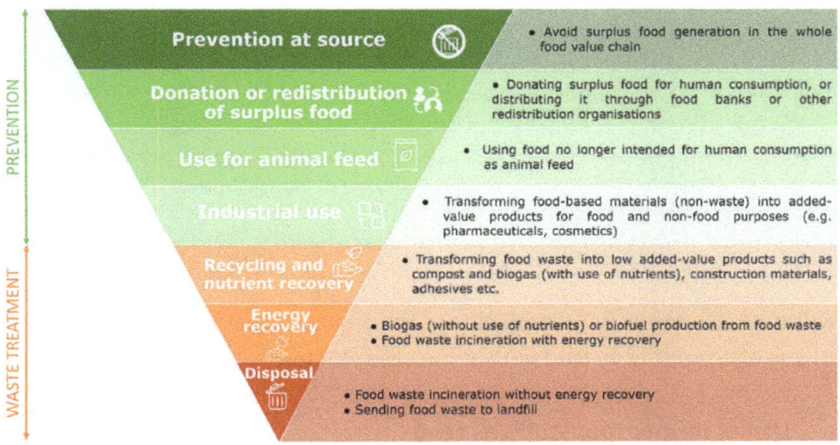

FIGURE 23.5 Food and waste use hierarchy (De Laurentiis et al., 2024)

'If a certain batch of a product is declared not intended for feed use, this declaration shall not be subsequently altered by an operator at a later stage of the chain.' This interpretation is consistent with the principle underpinning the prohibition of certain materials as feed in Annex III of Regulation (EC) No 767/2009, for the two following entries: i) faeces, urine and separated digestive tract content resulting from the emptying or removal of digestive tract, irrespective of any form of treatment or admixture and ii) all waste obtained from the various phases of the treatment of the urban, domestic and industrial waste water, as defined in Article 2 of Council Directive 91/271/EEC of 21 May 1991 concerning urban waste water treatment, irrespective of any further processing of that waste and irrespective of the origin of the waste waters. In its circular letter, the Belgian Federal Agency for the Safety of the Food Chain (FASFC) stresses that materials destined to be used as feed may carry a limited number of legal statuses, i.e. food, animal by-products or feed (FASFC, 2021).

So when considering the legal situation of a product and the relevant applicable legislation, the approach of Table 23.1 should be considered.

TABLE 23.1 Key steps to be followed to identify legal requirements when consididering the placing of a product on the feed market in the EU

Decision / action	Relevant legal texts
Finding whether a material qualifies a feed	Regulations (EC) No 178/2002, (EC) No 767/2009, (EC) No 1831/2003. Recommendation 2011/25.
Finding whether the material is subject to restrictions of use (animal-products, waste)	Regulation (EC) No 767/2009 (Annex III), (EC) No 1069/2009, (EC) No 999/2009
Applying safety legal requirements relevant to any feed (hygiene, contaminants)	Regulation (EC) No 183/2005, Directive 2002/32/EC
Finding whether a feed requires an authorisation (additive or genetically modified feed)	Regulations (EC) 68/2013, (EC) No 1831/2003 and (EC) No 1829/2003 Register of feed additives Recommendation 2011/25.
Applying requirements for 'regulated' feed products (additive, genetically modified feed)	Regulations (EC) No 1831/2003 and (EC) No 1829/2003
Applying requirements specific to feed materials	Regulations (EC) No 767/2009 and (EC) No 68/2013. Register of feed materials

We will now illustrate the complexity of the legal analysis of the conditions for a lawful use of a product as feed with some case studies.

3.2 Use of former foodstuffs as feed

The targeted reduction of food waste at several levels of the food chain and the aspiration of the feed industry to use resources not in direct competition with human consumption prompted a renewed interest in using as feed foodstuffs no longer intended for human consumption, so called 'former foodstuffs'.

Former foods are defined in the EU Catalogue of feed materials as 'foodstuffs, other than catering reflux (catering waste), which were manufactured for human consumption in full compliance with the Union food law but which are no longer intended for human consumption for practical or logistical reasons or due to problems of manufacturing or packaging defects or other defects and which do not present any health risks when used as feed'. The evolution of the legal status of the product (food, animal by-products, feed or waste) from the moment it is declared as non-longer intended to human consumption and the moment it is used as feed has evident implications with respect to the legislative framework governing the different kinds of products concerned.

The EU Commission issued in 2018 a notice 2018/C 133/02[22] providing guidance to

22 Commission notice 2018/C 133/02- Guidelines for the feed use of food no longer intended for human consumption. 2018. https://eur-lex.europa.eu/legal-content/EN/TXT/?uri=CELEX%3A52018XC0416%2801%29&qid=1775493857476

authorities and operators regarding the legal requirements applying to the different operators involved, i.e. the food business operator producing the former foodstuffs, the retailer of food products (if relevant), the operator collecting and processing the former foodstuffs and the user (see Figure 23.6). The critical questions being addressed in this guidance are in particular whether the operator producing or re-selling the former foodstuffs should qualify as feed business operator and whether operators processing former foodstuffs containing products of animal origin should be subject to the requirements of the ABP legislation. It should be noted that the Commission notice was released before the amendment of the Waste Framework Directive, so that the question of the need to justify the non-waste status of a material destined to feed use is no longer relevant. The EU Commission notice should therefore be read in conjunction with subsequent guidance provided by national authorities taking into account this modification, such as the FASFC circular letter (FASFC, 2021).

The EU Commission notice clarifies that the food business operator delivering the former foodstuffs for feed use qualifies as feed business operators, whereas the food retailers do not since the product still qualifies as food when leaving the control of the retailer and the feed chain starts with the operator receiving the food. In

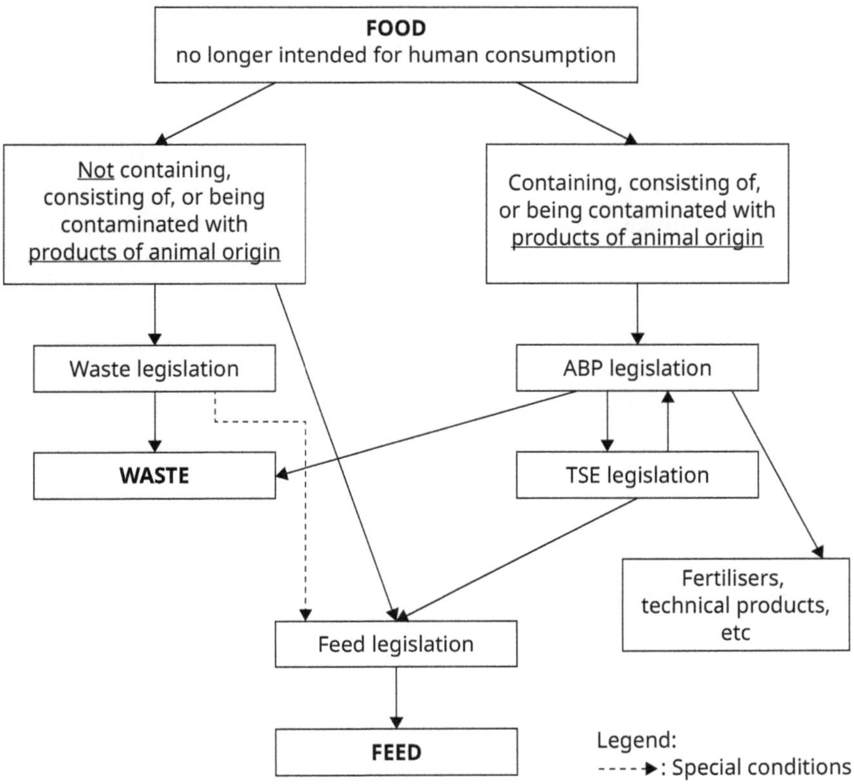

FIGURE 23.6 Flow chart from food to feed (EU Commission, 2018)

practice, food retailers who make former foodstuffs available to former foodstuff processors, typically use a collection point (normally their distribution centre) to re-transport certain unsold food, at which stage the material is categorised as feed. This facility owned by the food retailer is therefore responsible for the compliance with the feed law.

As regards food containing products of animal origin, the Commission notice clarifies that food no longer intended for human consumption and consisting of, containing or being contaminated with products of animal origin may not be directly used in the manufacturing of feed but is always subject first to the provisions of the ABP legislation and operators transporting, handling, storing, processing placing on the market must be registered in accordance with the ABP legislation.

3.3 Use of insect-derived products in feed

The use of insect-derived products was authorised for feeding fish in 2017 and pig and poultry in 2021. The legal pathway for such an authorisation involved dealing with and sometimes adapting several legislations. The EU Commission issued a background paper[23] summarising the different issues at stake shortly before the authorisation of insect Processed Animal Proteins for use in fish. The International Platform for Insects for Food and Feed (IPIFF) provides a more updated insight in the different legal aspects governing the production and use of insect products[24]:

- Insects being farmed animals shall comply with the environmental legislation applying to farming and more specifically the list of insect species that are eligible for farming purposes.
- Insect farmers shall comply with the Animal Health law but are exempted from the animal welfare legislation.
- As farmed animals, their feeding shall comply with feed-related legislation (Feed Hygiene, Feed marketing, etc.) including prohibitions established by the ABP Legislation (ban on intra-species recycling; feeding permitted only with Category 3 animal by-products; ban on use of catering waste) and the 'Feed ban' (ban on ruminant processed animal proteins).
- Insects destined for feed use qualify as animal by-products and shall therefore be handled, stored, processed and used in accordance with the ABP legislation.

3.4 Algae grown on manure

A potential interesting approach to reduce nitrogen leakages while reducing EU feed protein dependency is to recover nitrogen from e.g. urban waste water or manure via algae (Ivanova et al., 2024 – ILVO, 2022). The use of microalgae in Recircu-

23 Strategic safety concept for insects as feed. 2016. https://food.ec.europa.eu/document/download/6479a80a-92f3-42b2-b8a9-1dd63a9baf50_en?filename=animal-feed_marketing_concept-paper_insects_201703.pdf

24 IPIFF - EU Legislation – General Overview. 2025. https://ipiff.org/insects-eu-legislation-general/

lating Aquaculture Systems to recover nutrients surplus is a promising technology (Ende, 2024), although, at this point in time, the recovered biomass is not used as feed, due, beyond some uncertainties as regards its safety, to an unclear legal status.

The regulatory situation is detailed in a legal analysis ordered by the European Sustainable Phosphorous Platform (Love, 2024), which shows the high current regulatory ambiguity concerning biomass grown using waste inputs (secondary nutrients).

- The ABP Regulation (EC) No 1069/2009 allows spreading of raw manure on fields, but does not indicate spreading of processed manure (e.g. after solid-liquid separation, acidification to reduce ammonia losses, digestion, composting). So, crops grown on a field on which raw manure was used as fertiliser are not concerned by waste status, but crops grown on a field on which acidified manure was used may be considered waste.
- Fodder crops grown on a field fertilised using sewage biosolids under a waste management plan are generally considered not to be waste, whereas algae grown in treated sewage in fishing ponds could be considered waste.

Waste status of biomass will generally also apply to any nutrient materials extracted or processed from this biomass.

The uncertain 'waste status' of such biomass may not by regulation directly prevent its use in animal feed, but discourages its uptake by industry, feed wholesalers or farmers. Also, the comments above on 'waste status' may apply similarly for ABP status where biomass production is fed by ABP inputs (raw or processed manure, food waste, food industry streams…) and to interpretation of Regulation (EC) No 767/2009, Annex III, points 1 and 5, effectively prohibiting the use in animal feed of nutrients recovered from sewage or manure 'irrespective of any form of treatment' or any processing.

3.5 Fermentation-derived proteins

The use of microbial biomass in animal feed is not really new. A number of food/drinks involve a fermentation step and a well-known co-product from such fermentation process is the so-called brewers-yeasts.

In a number of applications, the chemical synthesis of chemical substances is increasingly replaced by fermentation, when possible. This is in particular the case for amino acids like lysine, valine, threonine or tryptophan which are nowadays produced exclusively by fermentation. This is also the case of enzymes and some vitamins. The remaining microbial biomass after removal of the product (the targeted chemical substance) is typically rich in highly digestible proteins (ranging from 35 to 65%). Additionally, the cell walls of certain micro-organisms can offer functional benefits, such as supporting gut health (for example yeast cell walls). Nowadays, fermentation biomass is not only a co-product of the production of chemical substances but a product per se, designed for feed use.

Legally speaking, in the 80's, the placing on the feed market of so-called "bioproteins" was subject to an authorisation procedure via Directive 82/471/EEC[25]. This Directive was repealed in 2009 with Regulation (EC) No 767/2009 with no pre-authorisation requirement anymore, unless the micro-organism would be genetically modified, in which case it would require an authorisation in accordance with Regulation (EC) 1829/2003 on genetically modified food and feed. Otherwise, like any other feed material, the responsibility for safe production and use of microbial biomass rests with the operators and if a material is deemed intrinsically unsafe, it is listed in Annex III of Regulation (EC) No 767/2009. This is the case for Protein products obtained from yeasts of the Candida variety cultivated on n-alkanes.

Having said that, a number of living microorganisms (so-called probiotics) exert zootechnical functions as gut flora stabilisers or may be used as hygiene condition enhancers and therefore qualify as feed additives. So a key question for operators and authorities is how to differentiate microbial biomass as a feed material and microbial biomass as feed additive. The key parameters being considered are whether the microorganism is alive or not and what the intention of the producer of the feed material containing the micro-organism is. In the EU Catalogue of feed materials, a specific chapter 12 of part B of the Annex is dedicated to "Products and co-products obtained by fermentation using microorganisms". In this chapter, 3 categories of fermentation products / co-products are identified: i) Fermentation products obtained from whole micro-organisms or their parts; ii) fermentation co-products mainly consisting of microbial biomass and iii) other fermentation co-products (like for example what remains from the fermentation medium after removal of the microbial biomass). A key requirement for all products classified in chapter 12 is that "All microorganisms (including germinable spores) used for fermentation shall be inactivated resulting in absence of viable micro-organisms in the feed materials." With such a requirement, a clear difference can be established between inactivated biomass as feed material and living micro-organism as feed additive.

However, this is not the end of the story. Micro-organisms may also be used to modify the nutritional characteristics of certain products, for example the production of soy protein concentrates or fermented cereals. In this case, the microorganism is usually not separated from the main product and its inactivation may not be suitable for nutritional or sanitary reasons. In such case, the feed material may contain living micro-organism. The problem for the legislator is then to make the difference between a feed material containing an inevitable quantity of living micro-organisms from feed materials used as "carrier" of living micro-organisms that the operator would have added deliberately with the intention to by-pass the feed additives registration. For this scenario, the legislator identified criteria to help the

25 Council Directive 82/471/EEC of 30 June 1982 concerning certain products used in animal nutrition. https://eur-lex.europa.eu/eli/dir/1982/471/oj/eng

differentiation in paragraph 6 of part A of the annex of regulation (EU) No 68/2013 (catalogue of feed materials) which reads as follows:

> (6) Feed materials not listed in Chapter 12 of Part C which have been manufactured by fermentation and/or which have a natural presence of microorganisms may be placed on the market with live microorganisms as long as the intended use of the feed materials and compound feed containing them is
> (a) not the multiplication of the microorganisms; and
> (b) not linked to a function exerted by microorganism(s) according to Annex I of Regulation (EC) No 1831/2003.
> The presence of micro-organisms, as well as any function resulting thereof shall not be claimed on the feed materials and the compound feed containing them.

Back to the case of microbial biomass, another key requirement is that, if the biomass is produced from a genetically modified micro-organism, it must have received an authorisation in accordance with Regulation (EC) No 1829/2003. At this point in time, no genetically modified micro-organism (GMMO), parts thereof or product derived thereof has been authorised in the EU for feed use. Two applications for authorisation of biomass from GMMO were submitted in 2008, with opinions[26] delivered by EFSA in 2017. In its conclusions, EFSA did not identify risks to human and animal health or the environment from this biomass attributed to the genetic modification of the strain. They however were not in a position to conclude on the safety for consumers of products from animals fed with this biomass.

There is clearly a discrimination here with biomass from non-GM microorganisms, not subject to an authorisation procedure, and which is nowadays safely used as feed ingredient. The Chapter 12 of the EU Catalogue of feed materials (Regulation (EU) 68/2013) includes many different products from non-GM microorganisms which are used safely as feed without the backing of an EFSA opinion.

Another interesting case when it comes to single-cell proteins is the production of biomass from animal cells. Since 2023, a feed material named 'Cultivated cells of mammalian origin' has been notified in the feed materials register with the petfood market as main targeted outlet. The question is whether to consider this feed material as fermentation biomass or as a product of animal origin. In the first case, the

26 Safety and nutritional value of a dried killed bacterial biomass from *Escherichia coli* (FERM BP-10942) (PT73 (TM)) as a feed material for pigs, ruminants and salmonids (https://efsa.onlinelibrary.wiley.com/doi/10.2903/j.efsa.2017.4936)
Safety and nutritional value of a dried killed bacterial biomass from *Escherichia coli* (FERM BP-10941) (PL73 (LM)) as a feed material for pigs, ruminants and salmonids (https://efsa.onlinelibrary.wiley.com/doi/full/10.2903/j.efsa.2017.4935)

same rules as for microbial biomass would apply. But if considered as a product of animal origin, then the feed material would fall under the ABP legislation, would have to be processed in accordance with ABP treatment and, as a processed animal protein, would be subject to the 'feed ban provisions' of Regulation (EC) 999/2001.

Although we are not aware of a public clarification by the EU Commission of the status of cultured mammalian cells, it is our understanding that the status of fermented feed material prevails.

In such case, nothing would prevent the use of such cultivated mammalian cells in feed for food producing animals, even cattle since the feed ban provisions would not apply. Having said that, it is unlikely that any producer of feed for food producing animals would take the risk of a non-compliance because of the detection of mammalian DNA present in the mammalian cells.

3.6 Microalgae products

Microalgae are unicellular organism and therefore one may expect that microalgae and products derived thereof are classified as feed materials together with bacterial, yeast and fungal feed products and are bound to comply with the same requirements. However, the process of multiplication of microalgae does not qualify as fermentation *per se*. Therefore, they cannot be classified under chapter 12 of the EU Catalogue of feed materials which includes 'Products and co-products obtained by fermentation using microorganisms.' They are instead classified under chapter 7 'Other plants, algae, funghi and products derived thereof'. Actually, there is no differentiation in this chapter between macro- and microalgae for entries 7.1.1 'Algae', 7.1.2 'Algae, dried', 7.1.3 'Algae meal' or 7.1.4 'Algal oil'. Only for entries 7.1.6 'Seaweed meal' and 7.1.7 'Algae meal from *Asparagopsis*' the description clearly refers to macroalgae.

The other difference between micro-algae and bacteria or yeast is that they do not play a role in digestion and therefore they may not pretend, as live micro-organism, to the status of gut flora stabiliser and therefore there is no ambiguity in terms of legal status of feed material vs. feed additives. Therefore, the requirement from chapter 12 that 'All microorganisms (including germinable spores) used for fermentation shall be inactivated resulting in absence of viable micro-organisms in the feed materials' should not be applicable to microalgae and products derived thereof. However, the legislator required in the description of 'Algae, dried' and 'Algae meal' that the algae are inactivated.

Certain microalgae produce active substances in high concentration (example of *Haematococcus pluvialis* rich in astaxanthin, notified to the register of feed material). Depending on the process undergone and the level of the active substance in the feed product, the classification as feed material vs. feed additive may be questionable.

Beyond the question of the legal status, microalgae and products thereof are bound by the same legal requirements as other feed materials, i.e. registration of the producer as feed business operator and compliance with the legal standards on contaminants.

3.7 Camelina (*Camelina sativa*)

The case of camelina is a lot easier legally speaking: the seeds, the oil and the expeller are listed in the EU Catalogue of feed materials and the only constraints are compliance with standard legislation on GM feed and food if the seed is genetically modified (no case so far) and compliance with the Undesirable Substances Directive 2002/32/EC, in particular as regards the presence of Volatile Mustard Oil and glucosinolates.

4 Assessment of the contribution of the EU legislation to feed circularity and recommendations for improvement

The increasing political priority put on optimisation of resources, minimisation of food waste and protein autonomy makes it legitimate to evaluate the extent to which the current EU legal framework is supportive of feed circularity and if not, what policy options may be considered to improve the situation. The possibility offered to operators to detoxify a feed material exceeding the maximum limits is clearly an asset of the EU feed legislation compared to e.g. the food legislation. However, at this point, the costs associated to detoxification and the lack of real incentives means that only a limited number of such detoxification processes have been evaluated by EFSA and only a few are used in practice (e.g. removal of dioxin from fishoil).

Increasing feed circularity requires understanding the reasons why a by-product or a secondary material[27] with high nutritional value is not used today as animal feed. It must be assessed whether legal restrictions are still legitimate and proportionate, as in some cases these restrictions were decided a long time ago in different circumstances.

In 2025, eight organisations of the circular bioeconomy issued a Catalogue of measures to improve feed circularity (FEFAC et al., 2025). The objective was to map existing legal provisions that restrict the feed use of certain resources or the upgrade / recovery of nutrients from waste and to evaluate the extent to which a revision of the restrictions can be argued and how it could contribute to increased feed circularity whilst meeting the high EU standards for feed and food safety. Beyond the sustainability and circularity considerations (Meijer et al., 2023), a number of other parameters may indeed justify such a re-assessment, starting with new scientific

27 'Secondary raw materials' are recycled materials that can be used in manufacturing processes instead of or alongside virgin raw materials.

evidence regarding the nature of hazard and exposure, technological developments that may contribute to minimise the risk (e.g. feed additive, detoxification process, etc.), new control tools to secure an efficient supervision by authorities (methods of analysis, digitisation) and also evolutions in social perception of the risk. They have defined a number of principles, thereof the following would contribute the most to more feed circularity.

- **Coherence should be ensured between the different legal acts and policies and be based on the Waste Hierarchy and the Food & Waste Use Hierarchy:** preventing waste remains a fundamental objective, however, the concept whereby any material should be regarded as a waste unless it can be proven that it is a resource may discourage operators from taking steps to preserve the resource status. Likewise, the existing orientations under the renewable energy policies encouraging the use of residual biomass for production of biogas must provide safeguards that production practices are not in conflict with the (food) waste hierarchy and thereby avoid encouraging the use for biogas production of materials safely used as feed for many decades already.
- **Coherence should be secured between the General Food Law, waste legislation, ABP regulation and other sectorial legislations (including end-of-waste status, supervision by authorities):** a proper functioning of a circular-based policy should secure coherence between different sectoral policies involved, avoid overlaps and gaps and also establish clear rules to facilitate an appropriate transition from one legal framework to another under well-structured supervision by competent control authorities. A number of feed safety related crises in the past stem from border line between Waste Framework Directive and Feed Hygiene Regulation, with operators breaching the legislation on feed either deliberately or by ignorance, taking also benefits of unclear allocation of responsibilities across control authorities responsible for environmental legislation and for feed/food safety.
- **Existing food/feed and ABP legislation and other relevant legislations should be checked for their contribution to circularity and any impact assessment of new legal initiatives should include circularity as a key evaluation criterion:** the 'better regulation' concept has been in force at EU level for 20 years, involving an impact assessment of any new legislative initiative or a fitness check of existing legislation based on 5 criteria (effectiveness, efficiency, coherence, relevance and EU added value) and aspects such as social or economic impact. Circularity was until now not systematically considered when evaluating EU legislation, despite clear incoherences between policy objectives and sectorial legislation. A typical example is the EU renewable energy legislation, with incentives to use certain 'waste' and 'residues', which in practice can be biomass flows, traditionally used as feed. Another example is the multiple-status issue for substances (e.g. pesticide and/or biocide and/or feed additive etc.) leading sometimes to

non-compliance and disposal of consignments even when the substance at stake is naturally present at a safe level in the feed.

– **Discrepancies among Member States in the enforcement of the EU legislation should be minimised via EU guidance and administrative burden for operators to comply with EU feed legislation should be simplified without compromising feed safety:** although circularity is a concept that favours local sourcing, business models for the valorisation as feed of by-products requires often cross-border transport and experience shows differences in the administrative requirements for transport, including unharmonized format and requirements for the transport of e.g. former foodstuffs between neighbouring countries. This may affect in particular the conditions for transport of material whose destination may change (e.g. former foodstuffs or fish oil).

The Catalogue of feed circularity measures includes also a list of case studies, pointing to existing restrictions, most of them related to access to proteins from animal origin and waste, which may deserve a reassessment.

5 Conclusions

Bringing a new product on the feed market requires a good understanding of the existing feed legislation, adhesion to feed safety culture and facing legal obstacles. The planned EU Circular Economy Act is expected to give a new impulse to the EU economy and decarbonation, while supporting more autonomy / sovereignty. It is essential that this Circular Economy Act analyses carefully the current legal constraints and identifies areas where more circularity could be achieved in particular, in the ABP legislation which was designed at a time of abundant resources and in a TSE context (high risk, lacking information) which has changed considerably today. In other words, it is the right moment for a Feed Circularity Roadmap.

References

Belgian Federal Agency for the Safety of the Food Chain (FASFC), 2021. Circulaire relative aux flux valorisés en alimentation animale : interdiction d'utilisation des déchets, nouveaux statuts juridiques de ces produits, version 1.1. Available at https://favv-afsca.be/fr/news/circulaire-relative-aux-flux-valorises-en-alimentation-animale-interdiction-dutilisation-des-dechets.

Bikker, P., Jansman, A.J.M., 2023. Review: composition and utilisation of feed by monogastric animals in the context of circular food production systems. The international journal of animal biosciences, Volume 17, Supplement 3. https://doi.org/10.1016/j.animal.2023.100892.

Bouxin, A., 2023. Management of Safety in the Feed Chain. In Food Safety Management (Second Edition), Academic Press, p. 19-35, https://doi.org/10.1016/B978-0-12-820013-1.00012-7.

De Laurentiis, V., Casonato, C., Mancini, L., García Herrero, L., Valenzano, A. et al., 2024. Building evidence on food waste prevention interventions. Publications Office of the European Union. https://data.europa.eu/doi/10.2760/684291.

Draghi, M., 2024. The future of European competitiveness: Report by Mario Draghi. Available at https://commission.europa.eu/topics/eu-competitiveness/draghi-report_en.

Ende, S., Henjes, J., Spille, M., Elshobary, M., Hanelt, D., Abomohra, A., 2024. Recent advances in recirculating aquaculture systems and role of microalgae to close system loop. Bioresource Technology, Volume 407. https://doi.org/10.1016/j.biortech.2024.131107.

European Commission, 2021. The revised feed ban. EURL-AP Workshop. Available at https://www.eurl.craw.eu/wp-content/uploads/2021/09/03-DG-Sante_Feed-ban.pdf.

European Commission, 2025. Competitiveness compass for the EU. Available at https://commission.europa.eu/topics/eu-competitiveness/competitiveness-compass_en.

European Feed Manufacturers' Federation (FEFAC), 2019. Resource efficiency champions - Co-products, an essential part of animal nutrition. Available at https://fefac.eu/wp-content/uploads/2020/07/05362_co-products_brochure_003.pdf.

European Feed Manufacturers' Federation (FEFAC), 2022. Circular Feed – Optimised nutrient recovery through Animal Nutrition. Available at https://fefac.eu/wp-content/uploads/2022/07/FEFAC-Circular-Feed-Publication-13-June-2022.pdf.

European Feed Manufacturers' Federation (FEFAC), 2023. Sustainable Animal Feeding Strategies - Advanced animal feeding strategies for more sustainable and circular livestock & aquaculture production. Available at https://fefac.eu/pages/sustainable-animal-feeding-strategies/.

European Feed Manufacturers' Federation (FEFAC) et al., 2025. EU Feed Circularity Catalogue – Existing restrictions to enhanced animal feed circularity. Available at https://europeanpetfood.org/wp-content/uploads/2025/05/Circularity-Catalogue_version1.0.pdf.

Flanders Research Institute for Agriculture, Fisheries and Food (ILVO), 2022. Valorisation of mineral nitrogen from livestock barns as slow-release fertilizer or animal feed. Available at https://pure.ilvo.be/ws/portalfiles/portal/20091568/Final_Report_2022_valorisation_mineral_nitrogen.pdf.

Food and Agriculture Organisation (FAO), 2023. Contribution of terrestrial animal source food to healthy diets for improved nutrition and health outcomes – An evidence and policy overview on the state of knowledge and gaps. Rome, FAO editor, 296 pp. https://doi.org/10.4060/cc3912en.

Food and Agriculture Organisation (FAO). 2024. Guidelines on the role of livestock in circular bioeconomy systems. FAO editor, in-press. https://www.fao.org/partnerships/leap/news-and-events/news/detail/launch-of-public-review--guidelines-on-the-role-of-livestock-in-circular-bioeconomy-systems/en.

Ivanova, J., Gigova, L., Alexandrov, S., 2024. Successful practices for valorization of microalgal biomass for inclusion in circular economy. South African Journal of Botany, Volume 175, p 408-425. https://doi.org/10.1016/j.sajb.2024.10.039.

Love, B., 2024. Legal opinion on the Waste/Animal By-Product (ABP) status of waste-derived algae with particular reference to the Animal Feeds Regulation. Available at https://www.phosphorusplatform.eu/images/Regulatory%20activities/Love%20ESPP%20Legal%20Opinion%20algae%20from%20waste%20ABPs%20v2_11_24.pdf.

Meijer, N. van Raamsdonk, L.W.D., Gerrits, E.W.J. Appel, M.J., 2023. The use of animal by-products in a circular bioeconomy: Time for a TSE road map 3. Heliyon, Volume 9, Issue 3, e14021. https://doi.org/10.1016/j.heliyon.2023.e14021.

Peyraud, J.-L. and MacLeod, M., 2020. Future of EU livestock: how to contribute to a sustainable agricultural sector. Available at https://op.europa.eu/en/publication-detail/-/publication/b10852e8-0c33-11eb-bc07-01aa75ed71a1/language-en.

Van Zanten, H.H.E., Van Ittersum, M. K. & De Boere, I. J. M., 2019. The role of farm animals in a circular food system. Global Food Security, Volume 21, p 18-22. https://doi.org/10.1016/j.gfs.2019.06.003.

CHAPTER 24

Regulatory pathways for animal feed and pet food ingredients in the US and Canada

K. Valm

SGS Nutrasource, 120 Research Lane, Suite 101, Guelph, Ontario, N1G 0B4
Correspondence: kvalm@nutrasource.ca

Abstract

Novel protein and lipid ingredients for use in animal food (including livestock and poultry feed and pet food) in the US and Canada are required to undergo a thorough safety review based on scientific data before entering commerce. In both countries, livestock feeds must be shown to be safe and effective (nutritive) for the animal, as well as safe for the humans handling the material and the humans consuming any products from the livestock. Key safety data need to be developed as part of regulatory submissions in both countries. These data include documentation of raw materials and manufacturing processes, characterization and specifications of finished ingredients, certificates of analysis demonstrating manufacturing consistency, and safety narratives using published and proprietary studies to demonstrate the safety of the ingredient in the target species under specific conditions of use.

Multiple pathways to market exist for novel protein and lipid ingredients in animal food. In the US, these pathways include Food Additive Petition (FAP), Generally Recognized as Safe (GRAS) determination, Animal Food Ingredient Consultation (AFIC), an existing AAFCO Ingredient Definition, AAFCO Ingredient Definition through Safety Review of Ingredients Submission (SRIS), or recognition that the ingredient is a common food. In Canada, livestock are considered to be a defined set of animal species and feeds for livestock must follow approval and registration processes through the Canadian Food Inspection Agency (CFIA). Ingredients for use in pet food and food for all other species that are not considered livestock are required to meet import/export regulations and labelling standards.

Keywords

safety – nutrition – CFIA – FDA-CVM – AAFCO

1 Introduction

Food ingredients for animals (including livestock and poultry feed and pet food) in the US and Canada have several different pathways for regulatory review and/or approval. Key factors, such as the target species and intended effect of the ingredient, are used to determine the most appropriate regulatory pathway. In the US, feed for both companion animals and livestock (animals intended to be part of the human food chain) can use the same regulatory pathways. In Canada, livestock have a distinct set of regulations compared to companion animals. In all cases, it is critical to demonstrate the safety and efficacy (also referred to as utility) of the ingredient using validated scientific evidence. Specifically, for lipid and protein ingredients, the utility is expected to be a nutritional benefit. This chapter will review key data requirements that are expected across multiple regulatory pathways for both countries, followed by a description of the regulatory requirements in the US and Canada.

2 Key data requirements

For both the US and Canada, regulatory submissions for new or modified feed ingredients require key information on the ingredient to substantiate safety and efficacy/utility. As discussed in several other chapters in this book, the discovery and commercialization of novel protein and lipid ingredients is a lengthy process, generating substantial scientific documentation. For regulatory dossiers, it is essential that these data are well organized and of sufficient quantity and quality to support the safe use of the ingredient under specified conditions in the target species. The following summary provides an overview of the information generally required for animal food ingredient safety reviews and should not be considered an exhaustive list of requirements for any specific ingredient.

2.1 Raw materials

Raw materials used in the manufacturing of the animal feed ingredient should be well characterized to permit the assessment of risk factors, and the source of raw materials should be well documented. This data set can include the specific species and strain of source organisms, geographical origin, cultivation practices, post-harvest processing, and other details as necessary on a case-by-case basis. The raw materials must have acceptable specifications for potential contaminants and be suitable for use in foods. Evidence for this can be verified production processes that follow Good Agricultural Practice (GAP) guidelines, analytical reports or full certificates of analyses showing the absence of common or ingredient specific contaminants (e.g. heavy metals, pathogenic microbes, antinutritional factors) or region-specific risks, such as pesticide residues or radiation.

2.2 Manufacturing

The final product will be required to be produced under Good Manufacturing Practice (GMP) meeting the requirements for animal food as outlined in 21 CFR Part 507 (FDA, 2025) or Preventive Control Program (CFIA, 2025)). Critical control points should be identified in both cases. Excipients, processing aids, and food contact materials used in manufacturing must be suitable and compliant for use.

2.2.1 Products of fermentation

A special case to note is products of fermentation. In this case, the information provided should also include details on the production organism, any genetic modifications, demonstration that the production organism does not contain any risk factors such as antimicrobial resistance, and that the production organism strain is maintained in a way to prevent genetic drift and contamination. Additionally, all materials used in the culture and production media should be suitable for food and not contribute unacceptable levels of contaminants to the final product. Evidence such as whole genome sequencing, *in silico* analysis of the genetic coding, and demonstrated absence of the production organism in the final product may be required.

2.3 Specifications

A critical first step in preparation for a regulatory submission is development of product specifications that will be continually achievable based on the established manufacturing process. The specifications should clearly outline the ingredient's identity (chemical and physical characteristics), key constituents (such as nutritional components that may inform any subsequent label guarantees), and contaminant monitoring (at a minimum, heavy metals and microbial analysis but others may be required depending on the material). If the ingredient is a complex mixture such as a botanical extract, it may be prudent to conduct and report on further analyses to understand the full composition. Typically, analysis of a minimum of three non-consecutive lots of the ingredient should be available to demonstrate compliance with the specifications.

2.4 Stability and packaging

As discussed in the Oxidation and Shelf-Life Chapter, the stability of an ingredient must be demonstrated as part of the overall safety profile, since oxidative degradation can compromise both nutritional value and consumer safety. Evidence can include comprehensive stability studies under conditions reflective of the intended use of the final ingredient. Accelerated stability studies (at higher temperature and humidity) may provide long-term storage data in a shorter study time frame. Packaging should be of materials suitable for food contact.

2.5 Intended use and nutritional data

In the US, food and food additive ingredients in animal feed may be required to have a demonstrated utility that includes nutrition, taste, aroma or technical effect (FDA 2025a) and in Canada, efficacy of the feed ingredient is a requirement for approval (Section 7 of the Feed Regulations, 2024, CFIA 2025a). For the purposes of this book chapter, all protein and lipid ingredients are assumed to have a nutritional purpose. Other important information for the intended use includes the animal species and life stage and the amount of an ingredient that will be ingested. As part of a regulatory filing, these parameters must be clearly stated and adequate evidence provided to support safety of the proposed use. Nutritional effects can be demonstrated through *in vitro* digestibility studies, feeding trials, comparability with established protein or lipid sources, or other suitable evidence.

2.6 Safety data

For novel ingredients, specific safety studies may be required if the ingredient does not have sufficient existing data for safe use. These studies should be specific to the intended species and life stage (e.g., adult dogs, beef cattle, pregnant sows, etc.), although some data can be transferred between species, such as the determination of a No Observed Adverse Effect Level (NOAEL) from rodent studies. These data could include feeding trials demonstrating nutritional adequacy, studies to illustrate the metabolism of any non-nutritional components of the ingredient, or general tolerability studies in companion animals. For animals that enter the human food chain, tissue residue analysis for any products from those animals (e.g. eggs, milk, meat) will be required to show safety for the human consumer. In certain cases, an environmental assessment could be required to demonstrate that the ingredient is not detrimental in the event of a spill or through byproducts in the waste materials of the intended species.

3 United States

In the United States, feed for livestock animals and pet food are regulated through similar processes. The key regulatory body is the US Food and Drug Administration's Center for Veterinary Medicine (FDA-CVM), which oversees the submission of Generally Recognized as Safe (GRAS) submissions as well as Food Additive Petitions (FAP).

The Association of American Feed Control Officials (AAFCO) is an industry group composed of State members (typically represented by employees of the State Feed Control office) who are charged by law with enforcing the safety of animal feed ingredients. AAFCO is the central non-regulatory body involved with state-level approval of feed ingredients. AAFCO provides a document called the Official

Publication (OP) that outlines information including, but not limited to, feed ingredient definitions from multiple sources, model feed regulations, model labelling requirements for feed and pet food, and nutritional requirements for dogs and cats that are recognized as one of the global standards for companion animal nutrition.

An additional key industry organization is the National Animal Supplement Council (NASC). The NASC supports their members to bring Animal Health Supplements, Food Supplements, and treats to market. Animal supplements in the US are unapproved animal drugs allowable through enforcement discretion, creating a unique regulatory environment. Animal Food Supplements can be used to provide an additional amount of a nutritional substance (e.g., vitamins, minerals, fatty acids) where the benefit of the ingredient is derived from a nutritional pathway. Novel proteins and lipids could potentially be used as sources of key nutrients, such as peptides and fatty acids, and, therefore, could be considered an Animal Food Supplement. Ingredients in Animal Food Supplements and treats are required to follow AAFCO guidelines and as such should be listed in the AAFCO Official Publication. Detailed discussion of Animal Health Supplements (supplements that provide a non-nutritional benefit) from novel protein and lipids is outside the scope of this chapter.

3.1 Assessment of novelty

There are a number of established protein and lipid sources available for animal diets in the US. Determining if a protein or lipid source is novel, and therefore requiring a new regulatory/safety assessment, considers multiple factors. For both pet food and livestock feed, a proposed ingredient that differs from established ingredients in its source, manufacturing, or specifications is likely to require a regulatory filing. An ingredient that is intended for use in a species that was not previously approved will also require a regulatory filing (FDA, 2020)

Evidence of safety for the target species is required for all ingredients; therefore, it is recommended to compile a safety dossier and keep it updated, even if submission to a regulatory authority is not required.

3.2 Ingredient review and approval pathways

The key regulatory pathways for animal feed ingredients are Food Additive Petition (FAP), Generally Recognized as Safe (GRAS) conclusion, Animal Food Ingredient Consultation (AFIC), AAFCO Ingredient Definition (historical or through Safety Review of Ingredients Submission (SRIS)), or recognition that the ingredient is a common food. Generally, ingredients that have safety concerns may require an FAP and review by FDA-CVM scientists (Table 24.1).

TABLE 24.1 Pathways to market for feed ingredients in the United States

Pathway	Overseeing organization	Outcome	Status
Food Additive Petition	FDA-CVM	Regulation/Law	Regulation 21 CFR 571
Notified GRAS	FDA-CVM	No Questions Letter	Regulation 21 CFR 570.30
Independent GRAS	Company or Expert Panel	Internal Dossier	Regulation 21 CFR 570.30
Animal Food Ingredient Consultation	FDA-CVM	No Questions Letter	Enforcement discretion
AAFCO Ingredient Definition – historical	AAFCO, FDA-CVM	Listed Ingredient Definition	Enforcement discretion
AAFCO Ingredient Definition – SRIS	AAFCO	Listed Ingredient Definition	Enforcement discretion
Common Food Listing	AAFCO	Presence on List	Based on non-novel whole food

3.2.1 Food Additive Petition

The FAP pathway is a legislative process that, when successful, lists the ingredient in a regulation in the Code of Federal Regulations (CFR). The FAP pathway is typically followed for ingredients that may have a higher risk profile or a very specific use case. Examples of protein source ingredients include Fermented ammoniated condensed whey (21 CFR 573.450) and *Komagateaella pastoris* dried yeast (21 CFR 573.587). The petition, containing detailed safety and intended use information (see Key Data Requirements), is filed by an organization (such as a manufacturer or company or their representative) for review by the FDA. Due to the fact that this submission is an approval process and rulemaking that becomes codified in the Code of Federal Regulations, an FAP is typically a lengthy and costly process but may be required for certain ingredients. Requirements for an FAP are outlined in 21 CFR 571 (FDA, 2025b).

3.2.2 Animal food GRAS determinations

A GRAS determination states that qualified experts agree that the ingredient is safe for its intended use. Ingredients following this pathway must have a specific use identified (minimally, the target animal species/life stage, level and duration of use) with supporting evidence of safety in that specific use case. A GRAS determination is specific to the use of the ingredient and the conclusion can be based on either scientific procedures or history of use. The filing has several key requirements, including publication of all pivotal safety data and the submitter's assertion that other qualified experts would agree with their conclusion of safety.

The specific use of a substance in animal food can be determined to be GRAS by the company or organization that intends to market that product. The company or organization may voluntarily notify the FDA-CVM of their GRAS determination. Alternatively, the GRAS determination may be kept on file internally by the company or organization utilizing the ingredient, where it should be updated as necessary and available for review upon request by FDA-CVM (see comment under Regulatory Activities in Process). The GRAS process requires submission of the same standard of data as an FAP; however, the result, if successful, is a "No Questions Letter" from the FDA that is posted publicly in the Animal Food GRAS Notice Inventory (FDA, 2025c). Examples of protein and lipid ingredients include AGRN 26 for Dried *Methylobacterium extorquens* biomass for use as a protein source in finfish feeds and AGRN 30 *Euphausia superba* (krill) meal for use in adult dog foods as a source of protein and lipids.

The FDA does not approve GRAS notifications but will issue a letter stating that they do not have any questions on the company's determination at this time. Information obtained in the future could be used to justify the modification or removal of GRAS determination if the ingredient can no longer be considered safe under the conditions of use.

3.2.3 AFIC

In 2024, a memorandum of understanding (MOU) between the FDA-CVM and AAFCO that outlined their cooperative review of new feed ingredients came to an end after 17 years (FDA, 2025d). As part of the FDA-CVM's actions following this change, the AFIC process was established to allow companies to submit data supporting the safety of ingredients intended for use in animal feed. Ingredients for this pathway are expected to be low risk and intended to provide a significant portion of the complete diet for the target species. If successful, the FDA will issue a letter indicating their intention to allow the ingredient into commerce without enforcement but does not provide a regulatory approval for use. As part of the AFIC review, the FDA-CVM may determine that there are safety concerns and the ingredient will now need to be reviewed via the FAP pathway. The FDA-CVM has started reviewing information through the AFIC process and the intent is to post minimal information about AFICs in a public database (FDA, 2025e). The first submissions have posted, and the first "Consultation complete" letter for Rice Protein Concentrate (AFIC 0001) is publicly available.

3.2.4 AAFCO ingredient definition through SRIS

In April 2025, AAFCO announced the commencement of their new ingredient definition pathway, retitled to the 'Scientific Review of Ingredient Submissions' or SRIS (SRIS Program, 2025). The process documentation indicates this pathway follows the existing safety data requirements for GRAS ingredients (see Key Data Requirements) as well as a proposed non-proprietary ingredient definition. The SRIS program will convene a panel of experts to review submissions with the requirement to conclude safety of the ingredient prior to acceptance as an AAFCO ingredient definition. The SRIS program began accepting submissions as of October 2025 (AAFCO, 2025).

3.2.5 AAFCO ingredient definition (pre-SRIS)

AAFCO as an organization has been operational for more than 110 years (AAFCO, 2025a). The intended goal is harmonization of feed regulations across the states, including definitions for specific ingredient terms. Between 2007 and 2024, the FDA-CVM worked in partnership with AAFCO to review these ingredient submissions to ensure the safety of new ingredients; however, the AAFCO Official Publication (OP) includes ingredient definitions that precedes the FDA/AAFCO MOU and will include new definitions that have not been reviewed by the FDA (see AAFCO Ingredient Definition through SRIS). The FDA has issued a Guidance for Industry (GFI#293, FDA 2024) indicating that barring any evidence of safety concerns, they will continue to exercise enforcement discretion for AAFCO defined ingredients (FDA, 2024a). AAFCO ingredient definitions are incorporated into State laws. Once an ingredient definition is established, the ingredient manufacturer or company bringing the ingredient into commerce should maintain adequate information on file showing the safety of the ingredient in the target species under the conditions of use (see Key Information Section).

It is important to note that all successful submissions under FAP, Animal Food GRAS and AFIC will become part of the approved definitions in the AAFCO OP.

3.2.6 Recognition as a common food

The FDA defines the phrase *common use in food* as '…a substantial history of consumption of a substance by a significant number of animals of the species to which the substance is intended to be fed (and, for food-producing animals fed with such substance, also means a substantial history of consumption by humans consuming human foods derived from those food-producing animals), prior to January 1, 1958.' In 2023, AAFCO introduced a Common Food Index (AAFCO, 2025b), inviting industry to submit ingredients for recognition as a common food.

3.3 Ingredient regulatory tools

In advance of filing a regulatory submission with the FDA (specifically, FAP or GRAS), a pre-submission consultation can be conducted informally or through two formal tools, the *Investigational Food Additive File* (IFA) or a *Food Use Authorization* (FUA) request.

3.3.1 IFA file

Prior to submission of an ingredient safety package to the FDA, consultation on the ingredient can be made through an IFA. Questions including the extent of safety data required, review of experimental protocol, submission of FUA (see below) documents, and other general inquires can be made in writing or via email. This collaborative approach supports the development of a data set sufficient for the submitters' regulatory goals.

3.3.2 FUA request

For any animal that will be fed an unapproved ingredient for research purposes and whose tissues or products may enter the human food chain, an FUA is required before those products are permitted for sale as human food. An FUA submission requires less data than full regulatory submissions, but should indicate the studies to be conducted, the number of animals that are expected to be fed the new ingredient, any anticipated metabolites expected to be present in the animal tissues that will enter the human food chain, an explanation of why this is expected to be safe for human consumption, and further details on the timing and location of slaughter and disposition of the animal tissues. The FDA-CVM will review and provide written authorization for those tissues to be used for human food purposes if they agree with the safety conclusion.

3.4 State level regulations

Although regulation of the safety of feed ingredients is typically done at a Federal level through FAP, GRAS, and others as discussed above, the sale of feed and feed ingredients is regulated at the State level. Discussion of individual State-level regulations is outside the scope of this book chapter. A summary of the regulations can be found on the AAFCO website. With few exceptions, States adopt the AAFCO ingredient definition list into their state regulations (AAFCO State Regulatory Requirements Summary, 2023).

3.5 Regulatory activities in process

As of the publication of this chapter, there are regulatory activities in process in the United States that may impact the regulation of animal feed ingredients. The reader is advised to review for updates on these, and other, relevant activities.

3.5.1 Innovative Feed Enhancement and Economic Development (Innovative FEED) Act

In March 2025, the Innovative FEED Act was reintroduced (Innovative FEED Act, 2025). This act proposes to define a 'zootechnical additive' as a food additive substance that acts on the digestive tract and affects the byproducts of digestion, and/or reduces foodborne pathogens, and/or provides structure/function effects from a non-nutritive pathway. This would allow for expanded health claims for these substances without requiring the ingredient to be filed as a drug. This is a subset of the FAP process and if passed, ingredients submitted to this pathway will require the same type, quantity and quality of information to support safety and utility.

3.5.2 Pet Food Uniform Regulatory Reform (PURR) act

The PURR act proposes changes to the regulatory authority over pet food specifically (US Congress, 2025). It is intended to harmonize the requirements for pet food across the States by creating a single standard at the federal level, as well as putting timelines for review of new ingredients and outlining several proposed allowable claims.

3.5.3 Requirement for notification of independent GRAS conclusions

In March 2025, the Secretary for Health and Human Services provided direction to review legislation for GRAS determinations, specifically the ability of a company to keep its GRAS determination confidential rather than submit a notification to the FDA (HHS, 2025). If successful, it is likely that any legislative change that occurs for human foods would be reflected in animal feeds as well.

3.6 Other Regulations

Labelling regulations for livestock feed and pet food are enforced at the state level, typically following the guidance documents developed by AAFCO. In 2024, a modernized labelling guidance for pet food specifically came into force with expected adoption by all states by 2030 and compliance by manufacturers at that time. Detailed review of labelling regulations is outside the scope of this chapter. When preparing data for a novel protein or lipid, consideration should be given to the 'common or usual name' that will be used on ingredient labels, as well as building a scientific data set that supports any additional claims to be made on the packaging regarding the ingredient, such as specific nutrient guarantees.

4 Canada

In Canada, livestock feeds are subjected to ingredient specific safety-related regulatory requirements. Livestock are specifically defined as cattle, sheep, goats, bison, water buffalo, cervids, llamas, alpacas, swine, poultry, ratites, pigeons, pheasants, partridges, quail, grouse, guinea fowl, pea fowl, horses, rabbits, bees, finfish intended for human consumption as food, and mollusks and crustaceans intended for human consumption as food. Feeds for livestock must follow approval and registration processes through the Canadian Food Inspection Agency (CFIA) and outlined in the Feed Regulations (Canadian Government, 2024), including the Canadian

Feed Ingredients Table (CFIT, CFIA 2025). Ingredients for use in pet food and food for all other species not considered livestock (such as foxes and mink) are required to meet import/export regulations, and finished goods are required to meet labelling standards.

Feed for livestock can be classified as single ingredient feeds (SIF) or mixed feeds and can be combined as part of complete feeds that meet all nutritional requirements for the intended species. Other feed types, such as medicated feed and specialty feed, are also defined in the Feed Regulations (2024) but are outside the scope of this chapter.

4.1 Livestock feeds

Approved livestock SIFs can be found in the CFIT which has two main sections, Part 1 and Part 2. Each part is subdivided into seven classes of ingredients: forage and roughages, energy feeds, protein feeds, recycled food products, nutritional ingredients, non-nutritive ingredients, and other. Ingredients listed in Part 1 of the CFIT are generally lower risk than those in Part 2, and ingredients that exactly meet the requirements in Part 1 are exempt from registration and approvals. Ingredients that exactly meet the requirements of Part 2 require registration but do not require any further approvals. The determination of placement of an ingredient in Part 1 or Part 2 is completed as part of the feed ingredient approval process.

4.1.1 Feed approvals

A feed ingredient is required to be submitted to the CFIA for pre-market approval if it is not listed in the CFIT, or if it is listed in the CFIT but differs based on novel traits, purpose, composition, hazards, nutritional quality, structure, physiological effect, manufacturing process, or intended use (CFIA, 2024). Feeds intended for use in research must also be approved. The premarket notification process involves administrative processes to register with CFIA and submission of a data package demonstrating the safety of the ingredient (see Key Data Requirements).

After review by CFIA and where approval or amendment of an SIF definition is recommended, the definition and supporting information are posted for public review and comment. The CFIA reviews comments from this process before making a final decision to add a new SIF or amend an existing definition.

4.1.2 Feeds requiring registration

Feeds must be registered with CFIA with some exceptions, including, but not limited to, SIFs listed in Part 1 of the CFIT and feeds that are approved through different licensing pathways (e.g. pest control products). Feeds that must be registered include (but are not limited to):

- Imported SIFs not listed in Part 1 of the CFIT
- Imported SIFs that are listed in Part 1 of the CFIT but have additional label claims, guarantees or languages

- SIFs listed in Part 2 of the CFIT
- Imported mixed feeds
- Feeds manufactured in Canada that are not otherwise exempt.

4.1.3 Claims

The use of certain claims for livestock feed has been outlined and includes established wording and requirements for substantiation. For claims that are not already listed, the intended claim should be submitted with the approval or registration dossier (CFIA, 2024a).

4.2 Pet foods and other non-livestock species

Ingredients for use in feed for dogs, cats, and other non-livestock species are required to meet import/export regulations in Canada. The manufacturer or importer is responsible for ensuring that the ingredients used in pet food in Canada is allowable and is not considered a drug under the definition provided by the Food and Drugs Act.

By convention, most manufacturers producing pet food for sale in Canada follow the AAFCO requirements for ingredient definitions, nutrient requirements and labelling.

A key organization is the Pet Food Association of Canada (PFAC, 2025), which provides members with regulatory support and connection to global markets.

4.2.1 Labelling requirements

Labelling is a key requirement for pet food in Canada (Competition Bureau Canada, 2001). The requirements are outlined by the Consumer Packaging and Labelling Act and Regulations, the Competition Act, and the Food and Drugs Act. The requirements apply to a complete finished good, but any ingredient specific information (e.g. microbial species name) will be included in the list of ingredients and if constituents other than protein, fat, fibre, and moisture are desired in the label, this will be required to be included in the guaranteed analysis.

5 Highlighted ingredients

5.1 Algae-based ingredients

The AAFCO OP includes algae-based ingredients that have been submitted as ingredient definitions, are recognized as GRAS in the CFR, or are the subject of a notified Animal GRAS, and are intended for use as nutritive ingredients or color additives (Table 24.2). The full definition for each of these ingredients, including additional conditions or restrictions for use, is available through the AAFCO website. Similarly, the CFIT includes several algae-based ingredients for use in livestock feed. The full definition and conditions of use are available in the CFIT.

TABLE 24.2 Summary of approved algae-based ingredients for use in animal food

Reference	Name	Status	Source
AAFCO			
60.76	Dried Seaweed Meal	AAFCO Ingredient Definition	Products from specific families of macroscopic marine algae in the red algae (*Rhodophyta*); brown algae (*Phaeophyta*) and green algae (*Chlorophyta*) divisions.
60.19	Dried Kelp	AAFCO Ingredient Definition	Products from *Laminariaceae* and *Fucaceae*.
87.130 21 CFR 73.185	*Haematococcus* Algae Meal	exempt color additive	For use as a color additive in feed for salmonid fish.
87.1 21 CFR 73.275	Dried Algae Meal	exempt color additive	Sourced from *Spongiococcum* and intended for use as a color additive in chicken feed.
33.27 21 CFR 573.615	Marine Microalgae	Listed in the CFR as an approved food additive	From *Schizochytrium* sp. and intended for use in dry food for adult dogs (maintenance).
101.1 AGRN 36	Marine Microalgae oil	Animal GRAS pathway	From *Schizochytrium* sp. For use in canned and dry dog food as a source of fatty acids.
100 21 CFR 582.30 21 CFR 582.40	Flavoring extracts	GRAS for their intended use as a flavor	Includes brown algae (kelp), red algae, and dulse.
Canada feed ingredient tables			
1-503-090	Seaweed extract liquid	Nutritional Ingredient - Minerals	From *Alariaceae, Bangiaceae, Fucaceae, Gelidiaceae, Gigartinaceae, Gracilariaceae, Laminariaceae, Monostromataceae, Palmariaceae, Sargassaceae, Solieriaceae*, and *Ulvacea*.
1-503-091	Pelleted whole seaweed (or dehydrated seaweed pellets)		
1-503-092	Dehydrated seaweed extract (or seaweed extract powder)		
1-503-093	Seaweed meal whole dehydrated (or dried seaweed meal)		
1-612-001	Agar	Encapsulating ingredient	From the genus *Gracilaria.*
1-701-002	Irish moss (or *Chondrus crispus*)	Other – Flocculating agent	From the family *Gigartinaceae* (class *Rhodophyceae*, order *Gigartinales*)
2-503-003	Calcified red seaweed dried ground (or coralline red algae dried ground or calcareous red seaweed dried ground)	Nutritional Ingredient – Minerals	From the phylum *Rhodophyta* and the order *Corallinales;* usually *Phymatolithon calcareum, Lithothamnion corallioides,* and *Lithothamnion glaciale.*
2-505-001	Omega-3 long chain polyunsaturated fatty acids micro-algal oil (or long chain omega-3 micro-algal oil or n-3 LC–PUFAs micro-algal oil)	Essential fatty acid - omega-3 long chain polyunsaturated fatty acids (n-3 LC-PUFAs), docosahexaenoic acid (DHA) and eicosapentaenoic acid (EPA).	From non-pathogenic strains of microalgae.

TABLE 24.2 Summary of approved algae-based ingredients for use in animal food *(cont.)*

Reference	Name	Status	Source
2-609-001	Carrageenan (or Kappa-carrageenan)	Viscosity Modifier	From the species *Eucheuma cottonii.*
2-616-008	*Haematococcus* algae meal, comminuted dehydrated (or dried *Haematococcus* algae meal)	Food modifying ingredient – microbial. Source of carotenoid pigments.	From *Haematococcus pluvialis* (Phylum *Chlorophyta*, Class *Chlorophyceae*, Order *Chlamydomonadales*, Family *Haematococcaceae*).
2-616-010	Dried micro-algae fermentation product (or micro-algae fermentation product dehydrated)	Food modifying ingredient – microbial. Source of DHA.	From non-pathogenic strains of micro-algae.

TABLE 24.3 Summary of approved insect-based ingredients for use in animal food

Reference	Name	Status	Source
AAFCO			
60.117	Dried Black Soldier Fly Larvae	AAFCO Ingredient definition; source of protein and fat	Dried larvae of *Hermetia illucens* raised on feed-grade materials
60.119	Dried Crickets	AAFCO Ingredient definition; source of protein and fat	Nymph to adult stage *Acheta domesticus* raised on feed-grade materials
33.29	Black Soldier Fly Larvae Oil	AAFCO Ingredient definition; source of energy	Dried larvae of *Hermetia illucens* raised on feed-grade materials
60.121	Dried mealworm meal	AAFCO Ingredient definition; source of protein	Dried larvae of *Tenebrio molitor* raised on feed-grade material
Canada feed ingredient tables			
2-205-015	Black soldier fly larvae oil mechanical extracted (or black soldier fly larvae oil)	Energy feeds – Source of energy	Whole larvae from *Hermetia illucens*
2-301-003	Dried whole black soldier fly larvae	Protein feeds – Source of protein and energy	Whole larvae from *Hermetia illucens*
2-301-009	Defatted black soldier fly larvae meal (or defatted whole black soldier fly larvae meal or defatted whole *Hermetia illucens* larvae meal or defatted dried whole black soldier fly (*Hermetia illucens*) larvae meal)	Protein feeds – Source of protein	Whole larvae from *Hermetia illucens*

5.2 Insect proteins

Several insect-based ingredients have been reviewed for use in pet and livestock feed, both in the AAFCO OP and in the CFIT (Table 24.3). Typical limitations include that the source of the insect raw material must have been fed feed-grade materials. The current insect species allowable include Black Soldier Fly (*Hermetia illucens*), crickets (*Acheta domesticus*) and mealworms (*Tenebrio molitor*).

5.3 Camelina (*Camelina sativa*)

Camelina oilseed products are increasingly used in aquafeed and pet food and ingredient definitions are included in both AAFCO and the CFIT (Table 24.4). Products sourced from genetically modified camelina would likely require a separate ingredient definition submission with evidence outlining what modifications have been made and presenting evidence for the safety of those modifications.

6 Summary and conclusions

Novel protein and lipid ingredients for use in livestock feed and pet food in the US and Canada are required to undergo a thorough safety review based on scientific data before entering commerce. In both countries, livestock feeds must be shown to be safe and effective (nutritive) for the animal and the humans consuming any products from the livestock. For pet food, AAFCO is a key organization for both the US and Canada. Ingredients used in pet food should contribute to the nutritional profiles for dogs and cats described by AAFCO and follow the naming conventions of the ingredient definitions. Safety in the target species should be part of a regu-

TABLE 24.4 Summary of approved camelina-based ingredients for use in animal food

Reference	Name	Status	Source
AAFCO			
71.300	Camelina meal, extracted	AAFCO ingredient definition	Crushed camelina seed
Canada feed ingredient tables			
1-205-028	Camelina oil mechanical extracted (or camelina oil cold-pressed mechanical extracted or *Camelina sativa* (L.) Crantz oil cold-pressed mechanical extracted)	Animal and plant fats and derived products – dietary energy	Whole seeds of *Camelina sativa* (L.) Crantz
1-302-037	Camelina meal mechanical extracted (or *Camelina sativa* (L.)	Protein products and by-products of plant origin – source of protein	Whole seeds of *Camelina sativa* (L.) Crantz

latory submission or maintained on file by the manufacturer or company bringing the ingredient to market.

Multiple regulatory pathways exist for novel protein and lipid ingredients in animal food. Key safety and efficacy/utility data need to be developed as part of regulatory submissions in both countries. Manufacturers of novel protein and lipid ingredients for use in animal food should consider their chosen regulatory pathway during the research and development phase to ensure that a robust data set is developed in preparation for a regulatory submission.

References

AAFCO (2023) State Regulatory Requirement Summary. https://www.aafco.org/wp-content/uploads/2023/01/State_Regulatory_Requirement_Summary.pdf

AAFCO (2025) SRIS Now Accepting Applications for Animal Food Ingredients. https://www.aafco.org/news/sris-now-accepting-applications-for-animal-food-ingredients/

AAFCO (2025a) What is AAFCO? https://www.aafco.org/about/

AAFCO (2025b). Common Food Index. https://www.aafco.org/about/committees/ingredient-definitions//?v20260211

Canadian Government (2024). SOR/2024-132 Regulation. https://laws-lois.justice.gc.ca/eng/regulations/SOR-2024-132/page-1.html#h-1458793

CFIA (2024). Approval and Registration List. https://inspection.canada.ca/en/animal-health/livestock-feeds/regulatory-modernization/approval-registration-list

CFIA. (2024a) Permissible Claims on Feed Labels. https://inspection.canada.ca/en/tables-permissible-claims-feed-labels

CFIA (2025). Preventive Control Plan (PCP). https://inspection.canada.ca/en/animal-health/livestock-feeds/preventive-control-plans-and-licences/model-systems-feeds-regulations-2024

CFIA (2025a). Feed Regulations, 2024. https://laws-lois.justice.gc.ca/eng/regulations/SOR-2024-132/page-2.html#h-1458974

CFIA (2025b). Canadian Feed Ingredients Table (CFIT). https://inspection.canada.ca/en/animal-health/documents-incorporated-reference/canadian-feed-ingredients-table-cfit

Competition Bureau Canada (2001) Guide to Labelling and Advertising Pet Foods. https://competition-bureau.canada.ca/en/how-we-foster-competition/education-and-outreach/publications/guide-labelling-and-advertising-pet-foods#s1

FDA. (2020). CVM GFI #262: Pre-Submission Consultation Process for Animal Food Additive Petitions or GRAS Notices. https://www.fda.gov/regulatory-information/search-fda-guidance-documents/cvm-gfi-262-pre-submission-consultation-process-animal-food-additive-petitions-or-generally

FDA. (2024) CVM GFI #293: FDA Enforcement Policy for AAFCO-Defined Animal Feed Ingredients. https://www.fda.gov/regulatory-information/search-fda-guidance-documents/cvm-gfi-293-fda-enforcement-policy-aafco-defined-animal-feed-ingredients

FDA (2024a). Letter to Stakeholders: Acknowledgment of Expiring FDA-AAFCO MOU. https://www.fda.gov/animal-veterinary/animal-foods-feeds/fda-letter-stakeholders-acknowledgment-expiring-fda-aafco-mou

FDA (2025). Code of Federal Regulations, Chapter 21, Section 507. https://www.ecfr.gov/current/title-21/chapter-I/subchapter-E/part-507

FDA (2025a). Animal Food Ingredients. https://www.fda.gov/animal-veterinary/animal-foods-feeds/animal-food-ingredients

FDA (2025b). Code of Federal Regulations, Chapter 21, Section 507. https://www.ecfr.gov/current/title-21/chapter-I/subchapter-E/part-571

FDA. (2025c). Current Animal Food GRAS Notices Inventory. https://www.fda.gov/animal-veterinary/generally-recognized-safe-gras-notification-program/current-animal-food-gras-notices-inventory

FDA. (2025d). CVM GFI #294: Animal Food Ingredient Consultation (AFIC). https://www.fda.gov/regulatory-information/search-fda-guidance-documents/cvm-gfi-294-animal-food-ingredient-consultation-afic

FDA. (2025e). Current Animal Food Ingredient Consultation Inventory. https://www.fda.gov/animal-veterinary/animal-food-ingredient-consultation-afic/current-animal-food-ingredient-consultation-inventory

HHS (2025). Revising GRAS Pathway. https://www.hhs.gov/press-room/revising-gras-pathway.html

Innovative FEED Act (2025) Congressman Langworthy. Bill to Support American Farmers. https://langworthy.house.gov/media/press-releases/congressman-langworthy-introduces-bill-support-american-farmers-national

PFAC (2025). Pet Food Association of Canada. https://pfac.com/

SRIS Program (2025). Scientific Review of Ingredient Submissions. Kansas State University. https://sris.k-state.edu/

U.S. Congress (2025). H.R.597 - 119th Congress. https://www.congress.gov/bill/119th-congress/house-bill/597

CHAPTER 25

Navigating the regulatory landscape for novel feed, aquafeed and pet food ingredients in China

X. Li

dsm-firmenich, PO Box 1, 2600 MA Delft, The Netherlands
Correspondence: xun.li@dsm-firmenich.com

Abstract

This chapter provides a structured overview of Chinese regulatory framework governing current and novel ingredients in the feed, aquafeed, and pet food sectors. It explains how authorities classify materials – distinguishing between feed materials (饲料原料), feed additives (饲料添加剂) and novel ingredients (新饲料) – and how these classifications determine approval pathways and data requirements. The chapter outlines registration processes under the Ministry of Agriculture and Rural Affairs (MARA 农业农村部) and the General Administration of Customs (GACC 海关总署), summarising application dossier and testing requirements such as identity, composition, efficacy, animal safety and environmental safety. Particular attention is given to four arising ingredient categories: macro- and microalgae, agricultural by-products, insects and insect-derived products, and fermentation-derived proteins. The analysis highlights areas with established regulatory clarity and those where interpretation remains. Practical insights are offered for companies preparing regulatory submissions and navigating China's market access process. The goal is to equip readers with an informed understanding of regulatory expectations, procedural challenges, and strategies for successful dossier planning and stakeholder engagement.

Keywords

feed – novel feed ingredients – regulatory approval

1 Introduction

China's animal feed sector underwent major regulatory reform during 2018–2019, reflecting national priorities on food safety, innovation, and international harmonization (GAIN, 2024). The institutional restructuring that followed with significant changes in the oversight of feed and feed additives, as well as in the registration and import approval procedures. Following the reformation, the former General Administration of Quality Supervision, Inspection and Quarantine (AQSIQ) was dissolved. Its import and export responsibilities were transferred to the General Administration of Customs (GACC 海关总署), while domestic government functions moved to the State Administration for Market Regulation (SAMR 市场监管总局). As a result, GACC now manages the import and export of feed materials, feed additives, and pet food, maintaining registration of overseas production facilities and publishing official lists of approved manufacturers (so called "white list") eligible for market access (GACC). Meanwhile, Ministry of Agriculture and Rural Affairs (MARA 农业农村部) remains responsible for the domestic production licensing, import registration licensing, cataloguing, and safety/efficacy evaluation of feed and feed additives. In practice, the import of feed and feed additive products into China requires dual approval from MARA and GACC, combining technical registration (product-level) and facility registration (manufacturer-level).

China's feed control system is primarily catalogue-based, following a positive list approach: only substances explicitly included in the Catalogue of Feed Materials (饲料原料目录) and the Catalogue of Feed Additives (饲料添加剂目录) are considered authorized and can be used in feed including pet food (MARA, 2025a; MARA, 2025b). Ingredients that are not listed in these catalogues are treated as novel ingredients (新饲料) (MARA, 2022a). Such ingredients must obtain MARA approval before they can be used, produced, imported, or sold in China. Their approval process involves a full technical review by the National Feed Evaluation Committee (NFEC 全国饲料评审委员会), covering product identity, manufacturing process, compositional analysis, toxicology, efficacy, and environmental safety (State Council, 2011).

2 Authorities and regulations

China's feed regulation operates through a coordinated multi-authority framework. Together, these authorities form an integrated system covering the evaluation, import control, and market supervision of feed materials, feed additives, including pet food ingredients. The following table (Table 25.1) lists the key authorities.

TABLE 25.1 Key national authorities and their roles

Authority	Chinese name	Responsibilities	Role in registration
Ministry of Agriculture and Rural Affairs (MARA)	农业农村部	Oversees registration of feed, maintenance of positive catalogues	Product-based registration, focusing on the scientific substantiation of safety and efficacy.
National Feed Evaluation Committee (NFEC)	全国饲料评审委员会	Evaluates the safety, efficacy, and environmental impact of new ingredients	A technical body under MARA composed of multidisciplinary experts, providing scientific recommendations for MARA's final approval decisions.
General Administration of Customs (GACC)	海关总署	import inspection, quarantine, and management of overseas manufacturer registrations.	GACC enforces compliance at the border and maintains official lists of approved manufacturers
State Administration for Market Regulation (SAMR)	市场监管总局	Oversees market supervision of feed product quality and labelling once products are commercialised domestically.	Not involved.

China's feed and feed additive regulations are based on a set of related laws and administrative measures. The following are the most relevant instruments currently in force:

- Regulation on the Management of Feeds and Feed Additives (动物饲料和饲料添加剂管理条例,) – The foundational regulation governing feed materials, additives, and enterprises engaged in their production, import, and sale (State Council, 2011). It establishes the legal basis for catalogue management, product registration, and inspection.
- Catalogue of Feed Materials (饲料原料目录) – A positive list of authorized feed materials. Ingredients with description (MARA, 2025a)
- Catalogue of Feed Additives (饲料添加剂目录) – A positive list of authorized feed additives, specifying their function, permitted species, (MARA, 2025b)
- MARA Announcement No. 2625 – Sets maximum allowable levels (dosage limits) for each additive based on scientific risk assessments.
- Administrative Measures for the Registration and Management of Imported Feed and Feed Additives (进口饲料和饲料添加剂登记管理办法) – Sets out procedures for registration of imported feed materials and additives, including documentation, testing, and validity periods (MARA, 2014). It outlines that if the imported product or its ingredients are not listed in the positive catalogue, the product must first obtain novel ingredient approval before import registration can proceed.

- Measures for the Administration of Inspection and Quarantine for Imported and Exported Feed and Feed Additives (进出口饲料和饲料添加剂检验检疫监督管理办法) – Defines import inspection and quarantine requirements, including risk classification and supervision protocols (GACC, 2018).
- Measures for the Administration of New Feed and Feed Additives(新饲料和新饲料添加剂管理办法) – Outlines approval procedures for feed materials and additives not listed in the positive catalogues (MARA, 2022a). It defines application data, testing requirements, and NFEC review steps.
- Feed label GB 10648 (饲料标签标准) – Defines the official classification of different type of feed products (GB, 2013). Establishes mandatory labelling and compositional declaration requirements for feed and additives in China.
- Administrative Measures on Pet Food (宠物饲料管理办法) – Extends the regulatory framework to pet food, including hygiene, labelling, and additive conformity (MARA, 2018a).

3 Registration procedures for exporting feed and feed additives to China

Exporting feed materials and feed additives to China requires registration and pre-market approval from MARA and GACC. These dual approvals serve complementary purposes: MARA registration ensures the safety and efficacy of the product itself, while GACC registration confirms that the manufacturing facility meets China's quality and import compliance requirements.

3.1 MARA import product registration

Overseas companies intending to export feed or feed additives to China for the first time must obtain an import product registration license from MARA. This license is product-specific, valid for five years, and is issued after a technical review. MARA publishes approved products through official notices and maintains a searchable online database containing registration details such as product name, manufacturer, license number, and expiry date (MARA).

The first and most critical stage in the registration process is product classification, which determines the applicable dossier requirements and review pathway. According to national feed label GB 10648 (GB, 2013):

- Feed materials (饲料原料) – Substances derived from animals, plants, microorganisms, or minerals that are used in feed processing but are not considered feed additives.
- Feed additives (饲料添加剂) – Small or trace compounds incorporated into feed formulations to improve nutritional value, quality, or animal performance. These are further divided into nutritional and general additives.

– Novel ingredients(新饲料) – Substances are not listed in the Catalogue of Feed Materials nor Catalogue of Feed Additives

If an ingredient is already listed in one of the catalogues, the applicant may apply for an import registration license through the standard procedure. MARA reviews the dossier, may request supplementary information, and issues a registration license once all requirements are met. The typical review period is 6–9 months after dossier submission (MARA, 2014).

If the ingredient is not listed, it is treated as a novel ingredient and must follow the new ingredient registration procedure. In this pathway, the company submits a complete technical dossier – including data on identity, safety, efficacy, and environmental impact – to MARA, which then forwards the dossier to the NFEC for expert scientific assessment. The overall approval process typically takes up to 24 months or longer, depending on data completeness, study quality, and the complexity of the case (MARA, 2014; 2022a).

International applicants must apply through a Chinese subsidiary or authorized local agent, as MARA does not accept direct submissions from overseas companies. MARA's import registration is issued for a specific product and producing manufacturer site – it is not granted to traders or sellers and is not transferable. If the name of manufacturing site changes, a change registration is required. If there are changes in product quality standards, production processes, or scope of application, a new application for registration should be submitted. The required dossier contents are summarized in Table 25.2, and the process flow chart (Figure 25.1) illustrates the main stages of application, evaluation, and approval.

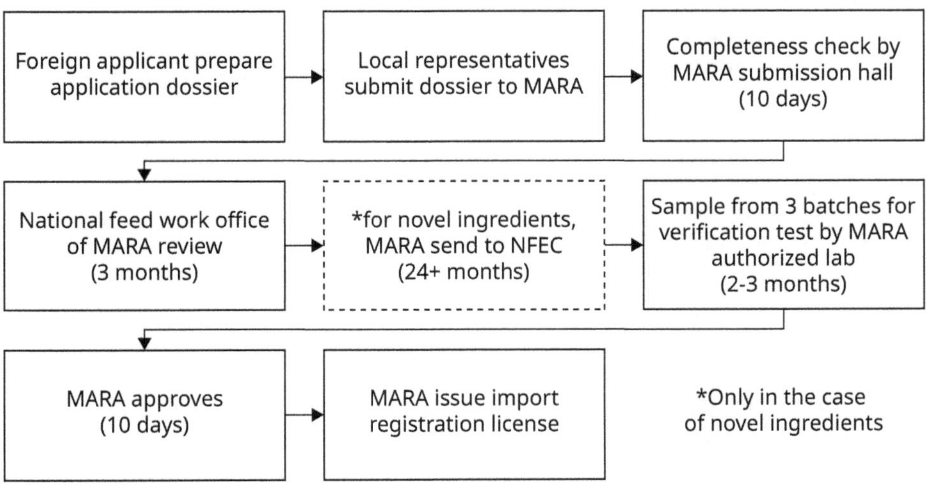

FIGURE 25.1 MARA import product registration process flow chart

TABLE 25.2 Outline of the dossier requirements and review pathway for normal and novel ingredient approval processes

List of documents	Approved feed additive or feed materials	Novel ingredients
Application form	√	√
Power of Attorney	√	√
Free sale certificate, approval documents and feed manufacturing license in origin countries	√	√
Product name, main ingredients, physical and chemical properties, product source, intended use, applicable scope, and usage instructions	√	√
Production process, quality standards, testing methods, and analysis reports.	√	√
Labels, trademarks, and Chinese label templates used at the place of production.	√	√
For microbial products or fermented products, a strain deposit certificate issued by an authoritative institution must also be provided.	√	√
Samples	√	√
A chemical structure identification report of the active ingredient or a taxonomic identification report of the animal, plant, or micro-organism.		√
A product efficacy evaluation test report and safety evaluation test report (including target animal tolerance evaluation report, toxicological safety evaluation report, metabolism, and residue evaluation report, etc.) issued by a testing institution designated by the Ministry of Agriculture. For applications for feed additive import registration, an analytical evaluation report on the potential impact of the feed additive residue in aquaculture products on human health must also be provided.		√
A stability test report and an environmental impact report.		√
If a maximum limit is required for the feed product, the maximum limit value, and the detection method for the active ingredient in the feed product must also be provided.		√

3.2 GACC facility registration

Once a MARA license has been granted, the manufacturing facility must also be registered with the General Administration of Customs (GACC) before export to China. Facility registration is handled via official diplomatic channels through the competent authority of the exporting country or region. Private companies cannot apply to GACC directly. In practice, the manufacturing facility submits its registration request and dossier to the competent authority, which then reviews and forwards the complete application package to GACC for approval (GACC, 2018).

This submission dossier must include:

- Basic company details (name, address, registration number)
- Factory layout and process flow
- Product information (composition, main raw materials, intended use)
- Quality system certifications or equivalent documents and production licenses
- Certificates of free sale

Facilities are expected to be able to demonstrate compliance with both domestic requirements and Chinese quality requirements. Depending on the risk category or product type, GACC may conduct on-site inspections or audits – either remotely or in person – to verify manufacturing conditions, traceability systems, and regulatory conformity. The list of approved overseas facilities is publicly available on the GACC website and is updated regularly (GACC).

The timeline for facility registration is influenced by bilateral coordination and typically takes 6–12 months. Because it depends on government-to-government communication, the schedule is less predictable than MARA registration.

4 Ingredient category status and case studies

china's regulation of novel feed ingredients continues to evolve as new types of proteins, algae, and other alternative sources appear in global feed markets. The main principle remains that only ingredients listed in the official catalogues are allowed to be used. When an ingredient is not listed, it is treated as new ingredient and must go through full registration and evaluation by MARA.

In April 2025, MARA issued an official notice reaffirming this policy direction and highlighting the state's focus on expanding the catalogue to include high-potential new feed sources:

"4. 增补开发潜力较大的新饲料原料。修订发布《饲料原料目录》，明确发酵饲料分类管理要求，增补微生物、微藻、昆虫等蛋白资源产品和食品、粮油、农产品加工副产品等地源性饲料原料品种。适时根据市场需求及时增补纳入《饲料原料目录》"

('Expand the inclusion of new feed materials with strong development potential. Revise and issue the Feed Materials Catalogue, clarify classification requirements for fermentation-based feeds, and add microbial, microalgal, insect-derived protein sources as well as feed materials derived from food, grain, and agricultural by-products. Adjust and supplement the catalogue in response to market demand.') (MARA, 2025c)

This policy confirms that the above-mentioned products are priority areas for catalogue expansion. The following four raising ingredient categories – algae, agricultural by-products, insects, and fermentation-derived proteins – illustrate how this evolving framework operates in practice.

4.1 Macro- and microalgae (藻类)

Algae are increasingly valued in animal feed for their rich nutritional profile (EC). They provide essential fatty acids such as EPA and DHA, along with vitamins, minerals, pigments, and antioxidants that contribute to animal growth and health performance.

In China, algae are regulated primarily as feed materials (饲料原料) under section 7.5 of Catalogue of Feed Materials (MARA, 2025a). However, the number of species officially included remains limited.

Macroalgae are generally accepted as feed materials without species-specific restrictions because of their long history of safe use. Microalgae, in contrast, are evaluated case by case, as their chemical composition and bioactive substances can vary between strains.

The Feed Materials Catalogue has been expanded over time to include additional algal species through MARA Announcement No. 22 and MARA Announcement No. 692 (MARA, 2018b;2023). At present, derived products from the following species are listed: *Schizochytrium* sp., *Spirulina platensis, Chlorella* sp., *Nannochloropsis* sp., *Isochrysis* sp., *Euglena* sp., *Haematococcus pluvialis, Tetraselmis* sp., and *Phaeodactylum* sp.

For microalgae species not listed in the catalogue, applications are treated as novel ingredients. In such cases, MARA and NFEC may require detailed data packages covering strain identification, toxicology, metabolic pathways, and environmental impact. Additional review is also required when chemical or solvent-based processing – such as extraction, purification, or modification – substantially alters the composition or active compounds of the algae, or when the strain is genetically modified.

If the production microorganism is genetically modified, the product is also subject to Regulations on Administration of Agricultural Genetically Modified Organisms Safety (农业转基因生物安全评价管理办法) (MARA, 2022b). This requires a separate biosafety assessment and approval prior to feed registration.

4.2 Agricultural by-products (农副产品)

China makes wide use of agricultural by-products in animal feed, such as distillers' grains, oilseed meals, bran, and molasses. These ingredients make efficient use of by-products from the food industry.

In China's regulatory system, agricultural by-products that are directly generated during food production – for example, oilseed meals from vegetable oil pressing or distillers' dried grains from alcohol production – are generally listed as feed materials in the Catalogue of Feed Materials (MARA, 2025a). Such products can be used directly in feed as long as they meet the corresponding catalogue definitions and specifications. Furthermore, feed ingredients derived from food by-products – such as corn oil, rapeseed oil, and similar materials – must originate from food-certified manufacturers (食品生产许可企业). This requirement ensures that the production facilities comply with food-grade hygiene, safety, and traceability standards before these by-products are permitted to enter the feed chain. However, when a by-product undergoes further processing beyond its original food-production stage, such as enzyme treatment, fermentation, extraction of bioactive components, concentration, or chemical modification, it is no longer covered by the existing catalogue entry. In these cases, MARA may require case-by-case evaluation.

4.3 Insects and insect-derived ingredients (昆虫及其衍生物)

Insects are becoming an important alternative protein source in global feed production, valued for their high protein content and sustainable production (A. van Huis, 2019). In China, this category is still developing, and the regulatory system is gradually adapting to new insect-based materials.

Silkworms (蚕蛹) and bees (蜂) have a long history of safe use, and their derived products – such as silkworm pupae meal and bee pollen – are explicitly listed in the Feed Materials Catalogue (Section 9.2) as authorized feed materials (MARA, 2025a). For other insect species, only simple dried or defatted meals (干燥昆虫粉) may be used under the general catalogue listing, whereas any further processed or derived products must undergo novel ingredient registration before they can be legally used in feed.

Overall, China's approach remains cautious. To date, no new insect species or products have been officially added to the Feed Materials Catalogue since 2012. However, with growing research and industrial interest, it is expected that more insect species and derived ingredients will gradually be evaluated and added to the feed catalogue in the coming years in line with MARA's stated plan to expand the scope of alternative protein sources.

4.4 Fermentation-derived and precision fermentation proteins (发酵蛋白和精准发酵产物)

Fermentation-derived ingredients – such as single-cell proteins (SCPs), microbial biomass, and fermentation by-products – represent one of the most dynamic and fast-growing categories in China's feed sector. Their classification depends mainly on the composition and degree of purification of the final product, rather than on the fermentation process itself.

Under the Catalogue of Feed Materials, Section 12 lists the following fermentation-related materials (MARA, 2025a):

- Meal or residue fermentation products (饼粕、糟渣发酵产品) – products obtained by fermenting agricultural or food by-products.
- Single-cell proteins (单细胞蛋白) – microbial biomass cultivated through fermentation and used as a protein source.
- Fermentation-derived product rich in protein (利用特定微生物和特定培养基培养获得的菌体蛋白类产品, 微生物细胞经休眠或灭活) – microbial biomass derived only from *Rhodobacter* sp. (类球红细菌), *Corynebacterium glutamicum* (谷氨酸棒杆菌), and *Clostridium autoethanogenum* (乙醇梭菌), with the bacteria inactivated (non-viable) before use.
- Fermentation by-products (糟渣类发酵副产物) – fermentation residues that retain nutritional value and can be incorporated directly into feed.

These mixture-type materials are regulated as feed materials when they comply with catalogue definitions and meet safety and quality requirements.

Direct-feed microorganisms (直接饲喂微生物) without fermentation residue, which contain live active microorganisms, are categorized as feed additives and only approved strains can be used in feed (MARA, 2025b).

When a fermentation-derived protein product is not listed in the catalogue – for example, a different microorganism species used for biomass production or other novel microbial source – MARA requires case-by-case registration as a novel ingredient. In such cases, MARA applies the Guidelines for the Identification and Safety Evaluation of Microbial Strains Used in Direct-Feed Microorganisms and Fermentation Products (直接饲喂微生物和发酵制品生产菌株鉴定及其安全性评价指南) (MARA, 2021). This guidance focuses on strain identity, safety (including AMR and toxin screening), and control of residues. It applies equally to live microbial additives and inactivated fermentation products. It is the current benchmark for new feed ingredient registration involving microorganisms.

If the production microorganism is genetically modified, the product is also subject to China's biosafety management regulations for agricultural GMOs (农业转基因生物安全评价管理办法) (MARA, 2022b). This requires separate biosafety assessment and approval prior to feed registration.

5 Differences between novel ingredients in feed, aquafeed, and pet food

Although feed, aquafeed, and pet food are governed under the same overarching legal framework, their evaluation priorities and testing requirements of novel ingredients differ in practice.

For livestock and poultry feed, regulatory procedures are the most mature. Evaluation criteria and technical requirements are clearly defined, including tolerance studies, residue testing, and efficacy trials (MARA, 2011a; 2011b). When a new ingredient is developed for pigs, poultry, or ruminants, applicants can rely on standardized study designs under MARA's established testing frameworks, ensuring a predictable and transparent approval process.

For aquafeed, the same registration pathway applies, but authorities place greater emphasis on aquatic environmental safety and species-specific metabolism. Reviews focus on water stability, leaching potential, and possible residue accumulation in fish or shrimp. When an ingredient has already been approved for livestock feed, those data may support an aquafeed submission if the applicant can justify cross-species relevance – for example, by showing comparable exposure or metabolism (MARA, 2012a). In addition, aquafeed evaluations follow national guidance documents and GB standards, such as GB/T 31812, GB/T 23186, GB/T 23389, GB/T 23388, GB/T 22488, and GB/T 22487, which define testing methods and study requirements for feed materials and additives in aquatic species (MARA, 2012b; GB, 2009a; 2009b; 2009c;2008a; 2008b).

Pet food is managed within the broader feed system but has separate administrative rules for finished products. The Administrative Measures on Pet Feed regulate the production, hygiene, labelling, and import supervision of complete and complementary pet food (MARA, 2018c). However, this regulation does not apply to individual ingredients.

At present, there is no dedicated official guidance for the evaluation or registration of novel pet food ingredients. In practice, novel pet food ingredients are assessed according to the same scientific principles applied to livestock feed; however, there is no specific evaluation criteria or technical guidelines – for example, on tolerance studies, residue testing, or efficacy trials – have been issued for pet food. Consequently, MARA conducts case-by-case reviews, and companies are expected to consult with MARA in advance to agree on the appropriate assessment approach and supporting data requirements.

6 Conclusions and outlook

China's regulatory system for feed, aquafeed, and pet food ingredients follows a catalogue-based approach that ensuring high safety standards with a growing openness to innovation. As China continues to modernize its feed, aquafeed and pet food sectors, authorities are expected to issue additional GB national standards and technical guidelines for feed and feed additives. Collaborative initiatives between regulators, industry, and academia are already emerging to harmonize testing methods and safety criteria. The rapid growth of the pet food market has already prompted the release of several specific regulations and standards, and further dedicated frameworks may be developed in the coming years to address ingredient evaluation and product oversight more comprehensively.

For international companies, the most effective strategy to register a novel ingredient is to:

– Engage early with MARA to confirm classification and data requirements.
– Leverage international safety and efficacy data but be prepared to generate local studies for novel ingredient approval.
– Work with experienced Chinese partners or agents who can manage communication, submission, and follow-up.

Overall, China's regulatory framework for feed, aquafeed and pet food ingredients is evolving toward greater clarity and harmonization with global standards. Companies that invest in robust science, regulatory transparency, and proactive engagement with authorities will be best positioned to succeed in this evolving market.

References

A. van Huis, 2019. Insects as food and feed, a new emerging agricultural sector: a review Journal of Insects as Food and Feed, 2020; 6(1): 27-44

China General Administration of Customs (GACC) Quarantine Registration List of Overseas Animals, Plants and Relevant Products (境外动植物及其产品检疫注册登记名单). Available at: https://www.gacc.app/quarantine-registration-list-of-overseas-animals-and-plants-and-relevant-products

China General Administration of Customs (GACC). 2018. Administrative Measures for the Inspection and Quarantine Supervision of Imported and Exported Feed and Feed Additives (进出口饲料和饲料添加剂检验检疫监督管理办法). Available at: https://www.gov.cn/zhengce/2018-11/23/content_5723442.htm

European Commission (EC). Algae for animal feed Available at: https://maritime-forum.ec.europa.eu/contents/algae-animal-feed_en

Global Agricultural Information Network (GAIN). 2024. China: Overview of Feed and Feed Additives and Pet Food Regulations for Export to the PRC. USDA Foreign Agricultural Service. Available at: https://www.fas.usda.gov/data/china-overview-feed-and-feed-additives-and-pet-food-regulations-export-prc

Ministry of Agriculture and Rural Affairs (MARA). Registration information for imported feed and feed additives (进口饲料和饲料添加剂登记信息查询) Available at: https://feedlicense.nahs.org.cn/publicsearch/dataQuery.jsp

National Standards of the People's Republic of China (GB). 2008a. Principle of aquafeed safety evaluation - Sub-acute toxicity test GB/T 22488 (水产饲料安全性评价 亚急性毒性试验规程 GB/T 22488) Available at: https://std.samr.gov.cn/gb/search/gbDetailed?id=71F772D75BC6D3A7E05397BE0A0AB82A

National Standards of the People's Republic of China (GB). 2008b. Principle of aquafeed safety evaluation – Acute toxicity test GB/T 22487 水产饲料安全性评价 急性毒性试验规程 GB/T

22487 Available at: https://std.samr.gov.cn/gb/search/gbDetailed?id=71F772D7F7B4D3A7E05397BE0A0AB82A

National Standards of the People's Republic of China (GB). 2009a. Principle of aquafeed safety evaluation - Chronic toxicity test GB/T 23186 (水产饲料安全性评价 慢性毒性试验规程 GB/T 23186) Available at: https://std.samr.gov.cn/gb/search/gbDetailed?id=71F772D7C99AD3A7E05397BE0A0AB82A

National Standards of the People's Republic of China (GB). 2009b. Principle of aquafeed safety evaluation - Reproduction test GB/T 23389 (水产饲料安全性评价 繁殖试验规程GB/T 23389) Available at: https://std.samr.gov.cn/gb/search/gbDetailed?id=71F772D7CED7D3A7E05397BE0A0AB82A

National Standards of the People's Republic of China (GB). 2009c. Principle of aquafeed safety evaluation - Residue and cumulation test GB/T 23388 (水产饲料安全性评价 残留和蓄积试验规程 GB/T 23388) Available at: https://std.samr.gov.cn/gb/search/gbDetailed?id=71F772D7CED6D3A7E05397BE0A0AB82A

Ministry of Agriculture and Rural Affairs (MARA). 2011a. Technical guidelines for tolerant studies of feed material and feed additive in livestock and poultry target animals (饲料和饲料添加剂畜禽靶动物 耐受性评价试验指南) Available at: https://view.officeapps.live.com/op/view.aspx?src=https%3A%2F%2Fwww.moa.gov.cn%2Fgovpublic%2FXMYS%2F201107%2FP020110720542595046799.doc&wdOrigin=BROWSELINK

Ministry of Agriculture and Rural Affairs (MARA). 2011b. Technical guidelines for efficacy studies of feed material and feed additive in livestock and poultry target animals GB/Z 31813 (饲料原料和饲料添加剂畜禽靶动物有效性评价试验技术指南 GB/Z 31813) Available at: https://view.officeapps.live.com/op/view.aspx?src=https%3A%2F%2Fwww.moa.gov.cn%2Fgovpublic%2FXMYS%2F201107%2FP020110720542594919927.doc&wdOrigin=BROWSELINK

Ministry of Agriculture and Rural Affairs (MARA). 2012a. Technical guidance for extrapolation of feed and feed additive evaluation data from major to minor species (饲料和饲料添加剂评价数据由主要畜禽物种向次要畜禽物种外推的技术指南) Available at: https://view.officeapps.live.com/op/view.aspx?src=https%3A%2F%2Fwww.moa.gov.cn%2Fgk%2Fzcfg%2Fnybz%2F201201%2FP020120130583473224606.doc&wdOrigin=BROWSELINK

Ministry of Agriculture and Rural Affairs (MARA). 2012b.Technical guidelines for efficacy studies of feed material and feed additive in aquaculture target animals GB/Z 31812 (饲料原料和饲料添加剂水产靶动物有效性评价试验技术指南 GB/Z 31812) Available at: https://view.officeapps.live.com/op/view.aspx?src=https%3A%2F%2Fwww.chinafeed.org.cn%2Fgkcx%2Fslfg%2F202310%2FW020231012583188571710.doc&wdOrigin=BROWSELINK

Ministry of Agriculture and Rural Affairs (MARA). 2014. Measures for the Administration of Registration for Imported Feed and Feed Additives (进口饲料和饲料添加剂登记管理办法) Available at: https://www.gov.cn/gongbao/content/2014/content_2654519.htm

Ministry of Agriculture and Rural Affairs (MARA). 2018a. Administrative Measures on Pet Food (宠物饲料管理办法) Available at: https://www.moa.gov.cn/gk/tzgg_1/gg/201805/t20180504_6141413.htm

Ministry of Agriculture and Rural Affairs (MARA). 2018 b. Announcement of the Ministry of

Agriculture and Rural Affairs No. 22(中华人民共和国农业农村部公告 第22号) Available at: https://www.moa.gov.cn/nybgb/2018/201805/201806/t20180620_6152776.htm

Ministry of Agriculture and Rural Affairs (MARA). 2018 c. Announcement of the Ministry of Agriculture and Rural Affairs No. 20 (中华人民共和国农业农村部公告 第20号) Available at: https://www.moa.gov.cn/nybgb/2018/201805/201806/t20180620_6152699.htm

Ministry of Agriculture and Rural Affairs (MARA). 2021. Guidelines for the Identification and Safety Evaluation of Microbial Strains Used in Direct-Feed Microorganisms and Fermentation Products (直接饲喂微生物和发酵制品生产菌株鉴定及其安全性评价指南) Available at: https://www.moa.gov.cn/nybgb/2021/202111/202112/t20211221_6385218.htm

Ministry of Agriculture and Rural Affairs (MARA). 2022a. Measures for the Administration of New Feeds and New Feed Additives (新饲料和新饲料添加剂管理办法) Available at: https://www.moa.gov.cn/govpublic/CYZCFGS/202201/t20220127_6387839.htm

Ministry of Agriculture and Rural Affairs (MARA). 2022b. Regulations on Administration of Agricultural Genetically Modified Organisms Safety (农业转基因生物安全评价管理办法) Available at: https://www.moa.gov.cn/ztzl/zjyqwgz/zcfg/202206/t20220607_6401864.htm

Ministry of Agriculture and Rural Affairs (MARA). 2023. Announcement of the Ministry of Agriculture and Rural Affairs No. 692 (中华人民共和国农业农村部公告 第692号) Available at: https://www.moa.gov.cn/govpublic/xmsyj/202307/t20230727_6433063.htm

Ministry of Agriculture and Rural Affairs (MARA). 2025a. Catalogue of Feed Materials (饲料原料目录). Available at: https://www.moa.gov.cn/ztzl/sqxzjcgs/xzjcbz/202506/t20250630_6475262.htm

Ministry of Agriculture and Rural Affairs (MARA). 2025b. Catalogue of Feed Additives (饲料添加剂目录). Available at: https://www.nahs.org.cn/xxcm/slpj/202109/P020250717557781046860.pdf

Ministry of Agriculture and Rural Affairs (MARA). 2025c. Implementation Plan for Grain-Saving Actions in Animal Husbandry (养殖业节粮行动实施方案) Available at: https://www.gov.cn/zhengce/zhengceku/202505/content_7022364.htm

National Standards of the People's Republic of China (GB). 2013. Feed label GB 10648 (饲料标签) Available at: https://www.chinafeed.org.cn/slbz/xxbz/202108/P020210820314575587282.pdf

State Council (2011) Regulation on the Management of Feeds and Feed Additives (饲料和饲料添加剂管理条例) Available at: https://www.gov.cn/gongbao/content/2011/content_2004716.htm

PART 6

Sustainability and decision-making strategies

CHAPTER 26

Sustainability perspectives on novel feed ingredients for feed and pet food production

S. Siddique[1] and N. Pelletier[1]*

[1]Food Systems PRISM Lab, University of British Columbia Okanagan, 3247 University Way, Kelowna, BC V1V 1V7, Canada
**Correspondence: shaiyan.siddique@ubc.ca*

Abstract

A major portion of the environmental impacts of today's agriculture can be attributed to livestock production. Feed production contributes a large share of these impacts. Many conventional animal feed ingredients could also be directly consumed by humans, representing a more efficient use of these resources. Feed production with conventional ingredients therefore incurs an opportunity cost. On the other hand, novel feed ingredients such as single-cell proteins, insects, and low opportunity-cost biomass resources such as agricultural residues and food waste, may be suitable for use as sources of protein, lipids, and energy for animal feed production. Some novel feed ingredients may offer resource and environmental benefits due to characteristics such as higher nutrient concentration and rapid proliferation. For low opportunity-cost biomass, conventional disposal methods are sometimes associated with significant environmental impacts, such as the common disposal of food waste in landfills, which exacerbates climate change impacts via methane emissions. Commercial-scale animal feed production using novel ingredients might, in some instances, have the potential to bring about multifaceted socio-economic, resource, and environmental benefits by improving nutrient circularity in food systems, diverting biomass away from landfills, and potentially improving livestock productivity and product quality. Improving process efficiencies and leveraging regional advantages can maximize these potential benefits. Revision of restrictive regulations based on current scientific knowledge, and the improvement of public perception and awareness regarding the benefits of such novel ingredients may be necessary to ensure commercial viability.

Keywords

novel feed ingredients – waste management – life cycle assessment – low opportunity-cost biomass – nutrient circularity

1 Introduction

The global food system faces an onerous task: feeding an increasingly large number of people while at the same time substantially reducing its resource and environmental impacts. Forcing the planetary systems that support food production to generate an increasingly larger output is contributing to their degradation (Poore & Nemecek, 2018). According to the Food and Agriculture Organization (FAO) of the United Nations (UN), approximately half of the planet's habitable land area is already utilized for agriculture (Ritchie & Roser, 2024), along with 70% of the total withdrawn freshwater (Fujs & Kashiwase, 2023). The world population is expected to continue growing for at least the next several decades, potentially peaking at nearly 10.5 billion people towards the end of the 21st century (Lam, 2025). Therefore, in the absence of substantial efficiency gains and/or changes in consumption patterns, these pressures on the global food system can reasonably be expected to increase commensurately over time.

Agricultural and food systems broadly focus on two types of activities, one being the production of plants, and the other the rearing of animals, for human use (FAO, 2021). Animal production is the single largest contributor to the overall environmental impacts of agriculture across many impact categories, accounting for around 80% of the land (Ritchie, 2021) and a third of the water (Gerbens-Leenes et al., 2013) used for all agricultural activities. Animal agriculture is also estimated to account for up to a fifth of the total greenhouse gas emissions worldwide (Blaustein-Rejto, 2023; X. Xu et al., 2021). These concerns are exacerbated by the rising consumption of animal products.

As developing countries achieve economic growth and higher purchasing power, their dietary preferences shift towards more animal products, substantially increasing the global demand for livestock animals (Komarek et al., 2021). While nutritionally balanced plant-based diets and lower-impact meat substitutes have been promoted in recent years due to health and environmental concerns, the trend of increasing meat consumption is likely to persist for some time in the future for several reasons. First, animal origin food products are generally considered to be of superior quality compared to plant-based foods across various subjective and objective aspects, such as amino acid profile, quality and digestibility of protein (Kaur et al., 2022), as well as taste and satisfaction (Vural et al., 2023). Second, meat consumption may have socio-cultural importance in some communities such as the Muslim community (Milford et al., 2019).

Animal production for human consumption is considered inefficient from a resource and environment perspective due to a large portion of the energy and nutrients from animal feed inputs going towards maintaining the animal's normal bodily functions (i.e. homeostasis). In fact, it is estimated that 90% of feed energy is lost with the addition of a trophic level in the food chain (Lindeman, 1942). Unsurprisingly, the environmental and economic footprint associated with feed use constitutes the lion's share of the overall footprint of livestock production (Sitepu et al., 2023; Turner et al., 2022; Vastolo et al., 2024). This is known as an "environmental impact hotspot." Many of the ingredients used to produce feed for animals, such as corn, soy, and wheat, can also be consumed by humans and could otherwise be utilized to meet human needs directly. Utilizing these human edible ingredients for animal feed therefore creates a "food-feed competition" (Palladini et al., 2024), where the opportunity cost of animal feed production is the diversion of these resources away from the potentially more efficient pathway of direct human consumption.

To further complicate the issue, in recent years some conventional animal feed ingredients, such as corn, have seen increased demand for non-food related uses with strong economic incentives such as biofuel production, creating a "food-feed-fuel" competition (Muscat et al., 2020). Therefore, substituting conventional animal feed ingredients with novel, "low opportunity-cost" ingredients may potentially offer both environmental and economic benefits.

2 Novel feed ingredients: a paradigm shift for sustainable feed production?

The definition of a novel feed ingredient is an ingredient that has not traditionally been utilized for producing animal feed in the past (CFIA, 2020). Novel feed ingredients can be categorized into several broad categories such as single-cell protein (e.g. yeasts, bacteria, microalgae), insect meal (e.g. black soldier fly larvae, mealworms), food waste and agri-food industrial waste. Commonly studied examples of novel feed ingredients include *Chlorella* and *Spirulina* microalgae (Kholif et al., 2020; Martins et al., 2021; Panaite et al., 2023), methanotrophic and methylotrophic bacteria such as *Methylococcus capsulatus* (Øverland et al., 2010; B. Xu et al., 2021), filamentous fungi (Sar et al., 2022), Black Soldier Fly Larvae (BSFL) (S. Lu et al., 2022), and processed grocery, food and agri-food industrial waste (Siddique et al., 2024, 2025). These ingredients may potentially be used, after necessary processing, to substitute conventional feed ingredients across different nutritional compartments of the feed. For example, BSFL meal may be used to substitute a portion of the protein component (usually soybean meal) and lipid component (usually animal or vegetable origin lipids) of poultry feed without affecting bird health (Facey et al., 2023; Heuel et al., 2021).

Governments in many regions across the world such as Canada and the European Union (European Commission, 2025; Government of Canada, 2025), as well as livestock industry stakeholders such as Egg Farmers of Canada (Egg Farmers of Canada, 2023) have set net-zero emission sustainability targets for 2050. Novel feed ingredients could potentially play a vital role as an emissions hotspot mitigator to achieve such targets by replacing higher impact conventional feed ingredients. Utilizing novel ingredients to substitute conventional feed ingredients may improve the resource and environmental sustainability of animal feed production in a number of ways. First, it can potentially free up human edible feed ingredients for direct human consumption, significantly improving the use efficiency of ingredients such as corn and soy due to the removal of a trophic level. Second, novel ingredients such as single-cell protein and insect meal could potentially have a smaller environmental footprint due to requiring less land, water, and other inputs compared to conventional feed ingredients (Fasihi et al., 2025; Van Huis & Oonincx, 2017). This is because they can potentially be more efficient from a resource and supply-chain perspective due to their shorter life cycles, rapid proliferation, lack of seasonal variation in productivity, and the entirety of their biomass being potentially usable (Nyyssölä et al., 2022).

Third, novel ingredients may potentially offer higher quantity and quality of nutrients compared to traditional sources. For example, it has been found that BSFL meal can have a protein content of up to 60% (Dalmoro et al., 2025), substantially higher than that of soybean meal, which can contain up to 48% crude protein after refining (Charles Bai et al., 2022). In the case of bacterial single-cell protein sources, a protein concentration of up to 80% on a dry weight basis can potentially be reached (Gundupalli et al., 2024). Studies have also found that the amino acid profile of BSFL meal and some bacterial single-cell protein sources were similar to conventional feed protein sources such as fish meal or soybean meal (Cheng et al., 2023; Gundupalli et al., 2024).

Fourth, the addition of novel feed ingredients could improve animal productivity and quality. Dao et al. (2023) found that the inclusion of processed food waste in layer hen diets can improve feed conversion efficiency. Salami et al. (2019) found that feeding oil and phytochemical-rich plant-based industrial residues to ruminants can improve meat quality and flavour by optimizing the fatty acid profile of the meat. Examples of such oil and phytochemical rich ingredients include apple and tomato pomace, which are rich in tannins, and palm oil mill effluent, which is rich in oil and fatty acids (Salami et al., 2019).

Fifth, studies have found that safe novel feed ingredients (notably single-cell protein and insect meal) could potentially be produced using low opportunity-cost biomass (LOCB) materials such as manure (Siva Raman et al., 2022), wastewater sludge (Filipigh et al., 2024), food waste (Hoek-van den Hil et al., 2023) and agricultural and forestry residues (Shirvani et al., 2023). Since many types of LOCB materials are currently disposed of as waste, their utilization to produce novel feed ingredients can establish a symbiotic relationship between the "grave" (i.e. waste management) and the "cradle" (i.e. feed ingredients production), potentially improving nutrient circularity in food systems. Thus, the use of LOCB to produce feed ingredients may be considered a step forward towards sustainable animal feed production due to its potential for multifaceted resource and environmental benefits.

3 Low opportunity-cost biomass for animal feed: a further step forward in sustainability

Low opportunity-cost biomass (LOCB) refers to those biomaterials for which there currently exists no large-scale, valued added competing use and which are instead currently downcycled or disposed of as waste. Many LOCB ingredients such as food waste, or underutilized agricultural, industrial, and forestry residues may potentially be upcycled into animal feed products (Gatto et al., 2024), to produce industrial chemicals (J. Lee et al., 2022), or to produce energy (Kakkar et al., 2024). This activity is known as "valorization", which is the addition of value to a low value material (such as food waste) and converting it into a high value product (e.g. animal feed). Various LOCB resources can potentially become novel sources for obtaining sustainable, low environmental impact ingredients such as lipids, proteins, and energy, for food and agricultural systems. For example, oil cakes, which are the residues remaining after pressing the oil out of oilseeds such as canola, sunflower, and mustard, are traditionally used as a protein-rich animal feed ingredient in many countries (Ramachandran et al., 2007). Similarly, black soldier flies can be raised on LOCB such as food waste and can potentially become a source of low environmental impact protein and lipids (H. Lu et al., 2023).

When it comes to the management of LOCB, it is important to decide between the available waste management options to minimize environmental impacts and maximize potential utility. A food waste management hierarchy has been developed to guide such decision-making for food waste, which may also potentially be applicable to other types of LOCB resources. The food waste management hierarchy is based on European policies regarding waste management starting from the 1970s and has since been adopted by many countries around the world (Papargyropoulou et al., 2014), including the United States (US EPA, 2025) and Canada (Environment and Climate Change Canada, 2024a).

According to the hierarchy (as shown in Figure 26.1), the best pathway by which to manage the negative impacts of food waste would be to minimize wasting food in the first place. When this is not possible, the best resource and environmental outcomes will be achieved by using the waste to feed animals. Non-food industrial uses such as producing chemicals and energy may yield lower overall benefits, while composting is even further down the list of priorities due to the loss of recoverable energy resources as well as associated environmental impacts such as global warming and eutrophication (Colón et al., 2010; Saer et al., 2013). Landfilling is considered as the poorest pathway for disposing of organic waste, as all potentially recoverable resources are lost.

While producing animal feed is one of the preferred choices from a resource efficiency and circularity perspective, a key aspect to consider in this regard would be the amount and type of energy and other inputs needed to convert the food waste into safe animal feed (Kobayashi et al., 2023), along with their associated emissions, as these would determine whether such pathways actually provide the best outcome. An efficient and well-designed valorization pathway will ideally maximize both resource/environmental and economic benefits.

Valorizing LOCB resources to produce novel animal feed ingredients may potentially reduce the environmental impacts of livestock production and disposal of LOCB, as well as potentially reducing the economic costs of animal feed and livestock production (Meza Ocampos et al., 2025; Nath et al., 2023). A large portion of LOCB such as food waste is currently disposed of in landfills, where it generates the potent greenhouse gas (GHG) methane (Siddique et al., 2024). Nearly 15% of

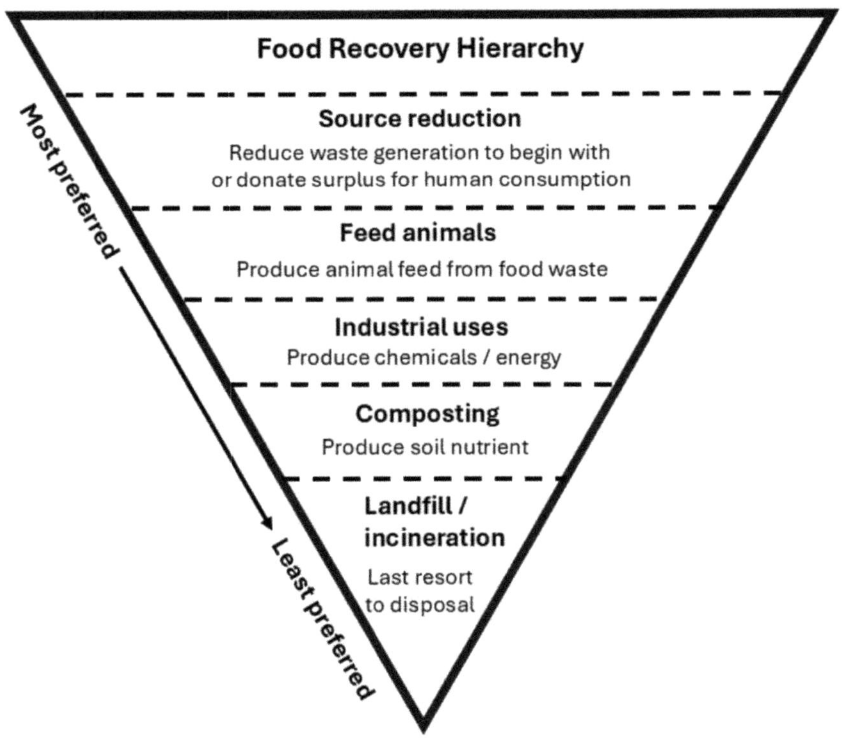

FIGURE 26.1 The food recovery hierarchy. Adapted from Papargyropoulou et al. 2014; and US EPA, 2021.

the total methane emissions in the United States in 2022 could be attributed to landfills (US EPA, 2024), while the share was higher at 17% in Canada (Environment and Climate Change Canada, 2024b). Globally, landfilling remains the most popular method for disposing of waste, and their use is predicted to increase as developing countries prosper economically and transition to landfilling from open disposal (Lou & Nair, 2009). Therefore, mitigating landfill GHG emissions is a priority for combatting climate change, and diverting LOCB away from landfills and utilizing them to create high value products such as animal feed might contribute substantially towards that end.

Feed accounts for around 70% of the economic costs of livestock production (Begna & Masho, 2024; Heidari et al., 2021). Protein-rich ingredients are considered to be the most expensive components of animal and aquaculture feed (Sitepu et al., 2023; Villaró-Cos et al., 2024). If novel feed ingredients can be produced at an affordable cost and can be used to substitute conventional protein feed ingredients, then feed costs could be reduced. For example, when considering substrates such as food waste, economic benefits across the stakeholder chain could be substantial, since large food waste generators such as grocery stores usually pay a tipping fee for transporting and disposing of their food waste in landfills. In contrast, grocery stores could donate grocery waste if a commercial-scale food waste valorization company collects it from them, saving the grocery stores tipping fees and potentially reducing the cost of valorization for the company in a symbiotic industrial relationship.

Table 26.1 summarizes some sources of LOCB suitable for potential use as animal and aquaculture feed, and the characteristics of the technologies associated with their valorization, as described in literature. As shown in Table 26.1, there are two major types of pathways for valorizing LOCB to feed: direct and indirect (via an intermediary). The direct pathway is where the substrate itself is consumed as the feed, either as it is, such as in the case of cactus cladodes (El Otmani et al., 2019; Mayer & Cushman, 2019) and fruits and vegetable peels (Onwuka et al., 1997), or after necessary processing, such as drying (Maia et al., 2023), grinding (Andersone et al., 2023), ensiling (Dou et al., 2022), and fermentation (Parchami et al., 2025). Indirect pathways involve feeding the LOCB substrate to an intermediary species, such as black soldier flies (Hoek-van den Hil et al., 2023; Veldkamp et al., 2021), microalgae (Villaró-Cos et al., 2024), yeast cells (Shirvani et al., 2023; Sitepu et al., 2023), or bacteria (Mahan et al., 2018), and then using the biomass of the intermediary species as feed for livestock or aquaculture.

While the addition of an intermediary species can result in substantial loss of efficiency with regards to energy and nutrient transfer, such a pathway may potentially improve the quality of or open up additional LOCB sources which might otherwise not be valorizable at all. For example, substrates such as urban wastewater, pig slurry and other manure might not have a viable direct valorization pathway at

TABLE 26.1 Examples of low opportunity-cost biomass sources and their valorization pathways

Biomass source	Valorization technology	Valorization type	Target species	Case study scale	References
Almond hulls	Yeast cell production	Indirect	Ruminants	Laboratory-scale	Sitepu et al., 2023
Kitchen/fast food waste	Black Soldier fly larvae production	Indirect	Monogastrics	Laboratory-scale	Hoek-van den Hil et al., 2023, Veldkamp et al., 2021
Urban wastewater, pig slurry	Microalgae production	Indirect	Monogastrics, ruminants, aquaculture	Pilot-scale	Villaró-Cos et al., 2024
Apple pomace, potato protein liquor	Acidogenic fermentation	Direct	Ruminants	Laboratory-scale	Parchami et al., 2025
Lupin straw and pod shells	Drying in oven	Direct	Ruminants	Laboratory-scale	Maia et al., 2023
Sea buckthorn twigs	Grinding using knife mill	Direct	Ruminants	Laboratory-scale	Andersone et al., 2023
Softwood pellets, tree bark, chicken manure	Steam gasification, Yeast cell production	Indirect	Monogastrics, ruminants, aquaculture	Laboratory-scale	Shirvani et al., 2023
Fruit and vegetable waste, corn cobs and stalks, mushroom stumps and spent compost, wet brewer's grains	Ensiling	Direct	Ruminants	Laboratory-scale	Dou et al., 2022
Chicken manure biogas slurry	Yeast cell production	Indirect	Monogastrics, ruminants, aquaculture	Laboratory-scale	Zhou et al., 2022
Cactus cladodes, prickly pear cactus, Halophytes	Direct feeding	Direct	Ruminants	Laboratory-scale	Mayer & Cushman, 2019, El Otmani et al., 2019, Nizar et al., 2016, El Shaer, 2010
Orange and lemon waste, corn stover effluent	Bacteria production	Indirect	Monogastrics, ruminants, aquaculture	Laboratory-scale	Mahan et al., 2018
Fruit and vegetable peel	Direct feeding	Direct	Ruminants	Not applicable	Onwuka et al., 1997

all, mandating the use of intermediaries such as microalgae to safely valorize them as feed (Villaró-Cos et al., 2024). Even for LOCB substrates that may directly be valorized, passing them through an intermediary species might, in some cases, turn out to be beneficial overall despite the efficiency loss associated with the addition of a trophic level, since this process may improve the nutritional properties and digestibility of the feed, as well as enhance species suitability for some substrates. For example, almond hulls can be used directly as a comparatively lower quality feed for ruminants only, however, using it as a substrate for growing protein rich yeast cells can produce a higher quality nutrition source for ruminants as well as non-ruminants (Sitepu et al., 2023). Similarly, apple pomace can be used directly as animal feed, but low protein concentration of fresh pomace (Yang et al., 2022) and high alcohol content of apple pomace silage (Alibes et al., 1984) have been identified as some of the key challenges that limit their utilization. However, controlled microbial treatment of apple pomace such as through solid state fermentation was found to have significantly improved crude protein concentrations (between 138 to 300%) along with other nutrients (Joshi & Sandhu, 1996; Yang et al., 2022), and the resultant feed was found to have improved flavour and taste (Yang et al., 2022).

4 Key contemporary challenges and concerns

While studies of novel feed ingredients and LOCB valorization might indicate significant potential for reducing the resource and environmental burdens of animal feed production, just like any new ideas at the beginning of their journey, these novel concepts are often not without formidable challenges and concerns. One of the key challenges associated with evaluating the impacts and benefits of such systems is their context dependent nature. Determining whether novel feed ingredients and/or LOCB valorization would result in net resource, environmental, and economic benefits for any given feed production system requires a detailed assessment of a number of aspects. These aspects include, but are not limited to, the type of outputs produced by the novel systems, their impacts compared to the conventional feed resources that they might substitute, how much and which of the conventional ingredients can be replaced, local geographic and infrastructural contexts, economic and supply chain aspects such as production and distribution costs, substrate availability, as well as marketing and public acceptance of the product (Carter & Codabaccus, 2022).

The requirement to consider such a multitude of interdisciplinary aspects can increase the complexity of conducting accurate assessments, and their contextualized nature can make it difficult to compare the findings between different studies. Due to the novel nature of these pathways, many of the studies available in the literature are laboratory-scale, as highlighted in Table 26.1. There might be concerns as to whether the benefits indicated by the laboratory-scale studies would hold when scaled up to commercial levels. Some studies have found that environmental benefits generally tend to increase for many types of novel ingredients, including single-cell protein (Martínez-Ibáñez et al., 2025), microalgae (Taelman et al., 2013), and BSF larvae meal (Domínguez Aldama, 2023) when scaled up from laboratory or pilot-scale to commercial-scale production due to increased efficiencies and yields (Martínez-Ibáñez et al., 2025).

The various contextual aspects of such systems need to be considered on a case-by-case basis when conducting assessments. For instance, different conventional feed ingredients will have different environmental footprints. Animal origin products (such as organ meats and slaughterhouse co-products) generally contribute more environmental impacts compared to ingredients such as corn and soy (Pelletier, 2017; Turner et al., 2022). Novel feed ingredients that substitute animal origin products might therefore potentially offer more sustainability benefits. How much of the novel ingredients can be added to the feed without negative side effects can also influence the net impacts and benefits (Siddique et al., 2025). There may also be concerns regarding taste, texture, and palatability of the feed (Pilmer et al., 2022). Studies into making novel feed ingredients more palatable for animals are already being conducted (Pilmer et al., 2022; Yang et al., 2022).

The net benefits of the novel ingredients might also be dependent on local contexts. Siddique et al. (2025) found through life cycle assessment that the environmental footprint of the local electricity grid would substantially influence the net environmental impacts and benefits of valorizing grocery waste for poultry feed in Canada. Canadian provinces and territories with an already established green electricity infrastructure, such as British Columbia and Quebec, were thus suggested as potential locations for setting up such an enterprise (Siddique et al., 2025). This is one example of how underlying local infrastructure contexts may influence the viability of novel feed ingredients and LOCB valorization initiatives from a resource and environmental perspective. Another important example of local geographic contexts could be the availability and dispersion of substrates. Sebastian et al. (2025) indicated that, despite forestry residues being generated in non-trivial quantities across Canada, their generation was most concentrated in the province of British Columbia. Setting up a forestry residue valorization facility in that province would reduce transportation related impacts and might mitigate some supply chain issues.

Economic viability is another major concern when deciding whether to invest in scaling up the production of novel feed ingredients (Carter & Codabaccus, 2022). Commercial-scale rollout of novel ingredient production may involve large start-up investment, infrastructure, and logistics requirements (Buccaro et al., 2023). For example, it has been reported that the equipment and infrastructure-related expenses associated with setting up an industrial-scale insect and algae based novel poultry feed ingredient manufacturing facility in Italy was more than 18 million Euros in 2023 (Buccaro et al., 2023). Similarly, it has been estimated that the set up of a raceway bioreactor for growing microalgae can cost hundreds of thousands of US dollars, reducing the economic viability of their widespread rollout in developing countries (Musa et al., 2025). Hence, collaborating with existing feed manufacturers to innovate their product lines with novel ingredients might reduce start-up investment requirements and improve chances for commercial success (Carter & Codabaccus, 2022).

Aside from capital expenditures, if novel feed ingredient production involves complicated resource and energy intensive technical processing, it may increase both the environmental impact and the economic cost of the product. Therefore, when conducting assessments the full contextual background needs to be considered on a case-by-case basis to determine whether there would remain a net environmental and economic benefit for the novel and/or valorized product after accounting for production and processing impacts and expenses. Nevertheless, when discussing the economic and market viability of novel feed ingredients, it is also important to consider the current state of sustainability of the market and how it influences the prices of conventional feed ingredients.

For example, it has been reported that BSF larvae meal might struggle to be economically competitive against conventional ingredients such as soybean meal and fishmeal under existing market conditions (Selvaraj & Won, 2024). However, it is important to note that existing market prices for the conventional ingredients may not reflect their true cost since they do not account for the substantial environmental and social costs associated with their production – also known as "externalities" (Reyes-García et al., 2025). There may be substantial externalities associated with the production of soybean and fish meals such as land use and transformation (Pexas et al., 2023) and depletion of fish stocks that may not be appropriately reflected in their market price (Reyes-García et al., 2025). In fact, it has been estimated that the global economies incur up to approximately US$ 23 trillion in environmental externalities alone each year (Reyes-García et al., 2025). These amounts do not get accounted for in the price of products in the economy, which contributes towards keeping prices artificially low and encourages unsustainable levels of consumption and sub-optimal consumer choice. Therefore, along with improving technical and process efficiencies to reduce production cost, it is also important to advocate for the internalization of externalities to allow for fair competition between conventional and novel feed ingredients.

Another crucial challenge for novel feed ingredients and LOCB valorization is to address pathogen and food safety concerns. The use of novel feed ingredients are severely restricted by legislation in many jurisdictions such as Canada and the European Union (Renna et al., 2023; Siddique et al., 2024). This is due to concerns such as prion diseases (e.g. mad-cow disease, scrapie) (Renna et al., 2023), viral and bacterial contamination (e.g. foot and mouth disease) (Iriarte et al., 2023), accumulation of heavy metals (Schrögel & Wätjen, 2019), pesticides (Siddiqui et al., 2024), and toxins (Ritala et al., 2017; Schrögel & Wätjen, 2019). Considering the potential resource and environmental benefits of novel ingredients, many of these concerns have been studied in detail to contribute towards developing strategies for their mitigation.

The concerns surrounding prion diseases and foot and mouth disease have garnered considerable research attention and scrutiny over the last few decades due to the severity of outbreaks across the United Kingdom and Europe in the 1990s and 2000s (Alarcon et al., 2023; Shanafelt et al., 2018). The mad-cow disease (Bovine Spongiform Encephalopathy or BSE) outbreak has been linked to the feeding of animal-origin feed products (particularly from mammals) to ruminants (European Commission, 2001). To mitigate the risk of spreading prion diseases via novel feed ingredients, several strategies have been described over the years. Novel feed ingredients and valorized LOCB may be processed as feed for animal species that are considered to be unaffected by prion diseases, such as poultry (Moore et al., 2011), dogs (Vidal et al., 2020), and horses (Sola et al., 2023). For waste valorization pathways such as grocery waste to feed valorization, materials of non-animal origin (e.g. fruits and vegetable, bread) may be valorized separately (Siddique et al., 2024). With regards to mitigating viral and bacterial contamination, heat treatment (e.g. cooking swill before feeding to swine) has been identified as an effective strategy to minimize risk from microbes (Dame-Korevaar et al., 2021), although this creates additional resource/environmental burdens.

Concerns surrounding the accumulation of heavy metals, pesticides, toxins, and other chemical contaminants in novel feed ingredients have also been studied. It has been found that the accumulation of heavy metals such as lead, cadmium, and mercury, insecticides, and drug residues in insects raised for feed production was primarily caused by raising insects on agricultural residues (Malematja et al., 2023). Strategies to mitigate this problem are still evolving, with strict monitoring of insect farms and their outputs, utilizing substrates with a lower risk of contamination, and research into insect species that accumulate lower levels of contaminants being some of the recommendations (Malematja et al., 2023). Aside from insects, the accumulation of contaminants was found to vary based on the type of contaminant and the species of microalgae (Mahlangu et al., 2024) and yeast (Ritala et al., 2017). As such, there might be an option to design the combination of substrates and novel ingredient species to minimize the accumulation of contaminants (Schrögel & Wätjen, 2019).

Another strategy for reducing the impact of heavy metal accumulation in novel feed ingredients may be to incorporate chelating agents in the feed, which can reduce the absorption of the heavy metals in the gut of the animals (Papaioannou et al., 2005). For BSF larvae meal production, a period of starvation for the larvae before being harvested for feed production may reduce the concentration of toxins, since the gut content of the larvae was found to contribute a significant portion of the contamination burden (Schrögel & Wätjen, 2019).

Last but not least, novel feed ingredient manufacturers must also address concerns surrounding public perception and acceptance of these ingredients in mainstream livestock and pet food production (Scheibenzuber et al., 2025). It has been found that communities with higher trust in science and innovation may have higher acceptance of novel ingredients (Beghin & Gustafson, 2021). Effective communication of potential sustainability and other benefits was also found to help improve public acceptance of novel ingredients (Laureati et al., 2024). Effective and transparent advertisement, public awareness strategies tailored to the target demography, transparent supply chain management, and certification by international organizations such as the International Organization for Standardization (ISO) might raise trust and acceptance of these ingredients.

5 Methods of assessment of sustainability benefits and tradeoffs

In recent years, with the development of various strategies to reduce the impacts of agriculture, it has become increasingly important to be able to measure the potential net benefits of such interventions for informed policymaking and decision support. To facilitate this, approaches such as Life Cycle Assessment (LCA) and Material Flow Analysis (MFA) have been developed under the umbrella of the field of "Industrial Ecology" (Graedel & Lifset, 2016). Industrial ecology takes inspiration from the field of traditional / natural ecology and aims to optimize the resource and environmental efficiency of industrial systems through biomimicry and symbiotic relationships. In recent years, social and economic aspects have also been increasingly incorporated into such analyses (Toniolo et al., 2020).

LCA is a multicriteria, systems-based approach that breaks down and evaluates resource inputs, emissions, and impacts associated with the sequence of activities involved in the production of any product or service into life cycle stages, starting from the collection of raw materials (also known as the "cradle") and ending with the final disposal of the product (also known as the "grave"). This concept is based on Life Cycle Thinking (LCT), of which LCA is a part. The methodology for LCA is internationally standardized by the ISO 14044 standard (ISO, 2006). Taking a holistic "cradle-to-grave" perspective, along with its multicriteria nature, means that information derived from LCA studies can help minimize the possibility of burden-shifting.

Burden-shifting occurs when any intervention appears to reduce the environmental impacts of one life cycle stage or one type of impact but increases it elsewhere. For example, a sustainability initiative to switch from single use plastic drinking cups to paper alternatives may result in burden-shifting between life cycle stages. This is because a large share of the environmental impacts of paper cups may be found at the cradle or raw material collection stage (i.e. deforestation), while for the plastic cups it may be found at the grave / disposal stage (i.e. plastic waste management). Similarly, an intervention that reduces climate change burdens may increase other types of environmental burden such as eutrophication, water scarcity, or ozone layer depletion. Due to the possibility of minimizing burden shifting and conducting nuanced, data-driven assessments, LCA has become one of the most common approaches to evaluate the environmental impacts and benefits of agri-food sustainability initiatives (Notarnicola et al., 2017).

While LCA is widely considered as a useful tool for sustainability assessment, it is not free of challenges and limitations. As such, since its inception, LCA has undergone methodological improvements and revisions to accommodate more complex systems (Sohn et al., 2020). The requirement to collect a large quantity of reliable and granular data to construct life cycle inventories is often highlighted as a key challenge (Lackner et al., 2023). Purchasing specialized software and obtaining access to secondary data from commercial databases may also be very expensive (Mba Wright et al., 2024). The results obtained from LCA studies may also vary substantially depending on practitioner choices, including system boundary definition, co-product allocation method, the selection of impact categories, and weighing and normalization methods (Mendoza Beltran et al., 2016; Ylmén et al., 2020). Such variations often make it difficult to compare the results between different studies. Conducting accurate uncertainty assessment is another challenge that may arise out of shortcomings in existing methodology (Bamber et al., 2020). The lack of consideration of spatial, temporal, and technological dynamism is also highlighted as a limitation of LCA (ISO, 2006), although some of these aspects are actively being developed at this time under the umbrella of "Dynamic LCA" (Levasseur et al., 2010; Pehnt, 2006).

In addition to the measurement of environmental impacts through LCA, the LCT concept has also been applied to evaluate the socio-economic impacts of products, services, and interventions using Social LCA (SLCA) and Life Cycle Costing (LCC) (Toniolo et al., 2020). While still in their infancy in terms of methodological development, SLCA and LCC are often used in conjunction with LCA to conduct a holistic assessment covering environmental, economic, and social impacts (Toniolo et al., 2020).

6 Case studies assessing novel ingredients using LCA

In recent years, the production of novel feed ingredients and the valorization of low opportunity-cost biomass have been studied more frequently through the lens of

LCA, and case studies of their environmental impacts and benefits have been published in the literature. Although many of these case studies are laboratory-scale and their results might differ from impacts seen during commercial-scale production, they can still provide important indications regarding the potential impacts and benefits of such pathways.

Comparative LCA studies investigating the production of various novel feed ingredients have reported that certain types of ingredients might be more environmentally burdensome to produce compared to others. For example, microalgae and yeast-based single-cell protein ingredients were considered to have more environmental burden across many impact categories including global warming, water consumption, acidification, and eutrophication compared to methanotrophic bacterial and insect-based pathways (Kobayashi et al., 2023). These higher impacts were found to be driven by energy consumption and the drying process, which were identified as major hotspots (Taelman et al., 2013). For BSFL systems, along with energy use, a major hotspot was the conventional diet used to grow the larvae (Beyers et al., 2023). Strategies such as improvements in the production process, utilizing green energy sources, and leveraging regional advantages can be important steps to minimize burden and maximize sustainability benefits (Francis et al., 2025; Koukoumaki et al., 2024).

Several LCA case studies of direct LOCB valorization without intermediaries have also been published, particularly for food waste. Ogino et al. (2007) conducted a case study on the climate change impacts and benefits of five commercial-scale mixed food waste to animal feed manufacturing facilities in Japan. This study found that producing wet animal feed from food waste can potentially result in substantial environmental benefits compared to other disposal options such as incineration or landfilling. S.-H. Lee et al. (2007) conducted a multi-criteria LCA study on a similar facility in South Korea and found that producing animal feed had a better outcome for the climate change impact category compared to landfilling, incineration, and composting. However, in their case study, feed production increased impacts in some other impact categories such as acidification, ecotoxicity, and eutrophication, indicating a potential for burden-shifting.

A later study conducted in South Korea by Kim & Kim (2010) on a commercial-scale food waste to pig feed manufacturing facility found similar results to Ogino et al. (2007), namely that producing wet feed generated the most climate change benefits, followed by composting, and producing dry feed. While landfilling was confirmed as having the worst climate change outcomes, the results for the other impact categories studied by S.-H. Lee et al. (2007) could not be compared as Kim & Kim (2010) only assessed climate change impacts. This highlights the importance of conducting multi-criteria assessments, since one of the fundamental aims of conducting an LCA is to minimize burden-shifting.

In Europe, Vandermeersch et al. (2014) conducted a case study on a commercial-scale facility in Belgium valorizing supermarket bread waste into animal feed. This study found that producing animal feed resulted in climate burden reductions of between 25%-32% compared to anaerobic digestion, a common disposal method in the European Union. Martín et al. (2016) conducted a laboratory-scale case study of producing animal feed from vegetable waste originating from food industries in the Basque region of Spain. This multi-criteria study found food waste to animal feed production to have the best environmental outcomes compared to anaerobic digestion, incineration, composting, and landfilling for not only the climate change impact category, but also across water and abiotic resources depletion, eutrophication, and human toxicity cancer impacts. Mosna et al. (2021) conducted a case study of reformulating two types of pet food to incorporate meat waste from packaged food waste in Italy. This study found that the substitution of fresh meat with meat waste from packaged food waste substantially reduced the environmental impacts of pet food production, with the land use footprint being reduced by 87.5%, climate burden being reduced by 56.4%, water consumption by ~23%, and fossil resource depletion by ~22%.

Some notable trends were seen among the case studies in the literature. For food waste to feed production, a few common technical steps were noted such as sorting and grinding of the substrate, heat treatment, and drying. The majority of the studies identified the drying process as the largest impact hotspot for such a system (Siddique et al., 2024). As a result, liquid feed production consistently showed lower climate change impacts due to avoiding the drying process. If appropriate for local use contexts, producing liquid valorized feed may potentially be a strategy for hotspot mitigation.

Asian countries like Japan, South Korea, and Taiwan were found to be more advanced in the field of food waste valorization due to a conducive socio-cultural and legislative environment (Siddique et al., 2024; Zu Ermgassen et al., 2016). As discussed previously in the challenges and concerns section, feeding animal origin food waste to ruminants (such as cows, goats, and sheep) is prohibited in the European Union and Canada due to concerns surrounding the spread of diseases. Therefore, most of the case studies conducted within the context of European countries were found to be theoretical or laboratory-scale, while Asian countries had more case studies of commercial-scale systems. Due to the legislative restrictions, European case studies often involved streams of segregated food waste substrates such as fruits, vegetable, and bread waste, while the Asian case studies involved mixed food waste (Siddique et al., 2024). Studies such as Mosna et al. (2021) discuss pathways that can potentially work around the legislative restrictions by producing pet food instead of livestock feed.

7 Conclusion

The commercial-scale production of novel feed ingredients and the utilization of low opportunity-cost biomass such as manure, agricultural residues and food waste may have the potential to reduce climate change burdens, land and water use, as well as other resource and environmental impacts of global food systems. However, there remains non-trivial challenges to overcome for this concept to be successful. To maximize the socio-economic and environmental benefits of such pathways, it is important to determine the net benefits achievable from such a system by accounting for the specific technical, regional, social, and economic contexts that might apply. Ensuring internal efficiency by addressing identified hotspots (such as substrate drying) can also help such a system succeed. It is important for regulatory authorities to update and revise restrictive legislation surrounding novel feed ingredients at a regular interval as new scientific knowledge become available regarding their safety and efficacy. Standardization of such legislation across countries and regions based on the latest scientific evidence may encourage more investments. Lastly, public perception and trust about novel feed ingredients needs to be improved by employing proven strategies for public engagement, increasing supply chain transparency, and raising awareness about the potential resource and environmental benefits of these ingredients.

References

Alarcon, P., Wall, B., Barnes, K., Arnold, M., Rajanayagam, B., & Guitian, J. (2023). Classical BSE in Great Britain: Review of its epidemic, risk factors, policy and impact. *Food Control*, *146*, 109490. https://doi.org/10.1016/j.foodcont.2022.109490

Alibes, X., Muñoz, F., & Rodriguez, J. (1984). Feeding value of apple pomace silage for sheep. *Animal Feed Science and Technology*, *11*(3), 189–197. https://doi.org/10.1016/0377-8401(84)90062-2

Andersone, A., Janceva, S., Lauberte, L., Zaharova, N., Chervenkov, M., Jurkjane, V., Jashina, L., Rieksts, G., & Telysheva, G. (2023). Granulated Animal Feed and Fuel Based on Sea Buckthorn Agro-Waste Biomass for Sustainable Berry Production. *Sustainability*, *15*(14), 11152. https://doi.org/10.3390/su151411152

Bamber, N., Turner, I., Arulnathan, V., Li, Y., Zargar Ershadi, S., Smart, A., & Pelletier, N. (2020). Comparing sources and analysis of uncertainty in consequential and attributional life cycle assessment: Review of current practice and recommendations. *The International Journal of Life Cycle Assessment*, *25*(1), 168–180. https://doi.org/10.1007/s11367-019-01663-1

Beghin, J. C., & Gustafson, C. R. (2021). Consumer Valuation of and Attitudes towards Novel Foods Produced with New Plant Engineering Techniques: A Review. *Sustainability*, *13*(20), 11348. https://doi.org/10.3390/su132011348

Begna, R., & Masho, W. (2024). Valuation of livestock population and national feed security to enhance livestock productivity in Ethiopia. *Veterinary Medicine and Science*, *10*(3), e1415. https://doi.org/10.1002/vms3.1415

Beyers, M., Coudron, C., Ravi, R., Meers, E., & Bruun, S. (2023). Black soldier fly larvae as an alternative feed source and agro-waste disposal route – A life cycle perspective. *Resources, Conservation and Recycling*, *192*, 106917. https://doi.org/10.1016/j.resconrec.2023.106917

Blaustein-Rejto, D. (2023). *Livestock Don't Contribute 14.5% of Global Greenhouse Gas Emissions*. https://thebreakthrough.org/issues/food-agriculture-environment/livestock-dont-contribute-14-5-of-global-greenhouse-gas-emissions

Buccaro, M., Toscano, A., Balzarotti, M., Re, I., Bosco, D., & Bettiga, M. (2023). Techno-Economic Assessment of APS-Based Poultry Feed Production with a Circular Biorefinery Process. *Sustainability*, *15*(3), 2195. https://doi.org/10.3390/su15032195

Carter, C. G., & Codabaccus, M. B. (2022). Assessing the value of single-cell ingredients in aquafeeds. *Current Opinion in Biotechnology*, *76*, 102734. https://doi.org/10.1016/j.copbio.2022.102734

CFIA. (2020). *Novel feeds*. https://inspection.canada.ca/en/animal-health/livestock-feeds/novel-feeds

Charles Bai, S., Hardy, R. W., & Hamidoghli, A. (2022). Diet analysis and evaluation. In *Fish Nutrition* (pp. 709–743). Elsevier. https://doi.org/10.1016/B978-0-12-819587-1.00010-0

Cheng, V., Shoveller, A. K., Huber, L.-A., & Kiarie, E. G. (2023). Comparative protein quality in black soldier fly larvae meal vs. Soybean meal and fish meal using classical protein efficiency ratio (PER) chick growth assay model. *Poultry Science*, *102*(1), 102255. https://doi.org/10.1016/j.psj.2022.102255

Colón, J., Martínez-Blanco, J., Gabarrell, X., Artola, A., Sánchez, A., Rieradevall, J., & Font, X. (2010). Environmental assessment of home composting. *Resources, Conservation and Recycling*, *54*(11), 893–904. https://doi.org/10.1016/j.resconrec.2010.01.008

Dalmoro, Y. K., De Godoy, G. L., Agilar, J. C., Raddatz, G. A., De Oliveira, F. D. C., Witt, N., & Stefanello, C. (2025). Dietary Black Soldier Fly Larvae Meal and Its Impact on the Growth Performance and Gut Health of Broilers Under an Intestinal Challenge. *Metabolites*, *15*(6), 347. https://doi.org/10.3390/metabo15060347

Dame-Korevaar, A., Boumans, I. J. M. M., Antonis, A. F. G., Van Klink, E., & De Olde, E. M. (2021). Microbial health hazards of recycling food waste as animal feed. *Future Foods*, *4*, 100062. https://doi.org/10.1016/j.fufo.2021.100062

Dao, H. T., Sharma, N. K., Swick, R. A., & Moss, A. F. (2023). Feeding recycled food waste improved feed efficiency in laying hens from 24 to 43 weeks of age. *Scientific Reports*, *13*(1), 8261. https://doi.org/10.1038/s41598-023-34878-2

Domínguez Aldama, D. (2023). *Evaluating the environmental performance of black soldier fly meal and its potential use as a sustainable feed input in the egg industry: A life cycle assessment*. https://doi.org/10.14288/1.0438335

Dou, Z., Toth, J. D., Pitta, D. W., Bender, J. S., Hennessy, M. L., Vecchiarelli, B., Indugu, N., Chen, T., Li, Y., Sherman, R., Deutsch, J., Hu, B., Shurson, G. C., Parsons, B., & Baker, L. D. (2022).

Proof of concept for developing novel feeds for cattle from wasted food and crop biomass to enhance agri-food system efficiency. *Scientific Reports*, *12*(1). https://doi.org/10.1038/s41598-022-17812-w

El Otmani, S., Chentouf, M., Hornick, J. L., & Cabaraux, J. F. (2019). Chemical composition and in vitro digestibility of alternative feed resources for ruminants in Mediterranean climates: Olive cake and cactus cladodes. *The Journal of Agricultural Science*, *157*(03), 260–271. https://doi.org/10.1017/S0021859619000558

El Shaer, H. M. (2010). Halophytes and salt-tolerant plants as potential forage for ruminants in the Near East region. *Small Ruminant Research*, *91*(1), 3–12. https://doi.org/10.1016/j.smallrumres.2010.01.010

Environment and Climate Change Canada. (2024a). *Food loss and waste*. https://www.canada.ca/en/environment-climate-change/services/managing-reducing-waste/food-loss-waste.html

Environment and Climate Change Canada. (2024b). *Waste and greenhouse gases: Canada's actions*. https://www.canada.ca/en/environment-climate-change/services/managing-reducing-waste/municipal-solid/waste-greenhouse-gases-canada-actions.html

European Commission. (2001). OPINION ON: HYPOTHESES ON THE ORIGIN AND TRANSMISSION OF BSE ADOPTED BY THE SCIENTIFIC STEERING COMMITTEE AT ITS MEETING OF *29-30* NOVEMBER *2001*. https://food.ec.europa.eu/document/download/2fc8838d-9c4b-4a69-8c01-7f15d043729f_en?filename=sci-com_ssc_out236_en.pdf

Facey, H., Kithama, M., Mohammadigheisar, M., Huber, L.-A., Shoveller, A. K., & Kiarie, E. G. (2023). Complete replacement of soybean meal with black soldier fly larvae meal in feeding program for broiler chickens from placement through to 49 days of age reduced growth performance and altered organs morphology. *Poultry Science*, *102*(1), 102293. https://doi.org/10.1016/j.psj.2022.102293

FAO. (2021). *The State of Food and Agriculture 2021*. FAO. https://doi.org/10.4060/cb4476en

Fasihi, M., Jouzi, F., Tervasmäki, P., Vainikka, P., & Breyer, C. (2025). Global potential of sustainable single-cell protein based on variable renewable electricity. Nature Communications, 16(1), 1496. https://doi.org/10.1038/s41467-025-56364-1

Filipigh, A. A., Rojo, E. M., Pila, A. N., & Bolado, S. (2024). Fractional recovery of proteins and carbohydrates from secondary sludge from urban wastewater treatment plants. *Chemical Engineering Journal Advances*, *20*, 100686. https://doi.org/10.1016/j.ceja.2024.100686

Francis, A., Schmitt, E., & Smetana, S. (2025). Making better Bugs: Improving black soldier fly production for a more sustainable future. *Journal of Cleaner Production*, *521*, 146240. https://doi.org/10.1016/j.jclepro.2025.146240

Fujs, T., & Kashiwase, H. (2023). *Strains on freshwater resources: The impact of food production on water consumption*. https://blogs.worldbank.org/en/opendata/strains-freshwater-resources-impact-food-production-water-consumption

Gatto, A., Kuiper, M., Van Middelaar, C., & Van Meijl, H. (2024). Unveiling the economic and environmental impact of policies to promote animal feed for a circular food system. *Resources, Conservation and Recycling*, *200*, 107317. https://doi.org/10.1016/j.resconrec.2023.107317

Gerbens-Leenes, P. W., Mekonnen, M. M., & Hoekstra, A. Y. (2013). The water footprint of poultry, pork and beef: A comparative study in different countries and production systems. *Water Resources and Industry*, *1–2*, 25–36. https://doi.org/10.1016/j.wri.2013.03.001

Graedel, T. E., & Lifset, R. J. (2016). Industrial Ecology's First Decade. In *Taking Stock of Industrial Ecology* (pp. 3–20). Springer International Publishing. https://doi.org/10.1007/978-3-319-20571-7_1

Gundupalli, M. P., Ansari, S., Vital Da Costa, J. P., Qiu, F., Anderson, J., Luckert, M., & Bressler, D. C. (2024). Bacterial single cell protein (BSCP): A sustainable protein source from methylobacterium species. *Trends in Food Science & Technology*, *147*, 104426. https://doi.org/10.1016/j.tifs.2024.104426

Heidari, M. D., Gandasasmita, S., Li, E., & Pelletier, N. (2021). Proposing a framework for sustainable feed formulation for laying hens: A systematic review of recent developments and future directions. *Journal of Cleaner Production*, *288*, 125585. https://doi.org/10.1016/j.jclepro.2020.125585

Heuel, M., Sandrock, C., Leiber, F., Mathys, A., Gold, M., Zurbrügg, C., Gangnat, I. D. M., Kreuzer, M., & Terranova, M. (2021). Black soldier fly larvae meal and fat can completely replace soybean cake and oil in diets for laying hens. *Poultry Science*, *100*(4), 101034. https://doi.org/10.1016/j.psj.2021.101034

Hoek-van den Hil, E. F., Meijer, N. P., Van Rozen, K., Elissen, H., van Wikselaar, P. G., Brust, H., Te Loeke, N. A. J. M., de Rijk, T., Tienstra, M., van de Schans, M. G. M., Wanrooij, J., Van der Weide, R., Veldkamp, T., & van der Fels-Klerx, H. J. (2023). Safety of black soldier fly (Hermetia illucens) larvae reared on waste streams of animal and vegetal origin and manure. *Journal of Insects as Food and Feed*, *10*(5), 771–783. https://doi.org/10.1163/23524588-20230080

Iriarte, M. V., Gonzáles, J. L., De Freitas Costa, E., Gil, A. D., & De Jong, M. C. M. (2023). Main factors associated with foot-and-mouth disease virus infection during the 2001 FMD epidemic in Uruguay. *Frontiers in Veterinary Science*, *10*, 1070188. https://doi.org/10.3389/fvets.2023.1070188

ISO. (2006). *Environmental management—Life cycle assessment—Requirements and guidelines (ISO 14044)*.

Joshi, V. K., & Sandhu, D. K. (1996). Preparation and evaluation of an animal feed byproduct produced by solid-state fermentation of apple pomace. *Bioresource Technology*, *56*(2–3), 251–255. https://doi.org/10.1016/0960-8524(96)00040-5

Kakkar, S., Dharavat, N., & Sudabattula, S. K. (2024). Transforming food waste into energy: A comprehensive review. *Results in Engineering*, *24*, 103376. https://doi.org/10.1016/j.rineng.2024.103376

Kaur, L., Mao, B., Beniwal, A. S., Abhilasha, Kaur, R., Chian, F. M., & Singh, J. (2022). Alternative proteins vs animal proteins: The influence of structure and processing on their gastro-small intestinal digestion. *Trends in Food Science & Technology*, *122*, 275–286. https://doi.org/10.1016/j.tifs.2022.02.021

Kholif, A. E., Hamdon, H. A., Kassab, A. Y., Farahat, E. S. A., Azzaz, H. H., Matloup, O. H., Mohamed, A. G., & Anele, U. Y. (2020). *Chlorella vulgaris* microalgae and/or copper supplementation enhanced feed intake, nutrient digestibility, ruminal fermentation, blood metabolites and lactational performance of Boer goat. *Journal of Animal Physiology and Animal Nutrition*, *104*(6), 1595–1605. https://doi.org/10.1111/jpn.13378

Kim, M.-N., & Kim, J. H. (2010). Comparison through a LCA evaluation analysis of food waste disposal options from the perspective of global warming and resource recovery. *Science of The Total Environment*, *408*(19), 3998–4006. https://doi.org/10.1016/j.scitotenv.2010.04.049

Kobayashi, Y., EL-Wali, M., Guðmundsson, H., Guðmundsdóttir, E. E., Friðjónsson, Ó. H., Karlsson, E. N., Roitto, M., & Tuomisto, H. L. (2023). Life-cycle assessment of yeast-based single-cell protein production with oat processing side-stream. *Science of The Total Environment*, *873*, 162318. https://doi.org/10.1016/j.scitotenv.2023.162318

Komarek, A. M., Dunston, S., Enahoro, D., Godfray, H. C. J., Herrero, M., Mason-D'Croz, D., Rich, K. M., Scarborough, P., Springmann, M., Sulser, T. B., Wiebe, K., & Willenbockel, D. (2021). Income, consumer preferences, and the future of livestock-derived food demand. *Global Environmental Change*, *70*, 102343. https://doi.org/10.1016/j.gloenvcha.2021.102343

Koukoumaki, D. I., Tsouko, E., Papanikolaou, S., Ioannou, Z., Diamantopoulou, P., & Sarris, D. (2024). Recent advances in the production of single cell protein from renewable resources and applications. *Carbon Resources Conversion*, *7*(2), 100195. https://doi.org/10.1016/j.crcon.2023.07.004

Lackner, K. S., Arcusa, S. H., Azarabadi, H., Sriramprasad, V., & Page, R. (2023). Carbon accounting without life cycle analysis. *Energy & Environmental Science*, *16*(11), 4968–4982. https://doi.org/10.1039/D3EE01138K

Lam, D. (2025). The Next 2 Billion: Can the World Support 10 Billion People? *Population and Development Review*, *51*(1), 63–102. https://doi.org/10.1111/padr.12685

Laureati, M., De Boni, A., Saba, A., Lamy, E., Minervini, F., Delgado, A. M., & Sinesio, F. (2024). Determinants of Consumers' Acceptance and Adoption of Novel Food in View of More Resilient and Sustainable Food Systems in the EU: A Systematic Literature Review. *Foods*, *13*(10), 1534. https://doi.org/10.3390/foods13101534

Lee, J., Chen, W.-H., & Park, Y.-K. (2022). Recent achievements in platform chemical production from food waste. *Bioresource Technology*, *366*, 128204. https://doi.org/10.1016/j.biortech.2022.128204

Lee, S.-H., Choi, K.-I., Osako, M., & Dong, J. I. (2007). Evaluation of environmental burdens caused by changes of food waste management systems in Seoul, Korea. *Science of The Total Environment*, *387*(1–3), 42–53. https://doi.org/10.1016/j.scitotenv.2007.06.037

Levasseur, A., Lesage, P., Margni, M., Deschenes, L., & Samson, R. (2010). Considering Time in LCA: Dynamic LCA and Its Application to Global Warming Impact Assessments. *ENVIRONMENTAL SCIENCE & TECHNOLOGY*, *44*(8), 3169–3174. https://doi.org/10.1021/es9030003

Lindeman, R. L. (1942). The Trophic-Dynamic Aspect of Ecology. *Ecology*, *23*(4), 399–417. https://doi.org/10.2307/1930126

Lou, X. F., & Nair, J. (2009). The impact of landfilling and composting on greenhouse gas emissions – A review. *Bioresource Technology*, *100*(16), 3792–3798. https://doi.org/10.1016/j.biortech.2008.12.006

Lu, H., Liu, Y., Shen, W., Zhou, Y., Ma, X., Sun, S., Dong, X., Ji, F., Tong, H., Xu, J., He, G., & Xu, W. (2023). Yeast enrichment facilitated lipid removal and bioconversion by black soldier fly larvae in the food waste treatment. *Waste Management*, *166*, 152–162. https://doi.org/10.1016/j.wasman.2023.04.003

Lu, S., Taethaisong, N., Meethip, W., Surakhunthod, J., Sinpru, B., Sroichak, T., Archa, P., Thongpea, S., Paengkoum, S., Purba, R. A. P., & Paengkoum, P. (2022). Nutritional Composition of Black Soldier Fly Larvae (Hermetia illucens L.) and Its Potential Uses as Alternative Protein Sources in Animal Diets: A Review. *Insects*, *13*(9), 831. https://doi.org/10.3390/insects13090831

Mahan, K. M., Le, R. K., Wells, T., Anderson, S., Yuan, J. S., Stoklosa, R. J., Bhalla, A., Hodge, D. B., & Ragauskas, A. J. (2018). Production of single cell protein from agro-waste using *Rhodococcus opacus*. *Journal of Industrial Microbiology and Biotechnology*, *45*(9), 795–801. https://doi.org/10.1007/s10295-018-2043-3

Mahlangu, D., Mphahlele, K., De Paola, F., & Mthombeni, N. H. (2024). Microalgae-Mediated Biosorption for Effective Heavy Metals Removal from Wastewater: A Review. *Water*, *16*(5), 718. https://doi.org/10.3390/w16050718

Maia, M. R. G., Monteiro, A., Valente, I. M., Sousa, C., Miranda, C., Castro, C., Cortez, P. P., Cabrita, A. R. J., Trindade, H., & Fonseca, A. J. M. (2023). Upcycling post-harvest biomass residues from native European Lupinus species: From straws and pod shells production to nutritive value and alkaloids content for ruminant animals. *Frontiers in Nutrition*, *10*. https://doi.org/10.3389/fnut.2023.1195015

Malematja, E., Manyelo, T. G., Sebola, N. A., Kolobe, S. D., & Mabelebele, M. (2023). The accumulation of heavy metals in feeder insects and their impact on animal production. *Science of The Total Environment*, *885*, 163716. https://doi.org/10.1016/j.scitotenv.2023.163716

Martín, D. S., Ramos, S., & Zufía, J. (2016). Valorisation of food waste to produce new raw materials for animal feed. *Food Chemistry*, *198*, 68–74. https://doi.org/10.1016/j.foodchem.2015.11.035

Martínez-Ibáñez, E., Laso, J., Pérez-Martínez, M. M., Martínez-Vazquez, R., Baptista De Sousa, D., Méndez, D., Olaya-Pérez, E., Marchisio, V., Aldaco, R., & Margallo, M. (2025). Environmental Insights into Single-Cell Protein Production: A Life Cycle Assessment Framework. *ACS Sustainable Chemistry & Engineering*, *13*(23), 8699–8710. https://doi.org/10.1021/acssuschemeng.5c02336

Martins, C. F., Pestana, J. M., Alfaia, C. M., Costa, M., Ribeiro, D. M., Coelho, D., Lopes, P. A., Almeida, A. M., Freire, J. P. B., & Prates, J. A. M. (2021). Effects of Chlorella vulgaris as a Feed Ingredient on the Quality and Nutritional Value of Weaned Piglets' Meat. *Foods*, *10*(6), 1155. https://doi.org/10.3390/foods10061155

Mayer, J. A., & Cushman, J. C. (2019). Nutritional and mineral content of prickly pear cactus: A highly water-use efficient forage, fodder and food species. *Journal of Agronomy and Crop Science*, *205*(6), 625–634. https://doi.org/10.1111/jac.12353

Mba Wright, M., Tan, E. C. D., Tu, Q., Martins, A., Parvatker, A. G., Yao, Y., Sunol, A., & Smith, R. L. (2024). Life Cycle Inventory Availability: Status and Prospects for Leveraging New Technologies. *ACS Sustainable Chemistry & Engineering*, *12*(34), 12708–12718. https://doi.org/10.1021/acssuschemeng.4c02519

Mendoza Beltran, A., Heijungs, R., Guinée, J., & Tukker, A. (2016). A pseudo-statistical approach to treat choice uncertainty: The example of partitioning allocation methods. *The International Journal of Life Cycle Assessment*, *21*(2), 252–264. https://doi.org/10.1007/s11367-015-0994-4

Meza Ocampos, G., Abdalla, A. L., González Reyna, A., & McAllister, T. A. (2025). Circular bioeconomy in livestock production: Harnessing crop by-products in MERCOSUR/MERCOSUL. *Animal Frontiers*, *15*(4), 44–53. https://doi.org/10.1093/af/vfaf032

Milford, A. B., Le Mouël, C., Bodirsky, B. L., & Rolinski, S. (2019). Drivers of meat consumption. *Appetite*, *141*, 104313. https://doi.org/10.1016/j.appet.2019.06.005

Moore, J., Hawkins, S. A., Austin, A. R., Konold, T., Green, R. B., Blamire, I. W., Dexter, I., Stack, M. J., Chaplin, M. J., Langeveld, J. P., Simmons, M. M., Spencer, Y. I., Webb, P. R., Dawson, M., & Wells, G. A. (2011). Studies of the transmissibility of the agent of bovine spongiform encephalopathy to the domestic chicken. *BMC Research Notes*, *4*(1), 501. https://doi.org/10.1186/1756-0500-4-501

Mosna, D., Bottani, E., Vignali, G., & Montanari, R. (2021). Environmental benefits of pet food obtained as a result of the valorisation of meat fraction derived from packaged food waste. *Waste Management*, *125*, 132–144. https://doi.org/10.1016/j.wasman.2021.02.035

Musa, M. N., Jirgi, G. M., Zango, Z. U., Isah, M. N., Abdurrazak, M., Adamu, A. A., Wadi, I. A., Adeleke, A. A., Garba, Z. N., Bello, U., Adamu, H., Hosseini-Bandegharaei, A., & Bokov, D. O. (2025). A review on techno-economic assessment of Spirulina for sustainable nutraceutical, medicinal, environmental, and bioenergy applications. *Bioresources and Bioprocessing*, *12*(1), 51. https://doi.org/10.1186/s40643-025-00888-3

Muscat, A., De Olde, E. M., De Boer, I. J. M., & Ripoll-Bosch, R. (2020). The battle for biomass: A systematic review of food-feed-fuel competition. *Global Food Security*, *25*, 100330. https://doi.org/10.1016/j.gfs.2019.100330

Nath, P. C., Ojha, A., Debnath, S., Sharma, M., Nayak, P. K., Sridhar, K., & Inbaraj, B. S. (2023). Valorization of Food Waste as Animal Feed: A Step towards Sustainable Food Waste Management and Circular Bioeconomy. *Animals*, *13*(8), 1366. https://doi.org/10.3390/ani13081366

Nizar, M., Hajer, G., & Kamel, H. (2016). *Potential Use of Halophytes and Salt Tolerant Plants in Ruminant Feeding: A Tunisian Case Study*.

Notarnicola, B., Sala, S., Anton, A., McLaren, S. J., Saouter, E., & Sonesson, U. (2017). The role of life cycle assessment in supporting sustainable agri-food systems: A review of the challenges. *Journal of Cleaner Production*, *140*, 399–409. https://doi.org/10.1016/j.jclepro.2016.06.071

Nyyssölä, A., Suhonen, A., Ritala, A., & Oksman-Caldentey, K.-M. (2022). The role of single cell protein in cellular agriculture. *Current Opinion in Biotechnology*, *75*, 102686. https://doi.org/10.1016/j.copbio.2022.102686

Ogino, A., Hirooka, H., Ikeguchi, A., Tanaka, Y., Waki, M., Yokoyama, H., & Kawashima, T. (2007). Environmental Impact Evaluation of Feeds Prepared from Food Residues Using Life Cycle Assessment. *Journal of Environmental Quality*, *36*(4), 1061–1068. https://doi.org/10.2134/jeq2006.0326

Onwuka, C. F. I., Adetiloye, P. O., & Afolami, C. A. (1997). Use of household wastes and crop residues in small ruminant feeding in Nigeria. *Small Ruminant Research*, *24*(3), 233–237. https://doi.org/10.1016/s0921-4488(96)00953-4

Øverland, M., Tauson, A.-H., Shearer, K., & Skrede, A. (2010). Evaluation of methane-utilising bacteria products as feed ingredients for monogastric animals. *Archives of Animal Nutrition*, *64*(3), 171–189. https://doi.org/10.1080/17450391003691534

Palladini, N. M., Gislon, G., Sandrucci, A., Zucali, M., Tamburini, A., & Bava, L. (2024). Assessment of food-feed competition for producing milk in cow dairy farms. *Agricultural Systems*, *218*, 103984. https://doi.org/10.1016/j.agsy.2024.103984

Panaite, T. D., Cornescu, G. M., Predescu, N. C., Cismileanu, A., Turcu, R. P., Saracila, M., & Soica, C. (2023). Microalgae (Chlorella vulgaris and Spirulina platensis) as a Protein Alternative and Their Effects on Productive Performances, Blood Parameters, Protein Digestibility, and Nutritional Value of Laying Hens' Egg. *Applied Sciences*, *13*(18), 10451. https://doi.org/10.3390/app131810451

Papaioannou, D., Katsoulos, P. D., Panousis, N., & Karatzias, H. (2005). The role of natural and synthetic zeolites as feed additives on the prevention and/or the treatment of certain farm animal diseases: A review. *Microporous and Mesoporous Materials*, *84*(1–3), 161–170. https://doi.org/10.1016/j.micromeso.2005.05.030

Papargyropoulou, E., Lozano, R., K. Steinberger, J., Wright, N., & Ujang, Z. B. (2014). The food waste hierarchy as a framework for the management of food surplus and food waste. *Journal of Cleaner Production*, *76*, 106–115. https://doi.org/10.1016/j.jclepro.2014.04.020

Parchami, M., Rustas, B.-O., Taherzadeh, M. J., & Mahboubi, A. (2025). An in vitro evaluation of partial energy replacement in a total mixed ration with volatile fatty acids derived from agro-industrial residues. *Systems Microbiology and Biomanufacturing*, *5*(2), 805–818. https://doi.org/10.1007/s43393-024-00278-4

Pehnt, M. (2006). Dynamic life cycle assessment (LCA) of renewable energy technologies. *Renewable Energy*, *31*(1), 55–71. https://doi.org/10.1016/j.renene.2005.03.002

Pelletier, N. (2017). Life cycle assessment of Canadian egg products, with differentiation by hen housing system type. *Journal of Cleaner Production*, *152*, 167–180. https://doi.org/10.1016/j.jclepro.2017.03.050

Pexas, G., Doherty, B., & Kyriazakis, I. (2023). The future of protein sources in livestock feeds: Implications for sustainability and food safety. *Frontiers in Sustainable Food Systems*, *7*, 1188467. https://doi.org/10.3389/fsufs.2023.1188467

Pilmer, L. W., Woolley, L. D., Lymbery, A. J., Salini, M., & Partridge, G. J. (2022). Using dietary additives to improve palatability of diets containing single-cell protein from methanotrophic bacteria in yellowtail kingfish (*Seriola lalandi*) diets. *Aquaculture Research*, *53*(14), 5006–5017. https://doi.org/10.1111/are.15986

Poore, J., & Nemecek, T. (2018). Reducing food's environmental impacts through producers and consumers. *Science*, *360*(6392), 987–992. https://doi.org/10.1126/science.aaq0216

Ramachandran, S., Singh, S. K., Larroche, C., Soccol, C. R., & Pandey, A. (2007). Oil cakes and their biotechnological applications – A review. *Bioresource Technology*, *98*(10), 2000–2009. https://doi.org/10.1016/j.biortech.2006.08.002

Renna, M., Rastello, L., Veldkamp, T., Toral, P. G., Gonzalez-Ronquillo, M., Jimenez, L. E. R., & Gasco, L. (2023). Are insects a solution for feeding ruminants? Legislation, scientific evidence, and future challenges. *Animal Frontiers*, *13*(4), 102–111. https://doi.org/10.1093/af/vfad026

Reyes-García, V., Villasante, S., Benessaiah, K., Pandit, R., Agrawal, A., Claudet, J., Garibaldi, L. A., Kabisa, M., Pereira, L., & Zinngrebe, Y. (2025). The costs of subsidies and externalities of economic activities driving nature decline. *Ambio*, *54*(7), 1128–1141. https://doi.org/10.1007/s13280-025-02147-3

Ritala, A., Häkkinen, S. T., Toivari, M., & Wiebe, M. G. (2017). Single Cell Protein—State-of-the-Art, Industrial Landscape and Patents 2001–2016. *Frontiers in Microbiology*, *8*, 2009. https://doi.org/10.3389/fmicb.2017.02009

Ritchie, H. (2021). *If the world adopted a plant-based diet, we would reduce global agricultural land use from 4 to 1 billion hectares*. https://ourworldindata.org/land-use-diets#:~:text=So%2C%20while%20the%20UN%20FAO,2018)%2C%20published%20in%20Science.

Ritchie, H., & Roser, M. (2024). *Half of the world's habitable land is used for agriculture*. Half of the World's Habitable Land Is Used for Agriculture. https://ourworldindata.org/global-land-for-agriculture

Saer, A., Lansing, S., Davitt, N. H., & Graves, R. E. (2013). Life cycle assessment of a food waste composting system: Environmental impact hotspots. *Journal of Cleaner Production*, *52*, 234–244. https://doi.org/10.1016/j.jclepro.2013.03.022

Salami, S. A., Luciano, G., O'Grady, M. N., Biondi, L., Newbold, C. J., Kerry, J. P., & Priolo, A. (2019). Sustainability of feeding plant by-products: A review of the implications for ruminant meat production. *Animal Feed Science and Technology*, *251*, 37–55. https://doi.org/10.1016/j.anifeedsci.2019.02.006

Sar, T., Larsson, K., Fristedt, R., Undeland, I., & Taherzadeh, M. J. (2022). Demo-scale production of protein-rich fungal biomass from potato protein liquor for use as innovative food and feed products. *Food Bioscience*, *47*, 101637. https://doi.org/10.1016/j.fbio.2022.101637

Scheibenzuber, S., Pucci, E., Presenti, O., Serafini, G., Nobili, C., Zoani, C., Duta, D. E., Mihai, A. L., Criveanu-Stamatie, G. D., Belc, N., Falch, E., Rustad, T., & Rychlik, M. (2025). Consumers acceptance of new food ingredients from the food industry's by-products—A focus group study. *Frontiers in Nutrition*, *12*, 1509833. https://doi.org/10.3389/fnut.2025.1509833

Schrögel, P., & Wätjen, W. (2019). Insects for Food and Feed-Safety Aspects Related to Mycotoxins and Metals. *Foods*, *8*(8), 288. https://doi.org/10.3390/foods8080288

Sebastian, R. M., Billal, M. M., & Kumar, A. (2025). The development of a framework to assess waste and biomass availability: A case study for Canada. *Resources, Conservation and Recycling*, *215*, 108170. https://doi.org/10.1016/j.resconrec.2025.108170

Selvaraj, V., & Won, E. (2024). Transforming aquaculture with insect-based feed: Restraining factors. *Animal Frontiers*, *14*(4), 24–27. https://doi.org/10.1093/af/vfae011

Shanafelt, D. W., Jones, G., Lima, M., Perrings, C., & Chowell, G. (2018). Forecasting the 2001 Foot-and-Mouth Disease Epidemic in the UK. *EcoHealth*, *15*(2), 338–347. https://doi.org/10.1007/s10393-017-1293-2

Shirvani, R., Bartik, A., Alves, G. A. S., Garcia De Otazo Hernandez, D., Müller, S., Föttinger, K., & Steiger, M. G. (2023). Nitrogen recovery from low-value biogenic feedstocks via steam gasification to methylotrophic yeast biomass. *Frontiers in Bioengineering and Biotechnology*, *11*. https://doi.org/10.3389/fbioe.2023.1179269

Siddique, S., Arulnathan, V., Turner, I., Sadiq, R., & Pelletier, N. (2025). Life cycle assessment of a commercial-scale valorized grocery food waste product and its potential use as a sustainable feed input for egg production. *Cleaner Environmental Systems*, *19*, 100364. https://doi.org/10.1016/j.cesys.2025.100364

Siddique, S., Grassauer, F., Arulnathan, V., Sadiq, R., & Pelletier, N. (2024). A review of life cycle impacts of different pathways for converting food waste into livestock feed. *Sustainable Production and Consumption*, *46*, 310–323. https://doi.org/10.1016/j.spc.2024.02.023

Siddiqui, S. A., Elsheikh, W., Ucak, İ., Hasan, M., Perlita, Z. C., & Yudhistira, B. (2024). Replacement of soy by mealworms for livestock feed—A comparative review between soy and mealworms considering environmental aspects. *Environment, Development and Sustainability*. https://doi.org/10.1007/s10668-024-04874-1

Sitepu, I. R., Hitomi, A., Wu, W., Wu, A., Jeoh, T., & Boundy-Mills, K. (2023). Production of high protein yeast using enzymatically liquefied almond hulls. *PLOS ONE*, *18*(11), e0293085. https://doi.org/10.1371/journal.pone.0293085

Siva Raman, S., Stringer, L. C., Bruce, N. C., & Chong, C. S. (2022). Opportunities, challenges and solutions for black soldier fly larvae-based animal feed production. *Journal of Cleaner Production*, *373*, 133802. https://doi.org/10.1016/j.jclepro.2022.133802

Sohn, J., Kalbar, P., Goldstein, B., & Birkved, M. (2020). Defining Temporally Dynamic Life Cycle Assessment: A Review. *Integrated Environmental Assessment And Management*, *16*(3), 314–323. https://doi.org/10.1002/ieam.4235

Sola, D., Artigas, R., Mediano, D. R., Zaragoza, P., Badiola, J. J., Martín-Burriel, I., & Acín, C. (2023). Novel polymorphisms in the prion protein gene (PRNP) and stability of the resultant prion protein in different horse breeds. *Veterinary Research*, *54*(1), 94. https://doi.org/10.1186/s13567-023-01211-8

Taelman, S. E., De Meester, S., Roef, L., Michiels, M., & Dewulf, J. (2013). The environmental sustainability of microalgae as feed for aquaculture: A life cycle perspective. *Bioresource Technology*, *150*, 513–522. https://doi.org/10.1016/j.biortech.2013.08.044

Toniolo, S., Tosato, R. C., Gambaro, F., & Ren, J. (2020). Life cycle thinking tools: Life cycle assessment, life cycle costing and social life cycle assessment. In *Life Cycle Sustainability Assessment for Decision-Making* (pp. 39–56). Elsevier. https://doi.org/10.1016/b978-0-12-818355-7.00003-8

Turner, I., Heidari, D., & Pelletier, N. (2022). Life cycle assessment of contemporary Canadian egg production systems during the transition from conventional cage to alternative housing systems: Update and analysis of trends and conditions. *Resources, Conservation and Recycling*, *176*, 105907. https://doi.org/10.1016/j.resconrec.2021.105907

US EPA. (2021). *Food Recovery Hierarchy*. https://19january2021snapshot.epa.gov/sustainable-management-food/food-recovery-hierarchy_.html

US EPA. (2024). *Basic Information about Landfill Gas*. https://www.epa.gov/lmop/basic-information-about-landfill-gas#:~:text=Municipal%20solid%20waste%20(MSW)%20landfills,the%20waste%20and%20generate%20methane.

US EPA. (2025). *From Field to Bin: The Environmental Impacts of U.S. Food Waste Management Pathways*. https://www.epa.gov/land-research/field-bin-environmental-impacts-us-food-waste-management-pathways

Van Huis, A., & Oonincx, D. G. a. B. (2017). The environmental sustainability of insects as food and feed. A review. Agronomy for Sustainable Development, 37(5). https://doi.org/10.1007/s13593-017-0452-8

Vandermeersch, T., De Alvarenga, R. F., Ragaert, P., & Dewulf, J. (2014). Environmental sustainability assessment of food waste valorization options. *Resources, Conservation and Recycling*, *87*, 57–64. https://doi.org/10.1016/j.resconrec.2014.03.008

Vastolo, A., Serrapica, F., Cavallini, D., Fusaro, I., Atzori, A. S., & Todaro, M. (2024). Editorial: Alternative and novel livestock feed: reducing environmental impact. *Frontiers in Veterinary Science*, *11*, 1441905. https://doi.org/10.3389/fvets.2024.1441905

Veldkamp, T., Van Rozen, K., Elissen, H., Van Wikselaar, P., & Van Der Weide, R. (2021). Bioconversion of Digestate, Pig Manure and Vegetal Residue-Based Waste Operated by Black Soldier Fly Larvae, Hermetia illucens L. (Diptera: Stratiomyidae). *Animals*, *11*(11), 3082. https://doi.org/10.3390/ani11113082

Vidal, E., Fernández-Borges, N., Eraña, H., Parra, B., Pintado, B., Sánchez-Martín, M. A., Charco, J. M., Ordóñez, M., Pérez-Castro, M. A., Pumarola, M., Mathiason, C. K., Mayoral, T., & Castilla, J. (2020). Dogs are resistant to prion infection, due to the presence of aspartic or glutamic acid at position 163 of their prion protein. *The FASEB Journal*, *34*(3), 3969–3982. https://doi.org/10.1096/fj.201902646R

Villaró-Cos, S., Cuaresma Franco, M., García-Vaquero, M., Morán, L., Alarcón, F. J., & Lafarga, T. (2024). Composition of microalgae produced using different types of water and nutrient sources. *Algal Research*, *78*, 103394. https://doi.org/10.1016/j.algal.2024.103394

Vural, Y., Ferriday, D., & Rogers, P. J. (2023). Consumers' attitudes towards alternatives to conventional meat products: Expectations about taste and satisfaction, and the role of disgust. *Appetite*, *181*, 106394. https://doi.org/10.1016/j.appet.2022.106394

Xu, B., Liu, Y., Chen, K., Wang, L., Sagada, G., Tegomo, A. F., Yang, Y., Sun, Y., Zheng, L., Ullah, S., & Shao, Q. (2021). Evaluation of Methanotroph (Methylococcus capsulatus, Bath) Bacteria Meal (FeedKind®) as an Alternative Protein Source for Juvenile Black Sea Bream, Acanthopagrus schlegelii. *Frontiers in Marine Science*, *8*, 778301. https://doi.org/10.3389/fmars.2021.778301

Xu, X., Sharma, P., Shu, S., Lin, T.-S., Ciais, P., Tubiello, F. N., Smith, P., Campbell, N., & Jain, A. K. (2021). Global greenhouse gas emissions from animal-based foods are twice those of plant-based foods. *Nature Food*, *2*(9), 724–732. https://doi.org/10.1038/s43016-021-00358-x

Yang, Z., Jiang, L., Zhang, M., Deng, Y., Suo, W., Zhang, H., Wang, C., & Li, H. (2022). Bioconversion of Apple Pomace into Microbial Protein Feed Based on Extrusion Pretreatment. *Applied Biochemistry and Biotechnology*, *194*(4), 1496–1509. https://doi.org/10.1007/s12010-021-03727-1

Ylmén, P., Berlin, J., Mjörnell, K., & Arfvidsson, J. (2020). Managing Choice Uncertainties in Life-Cycle Assessment as a Decision-Support Tool for Building Design: A Case Study on Building Framework. *Sustainability*, *12*(12), 5130. https://doi.org/10.3390/su12125130

Zhou, P., Zhang, L., Ding, H., Gao, X., Chen, Y., & Li, D. (2022). Optimization of culture conditions of screened *Galactomyces candidum* for the production of single cell protein from biogas slurry. *Electronic Journal of Biotechnology*, *55*, 47–54. https://doi.org/10.1016/j.ejbt.2021.11.006

Zu Ermgassen, E. K. H. J., Phalan, B., Green, R. E., & Balmford, A. (2016). Reducing the land use of EU pork production: Where there's swill, there's a way. *Food Policy*, *58*, 35–48. https://doi.org/10.1016/j.foodpol.2015.11.001

CHAPTER 27

Holistic decision modeling for sustainable feed and pet food systems: integrating predictive tools and a human ecology perspective

C. Whitney[1,2*] *and E. Luedeling*[1]

[1]*University of Bonn, The Institute of Crop Science and Resource Conservation (INRES)*
[2]*Center for Development Research (ZEF), Bonn, Germany*
**Correspondence: cory.whitney@uni-bonn.de*

Abstract

Transforming global food systems requires a radical overhaul of how we feed animals. Novel ingredients like insect meal and algae offer a sustainable path forward, yet their adoption is paralyzed by complex trade-offs between cost, scalability, environmental impact and social acceptance. This chapter cuts through this paralysis by synthesizing principles from decision theory, behavioral economics and human ecology into a single, practical framework. We present a four-step process that moves beyond narrow techno-economic analysis, enabling stakeholders to define problems rigorously, model uncertainties transparently, account for behavioral biases and anticipate system-wide consequences. The result is a structured approach for navigating the transition to feed systems that are not only efficient but also resilient, just and aligned with planetary boundaries.

Keywords

sustainable feed – decision analysis – novel ingredients – circular economy – behavioral economics – multi-criteria decision analysis – social-ecological systems – animal nutrition

1 Introduction

To transform our global food systems, we must fundamentally rethink animal feed. The promise of novel protein sources is clear, but the path to their adoption is fraught with uncertainty and competing priorities. The solution is not to search for a perfect ingredient, but to implement a superior decision-making process.

This chapter introduces a structured framework to guide policymakers, corporate leaders and farmers in evaluating sustainable feed alternatives. By integrating the analytical rigor of decision theory with the real-world insights of behavioral economics and human ecology, this approach provides a clear methodology to replace confusion with clarity. It empowers stakeholders to make choices that are not only economically viable but also environmentally sound and socially robust, turning the promise of sustainable agri-food systems into a practiced outcome.

2 The decision-making crisis in sustainable feed and pet food

The field of novel feed ingredients, from insect meal to single-cell proteins, shows technical promise for creating more sustainable feed systems (Chamodi et al. 2025; Rossi et al. 2025). Yet, these innovations have created a parallel crisis for decision-makers in industry and policy: paralysis by choice. Faced with competing claims about sustainability, cost and risk, the prevailing approach, relying on a single metric, such as carbon footprint, is not only inadequate but also dangerous. It risks burden-shifting, unintended social consequences and investment in solutions that fail at scale. The complex realities of the circular economy exacerbate industry-level decision-making challenges (Chhimwal et al. 2022; Lotfian Delouyi et al. 2023). While novel ingredients can upcycle waste streams and reduce land use, addressing critical issues like the use of agricultural land for livestock, adoption hinges on more than nutritional adequacy. Decision-makers must navigate a web of uncertain trade-offs. These include economic viability and supply chain risks, as many novel ingredients remain niche due to high costs and infrastructure limitations (Dierenfeld 2023; Malenica et al. 2023). Decision-makers face stringent regulatory barriers, particularly in regions such as the EU (Focker et al. 2022; Van Raamsdonk et al. 2023). Furthermore, choices are fraught with ethical considerations, such as ensuring animal welfare is not compromised for circularity (Bracke et al. 2023) and managing the food safety hazards introduced by new feed sources (Van Raamsdonk et al. 2023).

The key bottleneck in sustainable change is no longer a lack of technical options, but a lack of robust decision-making frameworks (Hersh 1999; Weaver et al. 2013). There is a need to move beyond cataloging alternatives toward providing practical, transdisciplinary evaluation frameworks. Integrating the principles of well-established fields of decision theory, behavioral economics and human ecology can equip stakeholders to systematically navigate the complex decision landscape in the pet food and animal feed industry. This chapter details a structured, four-step process for making decisions that are not just computationally sound but also socially robust and strategically wise. The goal is not to prescribe a single 'best' ingredient, but to provide a systematic way to think through the complex, value-laden decisions required to create truly sustainable, ethical and resilient feed systems.

3 A practical framework for decision-making on feed alternatives

Identifying sustainable alternatives to conventional pet food and animal feed requires more than technical innovation, it demands structured thinking in the face of uncertainty (Suckling et al. 2020; Akintan et al. 2024). Stakeholders must navigate ecological risks, economic trade-offs and behavioral constraints, often in data-scarce environments. Decision theory provides a systematic approach to addressing these challenges (Howard and Abbas 2015). At its core, it supports making rational choices under uncertainty by defining clear objectives, identifying alternatives, quantifying trade-offs and selecting actions that maximize expected value or align with stakeholder preferences (Howard and Abbas 2015; Keeney 2021).

However, purely rational models fall short if they ignore human behavior (Kahneman and Egan 2011). Behavioral economics highlights how decisions are shaped by framing effects, status quo bias, loss aversion and time inconsistency, biases that frequently influence consumer, investor and policy choices in the pet food and animal feed sectors. For example, aversion to novel ingredients like insects or algae may be rooted more in social norms and perceived naturalness than in actual safety or sustainability concerns (Rozin et al. 2009; Siegrist and Hartmann 2020). Accounting for biases that influence behavior can enhance the adoption of sustainable options and inform the communication of alternatives to different stakeholders.

Movement toward sustainable pet food and animal feed systems also requires attention to broader social-ecological contexts. Human ecology is a transdisciplinary approach to understanding how human social systems relate to and interact with the ecological systems on which they depend (Marten 2001; Kofinas and Chapin 2009). Human ecology offers a lens for examining how feed choices intersect with ecosystems, cultural practices, labor relations and governance structures. A circular ingredient may score well on environmental metrics but raise ethical or equity concerns depending on who benefits and who bears the burden (Liu 2024). For instance, diverting food waste to animal feed may undermine informal food recovery networks or conflict with cultural taboos. Understanding the full landscape of impacts is essential to avoiding unintended consequences. A narrow focus on selected impacts may risk presenting an incomplete picture. Objective decision-making must acknowledge that other equally vital outcome values exist, for which established proteins may hold distinct advantages. Other outcome values may include economic stability within existing supply chains, the scalability of novel production methods, nutritional bioavailability for specific species, food safety protocols, consumer acceptance and the palatability of pet food.

To support decision-making for novel pet food and animal feed, this chapter will outline a structured, four-step approach (Figure 27.1) grounded in decision analysis, enriched by insights from behavioral economics and human ecology. The approach includes four steps: (1) defining the decision problem and objectives, (2) modeling the outcomes and uncertainties of available alternatives, (3) identifying cognitive biases and behavioral barriers and (4) embedding the analysis in its broader socio-environmental context. This framework is intended to guide both public and private sector actors in evaluating feed alternatives through a structured, four-step process (Figure 27.1).

To see this framework in action, consider an example case we are calling 'VitaliFeed,' a hypothetical multinational pet food company. The company faces a critical strategic choice: selecting a primary protein source for a new line of sustainable dog food. The contenders are Black Soldier Fly Larvae (BSFL), Single-Cell Protein (SCP) from microorganisms and a novel, high-yield soybean variant. The path to a sound decision begins not with analyzing these options, but with a careful definition of the problem itself.

3.1 Define the problem

The first and most crucial step in any structured decision process is to frame the decision precisely and accurately (Keeney 1992). This foundational stage, often overlooked, involves clarifying four key elements: the decision-maker, the alternatives, the objectives and the constraints (Howard and Abbas 2015).

First, we must ask, who is the decision-maker? The answer fundamentally shapes the analysis. A multinational corporation like VitaliFeed will be driven by concerns

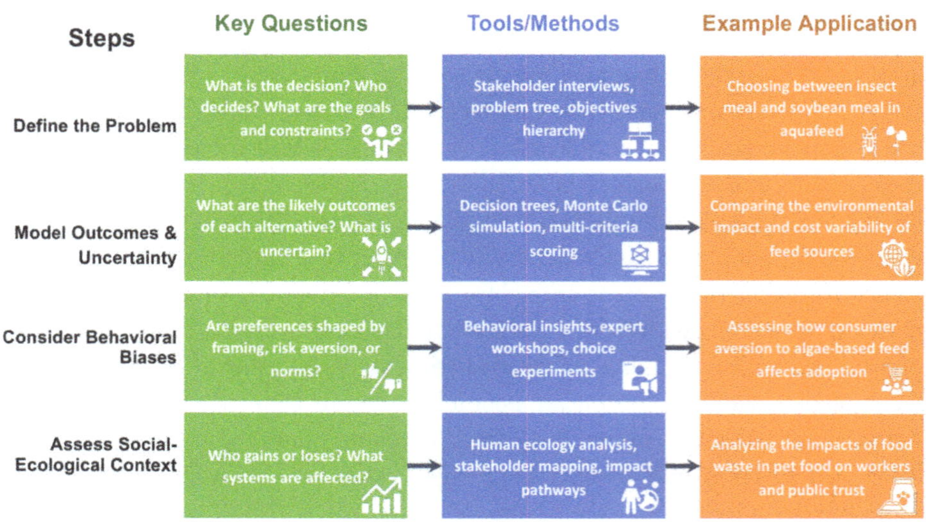

FIGURE 27.1 Four-step approach based on principles of decision analysis, enriched by insights from behavioral economics and human ecology

over global supply chain stability, shifting consumer preferences in key markets and compliance with international trade regulations. Its priorities would differ sharply from those of a small-scale farmer, who might focus on immediate costs and local availability, or from those of a government policymaker concerned with national food security. Defining the decision-maker acts as a filter, ensuring that only the relevant criteria are carried forward.

Next, the alternatives must be clearly specified. A decision model cannot function with vague possibilities. For VitaliFeed, the choice is between three distinct, tangible ingredients. In other contexts, the choice might be between broader strategies, such as investing in proprietary production versus sourcing from third-party suppliers. Clarity at this stage prevents false comparisons and grounds the analysis in reality.

The third element involves articulating the fundamental objectives, what VitaliFeed truly hopes to achieve. This moves beyond simple profit to encompass a hierarchy of goals. Fundamental objectives might include maximizing long-term economic performance, minimizing total environmental harm and safeguarding animal health. These are supported by more specific means objectives[28], such as improving protein digestibility or enhancing brand reputation through ethical sourcing. Explicitly stating these goals ensures the decision serves a purpose beyond mere selection.

Finally, the process requires identifying the constraints that bound the set of feasible options. VitaliFeed does not operate in a vacuum. It must navigate regulatory limits, such as EU regulations on novel feed ingredients (EU 2015), technical hurdles, including the shelf-life and processing requirements of each protein, and the powerful, intangible force of cultural acceptance, for instance, consumer willingness to feed insects to their pets. These constraints are not objectives to be optimized; they are hard limits that define the playing field.

Practical tip: a useful tool at this stage is an outline of the *decision context* and *problem definition.* Generate a list or table that clearly lists the main points related to your decision:

- Decision-maker
- Alternatives
- Objectives (and their relative importance if known)
- Constraints and external conditions

Such a list keeps the rest of the decision process anchored in the real-world situation.

28 'Means objectives' are the steps or actions taken to achieve larger, fundamental objectives. Means objectives are the "how" to the "why" of a fundamental objective.

3.2 Model the outcomes and uncertainty

With the decision problem clearly framed for VitaliFeed, the next step is to model the potential outcomes of each protein source, embracing the inherent uncertainty that accompanies such a choice. A model is a powerful tool for exploration, but it is not an impartial judge; it is a mirror that reflects the priorities embedded in its design (Sterman 1989). If the analysis focuses narrowly on techno-economic metrics, novel alternatives often show a compelling advantage. Our framework, however, requires a more comprehensive interrogation that includes economic stability, scalability and social acceptance.

The purpose here is not to predict the future perfectly, but to make the key trade-offs and unknowns visible and explicit. A useful starting point is a multi-criteria table (Table 27.1) that synthesizes the best available estimates for each alternative against VitaliFeed's core objectives. This provides a snapshot for initial comparison.

This table immediately illuminates the central dilemma: the novel soybean offers a low-cost, scalable, and consumer-friendly option, but at a significant environmental cost. In contrast, Black Soldier Fly Larvae (BSFL) and Single-Cell Protein (SCP) present a compelling environmental case, particularly on land use and emissions, but are hampered by higher costs and questions of scale and acceptance.

The point estimates in Table 27.1, however, tell only part of the story. A foundational principle of decision science is to move beyond averages and confront uncertainty directly (Hubbard 2014). For VitaliFeed, the future is not a single number but a range of possibilities. Probabilistic models, such as Monte Carlo simulations, can explicitly incorporate and account for variability. Key uncertainties include *supply chain stability* (How vulnerable is the SCP supply to price volatility or production hiccups? How might climate events affect soybean yields?), *social license* (what is the plausible range for consumer acceptance of insect-based pet food? Could a marketing campaign shift these probabilities?), *technological evolution* (Could breakthroughs in bioreactor design drastically lower the cost of SCP within five

TABLE 27.1 Hypothetical performance profile of alternative protein sources for VitaliFeed's sustainable dog food line

Novel ingredient	Black Soldier Fly Larvae (BSFL)	Single-Cell Protein (SCP)	Novel soybean variant
Cost (USD/ton protein)	High ($4,200)	Very high ($5,500)	Low ($1,100)
GHG emissions (kg CO_2e/ton)	Low (800)	Very low (500)	Moderate (1,500)
Land use (hectares/ton)	Very low (0.5)	Negligible (0.1)	High (3.5)
Scalability (5-yr outlook)	Moderate	Low	Very high
Protein digestibility (%)	High (90%)	High (92%)	Moderate (85%)
Consumer acceptance	Low/moderate	Low	Very high

years?). Where quantitative data is scarce, structured expert judgment can be used to bracket plausible ranges to express these uncertainties, a common practice for complex, emerging topics (Aspinall 2010; Whitney et al. 2024).

Finally, these alternatives do not exist in isolation; they are part of a dynamic system. A full model must consider interdependencies. For instance, a market feedback loop could occur: if BSFL is widely adopted, might economies of scale drive its cost down toward competitiveness? Conversely, a policy signal, such as the introduction of a carbon tax, could instantly reshape the economic landscape, penalizing soybeans for their higher emissions. Mapping these links, even qualitatively, helps VitaliFeed anticipate how the decision landscape might evolve, ensuring the choice made today is robust to the uncertainties of tomorrow.

Practical tip: a range of software options exist, from add-ins for common spreadsheets (@Risk) to open-source programming environments (R with the decisionSupport package) and dedicated simulation tools (Simul8, Analytica). The choice depends on the decision's complexity and the team's technical capacity.

- Luedeling E, Goehring L, Schiffers K, Whitney C, Fernandez E (2024). decisionSupport: Quantitative Support of Decision Making under Uncertainty. http://doi.org/10.32614/CRAN.package.decisionSupport
- Palisade Corporation (2020). @Risk for Microsoft Excel (Version 8.0) [Computer software]. https://www.palisade.com/risk/
- SIMUL8 Corporation (2021). SIMUL8 (Version 2021) [Computer software]. https://www.simul8.com/
- Lumina Decision Systems (2020). Analytica (Version 6.3) [Computer software]. https://www.lumina.com/

3.3 Consider behavioral biases

Even the most carefully modeled decision can fail in practice if it ignores the predictable ways in which human judgment deviates from pure rationality. For VitaliFeed, the choice between protein sources is not merely a technical calculation but a challenge in navigating the cognitive biases of consumers, retailers and their own decision-makers. Behavioral economics demonstrates that people rely on mental shortcuts, are powerfully influenced by how options are framed, and often prioritize immediate, certain outcomes over larger, future benefits (Tversky and Kahneman 1974; Thaler and Sunstein 2008).

The very language used to describe an alternative can predetermine its acceptance through framing effects. For VitaliFeed, an ingredient derived from "recycled food waste" may trigger immediate consumer rejection, whereas the same ingredient described as "upcycled nutrients" could be welcomed. This extends to internal decision-making, where present bias may cause a procurement manager to favor the novel soybean variant solely for its immediate cost savings, anchoring on this familiar metric while discounting the long-term strategic advantage of investing in a more resilient, lower-emission supply chain, such as Black Soldier Fly Larvae.

Aversion to the unfamiliar and perceived risk presents another significant hurdle, particularly for novel ingredients. Despite robust evidence of nutritional adequacy and safety, both retailers and consumers may exhibit a strong "ick factor" toward insect-based proteins (Higa et al. 2021). This risk aversion is not a reflection of objective danger but of aesthetic discomfort and the fear of reputational damage, leading stakeholders to reject an option that models suggest is superior. Similarly, Single-Cell Protein (SCP) may face rejection due to a naturalness bias, in which consumers perceive a "bacteria-derived" ingredient as synthetic or unnatural, even when it outperforms traditional options on key sustainability and purity metrics (Siegrist and Hartmann 2020).

These behavioral challenges are not confined to corporate or consumer contexts. The difficulties of transitioning to sustainable systems are vividly illustrated by research on Dutch farmers, where the adoption of Circular Agriculture practices was strongly influenced not only by objective data, but also by whether farmers felt capable of implementing the changes and how much they thought they knew (Lauwere et al. 2022). This case demonstrates that without confidence and capability, intention rarely translates into action. Furthermore, social pressure from conventionally operating peers created a powerful external barrier, showing that social dynamics can thwart technical superiority.

For VitaliFeed, this underscores a critical lesson: a technically optimal solution can be rejected due to cognitive biases, social norms, or consumer perceptions. Success, therefore, requires designing choice architectures that mitigate these biases. This involves pre-testing messages to defuse resistance, using pilot projects with guaranteed buyback schemes to reduce perceived risk, and crafting narratives that align with intrinsic values such as stewardship and naturalness. Explicitly incorporating these behavioral dynamics ensures that sustainable solutions are not only modeled but also adopted. Yet, even this is not enough if the solution fails to align with the broader social and ecological landscape.

3.4 Assess social-ecological context

Understanding the cognitive biases of individuals is crucial, but these biases themselves are shaped by, and shape, a broader context. A human ecology perspective emphasizes that feed choices are not made in a psychological vacuum any more than they are in a technical one; they are embedded within complex social-ecological systems comprising people, institutions, ecosystems and deeply held cultural values (Marten 2001; Kofinas and Chapin 2009). A feed ingredient is more than a nutritional input; it is a node within a web of decisions about land, labor, ethics and environmental relationships. For VitaliFeed, this means evaluating how each protein source performs within the human ecological system.

The land-use implications of each alternative reveal critical environmental trade-offs. While the novel soybean variant may offer cost advantages, its cultivation could perpetuate deforestation pressures and biodiversity loss in sensitive biomes. Conversely, the production of Black Soldier Fly Larvae (BSFL) can valorize organic waste streams and dramatically reduce land use, but scaling this system introduces new questions about the feedstock sourcing, processing requirements and the robustness of biosafety protocols to prevent contamination (Van Huis and Oonincx 2017). Single-Cell Protein (SCP), while land-efficient, may require significant water and energy resources, potentially competing with local community needs if production is sited in water-stressed regions.

Beyond environmental impacts, feed decisions must navigate cultural norms and ethical values. What is considered a sustainable innovation in one context may be unacceptable in another. For VitaliFeed, marketing a BSFL-based product in European markets may align with growing consumer interest in circular economies. Still, the same product could face resistance in other regions where insects are not considered appropriate animal feed due to cultural or religious norms. Furthermore, ethical considerations arise regarding the framing of animals within these systems, questioning whether they should be viewed as partners in waste processing or as sentient beings whose welfare is paramount (Bracke et al. 2023).

Finally, the distribution of benefits and burdens from feed innovation is a critical equity issue. A shift to these novel proteins will create winners and losers. A decentralized model for BSFL production could empower smallholder farmers by providing a new source of income, but only if supported by training and market access. Conversely, if SCP production becomes dominated by a few large corporations controlling proprietary microbial strains, it could marginalize smaller actors and lead to vertical integration that concentrates power (Hermans et al. 2017). For VitaliFeed, this means mapping its supply chain to understand who controls production and who is most vulnerable to shifts in policy or price.

Integrating this social-ecological thinking requires VitaliFeed to look beyond efficiency. It involves actively mapping stakeholder networks and power dynamics, considering impacts on ecosystem services like biodiversity and soil health, and engaging local knowledge to understand unintended cultural or labor consequences. This broader assessment helps ensure that the company's sustainable feed innovation contributes to a system that is resilient and just, not merely efficient (Davila and Dyball 2017).

4 Embedded decision-making for novel feed ingredients

To fully appreciate the interplay of factors outlined in the previous sections, we can visualize the decision process as a set of nested, interacting systems, as shown in the Venn diagram in Figure 27.2. This conceptual model moves beyond a linear checklist and instead posits that any decision regarding a novel feed ingredient, VitaliFeed's choice of protein source being a prime example, is embedded within and constantly influenced by a series of overlapping contexts.

At the core of the system lies the specific *Novel Feed Decision* itself, such as the adoption of insect meal. This core is where immediate technical and economic metrics, cost, protein content and lifecycle assessment results, are directly weighed. However, this core is not an isolated entity; it is immediately enveloped and constrained by the *Farm & Innovation System*. This ring represents the practical realities of implementation, encompassing the available production technology, the robustness of supply chains for sourcing and distribution, and the knowledge ecosystem of R&D and advisory services that support the innovation.

Expanding outward, the decision is further shaped by the *Social & Cultural System*, which governs its social license to operate. This domain includes the cultural, ethical, and religious values that define what is acceptable, conceptions of animal welfare, and the critical equity and power dynamics that determine how benefits and costs are distributed among actors, from smallholder farmers to multinational corporations. Simultaneously, regulations, subsidies and incentives within this system create a formal structure that can either accelerate or obstruct a novel alternative.

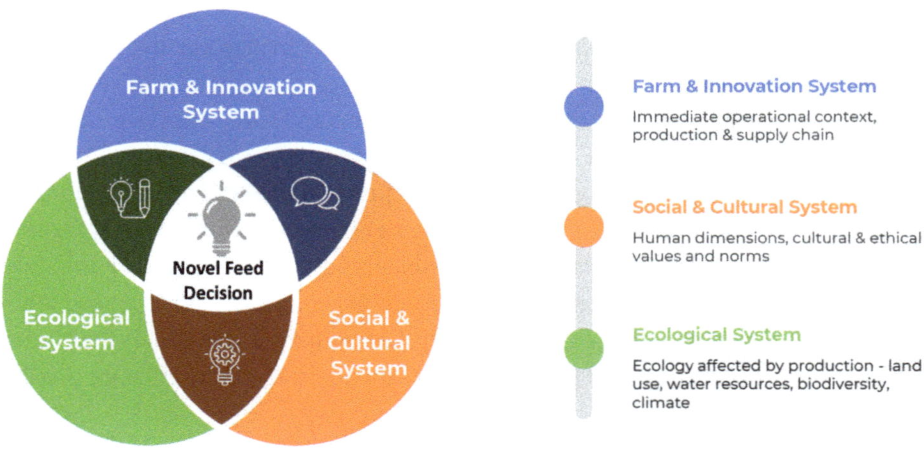

FIGURE 27.2 The nested contexts of a novel feed decision, illustrating how it is embedded within and influenced by farm/innovation, social/cultural and ecological systems.

Underpinning all else is the *Ecological System*, the ultimate foundation that provides essential ecosystem services such as climate regulation, biodiversity and freshwater. Decisions at the core inevitably impact this outer ring, and the health of this ring, in turn, feeds back to alter the context for all future decisions; for instance, the degradation of ecosystem services directly increases long-term production risks.

Crucially, influence in this model is not a one-way street from the outer rings inward. A decision made in the core creates feedback loops that alter all other systems. The widespread adoption of a novel feed can reshape supply chains, gradually shift cultural perceptions and trigger new policy developments, all while leaving a tangible footprint on local ecosystems. This continuous, multidirectional interaction defines the system as complex, dynamic and adaptive.

Using this framework ensures that decision-making moves beyond a narrow techno-economic analysis. It provides a structured way to map stakeholders and power dynamics, consider impacts on ecosystem services and incorporate context-specific knowledge. By visualizing these interactions, as VitaliFeed must, businesses, policymakers, and farmers can better anticipate the system's multifaceted responses, designing strategies that harness benefits while mitigating unintended consequences.

5 Conclusion

Novel protein sources, from insect meal to single-cell proteins, hold potential to steer our feed and pet food systems toward a more sustainable future. However, as the VitaliFeed case illustrates, their adoption is far from a straightforward technical calculation. It demands a careful navigation of complex, often conflicting, priorities, weighing ecological benefits against economic costs, ethical considerations against scalability, and scientific data against deep-seated social perceptions, all under conditions of significant uncertainty.

This chapter has argued that confronting this complexity requires a disciplined, multi-faceted approach. The framework presented here provides a path forward by insisting on a clearly defined decision problem, transparent modeling of outcomes and uncertainties, honest accounting of behavioral biases, and a thorough assessment of the broader social-ecological context. When used together, these steps transform an overwhelming dilemma into a structured inquiry.

The tools of decision theory and behavioral economics, combined with the systems-thinking of human ecology, do not generate perfect, one-time answers. Instead, they create a robust and adaptive process. They make values, assumptions, and trade-offs explicit, establishing a more informed and defensible basis for action. The key insights are clear: sustainable feeds involve inescapable trade-offs; public and stakeholder perceptions are as critical as protein scores; and every feed choice is embedded in a unique ecological and cultural context that determines its ultimate impact.

Supporting a true transition to sustainable feed systems will require more than technical innovation; it will necessitate structured thinking and context-sensitive governance. The framework presented in this chapter enables the learning and adaptation necessary for sustainable transitions. The framework provides a structure not just for making an initial choice, but for revisiting it as markets evolve, new science emerges and societal values shift. In this light, adopting such a structured decision-making process is not merely an analytical exercise, it is a foundational competency for any organization serious about navigating the coming decades of food system transformation.

6 Key insights

- *Not all sustainable feeds are equal*: trade-offs exist across cost, emissions and social values
- *Perceptions matter*: acceptance of novel ingredients depends on framing and trust
- *Context matters*: feed choices impact ecosystems, cultures and livelihoods
- *Tools help*: simple models + human-centered thinking improve outcomes

References

Akintan, Oreofeoluwa, Kifle G. Gebremedhin, and Daniel Dooyum Uyeh. 2024. Animal feed formulation—Connecting technologies to build a resilient and sustainable system. *Animals* 14. Multidisciplinary Digital Publishing Institute: 1497. https://doi.org/10.3390/ani14101497.

Aspinall, Willy. 2010. A route to more tractable expert advice. *Nature* 463. Nature Publishing Group: 294–295. https://doi.org/10.1038/463294a.

Bracke, M. B. M., I. J. M. M. Boumans, H. J. Nijland, and E. A. M. Bokkers. 2023. Review: connecting circularity to animal welfare calls for a 'novel' conceptual framework based on integrity. *animal* 17: 100694. https://doi.org/10.1016/j.animal.2022.100694.

Chamodi, Korale Kankanamge Dinuka, Nguyen Thanh Vu, Jose A. Domingos and Jiun-Yan Loh. 2025. Cellular Solutions: Evaluating Single-Cell Proteins as Sustainable Feed Alternatives in Aquaculture. *Biology* 14(7):764. https://doi.org/10.3390/biology14070764

Chhimwal, Madhukar, Saurabh Agrawal, and Girish Kumar. 2022. Challenges in the implementation of circular economy in manufacturing industry. *Journal of Modelling in Management* 17: 1049–1077. https://doi.org/10.1108/JM2-07-2020-0194.

Davila, Federico, and Robert Dyball. 2017. Food systems and human ecology: an overview. In *Sustainability science*. Routledge.

Dierenfeld, Ellen S. 2023. 94 sustainable feeding solutions across the animal kingdom: identifying and implementing. *Journal of Animal Science* 101: 116–117. https://doi.org/10.1093/jas/skad281.142.

EU. 2015. Regulation (EU) 2015/2283 of the European Parliament and of the Council of 25 November 2015 on novel foods, amending Regulation (EU) No 1169/2011 of the European Parliament and of the Council and repealing Regulation (EC) No 258/97 of the European Parliament and of the Council and Commission Regulation (EC) No 1852/2001 (Text with EEA relevance). *Official Journal of the European Union* OJ L 327,: 327. data.europa.eu. http://data.europa.eu/eli/reg/2015/2283/oj.

Focker, M., E. D. van Asselt, B. J. A. Berendsen, M. G. M. van de Schans, S. P. J. van Leeuwen, S. M. Visser, and H. J. van der Fels-Klerx. 2022. Review of food safety hazards in circular food systems in Europe. *Food Research International* 158: 111505. https://doi.org/10.1016/j.foodres.2022.111505.

Hermans, Frans, Murat Sartas, Boudy Schagen, Piet Asten, and Marc Schut. 2017. Social network analysis of multi-stakeholder platforms in agricultural research for development: Opportunities and constraints for innovation and scaling. *PLoS ONE* 12. https://doi.org/10.1371/journal.pone.0169634.

Hersh, M.A. 1999. Sustainable decision making: the role of decision support systems. *IEEE Transactions on Systems, Man, and Cybernetics, Part C (Applications and Reviews)* 29: 395–408. https://doi.org/10.1109/5326.777075.

Higa, Jennifer E., Matthew B. Ruby, and Paul Rozin. 2021. Americans' acceptance of black soldier fly larvae as food for themselves, their dogs, and farmed animals. *Food Quality and Preference* 90: 104119. https://doi.org/10.1016/j.foodqual.2020.104119.

Howard, Ronald A., and Ali E. Abbas. 2015. *Foundations of Decision Analysis*. NY, NY: Prentice Hall.

Hubbard, Douglas W. 2014. *How To Measure Anything: Finding the Value of Intangibles in Business*. 2nd ed. Vol. Second Edition. Hoboken, New Jersey: John Wiley & Sons.

Kahneman, Daniel, and Patrick Egan. 2011. *Thinking, fast and slow*. Vol. 1. New York: Farrar, Straus and Giroux.

Keeney, Ralph L. 1992. *Value-Focused Thinking: A Path to Creative Decisionmaking*. Harvard University Press. https://doi.org/10.2307/j.ctv322v4g7.

Keeney, Ralph L. 2021. Give yourself a nudge to make smarter business decisions. *MBR, Fuqua School of Business, Duke University* 01: 195–203.

Kofinas, Gary P., and F. Stuart Chapin. 2009. Sustaining livelihoods and human well-being during social-ecological change. In *Principles of ecosystem stewardship: resilience-based natural resource management in a changing world*, ed. Carl Folke, Gary P. Kofinas, and F. Stuart Chapin, 55–75. New York, NY: Springer. https://doi.org/10.1007/978-0-387-73033-2_3.

Lauwere, Carolien, Monique Slegers, and Marieke Meeusen. 2022. The influence of behavioural factors and external conditions on Dutch farmers' decision making in the transition towards circular agriculture. *Land Use Policy* 120: 106253. https://doi.org/10.1016/j.landusepol.2022.106253.

Liu, Kang. 2024. Circular economy and the separated yet inseparable social dimension: views from european circular city experts. *Sustainable Production and Consumption* 51: 474–483. https://doi.org/10.1016/j.spc.2024.09.016.

Lotfian Delouyi, Fahime, Meisam Ranjbari, and Zahra Shams Esfandabadi. 2023. A hybrid multi-criteria decision analysis to explore barriers to the circular economy implementation in the food supply chain. *Sustainability* 15: 9506. https://doi.org/10.3390/su15129506.

Malenica, Dunja, Marko Kass, and Rajeev Bhat. 2023. Sustainable management and valorization of agri-food industrial wastes and by-products as animal feed: for ruminants, non-ruminants and as poultry feed. *Sustainability* 15. Multidisciplinary Digital Publishing Institute: 117. https://doi.org/10.3390/su15010117.

Marten, Gerald G. 2001. *Human ecology: Basic concepts for sustainable development*. Sterling, VA: Earthscan.

Rossi, G., M. Psarianos, S. Ojha, and O. K. Schlüter. 2025. Review: insects as a novel feed ingredient: processing technologies, quality and safety considerations. *animal* 19. From Farm to Feed - Harnessing Insects for Sustainable Food and Feed Systems: 101495. https://doi.org/10.1016/j.animal.2025.101495.

Rozin, Paul, Claude Fischler, and Christy Shields-Argelès. 2009. Additivity dominance: additivites are more potent and more often lexicalized across languages than are "subtractives." *Judgment and Decision Making* 4: 475–478. https://doi.org/10.1017/S1930297500004010.

Siegrist, Michael, and Christina Hartmann. 2020. Consumer acceptance of novel food technologies. *Nature Food* 1: 343–350. https://doi.org/10.1038/s43016-020-0094-x.

Sterman, John D. 1989. Modeling managerial behavior: misperceptions of feedback in a dynamic decision making experiment. *Management Science* 35: 259-385. https://doi.org/10.1287/mnsc.35.3.321.

Suckling, J., A. Druckman, C. D. Moore, and D. Driscoll. 2020. The environmental impact of rearing crickets for live pet food in the UK, and implications of a transition to a hybrid business model combining production for live pet food with production for human consumption. *The International Journal of Life Cycle Assessment* 25: 1693–1709. https://doi.org/10.1007/s11367-020-01778-w.

Thaler, Richard H., and Cass R. Sunstein. 2008. *Nudge: improving decisions about health, wealth, and happiness*. New Haven (Conn.): Yale university press.

Tversky, Amos, and Daniel Kahneman. 1974. Judgment under uncertainty: Heuristics and biases. *Science, New Series* 185: 1124–1131.

Van Huis, Arnold, and Dennis G. A. B. Oonincx. 2017. The environmental sustainability of insects as food and feed. A review. *Agronomy for Sustainable Development* 37: 43. https://doi.org/10.1007/s13593-017-0452-8.

Van Raamsdonk, L.W.D., N. Meijer, E.W.J. Gerrits, and M.J. Appel. 2023. New approaches for safe use of food by-products and biowaste in the feed production chain. *Journal of Cleaner Production* 388: 135954. https://doi.org/10.1016/j.jclepro.2023.135954.

Weaver, Christopher P., Robert J. Lempert, Casey Brown, John A. Hall, David Revell, and Daniel Sarewitz. 2013. Improving the contribution of climate model information to decision making: the value and demands of robust decision frameworks. *WIREs Climate Change* 4: 39–60. https://doi.org/10.1002/wcc.202.

Whitney, Cory, Christine M. Schmitz, and Eike Luedeling. 2024. Learn how to make accurate estimates: Calibration training with an interactive online application. In *2024 Annual Conference of the Decision Sciences Institute Proceedings*, 849–864. Phoenix, Arizona: University of Arizona.

PART 7

Market insights in feed and pet food

CHAPTER 28

Innovation in pet food: era of new ingredients

E.A. Tretyakova

Innova Market Insights, Velperweg 18, 6824 BH Arnhem, Netherlands
Correspondence: ekaterina.tretyakova@innovami.com

Abstract

Driven by climate change and increasing environmental concerns among consumers, a new generation of innovative ingredients is emerging in the pet food sector. Algae, insect-based ingredients, fermentation-driven and plant-based proteins are gaining attention not only for their sustainability benefits, but also for their nutritional value, offering essential components such as proteins, fatty acids, vitamins, and more. However, the common usage of these novel ingredients in pet food is still in its early stages. Broader adoption depends on increased awareness among pet owners, who show interest but remain cautious, as well as a stronger innovation pipeline from manufacturers.

Keywords

sustainability – trends – protein – algae – insects – fermentation – plant-based

1 Introduction

The rise of innovative sustainable ingredients in cat and dog food and snacks – referred to in this article as pet food – reflects a broader transformation in how consumers approach both nutrition and environmental responsibility. As awareness of climate change and resource scarcity grows, pet owners are increasingly seeking food options that align with their values around sustainability, without compromising on health or quality for their pets. This evolving mindset is driving interest in alternative protein sources. Though they remain niche, they are gaining attention for their potential to reduce the environmental footprint of pet food production.

Leveraging data from Innova Market Insights' New Products Database, as well as the Trends Survey 2025 and Pet Owner Survey 2025, this article explores current developments in sustainable pet food innovation, examines shifting consumer attitudes, and highlights emerging trends in alternative proteins and lipids.

2 Sustainability awareness

2.1 Growing interest in sustainable ingredients

Climate change is increasingly gaining traction, and its consequences become more feasible for consumers. According to Innova Market Insights' Trends Survey 2025, conducted in July–August 2024, 80% of global consumers are aware of the impact of climate change. Consumers in Brazil and Mexico – countries more exposed to its consequences – feel the most severe impact of climate change. The survey covered 14,000 respondents across Brazil, Canada, China, France, Germany, India, Indonesia, Mexico, Spain, the UK, and the US.

Considering the accelerating concerns about climate change, there is growing interest in sustainable solutions. The survey shows that 46% of consumers consider the sourcing and production of meat 'very or extremely' impacted by climate change. These numbers are even higher for fish and seafood with every second consumer saying that the category is being severely impacted by climate change. It raises a question about alternative sources of protein in pet food and reveals a demand for sustainable sourcing.

Sustainable product features play an increasingly important role in purchase decisions. Globally, 55% of consumers say that they pay attention to nature protection-related claims in food and beverage packaging. This number is relatively low in the UK, the US or Germany, while being relatively higher in Global South – Brazil, China, India, and Indonesia. This pattern suggests that consumers in regions more affected by climate change tend to place greater importance on sustainability messaging. Reflecting this, consumers in these markets also report a higher willingness to pay more for brands and products that actively address sustainability issues. Overall, nearly one in two consumers states a willingness to pay extra for sustainable brands and products.

Sustainability has a significant influence on brand loyalty, as 44% of pet owners say they are likely to switch pet food brands based on the product's environmental impact. Almost every second consumer claims that he would stop or reduce consumption if he knew a food or beverage product was unsustainable.

2.2 Consumer demand for transparency

There is a clear demand for sustainably sourced and produced products. Meanwhile, often the problem is that consumers find it difficult to distinguish between sustainable and non-sustainable products. Globally, half of consumers report having trouble identifying which food and beverage products are truly sustainable.

Several barriers contribute to this challenge. The key one is the limited availability of sustainable options, making it the most significant obstacle to environmentally conscious choices. Confusing labels hinder 28% of consumers, and 26% worry about misinformation such as misleading advertising or unverified claims.

This reveals a key challenge for manufacturers: while demand for sustainable products is strong, it must be matched by improved availability and clear, transparent communication. To meet consumer expectations, brands need to ensure that sustainability claims are credible and that sourcing and production processes are communicated in a way that builds trust and enables informed purchasing decisions.

2.3 Shifts towards natural ingredients

In the pet food category, natural ingredients and health have consistently remained top priorities for pet owners. As per Innova Market Insights' Pet Owner Survey 2025, held in October-November 2024 in the US, Germany, UK, China and Brazil, clean label claims are particularly important. The survey reveals that 59% of 5,000 respondents prefer to avoid preservatives in pet food, while 64% actively avoid artificial colorants and flavors. Health aspects strongly influence purchasing decisions: two-thirds of pet owners say that they prefer buying pet food with health claims over cheaper products without them. When it comes to packaging, most pet owners pay attention to the ingredient list and nutritional facts, while sustainable sourcing attracts less attention (Table 28.1).

These findings clearly reflect strong consumer demand for clean-label, natural ingredients and health-focused products in pet food. This trend is contributing to the expansion of products that emphasize natural composition, nutritional value, and health benefits. While sustainable sourcing is of lesser importance compared to health and natural attributes, the market is gradually opening up to alternative sustainable ingredients, supported by transparent packaging communication on origin, processing, and environmental impact.

TABLE 28.1 The most important packaging information for pet owners

Category	% of respondents
Ingredient lists	49%
Nutrition facts/calories	35%
Clean label/natural	16%
Health benefits	15%
Sustainable farming/sourcing	9%

Question: What type of packaging information/elements do you pay most attention to? Only relevant answers are included. Source: Innova Market Insights' Pet Owner Survey 2025.

3 Expansion of novel ingredients

Pet owners are eager to experiment with new types of pet food with over 60% saying that they are open to trying new types of food for their pets. However, when it comes to truly novel ingredients, they still remain niche within the mainstream pet food market. Even among meat-free pet food products, the most preferred protein sources are still relatively conventional – soy protein, wheat protein, and pea protein continue to dominate.

3.1 Algae

Some other innovative alternatives, such as algae, are far less common. Only five and seven percent of dog and cat owners respectively prefer algae as a protein base for meat-free pet food.

However, the interest in algae as a protein source is emerging. Around one-third of pet owners say they would probably consider feeding their pet food that contains microalgae, and 1 in 10 state they would definitely do so. This openness is more common among dog owners, as cats are generally more selective with their food. The main reasons pet owners consider microalgae as a protein source include its natural ingredients, perceived health reasons, and recommendations by veterinarians.

As per Innova Market Insights data, among dog and cat food and snack innovations, the share of new launches featuring algae ingredients has grown from one percent in 2020 to five percent in 2024. Algae is most commonly used in dry dog food, and to a lesser extent in cat food, while algae-based innovations in pet snacks are emerging from a non-existence.

As a source of protein, antioxidants, and Omega-3, algae offer the opportunity to position pet products as both nutritious and sustainable. Nearly half of new pet products containing algae launched since 2020 include a 'high in protein' claim, while about one-third highlight skin health benefits and Omega-3. At the same time, sustainability-related claims remain rare. This reveals a clear opportunity for brands to better connect with environmentally conscious consumers, for whom sustainability was mentioned as the main reason to consider novel ingredients.

3.2 Insects

Another emerging source of protein in pet food is insects. According to Innova's Pet Owner Survey 2025, less than 10% of pet owners fed their pets with insect-based protein during the last year. This figure is twice as high in China, where acceptance of insect protein in pet food is generally higher than in other countries. Awareness of insect-based pet food also remains low. Overall, only 17% of pet owners are familiar with innovations in pet food involving alternative proteins such as insect protein, with Millennials and Gen Z being the most informed age groups.

When it comes to openness, 38% of pet owners say they would feed their pets food containing insects. The top reasons cited for considering insect protein include its natural origin, being a natural food source, and its reputation as a good source of nutrients such as fats, minerals, and vitamins.

Insects remain a niche ingredient, featured in less than one percent of new pet food and snack launches. They are most commonly used in dog snacks and dry food, with black soldier fly and insect protein being the primary ingredients within this category. Although consumer acceptance of insects as an ingredient is higher in Asia, it is Western Europe and North America that are currently driving innovation in this space.

Still considered a niche ingredient, insect protein is gradually gaining traction among consumers. While current usage remains low, higher adoption in markets like China and growing awareness among younger generations suggest that acceptance is likely to increase.

3.3 Plant-based proteins

Plant-based proteins are more established ingredients in pet food. 48% of pet owners say they are willing to include plant-based ingredients in their pet's diet. However, actual behavior tells a different story – only 1 in 5 pet owners reported buying pet food with plant-based proteins in the past 12 months. This gap suggests that while interest exists, it is not yet strongly supported by the factors that drive real purchasing decisions.

One key barrier is price. According to Innova's Trends Survey 2025, 54% of global consumers cite price as a barrier when choosing sustainable food and beverage products. This is likely to the pet food category as well, where plant-based or alternative protein products can carry a higher cost.

In addition, many pet owners are cautious about applying their own dietary restrictions to their animals. For example, only 23% of vegan pet owners feed their pets a meat-free diet – most believe that dogs and especially cats require meat for proper nutrition.

Plant-based proteins were featured in nearly 30% of new pet food launches in 2024. These ingredients are used across all pet food formats, with the highest relevance seen in dry dog and cat food. The most common plant-based proteins include vegetable protein, corn gluten, and wheat gluten, while corn and potato proteins are among the fastest-growing.

Although plant-based proteins are clearly used as a protein source, only 41% of pet food products containing these ingredients since 2020 have included a 'high protein' claim on the packaging. Products with potato, corn, and rice protein show the highest rate of high-protein positioning, with such claims appearing on over 60% of new launches in these categories. Rice protein also shows a relatively higher share of products featuring sustainable sourcing claims, though even in this case, the figure remains below two percent.

Mushroom proteins are fairly popular among consumers, with 26% of cat owners and 28% of dog owners considering them as a protein base in meat-free pet food. However, they remain rare in actual product innovation, appearing in the ingredient lists of less than one percent of new pet food launches. Mushroom-based ingredients are more frequently found in products launched in North America, with significantly lower presence in Europe.

3.4 Lab-grown meat

Lab-grown meat still requires increased awareness among pet owners. Nearly half of pet owners (47%) say they would prefer to avoid lab-grown meat in their pet's food. In the US and Brazil, this sentiment is even stronger, with 26% of pet owners in both countries saying they would definitely avoid such products.

Still, 36% of pet owners say they would consider feeding their pet lab-grown meat. The main reasons for considering lab-grown meat include recommendations from veterinarians, health benefits, and the perception that it is a good source of nutrients such as fats, minerals, and vitamins.

While awareness of lab-grown meat in pet food remains low – particularly in the US – and pet owners have concerns, the willingness to try such products suggests potential for growth if these concerns are addressed, the products become more available and accessible, and awareness improves. For now, however, strong avoidance and limited familiarity indicate that widespread acceptance is still a long way off.

3.5 Fermentation

The pet food segment with protein derived from microbial fermentation is still tiny but is showing signs of diversification beyond yeast-based ingredients. While yeast protein and yeast protein hydrolysates still dominate this space, novel microbial proteins produced via fermentation of bacteria and fungi are beginning to emerge. These innovations use biomass fermentation to produce sustainable protein alternatives with a low environmental footprint. Despite the niche status of these products, consumer openness suggests growth potential. 43% of pet owners say they would feed their pet food containing such proteins.

3.6 Comparison of innovative ingredients

Overall, innovations with novel ingredients remain niche, and awareness of their use in pet food is limited. However, potential interest is strong as over 40% of pet owners say they would consider feeding their pets algae or fermented proteins. Insects and lab-grown proteins show slightly lower appeal, with 36% of pet owners saying they would avoid feeding them to their pets.

Still, the cautious approach applies across various novel protein sources. Most pet owners are more comfortable trying ingredients like microalgae, insects, lab-grown meat, and fermentation-derived proteins in treats or supplements, rather than in their pet's main meals. This likely reflects a preference for lower-risk experimentation – trying new ingredients in smaller quantities before considering a broader dietary change.

This trend indicates that introducing novel proteins through treats and supplements may be a more effective entry point for consumer acceptance, especially for owners who are open to innovation but hesitant to shift their pet's core diet.

Looking at the share of pet owners who are willing to try novel ingredients, the US appears to be the most cautious market. No more than 25% of pet owners in the US say they would be willing to try insects, lab-grown meat, or algae in their pet's food. In contrast, China stands out as the most open market, where consumers are more willing to experiment with new types of pet food.

Among the four protein sources, microalgae seem to be the most appealing to pet owners overall. On the other hand, lab-grown meat and fermentation-derived proteins still face significant challenges and will require greater efforts in awareness-building and education to gain wider acceptance.

When comparing the main barriers mentioned by pet owners across these novel ingredients, the leading concern is lack of information (Figure 28.1). 21% of consumers say they would need to see more research and background before purchasing pet food with microalgae. The second most common concern is safety, again highlighting the need for clear communication to build trust in the safety of innovative ingredients. Price rounds out the top three barriers, reinforcing the need to make novel protein pet foods more accessible and affordable. For lab-grown meat, an additional key barrier is the perception of unnaturalness. As previously discussed, pet owners strongly value natural ingredients, making it harder for lab-grown meat to break through mainstream expectations.

When it comes to insects, consumer reluctance is often tied to discomfort with the idea of insects as food ingredients and doubts about palatability for pets. These concerns appear to outweigh even price as a barrier in this case.

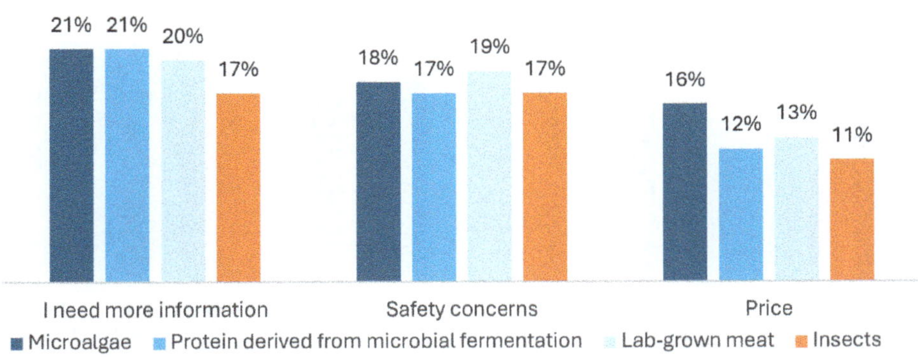

FIGURE 28.1 Key barriers for novel ingredients in pet food. Question: Why would you NOT consider giving the below protein sources to your pet? Source: Innova Market Insights' Pet Owner Survey 2025

Right positioning remains key for spreading awareness about ingredients. Products with microalgae and insects often highlight high-protein and gut-health benefits, while consumers are also seeking these ingredients for their naturalness (Figure 28.2). Companies could leverage this interest by using more natural-oriented language on packaging, emphasizing authenticity. At the same time, sustainability-related claims are largely absent from most packages featuring algae and insects. This represents a clear opportunity for companies to step in and differentiate.

4 Future outlook

The pet food industry is entering a new era of innovation, driven by increasing consumer awareness of climate change, sustainability, and pet health. Novel ingredients such as algae, insects, plant-based proteins, fermentation-derived proteins, and lab-grown meat are gaining attention for their dual benefits: reducing environmental impact while providing essential nutrition for pets.

However, the current adoption of these ingredients remains limited, and many innovations are still in early-stage product launches. Consumer interest is growing, but barriers such as lack of information, safety concerns, perceptions of unnaturalness, and price sensitivity continue to slow widespread acceptance.

Looking ahead, education and transparency will be essential. Many consumers remain unfamiliar with novel sources of protein, and uncertainty around their safety, nutritional adequacy, and environmental benefits continues to hold back broader acceptance. For these ingredients to gain traction, brands must take an active role in educating pet owners. They should highlight not only the nutritional value of such ingredients but also provide clear, credible information about sourcing, pro-

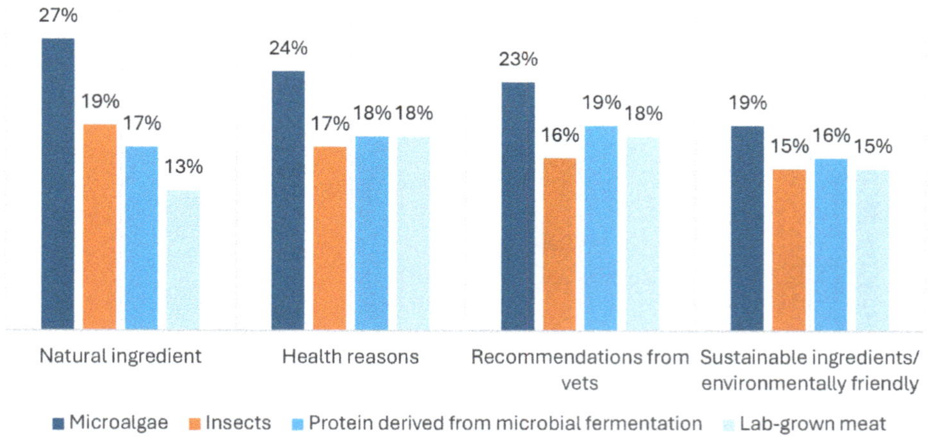

FIGURE 28.2 Selected reasons for pet owners to buy pet food with novel ingredients. Question: Why would you consider giving the below protein sources to your pet? (Select all that apply). Source: Innova Market Insights' Pet Owner Survey 2025

duction methods, and the science behind the innovation. Transparency will be especially important in building consumer trust, and messaging must be consistent, evidence-based, and easy to understand. While there is growing interest, particularly among younger pet owners, the market is still in its early stages, and a broader shift will take time. If awareness keeps rising and brands address barriers like price and accessibility, adoption of these ingredients is likely to grow – gradually, and supported by ongoing investment in consumer engagement and education.

4.1 What this means for the industry

Companies seeking to accelerate adoption of novel ingredients should focus on increasing awareness while building credibility. One effective approach is to highlight specific health benefit claims and integrate veterinary expertise, which can help pet owners feel more confident about the safety and value of these innovations.

Building this confidence is often easiest through lower-risk product formats. Treats and supplements appear to be the most effective entry point, as pet owners are more willing to test new ingredients in smaller, less disruptive formats. This strategy allows manufacturers to introduce novel proteins gradually, creating familiarity and paving the way for wider inclusion in core meals.

Alongside health-focused positioning, the strategic use of sustainability claims offers another pathway to differentiation. As highlighted at the beginning of this article, climate change and environmental responsibility are becoming powerful forces shaping consumer purchasing decisions. Yet, sustainability messages remain underrepresented in launches with algae, insects, or fermentation-based proteins. Framing these ingredients not only as nutritional innovations but also as climate-conscious solutions can help brands better connect with environmentally aware pet owners.

Equally important is ensuring authenticity in communication. Beyond underlining technical innovation, companies should emphasize the natural origins of alternative proteins to resonate emotionally with pet owners. Packaging and marketing that adopt a more consumer-friendly, natural-oriented language will strengthen perceptions of both functionality and naturality.

Although novel ingredients in pet food are still niche, consumer interest signals potential. To capture this opportunity, brands should invest in raising awareness and building credibility, turning curiosity into lasting adoption.

References

Innova Market Insights. (2020–2025). Innovation Database.

Innova Market Insights. (2024). Pet Owner Survey 2025.

Innova Market Insights. (2024). Trends Survey 2025.

CHAPTER 29

Global compound feed market overview

D.G. Yildiz

Feed & Additive Magazine, Istanbul, Türkiye
Correspondence: derya@feedandadditive.com

1 Introduction

Compound feeds are one of the cornerstones of modern livestock production. Formulated according to the species and life stage of the animal, they contain a balanced mix of cereals, proteins, vitamins, minerals, enzymes, and probiotics. With the right formulation, they support animal health, growth, reproduction, and performance, while also improving feed efficiency and providing economic advantages to producers. They offer ease of use on farms and eliminate the need to mix individual ingredients separately. Although the initial cost is higher, optimized nutrient utilization, reduced waste, and the preference for sustainable ingredients help lower costs in the long term, while also reducing environmental impacts and supporting animal welfare.

The growing demand for animal protein, the expansion of industrial livestock production, technological advancements, and government support are driving the global growth of the compound feed market. However, price fluctuations in raw materials, environmental regulations, disease outbreaks, and the traditional feed usage of small producers are factors that limit growth.

2 Global compound feed production

According to the 2024/25 Annual Report of International Feed Industry Federation (IFIF), world compound feed production will likely reach an estimated 1.4 billion tons annually in 2025 (IFIF, 2025). Additionally, global commercial feed manufacturing generates an estimated annual turnover in excess of USD 500 billion. According to the same report, the species distribution of world compound feed production is as follows:

- Poultry feed: 43%
- Pig feed: 26%
- Dairy feed: 11%
- Beef feed: 10%

- Aquafeed: 4%
- Pet food: 3%
- Other ruminant feed: 2%
- Equine feed: 0.6%
- Other species: 0.4%

Similar figures and additional details are provided by Alltech in the Agri-Food Outlook 2025 (Alltech, 2025b). According to this report, after a slowdown in 2023, global compound feed production recovered in 2024, reaching approximately 1.396 billion tons. This represents an increase of 16.7 million tons, or 1.2%, compared to 2023. Regionally, the largest producer is the Asia-Pacific region, with production of 533 million tons. It is followed by North America's 291 million tons, Europe's 268 million tons, Latin America's 198 million tons, Africa's 58 million tons, the Middle East's 38 million tons, and Oceania's 11 million tons. By country, the largest producer is China, with approximately 315 million tons of feed. It is followed by the USA's 270 million tons, Brazil's 87 million tons, India's 55 million tons, Mexico's 41 million tons, Russia's 38 million tons, Spain's 36 million tons, Vietnam's 26 million tons, Turkey's 25 million tons, and Japan's 24 million tons (Alltech, 2025b). The total compound feed production of these 10 countries is approximately 917 million tons, representing around 65.7% of global production.

3 Global market value and growth expectations

Despite periodic fluctuations in production, the value of the global compound feed market continues to increase year on year. According to a report published by Fortune Business Insights (Fortune Business Insights, 2025), the global compound feed market was valued at approximately USD 583 billion in 2024 and reached USD 611 billion in 2025. The company projects a compound annual growth rate (CAGR) of 5.4% for the 2025–2032 period, forecasting that the market will reach USD 881 billion by the end of 2032.

Similarly, a report released by Precedence Research, also points to a similar market size, though it presents a slightly higher projection for the future. The report estimates the size of the global compound feed market at around USD 621 billion in 2025 and anticipates that it will expand at a CAGR of 4.3%, reaching USD 907 billion by 2034 (Precedence Research, 2025).

Although differences in methodology and pricing parameters lead to some variation in these forecasts, the estimated size of the global compound feed market for 2025 generally falls within the range of USD 590–620 billion. Overall, these analyses indicate that the global compound feed industry – one of the cornerstones of the food system – will continue to experience strong growth, driven by the rising global demand for animal-based products such as meat, milk, and eggs.

4 Factors supporting market growth

The dynamics shaping the future of animal-based food production also stand out as the main drivers supporting the growth of the global compound feed market. The most important of these is the rising global population and demand for animal protein. The increasing world population requires a significant increase in food production to ensure global food security. At the same time, rising disposable incomes in developing countries are fuelling greater demand for animal proteins such as meat, milk, and eggs. This trend drives the expansion of livestock production and, consequently, the demand for compound feed. The expansion of industrial livestock farming is another factor driving up demand for compound feed. Worldwide, particularly in developing regions such as Asia-Pacific and Africa, there has been a transition from small-scale traditional farming to larger, more intensive operations. These modern farms naturally require specially formulated compound feeds to achieve efficiency and productivity. The need for feed efficiency and quality is increasing demand for specially formulated feeds. Because profitability in livestock production is directly linked to feed efficiency. Advances in feed formulation and animal nutrition science now enable the development of highly specialized feeds tailored to the specific needs of different species, breeds, and life stages. This approach, known as precision nutrition, maximizes feed efficiency and animal performance while encouraging the use of customized compound feeds. Another factor supporting growth is advancements in feed manufacturing technologies. Because these technologies are improving production efficiency, reducing labour costs, and strengthening consistency in feed quality.

Emerging technologies such as artificial intelligence (AI) and machine learning are increasingly being used to analyse data, optimize feed formulations, and enhance farm management practices. Meanwhile, developments in feed additives (enzymes, probiotics, vitamin-mineral premixes), processing technologies, and quality control systems are making it possible to produce more efficient, sustainable, and in-demand feeds. Additionally, sustainability and environmental factors can also be supportive for the compound feed market. It is clear that the livestock industry faces growing pressure to minimize its environmental footprint. This has accelerated the development of more sustainable feed ingredients, including alternative protein sources such as insects, algae, and plant-based proteins. In addition, the adoption of circular economy practices such as recycling and reusing agricultural by-products as feed ingredients helps reduce waste and improve resource efficiency. These solutions are increasingly integrated into compound feeds through proper formulation. The increasing importance of animal health and welfare also increases the demand for compound feed that offers functional benefits. The inclusion of prebiotics, probiotics, and other functional ingredients in feed helps improve gut health, strengthen immune function, and reduce dependency on antibiotics. The use of compound feeds that provide high-quality, nutritionally balanced mixtures contributes to improved animal health, welfare, and overall productivity. Policies that encourage sustainable livestock production, promote research and development in animal nutrition, and foster the adoption of innovative feeding practices can significantly positively influence market growth. In addition to all this, the growth of global trade also contributes to the expansion of the market. For example, the expansion of grain and oilseed trade facilitates access to raw materials for feed producers, thereby strengthening the feed production supply chain.

In conclusion, all of these factors combined translate into greater demand for compound feeds, which provide significant advantages in terms of animal health, performance, and welfare while supporting sustainable food production and driving the overall growth of the global feed market.

5 Factors constraining market growth

Although the global compound feed market benefits from numerous growth drivers, it is also constrained by several economic, environmental, and structural challenges. According to market analysts, volatility in raw material prices is the primary factor constraining this growth. The prices of key feed raw materials such as corn, soybean, barley, and wheat are subject to fluctuations driven by climate change, geopolitical tensions, trade policies, global supply-demand imbalances, and broader economic conditions. Such volatility increases production costs and feed prices, and makes compound feed less affordable, particularly for small-scale farmers. Particularly in developing countries, these farmers continue to rely on traditional feed sources such as straw, forage, and crop residues grown on their own farms or sourced locally, instead of using commercial compound feeds. Competition for resources is also a major problem. The growing demand from both food and feed producers for arable land and water resources creates a competitive environment. This competition can restrict the availability and affordability of essential feed ingredients, thereby increasing feed costs. Environmental impacts, sustainability concerns, and strict regulations are other factors restraining market growth. Industrial livestock production and associated compound feed production are linked to environmental issues such as greenhouse gas emissions, deforestation, water pollution, and biodiversity loss. As a result of increasing global pressure, governments and international institutions are tightening regulations on the sector, causing producers to turn to more costly methods. Consequently, both compliance costs are rising and the market's growth momentum is slowing down. It is known that disease outbreaks and animal health regulations have also caused declines in feed production in some regions in recent years. Outbreaks such as avian influenza, African swine fever, and bovine spongiform encephalopathy can reduce animal populations and lower consumer demand for animal-based foods, thereby reducing feed demand. In addition, strict regulations aimed at preventing and controlling these diseases may impose further restrictions on feed ingredients and production processes. Potential economic downturns are also a factor that could affect market growth, because periods of economic recession can lead to reduced consumer spending on animal proteins, negatively affecting feed demand. At the same time, consumers' dietary preferences are also changing. For example, vegan and vegetarian diets have become significantly more popular in recent years. This has increased interest in plant-based proteins offered as alternatives to animal proteins and in meat products produced in laboratory environments. The growing demand for these products could reduce demand for traditional animal products and, consequently, animal feed.

In summary, the growth of the compound feed market is constrained by multidimensional factors such as fluctuations in raw material prices, competition for resources, environmental regulations, animal health issues, and changing consumer preferences. These obstacles increase production costs and limit market access, particularly for small-scale producers. The market's sustainable growth depends on developing strategic solutions to these economic, environmental, and structural challenges.

6 Global compound feed market by species

The global compound feed market is a dynamic and diverse sector that caters to the specific nutritional needs of various animal species. The market has evolved to effectively address these needs. Poultry, due to their rapid growth rates and cost-efficient protein production, account for approximately 43% of the global compound feed market (Alltech, 2025b; IFIF, 2025). They are followed by pig feed with approximately 26%, ruminant feed with 21%, and aquafeed with 4%. Other animal species (pets, horses, etc.) share the remaining portion.

According to Alltech Agri-Food Outlook 2025, total poultry feed production, which was approximately 595.8 million tons in 2024 (Alltech, 2025b), is divided into broiler, layer, and other poultry feeds. Broiler feed production increased by 1.8% in 2024 to reach 385.4 million tons, while layer feed production increased by 1.4% to 173 million tons. Asia-Pacific has the largest share of broiler feed at 161.5 million tons. This region is followed by Latin America with 73.6 million tons, North America with 59.6 million tons, Europe with 56.0 million tons, Africa with 17.7 million tons, the Middle East with 13.0 million tons, and Oceania with 4.2 million tons. In layer feed, Asia-Pacific ranks first with 83.0 million tons, followed by Europe with 31.0 million tons, Latin America with 26.2 million tons, North America with 16.8 million tons, Africa with 8.5 million tons, and Oceania with 1.0 million tons. The growth of this segment is supported by the increasing global population, rising disposable income, demand for easily accessible protein sources alongside urbanization, and the expansion of poultry production in developing countries.

Global pig feed production declined by 0.6% to 369.3 million tons in 2024 (Alltech, 2025b). Africa and the Asia-Pacific region, which have been battling African swine fever (ASF) for a long time, account for the largest share of this decline. Despite this decline, Asia-Pacific remained the world's leading producer of pig feed in 2024, with 183.6 million tons. It is followed by Europe with 76.3 million tons, North America with 64.5 million tons, Latin America with 40.8 million tons, Africa with 2.8 million tons, and Oceania with 1.3 million tons. Unlike Asia-Pacific and Africa, the pig feed markets in Europe, North America, and Latin America have continued to grow during this period. Income growth and increased demand for processed meat products are the main factors supporting growth in these regions, but ASF continues to limit the expansion rate of this segment. Analysts attribute these growth differences between regions to the speed at which countries are recovering from ASF outbreaks.

The global total production of ruminant feed, generally classified into three different groups including dairy feed, beef feed and other ruminant feed, reached 323.2 million tons in 2024 (Alltech, 2025b). In 2024, dairy feed production reached 165.5 million tons, an increase of 3.2%, with North America leading the way at 48.3 million tons. It is followed by Europe with 47.5 million tons, Asia-Pacific with 29.9 million tons, Latin America with 22.7 million tons, Africa with 8.3 million tons, the Middle East with 7.2 million tons, and Oceania with 1.6 million tons. The most notable data in dairy feed production is the 25.7% increase in Africa. The growth of the urban population and economic prosperity, along with improvements in cold chain logistics, are among the main factors fueling this surge. The transformation of small-scale dairy farms in the region into more commercial operations is also increasing the need for formulated dairy feed and boosting feed tonnage. Similar reasons are supporting production growth in Asia-Pacific and Latin American countries. Beef feed production increased by 1.8% to 134 million tons during the same period. North America ranked first with 71.3 million tons, followed by Asia-Pacific with 19.8 million tons, Europe with 18 million tons, Latin America with 15.4 million tons, Africa with 6.8 million tons, the Middle East with 2.0 million tons, and Oceania with 0.8 million tons. Africa showed a significant jump of 32.2% in beef feed production, as mentioned above, as well as in dairy feed. In Asia-Pacific, difficulties experienced in 2024, primarily disease outbreaks (such as LSD and FMD) and economic factors, led to a notable 6% decline in production. Overall, growing consumer demand for meat and dairy products, the transition to modern and intensive production practices, and farmers' efforts to increase productivity continue to support growth in the ruminant segment. However, price fluctuations in feed raw materials, disease outbreaks, and environmental concerns regarding methane emissions are emerging as limiting factors for this segment.

Aquafeeds stand out in many analyses as the segment with the fastest growth potential in the long term. However, according to Alltech data, the downward trend in aquafeed production that began in 2023 continued in 2024 at 1.1%, with total global production reaching 53.0 million tons (Alltech, 2025b). Asia-Pacific is the clear leader with a production of 37 million tons. It is followed by Latin America with 6.8 million tons, Europe with 4.8 million tons, Africa with 1.7 million tons, North America with 1.3 million tons, the Middle East with 1.0 million tons, and Oceania with 0.2 million tons. The partial decline in 2024 is largely attributed to diseases, extreme weather events, cost pressures, and low market prices. However, the growing global population, declining wild fish stocks, and increasing consumer preference for seafood support the medium- and long-term growth of this segment.

7 Global compound feed market by region

The global compound feed market exhibits varying growth trends across different regions, influenced by factors such as economic development, dietary habits, and livestock production practices.

The Asia-Pacific region, with a compound feed production of 533.1 million tons and a 38.2% market share in 2024, is the largest and fastest-growing feed market in the world (Alltech, 2025b). The main growth drivers in the region are China, India, and Southeast Asian countries. China is the largest producer in both the region and the world, with a total compound feed production of 315.0 million tons in 2024. Among feed types, pig feed ranks first at 143.9 million tons, followed by broiler feed at 97.5 million tons, layer feed at 32.4 million tons, and aquafeed at 22.6 million tons (Alltech, 2025a). India's total production is approximately 55.2 million tons, with broiler feed accounting for 19.8 million tons, layer feed 16.9 million tons, and dairy feed 15.5 million tons. Vietnam, ranked among the top 10 global feed producers, stands out with 25.9 million tons of total production, with the highest output being pig feed at 11.9 million tons. Japan, producing 24.3 million tons, is another significant regional producer, with layer feed at 6 million tons and pig feed at 5.5 million tons. Key factors driving rapid market growth in the region include rising population, increasing income levels, urbanization, and changing dietary habits, which boost consumer demand for meat, milk, and eggs and support the expansion of the compound feed market. Additionally, Asia-Pacific hosts a large portion of global aquaculture production, further increasing demand for aquafeeds. According to Mordor Intelligence data, the Asia-Pacific compound feed market is expected to be valued at approximately USD 275 billion in 2025 and reach USD 345 billion by 2030, with a 4.6% CAGR (Mordor Intelligence, 2025b). However, growth may be partly constrained by animal diseases such as African swine fever and avian influenza, as well as increasing discussions on the environmental impacts of industrial livestock farming.

The Americas, particularly North America, have a strong compound feed market due to large-scale and technologically advanced livestock operations. According to Alltech data, North America ranked second globally in compound feed production in 2024, producing 290.7 million tons and accounting for a 20.8% market share (Alltech, 2025b). The leading producer in the region, the United States, produced 269.6 million tons in 2024, making it the second-largest feed producer globally on a national basis. Feed production in the U.S. includes 70.5 million tons of beef feed, 56.6 million tons of broiler feed, and 56.2 million tons of pig feed (Alltech, 2025a). Total compound feed production in Latin America is estimated to have reached 198.4 million tons (14.2% share) in 2024 (Alltech, 2025b). Brazil increased its total feed production to 86.6 million tons in 2024, of which 37.1 million tons were broiler feed and 21 million tons were pig feed (Alltech, 2025a). In Mexico, a total of 41.4 million tons of compound feed was produced, of which 11.8 million tons was broiler feed, 7.8 million tons was layer feed, and 7 million tons was pig feed. Key factors driving market growth in the Americas include increasing demand for high-quality animal protein, modern and expanding agricultural and livestock practices, focus on animal health and welfare, and the adoption of advanced feeding technologies. According to Market Data Forecast, the North American compound feed market is projected to be valued at USD 125.1 billion in 2025 and grow at a 1.3% CAGR to USD 138.7 billion by 2033 (Market Data Forecast, 2025). The Latin American market is valued at USD 55.1 billion in 2024 and is expected to reach USD 60.5 billion by 2029, growing at a 1.9% CAGR (Market Data Forecast, 2024). However, deforestation, and health concerns related to hormones and antibiotics in feed are key barriers to market expansion in the region.

Europe features a highly regulated, technology-driven compound feed market. In 2024, Europe produced 267.8 million tons of feed (19.2% share), ranking third globally. The top three producers in the region are Russia, Spain, and Turkey, producing 38.5 million tons, 36.0 million tons, and 24.5 million tons, respectively (Alltech, 2025b). In Russia, pig feed dominates production at 16.1 million tons, followed by 11.3 million tons of broiler feed (Alltech, 2025a). In Spain, pig feed leads at 17.6 million tons, while dairy feed, beef feed, broiler feed, and layer feed range from 2 to 5 million tons. Sustainability, animal welfare, and environmental protection are key focus areas in the European market. This approach is driving increased demand for organic and high-quality compound feed, while smart feeding systems, genetic improvements, and feed additives that enhance digestion are supporting market growth. According to Mordor Intelligence data, the value of the Europe compound feed market is estimated to be USD 116.9 billion in 2025 and is projected to reach USD 136.2 billion with a CAGR of 3.1% during the 2025–2030 period (Mordor Intelligence, 2025c). The main obstacles to growth in the region are low population growth, environmental regulations, animal diseases, and geopolitical tensions.

Africa and the Middle East present significant growth potential due to rising populations, increasing incomes, and urbanization. Growth in poultry and aquaculture sectors, along with rising demand for animal protein, supports the regional compound feed market. Alltech reports that Africa produced 57.8 million tons of feed in 2024, accounting for 4.1% of the global market (Alltech, 2025a). The Middle East produced approximately 37.7 million tons in 2024, representing 2.7% of the market. Mordor Intelligence estimates the African compound feed market at USD 22.9 billion in 2025, projected to reach USD 28.2 billion by 2030 with a 4.3% CAGR (Mordor Intelligence, 2025a). The GCC (Gulf Cooperation Council) countries' market is valued at USD 14.2 billion in 2025 and expected to reach USD 18.4 billion by 2030 with a 5.3% CAGR (Mordor Intelligence, 2025d). Key challenges include limited access to raw materials, inadequate transport and storage infrastructure, animal diseases, and food safety concerns. Nonetheless, countries such as Saudi Arabia, Egypt, and South Africa are expected to drive significant market growth in the coming years.

8 Conclusion – Key takeaways

The global compound feed market holds promising growth potential but also faces significant challenges in terms of sustainability, cost, and food security. In particular, the heavy reliance on traditional ingredients such as corn, soybean meal, fishmeal, and fish oil poses long-term risks. To meet the needs of a growing population and the rising demand for animal protein, it appears inevitable to move beyond current feeding models and turn toward alternative sources.

References

Alltech, 2025a. Agri-Food Outlook 2025, Map. Available at: https://www.alltech.com/agri-food-outlook/results

Alltech, 2025b. Agri-Food Outlook 2025. Available at: https://www.alltech.com/press-release/2025-alltech-agri-food-outlook-shares-global-feed-production-survey-data

Fortune Business Insights, 2025. Compound Feed Market Report. Available at: https://www.fortunebusinessinsights.com/industry-reports/compound-feed-market-100525

IFIF, 2025. International Feed Industry Federation Annual Report 2024/25. Available at: https://annualreport.ifif.org/#one-voice/feed-outlook

Market Data Forecast, 2024. Latin America Compound Feed Market Research Report. Available at: https://www.marketdataforecast.com/market-reports/latin-america-compound-feed-market

Market Data Forecast, 2025. North America Compound Feed Market. Available at: https://www.marketdataforecast.com/market-reports/north-america-compound-feed-market

Mordor Intelligence, 2025a. Africa Compound Feed Market Report. Available at: https://www.mordorintelligence.com/industry-reports/african-compound-feed-market-industry

Mordor Intelligence, 2025b. Asia-Pacific Compound Feed Market Report. Available at: https://www.mordorintelligence.com/industry-reports/asia-pacific-compound-feed-market-industry

Mordor Intelligence, 2025c. Europe Compound Feed Market Report. Available at: https://www.mordorintelligence.com/industry-reports/european-compound-feed-market-industry

Mordor Intelligence, 2025d. GCC Compound Feed Market Report. Available at: https://www.mordorintelligence.com/industry-reports/gcc-compound-feed-market

Precedence Research, 2025. Compound Feed Market Report. Available at: https://www.precedenceresearch.com/compound-feed-market

PART 8

Case studies from industry

CHAPTER 30

Success stories in integrating novel ingredients in pet food

Editorial note on industry case studies

The following contributions are industry-authored case studies describing practical experiences with the development, formulation, and market introduction of pet foods containing novel ingredients. They are included to provide real-world perspectives on innovation pathways, technological feasibility, regulatory navigation, and consumer response, complementing the scientific and regulatory chapters presented elsewhere in this volume.

These case studies may include company-specific examples, field observations, and customer or veterinarian testimonials reflecting the authors' professional experience. Such accounts are intended to illustrate applied practice and hypothesis generation, but they should not be interpreted as clinical evidence or as substitutes for controlled nutritional or veterinary studies.

Responsibility for product descriptions, brand-specific statements, and illustrative examples rests solely with the respective authors. The editor and publisher do not endorse individual products or companies and remain neutral with respect to commercial interests. Readers are encouraged to consult the cited scientific literature and regulatory sources when evaluating nutritional efficacy, safety, or health-related outcomes.

1 Introduction

Following the editorial note above, the case studies presented in this section illustrate how novel ingredients are being implemented in pet food systems across different market segments, technological approaches, and company cultures.

The pet food market is inherently more emotion-driven than feed sectors such as livestock production or aquaculture. Decisions are often shaped not only by technical performance and cost efficiency, but also by pet owner values, perceptions, and the human–animal bond. As a result, innovation pathways in pet food tend to reflect a broader diversity of voices, motivations, and communication styles than those typically observed in case studies from other feed sectors.

Accordingly, the contributions in this section show a wide spectrum of perspectives. Some are deeply narrative and experience-driven, reflecting the emotional

 HTTPS://DOI.ORG/10.1163/9789004692824-031

motivations and consumer-facing realities that often catalyze innovation, as illustrated in the Jiminy's cases. Others adopt a more technical and formulation-focused approach, emphasizing ingredient functionality, processing considerations, and product positioning, as seen in the Visán contribution. Additional cases highlight strategic decision-making at company level, integration of novel proteins into existing portfolios, and the balancing of innovation with regulatory, quality, and market constraints, as exemplified by Mera and Interquell.

Taken together, these case studies do not aim to provide definitive conclusions, but rather to offer complementary lenses through which innovation can be understood: from inspiration to implementation and laboratory insights to market reality. They are intended to stimulate reflection, dialogue, and hypothesis generation for researchers, practitioners, and students interested in the practical deployment of novel ingredients in pet food.

The editor is grateful for the willingness of the contributing pet food companies to share their experiences openly and directly within this volume. While case studies in the livestock feed and aquaculture sections rely primarily on market analyses and secondary data, reflecting the more limited participation of companies in those sectors, this section benefits from first-hand contributions made possible by the strong engagement and trust of stakeholders within the pet food community. These voices add practical depth and transparency to the discussion and provide insight into how novel ingredients are implemented under real commercial conditions.

CHAPTER 31

From inspiration to transformation: stories of change with insect-based dog food

A. Carlson

Founder and CEO at Jiminy's, Inc. 1831 Solano Ave #8510, Berkeley, CA 94707 USA
Correspondence: anne.carlson@jiminys.com

1 Introduction

Behind every nutritional breakthrough lies a collection of personal experiences that help contextualize scientific findings in real-world settings. While research papers can describe digestibility or environmental benefits of insect protein, individual owner-reported observations may help illustrate how such diets are perceived and applied in practice by pet owners, particularly in dogs with nutritional sensitivities.

These illustrative accounts highlight that the potential impact of nutrition may extend beyond measurable parameters, influencing daily routines and the human–animal bond. The following sections present two owner-reported experiences intended to complement, not replace, controlled scientific evidence.

2 When scratching stops: Savannah's allergy relief

Savannah got off to a rough start. Rescued from the euthanasia list by a Phoenix organization, the Border Collie/German Shepherd mix was visibly scared and sheepish. Lori, who was volunteering that day, later recalled that Savannah appeared to be trying to hide under the concrete, an image that stayed with her and prompted the immediate decision to foster the dog, convinced that Savannah would never leave her side again.

Savannah joined a loving household that day, and outside the usual canine agitators – mailmen, squirrels, hats – Savannah didn't have much to complain about. That is, until Lori noticed Savannah was always scratching at herself.

About eight years ago, Savannah developed seasonal allergies. Lori took her to the veterinarian and was prescribed steroids. Though not thrilled at the idea of giving her dog steroids, she reasoned it was only for a few months of the year. Fast forward, and these allergies became a year-round nuisance. Savannah was constantly itching and in visible discomfort.

Lori, not wanting to see her dog suffer, knew it was time for another veterinarian visit. This time around, Savannah was prescribed Apoquel, which led to a noticeable reduction in itching and scratching. Lori's main concern was that it treated the symptoms, not the cause. Lori later reflected that, while the medication helped manage symptoms, it did not appear to address the underlying cause and was perceived as a temporary solution.

When Lori heard about Jiminy's dog food being positioned as hypoallergenic and formulated for dogs with food sensitivities, she didn't think twice. She switched both of her pups to our Good Grub.

Within a couple of weeks, Lori noticed that Savannah was itching much less. Over time, Lori reported that the frequency with which Apoquel was administered decreased, as it was perceived to be less frequently required. Since Savannah is an aging dog, cutting back on medication rather than increasing it is a rare and welcomed phenomenon.

Her coat grew shinier, her energy returned, and for the first time in years, Lori felt she had her happy, playful dog back.

3 The science behind the relief

Insect protein is novel – meaning most dogs' immune systems don't recognize it as an allergen. Additionally, the natural antimicrobial and antioxidant compounds in insects help modulate immune responses, potentially contributing to immune modulation and inflammatory pathways (Lee et al., 2018, Xiao et al., 2018, Patyra and Kwiatek, 2023, Koutsos, 2022). Supporting studies confirm that dogs with food sensitivities can thrive on insect protein, offering similar impact to hydrolyzed proteins but without extensive processing (Böhm et al., 2018, Lee et al., 2021).

4 Finding stability: Mordecai's digestive transformations

Lauren first spotted Mordecai at the rescue event, and it was impossible to look away. The mostly white dog had a kind of striking appearance with piercing blue eyes, a confident stance, and distinctive markings. When Lauren approached and called his name, however, Mordecai did not respond, not out of disinterest or stubbornness, but because he was deaf. The volunteer later explained that Mordecai had been born that way, a condition that is relatively common in dogs with this type of coloring, while also noting his particularly expressive and engaging personality.

Lauren, who had been looking for a rescue companion, felt an immediate connection. As someone deeply committed to environmental sustainability, she had already been researching eco-friendly pet foods and was intrigued by the emerging science around insect protein. Adopting a special-needs dog who would thrive on sustainable nutrition felt like perfect alignment with her values.

What Lauren did not anticipate was the challenge that would test both her daily routines and her relationship with her new companion. Mordecai, despite his beauty and charm, exhibited pronounced digestive discomfort, including frequent and intense flatulence, which quickly became noticeable in social settings.

During the first weeks after adoption, Lauren observed that Mordecai's digestive issues were not transient. Initial assumptions that stress from the new environment might be responsible did not hold, as the symptoms persisted and, over time, intensified. The digestive disturbances extended beyond occasional inconvenience, affecting Lauren's willingness to invite visitors into her home and altering her social routines.

In response, Lauren explored a wide range of dietary approaches, including fresh food delivery services, premium dry diets, probiotics, digestive enzymes, and limited-ingredient formulations. Despite these efforts, none of the strategies resulted in consistent or meaningful improvement.

Beyond the practical challenges, Lauren observed changes in Mordecai's behavior. The dog, who had initially displayed a lively and playful demeanor, appeared increasingly subdued, which Lauren associated with ongoing digestive discomfort and her own visible frustration with the situation.

When Lauren ultimately decided to try an insect-based protein diet, the decision was not driven by expectations of resolving the digestive issues, but rather by her environmental convictions. She reasoned that, if dietary changes were to continue, they should at least align with her broader sustainability goals. She therefore transitioned Mordecai to a commercially available cricket-based dog food.

According to Lauren's observations, changes were not immediate but became noticeable over time. Within the first weeks, she reported a reduction in the intensity and frequency of digestive symptoms, including flatulence. As Mordecai's digestive comfort appeared to improve, Lauren felt more at ease inviting visitors into her home, and she observed that Mordecai's energy levels and playful behavior increased.

Over time, Lauren perceived that Mordecai was more relaxed and confident within the household. His deafness, which she had initially considered a potential limitation, became a minor aspect of a broader and more expressive personality that she felt was no longer overshadowed by digestive discomfort. While these observations are subjective and owner-reported, they illustrate how perceived improvements in digestive stability can influence both animal behavior and the human-animal relationship.

5 The science of gut health

Insect protein contains a unique fiber called chitin that feeds beneficial gut bacteria, making it a natural prebiotic for dogs. Insect protein is also naturally antimicrobial, helping control harmful bacteria such as Clostridium perfringens – a common cause of chronic diarrhea (Koutsos et al. 2022, Dong et al. 2021). Through their potential influence on gut microbiota composition and activity, insect-based diets may contribute to a more balanced intestinal ecosystem, which is increasingly recognized as relevant not only for digestive function but also for immune regulation and overall health (Jarett et al., 2021; Areerat et al., 2023).

6 Lessons learned and the road ahead

What these stories reveal is that innovation is more than new ingredients – it's about rethinking how we nourish our pets with respect for both their biology and the planet. Insect-based dog food isn't a miracle cure, but it is increasingly being explored as a practical tool that can support nutritional strategies and sustainability goals.

Each story teaches us something different. Savannah's journey reminds us that, in some cases, progress can involve stepping beyond conventional approaches and considering alternative protein sources within a broader management strategy.

These families did not just pursue better nutrition for their dogs – they reported renewed hope, stronger bonds, and the satisfaction that comes from seeing a beloved companion do well. In several cases, veterinarians were involved as partners in the process, helping guide decisions rather than managing symptoms. Daily routines shifted from continuous problem management toward greater confidence in feeding choices.

We learn from each dog, each pet parent, and each reported experience. While these accounts are anecdotal, they can help generate questions for further research and connect scientific evidence with real-world feeding contexts. Together, they point toward a constructive path forward – not just for our pets, but for the planet they'll inherit.

Every bite tells a story. Every story builds a future.

References

Areerat, S., Chundang, P., Lekcharoensuk, C., Patumcharoenpol, P., Kovitvadhi, A. 2023. Insect-based diets (house crickets and mulberry silkworm pupae): a comparison of their effects on canine gut microbiota. Vet. World. 16(8): 1627-1635. https://doi.org/10.14202/vetworld.2023.1627-1635

Böhm, T. M. S. A., Klinger, C. J., Gedon, N., Udraite, L., Hiltenkamp, K., Mueller, R.S. 2018. Effect of an insect protein-based diet on clinical signs of dogs with cutaneous adverse food reactions. Tierarztl. Prax. Ausg. K. Kleintiere Heimtiere. 46(5): 297-302. https://doi.org/10.15654/tpk-170833.

Di Stefano, M., Polizzi, A., Santonocito, S., Romano, A., Lombardi, T., Isola, G. 2022. Impact of Oral Microbiome in Periodontal Health and Periodontitis: A Critical Review on Prevention and Treatment. Int. J. Mol. Sci. 23(9): 5142. https://doi.org/10.3390/ijms23095142.

Koutsos, E., Modica, B., Freel, T. 2022. Immunomodulatory potential of black soldier fly larvae: applications beyond nutrition in animal feeding programs. Transl. Anim. Sci. 6(3): txac084. https://doi.org/10.1093/tas/txac084.

Lee, J., Kim, Y.-M., Park, Y.-K., Yang, Y.-C., Jung, B.-G., Lee, B.-J. 2018. Black soldier fly (*Hermetia illucens*) larvae enhances immune activities and increases survivability of broiler chicks against experimental infection of *Salmonella Gallinarum*. J. Vet. Med. Sci.; 80(5):736-740. https://doi.org/10.1292/jvms.17-0236.

Lee, K.-I., Chae, Y., Yun, T., Koo, Y., Lee, D., Kim, H., So, K.-M., Cho, W. J., Kim, H.-J., Yang, M.-P., Kang, B.-T. 2021. Clinical application of insect-based diet in canine allergic dermatitis. Korean J. Vet. Res. 61(4): e36. https://doi.org/10.14405/kjvr.2021.61.e36.

Okin, G.S., 2017. Environmental impacts of food consumption by dogs and cats. PLoS ONE 12(8): e0181301. https://doi.org/10.1371/journal.pone.0181301

Patyra, E., Kwiatek, K. 2023. Insect meals and insect antimicrobial peptides as an alternative for antibiotics and growth promoters in livestock production. Pathogens 12(6): 854. https://doi.org/10.3390/pathogens12060854

van Huis, A., Van Itterbeeck, J, Klunder, H, Martens, E., Halloran, A., Muir, G., Vantomme, P. 2013. Edible insects: Future prospects for food and feed security. Food and Agriculture Organization of the United Nations (FAO foresty paper; No. 171), Rome, Italy, 201 pp. https://edepot.wur.nl/258042

Xiao, X., Jin, P., Zheng, L., Cai, M., Ziniu, Y., Jeffrey, Y., Zhang, J. 2018. Effects of black soldier fly (*Hermetia illucens*) larvae meal protein as a fishmeal replacement on the growth and immune index of yellow catfish (*Pelteobagrus fulvidraco*). Aquacult. Res. 49(4): 1569-1577. https://doi.org/10.1111/are.13611.

CHAPTER 32

A journey toward sustainable pet nutrition: implementing microbial protein

S. Scherber and J. Wimmers*

Mera Tiernahrung GmbH, Industriestraße 16, 47623 Kevelaer, Germany
**Correspondence: kontakt@susannescherber.com*

Exploring new protein sources for dog and cat food has become a key strategic priority for today's pet food industry. It brings together the goals of sustainability, nutritional innovation, and market differentiation to deliver high-quality products that are fit for the future. Across Germany and Europe, the demand for premium pet food continues to rise as consumers become more discerning and better informed. They increasingly value transparency, environmental responsibility, and scientifically grounded nutrition qualities that define the next generation of successful pet food brands.

For Mera, a German family-owned pet food company, the decision to trial microbial protein from *Methylococcus capsulatus* was based on different core strategic considerations, such as sustainability, animal health and sensitive nutrition, innovation and differentiation, raw material quality and availability and market dynamics. Beyond these rational factors, there was also a genuine curiosity about the potential of this pioneering ingredient and what it could contribute to the company's mission and product portfolio.

1 Key motivations for implementing microbial protein in pet food development

1.1 Sustainability and environmental impact

As sustainability is one of Mera's company goals, the reduced environmental impact of microbial protein, in terms of less water use and lower land requirements, compared with traditional protein sources, has driven us to explore this topic. Alternative proteins, including insects, algae, plant-based ingredients, and cultivated microbial proteins, offer new avenues for resource diversification and supply resilience. They can strengthen sustainability further by reducing dependence on livestock production, while supporting cruelty-free, traceable, and circular sourcing. These sources align with EU sustainability and ESG standards and resonate with eco-conscious consumers. By incorporating microbial proteins, Mera contributes to lowering the environmental footprint of pet food production while meeting the expectations of a new generation of responsible pet owners.

1.2 Animal health and sensitive nutrition

Another central driver for innovation lies in addressing animal health and the rising incidence of food intolerances and allergies among pets. There is growing demand for hypoallergenic, highly digestible formulations that reduce the risk of sensitivities. Mera's pure sensitive range exemplifies this philosophy by providing tailored nutrition for animals with special dietary needs. The introduction of a variant containing microbial protein adds a new dimension to this product range, combining nutritional functionality with sustainability and reinforcing Mera's market position through a clear and credible unique selling proposition.

1.3 Innovation, differentiation, and technical feasibility

Novel protein sources enable companies to demonstrate technical expertise and forward-thinking leadership. By integrating innovative ingredients such as microbial proteins, Mera strengthens its reputation as a pioneer in sustainable and health-oriented pet nutrition. This approach ensures long-term competitiveness in a market that increasingly values transparency, science-based claims, and environmental responsibility.

Over recent years, the availability and scientific understanding of novel proteins have increased significantly. Mera was among the first German brands to introduce insect-based pet food using *Hermetia illucens* (black soldier fly larvae) and continues to produce it successfully. Building on this experience, the company turned its attention to fermented microbial protein, investing in research to understand application possibilities, inclusion rates, and safety aspects.

The comprehensive safety study with *Methylococcus capsulatus* played a decisive role in this decision-making process. The ingredient's amino acid profile is comparable to that of animal-derived proteins, making it nutritionally attractive for pet food formulations. This similarity addresses a key limitation of many plant-based proteins, which often lack sufficient levels of for example sulfur-containing amino acids such as methionine and cysteine essential for taurine synthesis and other metabolic functions in dogs and cats.

2 Challenges and practical implementation

During product development, several minor challenges emerged. As with any new raw material, the process began with detailed specification checks, analytical quality controls, availability assessments, and production trials. One key limitation arose from the safety study's recommendation to restrict inclusion levels to 8%. To achieve a nutritionally meaningful formulation and marketing claim such as "rich in, high in or with extra fermented protein," Mera targeted a 10% inclusion rate, supplementing the recipe with peas and pea protein concentrate to maintain total protein content.

Another challenge was the naturally higher copper content of microbial protein compared to conventional meat meals. To comply with legal requirements, the company adjusted its mineral premix accordingly, developing a new formulation tailored to this ingredient. A small-scale production test of 500 kilograms was conducted to validate manufacturability. The results were encouraging – no adjustments to machinery or processing steps were necessary, and the material integrated seamlessly into existing production workflows.

3 Market reception and communication

The market response to pet food containing novel proteins varies between business-to-business (B2B) and business-to-consumer (B2C) segments. Retail partners and distributors have shown strong enthusiasm. As Mera being among the first companies in Europe to commercialize microbial protein-based pet food, the company has earned recognition for innovation and reliability within the industry.

On the consumer side, adoption is more gradual. Unlike meat, plant-based, or even insect proteins, microbial proteins are still largely unknown to most pet owners. Initial skepticism often arises from concerns about safety, naturalness, and origin – frequently summarized by the question, "What exactly is this?" Additionally, many pet owners project their own dietary habits and preferences onto their pets, making it essential to provide clear, transparent information about the benefits and nutritional quality of microbial proteins. Another challenge was choosing a name that would be clearly understood and minimize potential confusion. In German, the term *fermentation* can sometimes evoke skepticism, whereas *microbial protein* in English conveys a more technical and scientific impression.

To address these challenges, Mera has invested considerable effort in consumer education. The company launched targeted social media campaigns, special promotions, and pre-sales events designed to spark interest and conversation among environmentally conscious "pet parents." These initiatives emphasize transparency, traceability, and the science behind fermentation, positioning microbial protein as a safe, natural, and future-oriented alternative to traditional meat ingredients.

4 Key learnings: courage, feedback, and iteration in product innovation

The development of Mera's *pure sensitive* dog food with microbial protein underscored an important lesson in innovation: progress begins with courage. The company adopted a "just do it" philosophy – starting small, testing prototypes, gathering authentic feedback from pet owners, and refining the product iteratively based on real-world insights. This agile, feedback-driven process allowed Mera to optimize everything from palatability to product communication before scaling up for wider distribution.

The main takeaway from this journey is clear: innovation in pet nutrition requires both scientific rigor and entrepreneurial courage. True progress does not stem from striving for perfection at the outset, but from learning through action, collaboration, and responsiveness to both market signals and consumer feedback. By combining sustainability, technological competence, and open-minded experimentation, Mera continues to position itself as a pioneer in shaping the future of responsible, high-quality pet food.

Note on the usage of AI tools

The authors of this chapter state that they used large language models (LLMs) strictly for English language correction and stylistic polishing of chapters, and not for content generation.

CHAPTER 33

Novel protein ingredients in pet food: a manufacturer's perspective on innovation, nutrition, and consumer acceptance

M.L. Rettenbeck

Interquell GmbH, Suedliche Haupstrasse 38, 86517 Wehringen, Germany
Correspondence: lucia.rettenbeck@interquell.de

1 Introduction

The projected shortages in meat and protein supply, driven by the rapid growth of the global population, demand innovative and sustainable solutions (Henchion et al., 2017). Traditional sources of animal protein are increasingly constrained by environmental, economic, and ethical limitations (Martin, 2024). This trend is also evident in the pet food sector, where the availability of conventional animal-derived proteins, in particular "rare" animal proteins, is declining. As a result, the industry must explore and adopt alternative protein sources that are not only resource-efficient but also ethically sound. The selection of such ingredients should consider factors such as land and water use, greenhouse gas emissions, and the broader implications of animal welfare and husbandry practices (Espinosa-Marrón et al., 2022). Insects, plant-based proteins, and microbial proteins represent promising alternatives that can contribute to a more sustainable and ethically responsible protein supply for both human and animal nutrition.

Therefore, manufacturers in the pet food industry are increasingly investing in novel and alternative protein sources due to a combination of environmental, market, and nutritional factors. The environmental and ethical cost of traditional animal-based proteins is substantial, however, due to its reliance on by-products and its close integration with livestock production and the human food system, the pet food industry occupies a unique position in discussions of sustainability (Swanson et al., 2013). Nevertheless, companies aiming to reduce their carbon footprints, meet sustainability and ethical goals are turning to alternatives like insect protein and recipes based merely on plant-based proteins. As indicated in many life cycle assessment studies insects require less land, water, and feed than conventional livestock, produce fewer greenhouse gas and therefore can have a lower environmental impact than traditional animal proteins from livestock when produced using renewable energy and fed with agricultural by-products (Oonincx and

HTTPS://DOI.ORG/10.1163/9789004692824-034

de Boer, 2012; Van Huis and Oonincx 2017). From a product development perspective, novel proteins enable formulations, which do not contain the rather common ingredients like chicken or beef and therefore can support clean-label innovation with limited, recognizable ingredients that appeal to health-conscious pet owners. These ingredients help brands connect with sustainability-minded and flexitarian consumers, a growing and influential segment. Nutritionally, many novel proteins offer high-quality amino acid profiles suitable for both dogs and cats. Black soldier fly larvae, for example, contain around 42–45% crude protein and a balanced amino acid composition comparable to traditional meat meals (Makkar et al., 2014). While dogs, as omnivores, generally adapt well to diets rich in plant proteins, cats – being obligate carnivores – require careful supplementation of nutrients such as taurine, arachidonic acid, and vitamin A when fed diets based solely on plant or insect proteins.

2 Challenges in using novel proteins as a pet food manufacturer

The use of novel proteins in pet food presents several challenges across supply, formulation, nutrition, and consumer acceptance. The integration of novel proteins into pet food formulations presents economic challenges that stem primarily from supply chain limitations and production costs. Insect farming, while promising in terms of sustainability, has yet to reach the scale necessary to ensure consistent availability, safety, and cost-efficiency. Industrial production is still maturing, and infrastructure investments remain high (Van Huis and Oonincx 2017). Similarly, plant-based proteins – such as pea, lentil, and soy isolates – often involve more complex processing and sourcing logistics, making them more expensive than conventional animal-derived proteins. As a result, pet food products incorporating these alternative protein sources currently face elevated production costs, which are reflected in higher retail prices. This cost barrier may limit widespread adoption unless economies of scale and technological advancements reduce input costs over time.

Regulatory frameworks complicate the adoption of insects as a sustainable alternative to conventional animal proteins. Within the European Union, regulatory constraints significantly limit its broader application. Insects are classified as food-producing animals under Regulation (EC) No 1069/2009, which subjects them to strict hygiene and feed regulations. Notably, the substrates permitted for insect rearing are tightly controlled: only plant-based materials and certain animal by-products such as milk, eggs, and honey are allowed. The use of slaughterhouse waste, catering leftovers, or supermarket waste containing meat or fish is strictly prohibited (Regulation (EU) 2017/893). These restrictions conflict with circular economy principles and reduce the ecological efficiency of insect farming. Furthermore, the approval process for new insect species intended for feed use is complex and time-consuming. Currently, only a few species – such as *Hermetia illucens*, *Tenebrio molitor*, and *Acheta domesticus* – are authorized for processing into pet food ingredients (European Pet Food Industry Federation, 2024). Although studies have demonstrated the environmental benefits of insect protein, including lower greenhouse gas emissions and reduced resource consumption (Dobermann, Swift, & Field, 2017), the existing regulatory framework hampers scalability and economic viability for manufacturers.

From a research and development perspective, formulating pet food with novel proteins requires careful attention to amino acid balance, particularly when using plant proteins that may be deficient in methionine or lysine. Palatability is another concern, especially for cats, who may reject diets with high levels of insect or legume-based ingredients (Bosch & Swanson, 2021). Additionally, insect protein often exhibits variability in nutrient composition depending on factors such as substrate, harvest age, and processing method, necessitating detailed ingredient specifications and sometimes pre-treatment methods like hydrolysis or fermentation to improve digestibility and taste (Malla, Nørgaard, & Roos, 2023).

Ensuring nutritional adequacy and digestibility is critical, with AAFCO and FEDIAF profiles serving as the baseline. Studies have shown that BSFL-based diets can be both digestible and nutritionally sufficient for adult dogs, although the bioavailability of certain minerals, such as calcium bound in the chitin exoskeleton, may be lower and require formulation adjustments (Freel, McComb, & Koutsos, 2021). Plant-based proteins, while effective, may contain anti-nutritional factors like phytates and trypsin inhibitors, which can reduce digestibility unless properly processed (Swanson, Carter, & Yount, 2013).

Finally, consumer perception remains a critical factor. Despite the environmental and health benefits, acceptance of insect-based pet food is mixed. While sustainability-minded consumers are increasingly open to insect-based or plant-based diets, the acceptance among the majority of customers is still lower. The results of a recent study showed that 52% of dog owners and 62% of cat owners expressed a positive attitude toward insect-based pet food, with cat owners demonstrating significantly higher acceptance. Acceptance increased notably when participants were informed about the nutritional and sustainability benefits of insect-based pet food. The findings suggest that emphasizing the environmental and nutritional advantages of insect-based pet food could enhance consumer acceptance, particularly within the feline segment, and offer valuable guidance for manufacturers and marketers in developing future products (Baptista da Silva et al., 2024).

2.1 Case report – use of novel proteins

As pet food manufacturer specializing in dry food for dogs and cats Interquell has successfully integrated plant-based proteins and insect protein into its formulations. The objective was to develop nutritionally balanced and sustainable products that comply with FEDIAF nutritional guidelines. Selected plant proteins, including pea, potato, and rice protein, were chosen for their functional properties and availability. The insect protein, derived from mealworm, was used in defatted form to improve oxidative stability and allow better usability in the extrusion process. To ensure a complete nutritional profile, supplementation with synthetic amino acids and taurine as well as other nutrients was applied.

Analytical testing of the finished products confirmed compliance with all nutritional requirements. Crude protein, fat, fiber, and mineral content were within target ranges, and the amino acid profile met species-specific needs. Microbiological analyses showed no irregularities. From a technological perspective, the extrusion process was stable and showed no limitations regarding the inclusion levels of either protein source. The defatted insect protein exhibited good flow properties and did not negatively affect texture, expansion, or process efficiency. Sensory trials with dogs and cats indicated high palatability, and feedback from pet owners was positive, particularly regarding the sustainability and innovation of the product concept.

The inclusion of plant-based proteins in recipes with traditional animal proteins or vegetable recipes and the inclusion of insect proteins proved to be both functionally effective and ethically beneficial. Although vegetarian and insect-based formulations can be produced efficiently and meet all nutritional and technological requirements, current consumer demand remains low. These products are currently considered niche offerings, with limited market penetration and acceptance. Despite their ecological advantages and compatibility with modern pet food processing technologies, widespread adoption is hindered by conservative purchasing behavior and a lack of familiarity among consumers. This highlights a gap between the technical feasibility of alternative protein sources and their commercial viability in the pet food sector.

2.2 Strategic outlook for pet food manufacturers and conclusion

The emergence of novel proteins represents a strategic inflection point for pet food manufacturers. Companies that invest in vertical integration – such as sourcing or co-owning insect farms – alongside advanced nutritional modeling and comprehensive palatability testing protocols are well positioned to lead in a segment poised for rapid expansion. Market forecasts indicate that the global insect protein market for pet food is expected to grow at a compound annual growth rate of 30 to 35 percent through 2030 (Allied Market Research, 2023). In addition to economic opportunity, novel proteins offer a pathway toward net-zero product lines, aligning with environmental, social, and governance objectives and contributing to the long-term resilience of brand portfolios.

Beyond commercial considerations, the adoption of novel proteins reflects a broader responsibility within the pet food industry to support both animal health and sustainable food systems. These ingredients are not merely a passing trend but signify a fundamental shift toward responsible innovation. By addressing challenges in formulation, supply chain logistics, and consumer perception, manufacturers can deliver high-quality, environmentally conscious nutrition that meets the evolving needs of pets, people, and the planet.

Note on the usage of AI tools

The author of this chapter states that large language models (LLMs) were used solely for English language correction and stylistic polishing, and not for content generation.

References

Allied Market Research. (2023). Insect Protein Market Size, Share, Competitive Landscape and Trend Analysis Report, by Insect Type and Application: Global Opportunity Analysis and Industry Forecast, 2023–2030. Retrieved from https://www.alliedmarketresearch.com/insect-protein-market-A14296

Baptista da Silva, C., dos Santos Neto, E. L., Broeckx, B. J. G., Ruiz-Suárez, N., Eeckhout, M., & Hesta, M. (2024). Assessing pet owners' willingness to embrace insects in pet food. Journal of Animal Physiology and Animal Nutrition, 108(6), 1833–1843. https://doi.org/10.1111/jpn.14019

Bosch, G., & Swanson, K. S. (2021). Protein quality of insects as potential ingredients for dog and cat foods. Journal of Animal Science, 99(3), skab007. https://doi.org/10.1093/jas/skab007

Bosch, G., Zhang, S., Oonincx, D.G.A.B., Hendriks, W.H. (2014). Protein quality of insects as potential ingredients for dog and cat foods. Journal of Nutritional Science, 3, e29. https://doi.org/10.1017/jns.2014.23

Dobermann, D., Swift, J. A., & Field, L. M. (2017). Opportunities and hurdles of edible insects for food and feed. Nutrition Bulletin, 42(4), 293–308. https://doi.org/10.1111/nbu.12291

European Pet Food Industry Federation. (2024). FEDIAF Fact Sheet: Insect-Based Ingredients in Pet Food. https://europeanpetfood.org

Freel, M. et al. (2021). Evaluation of black soldier fly larvae meal in adult dog diets. Animals, 11(8), 2330. https://doi.org/10.3390/ani11082330

Freel, T. A., McComb, B. L., & Koutsos, E. A. (2021). Evaluation of black soldier fly larvae meal as a protein source in adult dog diets. Translational Animal Science, 5(1), txaa234. https://doi.org/10.1093/tas/txaa234

Lähteenmäki-Uutela, A., Marimuthu, S., & Neubauer, P. (2022). Consumer acceptance of insect-based pet food in Europe. Foods, 11(3), 456. https://doi.org/10.3390/foods11030456

Makkar, H.P.S. et al. (2014). State-of-the-art on use of insects as animal feed. Animal Feed Science and Technology, 197, 1–33. https://doi.org/10.1016/j.anifeedsci.2014.07.008

Malla, S., Nørgaard, J. V., & Roos, N. (2023). Nutritional variability of insect meals: Influence of species, substrate, and processing. Animal Feed Science and Technology, 295, 115528. https://doi.org/10.1016/j.anifeedsci.2023.115528

Oonincx, D. G., & De Boer, I. J. (2012). Environmental impact of the production of mealworms as a protein source for humans–a life cycle assessment. PLoS ONE, 7(12), e51145. https://doi.org/10.1371/journal.pone.0051145

Regulation (EC) No 1069/2009. (2009). Laying down health rules as regards animal by-products and derived products not intended for human consumption. EUR-Lex. https://eur-lex.europa.eu/eli/reg/2009/1069/oj

Regulation (EU) 2017/893. (2017). Amending Annexes to Regulation (EC) No 999/2001 as regards the use of insect protein. EUR-Lex. https://eur-lex.europa.eu/eli/reg/2017/893/oj

Spranghers, T. et al. (2017). Nutritional composition of black soldier fly prepupae. Journal of the Science of Food and Agriculture, 97(8), 2594–2600. https://doi.org/10.1002/jsfa.8081

Swanson, K. S., Carter, R. A., & Yount, T. P. (2013). Nutritional sustainability of pet foods. Advances in Nutrition, 4(2), 141–150. https://doi.org/10.3945/an.112.003335

Van Huis, A., & Oonincx, D. G. (2017). The environmental sustainability of insects as food and feed. A review. Agronomy for Sustainable Development, 37, 43. https://doi.org/10.1007/s13593-017-0452-8 1

Zentek, J. et al. (2021). Plant-based diets for dogs and cats—risks and benefits. Veterinary Sciences, 8(5), 85. https://doi.org/10.3390/vetsci8050085

CHAPTER 34

Fermentation derived proteins in superpremium pet food: a step forward with *Corynebacterium glutamicum*

J. García Álvarez

VISAN Industrias Zootécnicas S.L., Ctra. M-208 km. 0,5 - 28500 Arganda del Rey, Madrid, Spain
Correspondence: tecnico@visan.es

1 Company overview

Visán Industrias Zootecnicas is a Spanish company with over 95 years of experience in animal nutrition. Over the past 25 to 30 years, its activity has focused on the production of dry and wet *superpremium* pet food for dogs and cats, as well as the manufacturing of specialized feeds for breeding centers and veterinary nutrition programs.

The core of Visán's strategy combines technological innovation, food safety, and environmental sustainability. In recent years, the company has intensified its investment in alternative ingredients and production processes that reduce environmental impact while preserving nutritional quality and palatability.

2 Motivation to incorporate fermentation derived proteins

The incorporation of fermentation-derived proteins responds to a dual challenge: securing long-term protein supply and providing more sustainable alternatives to traditional animal sources. The growth of the *superpremium* segment has increased the demand for high-quality proteins while also revealing the volatility of prices and availability of conventional raw materials.

Beyond their environmental advantages, fermentation proteins offer functional benefits. They show a balanced amino acid profile and high digestibility, which makes them suitable for hypoallergenic formulations and diets with a more technical focus. From a market standpoint, their use enables product differentiation supported by verifiable innovation and sustainability claims.

3 Type of ingredient tested

Visán has evaluated several alternative protein sources, including novel animal proteins, plant proteins, and microbial proteins obtained by fermentation, giving priority to those supported by technical validation and industrial traceability.

Among these, a protein fraction produced from *Corynebacterium glutamicum* was selected. This microorganism is well established in industrial biotechnology for its role in amino acid production.

The protein is obtained through controlled aerobic fermentation, followed by cell separation, drying, and standardization. Its protein content exceeds 70% on a dry matter basis, with a balanced profile of essential amino acids, particularly rich in lysine and threonine. As a microbial protein, it does not compete with agricultural feedstocks nor depend on livestock cycles, which contributes to the stability of supply chains.

4 Main challenges encountered

Incorporating a fermentation-derived protein into a complete diet required addressing several challenges. Technologically, its cell structure demanded adequate milling and dispersion to ensure uniform integration into the extruded matrix. From a formulation perspective, it was necessary to adjust the balance between animal, plant, and microbial proteins, as well as the total protein contribution from the recipe, to maintain optimal digestibility, palatability, and biological value, aligning with the amino acid requirements of dogs and cats at different life stages.

From the regulatory side, the main difficulty was the classification of such materials under existing European legislation and the preparation of appropriate toxicological and safety documentation for pet food applications, although this particular ingredient is already authorized and listed in the official EU Catalogue of Feed Materials.

Finally, consumer perception remains an ongoing challenge. While pet owners appreciate sustainability, the concept of "microbial protein" still requires communication efforts to be perceived as a scientific advancement rather than a lower-quality substitute.

5 Results and learnings

Factory-scale tests confirmed good thermal stability of the protein during extrusion, with no noticeable loss of functionality. Pilot batches maintained a homogeneous structure and stable kibble density. *In vivo* digestibility studies showed utilization comparable to animal protein sources used as reference.

Acceptance trials with adult dogs and cats revealed that moderate inclusion levels of fermentation protein did not negatively affect palatability. On the commercial side, distributors and final customers expressed particular interest in the ingredient's sustainability and traceability.

A key learning was the importance of complementing technical innovation with clear communication materials that explain the biotechnological origin of the ingredient and its tangible benefits for health and the environment.

6 Future outlook

Visán plans to expand the use of fermentation-derived proteins to additional dry and wet ranges, and to explore synergies with postbiotics and functional extracts. In parallel, new fermentation variants are being studied to optimize lipid profiles and nucleotide content, potentially enhancing digestibility and immune response.

In the medium term, the company sees the integration of biotechnology-based ingredients as a key strategy to improve the sustainability of the protein supply chain and reduce dependence on animal raw materials. This vision aligns with Visán's broader goal: to advance toward more precise, responsible, and sustainable nutrition that meets the expectations of both pet owners and veterinary professionals.

Note on the usage of AI tools

The author of this chapter states that large language models (LLMs) were used solely for English language correction and stylistic polishing, and not for content generation.

CHAPTER 35

The role of novel protein and lipid sources in the feed industry

D.G. Yildiz

Feed & Additive Magazine, Istanbul, Türkiye
Correspondence: derya@feedandadditive.com

Editor's note

This chapter offers a market-oriented overview of novel and emerging protein and lipid sources, drawing on publicly available information including company announcements, trade media, selected peer-reviewed literature, and regulatory sources. The practice examples illustrate implementation pathways and scaling dynamics; they do not constitute independent performance verification and should not be interpreted as efficacy or performance claims. Regulatory status and permitted uses vary by jurisdiction and may evolve over time. Readers are encouraged to consult primary scientific studies and competent authority documentation when evaluating safety, performance, and regulatory compliance.

1 Introduction

The livestock sector is undergoing a critical transformation driven by the need to increase sustainability, improve resource utilization, and reduce environmental impacts. The use of traditional sources such as soybean meal, fish meal, fish oil, and grains as feed ingredients has become increasingly questioned. However, global demand for animal protein and, consequently, livestock operations continue to grow. The use of traditional sources directly in human nutrition for food security, as well as the shift towards new ingredients in feed formulations for animal nutrition, is becoming more than just a necessity.

In this environment, insect- and algae-based ingredients, yeast and fermentation-derived single-cell proteins (SCPs), and circular feed materials from agricultural and food-chain by-products have emerged as prominent alternative feed ingredients. These ingredients have the potential to diversify feed inputs, improve nutritional value, align animal production with circular economy principles, and thus reduce environmental pressures. This section focuses on industrial collaborations showing how innovative feed ingredients are used in animal nutrition and addresses market development, trends in adoption, challenges, and future opportunities.

2 Industry collaborations and practice examples in novel feed ingredients

In recent years, many countries, particularly those in the European Union, have introduced regulatory frameworks for novel proteins in feed such as insects, algae, and SCP, and their applications have expanded (see the regulatory part of this volume). Initially prominent in aquaculture feeds, these new proteins are now spreading to poultry and pig feeds as well as pet foods. Large global feed producers are using these resources in commercial or pilot-scale feed for fish, shrimp, poultry, and other species through partnerships with novel ingredient producers. They are also accelerating R&D efforts to support scaling up in this area. Some of the current studies on the use of novel protein and lipid ingredients in animal feed include:

2.1 Skretting

Skretting, Nutreco's aquafeed brand, focuses on flexibility in feed ingredients and considers the use of novel ingredients that do not compete with human consumption, such as insect protein, microbial protein and algae oil, as responsible sourcing (Skretting, 2025). The company, which has been conducting R&D on the use of such novel ingredients as feed ingredients for many years, utilises this knowledge in commercial feed production. The company has various regional partnerships and agreements with leading representatives of the insect sector, such as Protix, InnovaFeed, Goterra and Volare. Some of these partnerships involve trial and research work, while others are aimed at securing supply.

Today, Skretting produces commercial feed using insect meal at its Averøy (Norway) factory (Skretting, 2018) and at Skretting China (International Aquafeed, 2023). In 2018, Nordlaks was the first company to test the commercial salmon feed containing insect meal produced by Skretting Norway, supplying this feed to 360,000 fry. According to statements made by Skretting Norway officials at the time, the results obtained from the trials showed that fish fed with feed containing insect meal exhibited the same growth performance as those fed with traditional protein sources, and they reported that insect meal could increase the fish's appetite (Skretting, 2018).

Skretting China used approximately 300 tonnes of insect meal in its feeds during the 2022–2023 period. Company officials highlight the high conversion efficiency and performance of black soldier fly meal, which has become an important component of shrimp feed, and report the following details: as a raw material, black soldier fly meal contains approximately 36% crude protein and up to 35% high-quality fat. Also, the antibacterial peptides and chitin present in this ingredient can improve animal health, while its unique flavour has been proven to be highly attractive to farmed species. Furthermore, Skretting company communications have reported reductions in fish meal and fish oil inclusion when insect meal is incorporated into specific shrimp formulations (International Aquafeed, 2023).

Skretting Ecuador currently uses locally sourced insect meal in its fish and shrimp feeds. The company has partnered with Protix and Veramaris to produce sustainable shrimp for a Dutch-based shrimp importer. As part of this collaboration, a new shrimp feed has been developed using insect meal and algae oil to partially replace fish meal and fish oil ingredients (International Aquafeed, 2023), and this feed has been used in shrimp farming by Cofimar, an Ecuador-based aquaculture farm. In 2024, a Dutch retail chain stocked its shelves with shrimp fed with this feed (Food Business Middle East & Africa, 2024).

As mentioned above, algae oil is among the novel liquid ingredients actively used by Skretting, and the company is working with Veramaris in this area. The two companies, together with InnovaFeed, have developed a feed containing algae oil and insect meal for trout to be marketed by the French retail group Auchan (Skretting, 2020a). Veramaris, which has collaborated with Skretting on many similar projects, also shared information from a joint commercial trial. In a trial conducted at a Chilean salmon farm, a salmon feed containing algae oil was tested on a commercial scale for the first time on more than 600,000 fish under the Optimum Omega Nutrition (OON) concept (Veramaris, 2025). The study, which examined the effects of higher EPA and DHA levels on fish immune health, performance and product quality, fed the fish two different diets: a standard diet containing 6% of total fatty acids as EPA and DHA, and an OON diet containing 10% of total fatty acids. Overall, the study showed that a diet containing 10% of total fatty acids as EPA and DHA improved biological outcomes, increased farm productivity and ultimately delivered a higher premium quality product (Veramaris, 2024; Carr et al., 2024).

Skretting, which uses insect and algae components, particularly in feeds for key commercial species such as salmon, trout, and shrimp, is also closely monitoring SCPs. The company is conducting performance trials with the PEKILO microbial protein developed by eniferBio (Skretting, 2021; Enifer, 2022), and the studies are reported to show promising results. For example, in a 9-week feeding trial with Atlantic salmon, it was demonstrated that PEKILO®Aqua can effectively replace 20% of the crude protein content of the diet without adversely affecting growth. According to the study, PEKILO®Aqua consumption was linked to significant improvements in feed conversion ratio and protein efficiency ratio (PER). The apparent digestibility coefficients (ADCs) remained high, around 85% (Keskitalo H., 2024; Hooft et al., 2024).

Skretting, which also tests bacterial proteins produced through carbon capture and usable gas fermentation, is collaborating with NovoNutrients on this (Skretting, 2020b). Another collaboration on SCP is ongoing with Unibio. Skretting is testing the effectiveness of Uniprotein, derived from methanotrophic bacteria, in fish and shrimp feeds (*AgTechNavigator*, 2020).

2.2 Cargill

As a diversified international company, Cargill's approach to innovative sources is aimed at expanding its portfolio through investment, commercial marketing agreements and customer pilots. This multi-dimensional strategy aims to reduce supply risk and offer customers a "sustainable option".

One of the most significant steps in the company's novel proteins portfolio is the strategic partnership established in 2019 with the France-based biotechnology company InnovaFeed (Cargill, 2019a). This partnership was initiated to market and scale the use of fish feed containing insect meal. The collaboration was subsequently expanded to include the use of insect oil in pig and poultry feed (InnovaFeed, 2021). This expanded partnership aims to enable the use of insect oil in the feeding of more than 20 million piglets by 2026. However, there is currently no clear information on commercial-scale feed production containing insect meal and insect oil. Sources indicate limited commercial sales and ongoing trial and development work. The partners state that they have conducted numerous feed trials in pig diets, comparing the impact of different oil sources. In these trials, they found that the nutritional profile of insect oil fits well with pig requirements, making it possible to supplement or replace vegetable oil without negative effects on performance (InnovaFeed, 2021). Additionally, company representatives shared the results of a research programme, where piglets were fed with feed containing insect oil, during a webinar. According to the statement, the research team observed an increased piglet body weight, a better stool quality on day 14 post-weaning and concluded that insect oil could be used to replace plant oils in piglet diets (Pig Progress, 2021).

Cargill, which develops formulations containing algae oil (microalgal omega-3), has been offering feed containing omega-3 derived from algae to interested customers since 2018. The company, which has entered into agreements to secure access to larger volumes, announced that it would add algae-derived omega 3 as part of the oil mix in all fish feed sold in Norway in 2022. Lerøy, a Norway-based company and one of the world's largest seafood producers and exporters, has taken on the role of pioneer and trial partner in this process (SalmonBusiness, 2022; Fishfarmingexpert, 2022). In 2024, Cermaq Canada committed to trialling Cargill's salmon feed containing Veramaris algae oil (Cermaq, 2024). This is seen as one of the concrete examples of Cargill beginning to offer algae-based omega-3s to customers.

Cargill also has various initiatives on the SCP side. By signing an agreement with White Dog Labs in 2019 to secure access to the single-cell protein ProTyton™, Cargill is focusing on using this product as a sustainable alternative to fish meal, particularly in salmon feed. According to the company's statement, in trials, salmon fed a diet containing ProTyton™ achieved a growth performance comparable to salmon on a conventional diet (Cargill, 2019b). In previous years, the company has also worked with Calysta on the production and commercialisation of FeedKind protein, produced through gas fermentation, in North America (Calysta, 2016). A trial at Japan's Kindai University has shown that FeedKind bacterial protein meal (BPM) can replace up to 30 percent of fish meal protein – or up to 20 percent of total feed – in the diet of yellowtail without any impact on growth rate, digestibility, daily feeding rate, or feed efficiency (Biswas A. et al., 2020).

Cargill's other studies on SCP typically focus on human food applications.

2.3 BioMar

BioMar, a Denmark-based aquafeed producer, stands out for its emphasis on using feed ingredients sourced from sustainable sources. The company aims to source 50% of its raw materials from circular and restoration economies and reduce the carbon footprint of its feeds by one-third by 2030 (BioMar, 2025b). In line with this vision, BioMar has launched its "Blue Impact" low-impact feed program. The program seeks to reduce dependence on wild fish by using novel ingredients such as algal oil, fish by-products, and insect meal (Fish Farming Expert, 2024). Within this framework, some farms (e.g., Caviar d'Aquitaine, France; Akaroa King Salmon, New Zealand; Loch Duart, UK) have started using Blue Impact feeds.

In April 2023, BioMar partnered with French insect producer Agronutris to begin R&D work on making black soldier fly (BSF) meal balanced and suitable for aquaculture (Agronutris, 2023). This collaboration is still in the product development phase. Another important development regarding the use of insects in aquafeeds is the collaboration between BioMar, Innovafeed, and retailer Auchan (BioMar, 2025a). On September 2, 2025, the partners announced the large-scale integration of insect protein in shrimp feed in Ecuador. While this development is noteworthy in terms of the commercialization of insect meal, global and high-volume supply is still progressing gradually.

BioMar is also pioneering work on novel alternatives to fish oil. As part of the Alga4Fuel&Aqua project, conducted in Greece in a research collaboration between BioMar Hellenic and the University of Thessaly, a dietary trial was carried out on sea bream over a period of 84 days. The trial investigated the complete replacement of fish oil in feed with a mix of microalgae of the species *Schizochytrium* sp. and *Nannochloropsis* sp. Results showed that the microalgae mixture can successfully replace fish oil in feed, while in fact slightly improving the specific growth rate (SGR) and the feed conversion rate (FCR) in sea bream. At the same time, the EPA and DHA content in fillets from the sea bream fed exclusively on lipids from microalgae was similar to that of fish fed exclusively on fish oil (AquaFeed, 2021).

Having demonstrated with similar trials that algal oil can completely replace fish oil, BioMar has begun implementing this in the field through programs such as Blue Impact. In a statement released in October 2024, BioMar reported that it uses algal oil in salmon, shrimp, and marine fish feeds, that more than 4 million tons of feed produced with this ingredient have been sold, and that prices remain competitive with other DHA/EPA sources (BioMar, 2024a). AlgaPrime DHA, developed in collaboration with Corbion, is actively used in these feeds (BioMar, 2021). BioMar has also commissioned a dedicated algal oil facility at its Wesley Vale plant in Northwest Tasmania, Australia, a first in the country (Hatchery Feed & Management, 2022).

BioMar is also turning to plant-based sources, focusing on omega-3-rich vegetable oils such as camelina oil. In 2024, the company entered into a long-term partnership with Yield10 Bioscience to commercialize camelina, which contains EPA and DHA levels equivalent to fish oil (WeAreAquaculture, 2023).

Another focus area for BioMar is single-cell proteins (SCPs). In May 2022, the company announced a long-term technical-commercial partnership with the Netherlands-based Aerbio (formerly Deep Branch) for Proton™ (Feed & Additive, 2022). This collaboration aims to use Aerbio's developed SCP as the primary protein source in salmon feed. BioMar is testing this new source at its own technology center, measuring performance and digestibility. The company, which conducts trials covering different SCPs (bacterial, yeast, fungus-based), has announced that including yeast derivatives in feed at rates of up to 20% has yielded positive results in growth and conversion rates. According to the company's statement, yeast protein has shown strong performance at 20% inclusion with no differences observed in growth or feed conversion. In addition to growth performance, these single-cell protein sources contain important cell wall components like b-glucans, mannan oligosaccharides, lipopolysaccharides, and peptidoglycan. These compounds can potentially stimulate immune responses over long-term feeding, influencing immune and physiological pathways. Advanced techniques such as transcriptomics, metabolomics, microbiota analysis, and histology have been used to assess these factors. So far, results have shown no unwanted upregulation of the immune system, and single-cell proteins are proving promising for long-term feeding (BioMar, 2024a). These findings demonstrate that BioMar is taking SCPs seriously not only in the research phase, but also in pilot and commercial evaluations.

2.4 Alltech & Alltech Coppens

Alltech, one of the global leaders in animal health and nutrition, has long been conducting research on novel feed ingredients for farm animals, particularly in aquaculture. The company's primary areas of focus include insect meal, algae-based ingredients, and single-cell proteins.

Alltech strengthened its algae production capacity and research capabilities by acquiring Martek Biosciences Corp.'s algae fermentation facility in Kentucky, USA, in 2010 (Biodiesel Magazine, 2010). This investment was a critical step in the company's shift towards algae-based products. In 2016, the company acquired fish feed producer Coppens International and restructured it as Alltech Coppens (Alltech Coppens, 2016). The company also took over Coppens' longstanding research and carried it forward. In 2017, Alltech Coppens announced both its own algal oil product, ForPlus™ DHA, and Neogreen-branded trout feed, which contains this oil source (eFeedlink, 2017). Although there were reports in 2018 that Alltech had closed its algae production pilot plant in Kentucky, the company emphasizes that it is continuing its algae-based research (The Irish Times, 2018).

Alltech Coppens is moving quickly, especially in commercializing new ingredients. The company explicitly states that the feed it produces specifically for koi fish and markets under the name Midori uses black soldier fly meal as well as algae as a source of EPA and DHA (Alltech, 2018). The company has also announced plans to add insect meal to more fish feeds in the future.

Alltech Fennoaqua, operating in Finland, has similarly turned to insect ingredients. The company has partnered with insect meal producer Volare and has begun using insect meal sourced from Volare in its feed formulations. These feeds are sent to the Kalankasvatus Vääräniemi trout farm (Alltech Fennoaqua, 2024). This collaboration in Finland introduces a new concept to the market, where a small portion of traditional feed ingredients is replaced with sustainable insect meal. It had been stated that approximately 150,000 kilograms of rainbow trout would be fed with this feed and made available to consumers during the winter of 2024–2025. An announcement made in April 2025 states that these fish are being sold in Finland (Alltech Fennoaqua, 2025), but there is still no detailed technical information available comparing them with traditional sources and regarding the effects of these feeds (digestibility, feed intake, palatability, etc.).

Single-cell proteins also hold an important place in Alltech's innovation portfolio. Work in this area aims to provide consistent, high-quality protein sources (feed additives) for animals and optimize performance, rather than feed containing SCP directly. One of the company's solutions in this field is its NuPro product, which is described as a high-quality, single-celled functional protein derived from yeast (Alltech, 2025c). The company states that it has obtained the following results in trials conducted with this product: in commercial trials of shrimp and fish fed NuPro, the results have included faster growth rates, improved growth performance and higher daily weight gain. NuPro is rich in amino acids and short-chain peptides, which are easily absorbed. Trials have also shown improved feed conversion ratio and improved feed palatability, which means improved feed intake.

Alltech aims to offer novel solutions to the feed industry by continuing its R&D activities on yeast, bacteria and fungus-based proteins in the SCP field.

2.5 Marubeni Nisshin Feed

Marubeni Nisshin Feed, a dominant player in the Japanese aquafeed market, formed a partnership with French insect protein producer Ÿnsect in March 2023 (Marubeni, 2023; SeafoodSource, 2023). As part of the partnership, an insect-based feed suitable for Japanese aquaculture was developed. This feed, which contains insect meal instead of fish meal, was used in the farming of yellowtail, and this study is considered to be the first example documenting the successful farming of yellowtail using insect meal-based feed (Japan 2 Earth, 2024). It was stated that fish fed with this feed would be served to consumers at an Italian restaurant in 2024 and that the project would be expanded based on consumer feedback regarding taste, but no results have been shared regarding this process yet.

Regarding algal oil and SCP, there is no evidence of a direct partnership between Marubeni Nisshin and any producer. However, the company's focus on sustainable proteins and its success in insect-based feed trials indicate that algal oil and SCP are logical and potential growth areas aligned with the company's strategy of developing "new business models".

2.6 De Heus Animal Nutrition

De Heus focuses on novel feed ingredients and sustainable aquaculture applications with the goal of reducing environmental impact and preserving biodiversity. In 2021, it acquired the Netherlands-based Coppens Diervoeding, thereby also acquiring that company's experience in insect- and algae-based feeds from previous years (World-Grain, 2021). Coppens previously processed Protix's insect protein and insect oil to produce commercial feeds in the Netherlands and launched a pig feed, Insecto Speen, containing insect oil in 2016 (Varkensbedrijf, 2016). Prior to this product, two companies researched how insect-based ingredients strengthen the immune system of piglets and reduce mortality rates. These two companies shared the following results from their Insecto Speen trials: trials conducted with Insecto Speen resulted in increased feed intake, improved gut health, and reduced diarrhoea. This is a natural feed for piglets and supports a more sustainable food chain. Protix LipidX contains 43% lauric acid, a naturally occurring antimicrobial that protects against bacteria, including *Streptococcus* sp. and *Clostridium* sp. Piglets' delicate intestinal villi and entire digestive system can develop optimally, resulting in a stronger immune system and reduced reliance on antibiotics, plus lower mortality rates (Protix, 2025).

De Heus, which has taken over this legacy from Coppens Diervoeding, has launched F3-Tom, a fish-free feed for shrimp, using insect meal and algae oil instead of amino acids derived from fish ingredients, starting in 2023 (De Heus, 2024). This step is considered a significant advancement in the development of fish-free feeds.

De Heus' sustainability strategy also encompasses the use of residual flows and food by-products. The company's units in the Netherlands and Poland are pioneers in this field; waste streams are also used in feed at its African business units and some farms in Spain (De Heus, 2025). These residual flows include wheat flour, cooked potato peels, liquid starch, stale bread, broken cookies, rice bran, malting barley from beer production and whey from milk and cheese processing. In addition to waste products from the human food industry, the company uses distiller's dried grains from biogas production (De Heus, n.d.).

3 Factors supporting use of novel protein and lipid sources in animal feed

The use of novel protein and lipid sources such as insects, algae, single-cell proteins, and agricultural by-products in animal feed is the result of a series of interrelated market pressures. These factors are driving the sector from an environmental, economic, and nutritional perspective.

3.1 Sustainability and environmental pressures

Traditional protein and energy sources such as soybean meal, fish meal, and fish oil are associated with high land use, intensive water consumption, and significant greenhouse gas emissions (FAO, 2024). This situation increases the pressure to reduce the environmental footprint of livestock farming. Alternative sources alleviate this pressure through environmentally friendly production methods. For example, insects and algae can offer high protein yields with much less water and land use, while the utilization of food waste and agricultural by-products also strengthens the circular economy concept. This increases resource efficiency and solves the waste management problem. This trend is also supported by regulations, corporate sustainability goals, and consumer expectations on a global scale.

3.2 Raw material prices and supply security

Traditional raw materials such as grains, oilseeds, fish meal, and fish oil are subject to constant price fluctuations due to supply-demand imbalances, trade restrictions, and geopolitical tensions. These fluctuations are driving the feed industry to seek more resilient and predictable supply chains. Novel protein sources play an important role in ensuring supply security by increasing diversity. In addition, the possibility of local insect farming or algae production offers a strategic advantage, especially in countries dependent on imports.

3.3 Nutritional and functional advantages

Novel feed ingredients are not only valued as basic sources of protein and energy. Their functional aspects, which directly affect animal health and product quality, are also noteworthy. For example, microalgae are rich in long-chain omega-3 fatty acids and are a valuable ingredient, especially for aquatic products. Insect meal contains compounds that support the immune system, such as chitosan and lauric acid, while single-cell proteins offer high digestibility with a balanced amino acid profile (Abenaim and Conti, 2025; Geofrey et al., 2024; Chamodi et al., 2025). Some alternatives also support gut health by creating antimicrobial peptides or prebiotic effects. These properties are considered important in terms of reducing antibiotic use and obtaining healthier animal products.

3.4 Technological advancements

Advancements in fermentation processes, bioreactor technologies, insect farming automation, and advanced extraction methods significantly reduce the production costs of novel ingredients. These technologies enable scalable production and standardize the quality of ingredients. For example, processing and disinfection techniques developed to convert food waste into safe feed ingredients reduce contamination risks and increase industry confidence (Chen et al., 2015; Nath et al., 2023). These developments are accelerating the competitiveness of novel ingredients against traditional sources.

3.5 Investment and market dynamics

In recent years, startups producing novel proteins and feed ingredients have received significant investment from both venture capital funds and large agriculture and food companies. These investments enable increased research and development activities, the rapid transition of pilot-scale projects to commercial applications, and the expansion of production capacity. Investor interest not only provides financing but also demonstrates confidence in the market. Market reports indicate that growth in this area will accelerate further in the coming years. For example, according to the latest report published by Grand View Research, the global novel protein ingredients (Plant Protein, Microbe-based Protein, Insect Protein) market size was estimated at USD 22.95 billion in 2024 and is anticipated to reach USD 50.22 billion by 2030, growing at a CAGR of 14.1% from 2025 to 2030 (Grand View Research, 2025). Furthermore, according to a report by IMARC Group, the global algae oil market size reached USD 2.1 billion in 2024. Looking forward, the company expects the market to reach USD 2.9 billion by 2033, exhibiting a growth rate (CAGR) of 3.6% during 2025–2033 (IMARC Group, 2025).

3.6 Regulations and policy incentives

The European Union and several other countries are updating their legislation on the use of insect protein, algae-based ingredients, and processed agricultural by-products in the feed sector (IPIFF, n.d.). These regulatory frameworks provide producers with the opportunity to safely use novel feed ingredients while also creating legal certainty for investors. Consumer demand for sustainable products, retailers' climate goals, and government environmental policies encourage the wider implementation of these regulations.

3.7 Cross-industry synergy and circular contribution

By-products from the food processing industry can become valuable ingredients for the feed industry. Vegetable processing waste, brewer's yeast, or distilled grain residues provide nutritious supplements for both ruminants and poultry while also reducing the waste burden of the food industry. This contributes to the circular economy by providing both cost advantages and cross-sector synergy. According to European Union legislation, by-products from the food processing industry may be used as feed materials provided that appropriate processing, hygiene and traceability conditions are met. This framework, defined by Regulations (EC) No 1069/2009 and 767/2009, aims to support circular economy objectives by safely incorporating waste into the feed chain. Furthermore, Commission Regulation (EU) No 68/2013 establishes the reference framework for the classification and definition of these materials through the Catalogue of feed materials (European Commission, 2025).

3.8 Pilot applications and gradual adoption

The process of introducing novel feed ingredients into the sector is generally gradual. Initial trials are mostly conducted in aquafeed, then transferred to the poultry and pig sectors. This step-by-step approach helps reduce risks and makes it easier for producers to accept new ingredients. In most cases, novel ingredients are included in formulations as partial additives rather than completely replacing existing feeds. This strategy offers a more economically manageable and safer solution in terms of production performance.

3.9 Increasing food security and population pressure

The rapid increase in the global population is constantly raising demand for animal products such as meat, milk, and eggs. This increase is directly reflected in feed demand and makes it difficult for traditional raw materials alone to meet this need. Novel sources play a strategic role in global food security by providing a solution to the increasing demand for feed.

4 Factors limiting use in animal feed

Novel raw materials such as insects, algae, single-cell proteins, agricultural by-products, and food waste offer significant opportunities but still face a number of economic, technical, regulatory, and social barriers. These challenges limit both the scaling of the sector and its adoption by farmers and consumers.

4.1 High production and processing costs

One of the biggest barriers to novel feed sources is that they cannot yet compete with traditional raw materials in terms of price. The production technologies used for insects, algae, and single-cell proteins require significant investment. Photobioreactors and fermenters used for microalgae production stand out for their high capital costs, while automation and mass production in insect farming are still in the development phase. Furthermore, processing steps such as drying, grinding, and sterilization, which are necessary for waste to be used as feed, result in additional energy consumption and increase costs. These conditions keep the cost per kilogram of novel ingredients higher than soybean meal or fish meal, limiting their widespread use. For example, while soybean meal costs approximately USD 353 per tonne and fish meal costs around USD 1,700 per tonne (IndexMundi, 2025), the price of insect protein ranges from €3,500 to €5,500 per tonne (approximately USD 4,000 to USD 6,400) (FAO, 2022). Such significant price differences limit the widespread use of novel proteins in terms of cost.

4.2 Regulatory uncertainties and approval processes

Clear and reliable regulatory frameworks are critical for new feed ingredients to enter the market. However, there are significant regulatory differences between countries. For example, in the European Union, insect proteins can only be used in certain sectors (mostly in aquafeed), while restrictions remain in place for poultry and ruminant feed. In the US and Asian countries, approval processes vary and require lengthy bureaucratic steps. The use of food waste as feed is also subject to strict legal restrictions due to the risk of pathogens and chemical residues. This increases the risk perception of investors and feed producers, slowing down the adoption process (EFSA, 2022; European Commission, 2025).

4.3 Food safety and contamination risks

The safe use of novel sources is one of the most debated issues in the industry. Agricultural by-products and food waste may pose risks in terms of mycotoxins, pesticide residues, or heavy metals. Insect-based proteins raise concerns about potential microbial contamination and possible allergenic reactions due to their similarity to crustaceans (Garino et al., 2019; Rodríguez et al., 2021). These risks complicate the regulatory approval process and limit consumer confidence. Therefore, widespread adoption does not seem possible without standardized processing technologies and strict control systems.

4.4 Nutritional composition inconsistency (variability) and quality standardization

The nutritional profiles of novel ingredients are not fixed. The protein, energy, or mineral content of agricultural by-products also varies depending on their source, processing method, and even season. This creates difficulties for feed producers seeking to formulate balanced feed. The lack of established standard quality parameters prevents the industry from using them confidently on a large scale.

4.5 Consumer perception and acceptance issues

Another factor complicating the adoption of novel ingredients is consumer perception. The idea of feeding animals insects, waste, or microorganisms creates psychological resistance in many consumers (Feed & Additive, 2025; Szendrő et al., 2020). This perception can also affect confidence in final animal products such as meat, milk, or eggs. Retailers and food brands may be reluctant to include these products in their supply chains without considering consumer reactions. Therefore, the environmental benefits and sustainability potential of novel sources must be communicated more strongly.

4.6 Technical and logistical challenges

The production and distribution of novel ingredients may not integrate seamlessly into existing feed supply chains. For example, microalgae production requires a high level of light, CO_2, and water efficiency, while scalable automation in insect farming is still limited. Collecting, sorting, transporting, and processing food waste is a complex and costly logistical process. Furthermore, appropriate storage infrastructure and protection strategies are needed to mitigate the risk of spoilage for such ingredients. These conditions make it difficult to establish a stable and reliable supply chain for novel ingredients.

4.7 Investment and financing risks

The novel feed market is still in its early stages, and uncertainties may cause investors to be reluctant to provide long-term financing. High initial costs, regulatory uncertainties, and difficulties in predicting the market's future prevent ventures in this field from achieving economies of scale. This situation reduces competitiveness against traditional raw materials.

4.8 Competitive traditional raw materials

Soybean meal and fish meal still have a strong price/performance advantage in the feed sector. Extensive production networks, established logistics systems, and long-standing reliability make these raw materials still attractive compared to alternatives. For novel sources to be widely adopted, either their costs must decrease or traditional raw materials must lose their appeal due to regulatory and environmental pressures.

5 Conclusion – Key takeaways

Novel protein and lipid sources are at the centre of a dynamic transformation driven by the need for sustainability, nutritional quality, and resource efficiency in animal feed. Real-world case studies demonstrate that insect meals and oils, algae-based oils, single-cell proteins, and agricultural by-products not only overcome current challenges but also hold the potential to create a more resilient and environmentally sustainable future for animal feed.

Aquaculture has been the first preferred area for testing and piloting new ingredients. Europe leads in regulations and pilot projects, while the US, Australia, and Asia are following this trend with increasing pilot applications. Most innovative ingredients are applied as partial contributions or functional uses rather than full replacements. Gradual integration is at the forefront. New ingredients are typically introduced in starter feeds, functional additives, or carbon reduction goals. This approach facilitates adoption by reducing risks and provides a reliable step for producers. Collaboration networks play a critical role. Partnerships between feed producers, biotechnology startups, farms, and academic institutions are vital for proving the effectiveness of novel ingredients, optimizing production processes, ensuring performance validation, commercializing innovations, and securing market acceptance. Sustainability-focused strategies not only provide environmental benefits but also offer marketing value and carbon credits. Early-stage pilot applications are yielding tangible gains in feed efficiency, animal health, and climate impacts, despite cost, scaling, and regulatory restrictions. Scalable production models such as urban insect farms, modular algae production facilities, and fermentation-based protein production support the spread of innovation across different sectors. The wide range of raw materials, from insects to algae and yeast, and from agricultural by-products to fermentation proteins, has the capacity to transform animal feeding systems. Reduced dependence on traditional feed sources will increase the sector's resilience and contribute to global sustainability goals. In the coming period, these innovative raw materials are expected to play a more prominent role in animal feed due to their functional and environmental benefits.

References

Abenaim, L., Conti, B., 2025. Harnessing Chitin from Edible Insects for Livestock Nutrition. Insects, 16(8), 799. https://doi.org/10.3390/insects16080799

Agronutris, 2023. BioMar and Agronutris partner on the next generation insect meal. Available at: https://www.agronutris.com/en/biomar-and-agronutris-partner-on-the-next-generation-insect-meal/

AgTechNavigator, 2020. Skretting testing Unibio's novel SCP, with purchase contract potential prize, Available at: https://www.agtechnavigator.com/Article/2020/05/07/skretting-testing-unibios-novel-single-cell-protein/

Alltech Fennoaqua, 2024. Alltech Fennoaqua partners with insect ingredient company Volare, a fish farming company and a fish wholesaler to launch a sustainable rainbow trout product. Available at: https://www.fennoaqua.fi/en/story/alltech-fennoaqua-partners-with-insect-ingredient-company-volare-a-fish-farming-company-and-a-fish-wholesaler-to-launch-a-sustainable-rainbow-trout-product/

Alltech Fennoaqua, 2025. Rainbow trout fed with insect meal arrived in shops. Available at: https://www.fennoaqua.fi/en/story/rainbow-trout-fed-with-insect-meal-arrived-in-shops/

Alltech Coppens, 2016. Coppens International announces rebrand to Alltech Coppens. Available at: https://www.alltechcoppens.com/de/pressebereich/rebrand-to-alltech-coppens

Alltech, 2018. Swimming art: Bringing koi feed back to nature. Available at: https://www.alltech.com/blog/swimming-art-bringing-koi-feed-back-nature

Alltech,2025c. Functional nutrients for aquaculture feed. Available at: https://www.alltech.com/en-in/blog/functional-nutrients-aquaculture-feed

AquaFeed, 2021. Microalgae mix can replace fish oil in seabream. Available at: https://www.aquafeed.com/newsroom/news/microalgae-mix-can-replace-fish-oil-in-seabream/

Biodiesel Magazine, 2010. Alltech acquires algae fermentation facility from Martek Bioscience. Available at: https://biodieselmagazine.com/articles/alltech-acquires-algae-fermentation-facility-from-martek-bioscience--6523

BioMar, 2021. BioMar reaches 1 million tonnes of salmon feed with microalgae. Available at: https://www.biomar.com/insights/insights-hub/biomar-reaches-1-million-tonnes-of-salmon-feed-with-microalgae

BioMar, 2024a. The Evolution of Nutrients in Aquafeeds. Available at: https://www.biomar.com/insights/insights-hub/the-evolution-of-nutrients-in-aquafeeds

BioMar, 2025a. BioMar, Innovafeed and Auchan announce landmark partnership to accelerate the commericialisation of insect meal. Available at: https://www.biomar.com/insights/insights-hub/biomar-innovafeed-and-auchan-announce-landmark-partnership-to-accelerate-the-commercialisation-of-insect-meal

BioMar, 2025b. A deep dive into restorative raw materials. Available at: https://www.biomar.com/insights/insights-hub/a-deep-dive-into-restorative-raw-materials

Biswas, A., Takakuwa, F., Yamada, S., Matsuda, A., Saville, R.M., LeBlanc, A, Silverman, J.A., Sato, N., Tanaka, H., 2020. Methanotroph (*Methylococcus capsulatus*, Bath) bacteria meal as an alternative protein source for Japanese yellowtail, *Seriola quinqueradiata*. Aquaculture, Volume 529, 2020. https://doi.org/10.1016/j.aquaculture.2020.735700

Calysta, 2016. Calysta completes $30 million Series C financing, accelerating the introduction of FeedKind protein at commercial scale. Available at: https://calysta.com/calysta-completes-30-million-series-c-financing-accelerating-the-introduction-of-feedkind-protein-at-commercial-scale/

Cargill, 2019a. Cargill and InnovaFeed partner to bring innovative, sustainable feed to animal producers, Available at: https://www.cargill.com/2019/cargill-and-innovafeed-partner-to-bring-innovative-sustainable

Cargill, 2019b. Cargill, White Dog Labs agreement to deliver scalable, sustainable feed for aquaculture customers. Available at: https://www.cargill.com/2019/cargill-white-dog-labs-agreement-to-deliver-scalable-sustainable

Carr, I., Santigosa, E., Chen, T., Costantino, J., 2024. Optimum eicosapentaenoic acid + docosahexaenoic acid levels for farmed Atlantic salmon: closing the gap between science and commercial practice. Frontiers in Marine Science, Volume 11 – 2024. https://doi.org/10.3389/fmars.2024.1379066

Cermaq, 2024. Cermaq Canada adopts Cargill salmon feed with Veramaris algal oil. Available at: https://www.cermaq.ca/news/cermaq-canada-adopts-cargill-salmon-feed-with-veramaris-algal-oil

Chamodi, K.K.D., Vu, N.T., Domingos, J.A., Loh, J.Y., 2025. Cellular Solutions: Evaluating Single-Cell Proteins as Sustainable Feed Alternatives in Aquaculture. Biology, 14(7), 764. https://doi.org/10.3390/biology14070764

Chen, T., Jin, Y., Shen, D., 2015. A safety analysis of food waste-derived animal feeds from three typical conversion techniques in China. Waste Management, Volume 45, 2015, Pages 42-50. https://doi.org/10.1016/j.wasman.2015.06.041.

De Heus, 2024. De Heus commitment to sustainable aquaculture: insights from Vietfish 2024. Available at: https://www.deheus.com/csr-stories/de-heus-commitment-to-sustainable-aquaculture-insights-from-vietfish-2024

De Heus, 2025. Pioneering the use of residual flows in animal feed. Available at: https://www.deheus.com/csr-stories/pioneering-the-use-of-residual-flows-in-animal-feed

eFeedlink, 2017. Coppens International unveils breakthrough aquatic feed innovations for 2017. Available at: https://www.efeedlink.com/contents/01-24-2017/8ff11fc1-141c-4c41-94fc-f51142f4fec8-a007.html

EFSA, 2022. James, K., Millington, A., Randall, N. Food and feed safety vulnerabilities in the circular economy. EFSA Supporting Publications 2022: 19(3):EN-7226. 112 pp. https://doi.org/10.2903/sp.efsa.2022.EN-7226

Enifer, 2022. Cooperation between Enifer and Skretting expanding. Available at: https://enifer.com/cooperation-between-enifer-and-skretting-expanding-2

European Commission, 2025. Regulation (EC) No 1069/2009, Regulation (EC) No 767/2009, Commission Regulation (EU) No 68/2013. Available at: https://eur-lex.europa.eu

FAO, 2022. Is the time ripe for using insect meal in aquafeeds?. Available at: https://openknowledge.fao.org/handle/20.500.14283/cb9023en

FAO, 2024. Greenhouse gas emissions from agrifood systems – Global, regional and country trends, 2000–2022. FAOSTAT Analytical Brief Series, No. 94. Rome, Available at: https://openknowledge.fao.org/handle/20.500.14283/cd3167en

Feed & Additive, 2022. Deep Branch, BioMar agree strategic partnership to enhance aquaculture industry. Available at: https://www.feedandadditive.com/deep-branch-biomar-agree-strategic-partnership-to-enhance-aquaculture-industry/

Feed & Additive, 2025. New Study: Insect-based foods unlikely to replace meat in Western diets. Available at: https://www.feedandadditive.com/new-study-insect-based-foods-unlikely-to-replace-meat-in-western-diets/

Fish Farming Expert, 2024. BioMar signs up first King Salmon farmer for Blue Impact feed. Available at: https://www.fishfarmingexpert.com/akaroa-king-salmon-biomar-blue-impact/biomar-signs-up-first-king-salmon-farmer-for-blue-impact-feed/1839507

Fishfarmingexpert, 2022. Cargill to use algal oil in all Norway fish feed. Available at: https://www.fishfarmingexpert.com/algal-oil-cargill-norway/cargill-to-use-algal-oil-in-all-norway-fish-feed/1162765

Food Business Middle East & Africa, 2024. Albert Heijn launches sustainable shrimp line in partnership with industry consortium. Available at: https://www.foodbusinessmea.com/albert-heijn-launches-sustainable-shrimp-line-in-partnership-with-industry-consortium

Garino, C., Zagon, J., Braeuning, A., 2019. Insects in food and feed – allergenicity risk assessment and analytical detection. EFSA Journal, Volume17, IssueS2, Special Issue:EU-FORA: Series 2, September 2019. https://doi.org/10.2903/j.efsa.2019.e170907

Geofrey, N., Zaina, N., Mugerwa, S., 2024. Insect-Based Diet, A Promising Protein Source for Monogastric, Affects Human Health and Poultry Performance: A Case Study on Black Soldier Fly Lavae. Biomedical Journal of Scientific & Technical Research, Volume 56- Issue 1. http://dx.doi.org/10.26717/BJSTR.2024.56.008786

Grand View Research, 2025. Alternative Protein Ingredients Market (2025 - 2030). Available at: https://www.grandviewresearch.com/industry-analysis/alternative-protein-market-report

Hatchery Feed & Management, 2022. BioMar Australia commissions algae oil facility. Available at: https://hatcheryfm.com/products/suppliers-news/biomar-australia-commissions-algae-oil-facility/

Hooft, J.M., Montero, R., Morales-Lange, B., Blihovde, V.F., Purushothaman, K., Press, C.M., Mensah, D.D., Agboola, J.O., Javed, S., Torunn, L., 2024. Paecilomyces variotii (PEKILO®) in novel feeds for Atlantic salmon: Effects on pellet quality, growth performance, gut health, and nutrient digestibility and utilization. Aquaculture, Volume 589, 2024, 740905. https://doi.org/10.1016/j.aquaculture.2024.740905.

IMARC Group, 2025. Algae Oil Market Report. Available at: https://www.imarcgroup.com/algae-oil-market

IndexMundi, 2025. Commodity Prices, Available at: https://www.indexmundi.com/commodities/

InnovaFeed, 2021. Cargill & InnovaFeed expand their partnership for sustainable feed beyond aquaculture, Available at: https://innovafeed.com/en/cargill-innovafeed-expand-their-partnership-for-sustainable-feed-beyond-aquaculture/

International Aquafeed, 2023. Insect meal in commercial Skretting feed. Available at: https://www.aquafeed.co.uk/insect-meal-in-commercial-skretting-feed/

IPIFF, n.d. Insects and EU Legislation: General Overview. Available at: https://ipiff.org/insects-eu-legislation-general/

Japan 2 Earth, 2024. Making Aquaculture More Sustainable: Marubeni Farms Yellowtail on Insect-Based Feed. Available at: https://featured.japan-forward.com/japan2earth/2024/03/6352/

Keskitalo, H., 2024. The role of PEKILO® Aqua mycoprotein in aquaculture to feed the increasing world population. Feed & Additive Magazine, Issue 43, August 2024. https://www.feedandadditive.com/digital-magazines/Issue43_August2024/44/

Marubeni, 2023. Marubeni Announce Collaboration for Ÿnsect, the World's Largest Insect Protein Manufacturer and Distributor, to Enter the Japanese Market. Available at: https://www.marubeni.com/en/news/2023/release/00028.html

Nath, P.C., Ojha, A., Debnath, S., Sharma, M., Nayak, P.K., Sridhar, K., Inbaraj, B.S., 2023. Valorization of Food Waste as Animal Feed: A Step towards Sustainable Food Waste Management and Circular Bioeconomy. Animals, 13(8), 1366. https://doi.org/10.3390/ani13081366

Pig Progress, 2021. On-demand: Webinar on insect oil in pig nutrition. Available at: https://www.pigprogress.net/health-nutrition/on-demand-webinar-on-insect-oil-in-pig-nutrition/

Protix, 2025. Case: Coppens Diervoeding and Protix. Available at: https://protix.com/discovery-centre/case-coppens-diervoeding

Rodríguez, B.S., Rivera, G.A., Valdés, A., Ibáñez, E., Cifuentes, A., 2021. Food by-products and food wastes: are they safe enough for their valorization? Trends in Food Science & Technology, Volume 114, 2021, Pages 133-147. https://doi.org/10.1016/j.tifs.2021.05.002.

SalmonBusiness, 2022. Cargill adds algae oil to all fish feed. Available at: https://www.salmonbusiness.com/cargill-algae-oil-fish-feed-omega

SeafoodSource, 2023. Marubeni, Sumitomo racing to market insect-based aquafeed in Japan. Available at: https://www.seafoodsource.com/news/premium/aquaculture/marubeni-and-sumitomo-leading-bettors-on-insect-protein-feed-in-japanese-market

Skretting, 2018. Insect meal now in commercial Skretting feed. Available at: https://www.skretting.com/en/news-and-stories/insect-meal-now-in-commercial-skretting-feed/

Skretting, 2020a. Skretting and Veramaris announce further collaboration as French supermarket Auchan launches sustainable super-nutritious trout. Available at: https://www.skretting.com/en/news-and-stories/skretting-and-veramaris-announce-further-collaboration-as-french-supermarket-auchan-launches-sustainable-super-nutritious-trout

Skretting, 2020b. Skretting to test NovoNutrients novel protein ingredients. Available at: https://www.skretting.com/en/news-and-stories/skretting-to-test-novonutrients-novel-protein-ingredients/

Skretting, 2021. Novel protein ingredient, Pekilo® by eniferBio, reaches Skretting ARC. Available at: https://www.skretting.com/en/news-and-stories/novelprotein-ingredient-pekilo-by-eniferbio-reaches-skretting-arc/

Skretting, 2025. Are our feeds sustainable? Available at: https://www.skretting.com/en-tr/transparency--trust/faqs/are-our-feeds-sustainable/

Szendrő, K., Nagy, M.Z., Tóth, K., 2020. Consumer Acceptance of Meat from Animals Reared on Insect Meal as Feed. Animals (Basel). 2020 Jul 30;10(8):1312. https://doi.org/10.3390/ani10081312

The Irish Times, 2018. Alltech closes algae fermentation plant in Kentucky. Available at: https://www.irishtimes.com/business/agribusiness-and-food/alltech-closes-algae-fermentation-plant-in-kentucky-1.3483465

Varkensbedrijf, 2016. Eerste levering insectenolie bij Coppens Diervoeding. Available at:

https://www.varkensbedrijf.nl/varkensvoer/mengvoer/eerste-levering-insectenolie-bij-coppens-diervoeding/

Veramaris, 2024. Commercial farm trial. Available at: https://www.veramaris.com/salmon-farm-proof-point

Veramaris, 2025. Optimum Omega Nutrition (OON) by Veramaris. Available at: https://www.veramaris.com/optimum-omega-nutrition

WeAreAquaculture, 2023. Is it possible to grow fish oil on land?. Available at: https://weare-aquaculture.com/news/feed/is-it-possible-to-grow-fish-oil-on-land

World-Grain, 2021. De Heus to acquire Coppens. Available at: https://www.world-grain.com/articles/15587-de-heus-to-acquire-coppens

All web-based sources were accessed on 28 December 2025.

PART 9

Conclusions and broader impact

CHAPTER 36

Conclusions: from innovation to application

C. Murcia

Kemin Nutrisurance, EMEA, Via della Tecnica 11, 37040, Veronella (VR), Italy

Throughout this book, we have seen how research, technology, and creativity are opening new ways to nourish animals while aiming to protect our planet. Yet progress remains uneven. Some ingredients are already part of commercial diets, while others are still navigating the complexities of science and regulation. What becomes clear is that sustainability in feed and pet food systems cannot rely on a single innovation; it requires a balanced transition where safety, functionality, affordability, and fairness evolve together.

Micro- and macroalgae are currently used primarily for their functional properties, including support of endogenous antioxidant defenses, immune response and product quality, rather than as full replacements for traditional protein or lipid sources. Their potential to reduce methane emissions in ruminants is scientifically promising, but large-scale use remains limited by production costs, scalability and variability in nutrient composition.

Insects stand out as the most actively explored and commercially implemented novel protein source. Their value lies not only in their nutritional quality but also in their capacity to contribute to circular production systems. However, unlocking their full sustainability potential requires continued progress in feed safety, processing technologies and energy efficiency. This is particularly relevant where food waste or agricultural by-products are used as substrates, and where reliance on renewable energy sources will be critical. Ensuring safety, traceability and process control remains essential to gain public confidence and regulatory trust.

Plant-based innovations also play an important role in this transition. Ingredients such as pulses in pet food and oilseed crops like camelina illustrate both the opportunities and the complexity of diversifying feed and pet food formulations. These examples highlight the importance of critically assessing popular or emerging ingredients through robust nutritional evaluation, digestibility, health outcomes and agronomic context, rather than relying on simplified sustainability narratives.

Agricultural by-products represent another important opportunity to close nutrient loops within feed systems. When valorized safely and responsibly, they can reduce dependence on crops with high opportunity costs. However, not all uses of by-products necessarily lead to improved sustainability outcomes. Their wider application requires careful, case-specific evaluation based on feed safety princi-

HTTPS://DOI.ORG/10.1163/9789004692824-037

ples, regulatory clarity and systemic environmental assessments. Trade-offs related to resource intensity, efficiency and the risk of perpetuating structurally inefficient production systems must be explicitly considered.

Across all novel ingredients discussed, the feed sector urgently needs more standardized and harmonized characterization. Reliable and comparable data are essential to formulators, researchers and regulators to identify which novel sources can truly contribute to specific nutrients, and under which conditions they perform best. The nutritional composition tables presented in this book are intended as a practical step toward improving transparency, comparability and evidence-based formulation.

Quality and sustainability are inseparable from stability and safety. As highlighted in the chapter on oxidation, lipid and protein degradation are key drivers of quality loss, reduced shelf life and rejection of feed and pet food products, with direct implications for food and feed waste. When oxidative processes are not adequately controlled, nutritional losses, sensory deterioration and safety concerns can negate potential environmental benefits. Managing oxidation is therefore critical to translating sustainability ambitions into real-world performance and reduced waste along the supply chain.

As highlighted in the chapter about decision-making and sustainability strategies, every innovation in feed ingredients represents a decision that carries consequences from discovery to market implementation. Building resilient and equitable feed and pet food systems requires these decisions to be guided not only by efficiency, but also by responsibility and fairness.

Achieving equilibrium between innovation and regulation, novelty and safety, economy and ecology, remains one of the central challenges ahead. However, it is also what makes this transition so compelling: the opportunity to feed animals better while sustaining the systems that feed us all.

We have aimed to make this book a space where science is accessible, knowledge is shared openly, and where innovation is not restricted to a few but available to many. By choosing an open-access format, we wanted to enable readers across regions and disciplines to engage with the evidence, reflect on its implications and apply it within their own contexts.

This volume does not claim to provide definitive answers to every question. Science is, by nature, iterative, and understanding evolves as evidence grows. Rather, this book offers a rigorous and transparent snapshot of where we currently stand, while opening a broader conversation about where the field may go next.

To our knowledge, it is the only resource that brings together novel feed and pet food ingredients and examines them across such a wide range of perspectives: nutritional, technological, regulatory, environmental and social. By combining scientific analysis with industry case studies, it connects sustainability and decision-making with real market experience and demonstrates that innovation in animal nutrition is already taking place in practice.

The purpose of this book is not to close the discussion, but to support informed engagement. Readers are invited to use this resource as a foundation for reflection, dialogue and further development. Progress in feed and pet food systems depends not on definitive answers, but on the continuous improvement of decisions guided by evidence and shared responsibility.

Perhaps this is the book's greatest contribution: to encourage critical thinking, collaboration and curiosity, and to remind us that the future of feed and pet food is not yet written, but is already being shaped, collectively through evidence-based decisions.

CHAPTER 37

Sustainability and the sustainable development goals: how this book contributes

C. Murcia

Kemin Nutrisurance, EMEA, Via della Tecnica 11, 37040, Veronella (VR), Italy

1 Introduction

Discussions around sustainability in feed, aquaculture and pet food are often complex, and sometimes uncomfortable. They touch on limits, trade-offs, and the reality that food and feed systems operate under growing environmental, economic and social pressures.

In this book, the Sustainable Development Goals (SDGs) are used as a shared framework to structure this discussion. They are not treated as a political checklist, nor as abstract aspirations, but as a useful lens to reflect on where current feed and pet food systems perform well, where pressure is accumulating, and where innovation may help. Sustainability in animal nutrition is no longer a distant objective; it is already shaping decisions across the entire value chain. Novel ingredients are not presented as miracle solutions, but as tools that, when applied carefully and based on evidence, can reduce pressure on critical resources, expand formulation options and contribute to more resilient systems. The SDGs provide a transparent and constructive way to articulate this transition, acknowledging both opportunities and limitations.

2 Alignment with the sustainable development goals

SDG 2 – Zero hunger

Feeding a growing global population of people, livestock and companion animals force us to look again at how we use the resources we have. It is not a comfortable discussion, but it is a necessary one. The truth is that our current food and feed systems are already feeling pressure, and it will be difficult to meet future demands if we depend only on the same traditional raw materials. By exploring insects, algae, microbial proteins and upcycled by-products among other ingredients, this book presents alternatives that can complement or partially replace, and in specific contexts even fully substitute conventional ingredients. When integrated responsibly, these alternatives can support more resilient and resource-efficient food and feed systems capable of nourishing a growing population while safeguarding planetary resources.

SDG 4 – Quality education

Access to reliable scientific knowledge is a prerequisite for sustainable transformation. By being published in an open-access format, this book contributes directly to SDG 4 by promoting equitable access to education and evidence-based information on the use of novel ingredients in feed and pet food systems. Students, researchers, professionals and decision-makers across regions and income levels can consult, use and build upon the content without financial or institutional barriers.

Open access is particularly relevant in the context of sustainability, where progress depends on shared understanding rather than exclusive expertise. By making current research, data and case studies openly available, this volume supports education for sustainable development and encourages informed decision-making across academia, industry and policy. In this way, the book aims not only to inform, but also to foster learning, dialogue and capacity building in animal nutrition worldwide.

SDG 9 – Industry, innovation and infrastructure

With the production of novel ingredients, this book also showcases the technologies that make them possible. Fermentation technologies for protein production, algae cultivation, insect farming, revalorization of agricultural by-products are tools that enable more efficient and circular infrastructures. These approaches are not idealized. Their potential is discussed alongside their current limitations and challenges including safety, precision, scalability and energy efficiency. Progress in this area requires robust science, long-term investment and a commitment to continuously improving industrial systems, so that they become smarter, fairer and better adapted to the demands of the future.

SDG 12 – Responsible consumption and production

Responsible use of resources requires greater awareness of the environmental costs associated with the production and consumption of feed materials, as well as a broader reflection on the role and future configuration of livestock and aquaculture systems. Diversifying feed ingredients can contribute to balancing nutritional demands with the way those materials are produced. Through the valorization of low-opportunity-cost biomass and underutilized side streams, and by presenting data transparently, this book encourages informed decisions along the food–feed value chain.

Novel ingredients are not shown as perfect solutions, but as promising components within broader efforts to develop production systems that are efficient, circular and respectful of planetary boundaries.

SDG 13 – Climate action

Meeting nutritional requirements with lower environmental footprints is central to climate mitigation. Novel ingredients can contribute to this goal, but only when evaluated through solid evidence and accurate life-cycle data. This volume examines where alternative proteins and lipids may realistically reduce greenhouse-gas emissions or land and water use, and where uncertainties remain. The message is clear: climate-smart feed systems will emerge not from replacing one raw material with another, but from combining innovation, transparency and continuous improvement.

SDG 14 – Life below water

Current fish-in:fish-out ratios continue to place pressure on marine ecosystems. Reducing this dependency is not a critique of aquaculture, but an acknowledgement of its rapid growth and its responsibility to evolve sustainably. The book presents several alternatives, such as algae (including those cultivated in controlled bioreactors), microbial proteins, insects and other emerging aquatic ingredients, that can complement marine resources while preserving their nutritional qualities. These options can help maintain performance in aquaculture feeds while easing extraction pressure on the oceans.

SDG 15 – Life on land

Land is a finite resource. By incorporating feed ingredients that require little or no arable land, such as insects, algae, fermentation-derived proteins or upcycled side streams, we can reduce competition between food and feed crops and support the recovery of terrestrial ecosystems. The aim is not to displace traditional agriculture, but to create space for natural carbon sinks to recover and biodiversity to flourish. Ingredient diversification represents one of several strategies that can help rebalance the relationship between agriculture and nature.

SDG 17 – Partnerships for the goals

Sustainability requires innovation, and innovation is rarely achieved in isolation. Harnessing the potential of novel ingredients responsibly depends on collaboration between academia, industry, policymakers and society. Education and open dialogue are essential to address misconceptions, increase acceptance and ensure that innovation serves the common good. This book embodies that spirit of cooperation: more than fifty authors from different disciplines, sectors and regions have contributed to a shared understanding of where we stand, what we know and how we can move forward together.

Ultimately, sustainability in animal nutrition is shaped not by individual ingredients, but by the collective decisions we make regarding knowledge, resources and responsibility.

Index

www.ingramcontent.com/pod-product-compliance
Ingram Content Group UK Ltd.
Pitfield, Milton Keynes, MK11 3LW, UK
UKHW050625101026
15624UKWH00017B/622

9 789004 692817